Differential Equations:
Techniques,
Theory,
and Applications

Differential Equations:
Techniques, Theory, and Applications

Barbara D. MacCluer | Paul S. Bourdon | Thomas L. Kriete

Cover image designed using Field Play software (https://anvaka.github.io/fieldplay/).
Use allowed under the MIT License. © 2017–2018 Andrei Kashcha.

2010 *Mathematics Subject Classification.* Primary 34-01, 35-01;
Secondary 97M10.

For additional information and updates on this book, visit
www.ams.org/bookpages/mbk-125

Library of Congress Cataloging-in-Publication Data
Names: MacCluer, Barbara D., author. | Bourdon, Paul, author. | Kriete, Thomas L., 1942- author.
Title: Differential equations : techniques, theory, and applications / Barbara D. MacCluer, Paul S. Bourdon, Thomas L. Kriete.
Description: Providence, Rhode Island : American Mathematical Society, [2019] | Includes bibliographical references and index.
Identifiers: LCCN 2019019108 | ISBN 9781470447977 (alk. paper)
Subjects: LCSH: Differential equations. | AMS: Partial differential equations – Instructional exposition (textbooks, tutorial papers, etc.). msc | Mathematics education – Mathematical modeling, applications of mathematics – Modeling and interdisciplinarity. msc
Classification: LCC QA371 .M1625 2019 | DDC 515/.35–dc23
LC record available at https://lccn.loc.gov/2019019108

Copying and reprinting. Individual readers of this publication, and nonprofit libraries acting for them, are permitted to make fair use of the material, such as to copy select pages for use in teaching or research. Permission is granted to quote brief passages from this publication in reviews, provided the customary acknowledgment of the source is given.

Republication, systematic copying, or multiple reproduction of any material in this publication is permitted only under license from the American Mathematical Society. Requests for permission to reuse portions of AMS publication content are handled by the Copyright Clearance Center. For more information, please visit www.ams.org/publications/pubpermissions.

Send requests for translation rights and licensed reprints to reprint-permission@ams.org.

© 2019 by the authors. All rights reserved.
Printed in the United States of America.
∞ The paper used in this book is acid-free and falls within the guidelines
established to ensure permanence and durability.
Visit the AMS home page at https://www.ams.org/
10 9 8 7 6 5 4 3 2 1   24 23 22 21 20 19

Contents

Preface		ix
Chapter 1.	Introduction	1
1.1.	What is a differential equation?	1
1.2.	What is a solution?	2
1.3.	More on direction fields: Isoclines	16
Chapter 2.	First-Order Equations	25
2.1.	Linear equations	25
2.2.	Separable equations	35
2.3.	Applications: Time of death, time at depth, and ancient timekeeping	45
2.4.	Existence and uniqueness theorems	65
2.5.	Population and financial models	83
2.6.	Qualitative solutions of autonomous equations	98
2.7.	Change of variable	112
2.8.	Exact equations	121
Chapter 3.	Numerical Methods	139
3.1.	Euler's method	139
3.2.	Improving Euler's method: The Heun and Runge-Kutta Algorithms	151
3.3.	Optical illusions and other applications	162
Chapter 4.	Higher-Order Linear Homogeneous Equations	171
4.1.	Introduction to second-order equations	171
4.2.	Linear operators	192
4.3.	Linear independence	209
4.4.	Constant coefficient second-order equations	216
4.5.	Repeated roots and reduction of order	228
4.6.	Higher-order equations	240

4.7.	Higher-order constant coefficient equations	245
4.8.	Modeling with second-order equations	254

Chapter 5. Higher-Order Linear Nonhomogeneous Equations — 265

5.1.	Introduction to nonhomogeneous equations	265
5.2.	Annihilating operators	275
5.3.	Applications of nonhomogeneous equations	288
5.4.	Electric circuits	303

Chapter 6. Laplace Transforms — 319

6.1.	Laplace transforms	319
6.2.	The inverse Laplace transform	330
6.3.	Solving initial value problems with Laplace transforms	335
6.4.	Applications	350
6.5.	Laplace transforms, simple systems, and Iwo Jima	357
6.6.	Convolutions	363
6.7.	The delta function	368

Chapter 7. Power Series Solutions — 381

7.1.	Motivation for the study of power series solutions	381
7.2.	Review of power series	383
7.3.	Series solutions	392
7.4.	Nonpolynomial coefficients	408
7.5.	Regular singular points	413
7.6.	Bessel's equation	430

Chapter 8. Linear Systems I — 443

8.1.	Nelson at Trafalgar and phase portraits	443
8.2.	Vectors, vector fields, and matrices	457
8.3.	Eigenvalues and eigenvectors	472
8.4.	Solving linear systems	482
8.5.	Phase portraits via ray solutions	495
8.6.	More on phase portraits: Saddle points and nodes	507
8.7.	Complex and repeated eigenvalues	524
8.8.	Applications: Compartment models	536
8.9.	Classifying equilibrium points	549

Chapter 9. Linear Systems II — 563

9.1.	The matrix exponential, Part I	563
9.2.	A return to the Existence and Uniqueness Theorem	580
9.3.	The matrix exponential, Part II	584
9.4.	Nonhomogeneous constant coefficient systems	595
9.5.	Periodic forcing and the steady-state solution	608

Chapter 10. Nonlinear Systems   615
    10.1. Introduction: Darwin's finches   615
    10.2. Linear approximation: The major cases   627
    10.3. Linear approximation: The borderline cases   647
    10.4. More on interacting populations   653
    10.5. Modeling the spread of disease   668
    10.6. Hamiltonians, gradient systems, and Lyapunov functions   683
    10.7. Pendulums   699
    10.8. Cycles and limit cycles   708

Chapter 11. Partial Differential Equations and Fourier Series   717
    11.1. Introduction: Three interesting partial differential equations   717
    11.2. Boundary value problems   719
    11.3. Partial differential equations: A first look   727
    11.4. Advection and diffusion   734
    11.5. Functions as vectors   745
    11.6. Fourier series   760
    11.7. The heat equation   777
    11.8. The wave equation: Separation of variables   792
    11.9. The wave equation: D'Alembert's method   804
    11.10. Laplace's equation   812

Notes and Further Reading   833

Selected Answers to Exercises   837

Bibliography   863

Index   867

Preface

A study of differential equations is crucial for anyone with an interest in mathematics, science, or engineering. While the traditional applications of differential equations have been in physics and engineering, recent years have seen their application greatly expand into diverse areas, consistent with a recent report from the U.S. National Academies that noted, "Mathematical sciences work is becoming an increasingly integral and essential component of a growing array of areas of investigation in biology, medicine, social sciences, business, advanced design, climate, finance, advanced materials, ..." We have been significantly influenced by this perspective in choosing applications to include in this text, and, by drawing our applications from as wide a variety of fields as possible, we hope to engage and motivate students with diverse interests and backgrounds. In particular, we have included a large number of examples and exercises involving differential equations modeling in the life sciences—biology, ecology, and medicine. This reflects both the growing importance of mathematics in these fields and the broad appeal of applications drawn from them.

We've subtitled the book "Techniques, theory, and applications", reflecting our perspective that these three components of the subject are intimately tied together and mutually reinforcing. "Techniques" include not just analytic or computational methods for producing solutions to differential equations, but also qualitative methods for extracting conceptual information about differential equations and the systems modeled by them. "Theory" helps to organize our understanding of how differential equations work and codify general and unifying principles, so that the "techniques" are not just a collection of individual tools. "Applications" show the usefulness of the subject as a whole and heighten interest in both solution techniques and theory. The organization of the book interweaves these three aspects, with each building on and complementing the others.

We seize all opportunities to choose techniques which promote conceptual understanding of the subject. This explains, for example, our strong preference for the beautiful "method of annihilating operators", rather than the "method of undetermined coefficients", to solve constant coefficient linear nonhomogeneous equations. Indeed, the method of annihilating operators could be described as "the method of undetermined coefficients, with justification". Our experience, at a variety of different types of institutions, is that students respond well to this approach, which reinforces the "operator" perspective we adopt throughout the text.

We've written the text to be read and conceive of it as a *conversation* with our readers, filling in background, motivation, and a bit of storytelling that builds a bridge to the core mathematical ideas of the text. While this approach doesn't necessarily lead to the most succinct presentation possible, it makes for an eminently readable one that should prepare students for lively discussions in both traditional and flipped classroom settings. Furthermore, we believe that learning to read technical material is a valuable goal of any mathematics course. We don't assume the typical student in a differential equations course has had a great deal of experience doing this, and one of our goals is to provide the means and opportunity to gain this experience.

Overview and core material. Chapters 1, 2 (with Sections 2.7 and 2.8.2 optional), 4, and the first two or three sections of Chapter 5 form the core material for a one-semester course, with some omissions possible from the applications. This leaves significant time to cover other topics chosen by the instructor. Chapter 1 is very short and Section 1.3 is optional; we usually cover this chapter in one lecture. In Chapter 2, there is flexibility in selecting which applications to discuss, though we highly recommend that the logistic model for population growth be included among the chosen applications. Chapter 3 (numerical methods) stands alone and can safely be omitted if desired. There are some exercises scattered elsewhere throughout the text that are intended to be done with some kind of computer algebra system (these are indicated by "(CAS)"), but with a few exceptions these problems do not require material from Chapter 3. As with Chapter 2, the choice of applications to cover in Chapter 4 (including a potpourri of introductory applications in the first section and mechanical vibrations in the last section) and in Chapter 5 (forced oscillations and pharmacology models in Section 5.3 and electrical circuits in Section 5.4) is at the discretion of the instructor.

An engineering-focused course would probably turn rather quickly to Chapter 6 on Laplace transforms and Chapter 7 on power series solutions, which may be covered in either order. Neither of these chapters is required for any later chapters. Although Laplace transforms are used briefly in solving some linear nonhomogeneous systems in Chapter 9, alternative approaches are provided there. The first four sections of Chapter 7 provide a good introduction to power series methods; this can be followed by a discussion of the Frobenius method in Section 7.5.

Linear systems are the main focus of Chapters 8 and 9. Chapter 8 concentrates on planar systems, and Sections 8.4–8.7 provide tools to solve any planar homogeneous system $\mathbf{X}' = \mathbf{AX}$ when $\mathbf{A}$ is a 2×2 real matrix. Applications based on compartment models are discussed in Section 8.8; these include models from pharmacology and ecology. The first section of Chapter 9 introduces the matrix exponential for 2×2 matrices. A novel aspect of the treatment here is the development of formulas for "$e^{t\mathbf{A}}$" that combine easily with elementary vector geometry, enabling the reader to sketch the corresponding phase portraits with confidence. In Section 9.3 matrix exponentials for larger matrices are discussed using Putzer's Algorithm, which gives an efficient way to discuss $e^{t\mathbf{A}}$ with minimal linear algebra prerequisites.

Chapter 10 on nonlinear systems is rich in applications, and one may pick and choose from these as time permits. For a course that has only limited time available for this material, Sections 10.1 and 10.2 and an application or two chosen from Sections 10.4 and 10.5 provide a coherent introduction that could immediately follow a study of linear systems in Sections 8.1–8.7.

An introduction to partial differential equations and Fourier series is found in Chapter 11. After sections presenting two-point boundary value problems and a first look at partial differential equations via the advection and diffusion equations, we discuss the elementary theory of Fourier series in Sections 11.5 and 11.6. Our treatment emphasizes not only the role of orthogonality, but also the analogy between vectors in Euclidean space and periodic functions on the line as well as that between the Euclidean dot product and the integral inner product of functions. In Sections 11.7–11.10 these ideas are applied to study the heat and wave equations in one space variable and Laplace's equation in two variables.

Exercises. There are a great number and a great variety of exercises. We have tried to provide exercises that will contain genuine appeal for students. In some cases, that means a bit of a "back story" is included in the exercise. The same applies to many of the examples in the text. While this adds some length to the text, we strongly believe the richness this provides well compensates for this additional length. When an exercise is referred to in the text, only its number is given if the exercise appears within the same section; otherwise both its number and the section in which it appears are specified.

In writing the exercises, we have taken to heart one of the lessons that came out of the recent MAA calculus study "Characteristics of successful calculus programs", namely, that asking students to work problems that require them to grapple with concepts (or even proofs) and do modeling activities is key to successful student experiences and retention in STEM programs. There are also exercises that are explicitly constructed to help students identify common misconceptions. Some of these (the "Sam and Sally problems") present two reasonable sounding, but contradictory, approaches to a question and ask which is right (and why).

Projects: Informal and formal. Throughout the text you will find exercises that are suitable for more in-depth study and analysis by students, working either individually or in groups. While the difficulty of these vary, all give scope for bringing the ideas of the course to bear on interesting questions of either an applied or theoretical nature.

The text website www.ams.org/bookpages/mbk-125 also includes a more formal collection of projects, identified with the prerequisite text material. Many of these provide students with opportunities (and motivation) to explore applications of differential equations in depth, and they will often include parts intended to be completed with the use of a computer algebra system. To facilitate this, each such project is accompanied by a "starter" *Mathematica* notebook or *Maple* worksheet that illustrates the relevant commands. Students can complete the CAS-based part of the project directly in the starter notebook or worksheet. Some of the web projects also provide an opportunity to test differential equations models against real-world data (employing, for example, least-squares analyses). We believe that most students benefit from working in groups and hence suggest that students be permitted to work with a partner from their class on these web projects (submitting for evaluation one project report for the partnership).

Prerequisites. The prerequisite for the vast majority of the text is a standard year-long course in single variable calculus. Some ideas from multivariable calculus make an appearance in a few places (e.g., partial derivatives), but any needed background is provided. We have taken pains to include a review of certain topics from single variable calculus—as they appear—that experience has told us some students will need. We have particularly wanted to minimize our expectations of student familiarity with linear algebra. In the material before Chapter 8, the only matrix algebra that makes an essential appearance is the determinant of a 2×2 matrix. In Chapter 8, where systems are introduced, we devote several subsections to presenting "from the ground up" relevant matrix algebra topics.

The role of "proof". A fair number of theorems are stated and discussed in the text. We carefully explain what each theorem says (and doesn't say), illustrating with examples that clarify the content. Our perspective on "proof" is that it is first and foremost an explanation of why something is true. When a formal proof will aid the student's understanding, it is included. At times we will instead give a "justification" of the theorem, which is a more informal argument that contains the ideas of the proof in a more conversational format. Similarly, while some exercises do ask the student to provide a proof of some statement, more are designed to draw out the *ideas* that would go into a formal proof.

Modularity and sample syllabi. After the core material, described above, from Chapters 1, 2, 4, and 5, the text offers a great deal of flexibility in terms of material that may be covered as well as the order in which it may be presented. For a semester-long course, some sample syllabi that have been used are:

- Chapters 1, 2, 4, and 5 (with some choices made as to applications included), Sections 8.1–8.7 on planar linear systems (possibly including an introduction to the matrix exponential in Section 9.1), some compartment model applications from Section 8.8, Sections 10.1 and 10.2 on nonlinear planar systems, followed by applications chosen from Sections 10.4 (interacting populations), 10.5 (epidemiology), and 10.7 (pendulums). As time permits, an introduction to power series solutions may be included (Sections 7.1–7.3).

- A course that is more focused on engineering students might begin as above with Chapters 1 and 2, then turn to numerical methods in Chapter 3, followed by Chapters 4, 5, and 6 (Laplace transforms). If Chapter 6 will be covered, then many omissions are possible in Chapter 5. There should also be time for some discussion of power series solutions and/or planar systems (e.g., Sections 7.1–7.3 and Sections 8.1–8.7).

- At the University of Virginia, we have occasionally offered a special section of our differential equations course for physics students. This includes Chapters 1, 2, 4, and 5, some discussion of planar linear systems (through Section 8.7), and power series solutions (through Section 7.3) and then culminates in an introduction to partial differential equations and Fourier series (the first part of Chapter 11, through Section 11.7 or 11.8).

For a two-quarter or two-semester sequence, there is, of course, scope for quite a bit of additional material to be covered. We want to emphasize that once the core material has been discussed, the remaining topics (numerical methods, Laplace transforms, power series solutions, linear and nonlinear systems, partial differential equations, and Fourier series) are largely independent of each other, and one can pick and choose from these in whatever order and with whatever emphasis are desired.

Supplementary resources. A complete and detailed solution manual, prepared by the authors, is available to instructors. It contains solutions to all exercises. A student solution manual with solutions to all odd numbered exercises is also available. Further information about the solution manuals can be found at the text website www.ams.org/bookpages/mbk-125. In collaboration with Cengage, many exercises are available through WebAssign. Details can be found at the text website.

To the student. This text is meant to be read. Reading the text will deepen your understanding of the material; moreover, learning to read technical material is a worthy goal with wide applicability beyond just this particular subject. We've written the text to be a conversation with you, the reader, and scattered throughout you will find some nonmathematical stories that set the stage for the mathematics that follows. If an exercise looks "long", it is probably because it is designed to talk you through the problem, much as we would if we were talking to you during office hours.

We hope you will come away from this text with the strong sense that differential equations are lurking around every corner in everyday life. There are dozens of examples and exercises here that are motivated by recent news articles, blog posts, movies and TV shows, and current pop culture. No matter what interests you have outside of mathematics, we hope you will find in the myriad applications of differential equations discussed here something that connects with these interests. And we'd love to hear from you if you find an interesting application of differential equations you'd like to see included in a future edition.

Acknowledgments. The active writing of this text has occupied its authors for more than twelve years, and during that time many people contributed to our efforts in a variety of ways. We thank Don Beville and Bill Hoffman for their early encouragement and Karen Saxe for making a crucial introduction late in the process. The editorial and production staff at the American Mathematical Society provided both encouragement and practical help. We particularly thank Steve Kennedy, Christine Thivierge, Arlene O'Sean, John Paul, and Brian W. Bartling. It was a pleasure to work with them.

Over a period of more than six years, many colleagues at the University of Virginia and elsewhere volunteered to teach from preliminary versions of the text, and we appreciate their willingness to do so. Particular thanks go to Juraj Foldes, Ira Herbst, John Imbrie, Tai Melcher at the University of Virginia and Christopher Hammond at Connecticut College. There was much benefit to us in having the text class-tested, and we are grateful to all the students at the University of Virginia, Washington and Lee University, Connecticut College, and area high schools St. Anne's-Belfield and Albemarle High School who were involved. We are particularly thankful for the student feedback we received, which both encouraged us and helped us improve the presentation in various places. Natalie Walther at Cengage facilitated the creation of a private WebAssign version of the text which one of the authors used several times in teaching a "flipped classroom" version of the course, giving us a different and useful perspective on the material.

We thank Rebecca Schmitz and Erin Valenti who, as graduate students at the University of Virginia in 2006, assisted one of us in producing a set of notes for a self-study course in differential equations offered to local high school students. We are indebted to many reviewers who provided helpful comments and suggestions at all stages of this project, including

William Green, Rose-Hulman Institute of Technology,

Lawrence Thomas, University of Virginia,

Richard Bedient, Hamilton College,

Michelle LeMasurier, Hamilton College,

Robert Allen, University of Wisconsin-La Crosse,

Stephen Pankavich, Colorado School of Mines,

and many anonymous reviewers commissioned by Pearson Education, Inc., early in the development of this project.

Chapter 1

Introduction

1.1. What is a differential equation?

A differential equation is an equation that contains one or more derivatives of a function of interest. For example

$$(1.1) \qquad \frac{dy}{dx} = 2e^x - 5xy$$

and

$$(1.2) \qquad \frac{d^2y}{dx^2} + x\frac{dy}{dx} + (\tan x)y = 0$$

are differential equations in which the "function of interest" is y, a function of the variable x.

An ordinary derivative is a derivative of a function of a single variable, and an **ordinary differential equation** is one in which only ordinary derivatives appear. By contrast, a **partial differential equation** involves one or more partial derivatives of a function of several variables. For example, equations (1.1) and (1.2) are ordinary differential equations. In each of these, x is the **independent variable** and y is the **dependent** variable. This means we are thinking of y as an (unknown) function of the single variable x, and roughly speaking, our goal is to determine y. The equation

$$(1.3) \qquad \frac{\partial^2 u}{\partial x^2} + \frac{\partial^2 u}{\partial y^2} = 0$$

is a partial differential equation. Here there are two independent variables, x and y, and $u = u(x, y)$ is the dependent variable (or unknown function). For most of this book, we'll be concerned with ordinary differential equations, so the term "differential equation" will always mean "ordinary differential equation" unless we say explicitly otherwise.

The **order** of an (ordinary or partial) differential equation is the order of the highest derivative that appears in the equation. Thus (1.1) is a first-order equation, while (1.2) is a second-order equation. Equation (1.3) is a second-order partial differential equation.

At times we'll use prime notation for ordinary derivatives, so for example

$$(1.4) \qquad y'' + 3y' = \sin t$$

1

is somewhat abbreviated notation for the differential equation
$$\frac{d^2y}{dt^2} + 3\frac{dy}{dt} = \sin t.$$
The disadvantage of this shorter notation is that the independent variable—in this case it's now t—is not as clearly displayed.

1.2. What is a solution?

A first-order equation with independent variable t and dependent variable y written in **normal form** looks like

(1.5) $$\frac{dy}{dt} = f(t,y).$$

In other words, on the left side is the derivative $\frac{dy}{dt}$ and on the right-hand side is some expression involving the independent variable t and the dependent variable y. What does it mean to solve such an equation? We might begin by seeking an explicit function $y = y(t)$ which satisfies (1.5) on some open interval $(a,b) = \{t : a < t < b\}$; this is called an **explicit solution**. It expresses the dependent variable y solely in terms of the independent variable t (and constants). For example,
$$y(t) = \frac{3}{7}e^{2t}$$
is an explicit solution of

(1.6) $$\frac{dy}{dt} + 5y = 3e^{2t}$$

on $(-\infty, \infty)$. You can verify this by computing
$$\frac{dy}{dt} = \frac{6}{7}e^{2t}$$
and substituting into the differential equation (1.6) to see that the equation is satisfied. Similarly,
$$y(t) = \frac{3}{7}e^{2t} - e^{-5t}$$
is also an explicit solution on $(-\infty, \infty)$ of the same differential equation. In fact, the functions

(1.7) $$y(t) = \frac{3}{7}e^{2t} + ce^{-5t},$$

where c is any real number, are all explicit solutions of equation (1.6). Checking whether a particular explicitly given function is a solution to a differential equation is usually easy—plug it in and see if it works.

When we give an explicit solution to a differential equation we may include a reference to the interval or intervals on which the solution function is defined and satisfies the equation. Such an interval is called an **interval of validity** or **interval of existence**. An interval of validity may be $(-\infty, \infty)$, or it may necessarily be restricted in some way. Intervals of validity are usually taken to be open intervals. (Recall that an interval is open provided that it does not include its endpoints.) Sometimes it's obvious from the differential equation itself that $(-\infty, \infty)$ is not the interval of validity of any solution. For example, consider the equation, with independent variable x,
$$\frac{dy}{dx} + \frac{1}{x}y = \ln x.$$

1.2. What is a solution?

The best we can hope for is a solution that is valid on $0 < x < \infty$, which is the domain of $\ln x$. You can verify by substitution that

$$y = \frac{x}{2}\ln x - \frac{x}{4} + \frac{c}{x}$$

is a solution valid on $(0, \infty)$ for any choice of constant c. Sometimes though, there is nothing obvious in the form of the differential equation to alert you to possible restrictions on intervals of validity for its solutions. For example, the function

$$y = \frac{1}{1-x}$$

solves

$$\frac{dy}{dx} = y^2$$

on the interval $(1, \infty)$ and also on $(-\infty, 1)$ but not on any interval containing $x = 1$. To make things even more mysterious, you can check that

$$y = \frac{1}{2-x}$$

also solves the same equation, but this time on any interval which does not include $x = 2$. We'll return to this issue later.

One-parameter families of solutions and general solutions. A **one-parameter family** of functions is a collection of functions, such as $f_c(x) = ce^x$ or $g_c(x) = x^2 + c$, whose definitions change according to the value of a single parameter, which is c in the given examples. Keep in mind that the parameter c is a constant. Choosing different values for c results in different functions in the family. Solutions of certain first-order differential equations may be found via a process that concludes with integration, which naturally yields a one-parameter family of solutions. For instance, integrating on both sides of the differential equation $\frac{dy}{dx} = \cos x$ yields $y = \sin x + c$, a one-parameter family of solutions on $(-\infty, \infty)$. In Section 2.1, we describe a procedure, concluding with integration, that would produce the one-parameter family

(1.8) $$y = \frac{x}{2}\ln x - \frac{x}{4} + \frac{c}{x}$$

of solutions to

$$\frac{dy}{dx} + \frac{1}{x}y = \ln x$$

on the interval $(0, \infty)$. As another example, you can verify by substitution that

$$y = \frac{1}{c-x}$$

is a one-parameter family of solutions to $\frac{dy}{dx} = y^2$ on any open interval not including the constant c.

For a first-order differential equation, when a one-parameter family of solutions captures the form of *all* possible solutions, then we call it **the general solution**. Using techniques introduced in Section 2.1, it's possible to show that (1.8) is the general solution of $\frac{dy}{dx} + \frac{1}{x}y = \ln x$ on $(0, \infty)$. However, $y = 1/(c-x)$ is not the general solution of $\frac{dy}{dx} = y^2$: Observe that the constant function $y(x) = 0$ for all x is a solution of this equation, but there is no choice of c for which $y = 1/(c-x)$ becomes the zero function.

General solutions of certain nth-order differential equations are expressed as n-parameter families of functions. For example, in Exercise 25, you will show that on any nonempty open interval, the two-parameter family $y(t) = c_1 x + c_2$ is the general solution of the second-order equation $y''(x) = 0$.

Implicit solutions. Sometimes we have to be content with an implicit solution. By an implicit solution to a first-order differential equation

(1.9) $$\frac{dy}{dx} = f(x, y)$$

we mean an equation of the form

(1.10) $$g(x, y) = 0$$

containing no derivatives, so that any differentiable function $y = y(x)$ that satisfies (1.10) will also satisfy (1.9). Furthermore, we require that there does actually exist at least one such differentiable function $y = y(x)$, defined on some open interval (a, b) and satisfying (1.10), that is, with $g(x, y(x)) = 0$ for $a < x < b$. As an example, we claim that

(1.11) $$x^2 + y^2 - 4 = 0$$

is an implicit solution to

$$\frac{dy}{dx} = -\frac{x}{y}.$$

To see this, we assume that y is implicitly determined as a function of x by equation (1.11) and use implicit differentiation in (1.11) to obtain

$$2x + 2y\frac{dy}{dx} = 0,$$

which we may rewrite as

$$\frac{dy}{dx} = -\frac{x}{y}.$$

The equation (1.11) is simple enough that we can algebraically solve it for y in terms of x:

(1.12) $$y = \sqrt{4 - x^2} \quad \text{and} \quad y = -\sqrt{4 - x^2}.$$

This says that the equation (1.11), which implicitly defines y as one or more functions of x, includes the two explicit relations (1.12). The graph of the points (x, y) satisfying (1.11) is a circle as shown in Fig. 1.1. This circle is not the graph of a single function, since corresponding to some single x values are two different y values. However, parts of this circle *are* graphs of functions $y(x)$. For example, the top half of the circle is the graph of the first function in (1.12), and the bottom half of the circle is the graph of the second function there, and both of these are (explicit) solutions of the differential equation $\frac{dy}{dx} = -\frac{x}{y}$ on the interval $(-2, 2)$.

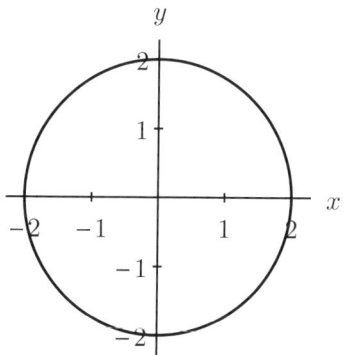

Figure 1.1. The circle $x^2 + y^2 = 4$.

1.2. What is a solution?

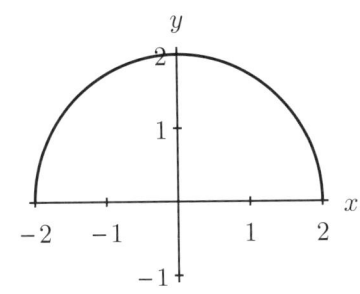

Figure 1.2. The top half of the circle is the graph of the function $y = \sqrt{4 - x^2}$.

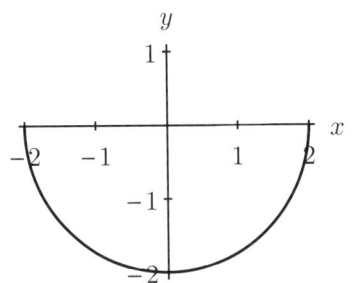

Figure 1.3. The bottom half of the circle is the graph of the function $y = -\sqrt{4 - x^2}$.

A theorem of advanced calculus, called the Implicit Function Theorem, says that under appropriate conditions an equation of the form $g(x, y) = 0$ will define y implicitly as a function of x. However, typically it is *not* possible to algebraically solve the equation $g(x, y) = 0$ explicitly for y in terms of x.

Example 1.2.1. We'll show that the equation

$$(1.13) \qquad 5y - y^5 + x^5 - 1 = 0$$

is an implicit solution to the differential equation

$$(1.14) \qquad \frac{dy}{dx} = \frac{x^4}{y^4 - 1}.$$

To verify this we differentiate (with respect to x) implicitly on both sides of equation (1.13) to obtain

$$5\frac{dy}{dx} - 5y^4\frac{dy}{dx} + 5x^4 = 0.$$

Solving this equation for $\frac{dy}{dx}$ gives

$$(5 - 5y^4)\frac{dy}{dx} = -5x^4$$

so that

$$\frac{dy}{dx} = \frac{x^4}{y^4 - 1},$$

which is equation (1.14). Although we cannot solve equation (1.13) explicitly for y as a function of x, for many purposes this implicit solution is an adequate substitute.

By looking at the (computer generated) graph of the relation $5y - y^5 + x^5 - 1 = 0$ in Fig. 1.4, we can see that sections of this graph do pass the "vertical line test" (for example, the thicker portion) and thus corresponding to each such section is *some* function y of x that is a solution to (1.14).

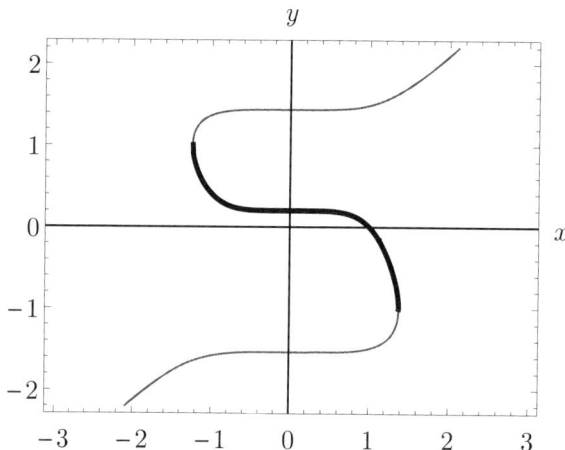

Figure 1.4. A portion of the graph of $5y - y^5 + x^5 - 1 = 0$.

Qualitative solutions. Sometimes our goal will be to produce a qualitative or graphical solution to the differential equation

(1.15)
$$\frac{dy}{dx} = f(x, y),$$

which emphasizes properties of solutions without finding explicit or even implicit formulas for them. One way to do this is by producing a **direction field** or **slope field** for the differential equation, which will allow us to sketch solutions to (1.15) even when solutions cannot be explicitly or implicitly described in terms of familiar functions. The idea is as follows: Equation (1.15) tells you that the slope of a solution at the point (x, y) is $f(x, y)$. Thus the graph of a solution—a **solution curve**—has to have the prescribed slope $f(x, y)$ at each point (x, y) it passes through. If we pick a point (x, y) in the plane, compute $f(x, y)$, and draw a short line segment (called a **field mark**) centered at (x, y) that has slope equal to the value $f(x, y)$, then a solution curve passing through (x, y) must be tangent to this field mark. If you choose a sampling of points in the plane and draw a field mark at each point, you produce a direction field for the differential equation. Solutions curves can then be sketched as they thread their way through the direction field.

Example 1.2.2. We'll sketch a direction field for the equation

(1.16)
$$\frac{dy}{dx} = 2 - \frac{1}{2}y$$

and show some solution curves in this picture. To construct a direction field for equation (1.16), notice that at any point with y-coordinate 4, the field marks will have slope 0, and thus at any point with y-coordinate 4 the field marks will be horizontal. At any point with y-coordinate 1, the field marks have slope $2 - \frac{1}{2} = \frac{3}{2}$, while at the points with y-coordinate 0, the field marks have slope 2, and so on. Continuing this process, we obtain a direction field as shown in Fig. 1.5. Fig. 1.6 shows how we use our direction field to sketch some solutions to equation (1.16). The field mark at any point is tangent to the solution curve passing through that point. Observe how Fig. 1.6 can be used to predict the qualitative behavior of solutions. For example, it appears that any solution $y = y(x)$ to equation (1.16) will have

$$\lim_{x \to \infty} y(x) = 4.$$

We will verify this prediction when we solve (1.16) analytically in Example 2.1.1 of Section 2.1.

1.2. What is a solution?

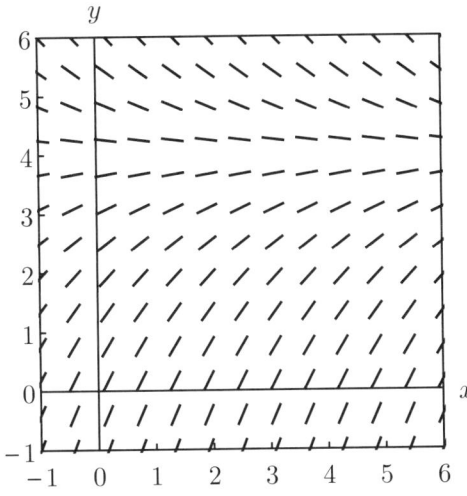

Figure 1.5. A portion of the direction field for $\frac{dy}{dx} = 2 - \frac{1}{2}y$.

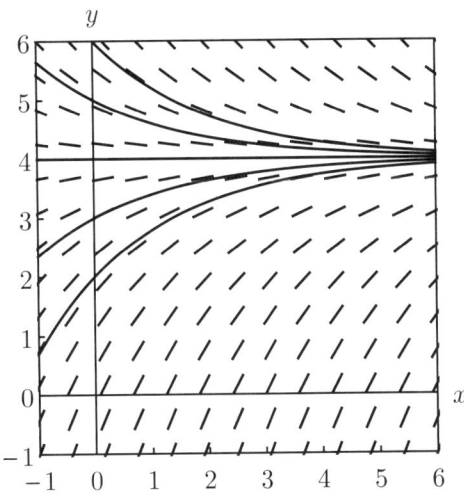

Figure 1.6. Direction field with some solution curves.

Equilibrium solutions. An equilibrium solution to a first-order differential equation of the form

$$\frac{dy}{dx} = f(x, y) \tag{1.17}$$

is a constant function $y = c$ that solves the equation. The corresponding solution curve is a horizontal line at height c. Since a constant function has derivative equal to 0 everywhere, we find equilibrium solutions of (1.17) by finding the constant values of y for which

$$0 = f(x, y).$$

In Example 1.2.2, the differential equation has equilibrium solution $y = 4$, and this appears in Fig. 1.6 as a horizontal line at height 4.

The next two examples further illustrate how equilibrium solutions are found.

Example 1.2.3. Find the equilibrium solutions to the differential equation

$$\frac{dy}{dx} = 4 - y^2.$$

Since $4 - y^2 = (2 - y)(2 + y)$, the constant values of y that make $4 - y^2$ equal to 0 are $y = 2$ and $y = -2$; these are the equilibrium solutions.

Example 1.2.4. The differential equation

$$\frac{dy}{dx} = 3xy - x,$$

or equivalently

$$\frac{dy}{dx} = x(3y - 1),$$

has one equilibrium solution, namely $y = \frac{1}{3}$.

When the independent variable in a differential equation is "time" (typically we would use the letter t instead of x in such a case), an equilibrium solution is one that doesn't change over time. While equilibrium solutions are in a sense "simple" solutions, we will see that they can sometimes have special significance for understanding the long-term behavior of other solutions. For example, Fig. 1.6 above illustrates solution curves approaching the equilibrium solution $y = 4$ as $x \to \infty$.

Initial value problems. When a differential equation $\frac{dy}{dx} = f(x, y)$ arises in an application, we'll often be looking for a solution whose graph passes through one specified point (x_0, y_0). In other words, we seek a solution satisfying $y = y_0$ when $x = x_0$. This requirement, written more briefly as $y(x_0) = y_0$, is called an **initial condition**, and the differential equation together with the initial condition is called an **initial value problem**.

Example 1.2.5. Solve the initial value problem
$$\frac{dy}{dx} + 5y = 3e^{3x}, \quad y(0) = 1.$$
We know from equation (1.7) that for any choice of the constant c, the function
$$y(x) = \frac{3}{7}e^{2x} + ce^{-5x}$$
solves the differential equation. In order that
$$y(0) = \frac{3}{7}e^{2(0)} + ce^{-5(0)} = 1$$
we choose $c = \frac{4}{7}$. Thus $y(x) = \frac{3}{7}e^{2x} + \frac{4}{7}e^{-5x}$ solves our initial value problem.

For an explicit solution $y = \varphi(x)$ of an initial value problem satisfying $y(x_0) = y_0$, *the interval of validity* is the largest open interval containing $x = x_0$ on which $\varphi(x)$ is a valid solution. For example, the interval of validity of the solution in Example 1.2.5 is $(-\infty, \infty)$.

Solving differential equations. Ideally, when we solve a differential equation we describe *all* of its solutions; however, providing an exact description of all solutions can be accomplished for only a limited variety of equations. Thus, especially in considering equations modeling real-world systems, our goal might be to solve an associated initial value problem exactly or approximately. In other situations, we may focus on developing an intuitive understanding of the behavior of solutions through qualitative means, or to find a family of solutions depending on one or more parameters. In all situations, we require the domain of a solution to a differential equation to be an open interval.

Recap. Differential equations are classified as ordinary differential equations or partial differential equations, and by their order.[1] We have seen that solutions of first-order differential equations of the form $dy/dx = f(x, y)$ may be described explicitly, implicitly, or, in a qualitative way, through direction fields.[2] While construction of a direction field is always possible, the nature of the differential equation will determine whether explicit or implicit solutions can be given in terms of the elementary functions that one encounters in a single-variable calculus course. (See Exercise 27.)

1.2.1. Differential equations and mathematical models.
Why study differential equations? One answer lies in the important role they play in constructing mathematics models for real-world systems. Mathematical modeling has long had an important role in physics and engineering. Its expanding importance in other fields is highlighted in a recent finding of the U.S. National Academies:

> Mathematical sciences work is becoming an increasingly integral and essential component of a growing array of areas of investigation in biology, medicine, social sciences, business, advanced design, climate, finance, advanced materials, and many more.[3]

---

[1] Later we will add a third kind of classification—linear or nonlinear.
[2] In Chapter 3 we will add a fourth type of solution—a numerical solution—to the list. Like a qualitative solution, this is an approximate solution and is typically produced with the aid of a computer.
[3] *The Mathematical Sciences in 2025*, The National Academies Press, 2013.

In this book we will model real-world problems from physics, engineering, biology, medicine, ecology, psychology, military science, and economics. It's not surprising that one or more differential equations frequently arise in mathematical models. A derivative is a rate of change and many fundamental relationships in the physical, biological, and social sciences ultimately involve certain quantities and their rates of change.

Mathematical modeling of a real-world system entails the following steps:

(I) **Set-up:** First we develop a mathematical description of the system—a mathematical model—that is, an equation or group of equations relating variables that provide information about the system.

(II) **Solution:** We solve the model, either exactly or approximately, and analyze our description, in particular how it suggests a change in one or more of its variables influences values of its other variable or variables.

(III) **Interpretation and Validity Assessment:** We use our results from the mathematical analysis of the model to answer questions and make predictions about the behavior of the system, and we test the model by comparing the way it suggests the system should function with experimental data or known facts.

(IV) **Refinement:** If the accuracy of the information it provides about the real-world system is deemed inadequate, we may repeat the process, creating a new model that we then solve, analyze, and interpret.

In the "Set-up" step, we often have to make many simplifying assumptions, extracting the most essential features from our original complex situation. We look for any relevant fundamental principles to guide the development of the model. In the physical sciences, there are often well-defined physical laws to be used. By contrast, in fields like biology and medicine we may have more freedom in choosing our starting assumptions, based on what seems reasonable to us. In the initial step, it may help to focus on a specific question we wish to answer and work to frame this question in mathematical form. For example, in Chapter 2 we will construct a mathematical model for the physiological effects of scuba diving, beginning with the specific question, "How long can a diver stay at a dive depth of d feet and still be able to ascend to the surface without risking the bends?"

The development and analysis phases are closely tied together. In particular, the process of analyzing a model should include preliminary testing:

- Do units of its equation or equations balance? If one side of an equation has, for example, units of Newtons per meter, so should the other.

- Are there general theorems that can be applied to confirm the existence of solutions of the model's equation(s)?

- For a situation in which the real-world system can behave in only one way, are there theorems that can be applied that ensure the system's mathematical model will produce a unique solution?

- Is the model simple enough that solving it (in some form) is feasible? Generally speaking, the more we want our model to accurately reflect the complexities of our real-world system, the more mathematically difficult it may be to solve. Even though the advent of powerful computer software for solving differential equations mitigates this to a certain extent, we still have to decide on the right balance of simplicity and realism in our model. If the model seems too complicated, various simplifying assumptions may be considered. For example, the time interval over which an accurate model is sought might be reduced in length, variables that are judged to be less influential than others might be eliminated, or higher powers of variables whose values are small might also be eliminated.

Ultimately, the real test of a mathematical model is how accurately its solutions, whether they are obtained approximately or exactly, describe the behavior of the system being modeled. Perhaps most satisfying is when our model allows us to make "unexpected" predictions about the system that turn out to be useful. A good model might influence future decision making; e.g., a model for tumor growth might have implications for future cancer treatment, a model of vaccination strategies might influence public health policies, or a model for retirement savings might influence decisions you make about investment practices.

As step (IV) indicates, the modeling process is often iterative. Having gone through the steps (I)–(III), we might then refine the model, perhaps making more sophisticated assumptions or including additional features, and repeat the steps. For example, if we first model the position of a falling object neglecting air resistance, on a second iteration we could include an air resistance term. Then too, we can make increasingly sophisticated assumptions about how this resistance term should be modeled. Any deficiencies we observe when we analyze our model can be addressed in subsequent refinements of the model, with each iteration of a model giving us new insights which guide these refinements.

In Chapter 2 we make a detailed study of first-order (ordinary) differential equations, devoting roughly equal energy to discussing

(a) methods for finding solutions of first-order equations (both exactly and approximately),

(b) underlying theory and conceptual understanding of such equations, and

(c) applications of first-order equations to a variety of modeling problems.

With regard to (c), we will see that different real-world problems can lead to the same differential equation based mathematical model. This has the fortunate consequence that the tools we develop to solve or analyze a particular differential equation may be relevant to modeling multiple real-world systems that initially appear to be quite different.

1.2.2. Exercises.

1. Verify that for each constant c, the function $y = ce^{2x}$ is a solution to the differential equation
$$\frac{dy}{dx} = 2y.$$
Is there a value of c so that the graph of $y = ce^{2x}$ passes through the point $(1, 3)$? Is there a value of c so that the graph passes through the point $(1, 0)$?

2. (a) The differential equation
$$\frac{dy}{dx} = \frac{1}{y^2(1-y)}$$
has implicit solution
$$4y^3 - 3y^4 = 12x.$$
Verify this by implicit differentiation.

   (b) Is $4y^3 - 3y^4 = 12x + c$ also an (implicit) solution for each constant c?

   (c) Is there a value of c so that the graph of the equation $4y^3 - 3y^4 = 12x + c$ passes through the point $(1, 2)$?

3. (a) The differential equation
$$(1.18) \qquad \frac{dy}{dx} = y(1-y)$$

1.2. What is a solution?

has implicit solution

(1.19)
$$\ln \frac{y}{1-y} = x,$$

provided $0 < y < 1$. Verify this by implicit differentiation.

(b) Solve equation (1.19) for y to find an explicit solution to (1.18).

(c) Show that for any constant c,
$$\ln \frac{y}{1-y} = x + c$$
is a solution to (1.18) provided $0 < y < 1$. What is the corresponding explicit solution?

4. (a) Verify that

(1.20)
$$\frac{1}{y} - x + c = 0$$

is an implicit solution to

(1.21)
$$\frac{dy}{dx} = -y^2$$

for each constant c.

(b) Equation (1.21) also has an equilibrium solution. Find it, and show that there is no choice for c in the one-parameter family in equation (1.20) that produces the equilibrium solution.

5. For what choices of c is $x^2 + y^2 - c = 0$ an implicit solution of $\frac{dy}{dx} = -\frac{x}{y}$?

6. (a) Show by substitution that $y = \sin x$ is a solution of the differential equation
$$\frac{dy}{dx} = \sqrt{1 - y^2}$$
on the interval $(-\pi/2, \pi/2)$.

(b) Is the function $y = \sin x$ a solution of the equation on the interval $(\pi/2, 3\pi/2)$? Explain. Hint: Remember that $\sqrt{a^2} = |a|$.

7. (a) Verify that for each real number c, the function

(1.22)
$$y = \frac{1}{c - x^2}$$

is a solution, on any interval in its domain, of the differential equation $\frac{dy}{dx} = 2xy^2$.

(b) What choice of c in (1.22) will give a solution to the initial value problem
$$\frac{dy}{dx} = 2xy^2, \quad y(0) = \frac{1}{4}?$$

(c) What is the interval of validity of the solution of the initial value problem you found in (b)?

8. There are two points on the graph in Fig. 1.4 where there is a vertical tangent line. Find them. Why can't the x-coordinates of either of these points be included in the interval of validity of any solution to the differential equation (1.14)?

9. (a) A group of equations and their corresponding direction fields appear below. Which picture goes with which equation?
   (i) $\frac{dy}{dx} = y^2 - 4$.
   (ii) $\frac{dy}{dx} = x - y$.
   (iii) $\frac{dy}{dx} = \sin^2 x$.
   (iv) $\frac{dy}{dx} = xy$.

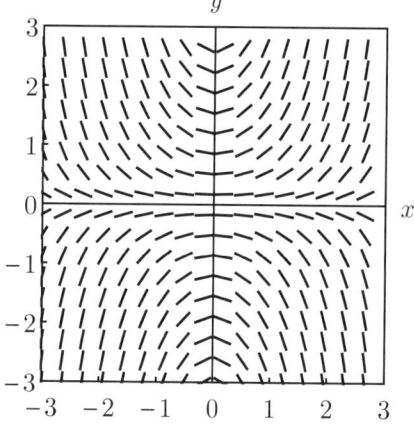

Figure 1.7. Direction field (A).

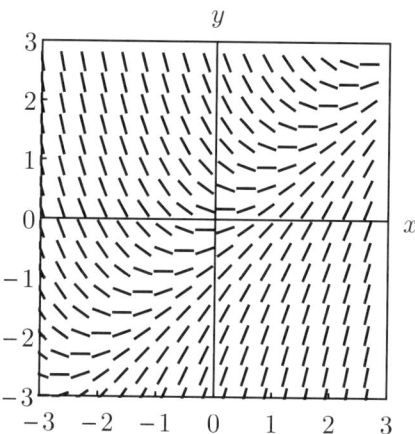

Figure 1.8. Direction field (B).

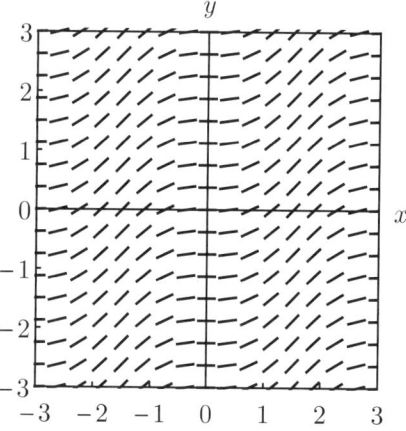

Figure 1.9. Direction field (C).

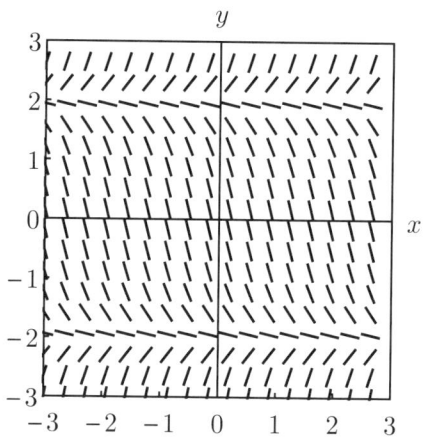

Figure 1.10. Direction field (D).

(b) The right-hand side of equation (i) above depends only on y. How do you see this fact reflected in the direction field you selected for (i)?

(c) The right-hand side of equation (iii) above depends only on x. How do you see this fact reflected in the direction field you selected for (iii)?

10. (a) Consider the differential equation

$$(1.23) \qquad \frac{dy}{dt} = ay - by^2.$$

Fig. 1.11 shows a portion of the direction field for equation (1.23) when $a = 0.2$ and $b = 0.005$. Using this, give an approximate sketch of the solution curve that passes through the point $(0, 10)$. Repeat for the solution curve passing through the point $(0, 50)$ and the solution through the point $(0, -1)$.

(b) Find all equilibrium solutions to equation (1.23) if $a = 0.2$ and $b = 0.005$. Does your answer fit with anything you can see in the direction field?

(c) The differential equation (1.23) will be used in Section 2.5 to model population growth; y is the population at time t. For such an interpretation, only the solution curves with $y \geq 0$ have any significance. What do you think happens to the solution curve passing through any point $(0, y_0)$ with $y_0 > 0$ as $t \to \infty$?

1.2. What is a solution?

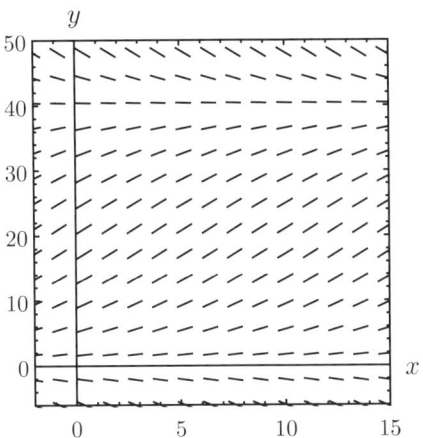

Figure 1.11. Direction field for $\frac{dy}{dx} = 0.2y - 0.005y^2$.

11. (a) The function $y = e^x$ has the property that its derivative is equal to itself. By writing the phrase "its derivative is equal to itself" in symbols, construct a first-order differential equation whose solutions include the function e^x.
    (b) Similarly, construct a first-order differential equation whose solutions include the function e^{5x}.
    (c) Construct a first-order differential equation whose solutions include e^{-2x}.

12. (a) The functions $y_1(t) = \sin t$ and $y_2(t) = \cos t$ have second derivatives equal to the negative of themselves. Use this fact to construct a second-order differential equation whose solutions include both y_1 and y_2.
    (b) Are $y(t) = -\sin t$ and $y(t) = -\cos t$ solutions to the differential equation you gave in (a)?
    (c) Is $y(t) = 2\sin t - 3\cos t$ a solution to the differential equation you gave in (a)?

13. (a) Verify that $3y^2 - 3x^2 + 2x^3 - 1 = 0$ is an implicit solution of

$$(1.24) \qquad \frac{dy}{dx} = \frac{x - x^2}{y}.$$

   (b) Substitute $x = 1$ into $3y^2 - 3x^2 + 2x^3 - 1 = 0$ and solve for y to see that both $(1, \sqrt{2/3})$ and $(1, -\sqrt{2/3})$ are points on the graph of $3y^2 - 3x^2 + 2x^3 - 1 = 0$, which is shown in Fig. 1.12.
   (c) Using Fig. 1.12, sketch a solution curve for the function $y = \varphi(x)$ that solves the initial value problem

$$\frac{dy}{dx} = \frac{x - x^2}{y}, \quad y(1) = \sqrt{\frac{2}{3}}.$$

   What is (approximately) the right-hand endpoint of the interval of validity of the solution to this initial value problem?

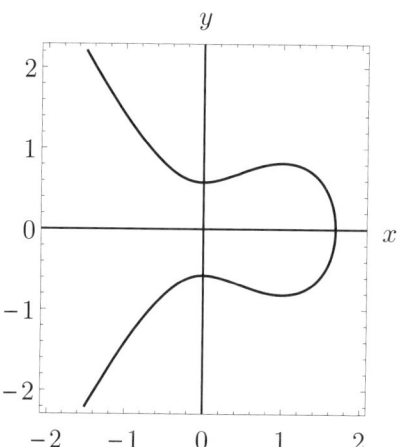

Figure 1.12. Graph of $3y^2 - 3x^2 + 2x^3 - 1 = 0$.

14. Two direction fields are shown in Figs. 1.13 and 1.14. One is for equation (1.14) (discussed in Example 1.2.1). The other is for equation (1.24) in Exercise 13. By comparing the direction fields with the implicit solutions shown in Figs. 1.4 and 1.12, decide which is which.

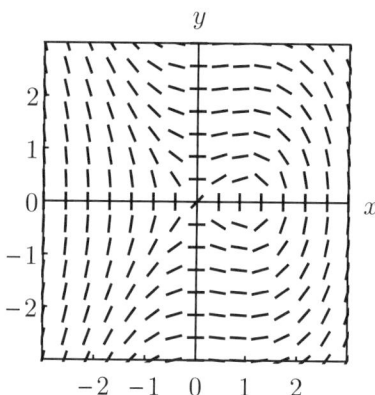

Figure 1.13. Direction field (A).

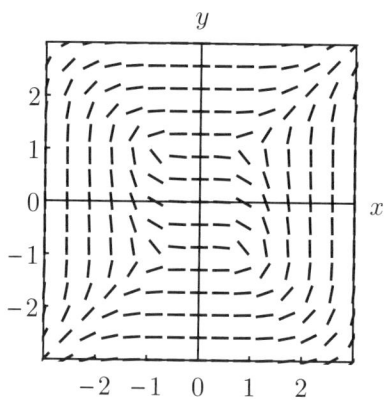

Figure 1.14. Direction field (B).

In Exercises 15–21, find all equilibrium solutions to the given differential equation.

15. $\frac{dy}{dx} = y^2 - y$.
16. $\frac{dy}{dx} = \cos y$.
17. $\frac{dy}{dx} = 2 - y$.
18. $\frac{dy}{dx} = 2 - x$.
19. $\frac{dy}{dx} = x - 2yx$.
20. $\frac{dy}{dx} = (y-4)(y-1) + 2$.
21. $\frac{dy}{dx} = \ln y$.

22. If $y = k$ is an equilibrium solution to $\frac{dy}{dx} = f(x,y)$, then how would this be reflected in a corresponding direction field?

23. In this problem, we look for equilibrium solutions to the equation
$$\frac{dy}{dx} = y(y-3) + k$$
where k is a constant.
    (a) Find all equilibrium solutions if $k = 0$.

1.2. What is a solution?

    (b) Find all equilibrium solutions if $k = \frac{5}{4}$.
    (c) For what values of k will there be no equilibrium solutions?

24. Show that $y = c_1 + c_2 e^x + c_3 e^{2x}$ is a three-parameter family of solutions to $y''' - 3y'' + 2y' = 0$ on $(-\infty, \infty)$.

25. Let $a < b$ and suppose that $y''(x) = 0$ for all x in (a, b).
    (a) Show that there is a constant c_1 so that $y'(x) = c_1$ for all x in (a, b).
    (b) Show that there is a constant c_2 so that $y(x) = c_1 x + c_2$ for all x in (a, b). Hint: Consider the function $g(x) = y(x) - c_1 x$ on (a, b).
    (c) Conclude that $y(x) = c_1 x + c_2$ is the general solution of $y''(x) = 0$ on (a, b).

26. Find a two-parameter family of solutions to $y'' = 6x + e^{3x}$.

27. Using the Fundamental Theorem of Calculus, verify that

$$(1.25) \quad y = \int_0^x e^{-t^2} \, dt$$

is a solution of the differential equation

$$\frac{dy}{dx} - e^{-x^2} = 0.$$

    Remarks: You may not view (1.25) as a satisfying solution to the preceding differential equation; however, one can show the equation does not have a solution that is an *elementary function*. Roughly speaking, an elementary function is one built by applying a *finite number* of elementary operations (addition, subtraction, multiplication, and division) and functional compositions using, as building blocks, the following types of functions: polynomial, exponential, trigonometric, and their inverses.
    That integration can be used as a tool for solving differential equations is not at all surprising because differentiation and integration are inverse processes. However, there are simple differential equations, such as $\frac{dy}{dx} + y^2 = x^2$, whose solutions cannot be expressed in terms of finite combinations of integrals of elementary functions.

28. (a) Let $a < b$. Show that the first-order equation

$$y^2 + \left(\frac{dy}{dx}\right)^2 = 0$$

       has one and only one solution on (a, b).
   (b) Modify the differential equation in (a) to give an example of a differential equation with no solution on any interval (a, b), $a < b$.

29. In 1798, T. R. Malthus proposed modeling human population growth by the following principle: If a population has size P at time t, then the rate of change of P with respect to t is proportional to P. Write Malthus's principle as a differential equation. What is the independent variable? The dependent variable?

30. Newton's law of cooling states that a heated object cools at a rate proportional to the difference between its temperature and the surrounding, or ambient, temperature.
    (a) Let T be the temperature of the object at time t and let A be the (constant) ambient temperature. Give a differential equation that expresses Newton's law of cooling.
    (b) Find an equilibrium solution of the differential equation obtained in (a).

31. In a rote memorization exercise, a student is asked to memorize a list of M three-digit numbers. We'll denote the number of items memorized by time t by $L(t)$. One model for memorization says that

$$\frac{dL}{dt} = k(M - L)$$

where k is a positive constant, so that the rate of memorization is proportional to the number of items still to be memorized.

(a) The value of the constant k is different for different people. If Sally memorizes more quickly than Sam, is her k value smaller or larger than Sam's?

(b) What initial condition corresponds to the assumption that initially no items have been memorized?

(c) Suppose we want to modify our model to include the fact that some items are forgotten after being initially memorized. If we assume that the "rate of forgetting" is proportional to the amount memorized, what new differential equation do we have for $L(t)$?

1.3. More on direction fields: Isoclines

To produce a direction field by hand can be a labor-intensive undertaking, and a computer is often used instead. However, by being a bit systematic we can often get a rough picture of the direction field by hand without a daunting amount of computation. We use the idea of isoclines. The **isoclines** for the differential equation

$$\frac{dy}{dx} = f(x, y)$$

are the curves with equations

$$f(x, y) = k$$

for different choices of the constant k. For a particular choice of k, at every point of the curve $f(x, y) = k$ the field marks have slope k; thus, any solution of the differential equation which intersects this isocline at a point P will have slope k at P. The isocline corresponding to the choice $k = 0$ is called a **nullcline**. The field marks at any point of a nullcline are horizontal.

Example 1.3.1. We'll find the isoclines for the differential equation

$$\frac{dy}{dx} = -2x$$

and use them to sketch a direction field. The isoclines have the form

$$-2x = k, \text{ or } x = -\frac{k}{2}.$$

Thus for each choice of k the isocline $x = -\frac{k}{2}$ is a vertical line. Choosing some specific values for k, we get Table 1.1.

Table 1.1. Some isoclines for $\frac{dy}{dx} = -2x$.	
value of k	isocline $x = -k/2$
-3	$x = \frac{3}{2}$
-2	$x = 1$
-1	$x = \frac{1}{2}$
0	$x = 0$
1	$x = -\frac{1}{2}$
2	$x = -1$
3	$x = -\frac{3}{2}$

1.3. More on direction fields: Isoclines

We can now graph an isocline and put field marks on it, all having the same slope k associated to that isocline. If we do this for some representative values of k as in Table 1.1, we produce a direction field as shown in Fig. 1.15. For example, with the choice $k = -3$ we draw the isocline $x = \frac{3}{2}$ and at points on this vertical line we show field marks all having slope -3. The nullcline is the vertical line $x = 0$ and all field marks at points along this line are horizontal. Using the field marks as a guide, we can then sketch some solution curves, as shown in Fig. 1.15. Three solution curves are also drawn so that they have the indicated slope as they cross the isoclines. These solution curves look parabolic, and indeed our differential equation $\frac{dy}{dx} = -2x$ can be solved explicitly by integrating both sides with respect to x to obtain $y = -x^2 + c$ for arbitrary constant c. The solution curves shown in Fig. 1.15 are parabolas from this family.

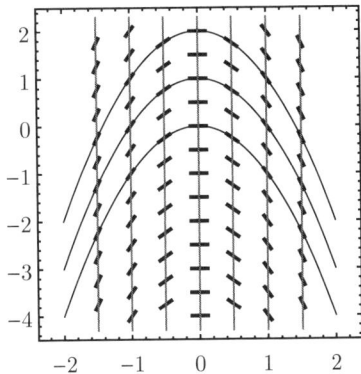

Figure 1.15. Isoclines, field marks, and three solution curves.

The next example illustrates the procedure for a somewhat more complicated differential equation.

Example 1.3.2. We'll sketch a direction field and some solution curves for the differential equation
$$\frac{dy}{dx} = y - x^2.$$
We begin by identifying some isoclines.

Table 1.2. Some isoclines for $\frac{dy}{dx} = y - x^2$.	
value of k	isocline curve $y - x^2 = k$
-1	$y = x^2 - 1$
0	$y = x^2$
$1/2$	$y = x^2 + 1/2$
1	$y = x^2 + 1$
$3/2$	$y = x^2 + 3/2$
2	$y = x^2 + 2$
$-1/2$	$y = x^2 - 1/2$

Fig. 1.16 shows these isoclines, all of which are parabolas, and some field marks. Notice that the parabola $y = x^2$ is the nullcline, and all field marks at points on this parabola are horizontal. Other isoclines are translates of this parabola, either up (if $k > 0$) or down (if $k < 0$). A few

solution curves are also sketched in Fig. 1.16. The slope of a solution curve matches the slope of the direction field at every point.

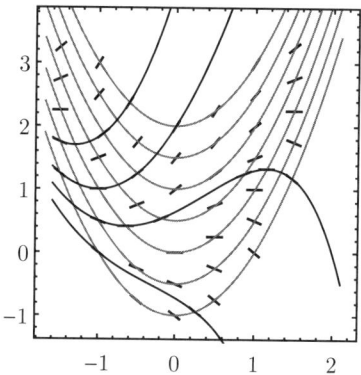

Figure 1.16. Isoclines, field marks, and three solution curves.

1.3.1. Computer-generated direction fields. To produce a direction field using a computer algebra system (CAS), one need only know the correct command or commands to enter. For example, the open-source CAS *SageMath* will plot a direction field for

$$\frac{dy}{dx} = \sin(xy), \tag{1.26}$$

over the rectangle $-4 \leq x \leq 4, -3 \leq y \leq 3$, upon execution of the commands

```
x,y=var('x y');
plot_slope_field(sin(x*y), (x,-4,4),(y,-3,3))
```

The result is

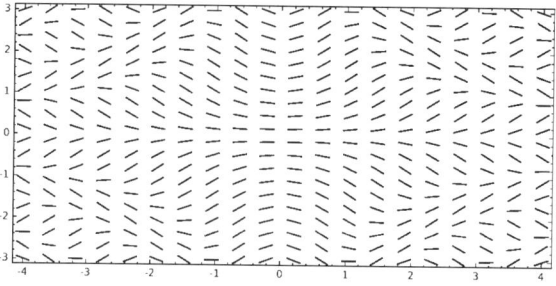

Figure 1.17. Direction field for $\frac{dy}{dx} = \sin(xy)$ produced by *SageMath*.

Commands for generating the same field in, respectively, *Maple*, *Mathematica*, and MATLAB's *MuPAD*, are displayed below.

CAS	Command or Commands Producing a Direction Field for Equation (1.26)
Maple	`with(DEtools);` `DEplot(diff(y(x),x)=sin(x*y(x)), y(x), x=-4..4,y=-3..3);`
Mathematica	`VectorPlot[{1, Sin[x y]}, {x, -4, 4}, {y, -3, 3}]`
MuPAD	`plot( plot::VectorField2d([1, sin(x*y)], x =-4..4,y=-3..3));`

The plots generated by the commands in the preceding table are shown in Fig. 1.18.

1.3. More on direction fields: Isoclines

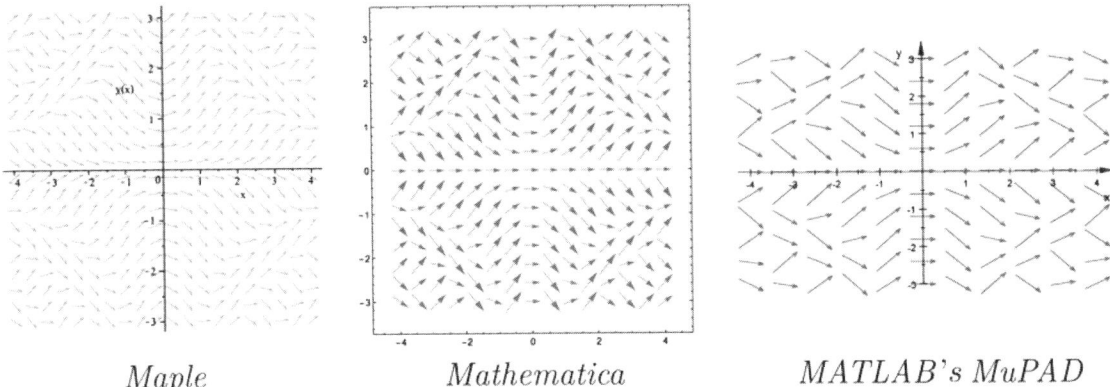

      Maple   *Mathematica*   *MATLAB's MuPAD*

Figure 1.18. Direction field for $\frac{dy}{dx} = \sin(xy)$.

Note that in these plots, the field marks are arrows rather than line segments. To generate segments as field marks, one needs to add an optional argument, e.g., in *MuPAD*, the optional argument is `TipLength = 0`. Optional arguments may allow you not only to modify the style of the field marks, but also their density and/or location. For, example Fig. 1.19 is a direction field plot for $\frac{dy}{dx} = \sin(xy)$ produced by the *MuPAD* command

```
plot( plot::VectorField2d([1, sin(x*y)], x =-4..4,y=-3..3, TipLength=0, Mesh=[30, 20]));
```

Here the two optional arguments specifying "`TipLength`" and "`Mesh`" produce a 20-by-30 array of equally spaced field marks that are segments.

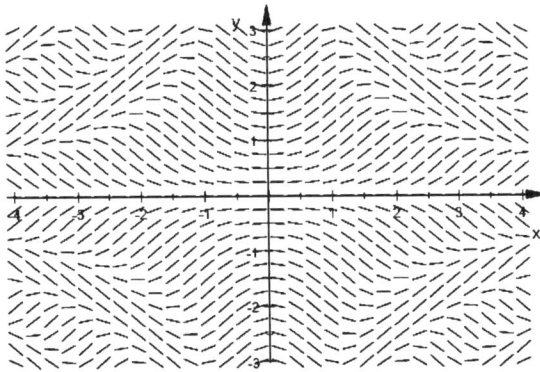

Figure 1.19. Direction field for $\frac{dy}{dx} = \sin(xy)$ produced by MATLAB's *MuPAD*.

The *Mathematica* command

```
VectorPlot[{1,y - x}, {x,-4,4}, {y,-4,4}, VectorStyle->{Black,Thick,"Segment"},
VectorScale->{.02,Automatic,None},VectorPoints->Flatten[Table[{n,n+k},{n,-4,4},{k,-7,7}],1]]
```

yields the following:

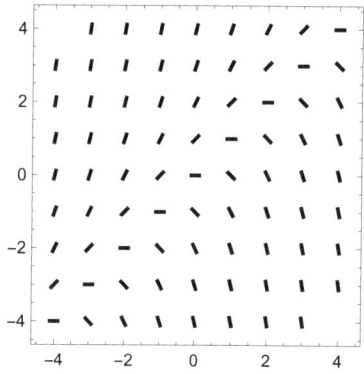

Figure 1.20. Direction field for $\dfrac{dy}{dx} = y - x$ produced by *Mathematica*.

In the command, the optional arguments "`VectorStyle`", "`VectorScale`", and "`VectorPoints`" are used to generate a direction field for $dy/dx = y - x$ consisting of field marks that are short segments (black, thick) centered at points on the isoclines $y = x + k$ for slopes of k, where k is an integer between -7 and 7 (including -7 and 7).

Your CAS may also provide optional arguments allowing you to add solution curves to your direction field plot. For example, Fig. 1.21 is a direction field plot for $\frac{dy}{dx} = \sin(xy)$ produced by the *Maple* commands

```
with(DEtools):
DEplot(diff(y(x),x)=sin(x*y(x)), y(x), x = -4 .. 4, y=-3..3, [[y(0)=-2],
                [y(0) = -1], [y(0) = 0], [y(0) = 1], [y(0)=2]],arrows=line);
```

The inclusion of the list `[[y(0)=-2], [y(0) = -1], [y(0) = 0], [y(0) = 1], [y(0)=2]]` in the command adds to the direction field plot solution curves[4] satisfying the initial conditions in the list.

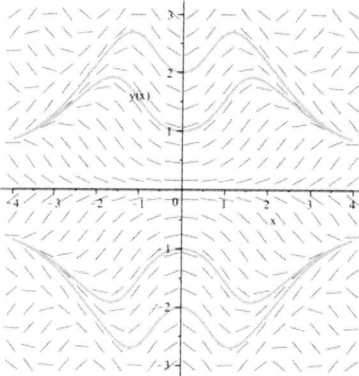

Figure 1.21. *Maple* plot of the direction field for $dy/dx = \sin(xy)$, along with some solution curves.

Computer algebra systems typically provide ways to combine plots produced by separate commands, allowing one to, e.g., add isoclines to a direction field plot:

---

[4]Generated by numerical methods—see Chapter 3.

1.3. More on direction fields: Isoclines

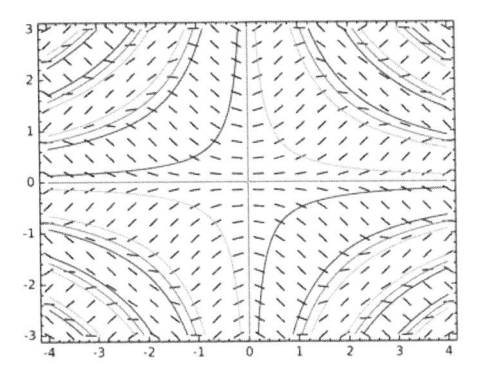

The preceding plot, featuring a direction field for $dy/dx = \sin(xy)$ together with isoclines for slopes of $k = -1 - 1/2, 0, 1/2,$ and 1, was produced via the following sequence of *SageMath* commands:

```
x,y=var('x y')
p1=plot_slope_field(sin(x*y), (x,-4,4),(y,-3,3))
p2=contour_plot(sin(x*y), (x, -4, 4), (y, -3, 3),contours=(-1,-1/2, 0, 1/2, 1),fill=false)
show(p1+p2)
```

Several exercises below invite you to use any computer algebra system to which you have ready access to generate direction fields.

1.3.2. Exercises.

1. Consider the differential equation

   $$\text{(1.27)} \qquad \frac{dy}{dx} = y.$$

   (a) Verify by substitution that any function

   $$y = Ce^x,$$

   where C is a constant, is a solution to equation (1.27).
   (b) Sketch the isoclines for equation (1.27) corresponding to the slopes $k = 0, \frac{1}{2}, 1, 2, -\frac{1}{2}, -1,$ and -2, and show the field marks on each of these isoclines.
   (c) Using your picture from (b), sketch some solution curves. Is your picture qualitatively consistent with the explicit solutions given in (a)?
   (d) Does equation (1.27) have any equilibrium solutions? If so, show them in your sketch in (c).

2. Consider the differential equation

   $$\text{(1.28)} \qquad \frac{dy}{dx} = \frac{x^2}{y}.$$

   (a) Sketch the isoclines for the equation in (1.28) corresponding to the slopes $k = 0, \frac{1}{2}, 1, 2, -\frac{1}{2},$ -1, and -2, and show the some field marks on each of these isoclines.
   (b) Using your sketch from (a), show several solution curves on the direction field.
   (c) If the field mark at point (a, b) has slope k, what is the slope of the field mark at point $(a, -b)$? What does this say about the relationship between the direction field above the x-axis and the direction field below the x-axis? How does this relationship translate into information about the solution curves?

3. Consider the differential equation

   $$\text{(1.29)} \qquad \frac{dy}{dx} + 2y = x.$$

(a) Sketch the isoclines corresponding to the slopes $k = -2, -1, -\frac{1}{2}, 0, \frac{1}{2}, 1, 2$, and show some field marks on each isocline.

(b) Using your work from (a), show several solution curves to equation (1.29) on the direction field.

(c) One of the isoclines is itself also a solution curve. Which one?

4. (a) Show that the isoclines for the differential equation
$$x\frac{dy}{dx} - 2y = 2$$
have equations
$$\frac{2}{x} + \frac{2y}{x} = k$$
for constant k.

(b) Sketch the isoclines corresponding to $k = 0, \frac{1}{2}, 1, 2, -\frac{1}{2}, -1$, and -2, and show the field marks at points on these isoclines.

(c) Using your sketch from (b), show several solutions curves.

(d) Verify by substitution that
$$y = -1 + cx^2$$
is a solution to
$$x\frac{dy}{dx} - 2y = 2$$
for each constant c. Does this fit with your picture in (c)?

5. A direction field for a first-order differential equation $\frac{dy}{dx} = f(x,y)$ is shown in Fig. 1.22, along with three isoclines labeled A, B and C.

(a) Which of these isoclines is a nullcline?

(b) Which isocline corresponds to a slope value $k > 0$?

(c) Which corresponds to a slope value $k < 0$?

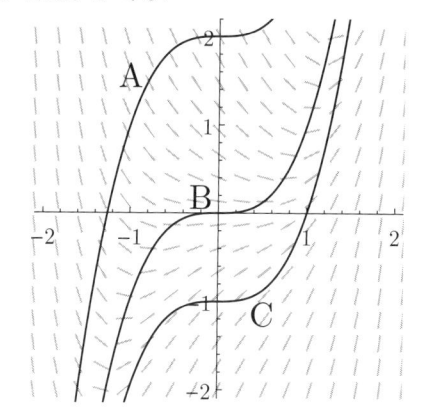

Figure 1.22. Direction field and three isoclines.

6. Consider the differential equation

(1.30) $$\frac{dy}{dx} = (y-x)^2 - 1.$$

(a) Plot the nullclines of (1.30) and notice that they separate the plane into three regions. In each region, determine whether the solution curves will be rising or falling.

(b) Add to your plot in (a) the isoclines corresponding to slopes $k = -1, 1$, and 3. Add field marks along these isoclines, as well as along the nullclines.

(c) Which of the isoclines that you have plotted are solution curves?

1.3. More on direction fields: Isoclines

(d) It's possible to show that solution curves of (1.30) do not intersect (see the No Interesection Theorem in Section 2.4). Assuming this, add to your plot a sketch of the solution of (1.30) satisfying the initial condition $y(0) = 0$.

(e) Recall that a slant asymptote of the graph of $f(x)$ is a line $y = mx + b$, $m \neq 0$, such that either
$$\lim_{x \to \infty} [f(x) - (mx + b)] = 0 \quad \text{or} \quad \lim_{x \to -\infty} [f(x) - (mx + b)] = 0.$$
What do you believe are slant asymptotes of the solution curve you plotted in (d)?

7. (CAS) Use a computer algebra system to plot a direction field for

(1.31) $$\frac{dy}{dx} = 1 + \sin(x + \pi - y).$$

Then use the field to obtain a one-parameter family of solutions to (1.31), where the parameter takes integer values.

8. (CAS) Consider the differential equation

(1.32) $$\frac{dy}{dx} = \ln(1 + (y - \sin x)^2) + \cos x.$$

(a) Use a computer algebra system to plot a direction field for (1.32) over, say, the rectangular region $-6 \leq x \leq 6$, $-4 \leq y \leq 4$.
(b) Use the field or inspection to obtain a solution to (1.32).
(c) Plot the direction field of (a) together with solution of (b).

9. (CAS) Suppose that

(1.33) $$\frac{dP}{dt} = \left(\frac{t}{t+4}\right) P(t)(3 - P(t))$$

models a population of bacteria in a certain culture at time $t \geq 0$, where t is measured in hours and P in thousands.
(a) Plot a direction field for (1.33) over, say, the rectangular region $0 \leq t \leq 10$, $0 \leq P \leq 4$.
(b) Based on the field, what do you believe to be the long-term population of bacteria in the culture assuming the initial population is positive?
(c) Learn how to use your computer algebra system to solve initial value problems and then have it solve
$$\frac{dP}{dt} = \left(\frac{t}{t+4}\right) P(t)(3 - P(t)), \quad P(0) = 1,$$
recording the solution given.
(d) Find $\lim_{t \to \infty} P(t)$, where P is your solution of (c). Is the answer consistent with that for part (b)?
(e) Plot the direction field of (a) together with the solution of (c).

Chapter 2

First-Order Equations

2.1. Linear equations

The goal of this section and the next is to develop two basic methods for finding solutions to certain types of first-order differential equations. After the classification of a differential equation as either an ordinary differential equation or a partial differential equation, and after deciding on the order of the equation, the next crucial step is to classify the equation as linear or nonlinear. A first-order ordinary differential equation is **linear** if it can be written in the form

(2.1) $$\frac{dy}{dt} + p(t)y = g(t);$$

we refer to this as the **standard form** of a first-order linear equation. Here t is the independent variable (we use the letter t frequently because in applications it often plays the role of time) and y is the dependent variable. If we compare this standard form for a first-order linear equation with the normal form $\frac{dy}{dt} = f(t,y)$ of *any* first-order equation, we see that

$$f(t,y) = -p(t)y + g(t),$$

so that $f(t,y)$ is a linear function of the variable y. The **coefficient functions** $p(t)$ and $g(t)$ are usually assumed to be continuous functions, at least on some open interval $a < t < b$. If the function $g(t)$ happens to be the constant 0, equation (2.1) is said to be **homogeneous**. A first-order equation that cannot be written in the form (2.1) is called **nonlinear**. For example, the equation

$$\frac{dy}{dt} = \cos t + 3y$$

is linear (with $p(t) = -3$ and $g(t) = \cos t$ when we put the equation in standard form), while the equation

$$\frac{dy}{dt} = \cos y + 3t$$

is nonlinear. The equation

$$2ty + t\cos t + (t^2+1)\frac{dy}{dt} = 0$$

is also linear, since upon dividing by t^2+1 it can be rewritten as

$$\frac{dy}{dt} + \frac{2t}{t^2+1}y = -\frac{t\cos t}{t^2+1},$$

and we see that it has the form of equation (2.1) with
$$p(t) = \frac{2t}{t^2+1} \quad \text{and} \quad g(t) = -\frac{t\cos t}{t^2+1}.$$
Notice that p and g are continuous on $(-\infty, \infty)$. The equations
$$\frac{dy}{dt} = y^2$$
and
$$\frac{dy}{dt} + t\sin y = t^3$$
are examples of nonlinear equations. We will generally assume that we can write a nonlinear equation in the normal form
$$\frac{dy}{dt} = f(t, y);$$
in other words, we assume we can isolate the derivative on one side of the equation.

There is a procedure for solving first-order *linear* differential equations. We'll illustrate this with an example before describing the method in general. This example uses the same differential equation we have already discussed qualitatively in Example 1.2.2 (except now we use t for the independent variable instead of x).

Example 2.1.1. Consider the equation

(2.2) $$2\frac{dy}{dt} = 4 - y,$$

which we rewrite in standard form as
$$\frac{dy}{dt} + \frac{1}{2}y = 2.$$
Now suppose we multiply both sides of this equation by $e^{t/2}$ (we'll soon know how this factor was chosen), obtaining

(2.3) $$e^{t/2}\frac{dy}{dt} + \frac{1}{2}e^{t/2}y = 2e^{t/2}.$$

Next we make the crucial observation that the sum on the left-hand side of (2.3) is exactly
$$\frac{d}{dt}(e^{t/2}y),$$
since by the product rule
$$\frac{d}{dt}(e^{t/2}y) = e^{t/2}\frac{dy}{dt} + \frac{1}{2}e^{t/2}y.$$
This means that we may rewrite (2.3) as

(2.4) $$\frac{d}{dt}\left(e^{t/2}y\right) = 2e^{t/2},$$

which tells us that
$$e^{t/2}y = \int 2e^{t/2}dt$$
or
$$e^{t/2}y = 4e^{t/2} + C.$$
The constant of integration C is an arbitrary real number. Solving for y we have an explicit solution to (2.2):

(2.5) $$y = 4 + Ce^{-t/2}.$$

Because we require the domain of a solution of a differential equation to be an interval and because every antiderivative of the function on the right-hand side of equation (2.4) has the form $4e^{t/2} + C$

2.1. Linear equations

for some constant C, every solution of (2.2) is given by equation (2.5) for some particular choice of C. Thus equation (2.5) gives the **general solution** of equation (2.2). The function $\mu(t) = e^{t/2}$ that got us started is called an **integrating factor**. We'll shortly see a method for how it was determined, but for right now notice what role it played: Upon multiplying the differential equation (written in standard form) by the integrating factor, we have an equation whose left-hand side is the derivative of the product of the integrating factor and the dependent variable y. Observe that we obtained not just one solution, but an infinite family of solutions with one arbitrary parameter C. Also notice that for any choice of a value for C,

$$\lim_{t \to \infty} 4 + Ce^{-t/2} = 4,$$

so that all solutions tend to 4 as $t \to \infty$. We had already guessed at this property when we discussed qualitative solutions to (2.2) in Section 1.1.

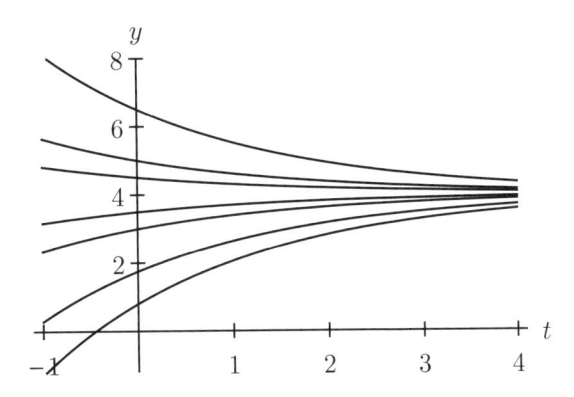

Figure 2.1. Some solutions of $\frac{dy}{dt} + \frac{1}{2}y = 2$.

Finding an integrating factor. How is an integrating factor found for a first-order linear differential equation? Consider the equation

$$\frac{dy}{dt} + p(t)y = g(t)$$

where $p(t)$ and $g(t)$ are continuous. Suppose we multiply both sides of this equation by some function $\mu(t)$, obtaining

(2.6) $$\mu(t)\frac{dy}{dt} + \mu(t)p(t)y = \mu(t)g(t).$$

How should $\mu(t)$ be chosen so that the left-hand side of (2.6) is exactly

$$\frac{d}{dt}(\mu(t)y)?$$

This requirement on $\mu(t)$ is precisely that we should have

(2.7) $$\mu(t)\frac{dy}{dt} + \mu(t)p(t)y = \frac{d}{dt}(\mu(t)y) = \mu(t)\frac{dy}{dt} + \mu'(t)y;$$

the last equality in (2.7) follows from applying the product rule to the derivative in the middle. Clearly, equation (2.7) will be satisfied exactly when

$$\mu'(t)y = \mu(t)p(t)y;$$

thus (2.7) will hold if

(2.8) $$\frac{\mu'(t)}{\mu(t)} = p(t).$$

Integrating both sides of (2.8) with respect to t gives

$$(2.9) \qquad \ln|\mu(t)| = \int p(t)dt + c$$

where c is an arbitrary constant. Upon exponentiating both sides of (2.9) we obtain

$$|\mu(t)| = e^{\int p(t)dt + c} = e^c e^{\int p(t)dt}.$$

If c is an arbitrary real number, e^c is an arbitrary positive number. Removing the absolute value bars on the left either does nothing (if $\mu(t) > 0$) or multiplies by -1 (if $\mu(t) < 0$). Thus

$$\mu(t) = (\pm e^c) e^{\int p(t)dt}.$$

If we replace $\pm e^c$ by C (an arbitrary nonzero real number), we get

$$\mu(t) = C e^{\int p(t)dt}.$$

Since we only need one integrating factor, we can choose $C = 1$ for convenience. Here's our conclusion:

$$(2.10) \qquad \text{An integrating factor for } \tfrac{dy}{dt} + p(t)y = g(t) \text{ is } \mu(t) = e^{\int p(t)dt}.$$

For instance, if we apply (2.10) in Example 2.1.1, where $p(t) = 1/2$, we see that it yields our integrating factor

$$\mu(t) = e^{\int \frac{1}{2} dt} = e^{\frac{1}{2}t}.$$

We now have the nice formula (2.10) for an integrating factor $\mu(t)$ for the equation

$$(2.11) \qquad \frac{dy}{dt} + p(t)y = g(t),$$

which we can use to describe the general solution to (2.11). With $\mu(t)$ given as in equation (2.10), we have

$$\frac{d}{dt}(\mu(t)y) = \mu(t)g(t).$$

Integrating both sides of this equation with respect to t gives

$$\mu(t)y = \int \mu(t)g(t)dt + C$$

for arbitrary constant C. We can solve for y to get

$$y(t) = \frac{1}{\mu(t)} \int \mu(t)g(t) + \frac{C}{\mu(t)}$$

$$(2.12) \qquad = e^{-\int p(t)dt} \int e^{\int p(t)dt} g(t)dt + Ce^{-\int p(t)dt}.$$

Faced with a specific first-order linear equation, you could in theory use (2.12) directly to solve it. However it is usually preferable to simply follow the *procedure* used to derive (2.12) to solve your linear differential equation, starting by multiplying on both sides by an integrating factor.

Example 2.1.2. We solve the initial value problem

$$ty' + 2y = t^2, \quad y(2) = 3$$

on the interval $0 < t < \infty$. Writing this linear equation in standard form

$$(2.13) \qquad y' + \frac{2}{t}y = t,$$

2.1. Linear equations

we see that
$$\mu(t) = e^{\int \frac{2}{t}dt} = e^{2\ln t} = e^{\ln(t^2)} = t^2$$
will serve as an integrating factor. Multiplying (2.13) by this integrating factor yields
$$t^2 y' + 2ty = t^3$$
or
$$\frac{d}{dt}(t^2 y) = t^3,$$
so that
$$t^2 y = \int t^3 dt = \frac{t^4}{4} + C$$
and
$$y = \frac{t^2}{4} + \frac{C}{t^2}.$$
To satisfy the initial condition we must have $y = 3$ when $t = 2$, so that
$$3 = 1 + \frac{C}{4}$$
and $C = 8$. Thus this initial value problem has solution
$$y = \frac{t^2}{4} + \frac{8}{t^2}.$$

A brief review of some calculus. Before proceeding to our next example we review a bit about the **Fundamental Theorem of Calculus**. Recall that the notation
$$\int x^2 dx$$
means "all antiderivatives of x^2"; this is the infinite family of functions
$$\frac{x^3}{3} + C$$
as C ranges over all real numbers. The notation
$$\int_0^x t^2 dt$$
denotes a *specific* function whose derivative is x^2 (the letter t used in the integrand is sometimes called a dummy variable; it is introduced because "x" has already been used as the upper limit of integration). If we set

(2.14) $$F(x) = \int_0^x t^2 dt,$$

then we know that
$$F'(x) = x^2.$$
This is precisely the Fundamental Theorem of Calculus, which says that for any constant a and continuous function g,
$$\frac{d}{dx}\left(\int_a^x g(t) dt\right) = g(x).$$
We also know that if $F(x)$ is as defined in (2.14), then $F(0) = 0$. In other words, F is the specific antiderivative of x^2 having value 0 at $x = 0$. The function
$$G(x) = \int_1^x t^2 dt$$

is another specific function having derivative x^2; it has the additional property that $G(1) = 0$. With this as preparation, we turn to another example.

Example 2.1.3. Solve the initial value problems

(a)
$$\frac{dy}{dx} - 2xy = 5, \quad y(0) = 2,$$

and

(b)
$$\frac{dy}{dx} - 2xy = 5, \quad y(1) = 8.$$

In both (a) and (b) the differential equation is a linear equation in standard form, with independent variable x, $p(x) = -2x$, and $g(x) = 5$. It has integrating factor
$$\mu(x) = e^{\int -2x\,dx} = e^{-x^2}.$$

Upon multiplying by this integrating factor the differential equation in (a) and (b) becomes
$$\frac{d}{dx}\left(e^{-x^2} y\right) = 5e^{-x^2}$$

or

(2.15)
$$e^{-x^2} y = \int 5e^{-x^2}\,dx.$$

Now we have a problem we didn't face in our earlier examples—we can't carry out the integration on the right-hand side, but we still need to bring the initial condition into play.[1] Our discussion about the Fundamental Theorem of Calculus tells us how to proceed. To implement the initial condition $y(0) = 2$, we write
$$\int_0^x 5e^{-t^2}\,dt$$

for the specific antiderivative of $5e^{-x^2}$ having value 0 at $x = 0$. By (2.15) we have
$$e^{-x^2} y = \int_0^x 5e^{-t^2}\,dt + C$$

where C is an arbitrary constant. To satisfy the condition $x = 0, y = 2$ we must have
$$2e^0 = \int_0^0 5e^{-t^2}\,dt + C$$

so that $C = 2$. Thus the initial value problem in (a) has solution

(2.16)
$$y = e^{x^2} \int_0^x 5e^{-t^2}\,dt + 2e^{x^2}.$$

We could use (2.16) to at least approximate the value of y at various values of x, for example
$$y(1) = e \int_0^1 5e^{-t^2}\,dt + 2e$$

where the definite integral can be estimated by a numerical method like the trapezoid rule or Simpson's Rule.

---

[1] In calculus, you may have learned that certain integrals, like $\int e^{-x^2}\,dx$, cannot be expressed as *elementary functions*. This means they cannot be built from power functions, exponential functions, logarithmic functions, and trigonometric and inverse trigonometric functions by taking finitely many sums, differences, products, quotients, and function compositions.

2.1. Linear equations

For the second initial value problem, with initial condition $y(1) = 8$ we choose

$$\int_1^x 5e^{-t^2}\,dt$$

as a specific antiderivative of $5e^{-x^2}$, this one having value 0 at $x = 1$. Now

$$e^{-x^2} y = \int_1^x 5e^{-t^2}\,dt + C$$

and setting $x = 1, y = 8$ yields $8e^{-1} = C$. Thus the initial value problem in (b) has solution

$$y = e^{x^2} \int_1^x 5e^{-t^2}\,dt + \frac{8}{e} e^{x^2}.$$

2.1.1. Exercises.

Solve the differential equation, giving an explicit solution. As in Example 2.1.1, your answer should contain an arbitrary constant and be the general solution to the given linear differential equation.

1. $\dfrac{dy}{dt} + 5y = 3e^{2t}$.

2. $\dfrac{dy}{dt} - 2y = 3te^{2t}$

3. $(x^2+1)\dfrac{dy}{dx} + 2xy = x^3 + x$.

4. $\dfrac{dy}{dx} + 2y = x$.

5. $t\dfrac{dy}{dt} - y = t^2 \sin t$.

6. $\dfrac{dy}{dx} + \dfrac{y}{x+2} = \dfrac{\sin x}{x+2}$.

7. $(t^2+1)\dfrac{dy}{dt} + ty = \sqrt{t^2+1}$.

8. $\sin t \dfrac{dy}{dt} - y \cos t = 2t \sin^2 t$.

9. $x\dfrac{dy}{dx} + y = \dfrac{1}{x^2-4}$.

10. $t\dfrac{dy}{dt} + \dfrac{1}{\ln t}y = t$.

Solve the initial value problems, and give the interval of validity of your solution.

11. $\dfrac{dy}{dt} + \dfrac{1}{t}y = \dfrac{e^{2t}}{t}$, $y(1) = 4$.

12. $\dfrac{dy}{dt} + \dfrac{2}{t}y = t^2$, $y(1) = 2$.

13. $\dfrac{dy}{dx} + y \tan x = x \cos x$, $y(0) = 1$.

14. $\dfrac{dr}{ds} + \dfrac{5r}{s-3} = 2$, $r(2) = 1$.

15. $\dfrac{dy}{dx} + 2xy = x$, $y(0) = -1$.

16. $\dfrac{dy}{dt} + 3y = e^{-3t} \ln t$, $y(1) = 1$.

17. Show by direct substitution that if some specific function $y = f(t)$ solves the homogeneous linear equation

(2.17) $$\dfrac{dy}{dt} + p(t)y = 0,$$

then any constant multiple $y = cf(t)$ also solves (2.17).

18. Use the method of Example 2.1.3 to solve the initial value problems
    (a)
    $$\dfrac{dy}{dx} + 2xy = 3, \quad y(2) = 4.$$
    (b)
    $$\dfrac{dy}{dt} + (\cos t)\, y = t, \quad y(\pi/2) = 1.$$

(c)
$$\frac{dy}{dt} = (1+y)\sqrt{1+\cos^3 t}, \quad y(0) = 1.$$

19. (a) Show that all solutions to
$$y' + ay = 0$$
and
$$y' + ay = be^{-at}$$
tend to 0 as $t \to \infty$ if a and b are constants and $a > 0$.
    (b) Solve $y' + ay = be^{ct}$ if $c \neq -a$. Show that if $a > 0$ and $c < 0$, then all solutions tend to 0 as $t \to \infty$.

20. Solve the equation

(2.18) $$\frac{d^2y}{dt^2} - \frac{dy}{dt} - 1 = 0$$

using the following two-step process. First set $v = \frac{dy}{dt}$ and rewrite (2.18) as a first-order equation with dependent variable v and independent variable t. Solve for v, and then use $v = \frac{dy}{dt}$ to solve for y. How many arbitrary constants are in your final answer?

21. Consider the homogeneous linear differential equation
$$\frac{dy}{dt} + p(t)y = 0,$$
where $p(t)$ is continuous.
    (a) Show that the solution to this equation is
$$y = Ce^{-\int p(t)dt}$$
for arbitrary constant C.
    (b) Show by substitution that if $y = f(t)$ is any specific function that solves
$$\frac{dy}{dt} + p(t)y = g(t),$$
then
$$y(t) = f(t) + Ce^{-\int p(t)dt}$$
also solves
$$\frac{dy}{dt} + p(t)y = g(t)$$
for any choice of C.
    (c) Find constants A and r so that $y = Ae^{rt}$ solves

(2.19) $$\frac{dy}{dt} - 5y = 2e^{3t}.$$

Do this by simply substituting into equation (2.19).
    (d) Use the results of (a), (b), and (c) to find an infinite list of solutions to (2.19) (containing one arbitrary parameter C). If an initial condition $y(t_0) = y_0$ is specified, will some function in your list satisfy this initial condition?

22. The differential equation $\frac{dy}{dx} + y = -x$ has a solution of the form $y = mx + b$.

(a) Using the direction field for $\dfrac{dy}{dx} + y = -x$ displayed in Fig. 2.2, find m and b by guessing and checking.

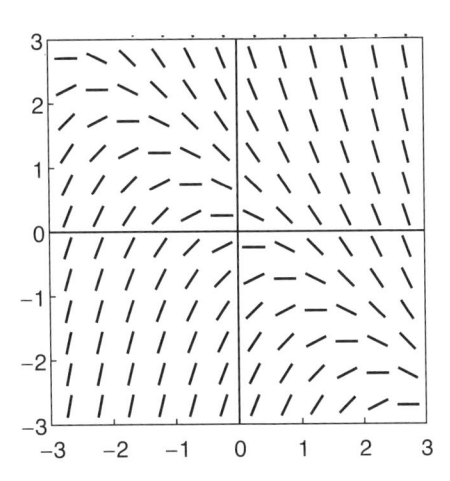

Figure 2.2. Direction field.

(b) Find m and b by solving the linear equation $\dfrac{dy}{dx} + y = -x$.

23. Two thousand fish are introduced into a large lake and their population p is projected to grow (for many years) at a rate proportional to the population present, satisfying $\dfrac{dp}{dt} = 0.1p$, where p is measured in thousands and t is measured in years. If fish are continuously harvested from the lake at a constant rate of 500 per year, then the following initial value problem models the number of fish present in the lake at time t:

$$\frac{dp}{dt} = 0.1p - 0.5, \quad p(0) = 2.$$

Solve this initial value problem, plot your solution, and describe in words the behavior of the population of fish in the lake. (See Section 2.5 for more on population modeling via differential equations.)

24. Various mathematical models have been proposed to study a sprinter running in a straight line (say, a 100-meter race).[2] The simplest of these uses the differential equation

$$m\frac{dv}{dt} = F_{prop} - F_{res}$$

where the runner's velocity is v and mass is m. The term F_{prop} is called the "propulsive force" of the runner, and "F_{res}" is the resistive force; the latter includes frictional losses within the body of the runner.

(a) As a simple first assumption, we set F_{prop} to be a constant (individualized to the specific runner being modeled) and assume F_{res} is proportional to the runner's velocity v. Upon dividing by the mass m we obtain the differential equation

$$\frac{dv}{dt} = \gamma - \sigma v$$

for positive constants γ and σ (both specific to the individual runner). Solve this equation for v, assuming that $v(0) = 0$.

---

[2] See the article *Physics of sprinting* by I. Alexandrov and P. Lucht in The Physics of Sport, A. Armenti, Jr., ed., American Institute of Physics, New York, 1992.

(b) Recalling that for an object moving along a straight line,
$$\frac{dp}{dt} = v$$
where $p(t)$ is the position of the object at time t and v is the velocity, use your work in (a) to determine the distance the sprinter has run as a function of elapsed time.

(c) Using data from the first 8 seconds of Usain Bolt's 100-meter final at the 2008 Beijing Olympics, a race that he ran in an astonishing 9.69 sec, and assuming the model in part (a), estimates for Bolt's values of γ and σ can be given as
$$\gamma \approx 8.5269 \text{ m/sec}^2, \quad \sigma \approx 0.6964 \text{ /sec}.$$
Using these values in your distance function from (b), determine his predicted position at his finishing time of $t = 9.69$; your answer should be greater than the actual position of 100 m at $t = 9.69$. This leads to an interesting feature of this particular race, which will be explored in the next part of this problem.

(d) (CAS) Bolt's time of 9.69 sec in the 100-meter final of the 2008 Beijing Olympics was a world record, but what was more remarkable is that about 8 seconds into the race, when about 20 meters remained, Bolt appears to begin to celebrate his victory, causing a significant drop in his speed for the last 20 meters or so. See the table in Fig. 2.3, which presents "100-meter splits" data often used to analyze this race.[3]

This drop in Bolt's speed led many commentators to speculate that he could have had a significantly faster time if he had not begun his celebration until the race was actually finished. His coach, for example, believed that his time could have been 9.52 sec. Based on your work in (b) and (c), you are in a position to predict what Bolt's finishing time might have been, had he not begun celebrating at around the 80-meter mark.

Let $d(t)$ be the distance function you derived in (c) based on your work from (b). This function represents Bolt's predicted position (distance from the start line) at a time t under the model of (a) based on data from the first 8 seconds of his run. The predicted distance given by $d(t)$, as shown in Fig. 2.3, fits the data from the table on the left of the figure quite well—especially over the first 8 seconds.

Usain Bolt Beijing Olympics 100 Meter Split Times		
Segment	Segment Time	Cummulative
0 – 10 m	1.85	1.85
10 – 20 m	1.02	2.87
20 – 30 m	.91	3.78
30 – 40 m	.87	4.65
40 – 50 m	.85	5.50
50 – 60 m	.82	6.32
60 – 70 m	.82	7.14
70 – 80 m	.82	7.96
80 – 90 m	.83	8.79
90 – 100 m	.90	9.69

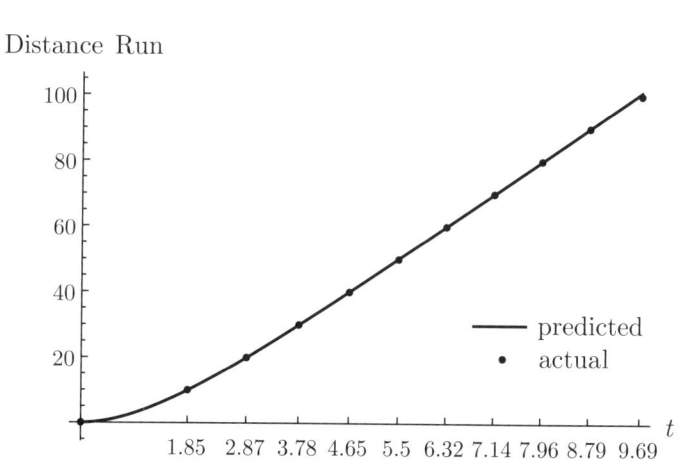

Figure 2.3. Modeling Bolt's 100-Meter Dash.

Using a computer algebra system, find the positive root of $d(t) = 100$ to two decimal places. (You might use the system's ability to solve equations numerically or its graphing

---
[3] http://speedendurance.com/2008/08/22/usain-bolt-100m-10-meter-splits-and-speed-endurance/.

capability.) The time you have found represents Bolt's predicted finishing time according to the model of (a), based on data from his run up to the 7.96 second mark.

(e) Compare your prediction of (d) to those of a group of physicists from the University of Oslo[4] who obtained detailed camera footage from the race and projected Bolt's potential time under two scenarios: for one their projected time is 9.61 ± 0.04 seconds and for the other it's 9.55 ± 0.04. Remark: In 2009, at the world championships in Berlin, Bolt ran the 100-meter race in 9.58 sec.

25. In this problem we continue the discussion from the previous exercise of Usain Bolt's 100-meter final in the 2008 Olympics. Physicists from the University of Oslo used camera footage from the race and tried to project Bolt's potential time under the following two hypothetical scenarios:

(A) Bolt maintains the same acceleration as the second place winner (Richard Thompson) from the 8 second mark until the end of the race.

(B) Bolt maintains an acceleration 0.5 m/sec^2 greater than Thompson's from the 8 second mark until the end. As justification for this assumption, they note that Bolt was actually considered to be a 200-meter specialist, not a 100-meter specialist, and thus would have been less likely to tire near the end.

Using numerical techniques they estimated a finishing time of 9.61 sec under scenario (A).

(a) What is Bolt's approximate position after 9.61 sec under scenario (B) (assume he keeps running even after crossing the finish line)?

(b) They predicted Bolt's position at $t = 9.55$ sec under scenario (A) to be approximately 99.4 m. Using this, show that under scenario (B) the approximate finishing time is 9.55 sec.[5]

Hints: Denote the acceleration, starting at $t = 8$, under scenario (A) by $a_1(t)$, and the acceleration under (B) starting at $t = 8$ by $a_2(t)$. You are told $a_2(t) = a_1(t) + 0.5$. Integrate (twice) to get a relationship between the position functions $p_1(t)$ and $p_2(t)$ under the two scenarios. Keep in mind that the positions and velocities at $t = 8$ are the same under both scenarios. For part (a), you have $p_1(9.61) = 100$ and you want $p_2(9.61)$. For (b), use $p_1(9.55) = 99.4$ and show that $p_2(9.55) \approx 100$.

Figure 2.4. Usain Bolt's 100-meter final run in the 2008 Olympics. Source: Wikimedia Commons, author: SeizureDog, licensed under Creative Commons Attribution 2.0 Generic (https://creativecommons.org/licenses/by/2.0/deed.en) license.

2.2. Separable equations

In this section we introduce **separation of variables**, which is a method for finding one-parameter families of solutions to certain first-order equations. Specifically, this method can be used for

---

[4] E. Erikson, J. Kristiansen, L. Langangen, and I. Wehus, *Velocity dispersion in a cluster of stars: How fast could Usian Bolt have run?*, arxiv.org/abs/0809.0209.
[5] In 2009, at the world championships in Berlin, Bolt ran the 100-meter race in 9.58 sec.

equations
$$\frac{dy}{dx} = f(x,y)$$
where $f(x,y)$ can be written as
$$f(x,y) = g(x)h(y).$$
Such an equation is termed **separable**. We'll illustrate with an example.

Example 2.2.1. Find a one-parameter family of solutions to

(2.20)
$$\frac{dy}{dx} = \frac{x^3}{y(1+x^4)}.$$

Notice that the right-hand side can be written
$$\frac{x^3}{1+x^4} \frac{1}{y}.$$
This allows us to separate the variables and rewrite our differential equation as

(2.21)
$$y\,dy = \frac{x^3}{1+x^4}\,dx.$$

Separating variables means putting all y's on one side times the dy and all x's on the other side times the dx (using valid algebra of course!). Integrating on both sides of (2.21) gives

$$\int y\,dy = \int \frac{x^3}{1+x^4}\,dx,$$
$$\frac{y^2}{2} = \frac{1}{4}\ln(1+x^4) + C.$$

Instead of the single constant term $+C$ on the right-hand side of the last line, we could have written a constant of integration on both the left-hand side and right-hand side, but the resulting equation would be no more general in that case. Thus by "separating variables and integrating" we have obtained a one-parameter family of implicit solutions

(2.22)
$$\frac{y^2}{2} = \frac{1}{4}\ln(1+x^4) + C$$

to equation (2.20).

Justification. To justify our work and explain the method of separation of variables in general consider any equation of the form

(2.23)
$$\frac{dy}{dx} = g(x)h(y).$$

Separating and integrating, we obtain
$$\int \frac{1}{h(y)}\,dy = \int g(x)\,dx,$$
an equation suggesting that if H is an antiderivative of $1/h$ and G is an antiderivative of g, then

(2.24)
$$H(y) - G(x) = C$$

is an implicit solution of (2.23). To see that this is correct, let $y = \varphi(x)$ be a function defined implicitly by equation (2.24) on some open interval I, so that

$$H(\varphi(x)) - G(x) = C \quad \text{for all } x \text{ in } I.$$

2.2. Separable equations

Differentiating the preceding equation, using the chain rule as needed, we obtain

$$\frac{1}{h(\varphi(x))}\varphi'(x) - g(x) = 0 \quad \text{for all } x \text{ in } I,$$

which may be rewritten

$$\varphi'(x) = g(x)h(\varphi(x)) \quad \text{for all } x \text{ in } I.$$

This says that $y = \varphi(x)$ is a solution of (2.23).

Example 2.2.2. Find a one-parameter family of solutions to

(2.25) $$\frac{dy}{dx} = (x-3)(y+1)^{2/3}.$$

We have

$$\frac{1}{(y+1)^{2/3}} dy = (x-3) dx,$$

$$\int \frac{1}{(y+1)^{2/3}} dy = \int (x-3) dx,$$

$$3(y+1)^{1/3} = \frac{x^2}{2} - 3x + C,$$

$$(y+1)^{1/3} = \frac{x^2}{6} - x + \frac{C}{3},$$

$$y = -1 + \left(\frac{x^2}{6} - x + \frac{C}{3}\right)^3.$$

Since C is an arbitrary real number, so is $\frac{C}{3}$, and we might prefer to write the final line in the equivalent but simpler form

(2.26) $$y = -1 + \left(\frac{x^2}{6} - x + C\right)^3$$

(C any real number).

Lost solutions. Notice that the original equation (2.25) has the constant function $y = -1$ as a solution (recall that a constant solution is called an equilibrium solution), but this solution is not present in the family (2.26), since no choice of C will produce it. It's easy to see why this solution got "lost", since the first step of our work was to divide by $(y+1)^{2/3}$ and we were tacitly assuming that $y \neq -1$ in doing this. *Other solutions have been lost too*: for instance, you can check that

$$y(x) = \begin{cases} -1 + \left(\frac{x^2}{6} - x\right)^3 & \text{for } x \leq 0, \\ -1 & \text{for } x > 0 \end{cases}$$

is a solution of (2.25) on the interval $(-\infty, \infty)$.

The moral of Example 2.2.2 is that separation of variables, unlike the linear equation method introduced in the preceding section, yields a one-parameter family of solutions that isn't *necessarily* the general solution. On the other hand, the family of solutions

$$\frac{y^2}{2} = \frac{1}{4}\ln(1+x^4) + C$$

obtained in Example 2.2.1 by separation of variables *is* the general solution to equation (2.20); see Exercise 21.

Initial value problems. Suppose we want to solve a separable equation

$$\text{(2.27)} \qquad \frac{dy}{dx} = g(x)h(y)$$

together with an initial condition $y(x_0) = y_0$. Let's assume that $g(x)$ and $h(y)$ are continuous functions, at least near x_0 and y_0, respectively. If $h(y_0) = 0$, then the constant function $y = y_0$ solves this initial value problem. If $h(y_0) \neq 0$, there are two slightly different ways to proceed. We can separate variables and integrate in (2.27) to obtain

$$\int \frac{1}{h(y)} \, dy = \int g(x) \, dx.$$

If we can carry out the integrations, this leads to an implicit solution containing an arbitrary constant C. Given our hypothesis on h, a value for C can be chosen so that the initial condition is satisfied.

For an alternate approach, we separate variables

$$\frac{1}{h(y)} \, dy = g(x) \, dx$$

and then integrate from y_0 to y on the left and from x_0 to x on the right:

$$\text{(2.28)} \qquad \int_{y_0}^{y} \frac{1}{h(v)} \, dv = \int_{x_0}^{x} g(u) \, du.$$

(notice how we've introduced the dummy variables of integration v and u), and finish by carrying out the integrations if possible[6].

Example 2.2.3. Impose the initial condition $y = 7$ when $x = 2$ on the differential equation in equation (2.25). If we use the solution produced in (2.26), together with the condition $y(2) = 7$, to obtain

$$7 = -1 + \left(\frac{2}{3} - 2 + C\right)^3,$$

we determine that $C = \frac{10}{3}$ and

$$\text{(2.29)} \qquad y = -1 + \left(\frac{x^2}{6} - x + \frac{10}{3}\right)^3.$$

Alternately, if we implement the initial condition as soon as the variables have been separated, we go directly from

$$\frac{1}{(y+1)^{2/3}} \, dy = (x - 3) \, dx$$

to

$$\int_{7}^{y} \frac{1}{(v+1)^{2/3}} \, dv = \int_{2}^{x} (u - 3) \, du.$$

Carrying out the integrations we have

$$3(y+1)^{1/3} - 3(8)^{1/3} = \frac{x^2}{2} - 3x - (2 - 6)$$

or

$$3(y+1)^{1/3} = \frac{x^2}{2} - 3x + 10.$$

You can easily check this agrees with the solution in (2.29).

---

[6]Tacitly, we are assuming that h is nonzero and continuous on the interval with endpoints y_0 and y and g is continuous on the interval with endpoints x_0 and x.

2.2.1. Exercises.

In Exercises 1–10, find a one-parameter family of solutions to the given differential equation, making the solution explicit if it is convenient to do so.

1. $\dfrac{dy}{dx} = 2xy^{2/3}$.

2. $\dfrac{dy}{dt} - \dfrac{y}{t} + \dfrac{y^2}{t} = 0$.

3. $xe^{x^2-y} = y\dfrac{dy}{dx}$.

4. $\dfrac{dy}{dx} = y^2(1+\sin x)$.

5. $\dfrac{dy}{dt} = -\dfrac{y}{y+1}$.

6. $\dfrac{dy}{dx} = \dfrac{x^2}{y^2+6y}$.

7. $\dfrac{dy}{dx} = \dfrac{e^{x+y}}{e^{x-y}}$.

8. $\dfrac{dy}{dx} - y\ln x = y$.

9. $\dfrac{dy}{dx} = 3y^{2/3}\sec^2 x$.

10. $(1+x^2)\dfrac{dy}{dx} = 1+y^2$. Hint: Use the identity
$$\tan(\alpha+\beta) = \dfrac{\tan\alpha + \tan\beta}{1 - \tan\alpha\tan\beta}$$
to simplify your explicit solution.

In Exercises 11–14, solve the initial value problem.

11. $\dfrac{dy}{dt} = \dfrac{t^3+t}{y^3}$, $y(1)=2$.

12. $\dfrac{ds}{dr} = s\sin r$, $s(\pi/2)=1$.

13. $\dfrac{dy}{dx} = e^{x-y}$, $y(0)=2$.

14. $\dfrac{dy}{dt} = \dfrac{y^3}{t^3+t}$, $y(1)=0$. (Hint: Think before you compute.)

15. Which of the following is a solution to the initial value problem
$$\dfrac{dy}{dt} = \dfrac{1}{3}y^{-2}\sin(t^2), \quad y(1)=2?$$
Indicate all correct answers.

(a)
$$y(t) = \left(\int_1^t \sin(u^2)\,du\right)^{1/3}.$$

(b)
$$y(t) = \left(\int_1^t \sin(u^2)\,du\right)^{1/3} + 2.$$

(c)
$$y(t) = \left(\int_1^t \sin(u^2)\,du + 8\right)^{1/3}.$$

(d)
$$y(t) = \left(\int_0^t \sin(u^2)\,du\right)^{1/3} + 8.$$

16. In Example 2.2.2 we decided that the solution could be written as
$$y = -1 + \left(\frac{x^2}{6} - x + \frac{C}{3}\right)^2$$
or as
$$y = -1 + \left(\frac{x^2}{6} - x + C\right)^2$$
(for arbitrary constant C) since the collection of functions obtained is the same in either case. In each of the following, two families of functions are given, each containing one or more arbitrary constants (C, k, a, b, and c). Except where indicated, these constants are allowed to have any real number value. Decide whether the two families are the same. If the families are not the same, give a specific example of a function in one family that is not in the other family.
   (a) $y(t) = Ce^{t+k}$, $y(t) = Ce^t$.
   (b) $y(t) = e^{t+k}$, $y(t) = Ce^t$.
   (c) $y(t) = C\ln(t^3)$, $y(t) = C\ln t$.
   (d) $y(t) = a + \ln(bt)$, $y(t) = a + \ln t$ ($b > 0$).
   (e) $y(t) = a + \ln(bt)$, $y(t) = \ln(bt)$ ($b > 0$).
   (f) $y(t) = (x+c)^2$, $y(t) = (c-x)^2$.

17. (a) Solve the equation
   (2.30) $$\frac{dy}{dx} = y^{3/2}$$
   by separation of variables. Obtain a one-parameter family of explicit solutions, and show that your family of solutions can be written as
   $$y = \frac{4}{(x+C)^2}.$$
   Also find all equilibrium solutions.
   (b) The direction field for (2.30) is shown in Fig. 2.5. There are no field marks below the x-axis; explain why this is.

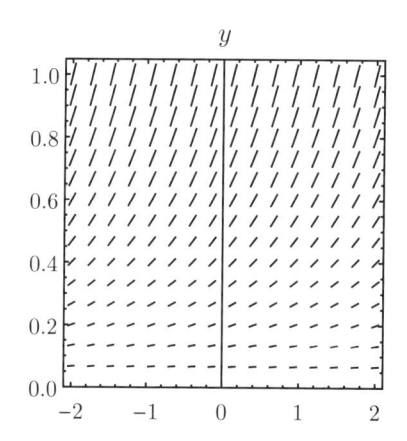

Figure 2.5. Direction field for (2.30).

   (c) One of the explicit solutions you found in (a) should be $y = 4/x^2$. What choice of the arbitrary constant C in your solution family will give this solution?
   (d) Does the graph of $y = 4/x^2$ for $x > 0$ fit with the direction field picture for (2.30)? Where in the work that you did in part (a) could you have predicted that the solution $y = 4/x^2$ would only be valid for $x < 0$?

2.2. Separable equations

(e) For what interval of x's is the solution $y = 4/(x+C)^2$ of (2.30) valid?

18. Find a family of solutions to

(2.31) $$\frac{d^2y}{dt^2} - \left(\frac{dy}{dt}\right)^2 - 1 = 0$$

using the following two-step process. First set $v = \frac{dy}{dt}$ and rewrite (2.31) as a first-order equation with dependent variable v and independent variable t. Solve for v, and then use $v = \frac{dy}{dt}$ to solve for y. How many arbitrary constants are in your final answer?

19. Find all equilibrium solutions (if any) of the following differential equations.
    (a) $\frac{dy}{dt} = (t-1)(y-2)$.
    (b) $\frac{dy}{dt} = \frac{y-2}{t-1}$.
    (c) $\frac{dy}{dt} = \frac{t-1}{y-2}$.

20. Consider a differential equation in the form $\frac{dy}{dx} = f(y)$, where f is a continuous function. Suppose that $f(3) = 2$ and $f(1) = -2$. Which of the following are correct?
    (a) There is an equilibrium solution $y = k$ for some value of k between 1 and 3.
    (b) There is an equilibrium solution $y = k$ for some value of k between -2 and 2.
    (c) We can't conclude anything about equilibrium solutions without more information.

21. In this problem you will verify that the family of solutions in equation (2.22) of Example 2.2.1 is the general solution of equation (2.20). Show that if $y = y(x)$ satisfies

$$\frac{dy}{dx} = \frac{x^3}{y(1+x^4)}$$

on an open interval (a,b), then the function

$$G(x) = \frac{y(x)^2}{2} - \frac{1}{4}\ln(1+x^4)$$

is constant on (a,b). Conclude that there is a constant C so that

$$\frac{y^2}{2} = \frac{1}{4}\ln(1+x^4) + C$$

for all x in (a,b).

22. (a) By separating variables, find a one-parameter family of solutions to

$$x\frac{dy}{dx} = \sqrt{1-y^2}.$$

Show that your solution can be given explicitly as

$$y = \sin(\ln|x| + c).$$

(b) Find all equilibrium solutions to the equation in (a).
(c) Care must be taken to analyze the open intervals on which $y = \sin(\ln|x| + c)$ is a solution. By analyzing the sign of the right- and left-hand sides of the differential equation, show that $y = \sin(\ln|x|)$ is not a solution on the interval $(e^{\pi/2}, e^{3\pi/2})$, even though the domain of $\sin(\ln|x|)$ includes this interval.

(d) The solutions from (a) and (b) do not comprise all solutions to the equation. To see this, explain why

$$y = \begin{cases} -1 & \text{for } x \leq e^{-\pi/2}, \\ \sin(\ln x) & \text{for } e^{-\pi/2} < x < e^{\pi/2}, \\ 1 & \text{for } x \geq e^{\pi/2} \end{cases}$$

is a solution to the differential equation on $(-\infty, \infty)$ but does not appear in the solutions from (a) or (b).

23. For a differential equation that contains a real number constant (we say the equation depends on a real parameter), a **bifurcation value** is a real number b_0 such that for parameter values near b_0 but on opposite sides, solutions to the corresponding equations behave quite differently, qualitatively speaking. Complete this exercise to see that $b_0 = 0$ is a bifurcation value of

(2.32) $$\frac{dy}{dx} = by^2 - 4.$$

(a) Show that (2.32) has two equilibrium solutions if b is a positive constant, and no equilibrium solutions if b is a nonpositive constant.

(b) Fill in the blanks, based on the representative direction fields below for equation (2.32) in the $b < 0$ and $b > 0$ cases.

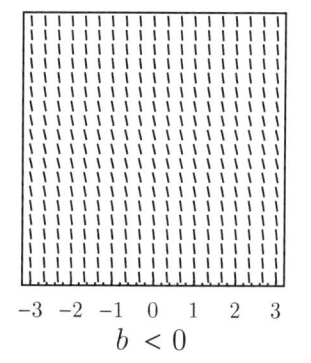

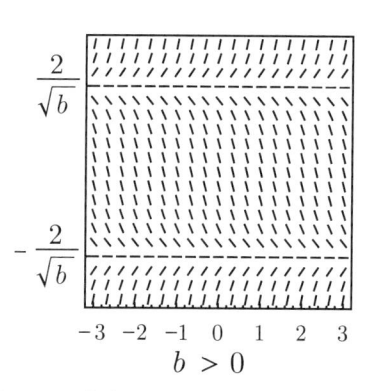

(i) Let $b < 0$. If y_0 is any real number and $y = y(x)$ solves the initial value problem

$$\frac{dy}{dx} = by^2 - 4, \quad y(0) = y_0,$$

then $y(x)$ approaches _____ as x increases (starting at 0).

(ii) Let $b > 0$. If $y_0 > \frac{2}{\sqrt{b}}$ and $y = y(x)$ solves the initial value problem

$$\frac{dy}{dx} = by^2 - 4, \quad y(0) = y_0,$$

then $y(x)$ approaches _____ as x increases (starting at 0).

(iii) Let $b > 0$. If $-\frac{2}{\sqrt{b}} < y_0 < \frac{2}{\sqrt{b}}$ and $y = y(x)$ solves the initial value problem

$$\frac{dy}{dx} = by^2 - 4, \quad y(0) = y_0,$$

then $y(x)$ approaches _____ as x increases (starting at 0).

(iv) Let $b > 0$. If $y_0 < -\frac{2}{\sqrt{b}}$ and $y = y(x)$ solves the initial value problem

$$\frac{dy}{dx} = by^2 - 4, \quad y(0) = y_0,$$

then $y(x)$ approaches _____ as x increases (starting at 0).

In the next exercise, for certain values of b, you will confirm your answers filling the blanks above are correct.

2.2. Separable equations

24. (a) Solve the initial value problem
$$\frac{dy}{dx} = -4y^2 - 4, \quad y(0) = 0.$$
What happens to your solution as x increases (starting at 0)? (This is the case $b = -4$ in Exercise 23.)

    (b) Solve the initial value problem
$$\frac{dy}{dx} = y^2 - 4, \quad y(0) = 0.$$
What happens to your solution as x increases (starting at 0)? (This is the case $b = 1$ in Exercise 23.)

25. A snowball melts so that the rate of change of its volume with respect to time is proportional to its surface area.

    (a) Give a formula for the snowball's surface area S in terms of its volume V. What differential equation models the volume of the melting snowball? Hint: If r is the radius of the snowball, then $V = \frac{4}{3}\pi r^3$ and $S = 4\pi r^2$.

    (b) Assume that V is measured in cubic centimeters and time t is measured in hours. If $V(0) = 27$ cm^3 and $V(2) = 8$ cm^3, when will $V(t) = 0$?

26. In this problem we outline an alternate method, called **variation of a parameter**, for solving the linear equation

    (2.33) $$\frac{dy}{dt} + p(t)y = g(t),$$

    where p and g are continuous. It foreshadows an important technique we will use in later chapters for higher-order equations.

    (a) The equation
$$\frac{dy}{dt} + p(t)y = 0$$
is both a homogeneous linear equation and a separable equation. Solve it as a separable equation, and show that it has explicit solution

    (2.34) $$y = Ce^{-\int p(t)dt}.$$

    (b) The method of variation of parameter gives a way to solve the nonhomogeneous equation (2.33) once you have solved the "associated" homogeneous equation in (a). The idea is to look for a solution of the form
$$y = h(t)e^{-\int p(t)dt}.$$
The terminology "variation of parameter" comes from the perspective that we have taken the constant C in (2.34) and "varied" it, by replacing C by a function $h(t)$. Show that if $y = h(t)e^{-\int p(t)dt}$ is substituted into equation (2.33), you obtain
$$h'(t) = g(t)e^{\int p(t)dt},$$
so that
$$h(t) = \int g(t)e^{\int p(t)dt}dt + c.$$

    (c) To see this method in action, find **one** solution to $\frac{dy}{dt} + y = 0$ by separation of variables, and call it $f(t)$. Next, solve $\frac{dy}{dt} + y = e^t$ by using the trial solution $y = h(t)f(t)$ and determining $h(t)$.

(d) Repeat (c) for the equations

$$\frac{dy}{dt} + \frac{2(1-t^2)}{t}y = 0$$

and

$$\frac{dy}{dt} + \frac{2(1-t^2)}{t}y = \frac{1}{t}.$$

27. (a) Find one solution to

$$\frac{dy}{dt} + y\tan t = 0$$

by separating variables.

(b) Use variation of parameter (see Exercise 26) to solve

$$\frac{dy}{dt} + y\tan t = \sin t.$$

28. In 2007 students in a Decision Analysis class in the Darden Graduate Business school at the University of Virginia participated in the following experiment, with cash provided by an anonymous donor. One student in the class would be randomly selected and given the opportunity to choose between two briefcases. One briefcase contained $18,500, and the other $1. (The actual dollar amounts have been slightly adjusted in this example; the figure of $18,500 corresponds to the approximate cost of one year of in-state tuition.) The student will also be offered an alternate cash payoff, which he or she can select instead of choosing one of the briefcases. However, before the dollar amount of this cash payoff is announced, each student in the class had to make a binding decision of their personal minimum acceptable cash payoff (how much they would have to be guaranteed to forgo a 50-50 chance at $18,500). When the actual payoff was announced, this would determine if the selected student would go on to choose a briefcase or would take the cash payoff instead.

In this problem we will analyze this game through the notions of utility functions and risk aversion.

(a) What would you personally choose as your minimum acceptable payoff amount? (There is no right or wrong answer to this question.) Call this C_0; you'll need it later in (f). Keep in mind that the **expected value** of this "choose a briefcase" game is

$$\frac{1}{2}(18{,}500) + \frac{1}{2}(1) = 9{,}250.50 \text{ dollars};$$

this is the average value of the gamble. This is computed by multiplying the probability $\frac{1}{2}$ of choosing each of the briefcases with the dollar amount in the briefcase and adding the results.

(b) Economists often use the notion of **expected utility** instead of expected value. They choose a simple function u, called a utility function, and multiply the probability of each outcome by the value of u at the outcome. For example, if the utility function is $u(x) = \sqrt{x}$, then the expected utility $E(u)$ in our briefcase game is

$$E(u) = \frac{1}{2}\sqrt{18{,}500} + \frac{1}{2}\sqrt{1} = 68.5.$$

The **certainty equivalent** allocation is the value C solving $u(C) = E(u)$. Show that for $u(x) = \sqrt{x}$, the certainty equivalent allocation is about $4,692. This means that a person operating under this particular utility function would choose a minimum acceptable payoff of $4,692; since this is smaller than the expected value, this utility function describes a **risk-averse** scenario.

(c) Economists often require that utility functions satisfy the differential equation

(2.35) $$-\frac{xu''(x)}{u'(x)} = k$$

where k is constant; these are called **constant relative risk aversion** utility functions. Determine solutions to this differential equation by first converting the equation to a separable first-order equation by setting $v(x) = u'(x)$. Among your solutions you should have the specific functions $u(x) = \ln x, u(x) = \sqrt{x}, u(x) = x$, and $u(x) = x^2$. What value of k corresponds to each of these?

(d) Show that with $u(x) = x$ the expected utility is the same as the expected value and these are equal to the certainty equivalent allocation. Show that if $u(x) = ax + b$ for positive constants a and b, the certainty equivalent allocation is the expected value.

(e) Suppose $u(x) = x^2$. What is the expected utility of the briefcase game with this choice, and for what payoff amount C is $u(C)$ equal to this expected utility? Does your answer change if the utility function is $u(x) = ax^2 + b$ for positive constants a and b? How does C compare to the expected value in this case? This is an example of a **risk-loving** scenario.

(f) Assume that $u(x) = x^b$. Using your answer to (a), *approximately* what value of b corresponds to

$$\frac{1}{2}u(18{,}500) + \frac{1}{2}u(1) = u(C_0)$$

where C_0 is your personal acceptable minimum payoff? Assume b is positive and large enough that $u(1)$ is negligible in comparison to $u(18{,}500)$.

(g) The student who was randomly selected to play this game at the University of Virginia chose his minimum acceptable payoff value as $8,000. Assuming a utility function of $u(x) = x^b$, determine his b approximately, as in (f).

Epilogue. Hideki Inoue was the student randomly chosen out of a class of 335 to participate in this contest. He had selected $8,000 as his personal minimum payoff value. The actual payoff offer (also randomly selected) was $5,769, so Mr. Inoue picked a briefcase. It was not the winning one.[7]

2.3. Applications: Time of death, time at depth, and ancient timekeeping

Mathematical models based on simple first-order linear or separable equations can be surprisingly useful in providing insight into diverse and complex real-world situations.

2.3.1. Newton's law of heating and cooling.

Example 2.3.1. A corpse is found at midnight with a body temperature of 93°F. Two hours later its temperature is 89°F. The surrounding room temperature is 70°F. When did death occur?

We'll approximate the time of death using Newton's law of cooling, which says that a heated object cools at a rate proportional to the difference between its temperature and the surrounding, or ambient, temperature. If we let T be the temperature of the object at time t and let A denote the (constant) ambient temperature, then

(2.36) $$\frac{dT}{dt} = -k(T - A).$$

Here k is a positive constant of proportionality. Because k is positive, equation (2.36) tells us that $\frac{dT}{dt}$ will be negative when $T > A$, consistent with the assumption that the object is cooling.

---

[7] See *A Jackpot, as Luck Would Have It*, by Susan Kinzie, The Washington Post, February 22, 2007.

This is a first-order equation that is both separable and linear. If we solve it by separating the variables t and T, we quickly obtain an implicit solution for T as follows:

(2.37) $$\int \frac{dT}{T-A} = \int -k\,dt,$$

so that

(2.38) $$\ln|T-A| = -kt + c.$$

Since our object (the corpse) is cooling, $T > A$ and we may rewrite our solution as

(2.39) $$\ln(T-A) = -kt + c.$$

We could go on to solve explicitly for T as a function of t, but to answer the question in our example it's just as easy to work with our implicit solution.

We are given that $A = 70°F$. Let's choose our timeline so that $t = 0$ corresponds to midnight and measure t in hours. We are told that when $t = 0$, $T = 93°F$ so that

$$\ln(93 - 70) = c$$

and $c = \ln 23$. Since $T = 89°F$ when $t = 2$, we have by (2.39)

$$\ln(89 - 70) = -k(2) + \ln 23$$

and

$$k = (\ln 23 - \ln 19)/2 \approx 0.09553.$$

What is the time of death? This is asking for the (last) time that $T = 98.6°F$. Solving

$$\ln(98.6 - 70) = -0.09553t + \ln 23$$

gives $t = -2.2811$, and we estimate the time of death to be 2.2811 hours before midnight or approximately 9:43 PM.

Newton's law (2.36) also applies when $T < A$, that is, for an object that is *warming* rather than cooling. Now $T - A$ will be negative, and the rate of change $\frac{dT}{dt}$ given by (2.36) will be positive. The solution (2.38) can be written as

(2.40) $$\ln(A - T) = -kt + c.$$

If we solve *either* (2.39) or (2.40) explicitly for T as a function of t, we obtain

(2.41) $$T(t) = A - (A - T(0))e^{-kt},$$

where $T(0)$ is the temperature of the object at $t = 0$ (see Exercise 5). This is an explicit solution for the temperature function in either the cooling or warming scenario.

In the next section, we apply the same differential equation (2.36) to a quite different problem in which we analyze decompression tables for scuba diving.

2.3.2. Differential equations and scuba diving. Decompression sickness, sometimes called "the bends", is one of several major medical problems potentially encountered in scuba diving. It occurs during ascent from deep dives and is caused when bubbles of nitrogen or other inert gases present are released from blood or tissues. (Oxygen does not pose a problem as it is continuously utilized by the body.) A diver underwater breathes air that is at a higher pressure than on land. This causes more nitrogen to be dissolved in the blood and tissues. As the diver ascends to a higher level, the rapid decrease in pressure can cause the formation of bubbles from the dissolved gas, much like the bubbles that can rapidly form when a bottle of soda is opened. No problem occurs as long as the decrease in pressure *in the tissues* is not greater than a factor of about 2.2, so that, for example, a change from a tissue pressure of 4 atm (atmosphere) to an (ambient) pressure of 2 atm is safe, while a change from 3 to 1 atm is not. The pressure on the earth's surface is 1

2.3. Applications: Time of death, time at depth, and ancient timekeeping   47

atm, and at depth d ft below the surface of the water it is $1 + \frac{d}{33}$ atm. So as long as the dive is to a depth of no more than $d = 1.2(33) \approx 39.6$ ft, a diver can spend any length of time there and be able to immediately ascend to the surface with no danger of the bends. At depths greater than this, the length of time spent at a depth determines whether it is safe to ascend to the surface all at once or whether intermediate stops are required. In this example, we investigate the mathematics behind "decompression tables". While no such table can guarantee complete safety, due to the highly individual reactions to varying pressures, they do provide reasonable guidelines, and when scrupulously followed the incidence of decompression illness is quite low. *Caution:* Because we make various simplifications in our model, the conclusions we draw should not be relied on for any actual dives!

No-stop dives. A "no-stop dive", also called a "bounce dive", is one in which the diver descends directly to fixed depth, stays there for some period of time, then ascends directly to the surface. Dive tables such as that shown in Table 2.1 (the most commonly used tables are provided by the US Navy) give the length of time allowed at the chosen depth with no decompression stops required on ascent. We will show how a differential equation model might be used to produce such a table.

Table 2.1. Bounce Dives (US Navy limits)									
depth d (ft)	40	50	60	70	80	90	100	110	120
time T (min)	200	100	60	50	40	30	25	20	15

We are interested in the pressure $P(t)$ in the tissues at varying times t. Let's first consider a situation where the ambient pressure is a fixed constant value P_a. The Merck Manual[8] describes how the pressure changes in the body tissues as follows:

"Gradients of partial pressure govern the uptake and elimination of the gas...."

The terminology "gradient of partial pressure" means the difference between the pressure P in the tissues and the ambient pressure P_a, and the statement is that the rate of change of P, $\frac{dP}{dt}$, is proportional to the difference between P and the ambient pressure P_a:

$$(2.42) \qquad \frac{dP}{dt} = k(P_a - P).$$

Notice how this has the same *form* as the differential equation in Newton's law of heating and cooling (equation (2.36)). The constant k depends on the tissue under consideration, and herein lies a bit of a complication. The body is composed of many different tissues, and we must design our dives to be safe for all of these different tissues. This means we will have to consider a variety of different values of k, over a large enough range that we are confident that we are protecting all different parts of the body. We'll return to this issue shortly.

Since we've already solved the differential equation for Newton's law of heating/cooling and (2.42) is really the same equation, we can use (2.41) to immediately write down its solution:

$$(2.43) \qquad P(t) = P_a - (P_a - P(0))e^{-kt},$$

where $P(0)$ is the pressure in the tissues at $t = 0$. This is applicable both to a situation where the tissue pressure is less than the ambient pressure P_a or when the reverse holds, so we can use (2.43)

---

[8] *The Merck Manual of Diagnosis and Therapy*, R. Berkow, Editor-in-Chief, 15th edition, Merck Sharp and Dohme Research Laboratories, 1987, p. 2382.

for both compression ($P_a > P$) and decompression ($P_a < P$) situations. We will measure pressure in atmospheres (atm); 1 atm is approximately 14 lbs/in^2.

Example 2.3.2. Now let's apply this to analyze a bounce dive. We assume that our diver moves *instantaneously*[9] from the surface to the dive depth, stays at that depth for a certain amount of time, and then instantaneously moves back to the surface. We set $P(0) = 1$ atm, the pressure on the surface and the pressure in the tissues at the start of the dive. If the depth of the dive is d feet and the time spent at this depth is T minutes, then at the end of the dive the pressure in the tissues is

$$P(T) = P_a - (P_a - 1)e^{-kT},$$

where $P_a = 1 + d/33$ atm is the pressure the diver experiences at depth d. Since the pressure on the surface is 1 atm and we don't want the pressure on ascent to decrease by more than a factor of 2.2, it will be safe for the diver to return (instantaneously) to the surface if $P(T) \leq 2.2$. Thus we determine the longest allowable time at depth d by solving the equation

$$2.2 = P_a - (P_a - 1)e^{-kT}$$

for T. Note that $P_a - 1$ is positive, and therefore, by the equation, so is $P_a - 2.2$. Solving for T gives

$$T = -\frac{1}{k} \ln \frac{P_a - 2.2}{P_a - 1} = \frac{1}{k} \ln \frac{P_a - 1}{P_a - 2.2}.$$

Since $P_a = 1 + \frac{d}{33}$, we can also determine this safe time T in terms of dive depth d as

(2.44) $$T = \frac{1}{k} \ln \frac{d/33}{(d/33) - 1.2} = \frac{1}{k} \ln \frac{d}{d - (1.2)(33)}.$$

As we have already noted, the body is composed of different sorts of tissues, with different corresponding "k-values", and to determine the safe time at depth d we must choose the smallest T in (2.44) calculated from these different k-values. It should be clear from (2.44) that this will be obtained by using the largest k-value in (2.44). To produce a bounce dive table, we use $k = 0.025$ min^{-1} and compute T for various depths as shown in Table 2.2. In Fig. 2.6 we have the graph of T as a function of d for this value of k. Note that this function of d is only defined for $d > (1.2)(33) = 39.6$ and its graph has the line $d = 39.6$ as a vertical asymptote. A comparison with a bounce dive table produced by the US Navy (Table 2.1 above) shows general qualitative agreement, although our dive table is generally more conservative, at least for shallower depths.

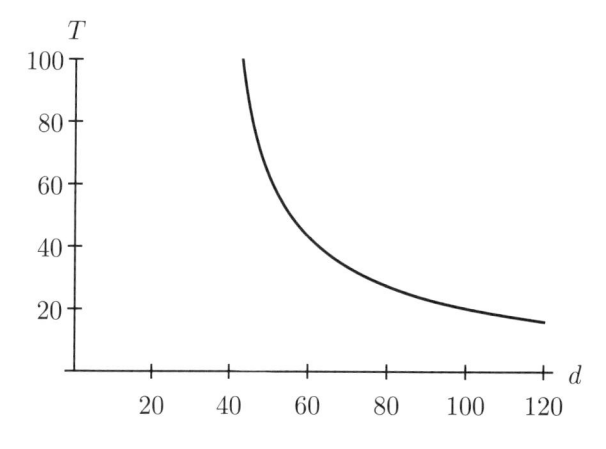

Figure 2.6. Maximum safe time T in terms of depth d.

---

[9]Exercise 10 replaces the assumption of instantaneous change of depth with the more realistic one of a steady rate of descent and ascent.

2.3. Applications: Time of death, time at depth, and ancient timekeeping

Table 2.2. Bounce Dives (Our Model)									
d (ft)	40	50	60	70	80	90	100	110	120
T (min)	184	63	43	33	27	23	20	17	16

Dives with stops. A more complex analysis is required for dives that involve decompression stops, although the analysis still depends on equation (2.42). We'll illustrate this in the next example.

Example 2.3.3. Suppose our diver instantaneously descends from the surface to 100 ft, spends 60 minutes there (pressure $\approx$ 4 atm). Since by Table 2.2 this is greater than the amount of time at this depth that would permit immediate ascent to the surface, one or more decompression stops must be planned. Let's assume we have three different tissues, with k-values

$$k_1 = 0.025, \quad k_2 = 0.01, \quad k_3 = 0.006$$

in units of min^{-1}. After 1 hour at 4 atm, the corresponding pressures in these three tissues are computed using (2.43) with $P_a = 4$, $t = 60$, and $P(0) = 1$, the latter since we are assuming the diver starts on the surface before instantaneously descending to 4 atm. We have

$$P(60) = 4 - 3e^{-(0.025)(60)} \approx 3.33,$$

$$P(60) = 4 - 3e^{-(0.01)(60)} \approx 2.35,$$

and

$$P(60) = 4 - 3e^{-(0.006)(60)} \approx 1.91.$$

It is safe to ascend to a depth where the pressure is $3.33/2.2 \approx 1.5$ atm. This corresponds to a depth d where $1 + \frac{d}{33} = 1.5$, or $d \approx 16.5$ ft. Most dive tables give stops at depths that are multiples of 10 ft, so let's plan our first stop at 20 ft. As before, we assume our ascent to the depth to be instantaneous, so that the pressure in the tissues when we start our ascent from 100 ft and when we stop at 20 ft is the same. How long must this stop be? The answer depends on whether and where a second stop is to be made.

Suppose we want to wait long enough at 20 ft so that no further decompression stops are needed and we can ascend directly to the surface. The ambient pressure at 20 ft is $1 + \frac{20}{33} = 1.6$ atm. Ascent from 20 ft should not take place until all tissues are at a pressure of at most 2.2 atm. The pressure in the third tissue is already lower than this at the beginning of the 20 ft stop (and will not be increasing during this stop), so we just must make sure we wait long enough at 20 ft so that the pressure in the first two tissues drops to at most 2.2 atm. Using equation (2.43), we see that we must determine a value of t that ensures that

$$1.6 - (1.6 - 3.33)e^{-0.025t} \leq 2.2$$

and

$$1.6 - (1.6 - 2.35)e^{-0.01t} \leq 2.2.$$

These two inequalities require that, respectively,

$$1.73e^{-0.025t} \leq 0.6$$

and

$$0.75e^{-0.01t} \leq 0.6.$$

Solving these last two inequalities gives $t \geq 42$ min and $t \geq 22$ min, where times are rounded to the nearest minute. This tells us that after a 42-minute stop at 20 feet, it will be safe to ascend to the surface.

We can decrease the total amount of time spent in decompression stops if we plan a second stop at 10 ft. The details are outlined in Exercise 17.

2.3.3. Mixing problems. Imagine a tank containing a given volume of water in which a certain amount of a second substance, say salt, is dissolved. At time $t = 0$ a faucet opens, letting additional water (also possibly containing some salt) into the tank. At the same time, a drain opens at the bottom of the tank, allowing the perfectly stirred mixture to flow out. We can model the amount of salt in the tank using a differential equation. If we let $S(t)$ denote the amount of salt (say, in pounds) dissolved in the tank at time t (in minutes), then $\frac{dS}{dt}$ is the net rate of change of dissolved salt at time t, expressed in units of lb/min. This net rate is the difference between the rate at which the salt comes into the tank from the faucet and the rate at which it is being removed through the drain at the bottom:

$$\frac{dS}{dt} = \text{rate in} - \text{rate out}.$$

We'll look at an example.

Example 2.3.4. Suppose the tank initially contains 10 pounds of dissolved salt in 100 gallons of water. At time $t = 0$ water containing $\frac{1}{5}$ lb of salt per gallon begins flowing in at a rate of 2 gallons per minute. At the same instant a drain opens allowing the perfectly stirred mixture to flow out at 2 gallons per minute.

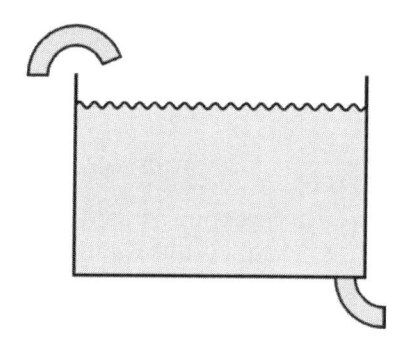

Figure 2.7.

Here the "rate in" term is

$$\text{rate in} = \left(\frac{1}{5} \text{ lb/gal}\right)(2 \text{ gal/min}) = \frac{2}{5} \text{ lb/min}.$$

The first factor, $\frac{1}{5}$ lb/gal, is the concentration of salt in the incoming fluid, while the second factor, 2 gal/minute, is the volume rate of flow in. The product has units lb/min, exactly as it should for the derivative $\frac{dS}{dt}$. Similarly, the rate out is computed as

$$\text{rate out} = (\text{concentration of salt in outgoing fluid in lb/gal})(2 \text{ gal/min}).$$

What is the expression for the concentration? The key phrase is "perfectly stirred". We assume that at each instant, all the salt in the tank is evenly dispersed throughout, so that its concentration at any time is given by

$$\frac{\text{total amount of salt in tank in lbs}}{\text{gal of mixture in tank}} = \frac{S(t)}{100}.$$

Notice how this expression involves our unknown function S. Thus our differential equation becomes

$$\frac{dS}{dt} = \text{rate in} - \text{rate out} = \left(\frac{1}{5}\right)(2) - \left(\frac{S(t)}{100}\right)(2),$$

or

$$\frac{dS}{dt} = \frac{2}{5} - \frac{S}{50}.$$

This equation is both separable and linear. From either perspective you can determine that it has solution

$$S(t) = 20 + Ce^{-\frac{t}{50}}.$$

To determine C, we use the initial condition $S(0) = 10$ lbs, which comes from the first sentence of the example. Thus $10 = 20 + C$, or $C = -10$, and

$$S(t) = 20 - 10e^{-\frac{t}{50}}.$$

Notice that $S(t) \to 20$ as $t \to \infty$, so that the limiting concentration in the tank is $\frac{20}{100} = \frac{1}{5}$ lb/gal, the same as the concentration in the incoming fluid. This is what we would expect, because when t is large, most of the fluid in the tank has been replaced by the fluid from the input faucet.

For an arbitrary initial value $S_0 = S(0)$ of salt in the tank, the corresponding solution is

$$S(t) = 20 + (S_0 - 20)e^{-\frac{t}{50}}.$$

If $S_0 = 20$, we have the equilibrium solution $S(t) = 20$; if $S_0 > 20$, the solution decreases to the equilibrum value 20; and if $S_0 < 20$, the solution increases to 20; see Fig. 2.8.

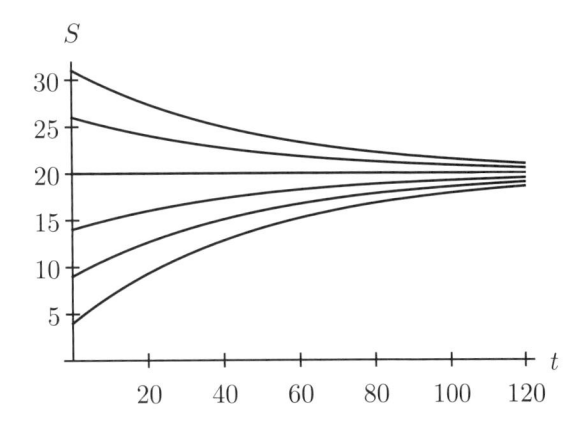

Figure 2.8. $S(t) = 20 + (S_0 - 20)e^{-t/50}$.

Example 2.3.5. We'll consider the same set-up as in the last example, but with one exception. Suppose the water (again containing $\frac{1}{5}$ lb/gal of salt) flows into the tank at 4 gal/min, while the perfectly mixed solution flows out of the tank at 2 gal/min. Thus, unlike in the previous example, the volume of salt solution in the tank does not stay constant. We'll still assume that at $t = 0$ there are 10 lbs of salt dissolved in the initial volume of 100 gal. Now our differential equation for the amount $S(t)$ of salt in the tank at time t is

$$\frac{dS}{dt} = \text{rate in} - \text{rate out}$$

where the "rate in" term is

$$(\text{concentration in lbs/gal})(\text{volume rate of flow in gal/min}) = \left(\frac{1}{5}\right)(4)$$

with units of lbs/min. Similarly the "rate out" term is computed as

$$(\text{concentration in lbs/gal})(\text{volume rate of flow in gal/min})$$

where the volume rate of flow out is 2 gal/min. The concentration in lbs/gal is determined as the ratio

$$\frac{S(t)}{\text{volume of fluid in tank at time } t}.$$

Since the fluid flows into the tank at 4 gal/min and out of the tank at only 2 gal/min, the volume in the tank increases at a rate of 2 gal/min. After t minutes, the volume is $100+2t$ gal. Therefore, the correct expression for concentration in lbs/gal is

$$\frac{S(t)}{100+2t}$$

and our equation for $\frac{dS}{dt}$ is

$$\frac{dS}{dt} = \frac{4}{5} - 2\frac{S(t)}{100+2t}.$$

This is a linear (but not separable) equation. Using the integrating factor $(100+2t)$, we can solve this equation to obtain

$$S(t) = \frac{1}{5}(100+2t) + \frac{C}{100+2t}.$$

With the initial condition $S(0) = 10$ we determine $C = -1{,}000$. If the tank will hold 200 gallons, how much salt does it contain when it starts to overflow? At this time, $100+2t = 200$, so $t = 50$ minutes at the instant it begins to overflow. The amount of salt at time $t = 50$ is

$$\begin{aligned} S(50) &= \frac{1}{5}(100+2(50)) - \frac{1{,}000}{100+2(50)} \\ &= 35 \end{aligned}$$

pounds.

2.3.4. The early measurement of time. The earliest devices for keeping track of time date from about 3500 BC and were obelisks or sundials which gave a rough idea of the time of day during sunlight hours. The first mechanical clock was the water clock, or clepsydra, from the Greek for "water thief". It was a standard device for measuring time from about 1500 BC until 1500 AD. The idea is very simple. A tank, cylindrical or another shape, is filled with water and a small opening at the bottom allows the water to flow out under gravity. We measure time by measuring the height of water remaining in the tank. Ultimately, the question is whether, by choosing the shape of the tank correctly, we can construct our clock so that the height of water changes by a fixed amount during a fixed period of time. This would mean that we can delineate the hours with equally spaced marks along the clepsydra.

Let's derive a differential equation for $\frac{dV}{dt}$ where $V(t)$ is the volume of water in the tank at time t. If the horizontal cross section at height y above the bottom of the tank has area $A(y)$ (see Fig. 2.9) and the tank is filled to height h, the basic volume principle of calculus says that the volume of water in the tank is given by integrating the cross-sectional area formula:

(2.45) $$V(h) = \int_0^h A(y)\,dy.$$

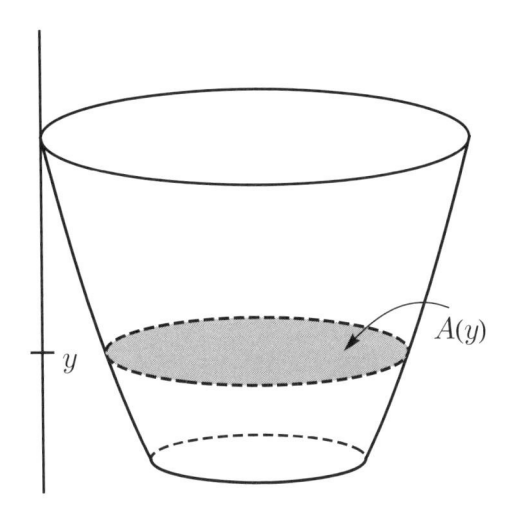

2.3. Applications: Time of death, time at depth, and ancient timekeeping

Figure 2.9. Cross section with area $A(y)$ at height y.

By the Fundamental Theorem of Calculus we see from (2.45) that

$$\frac{dV}{dh} = A(h).$$

The volume of water V is a function of the height of water h, and h is a function of time t, so that by the chain rule

(2.46) $$\frac{dV}{dt} = \frac{dV}{dh} \cdot \frac{dh}{dt} = A(h)\frac{dh}{dt}.$$

On the other hand, $\frac{dV}{dt}$, the rate of change of the volume with respect to time, will depend on how fast the water is flowing out of the hole at the bottom of the tank and how big the hole is. A reasonable assumption to make is that $\frac{dV}{dt}$ is the product of the exit speed of the fluid flow, or particle velocity, and the effective cross-sectional area of the hole:

(2.47) $$\frac{dV}{dt} = \text{particle velocity} \times \text{effective area of hole}.$$

According to Torricelli's law of fluid flow,[10] the exit speed of the water is the speed attained in free fall from a height equal to the height of the water in the tank. In Exercise 23 you are asked to show as a calculus exercise that the particle velocity is

(2.48) $$-\sqrt{2gh},$$

when the height of water in the tank is h. Here g is the acceleration due to gravity near the earth's surface in whatever units are appropriate to the situation at hand. For example, if we measure distance in feet and time in seconds, $g \approx 32$ ft/sec^2, while if we measure distance in centimeters, $g \approx 980$ cm/sec^2. The negative sign in the expression (2.48) for the velocity appears because we have chosen the positive direction to be upward.

Experimentally it is observed that as the water flows out of a circular hole at the bottom of the tank, the fluid stream has a smaller diameter than the aperture of the hole. This means that the effective cross-sectional area of the hole is $\alpha\pi r^2$, where r is the radius of the hole and α is a constant less than 1; for water we have $\alpha \approx 0.6$. By (2.47) we have

(2.49) $$\frac{dV}{dt} = -\sqrt{2gh}(0.6\pi r^2).$$

---

[10] Torricelli was a seventeenth-century Italian mathematician and physicist. He invented the barometer in 1643.

The units on this expression depend on our choice of time and distance units. For example, if we measure time in seconds and distance in centimeters, so $g \approx 980$ cm/sec^2, the units on dV/dt will be cm^3/sec. Notice that there are three variables in equation (2.49): V, h, and t. To obtain an equation with just two variables, equate the expression for dV/dt from (2.49) with our first expression for $\frac{dV}{dt}$ in (2.46). This gives the equation

$$A(h)\frac{dh}{dt} = -\sqrt{2gh}(0.6\pi r^2),$$

or

(2.50) $$\frac{dh}{dt} = \frac{-0.6\pi r^2 \sqrt{2gh}}{A(h)}.$$

This equation tells us the rate of change of the height of water in the tank with respect to time. We'll first use this to analyze a cylindrical tank.

Example 2.3.6. A cylindrical clepsydra has height 40 cm, radius 15 cm, and a circular hole on the bottom with a diameter of 1 mm. If the clepsydra is filled with water, how long does it take for the water level to drop 1 cm?

Figure 2.10. Cylindrical clepsydra.

Because the clepsydra is cylindrical with radius 15 cm, the cross-sectional area $A(h)$ at height h is constant $\pi(15)^2$, and equation (2.50) simplifies to

$$\frac{dh}{dt} = -\frac{(0.6)(0.05)^2}{15^2}\sqrt{2(980)h}.$$

Notice we wrote the radius of the hole as 0.05 cm to keep the units consistent. This is a separable equation. Using $(0.6)(0.05)^2\sqrt{2(980)}/(15^2) \approx 2.951 \times 10^{-4}$, we have the (approximate) implicit solution

$$2\sqrt{h} = (-2.951 \times 10^{-4})t + c.$$

We set $h = 40$ at time $t = 0$ to determine $c = 2\sqrt{40}$. Solving explicitly for h as a function of t we obtain

$$h(t) = (-1.475 \times 10^{-4}t + \sqrt{40})^2.$$

The time for the water level to drop from 40 to 39 cm is determined by solving

$$39 = (-1.475 \times 10^{-4}t + \sqrt{40})^2.$$

This gives $t \approx 539$ seconds, or almost 9 minutes.

2.3. Applications: Time of death, time at depth, and ancient timekeeping   55

To get some idea for how well a cylindrical-shaped clepsydra will work as a clock, let's compare how long it takes for the water level to fall the first 1 cm with the time required for the level to drop the final 1 cm. The time for the water level to fall from 1 to 0 cm is determined by solving

$$1 = (-1.475 \times 10^{-4} t + \sqrt{40})^2$$

and

$$0 = (-1.475 \times 10^{-4} t + \sqrt{40})^2$$

and subtracting the two values. The result is approximately $42{,}878 - 36{,}098 \approx 6{,}780$ sec. This means it takes about 113 min for the water level to drop the last centimeter. Thus a cylindrical clepsydra actually functions very poorly as a timekeeping device, since a fixed change in the water level doesn't correspond at all well to a fixed change in time. Of course, we could use the cylindrical clepsydra to measure a fixed interval of time (more like a stopwatch than a clock).[11] The problem with using the clepsydra as a clock was certainly recognized from the beginning, and even some ancient specimens had shapes that worked better than our cylindrical one (see Exercise 25).

A perfect clepsydra? Starting with the differential equation model we developed above, can we determine a shape for the clepsydra so that it functions correctly as a clock? This problem was posed and solved by the French physicist Edme Mariotte in about 1680, who described it as follows: *It is proper enough in this Place to resolve a pretty curious Problem which Toricelly [sic] has not undertaken to resolve, tho' he propos'd it; this Problem is to find a Vessel of such a Figure that being pierc'd at the Bottom with a small Hole the Water should go out, its upper Surface descending from equal Heights in equal Times....*[12] Mariotte did obtain a correct answer, but others later misunderstood it—no great surprise given Mariotte's convoluted reasoning—and repeated it incorrectly.

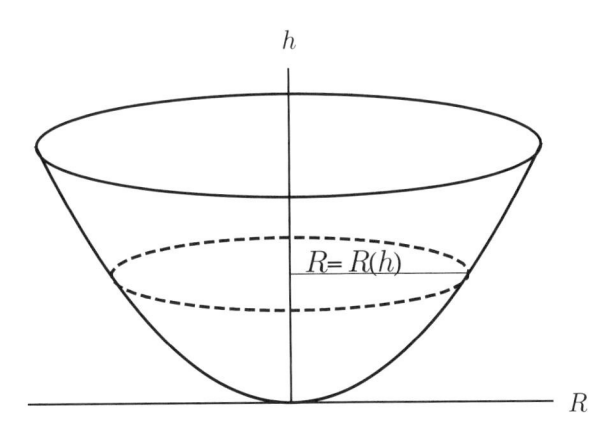

Figure 2.11. Revolving a curve to design a clepsydra.

Let's assume our desire is to design a "perfect" clepsydra as a surface of revolution (see Fig. 2.11) with circular cross sections of radius $R(h)$ at height h from the bottom of the tank. We want to determine $R(h)$ so that the corresponding clepsydra functions as an accurate timekeeper. The key observation is that this happens precisely when $\frac{dh}{dt}$ is a (necessarily negative) constant, that is,

---

[11] In Greece, clepsydras were used in this way in legal proceedings. The advocate for the accused was permitted to argue the case until the water ran out, with different size clepsydras used for different crimes. From this we get the expression, "Your time has run out."

[12] This is from an English translation by J. T. Desaguliers in 1717 of Mariotte's 1686 book *Traite du Mouvement des Eaux et des Autres Corps Fluides*, or *The Motion of Water and Other Fluids*.

when the rate of change of the height with respect to time is constant. By equation (2.50) we have

$$\frac{dh}{dt} = \frac{-0.6\pi r^2 \sqrt{2gh}}{A(h)},$$

where the cross-sectional area $A(h)$ is given by $\pi R(h)^2$ if we are assuming the cross section to be a circle of radius $R = R(h)$ at height h. Our problem is to determine R (as a function of h) so that $\frac{dh}{dt}$ is some constant C. Solving

$$C = \frac{-0.6\pi r^2 \sqrt{2gh}}{\pi R^2}$$

and remembering that C, r, and g are all constants while h and R are variables, we see that we want

$$cR^2 = \sqrt{h}$$

for c constant. Equivalently,

$$h = KR^4$$

for some (positive) constant K. Thus we construct our "perfect" clepsydra as a surface of revolution by revolving any curve of the form $h = KR^4$ in the Rh-plane around the h-axis; see Figs. 2.12–2.13.

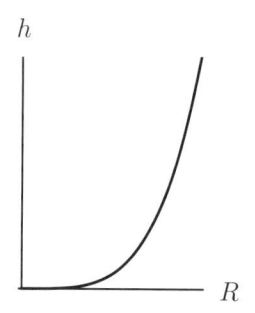

Figure 2.12. Creating a perfect clepsydra by revolving the curve $h = KR^4$ about the h-axis.

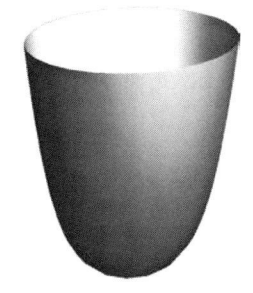

Figure 2.13. A perfect clepsydra.

2.3.5. Exercises.

1. A body is discovered with a temperature of 83°F in a room with ambient temperature 70°F. Three hours later the temperature is 78°F. About how long after death was the body discovered?

2. Suppose that a body with temperature 85°F is found at midnight and the ambient temperature is a constant 70°F. The body is removed quickly (assume instantly) to the morgue where the ambient temperature is 40°F. After one hour the body temperature is found to be 75°F. Estimate the time of death using Newton's law of cooling (note that the value of k doesn't change when the body is moved).

3. An unruly toddler is having his temperature checked. It is 95°F at the outset (time of insertion of thermometer), 99°F one minute later, and 100°F one minute after that, at which point the toddler stops cooperating. What is his temperature?

4. Professor Tarry always has a cup of coffee before his 9:00 AM class. Suppose the temperature of the coffee is 200°F when it is first poured at 8:30 AM and 15 minutes later it has cooled to 120°F in a room whose temperature is 70°F. However, Professor Tarry will never drink his coffee until it has cooled to 95°F. Will he be late to class?

2.3. Applications: Time of death, time at depth, and ancient timekeeping

5. Show that when you solve either equation (2.39) or (2.40) explicitly for T as a function of t, the result can be written as
$$T = A - (A - T(0))e^{-kt}$$
where $T(0)$ is the temperature of the object at time $t = 0$.

6. A 5 lb tenderloin of beef roast, initially at 45°F, is put into a 350°F oven at 5:00 PM. After 75 minutes its internal temperature is 120°F. When will the roast be medium rare (internal temperature 145°F)?

7. A time-of-death rule of thumb in some homicide departments is that when a body is found with temperature within 1% of the room temperature, the person has been dead (at least) 18 hours. In a room with ambient temperature of 70°, what does this rule of thumb principle give for the constant k in Newton's law of cooling? Assume normal body temperature is 98.6°, and be sure to give units on your answer.

8. President Donald Trump and Chinese President Xi Jinping meet to discuss trade issues. They order coffee and receive cups of fine Columbian (9 ounces each) served at 170°F in identical cups. President Xi immediately adds 1 ounce of cream (kept chilled at 40°F) to his cup. Mr. Trump, however, at the earlier suggestion of his press secretary Sarah Huckabee Sanders, does not add his ounce of cream until the two are ready to drink their coffee sometime later, for the White House expects a stalemate on issues of substance, and Ms. Sanders knows that in this event, the media will scrutinize the tiniest trivia for symbolic winners and losers. Ms. Sanders plans to spin the media on the president's savvy in conserving heat in his coffee and thus drinking the hotter cup. Question: Does Ms. Sanders deserve
   (a) a raise or
   (b) compulsory enrollment in a night school differential equations course at George Washington University?

   To answer this intelligently, you need Newton's law of cooling to calculate the temperature of each cup of coffee at the time it is drunk and so decide which man (if either) drinks the hotter coffee. Assume the cream is kept at a constant temperature of 40°F until it is poured.

   In your calculations, assume that if you have b ounces of a liquid at temperature T_1 and c ounces of another liquid at temperature T_2, then mixing them together yields a mixture whose temperature is the weighted average
   $$\frac{bT_1 + cT_2}{b + c}.$$

9. Repetitive dives carry special decompression dangers (repeated dives carried out in a 12-hour period are considered "repetitive dives"). This is because on the second, or subsequent, dive the tissues are already at higher than 1 atm pressure at the outset. Suppose the first dive is made at 66 ft for 20 min, followed by a direct ascent to the surface and an hour spent at the surface. How long can this diver then spend at 80 ft if he wants to make a "no-stop" second dive? Use a two-tissue model with $k_1 = 0.025$ min^{-1} and $k_2 = 0.01$ min^{-1}.

10. (a) The model discussed in Example 2.3.2 assumes *instantaneous* descent and ascent from one depth to the next "stopping depth". Suppose instead that we assume the diver makes a steady descent at a rate of r ft/min from the surface. The differential equation
$$\frac{dP}{dt} = k(P_a - P)$$
should be modified so that the external pressure P_a is not constant but reflects the changing pressure as the diver swims down:
$$P_a = 1 + r\frac{t}{33}.$$

Solve the resulting differential equation
$$\frac{dP}{dt} = k\left(1 + r\frac{t}{33} - P\right).$$

(b) Next assume that starting from a depth of D ft, the diver makes a steady ascent at r ft/min. Give an expression for the ambient pressure P_a after t minutes, and then give the corresponding differential equation for $\frac{dP}{dt}$ during the ascent.

11. Recently *The Washington Post* ran a front-page story about the deaths of two Navy divers and subsequent court-martial trial of a Navy supervisor.[13] This brought the number of military divers who had died in operations at the Aberdeen Proving Ground "super pond" since January 2012 to five. Extreme decompression sickness had been implicated in at least some of the earlier deaths. The dive in February 2013 was to be a bounce dive (no decompression stops) to a depth of 150 feet and was to take 4 minutes for the descent and 5 minutes for the ascent, with only enough time spent at the bottom to make a visual inspection. One diving supervisor had reportedly said beforehand, "Not a good idea. Do the math." The purpose of this problem is to do some of the math, assuming perfect execution of the plan.

(a) Using the differential equation (see Exercise 10)
$$\frac{dP}{dt} = k\left(1 + \frac{150}{4}\frac{t}{33} - P\right)$$
for divers taking 4 minutes to make a steady descent from the surface to 150 feet, with the single k-value $k = 0.1$, find the pressure in the tissues as the divers reach the bottom. Assume $P(0) = 1$.

(b) Assume the divers spend 2 minutes at the depth of 150 feet making their inspection. Find their tissue pressure at the end of this 2 minutes.

(c) Find an initial value problem to model the ascent, assuming 5 minutes to go 150 ft. You'll need first to find an expression for the ambient pressure t minutes into the ascent phase; then use this to modify the differential equation in (a).

(d) Still using $k = 0.1$, find the tissue pressure as the divers surface. Is it more than the "safe" value of 2.2?

12. This problem is inspired by the Austrian diver Herbert Nitsch who in 2007 successfully completed a 702-ft "breath-hold" dive (no air tank!) to earn the title of "Deepest Man on Earth". He competed in the "no-limits" category of breath-hold diving, which permits the use of certain equipment to speed his descent and ascent. The total time of his dive was 4.25 min.

Assume a scuba diver descends to a depth of 702 ft at a constant rate of $1{,}404/4.25 \approx 330$ ft/min and then immediately begins an ascent to the surface at the same rate. Using the differential equations of Exercise 10, with constant descent/ascent rates of 330 ft/min and only one tissue with $k = 0.1$, what would the pressure in this tissue be at 702 ft, and what would the pressure be upon return to the surface? Do you expect this to cause decompression sickness?

In actuality, Nitsch took 90 sec to reach the bottom, spent 7 sec on the bottom, ascended to 31 ft in 104 sec where he had a decompression stop of 30 sec, and then finished his ascent in 24 sec. At one time it was believed that breath-hold divers could not get decompression sickness since they are not breathing air at high pressure. However, this is now known to be false. Breath-hold divers who make repeated dives in a short period of time are particularly at risk.[14]

---

[13] Michael Ruane, *After deaths of 2 Navy divers, a court-martial*, The Washington Post, January 13, 2014.

[14] Several years later, Nitsch attempted another no-limits breath-hold dive off the coast of Greece, intending to descend 800 feet at a rate of 10 ft/sec. He planned for a 1 minute decompression stop on the ascent, at a depth of 30 feet. He actually descended 830 feet, but at a slower rate. On the ascent, he blacked out and was rescued by safety divers who brought him immediately to the surface. Briefly regaining consciousness, Nitsch attempted unsuccessfully to return to 30 feet (with a mask

13. Diving literature tends to refer to "half-times" more than k-values. Using the model discussed in Example 2.3.2, the connection between the two is simple.
    (a) When is the difference between the ambient pressure and the tissue pressure, as given by equation (2.43), exactly half of the original difference between ambient pressure and tissue pressure, $P_a - P(0)$? This is the half-time, and your answer should only involve k, not the original tissue pressure $P(0)$ or the ambient pressure P_a.
    (b) What are the half-times corresponding to $k = 0.025$ min^{-1}, $k = 0.01$ min^{-1}, and $k = 0.006$ min^{-1}? Original researchers in decompression sickness used a range of half-times varying from 5 min to 120 min when constructing decompression tables.

14. This problem is inspired by a 2007 episode of Fox TV's *House*. Dr. House and Dr. Cuddy are flying back from Singapore when a man on board the plane gets violently ill. Initially they suspect meningococcus meningitis, as others on the plane start to show similar symptoms. After House performs a spinal tap with ad hoc supplies, it appears this diagnosis is not correct. In the last 5 minutes of the show House finds a receipt from a scuba rental shop and diagnoses the bends. The point of this problem is to investigate whether flying soon after diving can cause decompression sickness.
    (a) Suppose a diver spends 22 min in a dive at 99 ft and then (instantaneously) ascends to the surface. What are the pressures in his tissues after 22 minutes at 99 ft if you use a two-tissue model with $k_1 = 0.025$ min^{-1} and $k_2 = 0.01$ min^{-1}?
    (b) What are the tissue pressures after 30 minutes at sea level? (Use your answers from (a) for $P(0)$ at the start of the time at sea level.)
    (c) Is it safe to then be instantaneously flying in an airplane where the pressurized cabin is at 0.75 atm?
    (d) If only 20 minutes are spent at sea level instead, is it safe to be instantaneously flying at 0.75 atm?

15. Extreme sports enthusiasts may find themselves flying in the same 12-hour period in which they have been diving. This increases the risk of decompression sickness. Suppose a diver spends 30 min in a dive at 100 ft, (instantaneously) ascends to the surface, spends 15 min on the surface, and then is (instantaneously) flying in an airplane where the pressurized cabin is at 0.75 atm. Is there a danger of decompression sickness? Assume a two-tissue model, with $k_1 = 0.01$ min^{-1} and $k_2 = 0.006$ min^{-1}.

16. When a diver is injured and must be medically evacuated by plane, the risk of decompression illness is increased due to the additional decrease in pressure caused by the flight. An injured diver has spent 22 min at 90 ft, instantaneously ascended to the surface, and been placed on a rescue plane 15 min later. Is it safe for the plane to (instantaneously) ascend to an altitude with pressure 0.75 atm? Use a single tissue with $k = 0.025$ min^{-1}.

17. Suppose in Example 2.3.3 we plan a second stop at 10 ft, which corresponds to a pressure of 1.3 atm. The ascent from 20 ft to 10 ft should not take place until all tissues are at a pressure of less than or equal to $(1.3)(2.2) = 2.86$ atm. The pressures in the second and third tissues are already lower than this at the beginning of the 20-ft stop, so we must just make sure we wait long enough at 20 ft so that the pressure in the first tissue drops to 2.86 atm.
    (a) How long should the stop at 20 ft be so that the pressure in the first tissue ($k_1 = 0.025$ min^{-1}) is at most 2.86 atm at the end of the stop? What are the pressures in the three tissues at the end of this stop?
    (b) After the required time at the 20-ft stop, we ascend (instantaneously) to 10 ft. How long must we wait there until we can safely ascend to the surface?

---

and an oxygen tank) for decompression but blacked out again. Transported to a hospital, he suffered a cardiac arrest and multiple strokes from decompression sickness. He lived but has not fully recovered.

(c) Compare the total time spent in decompression stops under this scenario, with the time in decompression with a single stop at 20 ft as in Example 2.3.3.

18. A tank contains 60 gallons of a saltwater solution. At time $t = 0$ the solution contains 30 lbs of dissolved salt. At that instant, water containing $\frac{1}{4}$ lb of salt per gallon begins flowing in from a faucet at a rate of 3 gallons per minute, and the perfectly stirred solution begins flowing out, also at a rate of 3 gallons per minute.
    (a) Find the amount of salt $S(t)$ in pounds in the tank at time t.
    (b) Suppose that after 20 minutes have elapsed, the concentration of salt in the inflow is doubled. Find $S(t)$ for all $t > 20$.
    (c) Sketch the graph of $S(t)$, showing the part corresponding to $0 \leq t \leq 20$ and the part corresponding to $t \geq 20$. What is the limiting value of $S(t)$ as $t \to \infty$?

19. At time $t = 0$ there are 4 lbs of salt dissolved in 40 gallons of water in a tank. At that instant, water containing $\frac{1}{2}$ pound of salt per gallon begins flowing into the tank at a rate of 1 gal/min. At the same time a drain opens and the well-stirred mixture begins flowing out at a rate of 2 gal/min.
    (a) Find the amount of salt in pounds in the tank for $t > 0$.
    (b) When is the amount of salt in the tank a maximum?
    (c) Find the concentration of salt in the tank at any time $t > 0$.
    (d) What is the concentration of salt in the tank at the instant the tank empties?

20. A tank initially contains 1 gram of blue dye dissolved in 10 liters of water. Pure water will be added to the tank at a certain rate, and the well-mixed solution will be drained from the tank at the same rate. The tank is considered "cleaned" when the dye concentration in the tank is no more than 0.001 gr/liter.
    (a) If the rate at which pure water is added (and the tank is drained) is 2 liters/min, how long does it take until the tank is cleaned?
    (b) If we want to clean the tank in 5 minutes, how quickly must the pure water be poured in?

21. Water from a chemical company's containment pond, containing 10g of the insecticide Kepone per liter, leaks into a small lake at a rate of 1 liter per hour. Assume the lake contains 50,000,000 liters of water, has a fresh water inflow of 9,999 liters per hour (containing no Kepone), and has a well-mixed outflow of 10,000 liters per hour. Based on these modeling assumptions, roughly how many days do authorities have to find the containment pond leak before the concentration of Kepone in the lake reaches 0.0003 g/liter? (Remark: Converted to parts per million, 0.0003 g/liter is equivalent to 0.3 ppm, which is the level at which fish becomes unsafe to eat according to the FDA.)

22. The five Great Lakes form the largest collection (by area) of fresh water lakes in the world, and the second largest by volume. The issue of pollution in the Great Lakes caught the nation's attention in June of 1969 when the Cuyahago River, which flows into Lake Erie, caught fire because of high levels of pollutants, and Lake Erie was declared "dead". Subsequently, a variety of steps, including the Clean Water Act of 1972, were taken to address pollution in these lakes.

    The purpose of this problem is to present a greatly simplified model of a lake and investigate the possibilities of recovery from pollution, applying our model to two of the Great Lakes. If we assume that precipitation into the lake is exactly balanced by evaporation and that the flow of water r in km^3/year into the lake is equal to the flow of water out of the lake, then the volume V (in km^3) of the lake will remain constant. We also assume (unrealistically) that pollutants are uniformly distributed throughout the lake and that *at time $t = 0$ all pollution into the lake is stopped*.
    (a) Find a first-order initial value problem for the amount $x(t)$ of pollutants in the lake at time t in years, if $x(0) = x_0$.

2.3. Applications: Time of death, time at depth, and ancient timekeeping 61

(b) For Lake Superior, at the top of the Great Lakes (see Fig. 2.14), V is approximately 12,220 km^3 and r is approximately 65 km^3/year. Using these figures, solve the initial value problem in (a) for $x(t)$, the amount of pollution in Lake Superior, in terms of x_0.

(c) How long does it take for the pollution level in Lake Superior to drop to 10% of its original values, assuming all new pollution is stopped?

(d) For Lake Michigan, the other lake at the "top" of the Great Lakes chain, $V = 4920$ km^3 and $r = 160$ km^3/year. How long would it take for pollution levels in Lake Michigan to drop to 10% of their original value, assuming all pollution is stopped?

For the lower lakes—Huron, Erie, and Ontario—a more complicated model is required, as they receive a significant portion of their water inflow from one (or more) of the other Great Lakes, so that even if all pollution into the lakes as a whole is stopped, pollutants continue to flow into the lower lakes from the upper ones. This issue will be considered later, in Section 8.8.

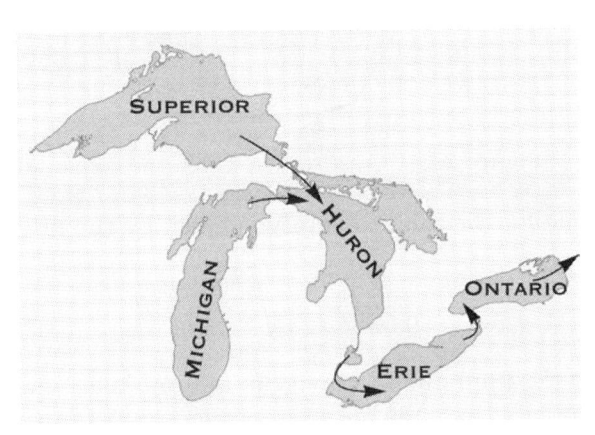

Figure 2.14. The Great Lakes. Source: Wikimedia Commons, author: Phizzy, licensed under the GNU Free Documentation license and the Creative Commons Attribution-ShareAlike 3.0 Unported (https://creativecommons.org/licenses/by-sa/3.0/deed.en) license.

23. Torricelli's law of fluid flow says that when water in a tank (with an open top) flows out through a hole in the bottom of the tank, it does so with velocity equal to that of an object falling freely from a height equal to the water level to the hole. In this problem we determine an expression for this exit velocity. When an object falls near the surface of the earth and we neglect air resistance, its acceleration $a(t)$ is constant

$$a(t) = -g, \qquad (2.51)$$

where the numerical value of the constant g depends on the units we wish to use.

(a) Recalling that for motion along a straight line, acceleration is the derivative of velocity, show that the velocity function is given by

$$v(t) = -gt + v(0) = -gt \qquad (2.52)$$

and then determine the position function for the freely falling object if it starts at height h.

(b) How long does it take to fall from height h to position 0?

(c) Determine the velocity at position 0; this is the exit velocity of the water as it leaves the tank.

(d) Show that the exit velocity v_e can also be determined from a conservation of energy principle by solving

$$mgh = \frac{1}{2}mv_e^2,$$

which equates the potential energy of a particle of mass m at height h (potential energy on the surface of the water) with the kinetic energy of a particle of mass m and velocity v_e (kinetic energy of water as it leaves through the hole on the bottom).

24. A clepsydra is designed so that there are no depressions that would prevent some of the water from draining out. Suppose the cross section of the clepsydra tank at height h above the bottom has area $A(h)$ (see Fig. 2.15) and the total height of the tank is H. Assume the hole in the bottom is circular, with radius r, and that the tank is initially filled with water.

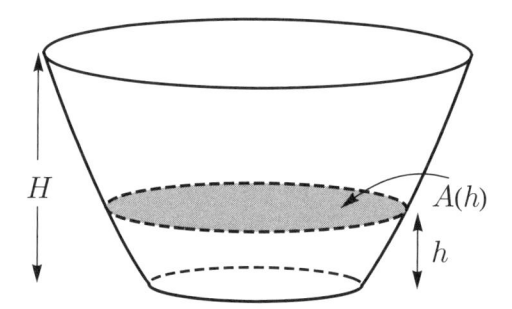

Figure 2.15. Cross section with area $A(h)$ at height h.

(a) Show that the time for the tank to empty is given by the integral

$$\frac{1}{0.6\pi r^2 \sqrt{2g}} \int_0^H \frac{A(h)}{\sqrt{h}} dh$$

where g is the acceleration due to gravity in appropriate units.

(b) Notice that the integral in (a) is an improper integral. Must it converge? In other words, will the tank always drain in a finite amount of time, even though the rate of flow through the hole in the bottom decreases to 0 as it drains? (Hint: Does the improper integral

$$\int_0^H \frac{1}{\sqrt{h}} dh$$

converge or diverge?)

(c) Two identical cylindrical tanks, with same size circular hole of diameter r at their bottoms, are placed as shown. The height H of the tank on the left is equal to twice the radius R of its circular base. When the tanks are filled with water, which takes longer to completely drain?

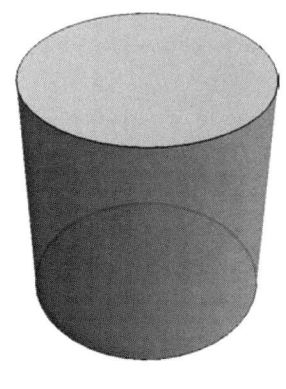

Figure 2.16. $H = 2R$.

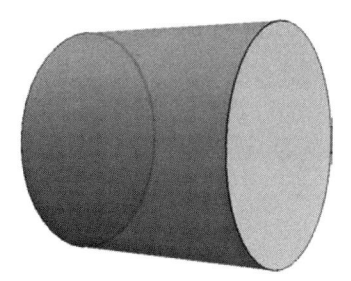

Figure 2.17. Same tank, turned on its side.

2.3. Applications: Time of death, time at depth, and ancient timekeeping   63

25. One of the oldest surviving clepsydras was used in the Temple of Amun-Re in Egypt c. 1400 BC. It was found in pieces in the ruins of Karnak and reconstructed. Shaped like a section of a cone, its interior dimensions are as shown in Fig. 2.18. In writing about this clepsydra, the French archaeologist Georges Daressy concludes that the conical shape must have been decorative rather than functional: *"...the container is very pretty, but without real use since its markings are whimsical.... the gradations are obviously equidistant from top to bottom, and given the shape of the container, the volume of water increases steadily for a given vertical distance, and so the hour marks should gradually get closer and closer as one goes up the tank. The engraver seems to have simply repeated the markings on a cylindrical clepsydra out of ignorance, rendering it useless"*.[15] The point of this problem is to show that he is wrong about this and that its shape, with equally spaced marks for the hours, makes it function more accurately as a clock than a cylindrical clepsydra.

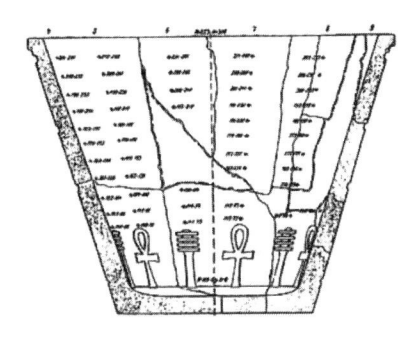

Figure 2.18. Height: 12.6 in; diameter at top: 17.4 in; diameter at base: 8.5 in.

(a) Show that the rate of change $\frac{dh}{dt}$ of the water in the Karnak clepsydra when the tank is full and the rate of change when there is only 1 inch of water left in the tank are almost equal, unlike the case of the cylindrical tank analyzed in the text. Assume that the radius of the hole at the bottom of the tank is 0.018 in.
(b) Suppose you have a cylindrical clepsydra with height 12.6 in and circular hole in the bottom of radius 0.018 in. What should its radius be so that it drains completely in 12 hours?
(c) Figs. 2.19 and 2.20 show the graph of h vs. $\frac{dh}{dt}$, $1 \leq h \leq 12.6$, for the Karnak clepsydra and for the clepsydra described in (b) (units on dh/dt are inches/sec). Which is which? How can you tell from these graphs that the Karnak design functions better as a clock?

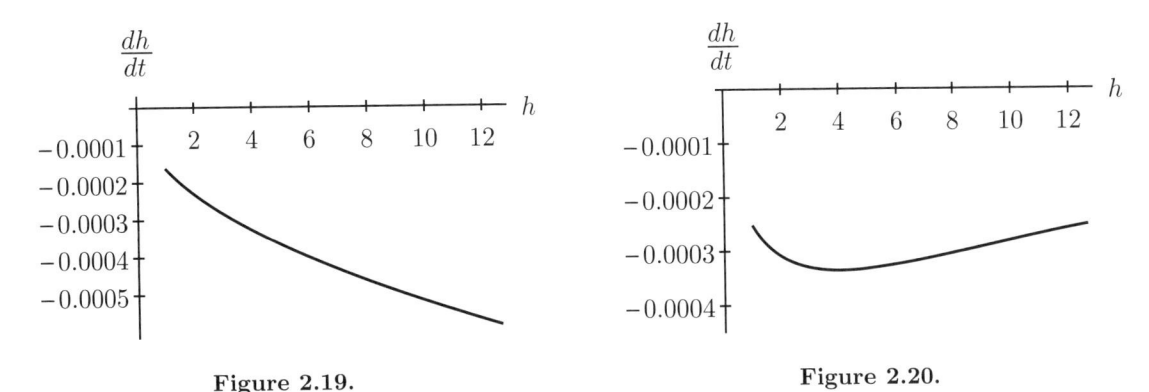

Figure 2.19.

Figure 2.20.

---

[15]From the article *Deux Clepsydras Antiques* by M. Georges Daressy, Bulletin de l'Institut egyptien, serie 5,9, 1915 pp. 5–16. Translation is by the authors.

26. In the clepsydra from Karnak described in the last exercise, the markings on the sides indicate it was used to measure 12 hours. Show how to use this information and the dimensions as shown in Fig. 2.18 to estimate the radius of the hole at the bottom to be approximately 0.018 in. Since the clepsydra was found broken up and not all pieces were recovered, this may be the only way to determine this crucial dimension.

27. The clepsydra of Edfu (another ancient specimen) is cylindrical in shape, with interior depth 10.8 inches and interior diameter 6.6 inches. If it is designed to measure 12 hours, what should the diameter of the circular hole at the bottom be?

28. A clepsydra as shown is obtained by revolving the portion of the graph of $y = cx^4$ for $0 \leq x \leq 1$ ft around the y-axis, and a circular hole is made at the bottom with radius $r = 0.0024$ ft. Determine c if the water level in the clepsydra is to fall at a constant rate of 3 in/hr. How many hours does this clock measure?

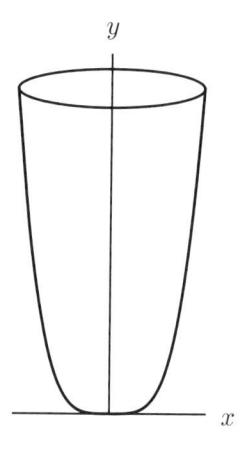

Figure 2.21. Revolving $y = cx^4$, $0 \leq x \leq 1$, about the y-axis.

29. One theory of metabolism asserts that for any animal, from a mouse to an elephant, the metabolic rate is proportional to the 3/4 power of the weight of the animal. Furthermore, when an animal is not fed, it will lose weight at a rate proportional to its metabolic rate. This leads to the differential equation
$$\frac{dW}{dt} = -kW^{3/4},$$
where W is the weight (in kg) at time t and k is a positive constant.
   (a) Determine W as a function of t (in days) if $k = 0.03$ and the animal's weight at time $t = 0$ is W_0.
   (b) How long does it take for the animal's weight to drop by 50 percent? The Law of Chossat says this is the approximate time to death by starvation.

30. The von Bertalanffy growth model is often applied to describe the average length L of a fish of a given species at time t. The length function L is assumed to obey the differential equation
$$\frac{dL}{dt} = k(L_\infty - L)$$
where k is the "growth rate constant" and L_∞ is the theoretical maximum of the length.
   (a) Before actually solving this differential equation, explain why it tells you that the length function will be increasing more slowly the older the fish is. What does this say about the concavity of the solution function?
   (b) Solve the von Bertalanffy growth differential equation assuming $L(0) = 0$.
   (c) Check that your solution does indeed have the kind of concavity you described in (a).

(d) What is $\lim_{t\to\infty} L(t)$?

(e) The value of k differs for different species of fish. For a certain group of tropical water fish, the mean value of k is approximately 0.8/yr. About how many years does it take for these fish to reach half of their limiting length value L_∞?

(f) For an individual species, the growth rate constant can also vary dramatically with the habitat. Suppose, for example, that a particular species of fish in one location has a growth constant of approximately $k = 3.6$ and limiting length value of $L_\infty = 13$ cm, while for the same species of fish in a different location k is about 0.12 and L_∞ is about 40 cm. What is the approximate length of a one-year-old fish of this species in each of these locations?

31. The differential equation for the von Bertalanffy growth model (see the previous problem) has also been used to model the development of human cognition. If we let $b(t)$ measure cognition t years after birth and set the theoretical maximum value of b at 1, we have

$$\frac{db}{dt} = k(1-b).$$

Solving this differential equation for $b(t)$ assuming that $b(0) = 0$ will give us a function that tells us what proportion of "total cognitive development" an individual has reached by age t. If an individual has reached 90% of his cognitive development by age 21, what is the value of k for this person, and at what age did he reach 50% of his cognitive development?

32. A tank contains 50 gallons of pure water. At time $t = 0$, water containing $1/(t+1)$ lb of salt per gallon begins flowing in at a rate of 2 gallons per minute, and the perfectly stirred solution begins flowing out, also at a rate of 2 gallons per minute.

(a) Use the method of Example 2.1.3 to find an expression for the amount $S(t)$ of salt in the tank at time t.

(b) (CAS) Plot your solution of part (a) and then estimate the approximate maximum value of S and approximately when it occurs.

2.4. Existence and uniqueness theorems

In this section we consider initial value problems

(2.53) $$\frac{dy}{dt} = f(t, y), \quad y(t_0) = y_0,$$

with two questions in mind:

(Q_1) Does this initial value problem have a solution?

(Q_2) If the answer to (Q_1) is yes, is the solution *unique*, so that (2.53) has only one solution?

Thus we are concerned with the *existence* (Q_1) and *uniqueness* (Q_2) of solutions to (2.53). Consideration of these questions will highlight some contrasts between linear and nonlinear equations.

We begin with several examples.

Example 2.4.1. Consider the initial value problem

(2.54) $$\frac{dy}{dt} = -2\sqrt{y}, \quad y(1) = 0.$$

Recall from our discussion in Section 2.3.4 (see Example 2.3.6 in particular) that the differential equation in (2.54) is basically the differential equation for the height y of water in a draining cylindrical tank at time t. We've chosen the constant 2 in this equation for convenience. With this interpretation, the initial condition in (2.54) says that "at time $t = 1$ the tank is empty".

Now (2.54) has equilibrium solution $y(t) = 0$ by "inspection". This corresponds to the tank being empty for all time. If we solve the differential equation by separating variables, we see that

$$\frac{1}{2\sqrt{y}} dy = -dt,$$

or

(2.55) $$\sqrt{y} = -t + c.$$

We can go one step further and solve for y explicitly as $y = (c - t)^2$; however, this requires one bit of caution. To see why, let's substitute the function $y = (c - t)^2$ into the differential equation of (2.54). The result is

$$\begin{aligned} -2(c-t) &= -2\sqrt{(c-t)^2} \\ &= -2|c-t| \end{aligned}$$

since $\sqrt{a^2} = |a|$. This is satisfied only if $c - t \geq 0$, or $t \leq c$. Thus we have solutions to $\frac{dy}{dt} = -2\sqrt{y}$ of the form

$$y(t) = (c - t)^2 \quad \text{for } t \leq c.$$

Imposing the initial condition $y(1) = 0$ gives $c = 1$, and $y(t) = (1 - t)^2$ for $t \leq 1$ solves our initial value problem. We can extend this solution to be defined for all times $-\infty < t < \infty$ by setting

$$y(t) = \begin{cases} (1-t)^2 & \text{if } t \leq 1, \\ 0 & \text{if } t > 1. \end{cases}$$

The graph of this function is shown in Fig. 2.22.

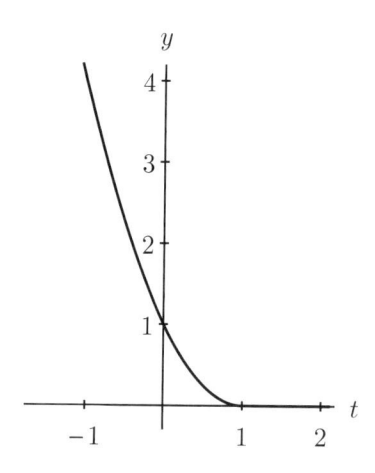

Figure 2.22. Graph of $y(t)$.

In Exercise 12 you are asked to verify that this extension is in fact differentiable at every point and satisfies the differential equation everywhere. In our physical set-up, this solution gives the height of the water in a tank that *just empties* at time $t = 1$ and stays empty for all times $t > 1$. This perspective suggests how we might find many other solutions to (2.54), namely

(2.56) $$y = \begin{cases} (c-t)^2 & \text{if } t \leq c, \\ 0 & \text{if } t > c, \end{cases}$$

for *any choice* of $c \leq 1$. Now the tank just empties at time $t = c$ and stays empty for later times. As long as $c \leq 1$, the height of water at time $t = 1$ is 0, so that the initial condition $y(1) = 0$ is met. Fig. 2.23 shows the graphs of several of these solutions to (2.54).

2.4. Existence and uniqueness theorems 67

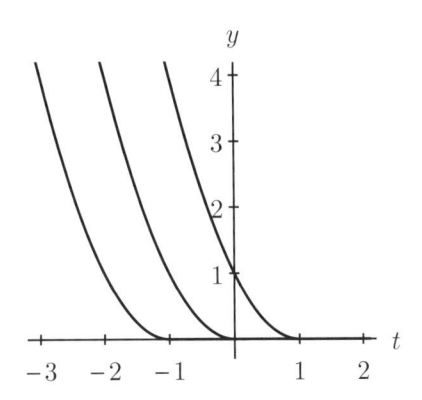

Figure 2.23. Several solutions to the initial value problem in (2.54).

Summarizing, the initial value problem in (2.54) does not have a *unique* solution. In fact we have found infinitely many solutions: the equilibrium solution and the solutions in (2.56) for any $c \leq 1$. We even gave a physical interpretation to the problem in which these infinitely many solutions are each meaningful.

Existence and uniqueness for linear equations. We begin with an example of a *linear* initial value problem.

Example 2.4.2. Consider the equation

$$\frac{dy}{dx} + y = xe^{-x}. \tag{2.57}$$

This is a linear equation with independent variable x, in standard form, having coefficient functions $p(x) = 1$ and $g(x) = xe^{-x}$. These functions are continuous on $-\infty < x < \infty$. Multiplying by the integrating factor e^x we can rewrite the equation equivalently as

$$\frac{d}{dx}(e^x y) = x \tag{2.58}$$

so that

$$e^x y = \frac{x^2}{2} + c,$$

or

$$y = \frac{x^2 e^{-x}}{2} + ce^{-x}, \tag{2.59}$$

for arbitrary constant c.

This is a solution to (2.57) for all x in $(-\infty, \infty)$. Before we impose an initial condition, let's pause to observe that the functions in (2.59) form the *complete* family of all solutions to (2.57). How do we know this? Equation (2.57) is equivalent to equation (2.58), meaning that they have exactly the same solutions. From calculus (or more precisely, from the Mean Value Theorem) we know that the only functions with derivative x are the functions $x^2/2 + c$ for c an arbitrary constant, so that if y solves (2.58), it must appear in the family (2.59).

Now suppose we add an initial condition:

$$y(x_0) = y_0, \tag{2.60}$$

where x_0 and y_0 can each be any real number. There will be exactly one choice of the constant c in (2.59) so that the resulting function y satisfies our initial condition. You can see this by substituting the values and solving for c:

$$y_0 = \frac{x_0^2 e^{-x_0}}{2} + ce^{-x_0}$$

if and only if

$$c = y_0 e^{x_0} - \frac{x_0^2}{2}.$$

In summary, the initial value problem consisting of (2.57) and (2.60) has *exactly one* solution, given by (2.59) with c as just determined. This solution is valid for $-\infty < x < \infty$.

The next example also uses a linear differential equation, but unlike the last example, the coefficient function $p(x)$ has a discontinuity.

Example 2.4.3. Every solution to the linear equation

$$\frac{dy}{dx} + \frac{1}{x}y = 0$$

is a solution to the equation obtained by multiplying both sides by x:

$$x\frac{dy}{dx} + y = 0.$$

Since this latter equation is exactly

$$\frac{d}{dx}(xy) = 0,$$

its complete family of solutions is given by

$$xy = c$$

for c an arbitrary constant. So long as we choose a value of $x_0 \neq 0$ we may find a function in the family

$$y = \frac{c}{x}$$

that satisfies any desired initial condition $y(x_0) = y_0$; just choose $c = x_0 y_0$. Summarizing, for $x_0 \neq 0$, the initial value problem

$$\frac{dy}{dx} + \frac{1}{x}y = 0, \quad y(x_0) = y_0,$$

has unique solution

$$y = \frac{x_0 y_0}{x}.$$

The interval of validity for this solution is the largest (open) interval containing x_0 and not containing 0.

What we saw in the last two examples—the existence and uniqueness of a solution to a *linear* first-order initial value problem on an interval on which the coefficient functions $p(x)$ and $g(x)$ are continuous—typifies all linear initial value problems. The next result summarizes this situation.

Theorem 2.4.4. *Consider the linear equation*

$$\frac{dy}{dx} + p(x)y = g(x)$$

with initial condition $y(x_0) = y_0$. If the functions p and g are continuous on an open interval (a, b) containing the point $x = x_0$, then there exists a unique function $y = \varphi(x)$ defined for $a < x < b$ that solves the differential equation

$$\frac{dy}{dx} + p(x)y = g(x)$$

2.4. Existence and uniqueness theorems

and also satisfies the initial condition

$$y(x_0) = y_0,$$

where y_0 is an arbitrary prescribed value.

One way to structure a proof of Theorem 2.4.4 is to think through the algorithm we have for solving first-order linear initial value problems and see that under the hypotheses of the theorem this algorithm will always produce a solution, and no other solution is possible. This approach is outlined in Exercise 30.

Existence and uniqueness for nonlinear equations. An important feature of Theorem 2.4.4 is the upfront guarantee that the solution exists for all x in the interval (a, b) on which p and g are continuous. Things are not so nice when we look at *nonlinear* initial value problems. The next result is an "existence and uniqueness" theorem that can be used for nonlinear initial value problems. The notion of a partial derivative for a function of more than one variable appears in the statement. If $f(x, y)$ is a function of the two variables x and y, the partial derivative $\frac{\partial f}{\partial y}$ is computed by holding x constant and differentiating "with respect to y". For example, if $f(x, y) = x^2 y + 2y^3 e^x + x$, then $\frac{\partial f}{\partial y} = x^2 + 6y^2 e^x$. You'll find more examples below, and you can find further discussion of partial derivatives in Section 2.8.2.

Theorem 2.4.5. *Consider the initial value problem*

$$\frac{dy}{dx} = f(x, y), \; y(x_0) = y_0. \tag{2.61}$$

(a) **Existence.** *Suppose that the function f is continuous in some open rectangle $\alpha < x < \beta$, $\gamma < y < \delta$ containing the point (x_0, y_0). Then in some open interval containing x_0 there exists a solution $y = \varphi(x)$ of the initial value problem (2.61).*

(b) **Uniqueness.** *If in addition, the partial derivative $\frac{\partial f}{\partial y}$ exists and is continuous in the rectangle $\alpha < x < \beta$, $\gamma < y < \delta$, then there is an open interval I containing x_0 and contained within (α, β) on which the solution to (2.61) is uniquely determined. (Thus if $y_1(x)$ and $y_2(x)$ are solutions to (2.61) on I, then $y_1(x) = y_2(x)$ for all x in I.)*

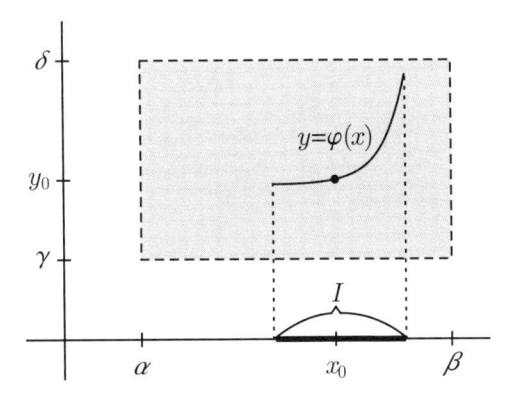

Figure 2.24. Theorem 2.4.5: If $f = f(x, y)$ and $\frac{\partial f}{\partial y}$ are continuous on the shaded region, then the initial value problem (2.61) has, on some open interval I containing x_0, a solution $y = \varphi(x)$ that is uniquely determined.

The next example illustrates some of the features of this theorem.

Example 2.4.6. Consider the nonlinear initial value problem
$$\frac{dy}{dx} = y^2, \ y(0) = y_0$$
for any real number y_0. Note the initial value of x here is $x_0 = 0$. To apply Theorem 2.4.5 we first find an open rectangle containing $(0, y_0)$ on which both $f(x, y) = y^2$ and $\frac{\partial f}{\partial y} = 2y$ are continuous functions. Since these functions have no discontinuities, the rectangle can be chosen to be the whole plane, $-\infty < x < \infty$, $-\infty < y < \infty$. Theorem 2.4.5 allows us to conclude that once we choose a value y_0, there will exist *some* interval containing $x_0 = 0$ and contained in $(-\infty, \infty)$ on which a unique solution exists. It makes no a priori guarantee about the size of this interval. To further illustrate this, observe that if $y_0 = 0$, this unique solution is the equilibrium solution $y(x) \equiv 0, -\infty < x < \infty$. For other values of y_0 we first solve the differential equation by separating variables:
$$y^{-2} \, dy = dx,$$
so that
$$\int y^{-2} dy = \int dx.$$
Integrating gives $-y^{-1} = x + c$, or
$$y = -\frac{1}{x+c}.$$
Solving for c by plugging in $x = 0$ and $y = y_0$, we get that $c = -\frac{1}{y_0}$. So for each initial value y_0, there is a unique solution
$$y = \frac{1}{\frac{1}{y_0} - x} = \frac{y_0}{1 - y_0 x}.$$
Since the denominator $1 - y_0 x$ is 0 at $x = 1/y_0$, the interval of validity for this solution is the largest open interval containing $x_0 = 0$ but not containing $x = 1/y_0$.

Let's examine our solution for some different initial values y_0.

y_0	$y(x)$	interval of validity
1	$\frac{1}{1-x}$	$x < 1$
2	$\frac{2}{1-2x}$	$x < \frac{1}{2}$
-1	$\frac{-1}{1+x}$	$x > -1$
-2	$\frac{-2}{1+2x}$	$x > -\frac{1}{2}$
0	0	$(-\infty, \infty)$
$\frac{1}{2}$	$\frac{1}{2-x}$	$x < 2$

Thus even though the functions
$$f(x, y) = y^2$$
and
$$\frac{\partial f}{\partial y} = 2y$$
are defined and continuous for all x and y, the solution $y(x)$ to our initial value problem may have its interval of validity restricted to some subinterval of the x-axis, and in fact the subintervals change as the value of y_0 changes. Furthermore, suppose we pick one of the solutions we found, for example,
$$y(x) = \frac{1}{1-x}.$$

2.4. Existence and uniqueness theorems

This solution was determined by the initial condition $y(0) = 1$, so its graph passes through the point $(x_0, y_0) = (0, 1)$. We say the domain of this solution is the interval $-\infty < x < 1$, even though it's also defined on $1 < x < \infty$ and satisfies the differential equation there, because the x-coordinate (namely 0) of our initial value point $(x_0, y_0) = (0, 1)$ lies in $-\infty < x < 1$ but not in $1 < x < \infty$; see Fig. 2.25. Note that $x = 1$ is a vertical asymptote of the solution passing through $(0, 1)$. In general, $x = \frac{1}{y_0}$ is a vertical asymptote of the solution passing through $x = 0, y = y_0$, provided $y_0 \neq 0$.

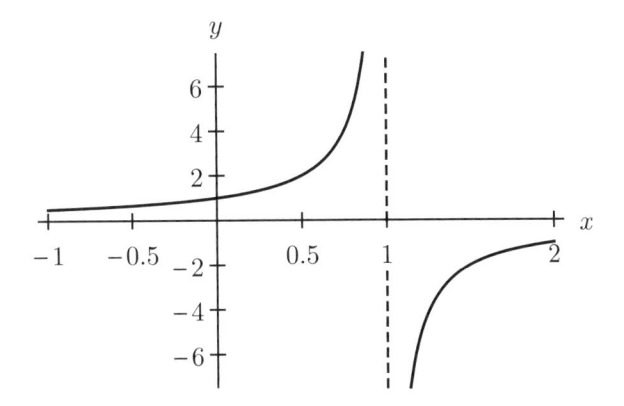

Figure 2.25. Graph of $y(x) = (1-x)^{-1}$.

Example 2.4.7. Consider the initial value problem

$$\frac{dy}{dx} = 2x - 2\sqrt{x^2 - y}, \ y(1) = 1.$$

You can verify by substitution that both $y_1(x) = x^2$ and

$$y_2(x) = \begin{cases} x^2 & \text{if } x < 1, \\ 2x - 1 & \text{if } x \geq 1 \end{cases}$$

are solutions to this initial value problem on $(-\infty, \infty)$. (For more on where these solutions came from, see Exercise 31.) Thus this initial value problem does not have a unique solution. Why does this not contradict Theorem 2.4.5? Let's check the hypotheses of the theorem. Set

$$f(x, y) = 2x - \sqrt{x^2 - y},$$

and notice that $f(x, y)$ is continuous only whenever $x^2 - y \geq 0$ (it is not even defined if $x^2 - y < 0$). Furthermore,

$$\frac{\partial f}{\partial y} = \frac{1}{2\sqrt{x^2 - y}},$$

which is continuous in that part of the xy-plane where $x^2 - y > 0$. Fig. 2.26 shows the part $\mathcal{S}$ of the xy-plane on which both f and $\frac{\partial f}{\partial y}$ are continuous. There is no open rectangle containing our initial condition point $(1, 1)$ and contained in $\mathcal{S}$. Therefore the theorem is not applicable, and there is no contradiction.

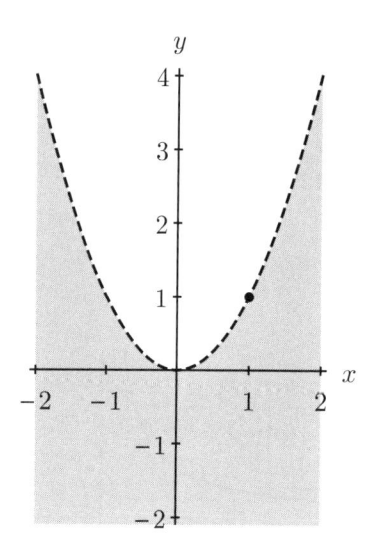

Figure 2.26. The region $\mathcal{S}$ in Example 2.4.7.

Solution curves can't intersect. As a consequence of Theorem 2.4.5, we have the following important observation.

Theorem 2.4.8 (No Intersection Theorem). *Suppose that f and $\partial f/\partial y$ are continuous in an open rectangle $\mathcal{R} = \{(x,y) : \alpha < x < \beta, \gamma < y < \delta\}$ and that I is a open subinterval of (α, β). Suppose also that $y_1(x)$ and $y_2(x)$ are two solutions of $\frac{dy}{dx} = f(x, y)$ having common domain I whose graphs are both contained in $\mathcal{R}$. If $y_1(x)$ and $y_2(x)$ are not identical, then their graphs cannot intersect.*

For example, consider the differential equation $\frac{dy}{dx} = y(1-y)$. Here $f(x,y) = y(1-y)$, and this function as well as $\partial f/\partial y = 1 - 2y$ are continuous in $-\infty < x < \infty, -\infty < y < \infty$. The constant functions $y(x) = 0$ and $y(x) = 1$ are solutions to this differential equation on the interval $-\infty < x < \infty$; their graphs are horizontal lines as shown in Fig. 2.27. The graph of any other solution cannot cross either of these lines, since if (x_0, y_0) were such an intersection point, the initial value problem

$$\frac{dy}{dx} = y(1-y), \; y(x_0) = y_0$$

would have more than one solution on some interval containing x_0, in contradiction to Theorem 2.4.5.

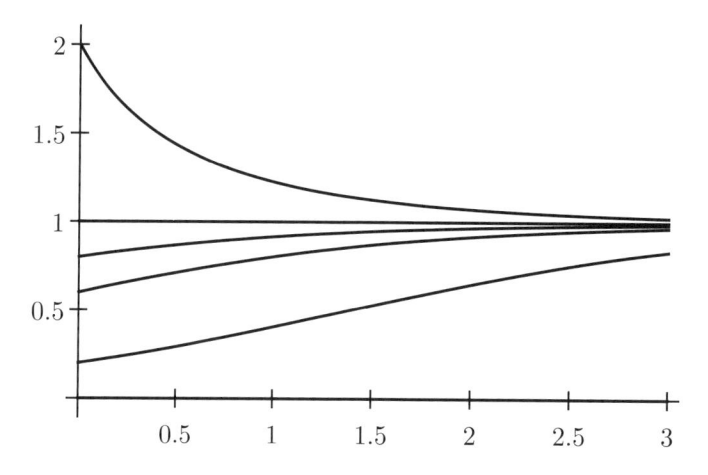

Figure 2.27. Some solutions of $\frac{dy}{dx} = y(1-y)$.

2.4. Existence and uniqueness theorems

2.4.1. Comparing linear and nonlinear differential equations. Here we summarize several key qualitative differences between first-order linear and nonlinear differential equations.

Existence and uniqueness for initial value problems. Theorem 2.4.4 assures us that if $p(x)$ and $g(x)$ are continuous on an open interval containing x_0, then the *linear* initial value problem

$$\frac{dy}{dx} + p(x)y = g(x), \quad y(x_0) = y_0,$$

has a unique solution on that interval. Any restriction on the interval of validity is apparent from the discontinuities of the coefficients $p(x)$ and $g(x)$ in the equation. By contrast, for a nonlinear initial value problem

(2.62) $$\frac{dy}{dx} = f(x,y), \quad y(x_0) = y_0,$$

even when $f(x,y)$ is continuous for all x and y, there may be a restriction on the interval of validity (see Example 2.4.6), or there may be more than one solution (see Exercise 13). While Theorem 2.4.5 gives conditions under which a unique solution to (2.62) exists, we do not have a simple way of determining the interval of validity from looking at the differential equation itself.

General solutions. The method we learned in Section 2.1 for solving a linear differential equation

(2.63) $$\frac{dy}{dx} + p(x)y = g(x)$$

on an interval I on which p and g are continuous produces an explicit solution with one arbitrary constant. This one-parameter family of solutions to (2.63) on I is the **general solution**: Every solution to the equation on I is included in the family.

For nonlinear equations

$$\frac{dy}{dx} = f(x,y),$$

even when we are able to produce a one-parameter family of solutions, it may not be complete. For example, we saw in Example 2.2.2 that separation of variables leads to the solutions

(2.64) $$y = -1 + \left(\frac{x^2}{6} - x + C\right)^3$$

for the equation

$$\frac{dy}{dx} = (x-3)(y+1)^{2/3}.$$

However, we also have (for example) the equilibrium solution $y = -1$, which is not obtained from (2.64) by any choice of C. For nonlinear equations it is also possible for nonequilibrium solutions to be missing from a one-parameter family of solutions; you can see a number of examples in the exercises, such as Exercise 13 or Exercise 31. Thus, in comparison with linear equations, knowing when we have obtained all solutions to a first-order nonlinear differential equation is a more difficult problem. Some positive results when the nonlinear equation is separable are discussed below, in Section 2.4.2. Finally, we note that the methods we have for solving nonlinear equations (separation of variables in Section 2.2 and other methods we will develop in Sections 2.7 and 2.8) typically produce *implicit* solutions, while linear equations can be solved explicitly.

2.4.2. Revisiting separable equations. We will use the No Intersection Theorem to help answer the following question: When can we know we have found *all* solutions of a separable equation

$$\frac{dy}{dx} = g(x)h(y)? \tag{2.65}$$

Recall that we have procedures to produce two general types of solutions:

- Separate variables to write $\frac{dy}{h(y)} = g(x)\,dx$ and obtain implicit solutions of the form

$$H(y) = G(x) + c$$

where $G'(x) = g(x)$, $H'(y) = \frac{1}{h(y)}$, and c is an arbitrary constant.

- Find the equilibrium solutions $y = b_j$ by determining the constants b_j satisfying $h(b_j) = 0$.

Now we ask when we can be assured that *any* solution of equation (2.65) will be one of these two types. An answer is provided by the next result.

Theorem 2.4.9. *Suppose that $g = g(x)$ is continuous on $(-\infty, \infty)$ and that $h = h(y)$ is differentiable with continuous derivative on $(-\infty, \infty)$. Suppose further that $h(y) = 0$ precisely for the values $y = b_1, b_2, \ldots, b_n$. Let G be an antiderivative of g on $(-\infty, \infty)$ (so that $G'(x) = g(x)$) and let H be an antiderivative of $\frac{1}{h}$ at all points except where $h(y) = 0$ (so that $H'(y) = 1/h(y)$ for all y not in the set $\{b_1, b_2, \ldots, b_n\}$). Then if $y = \varphi(x)$ is any solution whatsoever of*

$$\frac{dy}{dx} = g(x)h(y),$$

then either y is an equilibrium solution $y(x) = b_j$ for some j in $\{1, 2, \ldots, n\}$ or there is a constant c such that

$$H(\varphi(x)) - G(x) = c \quad \text{for all } x \text{ in the domain of } y = \varphi(x).$$

To see this theorem in action, we revisit two earlier examples. First, recall the separable equation $\frac{dy}{dx} = y^2$, which we solved in Example 2.4.6. Note that the hypotheses of Theorem 2.4.9 apply, since $g(x) = 1$ is continuous on $(-\infty, \infty)$ and $h(y) = y^2$ is differentiable with continuous derivative $2y$ on $(-\infty, \infty)$. The function $h(y) = y^2$ is 0 only at the point $b_1 = 0$, so that $y = 0$ is the only equilibrium solution. An antiderivative for g is $G(x) = x$, and an antiderivative for $\frac{1}{h}$ is $H(y) = -y^{-1}$. Separating variables gives the implicit solutions $H(y) - G(x) = c$ or $-y^{-1} - x = c$, for any constant c. Some algebra lets us write these implicit solutions as $y = -\frac{1}{x+c}$, a function having domain all real numbers except $-c$. Theorem 2.4.9 guarantees that we have found all solutions of $\frac{dy}{dx} = y^2$.

By contrast, let's look again at the equation $\frac{dy}{dx} = (x-3)(y+1)^{2/3}$ considered in Example 2.2.2. Here $g(x) = x - 3$ is continuous on $(-\infty, \infty)$ with antiderivative $G(x) = \frac{x^2}{2} - 3x$, but $h(y) = (y+1)^{2/3}$, while having a continuous derivative on $(-\infty, -1)$ and $(-1, \infty)$, fails to be differentiable at $y = -1$. Thus Theorem 2.4.9 does not apply. Moreover, while we have equilibrium solution $y = -1$ and the separation of variables solutions $3(y+1)^{1/3} - (\frac{x^2}{2} - 3x) = c$ (or equivalently $y = -1 + (\frac{x^2}{6} - x + c)^3$), we have many other solutions as well, including

$$y(x) = \begin{cases} -1 + \left(\frac{x^2}{6} - x\right)^3 & \text{if } x \leq 0, \\ -1 & \text{if } x > 0 \end{cases}$$

and

$$y(x) = \begin{cases} -1 + \left(\frac{x^2}{6} - x + \frac{5}{6}\right)^3 & \text{if } x \leq 1, \\ -1 & \text{if } x > 1. \end{cases}$$

These are nonequilibrium solutions which are not simply obtained by some choice of constant in the implicit solution $3(y+1)^{1/3} - (\frac{x^2}{2} - 3x) = c$.

Proof of Theorem 2.4.9. The hypotheses on the functions g and h guarantee that we can apply the No Intersection Theorem, Theorem 2.4.8, on the rectangle $\mathcal{R}$ consisting of the entire plane $-\infty < x < \infty$, $-\infty < y < \infty$, since $f(x,y) = g(x)h(y)$ and $\frac{\partial f}{\partial y} = g(x)h'(y)$ are continuous in $\mathcal{R}$. The antiderivatives $G(x)$ and $H(y)$ exist by the Fundamental Theorem of Calculus.

Now suppose that $y = \varphi(x)$ is a nonequilibrium solution to $\frac{dy}{dx} = g(x)h(y)$ on some open interval I. Note that this means $\varphi'(x) = g(x)h(\varphi(x))$ on I. The No Intersection Theorem tells us that φ never takes the values $b_1, b_2, \ldots, b_n$, since the graph of φ cannot intersection the graph of any equilibrium solution $y = b_j$. This means that for x in I, the values of $\varphi(x)$ are in the domain of $H(y)$ and we compute by the chain rule that

$$\frac{d}{dx}[H(\varphi(x)) - G(x)] = H'(\varphi(x))\varphi'(x) - G'(x) = \frac{1}{h(\varphi(x))}\varphi'(x) - g(x) = \frac{\varphi'(x) - g(x)h(\varphi(x))}{h(\varphi(x))} = 0$$

for all x in I. Thus there is a constant c so that

$$H(\varphi(x)) - G(x) = c$$

for all x in I, as desired.

2.4.3. Exercises.

For the equations in Exercises 1–4, find the largest open interval on which the given initial value problem is guaranteed, by Theorem 2.4.4, to have a unique solution. Do not solve the equation.

1. $\dfrac{dy}{dx} + y = \tan x$, $y(1) = 2$.

2. $\dfrac{dy}{dx} + y = \sqrt{\sin^2 x + 1}$, $y(3) = 2$.

3. $x\dfrac{dy}{dx} + (\cos x)y = \ln(x+2)$, $y(-1) = 2$.

4. $\ln(x+3)\dfrac{dy}{dx} + (\sin x)y = e^x$, $y(0) = 2$.

In Exercises 5–10, determine whether Theorem 2.4.5 ensures existence of a solution to the initial value problem. If so, determine whether the theorem guarantees that there is some open interval containing the initial value of the independent variable on which the solution is unique.

5. $\dfrac{dy}{dx} = x^2 + y^2$, $y(1) = 2$.

6. $\dfrac{dy}{dx} = y^{1/3}$, $y(2) = 0$.

7. $\dfrac{dy}{dx} = \sqrt{y}$, $y(1) = 0$.

8. $\dfrac{dy}{dx} = \dfrac{x}{y^2 - 4}$, $y(2) = 1$.

9. $\dfrac{dy}{dx} = \dfrac{x}{y^2 - 4}$, $y(1) = 2$.

10. $\dfrac{dP}{dt} = P - 2P^2$, $P(0) = 1$.

11. Show that the initial value problem

$$x\frac{dy}{dx} = 2y, \quad y(0) = 0,$$

has infinitely many solutions. Note that the differential equation is linear. Why does this example not contradict Theorem 2.4.4?

12. Consider the initial value problem

(2.66) $$\frac{dy}{dt} = -2\sqrt{y}, \quad y(1) = 0,$$

which we discussed in Example 2.4.1.

(a) Pick any number c and show that the function

$$y(t) = \begin{cases} (c-t)^2 & \text{if } t \le c, \\ 0 & \text{if } t > c \end{cases}$$

is differentiable at every real number t. Pay particular attention to the point $t = c$; you may find it helpful to talk about "left-hand" and "right-hand" derivatives at this point.

(b) The direction field for the equation

$$\frac{dy}{dt} = -2\sqrt{y}$$

is shown in Fig. 2.28. Explain how you can tell from this picture that the function $y = (c-t)^2$ cannot solve this equation for all $-\infty < t < \infty$.

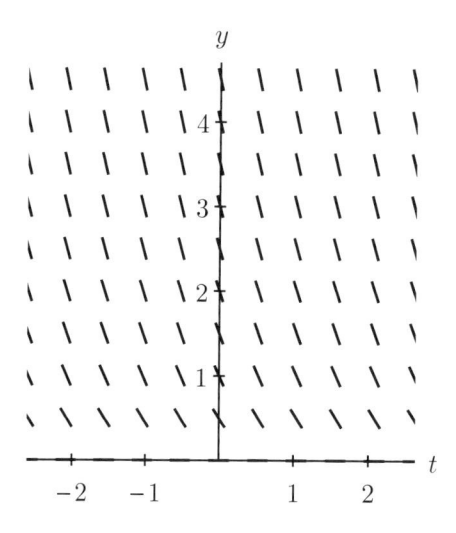

Figure 2.28. Direction field.

13. (a) If the equation $\frac{dy}{dx} = 3y^{2/3}$ is solved by separating variables, what one-parameter family of solutions is obtained?

    (b) Consider the function

    $$y = \begin{cases} (x-a)^3 & \text{if } x \le a, \\ 0 & \text{if } a < x < b, \\ (x-b)^3 & \text{if } x \ge b, \end{cases}$$

    where a is an arbitrary negative number and b is an arbitrary positive value. Sketch the graph of this function. Is y differentiable at $x = a$? Is it differentiable at $x = b$? If your answer is yes, what are $y'(a)$ and $y'(b)$?

    (c) Show that the function y defined in (b) solves the initial value problem $\frac{dy}{dx} = 3y^{2/3}$, $y(0) = 0$. Notice that as a ranges over all negative numbers and b ranges over all positive numbers, this gives infinitely many solutions (in fact, a two-parameter family's worth) to this initial value problem. Why does this not contradict Theorem 2.4.5?

    (d) Does Theorem 2.4.9 apply to the differential equation $\frac{dy}{dx} = 3y^{2/3}$?

14. Find two different solutions to the initial value problem

$$4\frac{dy}{dx} = 5y^{1/5}, \quad y(1) = 0,$$

on $(-\infty, \infty)$. Why doesn't this contradict Theorem 2.4.5? Hint: You may want to define one of your solutions in a piecewise manner.

2.4. Existence and uniqueness theorems

15. The two initial value problems

$$\frac{dy}{dt} = 1 - y^2, \quad y(0) = 0,$$

and

$$\frac{dy}{dt} = 1 + y^2, \quad y(0) = 0,$$

look superficially quite similar. Solve each of these initial value problems, showing that the first has solution

$$y = \frac{e^{2t} - 1}{e^{2t} + 1}$$

and the second has solution $y = \tan t$. What is the interval of validity for each of these solutions?

16. Various solution curves of the differential equation

$$\frac{dy}{dx} = 2x^2 - xy^2$$

are shown in Fig. 2.29. Which of the following statements is correct? Note: If two curves are sometimes closer together than the width of the instrument that draws them, the curves may appear to be the same in that part of the picture.

(a) Any two distinct solution curves shown have exactly one point of intersection.
(b) No two distinct solution curves shown have a point of intersection.
(c) Any two of the distinct solution curves shown have more than one point of intersection.
(d) It is impossible to know anything about points of intersection without first solving the differential equation.

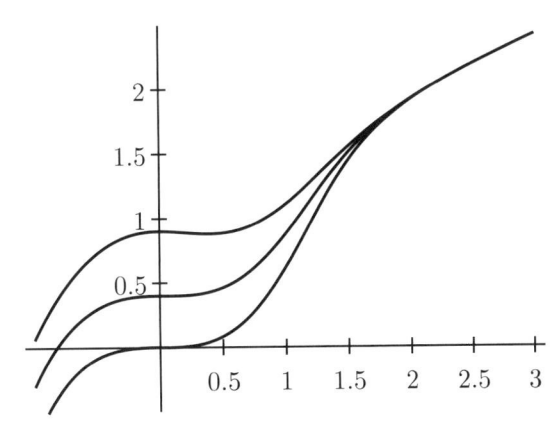

Figure 2.29. Some solutions of $\frac{dy}{dx} = 2x^2 - xy^2$.

17. (a) Show that

$$y(x) = 1 \quad \text{and} \quad y(x) = \frac{(x-1)^3}{27} + 1$$

both solve the initial value problem

$$\frac{dy}{dx} = (y-1)^{2/3}, \quad y(1) = 1.$$

(b) Suppose we define

$$g(x) = \begin{cases} \frac{(x-1)^3}{27} + 1 & \text{for } x < 1, \\ 1 & \text{for } x \geq 1 \end{cases}$$

and
$$h(x) = \begin{cases} 1 & \text{for } x < 1, \\ \frac{(x-1)^3}{27} + 1 & \text{for } x \geq 1. \end{cases}$$
Do g and h also solve the initial value problem in (a)? In answering this, check whether g and h are differentiable at $x = 1$.

18. Show that the differential equation
$$\left(\frac{dy}{dt}\right)^2 + y^2 + 1 = 0$$
has no solutions. Why doesn't this contradict Theorem 2.4.5?

19. Though we think of Theorem 2.4.5 as an existence and uniqueness theorem for *nonlinear* initial value problems, the hypotheses do not *require* this.
    (a) If a linear equation
    $$\frac{dy}{dx} + p(x)y = g(x)$$
    is written in the form
    $$\frac{dy}{dx} = f(x, y),$$
    what is $f(x, y)$?
    (b) Show that if $p(x)$ and $g(x)$ are continuous on an interval (a, b), then $f(x, y)$ and $\frac{\partial f}{\partial y}$ are continuous on the infinite rectangle $a < x < b$, $-\infty < y < \infty$.
    (c) What conclusion does Theorem 2.4.5 let you draw about solutions to an initial value problem $\frac{dy}{dx} + p(x)y = g(x)$, $y(x_0) = y_0$, and how does this compare with the conclusion if you instead apply Theorem 2.4.4?

20. Suppose $f(x, y)$ and $\frac{\partial f}{\partial y}$ are continuous in $-\infty < x < \infty$, $-\alpha < y < \alpha$ for some $\alpha > 0$, and suppose that $f(x, 0) = 0$ for all x. Can there be a solution $y = \varphi(x)$ (defined on an interval) to
$$\frac{dy}{dx} = f(x, y)$$
which is positive for some values of x and negative for other values of x? Explain.

21. The purpose of this problem is to explain why we focus on the existence and uniqueness of solutions on *intervals* rather than some other kind of set. Suppose we wish to solve the initial value problem
$$\frac{dy}{dt} + \frac{1}{t}y = 0, \quad y(1) = 2,$$
on the set $(-\infty, 0) \cup (0, \infty)$ (this is the set of all real numbers not equal to 0).
    (a) Show that
    $$y(t) = \begin{cases} 2/t & \text{if } t > 0, \\ c/t & \text{if } t < 0, \end{cases}$$
    where c is an arbitrary constant solves this initial value problem. Conclude that relative to the set $(-\infty, 0) \cup (0, \infty)$ this initial value problem has infinitely many solutions.
    (b) Show that if we restrict to any open *interval* I containing $t = 1$ on which the function $p(t) = \frac{1}{t}$ is continuous, then the initial value problem has the unique solution $y(t) = \frac{2}{t}$.

22. (CAS) Sally relies on the computer algebra system *Maple* to solve initial value problems numerically. When she used *Maple*'s default numerical solution command to solve
$$\frac{dy}{dx} = x^2 y + \frac{x+1}{x+2}, \quad y(1) = 1,$$
and then evaluate the solution at $x = 13$, this is what happened:

2.4. Existence and uniqueness theorems

$$F := \text{dsolve}\left(\left\{\text{diff}(y(x),x) = x^2 \cdot y(x) + \frac{(x+1)}{x+2}, y(1) = 1\right\}, \text{type} = \text{numeric}\right);$$

$$\text{proc}(x\_rkf45) \ldots \text{end proc}$$

$F(13)$

Error, (in F) cannot evaluate the solution further right of 12.788291, probably a singularity

When *Maple* says the solution probably has a singularity, it is suggesting there is a number $x_0 \approx 12.788291$ such that the solution of Sally's initial value problem is not defined for any x exceeding x_0. Sally knows that *Maple* is wrong. That is, she knows that the solution to her initial value problem should exist at $x = 13$. Give a thorough explanation of why Sally is right and thus *Maple* is wrong.

In Exercises 23–26, (a) Find all equilibrium solutions. (b) By separating variables, find a one-parameter family of implicit solutions, and then write the solutions explicitly. (c) Does Theorem 2.4.9 apply to guarantee that you have found all solutions? (d) If your answer to (c) is no, try to find a solution not obtained in (a) or (b).

23. $\frac{dy}{dx} = e^{2x-y}$.

24. $\frac{dy}{dx} = (y+3)^2 \cos x$.

25. $\frac{dy}{dx} = 1 + y^2$.

26. $\frac{dy}{dx} = xy^{1/3}$.

27. Consider the initial value problem

    (2.67) $$\frac{dy}{dx} = \frac{-\epsilon x}{\sqrt{x^2 + y^2 + \epsilon y}}, \quad y(0) = 1,$$

    where ϵ is a (small) positive constant.
    (a) Suppose that g is a positive function such that $y = g(x)$ satifies the initial value problem (2.67) on some interval centered at the origin, say $-2 < x < 2$. Show that $y = g(-x)$ also satisfies the same initial value problem on $(-2, 2)$.
    (b) Show how the No Intersection Theorem, Theorem 2.4.8, can be applied to conclude that g must be an even function ($g(-x) = g(x)$ for all x in $(-2, 2)$).
    Remark: We'll revisit the initial value problem (2.67) in Section 3.3, where it will be used to model the "apparent curve" for an optical illusion called the Hering illusion. The symmetry present in the illusion requires that any such apparent curve be the graph of an even function. Thus, this exercise may be regarded as a plausibility check on the model (2.67) for the Hering illusion.

28. Based on your work for Exercise 27, describe properties of a function f of x and y that will ensure that there is a open interval $(-\kappa, \kappa)$ containing 0 such that

    (2.68) $$\frac{dy}{dx} = f(x, y), \quad y(0) = y_0,$$

    has a unique solution on $(-\kappa, \kappa)$ that is necessarily an even function (on $(-\kappa, \kappa)$).

29. Let $f(x, y) = -6xy^{2/3}$ and observe $f(-x, y) = -f(x, y)$ for all real numbers x and y. Find a function φ such that
    (i) $y = \varphi(x)$ is a solution of
    $$\frac{dy}{dx} = -6xy^{2/3}, \quad y(0) = 1,$$
    on $(-\infty, \infty)$ and
    (ii) φ is not an even function on $(-\infty, \infty)$.

30. In this problem we outline a proof of Theorem 2.4.4. You should keep in mind the Fundamental Theorem of Calculus, which says that if f is continuous on an open interval I and if x_0 is any point in I, then for every x in I, the function F defined by
$$F(x) = \int_{x_0}^{x} f(u)\, du$$
exists and is a differentiable function, with $F'(x) = f(x)$ (for any x in I).

(a) First consider a homogeneous linear equation

(2.69) $$\frac{dy}{dx} + p(x)y = 0$$

with initial condition $y(x_0) = y_0$, where p is continuous on some open interval I containing x_0. Show that a solution to this initial value problem *exists* by showing that

(2.70) $$y(x) = Ce^{-P(x)}$$

solves (2.69) if P is defined on I by

$$P(x) = \int_{x_0}^{x} p(u)\, du$$

and checking that there is a choice of C so that the initial condition is satisfied. Show that the solution is *unique* by showing that the original differential equation is equivalent to

$$\frac{d}{dx}(e^{P(x)} y) = 0$$

and that every solution to this equation appears in the family (2.70) and finally that there is only one choice of C so that (2.70) satisfies a given initial condition.

(b) Now consider a nonhomogeneous equation

(2.71) $$\frac{dy}{dx} + p(x)y = g(x)$$

with initial condition $y(x_0) = y_0$, where p and g are continuous on an open interval I containing x_0. Show that there *exists* a solution to this initial value problem on the interval I by verifying that

$$y = e^{-P(x)} \left(\int_{x_0}^{x} e^{P(u)} g(u)\, du + y_0 \right),$$

where P is as defined in (a), satisfies (2.71), and has $y(x_0) = y_0$. Finally, show this is the unique solution by arguing that if y_1 and y_2 solve this initial value problem, then $y_1 - y_2$ solves the initial value problem

$$\frac{dy}{dx} + p(x)y = 0, \quad y(x_0) = 0.$$

Use part (a) to explain why this tells you that $y_1 = y_2$.

31. In this problem we will work with the family of lines having equations

(2.72) $$y = 2cx - c^2$$

for c an arbitrary constant.

(a) Show that any line in the family (2.72) is a tangent line to the parabola $y = x^2$ at some point of the parabola. Describe this point in terms of c. Fig. 2.30 illustrates this relationship.

(b) Verify that each of the functions in the one-parameter family in (2.72) solves the differential equation

$$4y = 4x\frac{dy}{dx} - \left(\frac{dy}{dx}\right)^2.$$

2.4. Existence and uniqueness theorems

(c) Does the function $y = x^2$ also solve this differential equation? Is it obtained by any choice of c in (2.72)? This "missing solution" is called a **singular solution**.

(d) Show that if
$$\frac{dy}{dx} = 2x - 2\sqrt{x^2 - y},$$
then
$$\left(\frac{dy}{dx}\right)^2 + 4y = 4x\frac{dy}{dx}.$$

This shows how the solutions in Example 2.4.7 can be obtained.

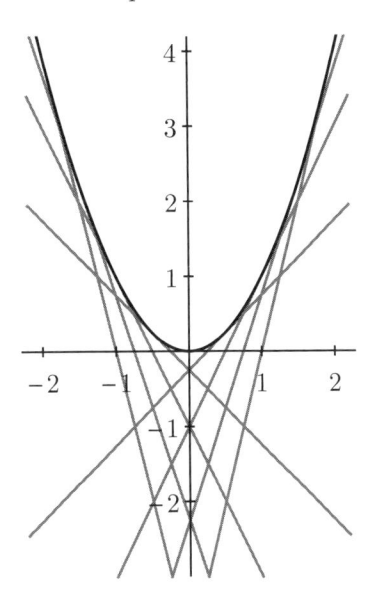

Figure 2.30. The parabola $y = x^2$ and some lines $y = 2cx - c^2$.

32. The differential equation

(2.73) $$y = x\frac{dy}{dx} - \frac{1}{4}\left(\frac{dy}{dx}\right)^2$$

from Exercise 31(b) is an example of a **Clairaut equation**. We saw in that exercise that it had the family of lines $y = 2cx - c^2$ as well as the parabola $y = x^2$ as solutions. The family of lines is the collection of tangent lines to the parabola, and the parabola is called the envelope of the family of lines. More generally, a differential equation which can be written as

(2.74) $$y = x\frac{dy}{dx} + f\left(\frac{dy}{dx}\right),$$

for some nonconstant twice-differentiable function f, is called a Clairaut equation.

(a) If equation (2.73) is written in the form of (2.74), what is the function f? In (2.74), the notation $f\left(\frac{dy}{dx}\right)$ means f evaluated at $\frac{dy}{dx}$.

(b) Show that equation (2.74) has the line solutions $y = mx + b$ for any constant m for which $f(m)$ is defined and an appropriate choice of the constant b. Identify b in terms of m and f.

(c) Show that the curve described by the parametric equations

(2.75) $$x = -f'(t), \quad y = f(t) - tf'(t)$$

is also a solution to (2.74). Hint: Use the fact that
$$\frac{dy}{dx} = \frac{dy/dt}{dx/dt}.$$

(d) Find the equation of the tangent line to the curve described by the parametric equations (2.75) at the point on the curve with coordinates $(-f'(m), f(m) - mf'(m))$. Check that this is one of the line solutions to (2.74), as described in (b).

(e) Find the solutions to the Clairaut equation
$$y - 2 = (x+1)\frac{dy}{dx} - \left(\frac{dy}{dx}\right)^2$$
described in (b) and (c).

33. The uniqueness part of Theorem 2.4.5 says that if two solutions to
$$\frac{dy}{dt} = f(t, y)$$
start out the same (with value y_0 at time t_0), then they stay the same, at least on some interval of t's containing t_0. There is an extension of this statement which quantifies how "close" two solutions stay to each other if they start out "close" at time t_0. In this problem we'll explore this extension.

Suppose $\mathcal{R}$ is a rectangle as in the hypothesis of Theorem 2.4.5, and suppose we have a number M so that
$$-M \le \frac{\partial f}{\partial y} \le M$$
at all points of $\mathcal{R}$. Suppose further that y_1 and y_2 are solutions to
$$\frac{dy}{dt} = f(t, y)$$
with
$$y_1(t_0) = a \quad \text{and} \quad y_2(t_0) = b,$$
where (t_0, a) and (t_0, b) are in $\mathcal{R}$. The following fact can be shown: As long as $(t, y_1(t))$ and $(t, y_2(t))$ continue to belong to $\mathcal{R}$, we have the estimate

(2.76) $$|y_1(t) - y_2(t)| \le |a - b|e^{M|t-t_0|}.$$

(a) Explain how the uniqueness part of Theorem 2.4.5 is a special case of this result.

(b) Now the exponential function
$$e^{M|t-t_0|}$$
can certainly grow quickly as t moves away from t_0, so by (2.76) the values of our two functions $y_1(t)$ and $y_2(t)$ can be far apart when $t \ne t_0$ even if a and b are very close. To see this with a concrete example, solve the initial value problem
$$\frac{dy}{dt} = 3y, \quad y(0) = a,$$
and call the solution y_1, and solve
$$\frac{dy}{dt} = 3y, \quad y(0) = b,$$
and call the solution y_2. Show directly that
$$|y_1(t) - y_2(t)| = |a - b|e^{3t}.$$
Explain how this gives an example where we have equality in the inequality (2.76).

(c) In the initial value problems of (b), how large must t be so that $|y_1(t) - y_2(t)|$ is more than 100 times larger than $|y_1(0) - y_2(0)|$?

2.5. Population and financial models

The problem of estimating population is fundamental. For example, accurate models for population growth allow for rational public policy decisions of many sorts.

In the simplest situation, we are interested in population models for a single species (human, animal, bacterial, etc.) in isolation, and we think of the population as homogeneous (we ignore individual variations like gender, age, etc.). More complicated models might look at two interacting populations (for example, a predator species and its prey), or divide the population into subgroups (by age, for instance) and model each separately.

2.5.1. Malthusian model. Given all the potential complexities in population growth, it may be surprising that we can give a useful model based on one of the simplest first-order differential equations. This goes back to the English economist Thomas Malthus in 1798. He proposed that if the population under study had size $P(t)$ at time t, then the rate of change of P with respect to t should be proportional to P itself:

$$(2.77) \qquad \frac{dP}{dt} = kP$$

for some positive constant k. Notice that if we measure population in numbers of individuals, P will always be integer-valued, but when we write a differential equation for P we're assuming it's a differentiable function and therefore continuous. So we are tacitly assuming that we "smooth out" our population function to be differentiable. This is a reasonable assumption if the population P is large, for then when we look at its graph as a function of time, we do so "from a distance" so as to see the whole picture. From this vantage point, individual changes in P (by a birth or death) are barely noticeable, leaving the impression of a smooth function.

Equation (2.77) is readily solved (by separating variables, for example) and we can write the solution as

$$P(t) = Ce^{kt}.$$

Since $P(0) = C$, we'll rewrite this as

$$P(t) = P_0 e^{kt}$$

where P_0 denotes the population at whatever we call time $t = 0$. Given the simplicity of this model and the complexity of population demographics in general, it is perhaps surprising that the Malthusian equation $P(t) = P_0 e^{kt}$ is of any use. Notice that if we want to fit this model to an actual population, we need two data points to start with: one to determine P_0 and one to determine k. Let's try this with the US population. The population in 1910 was 91.972 million. In 1920 it was 105.711 million. If we set our timeline so that $t = 0$ corresponds to 1910 and measure t in years, then when $t = 10$ we have $P(10) = 105.711$, in units of millions of people. We determine k from

$$105.711 = 91.972 e^{10k},$$

or

$$1.149 = e^{10k}.$$

Taking the natural logarithm gives $\ln 1.149 = 10k$ so that $k = 0.0139$. What does this model predict for the populations in 1930, 1940, and 2000? We compute

$$P(20) = 91.972 e^{(0.0139)(20)} = 121.448,$$

$$P(30) = 91.972 e^{(0.0139)(30)} = 139.559,$$

and
$$P(90) = 91.972 e^{(0.0139)(90)} = 321.335$$
as estimates for the US population, in millions of individuals, for these three dates. The actual population in these years was, respectively, 122.775, 131.669, and 282.434 million. So we can see that error in our 1930 prediction was approximately 1 percent, and our 1940 prediction is about 6 percent above the actual value. The prediction for 2000 is not so good; the predicted value is more than 13 percent higher than the actual value of 282.434 million. By the year 2914, this model gives a US population of approximately 105,800,000 million, which is about the same as the total surface area in square feet of the US. So clearly this model eventually becomes unrealistic.

Doubling time in the Malthusian model. One noteworthy feature of a population that grows according to a Malthusian model is the concept of **doubling time**: It takes a fixed amount of time for the population to go from any specific value to twice that value. To see this, let's determine the time for the population to increase from the value P_1 to the value $2P_1$ by solving the equations
$$P_1 = P_0 e^{kt}$$
and
$$2P_1 = P_0 e^{kt}$$
for t. In the first equation we get

(2.78) $$t = \frac{1}{k} \ln \frac{P_1}{P_0}$$

and in the second we have

(2.79) $$t = \frac{1}{k} \ln \frac{2P_1}{P_0}.$$

Using properties of the logarithm we rewrite the second solution as
$$t = \frac{1}{k} \left(\ln 2 + \ln \frac{P_1}{P_0} \right)$$
so that the difference between the values in (2.78) and (2.79), which is the time to double the population from P_1 to $2P_1$, is
$$\frac{\ln 2}{k}.$$
For example, according to our Malthusian model of US population growth, the population of the United States should double every
$$\frac{\ln 2}{0.0139} \approx 50 \text{ years.}$$

2.5.2. The logistic model. A difficulty with the Malthusian model for population growth is obvious—no population can grow exponentially indefinitely. Eventually competition for available resources needed to support the population must begin to limit the rate of growth. To develop a model that allows for such consideration, we consider a differential equation of the form
$$\frac{dP}{dt} = rP$$
where we now think of r as not necessarily a constant (as in the Malthusian model) but rather a *function* $r(P)$ of P. What sort of properties should this function (called the **proportional growth rate**) have? The basic idea is that, at least when the population is large enough, it should decrease as the population increases. This reflects the fact that as the population increases, there

2.5. Population and financial models

is increased competition for resources. The simplest form we might assume for the function $r(P)$ that fits this desired property is

$$r(P) = a - bP,$$

for positive constants a and b, so that $r(P)$ is a linear function that decreases as P increases; see Fig. 2.31. The constant a is sometimes called the intrinsic growth rate. This leads to the differential equation

$$\frac{dP}{dt} = (a - bP)P = aP - bP^2 \qquad (2.80)$$

for the population. This should make a certain amount of intuitive sense, especially if we assume that the constant a is much larger than the constant b, so that when the population is small, relatively speaking, the term aP on the right-hand side is more significant, allowing something like Malthusian growth, but as the population gets larger, the term $-bP^2$ becomes progressively more important and acts to slow the growth of the population. We can think of the term $-bP^2$ as representing the interaction or competition among members of the population. When P is small, the population density is small and interaction is limited. As P increases, population density increases and an individual competes with others for resources. To see why the competition term should be proportional to P^2, imagine rhinos gathering at a watering hole in a time of drought, competing for space to drink. For each of the P animals in the population, there are $P - 1$ potential competitors to pick a fight with, leading to $P(P-1)/2$ possible hostile encounters. For a large population, $P \approx P - 1$, so the number of deaths resulting from hostile encounters should be (approximately) proportional to P^2 and thus having form $-bP^2$.

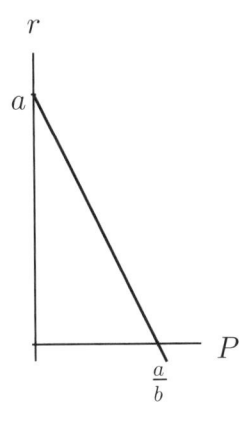

Figure 2.31. Graph of $r(P) = a - bP$.

Solving the logistic equation. Equation (2.80), which is called the **logistic equation**, is nonlinear. It is separable, and shortly we will solve it analytically. But first, we discuss some qualitative properties of its solutions and use these to give a rough sketch of the graphs of these solutions. The logistic equation is an example of an **autonomous equation**; this means that the independent variable t does not explicitly appear on the right-hand side of (2.80), so that the dependent variable P alone governs its own rate of change.

We first ask whether the logistic equation has any *equilibrium solutions*. In other words, does (2.80) have any *constant* solutions? If $P = c$ is a constant solution, then $\frac{dP}{dt}$ is 0, so we have

$$0 = (a - bP)P.$$

This tells us that there are two equilibrium solutions, $P = 0$ and $P = a/b$; shown in Fig. 2.32.

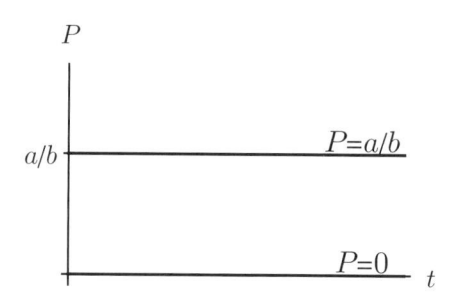

Figure 2.32. Equilibrium solutions to the logistic equation (2.80).

The meaning of the first of these is obvious, although rather uninteresting. If the population is 0, it stays 0 for all future times. The meaning of the second equilibrium solution a/b will become clear shortly. Notice that if a is much larger than b, as we assume is the typical case, a/b is large. Whenever P is between 0 and a/b, equation (2.80) tells us that $\frac{dP}{dt}$ is positive (why?) which in turn tells us that P is an increasing function of t.[16] This means that if we sketch the graph of t vs. P for solutions of (2.80), any curve in the horizontal strip between $P = 0$ and $P = a/b$ will be increasing from left to right. Whenever $P > a/b$, $\frac{dP}{dt}$ is negative (see Fig. 2.33), and the solution curves will be decreasing in this range. We can also determine the concavity of the solution curves by analyzing the second derivative

$$\frac{d^2P}{dt} = \frac{d}{dt}\left(\frac{dP}{dt}\right) = \frac{d}{dt}\left(aP - bP^2\right) = a\frac{dP}{dt} - 2bP\frac{dP}{dt} = (a - 2bP)(a - bP)P.$$

The second derivative is 0 when $P = 0, P = a/(2b)$, or $P = a/b$. Fig. 2.34 shows the sign for this second derivative in the ranges $0 < P < a/(2b)$, $a/(2b) < P < a/b$, and $a/b < P$; we ignore $P < 0$ since this has no physical meaning for our model. Thus solution curves are concave up for $0 < P < a/(2b)$ and for $P > a/b$, while they are concave down for $a/(2b) < P < a/b$. The No Intersection Theorem, Theorem 2.4.8, says that no two solution curves intersect (because $f(t, P) = aP - bP^2$ and $\frac{\partial f}{\partial P}(t, P) = a - 2bP$ are everywhere continuous). We use all of this information to sketch some of the solutions to the logistic equation in Fig. 2.35. The equilibrium solution $P = 0$ is said to be **unstable** since if $P(0)$ is close to but not equal to 0, it does *not* stay close to 0 for all later times. On the other hand, the equilibrium solution $P = a/b$ is **stable**. If $P(0)$ is near a/b (either larger or smaller), then for all later times $P(t)$ will be near a/b.

Figure 2.33. Sign of dP/dt.

Figure 2.34. Sign of d^2P/dt^2.

---

[16]Recall that a function $g(t)$ is increasing on the interval $a < t < b$ if whenever $a < t_1 < t_2 < b$ we have $g(t_1) \leq g(t_2)$. The function is strictly increasing on (a, b) if we have $g(t_1) < g(t_2)$ whenever $a < t_1 < t_2 < b$. Decreasing/strictly decreasing are defined analogously.

2.5. Population and financial models

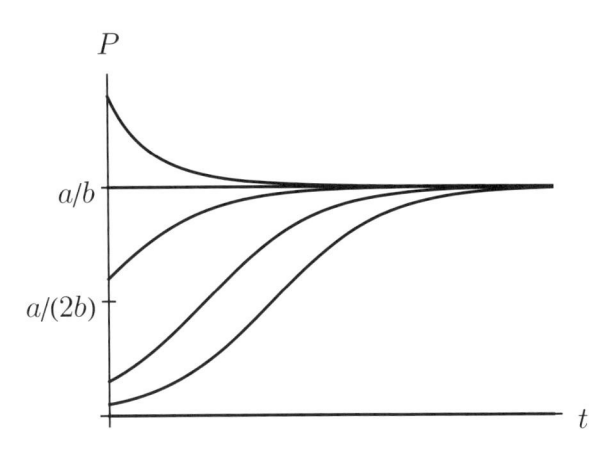

Figure 2.35. Some solutions to the logistic equation.

To solve (2.80) analytically we begin by separating the variables:

$$(2.81) \qquad \int \frac{dP}{P(a - bP)} = \int dt.$$

To proceed, we determine a partial fraction decomposition on the left-hand side by writing

$$(2.82) \qquad \frac{1}{P(a - bP)} = \frac{c_1}{P} + \frac{c_2}{a - bP}.$$

We determine values for c_1 and c_2 so that this decomposition holds. If we find a common denominator on the right-hand side of equation (2.82) and then equate numerators on the left and right, we find that

$$1 = c_1(a - bP) + c_2 P.$$

If we set $P = 0$ and $P = a/b$ in the last equation, we determine that $c_1 = \frac{1}{a}$ and $c_2 = \frac{b}{a}$. Thus

$$\int \frac{dP}{P(a - bP)} = \frac{1}{a} \int \frac{1}{P} dP + \frac{b}{a} \int \frac{1}{a - bP} dP$$

and from equation (2.81) we have the implicit solution

$$\frac{1}{a} \ln |P| - \frac{1}{a} \ln |a - bP| = t + k$$

for an arbitrary constant of integration k. Because of the physical meaning of P as population, we can replace $\ln |P|$ with $\ln P$. Multiplying by a and using properties of logarithms we write

$$(2.83) \qquad \ln \frac{P}{|a - bP|} = at + k'$$

where k' is an arbitrary constant.

It will be useful to have an explicit solution for P in terms of t. If we exponentiate both sides of equation (2.83), we obtain

$$(2.84) \qquad \frac{P}{|a - bP|} = Ce^{at}$$

where $C = e^{k'}$ is an arbitrary *positive* constant. We can remove the absolute value bars on the left-hand side of (2.84) by allowing C to be an arbitrary *real* constant, so that

$$(2.85) \qquad \frac{P}{a - bP} = Ce^{at},$$

for C real. The choice $C = 0$ corresponds to the equilibrium solution $P = 0$ that we've already noted. Notice that C is related to the population P_0 at time $t = 0$ by

(2.86) $$\frac{P_0}{a - bP_0} = C.$$

We can solve (2.85) explicitly for P as follows:
$$P = Ce^{at}(a - bP),$$
$$P + Cbe^{at}P = aCe^{at},$$
$$P = \frac{aCe^{at}}{1 + bCe^{at}} = \frac{aC}{e^{-at} + Cb}.$$

Recalling (2.86) we can write the population function as

(2.87) $$P(t) = \frac{a\frac{P_0}{a-bP_0}}{e^{-at} + \frac{P_0}{a-bP_0}b} = \frac{aP_0}{(a - bP_0)e^{-at} + bP_0}.$$

We can analyze (2.87) to draw some interesting conclusions about populations that grow according to the logistic equation. First, observe that (2.87) does give the expected equilibrium solutions in case $P_0 = 0$ or $P_0 = a/b$. Next, let's assume $P_0 > 0$ and look at long-time behavior; that is, we consider
$$\lim_{t \to \infty} \frac{aP_0}{(a - bP_0)e^{-at} + bP_0}.$$

Since $a > 0$,
$$e^{-at} \to 0 \text{ as } t \to \infty.$$

This means
$$P(t) \to \frac{aP_0}{bP_0} = \frac{a}{b}$$

as $t \to \infty$. This represents a significant qualitative difference from the Malthusian population model. No longer can the population grow without bound. The ratio a/b is called the **carrying capacity** of the environment. Our remark that a is typically much larger than b is consistent with the ratio a/b being large.

The logistic model for population growth was first proposed by the Belgian sociologist and mathematician Pierre-Francois Verhulst in 1837. Verhulst lacked access to population data that could have provided support for this theory, so his work did not get much attention until it was rediscovered 80 years later by Raymond Pearl and Lowell Reed, who are acknowledged as being the first to apply the logistic model to study the population growth of the United States.

Example 2.5.1. Let's use the logistic equation to model world population. The natural value of a, or the intrinsic growth rate for humans when there are plenty of resources and thus no competition, is about $a = 0.029$. In 1979, the world population was 4.38×10^9 and was increasing at 1.77% per year. The population increasing at 1.77% per year translates into $\frac{1}{P}\frac{dP}{dt} = 0.0177$ because $\frac{dP}{dt}$ is the population's rate of increase, and dividing by the population P gives the fractional rate of change for the population. This gives
$$0.0177 = 0.029 - b(4.38 \times 10^9),$$
and therefore
$$b = 2.580 \times 10^{-12}.$$

The carrying capacity is therefore $\frac{a}{b} = 11.24 \times 10^9$, or 11.24 billion. So our model predicts that the population of earth over time will approach 11.24 billion people. The world population in July

2005 was 6.45 billion people, which is larger than half of the carrying capacity. Thus we are above the inflection point and on the stressful "concave down" part of the population curve.[17]

In principle if we know the population at three points in time, we can determine the unknowns a, b, and P_0 in (2.87).[18] The computations are somewhat involved algebraically and are simplest if these three points are equally spaced in time. Exercise 15, which analyzes the post-Katrina population of New Orleans, shows how to manage the algebra in this case.

Other differential equation models that have features in common with the logistic model are often used. One of these, called the Gompertz equation, is discussed in the exercises. However, none of the models described in this section allow for oscillating populations, a phenomenon that is quite common in nature. For example, oscillations may occur if a population is subject to epidemics when it gets too large, or if the environment varies seasonally in its ability to support the population. A modification of the logistic model to allow for such seasonal fluctuations is given in Exercise 23 of Section 2.7.

Logistic models can be made more accurate over longer periods of time by periodically re-evaluating the coefficients a and b in the logistic equation. In particular, as technological advances allow the population to use resources more productively, we should decrease the value of b.

2.5.3. Financial models. When money is invested in a savings account that pays interest at a rate of $r\%$ per year, compounded continuously, the amount A in the account after t years is determined by the differential equation

$$(2.88) \qquad \frac{dA}{dt} = rA$$

where r is expressed as a decimal. This is analogous to a Malthusian population model; now our "population" is the amount of money in the account, and the money grows at a rate proportional to the current size of the account. Equation (2.88) has solution $A(t) = A_0 e^{rt}$ where A_0 is the account balance at time $t = 0$. This simple model assumes a constant interest rate r, which may not be realistic. It also assumes no withdrawals or deposits during the time period under consideration. In the next examples we show how to adjust for steady withdrawals or deposits; this is analogous to allowing for constant emigration or immigration in a population model.

Example 2.5.2. You open a savings account that pays 4% interest per year, compounded continuously, with an initial deposit of $1,000. If you make continuous deposits at a constant rate that totals $500 per year, what is your balance after 5 years? Notice that your total contribution is $1,000 + 5(\$500) = \$3,500$. The linear differential equation for the dollar amount A in the account after t years is

$$\frac{dA}{dt} = 0.04A + 500, \quad A(0) = 1{,}000.$$

We solve this using the integrating factor $e^{-0.04t}$ and obtain

$$Ae^{-0.04t} = -\frac{500}{0.04}e^{-0.04t} + C.$$

The initial condition $A(0) = \$1{,}000$ determines $C = 13{,}500$, so that

$$A(t) = -12{,}500 + 13{,}500 e^{0.04t}.$$

After 5 years the account balance is $A(5) = \$3{,}989$.

---

[17] A recent United Nations report gave a prediction of a world population of 10.1 billion by the year 2100, considerably higher than earlier predictions. One factor cited was a lessened *demographic* significance of the AIDS epidemic in Africa. The new report predicts the population in Africa to triple in this century.

[18] More realistically, we might have a set of data points that we conjecture lie on a logistic curve. Choosing different sets of 3 points from our data would typically give different parameters for the corresponding logistic equation. The challenge then is to determine a logistic curve that "best fits" the entire set of data.

Example 2.5.3. Suppose a retirement account earns 5% per year, continuously compounded. What should the initial balance in the account be to permit steady withdrawals of $50,000 per year for exactly 35 years? If we let $A(t)$ be the account balance in thousands of dollars at time t measured in years, we interpret this question as: What should $A(0)$ be so that $A(35) = 0$, assuming

$$\frac{dA}{dt} = 0.05A - 50?$$

Check that this linear (and separable) equation has solutions

$$A(t) = 1{,}000 + Ce^{0.05t}.$$

We want $A(35) = 0$ so that

$$1{,}000 + Ce^{1.75} = 0$$

and $C \approx -173.77$. This tells us that

$$A(t) = 1{,}000 - 173.77e^{0.05t}$$

and $A(0) \approx 826.23$. Thus we need an initial account balance of approximately $826,230. The next example assesses how easy this will be to achieve.

Example 2.5.4. Suppose that starting at age 25, you make steady contributions to a retirement account (with initial balance 0). What should your yearly contribution be if you want to have a balance of $826,230 after 40 years? Assume your account will earn 5% interest, compounded continuously. Denote by $R(t)$ the amount, in thousands of dollars, in the retirement account at time t in years. We have

$$\frac{dR}{dt} = 0.05R + d$$

where d is our deposit rate (in thousands of dollars). We have $R(0) = 0$ and we want $R(40) = 826.23$. The differential equation has solution

$$R(t) = -\frac{d}{0.05} + \frac{d}{0.05}e^{0.05t}$$

and upon setting $t = 40$ and $R = 826.23$ we have

$$826.23 = -\frac{d}{0.05} + \frac{d}{0.05}e^2.$$

This says

$$d = \frac{(826.23)(0.05)}{e^2 - 1} \approx 6.466,$$

so our desired deposit rate is about $6,466 dollars per year.

More realistic models of long-term retirement planning usually assume that you will increase your withdrawal rate to take inflation into account. Suppose we plan to increase our withdrawal continuously by 3% per year, so

$$\frac{dW}{dt} = 0.03W.$$

The withdrawal rate at time t (in years) is then

$$W(t) = W_0 e^{0.03t},$$

where W_0 is the initial withdrawal rate. Let's take W_0 to be $50,000 as in Example 2.5.3, so that our withdrawal rate in thousands of dollars is

$$W(t) = 50e^{0.03t}.$$

Now the balance A in our account earning 5% per year (continuously compounded) is determined by
$$\frac{dA}{dt} = 0.05A - 50e^{0.03t},$$
where we continue to measure A in thousands of dollars. Solving this linear equation gives
$$A(t) = 2{,}500e^{0.03t} + Ce^{0.05t}.$$
We can describe C in terms of the initial account balance A_0 as
$$C = A_0 - 2{,}500,$$
so that

(2.89) $$A(t) = 2{,}500e^{0.03t} + (A_0 - 2{,}500)e^{0.05t}.$$

What should our initial account balance be to permit 35 years of withdrawals? We want $A(35) = 0$; substituting this into (2.89) gives $A_0 = 1{,}258.5$, or approximately \$1,258,500, more than one and a half times the amount given by the model of Example 2.5.3.

2.5.4. Exercises.

1. In 2000 a man living in Maryland purchased two live adult Northern snakehead fish from a market in Chinatown, NY, initially intending to prepare an Asian dish with them. Snakeheads are an invasive species whose introduction into lakes and rivers in the US is much feared because of the potential negative impact on native species.[19] However instead of cooking the fish, the man kept them as pets for a while and later released the pair into a Crofton, Maryland, pond. Beginning in early summer 2002, snakeheads started being seen in the pond by fishermen. In September of 2002, in a desperate attempt to eradicate the population from this pond before it spread to the nearby Little Patuxent River, the pond was poisoned with rotenone and 1,006 dead snakeheads were recovered. Assuming the snakehead population grew according to the Malthusian model
$$\frac{dP}{dt} = kP$$
and that the initial pair was introduced into the pond in June 2000, estimate the value of k and predict what the population would have been 1 year later (in September 2003) had not the State biologists intervened. The SciFi Network has aired two movies inspired by this incident: *Snakehead Terror* and *Frankenfish*.

    A possible criticism of this model is the fact that, since we approximate the (integer-valued) population by a differentiable function, it should be used only for populations that are "large".

2. In the period 1790–1840 US population doubled about every 25 years. If population growth during this period is modeled by a Malthusian equation $dP/dt = kP$, what should k be?

3. Yasir Arafat, a long-time Palestinian leader, died in November 2004 under mysterious circumstances. Traces of the toxic radioactive isotope polonium 210 were found on his toothbrush and on clothes he wore shortly before falling ill. In November 2012 Arafat's body was exhumed to search for further evidence of ^{210}Po poisoning. What percentage of ^{210}Po would be present after 8 years, given that it has a half-life of 138 days? (This means that it decays at a rate proportional to the amount present, and the proportionality constant is such that after 138 days only half the initial amount is left.)

---

[19] In addition to their voracious appetites, the fact that they can breath air and crawl considerable distances across land adds to their "monster fish" reputation.

4. A bacterial culture is growing according to the Malthusian equation
$$\frac{dM}{dt} = kM$$
where M is the mass of the culture in grams. Between 1:00 PM and 2:00 PM the mass of the culture triples.
   (a) At what time will the mass be 100 times its value at 1:00 PM?
   (b) If the mass at 3:00 PM is 11 grams, what was the mass at 1:00 PM?

5. Suppose a population is growing according to the Malthusian model
$$\frac{dP}{dt} = 0.015P$$
where we measure P in millions of people and t in years. Modify this equation to include a constant emigration rate of 0.2 million people per year. Assume the population on January 1 is 8.5 million. Solve your modified equation and estimate the population 6 years later.

6. You borrow $25,000 to buy a car. The interest rate compounded continuously on your loan is 6%. Assume that payments of k dollars per year are made continuously. What should k be if your loan is for 48 months? How much do you pay in total over the 48-month period of the loan?

7. (a) You borrow $10,000 at an interest rate of 4%. You will make payments of k dollars per year continuously, and the period of the loan is 4 years. What is the total amount of interest paid (above the $10,000 principal) over the course of the loan?
   (b) If you borrow $20,000 instead, but the interest rate and loan period are unchanged, do you end up paying exactly twice as much in total interest?
   (c) If you borrow $10,000, but now at an interest rate of 8%, still with a loan period of 4 years, do you end up paying exactly twice as much in total interest as in (a)?

8. In a 2009 article in *The Washington Post*[20] regarding depleted 401(k) accounts, the following statement is made: "'A $100,000 account that has declined by roughly a quarter can return to its original level in two years, if a worker contributes as much as he or she can to the 401(k) while the employer matches half of that amount and the market gains 8% each year,' said Steve Utkus, director of Vanguard's Center for Retirement Research." Determine the dollar amount of the (continuously made) contributions of the worker and the employer if this statement is correct.

9. (a) If $1,000 is invested at an interest rate of 5%, how much is in the account after 8 years?
   (b) How much should be invested initially if you want to have an account balance of $20,000 after 10 years?
   (c) Suppose we start with an initial deposit of $5,000 and make deposits of $1,000 per year in a continuous fashion. What is the account balance after 8 years? Assume an interest rate of 5%.
   (d) Starting with an initial deposit of $5,000, we plan to make withdrawals of k dollars per year in a continuous fashion. If $k = \$750$, how long will the money last? Is there a value of $k > 0$ so that the account balance never gets to 0? If so, what is the maximum such k? Assume an interest rate of 5%.

10. Redo Example 2.5.4 assuming an interest rate of 8%.

11. A recent front-page story in the New York Times entitled *In the Latest London Blitz, Pets Turn Into Pests*[21] described the rapidly increasing population of rose-ringed parakeets in London and the surrounding suburbs. It is believed that escapees from pet cages fueled this burgeoning

---

[20] Nancy Trejos, *Toll on 401(k) Savings Adds Years More of Toil*, The Washington Post, January 29, 2009.
[21] Elisabeth Rosenthal, The New York Times, May 14, 2011.

population of tropical birds and that perhaps slightly warmer temperatures have facilitated their breeding. Though they are beautiful birds with exuberant personalities, their presence in large numbers in disruptive.

Assume that the wild parakeet population in Britain follows a Malthusian model. The article gives a population in 1995 of 1,500 birds, and of 30,000 "a few years ago"; interpret this as being the 2010 population. What is the projected population for 2020? When was the population 100?

12. A fish population grows according to the Malthusian model
$$\frac{dP}{dt} = 0.05P.$$
When fishing is allowed, the population is determined by
$$\frac{dP}{dt} = 0.05P - 0.65P = -0.6P.$$
Suppose when the fish population is P_0, a year of fishing takes place. Assuming that fishing is then suspended, how long does it take for the population to recover from the fishing (that is, how long to return to a population of P_0)?

13. An intuitive approach to the logistic equation is to suppose that the rate of change $\frac{dP}{dt}$ is jointly proportional to (i.e., proportional to the product of) the population and to the "unused fraction of the population resource" which is $1 - \frac{P}{K}$ where K is the upper limit of sustainable population. Show that this assumption does lead to the logistic equation.

14. In both the Malthusian and logistic models, we set the timeline so that the initial value of $P(t)$ occurs at $t = 0$ (so $P(0) = P_0$ is the initial value). Sometimes it is convenient to have the initial value P_0 occur at another time $t = t_0$, so that $P_0 = P(t_0)$.

    (a) Show that the solution of the initial value problem
    $$\frac{dP}{dt} = kP, \quad P(t_0) = P_0,$$
    is
    $$P(t) = P_0 e^{k(t-t_0)}.$$

    (b) Show that the solution of the initial value problem
    $$\frac{dP}{dt} = aP - bP^2, \quad P(t_0) = P_0,$$
    is given by
    $$P(t) = \frac{aP_0}{(a - bP_0)e^{-a(t-t_0)} + bP_0}.$$

15. The population of New Orleans in July 2005, just before Hurricane Katrina, was 455,000. The population at the first, second, and third year anniversaries of Katrina was, respectively, 50,200, 273,600, and 327,600. In this problem we will show how to fit this data to a logistic model and determine the coefficients a and b in equation (2.80). Recall that the solution to
    $$\frac{dP}{dt} = aP - bP^2$$
    is
    $$P(t) = \frac{1}{\frac{b}{a} + \left(\frac{1}{P_0} - \frac{b}{a}\right)e^{-at}}$$
    where $P_0 = P(0)$. We'll measure t in years and take $t = 0$ to correspond to July 2006, so that $P_0 = 50{,}200$.

(a) Set

$$r = \frac{b}{a}, \quad s = \frac{1}{P_0} - \frac{b}{a}, \quad \text{and} \quad x = e^{-a}.$$

Show that if you know P_0 and have determined x and s, then you can easily compute a and b. Also verify that

$$\frac{1}{P(t)} = r + sx^t.$$

(b) If we set $P_1 = P(1)$ and $P_2 = P(2)$, check that

$$\frac{1}{P_0} = r + s, \quad \frac{1}{P_1} = r + sx, \quad \frac{1}{P_2} = r + sx^2.$$

From the given data, we have $P_1 = 273{,}600$ and $P_2 = 327{,}600$, but we'll defer using these numerical values until we have carried out some algebraic simplifications.

(c) Show that

$$\frac{\frac{1}{P_1} - \frac{1}{P_2}}{\frac{1}{P_0} - \frac{1}{P_1}} = x$$

and

$$\frac{\frac{1}{P_0} - \frac{1}{P_1}}{1 - x} = s.$$

(d) Now use the numerical values of P_0, P_1, and P_2 and determine x, s, r, a, and b.

(e) What is the predicted population of New Orleans in July 2009?

(f) What does this model give for the carrying capacity of New Orleans? As of July 2008, had the inflection point on the logistic curve been passed?

16. Total sales of a new product sometimes follow a logistic model. If $S(t)$ denotes the total number of a product sold by time t, then $dS/dt = aS - bS^2$ for some constants a and b. In thousands of units, the total number of iPods sold by Apple by the end of 2005 was 4,580, by the end of 2006 it was 18,623, and by the end of 2007 it was 39,689. Using the results of the last exercise, use these three data points to determine a and b. At the start of 2008 were iPod sales on the concave up or concave down portion of the logistic sales curve?

17. Two populations with the same initial size P_0 and same carrying capacity $K = a_1/b_1 = a_2/b_2$ are modeled by the differential equations

$$\frac{dP}{dt} = a_1 P - b_1 P^2$$

and

$$\frac{dP}{dt} = a_2 P - b_2 P^2.$$

The graphs of the two populations are shown in Figs. 2.36 and 2.37. Which is larger, a_1 or a_2? Hint: For any solution $P(t)$ to the logistic equation, determine the slope of the graph of $P(t)$ at its inflection point in terms of a and the carrying capacity $\frac{a}{b}$. You do not need to solve the differential equation to do this!

2.5. Population and financial models

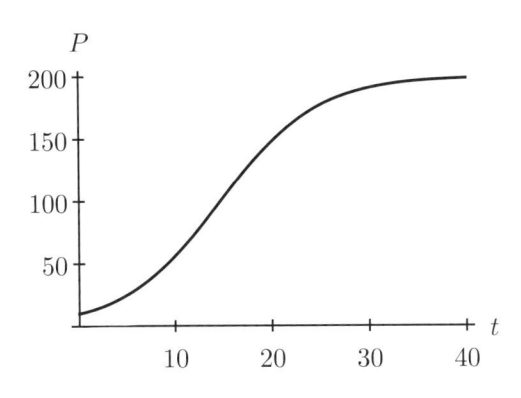

Figure 2.36. Coefficients a_1, b_1.

Figure 2.37. Coefficients a_2, b_2.

18. Suppose that in the logistic model we choose $t = 0$ to correspond to the time when the population is one-half of its carrying capacity a/b. Show that then we can write (2.87) as

$$P(t) = \frac{a/b}{1 + e^{-at}}.$$

19. For positive constants a and b, an equation of the form

$$\frac{dP}{dt} = aP^2 - bP$$

is sometimes called an explosion/extinction equation. This model is appropriate for a situation where births occur at a rate proportional to the square of the population size, while deaths occur at a rate proportional to the population.
   (a) Solve this differential equation in the case that $a = b = 1$. Since $P(t)$ represents a population, think of units of P as millions of individuals, so that the equilibrium at $P = 1$ represents one million individual organisms (whatever they are).
   (b) Find the solution with $P(0) = 2$. At what time t does the population "explode" (become infinite)?
   (c) Find the solution with $P(0) = \frac{1}{2}$. What happens to the population as time tends to infinity?

20. Suppose we propose to model population growth by the equation

$$\frac{dP}{dt} = aP + bP^2$$

where a and b are positive constants.
   (a) Explain how you know without doing any calculations at all that the population will grow without bound, assuming $P(0) > 0$.
   (b) Show, by solving the equation exactly, that there is a finite time t_0 so that

$$\lim_{t \to t_0} P(t) = \infty.$$

This model is sometimes called the doomsday model, with t_0 being doomsday.

21. The **Gompertz differential equation** is

$$\frac{dP}{dt} = aP - bP \ln P$$

where a and b are positive constants. Show that it has the form

$$\frac{dP}{dt} = r(P)P$$

for some function $r(P)$ (which you should identify explicitly). Sketch the graph of P vs. $r(P)$. The Gompertz equation is used to model population growth and also the closely related problem of tumor growth. Tumor cells grow in a confined environment with restricted availability of nutrients, a situation that has parallels with a population growing in a fixed space with limited resources.

22. Solve the Gompertz equation (see Exercise 21) as a separable equation. A substitution will be helpful to carry out the necessary integration.

    What is the long-term behavior of a (positive) population that grows according to this equation? That is, determine
    $$\lim_{t \to \infty} P(t).$$

23. Suppose the population P at time t (in years) of a city is modeled by a logistic equation
    $$\frac{dP}{dt} = \frac{1}{3}P - \frac{1}{18 \times 10^4}P^2.$$
    How would you modify this equation if additionally 5,000 people per year move out of the city (in a continuous fashion)? Find any equilibrium solutions of your modified equation.

24. Sally is cramming for a history exam. She has an amount of material M to memorize. Let $x(t)$ denote the amount memorized by time t (in hours).
    (a) Assume that the rate of change of the amount of material memorized is proportional to the amount that remains to be memorized. The proportionality constant is a measure of natural learning ability, and in Sally's case set it to be 0.2. Suppose that $x(0) = 0$. Solve the resulting initial value problem, and determine how long it takes her to memorize half of the material.
    (b) How long does it take to memorize 90 percent of the material?
    (c) More realistically, while she is memorizing new material, she is also forgetting some of what she has already memorized. Assume that the rate of forgetting is proportional to the amount already learned, with proportionality constant 0.02. Modify your differential equation from (a) to take forgetting into account, and determine the value of t for which $x(t) = M/2$. Can she get everything memorized? If not, what's the best she can do?

25. A spherical mass of bacteria is growing in a Petri dish. The mass grows uniformly in all directions (so it continues to be spherical in shape) and only the bacteria on the surface of the mass reproduce.
    (a) The differential equation
    $$\frac{dN}{dt} = kN^{2/3}$$
    (k a positive constant) is proposed to model the number N of bacteria at time t. Solve this equation with initial condition $N(0) = N_0$.
    (b) Using the formulas
    $$S = 4\pi r^2 \quad \text{and} \quad V = \frac{4}{3}\pi r^3$$
    for the surface area S and the volume V of a sphere in terms of the radius r, explain where the power $2/3$ comes from in the equation in (a).

26. **Retrieval from long-term memory.** Suppose you are asked to list all items you can think of in a particular category; for example list animals, or types of flowers. If we assume you have some finite number M of items in the specified category stored in your long-term memory, one model of how these items are retrieved hypothesizes that the rate of recollection is proportional to the difference between M and the number you have already named. Translate this into a differential equation for $n(t)$, the number of items you have listed by time t, and then solve the differential equation assuming $n(0) = 0$. What happens to $n(t)$ as $t \to \infty$?

27. A simple blood test measuring β-HCG (human chorionic gonadotropin) levels in the blood is used to assess the health of a pregnancy in the earliest weeks, before ultrasound technology

becomes useful. The rate of increase of β-HCG is more predictive than the absolute numbers. In the early weeks of a normally progressing pregnancy β-HCG doubles about every 48 hours, and a doubling time of greater than 72 hours is considered abnormal. Serial β-HCG measurement is routine in couples undergoing any kind of infertility treatment, as well as in many other cases.

(a) If β-HCG is 80 units when measured at 14 days past ovulation (DPO) and 195 at 18 DPO, is this within normal limits? Assume the blood is drawn at the same time each day and that the amount A of HCG grows according to the differential equation $\frac{dA}{dt} = kA$. Hint: If a normal doubling time is anything less than or equal to 72 hours, what does this tell you about a "normal" value of k?

(b) If β-HCG is 53 units when measured at 9:00 AM on 13 DPO and 100 when measure at 5:00 PM on 15 DPO, is this within normal limits?

(c) In cases of in-vitro fertilization (IVF) where multiple embryos are implanted, it is typical to measure β-HCG 4 or 5 times over a period of several weeks before the first ultrasound can be done at approximately 40 DPO. Here the phenomenon known as a "vanishing twin" is not uncommon. This means one or more embryos that begin early development (and hence initially contribute to the production of β-HCG) cease development before they are visualized on ultrasound. The typical β-HCG pattern in such a situation may show some measurements that do not meet the "normal" doubling time of at most 3 days before settling into a normal rate of increase (indicating healthy development of the remaining embryos). Suppose the β-HCG measurements for an IVF pregnancy are as shown below. Assume the blood is drawn at the same time each day. Do any two consecutive measurements fall within normal limits?

DPO	14	16	19	26	38
β-HCG	227	310	570	2,800	45,000

28. Suppose we have a collection of data points for population as shown below. We want to decide if this set of data is reasonably "logistic", and if so, to give estimates for the parameters a and b in the logistic equation $\frac{dP}{dt} = aP - bP^2$.

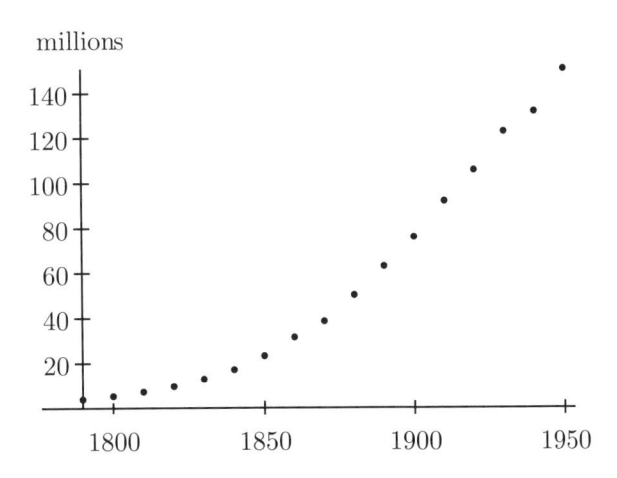

Figure 2.38. US population 1790–1950.

(a) Imagine that the points are spaced in time by 1 unit; in our example one unit of time will be a decade, and $t = 0$ will correspond to the year 1790. For $t = 1, 2, 3, 4, \ldots$, imagine calculating the ratios
$$\Delta_t = \frac{P(t+1) - P(t-1)}{2}.$$
Explain why this is a (very) rough approximation to $P'(t)$.

(b) Now imagine computing the ratios $\dfrac{\Delta_t}{P(t)}$ for $t = 1, 2, 3, 4, \ldots$. Explain why this is an approximation for the proportional growth rate

$$\dfrac{\frac{dP}{dt}}{P}.$$

(c) If you make a graph of P (horizontal axis) vs. $\dfrac{\Delta_t}{P(t)}$ (using your calculations from (b)), what should this graph look like if the population growth is logistic? If it is logistic, how would you estimate the parameters a and b from this graph?

(d) Using any computer software at your disposal, calculate Δ_t and $\Delta_t/P(t)$ as in parts (a)–(b) for $t = 1, 2, \ldots, 15$ using the US population data given below. Graph the points $(P(t), \Delta_t/P(t))$ you obtained. They should (roughly!) look like they lie on a line. Estimate the parameters a and b from this line.

(e) Plot on the same coordinate grid both the raw population data and the solution of the logistic equation with the a and b values you found in part (d) and $P(0) = 3{,}930{,}000$.

t	Year	Population
0	1790	3,930,000
1	1800	5,310,000
2	1810	7,240,000
3	1820	9,640,000
4	1830	12,870,000
5	1840	17,070,000
6	1850	23,190,000
7	1860	31,440,000
8	1870	38,560,000
9	1880	50,160,000
10	1890	62,950,000
11	1900	75,990,000
12	1910	91,970,000
13	1920	105,710,000
14	1930	122,780,000
15	1940	131,670,000
16	1950	150,700,000

2.6. Qualitative solutions of autonomous equations

Recall that the logistic differential equation is an example of an *autonomous* differential equation. A first-order equation is said to be autonomous if it can be written in the form

(2.90) $$\dfrac{dy}{dt} = f(y),$$

so that the independent variable t does not appear explicitly on the right-hand side of the equation. Autonomous equations have several noteworthy features that make them particularly amenable to qualitative analysis. The field marks for the direction field of any autonomous equation have the same slope along any *horizontal* line; see Fig. 2.39.

2.6. Qualitative solutions of autonomous equations

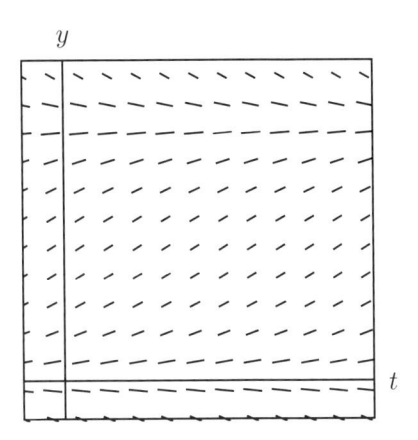

Figure 2.39. Direction field for an autonomous equation.

So if we calculate the field marks at points along one *vertical* line, we can fill out a fuller picture of the direction field by just shifting these field marks left and right.

This should make the following assertion believable: If you have a solution $y_1(t) = \varphi(t)$ of equation (2.90) and you translate this solution (to the right or the left) by constructing the function

$$y_2(t) = \varphi(t-c)$$

for some constant c, then the resulting new function is also a solution of (2.90). For example, the function $y(t) = 2e^{3t}$ solves the autonomous equation $\frac{dy}{dt} = 3y$, and so does the function $y(t) = 2e^{3(t-2)} = 2e^{-6}e^{3t}$.

Short of sketching solutions using the direction field, we can often get a useful picture of the solution curves to an autonomous equation $\frac{dy}{dt} = f(y)$ by determining the equilibrium solutions and the increasing/decreasing behavior of the other solutions. We illustrate how this goes with the equation

$$\frac{dy}{dt} = f(y) = (y-2)(y-3)(y+1)^2. \tag{2.91}$$

Equilibrium solutions. To find the equilibrium solutions we determine all constants k for which $f(k) = 0$. Corresponding to each such constant k is a horizontal-line solution $y = k$ in the ty-plane, and we begin our qualitative solution by sketching these solutions. In equation (2.91), the equilibrium solutions are $y = 2$, $y = 3$, and $y = -1$, as shown in Fig. 2.40.

Figure 2.40. Equilibrium solutions for $dy/dt = (y-2)(y-3)(y+1)^2$.

Increasing/decreasing properties. Continuing with equation (2.91), we next determine for each of the ranges of values $-\infty < y < -1$, $-1 < y < 2$, $2 < y < 3$, and $3 < y < \infty$ whether

$\frac{dy}{dt}$ is positive or negative. Continuity of f will guarantee that $f(y)$ is either always positive or always negative in each of these ranges. We'll keep track of the positive or negative sign of $f(y)$ on a number line, which for later convenience we orient vertically; see Fig. 2.41. For example, when $y > 3$, each factor $(y - 2)$, $(y - 3)$, and $(y + 1)^2$ is positive, ensuring that $\frac{dy}{dt} > 0$ whenever $y > 3$.

What does this number line in Fig. 2.41 tell us? Since the sign of $\frac{dy}{dt}$ determines the increasing/decreasing property of a solution $y = y(t)$, we see that solution curves in the horizontal strips $y > 3$, $-1 < y < 2$, or $y < -1$ are increasing (since $\frac{dy}{dt}$ is positive) while any solution curve in the strip $2 < y < 3$ is decreasing. We summarize this information in Fig. 2.42 using "up" and "down" arrows along our vertical line to indicate whether solution curves in the corresponding horizontal strips increase or decrease. This up/down line is called a **phase line** for the differential equation (2.91).

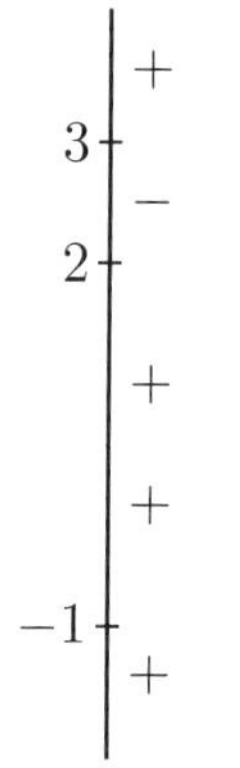

Figure 2.41. Sign of $dy/dt = (y-2)(y-3)(y+1)^2$.

Figure 2.42. Phase line for $dy/dt = (y-2)(y-3)(y+1)^2$.

No two distinct solution curves can intersect (by the No Intersection Theorem, Theorem 2.4.8). Thus, in particular, none can cross the horizontal lines $y = 2$, $y = 3$, and $y = -1$ (the equilibrium solutions). So, as suggested by the phase line, the equilibrium solutions break the ty-plane up into horizontal strips on which the solution curves are either increasing or decreasing.

Moreover, any solution to an autonomous differential equation $dy/dt = f(y)$, with f and f' continuous, either

- tends to ∞ as t increases or
- tends to $-\infty$ as t increases or
- tends to an equilibrium solution as t increases.

A key ingredient in a formal proof of these claims is the topic of Exercise 23, which outlines how to show that if $y(t)$ is any solution to an autonomous equation $dy/dt = f(y)$ and if

$$\lim_{t \to \infty} y(t) = L$$

for some finite value L, then L must be an equilibrium value: $f(L) = 0$. Moreover, if at time $t = t_0$ a solution to $dy/dt = f(y)$ (with f and f' continuous) is between two consecutive equilibrium values k_1 and k_2 with $k_1 < k_2$, then either

- the solution tends to the larger equilibrium value k_2 as $t \to \infty$ if the solution curves are increasing in the range $k_1 < y < k_2$ or
- the solution tends to the smaller equilibrium value k_1 as $t \to \infty$ if the solution curves are decreasing in the range $k_1 < y < k_2$.

2.6. Qualitative solutions of autonomous equations

Using these ideas, we show qualitatively some solutions for the equation

$$dy/dt = (y-2)(y-3)(y+1)^2.$$

Figure 2.43. Qualitative solutions for $dy/dt = (y-2)(y-3)(y+1)^2$.

Stability of equilibrium solutions. Fig. 2.43 also illustrates the notion of stability for an equilbrium solution. The equilibrium solution $y = 2$ is said to be **(asymptotically) stable** since any solution $y(t)$ whose initial value $y(0)$ is sufficiently close to the equilibrium value 2 tends to this equilibrium value as $t \to \infty$. By contrast, the equilibrium solution $y = 3$ is **unstable**. Any solution $y(t)$ whose initial value is close to 3, but not equal to 3, will ultimately move away from 3, either approaching 2 as $t \to \infty$ or tending to ∞. A third possibility is illustrated by the equilibrium solution $y = -1$, which is called **semistable**. Any solution whose initial value is a little bigger than -1 will move away from $y = -1$, eventually approaching 2 as $t \to \infty$. On the other side though, a solution whose initial value is less than -1 approaches the equilibrium value -1 as t increases. In terms of the phase line, an equilibrium solution $y = k$ is stable if both arrows adjacent to k point towards k, and it is unstable if the adjacent arrows both point away from k. The semistable classification occurs when one adjacent arrow points towards k and the other points away from k.

Blowing up in finite time. One caution in interpreting the "solution tends to ∞" or "solution tends to $-\infty$" conditions above: These can happen in finite time! For example, let's look at the equation

$$\frac{dy}{dt} = y + y^2, \qquad (2.92)$$

whose phase line is shown in Fig. 2.44.

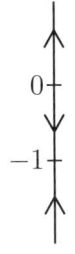

Figure 2.44. Phase line for $dy/dt = y + y^2$.

Explicit solutions to (2.92) can be found by separating variables, and we find solutions

(2.93) $$y = \frac{Ce^t}{1 - Ce^t}$$

for arbitrary constant C (see Exercise 9). You can check that the value of C is related to the value y_0 of y at $t = 0$ by

(2.94) $$C = \frac{y_0}{1 + y_0}.$$

In particular, the solution with value $y = 1$ at $t = 0$ has $C = \frac{1}{2}$:

(2.95) $$y = \frac{\frac{1}{2}e^t}{1 - \frac{1}{2}e^t} = \frac{e^t}{2 - e^t}.$$

By the phase line sketch, we expect this solution to be increasing, and since there are no equilibrium values larger than 0, any solution with $y(0) > 0$ must tend to infinity. Notice that our solution in (2.95) has (left-hand) limit ∞ at $t = \ln 2$:

$$\lim_{t \to \ln 2^-} \frac{e^t}{2 - e^t} = \infty,$$

so that this solution tends to infinity in finite time, not as $t \to \infty$. Other solutions with $y(0) > 0$ also blow up in finite time, since

$$\lim_{t \to \ln \frac{1}{C}^-} \frac{Ce^t}{1 - Ce^t} = \infty.$$

Equation (2.94) says that if $y(0) = y_0$ is positive, then C is less than 1, so that $\ln \frac{1}{C} > 0$. We cannot tell from the phase line alone if a solution tends to infinity in finite time.

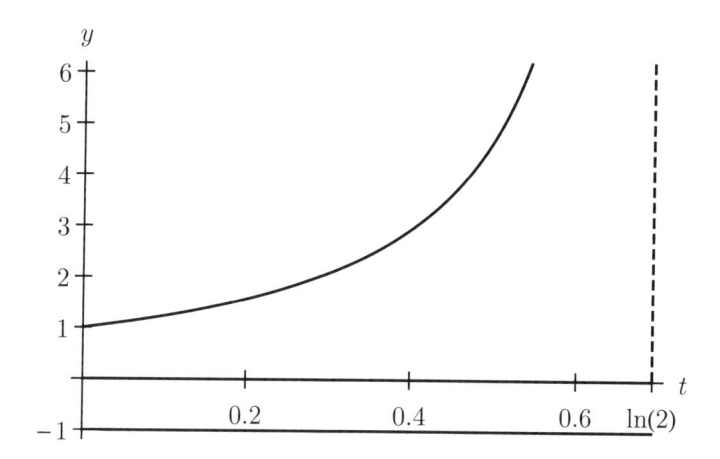

Figure 2.45. $y = e^t/(2 - e^t)$ and equilibrium solutions $y = 0$, $y = -1$.

Sometimes the information at our disposal will only allow us to find the equilibrium solutions and phase line *qualitatively*. The next example shows how this might be done.

Example 2.6.1. We will sketch some solutions to the equation

(2.96) $$\frac{dy}{dt} = f(y),$$

where the graph of $f(y)$ is as shown in Fig. 2.46.

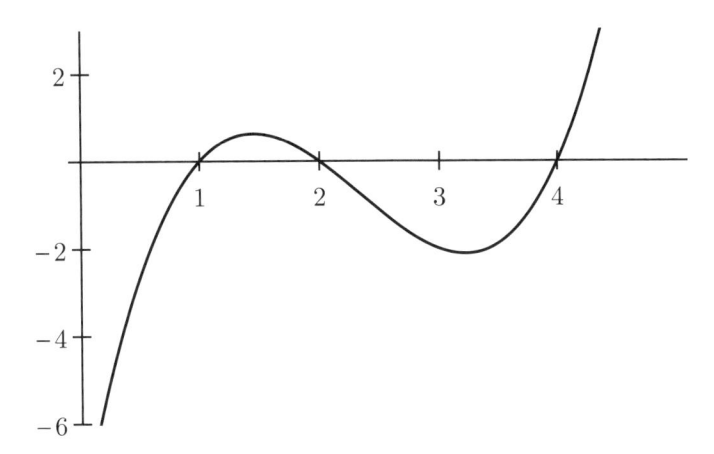

2.6. Qualitative solutions of autonomous equations

Figure 2.46. Graph of $f(y)$.

The equilibrium solutions take the form $y = k$ where $f(k) = 0$ (and hence $\frac{dy}{dt} = 0$). From the graph, we see the equilibrium solutions are approximately $y = 1$, $y = 2$, and $y = 4$. The phase line shows the equilibrium values 1, 2, and 4. Next we want to decide if $\frac{dy}{dt}$ is positive or negative in each of the intervals $-\infty < y < 1$, $1 < y < 2$, $2 < y < 4$, and $y > 4$. From the graph of $f(y)$ we see that for $y < 1$, $f(y)$ is negative. For y between 1 and 2, $f(y)$ is positive; between 2 and 4, $f(y)$ is negative; and finally for $y > 4$, $f(y)$ is positive. The phase line for equation (2.96) is as shown in Fig. 2.47.

Figure 2.47. Phase line for $dy/dt = f(y)$.

Using the phase line, we give a rough sketch in Fig. 2.48 of some solutions curves for (2.96). Begin the sketch by showing the equilibrium solutions, $y = 1$, $y = 2$, and $y = 4$, as horizontal lines. Next use the phase line to determine the increasing and decreasing behavior of solution curves between the equilibrium solutions. For example, solutions in the horizontal strip between $y = 1$ and $y = 2$ are increasing, while those that lie in the strip between $y = 2$ and $y = 4$ are decreasing. Our discussion above also tells us about long-term behavior: If $y(0)$ is between 1 and 2 or between 2 and 4, then $y(t)$ tends to 2 as $t \to \infty$. Solutions with $y(0) > 4$ tend to infinity, while those with $y(0) < -1$ tend to $-\infty$.

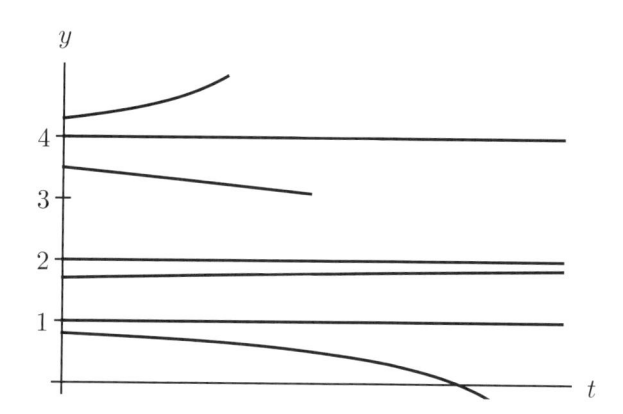

Figure 2.48. Some approximate solution curves.

2.6.1. Bifurcations. Many of the differential equations we have used in applications contain one or more **parameters**; these are *constants* which may take different values in different particular situations. The Malthusian equation $\frac{dP}{dt} = kP$ has one parameter k while the logistic equation $\frac{dP}{dt} = aP - bP^2$ has two parameters a and b. We expect the value of these parameters to depend on the particular population we wish to study; moreover, we typically have only approximate values for the parameters.

In this section we will consider autonomous first-order differential equations containing one real parameter b. The equilibrium solutions and the qualitative behavior of the nonequilibrium solutions may change as the value of this parameter changes. We might hope that a small change in b corresponds to a small change in the solution curves, but this is not always the case. The next example illustrates this.

Example 2.6.2. Consider the differential equation

$$\frac{dy}{dt} = by - y^3 \tag{2.97}$$

which depends on the parameter b. The equilibrium solutions are found by solving $y(b - y^2) = 0$, and when $b > 0$ we have three equilibrium solutions: $y = 0$, $y = \sqrt{b}$, and $y = -\sqrt{b}$. The phase line for the case $b > 0$ is shown in Fig. 2.49. Solutions with $y(0) < -\sqrt{b}$ are increasing to $-\sqrt{b}$ as $t \to \infty$ and solutions with $-\sqrt{b} < y(0) < 0$ are decreasing to $-\sqrt{b}$ as $t \to \infty$. If $0 < y(0) < \sqrt{b}$, then the solution increases to $\sqrt{b}$ as $t \to \infty$, and if $\sqrt{b} < y(0)$, the solution decreases to $\sqrt{b}$ as $t \to \infty$. The equilibrium solution $y = 0$ is unstable, while the solutions $y = \sqrt{b}$ and $y = -\sqrt{b}$ are stable.

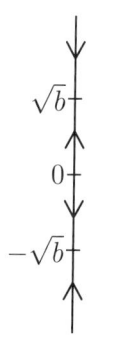

Figure 2.49. Phase line for $dy/dt = by - y^3$ when $b > 0$.

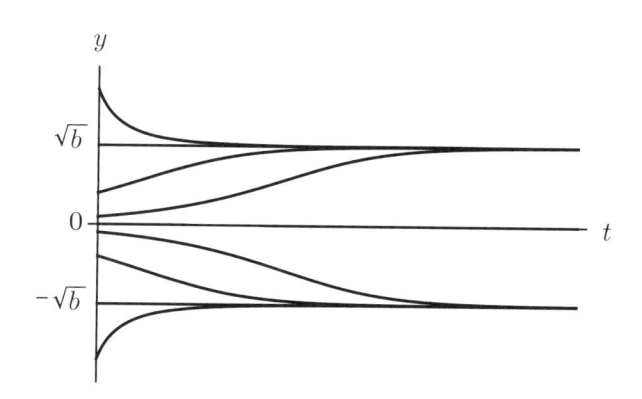

Figure 2.50. Typical solution curves for $\frac{dy}{dt} = by - y^3$ when $b > 0$.

2.6. Qualitative solutions of autonomous equations

By contrast, if $b \leq 0$, there is only one equilibrium solution, $y = 0$, which is stable. Solutions with $y(0) > 0$ decrease to 0 as $t \to \infty$, while solutions with $y(0) < 0$ increase to 0 as $t \to \infty$.

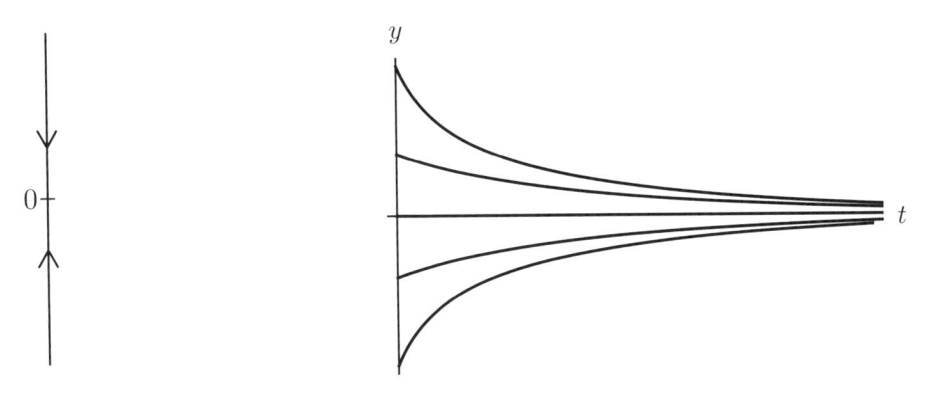

Figure 2.51. Phase line for $dy/dt = by - y^3$ when $b < 0$.

Figure 2.52. Typical solution curves for $\frac{dy}{dt} = by - y^3$ when $b < 0$.

Thus the number of equilibrium solutions and the long-term behavior of nonequilibrium solutions are dramatically different for $b > 0$ than for $b < 0$ and we say that a **bifurcation** occurs at the parameter value $b = 0$.

The graph of the equation $by - y^3 = 0$ in the by-plane (with the b-axis horizontal) is called a **bifurcation diagram**. Shown in Fig. 2.53, it consists of the line $y = 0$ and the parabolic curve $b - y^2 = 0$. This particular bifurcation diagram is sometimes called a **pitchfork**. A point (b, k) lies on the bifurcation diagram if and only if $y = k$ is an equilibrium solution of equation (2.97).

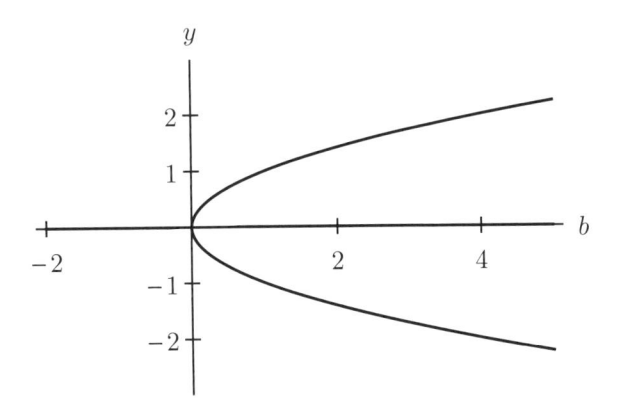

Figure 2.53. Bifurcation diagram for $\frac{dy}{dt} = by - y^3$.

You can see from the bifurcation diagram that for each value of b less than the bifurcation value 0 there is only one equilibrium solution (each vertical line to the left of 0 intersects the pitchfork in only one point), while for each value of b greater than 0 there are three equilibrium solutions (each vertical line to the right of 0 intersects the pitchfork in three points).

In general, for a differential equation that depends on a real parameter, a bifurcation value is a real number b_0 such that for parameter values near b_0 but on opposite sides, solutions to the corresponding equations behave quite differently, qualitatively speaking. When a differential equation with a parameter appears in an application, the value of the parameter is often controlled by external conditions. Its size relative to any bifurcation values may have a dramatic influence on the long-term behavior of the solutions. Exercise 20 explores this in the context of a fish farm.

2.6.2. Exercises.

For the autonomous equations in Exercises 1–8, (a) find all equilibrium solutions, (b) draw a phase line, (c) based on the phase line, describe the stability of equilibrium solutions, and (d) using the phase line, determine $\lim_{t \to \infty} y(t)$, given that y satisfies the equation and $y(0) = 1$.

1. $\frac{dy}{dt} = y^2 - 3y - 4$.
2. $\frac{dy}{dt} = (2-y)^2$.
3. $\frac{dy}{dt} = y^2(3-y)$.
4. $\frac{dy}{dt} = k(A-y)$, where $k > 0$ and $A > 1$.
5. $\frac{dy}{dt} = y(y-2)$.
6. $\frac{dy}{dt} = -y^3 + 3y^2 - 2y$.
7. $\frac{dy}{dt} = (2-y)\ln(1+y^2)$.
8. $\frac{dy}{dt} = y(e^y - 3)$.

9. Solve
$$\frac{dy}{dt} = y + y^2$$
by separating variables, obtaining (2.93) as a one-parameter family of solutions.

10. Consider the differential equation
$$\frac{dy}{dt} = k(1-y)^2$$
where k is a positive constant.
   (a) What is the equilibrium solution?
   (b) Explain why you know from the differential equation that every (nonequilibrium) solution is an increasing function of t.
   (c) Show that
$$\frac{d^2y}{dt^2} = -2k^2(1-y)^3,$$
   and describe the concavity of the solution curves.
   (d) Set $k = 1$ and solve the differential equation with the initial condition $y(0) = 1/2$ and with the initial condition $y(0) = 3/2$. Sketch the first solution for $t > -2$, and the second solution for $t < 2$.

11. In this problem you will give a qualitative solution to
$$(2.98) \qquad \frac{dy}{dx} = (y+2)(y-3).$$
   (a) Find all equilibrium solutions, and sketch them in the xy-plane.
   (b) Draw the phase line, showing where solutions are increasing and decreasing.
   (c) What is $\frac{d^2y}{dx^2}$? Determine where this second derivative is zero, where it is positive, and where it is negative. Explain what this tells you about the concavity of solution curves for equation (2.98).
   (d) Using the information you have just determined, sketch solution curves with $y(0) = 0$, $y(0) = 1$, $y(0) = 2$, $y(0) = 4$, $y(0) = -1$, and $y(0) = -3$. Show a portion of each of these solution curves with both positive and negative values of x.
   (e) Classify each equilibrium solution as either stable, unstable, or semistable.

12. Consider the differential equation
$$(2.99) \qquad \frac{dy}{dt} = 4 - y^2.$$
   (a) What are the equilibrium solutions of (2.99)?
   (b) Draw a phase line for (2.99) and describe the long-term behavior of the solution of (2.99) satisfying $y(0) = 1$.

2.6. Qualitative solutions of autonomous equations

(c) Solve the initial value problem consisting of (2.99) and $y(0) = 1$ exactly and check that your analysis of its long-term behavior in part (b) is correct.

13. A differential equation of the form

$$\frac{dP}{dt} = P \cdot f(P),$$

where the graph of $f(P)$ is as shown in Fig. 2.54, is proposed to model the population of a certain species.

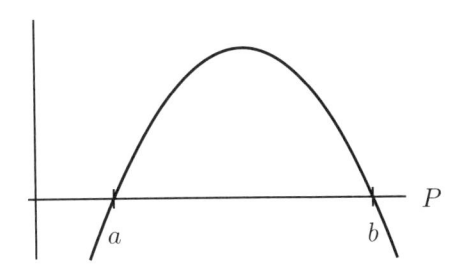

Figure 2.54. Graph of $f(P)$.

(a) Show that there are 3 equilibrium populations.
(b) Are either of the positive equilibrium populations stable? If so, which one(s)?
(c) Show that the limiting value of the population can be different depending on the initial (positive) value of the population. Explain why this is different from a logistic population model.

14. Consider the differential equation

$$\frac{dy}{dt} = f(y)$$

where the graph of $f(y)$ is as shown in Fig. 2.55.
(a) What are the equilibrium solutions?
(b) Sketch the phase line for $\frac{dy}{dt}$.
(c) Using the phase line, sketch some solution curves in the ty-plane.

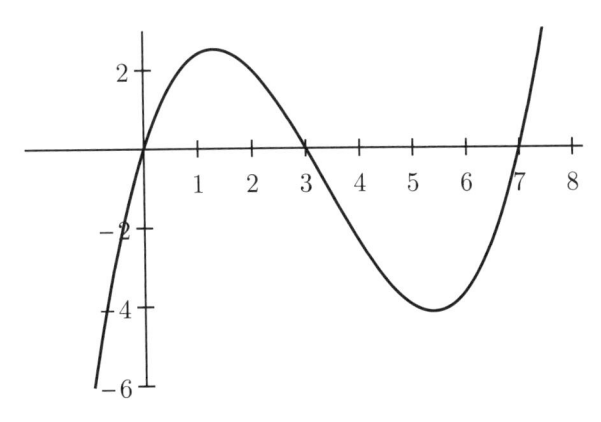

Figure 2.55. Graph of $f(y)$.

15. Consider the initial value problem

$$\frac{dy}{dt} = \frac{(y-5)^2(y-14)}{y^2+1}, \quad y(0) = 10.$$

Identify each of the following statements as true or false, and explain your reasoning. Hint: Start by determining all equilibrium solutions of the differential equation.
(a) $y(10)$ must be between 5 and 14.
(b) $y(15)$ cannot be equal to 0.
(c) $y(15)$ cannot be equal to 15.

16. Consider the initial value problem

$$\frac{dy}{dt} = y(y-1)^2(y-2)(y-3), \quad y(0) = y_0.$$

Describe the behavior of the solution $y(t)$ as t increases if
(a) $y_0 = 1/2$,
(b) $y_0 = 3/2$,
(c) $y_0 = 5/2$,
(d) $y_0 = 7/2$,
(e) $y_0 = 3$.

17. Suppose we have a differential equation

$$\frac{dy}{dt} = f(y)$$

where f is a continuous function. If we also know that $f(1) = 2$ and $f(3) = -3$, which of the following statements is correct?
(a) There is an equilibrium solution $y = k$ for some value of k between 1 and 3.
(b) There is an equilibrium solution $y = k$ for some value of k between -3 and 2.
(c) You can't conclude anything about equilibrium solutions without further information.
Suggestion: If you don't remember the Intermediate Value Theorem, look it up in a first-year calculus book.

18. Based on a phase line for

(2.100) $$\frac{dy}{dt} = y^{2/3}$$

Sally says that if y satisfies (2.100) and $y(0) = -1$, then $\lim_{t \to \infty} y(t)$ must be 0. Sam disagrees. Who is right?

19. The fish population $P(t)$ in a large fish farm changes according to the differential equation

$$\frac{dP}{dt} = 3P - P^2 - 2,$$

where P is in millions of fish and t is in years. Notice that we have modified the logistic equation by assuming that the fish are being harvested at a constant rate of 2 million per year.
(a) Find all equilibrium solutions of the differential equation.
(b) For what range of P values is the population increasing with time, and for what P values is it decreasing?
(c) If the initial population is $P(0) = 1.5$, can we continue to harvest at a rate of 2 million fish per year indefinitely? Answer the same question if $P(0) = 0.8$.
(d) What is the smallest initial population which permits harvesting 2 million fish per year indefinitely?
(e) Find the solution $P(t)$ with $P(0) = 1.5$.

(f) Using your solution in (e), determine approximately how many years must pass before the fish population reaches 1.8 million, which is 90% of the fish farm's carrying capacity.

20. In this exercise we consider the fish farm population model of Exercise 19, assuming a constant harvesting rate h, the value of which will vary. Thus, we have

(2.101) $$\frac{dP}{dt} = 3P - P^2 - h$$

where P is in millions of fish, t is in years, and $h \geq 0$ is the constant harvest rate in millions per year.
   (a) If $h = 0$, what is the fish farm's carrying capacity?
   (b) For $h = 0, h = 1, h = 9/4$, find equilibrium solutions of (2.101) and describe their stability.
   (c) Show that if $h > 9/4$, the equation (2.101) has no equilibrium solutions.
   (d) What value of h is a bifurcation value of (2.101)? For positive values of h less than the bifurcation value, describe the long-term behavior of the fish population assuming it is initially at least 1.5 million. Do the same for parameter values exceeding the bifurcation value.
   (e) Give the bifurcation diagram for (2.101); that is, graph the equation

(2.102) $$3P - P^2 - h = 0$$

   in the hP-plane (the horizontal axis is the h-axis).
   (f) Recall that a point (h, k) lies on the graph of (2.102) if and only if $y = k$ is an equilibrium solution of (2.101). Your bifurcation diagram of part (e) is a parabola. Which part of the parabola corresponds to stable equilibrium solutions? What terminology applies to the first coordinate of the parabola's vertex?

21. **Harvesting renewable resources.** In this problem we look at the effect of fishing on a fish population that grows logistically.
   (a) Suppose our fish population P is growing according to the equation

$$\frac{dP}{dt} = aP - bP^2 \quad (a > 0, b > 0)$$

   and we begin to remove fish from the population. Suppose first that the fish are removed at a rate proportional to the size of the population, so now

(2.103) $$\frac{dP}{dt} = aP - bP^2 - cP$$

   for some positive constant c (sometimes called the effort coefficient). Show that if $a > c$, this new equation has two equilibrium values, $p_1 = 0$ and $p_2 = \frac{a-c}{b}$.
   (b) Still assuming $a > c$, what happens to the solutions to (2.103) as $t \to \infty$, if $P(0) > 0$?
   (c) The product of the stable equilibrium p_2 and the effort coefficient c is a *sustainable yield*, $Y = cp_2$. When fishing occurs at this level it can be continued indefinitely. Give a formula for Y in terms of a, b, and c. What choice of c will maximize Y (assuming a and b are constant)? What is the corresponding maximal sustainable yield?
   (d) Alternately, assume that fish are harvested at a constant rate so that

(2.104) $$\frac{dP}{dt} = aP - bP^2 - h$$

   for some positive constant h. Show that if

$$0 < h < \frac{a^2}{4b},$$

   this differential equation has two positive equilibrium solutions. Classify each as stable or unstable.

(e) If $h = \frac{a^2}{4b}$ in (2.104), show there is one equilibrium value $p = \frac{a}{2b}$. Explain why if $P(0) > \frac{a}{2b}$, the population will stay above $\frac{a}{2b}$ for all $t > 0$, so that this level of fishing can be maintained indefinitely, and there is a sustainable yield of $\frac{a^2}{4b}$.

(f) Show that when $h > \frac{a^2}{4b}$ there is no equilibrium solution to (2.104).

22. In this problem,[22] we take the simplest possible view of the average temperature $T(t)$ of the earth at time t and assume that the two factors that control $T(t)$ are radiation from the sun and radiation leaving the earth. We model this with the differential equation

$$(2.105) \qquad \frac{dT}{dt} = r_i(T) - r_o(T)$$

where $r_i(T)$ and $r_o(T)$ are functions of the temperature T. For example, a higher average temperature T may cause a decrease in the amount of ice and snow near the poles. Since ice and snow are more reflective than, say, water, this causes an increase in the absorption of solar energy. Also, as T increases, the radiation leaving the earth increases too, as a warm body radiates more than a cold one. Equation (2.105) is an autonomous differential equations since the right-hand side is a function of T but not explicitly of t. Suppose the functions $r_i(T)$ and $r_o(T)$ have graphs as shown in Fig. 2.56.

(a) On what intervals of T is $r_i(T) - r_o(T)$ positive? Negative?

(b) What are the equilibrium solutions of equation (2.105) if $r_i(T)$ and $r_o(T)$ are as shown? Sketch these in a t vs. T graph as horizontal lines.

(c) Using your answers to (a) and (b), give a qualitative solution to equation (2.105). Just concentrate on the increasing and decreasing behavior of the solution curves.

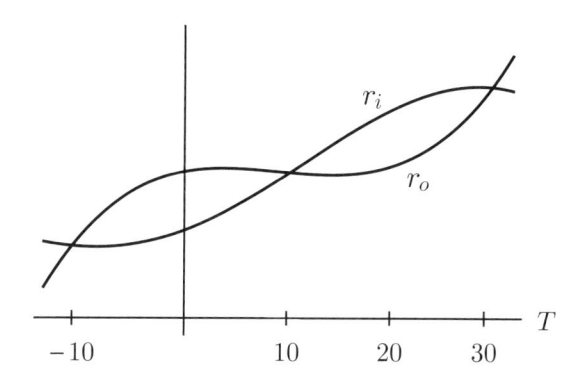

Figure 2.56. Graphs of $r_i(T)$ and $r_o(T)$.

23. Suppose that $y_1(t)$ is a solution on $[0, \infty)$ to the autonomous differential equation

$$(2.106) \qquad \frac{dy}{dt} = f(y),$$

where f is a continuous function. Show that if the limit

$$\lim_{t \to \infty} y_1(t) = L$$

exists (where L is a finite number), then $y = L$ must be an equilibrium solution of the equation (2.106). Hints:

(a) Explain why $y_1(t+1) - y_1(t) \to 0$ as $t \to \infty$.

(b) Using a result from calculus, show that for each $t \geq 0$ there is a number $s(t)$ between t and $t+1$ satisfying $y_1(t+1) - y_1(t) = y_1'(s(t))$.

---

[22] Adapted from Clifford Taubes, *Modeling Differential Equations in Biology*, 2nd edition, Cambridge University Press, New York, 2008.

2.6. Qualitative solutions of autonomous equations

(c) Using parts (a) and (b) show that $f(L) = 0$. Conclude that $y = L$ is an equilibrium solution.

24. The following differential equation has been proposed to model grazing by a herd of cows on vegetation in a certain field:[23]

$$\text{(2.107)} \qquad \frac{dV}{dt} = \frac{1}{3}V - \frac{1}{75}V^2 - H\frac{0.1V^2}{9+V^2}.$$

In this equation time t is measured in days, $V(t)$ is the amount of vegetation in the field, in tons, at time t, H is the (constant) size of the herd, and

$$\frac{0.1V^2}{9+V^2}$$

is the amount of vegetation eaten per cow in one day. An initial condition for this model specifies the amount of vegetation at time 0, $V(0) = V_0$. The term $H\frac{0.1V^2}{9+V^2}$, which is sometimes referred to as the "predation term" has a graph whose general shape is shown below. It starts out small, grows slowly up to a certain point, then increases sharply until it begins to level off. This leveling off represents a saturation—the cows can only eat so much per day, no matter how abundant the vegetation.

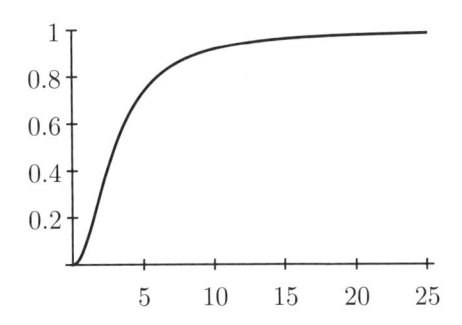

Figure 2.57. Graph of $V^2/(V^2+9)$.

(a) Show that if $H = 0$, the vegetation grows logistically, and find the carrying capacity.
(b) Figs. 2.58–2.60 show the graphs of

$$\text{(2.108)} \qquad y = \frac{1}{3}V - \frac{1}{75}V^2 \text{ (dashed)} \quad \text{and} \quad y = H\frac{0.1V^2}{9+V^2} \text{ (solid)}$$

for three different herd sizes: $H = 10$, $H = 20$, and $H = 30$. For each of these three herd sizes, use the information available in these graphs to predict that there is one positive equilibrium solution to equation (2.107) when $H = 10$, three positive equilibrium solutions when $H = 20$, and one positive equilibrium solution when $H = 30$.

(c) For the smallest herd size, $H = 10$, approximately what is the positive equilibrium solution? Predict what happens as $t \to \infty$ to those solutions that have $V(0)$ greater than this equilibrium value. Predict what happens to those solutions with $V(0)$ less than the equilibrium value.

(d) Now focus on the intermediate herd size $H = 20$. Using the information in the graphs in Figs. 2.58–2.60, give a rough sketch of several solutions to the differential equation, showing V as a function of t, with various initial conditions. Begin by showing the equilibrium solutions (as horizontal lines) in your sketch. Show that if $V(0) = 6$, then the amount of vegetation decreases for $t > 0$, but if $V(0) = 9$, the amount of vegetation increases for $t > 0$.

---
[23] R. M. May, *Stability and Complexity in Model Ecosystems*, Princeton University Press, Princeton, NJ, 1974.

(e) If your goal is to ensure that V eventually stays above 10, is it better to have $H = 20, V(0) = 9$ or $H = 30, V(0) = 16$?

(f) (CAS) Using your computer algebra system's ability to plot implicitly, plot a bifurcation diagram for equation (2.107) in the region $0 \leq H \leq 40$, $-5 \leq V \leq 30$ of the HV-plane. Approximately what value of H in the interval $[0, 40]$ is a bifurcation value?

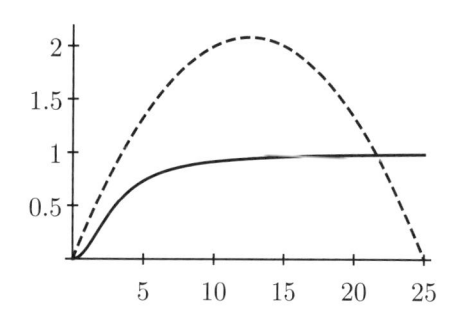

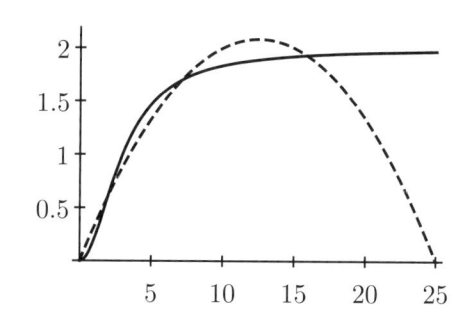

Figure 2.58. $H = 10$.

Figure 2.59. $H = 20$.

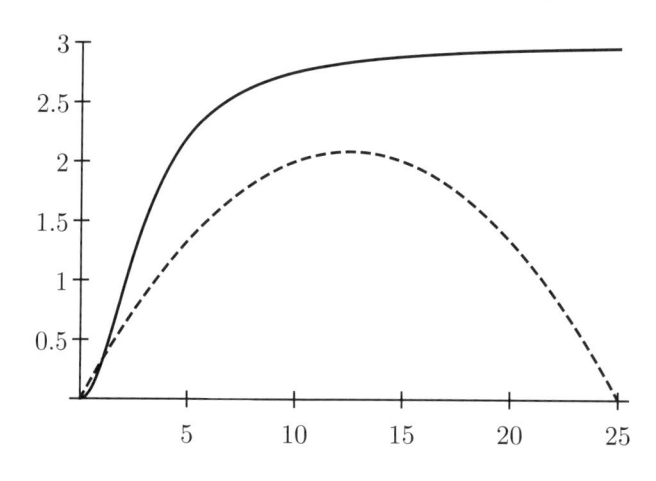

Figure 2.60. $H = 30$.

2.7. Change of variable

In this section we will look at some first-order differential equations for which a change of variable can be used to convert the equation to one that is either separable or linear. Choosing useful substitutions is something of an art form; nevertheless, we can single out certain types of equations for which there are general principles.

2.7.1. Homogeneous equations.

Example 2.7.1. Consider the equation

$$\frac{dy}{dx} = \frac{y^3 - x^3}{y^2 x}$$

and convince yourself that this equation is neither separable nor linear. However, notice that the right-hand side can be written as

$$\frac{y^3 - x^3}{y^2 x} = \frac{y^3}{y^2 x} - \frac{x^3}{y^2 x} = \frac{y}{x} - \left(\frac{x}{y}\right)^2,$$

2.7. Change of variable

and this suggests we try making the substitution

(2.109) $$v = \frac{y}{x},$$

so that the right-hand side becomes

$$v - \left(\frac{1}{v}\right)^2$$

in terms of the new dependent variable v. What is the left-hand side $\frac{dy}{dx}$ in terms of v and x? By (2.109) we have $y = vx$, and differentiating this equation with respect to x gives

$$\frac{dy}{dx} = v + x\frac{dv}{dx}.$$

So with the substitution (2.109), the equation becomes

$$v + x\frac{dv}{dx} = v - \left(\frac{1}{v}\right)^2.$$

This equation *is* separable:

$$v^2 dv = -\frac{1}{x}dx.$$

Solving, we obtain $v^3 = -3\ln|x| + c$. In terms of the original dependent variable y this becomes

$$y^3 = -3x^3 \ln|x| + cx^3,$$

or solving explicitly for y,

$$y = \left(-3x^3 \ln|x| + cx^3\right)^{1/3}.$$

The homogeneous equation substitution in general. We can generalize the last example as follows. Suppose we start with an equation

(2.110) $$\frac{dy}{dx} = f(x, y),$$

where the right-hand side $f(x, y)$ can be written as a function $g(v)$ of $v = \frac{y}{x}$ alone. The substitution $v = \frac{y}{x}$ will convert (2.110) to the equation

$$v + x\frac{dv}{dx} = g(v).$$

This is always a separable equation

$$\frac{1}{g(v) - v}dv = \frac{1}{x}dx.$$

The differential equations of the type just described are called **homogeneous** equations.[24]

---

[24] Recall that earlier "homogeneous" was used to describe linear equations that have zero on the right when written in standard form. Thus, the terminology "homogeneous" has two unrelated meanings in the study of differential equations. Fortunately, there is little chance for confusion between the two.

2.7.2. Bernoulli Equations.

A change of variable can be prescribed to convert any equation of the form

$$(2.111) \qquad \frac{dy}{dt} + p(t)y = q(t)y^b,$$

where b is a real number, into a linear equation. When $b = 0$ or $b = 1$ the equation is already linear (see Exercise 11); so we're interested only in values of b different from 0 and 1. Equations of the form (2.111) are called **Bernoulli equations**.

Let's look at a concrete example first.

Example 2.7.2. The logistic equation

$$(2.112) \qquad \frac{dP}{dt} = aP - bP^2,$$

which can be rewritten as

$$\frac{dP}{dt} - aP = -bP^2,$$

is a Bernoulli equation (with $b = 2$). We make the substitution

$$(2.113) \qquad v = P^{-1}.$$

Differentiating both sides of (2.113) with respect to t, we obtain

$$\frac{dv}{dt} = -P^{-2}\frac{dP}{dt}.$$

Since $P^{-2} = v^2$, this tells us that

$$\frac{dP}{dt} = -\frac{1}{v^2}\frac{dv}{dt},$$

so that equation (2.112) becomes

$$-\frac{1}{v^2}\frac{dv}{dt} = a\frac{1}{v} - b\frac{1}{v^2},$$

or more simply,

$$\frac{dv}{dt} + av = b.$$

This is a linear equation, with integrating factor e^{at}. Solving first for v we obtain

$$v = \frac{b}{a} + ce^{-at}$$

for an arbitrary constant c. Since $v = \frac{1}{P}$, we have

$$P(t) = \frac{1}{\frac{b}{a} + ce^{-at}}.$$

If $P(t)$ is interpreted as population, we can relate the constant c to the initial population $P_0 = P(0)$ (as we did in Section 2.5):

$$P_0 = \frac{1}{\frac{b}{a} + c}$$

so that

$$c = \frac{1}{P_0} - \frac{b}{a}$$

and our population function is

$$P(t) = \frac{1}{\frac{b}{a} + \left(\frac{1}{P_0} - \frac{b}{a}\right)e^{-at}}.$$

Some simple algebra shows this is the same as the solution we obtained in Section 2.5.

2.7. Change of variable

The Bernoulli substitution in general. We have the following principle:

> When b is a real number different from 0 and 1, the substitution $v = y^{1-b}$ will convert any equation of the form
>
> (2.114) $$\frac{dy}{dt} + p(t)y = q(t)y^b$$
>
> into a linear equation.

To see why this is so, notice first that $v = y^{1-b}$ implies
$$\frac{dv}{dt} = (1-b)y^{-b}\frac{dy}{dt}.$$

Multiplying both sides of equation (2.114) by y^{-b} gives
$$y^{-b}\frac{dy}{dt} + y^{1-b}p(t) = q(t),$$

and upon implementing our substitution we obtain
$$\frac{1}{1-b}\frac{dv}{dt} + p(t)v = q(t).$$

This is clearly a linear equation with dependent variable v.

2.7.3. Change in time-scale.
It is also possible to make a change of variable in the *independent* variable.

Example 2.7.3. Consider the equation
$$t\frac{dy}{dt} - y = t\ln^2 t$$
in which the independent variable is t. We will make the substitution $s = \ln t$ to obtain a simpler linear equation with new independent variable s and dependent variable y. To implement this substitution, recall that by the chain rule
$$\frac{dy}{dt} = \frac{dy}{ds}\frac{ds}{dt},$$
so that if $s = \ln t$,
$$\frac{dy}{dt} = \frac{1}{t}\frac{dy}{ds}.$$
Since $s = \ln t$ implies $e^s = t$, the substitution produces the new equation
$$\frac{dy}{ds} - y = e^s s^2$$
which is linear and is easily solved (using the integrating factor e^{-s}) to obtain
$$y(s) = \frac{s^3}{3}e^s + ce^s,$$
or in terms of the original variable
$$y(t) = \frac{t\ln^3 t}{3} + ct.$$
Even though the original equation was linear, this substitution simplifies its solution by simplifying the required integrations—solve the original equation to see the difference.

A change in the independent variable is sometimes referred to as a **change in time-scale**. It is sometimes used to reduce the number of parameters in the differential equation or to simplify constants in the original equation. Exercises 20, 21, and 22 provide examples.

Other examples of equations that can be solved by a substitution are explored in the exercises.

2.7.4. Exercises.

In Exercises 1–10, solve the differential equation by making an appropriate change of variable.

1. $\dfrac{dy}{dt} - 4y = -2y^3.$

2. $\dfrac{dy}{dx} = \dfrac{x+y}{x}.$

3. $\dfrac{dy}{dt} = \dfrac{y^2 + 2ty}{3y^2}.$

4. $\dfrac{dy}{dx} = \dfrac{x^2 + y^2}{xy - x^2}.$

5. $\dfrac{dy}{dx} = y \tan x + 3y^3 \cos^2 x.$

6. $\dfrac{dy}{dt} + 3y = y^3 t.$

7. $\dfrac{dy}{dx} = \dfrac{x+y}{x-y}.$

8. $\dfrac{dy}{dt} - y = y^3.$

9. $\dfrac{dy}{dx} = \dfrac{2y^4 + x^4}{xy^3}.$

10. $\dfrac{dy}{dt} = \dfrac{-y}{t} + \dfrac{t-1}{2y}.$

11. Show that when $b = 0$ or $b = 1$, the Bernoulli equation (2.111) is linear.

12. Show that the substitution $v = xy$ converts the equation
$$\frac{dy}{dx} = \frac{y - xy^2}{x + x^2 y}$$
into a separable equation. Solve the equation, giving an implicit solution.

13. Solve the equations in (a) and (b) by making the substitution $v = x - y$. Give an implicit solution.
    (a)
    $$\frac{dy}{dx} = x - y + 1 + \frac{1}{x - y + 2}.$$
    (b)
    $$\frac{dy}{dx} = \sin^2(x - y).$$
    (c) Show that the substitution $v = ax + by$ will convert any equation in the form
    $$\frac{dy}{dx} = f(ax + by)$$
    into a separable equation with dependent variable v.

14. Solve the equation
$$\frac{dy}{dx} = \frac{y}{x}(x^2 y - x^4 y^2)$$
by making the substitution $v = x^2 y$.

15. The von Bertalanffy growth model proposes the differential equation
$$\frac{dW}{dt} = aW^{2/3} - bW,$$
where a and b are positive constants, for the weight $W(t)$ of a fish at time t.
    (a) Find any equilibrium solutions, and then solve the equation as a Bernoulli equation.

(b) From the Bernoulli solutions you found in (a), find the solution that satisfies the initial condition $W = 0$ at $t = 0$. Show that for this solution, $W \to (\frac{a}{b})^3$ as $t \to \infty$ and that we may write the solution as
$$W = W_\infty \left(1 - e^{-rt}\right)^3,$$
where $W_\infty = (a/b)^3$ is the limiting value of W as $t \to \infty$ and $r = b/3$.

(c) In Exercise 30 of Section 2.3, we had the equation

(2.115) $$\frac{dL}{dt} = k(L_\infty - L)$$

for the length L of a fish at time t. Both k and L_∞ are constants, and L_∞ can be interpreted as the theoretical maximum length. Suppose we make the (reasonable) assumption that the weight and length are related by

(2.116) $$W = sL^3$$

for some constant s. Show that with the change of variable (2.116), the differential equation (2.115) becomes
$$\frac{dW}{dt} = aW^{2/3} - bW$$
for
$$a = 3ks^{1/3}L_\infty \quad \text{and} \quad b = 3k.$$

(d) Is $W_\infty = sL_\infty^3$?

16. The Gompertz differential equation,
$$\frac{dP}{dt} = aP - bP \ln P,$$
was used in Exercise 21 of Section 2.5 to model population growth or tumor growth. Show that the substitution $Q = \ln P$ converts this to a linear equation with new dependent variable Q. Solve the resulting equation.

17. Make the change of independent variable $t = x^2$ in the equation
$$\frac{dy}{dx} = 2x^3 + 4xy + 2x.$$
Solve the new equation, and then give the solution in terms of the original variables.

18. Show that when you make the substitution $s = t^3$ in the equation
$$\frac{dy}{dt} = 3t^2 y + 3t^5$$
the resulting equation, with new independent variable s, is
$$\frac{dy}{ds} = y + s.$$
Use this to solve the original equation.

19. (a) Make the substitution $v = y - t$ in the differential equation
$$\frac{dy}{dt} = y^2 - 2yt + t^2 - y + t - 1.$$
Check that the resulting equation, with new dependent variable v, is autonomous.

(b) Give a qualitative solution to the new equation by sketching some solution curves in the tv-plane, showing the concavity of these curves. Begin by finding all equilibrium solutions.

(c) What solutions $y = y(t)$ correspond to your equilibrium solutions $v = c$ in (b)?

(d) Part of the direction field for the original differential equation is shown in Fig. 2.61. Sketch some solution curves on it and compare with your graph in (b). Comments?

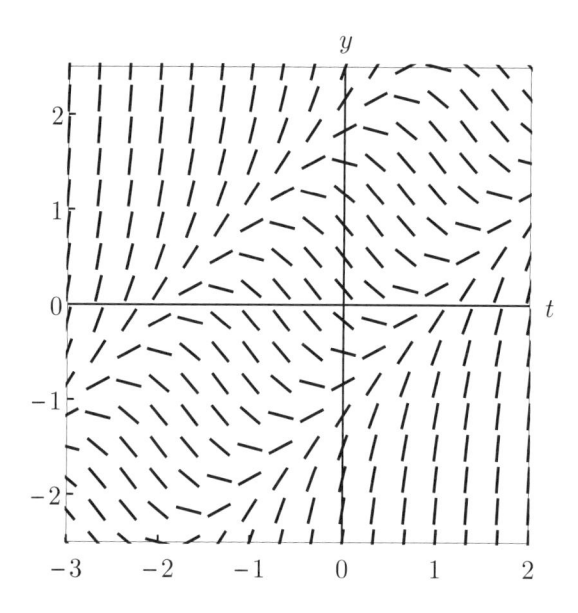

Figure 2.61. Direction field.

20. Consider the logistic equation

$$\frac{dP}{dt} = aP - bP^2$$

where P is the number of people at time t measured in years.
    (a) Note that this equation contains two parameters, a and b. What are the units on these parameters?
    (b) Make the simple substitution $Y = \frac{b}{a}P$ to convert the logistic equation to a differential equation with independent variable t and new dependent variable Y. Show that the new variable Y is "dimensionless"; that is, it has no units. Since $\frac{a}{b}Y = P$ and $\frac{a}{b}$ is the carrying capacity, you can think of Y as telling you the population as a fraction of the carrying capacity; for example $Y = \frac{1}{2}$ means the population is half the carrying capacity.
    (c) Starting with the differential equation you obtained in (b), make another change of variable $s = at$, to rewrite the equation again, this time with new independent variable s and dependent variable Y. Is s dimensionless?
    (d) Solve your differential equation in (c) (for Y as a function of s) as a Bernoulli equation, and then use your answer to give P as a function of t.

21. The equation

$$\frac{dv}{dt} = g - \frac{c}{m}v$$

will be used in Section 4.1 to describe the velocity v of a falling object of mass m subject to a drag force. Here g is the acceleration due to gravity and c is a positive constant (with units mass/time). In this problem we make the two simple substitutions

$$w = \alpha v \quad \text{and} \quad s = \beta t$$

where α and β are constants to be chosen with the goal of reducing the number of parameters in the equation and making the new variables w and s dimensionless (see Exercise 20 for an explanation of what this means).

2.7. Change of variable

(a) Show that with the substitution $w = \alpha v$ the equation becomes
$$\frac{dw}{dt} = \alpha g - \frac{c}{m}w.$$

(b) Show that the further substitution $s = \beta t$ gives the equation
$$\frac{dw}{ds} = \frac{\alpha g}{\beta} - \frac{c}{m\beta}w.$$

(c) Determine α and β (in terms of c, m, and g) so that
$$\frac{\alpha g}{\beta} = 1 \quad \text{and} \quad \frac{c}{m\beta} = 1.$$

(d) Check that, with the choices of α and β as in (c), the variables w and s are dimensionless.

22. The equation
$$\frac{dA}{dt} = -3.5 \times 10^{-8} A + 2.7 \times 10^{-8}$$
is simple enough to solve, but the extremely small coefficients can be awkward to work with. Make the change in time-scale
$$s = \frac{t}{10^8}$$
and solve the resulting equation.

23. The logistic model for population growth is modified to allow seasonal fluctuations in the carrying capacity. Suppose the new differential equation for the population P at time t is
$$\frac{dP}{dt} = aP - bP^2(1 + \beta \cos(\gamma t)).$$
Here a, b, γ, and β are positive constants. Solve this equation using the substitution
$$v = \frac{1}{P}$$
and the initial condition $P(0) = P_0$. At some point you'll need to integrate $e^{at} \cos(\gamma t)$; use integration by parts twice, or the integration formula
$$\int e^{at} \cos(\gamma t)\, dt = \frac{e^{at}}{a^2 + \gamma^2}(a \cos(\gamma t) + \gamma \sin(\gamma t)).$$

24. This problem outlines an alternate method for solving a Bernoulli equation

(2.117) $$\frac{dy}{dt} + p(t)y = q(t)y^b,$$

using a variation of parameter technique.

(a) Suppose that $f(t)$ is a nontrivial solution of the homogeneous linear equation
$$\frac{dy}{dt} + p(t)y = 0.$$
We'll look for a solution to (2.117) in the form

(2.118) $$y = h(t)f(t).$$

Using (2.118) as a trial solution for (2.117) show that h must satisfy
$$f(t)\frac{dh}{dt} = q(t)[h(t)f(t)]^b,$$
and check that this is a separable equation for the unknown function h.

(b) Apply the method in (a) to solve
$$\frac{dy}{dt} = ty^2 - y.$$

25. An ultralight airplane leaves an airport, located 100 km east of its destination airport. The plane moves at constant speed s km/hr relative to the air current within which it travels. During the plane's flight, there is a constant windspeed of w km/hr, south to north. Assuming the destination airport is located at the origin of an xy-coordinate system, with east as the positive x-direction and north as the positive y-direction, we conclude that the wind's velocity vector is $(0, w)$. Assuming that at any instant the plane's velocity vector relative to the air points from the plane's position towards the origin, we see that when the plane's location is (x, y), its velocity vector relative to the air is $s(-x, -y)(x^2 + y^2)^{-1/2}$, for $(x, y) \neq (0, 0)$. During its flight, the plane's velocity vector relative to the ground is thus

$$\left(\frac{-sx}{\sqrt{x^2 + y^2}}, \frac{-sy + w\sqrt{x^2 + y^2}}{\sqrt{x^2 + y^2}} \right).$$

Hence, if $y = y(x)$ is the curve along which the plane travels during its fight, we see that

(2.119) $$\frac{dy}{dx} = \frac{y}{x} - \alpha \sqrt{1 + \left(\frac{y}{x}\right)^2},$$

where $\alpha = w/s$.

(a) Make the substitution $v = y/x$ and use $\frac{d}{dv}[\sinh^{-1}(v)] = \frac{1}{\sqrt{1+v^2}}$ to show that for any constant C

$$y(x) = x \sinh(C - \alpha \ln x)$$

is a solution of (2.119).

(b) Using the solution from part (a), solve (2.119) subject to the initial condition $y(100) = 0$.

(c) Using the definition of $\sinh$, $\sinh(t) = \frac{e^t - e^{-t}}{2}$, show your solution to part (b) may be written as

$$y(x) = 50 \left(\left(\frac{x}{100}\right)^{1-\alpha} - \left(\frac{x}{100}\right)^{\alpha+1} \right).$$

(d) If the plane's top airspeed is $s = 40$ km/hr and the windspeed $w = 40$ km/hr, then what is the closest point due north of the destination airport that the plane can attain?

(e) (CAS) Plot the plane's flight path (in the first quadrant) for $\alpha = 0.2, 0.4, 0.6, 0.8, 1$, and 1.2.

26. The spruce budworm is a destructive insect that feeds on northern spruce and fir trees in the Eastern United States and Canada. It can cause severe defoliation, turning the needles brown and possibly killing the affected trees. The budworm is preyed on by birds. The simplest model of budworm population starts with the logistic equation and modifies it to include this predation by birds. One differential equation proposed for the budworm population P is

$$\frac{dP}{dt} = aP - bP^2 - \frac{AP^2}{B^2 + P^2}$$

where a, b, A, and B are positive constants.

(a) Show that with the substitutions

$$v = \frac{P}{B} \quad \text{and} \quad s = \frac{At}{B}$$

the equation can be rewritten with variables v and s and only two parameters, $\rho = \frac{aB}{A}$ and $\kappa = \frac{a}{bB}$.

(b) Clearly, $v = 0$ is an equilibrium solution of the equation you obtained in (a). Show that there are at most 3 other equilibrium solutions.

(c) Fig. 2.62 shows the graphs of the functions
$$f(v) = \frac{v}{1+v^2}$$
and
$$g(v) = \rho\left(1 - \frac{v}{\kappa}\right)$$
for certain values of the parameters ρ and κ, and Fig. 2.63 shows the graphs of the same functions for different values of ρ and κ. In each case, say how many nonzero equilibrium solutions there are, and classify each as a stable or unstable equilibrium.

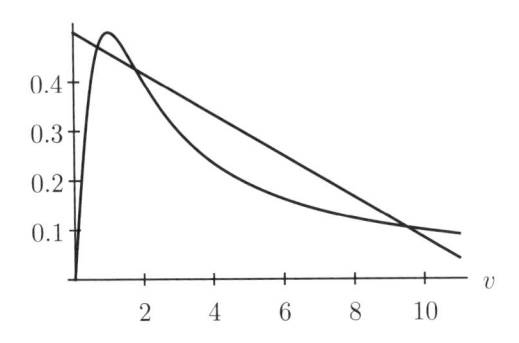

Figure 2.62. Possible graphs of $f(v)$ and $g(v)$.

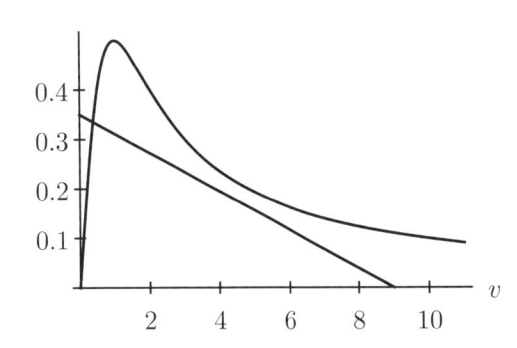

Figure 2.63. Other possible graphs of $f(v)$ and $g(v)$.

2.8. Exact equations

If you are an optimist, you'll think of the results in this section as expanding the list of nonlinear first-order equations we can solve. If you are a pessimist, the results of this section will tell you why we'll never be able to solve analytically many nonlinear equations.

The basic idea behind the notion of an "exact equation" is contained in the following example.

Example 2.8.1. Suppose y is a differentiable function of x satisfying the equation
$$(2.120) \qquad x^3 + y^2 x = c$$
on some open interval I, where c is a constant. Differentiating implicitly with respect to x, we find that
$$(2.121) \qquad 3x^2 + y^2 + 2yx\frac{dy}{dx} = 0.$$
Thus $x^3 + y^2 x = c$ is an implicit solution of the differential equation in (2.121); note that this equation is neither linear nor separable.

Can we reverse our steps? In other words, if we *start* with a differential equation
$$(2.122) \qquad M(x,y) + N(x,y)\frac{dy}{dx} = 0$$
(a quite general first-order equation), can we find a function $\varphi(x,y)$ so that
$$\frac{d}{dx}(\varphi(x,y)) = M(x,y) + N(x,y)\frac{dy}{dx}?$$
If so, then our equation (2.122) looks like
$$\frac{d}{dx}(\varphi(x,y)) = 0,$$

which has implicit solution
$$\varphi(x,y) = c.$$

If φ is a function of x and y and y in turn is a function of x, say, $y = y(x)$, then we can think of φ as a function of the single variable x. From this point of view, it makes sense to talk about the ordinary derivative $\frac{d\varphi}{dx}$, keeping in mind that it is an abbreviation for $\frac{d}{dx}(\varphi(x,y(x)))$. One form of the chain rule from multivariable calculus says that
$$\frac{d}{dx}(\varphi(x,y(x))) = \frac{\partial \varphi}{\partial x}(x,y(x)) + \frac{\partial \varphi}{\partial y}(x,y(x))\frac{dy}{dx},$$
or in more abbreviated language,

(2.123) $$\frac{d\varphi}{dx} = \frac{\partial \varphi}{\partial x} + \frac{\partial \varphi}{\partial y}\frac{dy}{dx}.$$

We will look at this more carefully in Section 2.8.2 below. For example, if $\varphi(x,y) = x^3 + xy^2$ where y is in turn an (unspecified) function of x, then
$$\frac{d\varphi}{dx} = \frac{\partial \varphi}{\partial x} + \frac{\partial \varphi}{\partial y}\frac{dy}{dx} = (3x^2 + y^2) + (2xy)\frac{dy}{dx}.$$

The preceding discussion shows we can rephrase our "reversal-of-steps" question as follows: Starting with a differential equation as in (2.122), can we find $\varphi(x,y)$ with

(a) $\dfrac{\partial \varphi}{\partial x} = M(x,y)$ and

(b) $\dfrac{\partial \varphi}{\partial y} = N(x,y)$?

If such a φ exists, then via (2.123) we see that equation (2.122) takes the form $\frac{d\varphi}{dx} = 0$, and thus $\varphi(x,y) = c$ is an implicit solution of the equation. We will say that (2.122) is **exact** when such a function $\varphi(x,y)$ exists.

The basic question of whether, given functions $M(x,y)$ and $N(x,y)$, we can find $\varphi(x,y)$ satisfying properties (a) and (b) above can be put in other language, which may be familiar to you from a physics or multivariable calculus course. For example, we can ask

- whether (M,N) is equal to $\nabla \varphi$, the gradient of some function $\varphi(x,y)$, or
- whether (M,N) is a conservative vector field or
- if the vector field (M,N) has a potential function.

No matter what terminology we use, the problem comes down to two issues:

(1) Can we tell in some simple way when
$$M(x,y) + N(x,y)\frac{dy}{dx} = 0$$
is exact?

(2) If it is exact, how do we find $\varphi(x,y)$ with properties (a) and (b) above?

The Test for Exactness. A basic fact from multivariable calculus provides a *necessary condition* for exactness. If there is a $\varphi(x,y)$ satisfying

(2.124) $$\frac{\partial \varphi}{\partial x} = M(x,y)$$

2.8. Exact equations

and

(2.125) $$\frac{\partial \varphi}{\partial y} = N(x,y),$$

then differentiating (2.124) with respect to y and (2.125) with respect to x says

$$\frac{\partial^2 \varphi}{\partial y \partial x} = \frac{\partial M}{\partial y} \quad \text{and} \quad \frac{\partial^2 \varphi}{\partial x \partial y} = \frac{\partial N}{\partial x}.$$

As long as φ is a "nicely behaved" function,[25] these two mixed second-order partial derivatives of $\varphi(x,y)$ must be equal, so that we must have

(2.126) $$\frac{\partial M}{\partial y} = \frac{\partial N}{\partial x}.$$

We will call (2.126) the **Test for Exactness**. When we say that the condition (2.126) is a necessary condition for exactness we mean that (2.126) must be satisfied in order for there to be a chance for (2.122) to be exact. If we check this condition for some particular $M(x,y)$ and $N(x,y)$ and find it is **not** satisfied, then we immediately conclude that (2.122) is not exact.

Example 2.8.2. Is the equation

$$\frac{y^2}{2} + 2ye^x + (y + e^x)\frac{dy}{dx} = 0$$

exact? Here, $M(x,y) = \frac{y^2}{2} + 2ye^x$ and $N(x,y) = y + e^x$ so

$$\frac{\partial M}{\partial y} = y + 2e^x \quad \text{while} \quad \frac{\partial N}{\partial x} = e^x.$$

These are not equal, so the equation is not exact, and for now there is nothing more to do.

If the necessary condition (2.126) is satisfied, can we conclude that (2.122) is exact? In other words, is this necessary condition also *sufficient* to conclude exactness? Almost; it is sufficient as long as $M(x,y)$ and $N(x,y)$ have the property that they and their derivatives $\frac{\partial M}{\partial y}$ and $\frac{\partial N}{\partial x}$ are continuous on the whole plane $\mathbb{R}^2$, or even just in some open rectangle $\mathcal{R}$ defined by $a < x < b, c < y < d$:

> Suppose that M, N, $\frac{\partial M}{\partial y}$, and $\frac{\partial N}{\partial x}$ are continuous on an open rectangle $\mathcal{R}$. If
>
> (2.127) $$\frac{\partial M}{\partial y} = \frac{\partial N}{\partial x}$$
>
> on $\mathcal{R}$, then the differential equation
>
> (2.128) $$M(x,y) + N(x,y)\frac{dy}{dx} = 0$$
>
> is exact in $\mathcal{R}$, meaning there is a function φ of x and y such that for all (x,y) in $\mathcal{R}$
>
> $$\frac{\partial \varphi}{\partial x} = M(x,y) \quad \text{and} \quad \frac{\partial \varphi}{\partial y} = N(x,y),$$
>
> and an implicit solution to (2.128) is given by
>
> $$\varphi(x,y) = c$$
>
> for any constant c. Moreover, equation (2.127) must hold if (2.128) is exact in $\mathcal{R}$.

---

[25] The condition here for φ to be nicely behaved is that the second-order mixed partial derivatives of φ are continuous.

Example 2.8.3. Consider the equation

(2.129) $$5y + e^x + (5x + \cos y)\frac{dy}{dx} = 0.$$

Here $M(x,y) = 5y + e^x$ and $N(x,y) = 5x + \cos y$. These functions have first-order partial derivatives that are continuous in all of $\mathbb{R}^2$. We check the exactness condition by computing

$$\frac{\partial M}{\partial y} = 5, \quad \frac{\partial N}{\partial x} = 5,$$

and thus we can conclude that (2.129) is exact. To solve (2.129) we now want to find a function $\varphi(x,y)$ satisfying

(a)
$$\frac{\partial \varphi}{\partial x} = M(x,y) = 5y + e^x$$

and

(b)
$$\frac{\partial \varphi}{\partial y} = N(x,y) = 5x + \cos y.$$

To recover φ from $\frac{\partial \varphi}{\partial x}$ we need to "undo" the derivative with respect to x; that is, we integrate $\frac{\partial \varphi}{\partial x}$ with respect to x. Thus condition (a) tells us that φ should have the form

(2.130) $$\varphi(x,y) = \int M(x,y)\,dx = \int (5y + e^x)\,dx = 5yx + e^x + h(y).$$

Note that in integrating with respect to x we treat y as a constant. This is because y was also treated as a constant when computing the partial derivative $\frac{\partial \varphi}{\partial x}$, the operation we seek to "undo". The quantity $h(y)$ in (2.130) is the "constant of integration" needed to make the right-hand side equal to $\varphi(x,y)$. Notice that $h(y)$, while constant with respect to the variable of integration x, can be a function of the other variable y. Working from this information about φ, we compute

$$\frac{\partial \varphi}{\partial y} = 5x + h'(y).$$

But by condition (b) we also want $\frac{\partial \varphi}{\partial y}$ to be equal to $5x + \cos y$. Upon comparing these two expressions for $\frac{\partial \varphi}{\partial y}$, we see that $h'(y) = \cos y$. The choice $h(y) = \sin y$ will do (as will $\sin y + c$ for any constant c), and

$$\varphi(x,y) = 5xy + e^x + \sin y$$

satisfies both properties (a) and (b). We conclude that

$$5xy + e^x + \sin y = c$$

is an implicit solution to (2.129).

The process carried out in the last example to find $\varphi(x,y)$ satisfying conditions (a) and (b) might be referred to as a kind of "partial integration"; the notation $\int (5y+e^x)dx$ refers to integrating $5y + e^x$ with respect to x while holding y constant. We could have also solved equation (2.129) by a similar process beginning with the computation of the partial integral of $5x + \cos y$ with respect to y; see Exercise 13.

Solving an exact equation in the manner of the last example produces an implicit solution of the form

(2.131) $$\varphi(x,y) = c.$$

2.8. Exact equations

If we impose an initial condition $y(x_0) = y_0$, we can use it to determine the value of c. Sometimes you can solve the equation (2.131) explicitly for y in terms of x. Even when you can't, equation (2.131) is a perfectly good implicit solution, and you can still substitute the initial condition $x = x_0$, $y = y_0$ to find c.

2.8.1. Integrating factors. Let's return to the differential equation of Example 2.8.2,

$$(2.132) \qquad \frac{y^2}{2} + 2ye^x + (y + e^x)\frac{dy}{dx} = 0,$$

and recall that we showed in that example that this equation is not exact. However, suppose we multiply both sides of (2.132) by e^x, obtaining the equivalent equation

$$\left(\frac{e^x y^2}{2} + 2ye^{2x}\right) + (ye^x + e^{2x})\frac{dy}{dx} = 0.$$

If we now take

$$M(x, y) = \frac{e^x y^2}{2} + 2ye^{2x}$$

and

$$N(x, y) = ye^x + e^{2x},$$

it is easy to see that

$$\frac{\partial M}{\partial y} = \frac{\partial N}{\partial x},$$

so that the new equation is exact. We say that the function $\mu(x) = e^x$ is an **integrating factor** for (2.132); multiplying by this integrating factor converted the nonexact equation (2.132) into one that is exact, and moreover the two equations have the same solutions. This new equation is now easily solved. We have

$$\varphi(x, y) = \int M(x, y)\, dx = \int \left(\frac{e^x y^2}{2} + 2ye^{2x}\right) dx = \frac{e^x y^2}{2} + ye^{2x} + h(y).$$

Differentiating this with respect to y gives

$$\frac{\partial \varphi}{\partial y} = e^x y + e^{2x} + h'(y).$$

On comparing this with

$$\frac{\partial \varphi}{\partial y} = N(x, y) = ye^x + e^{2x}$$

we see that

$$e^x y + e^{2x} + h'(y) = ye^x + e^{2x},$$

so that $h'(y) = 0$, and we can choose $h(y) = 0$ and obtain the solution to (2.132) as

$$\frac{e^x y^2}{2} + ye^{2x} = c.$$

In general, an integrating factor may be a function $\mu(x, y)$ of both x and y. For example, the equation

$$4xy^2 + 3y + (3x^2 y + 2x)\frac{dy}{dx} = 0$$

has integrating factor $\mu(x, y) = x^2 y$ (verify this). However, finding an integrating factor may be as hard as solving the original differential equation. Nevertheless, there are a few situations where an integrating factor can be determined relatively easily. The next example illustrates how to tell if there is an integrating factor that is a function of x alone and how to then find this integrating factor.

Example 2.8.4. We'll see whether there is an integrating factor $\mu(x)$ that is a function of x alone for the differential equation

(2.133) $$(x - y^3) + (3xy^2 + x^2)\frac{dy}{dx} = 0.$$

If we multiply this equation by $\mu(x)$, we obtain

$$\mu(x)(x - y^3) + \mu(x)(3xy^2 + x^2)\frac{dy}{dx} = 0.$$

Next we impose the Test for Exactness:

$$\frac{\partial}{\partial y}(\mu(x)(x - y^3)) = \frac{\partial}{\partial x}(\mu(x)(3xy^2 + x^2)).$$

This says that $\mu(x)$ should satisfy

$$-\mu(x)(3y^2) = \mu(x)(3y^2 + 2x) + \mu'(x)(3xy^2 + x^2).$$

Rearranging the last line gives

$$-\mu(x)(6y^2 + 2x) = \mu'(x)(3xy^2 + x^2).$$

Now a crucial bit of algebra must take place. If we divide both sides of the last equation by $3xy^2 + x^2$, the result is

(2.134) $$-\mu(x)\frac{6y^2 + 2x}{3xy^2 + x^2} = \mu'(x).$$

For this to be possible, we must be able to simplify the left-hand side so that the variable y disappears. But notice that the left side is just

$$-\mu(x)\frac{2(3y^2 + x)}{x(3y^2 + x)} = -\frac{2}{x}\mu(x),$$

so equation (2.134) is simply

$$-\frac{2}{x}\mu(x) = \mu'(x) = \frac{d\mu}{dx}.$$

Solving this either as a separable or as a linear equation we see that $\mu(x) = \frac{C}{x^2}$. Since we just need one integrating factor, we choose $C = 1$ for convenience and conclude that $\mu(x) = \frac{1}{x^2}$ is an integrating factor for our original equation. In Exercise 15 you are asked to use this integrating factor to solve equation (2.133).

An integrating factor of the form $\mu(x)$. The generalization of the last example is as follows.

> The equation
> $$M(x,y) + N(x,y)\frac{dy}{dx} = 0$$
> will have an integrating factor $\mu(x)$ that is a function of x alone if
> $$\frac{\frac{\partial M}{\partial y} - \frac{\partial N}{\partial x}}{N}$$
> is a continuous function of x alone. We can then find $\mu(x)$ by solving the separable (and linear) equation
> $$\frac{d\mu}{dx} = \frac{\frac{\partial M}{\partial y} - \frac{\partial N}{\partial x}}{N}\mu.$$

2.8. Exact equations

To see why these assertions are true, multiply the equation

$$M(x,y) + N(x,y)\frac{dy}{dx} = 0$$

by $\mu(x)$:

$$\mu(x)M(x,y) + \mu(x)N(x,y)\frac{dy}{dx} = 0.$$

The Test for Exactness requires that

$$\frac{\partial}{\partial y}(\mu(x)M(x,y)) = \frac{\partial}{\partial x}(\mu(x)N(x,y)),$$

or equivalently,

$$\mu(x)\frac{\partial M}{\partial y} = \mu'(x)N(x,y) + \mu(x)\frac{\partial N}{\partial x}$$

(notice that we have used the product rule on the right-hand side). Rearranging the last line, we have

(2.135) $$\frac{\left(\frac{\partial M}{\partial y} - \frac{\partial N}{\partial x}\right)}{N(x,y)} \mu(x) = \mu'(x).$$

Thus

$$\frac{\left(\frac{\partial M}{\partial y} - \frac{\partial N}{\partial x}\right)}{N(x,y)} = \frac{\mu'(x)}{\mu(x)}$$

is a function $R(x)$ of x alone (this is the "crucial algebra" step as we saw in the last example), and the differential equation (2.135) for the unknown function μ is just

$$\mu'(x) = R(x)\mu(x)$$

(both linear and separable). Solving for $\mu(x)$ we obtain

$$\mu(x) = e^{\int R(x)dx}.$$

An analogous statement holds when there is an integrating factor $\mu(y)$ that is a function of y alone; see Exercise 20.

What happens when there is no integrating factor that is a function of x or y alone? Let's look at the reasoning in this case. Suppose $\mu(x,y)$ is an integrating factor for the differential equation

(2.136) $$M(x,y) + N(x,y)\frac{dy}{dx} = 0,$$

so that

(2.137) $$\mu(x,y)M(x,y) + \mu(x,y)N(x,y)\frac{dy}{dx} = 0$$

is exact. By the Test for Exactness we have

(2.138) $$\frac{\partial(\mu N)}{\partial x} = \frac{\partial(\mu M)}{\partial y}.$$

Computing with the product rule gives

(2.139) $$N\frac{\partial \mu}{\partial x} + \mu\frac{\partial N}{\partial x} = M\frac{\partial \mu}{\partial y} + \mu\frac{\partial M}{\partial y}.$$

Here M, N, and their partial derivatives are known and $\mu = \mu(x,y)$ is the unknown function. Equation (2.139) is a *partial differential equation*, and it is as hard to solve for μ as the original differential equation (2.136) is to solve for $y = y(x)$. Nevertheless, there are some situations where (2.139) can be solved, even when μ is a function of both x and y; see Exercise 26.

2.8.2. A deeper look at differentials. Let's review a bit more about partial derivatives of a function $\varphi(x,y)$ of two variables. To find the partial derivative of φ with respect to x, hold y constant, think of $\varphi(x,y)$ as a function of x only, and differentiate with respect to x. The result is the partial derivative $\frac{\partial \varphi}{\partial x}$, which we can abbreviate as φ_x. The quantity φ_x is the rate of change of φ in the x-direction. Similarly, the partial derivative of φ with respect to y is computed by holding x constant, thinking of $\varphi(x,y)$ as a function of y only, and differentiating $\varphi(x,y)$ with respect to y. This gives you the partial derivative $\frac{\partial \varphi}{\partial y}$ (we abbreviate it as φ_y), which represents the rate of change of φ in the y-direction.

We can draw pictures to explain the geometric meaning of the partial derivatives of a function $\varphi(x,y)$. The graph of $\varphi(x,y)$ is the *surface* $z = \varphi(x,y)$ in three-dimensional space, where (x,y) ranges over the domain D of φ. In computing $\frac{\partial \varphi}{\partial y} = \varphi_y$, we hold x constant and let y vary. This corresponds to looking at the cross section of the graph lying in the vertical plane containing the point (x,y) as shown and perpendicular to the x-axis; see Fig. 2.64. Fig. 2.65 shows a front view of this cross section. The curve is the graph of $z = \varphi(x,y)$ with x held constant, and the slope of the tangent line above y is $\frac{\partial \varphi}{\partial y}(x,y)$.

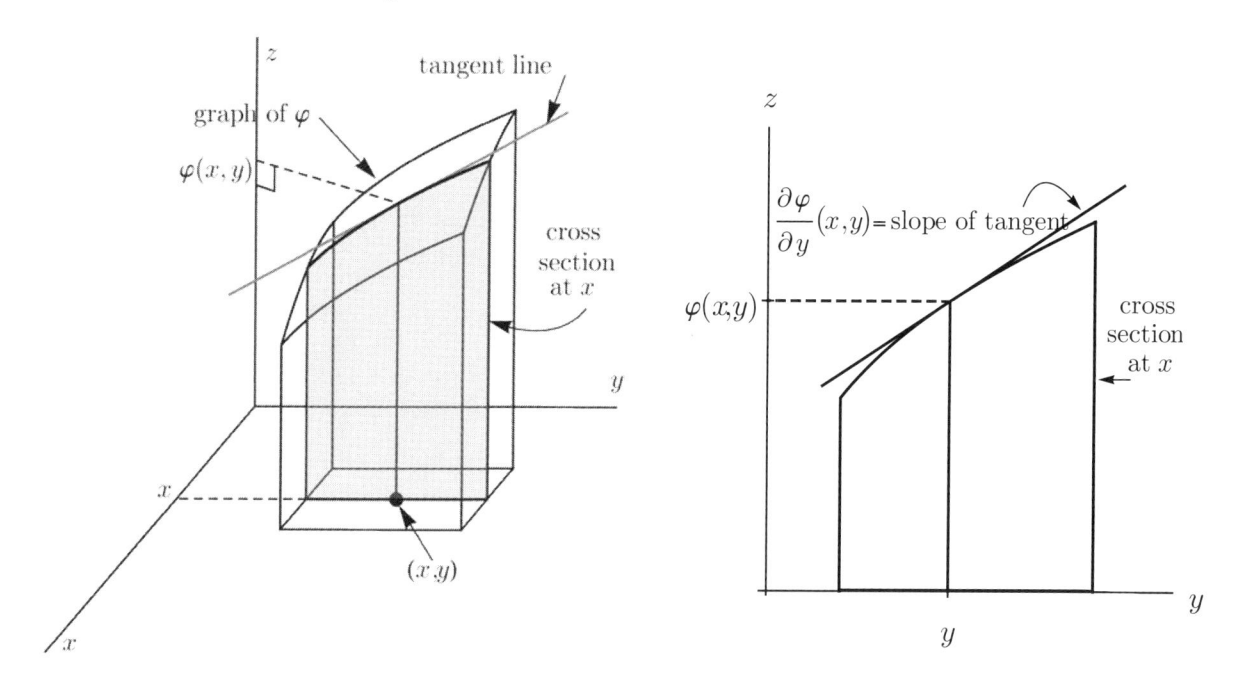

Figure 2.64. A cross section perpendicular to the x-axis.

Figure 2.65. Front view.

Now consider $\frac{\partial \varphi}{\partial x}$. Holding y constant and letting x vary gives rise to the cross section of the graph as shown in Fig. 2.66, in the vertical plane perpendicular to the y-axis and passing through the point (x,y) as shown. Fig. 2.67 shows a view of this cross section as seen from the left side in Fig. 2.66. The curve in Fig. 2.67 is the graph of $z = \varphi(x,y)$ with y held constant. The slope of the tangent line above x is $\frac{\partial \varphi}{\partial x}(x,y)$.

2.8. Exact equations

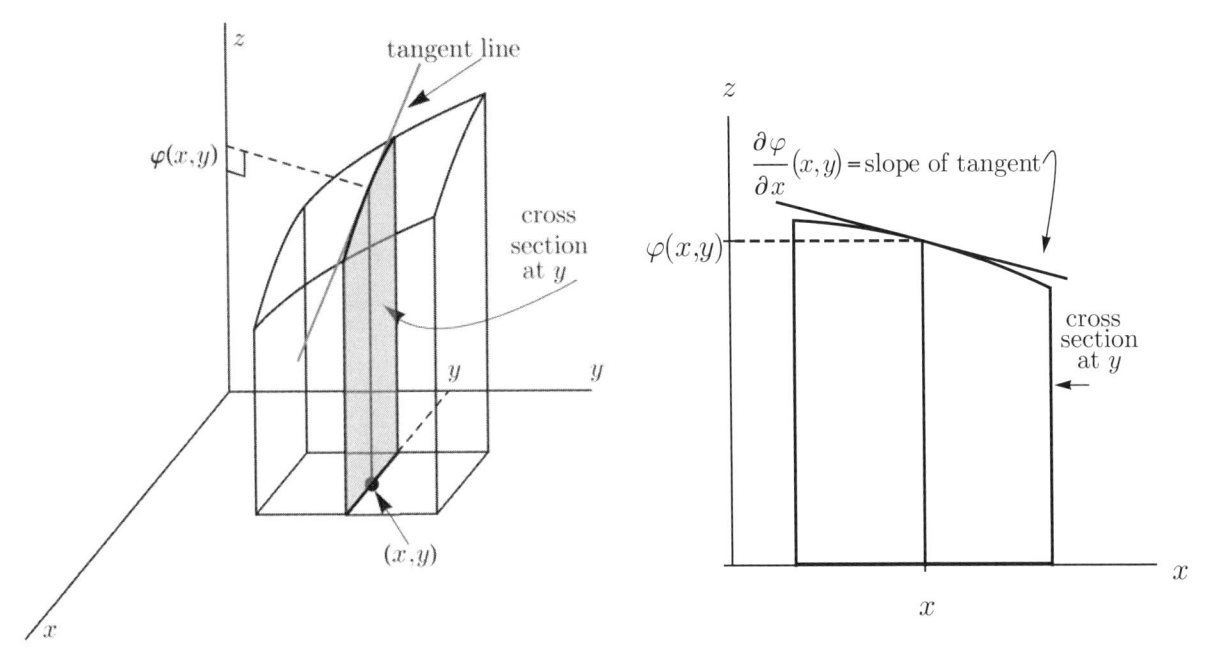

Figure 2.66. A cross section perpendicular to the y-axis.

Figure 2.67. Side view.

In Fig. 2.68 we combine Figs. 2.64 and 2.66 into one picture. The plane that contains the two tangent lines shown in Figs. 2.64 and 2.66 is the **tangent plane** to the graph of the surface $z = \varphi(x, y)$ at the point $(x, y, \varphi(x, y))$. The *tangent plane* at a point on the graph of a function of two variables is the analogue of the *tangent line* at a point on the graph of a function of one variable. In particular, the tangent plane is the plane which most closely approximates the surface $z = \varphi(x, y)$ near the point $(x, y, \varphi(x, y))$ on the surface.

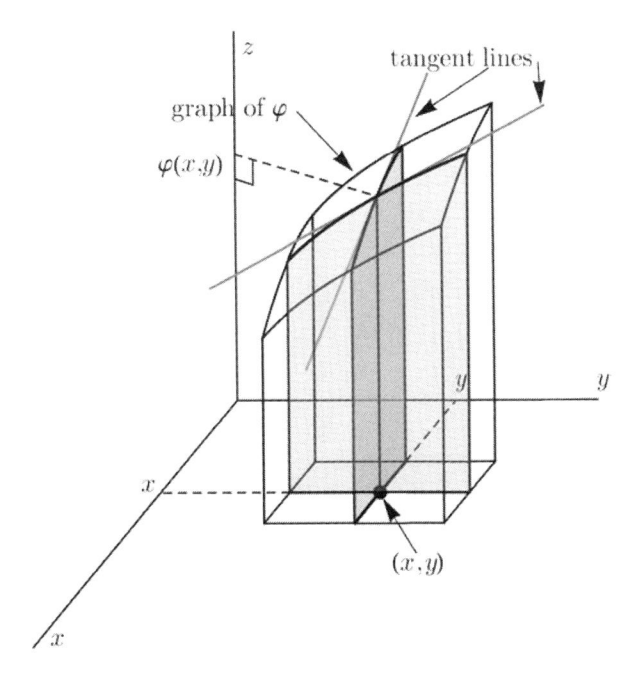

Figure 2.68. Combining Figs 2.64 and 2.66 to find the tangent plane.

The differential of φ: Starting at (x, y) in the domain D of φ, let x change by a small amount dx (while holding y constant). By the above discussion of partial derivatives, φ changes by an amount that is approximately $\varphi_x dx$. Now hold x fixed and let y change by a small amount dy. This makes φ change by approximately $\varphi_y dy$. If you let *both* x and y change by the respective amounts dx and dy, then the total change in φ will be approximately

(2.140) $$d\varphi = \varphi_x dx + \varphi_y dy.$$

This expression is called the **differential** of φ corresponding to the differentials dx and dy. Geometrically, it represents the change in height (that is, in the z-coordinate) along the tangent plane when you move from an input of (x, y) to a "nearby" input $(x + dx, y + dy)$. If we think of dx and dy as infinitesimally small, then we can consider $d\varphi$ to be exactly the resulting (also infinitesimal) total change in φ. Notice that formally dividing equation (2.140) by dx yields the two-variable chain rule

$$\frac{d\varphi}{dx} = \frac{\partial \varphi}{\partial x} + \frac{\partial \varphi}{\partial y}\frac{dy}{dx}$$

from equation (2.123). Here we are thinking of y as a function of x, and $\frac{d\varphi}{dx}$ means $\frac{d}{dx}\varphi(x, y(x))$. More generally, if x and y are functions of a third variable t, we can formally divide equation (2.140) by dt to obtain the full two-variable chain rule

$$\frac{d}{dt}\varphi(x(t), y(t)) = \frac{d\varphi}{dt} = \frac{\partial \varphi}{\partial x}\frac{dx}{dt} + \frac{\partial \varphi}{\partial y}\frac{dy}{dt}.$$

Now consider a first-order differential equation in the form

(2.141) $$M(x, y) + N(x, y)\frac{dy}{dx} = 0.$$

If we multiply by the differential dx, we get

(2.142) $$M(x, y)dx + N(x, y)dy = 0,$$

which is called the **symmetric form** of the differential equation.

Just as an implicit solution $\varphi(x, y) = c$ treats both variables x and y symmetrically, writing the differential equation in the form (2.142) also treats the variables symmetrically. The left side of (2.142) looks a bit like a differential $d\varphi$. It may or may not actually be a differential; it *is* precisely in the case that the differential equation is exact, since then there exists a function $\varphi(x, y)$ such that $\varphi_x = M$ and $\varphi_y = N$. In this case the implicit solution is $\varphi(x, y) = c$.

One method for finding integrating factors is **recognizing differentials**. This involves putting the differential equation in a form where you can see some known differentials. Here is a brief table of common differentials.

2.8. Exact equations

	Table 2.3. Some Common Differentials.
1	$d\varphi = \varphi_x dx + \varphi_y dy$
2	$d(xy) = ydx + xdy$
3	$d(\frac{y}{x}) = \dfrac{-ydx + xdy}{x^2}$
4	$d(\frac{x}{y}) = \dfrac{ydx - xdy}{y^2}$
5	$d(\frac{1}{2}\ln(x^2 + y^2)) = \dfrac{xdx + ydy}{x^2 + y^2}$
6	$dh(x) = h'(x)dx$
7	$dk(y) = k'(y)dy$
8	$d\left(\tan^{-1}(\frac{y}{x})\right) = \dfrac{-ydx + xdy}{x^2 + y^2}$
9	$dh(\varphi(x,y)) = h'(\varphi(x,y))d\varphi = h'(\varphi(x,y))(\varphi_x dx + \varphi_y dy)$

The next examples illustrate how being able to recognize these common differentials can help in finding integrating factors.

Example 2.8.5. Here's a differential equation in symmetric form:
$$(x^2 + y^2 - x)dx - ydy = 0.$$
Notice that we have a $-(xdx + ydy)$ and a $(x^2 + y^2)dx$ term. So let us try to use common differential 5 from the table and first divide our equation by $x^2 + y^2$. This gives us
$$dx - \frac{xdx + ydy}{x^2 + y^2} = 0.$$
By common differential 5, this is the same as
$$d\left(x - \frac{1}{2}\ln(x^2 + y^2)\right) = 0,$$
so our general solution is
$$x - \frac{1}{2}\ln(x^2 + y^2) = c,$$
c being any constant. The key step was multiplying our differential equation by $\frac{1}{x^2+y^2}$ to make it exact, so
$$\mu(x,y) = \frac{1}{x^2 + y^2}$$
is our integrating factor.

Example 2.8.6. Consider $ydx + (x^2y^3 + x)dy = 0$, or
$$ydx + xdy + x^2y^3 dy = 0.$$
Using common differential 2, this is just
$$d(xy) + x^2y^3 dy = 0.$$
Dividing each term by x^2y^2 gives
$$\frac{d(xy)}{(xy)^2} + ydy = 0.$$

Using common differential 9 with $h(t) = -\dfrac{1}{t}$ and $\varphi(x,y) = xy$, as well as common differential 7, we can rewrite this equation as

$$d\left(-\frac{1}{xy} + \frac{y^2}{2}\right) = d\left(-\frac{1}{xy}\right) + d\left(\frac{y^2}{2}\right) = 0.$$

Therefore we have a one-parameter family of solutions

$$-\frac{1}{xy} + \frac{y^2}{2} = c,$$

for any constant c. Recognizing differentials suggested that we divide our differential equation by $x^2 y^2$, so

$$\mu(x,y) = \frac{1}{x^2 y^2}$$

is our integrating factor.

Example 2.8.7. Consider

$$xdy - ydx = (x^2 + y^2)dy.$$

Looking at the table, we see that we can put this in the form of common differential 8 by dividing by $x^2 + y^2$. Doing this gives us

$$\frac{xdy - ydx}{x^2 + y^2} = dy,$$

or

$$d\left(\tan^{-1}\left(\frac{y}{x}\right) - y\right) = d\left(\tan^{-1}\left(\frac{y}{x}\right)\right) - dy = 0.$$

This gives a one-parameter family of solutions

$$\tan^{-1}\left(\frac{y}{x}\right) - y = c,$$

for any constant c. In this case, our integrating factor was

$$\mu(x,y) = \frac{1}{x^2 + y^2}.$$

Example 2.8.8. For our final application in this chapter, we'll design a spotlight. Imagine a point source of light (a high-intensity filament) located at the origin O in the xy-plane. A mirror-finished surface is to be constructed by revolving a curve in the xy-plane about the x-axis. We want to choose the curve in such a way that the reflected light rays will all be parallel to the x-axis. Light is reflected according to the **Law of Reflection**, which says that the angle of incidence is equal to the angle of reflection. For a flat mirror, this gives the situation depicted in Fig. 2.70. For a curved mirror, these angles are measured with respect to the tangent plane to the mirror at the point of reflection P; the cross section in the xy-plane is shown in Fig. 2.71.

2.8. Exact equations

Figure 2.69. A spotlight.

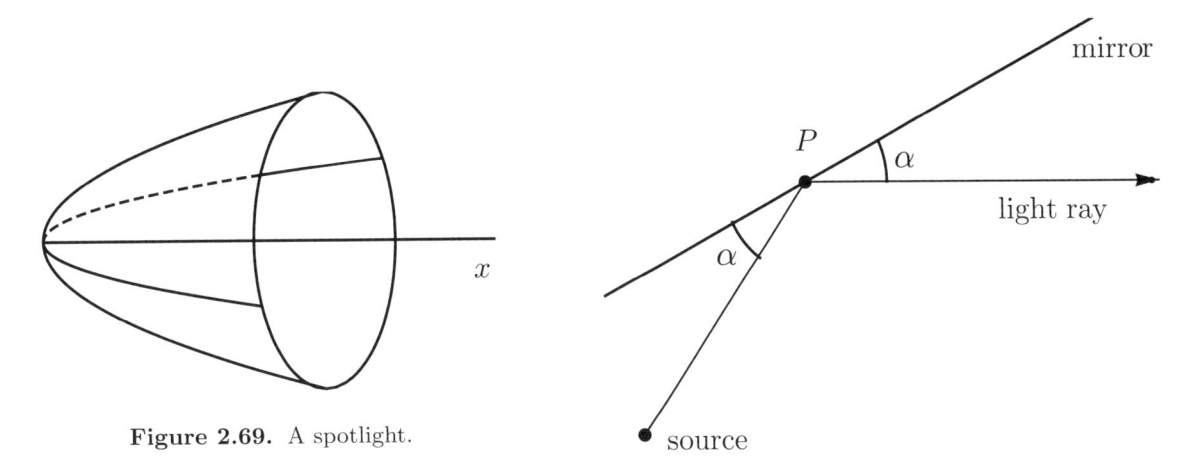

Figure 2.70. Light reflected by a flat mirror.

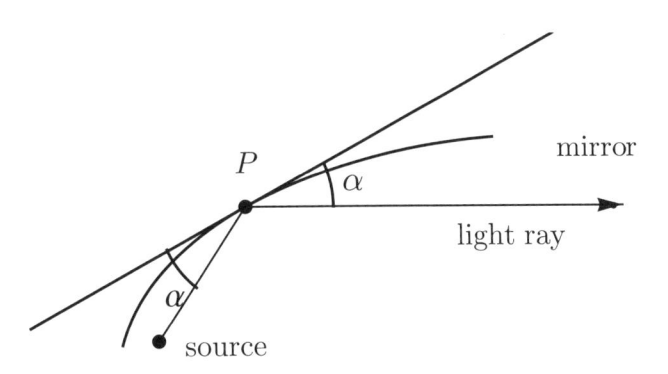

Figure 2.71. Light reflected by a curved mirror.

Fig. 2.72 shows our unknown curve in the xy-plane and an arbitrary point P on the curve. Our goal is to determine this curve so that the reflected ray of light is parallel to the x-axis. By symmetry, when this curve is revolved about the x-axis, the resulting mirrored surface will focus all reflected light rays in a beam parallel to the x-axis.

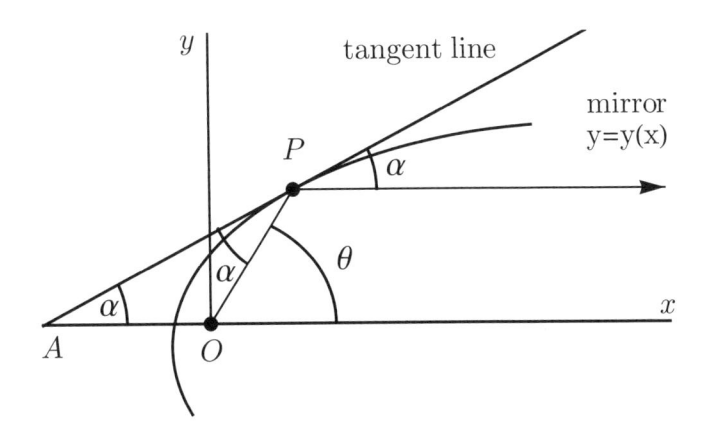

Figure 2.72. Cross section in the xy-plane with light source at origin.

If PA is the tangent line to the curve, by elementary geometry the angle PAO has measure α. Since the three angles in triangle PAO sum to π, we have

$$\alpha + \alpha + (\pi - \theta) = \pi,$$

or $\theta = 2\alpha$, where θ is the angle as shown in Fig. 2.72. Since $\theta = 2\alpha$, a trigonometric identity tells us

$$\tan \theta = \tan(2\alpha) = \frac{2 \tan \alpha}{1 - \tan^2 \alpha}.$$

But

$$\tan \theta = \frac{y}{x}$$

where P has coordinates (x, y). Since $\tan \alpha$ is the slope of the tangent line at P, this says

$$\frac{y}{x} = \frac{2 \frac{dy}{dx}}{1 - \left(\frac{dy}{dx}\right)^2},$$

so that

$$y \left(\frac{dy}{dx}\right)^2 + 2x \frac{dy}{dx} - y = 0.$$

Using the quadratic formula we can solve this for $\frac{dy}{dx}$, obtaining

$$\frac{dy}{dx} = \frac{-x \pm \sqrt{x^2 + y^2}}{y}.$$

We write this in symmetric form as

$$x \, dx + y \, dy = \pm \sqrt{x^2 + y^2} \, dx.$$

In Exercise 27 you are asked to solve this for the unknown function $y = y(x)$ which describes the cross section of the mirror in the xy-plane.

2.8.3. Modeling recap.

In this chapter we have seen a variety of models based on first-order differential equations. Using the perspective we've gained from this experience, let's briefly revisit the general description of the modeling process at the end of Section 1.2 in Chapter 1. There, our first step was to develop a mathematical description of the real-world system being modeled. Sometimes, such as in heating/cooling problems or in the draining of a tank or in the spotlight example above, a well-defined physical law (Newton's law of heating and cooling, Torricelli's law, the Law of Reflection) gave us the starting point for developing the model. In other applications, such as the various population models we studied in Section 2.5, we began with a "reasonable" assumption about how a quantity changed over time that led to a first-order differential equation. Our population models illustrate the iterative feature of the modeling process. The deficiences we observed in the simple Malthusian model led us to consider next the more sophisticated logistic model (as well as other variations that appeared in exercises). We also saw how the same differential equation can model more than one real-world system (for example, the same differential equation told us how a heated object cools and how to construct a dive table).

In some models, like the financial models in Section 2.5, we wanted, and were able to obtain, exact solutions to the relevant differential equation. In other cases (see especially Section 2.6) we learned how to analyze qualitatively our model without providing an exact solution. In the next chapter we will develop tools for producing approximate numerical solutions to first-order initial value problems. Being able to solve differential equations qualitatively or numerically gives us more flexibility in constructing mathematical models, as we are not limited to models whose equations can be solved analytically.

2.8. Exact equations

2.8.4. Exercises.

In Exercises 1–8, determine whether the given equation is exact. If it is, solve the equation, giving a one-parameter family of solutions.

1. $2xy^3 + 3x^2y^2 \frac{dy}{dx} = 0$.
2. $y + (x - \sin y)\frac{dy}{dx} = 0$.
3. $3x^2y^3 + y + (3x^3y^2 + 1)\frac{dy}{dx} = 0$.
4. $2xy^2 + y\cos(xy) + [x\cos(xy) + 2x^2y]\frac{dy}{dx} = 0$.
5. $2y + \cos x + (2x + \cos x)\frac{dy}{dx} = 0$.
6. $ye^{xy} + (xe^{xy} + 2y)\frac{dy}{dx} = 0$.
7. $\frac{2x}{1+x^2+y^2} + \frac{2y}{1+x^2+y^2}\frac{dy}{dx} = 0$.
8. $3xy + y^2 + (x^2 + xy)\frac{dy}{dx} = 0$.

In Exercises 9–12, solve the initial value problem.

9. $2xy^3 + (3x^2y^2 + y)\frac{dy}{dx} = 0$, $y(2) = 1$.
10. $2xye^{x^2y} + y + (x^2e^{x^2y} + x)\frac{dy}{dx} = 0$, $y(0) = 1$.
11. $y\cos(xy) + (x\cos(xy) + 1)\frac{dy}{dx} = 0$, $y\left(\frac{\pi}{6}\right) = 1$.
12. $\sec^2 x + 2xy + x^2\frac{dy}{dx} = 0$, $y\left(\frac{\pi}{4}\right) = 0$.

13. Solve the equation
$$5y + e^x + (5x + \cos y)\frac{dy}{dx} = 0$$
of Example 2.8.3 via the following steps:
   (a) Evaluate the "partial integral" $\int (5x + \cos y)\, dy$. Here, playing the role of the constant of integration will be a function h depending only on the variable x.
   (b) Compute the partial derivative with respect to x of your answer in (a) and equate it with $5y + e^x$ to determine h.

14. Find a function $f(x)$ so that
$$y^2 \cos x + yf(x)\frac{dy}{dx} = 0$$
is exact. Solve the resulting equation.

15. Use the integrating factor found in Example 2.8.4 to solve the differential equation given there.

16. Show that if you write any separable equation
$$\frac{dy}{dx} = g(x)h(y)$$
in the form
$$M(x,y) + N(x,y)\frac{dy}{dx} = 0$$
where $N(x,y) = \frac{1}{h(y)}$, the resulting equation is exact on any rectangle $\alpha < x < \beta$, $\gamma < y < \delta$ such that g is continuous on (α, β) and h is continuous and nonzero on (γ, δ). Are you doing essentially the same work or different work when you solve it first as a separable equation and then as an exact equation?

17. Consider the differential equation
$$\frac{dy}{dx} + p(x)y = g(x).$$
   (a) If you solve this as a linear equation, what do you use as an integrating factor?
   (b) If you rewrite the differential equation in the form
$$M(x,y) + N(x,y)\frac{dy}{dx} = 0$$
   where $N(x,y) = 1$, what is $M(x,y)$?

(c) Writing the equation as in (b), show there is an integrating factor that is a function of x alone. How does it compare with your integrating factor in (a)?

18. Solve $xy + y + 2x\frac{dy}{dx} = 0$ by using the integrating factor ye^x.

19. If you are given that the equation
$$\sin x + y + f(x)\frac{dy}{dx} = 0$$
has $\mu(x) = x$ as an integrating factor, determine all possible choices for $f(x)$.

20. Show that $M(x,y) + N(x,y)\frac{dy}{dx} = 0$ has an integrating factor that is a function of y alone provided
$$\frac{\frac{\partial N}{\partial x} - \frac{\partial M}{\partial y}}{M}$$
is a function of y alone. What first-order linear (or separable) equation do you solve to find this integrating factor?

21. Consider the equation
$$r^2 ds = (4r^3 - 3rs)dr.$$
By writing the equation in this symmetric form, it is not clear which variable—r or s—we intend to be the independent variable.
(a) Show that if r is the independent variable, the equation is linear, but it is not linear if we choose s as the independent variable.
(b) Solve it in the linear case, obtaining $s = \varphi(r)$ explicitly.
(c) Solve the equation
$$(x^2 - y^2)dy = 2xy\,dx$$
by thinking of y as the independent variable and x as the dependent variable.

22. By recognizing differentials, solve the initial value problem
$$y\,dx - x\,dy = x^3 y^5 (y\,dx + x\,dy), \quad y(4) = \frac{1}{2}.$$
(An implicit solution will do.)

23. Find a one-parameter family of solutions to the differential equation
$$xy\,dx - x^2\,dy = x^5 y^6 (y\,dx + x\,dy).$$

24. **Lost or extraneous solutions when using integrating factors.**
(a) Check that the equation
$$(2.143) \qquad y^2 + y + xy\frac{dy}{dx} = 0$$
is not exact. Show that upon multiplying this equation by $\mu(y) = 1/y$, the new equation
$$(2.144) \qquad y + 1 + x\frac{dy}{dx} = 0$$
is exact. Solve equation (2.144).
(b) Find a solution to (2.143) that is not a solution to (2.144). Thus solutions can be lost when we multiply by an integrating factor.
(c) Check that
$$(2.145) \qquad y + \frac{1}{y} + \left(3x + \frac{x}{y^2}\right)\frac{dy}{dx} = 0$$

2.8. Exact equations

is not exact, but the equation obtained by multiplying by $\mu(y) = y^2$, which is

(2.146) $$y^3 + y + (3xy^2 + x)\frac{dy}{dx} = 0,$$

is exact. Solve the new equation, and find a solution to (2.146) that is not a solution to (2.145). Thus solutions can be gained upon multiplying by an integrating factor.

25. (a) Show that the equation
$$xy^2 - y + (x^2y - x)\frac{dy}{dx} = 0$$
is exact, and then solve it.

(b) Take the differential equation in (a), multiply it by
$$\frac{1}{xy},$$
and check that this new equation is exact. Solve it as an exact equation.

(c) Show that the equation in (a) can also be written as
$$\frac{dy}{dx} = \frac{y - xy^2}{x^2y - x} = -\frac{y}{x},$$
which is linear. Solve this linear equation.

(d) Your solutions from (a), (b), and (c) should look rather different. Can you reconcile these different looking solutions?

26. (a) Show that $\mu(x, y) = xy$ is an integrating factor for
$$(2y^2 - 6xy) + (3xy - 4x^2)\frac{dy}{dx} = 0$$
and then solve the equation.

(b) The integrating factor in (a) is a simple example of an integrating factor that is a function of the product xy. We'll determine a condition under which
$$M(x, y) + N(x, y)\frac{dy}{dx} = 0$$
has an integrating factor of the form $\mu(x, y) = h(xy)$ for some single-variable function h. By the chain rule, $\mu_x = h'(xy)y$ and $\mu_y = h'(xy)x$. Suppose that the function
$$\frac{N_x - M_y}{xM - yN}$$
is a function of the product xy; call it $R(xy)$. Show that $\mu(x, y) = h(xy)$ is an integrating factor if
$$h'(xy) = R(xy)h(xy).$$
Notice that what is plugged into h', R, and h is the single quantity xy. Replace xy by a single variable t and obtain a simple first-order differential equation satisfied by the function $h(t)$. Solve it to get a formula for $h(t)$ and thus also for $\mu(x, y) = h(xy)$.

(c) Apply the method of part (b) to the differential equation in part (a) to produce an integrating factor for that equation. Is it the same one given in part (a)?

27. Solve the differential equation
$$\frac{dy}{dx} = \frac{-x + \sqrt{x^2 + y^2}}{y}, \quad \text{or} \quad x\,dx + y\,dy = \sqrt{x^2 + y^2}\,dx,$$
from Example 2.8.8 by "recognizing differentials". What kind of curve do you get?

Chapter 3

Numerical Methods

3.1. Euler's method

In spite of our efforts in the preceding chapter, the sad fact remains that most first-order initial value problems

(3.1) $$\frac{dy}{dt} = f(t,y), \ y(t_0) = y_0$$

cannot be solved exactly. In practice, we often have to be content with a graphical or approximate numerical solution, typically produced with the aid of a computer. In this section we discuss Euler's method, which is one way that a numerical solution is obtained, and investigate the expected error in this method. This is a small part of a big subject, the study of numerical methods in differential equations. Euler's method is conceptually simple and illustrates some of the basic ideas common to all numerical solution methods. Improvements of Euler's method are discussed in Section 3.2. In Section 3.3 we will use numerical solution techniques in a number of modeling applications, including an analysis of certain optical illusions and a model for phosphorus accumulation in a lake (see Exercise 7 in Section 3.3).

Euler's method. Euler's method is based on the same idea used to produce a graphical solution from a direction field. Recall that a direction field for the equation $\frac{dy}{dt} = f(t,y)$ shows short segments ("field marks") of slope $f(t,y)$ at various points in the ty-plane, as in Fig. 3.1. Euler's method is a numerical implementation of the following procedure: Pick a starting point (t_0, y_0) (i.e., an initial condition), and move a short way from this point in the direction of the field mark at (t_0, y_0), stopping at a new point (t_1, y_1). We assume that movement along the field marks is in the direction of increasing time, so $t_1 > t_0$, and we are moving to the right. From the point (t_1, y_1), move in the direction of the field mark at that point for a short distance stopping at a point that we denote (t_2, y_2). Continue this process, each time using the field mark at the stopping point to determine in what direction to move next. This is illustrated in Fig. 3.2. In practice, t_k's will be evenly spaced, and the distance between any two consecutive t-values is a positive constant h called the **step size**:

$$h = t_1 - t_0 = t_2 - t_1 = t_3 - t_2 = \cdots.$$

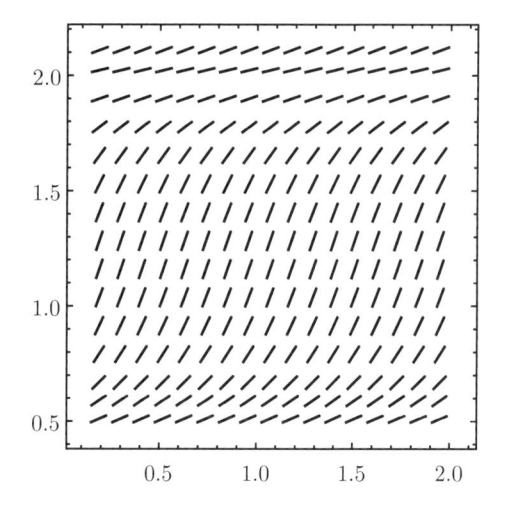

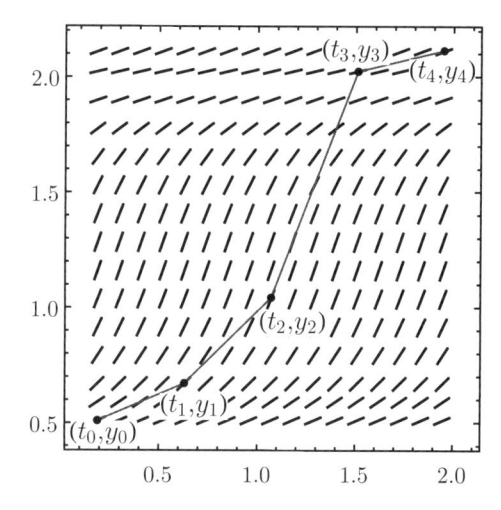

Figure 3.1. A direction field.

Figure 3.2. The idea of Euler's method.

We want to make this precise and in the process give an iterative formula for the points (t_k, y_k). Having chosen a step size h, we begin Euler's method by approximating the solution y to (3.1) at $t_1 = t_0 + h$. How should we do this? Look at Fig. 3.3, which illustrates this first step via the segment connecting (t_0, y_0) to (t_1, y_1). We know the actual solution curve passes through the point (t_0, y_0) (by our initial condition) and at this point the slope of the solution curve is the *number* $f(t_0, y_0)$ (by the differential equation). Thus

(3.2) $$L(t) = y_0 + f(t_0, y_0)(t - t_0)$$

is the linear function whose graph is the tangent line to the solution curve at (t_0, y_0). If t_1 is close to t_0 (i.e., if the step size is small), we expect this tangent line to still be close to the actual solution curve when $t = t_1$; so $L(t_1)$ then approximates $y(t_1)$, the solution at t_1. Setting $t = t_1$ in the right-hand side of (3.2) gives

$$y_0 + f(t_0, y_0)(t_1 - t_0) = y_0 + f(t_0, y_0)h.$$

This is illustrated in Fig. 3.3 (where we have made the step size not so small, to make the picture clearer): The point with coordinates $(t_1, y(t_1))$ lies on the actual solution curve, while the point $(t_1, L(t_1)) = (t_1, y_1)$ lies on the tangent line; y_1 is our desired approximation to $y(t_1)$. The error in this approximation is $y(t_1) - y_1$. We'll say more about errors below.

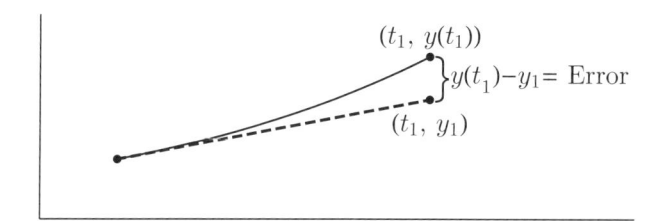

Figure 3.3. The first approximation $y_1 \approx y(t_1)$.

Let's move on to estimating the value of the solution at $t = t_2$. Starting at the point (t_1, y_1), we next move in the direction of the field mark at this point. The field mark has slope $f(t_1, y_1)$, and the corresponding tangent line function having slope $f(t_1, y_1)$ and passing through the point (t_1, y_1) has equation

$$L(t) = y_1 + f(t_1, y_1)(t - t_1).$$

3.1. Euler's method

Since our step size is h, we move along this line until we get to the point with t-coordinate $t_2 = t_1 + h = t_0 + 2h$. The corresponding point on the tangent line is
$$y_2 = y_1 + f(t_1, y_1)h.$$
The just computed value y_2 is our approximation to $y(t_2)$, the value of the solution at $t = t_2$. We repeat this process, computing in turn the values y_3 (approximating $y(t_3)$), y_4 (approximating $y(t_4)$), and so on. In general,
$$t_k = t_0 + kh$$
and

(3.3) $$y_k = y_{k-1} + f(t_{k-1}, y_{k-1})h.$$

Example 3.1.1. In this example we apply Euler's method to approximate the solution to the initial value problem

(3.4) $$\frac{dy}{dt} = \frac{t}{y}, \quad y(1) = 2,$$

at $t = 2$, and compare it to the exact value of the solution there. We begin by choosing a step size. In general, we expect that the smaller the step size, the better the approximation. However, the smaller the step size, the more steps are needed to get us from our starting point $t_0 = 1$ to our desired endpoint $t = 2$. Here we'll use a step size of $h = 0.25$, so we'll be making four steps from $t_0 = 1$: $t_1 = 1.25$, $t_2 = 1.5$, $t_3 = 1.75$, and $t_4 = 2$. From the initial condition we have $y_0 = 2$ at $t_0 = 1$, so by equation (3.3), with $f(t, y) = \frac{t}{y}$, our first computation is

$$y_1 = y_0 + \frac{t_0}{y_0}h = 2 + \frac{1}{2}(0.25) = 2.125.$$

Next we compute y_2, y_3, and y_4 in turn, using (3.3) with $k = 2, 3$, and 4, to get

$$y_2 = y_1 + \frac{t_1}{y_1}h = 2.125 + \frac{1.25}{2.125}(0.25) = 2.272,$$

$$y_3 = y_2 + \frac{t_2}{y_2}h = 2.272 + \frac{1.5}{2.272}(0.25) = 2.437,$$

and

$$y_4 = y_3 + \frac{t_3}{y_3}h = 2.437 + \frac{1.75}{2.437}(0.25) = 2.617.$$

Since we can solve (3.4) as a separable equation, we have
$$y = \sqrt{t^2 + 3},$$
and the exact value of y when $t = 2$ is $y = \sqrt{7} \approx 2.6457$. Our error is approximately 0.03. We might also compute the **relative percent error**; this is
$$\left|\frac{\text{estimated value} - \text{actual value}}{\text{actual value}}\right| \times 100.$$

Here the relative percent error is about 1.1 percent. Our error arises from two sources: a round-off error in the computations and what is called a **truncation error** that is inherent in our method of computing the approximations y_k by moving along field marks, not the solution curve itself. Types of errors will be discussed more fully below.

Fig. 3.4 shows the approximation points (t_k, y_k) for $k = 0, 1, 2, 3$, and 4. In Fig. 3.5 we have joined these approximation points with line segments to produce an approximate solution curve on the interval $[1, 2]$. The line segment joining the point (t_{k-1}, y_{k-1}) to (t_k, y_k) points in the direction of the field mark at (t_{k-1}, y_{k-1}), by equation (3.3). In Fig. 3.6 we show the actual solution curve $y = \sqrt{t^2 + 3}$ on the interval $[1, 2]$, together with our approximation points (t_k, y_k).

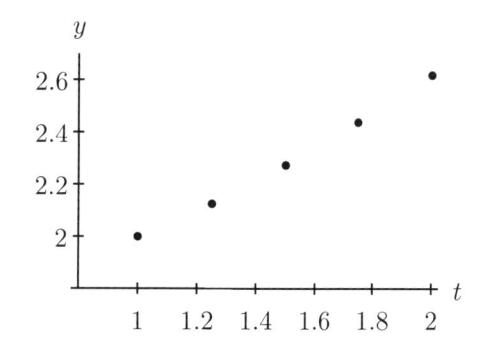

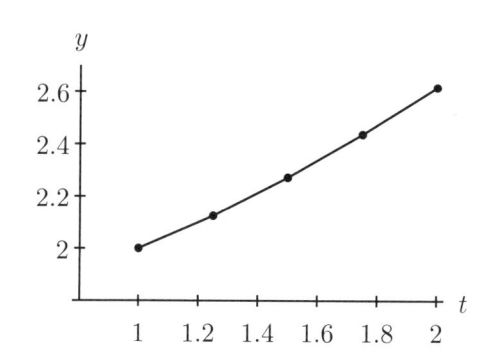

Figure 3.4. Approximations with step size $1/4$.

Figure 3.5. Approximate solution curve.

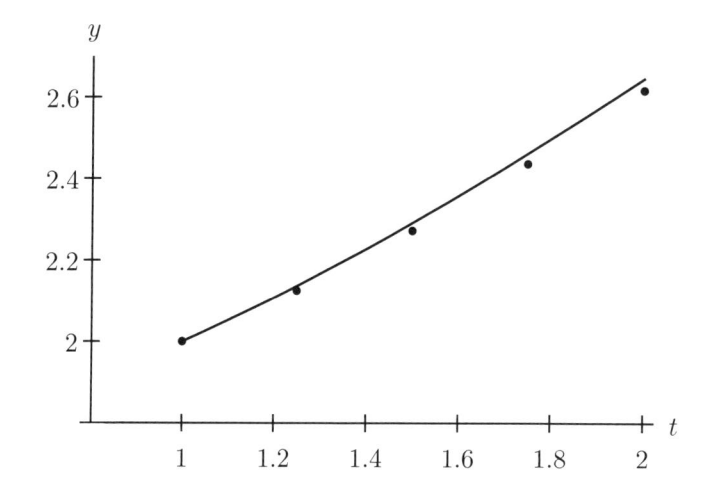

Figure 3.6. Actual solution curve and approximation points (t_k, y_k).

As we have noted, the smaller the step size, the better we expect the approximations obtained from Euler's method to be. If we return to the initial value problem in (3.4) but decrease the step size to $\frac{1}{10}$, Euler's method gives the estimates in Table 3.1. Of course, we need more steps of computation to estimate $y(2)$ with a step size of $\frac{1}{10}$ than with a step size of $\frac{1}{4}$. The values for y_k in Table 3.1 were calculated via a simple computer program whose structure is outlined below.

Euler's method

(1) **Initialization:** $h := 0.1$; $t_0 := 1$; $y_0 := 2$; $f(t, y) = t/y$.

(2) **Loop defining y_k recursively and incrementing time:**
   for k from 1 to 10 (k an integer) do
   $$y_k := y_{k-1} + f(t_{k-1}, y_{k-1})h;$$
   $$t_k := t_{k-1} + h;$$
   end for.

(3) **Output desired data:** Display sequence $(y_k, k = 0, \ldots, 10)$.

Translating the preceding outline into a particular programming language should be routine.

3.1. Euler's method

Table 3.1. Computations for $y' = t/y$, $y(1) = 2$, step size 0.1.

k	t_k	y_k
0	1	2
1	1.1	2.05000
2	1.2	2.10366
3	1.3	2.16070
4	1.4	2.22087
5	1.5	2.28391
6	1.6	2.34958
7	1.7	2.41769
8	1.8	2.48801
9	1.9	2.56034
10	2	2.63455

With a step size of $\frac{1}{10}$, Euler's method gives $y(2) \approx 2.63455$. Comparing this with the actual value of $y(2) = \sqrt{7}$, we now have a relative percent error of approximately 0.4 percent.

Sources of errors in the Euler method. Each step of Euler's method contributes some error to our numerical solution. We can identify different sources for the error. As we do the computations to implement Euler's method, we typically have round-off errors. Round-off error is the error you make when, for example, in a calculation you approximate 1/3 by 0.3333, or π by 3.14159. Though round-off error can accumulate when many calculations are made, computers can do calculations with a high degree of accuracy, so we will assume that our round-off errors are less significant that the errors that appear in Euler's method by virtue of the procedure itself. In the rest of this section we will generally ignore round-off errors.

To understand the errors made by the method itself, let's suppose we use Euler's method to approximate the solution of the initial value problem (3.1) at $t = t_*$ for some $t_* > t_0$. We chose a positive integer N representing the number of steps of the Euler method we decide to use to obtain an estimate y_N for $y(t_*)$; thus, we have a step size given by

$$(3.5) \qquad h = (t_* - t_0)/N.$$

Hence, $t_* = t_0 + Nh = t_N$. Consider the first step of Euler's method. We compute y_1, our approximation to the exact solution at t_1, by the formula

$$(3.6) \qquad y_1 = y_0 + f(t_0, y_0)h.$$

Graphically, we are replacing the solution $y(t)$ by the tangent line approximation to $y(t)$ at the point (t_0, y_0).

In calculus you learn that if $y(t)$ is a nice enough function (specifically, if it can be differentiated twice and has continuous first and second derivatives), then its exact value at t_1 is

$$(3.7) \qquad y(t_1) = y(t_0 + h) = y(t_0) + y'(t_0)h + R(h)$$

where $R(h)$, called the Taylor remainder, is *estimated* by

$$(3.8) \qquad |R(h)| \leq \frac{M}{2}h^2$$

for some constant M (where M can be chosen to be the maximum of $|y''(t)|$ on the interval $[t_0, t_0 + h]$). Since $y_0 = y(t_0)$ and $f(t_0, y_0) = y'(t_0)$, equations (3.6) and (3.7) say that the error we make at the first step when we estimate $y(t_1)$ by y_1 is the Taylor remainder $R(h)$, and the inequality (3.8) tells us that $|R(h)|$ is bounded above by a quantity proportional to h^2. It's called a **truncation error** since the right-hand side of (3.7) gets truncated by discarding the term $R(h)$. Notice that, as we expected, this first term error gets smaller as the step size h gets smaller.

In computing the second estimate y_2 and all subsequent y_k, we make a similar truncation error, but with an extra wrinkle. For example, Euler's method gives

$$y_2 = y_1 + f(t_1, y_1)h$$

as our estimate for $y(t_2)$. Remember that this arises by setting $t = t_2$ in the function

$$(3.9) \qquad L(t) = y_1 + f(t_1, y_1)(t - t_1).$$

This is a tangent line function, but (unlike when we computed y_1) its graph is *not* the tangent line to the actual solution curve when $t = t_1$. Instead, it is the tangent line to a *nearby* solution curve passing through (t_1, y_1); see Fig. 3.7. So there are two factors contributing to the error $y(t_2) - y_2$: the error that comes from using the tangent line to a *nearby* solution curve, rather than to the actual solution curve, and the truncation error that comes from the tangent line approximation itself. In Fig. 3.7, the nearby solution curve through (t_1, y_1) is shown dotted. The cumulative error, $y(t_2) - y_2$, is the sum of these two effects.

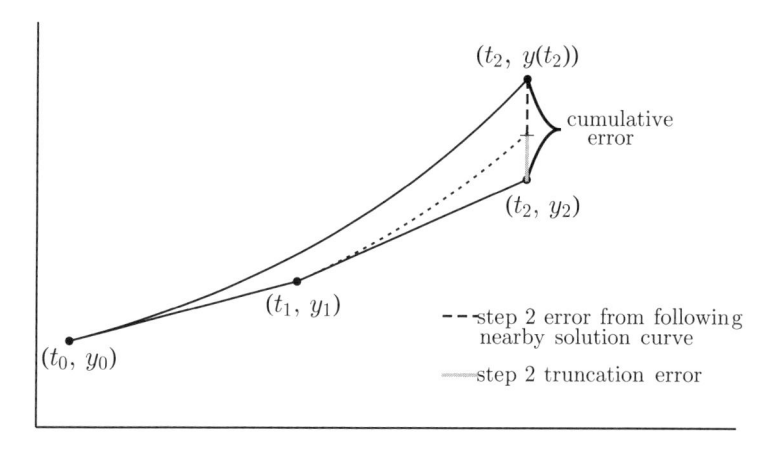

Figure 3.7. Error at step (2).

There is an analogous two-part contribution to the error at each of the remaining steps in the computations of the Euler estimates. **If** we had only the truncation error to worry about at each step, then we could bound the total error in going from (t_0, y_0) to $(t_N, y_N) = (t_*, y_N)$ by the sum of our N truncation error-bounds at each step

$$\frac{M_1}{2}h^2 + \frac{M_2}{2}h^2 + \cdots + \frac{M_N}{2}h^2,$$

where M_j is the maximum of $|y''(t)|$ on $[t_{j-1}, t_j]$. Moreover, since each M_j is no bigger than the maximum of $|y''(t)|$ on the *whole* interval $[t_0, t_N]$, this would give a total "truncation error" in going from t_0 to $t_N = t_*$ of no larger than

$$\frac{M}{2}h^2 N = \frac{M}{2}(t_N - t_0)h,$$

where M is the maximum of $|y''(t)|$ on $[t_0, t_N]$ and $h^2 N = (t_N - t_0)h$ by (3.5). This *intuitive* argument suggesting $|y(t_*) - y_N| \leq Ch$, where C is the constant $M(t_* - t_0)/2$, ignores the other factor contributing to the cumulative error in each step. Nevertheless, it is possible, although harder, to show that the **total cumulative error** in an Euler method approximation on the

3.1. Euler's method

interval $[t_0, t_*]$ with step size h is no larger than a constant times h; that is, there is a constant C_1 such that if $h > 0$ and $t_k = t_0 + kh \leq t_*$, then

(3.10) $$|y(t_k) - y_k| \leq C_1 h,$$

provided we assume that equation (3.1) has a unique solution $y(t)$, with continuous second derivative on the interval $[t_0, t_*]$. This justifies our intuition that the smaller the step size, the better the approximations provided by Euler's method. In particular, if h is chosen to be $(t_* - t_0)/N$, we see from (3.10) that

$$|y(t_*) - y_N| \leq C_1 \frac{(t_* - t_0)}{N}$$

so that y_N converges to the solution $y(t_*)$ as $N \to \infty$, assuming no round-off error. The constant C_1 on the right-hand side of (3.10) doesn't depend on the step size, but it may get larger as the interval $[t_0, t_*]$ on which we do our approximation grows. Example 3.1.2 below illustrates this. In practice, those who use a numerical solution routine whose accuracy improves as step size decreases typically compute approximations corresponding to smaller and smaller step sizes, hoping that the approximations will stabilize to the desired number of decimal places.

Example 3.1.2. The initial value problem

$$\frac{dy}{dt} = y \cos t, \quad y(0) = 1,$$

has exact solution

$$y = e^{\sin t}.$$

Because the function $\sin t$ oscillates periodically between 1 and -1, the function $e^{\sin t}$ oscillates between e and $1/e$. Using Euler's method with a step size of $\frac{1}{10}$ on the interval $0 \leq t \leq 30$, we obtain 300 approximation points. These are shown in Fig. 3.8, along with the graph of the solution $y = e^{\sin t}$ for $0 \leq t \leq 30$. Our approximation looks pretty good for the first cycle or so, but Euler's method is unable to keep up with the oscillations because of the effect of the cumulative errors. In fact, the "long-term" behavior of an approximate Euler solution can be considerably worse than what we see in this example.

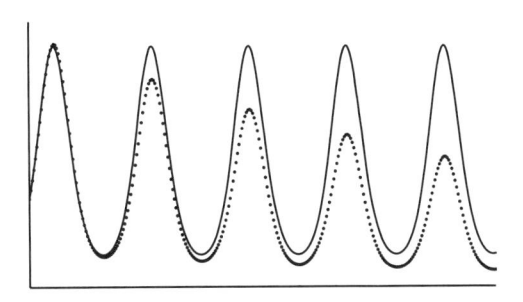

Figure 3.8. Actual solution and Euler approximation points.

3.1.1. Exercises. Use whatever technology is at your disposal to help with the Euler method computations in these exercises.

1. Use Euler's method with a step size of $\frac{1}{4}$ to estimate $y(2)$, where y is the solution to

$$\frac{dy}{dt} = \frac{1}{1+t^2}, \quad y(1) = \frac{\pi}{4}.$$

Explain why your answer is an approximation to $\arctan 2$.

2. Use Euler's method with a step size of $\frac{1}{10}$ to estimate $y(0.5)$, where y solves the initial value problem

$$\frac{dy}{dt} = e^{-t^2}/y, \quad y(0) = 1.$$

3. In this problem we consider the initial value problem

$$\frac{dy}{dt} = ty, \quad y(1) = \frac{1}{2}.$$

   (a) Approximate $y(2)$ using Euler's method with a step size of $\frac{1}{4}$.
   (b) Approximate $y(2)$ by Euler's method with a step size of $\frac{1}{10}$.
   (c) Give an explicit solution of the initial value problem, and compute $y(2)$ exactly.
   (d) Compute the relative percentage error in your approximations from (a) and (b).
   (e) In applying Euler's method, we may step to the left instead of to the right, so that $t_k = t_0 - kh$ for positive integers k. Use Euler's method to estimate $y(\frac{1}{2})$, using the step size $h = \frac{1}{10}$. What is your relative percent error?

4. (a) Solve the initial value problem

$$\frac{dy}{dt} = -100y, \quad y(0) = 1,$$

   and determine $y(1)$.
   (b) Show that if you use Euler's method with step size h to estimate $y(1)$, the resulting estimate is

$$(1 - 100h)^{1/h}.$$

   (You may assume $1/h$ is an integer.)
   (c) Using the result in (b), what is the Euler estimate for $y(1)$ if $h = 0.001$? If $h = 0.0001$? Compute the relative percent error in the second case.
   (d) Compute

$$\lim_{h \to 0}(1 - 100h)^{1/h}.$$

5. Suppose we model US population $P(t)$, in millions of people at time t (in years), by the logistic equation

$$\frac{dP}{dt} = aP - bP^2$$

where $a = 0.0313$ and $b = 1.589 \times 10^{-10}$. Using Euler's method, with a step size of 10, estimate the population in 1900 if the population in 1800 was 5.308 million.

6. Euler's method is carried out with a step size of h, and the estimated solution values y_k are computed with the formula

$$y_k = y_{k-1} + 3y_{k-1}h + 5h.$$

If the differential equation has the form $\frac{dy}{dt} = f(t,y)$, what could $f(t,y)$ be?

7. Three direction fields, for three first-order equations $\frac{dy}{dt} = f(t,y)$, are given below. We are interested in the solution passing through the point $P = (1, y_0)$ as shown. With a step size of $\frac{1}{2}$, sketch (approximately) the Euler approximation for $y(2)$ (using two steps in Euler's method). Do you expect your approximation for $y(2)$ to be larger or smaller than the actual value?

3.1. Euler's method

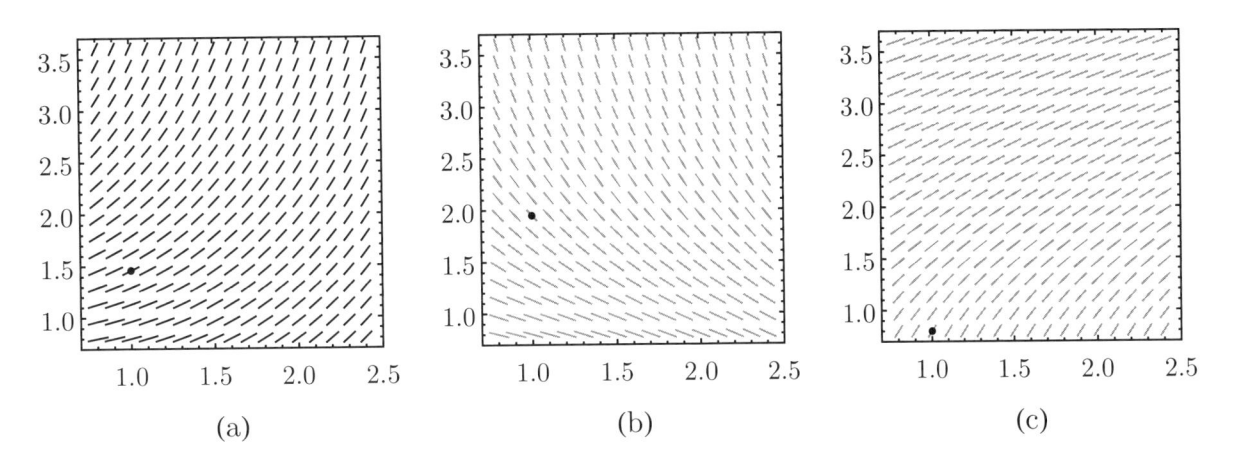

Figure 3.9. Direction fields and $P = (1, y_0)$.

8. (a) Check that the solution to
$$\frac{dy}{dt} = y, \quad y(0) = 1,$$
is $y(t) = e^t$, so that in particular $y(1) = e$.

(b) Show that if we use Euler's method with step size $\frac{1}{n}$ for some positive integer n to approximate $y(1)$, we will have
$$y_1 = y_0 \left(1 + \frac{1}{n}\right) = 1 + \frac{1}{n},$$
$$y_2 = y_1 \left(1 + \frac{1}{n}\right) = \left(1 + \frac{1}{n}\right)^2,$$
$$y_3 = y_2 \left(1 + \frac{1}{n}\right) = \left(1 + \frac{1}{n}\right)^3,$$
and so on, so that
$$y_n = \left(1 + \frac{1}{n}\right)^n.$$
Thus Euler's method gives
$$\left(1 + \frac{1}{n}\right)^n$$
as an approximation for e.

9. (a) Use Euler's method with a step size of $h = \frac{1}{10}$ to estimate $y(1)$ for
$$\frac{dy}{dt} = t + y, \quad y(0) = 1.$$

(b) Find the exact solution to this initial value problem, and compute your relative percent error in (a).

(c) Repeat (a) with step sizes $h = \frac{1}{100}$ and $h = \frac{1}{200}$, and compute the relative percent error.

(d) The error estimate for Euler's method says, *roughly*, if you divide the step size by d, the error is divided by d. For this problem, with $h = 0.1$, $h = 0.1/10 = 0.01$, and $h = 0.01/2 = 0.005$, do you see this effect?

(e) Sketch a graph showing the actual solution for $0 \leq t \leq 1$ and the approximate solution from (a).

10. (a) Use Euler's method with a step size $h = \frac{1}{10}$ to estimate $y(1)$, where y is the solution to
$$\frac{dy}{dt} = \frac{t}{\sqrt{t^2 + y^2}}, \quad y(0) = 1.$$

(b) As mentioned in Exercise 3(e), we may step leftward in applying Euler's method ($t_k = t_0 - kh$, $k > 0$). Suppose you use a step size of $\frac{1}{10}$ to estimate $y(-1)$. Can you say what your answer will be, without doing any new computations, based on the symmetry in the direction field (see Fig. 3.10)?

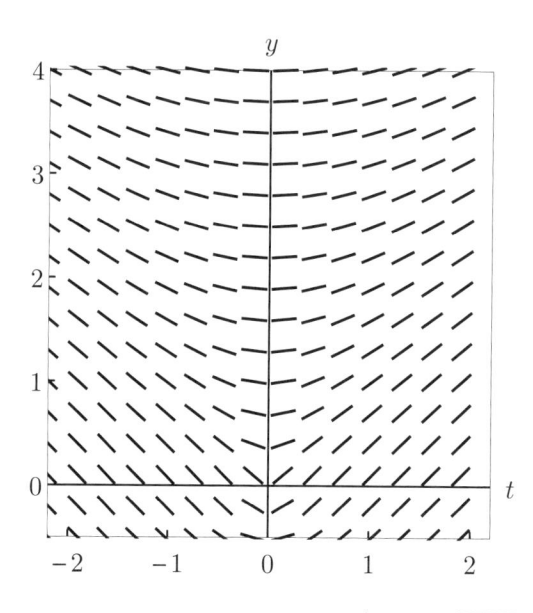

Figure 3.10. Direction field for $\frac{dy}{dt} = t/\sqrt{t^2 + y^2}$.

11. Caution: Euler's method can appear to give a solution when no solution exists!
    (a) Estimate $y(1)$ by Euler's method with step size $h = \frac{1}{10}$ if
$$\frac{dy}{dt} = y^2, \quad y(0) = 1.$$

    (b) Repeat (a) with step size $h = \frac{1}{100}$.
    (c) Find an explicit solution to the initial value problem in (a). What happens if you try to compute $y(1)$ exactly using your solution?

12. We know from Section 2.4 that (under the hypotheses of Theorem 2.4.5) two distinct solution curves to the differential equation $\frac{dy}{dt} = f(t,y)$ cannot intersect. In particular, no solution curve can intersect an equilibrium solution curve. However, can an *approximate solution curve* constructed by Euler's method intersect an equilibrium solution curve? This exercise investigates this question.
    (a) Find all equilibrium solutions to
$$\frac{dy}{dt} = y(8 - y).$$

    (b) Using a step size of $h = 0.25$ and initial point $t_0 = 0, y_0 = 7$, compute $y_1, y_2, \ldots, y_6$ by Euler's method.
    (c) Sketch the equilibrium solutions and the approximate solution obtained by joining the points (t_k, y_k), $k = 0, 1, 2, \ldots, 6$ from (a). Do they intersect?

3.1. Euler's method

13. Consider the differential equation

$$\frac{dy}{dt} = y^3 - y^2 - 1.2y + 0.7.$$

Part of the direction field for this equation is shown in Fig. 3.11.
(a) What is the connection between roots of the polynomial

$$p(y) = y^3 - y^2 - 1.2y + 0.7$$

and (certain) solutions to the differential equation?
(b) Roughly approximate the roots of $p(y)$ by looking at the direction field.
(c) If you use Euler's method with $t_0 = 0$, $y_0 = 0.4$, and step size $h = \frac{1}{100}$, do you expect the approximations $y_1, y_2, y_3, \ldots$ to converge to one of the roots of $p(y)$? If not, why not? If so, which one?
(d) If you use Euler's method with $t_0 = 0$, $y_0 = -1.1$, and a step size $h = \frac{1}{100}$, do you expect the approximations $y_1, y_2, y_3, \ldots$ to converge to one of the roots of $p(y)$? If not, why not? If so, which one?
(e) Answer (d) if we use *leftward* steps (see Exercise 3(e)) with step size is $h = \frac{1}{100}$.

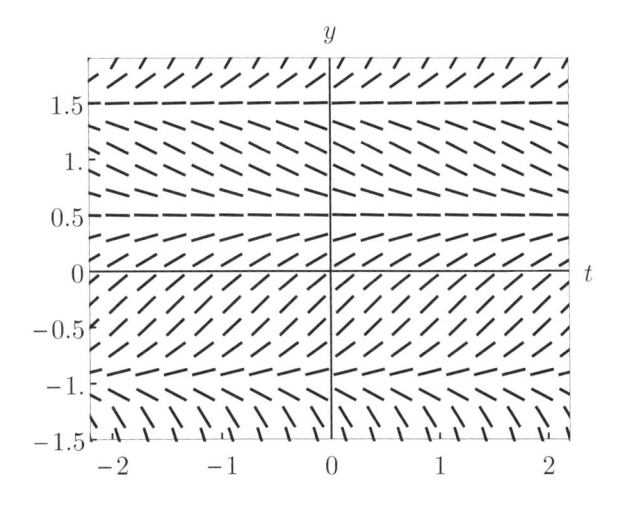

Figure 3.11. Direction field for $\frac{dy}{dt} = y^3 - y^2 - 1.2y + 0.7$.

14. (a) Show that one step of the Euler method with $h = \frac{1}{4}$ applied to the problem

$$\frac{dy}{dt} = t + y, \quad y(0) = 1,$$

gives the approximate solution point $(0.25, 1.25)$.
(b) Suppose we use Euler's method with *leftward* steps of size $h = \frac{1}{4}$ (see Exercise 3(e)) to approximate the solution to

$$\frac{dy}{dt} = t + y, \quad y(0.25) = 1.25,$$

at $t = 0$. Do you get back to the point $t = 0, y = 1$ as in (a)?
(c) Use the direction field pictured in Fig. 3.12 to explain your answer to (b).

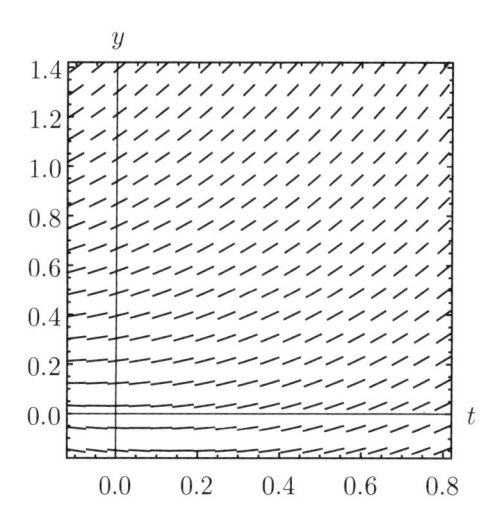

Figure 3.12. Direction field for $\frac{dy}{dt} = t + y$.

15. In this problem we introduce a modification of Euler's method (called the Heun method), which will be discussed further in the next section. Like the Euler method, it begins by choosing a step size h and produces approximate solutions at the points $t_1 = t_0 + h$, $t_2 = t_1 + h$, $t_3 = t_2 + h, \ldots$ for an initial value problem

$$\frac{dy}{dt} = f(t, y), \quad y(t_0) = y_0.$$

The approximate solution at $t = t_1$ is

(3.11) $$y_1 = y_0 + \frac{1}{2}[f(t_0, y_0) + f(t_1, z_1)]h$$

where z_1 is the Euler approximation point

$$z_1 = y_0 + f(t_0, y_0)h.$$

Notice that on the right-hand side of (3.11) we are averaging the slopes $f(t_0, y_0)$ and $f(t_1, z_1)$, the slopes of the field marks at the initial condition point (t_0, y_0) and the first Euler point (t_1, z_1). In general, our improved approximate solution y_k at t_k is obtained by the formula

$$y_k = y_{k-1} + \frac{1}{2}[f(t_{k-1}, y_{k-1}) + f(t_k, z_k)]h,$$

where

$$z_k = y_{k-1} + f(t_{k-1}, y_{k-1})h$$

is obtained by taking one "Euler step" of size h from the previous point (t_{k-1}, y_{k-1}).

In this problem we will implement this improved Euler method for the initial value problem

$$\frac{dy}{dt} = t + y, \quad y(0) = 1.$$

(a) Using a step size of $h = \frac{1}{10}$, show that

$$y_1 = 1 + \frac{1}{20}\left(1 + \frac{12}{10}\right) = 1.11.$$

(b) Show that for $k \geq 0$,

$$t_k = \frac{k}{10},$$

$$y_{k+1} = y_k + \frac{1}{20}\left(t_k + y_k + t_{k+1} + y_k + \frac{1}{10}(t_k + y_k)\right),$$

so that
$$y_{k+1} = y_k + \frac{1}{20}\left(\frac{k}{10} + y_k + \frac{k+1}{10} + y_k + \frac{1}{10}(\frac{k}{10} + y_k)\right).$$

(c) Using the result of (b), confirm that the value of y_1 computed in (a) is correct and then find y_k for $k = 2, 3, \ldots, 10$.

(d) Compare the relative percent error in the Heun estimate y_{10} for $y(1)$ with the relative percent error obtained in Exercise 9.

Remark: Assuming that the initial value problem (3.1) has a unique solution with continuous third derivative, it can be shown that the total cumulative error using the Heun method with step size h on a fixed bounded interval satisfies an estimate of the form (3.10) with h replaced by h^2.

16. (a) Use Euler's method with $h = \frac{1}{10}$ to estimate the value of the solution to
$$\frac{dy}{dt} = t(1+y), \quad y(0) = 1,$$
at $t = 2$.

(b) Redo the estimate from (a), this time using the Heun method (see Exercise 15).

(c) Solve the initial value problem in (a) exactly, and compute $y(2)$. What is the relative percent error of your estimates in (a) and (b)?

3.2. Improving Euler's method: The Heun and Runge-Kutta Algorithms

In the last section we saw that when we apply Euler's method, with step size h, to the initial value problem

(3.12) $$\frac{dy}{dt} = f(t, y), \quad y(t_0) = y_0,$$

we obtain a sequence of approximations $y_1, y_2, \ldots, y_{N-1}, y_N$ for, respectively, $y(t_1), y(t_2), \ldots, y(t_{N-1}), y(t_N)$, where $y(t_k)$ is the (exact) solution of (3.12) at $t = t_k$ and $t_k = t_0 + kh$ for $k = 1, 2, \ldots, N$. The approximations are generated from the recurrence relation (3.3), which may be written
$$y_k - y_{k-1} = f(t_{k-1}, y_{k-1})h.$$

Thus, keeping in mind that y_k approximates $y(t_k)$ while y_{k-1} approximates $y(t_{k-1})$, we see that Euler's method is based on the estimate
$$y(t_k) - y(t_{k-1}) \approx f(t_{k-1}, y_{k-1})h.$$

In this section we will explore different ways of estimating $y(t_k) - y(t_{k-1})$ and thus different numerical methods for solving the initial value problem (3.12). We derive these methods from the same starting point: If y is a solution of

(3.13) $$\frac{dy}{dt} = f(t, y(t)),$$

then by integrating both sides of (3.13) from t_{k-1} to t_k we obtain

(3.14) $$y(t_k) - y(t_{k-1}) = \int_{t_{k-1}}^{t_k} f(t, y(t))\, dt.$$

We'll estimate $y(t_k) - y(t_{k-1})$ by estimating the integral on the right-hand side. Different methods of estimating the integral will lead to different numerical solution methods for the initial value problem (3.12). For example, we will show that a "left-endpoint approximation" of the integral yields Euler's method, while the "trapezoidal rule" yields Heun's method (introduced in Exercise

15 of the preceding section). Using Simpson's Rule to approximate the integral will lead to yet another technique for solving (3.12) numerically, called the Runge-Kutta method.

Throughout the remainder of this section, we assume that (3.12) has a unique solution $y = y(t)$ on the interval $[t_0, t_*]$ over which we find its approximate values $y_1, y_2, \ldots, y_N$ at equally spaced points $t_1, t_2, \ldots, t_N$, with $t_k - t_{k-1} = h$ for $1 \leq k \leq N$, where h is a positive step size for which $t_N \leq t_*$. We also assume that the function f of two variables appearing in (3.12) is continuous on a rectangle large enough to contain all the ordered pairs $(t, y(t))$ for $t_0 \leq t \leq t_*$. In particular, the integrand on the right-hand side of (3.14) is a continuous function of t on $[t_{k-1}, t_k]$, and thus the integral exists.

Euler's method: Left-endpoint approximation. Consider first equation (3.14) with $k = 1$:

$$y(t_1) - y(t_0) = \int_{t_0}^{t_1} f(t, y(t))\, dt.$$

If we use a *left-endpoint approximation* (see Fig. 3.13), then

$$\int_{t_0}^{t_1} f(t, y(t))\, dt \approx f(t_0, y(t_0))(t_1 - t_0) = f(t_0, y_0)h,$$

and we obtain the estimate

$$y(t_1) - y(t_0) \approx f(t_0, y_0)h,$$

or

$$y(t_1) \approx y_0 + f(t_0, y_0)h,$$

where y_0 is given in the initial condition. This is exactly the Euler method approximation for $y(t_1)$. In general, for the k-th step, which produces y_k from y_{k-1}, we use (3.14) and the left-endpoint estimate for the integral, obtaining

$$\begin{aligned} y(t_k) - y(t_{k-1}) &= \int_{t_{k-1}}^{t_k} f(t, y(t))\, dt \\ &\approx f(t_{k-1}, y(t_{k-1}))(t_k - t_{k-1}) \\ &= f(t_{k-1}, y(t_{k-1}))h. \end{aligned}$$

Replacing $y(t_{k-1})$ with its approximate value y_{k-1}, we have

(3.15) $$y(t_k) \approx y_{k-1} + f(t_{k-1}, y_{k-1})h.$$

Setting the left-hand side equal to y_k, we have obtained the recurrence relation of Euler's method.

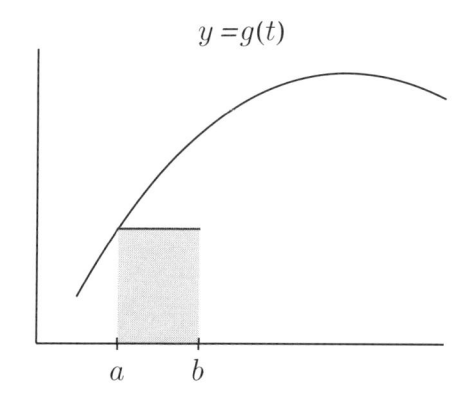

Figure 3.13. Left-endpoint approximation: $\int_a^b g(t)\, dt \approx g(a)(b - a)$.

3.2. Improving Euler's method: The Heun and Runge-Kutta Algorithms

Heun's method: Trapezoid rule approximation. Suppose now that we use the trapezoid rule (see Fig. 3.14) to estimate the integral on the right-hand side of (3.14). For $k = 1$ this looks like

$$\begin{aligned} y(t_1) - y(t_0) &= \int_{t_0}^{t_1} f(t, y(t)) \, dt \\ &\approx \frac{1}{2} \left[f(t_0, y(t_0)) + f(t_1, y(t_1)) \right] (t_1 - t_0) \\ &= \frac{h}{2} \left[f(t_0, y(t_0)) + f(t_1, y(t_1)) \right]. \end{aligned}$$

In the last line, we know that $y(t_0) = y_0$ (from the specified initial condition) but we don't know $y(t_1)$. However, we can approximate $y(t_1)$ by the estimate provided by Euler's method: $y(t_1) \approx y_0 + f(t_0, y_0)h$. Thus the first step of the Heun method gives the estimate y_1 for $y(t_1)$ as

$$y_1 = y_0 + \frac{h}{2} \left(f(t_0, y_0) + f(t_1, z_1) \right),$$

where $z_1 = y_0 + f(t_0, y_0)h$. In the k-th step of the Heun method we have

$$\begin{aligned} y(t_k) - y(t_{k-1}) &= \int_{t_{k-1}}^{t_k} f(t, y(t)) \, dt \\ &\approx \frac{1}{2} \left(f(t_{k-1}, y(t_{k-1})) + f(t_k, y(t_k)) \right) (t_k - t_{k-1}) \\ &= \frac{h}{2} \left(f(t_{k-1}, y(t_{k-1})) + f(t_k, y(t_k)) \right). \end{aligned}$$

Replacing $y(t_{k-1})$ with its approximate value y_{k-1} (computed in the $(k-1)$-st step) and using the approximation of $y(t_k)$ supplied by Euler's method, we obtain

$$\int_{t_{k-1}}^{t_k} f(t, y(t)) \, dt \approx \frac{h}{2} \left(f(t_{k-1}, y_{k-1}) + f(t_k, z_k) \right)$$

where $z_k = y_{k-1} + f(t_{k-1}, y_{k-1})h$. We thus are led to the following recurrence relation for the approximation y_k to the solution (3.12) at time $t_k = t_0 + kh$ for $k = 1, 2, \ldots, N$:

$$(3.16) \quad y(t_k) \approx y_k = y_{k-1} + \frac{h}{2} \left(f(t_{k-1}, y_{k-1}) + f(t_k, z_k) \right),$$
$$\text{where } z_k = y_{k-1} + f(t_{k-1}, y_{k-1})h.$$

Equation (3.16) describes Heun's method of numerical solution, which is also called the "Improved Euler method". It is referred to as a **predictor-corrector** method. We first "predict" the approximation z_k of $y(t_k)$ by Euler's method; then we use this predicted value to obtain via equation (3.16) an approximation y_k that is typically closer to the "correct" value $y(t_k)$.

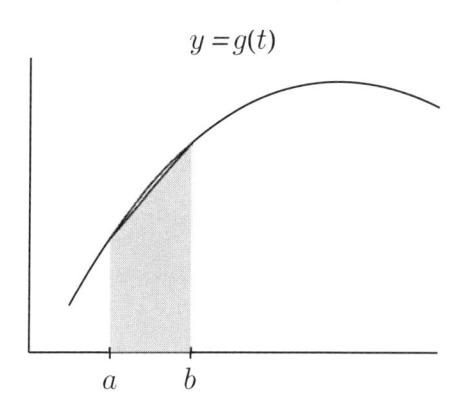

Figure 3.14. The Trapezoid Rule: $\int_a^b g(t)\,dt \approx$ area of shaded trapezoid $= \frac{1}{2}\left(g(a) + g(b)\right)(b - a)$.

Example 3.2.1. We rework Example 3.1.1 using Heun's method instead of Euler's method. Our goal is to approximate the value of the solution of

(3.17) $$\frac{dy}{dt} = \frac{t}{y}, \quad y(1) = 2,$$

at $t = 2$. We choose, just in as Example 3.1.1, a step size $h = \frac{1}{4} = 0.25$, making y_4 our approximation of $y(2)$. We have $t_0 = 1$, $t_1 = 1.25$, and $y_0 = 2$. As we compute, via Heun's method, y_k and z_k for $k = 1, 2, 3$, and 4, we round to three decimal places:

$$z_1 = y_0 + \frac{t_0}{y_0}h = 2 + \frac{1}{2}(0.25) = 2.125,$$

$$y_1 = y_0 + \frac{h}{2}\left(\frac{t_0}{y_0} + \frac{t_1}{z_1}\right) = 2 + \frac{0.25}{2}\left(\frac{1}{2} + \frac{1.25}{2.125}\right) = 2.136;$$

$$z_2 = y_1 + \frac{t_1}{y_1}h = 2.136 + \frac{1.25}{2.136}(0.25) = 2.282,$$

$$y_2 = y_1 + \frac{h}{2}\left(\frac{t_1}{y_1} + \frac{t_2}{z_2}\right) = 2.136 + \frac{0.25}{2}\left(\frac{1.25}{2.136} + \frac{1.5}{2.282}\right) = 2.291;$$

$$z_3 = y_2 + \frac{t_2}{y_2}h = 2.291 + \frac{1.5}{2.291}(0.25) = 2.455,$$

$$y_3 = y_2 + \frac{h}{2}\left(\frac{t_2}{y_2} + \frac{t_3}{z_3}\right) = 2.291 + \frac{0.25}{2}\left(\frac{1.5}{2.291} + \frac{1.75}{2.455}\right) = 2.462;$$

$$z_4 = y_3 + \frac{t_3}{y_3}h = 2.462 + \frac{1.75}{2.462}(0.25) = 2.640,$$

$$y_4 = y_3 + \frac{h}{2}\left(\frac{t_3}{y_3} + \frac{t_4}{z_4}\right) = 2.462 + \frac{0.25}{2}\left(\frac{1.75}{2.462} + \frac{2}{2.640}\right) = 2.646.$$

Thus, we obtain the approximation $y(2) \approx 2.646$. Recall that the exact value of $y(2)$, calculated using the solution $y(t) = \sqrt{t^2 + 3}$ of (3.17), is $\sqrt{7} \approx 2.64575$. Thus, our approximation is correct to three decimal places. Compare our estimate of 2.646 for $y(2)$, computed with step size 0.25 via the Heun method, to the approximation 2.635 from Table 3.1, obtained via the Euler method with step size $1/10$. Heun's method with just 4 steps produces a more accurate result than Euler's method with 10 steps. This is not particularly surprising because the trapezoidal rule for integral approximation, which underlies the Heun method, is typically more accurate than the left-endpoint approximation, which underlies the Euler method. Observe that the increased accuracy of Heun's method has a price: Heun's method requires more computations per iteration than does Euler's.

3.2. Improving Euler's method: The Heun and Runge-Kutta Algorithms

Error estimates for the Heun method. Suppose that on $[t_0, t_*]$ the solution $y = y(t)$ to the initial value problem (3.12) is unique and has continuous third-order derivative. It is possible to show that there is a constant C_2 such that for any step size h and any positive integer k for which $t_k = t_0 + kh \leq t_*$, the total cumulative error $|y(t_k) - y_k|$ satisfies

(3.18) $$|y(t_k) - y_k| \leq C_2 h^2,$$

where y_k is the Heun approximation of $y(t_k)$ with step size h. Comparing this result to the corresponding estimate "$C_1 h$" for Euler's method (see (3.10)), we again have reason to believe that for small values of h, Heun's method with step size h should be more accurate than Euler's method with step size h.

When an approximation method has the property that

$$\text{Error} \leq Ch^p,$$

for some positive real number p and constant C independent of h, we say the method is *of order p*. Thus, Euler's method is of order 1 while Heun's is of order 2. We conclude this section by describing the Runge-Kutta method, which is sometimes referred to as "RK4" because it is a method of order 4.

The Runge-Kutta method: Simpson's Rule approximation. The Runge-Kutta method arises from using Simpson's Rule for approximating the integral in (3.14). Simpson's Rule estimates an integral by replacing the integrand by an approximating quadratic polynomial; see Fig. 3.15 for the details.

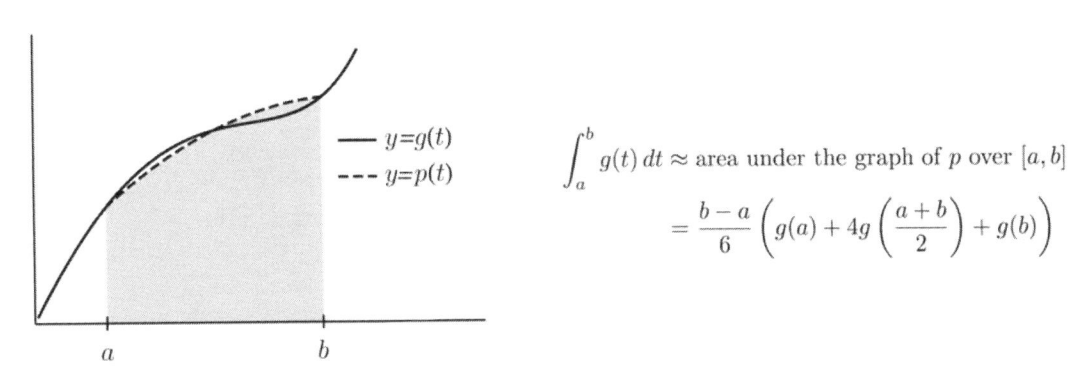

$$\int_a^b g(t)\,dt \approx \text{area under the graph of } p \text{ over } [a, b]$$
$$= \frac{b-a}{6}\left(g(a) + 4g\left(\frac{a+b}{2}\right) + g(b)\right)$$

Figure 3.15. Simpson's Rule. Here p is the quadratic polynomial whose values agree with those of g at a, $(a+b)/2$, and b. (See Exercise 2.)

Instead of $g(t)$, our integrand is $f(t, y(t))$, and to employ the integral estimate of Simpson's Rule, illustrated in the figure above, we see that our integrand's values at the endpoints, as well as the midpoint, of the interval of integration are needed. For the Runge-Kutta method, the values of $f(t, y(t))$ at these points of the interval of integration $[t_{k-1}, t_k]$ are approximated as follows. For the left endpoint, we use

$$s_1 \equiv f(t_{k-1}, y_{k-1});$$

for the midpoint, we use the average of

$$s_2 \equiv f\left(t_{k-1} + \frac{1}{2}h, y_{k-1} + \frac{1}{2}s_1 h\right) \quad \text{and} \quad s_3 \equiv f\left(t_{k-1} + \frac{1}{2}h, y_{k-1} + \frac{1}{2}s_2 h\right);$$

finally, for the right endpoint, we use

$$s_4 \equiv f(t_k, y_{k-1} + s_3 h).$$

Using these estimates, we obtain from Simpson's Rule

$$\int_{t_{k-1}}^{t_k} f(t, y(t))\, dt \approx \frac{t_k - t_{k-1}}{6} \left(s_1 + 4 \left(\frac{s_2 + s_3}{2} \right) + s_4 \right)$$

$$= \frac{t_k - t_{k-1}}{6} (s_1 + 2s_2 + 2s_3 + s_4).$$

Replacing $t_k - t_{k-1}$ with h and substituting this result into (3.14), we arrive at the following recurrence relation for the approximate values of the solution of (3.12) at times $t_k = t_0 + kh$:

$$(3.19) \qquad y_k = y_{k-1} + \frac{h}{6}(s_1 + 2s_2 + 2s_3 + s_4).$$

The preceding equation, with s_1, s_2, s_3, and s_4 defined as above, constitutes the Runge-Kutta method.

To illustrate the method, we once again consider the initial value problem $dy/dt = t/y$, $y(1) = 2$, of (3.17) and seek to approximate $y(2)$. We choose a step size $h = 1$, finding our approximation in a single iteration. This single iteration requires estimates of four values of $f(t, y(t)) = t/y(t)$, as opposed to one in Euler's method and two in Heun's. We have $t_0 = 1$, $y_0 = 2$, $h = 1$,

$$s_1 = \frac{t_0}{y_0} = \frac{1}{2}, \quad s_2 = \frac{t_0 + \frac{h}{2}}{y_0 + \frac{1}{2}s_1 h} = \frac{2}{3}, \quad s_3 = \frac{t_0 + \frac{h}{2}}{y_0 + \frac{1}{2}s_2 h} = \frac{9}{14}, \quad s_4 = \frac{t_1}{y_0 + s_3} = \frac{28}{37}.$$

Thus by equation (3.19),

$$y_1 = 2 + \frac{1}{6}\left(\frac{1}{2} + 2\frac{2}{3} + 2\frac{9}{14} + \frac{28}{37}\right) = \frac{24{,}671}{9{,}324} \approx 2.64597.$$

Comparing this estimate to the exact value $y(2) = \sqrt{7} \approx 2.64575$, we have achieved three decimal place accuracy with one iteration of the Runge-Kutta method.

Example 3.2.2. In this example we'll compare the numerical solutions of the initial value problem

$$(3.20) \qquad \frac{dy}{dt} = 4t^2 \sqrt{y} - \frac{2}{t}y, \quad y(2) = 1,$$

over the interval $[2, 3]$, computed by the Euler, Heun, and Runge-Kutta methods. We choose a step size $h = \frac{1}{10}$ and display approximate values of the solution at $2 + k/10$ for $k = 1$ to 10 for these three methods. We also give the values of the exact solution,

$$y(t) = \frac{(t^4 - 12)^2}{4t^2},$$

to six digits, where the exact solution may be obtained via a Bernoulli equation substitution. The results are shown in Table 3.2.

3.2. Improving Euler's method: The Heun and Runge-Kutta Algorithms

Table 3.2. Comparing Euler, Heun, and Runge-Kutta for (3.20).

k	t_k	y_k (Euler)	y_k (Heun)	y_k (RK4)	Exact Solution (to 6 digits)
0	2	1	1	1	1
1	2.1	2.50000	3.02552	3.14092	3.14480
2	2.2	5.05103	6.48410	6.73621	6.74299
3	2.3	8.94291	11.6645	12.0649	12.0743
4	2.4	14.4931	18.8967	19.4539	19.4657
5	2.5	22.0566	28.5593	29.2808	29.2952
6	2.6	32.0332	41.0843	41.9777	41.9944
7	2.7	44.8732	56.9615	58.0342	58.0534
8	2.8	61.0828	76.7427	78.0028	78.0244
9	2.9	81.2293	101.047	102.502	102.526
10	2	105.946	130.563	132.223	132.250

The approximations populating the preceding table were obtained from computer programs implementing the three different numerical methods. The programs for the Heun and Runge-Kutta methods have the same basic structure as that used to implement Euler's method. Here's the outline for Runge-Kutta.

> **Runge-Kutta method**
> (1) **Initialization:** $h := 0.1$; $t_0 := 2$; $y_0 := 1$; $f(t,y) = 4t^2\sqrt{y} - \frac{2}{t}y$.
> (2) **Loop defining y_k recursively and incrementing time:**
> for k from 1 to 10 (k an integer) do
>
> $t_k := t_{k-1} + h$;
>
> $s_1 := f(t_{k-1}, y_{k-1})$;
>
> $s_2 := f\left(t_{k-1} + \frac{1}{2}h, y_{k-1} + \frac{1}{2}s_1 h\right)$;
>
> $s_3 := f\left(t_{k-1} + \frac{1}{2}h, y_{k-1} + \frac{1}{2}s_2 h\right)$;
>
> $s_4 := f(t_k, y_{k-1} + s_3 h)$;
>
> $y_k = y_{k-1} + \frac{h}{6}(s_1 + 2s_2 + 2s_3 + s_4)$;
>
> end for.
> (3) **Output desired data:** Display sequence $(y_k, k = 0, \ldots, 10)$.

As we have indicated, the Runge-Kutta method is of order 4. Under the assumption that on $[t_0, t_*]$ the solution $y = y(t)$ to the initial value problem (3.12) is unique and has continuous fifth-order derivative, it is possible to show that there is a constant C_3 such that for any step size h and any positive integer k for which $t_k = t_0 + kh \leq t_*$, the total cumulative error $|y(t_k) - y_k|$ satisfies

(3.21) $$|y(t_k) - y_k| \leq C_3 h^4,$$

where y_k is the Runge-Kutta approximation of $y(t_k)$ with step size h.

Further modifications. There are a number of ways that the numerical methods we have described might be modified. For example, each of the numerical methods of solution we have considered—Euler, Heun, Runge-Kutta—are similar in that y_k is obtained using only y_{k-1} and the step size. Of these and other similar "one-step" methods, RK4 is often considered to be the standard general method, optimizing the trade-off between maximizing accuracy and minimizing computational work. *Multistep* numerical methods consider several preceding estimates (perhaps y_{k-1}, y_{k-2}, and y_{k-3}, assuming $k \geq 3$). Another possible modification would be to allow the step size h to vary in the course of obtaining the approximations $y_1, \ldots, y_N$ of the exact solution over $[t_0, t_*]$. More sophisticated methods adjust the size of the step as the calculation proceeds, using smaller steps over subintervals of $[t_0, t_*]$ where initial analysis suggests errors are relatively large (owing to rapid variation of the solution). These *adaptive methods* might estimate error over a subinterval of $[t_0, t_*]$ by comparing approximations generated by methods of different orders, say, p and $p + 1$, and when these approximations differ by an amount greater than a tolerance factor, then the step size is decreased until the difference is within tolerance.

3.2.1. Exercises. Use whatever technology is at your disposal to help with the computations in these exercises.

1. Solve the initial value problem (3.20) by, e.g., using a Bernoulli equation substitution.

2. Suppose $g = g(t)$ is a continuous, real-valued function on a finite interval $[a, b]$, and let $m = (a + b)/2$ be the midpoint of this interval.
   (a) Verify that
   $$p(t) = \frac{g(a)}{(a-m)(a-b)}(t-m)(t-b) + \frac{g(m)}{(m-a)(m-b)}(t-a)(t-b)$$
   $$+ \frac{g(b)}{(b-a)(b-m)}(t-a)(t-m)$$
   is a polynomial of degree at most 2 whose values agree with those of g at $a, (a+b)/2$, and b.
   (b) Show that for $p(t)$ as described in (a),
   $$\int_a^b p(t)\, dt = \frac{b-a}{6}\left(g(a) + 4g\left(\frac{a+b}{2}\right) + g(b)\right),$$
   which is the formula of Simpson's Rule, illustrated in Fig. 3.15.

3. Use Heun's method, as well as the Runge-Kutta method, both with step size $h = 0.25$, to approximate $y(1)$ given
   $$\frac{dy}{dt} = y, \quad y(0) = 1.$$

4. Use Heun's method, as well as the Runge-Kutta method, both with step size $h = 0.25$, to approximate $y(1)$ given
   $$\frac{dy}{dt} = \frac{\ln(1+t)}{\sqrt{1+y^5}}, \quad y(0) = 1.$$

5. Let $y = y(t)$ solve the initial value problem
   $$\frac{dy}{dt} = 2ty, \quad y(0) = 1.$$
   For each $j \geq 1$, let $h_j = 1/2^j$ and let a_j be the Heun method approximation of $y(1)$ with step size h_j. Find the least j such that a_j agrees with a_{j-1} to three decimal places.

6. Repeat the preceding exercise with the Runge-Kutta method replacing the Heun method. Find the exact solution $y(1)$ to three decimal places and compare it to your numerical solution.

3.2. Improving Euler's method: The Heun and Runge-Kutta Algorithms

7. This problem is a continuation of Exercise 2 of Section 3.1.1, which concerns the initial value problem

   (3.22)
   $$\frac{dy}{dt} = e^{-t^2}/y, \quad y(0) = 1.$$

   (a) Use Heun's method with a step size of $\frac{1}{10}$ to estimate $y(0.5)$, where y solves the initial value problem (3.22).
   (b) Use the Runge-Kutta method with step size of 0.25 to estimate $y(0.5)$, where y solves the initial value problem (3.22).
   (c) Compare your answers to (a) and (b) to that obtained for Exercise 2 of Section 3.1.1.
   (d) Solve (3.22) using your computer algebra system's default numerical solution method and compare the result to those examined in part (c). Of those examined in (c), which is closest to your computer algebra system's estimate?

8. (a) Explain why a numerical solution to the initial value problem
   $$\frac{dy}{dt} = \frac{4}{1+t^2}, \quad y(0) = 0,$$
   at $t = 1$ is an approximation for π.
   (b) Explain why in the Runge-Kutta method for this initial value problem, $s_2 = s_3$ at every step.
   (c) What approximation for π is obtained using a step size of $h = 0.25$ in the Runge-Kutta method?

9. Find a choice of function $f(t, y)$ and value y_0 so that the initial value problem
   $$\frac{dy}{dt} = f(t, y), \quad y(1) = y_0,$$
   has solution whose value at $t = 2$ is exactly $\ln 2$. Then use Runge-Kutta with step size $h = 1/10$ to estimate $\ln 2$.

10. When round-off error is also considered, inequality (3.18) estimating the error in Heun's method over $[t_0, t_*]$ with step size h becomes
    $$|y_k - y(t_k)| \leq Ch^2 + \frac{b}{h},$$
    where b is a small constant depending on machine precision and the length of $[t_0, t_*]$, but neither b nor the constant C depends on h. What value of h (in terms of C and b) minimizes the error estimate $Ch^2 + \frac{b}{h}$?

11. The initial value problem
    $$\frac{dy}{dt} = 1 - 2ty, \quad y(0) = 0,$$
    has exact solution
    $$y(t) = e^{-t^2} \int_0^t e^{u^2} du.$$

    (a) Use the Heun method with step size $h = 0.25$ to estimate $y(1)$.
    (b) Explain how your answer to (a) provides an estimate for $\int_0^1 e^{u^2} du$.

12. Part of the direction field for an equation $dy/dt = f(t, y)$ is shown below. If Euler's method and the Heun method are both used to approximate the value $y(2)$ for the initial value problem
    $$\frac{dy}{dt} = f(t, y), \quad y(1) = 3,$$
    with step size $h = 1$, which will be larger, the Euler approximation or the Heun approximation? Explain how you know.

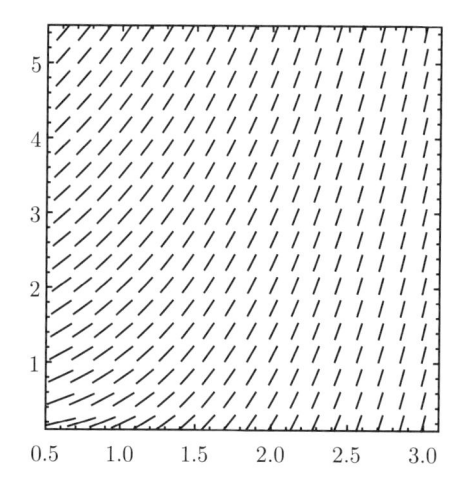

13. If we count the number of times the function $f(t,y)$ must be evaluated in one step (getting us from y_{k-1} to y_k) of a numerical method for solving

$$\frac{dy}{dt} = f(t,y), \quad y(t_0) = y_0,$$

we get a measure of the "computational cost" of the method. For example, one step of Euler's method requires only one such computation since

$$y_k = y_{k-1} + f(t_{k-1}, y_{k-1})h$$

so that f is evaluated once, at (t_{k-1}, y_{k-1}). How many evaluations of f are needed for one step of the Heun method? For one step of the Runge-Kutta method?

14. In this problem we investigate how the values s_1, s_2, s_3, and s_4 in the Runge-Kutta method each represent the slope of a particular field mark in the direction field for the equation $dy/dt = f(t,y)$. By "one Euler step from a point P with slope s" we mean "starting at P and moving along the line with slope s, go to the point at horizontal distance h from P"; see Fig. 3.16. Each of the following describes one of the numbers s_j, $j = 1, 2, 3$, or 4. Which one?
    (i) The slope of the field mark at (t_{k-1}, y_{k-1}).
    (ii) The slope of the field mark one Euler step from (t_{k-1}, y_{k-1}) with slope s_3.
    (iii) The slope of the field mark one-half Euler step from (t_{k-1}, y_{k-1}) with slope s_1.
    (iv) The slope of the field mark one-half Euler step from (t_{k-1}, y_{k-1}) with slope s_2.

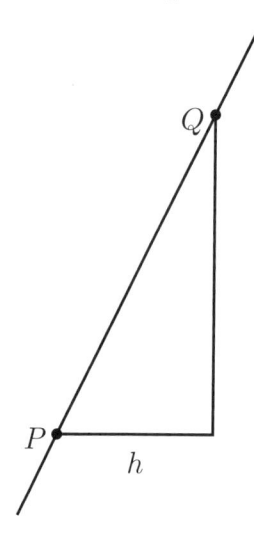

Figure 3.16. Point Q is one Euler step from point P, with slope s.

15. Example 3.1.2 featured the initial value problem
$$\frac{dy}{dt} = y\cos t, \quad y(0) = 1,$$
having exact solution $y = e^{\sin t}$.
   (a) Use the Heun method with a step size of $\frac{1}{10}$ on the interval $0 \leq t \leq 30$ to approximate $y(30)$ and compare the exact solution $y(30) = e^{\sin(30)}$ to your approximation.
   (b) Plot the entire Heun method approximation sequence $\{(t_k, y_k) : k = 0, 1, \ldots, 300\}$ along with the exact solution. Compare your plot to Fig. 3.8.

16. A water tank, filled with water, has the shape pictured below.

The outer surface of the tank has the shape generated by the curve $x = 1 + 2^{y-2}$, $0 \leq y \leq 3$, as it is rotated about the y-axis. Assume units along the coordinate axes are in meters. The cross-sectional area of the tank at level h meters, $0 \leq h \leq 3$, is thus given by $A(h) = \pi(1+2^{h-2})^2$ m². At time $t = 0$ seconds, nearly the entire bottom of the tank is removed, leaving a hole of radius $r = 1$ m. According to Torricelli's law the level of water $h(t)$ in the tank t seconds after the plug is removed satisfies
$$\frac{dh}{dt} = -\frac{0.6\sqrt{2(9.8)h}}{(1+2^{h-2})^2}, \quad h(0) = 3.$$
Attempt to find an approximate numerical solution of the preceding initial value problem over the interval $0 \leq t \leq 4$, using the Runge-Kutta method with step size $h = 1/10$. Identify a time interval of length $1/10$ second during which you have reason to believe the tank becomes empty (according to the model). Reduce the step size to $1/20$ and repeat the approximate solution attempt over $[0, 4]$. Does the interval you identified using step size $1/10$ still seem plausible?

17. You are to devise an "adaptive method" for evaluating numerically the integral
$$\int_0^1 \sqrt{1-t^2}\, dt.$$
You will subdivide the interval of integration $[0, 1]$ into 4 subintervals of varying length, with endpoints $0 = t_0 < t_1 < t_2 < t_3 < t_4 = 1$, and will use a simple, left-endpoint approximation for integration over the subintervals:
$$\int_{t_{k-1}}^{t_k} \sqrt{1-t^4}\, dt \approx \sqrt{1-t_{k-1}^2}\,(t_k - t_{k-1}), \quad k = 1, 2, \ldots, 4.$$
   (a) Which interval should be longer, $[t_0, t_1]$ or $[t_3, t_4]$?
   (b) Make what you believe to be reasonable choices for t_1, t_2, and t_3; then compute your numerical approximation based on these choices.
   (c) Compute $\int_0^1 \sqrt{1-t^2}\, dt$ numerically assuming a uniform step size of $1/4$ (with left-endpoint approximation over subintervals).
   (d) Compare the exact value of $\int_0^1 \sqrt{1-t^2}\, dt$ to your answers to (b) and (c).

3.3. Optical illusions and other applications

Optical illusions like those shown in Fig. 3.17 and Fig. 3.18 are called stable illusions of angle. Straight lines appear to be curved in these illusions. One explanation of these types of optical illusions comes from the hypothesis that we perceive angles as being closer to 90° than they actually are; this is an acquired trait which is a consequence of living in a "rectangular" world. Observe, for example, that if the line segment forming the top side of the square in Fig. 3.18 were bowed downward, all of the acute angles at the intersections with the circles would be slightly closer to 90° than they are with a straight line segment. Various models have been given that use this principle to determine a differential equation for the **apparent curves** in the illusion (the curves that you think you're seeing instead of a line). In this section we will solve some of these differential equation models, both analytically and numerically. For each illusion we consider, we will discuss two differential equation models for the apparent curves. For the most part we will concentrate on the *solution* of the various models, although we will briefly discuss how the models arise below. You can find more details on the *derivations* of the models, which use some elementary ideas from multivariable calculus, in the excellent article [**46**], on which this section is based.

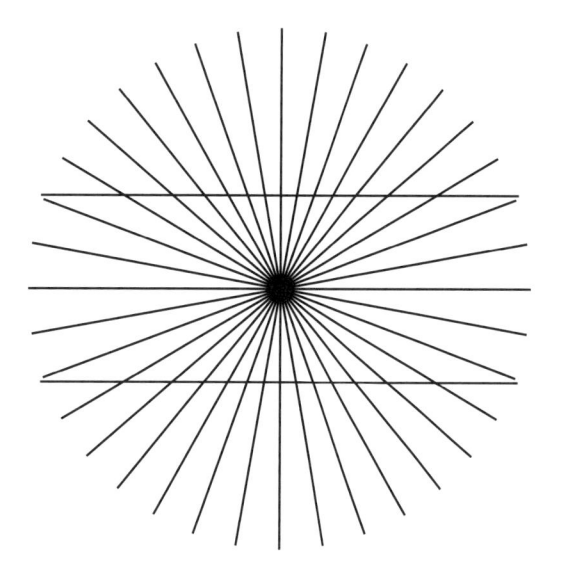

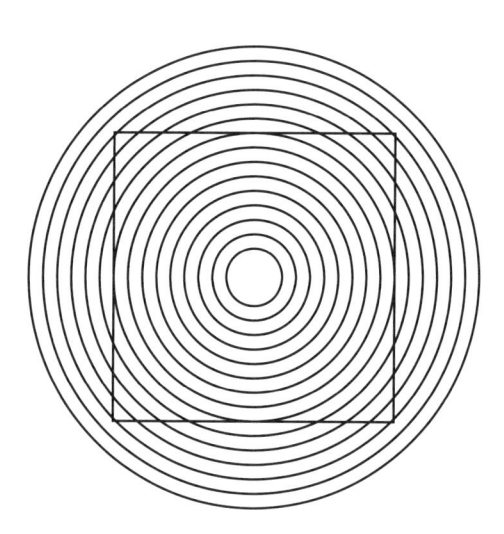

Figure 3.17. The Hering illusion.

Figure 3.18. The Orbison illusion.

The Hering illusion, first model. Fig. 3.17 shows the Hering illusion. The background family of lines passing through the origin is called the **distortion pattern**. The other curves in the figure are a pair of horizontal lines, which *appear* to be curves which bend away from the origin. According to one model, the differential equation for the "upper" apparent curve is

$$\frac{dy}{dx} = \frac{-\epsilon x}{1 + \epsilon y},$$

where ϵ is a small positive constant. This separable equation is easily solved to give

$$x^2 + y^2 + \frac{2}{\epsilon}y = C$$

for a constant C. Completing the square on y we can rewrite this as

$$x^2 + \left(y + \frac{1}{\epsilon}\right)^2 = C + \frac{1}{\epsilon^2}.$$

This is the equation of a circle centered at $(0, -\frac{1}{\epsilon})$ with radius $\sqrt{C + 1/\epsilon^2}$. As ϵ gets smaller and smaller, the center moves far down the y-axis and the radius increases. Notice that this does have the general features expected for the upper apparent curve.

The Hering illusion, second model. Another model for the Hering illusion gives rise to

(3.23) $$\frac{dy}{dx} = \frac{-\epsilon x}{\sqrt{x^2 + y^2} + \epsilon y}$$

as the differential equation of the apparent curves in Fig. 3.17. Here, ϵ is a (small) positive constant if we are describing the "upper" apparent curve in the figure and a negative constant if we are describing the lower one. While this is a homogeneous equation, trying to solve it that way quickly leads to a difficult integration. Instead, we'll solve it below numerically. Our plan is as follows: We'll set our scale so that the point $(0, 1)$ is the point of intersection of the top curve with the y-axis (and $(0, -1)$ for the lower curve). Using the Runge-Kutta method with $(x_0, y_0) = (0, 1)$, step size $h = 0.02$, and parameter value $\epsilon = 0.025$, we'll compute approximate solution points (x_k, y_k) for $k = 1, 2, \ldots, 100$. This will give 100 approximate solution points along the (approximate) apparent curve. Joining these closely spaced points by line segments gives the approximate apparent curve shown in Fig. 3.19. Reflecting this curve first across the y-axis and then reflecting across the x-axis fills out our picture; see Fig. 3.20. The relevant equations for implementing Runge-Kutta here are

$$y_k = y_{k-1} + \frac{h}{6}(s_1 + 2s_2 + 2s_3 + s_4)$$

where

$$s_1 \equiv f(x_{k-1}, y_{k-1}),$$

$$s_2 \equiv f\left(x_{k-1} + \frac{1}{2}h, y_{k-1} + \frac{1}{2}s_1 h\right),$$

$$s_3 \equiv f\left(x_{k-1} + \frac{1}{2}h, y_{k-1} + \frac{1}{2}s_2 h\right),$$

$$s_4 \equiv f(x_k, y_{k-1} + s_3 h)$$

for

$$f(x, y) = \frac{-\epsilon x}{\sqrt{x^2 + y^2} + \epsilon y}$$

and $x_k = x_0 + hk = (0.02)k$.

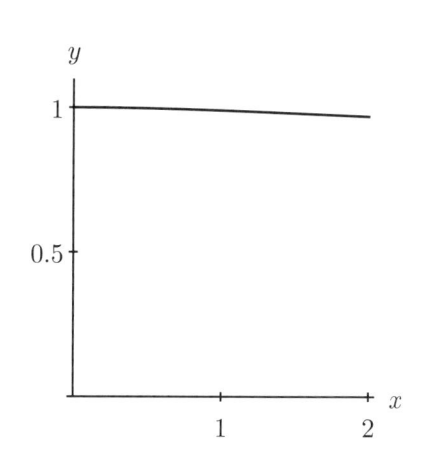

Figure 3.19. Approximate apparent curve for $x, y \geq 0$, $\epsilon = 0.025$.

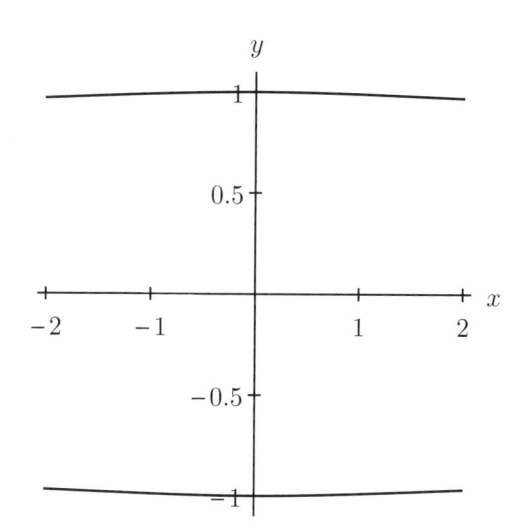

Figure 3.20. Approximate apparent curves, $\epsilon = 0.025$.

The Orbison illusion, first model. In one model, the apparent curves corresponding to the top side of the square (and it is a square) in Fig. 3.18 are solutions to the differential equation

$$(3.24) \qquad \frac{dy}{dx} = \frac{\epsilon y}{1 + \epsilon x}$$

where ϵ is a small constant, which is taken to be positive if $x > 0$ and negative if $x < 0$. Here we're assuming that the background circles forming the distortion pattern are centered at the origin and the sides of the square are parallel to the axes. Let's also specify a scale on the figure by assuming the top apparent curve crosses the y-axis at $y = 1$. Equation (3.24) is separable, and solving it with the initial condition $x = 0$, $y = 1$ gives $y = 1 + \epsilon x$ as the equation of the apparent curve, that is, a line with small positive slope if $\epsilon > 0$ is small and small negative slope if $\epsilon < 0$.

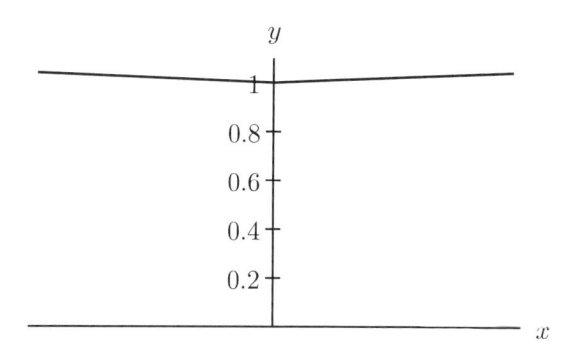

Figure 3.21. The apparent curve in the first model for the Orbison illusion.

The Orbison illusion, second model. Another model for the same illusion gives the equation

$$(3.25) \qquad \frac{dy}{dx} = \frac{\epsilon y}{\sqrt{x^2 + y^2} + \epsilon x}$$

for the apparent curves in Fig. 3.18. You can solve this as a homogeneous equation (see Exercise 1) and see that it gives a different apparent curve than the first Orbison model.

3.3. Optical illusions and other applications

The correction curves. So far our only method of evaluating the solutions produced by our models has been to subjectively ask if they give apparent curves that look like what we think we see in the illusion. There is a way to quantify this that also explains the role of the parameter ϵ that appears in the differential equation for the apparent curve. Recall that ϵ is a small constant, which could be positive or negative depending on the part of the figure. A **correction curve** for one of our illusions is the curve that results when we change the sign of ϵ in our model and then solve the resulting differential equation. If the graph of the correction curve is plotted against the distortion pattern, the illusion should disappear for some particular numerical value of ϵ. That is, the lines that appeared curved in the illusion are now replaced by curves (the correction curves) that appear straight. By experimenting with different numerical choices for ϵ we can try to find a choice for which the corrected curve best appears to be straight; most people come to general agreement on this numerical value. The corresponding value of $|\epsilon|$ is sometimes called the **strength** of the illusion.

For example, if we use the second model for the Hering illusion and replace ϵ by $-\epsilon$, we have the differential equation

$$(3.26) \qquad \frac{dy}{dx} = \frac{\epsilon x}{\sqrt{x^2 + y^2} - \epsilon y}.$$

Using $\epsilon = 0.035$, a Runge-Kutta solution to equation (3.26) with initial condition $y(0) = 1$ gives the top curve in Fig. 3.22 (the lower curve is then obtained by reflecting across the x-axis). For most people, these correction curves appear as straight lines in this picture; however if you put a straightedge along them you will see that this is not so.

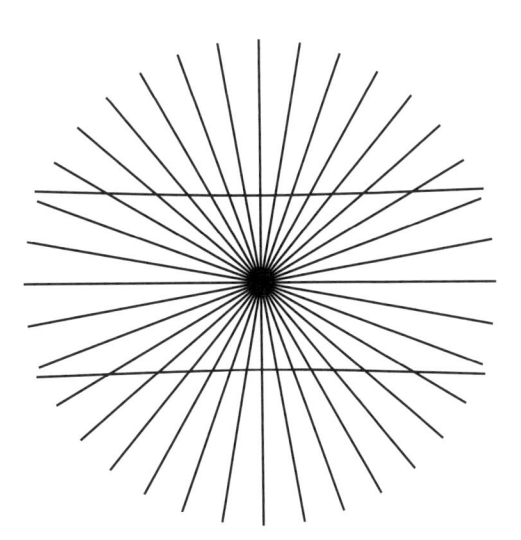

Figure 3.22. The correction curves, $\epsilon = 0.035$.

Derivation of the Hering models. We will outline how the two models discussed above for the Hering illusion (Fig. 3.17) arise. This requires the notion of normal and tangent vectors to curves in the plane. Focus on a point P of intersection between one of the lines in the distortion pattern and the top horizontal line. Fig. 3.23 shows such a point, with a normal vector $N = (0, 1)$ to the horizontal line and a tangent vector T to the distortion line through P. Notice that if the coordinates of P are (x, y), then the vector (x, y) can serve as a tangent vector T to the distortion line through P.

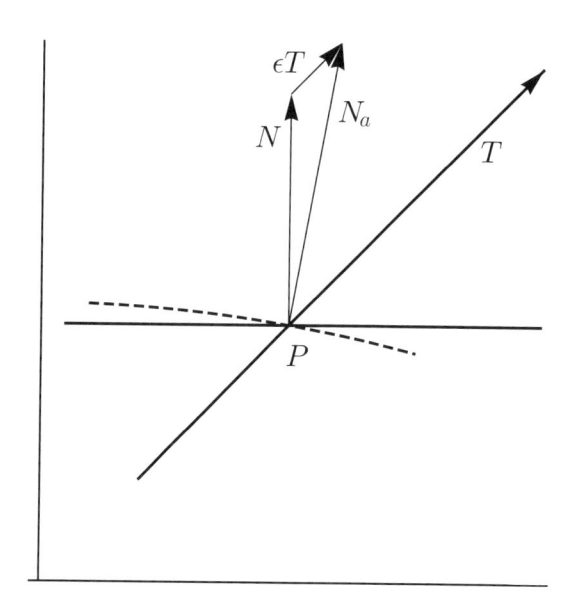

Figure 3.23. The apparent normal vector N_a and the apparent curve (dashed) at P.

The basic principle is that the apparent curve at P, which is dashed in Fig. 3.23, makes an angle a little closer to $90°$ with the distortion line passing through P than the actual curve (the horizontal line) does. This is the same thing as saying that a normal vector N_a to the apparent curve at P is $N_a = N + \epsilon T$ for some small positive ϵ; see Fig. 3.23. Thus the slope of the apparent curve at P is the negative reciprocal of the slope of $N_a = (0,1) + \epsilon(x,y)$. Writing this last sentence as an equation gives

$$\frac{dy}{dx} = -\frac{\epsilon x}{1+\epsilon y}.$$

This is the differential equation for the "upper" apparent curve (first model), where ϵ is a small positive constant.

In the model just discussed, we used $T = (x, y)$ for the tangent vector to the distortion line at the point $P(x,y)$. If we multiply T by a positive scalar, we get another choice for a tangent vector. Different choices give different models. A "natural choice" would be to choose the tangent vector to have length one; this gives

$$\left(\frac{x}{\sqrt{x^2+y^2}}, \frac{y}{\sqrt{x^2+y^2}} \right)$$

as our new choice for T. In Exercise 4 you are asked to check that this yields the differential equation in the second model of the Hering illusion. Starting with Exercise 5 you are invited to use numerical-solution techniques in applications unrelated to optical illusions.

3.3.1. Exercises.

1. (a) Find an implicit solution of the differential equation

$$\frac{dy}{dx} = \frac{\epsilon y}{\sqrt{x^2+y^2} + \epsilon x}$$

by means of the substitution $v = y/x$. The integration formula

$$\int \frac{1}{v\sqrt{v^2+1}} \, dv = -\ln\left|\frac{1+\sqrt{1+v^2}}{v}\right|$$

may be useful.

(b) (CAS) Give, implicitly, the solution to the differential equation in (a) with $\epsilon = \frac{1}{20}$ that satisfies the initial condition $y(0) = 1$. Using a computer algebra system, sketch the graph of this solution for $0 < x < 2$, $0 < y < 2$.

2. (CAS) Give the Runge-Kutta solution to equation (3.25) using $(x_0, y_0) = (0, 1)$, step size $h = \frac{1}{10}$, and parameter value $\epsilon = 0.05$, computing (x_k, y_k) for $k = 1, 2, \ldots, 20$. Sketch the resulting approximate solution points joined by line segments to get an approximate sketch of the solution curve for $0 \leq x \leq 2$.

3. (a) Show that the solutions to
$$\frac{dy}{dx} = -\frac{x}{y}$$
are the family of circles centered at $(0, 0)$, the distortion pattern in the Orbison illusion. Explain why $T = (-y, x)$ is a tangent vector to the circle passing through the point $P = (x, y)$.

(b) Choosing $N = (0, 1)$ as a normal vector to a horizontal line and
$$N_a = N + \epsilon T = (0, 1) + \epsilon(-y, x)$$
as an apparent normal to the upper apparent curve ($x \geq 0$) in the Orbison illusion, give a differential equation for this apparent curve.

4. Show that using the tangent vector
$$T = \left(\frac{x}{\sqrt{x^2 + y^2}}, \frac{y}{\sqrt{x^2 + y^2}}\right)$$
to derive the differential equation of the apparent curve in the Hering illusion gives equation (3.23).

5. (CAS) A population whose growth varies seasonally is modeled by the initial value problem
$$\frac{dP}{dt} = P - 0.1P^2(1 + \cos(2\pi t)), \quad P(0) = 3.$$
Using the Runge-Kutta method with $(x_0, y_0) = (0, 3)$ and a step size $h = \frac{1}{10}$, compute (x_k, y_k) for $k = 1, 2, \ldots, 40$. Using the computed approximate solution points, sketch the approximate solution curve for $0 \leq t \leq 4$.

6. (CAS) Differential equation models are used to estimate blood alcohol concentration $x(t)$ after a person consumes alcoholic drinks. One important model, called the Wagner model, uses the equation
$$\frac{dx}{dt} = u - \frac{1}{V}\frac{cx}{k + x}$$
during the "drinking" phase (assuming the drinking occurs at a constant rate) and
$$\frac{dx}{dt} = -\frac{1}{V}\frac{cx}{k + x}$$
during the "recovery" phase (after drinking is stopped). In these equations, u, V, c, and k are constants, described as follows:

- The constants u and V are individualized to the person and situation: u is the (constant) rate at which alcohol is being consumed (in grams per hour) divided by the number of deciliters (dL) of body water V in the drinker's body. The amount of body water depends on the person's weight and gender.[1] Typical formulas used for body water volume (in dL) is $V = 6.8W$ for men and $V = 5.5W$ for women, where W is the person's mass in kg.

---
[1] Women have a higher percentage of body fat than men do, and since alcohol does not dissolve into body fat, the "body water" value per kg of body mass is lower for women than for men.

- The constants c and k are related to kidney and liver function, and we will use the values $c = 10.2$ g/hr and $k = 0.0045$ g/dL.[2]

(a) Sally, a 120 lb woman (54 kg) drinks 4 glasses of wine, each containing 10 g of alcohol, at a steady rate during a two-hour period. Using either a Runge-Kutta approximation or the built-in numerical solver of your computer algebra system, sketch the graph of Sally's blood alcohol concentration (BAC) for $0 \leq t \leq 2$ and for $2 \leq t \leq 8$. What is her BAC at the end of the drinking phase? How long a recovery period is needed for Sally's BAC to drop below 0.02? Hint: $V = (5.5)(54) = 297$ dL and $u = 20/297$ g/(hr-dL).

(b) Sam, a 160 lb man (72 kg) consumes 6 ounces of whiskey (10 g of alcohol in each ounce) during a two-hour period. What is his BAC at the end of this drinking phase, and how long a recovery period is needed for his BAC to drop below 0.02? As in part (a) produce a graph of Sam's BAC for $0 \leq t \leq 2$ and for $2 \leq t \leq 8$ to answer these questions.

7. (CAS) Excessive input of phosphorus into a lake (from agricultural runoff, for example) is a primary cause of lake eutrophication, an undesirable state of murky water, oxygen depletion, and toxicity. Even if phosphorus input into the lake can be decreased, eutrophication cannot always be reversed, due to the recycling of phosphorus within the lake (from phosphorus accumulation in lake sediment). The differential equation[3]

$$(3.27) \qquad \frac{dP}{dt} = L - sP + \frac{rP^q}{m^q + P^q}$$

is used to model phosphorus levels P at time t. The parameters s, r, q, and m are positive constants. The first term on the right-hand side, L, is the input of phosphorus to the lake; we will assume this to be constant. The second term, $-sP$, represents the loss of phosphorus through outflow, sedimentation, and other processes. The term $rP^q/(m^q + P^q)$ represents the recycling of phosphorus back into the lake.

(a) The equilibrium solutions of equation (3.27) are found by solving

$$L - sP + \frac{rP^q}{m^q + P^q} = 0.$$

With specific values for the constants r, m, q, s, and L, this can be done numerically using a CAS or graphically. Fig. 3.24 shows the graphs of the curve with equation

$$(3.28) \qquad y = L + \frac{rP^q}{m^q + P^q}$$

and the line

$$y = sP$$

for typical values of r, m, q, s, and L. How many equilibrium solutions are there? Classify each as stable or unstable.

---

[2] These values, as well as the differential equations for the model, are from *Modeling intake and clearance of alcohol in humans* by Andre Heck, Electronic Journal of Mathematics and Technology, Vol. 1, No. 3, 232–244.

[3] The differential equation and parameter values in this problem are taken from *Management of eutrophication for lakes subject to potentially irreversible change* by S. Carpenter, D. Ludwig, and W. Brock, Ecological Applications, 9(3), 1999, 751–771.

3.3. Optical illusions and other applications

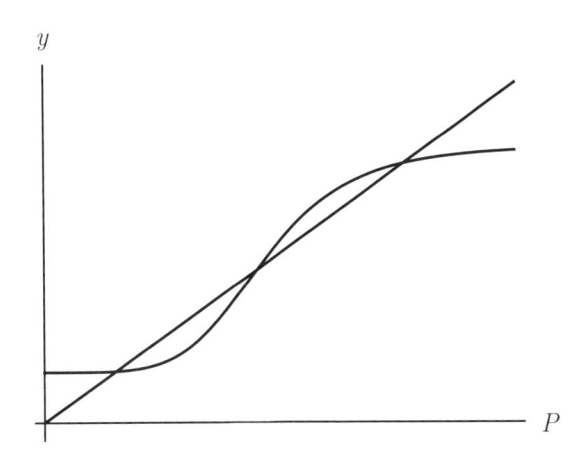

Figure 3.24. Finding the equilibrium solutions graphically.

(b) As the constant L is varied, the graph of (3.28) is translated up or down. Using Fig. 3.24, show that for either a sufficiently small value of L or a sufficiently large L there may be only one equilibrium solution. Classify it (in both the small L and large L cases) as stable or unstable.

(c) Phosphorus levels in Lake Mendota (located on the edge of the University of Wisconsin campus) have been much studied. Estimates for the relevant parameters are

$$s = 0.817, \quad r = 731{,}000 \text{ kg/yr}, \quad m = 116{,}000 \text{ kg}, \quad q = 7.88.$$

Using a computer algebra system, sketch on the same axes the curves $y = 0.817P$ and

$$y = 4{,}000 + \frac{731{,}000 P^{7.88}}{116{,}000^{7.88} + P^{7.88}}$$

(this is equation (3.28) with the specified values for the parameters and $L = 4{,}000$ kg/yr) for $0 < P < 1{,}000{,}000$. By zooming in on the relevant portions of this graph, estimate the three equilibrium solutions for the differential equation

$$\frac{dP}{dt} = 4{,}000 - 0.817P + \frac{731{,}000 P^{7.88}}{116{,}000^{7.88} + P^{7.88}}.$$

(d) By using either a Runge-Kutta approximation or the built-in numerical solver of your computer algebra system, sketch the solution to the initial value problem

$$(3.29) \qquad \frac{dP}{dt} = 4{,}000 - 0.817P + \frac{731{,}000 P^{7.88}}{116{,}000^{7.88} + P^{7.88}}, \quad P(0) = P_0,$$

on the interval $0 < t < 10$, for the three values $P_0 = 900$ kg, $P_0 = 9{,}000$ kg, and $P_0 = 90{,}000$ kg. In each case, the values of P should appear to approach one of the stable equilibrium values you found in (c). The larger stable equilibrium value corresponds to the undesirable eutrophic state, while the smaller equilibrium value corresponds to a healthier state.

(e) The values of the parameters s, m, r, and q can be difficult to estimate accurately. The parameter q typically ranges from about $q = 2$ for deep, cold lakes, to $q = 20$ for warm, shallow lakes. If the q-value in equation (3.29) is changed to $q = 2$, how does this change the equilibrium solutions, and what *practical* effect does this have?

8. (CAS) In this problem, we consider the cow-grazing model of Exercise 24 of Section 2.6 with a herd size $H = 20$:

$$(3.30) \qquad \frac{dV}{dt} = \frac{1}{3}V - \frac{1}{75}V^2 - 20\frac{(0.1)V^2}{9 + V^2}.$$

In this equation time t is measured in days and $V(t)$ is the amount of vegetation in the field, in tons, at time t.

(a) Estimate the values of the three positive equilibrium solutions $e_1 < e_2 < e_3$ of equation (3.30) by either solving the equation
$$\frac{1}{3}V - \frac{1}{75}V^2 - 20\frac{(0.1)V^2}{9+V^2} = 0$$
numerically or by zooming in on appropriate portions of the graph of its left-hand side over the interval $0 < V < 20$.

(b) Draw a phase line for equation (3.30). Using the phase line, confirm that if V solves (3.30) and satisfies $V(0) = 10$, then
$$\lim_{t \to \infty} V(t) = e_3.$$

(c) Assume, as in part (b), that $V(0) = 10$ and that V solves (3.30). Use the Runge-Kutta method with step size $h = \frac{1}{4} = 0.25$ to find approximate values of V over the interval $0 \leq t \leq 100$, thus at times $t_k = kh$ for $k = 0$ to 400. Plot the resulting approximation sequence. Based on the approximate value of e_3 from part (a) as well as the phase line and limit from part (b), do you believe the approximation sequence for V obtained by the Runge-Kutta method with $h = 0.25$ appears plausible? Explain.

(d) Based on your Runge-Kutta approximation of part (c), estimate the time when the vegetation level in the field reaches 15 tons.

(e) Again assume that V solves the initial value problem consisting of (3.30) with $V(0) = 10$. Use the Euler method with step size $h = 0.25$ to find approximate values of V over the interval $0 < t < 100$, thus at times $t_k = kh$ for $k = 0$ to 400. Compare the Euler approximation sequence to that obtained by the Runge-Kutta method, e.g., by plotting $(t_k, V_k - \tilde{V}_k)$ for $0 \leq k \leq 400$, where V_k is produced by the Runge-Kutta method while $\tilde{V}_k$ is produced by the Euler method (both with step size 0.25). Based on this comparison, would you expect the Euler approximation sequence to change significantly if the step size were reduced from $h = 1/4$ to $h = 1/8$?

Chapter 4

Higher-Order Linear Homogeneous Equations

4.1. Introduction to second-order equations

We introduce the topic of this chapter with several modeling problems that lead to second-order differential equations.

In the first example, we will be looking at an object falling in a straight line to the earth and subject to two forces: its weight (the force due to gravity) and a drag force (or force due to air resistance). We make two assumptions about these forces. First, we assume that the object is close enough to the surface of the earth that its weight W is constant:

$$W = mg,$$

where m is the mass of the object and g is the (constant) acceleration due to gravity. The gravitational force acts downward, that is, towards the center of the earth. Second, we assume that the drag force is proportional to the speed of the object and acts opposite to the direction of motion. This is a reasonable assumption for low speeds only; for higher speeds we might need to assume the drag force is proportional to some higher power of the speed. For sufficiently large speeds, a drag force proportional to the square of the speed gives a better, although somewhat more complicated, model. Thus, in general, we assume that the magnitude of the drag force takes the form $c|v|^p$, where v is the object's velocity, $1 \leq p \leq 2$, and c is a positive constant that we call the *drag coefficient*. The drag coefficient typically depends on the object's shape and size and the roughness of its surface,[1] but not its mass. A skydiver will have a low drag coefficient until her parachute opens, at which point the drag coefficient may increase by a factor of 10 or more. With these basic principles in mind, we turn to an example.

Example 4.1.1. According to the Guinness Book of World Records, Paul Tavilla holds the record for catching in his mouth grapes dropped from great heights (see http://thegrapecatcher.com). In 1988 he caught a grape dropped from the top of the 60-story John Hancock Building in Boston, a distance of 782 feet, beating his previous record of a 660 foot catch from a skyscraper in Tokyo. Just how dangerous a stunt was this? We'd like to determine the speed of the grape as Mr. Tavilla catches it in his mouth (or as it hits him somewhere else).

---

[1] Perhaps contrary to intuition, roughness of the surface can sometimes reduce drag. This is why golf balls are dimpled. At speeds that a golf ball is driven, the drag is less with the standard dimpled ball than it would be with a smooth ball of the same diameter.

By Newton's second law, the total force acting on our object is equal to the product of the mass of the object and its acceleration due to the force: $F = ma$. We're going to be assuming that our object, one of Mr. Tavilla's black ribier grapes, is falling in a straight line, and we'll introduce a coordinate system so that the object moves along a number line, oriented so that the origin is located at the point from which the grape is dropped and the positive direction is downward. This means that the gravitational force, mg, will be positive, while the drag force, acting in the upward direction, will be negative. By our discussion above, we model the drag force as $-cv$, where c is the drag coefficient and v is the (positive) velocity of the falling grape. By Newton's law we have

(4.1) $$ma = mg - cv$$

where a is the acceleration of the grape.

Figure 4.1. The grape catcher.

We let $y(t)$ denote the position of the grape along our axis at time t. Keep in mind that velocity is the rate of change of position with respect to time, so that $v = \frac{dy}{dt}$, and acceleration is the rate of change of velocity with respect to time, so that

$$a = \frac{dv}{dt} = \frac{d^2y}{dt^2}.$$

Thus equation (4.1) becomes

$$m\frac{d^2y}{dt^2} = mg - c\frac{dy}{dt}.$$

Dividing by the mass m we have

$$\frac{d^2y}{dt^2} + \frac{c}{m}\frac{dy}{dt} = g,$$

or more compactly,

(4.2) $$y'' + \frac{c}{m}y' = g.$$

This is a *second-order* differential equation for the position function $y(t)$. We will use it, together with information about the initial position and velocity of the grape, to answer the question, "What is $y'(t)$ when $y(t) = 78$?"

Notice that we can rewrite equation (4.2) using the substitution $v = y'(t)$, and we're then left with a *first*-order linear equation:

$$v' + \frac{c}{m}v = g,$$

with unknown function $v = v(t)$. If we use the British system of units and measure distance in feet and time in seconds, the approximate value of g is 32 ft/sec^2 (see Table 4.1 for a comparison of various systems of units).

4.1. Introduction to second-order equations

Table 4.1. Comparing Systems of Units.					
	Time	Distance	Mass	Force	g
British	sec	foot (ft)	slug	pound (lb)	32 ft/sec^2
MKS (SI)	sec	meter (m)	kilogram (kg)	Newton (N) kg-m/sec^2	9.8 m/sec^2
CGS	sec	centimeter (cm)	gram (gm)	dyne gm-cm/sec^2	980 cm/sec^2

Reasonable choices for the weight of the grape and the drag coefficient c are

$$\text{weight } = 0.02 \text{ lbs}, \quad c = 1.2 \times 10^{-4} \text{ lb-sec/ft}.$$

Therefore its mass, m, is $0.02/32$ (slugs), and $c/m = 0.192$. Thus our differential equation now looks like

(4.3) $$v' + 0.192v = 32.$$

Using the integrating factor $e^{0.192t}$, we solve equation (4.3) and see that (approximately)

$$v(t) = 167 + c_1 e^{-0.192t}.$$

If we assume the grape is dropped at time $t = 0$, then $v(0) = 0$, so that $c_1 = -167$. Since $y'(t) = v(t)$ we have

$$y'(t) = 167 - 167 e^{-0.192t}$$

so that an integration gives

$$y(t) = 167t + \frac{167}{0.192} e^{-0.192t} + c_2.$$

Since we have $y(0) = 0$ by the orientation of our number line,

$$\frac{167}{0.192} + c_2 = 0$$

and $c_2 = -167/0.192 \approx -870$. Our position function $y(t)$ is (approximately)

(4.4) $$y(t) = 167t + 870 e^{-0.192t} - 870.$$

We want to determine the velocity of the grape at the instant when $y(t) = 782$. Setting $y(t) = 782$ in equation (4.4), we have

$$782 = 167t + 870 e^{-0.192t} - 870,$$

or

(4.5) $$167t + 870 e^{-0.192t} - 1{,}652 = 0.$$

Instead of trying to solve this exactly for t, we will be content with an approximate solution for t, the time the grape reaches Mr. Tavilla's mouth. This can be done graphically (see Fig. 4.2) or by calculating that when $t = 8.9$ the left-hand side of equation (4.5) is negative and when $t = 9.0$ it is positive, so by the Intermediate Value Theorem, there is a solution to equation (4.5) between $t = 8.9$ and $t = 9.0$. Thus the velocity of the grape when he catches it is approximately $v(8.95) = 137$ ft/sec, or approximately 93 miles per hour.[2] This is about the speed (crossing home plate) of the fastest baseballs thrown by top pitchers.

---

[2] The rough approximations 1 m/sec ≈ 2 miles per hour and 3 ft/sec ≈ 2 miles per hour may give a more intuitive feel for speeds in this section.

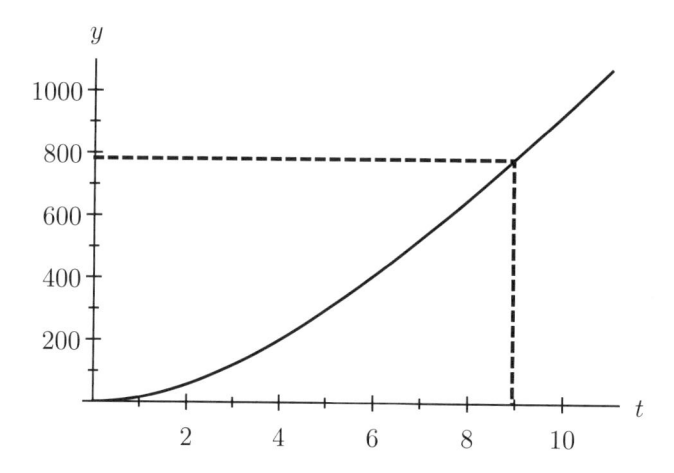

Figure 4.2. $y = 167t + 870e^{-0.192t} - 870$.

Notice that no matter what the height from which the grape is dropped, our model says that its speed never exceeds 167 ft/sec:

$$v(t) = 167 - 167e^{-0.192t} \leq 167$$

for all $t > 0$. In fact,

$$\lim_{t \to \infty} \left(167 - c_1 e^{-0.192t} \right) = 167.$$

This limiting value is called the **terminal velocity**; see Exercise 5 for further discussion of this.

Some basic terminology. With Example 4.1.1 in hand, let's look at some relevant terminology. A second-order differential equation has the "normal form"

(4.6) $$y''(t) = F(t, y(t), y'(t))$$

for an unknown function $y(t)$; the right-hand side of (4.6) is an expression in terms of t, y, and y'. The equation is classified as **linear** if it can be written in the form

(4.7) $$P(t)y'' + Q(t)y' + R(t)y = G(t).$$

This is the same as saying in (4.6) that

$$F(t, y(t), y'(t)) = \frac{G(t) - Q(t)y'(t) - R(t)y(t)}{P(t)}.$$

Equation (4.7) is said to be a **homogeneous linear** equation if $G(t) \equiv 0$. In Example 4.1.1, equation (4.2) is linear, but not homogeneous.

We were able to solve the second-order differential equation in (4.2) because the absence of a "$y(t)$" term made it possible to convert the equation to a *first*-order equation that we could solve because it is linear. We describe this type of equation as one with "the *dependent* variable missing". In the next example we'll have a *nonlinear* second-order equation with the dependent variable missing.

Example 4.1.2. It's early morning at the shore. A smuggler is offloading contraband onto the beach at the origin in the xy-plane. The x-axis is the shoreline. One mile to the right along the shoreline waits a Coast Guard boat. At time $t = 0$, the fog lifts. The smuggler and the Coast Guard spot each other. Instantly, the smuggler heads for open seas, fleeing up the y-axis. Also, instantly, the Coast Guard pursues. The speed of both boats is a mph. What path does the Coast Guard take to try to catch the smuggler? This path is called a **pursuit curve**, and the understanding is that the Coast Guard helmsman always aims his boat *directly at* the smuggler.

4.1. Introduction to second-order equations

Note that at time t, the distance s traveled by the Coast Guard and the distance at traveled by the smuggler are the same.

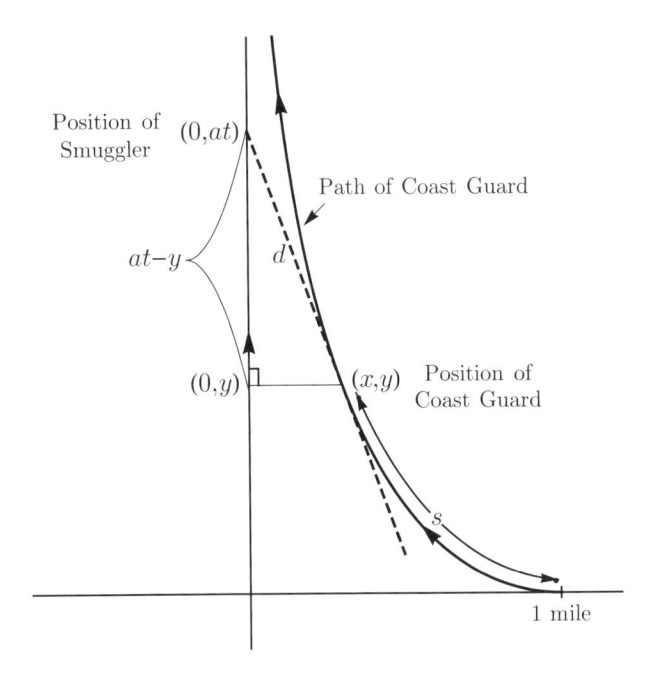

Figure 4.3. The smuggler and the Coast Guard.

It is apparent from Fig. 4.3 that the slope of the dashed tangent line can be written two ways:
$$\frac{dy}{dx} = \frac{y - at}{x},$$
where the graph of $y = y(x)$ is the path of the Coast Guard. This is a first-order differential equation but with too many variables! Instead, write the equation as
$$x\frac{dy}{dx} = y - at,$$
and differentiate this equation with respect to x to get
$$x\frac{d^2y}{dx^2} + \frac{dy}{dx} = \frac{dy}{dx} - a\frac{dt}{dx},$$
which simplifies to

(4.8) $$x\frac{d^2y}{dx^2} = -a\frac{dt}{dx}.$$

Now, the arc-length formula from calculus tells us that the distance s that the Coast Guard has traveled along its path is given by
$$s = \int_x^1 \sqrt{1 + y'(u)^2}\, du$$
where y' denotes the derivative $\frac{dy}{dx}$. Since both boats have the same speed a, they have both traveled the same distance at time t: $at = s$. Thus,
$$at = s = \int_x^1 \sqrt{1 + y'(u)^2}\, du = -\int_1^x \sqrt{1 + y'(u)^2}\, du.$$

Differentiate this equation with respect to x to get

(4.9) $$a\frac{dt}{dx} = \frac{ds}{dx} = -\sqrt{1 + y'(x)^2};$$

note that we've used the Fundamental Theorem of Calculus to compute

$$\frac{d}{dx}\int_1^x \sqrt{1 + y'(u)^2}\, du = \sqrt{1 + y'(x)^2}.$$

Plugging (4.9) back into equation (4.8) gives

$$x\, y''(x) = \sqrt{1 + y'(x)^2},$$

a nonlinear second-order differential equation.

Pause to consider what we have done. Our first differential equation had too many variables: y, x, and t. We tried differentiating to see if we could get rid of t. We did (using the arc-length formula). This was just trial and error, but it worked, at the expense, however, of ending up with a *second-order* equation. Luckily, it fits a situation we can handle—a second-order equation with the dependent variable ("y") missing. Let's solve it. Letting $w(x) = y'(x)$, we get $w'(x) = \frac{dw}{dx} = \frac{d^2y}{dx^2}$, so

$$xw' = \sqrt{1 + w^2},$$

a first-order differential equation with unknown $w(x)$. We separate variables and integrate to get

$$\int \frac{dw}{\sqrt{1 + w^2}} = \int \frac{dx}{x}.$$

For the integral on the left, use the trigonometric substitution $w = \tan\theta$, $dw = \sec^2\theta\, d\theta$. Then, by the trigonometric identity $1 + \tan^2\theta = \sec^2\theta$,

$$\sqrt{1 + w^2} = \sqrt{1 + \tan^2\theta} = \sqrt{\sec^2\theta} = |\sec\theta| = \sec\theta.$$

We find

$$\int \frac{dw}{\sqrt{1+w^2}} = \int \frac{\sec^2\theta}{\sec\theta}\, d\theta = \int \sec\theta\, d\theta = \ln|\sec\theta + \tan\theta| + c.$$

Substituting back with $\sec\theta = \sqrt{1 + w^2}$, $\tan\theta = w$, we get

(4.10) $$\ln\left(\sqrt{1+w^2} + w\right) = \ln x + c_1.$$

What about initial conditions? At time $t = 0$, the position (x, y) of the Coast Guard is $(1, 0)$ (one mile up the beach, right on the shore's edge). At the first instant of pursuit, it heads horizontally toward the smuggler, who is at the origin $(0, 0)$. Hence, $w(0) = y'(0) = 0$. Thus our initial conditions are $t = 0$, $x = 1$, $w = y' = 0$, and we have

$$\ln\left(\sqrt{1+0} + 0\right) = \ln 1 + c_1,$$

so $c_1 = 0$. Using this in (4.10) we see that $\sqrt{1 + w^2} + w = x$, so

$$1 + w^2 = (x - w)^2 = x^2 - 2xw + w^2,$$

or $2xw = x^2 - 1$. Solving for w we get

$$y'(x) = w(x) = \frac{x^2 - 1}{2x} = \frac{1}{2}\left(x - \frac{1}{x}\right).$$

Integrate this to get

$$y(x) = \frac{x^2}{4} - \frac{1}{2}\ln x + c_2.$$

4.1. Introduction to second-order equations

Using the initial condition $y(1) = 0$, we get $0 = \frac{1}{4} - \frac{1}{2}\ln 1 + c_2$, or $c_2 = -\frac{1}{4}$. So

$$y(x) = \frac{x^2}{4} - \frac{1}{2}\ln x - \frac{1}{4}.$$

This is the equation of the *pursuit curve*, the path the Coast Guard follows. Its graph is shown in Fig. 4.4.

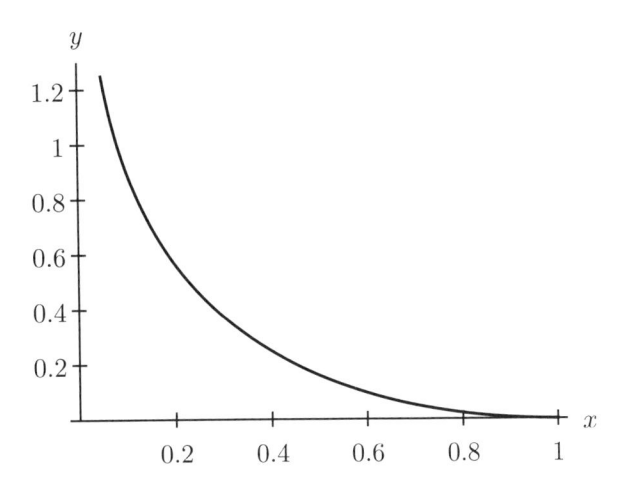

Figure 4.4. The pursuit curve.

Since the two boats in this example have the same speed, it should be clear that the Coast Guard never catches the smuggler. How close does it get? You're asked about this in Exercise 15. Then in Exercise 14 you'll modify this model to handle the case that the Coast Guard is faster than the smuggler.

In the preceding two examples, the dependent variable y did not explicitly appear in the differential equation for y'', allowing us to reduce to a first-order differential equation. Next we consider second-order equations with the special form

(4.11) $$y'' = f(y),$$

in which neither y' nor the independent variable t explicitly appear and f is a continuous function. Again the key is to reduce to a first-order equation. To see how this goes, multiply both sides of equation (4.11) by $2y'$ to obtain

(4.12) $$2y'y'' = 2y'f(y).$$

Keep in mind that the primes denote differentiation with respect to t:

$$y' = \frac{dy}{dt} \quad \text{and} \quad y'' = \frac{d^2y}{dt^2}.$$

Now on the left-hand side of (4.12), $2y'y''$ is the derivative of $(y')^2$ by the chain rule:

$$\frac{d}{dt}((y')^2) = 2y'y''.$$

If F is an antiderivative of the function f (meaning $\frac{dF}{dy} = f(y)$), then the right-hand side of (4.12) is the derivative of $2F(y)$:

$$\frac{d}{dt}(2F(y)) = 2F'(y)y' = 2f(y)y'.$$

Again, we used the chain rule in this computation. Thus equation (4.12) says

$$\frac{d}{dt}((y')^2) = \frac{d}{dt}(2F(y)),$$

and since the derivatives are equal, the functions must differ by a constant:

(4.13) $$(y')^2 = 2F(y) + c.$$

Notice that we've reduced the second-order equation (4.11) to a first-order equation. Let's look at a concrete example.

Example 4.1.3. A mass m is attached to a horizontal spring lying on a frictionless table as shown in Fig. 4.5. The other end of the spring is attached to a stationary wall. We set up a horizontal number line so that the mass is at the origin when the system is at rest (with the spring at its natural length) and the positive direction is to the right. We put the mass in motion by pulling it out and letting it go. We ignore any friction or air resistance. The force acting on the mass is the **restoring force** due to the spring; the term "restoring" is used as this force acts to try to restore the spring to its natural length. At any moment when the mass is to the right of its rest position, so that the spring is stretched from its natural length, the restoring force acts to the left (or in the negative direction) and when the mass is to the left of its rest position (the spring is compressed from its natural length), the restoring force acts to the right, in the positive direction.

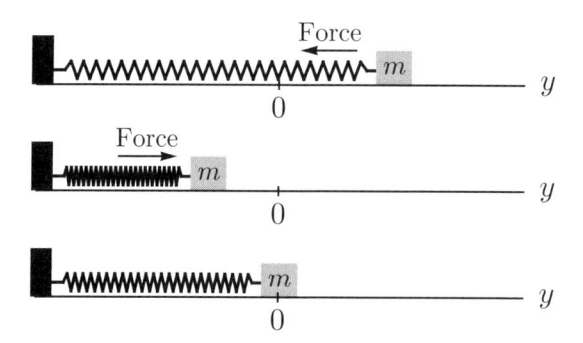

Figure 4.5. A horizontal mass-spring system.

For an "ideal" spring, Hooke's law[3] states that the magnitude of the restoring force is proportional to the displacement of the spring from its natural length. If we call the (positive) proportionality constant k and if we let $y(t)$ denote the position of the mass at time t (remembering that a positive value of y means the mass is to the right of its rest position), then an expression for the restoring force is $-ky(t)$. You should check that the minus sign gives the correct direction for the restoring force whether the spring is compressed or stretched. This is the only force at play in our simplified model; so by Newton's second law we have

(4.14) $$my'' = -ky$$

to model this mass-spring system. This is sometimes called the equation of an (undamped) harmonic oscillator; the meaning of this terminology should become clearer soon. If we write equation (4.14) as

$$y'' = -\frac{k}{m}y,$$

---

[3]Experimentally observed by the seventeenth-century British physicist Robert Hooke, who first stated this law as a Latin anagram "ceiiinosssstttuu". Two years later, in 1678, Hooke published the solution to this anagram, "Ut tensio, sic vis," which translates as, "As the displacement, so the force."

4.1. Introduction to second-order equations

we see that this equation has the form of equation (4.11) with

$$f(y) = -\frac{k}{m}y,$$

and for now we will solve this using equation (4.13). Later, in Section 4.4, we will see another, admittedly quicker, way to solve (4.14). Since the function $f(y) = -\frac{k}{m}y$ has antiderivative

$$F(y) = -\frac{k}{2m}y^2,$$

equation (4.13) tells us that

$$(y')^2 = -\frac{k}{m}y^2 + C.$$

Suppose that you pull the mass out to position y_0 and *release* it at time $t = 0$. This gives us the initial condition

$$y(0) = y_0, \quad y'(0) = v(0) = 0,$$

where v denotes the velocity of the mass. From this we determine C:

$$0 = -\frac{k}{m}y_0^2 + C$$

so

$$C = \frac{k}{m}y_0^2.$$

Thus

$$(y')^2 = \frac{k}{m}(y_0^2 - y^2),$$

or

$$y' = \pm\sqrt{\frac{k}{m}(y_0^2 - y^2)}.$$

Choosing either the plus or minus sign, this is a separable first-order differential equation. Let's solve it (using the $+$ sign):

$$\frac{dy}{\sqrt{y_0^2 - y^2}} = \sqrt{\frac{k}{m}}\,dt,$$

so

(4.15) $$\int \frac{dy}{\sqrt{y_0^2 - y^2}} = \int \sqrt{\frac{k}{m}}\,dt.$$

On the left-hand side, use the trigonometric substitution

(4.16) $$y = y_0 \sin\theta, \quad dy = y_0 \cos\theta\, d\theta,$$

along with the identity $\cos^2\theta = 1 - \sin^2\theta$, to get

$$\int \frac{dy}{\sqrt{y_0^2 - y^2}} = \int \frac{y_0 \cos\theta}{y_0 \cos\theta}\,d\theta = \int d\theta = \theta + \text{constant}.$$

From (4.15) we have

$$\theta = \sqrt{\frac{k}{m}}\,t + c.$$

Substituting this into (4.16) gives us

$$y = y_0 \sin\left(\sqrt{\frac{k}{m}}\,t + c\right).$$

Since our initial condition tells us that when $t = 0$, $y = y_0$, we must have $1 = \sin c$, and we choose $c = \frac{\pi}{2}$. Our solution is then

$$y = y_0 \sin\left(\sqrt{\frac{k}{m}}\, t + \frac{\pi}{2}\right)$$

$$= y_0 \cos\left(\sqrt{\frac{k}{m}}\, t\right),$$

a periodic function with amplitude y_0 and period $2\pi\sqrt{\frac{m}{k}}$; see Fig. 4.6.

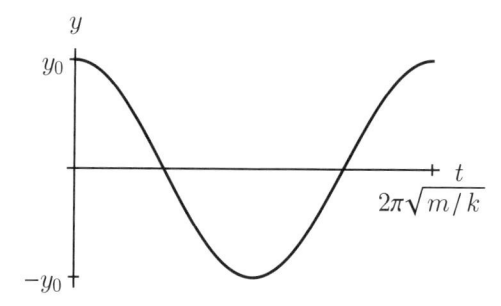

Figure 4.6. Position y of mass at time t.

In Exercise 20 you're asked to modify the model of Example 4.1.3 to include a drag force.

The three modeling examples just discussed resulted in second-order differential equations, which we were able to solve by reducing them to first-order equations. In the later sections of this chapter, we will want to develop techniques that will allow us to move beyond this special situation. Our focus will be on linear equations, because for second- and higher-order equations it is the linear equations that are most tractable, at least if the goal is to produce an analytic solution (as opposed to a graphical or numerical one). We also concentrate first on homogeneous linear equations, turning to the nonhomogeneous case in Chapter 5.

4.1.1. Existence and uniqueness. A fundamental result that underlies our study of second-order linear differential equations is an existence and uniqueness theorem. Like the analogous theorems we discussed in Chapter 2 for first-order equations, it deals with initial value problems, that is, a differential equation together with a suitable type of initial condition, and it addresses two related issues: Does a solution *exist* for this initial value problem, and if so, is the solution *unique*? Looking back at the examples we have just worked through, we can predict the appropriate form of the initial conditions for a second-order equation. They should look like

(4.17) $\qquad\qquad\qquad y(t_0) = \alpha, \quad y'(t_0) = \beta;$

that is, they specify the value of our unknown function $y(t)$ at some point t_0, as well as the value of the first derivative $y'(t_0)$ at the *same* point t_0. While you might imagine other kinds of data points (see Exercise 24), by initial conditions for a second-order differential equation we always mean data of the form given in (4.17). Roughly speaking, solving a second-order equation requires two steps of integration (as we saw in the examples above), each introducing a constant of integration, and the two pieces of data in equation (4.17) allow us to determine these two constants.

Our Existence and Uniqueness Theorem, which we state next, applies to homogeneous and nonhomogeneous *linear* equations alike. In its statement, recall that an *open* interval is one that

4.1. Introduction to second-order equations

doesn't contain any endpoints. It might be a finite interval $a < t < b$ or an unbounded interval like $a < t < \infty$ or $-\infty < t < \infty$.

Theorem 4.1.4 (Existence and Uniqueness Theorem). *Suppose that $p(t)$, $q(t)$, and $g(t)$ are continuous on an open interval I containing the point t_0. Then for any pair of constants α and β, the initial value problem*
$$y'' + p(t)y' + q(t)y = g(t), \qquad y(t_0) = \alpha, \quad y'(t_0) = \beta,$$
has a unique solution $y = \varphi(t)$ valid on the entire interval I.

On the one hand, this theorem may seem a bit theoretical, since it doesn't give any information about how you might *find* this solution whose existence (and uniqueness) is guaranteed. But we will soon put the theorem to practical use; for example, in the next section we apply it to characterize the form of the collection of all solutions to $y'' + p(t)y' + q(t)y = 0$ on any interval where p and q are continuous.

4.1.2. Presenting solutions graphically. How might we present the solution to some initial value problem

(4.18) $$y'' + p(t)y' + q(t)y = g(t), \qquad y(t_0) = \alpha, \quad y'(t_0) = \beta,$$

graphically? The most obvious way to do this is to sketch the graph of y vs. t, as we did in Examples 4.1.1 and 4.1.3. The graph of y' vs. t is sometimes also of interest, especially when it has some physical meaning for the problem. For instance, in Example 4.1.1 and Example 4.1.3, y' is the velocity of our falling grape or our moving mass.

Another type of graphical information will also prove to be quite useful in some circumstances. Suppose in equation (4.18) we set $v(t) = y'(t)$, so that the second-order equation

$$y'' + p(t)y' + q(t)y = g(t)$$

can be rewritten as a pair, or system, of first-order equations

(4.19) $$y' = v,$$
(4.20) $$v' = g(t) - q(t)y - p(t)v$$

for the two unknown functions y and v. The initial conditions in (4.18) translate to

(4.21) $$y(t_0) = \alpha, \quad v(t_0) = \beta.$$

When the functions $p(t), q(t)$, and $g(t)$ are *constants* the system (4.19)–(4.20) is called **autonomous**, meaning that the right-hand sides of these equations have no explicit dependence on t. Imposing the initial conditions (4.21), the solution functions

$$y = y(t), \quad v = v(t)$$

for an autonomous system give parametric equations for a curve in the yv-plane: for each t we imagine plotting the point $(y(t), v(t))$; as t varies, this point sketches out a curve in the yv-plane. If the initial conditions (4.21) are changed, the new values give rise to other solution curves. As the initial conditions are varied, the parametric solution curves just described constitute a **phase portrait** of the system of equations in (4.19) and (4.20), and the yv-plane is called the **phase plane**.

Example 4.1.5. When the second-order initial value problem

(4.22) $$y'' + y' + \frac{1}{2}y = 0, \qquad y(0) = 1, \quad y'(0) = 1,$$

is converted to a system of two first-order equations by the substitution $v = y'$, the result is the pair of equations

(4.23) $$y' = v, \quad v' = -\frac{1}{2}y - v,$$

with initial conditions $y(0) = 1, v(0) = 1$. Referring to Theorem 4.1.4, we see that because $p(t) = 1$, $q(t) = 1/2$, and $g(t) = 0$ are all constants and thus are continuous on the real line, the initial value problem (4.22) has a unique solution $y = y(t)$ defined on $(-\infty, \infty)$. Hence, the corresponding autonomous system (4.23) has a unique solution as well: $y = y(t)$, $v = y'(t)$. The graph in Fig. 4.7 shows the solution curve $(y(t), v(t)), t \geq 0$, in the phase plane of this system; the arrows on the curve indicate how it is traced out for increasing times t. Figs. 4.8 and 4.9 give sketches of the corresponding graphs of y vs. t and of v vs. t. In Section 4.4, we show how the solution

$$y(t) = e^{-t/2} \cos(t/2) + 3e^{-t/2} \sin(t/2)$$

of (4.22) can be computed. For now, you should focus on how the solution curve in the phase plane is at least consistent with the qualitative features of the two other graphs. For example, starting at $t = 0$ and going forward in time, Fig. 4.7 says that y, which is 1 when $t = 0$, initially increases to about 1.4, then decreases, eventually becoming negative for a time, and appears to eventually approach 0, all of which we see in Fig. 4.8. Similarly, starting at $t = 0$, v decreases initially from its starting value of 1, becoming negative before it starts to increase, as we see in Fig. 4.9.

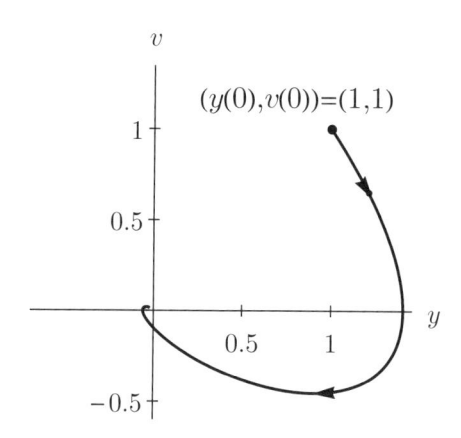

Figure 4.7. Solution curve in the phase plane.

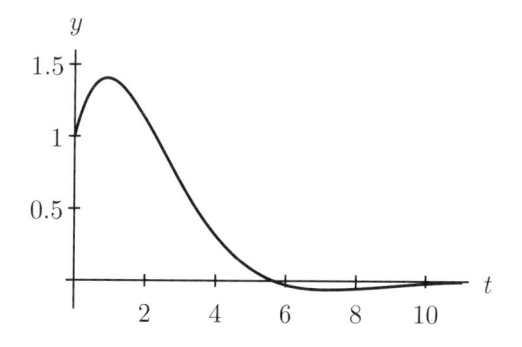

Figure 4.8. Graph of y vs. time t.

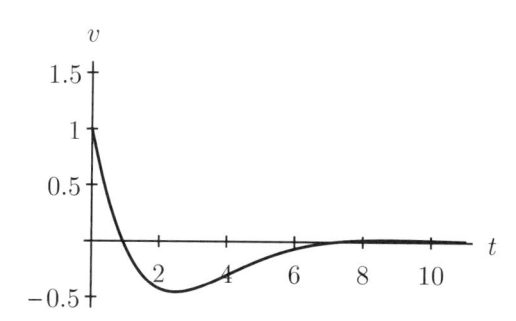

Figure 4.9. Graph of v vs. time t.

The graph below is a phase portrait of the system (4.23) in which we have included (in bold) the system's solution depicted in Fig. 4.7.

4.1. Introduction to second-order equations

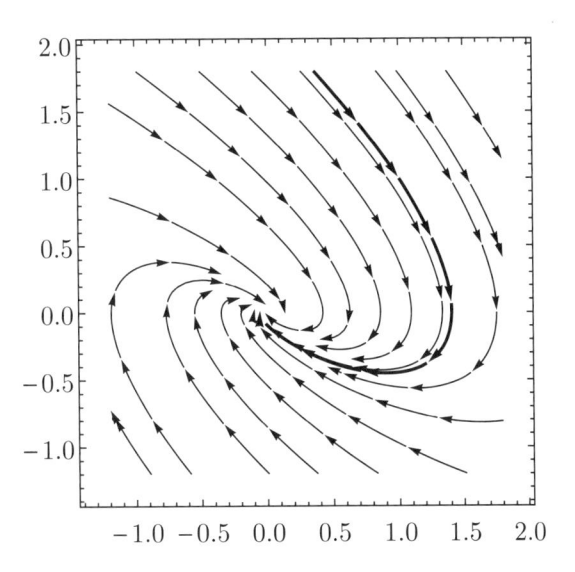

Figure 4.10. A phase portrait for (4.23).

4.1.3. Exercises.

1. For each of the following second-order differential equations, classify the equation as linear or nonlinear. For the linear equations, decide whether the equation is homogeneous or nonhomogeneous.
    (a) $y'' + 5y' + 6y = 2\cos t$.
    (b) $y'' + 5y = 0$.
    (c) $y'' + 5 = 0$.
    (d) $(1 - t^2)y'' = 3y$.
    (e) $(1 - y)y'' = 2y$.
    (f) $t^2 y'' + ty' = y$.
    (g) $t^2 y'' + ty' = y^2$.

2. Consider the second-order linear equation
$$y'' + 3y = 12$$
where $y(t)$ measures position in inches and time t is measured in seconds.
    (a) What are the units on 12? What are the units on the coefficient 3?
    (b) If we want to measure the position function y in feet rather than inches, how does the differential equation change?

3. For the equations below, find the largest open interval containing $t = 1$ on which the given initial value problem is guaranteed, by Theorem 4.1.4, to have a unique solution.
    (a) $y'' + \ln t \, y' + 3y = \tan t$, $y(1) = 2$, $y'(1) = -1$.
    (b) $y'' + \dfrac{1}{9 - t^2} y' + \dfrac{1}{2 - t} y = \ln(t^2)$, $y(-1) = 2$, $y'(-1) = -1$.
    (c) $t^2 y'' + ty' + y = t^3$, $y(3) = 2$, $y'(3) = -1$.
    (d) $y'' + 2y' + \cos(t)y = e^t$, $y(-2) = 2$, $y'(-2) = -1$.

4. In 1908, Washington Senators' catcher Gabby Street caught a baseball dropped from the top of the Washington Monument, a height of 555 feet. A baseball weighs 5.125 ounces. Assume that the drag force is proportional to the velocity and that the drag coefficient is $c = .0018$ lb-sec/ft. How fast was the baseball going when Street caught it 7 feet above ground level?

5. If an object falls in a straight line according to the differential equation

$$v' + \frac{c}{m}v = g$$

where v, c, m, and g are as in Example 4.1.1, show that the *terminal velocity* of the object is mg/c. By the terminal velocity we mean $\lim_{t \to \infty} v(t)$, the limiting value of the velocity if it could continue to fall forever.

6. In an unsolved airplane hijacking case from 1971, the hijacker (given the alias of D. B. Cooper) parachuted out of the back of a Northwest Boeing 727 over the southwest part of Washington State with $200,000 in $20 bills strapped to his body. The money and four parachutes had been provided by the FBI after Cooper took over the plane, claiming to have a bomb. Despite an intensive search by the FBI and 400 troops from Fort Lewis near the projected landing spot, Cooper was never found and his fate remains unknown.[4] The combined weight of Cooper, the money, and his parachutes was approximately 230 lbs. He jumped from an altitude of 10,000 feet. Assume a linear drag term as in Example 4.1.1, and assume that he opened his parachute instantaneously 30 seconds after he stepped out of the plane and that his drag coefficient was $c = 1.1$ lbs-sec/ft before the chute opened and $c = 12$ lbs-sec/ft after. Estimate the speed with which he hit the ground.

7. For both recreational and military parachute jumps, there are numerous stories of jumpers surviving long falls in spite of parachutes that failed to fully deploy. In most of these cases, there was at least partial opening of the main or backup chute. Assume that the combined weight of a jumper and his gear is 100 kg and that he falls far enough to reach 95% of his terminal velocity. If an impact velocity of 50 miles per hour (approximately 22 m/sec) is the practical limit of "survivability" (and requires a good landing position), what drag coefficient c for a partially deployed chute gives the jumper a chance of surviving? Compare this with a typical value of $c/m = 1.6$ for a fully open chute and $c/m = 0.16$ for a person with a completely unopened chute.

8. **Feline high-rise syndrome.** Various veterinary journals report on the injury and survival rates for cats falling out of windows on upper stories of apartment buildings. Curiously, the severity of injury is lower for heights substantially above 7 stories than it is at around the 7-story height. One explanation given is that falling from the greater height, the cat gets close enough to terminal velocity so that it no longer perceives a downward acceleration (the feeling of falling). This effects the balance sensing mechanism in the cat's inner ear, so that it no longer maintains its legs in a downward pointing position but rather tends to move into a "spread-eagle" position. This causes a greater drag force, and the result is a lower velocity landing. In fact, the survival rate for cats falling from 9 stories is 95% and, amazingly, holds steady at 95% up to 32 stories.[5] Suppose that a cat weighing 12 lbs falls from an upper floor and reaches a velocity of 90 ft/sec as it passes the ninth story. At this point, it relaxes into a spread-eagle position, increasing its drag coefficient to $c = 0.6$. What is its approximate speed when it hits the ground? Assume the drag is proportional to the speed and that 9 stories is 120 ft.

9. Assume a spherical raindrop[6] with diameter 1 mm and mass $\frac{4}{3}\pi(0.05)^3$ gm has drag coefficient $c = 1.2 \times 10^{-3}$ gm/sec.
   (a) What is its terminal velocity?

---

[4]D. B. Cooper was back in the news in 2011, with the FBI investigating "new evidence" in the case, which apparently suggested Cooper might have survived, dying of natural causes around 2001.
[5]For further discussion of feline high-rise syndrome, see M. Memmott, *Cat Falls 19 Floors, Lands Purrfectly*, NPR news, March 22, 2012.
[6]Some of the data in this problem is taken from W. J. Humphreys, *Physics of the Air*, Dover, New York, 1964.

(b) If the rain is falling from a cloud at height 500 m, with initial velocity 0, when is it within 10% of its terminal velocity? How far, approximately, has it fallen by this point? (Caution: Watch units!)

10. Suppose an object falls from rest along a vertical line oriented as in Fig. 4.1. In this problem, we'll assume that the drag force is proportional to v^2, the square of the velocity.

   (a) Show that this leads to the first-order equation

   $$\text{(4.24)} \quad \frac{dv}{dt} = g - \frac{c}{m}v^2$$

   where c is a positive constant.

   (b) Give a qualitative solution to equation (4.24); that is, sketch some solution curves in the tv-plane (for $v, t \geq 0$) using the techniques of Section 2.6.

   (c) What is the limiting value of $v(t)$ as $t \to \infty$? Compare your answer here with the terminal velocity computed in Exercise 5.

   (d) Why does equation (4.24) have to be adjusted if $v(t)$ is ever negative (for example, if the object is initially thrown upward)? Hint: The drag force should act opposite to the direction of motion. Argue that

   $$\frac{dv}{dt} = g - \frac{c}{m}v|v|$$

   (for positive constant c) is an appropriate equation if the velocity assumes both positive and negative values.

11. The equation

    $$\frac{dv}{dt} = g - \frac{c}{m}v^2$$

    of Exercise 10 can be made to look cleaner by a simple *rescaling* of the variables. In the original equation the independent variable is t and the dependent variable is v. Suppose we set

    $$s = \sqrt{\frac{cg}{m}}\, t, \quad w = \sqrt{\frac{c}{mg}}\, v,$$

    so that s is "rescaled time" and w is "rescaled velocity".

    (a) Use the chain rule

    $$\frac{dw}{ds} = \frac{dw}{dt} \cdot \frac{dt}{ds}$$

    to show that

    $$\frac{dw}{ds} = \frac{1}{g}\frac{dv}{dt}.$$

    (b) Show that this rescaling converts the original differential equation into

    $$\frac{dw}{ds} = 1 - w^2.$$

    (c) Solve the equation in (b) by separating variables, using the initial condition $s = 0$, $w = 0$. What is $\lim_{s \to \infty} w(s)$? Can you reconcile your answer here to the answer in (c) of Exercise 10?

12. In Example 4.1.1 we can replace the second-order differential equation in (4.2) by a system of two first-order equations

    $$\frac{dy}{dt} = v,$$
    $$\frac{dv}{dt} = g - \frac{c}{m}v.$$

A solution to this system is a pair of functions

(4.25) $$y = y(t), \quad v = v(t)$$

specifying the position and velocity of our falling object at time t (for the duration of its fall). A computer-produced solution to this system is shown below, with $g = 32$ ft/sec^2, $c = 1.2 \times 10^{-4}$ lb-sec/ft, and $m = 0.02/32$; these are the curves in the yv-plane described by the *parametric equations* (4.25). Different curves correspond to different initial conditions $y(0) = \alpha$, $v(0) = \beta$. Use this picture to answer (approximately) the question we asked in Example 4.1.1: How fast was the grape going when Mr. Tavilla caught it?

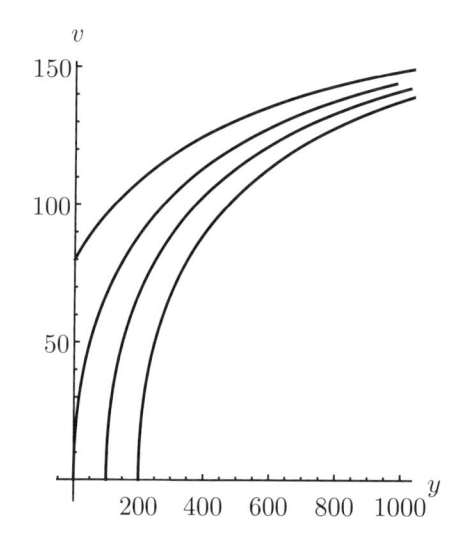

Figure 4.11. The parametric curves $y = y(t)$, $v = v(t)$ with different initial conditions.

13. (a) Is there a solution to the differential equation $y'' + 6y' = 32$ in which y' is a constant function k? If so, find k.
    (b) If we interpret the equation in (a) as giving the position of a falling object, what is the terminal velocity? Is the terminal velocity reached in finite time? How does this compare to your answer to (a)?

14. Suppose we have the same set-up as in Example 4.1.2, except the Coast Guard is twice as fast as the smuggler. At what point on the y-axis will the Coast Guard intercept the smuggler (i.e., the point where they are both in the same place at the same time)? As in Example 4.1.2, you will need to find a formula for the path of the Coast Guard boat. Unlike the case of equal speeds, this path will cross the y-axis (at the point of interception).

15. In Example 4.1.2, both boats have the same speed a, so the Coast Guard will never catch the smuggler. But suppose the Coast Guard boat has a small cannon with a maximum range of $\frac{1}{2}$ mile. Will the Coast Guard boat get within range? The skipper quickly works through Example 4.1.2 and orders the first mate (you) to determine whether the chase is worthwhile; that is, will you get within $\frac{1}{2}$ mile of the smuggler? Look at Fig. 4.3 again. The distance d from Coast Guard to smuggler is (by the Pythagorean Theorem—notice the right triangle with hypotenuse d) is

$$d = \sqrt{x^2 + (at - y)^2}.$$

Both boats have the same speed, so

$$at = s = \int_x^1 \sqrt{1 + y'(u)^2}\, du.$$

4.1. Introduction to second-order equations

Plug your known formula for y' into this integral, evaluate the integral, also use the formula we found for y, and thus find d as a function of x. Do you ever get within range? Find
$$\lim_{t \to \infty} d = \lim_{x \to 0} d.$$
(Note that $x \to 0$ as $t \to \infty$ and vice versa.)

16. In Example 4.1.3, suppose we use the MKS system of units (see Table 4.1). What are the units on the spring constant k? Answer the same question for the British and CGS systems of units.

17. The functions graphed below are solutions to the differential equation $my'' + ky = 0$ of Example 4.1.3 for a fixed m but differing values of k. Which one corresponds to the largest value of k?

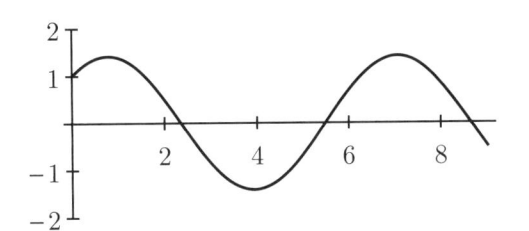

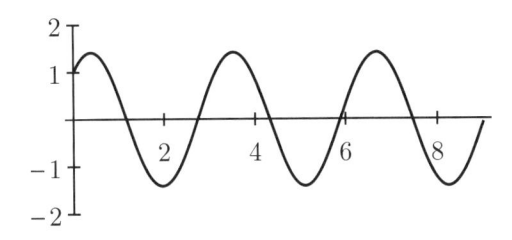

Figure 4.12. (A)

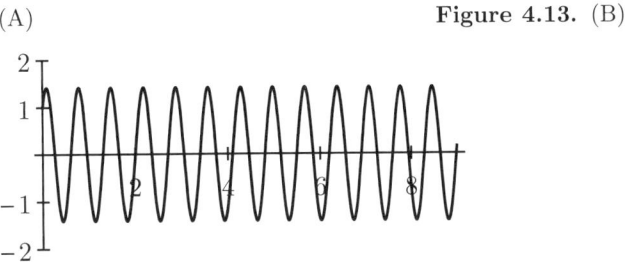

Figure 4.13. (B)

Figure 4.14. (C)

18. Suppose that in Example 4.1.3 we want to use our frictionless mass-spring set-up as a clock, so that each extension of the mass to its maximum displacement ticks off one second.
    (a) If the spring constant $k = 4$ gm/sec^2, what mass m will make our desired ticks of one second?
    (b) If the mass is slightly larger than what you obtained as your answer in (a), will the clock run fast or slow?
    (c) If the spring constant decreases slightly over time, will the clock run fast or slow?

19. Which of the following cannot be a solution of $y'' + ay = 0$ for any value of a?
    (a) $y = 2\sin(2t)$.
    (b) $y = 4\cos(3t)$.
    (c) $y = 2e^{3t}$.
    (d) $y = 5$.
    (e) $y = t^2$.

20. Suppose we want to include a drag or damping term in Example 4.1.3 by assuming a damping force proportional to the speed of the mass, with (positive) constant of proportionality c, and acting opposite to the direction of motion.
    (a) If the mass is moving to the right at some instant, is $y'(t)$ positive or negative? Is the damping force term $cy'(t)$ or $-cy'(t)$? Recall that we have oriented our number line so that the positive direction is to the right.
    (b) If the mass is moving to the left at some instant, is the damping force term $cy'(t)$ or $-cy'(t)$?

(c) Using Newton's second law, $F = ma$, where F is the total force (the sum of the restoring force and the damping force) and $a = y''$ is the acceleration of the mass due to that force, find a differential equation to model the motion of our horizontal mass-spring system.

21. In Example 4.1.3 we solved $my'' + ky = 0$ to obtain
$$(y')^2 = -\frac{k}{m}y^2 + c.$$
This shows that the function
$$\frac{1}{2}m(y')^2 + \frac{1}{2}ky^2$$
is constant C (as a function of time, t). The term $\frac{1}{2}m(y')^2$, or $\frac{1}{2}mv^2$, where v is velocity, is the kinetic energy of the mass. The second term, $\frac{1}{2}ky^2$, is the potential energy, and the statement
$$\text{kinetic energy } + \text{ potential energy } = C$$
is a **conservation of energy** principle. Show that if we modify the differential equation to include a damping term, so that $my'' + by' + ky = 0$, then the total energy
$$E(t) = \frac{1}{2}m(y')^2 + \frac{1}{2}ky^2$$
is a decreasing function of t. (Hint: What is $\frac{d}{dt}(E(t))$?)

22. We saw that the second-order equation of the form $y'' = f(y)$ can be reduced to a first-order equation
$$(y')^2 = 2F(y) + c$$
where F is an antiderivative of f. Solving for y' gives
$$(4.26) \qquad y' = \pm\sqrt{2F(y) + c}.$$

(a) Show that separating variables in equation (4.26) gives the solutions
$$\pm \int \frac{dy}{\sqrt{2F(y) + c}} = t + k.$$

(b) Even for simple choices of f, the integral
$$\int \frac{dy}{\sqrt{2F(y) + c}}$$
may not be computable in terms of familiar functions. However, we may be able to produce *some* (explicit) solutions by choosing $c = 0$. Do so for the following equations:

(i)
$$y'' = \frac{3}{4}\sqrt{y}.$$

(ii)
$$y'' = \frac{3}{2}y^2.$$

(iii)
$$y'' = 24y^{\frac{1}{3}}.$$

(c) Consider two objects having, respectively, masses m_1 and m_2, such that distance between their centers of mass at time t is $y(t)$. If the only force the objects experience is that of their mutual gravitational attraction, then Newton's law of gravitation yields
$$(4.27) \qquad \frac{d^2y}{dt^2} = -\frac{G(m_1 + m_2)}{y^2},$$

4.1. Introduction to second-order equations 189

where G is the universal gravitational constant. Solve the initial value problem

$$\frac{d^2y}{dt^2} = \frac{-2}{y^2}, \quad y(1) = 1, \quad y'(1) = 2,$$

noting this corresponds to (4.27) applied to two (very massive) objects each with mass numerically equal to $1/G$.

23. A particle of mass $m = 1$ moves along the positive y-axis under the influence of a force F given by

$$F = \frac{v^2}{y} - vy$$

where $y = y(t)$ is the position of the particle at time t and $v = v(t) = y'(t)$ is the velocity (assume no gravity is present). The acceleration $a(t)$ of the particle at time t is $a(t) = v'(t) = y''(t)$, so that by Newton's second law we have

$$y'' = \frac{v^2}{y} - vy.$$

  (a) Convert this second-order equation to a first-order equation with independent variable y and dependent variable v, and then solve this first-order equation. Hint: Since $v = y'$, we have

$$y'' = v' = \frac{dv}{dt} = \frac{dv}{dy}\frac{dy}{dt}$$

  by the chain rule.
  (b) Suppose that at time $t = 0$ the particle has position $y(0) = 1$ and velocity $y'(0) = -1$. Find $y(t)$ for all $t > 0$. What happens to y as t tends to infinity?
  (c) Keep the initial position the same as in (a), namely $y(0) = 1$, but make the initial velocity be $v(0) = -\frac{1}{2}$. Thus the particle still starts off moving downward towards 0, but not as fast. Find $y(t)$. What happens to y as t tends to infinity?

24. Initial conditions for the equation

(4.28) $$y'' + y = 0$$

specify y and y' at a single value of t. Theorem 4.1.4 then guarantees the resulting initial value problem has a unique solution on $(-\infty, \infty)$. If we instead specify the value of y at two different values of t, we have a **boundary value problem**. In this problem, you will consider the fate of "existence" and "uniqueness" for four boundary value problems, using the fact that all solutions of equation (4.28) have the form $c_1 \cos t + c_2 \sin t$ for some constants c_1 and c_2.
  (a) Show that the boundary value problem

$$y'' + y = 0, \quad y(0) = 1, \quad y(\pi/2) = 0,$$

  has exactly one solution, and find this solution.
  (b) Show that the boundary value problem

$$y'' + y = 0, \quad y(0) = 1, \quad y(\pi) = -1,$$

  has infinitely many solutions. Describe all of them.
  (c) Show that the boundary value problem

$$y'' + y = 0, \quad y(0) = 1, \quad y(\pi) = 0,$$

  has no solution.
  (d) Find all solutions to the boundary value problem

$$y'' + y = 0, \quad y(0) = 0, \quad y(\pi) = 0.$$

25. (a) The graph below shows several solutions to the first-order differential equation $y'(t) = y(t)\sin y(t) + t$. Could these solution curves intersect?

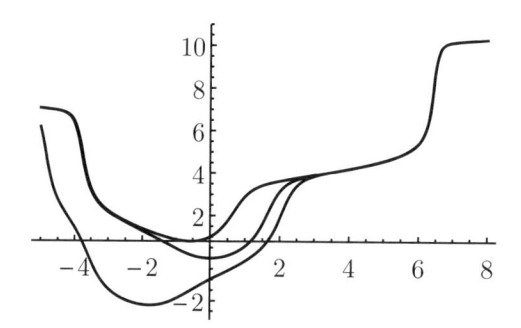

(b) Two different functions are graphed below. Could these both be solutions to the same second-order linear differential equation $y''(t) + p(t)y'(t) + q(t)y(t) = g(t)$ where $p(t), q(t)$, and $g(t)$ are continuous functions on $-\infty < t < \infty$? If not, why not?

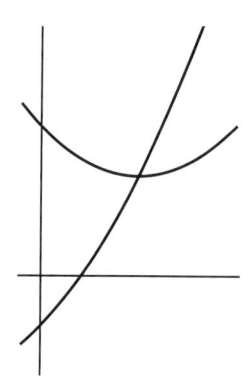

(c) The two functions graphed below intersect at $t = 1$ and have the same tangent line at that point. Could these both be solutions to the same second-order linear differential equation $y''(t) + p(t)y'(t) + q(t)y(t) = g(t)$ where $p(t), q(t)$, and $g(t)$ are continuous functions on $-\infty < t < \infty$? If not, why not?

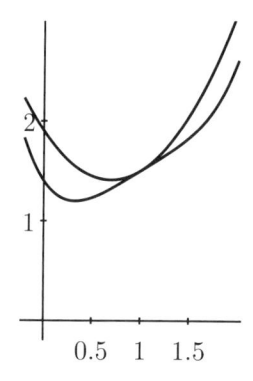

Hint: Think about the existence and uniqueness theorems for the appropriate type of differential equation.

26. True or false? If $y(t)$ is a solution of
$$y'' + (\sin t)y' + t^2 y = 0$$
whose graph (in the ty-plane) touches and is tangent to the t-axis at $t_0 = 0$, then $y(t)$ is the constant zero.

4.1. Introduction to second-order equations

27. (a) Fig. 4.15 shows the graphs of $y(t)$ for three solutions of $y'' + p(t)y' + q(t)y = 0$ for certain functions $p(t)$ and $q(t)$. Which of these solutions also satisfies the initial conditions $y(0) = 1$, $y'(0) = 1$?

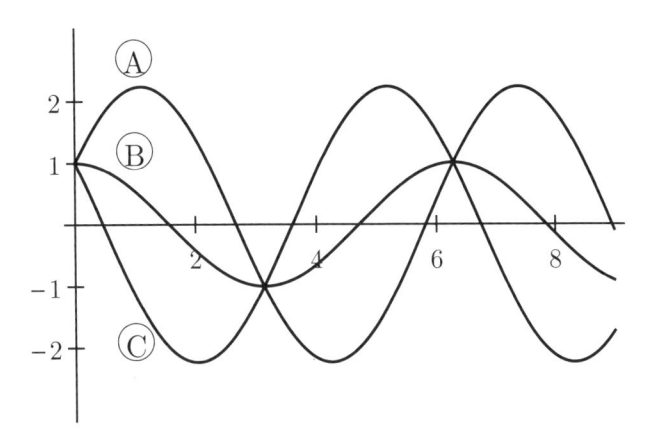

Figure 4.15. Graph of three solutions $y(t)$.

(b) Suppose we replace the second-order equation $y'' + ay' + by = 0$ by the system

$$\frac{dy}{dt} = v,$$
$$\frac{dv}{dt} = -by - av.$$

Several curves in the phase portrait for this system are shown in Fig. 4.16. Which solution curve could correspond to the initial conditions $y(0) = 0, v(0) = 1$?

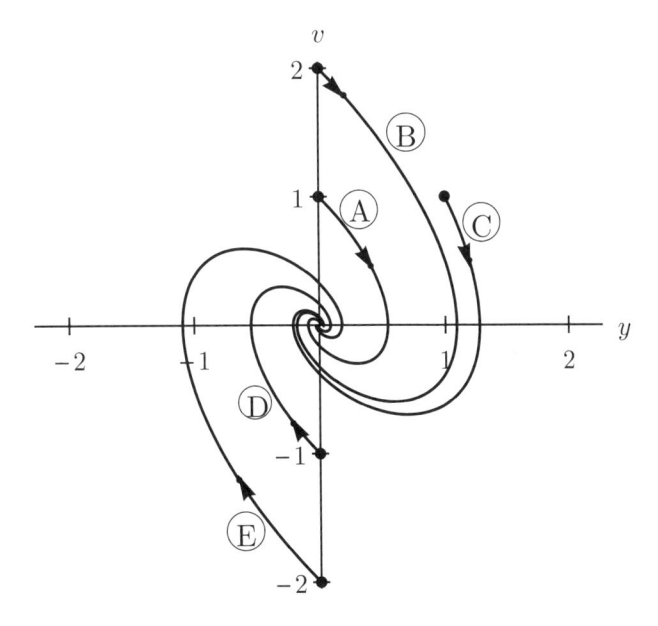

Figure 4.16. Curves in the phase portrait.

(c) Fig. 4.17 gives the graph of y vs. t for one of the solution curves shown above in the phase portrait. Which one?

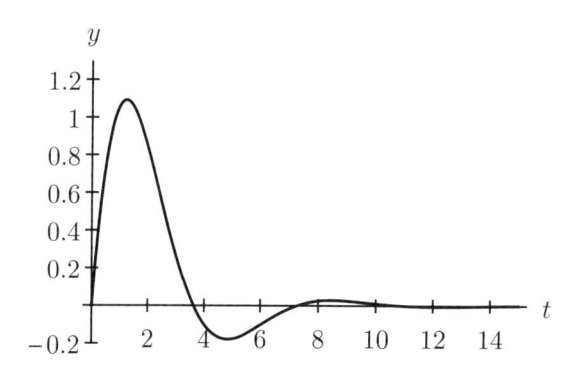

Figure 4.17. y vs. t.

28. Consider the homogeneous equation $y'' + p(t)y' + q(t)y = 0$ where $p(t)$ and $q(t)$ are continuous in some open interval I containing 0, with the initial conditions $y(0) = 0$, $y'(0) = 0$.
    (a) Find an "obvious" solution to this initial value problem, no matter what the specific functions $p(t)$ and $q(t)$ are. Hint: An "obvious" solution is often a constant solution.
    (b) Explain why the function $y(t) = t^3 \sin t$ cannot solve the equation $y'' + p(t)y' + q(t)y = 0$ for any choice of $p(t)$ and $q(t)$ that are continuous in an open interval containing 0. Hint: For $y(t) = t^3 \sin t$, what are $y(0)$ and $y'(0)$?

29. (CAS) In Example 4.1.5 we saw how to convert a second-order initial value problem into a system of two first-order equations with corresponding initial conditions. Such a conversion facilitates numerical approximation of the solution. Suppose that $y = y(t)$, $v = v(t)$ is the solution of an initial value problem of the form

    (4.29) $\quad y'(t) = F(t, y(t), v(t)), \quad v'(t) = G(t, y(t), v(t)), \quad y(t_0) = y_0, \; v(t_0) = v_0,$

    where t is time, measured in seconds. View $(y(t), v(t))$ as the horizontal and vertical coordinates at time t of a particle traveling in a plane. View the equations for y' and v' of (4.29) as specifying, respectively, the horizontal and vertical components of the particle's velocity. Euler's method with step size h for (4.29) provides approximate values y_k and v_k for, respectively, $y(t_0 + kh)$ and $v(t_0 + kh)$ via the iterative formula

    $$(y_k, v_k) = (y_{k-1}, v_{k-1}) + (F(t_{k-1}, y_{k-1}, v_{k-1}), G(t_{k-1}, y_{k-1}, v_{k-1}))\, h,$$

    where $t_k = t_0 + kh$, $y_0 = y(t_0)$, and $v_0 = v(t_0)$.
    (a) Verify that (y_1, v_1) is obtained from (y_0, v_0) by starting at (y_0, v_0) and then following, for h seconds, the velocity vector $(y'(t_0), v'(t_0))$. Then confirm that for $k > 1$, (y_k, v_k) is obtained from (y_{k-1}, v_{k-1}) by starting at (y_{k-1}, v_{k-1}) and then following, for h seconds, the approximation of the velocity vector $(y'(t_{k-1}), v'(t_{k-1}))$ given by

    $$(F(t_{k-1}, y_{k-1}, v_{k-1}), G(t_{k-1}, y_{k-1}, v_{k-1})).$$

    (b) For the initial value problem of Example 4.1.5, use the iterative formula given above and a computer algebra system to produce the sequence of approximations (y_k, v_k), for $k = 0, 1, \ldots, 10$ and step size $h = \frac{1}{10}$. Compare a plot of your approximations with Fig. 4.7.

4.2. Linear operators

In this section we look at second-order linear homogeneous differential equations and focus on understanding the basic structure of their solutions.

Let's start with an example. Consider the equation

$$y'' - 6y' + 8y = 0;$$

4.2. Linear operators

this is a second-order, linear, homogeneous equation with *constant* coefficients. Suppose we try to find a solution of this equation in the form

$$y = e^{rt},$$

where r is to be determined. Since $y' = re^{rt}$ and $y'' = r^2 e^{rt}$, when we substitute into the differential equation we obtain

(4.30) $$r^2 e^{rt} - 6re^{rt} + 8e^{rt} = 0.$$

Do any values of r produce a solution? Factoring in (4.30) gives

$$e^{rt}(r^2 - 6r + 8) = 0.$$

Since the exponential factor is never 0, equation (4.30) holds if and only if

$$r^2 - 6r + 8 = 0.$$

Thus e^{rt} is a solution of $y'' - 6y' + 8y = 0$ exactly when r is a root of the polynomial

$$p(z) = z^2 - 6z + 8.$$

By either factoring

$$z^2 - 6z + 8 = (z-2)(z-4)$$

or using the quadratic formula, we find that the roots are 2 and 4. Thus the functions

$$y_1 = e^{2t} \quad \text{and} \quad y_2 = e^{4t}$$

are both solutions to the equation $y'' - 6y' + 8y = 0$, and these are the only solutions of the form e^{rt} for some constant r. Are there other solutions? Can we produce the family of all solutions to this equation?

To answer these questions, it will be helpful to discuss some general principles for a second-order linear, homogeneous differential equation

$$y'' + p(x)y' + q(x)y = 0.$$

These general principles will be most easily described by using some operator notation. Think of an operator as a "black box" into which we input functions. Each inputted function is then converted by some "rule" into an output function.

For example, the differentiation operator, which we denote by D in the schematic below, takes a (differentiable) function f as input and produces the function f' as output: Two other examples of operators are also shown below. The operator S (for squaring) converts f to f^2, and the multiplication operator $5\times$ takes the input function and multiplies it by 5 to produce the output.

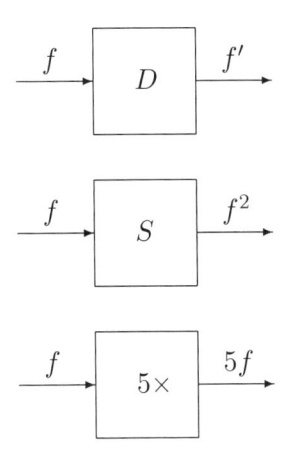

Linear operators. We will be primarily interested in operators that have two special properties, collectively called the linearity properties. These two properties tell us how the output for $f + g$, a sum of two functions, is related in a natural way to the outputs for f and g individually, and how the output for cf, a constant multiple of f, is related to the output of f. If L denotes our operator and f is an input function, we write $L[f]$ for the corresponding output function. Here is the definition:

Definition 4.2.1. An operator L is said to be **linear** if both

(a) $L[f + g] = L[f] + L[g]$ and

(b) $L[cf] = cL[f]$

for all functions f and g and scalars c.

In other words, the operator is linear if the output for $f + g$ is just the sum of the outputs for f and g and if the output for cf is just c times the output for f.

Let's look at our examples above of operators and decide which are linear.

Example 4.2.2. The operator D defined by $D[f] = f'$ is linear. This is just another way of saying that the derivative of a sum is the sum of the derivatives and the derivative of a constant multiple of a function is the constant times the derivative of the function:

$$D[f + g] = (f + g)' = f' + g' = D[f] + D[g]$$

and

$$D[cf] = (cf)' = cf' = cD[f].$$

Example 4.2.3. By contrast, the squaring operator S is not linear, since it is **not** true that $(f + g)^2 = f^2 + g^2$ for all choices of functions f and g. It is also **not** true that $(cf)^2 = cf^2$ for all scalars c and all functions f; so the squaring operator S fails both criteria for linearity in Definition 4.2.1.

Example 4.2.4. The "5×" operator is linear, since $5(f + g) = 5f + 5g$ and $5(cf) = c(5f)$.

We are especially interested in a **linear differential operator** L whose black-box rule is described by

(4.31) $$L[f] = f'' + p(t)f' + q(t)f,$$

where $p(t)$ and $q(t)$ are chosen functions that are continuous on some interval $\alpha < t < \beta$. Let's check that this operator is linear. When we input a function f, the output is $f'' + p(t)f' + q(t)f$, and when we input g, the output is $g'' + p(t)g' + q(t)g$. What is $L[f + g]$, the output when we input $f + g$? By the rule for L, it is

$$L[f + g] = (f + g)'' + p(t)(f + g)' + q(t)(f + g).$$

Since $(f + g)'' = f'' + g''$ and $(f + g)' = f' + g'$ we can rewrite $L[f + g]$ as

$$f'' + g'' + p(t)(f' + g') + q(t)(f + g)$$

and rearrange this as

$$(f'' + p(t)f' + q(t)f) + (g'' + p(t)g' + q(t)g),$$

which we recognize as $L[f] + L[g]$. So the first of the two requirements for linearity of L is satisfied. Checking the second requirement, that $L[cf] = cL[f]$, is done similarly and we leave it to the reader to write down the details (see Exercise 2).

Let's go back to our example $y'' - 6y' + 8y = 0$ and think about this in operator notation. This differential equation can be written as $L[y] = 0$ where L is the linear operator

$$L[f] = f'' - 6f' + 8f,$$

4.2. Linear operators

so that we are choosing $p(t) = -6$ and $q(t) = 8$ in equation (4.31). The work at the beginning of this section, which showed that $y_1 = e^{2t}$ and $y_2 = e^{4t}$ solve $y'' - 6y' + 8y = 0$, can be summarized by

$$L[e^{2t}] = 0 \quad \text{and} \quad L[e^{4t}] = 0.$$

Since L is linear,

$$L[c_1 e^{2t}] = c_1 L[e^{2t}] = c_1(0) = 0$$

for any constant c_1 and

$$L[c_2 e^{4t}] = c_2 L[e^{4t}] = c_2(0) = 0$$

for any constant c_2. This says that $c_1 e^{2t}$ and $c_2 e^{4t}$ are also solutions to $y'' - 6y' + 8y = 0$, for arbitrary constants c_1 and c_2. Linearity also says that

$$L[c_1 e^{2t} + c_2 e^{4t}] = L[c_1 e^{2t}] + L[c_2 e^{4t}] = 0 + 0 = 0,$$

so that any function of the form

(4.32) $$y = c_1 e^{2t} + c_2 e^{4t}$$

is a solution to $y'' - 6y' + 8y = 0$. We call the function in (4.32) a **linear combination** of e^{2t} and e^{4t}. Notice how the fact that the differential equation was homogeneous was crucial to the calculations we just did. The next result just states that what we were able to do in this particular example holds equally well for any second-order linear homogeneous differential equation.

Theorem 4.2.5 (The Superposition Principle). *If y_1 and y_2 are solutions to the second-order linear, homogeneous differential equation*

$$y'' + p(t)y' + q(t)y = 0,$$

then any linear combination

$$c_1 y_1 + c_2 y_2$$

is also a solution to this differential equation.

With linear differential operator notation it is easy to see why this theorem is true. If we set

$$L[y] = y'' + p(t)y' + q(t)y,$$

then we are given that $L[y_1] = 0$ and $L[y_2] = 0$. By linearity,

$$L[c_1 y_1 + c_2 y_2] = L[c_1 y_1] + L[c_2 y_2] = c_1 L[y_1] + c_2 L[y_2] = 0,$$

which is exactly the statement that $c_1 y_1 + c_2 y_2$ is a solution to the differential equation.

Returning to our example $y'' - 6y' + 8y = 0$ one more time, we are now led to ask two more questions:

(Q_1) Is $\{c_1 e^{2t} + c_2 e^{4t}\}$ the *complete* family of all solutions to $y'' - 6y' + 8y = 0$ as we let c_1 and c_2 range over all real numbers?

(Q_2) Can we find a function $y(t)$ in the family

(4.33) $$\{c_1 e^{2t} + c_2 e^{4t}\}$$

that will satisfy any initial conditions

$$y(t_0) = \alpha, \quad y'(t_0) = \beta$$

we'd like?

These two questions are related, and we'll tackle the second one, Q_2, first. Pick any numbers t_0, α, and β. We want to know if we can solve the pair of equations

$$(4.34) \qquad c_1 e^{2t_0} + c_2 e^{4t_0} = \alpha,$$

$$(4.35) \qquad 2c_1 e^{2t_0} + 4c_2 e^{4t_0} = \beta$$

where the "unknowns" are c_1 and c_2. If we multiply the first equation by 4 and subtract the second equation from the result, we determine

$$c_1 = \frac{4\alpha - \beta}{2e^{2t_0}}.$$

Once c_1 is known, we can substitute into either equation to determine a unique solution for c_2. Instead of carrying out this algebra, let's think geometrically about what we are doing.

The equations (4.34) and (4.35) are the equations of two lines in the $c_1 c_2$-plane. When we solve this system of equations, we are determining any point(s) of intersection of these two lines. Now given any pair of lines, there are three possibilities: They intersect in a unique point, they are parallel and don't intersect at all, or they are in fact the same line (and hence intersect in infinitely many points). We recall how, in general, to distinguish these possibilities using matrix notation. Recall that a 2×2 matrix is just an array with two rows and two columns. A pair of lines in the xy-plane,

$$(4.36) \qquad \begin{aligned} ax + by &= k_1, \\ cx + dy &= k_2, \end{aligned}$$

has a unique point of intersection if and only if the determinant of the matrix of coefficients

$$\begin{pmatrix} a & b \\ c & d \end{pmatrix}$$

is not equal to 0. To see why, recall that the determinant of this 2×2 matrix is given by

$$(4.37) \qquad \det \begin{pmatrix} a & b \\ c & d \end{pmatrix} = ad - bc.$$

This determinant is 0 if and only if the lines are either parallel or they coincide, since $ad - bc = 0$ if and only if the slopes of the two lines are the same or $b = d = 0$ (making both lines vertical).[7] The lines coincide (which is the same as saying the two equations are in fact equations of the same line) exactly when one equation is just a multiple of the other. Thus if $ad - bc = 0$, there will always be a choice of k_1 and k_2 so that the two lines are distinct and parallel and the system (4.36) has no solution.

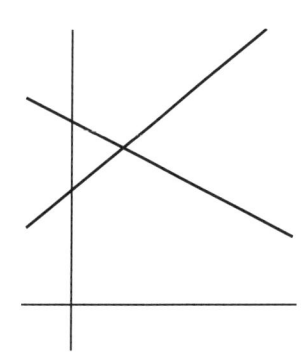

Figure 4.18. Intersecting; determinant in (4.37) is nonzero.

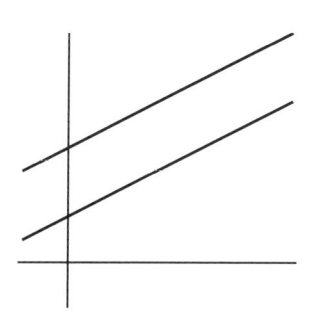

Figure 4.19. Parallel; determinant in (4.37) equals zero.

---

[7]Note that $b = 0$ if and only if $d = 0$ because if one coefficient in the equation of a line is zero, the other must be nonzero.

4.2. Linear operators

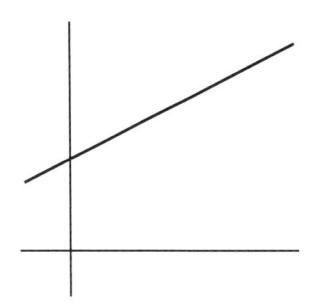

Figure 4.20. Coincident; determinant in (4.37) equals zero.

In our example (equations (4.34) and (4.35)), the coefficients are $a = e^{2t_0}, b = e^{4t_0}, c = 2e^{2t_0}$, and $d = 4e^{2t_0}$, and we compute

$$\det \begin{pmatrix} e^{2t_0} & e^{4t_0} \\ 2e^{2t_0} & 4e^{4t_0} \end{pmatrix} = 4e^{6t_0} - 2e^{6t_0} = 2e^{6t_0}, \tag{4.38}$$

and this is nonzero no matter what t_0 is. This tells us geometrically that the two lines in equations (4.34), and (4.35) intersect in a unique point, regardless of the values of α, β, and t_0. In short, we can satisfy any desired initial conditions with a function from the family of solutions in (4.33). Observe that the matrix in equation (4.38) is equal to

$$\begin{pmatrix} y_1(t_0) & y_2(t_0) \\ y_1'(t_0) & y_2'(t_0) \end{pmatrix},$$

where $y_1(t) = e^{2t}$ and $y_2(t) = e^{4t}$, the two solutions to $y'' - 6y' + 8y = 0$ we started with. We'll return to this observation shortly.

Now that we know that for our example the answer to (Q_2) is "yes", we tackle our other question, (Q_1): Is $\{c_1 e^{2t} + c_2 e^{4t}\}$ the complete family of all solutions to $y'' - 6y' + 8y = 0$? Suppose we have a specific function f that solves this differential equation. Pick any point t_0 in the set of real numbers $\mathbb{R}$, and compute the two numbers $f(t_0)$ and $f'(t_0)$. If we set $\alpha = f(t_0)$ and $\beta = f'(t_0)$, then we have *manufactured* an initial value problem

$$y'' - 6y' + 8y = 0, \quad y(t_0) = \alpha, \quad y'(t_0) = \beta, \tag{4.39}$$

having f as a solution. We also know, from the work that we just did, that there is a solution to this same initial value problem (4.39) from the family $\{c_1 e^{2t} + c_2 e^{4t}\}$. The Existence and Uniqueness Theorem from the last section tells us that our initial value problem has a *unique* solution, so we conclude that $f(t) = c_1 e^{2t} + c_2 e^{4t}$ for some choice of the constants c_1 and c_2, and thus the answer to our question (Q_1) is "yes".

The ideas in this example carry over to all second-order linear homogeneous differential equations. We summarize these next.

A fundamental set of solutions and the general solution. Suppose we have found two solutions y_1 and y_2 to the linear homogeneous equation

$$y'' + p(t)y' + q(t)y = 0 \tag{4.40}$$

on some open interval $I = (a, b)$, where $p(t)$ and $q(t)$ are continuous on I. We say that this pair of functions $\{y_1, y_2\}$ constitutes a **fundamental set of solutions** to the homogeneous linear differential equation (4.40) in the interval I if the set of **linear combinations** of this pair of functions,

$$c_1 y_1 + c_2 y_2,$$

for arbitrary choice of the *constants* c_1 and c_2, is the family of **all** solutions to (4.40) on I. The term **general solution** (on I) is used for

$$y = c_1 y_1 + c_2 y_2,$$

and we think of a fundamental set of solutions as being "building blocks" for a general solution.

The Wronskian. If we have two solutions y_1 and y_2 to (4.40), we can test to see if they form a fundamental set of solutions by computing the following:

$$W(y_1, y_2) \equiv \det \begin{pmatrix} y_1 & y_2 \\ y_1' & y_2' \end{pmatrix} = y_1 y_2' - y_1' y_2.$$

If this determinant is nonzero at some point t_0 in our interval I, then by the same reasoning that provided answers of "yes" to questions (Q_1) and (Q_2) above, we are guaranteed that $\{y_1, y_2\}$ is a fundamental set of solutions to (4.40). This 2×2 determinant W is called the **Wronskian determinant**, or just the Wronskian, of the pair of functions y_1, y_2. Notice that since y_1 and y_2 are functions of the independent variable t, so is W, and we occasionally write $W(y_1, y_2)(t)$, or even just $W(t)$, instead of $W(y_1, y_2)$ if we want to emphasize this point of view.

We'll see shortly (see Theorem 4.2.12) that as long as the functions y_1 and y_2 came to us as solutions on some open interval I to a differential equation as in (4.40), their Wronskian will be either *never* equal to 0 on I or *always* equal to 0 on I; no other scenario is possible. The first option, $W \neq 0$ on I, is the one that tells us that $\{y_1, y_2\}$ is a fundamental set of solutions and that the equation (4.40) has general solution

$$y = c_1 y_2 + c_2 y_2.$$

We illustrate these general principles with several examples.

Example 4.2.6. Is $\{e^t, e^{2t}\}$ a fundamental set of solutions for $y'' - 3y' + 2y = 0$ on $(-\infty, \infty)$?

Check by substitution that both $y_1(t) = e^t$ and $y_2(t) = e^{2t}$ do indeed solve the differential equation. Their Wronskian is

$$W(e^t, e^{2t}) = \det \begin{pmatrix} e^t & e^{2t} \\ e^t & 2e^{2t} \end{pmatrix} = e^{3t},$$

which is never equal to 0. Thus we conclude that $\{e^t, e^{2t}\}$ is a fundamental set of solutions and

$$y = c_1 e^t + c_2 e^{2t}$$

is the general solution.

Example 4.2.7. We keep the same differential equation as in the preceding example and now ask if $\{e^t, 2e^t - 3e^{2t}\}$ is a fundamental set of solutions. Again, $y_1 = e^t$ and $y_2 = 2e^t - 3e^{2t}$ are both solutions to the differential equation. Computing

$$W(y_1, y_2) = \det \begin{pmatrix} e^t & 2e^t - 3e^{2t} \\ e^t & 2e^t - 6e^{2t} \end{pmatrix} = -3e^{3t},$$

we see that this is nonzero on $(-\infty, \infty)$. Thus $\{e^t, 2e^t - 3e^{2t}\}$ is also a fundamental set of solutions to this differential equation. This leads to a description of the general solution as

$$y = c_1 e^t + c_2(2e^t - 3e^{2t});$$

you should convince yourself that this is equivalent to the general solution produced in Example 4.2.6.

4.2. Linear operators

Example 4.2.8. We'll generalize what we saw in the preceding examples. Consider any differential equation
$$ay'' + by' + cy = 0.$$
When we substitute the trial solution $y = e^{rt}$ into this differential equation we are led to the equation
$$ar^2 + br + c = 0.$$
Suppose this equation has two **different** real numbers solutions; call them r_1 and r_2. Is $\{e^{r_1 t}, e^{r_2 t}\}$ a fundamental set of solutions? Since $y_1 = e^{r_1 t}$ and $y_2 = e^{r_2 t}$ are solutions to our differential equation and their Wronskian is
$$W(e^{r_1 t}, e^{r_2 t}) = \det \begin{pmatrix} e^{r_1 t} & e^{r_2 t} \\ r_1 e^{r_1 t} & r_2 e^{r_2 t} \end{pmatrix} = (r_2 - r_1) e^{(r_1 + r_2)t},$$
which is never 0 (since we're assuming $r_1 \neq r_2$), the answer is *yes*. Furthermore, the general solution is
$$y = c_1 e^{r_1 t} + c_2 e^{r_2 t}.$$

Example 4.2.9. You can easily check that
$$y_1(t) = \cos(2t) \quad \text{and} \quad y_2(t) = \sin(2t)$$
are solutions to the equation $y'' + 4y = 0$. Are they a fundamental set of solutions on $(-\infty, \infty)$? We compute their Wronskian to answer this:
$$W(\cos(2t), \sin(2t)) = \det \begin{pmatrix} \cos(2t) & \sin(2t) \\ -2\sin(2t) & 2\cos(2t) \end{pmatrix} = 2\cos^2(2t) + 2\sin^2(2t) = 2,$$
so $\{\cos(2t), \sin(2t)\}$ is a fundamental set of solutions for $y'' + 4y = 0$.

Example 4.2.10. In this example, the coefficients of our homogeneous differential equation are not constants. Consider

(4.41) $$2t^2 y'' + 3t y' - y = 0,$$

and check that $\sqrt{t}$ and $1/t$ are both solutions to this equation. Is $\{\sqrt{t}, \frac{1}{t}\}$ a fundamental set of solutions on $(0, \infty)$? Before we answer this, recall why we restrict ourselves to the interval $(0, \infty)$. The standard form of the equation in (4.41) is
$$y'' + \frac{3}{2t} y' - \frac{1}{2t^2} y = 0$$
and the coefficient functions
$$p(t) = \frac{3}{2t} \quad \text{and} \quad q(t) = -\frac{1}{2t^2}$$
are not defined at $t = 0$ but are continuous everywhere else. This means we need to work on an open interval not containing 0; $(0, \infty)$ is the obvious choice since $\sqrt{t}$ is defined only for $t > 0$. Computing the Wronskian
$$W\left(\sqrt{t}, \frac{1}{t}\right) = W(\sqrt{t}, t^{-1}) = \det \begin{pmatrix} \sqrt{t} & t^{-1} \\ \frac{1}{2\sqrt{t}} & -t^{-2} \end{pmatrix} = -\frac{3}{2} t^{-\frac{3}{2}}$$
we observe this is nonzero on $(0, \infty)$. This tells us that $\{\sqrt{t}, \frac{1}{t}\}$ is a fundamental set of solutions, and the general solution to equation (4.41) on $(0, \infty)$ is
$$y(t) = c_1 \sqrt{t} + \frac{c_2}{t}.$$

Notice that the constant function $y = 0$ is a solution of (4.41). If we ask whether $\{0, \sqrt{t}\}$ is a fundamental set of solutions to this same differential equation on the same interval $(0, \infty)$, the answer is no, since

$$W(0, \sqrt{t}) = \det \begin{pmatrix} 0 & \sqrt{t} \\ 0 & \frac{1}{2\sqrt{t}} \end{pmatrix} = 0$$

for all t in $(0, \infty)$. This also means that not every solution to (4.41) can be produced by making some choice of constants c_1 and c_2 in the expression

$$c_1(0) + c_2(\sqrt{t});$$

for example the solution $\frac{1}{t}$ cannot be so obtained.

We summarize our work so far with the following useful result:

Theorem 4.2.11 (General Solutions Theorem). *Suppose that y_1 and y_2 are solutions to*

(4.42) $$y'' + p(t)y' + q(t)y = 0$$

on an open interval I on which p and q are continuous and that $W(y_1, y_2)$ is nonzero at a point of I. Then $\{y_1, y_2\}$ is a fundamental set of solutions to equation (4.42) on I. For each solution y of (4.42) on I, there are constants c_1 and c_2 such that

$$y = c_1 y_1 + c_2 y_2$$

on I.

Abel's Theorem and the Wronskian dichotomy. In Examples 4.2.6–4.2.10, either $W(y_1, y_2)$ was identically equal to 0 on the interval I or $W(y_1, y_2)$ was never equal to 0 on I. The next result says that this dichotomy is always present. It will give a formula, called **Abel's formula**, for the Wronskian directly in terms of information present in the differential equation itself (in other words, it will tell us what the Wronskian looks like without having to first find solutions to the differential equation).

Theorem 4.2.12 (Abel's Theorem). *If y_1 and y_2 are solutions to*

$$y'' + p(t)y' + q(t)y = 0$$

on an open interval I on which p and q are continuous, then

(4.43) $$W(y_1, y_2)(t) = Ce^{-\int p(t)dt}$$

for some constant C. In particular, the Wronskian is either never 0 (if $C \neq 0$) or identically equal to 0 (if $C = 0$) in I.

Before we see why this theorem is true, let's look back at Examples 4.2.6 and 4.2.7 and relate what we did there to the statement of Theorem 4.2.12. The differential equation in both of these examples was

$$y'' - 3y' + 2y = 0$$

and so $p(t) = -3$, $q(t) = 2$, and we may choose $I = (-\infty, \infty)$. In Abel's formula,

$$e^{-\int p(t)dt} = e^{-\int (-3)dt} = e^{3t}$$

if we choose $-3t$ as our antiderivative of -3. According to Abel's Theorem, the Wronskian of any pair of solutions to this differential equation should be Ce^{3t} for some constant C. Looking back at Example 4.2.6 we see that the constant C for the solution pair there is $C = 1$, while the different pair of solutions in Example 4.2.7 corresponds to the choice $C = -3$.

4.2. Linear operators

Verifying Abel's formula. Why is Abel's formula (4.43) true? We are given two functions y_1 and y_2, both of which solve the equation $y'' + p(t)y' + q(t)y = 0$ on some open interval I. This tells us that

(4.44) $$y_1'' + p(t)y_1' + q(t)y_1 = 0$$

and

(4.45) $$y_2'' + p(t)y_2' + q(t)y_2 = 0.$$

If we multiply equation (4.44) by $-y_2$, multiply equation (4.45) by y_1, and add the resulting equations together, we obtain

(4.46) $$-y_2 y_1'' + y_1 y_2'' + p(t)[y_1 y_2' - y_1' y_2] + q(t)[y_2 y_1 - y_1 y_2] = 0.$$

The coefficient of $q(t)$ in equation (4.46) is 0 and the coefficient of $p(t)$ is $W(y_1, y_2)$. Check that the expression $-y_2 y_1'' + y_1 y_2''$ is equal to W', the derivative of the Wronskian $y_1 y_2' - y_1' y_2$. Thus we can rewrite equation (4.46) as

(4.47) $$W' + p(t)W = 0.$$

This is a first-order linear (or separable) equation. Solving it as a linear equation, we use the integrating factor

$$e^{\int p(t)dt}$$

and obtain the solutions

$$W = Ce^{-\int p(t)dt},$$

which is Abel's formula in Theorem 4.2.12.

4.2.1. Revisting $ay'' + by' + cy = 0$: Solving by factoring operators. At the start of this section we solved the constant coefficient homogeneous equation $y'' - 6y' + 8y = 0$ by "guessing" that it has solutions in the form $y = e^{rt}$ and then determining by substitution for which values of r this is the case. We will close this section by returning to the same example and use it to show how an operator-factoring perspective could produce the solution with no guessing involved. This will also illustrate how our knowledge of solving first-order equations can be used to solve a second-order equation.

Writing D for the linear operator sending y to y' and D^2 for the linear operator sending y to y'', we have $y'' - 6y' + 8y = (D^2 - 6D + 8)[y]$ (notice that the third term $+8$ is the operator of multiplication by 8). Next observe that we can "factor" the operator $D^2 - 6D + 8$ as $(D-4)(D-2)$. By this we mean

$$(D^2 - 6D + 8)[y] = (D-4)(D-2)[y]$$

for any function y. To check this, we compute

$$\begin{aligned}(D-4)(D-2)[y] &= (D-4)[y' - 2y] = (y' - 2y)' - 4(y' - 2y) \\ &= y'' - 2y' - 4y' + 8y = y'' - 6y' + 8y \\ &= (D^2 - 6D + 8)[y].\end{aligned}$$

(It is also true that $D^2 - 6D - 8 = (D-2)(D-4)$, which you can similarly check.)

Now suppose that $y(t)$ is any solution whatsoever to the equation $y'' - 6y' + 8y = 0$ on $(-\infty, \infty)$, so that $(D-4)(D-2)[y] = 0$. Set

(4.48) $$h(t) = (D-2)[y] = y' - 2y$$

and observe that $(D-4)[h] = 0$. This last equation says $h' - 4h = 0$, and our work in Section 2.1 on solving first-order linear equations says that $h(t) = ce^{4t}$ for constant c. Using this together with the definition of h in equation (4.48) gives
$$y' - 2y = ce^{4t},$$
another first-order linear equation we can solve, using the integrating factor $\mu(t) = e^{-2t}$:
$$\frac{d}{dt}(ye^{-2t}) = ce^{2t},$$
$$ye^{-2t} = \frac{c}{2}e^{2t} + c_2,$$
$$y = c_1 e^{4t} + c_2 e^{2t},$$
where we have written c_1 for $\frac{c}{2}$. Thus any solution to $y'' - 6y' + 8y = 0$ must be in the family $c_1 e^{4t} + c_2 e^{2t}$, and conversely, any function in this family solves the differential equation, as is easily verified by substitution. This gives our general solution $y = c_1 e^{4t} + c_2 e^{2t}$. Effectively, we replaced the problem of solving the original second-order equation by the problem of solving two first-order equations in succession.

In Exercises 29–32 you can practice this operator-factoring technique to solve some other constant coefficient equations.

4.2.2. Exercises.

1. For each of the operators described below, decide whether or not the operator is linear.
   (a) The operator D^2, which takes any twice-differentiable function f to f'', its second derivative.
   (b) The operator R, which takes any nonzero function f to its reciprocal $\frac{1}{f}$.
   (c) The operator M_{t^2}, which takes any function $f(t)$ to the function $t^2 f(t)$.
   (d) The operator M_h, which takes any function f to the function hf, where h is some chosen fixed function. (In part (c) we used the choice $h(t) = t^2$.)
   (e) The operator C, which takes any function f to the function $\cos(f)$.
   (f) The operator A, which takes any continuous function f to $F(t) = \int_0^t f(s)\,ds$.

2. Verify the assertion made in the text that for the operator L defined by
$$L[y] = y'' + p(t)y' + q(t)y,$$
we have $L[cf] = cL[f]$, for every constant c and every twice-differentiable function f.

3. If $L[y] = y'' + 3t^2 y' - 2ty$, compute the following:
   (a) $L[t^3]$.
   (b) $L[\sin t]$.
   (c) $L[t^3 + \sin t]$.
   (d) $L[5t^3 - 3\sin t]$.
   (e) $L[-2t^3 + 6\sin t]$.

In Exercises 4–7, use Example 4.2.8 to record the general solution (assuming t is the independent variable).

4. $y'' - 9y = 0$.
5. $y'' + 3y' = 0$.
6. $y'' - 5y' + 6y = 0$.
7. $2y'' - y' - 3y = 0$.

In Exercises 8–11, use Example 4.2.8 to solve the initial value problem (assuming t is the independent variable).

4.2. Linear operators

8. $y'' - 4y = 0, y(0) = 2, y'(0) = 0.$
9. $y'' - 3y' - 4y = 0, y(0) = -1, y'(0) = 6.$
10. $y'' + y' = 0, y(0) = 0, y'(0) = 2.$
11. $y'' - 3y' + 2y = 0, y(0) = 1, y'(0) = 2.$

12. Show that $y_1(t) = \frac{1}{t}$ is a solution to the differential equation $y'' - 2y^3 = 0$. Determine all values of c such that $y_2 = cy_1$ is a solution to this same equation. Why doesn't this contradict Theorem 4.2.5?

13. Show that both $y_1(t) = 2$ and $y_2(t) = \sqrt{t}$ are solutions to the differential equation $yy'' + (y')^2 = 0$. Is $y_1 + y_2$ also a solution? Why doesn't this contradict Theorem 4.2.5?

14. Consider the differential equation
$$(4.49) \qquad y'' + 5y' + 6y = 0.$$
   (a) Suppose that $y_1(t) = e^{-2t} + 2e^{-3t}$ and $y_2(t) = 4e^{-2t} + be^{-3t}$. Show y_1 and y_2, for any choice of b, are solutions to equation (4.49).
   (b) Find a value of b such that $\{y_1, y_2\}$ is a fundamental set of solutions to equation (4.49).
   (c) Find a value of b such that $\{y_1, y_2\}$ is not a fundamental set of solutions to equation (4.49).
   (d) How many correct answers to (b) are there? How many correct answers to (c)?

15. (a) Show that $\{e^t, e^{-t}\}$ is a fundamental set of solutions to $y'' - y = 0$ on $(-\infty, \infty)$.
    (b) Show that $\{(e^t + e^{-t})/2, (e^t - e^{-t})/2\}$ is also a fundamental set of solutions to this same equation on $(-\infty, \infty)$.
    (c) The hyperbolic sine and cosine functions are defined by
    $$\sinh t = \frac{e^t - e^{-t}}{2} \quad \text{and} \quad \cosh t = \frac{e^t + e^{-t}}{2}.$$
    Verify that
    $$\sinh 0 = 0, \quad \cosh 0 = 1$$
    and
    $$\frac{d}{dt} \sinh t = \cosh t, \quad \frac{d}{dt} \cosh t = \sinh t.$$
    (d) Using the hyperbolic sine and cosine functions and the result of (b), solve the initial value problem
    $$y'' - y = 0, \quad y(0) = 5, \quad y'(0) = 8.$$
    Can you see why the computations are easier than if we had used the fundamental set of solutions from (a)?

16. (a) Show that if e^{rt} is a solution to
    $$ay'' + by' + cy = 0$$
    on $(-\infty, \infty)$, then so is $e^{r(t-t_0)}$ for any choice of a constant t_0.
    (b) Find two values of r so that e^{rt} solves
    $$y'' - 2y' - 3y = 0.$$
    (c) Solve the initial value problem
    $$y'' - 2y' - 3y = 0, \quad y(4) = 1, \quad y'(4) = -5,$$
    using your answer to (a) and (b) and making a judicious choice of t_0 (chosen, on the basis of the initial conditions, to simplify the calculations).

17. Suppose the initial value problem
    $$y'' + p(t)y' + q(t)y = 0, \quad y(t_0) = \alpha, \quad y'(t_0) = 0,$$
    has solution y_1 and the initial value problem
    $$y'' + p(t)y' + q(t)y = 0, \quad y(t_0) = 0, \quad y'(t_0) = \beta,$$

has solution y_2. Can you solve
$$y'' + p(t)y' + q(t)y = 0, \quad y(t_0) = \alpha, \quad y'(t_0) = \beta,$$
without further computation? Explain.

18. Construct a second-order linear homogeneous differential equation that has the given pair of functions as a fundamental set of solutions.
    (a) $\{e^{2t}, e^{-t}\}$.
    (b) $\{e^{4t}, e^{-2t}\}$.
    (c) $\{2e^{4t} + 3e^{-2t}, e^{4t} - 5e^{-2t}\}$.
    Hint: In each case there is a choice of constants b and c so that $y'' + by' + cy = 0$ has the desired fundamental set of solutions.

19. Without solving the equation
$$2t^2 y'' + 3ty' - y = 0$$
show that for any pair of solutions y_1, y_2 on $(0, \infty)$ the Wronskian $W(y_1, y_2)$ will be equal to $Ct^{-3/2}$ for some constant C.

20. Sally finds two solutions to the homogeneous equation
$$y'' + p(t)y' + q(t)y = 0$$
on an open interval I on which p and q are continuous and calculates that the Wronskian of her two solutions is e^{5t}. Sam finds two solutions to the same equation and calculates the Wronskian of his solution to be $3e^{2t}$. Can both be correct? Answer the same question if instead the Wronskian of Sam's solutions is 0.

21. (a) Suppose the function $y_1(t)$ solves the initial value problem
$$y'' + p(t)y' + q(t)y = 0, \quad y(t_0) = 0, \quad y'(t_0) = 1,$$
and $y_2(t)$ solves the initial value problem
$$y'' + p(t)y' + q(t)y = 0, \quad y(t_0) = 1, \quad y'(t_0) = 0,$$
on an open interval I containing t_0 and that $p(t)$ and $q(t)$ are continuous on I. Is the pair $\{y_1, y_2\}$ a fundamental set of solutions for the equation $y'' + p(t)y' + q(t)y = 0$ on I?
(b) Answer the same question if instead y_2 satisfies $y_2(t_0) = 0, y_2'(t_0) = 0$.

22. Using $y(t) = t^r$ as a trial solution, find a fundamental set of solutions for
$$t^2 y'' - 2ty' + 2y = 0$$
of the form $\{t^{r_1}, t^{r_2}\}$ for particular real numbers r_1, r_2. Compute the Wronskian of the functions in your fundamental set and notice that the Wronskian is 0 at $t = 0$ and nonzero elsewhere. Why doesn't this contradict the second assertion in Abel's Theorem?

23. (a) Show that both $y_1(t) = e^{-t^2}$ and $y_2(t) = e^{-t^2} \int_0^t e^{s^2} ds$ are solutions to $y'' + 2ty' + 2y = 0$.
    (b) Is $\{y_1, y_2\}$ a fundamental set of solutions to this equation on the interval $(-\infty, \infty)$?
    (c) Solve the initial value problem
$$y'' + 2ty' + 2y = 0, \quad y(0) = 1, \quad y'(0) = 1.$$

24. Suppose that y_1 and y_2 are solutions to the differential equation
$$y'' + p(t)y' + q(t)y = 0$$
on the interval $a < t < b$, where $p(t)$ and $q(t)$ are continuous on this interval. If there is a point t_0 in this interval with
$$y_1(t_0) = y_2(t_0) = 0,$$
show that $\{y_1, y_2\}$ is **not** a fundamental set of solutions to the differential equation on $a < t < b$.

4.2. Linear operators

25. Suppose that y_1 and y_2 are solutions to the differential equation
$$y'' + p(t)y' + q(t)y = 0$$
on the interval $a < t < b$, where p and q are continuous on (a, b). If there is a point t_0 in this interval such that both y_1 and y_2 have either a maximum or a minimum at t_0, show that $\{y_1, y_2\}$ is **not** a fundamental set of solutions to the differential equation on $a < t < b$.

26. Follow the outline given below to show that if y_1 and y_2 are solutions to the differential equation
$$y'' + p(t)y' + q(t)y = 0$$
on the interval $a < t < b$ (on which p and q are continuous) and each has an inflection point at t_0, then $\{y_1, y_2\}$ cannot be a fundamental set of solutions on $a < t < b$ unless both p and q are zero at t_0.
    (a) Explain why y_1'' and y_2'' are necessarily continuous on the interval (a, b).
    (b) You are given that
    $$y_1'' + py_1' + qy_1 = 0$$
    and
    $$y_2'' + py_2' + qy_2 = 0$$
    on $a < t < b$. Plug in t_0 and simplify, using what you know about $y_1''(t_0)$ and $y_2''(t_0)$ (since there is an inflection point at t_0).
    (c) If not both p and q are zero at t_0, at least one of them is nonzero there. If $p(t_0)$ is not equal to zero, solve the equations from (a) for $y_1'(t_0)$ and $y_2'(t_0)$, and compute $W(y_1, y_2)(t_0)$. What do you conclude about W?
    (d) Now handle the case where $q(t_0)$ is nonzero.

27. Each of the pictures below show the graphs of two functions. Explain why, in each case, the two functions cannot be a fundamental set of solutions to a second-order linear homogeneous differential equation with continuous coefficient functions.

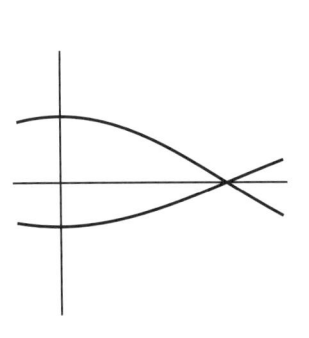

Figure 4.21.

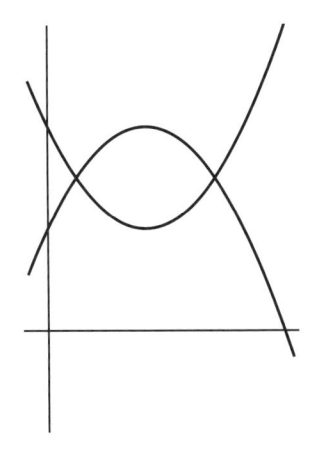

Figure 4.22.

28. Suppose that $p(t)$ and $q(t)$ are continuous on $(-\infty, \infty)$, and suppose we have a fundamental set of solutions $\{y_1, y_2\}$ of
$$y'' + p(t)y' + q(t)y = 0.$$
Suppose that a and b are two consecutive points where the graph of y_2 crosses the t-axis (with $a < b$), so that the graph of $y_2(t)$ looks like one of the two pictures shown below.

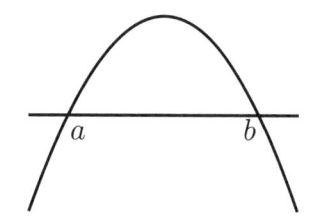

Figure 4.23. Graph of $y_2(t)$.

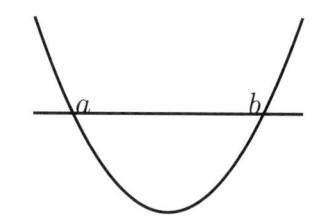

Figure 4.24. Graph of $y_2(t)$.

Calculate $W(y_1, y_2)(a)$ and $W(y_1, y_2)(b)$ and use this to show that $y_1(a)$ and $y_1(b)$ must have opposite signs. Conclude that y_1 must be equal to zero somewhere between $t = a$ and $t = b$. This conclusion is sometimes described by saying "the zeros of y_1 and y_2 interlace".

29. This exercise explores some properties of linear operator notation. We will look at three operators, denoted $(D+3)$, $(D+4)$, and $(D^2 + 7D + 12)$, whose rules are indicated below.

$$f \longrightarrow \boxed{D+3} \longrightarrow f' + 3f$$

$$f \longrightarrow \boxed{D+4} \longrightarrow f' + 4f$$

$$f \longrightarrow \boxed{D^2 + 7D + 12} \longrightarrow f'' + 7f' + 12f$$

(a) Check that each of these operators is linear.

(b) Now imagine hooking the operators $(D+3)$ and $(D+4)$ together, in each of two ways, as illustrated below.

$$f \longrightarrow \boxed{D+3} \xrightarrow{f'+3f} \boxed{D+4} \longrightarrow (f'+3f)' + 4(f'+3f)$$

$$f \longrightarrow \boxed{D+4} \xrightarrow{f'+4f} \boxed{D+3} \longrightarrow ?$$

In the first schematic, f is inputted into the operator $(D+3)$ and the resulting output, $f' + 3f$, is then used as input for the second operator $(D+4)$; this is sometimes written $(D+4)(D+3)[f]$. The net result of these two operators in succession is to produce a final output of

$$(f' + 3f)' + 4(f' + 3f) = f'' + 7f' + 12f.$$

What is the final output in the case of the second schematic,

$$(D+3)(D+4)[f]?$$

How do both of these compare with $(D^2 + 7D + 12)[f]$? Comments?

(c) Suppose that y is any solution to $y'' + 7y' + 12y = 0$ on $(-\infty, \infty)$, so that

$$(D^2 + 7D + 12)[y] = 0.$$

4.2. Linear operators

Set $h(t) = y'(t) + 4y(t)$ for all t. Show that
$$(D+3)[h] = 0.$$
Solve this last (first-order) equation for h; then solve $y'(t) + 4y(t) = h(t)$ for $y(t)$ to obtain the general solution of $y'' + 7y' + 12y = 0$.

30. In this problem you will use operator-factoring to solve $y'' + y' - 6y = 0$.
    (a) Write the operator $D^2 + D - 6$ in factored form as $(D+a)(D+b)$ for appropriate constants a and b. Verify that your factorization is correct by computing $(D^2 + D - 6)[y]$ and $(D+a)(D+b)[y]$.
    (b) Use the method of Section 4.2.1 to find the general solution to $y'' + y' - 6y = 0$.

31. Follow the same outline as in Exercise 30 to solve $y'' - 4y = 0$.

32. (a) Write the operator $6D^2 + D - 2$ in factored form as $(a_1 D + b_1)(a_2 D + b_2)$ for appropriate constants a_j and b_j.
    (b) Use the method of Section 4.2.1 to find the general solution to $6y'' + y' - 2y = 0$.

33. The ideas in this problem and the next two will appear again in later sections.
    (a) Show that any function that solves $y' = \frac{1}{2}y$ also solves $y'' = \frac{1}{4}y$.
    (b) Find a function of the form $y(t) = e^{rt}$ that solves $y'' = \frac{1}{4}y$ but does not solve $y' = \frac{1}{2}y$.
    (c) Show that any function that solves $y'' + y' = 5$ also solves $y''' + y'' = 0$.
    (d) Find a function of the form $y = c_1 t + c_2 t^2$ that solves $y''' + y'' = 0$ but does not solve $y'' + y' = 5$.

34. Consider the operators $(D+2)$ and $(D^2 + 2D)$ as shown below.

    $$f \longrightarrow \boxed{D+2} \xrightarrow{f' + 2f}$$

    $$f \longrightarrow \boxed{D^2 + 2D} \xrightarrow{f'' + 2f'}$$

    (a) If $(D+2)[g] = 0$, is $(D^2 + 2D)[g] = 0$?
    (b) If $(D^2 + 2D)[g] = 0$, is $(D+2)[g] = 0$?

35. Consider the operators $(D+2)$ and $(D^2 + 4D + 4)$ as shown below.

    $$f \longrightarrow \boxed{D+2} \xrightarrow{f' + 2f}$$

    $$f \longrightarrow \boxed{D^2 + 4D + 4} \xrightarrow{f'' + 4f' + 4f}$$

    (a) If $(D+2)[f] = 0$, is $(D^2 + 4D + 4)[f] = 0$?
    (b) If $(D^2 + 4D + 4)[f] = 0$, must $(D+2)[f] = 0$? Hint: Try $f(t) = te^{bt}$ for appropriate choice of b.
    (c) Verify that $(D+2)(D+2) = D^2 + 4D + 4$.
    (d) Suppose that y is a solution to $y'' + 4y' + 4y = 0$ on $(-\infty, \infty)$. Set $h = y' + 2y$ and show that $(D+2)[h] = 0$. Finally, solve $(D+2)[h] = 0$ for h and then $y' + 2y = h$ to obtain the general solution to $y'' + 4y' + 4y = 0$.

36. Cramer's Rule says that the system of equations

$$ax + by = k_1,$$
$$cx + dy = k_2$$

has exactly one solution (x, y) if and only if

$$W = \det \begin{pmatrix} a & b \\ c & d \end{pmatrix} \neq 0$$

and that the unique solution is then

$$x = \frac{\det \begin{pmatrix} k_1 & b \\ k_2 & d \end{pmatrix}}{W} = \frac{k_1 d - k_2 b}{ad - bc}$$

and

$$y = \frac{\det \begin{pmatrix} a & k_1 \\ c & k_2 \end{pmatrix}}{W} = \frac{k_2 a - k_1 c}{ad - bc}.$$

Use Cramer's Rule to solve

$$2x + 3y = 1,$$
$$x + 4y = -1.$$

37. Suppose that y_1 and y_2 are solutions to

(4.50) $$y'' + p(t)y' + q(t)y = 0$$

on an open interval I on which p and q are continuous. This exercise outlines an argument that uses the Existence and Uniqueness Theorem, rather than Abel's Theorem, to establish that if $W(y_1, y_2)$ is nonzero at one point of I, then it is nonzero at every point of I.

Suppose that $W(y_1, y_2)(t_0) \neq 0$ for some t_0 in I. Let t_1 in I be arbitrary, and let k_1 and k_2 be arbitrary constants. By Theorem 4.1.4, the initial value problem

$$y'' + p(t)y' + q(t)y = 0, \qquad y(t_1) = k_1, \quad y'(t_1) = k_2,$$

has a unique solution $y = \varphi(t)$ on I. Let $\alpha = \varphi(t_0)$ and $\beta = \varphi'(t_0)$.

(a) Show that there are constants c_1 and c_2 such that $y = c_1 y_1 + c_2 y_2$ solves, on I, the initial value problem

$$y'' + p(t)y' + q(t)y = 0, \qquad y(t_0) = \alpha, \quad y'(t_0) = \beta.$$

(b) Explain why $\varphi(t) = c_1 y_1(t) + c_2 y_2(t)$ for every t in I, and conclude that

$$c_1 y_1(t_1) + c_2 y_2(t_1) = k_1, \quad c_1 y_1'(t_1) + c_2 y_2'(t_1) = k_2.$$

(c) Using facts about determinants and linear systems discussed in this section, explain why

$$W(y_1, y_2)(t_1) \neq 0.$$

38. Suppose both y_1 and y_2 solve a nonhomogeneous linear equation

$$y'' + p(t)y' + q(t)y = g(t).$$

Use operator notation to show that $y_1 - y_2$ solves the associated homogeneous equation

$$y'' + p(t)y' + q(t)y = 0.$$

39. Suppose that y_1 solves
$$y'' + p(t)y' + q(t)y = g_1(t)$$
on an interval I while y_2 solves
$$y'' + p(t)y' + q(t)y = g_2(t)$$
on I. The linear combination $c_1y_1 + c_2y_2$ is a solution on I of
$$y'' + p(t)y' + q(t)y = G(t)$$
for what function G?

40. Suppose that y_1 and y_2 are solutions of $ty'' + 3y' + (t\sin t)y = 0$ on $0 < t < \infty$, and suppose that the Wronskian $W(y_1, y_2)(1) = 2$. What is $W(y_1, y_2)(t)$?

41. Suppose that y_1 and y_2 are solutions of $y'' + 2ty' + 2y = 0$ on $-\infty < t < \infty$, and suppose that the Wronskian $W(y_1, y_2)(0) = 3$. What is $W(y_1, y_2)(t)$?

42. Suppose that y_1 and y_2 are any two functions with $W(y_1, y_2) \neq 0$ on an open interval I. Construct a linear homogeneous differential equation
$$y'' + p(t)y' + q(t)y = 0$$
with fundamental set of solutions $\{y_1, y_2\}$ on I. Hint: We want
$$py_1' + qy_1 = -y_1'',$$
$$py_2' + qy_2 = -y_2''.$$
Solve for the "unknowns" p and q.

43. Construct a differential equation of the form $y'' + p(t)y' + q(t)y = 0$ that has fundamental set of solutions $\{t, t^3\}$ on the interval $(0, \infty)$. Hint: See Exercise 42.

44. Construct a differential equation of the form $y'' + p(t)y' + q(t)y = 0$ that has fundamental set of solutions $\{t^{1/2}, t^{3/2}\}$ on the interval $(0, \infty)$. Hint: See Exercise 42.

4.3. Linear independence

A fuller understanding of when we have a fundamental set of solutions to a linear homogeneous differential equation (and thus the building blocks for a general solution) comes from the idea of linear independence. This is a concept that appears in many areas of mathematics; our interest here will be in how it relates to linear differential equations.

We'll begin by discussing linear independence for a pair of functions; this is the scenario relevant to second-order differential equations. Later, in Section 4.6.2, we will expand the notion of linear independence to sets of more than two functions in order to study differential equations of order greater than two.

Definition 4.3.1. We say that two functions f and g are **linearly dependent** on an interval I provided that we can find two constants c_1 and c_2, **not both equal to** 0, with
$$c_1 f + c_2 g = 0$$
on I.

In words, this is expressed as "some nontrivial linear combination of f and g is the zero function". Notice that the "not both equal to 0" restriction on the constants c_1 and c_2 is crucial. The functions f and g are **linearly independent** on I precisely when they are **not** linearly dependent on I.

Example 4.3.2. Are the functions $\{e^x, e^{x-2}\}$ linearly dependent or linearly independent on $(-\infty, \infty)$? According to the definition, we are asking if we can fill in the boxes

$$\boxed{} e^x + \boxed{} e^{x-2} = 0$$

with constants, at least one of which is not equal to 0, to make an identity that is true for all real numbers x. The key idea is to recall that

$$e^{x-2} = e^x e^{-2}.$$

Thus we have

$$\boxed{-1} e^x + \boxed{e^2} e^{x-2} = 0,$$

and the functions e^x and e^{x-2} are linearly dependent on $(-\infty, \infty)$. They are also linearly dependent on any other interval, by the same reasoning. Other "fill in the boxes" choices are possible too; for example,

$$\boxed{e^{-2}} e^x + \boxed{-1} e^{x-2} = 0$$

is another natural one. We only need to find one allowable choice to conclude linear dependence.

Example 4.3.3. Are the functions $f(x) = \sin x$ and $g(x) = 0$ linearly dependent or linearly independent on $(-\infty, \infty)$? Since

$$\boxed{0} \sin x + \boxed{1} \, 0 = 0,$$

we conclude these functions are linearly dependent.

The idea in the last example generalizes to show that for any function f, the pair of functions $\{f, 0\}$ is linearly dependent; see Exercise 5.

Example 4.3.4. Show that the functions $f(x) = e^x$ and $g(x) = \sin x$ are linearly independent on $(-\infty, \infty)$. We must argue that if

(4.51) $$c_1 e^x + c_2 \sin x = 0$$

for all real numbers x, then **both** c_1 and c_2 must equal 0. If c_1 is not equal to 0, then we may rearrange equation (4.51) to say

$$e^x = -\frac{c_2}{c_1} \sin x,$$

which is clearly not true, since, for example, the function $\sin x$ is 0 at $x = 0$ but e^x is never 0. If c_1 is equal to 0, equation (4.51) becomes $c_2 \sin x = 0$ (for all x), which forces $c_2 = 0$ as well.

These examples give a clue as to a useful way to think about linear independence for a pair of functions:

> The pair of functions $\{f, g\}$ is linearly dependent on an interval I if and only if one function is a constant multiple of the other on that interval. They are linearly independent on I if and only if neither function is a constant multiple of the other on I.

To see why, notice that if $f = cg$ on I for some constant c, then

$$\boxed{1} f + \boxed{-c} g = 0$$

on I. This says f and g are linearly dependent on I. On the other hand, if $c_1 f + c_2 g = 0$ on I, where at least one of c_1, c_2 is not zero, we may solve this equation as in Example 4.3.4 to express one of the functions f, g as a constant multiple of the other. For example, if $c_1 \neq 0$, then $f = -\frac{c_2}{c_1} g$.

Looking back at Example 4.3.2 from this perspective, we see that $e^{x-2} = e^{-2} e^x$, so that e^{x-2} is a constant multiple of e^x on $(-\infty, \infty)$, and thus the two functions are linearly dependent there.

4.3. Linear independence

In Example 4.3.3 we have $0 = 0 \cdot \sin x$, and again one function in the pair is a constant multiple of the other, leading to linear dependence. Notice that the property "f is a constant multiple of g" need not imply that "g is a constant multiple of f". For instance, 0 is a constant multiple of $\sin x$ since $0 = 0 \cdot \sin x$, but $\sin x$ is not a constant multiple of 0.

The next example shows the importance of paying attention to the interval on which the problem is set.

Example 4.3.5. Suppose $f(x) = 2x^2$ and $g(x) = x|x|$. On the interval $I_1 = (0,1)$, f and g are linearly dependent, since $f = 2g$ for all points in I_1. On the interval $I_2 = (-1,0)$, $f = -2g$, so f and g are also linearly dependent on I_2. However, if we ask if f and g are linearly dependent on $I_3 = (-1,1)$, the answer is **no**, since there is no *single* constant c with $2x^2 = cx|x|$ (or with $c(2x^2) = x|x|$) for all x in I_3.

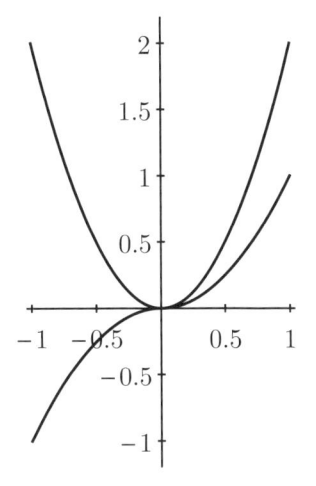

Figure 4.25. Graphs of $y = 2x^2$ and $y = x|x|$.

4.3.1. Connection with the Wronskian. We are mainly interested in linear independence as it applies to functions that are solutions of a linear homogeneous differential equation. The next theorem relates linear independence to the Wronskian in this situation. We already know from Theorem 4.2.12 that the Wronskian of a pair of solutions to $y'' + p(t)y' + q(t)y = 0$ is either identically equal to 0 or is never equal to 0 on an interval where $p(t)$ and $q(t)$ are continuous. The point of the next result is to relate these two options to linear dependence/independence of the solution pair.

Theorem 4.3.6. *Suppose that y_1 and y_2 are solutions to the differential equation*

$$y'' + p(t)y' + q(t)y = 0$$

on an open interval I on which p and q are continuous.

(a) *If y_1 and y_2 are linearly dependent on I, then their Wronskian $W(y_1, y_2)$ is identically equal to 0 in I.*

(b) *If the Wronskian $W(y_1, y_2)$ is identically equal to 0 on I, then y_1 and y_2 are linearly dependent on I.*

More compactly, the theorem says: "The solutions y_1 and y_2 are linearly dependent on I if and only if their Wronskian $W(y_1, y_2)$ is identically equal to 0 on I." We can also phrase the content of Theorem 4.3.6 as a statement about linear *independence*:

> If y_1 and y_2 are two solutions to $y'' + p(t)y' + q(t)y = 0$ on an open interval I on which $p(t)$ and $q(t)$ are continuous, then y_1 and y_2 are linearly independent on I if and only if their Wronskian is never equal to 0 on I.

Justifying Theorem 4.3.6. It is easy to see why (a) is true. Linear dependence of y_1 and y_2 means one function is a constant multiple of the other on I, say, $y_2 = cy_1$ for some constant c. We then have

$$W(y_1, y_2) = \det \begin{pmatrix} y_1 & y_2 \\ y_1' & y_2' \end{pmatrix} = \det \begin{pmatrix} y_1 & cy_1 \\ y_1' & cy_1' \end{pmatrix} = cy_1 y_1' - cy_1 y_1' = 0.$$

Notice that this part of the argument did not actually make use of the hypothesis that y_1 and y_2 are solutions to our differential equation, and so (a) holds for *any* pair of functions.

Let's see why (b) is true. We are given that $W(y_1, y_2) = 0$ on I, where y_1 and y_2 solve $y'' + p(t)y' + q(t)y = 0$ on I. Our goal is to show that y_1 and y_2 are linearly dependent. This is easy if one of the two functions, say, y_1, is never equal to 0 on I. In this case, we can form the quotient y_2/y_1 and compute its derivative:

$$\frac{d}{dt}\left(\frac{y_2}{y_1}\right) = \frac{y_1 y_2' - y_2 y_1'}{y_1^2} = \frac{W(y_1, y_2)}{y_1^2} = 0$$

on the interval I. Since the derivative is equal to 0, the function y_2/y_1 is constant on I, and we conclude that $y_2 = cy_1$ for some constant c. Notice that still so far we haven't used the fact that the functions y_1 and y_2 came to us as solutions to a differential equation.

To handle the case that y_1 has a zero at some point in I we have to work a little harder and use the fact that y_1 and y_2 solve our differential equation. We'll give one approach here, and an alternate argument is outlined in Exercise 19. Pick a point t_0 in I and consider the pair of equations

(4.52) $$y_1(t_0)X + y_2(t_0)Y = 0,$$
(4.53) $$y_1'(t_0)X + y_2'(t_0)Y = 0,$$

where X and Y are variables. These are equations of lines in the XY-plane. Since $W(y_1, y_2)(t_0) = 0$ by assumption,

$$\det \begin{pmatrix} y_1(t_0) & y_2(t_0) \\ y_1'(t_0) & y_2'(t_0) \end{pmatrix} = 0.$$

This tells us that the lines in the XY-plane with equations (4.52) and (4.53) have the same slope and are thus parallel. But it's also clear from equations (4.52) and (4.53) that both lines pass through the origin $(0,0)$ (why?). Thus they are the *same* line in the XY-plane. Pick any point (c_1, c_2) on this line other than the point $(0,0)$, and consider the function $y = c_1 y_1 + c_2 y_2$. Check that this function solves the initial value problem

$$y'' + p(t)y' + q(t)y = 0, \qquad y(t_0) = 0, \quad y'(t_0) = 0.$$

But the constant function $y(t) \equiv 0$ also solves this same initial value problem, and the uniqueness part of the Existence and Uniqueness Theorem (Theorem 4.1.4) tells us that

$$c_1 y_1 + c_2 y_2 = y = 0$$

on I, which is exactly the statement that y_1 and y_2 are linearly dependent on I.

Example 4.3.7. You can check by direct substitution that $y_1(t) = e^t$ and $y_2(t) = te^t$ are solutions to the differential equation $y'' - 2y' + y = 0$ on $(-\infty, \infty)$. It is also clear by inspection that these two functions are linearly independent on $(-\infty, \infty)$, since neither is a *constant* multiple of the

4.3. Linear independence

other. According to Theorem 4.3.6 this means that the Wronskian of e^t and te^t is never 0, and therefore $\{e^t, te^t\}$ is a fundamental set of solutions to $y'' - 2y' + y = 0$ on $(-\infty, \infty)$. Of course, we could have verified this by directly calculating $W(e^t, te^t)$, but this would be more computational work.

Summary. We can combine the results of Theorems 4.2.12, 4.2.11, and 4.3.6 to summarize our main conclusions about fundamental sets of solutions as follows:

Theorem 4.3.8. *Suppose that y_1 and y_2 are solutions of the differential equation*

(4.54) $$y'' + p(t)y' + q(t)y = 0$$

on an open interval I on which p and q are continuous. Then the following conditions are equivalent, so that if any one of them holds, so do all of the others.

(a) *The pair $\{y_1, y_2\}$ is a fundamental set of solutions of equation (4.54) on I; that is,* **every** *solution on I has the form $c_1 y_1 + c_2 y_2$ for some constants c_1 and c_2.*

(b) *The Wronskian $W(y_1, y_2)$ is nonzero at some point of I.*

(c) *The Wronskian $W(y_1, y_2)$ is nonzero at every point of I.*

(d) *The functions y_1 and y_2 are linearly independent on I.*

A natural question to ask is whether a fundamental set of solutions must always exist for equation (4.54) on an interval I on which its coefficient functions p and q are continuous. The answer is yes: Pick a point t_0 in I. By the Existence and Uniqueness Theorem, there are solutions y_1 and y_2 on I satisfying the initial conditions $y_1(t_0) = 0, y_1'(t_0) = 1$ and $y_2(t_0) = 1, y_2'(t_0) = 0$. In Exercise 16 you are asked to verify that then $\{y_1, y_2\}$ is a fundamental set of solutions.

4.3.2. Exercises.

In Exercises 1–3, use the *definition* of linear independence/dependence.

1. Show that the functions $f(t) = 2\sin^2 t$ and $g(t) = 1 - \cos^2 t$ are linearly dependent on $(-\infty, \infty)$.
2. Show that the functions $\ln t$ and $\ln t^2$ are linearly dependent on $(0, \infty)$.
3. Show that the functions $f(t) = \sin(t + \frac{\pi}{2})$ and $g(t) = 2\cos t$ are linearly dependent on $(-\infty, \infty)$.
4. In each of the following, decide whether the given pair of functions is linearly dependent or linearly independent on $(-\infty, \infty)$.
   (a) $\{\sin(2t), \sin t \cos t\}$.
   (b) $\{\sin(2t), \sin(3t)\}$.
   (c) $\{\sin(2t), \sin(-2t)\}$.
   (d) $\{\sin(2t), \cos(2t - \pi/2)\}$.
   (e) $\{e^{2t}, te^{2t}\}$.
   (f) $\{e^{2t}, e^{3t}\}$.
   (g) $\{e^{2t}, -e^{2t}\}$.
   (h) $\{e^{2t}, 1\}$.
   (i) $\{e^{2t}, 0\}$.
   (j) $\{e^{2t}, e^{-2t}\}$.
5. Show that for any function f, the pair of functions $\{f, 0\}$ is linearly dependent.
6. In each part, decide whether the given function is a linear combination of the functions $f(t)$ and $g(t)$.
   (a) $2f(t) + 3g(t)$.
   (b) $2f(t) + 3g(t) + 5$.
   (c) $f(t) + 3g(t) + t$.

(d) $f(t) - g(t)$.
(e) $f(t)g(t)$.

7. Show that the function $\cos(t - \pi/3)$ is a linear combination of the functions $\cos t$ and $\sin t$.

8. Suppose $f(t) = (1+t)^2$ and $h(t) = (1-t)^2$ on $(-\infty, \infty)$.
   (a) Show that $g_1(t) = t$ is a linear combination of f and h.
   (b) Show that $g_2(t) = t^2$ is not a linear combination of f and h.

9. True or false?
   (a) The constant function 0 is a linear combination of $f(t) = t$ and $g(t) = t^2$.
   (b) There is a nontrivial linear combination of $f(t) = t$ and $g(t) = t^2$ that is equal to 0 on $(-\infty, \infty)$. (A *nontrivial* linear combination is one in which not both coefficients are zero.)

10. Each of the functions in the three pairs shown below solves $y'' - y = 0$. For each pair, use the notion of linear dependence/independence (without computing any Wronskians) to decide whether or not the pair is a fundamental set of solutions to $y'' - y = 0$.
    (a) $\{e^t, e^{-t}\}$.
    (b) $\{e^t, e^{1-t}\}$.
    (c) $\{e^{-t}, e^{1-t}\}$.

11. The differential equation $y'' + y' - 6y = 0$ has fundamental set of solutions $\{e^{2t}, e^{-3t}\}$ on $(-\infty, \infty)$. For each of the following, decide whether or not it is also a fundamental set of solutions.
    (a) $\{2e^{2t}, -e^{-3t}\}$.
    (b) $\{e^{2t}, e^{2t} + e^{-3t}\}$.
    (c) $\{3e^{2t}, 5e^{2t}\}$.
    (d) $\{e^{2t} - e^{-3t}, e^{2t} + e^{-3t}\}$.

12. If f_1 and f_2 are linearly independent on an interval I, are they necessarily also linearly independent on any smaller interval contained in I?

13. If f_1 and f_2 are linearly dependent on an interval I, are they necessarily also linearly dependent on any smaller interval contained in I?

14. Recall that in Example 4.2.9 we saw that $\{\cos(2t), \sin(2t)\}$ is a fundamental set of solutions to $y'' + 4y = 0$.
    (a) Show that
    $$\cos\left(2t + \frac{\pi}{4}\right)$$
    also solves $y'' + 4y = 0$.
    (b) Write $\cos(2t + \frac{\pi}{4})$ as a linear combination of $\cos(2t)$ and $\sin(2t)$.
    (c) For what values of c is
    $$\{\cos(2t + c), \sin(2t + c)\}$$
    a fundamental set of solutions to $y'' + 4y = 0$?

15. Let $y_1(t) = t^2$ and $y_2(t) = t|t|$.
    (a) Show that y_1 and y_2 are linearly independent on $(-\infty, \infty)$.
    (b) Compute $W(y_1, y_2)$ and show that it is identically equal to 0 on $(-\infty, \infty)$. Notice that you'll need to compute the derivative of the function $t|t|$; work separately for $t > 0$, $t < 0$, and $t = 0$. For the last, use the definition of the derivative as a limit of difference quotients.
    (c) Why doesn't this example contradict Theorem 4.3.6?

16. Suppose that y_1 solves the initial value problem
    $$y'' + p(t)y' + q(t)y = 0, \quad y(t_0) = 0, \quad y'(t_0) = 1,$$
    and y_2 solves the initial value problem
    $$y'' + p(t)y' + q(t)y = 0, \quad y(t_0) = 1, \quad y'(t_0) = 0,$$

4.3. Linear independence

on some interval I, where p, q are continuous on I and t_0 is in I. Show that y_1 and y_2 are linearly independent on I, and conclude that $\{y_1, y_2\}$ is a fundamental set of solutions to $y'' + p(t)y' + q(t)y = 0$ on I.

17. Suppose you know that f and g are linearly independent on some interval I.
    (a) Show that $f + g$ and $f - g$ are linearly independent on I.
    (b) Show that $2f + 2g$ and $3f + 3g$ are linearly dependent on I.
    (c) Are $f + 2g$ and $2f + g$ linearly independent or linearly dependent on I?

18. Suppose you know that f and g are linearly independent on some interval I. Find a condition on constants b_1, b_2, c_1, c_2 that will guarantee that

    $$b_1 f + b_2 g \quad \text{and} \quad c_1 f + c_2 g$$

    are linearly independent on I.

19. Suppose that y_1 and y_2 solve $y'' + p(t)y' + q(t)y = 0$ on an open interval I on which p and q are continuous. Suppose further that $W(y_1, y_2) = 0$ on I. The goal of this problem is to show that y_1 and y_2 are linearly dependent on I, by a different argument than that in the text. If $y_1 \equiv 0$, we can just appeal to Exercise 5, so assume that there is a point t_0 in I with $y_1(t_0) \neq 0$. Since y_1 is continuous, this means there is an open interval J containing t_0 and contained in I with y_1 never equal to 0 on J.
    (a) Show that

    $$\frac{d}{dt}\left(\frac{y_2}{y_1}\right) = 0$$

    on J, and conclude that $y_2 = cy_1$ on J, for some constant c.
    (b) We want to conclude that $y_2 = cy_1$ on all of I. Use the Existence and Uniqueness Theorem to do this. Hint: Let $y_1(t_0) = \alpha$ and $y_1'(t_0) = \beta$. On the interval I, both y_1 and cy_2 solve the initial value problem

    $$y'' + p(t)y' + q(t)y = 0, \quad y(t_0) = \alpha, \quad y'(t_0) = \beta.$$

20. If you have had a linear algebra course, how might you fill in the blank in the following: A fundamental set of solutions $\{y_1, y_2\}$ of a second-order linear homogeneous equation is a _____ for the space of all solutions of the equation.

21. (a) Show that

    $$\{e^{2t}, e^t\} \quad \text{and} \quad \{e^{2t} + e^t, e^{2t} - e^t\}$$

    are both fundamental sets of solutions to $y'' - 3y' + 2y = 0$. Notice that each function in the second pair is a linear combination of the functions in the first pair; explain why this must be the case.
    (b) If $\{f_1, f_2\}$ is *any* fundamental set of solutions to $y'' - 3y' + 2y = 0$, explain why there must be constants a_1, a_2, b_1, b_2 with

    $$f_1 = a_1 e^{2t} + a_2 e^t, \quad f_2 = b_1 e^{2t} + b_2 e^t.$$

    (c) Show that the result of part (b) can be written using matrix multiplication as

    $$\begin{pmatrix} f_1 \\ f_2 \end{pmatrix} = \begin{pmatrix} a_1 & a_2 \\ b_1 & b_2 \end{pmatrix} \begin{pmatrix} e^{2t} \\ e^t \end{pmatrix}$$

    (d) Suppose that $\{f_1, f_2\}$ and $\{g_1, g_2\}$ are fundamental sets of solution to $L[y] = 0$, where L is the linear differential operator

    $$L[y] = y'' + p(t)y' + q(t)y.$$

Show that there are 2×2 matrices

$$\begin{pmatrix} a_1 & a_2 \\ b_1 & b_2 \end{pmatrix}$$

and

$$\begin{pmatrix} c_1 & c_2 \\ d_1 & d_2 \end{pmatrix}$$

with

$$\begin{pmatrix} f_1 \\ f_2 \end{pmatrix} = \begin{pmatrix} a_1 & a_2 \\ b_1 & b_2 \end{pmatrix} \begin{pmatrix} g_1 \\ g_2 \end{pmatrix}$$

and

$$\begin{pmatrix} g_1 \\ g_2 \end{pmatrix} = \begin{pmatrix} c_1 & c_2 \\ d_1 & d_2 \end{pmatrix} \begin{pmatrix} f_1 \\ f_2 \end{pmatrix}.$$

How are

$$\begin{pmatrix} a_1 & a_2 \\ b_1 & b_2 \end{pmatrix} \text{ and } \begin{pmatrix} c_1 & c_2 \\ d_1 & d_2 \end{pmatrix}$$

related? (Hint: If you have not worked with matrices before, you might want to postpone this problem until after we review matrix algebra in Chapter 8).

4.4. Constant coefficient second-order equations

In this section and the next we will tell the full story on solving second-order linear homogeneous equations with constant coefficients

(4.55) $$ay'' + by' + cy = 0,$$

where $a, b,$ and c are real numbers. As already discussed, we look for solutions of the form $y = e^{rt}$. By substitution, this leads to the equation

$$e^{rt}(ar^2 + br + c) = 0,$$

so e^{rt} is a solution precisely when

(4.56) $$ar^2 + br + c = 0,$$

that is, when r is a root of the quadratic polynomial $p(z) = az^2 + bz + c$. Equation (4.56) is called the **characteristic equation** of equation (4.55). We can always solve it with the quadratic formula to obtain

$$r = \frac{-b \pm \sqrt{b^2 - 4ac}}{2a}.$$

If this computation leads to two *different* real roots r_1 and r_2, then we know (see Example 4.2.8) that $\{e^{r_1 t}, e^{r_2 t}\}$ is a fundamental set of solutions to equation (4.55) and $c_1 e^{r_1 t} + c_2 e^{r_2 t}$ is the general solution.

For example, the characteristic equation for $y'' + 3y' = 0$ is $z^2 + 3z = 0$, which factors as $z(z+3) = 0$. There are two real roots, $z = -3$ and $z = 0$, which leads to the fundamental set of solutions $\{e^{-3t}, 1\}$ (since $e^{0t} = 1$) and general solution

$$y = c_1 e^{-3t} + c_2.$$

The case of two different real roots for equation (4.56) occurs when $b^2 - 4ac > 0$. However, two other possibilities exist:

(i) If $b^2 - 4ac < 0$, the solutions in equation (4.56) are no longer real numbers, and we must determine the significance of this for our ultimate goal of solving equation (4.55).

(ii) If $b^2 - 4ac = 0$, then equation (4.56) has only one (real) solution, namely $r_1 = -\frac{b}{2a}$. This is sometimes described by saying that the characteristic equation has a "repeated root", or a "root of multiplicity two", since when $b^2 - 4ac = 0$ the equation $ar^2 + br + c = 0$ factors as

$$a\left(r + \frac{b}{2a}\right)\left(r + \frac{b}{2a}\right) = 0.$$

This leads to a solution $e^{r_1 t}$ to equation (4.55), but as yet we do not have enough information to find a fundamental set of solutions and produce a general solution.

In this section we will discuss how to proceed in case (i), when our characteristic equation has nonreal roots. We'll take up the matter of repeated roots in Section 4.5.

4.4.1. Complex arithmetic. A complex number is an expression of the form $z = a + bi$ (or $z = a + ib$), where a and b are real numbers and $i^2 = -1$. We call a the **real part** of z and b the **imaginary part** of z and write $a = \text{Re } z, b = \text{Im } z$. When $a = 0$, the number is said to be purely imaginary. When $b = 0$, z is real, and the real numbers are a subset of the complex numbers. It is useful to visualize complex numbers as points in the **complex plane** by identifying the number $a + bi$ with the point (a, b); see Fig. 4.26. In this context, the x-axis is called the **real axis** and the y-axis is called the **imaginary axis**.

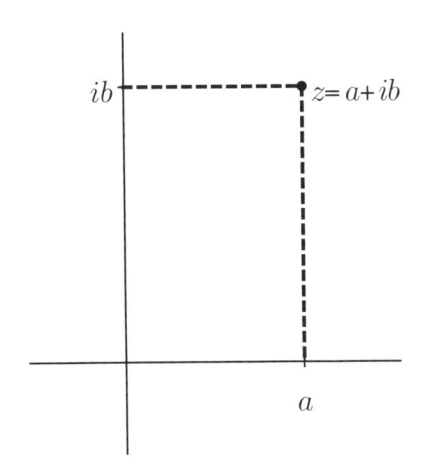

Figure 4.26. The point $z = a + bi$ in the complex plane.

Basic arithmetic operations on the complex numbers are defined by
$$(a + bi) + (c + di) = (a + c) + (b + d)i,$$
$$(a + bi) - (c + di) = (a - c) + (b - d)i,$$
and
$$(a + bi)(c + di) = (ac - bd) + (bc + ad)i.$$
Notice that in the definition of multiplication we multiply the expressions $a + bi$ and $c + di$ as if they were binomials and replace i^2 by -1.

Thus the complex numbers occupy what seems merely to be a copy of the xy-plane. However, they possess one new operation: multiplication, based on the fact that $i^2 = -1$. It is interesting to look at the list of powers of i as well:
$$i^0 = 1,\ i^1 = i,\ i^2 = -1,\ i^3 = -i,\ i^4 = 1,\ i^5 = i,\ i^6 = -1,\ i^7 = -i,\ i^8 = 1,\ \text{etc.}$$
Notice how the list repeats with period 4, so that
(4.57)
$$i^{4n+k} = i^k$$
for $k = 0, 1, 2, 3 \ldots$ and any integer n.

Example 4.4.1. The product $(-1+3i)(2-5i)$ is computed as

$$\begin{aligned}(-1+3i)(2-5i) &= (-1)(2)+(3i)(-5i)+(3i)(2)+(-1)(-5i)\\ &= -2-15i^2+6i+5i\\ &= 13+11i.\end{aligned}$$

The **conjugate** of a complex number $z = a+bi$ is defined to be the complex number $\bar{z} = a-bi$. Geometrically, $\bar{z}$ is the reflection of z across the real axis; see Fig. 4.27. Note that z is a real number precisely when $z = \bar{z}$. Here is a short list of properties of conjugation; you are asked to verify these in Exercise 14. For any complex numbers z and w and any positive integer n,

$$\begin{aligned}\overline{z+w} &= \bar{z}+\bar{w},\\ \overline{z-w} &= \bar{z}-\bar{w},\\ \overline{zw} &= \bar{z}\,\bar{w},\\ \overline{z^n} &= (\bar{z})^n.\end{aligned}$$

Suppose that the polynomial $az^2 + bz + c$ has real coefficients a, b, and c. If $b^2 - 4ac < 0$, the roots of this polynomial will be a pair of *complex conjugates*

$$-\frac{b}{2a}+i\frac{\gamma}{2a},\quad -\frac{b}{2a}-i\frac{\gamma}{2a}$$

where

$$\gamma = \sqrt{|b^2-4ac|}$$

is the positive square root of the positive number $|b^2 - 4ac|$.

Something like this is true for polynomials of degree greater than 2 as well. For example, consider the equation

$$az^3 + bz^2 + cz + d = 0$$

where a, b, c, and d are real numbers. Suppose we determine that the complex number w is a solution to this equation. Its conjugate, $\bar{w}$, will also be a solution. To see why, we use the properties of conjugation to write

$$\begin{aligned}a(\bar{w})^3+b(\bar{w})^2+c\bar{w}+d &= \overline{aw^3}+\overline{bw^2}+\overline{cw}+d\\ &= \bar{a}\overline{w^3}+\bar{b}\,\overline{w^2}+\overline{cw}+\bar{d}\\ &= \overline{aw^3}+\overline{bw^2}+\overline{cw}+\bar{d}\\ &= \overline{aw^3+bw^2+cw+d} = \bar{0} = 0.\end{aligned}$$

In short, when a polynomial has real coefficients, its roots come in complex conjugate pairs, w and $\bar{w}$.

The **modulus** of a complex number $z = a + bi$ is defined to be the nonnegative number

$$|z| \equiv \sqrt{a^2+b^2}.$$

You can check that $|z|^2 = z\bar{z}$. Geometrically, $|z|$ is the distance from the point z to $0 + i0 = 0$ (corresponding to the origin in the xy-plane). Notice that $|z| = |\bar{z}|$ and that $|z| = 0$ if and only if $z = 0$.

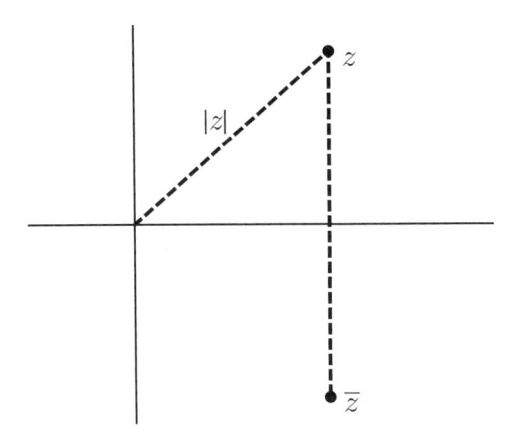

4.4. Constant coefficient second-order equations

Figure 4.27. The modulus and conjugate of z.

The notions of conjugate and modulus give an easy way to describe division of complex numbers: If $z = a + bi$ and $w = c + di$, we compute z/w by multiplying by $\overline{w}/\overline{w}$ as shown below:

$$\begin{aligned}\frac{z}{w} &= \frac{z}{w}\frac{\overline{w}}{\overline{w}} \\ &= \frac{z\overline{w}}{|w|^2} \\ &= \frac{(a+bi)(c-di)}{c^2+d^2} = \frac{ac+bd}{c^2+d^2} + \frac{bc-ad}{c^2+d^2}i.\end{aligned}$$

Example 4.4.2. We compute

$$\begin{aligned}\frac{1-3i}{5+2i} &= \frac{1-3i}{5+2i} \cdot \frac{5-2i}{5-2i} = \frac{5+6i^2-15i-2i}{5^2+2^2} \\ &= -\frac{1}{29} - \frac{17}{29}i.\end{aligned}$$

Complex roots for the characteristic equation. If the characteristic equation of the differential equation
$$ay'' + by' + cy = 0$$
(where a, b, and c are real) has nonreal roots, our discussion above shows that these roots will have the form
$$r_1 = \lambda + i\mu, \quad r_2 = \lambda - i\mu$$
for some real numbers λ and $\mu \neq 0$. By analogy with the case of real roots, we ask if the functions
$$y_1(t) = e^{(\lambda+i\mu)t} \quad \text{and} \quad y_2(t) = e^{(\lambda-i\mu)t}$$
solve our differential equation. However, we are immediately faced with a serious difficulty; namely, how do we even *define* these functions, as they involve *complex* powers of e? That is, how we should define
$$e^{(\lambda+i\mu)t}$$
for real λ and μ?

Let's back up a bit and first propose a definition for e^z, where z is any complex number. Recall the Taylor series expansion about the origin for the function e^x, x any real number:

(4.58) $$e^x = 1 + x + \frac{x^2}{2!} + \frac{x^3}{3!} + \cdots + \frac{x^n}{n!} + \cdots, \quad -\infty < x < \infty.$$

The meaning of this identity is that the infinite series converges for all real numbers x, and for each x, the value of this convergent sum is e^x. Motivated by this, we define

$$(4.59) \qquad e^z = 1 + z + \frac{z^2}{2!} + \frac{z^3}{3!} + \cdots + \frac{z^n}{n!} + \cdots = \sum_{n=0}^{\infty} \frac{1}{n!} z^n$$

for any complex number z. The series on the right-hand side of (4.59) converges for all z in the complex plane.

Is this a reasonable definition? Equation (4.59) is a *definition*, and as such it doesn't require "proof". However, we'd like to provide evidence that it is a reasonable definition. In particular, we hope that the familiar property for the product of exponentials still holds with our expanded definition:

$$(4.60) \qquad e^z \cdot e^w = e^{z+w}$$

for all complex numbers z and w. You can check this using power series as follows. By equation (4.59),

$$e^{z+w} = \sum_{n=0}^{\infty} \frac{1}{n!} (z+w)^n,$$

and by the Binomial Theorem,

$$(z+w)^n = \sum_{k=0}^{n} \frac{n!}{k!(n-k)!} z^k w^{n-k},$$

so that

$$e^{z+w} = \sum_{n=0}^{\infty} \frac{1}{n!} \left(\sum_{k=0}^{n} \frac{n!}{k!(n-k)!} z^k w^{n-k} \right).$$

Cancel the $n!$'s and we have

$$(4.61) \qquad e^{z+w} = \sum_{n=0}^{\infty} \left(\sum_{k=0}^{n} \frac{1}{k!(n-k)!} z^k w^{n-k} \right).$$

We want to compare the result in (4.61) with

$$(4.62) \qquad \begin{aligned} e^z e^w &= \left(\sum_{k=0}^{\infty} \frac{z^k}{k!} \right) \left(\sum_{j=0}^{\infty} \frac{w^j}{j!} \right) \\ &= \left(1 + z + \frac{z^2}{2!} + \frac{z^3}{3!} + \cdots \right) \left(1 + w + \frac{w^2}{2!} + \frac{w^3}{3!} + \cdots \right). \end{aligned}$$

We'll expand the product in (4.62) by multiplying the two factors as if they were "big polynomials". (The general principle here, which does require justification, is discussed in Section 7.2.) Carrying out this multiplication and grouping together all terms

$$\frac{z^k}{k!} \frac{w^j}{j!}$$

where $k + j = n$, we can write (4.62) as

$$(4.63) \qquad \sum_{n=0}^{\infty} \left(\sum_{k+j=n} \frac{z^k}{k!} \frac{w^j}{j!} \right).$$

4.4. Constant coefficient second-order equations

Since $k + j = n$ says $j = n - k$, we can rewrite the inner sum in equation (4.63) as

$$\sum_{k+j=n} \frac{z^k}{k!} \frac{w^j}{j!} = \sum_{k=0}^{n} \frac{z^k}{k!} \frac{w^{n-k}}{(n-k)!}.$$

Thus

(4.64) $$e^z e^w = \sum_{n=0}^{\infty} \left(\sum_{k=0}^{n} \frac{z^k}{k!} \frac{w^{n-k}}{(n-k)!} \right)$$

which by equation (4.61) is equal to e^{z+w}. We've verified equation (4.60).

Using definition (4.59). If we replace z by it in equation (4.59), where t is real, and use equation (4.57), we have

$$\begin{aligned} e^{it} &= 1 + it + \frac{(it)^2}{2!} + \frac{(it)^3}{3!} + \cdots + \frac{(it)^n}{n!} + \cdots \\ &= 1 + it - \frac{t^2}{2!} - \frac{t^3}{3!}i + \frac{t^4}{4!} + \frac{t^5}{5!}i - \frac{t^6}{6!} + \cdots. \end{aligned}$$

Rearrange the last line as

$$\left(1 - \frac{t^2}{2!} + \frac{t^4}{4!} - \frac{t^6}{6!} + \cdots \right) + i\left(t - \frac{t^3}{3!} + \frac{t^5}{5!} - \frac{t^7}{7!} + \cdots \right),$$

and recall that the sine and cosine functions have Taylor series expansions

$$\cos t = 1 - \frac{t^2}{2!} + \frac{t^4}{4!} - \frac{t^6}{6!} + \cdots + (-1)^n \frac{t^{2n}}{(2n)!} + \cdots$$

and

$$\sin t = t - \frac{t^3}{3!} + \frac{t^5}{5!} - \frac{t^7}{7!} + \cdots + (-1)^n \frac{t^{2n+1}}{(2n+1)!} + \cdots$$

valid for all real numbers t. Putting this all together we have

$$e^{it} = \cos t + i \sin t$$

for all real numbers t. More generally, for real μ we have the function

$$e^{i\mu t} = \cos(\mu t) + i \sin(\mu t),$$

with domain $-\infty < t < \infty$. Using equation (4.60), we see that for real λ and μ

$$e^{(\lambda + i\mu)t} = e^{\lambda t} e^{i\mu t} = e^{\lambda t} \cos(\mu t) + i e^{\lambda t} \sin(\mu t)$$

and we have resolved the issue of the meaning of the function

$$e^{(\lambda + i\mu)t}.$$

The identity $e^{it} = \cos t + i \sin t$ is known as Euler's formula,[8] after the great eighteenth-century Swiss mathematician Leonard Euler. This formula appeared on a Swiss postage stamp commemorating the 250th anniversary of Euler's birth.

Example 4.4.3. According to our definition,

$$e^{i\pi/3} = \cos\frac{\pi}{3} + i\sin\frac{\pi}{3} = \frac{1}{2} + i\frac{\sqrt{3}}{2}$$

and

$$e^{2+i\frac{\pi}{2}} = e^2\left(\cos\frac{\pi}{2} + i\sin\frac{\pi}{2}\right) = e^2(0 + i) = e^2 i.$$

Differentiation of complex exponential functions. It is important to check that the expected differentiation formulas hold for our complex exponential functions. What we mean by this is that

(4.65) $$\frac{d}{dt}e^{rt} = re^{rt}$$

even when r is a complex number. Verifying this is not hard. If $r = \lambda + i\mu$, so that

$$e^{rt} = e^{\lambda t}\cos(\mu t) + ie^{\lambda t}\sin(\mu t),$$

then by the product rule

$$[e^{\lambda t}\cos(\mu t)]' = \lambda e^{\lambda t}\cos(\mu t) - \mu e^{\lambda t}\sin(\mu t)$$

and

$$[e^{\lambda t}\sin(\mu t)]' = \lambda e^{\lambda t}\sin(\mu t) + \mu e^{\lambda t}\cos(\mu t)$$

and hence

(4.66) $$\frac{d}{dt}e^{rt} = e^{\lambda t}\left[(\lambda\cos(\mu t) - \mu\sin(\mu t)) + i(\mu\cos(\mu t) + \lambda\sin(\mu t))\right].$$

On the other hand, complex multiplication shows that

$$(\lambda + i\mu)(\cos(\mu t) + i\sin(\mu t)) = (\lambda\cos(\mu t) - \mu\sin(\mu t)) + i(\mu\cos(\mu t) + \lambda\sin(\mu t))$$

so that the derivative of e^{rt} in equation (4.66) can be written as

$$\begin{aligned}(e^{rt})' &= (\lambda + i\mu)e^{\lambda t}(\cos(\mu t) + i\sin(\mu t)) \\ &= re^{rt},\end{aligned}$$

for $r = \lambda + i\mu$, exactly as we hoped. As a practical matter, this tells us that even when r is complex, the function e^{rt} will be a solution to $ay'' + by' + cy = 0$ if r is a root of the associated characteristic equation.

It's time for an example.

Example 4.4.4. Let

$$y'' + y = 0.$$

The associated characteristic equation is $r^2 + 1 = 0$, which has roots $r = \pm i$. Thus the differential equation has solutions

$$y_1(t) = e^{it} = \cos t + i\sin t$$

and

$$y_2(t) = e^{-it} = \cos(-t) + i\sin(-t) = \cos t - i\sin t$$

where the last identity follows from the fact that the cosine function is an even function ($\cos(-t) = \cos t$) and the sine function is odd ($\sin(-t) = -\sin t$). So far, so good, but notice that our solutions

---

[8] The particular choice $t = \pi$ in Euler's formula leads to the statement $e^{i\pi} + 1 = 0$, relating the five fundamental numbers $1, 0, e, \pi$, and i.

4.4. Constant coefficient second-order equations 223

are complex-valued functions. We'd prefer our answer to involve only real-valued functions (for example, in our modeling work, only real-valued functions will typically be of interest). We can use the Superposition Principle (Theorem 4.2.5) to produce some real-valued solutions. Notice that

$$u_1(t) \equiv y_1(t) + y_2(t) = \cos t + i \sin t + \cos t - i \sin t = 2\cos t,$$

so that $2\cos t$, or any (real) scalar multiple of it, is a (real-valued) solution of $y'' + y = 0$. Similarly,

$$u_2(t) \equiv y_1(t) - y_2(t) = 2i \sin t$$

is a solution, and if we multiply it by the complex scalar $\frac{1}{2i}$, we obtain a real-valued solution $\sin t$. The pair of real-valued functions $\{\cos t, \sin t\}$ forms a fundamental set of solutions to our differential equation (since their Wronskian is $\cos^2 t + \sin^2 t = 1$ for all t) and the general solution to $y'' + y = 0$ is $y = c_1 \cos t + c_2 \sin t$.

Producing real-valued solutions. We generalize this example as follows. Suppose the roots of the characteristic equation for the differential equation $ay'' + by' + cy = 0$ are a pair of complex conjugates $\lambda + i\mu, \lambda - i\mu$ with $\mu \neq 0$. These complex roots give complex-valued solutions

(4.67)
$$e^{(\lambda+i\mu)t} = e^{\lambda t}(\cos(\mu t) + i \sin(\mu t))$$

and

(4.68)
$$e^{(\lambda-i\mu)t} = e^{\lambda t}(\cos(\mu t) - i \sin(\mu t)).$$

By the Superposition Principle, the real-valued function

(4.69)
$$e^{\lambda t} \cos(\mu t) = \frac{1}{2} e^{(\lambda+i\mu)t} + \frac{1}{2} e^{(\lambda-i\mu)t},$$

being a linear combination of solutions, is also a solution. Similarly, the linear combination

(4.70)
$$e^{\lambda t} \sin(\mu t) = \frac{1}{2i} e^{(\lambda+i\mu)t} - \frac{1}{2i} e^{(\lambda-i\mu)t}$$

is also a real-valued solution to the differential equation. The solution functions $e^{\lambda t} \cos(\mu t)$ and $e^{\lambda t} \sin(\mu t)$ are linearly independent on $(-\infty, \infty)$, as can be verified either by noting that neither is a constant multiple of the other or by computation of the Wronskian. We conclude that the general solution is

$$y(t) = c_1 e^{\lambda t} \cos(\mu t) + c_2 e^{\lambda t} \sin(\mu t).$$

The real-valued functions in equations (4.69) and (4.70) can be described as the *real and imaginary parts* of the single complex-valued solution in equation (4.67). Taking the real and imaginary parts of the complex-valued solution in equation (4.68) contributes nothing essentially new, since the functions produced are scalar multiples of the two real-valued solutions we had already identified.

A generalization. In fact, whenever our differential equation

$$y'' + p(t)y' + q(t)y = 0$$

has real-valued continuous coefficients $p(t)$ and $q(t)$ and $y(t) = u(t) + iv(t)$ is a complex-valued solution to this equation, where $u(t)$ and $v(t)$ are real-valued functions, then $u(t)$, the real part of y, and $v(t)$, the imaginary part of y, are also solutions to the differential equation. This is easy to verify with linear operator notation. We are given that $L[u + iv] = 0$, where

$$L[y] = y'' + p(t)y' + q(t)y.$$

By the linearity of the operator L,

$$0 = L[u + iv] = L[u] + iL[v].$$

Since a complex number is equal to 0 only if its real and imaginary parts are both 0, we conclude
$$L[u] = 0 \quad \text{and} \quad L[v] = 0,$$
which is just another way of saying that u and v are solutions to
$$y'' + p(t)y' + q(t)y = 0.$$

Example 4.4.5. Solve the initial value problem
$$y'' + 2y' + 3y = 0, \qquad y(0) = 1, \quad y'(0) = 5.$$
The characteristic equation $r^2 + 2r + 3 = 0$ has roots $-1 \pm i\sqrt{2}$. Since the real part of $e^{(-1+i\sqrt{2})t} = e^{-t}(\cos(\sqrt{2}\,t) + i\sin(\sqrt{2}\,t))$ is
$$e^{-t}\cos(\sqrt{2}\,t)$$
and the imaginary part is
$$e^{-t}\sin(\sqrt{2}\,t),$$
the general solution to the differential equation is

(4.71) $$y(t) = c_1 e^{-t}\cos(\sqrt{2}\,t) + c_2 e^{-t}\sin(\sqrt{2}\,t).$$

To solve the initial value problem, compute
$$y'(t) = c_1\left(-e^{-t}\cos(\sqrt{2}\,t) - e^{-t}\sqrt{2}\sin(\sqrt{2}\,t)\right)$$
(4.72) $$+ c_2\left(-e^{-t}\sin(\sqrt{2}\,t) + e^{-t}\sqrt{2}\cos(\sqrt{2}\,t)\right).$$

Plugging the initial conditions into equations (4.71) and (4.72) gives
$$1 = y(0) = c_1 \quad \text{and} \quad 5 = y'(0) = -c_1 + \sqrt{2}c_2 = -1 + \sqrt{2}c_2,$$
so that $c_2 = 6/\sqrt{2}$. The solution to the initial value problem is
$$y(t) = e^{-t}\cos(\sqrt{2}\,t) + \frac{6}{\sqrt{2}}e^{-t}\sin(\sqrt{2}\,t).$$

4.4.2. Amplitude and phase shift. When the characteristic equation has complex conjugate roots $\lambda \pm i\mu$, the general solution to the associated differential equation is
$$y(t) = c_1 e^{\lambda t}\cos(\mu t) + c_2 e^{\lambda t}\sin(\mu t).$$
To better understand these solution functions, we can use a trigonometric identity to write them in a more "transparent" form. The idea is to use the identity
$$\cos(A - B) = \cos A \cos B + \sin A \sin B$$
to obtain a nonnegative number R and some angle δ such that
$$c_1 e^{\lambda t}\cos(\mu t) + c_2 e^{\lambda t}\sin(\mu t) = R e^{\lambda t}\cos(\mu t - \delta).$$
Dividing by $e^{\lambda t}$, this means we want
$$R\cos(\mu t - \delta) = c_1\cos(\mu t) + c_2\sin(\mu t).$$
Comparing this with the expansion
$$R\cos(\mu t - \delta) = R\cos(\mu t)\cos\delta + R\sin(\mu t)\sin\delta,$$
we see that R and δ should be chosen so that
$$R\cos\delta = c_1 \quad \text{and}$$
$$R\sin\delta = c_2;$$

4.4. Constant coefficient second-order equations

that is, R and δ are polar coordinates for the point (c_1, c_2) in the xy-plane. Since

$$R^2 \cos^2 \delta + R^2 \sin^2 \delta = R^2,$$

we have $c_1^2 + c_2^2 = R^2$ and we choose

$$R = \sqrt{c_1^2 + c_2^2}.$$

Assuming c_1 and c_2 are not both 0, the angle δ is chosen so that

(4.73) $$\cos \delta = \frac{c_1}{R} \quad \text{and} \quad \sin \delta = \frac{c_2}{R}.$$

Although R is positive, c_1 and c_2 may not be, so we'll have to pay attention to their signs to determine the correct quadrant for δ. Notice that δ is not unique, since if δ solves the equations in (4.73), then so does $\delta + 2\pi k$ for any integer k. We will have a unique solution if we require that $0 \leq \delta < 2\pi$, for example.

To summarize, we have shown that if
$$f(t) = c_1 \cos(\mu t) + c_2 \sin(\mu t),$$
then
(4.74) $$f(t) = R \cos(\mu t - \delta),$$
where $R = \sqrt{c_1^2 + c_2^2}$ is called the amplitude of f, while δ, called a **phase shift**, or phase angle, satisfies (4.73). The **period** T of f is given by $T = \frac{2\pi}{\mu}$, while its **frequency** ν is given by $\nu = \frac{\mu}{2\pi}$ and its **circular frequency** is μ.

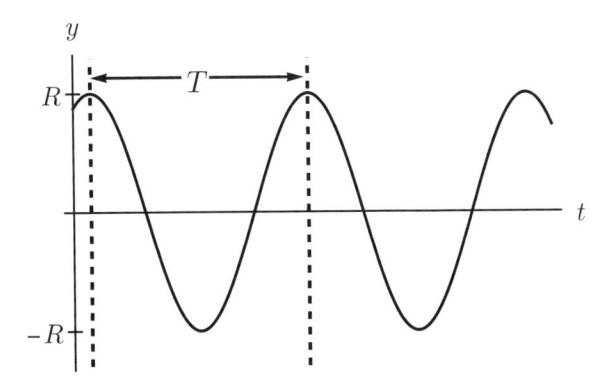

Figure 4.28. $f(t) = R\cos(\mu t - \delta)$ has amplitude R and period $T = 2\pi/\mu$.

For applications, where $f(t) = R\cos(\mu t - \delta)$ models an oscillating object's displacement from equilibrium position at time t, the amplitude R is the maximum displacement, the period T is a time period—that required for one complete cycle of the motion—and the frequency ν represents the number of cycles completed per unit of time.

Our work above shows how a general solution $y(t) = c_1 e^{\lambda t} \cos(\mu t) + c_2 e^{\lambda t} \sin(\mu t)$ of a second-order equation can also be written in the form

$$y(t) = e^{\lambda t} R \cos(\mu t - \delta).$$

Here, the quantity $e^{\lambda t} R$ is sometimes called a "time-varying amplitude" (when $\lambda \neq 0$). As the next example illustrates, such a function y may model a "damped oscillation".

Example 4.4.6. Consider the initial value problem

$$y'' + 2y' + 10y = 0, \quad y(0) = -2, \quad y'(0) = 8.$$

The roots of the characteristic equation, $r^2 + 2r + 10 = 0$, are $-1 \pm 3i$, so the general (real-valued) solution to the differential equation is

$$y(t) = c_1 e^{-t} \cos(3t) + c_2 e^{-t} \sin(3t).$$

We use the initial condition to determine that $c_1 = -2$ and $c_2 = 2$. Thus the solution to the initial value problem is

(4.75) $$y(t) = -2e^{-t} \cos(3t) + 2e^{-t} \sin(3t).$$

This solution is easier to understand if we write it in the form $Re^{-t}\cos(3t - \delta)$. From our discussion above we choose

$$R = \sqrt{(-2)^2 + (2)^2} = \sqrt{8} = 2\sqrt{2}.$$

The angle δ must be chosen so that

$$\cos \delta = \frac{-2}{2\sqrt{2}} = -\frac{\sqrt{2}}{2} \quad \text{and} \quad \sin \delta = \frac{2}{2\sqrt{2}} = \frac{\sqrt{2}}{2}.$$

Thus we can choose $\delta = \frac{3\pi}{4}$. This tells us that we can rewrite the function in (4.75) as

(4.76) $$y(t) = 2\sqrt{2} e^{-t} \cos\left(3t - \frac{3\pi}{4}\right).$$

Writing it in this form lets us quickly sketch its graph; see Fig. 4.29. Notice how we have used the graphs of $y = 2\sqrt{2} e^{-t}$ and $y = -2\sqrt{2} e^{-t}$ as guidelines (shown dashed). The oscillation pattern is due to the presence of the periodic factor $\cos(3t - \frac{3\pi}{4})$. The solution function (4.76) is sometimes described as exhibiting "damped oscillations", and the factor $2\sqrt{2} e^{-t}$ is its time-varying amplitude. The solution crosses the t-axis at regular intervals; two consecutive crossings are separated by $t = \frac{\pi}{3}$ units.

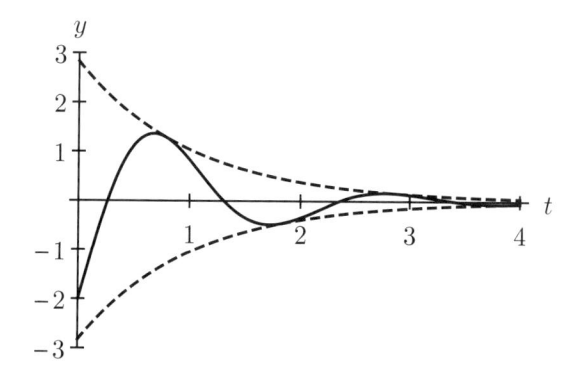

Figure 4.29. Graph of $y(t) = 2\sqrt{2} e^{-t} \cos\left(3t - \frac{3\pi}{4}\right)$, modeling a damped oscillation.

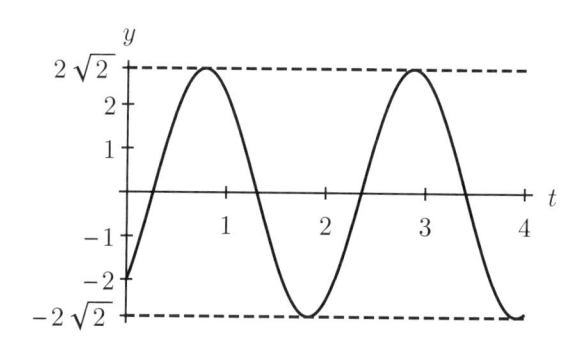

Figure 4.30. Graph of $y(t) = 2\sqrt{2} \cos\left(3t - \frac{3\pi}{4}\right)$, modeling an undamped oscillation of amplitude $2\sqrt{2}$ and period $2\pi/3$.

4.4.3. Exercises.

In Exercises 1–8, give the general solution.

1. $y'' + y' + y = 0$.
2. $y'' - 2y' + 6y = 0$.
3. $y'' - y' - 6y = 0$.
4. $4y'' + 9y = 0$.
5. $y'' + 4y' + 7y = 0$.
6. $y'' + 2y' = 0$.
7. $y'' + 2y = 0$
8. $2y'' + 6y' + 5y = 0$.

In Exercises 9–12, write the given expression in the form $R\cos(\mu t - \delta)$, and give the amplitude R, period T, and frequency ν of the oscillation.

9. $\cos t + \sin t$.
10. $\cos(2t) - \sin(2t)$.
11. $\cos(2t) + \sqrt{3}\sin(2t)$.
12. $3\cos(2t) - 4\sin(2t)$.

13. Solve the initial value problem
$$y'' + 2y' + 4y = 0, \quad y(0) = 1, \quad y'(0) = -1 - \sqrt{3}.$$
Write your answer in the form $y(t) = Re^{\lambda t}\cos(\mu t - \delta)$ and use this to sketch the solution. What is $\lim_{t \to \infty} y(t)$?

14. For complex numbers z and w, verify the following.
    (a) $\overline{z + w} = \overline{z} + \overline{w}$.
    (b) $\overline{z - w} = \overline{z} - \overline{w}$.
    (c) $\overline{zw} = \overline{z}\,\overline{w}$.
    (d) $\overline{\left(\frac{z}{w}\right)} = \frac{\overline{z}}{\overline{w}}$.
    (e) Show by induction that $\overline{z^n} = \overline{z}^n$ for each positive integer n.

15. (a) For the function
$$y(t) = 2\sqrt{2}e^{-t}\cos\left(3t - \frac{3\pi}{4}\right)$$
of Example 4.4.6, find the first time $t_0 > 0$ at which the graph of $y(t)$ intersects the graph of $g(t) = 2\sqrt{2}e^{-t}$.
    (b) Is $y'(t_0) = 0$?
    (c) Are the times at which the graphs of $y(t)$ and $g(t)$ intersect the same as the times at which $y(t)$ has a relative maximum?
    (d) Find the first time $t > 0$ at which $y(t)$ has a relative maximum.

16. Suppose that $c_1\cos(\mu t) + c_2\sin(\mu t) = R\cos(\mu t - \delta)$, where at least one of c_1 and c_2 is nonzero. We know that $R = \sqrt{c_1^2 + c_2^2}$ and $\cos\delta = c_1/R$ while $\sin\delta = c_2/R$. Suppose we insist that $-\frac{\pi}{2} < \delta \leq \frac{3\pi}{2}$.
    (a) If $c_1 = 0$ and $c_2 > 0$, then what's the value of δ?
    (b) If $c_1 = 0$ and $c_2 < 0$, then what's the value of δ?
    (c) If $c_1 > 0$, express δ in terms of the inverse tangent function.
    (d) If $c_1 < 0$, express δ in terms of the inverse tangent function.

17. Consider the initial value problem
$$y'' + 2y' + 10y = 0, \quad y(0) = -2, \quad y'(0) = 8,$$
which was solved in Example 4.4.6.
    (a) Find the first value of $t > 0$ with $y(t) = 0$.
    (b) Is the time interval between two successive zeros of the solution function always the same?

18. Solve the initial value problem
$$y'' + 4y' + 13y = 0, \quad y(0) = 4, \quad y'(0) = 1.$$
What is the time-varying amplitude of the solution?

19. (a) Solve the equations $y'' - k^2 y = 0$ and $y'' + k^2 y = 0$ for any real constant k, giving the general solutions.
    (b) Solve each of the equations in (a) subject to the initial condition $y(0) = 1$, $y'(0) = 2k$. Sketch each of these solutions with $k = 1$, and describe their behavior for $t > 0$.
    (c) Do the functions $y(t) = c_1 \cosh(kt) + c_2 \sinh(kt)$ solve either of the differential equations in (a)? (The hyperbolic trigonometric functions $\cosh t$ and $\sinh t$ are defined in Exercise 15 of Section 4.2). If so, they must be contained in the general solution you found in (a); explain.

20. Solve the equation $y'' - 4y' + (4 + k^2)y = 0$.

21. In this exercise we explore constant coefficient homogeneous differential equations where the coefficients are not necessarily real numbers.
    (a) Verify that $(z - 3i)(z + i) = z^2 - 2iz + 3$.
    (b) Find two linear independent (complex-valued) solutions of the form e^{rt} with r complex for the equation $y'' - 2iy' + 3y = 0$. Are the real and imaginary parts of your solutions separately solutions to the equation?
    (c) Give a general (complex-valued) solution to $y'' - 2iy' + 3y = 0$.
    (d) Repeat steps (b) and (c) for the equation $y'' - iy' + 2y = 0$. You'll need a factorization of $z^2 - iz + 2$ of the form $(z + ai)(z + bi)$ for some real numbers a and b.

22. Describe (in terms of m, k, y_0, v_0) the maximum displacement from the rest position for a mass-spring system as in Example 4.1.3 modeled by
$$my'' + ky = 0, \quad y(0) = y_0, \quad y'(0) = v_0.$$

4.5. Repeated roots and reduction of order

When the characteristic equation $ar^2 + br + c = 0$ for the constant coefficient differential equation $ay'' + by' + cy = 0$ has only one (repeated) root r_1, we know that $e^{r_1 t}$ solves the differential equation, but we need a second linearly independent solution to obtain the general solution. The next example shows one way to proceed.

Example 4.5.1. Our goal is to find the general solution to
(4.77) $$y'' + 8y' + 16y = 0.$$
The characteristic equation, $r^2 + 8r + 16 = 0$, factors as $(r+4)^2 = 0$ and thus has one root, $r = -4$. Thus $y_1(t) = e^{-4t}$ solves the differential equation, as does ce^{-4t} for any constant c (why?), but any two functions from this family are linearly dependent. To find a second linearly independent solution we will look for a solution in the form
$$y_2(t) = v(t)e^{-4t}$$
for some "unknown" function $v(t)$. Computing, we see that
$$\begin{aligned} y_2'(t) &= v'(t)e^{-4t} - 4v(t)e^{-4t}, \\ y_2''(t) &= v''(t)e^{-4t} - 8v'(t)e^{-4t} + 16v(t)e^{-4t}. \end{aligned}$$
Substituting into equation (4.77) we see, after the dust clears, that $v(t)e^{-4t}$ will solve this differential equation precisely when
$$v''(t)e^{-4t} = 0.$$
This says $v''(t) = 0$, so that $v(t) = k_1 t + k_2$ for some constants k_1 and k_2. Since we're after one new solution that is linearly independent from e^{-4t}, we make a convenient choice for $k_1 \neq 0$ and

4.5. Repeated roots and reduction of order

k_2, say, $k_1 = 1$ and $k_2 = 0$. This gives $v(t) = t$ and $y_2(t) = te^{-4t}$. We claim that $\{e^{-4t}, te^{-4t}\}$ is a fundamental set of solutions. To see this, either note that the functions e^{-4t} and te^{-4t} are linearly independent because neither is a constant multiple of the other or compute their Wronskian

$$W(e^{-4t}, te^{-4t}) = \det \begin{pmatrix} e^{-4t} & te^{-4t} \\ -4e^{-4t} & e^{-4t} - 4te^{-4t} \end{pmatrix} = e^{-8t},$$

and observe that this is never equal to 0.

The general case. Whenever the characteristic equation for $ay'' + by' + cy = 0$ has a repeated root $r = m$, the functions

$$y_1(t) = e^{mt} \quad \text{and} \quad y_2(t) = te^{mt}$$

form a fundamental set of solutions, and the equation has general solution

$$c_1 e^{mt} + c_2 t e^{mt}.$$

You are asked to verify this in Exercise 15.

4.5.1. Reduction of order. The method we used in Example 4.5.1 applies more generally to any second-order linear homogeneous equation

$$y'' + p(t)y' + q(t)y = 0,$$

whether the coefficients $p(t)$ and $q(t)$ are constant or not, to produce a second linearly independent solution, when one solution (other than the constant function 0) is known. This method is called **reduction of order**, since, as we'll show below, the problem of finding a second solution is reduced to that of solving a first-order differential equation.

Consider any second-order linear homogeneous differential equation

$$(4.78) \qquad y'' + p(t)y' + q(t)y = 0$$

with continuous coefficient functions $p(t)$ and $q(t)$ on some interval I. Suppose you have found one (nontrivial) solution $y_1(t)$ on I and you look for a second solution in the form

$$y_2(t) = v(t)y_1(t),$$

where you hope to find a choice for $v(t)$ that is *not* a constant, so that the set $\{y_1(t), v(t)y_1(t)\}$ will be a fundamental set of solutions. Compute

$$y_2' = vy_1' + v'y_1 \quad \text{and} \quad y_2'' = vy_1'' + 2v'y_1' + v''y_1$$

and substitute these into equation (4.78). The resulting equation is

$$(4.79) \qquad v''y_1 + v'[2y_1' + p(t)y_1] + v[y_1'' + p(t)y_1' + q(t)y_1] = 0.$$

Since y_1 is assumed to be a solution to the differential equation, the coefficient of v in (4.79) is equal to 0. This is crucial, since then equation (4.79) simplifies to

$$(4.80) \qquad v''y_1 + v'[2y_1' + p(t)y_1] = 0$$

and the substitution $z = v'$ reduces equation (4.80) to a *first-order* equation for z:

$$(4.81) \qquad z'y_1 + z[2y_1' + p(t)y_1] = 0.$$

Equation (4.81) is both linear and separable. It can be solved for z, provided we can carry out the necessary integration, and then we can determine v from the relation $v' = z$. We only need one specific, nonconstant, solution for v, so any constants that appear in the computation of z and v can be chosen for convenience.

Example 4.5.2. Given that $y_1(t) = \frac{1}{t}$ is a solution to

(4.82) $$t^2 y'' + 3t y' + y = 0$$

on $(0, \infty)$, we will find a general solution. We look for a second linearly independent solution in the form
$$y_2(t) = v(t)\frac{1}{t}.$$

Computing $y_2' = -vt^{-2} + v't^{-1}$ and $y_2'' = 2vt^{-3} - 2v't^{-2} + v''t^{-1}$ and substituting into equation (4.82) we obtain the equation
$$tv'' + v' = 0.$$

Making the substitution $z(t) = v'(t)$, we obtain
$$tz' + z = 0,$$

a first-order equation that is both separable and linear. Solving it as a linear equation, with integrating factor t, leads to
$$\frac{d}{dt}(tz) = 0.$$

Thus $tz = k$, or
$$z(t) = \frac{k}{t}.$$

We make a convenient choice for the constant k, say, $k = 1$, which gives us $v'(t) = z(t) = 1/t$, or
$$v(t) = \ln t + c.$$

Choosing $c = 0$ for convenience, we have $v(t) = \ln t$ and
$$y_2(t) = \frac{\ln t}{t}.$$

Since $y_1(t) = 1/t$ and $y_2(t) = \ln t / t$ are linearly independent,
$$y(t) = c_1 \frac{1}{t} + c_2 \frac{\ln t}{t}$$

is the general solution on $(0, \infty)$.

We can use equation (4.81) to give a general formula for a second linearly independent solution to $y'' + p(t)y' + q(t)y = 0$ on any interval I on which p is continuous and a nonzero solution y_1 is known. Set
$$Q(t) = \frac{2y_1'}{y_1} + p(t)$$

so that equation (4.81) becomes

(4.83) $$z' + Q(t)z = 0.$$

An integrating factor for this first-order linear equation is
$$\mu(t) = e^{\int Q(t) dt}.$$

Since
$$\int Q(t) dt = \int \left(\frac{2y_1'}{y_1} + p(t) \right) dt = 2\ln|y_1| + \int p(t)\, dt,$$

the integrating factor is
$$\mu(t) = y_1^2 \, e^{\int p(t)\, dt}.$$

Using this integrating factor to solve (4.83) gives
$$\frac{d}{dt}\left(z y_1^2 \, e^{\int p(t)\, dt} \right) = 0$$

4.5. Repeated roots and reduction of order

so that
$$zy_1^2 e^{\int p(t)\,dt} = c,$$
or
$$z = \frac{c}{y_1^2} e^{-\int p(t)\,dt}.$$

We can choose $c = 1$, and then
$$v'(t) = z(t) = \frac{1}{(y_1(t))^2} e^{-\int p(t)\,dt}$$

which leads to the formula

(4.84)
$$y_2(t) = y_1(t) \int \frac{e^{-\int p(t)\,dt}}{y_1(t)^2}\,dt.$$

The next example illustrates using the formula (4.84) directly.

Example 4.5.3. Solve the initial value problem
$$t^2 y'' + 2t y' - 2y = 0, \qquad y(1) = 3, \quad y'(1) = 9,$$
for $0 < t < \infty$, given that $y_1(t) = t$ solves the differential equation. We begin by putting the equation in standard form as

(4.85)
$$y'' + \frac{2}{t} y' - \frac{2}{t^2} y = 0.$$

From this, the restriction to the interval $(0, \infty)$ becomes apparent, since this is the largest open interval that contains $t = 1$ (where the initial condition is specified) and on which the coefficient functions $p(t) = 2/t$ and $q(t) = -2/t^2$ are continuous. From equation (4.84) we see that

$$\begin{aligned}
y_2(t) &= t \int \frac{e^{-\int (2/t)\,dt}}{t^2}\,dt = t \int \frac{t^{-2}}{t^2}\,dt = t \int t^{-4}\,dt \\
&= t\left(-\frac{1}{3} t^{-3} + C\right) = -\frac{1}{3} t^{-2} + Ct,
\end{aligned}$$

where in the second equality we used the calculation
$$e^{-\int 2/t\,dt} = e^{-2\ln t} = e^{\ln(t^{-2})} = t^{-2}.$$

We can choose C as convenient; the simplest choice is $C = 0$. At this point we have the general solution
$$y(t) = c_1 t + c_2 \frac{1}{t^2},$$
where we have incorporated the $-\frac{1}{3}$ into the arbitrary constant c_2. To meet our initial condition we want $c_1 + c_2 = 3$ and $c_1 - 2c_2 = 9$. Solving this pair of equations gives $c_2 = -2$ and $c_1 = 5$.

Repeated roots and operator-factoring. We'll revisit the equation $y'' + 8y' + 16y = 0$ in Example 4.5.1 to see an alternative approach to the method of reduction of order.

Example 4.5.4. Notice that $y'' + 8y' + 16y = (D^2 + 8D + 16)[y]$, and we can factor the linear operator $D^2 + 8D + 16$ as $(D+4)(D+4)$, since
$$(D+4)(D+4)[y] = (D+4)[y' + 4y] = (y' + 4y)' + 4(y' + 4y) = y'' + 8y' + 16y.$$

If y is any solution of $y'' + 8y' + 16y = 0$ on $(-\infty, \infty)$, set

(4.86)
$$h = (D+4)[y] = y' + 4y$$

and observe that $(D+4)[h] = 0$. This first-order equation for h tells us that $h(t) = ce^{-4t}$, and by equation (4.86),
$$y' + 4y = ce^{-4t}.$$
Solving this first-order equation for y, using the integrating factor e^{4t}, we have
$$\begin{aligned} \frac{d}{dt}(ye^{4t}) &= ce^{-4t}e^{4t} = c, \\ ye^{4t} &= ct + c_1, \\ y &= c_1 e^{-4t} + c_2 t e^{-4t}, \end{aligned}$$
writing c_2 for c. This is the same solution we produced in Example 4.5.1 by reduction of order.

4.5.2. Long-term behavior. We turn next to an example that will be important for applications later in this chapter.

Example 4.5.5. Suppose we have any constant coefficient homogeneous equation
$$(4.87) \qquad ay'' + by' + cy = 0,$$
where the constants a, b, and c are *strictly positive*. The goal of this example is to show that every solution y of this equation approaches the equilibrium solution $y = 0$ in the long run:
$$\lim_{t \to \infty} y(t) = 0.$$
We can write down a general solution to our differential equation once we have found the roots of the characteristic equation $az^2 + bz + c = 0$, and it is convenient to distinguish three cases.

Case (i). If $b^2 - 4ac = 0$, there is only one root, $z = -\frac{b}{2a}$, which is negative. The general solution in this case is
$$y(t) = c_1 e^{-\frac{b}{2a}t} + c_2 t e^{-\frac{b}{2a}t}.$$
Now the exponential
$$e^{-\frac{b}{2a}t} \to 0 \quad \text{as } t \to \infty$$
since $-\frac{b}{2a}$ is negative. Moreover, this exponential function decays so rapidly that we also have
$$\lim_{t \to \infty} t e^{-\frac{b}{2a}t} = 0;$$
you should verify this using l'Hôpital's Rule, for example. Thus when $b^2 - 4ac = 0$, every solution to equation (4.87) tends to 0 as $t \to \infty$.

Case (ii). If $b^2 - 4ac > 0$, then the two roots
$$\frac{-b \pm \sqrt{b^2 - 4ac}}{2a}$$
are real and different. It's clear that the choice of the minus sign will result in a negative value. But it's also true that the choice of the plus sign in this expression will result in a negative value, since $b^2 - 4ac < b^2$ (remember, we're assuming a, b, and c are all positive), so $-b + \sqrt{b^2 - 4ac} < 0$. If we denote these two *negative* roots of our characteristic equation by r_1 and r_2, this gives as a general solution
$$y(t) = c_1 e^{r_1 t} + c_2 e^{r_2 t},$$
which, for every choice of constants c_1 and c_2, tends to 0 as $t \to \infty$.

Case(iii). If $b^2 - 4ac < 0$, our general solution looks like

(4.88) $$y(t) = c_1 e^{\lambda t} \cos(\mu t) + c_2 e^{\lambda t} \sin(\mu t)$$

where
$$\lambda = -\frac{b}{2a}$$

is negative and
$$\mu = \frac{\sqrt{|b^2 - 4ac|}}{2a}.$$

For any constants c_1 and c_2, we can use the technique of Section 4.4.2 to rewrite $y(t)$ as

$$e^{\lambda t} R \cos(\mu t - \delta)$$

for some positive number R and some phase shift δ. Since $R\cos(\mu t - \delta)$ stays bounded between R and $-R$ and $e^{\lambda t} \to 0$ as $t \to \infty$, we see that any solution in equation (4.88) tend to 0 as $t \to \infty$.

Exercise 29 explores the question of long-term behavior when a, b, and c are nonnegative and either b or c is 0.

4.5.3. Cauchy-Euler equations. We got started on solving a constant coefficient equation

(4.89) $$ay'' + by' + cy = 0$$

by using $y = e^{rt}$ as a trial solution. With this choice, each term of (4.89) has the form

$$\text{constant} \times e^{rt}.$$

By substituting the trial solution into a particular constant coefficient equation we learn the specific values of r that make e^{rt} a solution to that equation. Another type of linear homogeneous differential equation for which a similar approach is applicable is a **Cauchy-Euler** equation

(4.90) $$b_2 t^2 y'' + b_1 t y' + b_0 y = 0,$$

where b_0, b_1, and b_2 are constants. If you look back at Examples 4.5.2 and 4.5.3, you'll see that the equations there are Cauchy-Euler equations. Writing (4.90) in the standard form for a second-order linear equation we have

$$y'' + \frac{b_1}{b_2}\frac{1}{t}y' + \frac{b_0}{b_2}\frac{1}{t^2}y = 0,$$

and we look for solutions on an interval not containing 0. Here a good choice for a trial solution is $y = t^m$ for some constant m. Can you see why we make this choice? Since $y' = mt^{m-1}$ and $y'' = m(m-1)t^{m-2}$, the presence of the coefficients t^2 and t in (4.90) mean that each term of (4.90) will have the form

$$\text{constant} \times t^m$$

after substituting our trial solution. Let's look at an example.

Example 4.5.6. We'll find the general solution to

$$t^2 y'' - ty' - 3y = 0$$

for $t > 0$. Upon substituting $y = t^m$ we have

$$m(m-1)t^m - mt^m - 3t^m = 0,$$

or

$$t^m(m^2 - 2m - 3) = 0.$$

Thus we must choose m so that

$$m^2 - 2m - 3 = 0.$$

Since $m^2 - 2m - 3 = (m-3)(m+1)$, we have $m = 3$ or $m = -1$, and $y_1(t) = t^3$ and $y_2(t) = t^{-1}$ are solutions to the differential equation on $(0, \infty)$ (notice that y_2 is not defined at $t = 0$, so our restriction to an interval not containing 0 really is necessary). This pair of functions is linearly independent on $(0, \infty)$, so by Theorem 4.3.8, $y = c_1 t^3 + c_2 t^{-1}$ is the general solution there.

In this example, substituting the trial solution $y = t^m$ led to two different real values of m from which we built the general solution. In the case that only one (real) value of m is obtained, reduction of order can be used to find a second linearly independent solution, as was done in Example 4.5.2. It's also possible that the allowable values of m will be complex numbers $\lambda \pm i\mu$. This possibility will be discussed later, in Section 7.5.

The parallel between solving a constant coefficient homogeneous equation and a Cauchy-Euler equation is more than superficial. In Exercise 25 you are asked to show that a substitution will convert any Cauchy-Euler equation to a differential equation with constant coefficients.

Cauchy-Euler equations and operator-factoring. A third approach to solving a Cauchy-Euler equation $b_1 t^2 y'' + b_2 t y' + b_0 y = 0$, which does not require "guessing" a trial solution $y = t^r$, uses operator-factoring. However, the operators involved in this factoring will not, in general, be constant coefficient operators, and as we will see, this requires a bit more care. We'll look at two examples.

Example 4.5.7. We'll find the general solution to
$$(4.91) \qquad t^2 y'' - t y' - 3y = 0$$
for $t > 0$. To get started, write the equation as $y'' - \frac{1}{t} y' - \frac{3}{t^2} y = 0$ and look for a factorization of the operator
$$D^2 - \frac{1}{t} D - \frac{3}{t^2}$$
in the form
$$\left(D + \frac{a_1}{t}\right)\left(D + \frac{a_2}{t}\right)$$
for constants a_1 and a_2 to be determined. We check that
$$\begin{aligned}
\left(D + \frac{a_1}{t}\right)\left(D + \frac{a_2}{t}\right)[y] &= \left(D + \frac{a_1}{t}\right)[y' + \frac{a_2}{t} y] \\
&= y'' + \left(\frac{a_2}{t} y\right)' + \frac{a_1}{t} y' + \frac{a_1 a_2}{t^2} y \\
&= y'' + \frac{a_2}{t} y' - \frac{a_2}{t^2} y + \frac{a_1}{t} y' + \frac{a_1 a_2}{t^2} y \\
&= y'' + \frac{a_1 + a_2}{t} y' + \frac{a_1 a_2 - a_2}{t^2} y,
\end{aligned}$$
where we have used the product rule to compute $(\frac{a_2}{t} y)' = \frac{d}{dt}(\frac{a_2}{t} y)$. We want $a_1 + a_2 = -1$ and $a_1 a_2 - a_2 = -3$, and we can satisfy this with $a_1 = -2$ and $a_2 = 1$. Thus to solve equation (4.91) we want to solve
$$\left(D - \frac{2}{t}\right)\left(D + \frac{1}{t}\right)[y] = 0.$$
If $y(t)$ is any solution on $(0, \infty)$, set
$$(4.92) \qquad h = \left(D + \frac{1}{t}\right) y = y' + \frac{y}{t}.$$
Now $h(t)$ must solve
$$\left(D - \frac{2}{t}\right)[h] = 0,$$

4.5. Repeated roots and reduction of order

or equivalently $h' - \frac{2}{t}h = 0$. This first-order equation may be solved using the integrating factor $e^{-2\ln t} = t^{-2}$ to obtain $h = ct^2$ for constant c. By the definition of h in equation (4.92),

$$y' + \frac{y}{t} = h(t) = ct^2.$$

We solve for y (using the integrating factor t), obtaining $\frac{d}{dt}(ty) = ct^3$, or $y = c_1 t^3 + c_2 t^{-1}$, where we have relabeled $c/4$ as c_1.

The next example is a Cauchy-Euler equation in which there is only one value of r for which t^r as a solution. Our previous method would have thus required a reduction of order computation to produce the general solution. As we will see, operator-factoring bypasses this step.

Example 4.5.8. We'll find the general solution to

$$t^2 y'' - t y' + y = 0$$

for $t > 0$. Writing the equation as $y'' - \frac{1}{t}y' + \frac{1}{t^2}y = 0$, we factor the operator $D^2 - \frac{1}{t}D + \frac{1}{t^2}$ as $D\left(D - \frac{1}{t}\right)$; verify that this factorization is correct by computing

$$D\left(D - \frac{1}{t}\right)[y] = D\left[y' - \frac{1}{t}y\right] = y'' - \left(\frac{1}{t}y\right)' = y'' - \frac{1}{t}y' + \frac{1}{t^2}y.$$

Thus we want to solve

$$D\left(D - \frac{1}{t}\right)[y] = 0.$$

If $y(t)$ is any solution on $(0, \infty)$, set $h(t) = (D - \frac{1}{t})[y] = y' - \frac{y}{t}$. Now $h(t)$ must solve

$$D[h] = 0,$$

or equivalently $h' = 0$, so that $h(t) = c$ for any constant c. By the definition of h,

$$y' - \frac{y}{t} = h = c.$$

Solve this first-order equation for y, using the integrating factor $\frac{1}{t}$, to obtain

$$\frac{d}{dt}\left(\frac{1}{t}y\right) = c\frac{1}{t},$$

or

$$\frac{1}{t}y = c_1 \ln t + c_2$$

(where $c_1 = c$), so that $y = c_1 t \ln t + c_2 t$.

Some cautions about operator-factoring for linear differential operators having non-constant coefficients. When a and b are constants, the operator product $(D + a)(D + b)$ is well-behaved in the following senses:

- $(D + a)(D + b) = (D + b)(D + a)$, and
- $(D + a)(D + b) = D^2 + (a + b)D + ab$,

so that $D+a$ and $D+b$ commute, and their product is computed as if multiplying two polynomials. When a and b are replaced by functions of t (as in the last two examples), neither of these nice properties may hold. For example, $D(D - \frac{1}{t}) \neq (D - \frac{1}{t})D$, and $D(D - \frac{1}{t}) \neq D^2 - \frac{1}{t}D$ since

$$D\left(D - \frac{1}{t}\right)[y] = y'' - \frac{1}{t}y' + \frac{1}{t^2}y \quad \text{and} \quad \left(D - \frac{1}{t}\right)D[y] = y'' - \frac{1}{t}y'.$$

In general, for a second-order linear equation $y''+p(t)y'+q(t)y = 0$ with nonconstant coefficients there is no systematic method for obtaining a solution via integrating factors and integration (to yield an integral-based solution formula analogous to that we found for first-order equations (equation (2.12) of Section 2.1). In Chapter 7 we explore a systematic method for solving nonconstant coefficient equations based on power series.

4.5.4. Exercises.

In Exercises 1–6, give the general solution.

1. $y'' - 4y' + 4y = 0$.
2. $2y'' + 4y' + 2y = 0$.
3. $y'' - 2y = 0$.
4. $y'' + 4y' + 7y = 0$.
5. $y'' + 2y' + y = 0$.
6. $6y'' + y' - 2y = 0$.

In Exercises 7–10, solve the initial value problem.

7. $y'' + 6y' + 9y = 0$, $y(0) = 3$, $y'(0) = -5$.
8. $4y'' - 12y' + 9y = 0$, $y(0) = 1$, $y'(0) = -1$.
9. $y'' + 2y' + y = 0$, $y(0) = 1$, $y'(0) = -3$.
10. $y'' - 4y' + 4y = 0$, $y(1) = 2$, $y'(1) = 2$.

11. Find a constant coefficient homogeneous linear differential equation with fundamental set of solutions $\{e^{2t}, 3e^{-t}\}$; then give an initial value problem whose solution is $y = e^{2t} - 6e^{-t}$.

12. Repeat Exercise 11 for the fundamental set of solutions $\{e^{3t}, te^{3t}\}$ and the solution $y = e^{3t} - te^{3t}$.

13. Repeat Exercise 11 for the fundamental set of solutions $\{\cos(2t), -\sin(2t)\}$ and the solution $y = \cos(2t) - \sin(2t)$.

14. Repeat Exercise 11 for the fundamental set of solutions $\{1, 2t\}$ and the solution $y = 1 + t$.

15. The purpose of this exercise is to verify that when the characteristic equation for $ay'' + by' + cy = 0$ factors as
$$a(r-m)^2 = 0$$
(so that there is a root $r = m$ of multiplicity two), then $\{e^{mt}, te^{mt}\}$ is a fundamental set of solutions to the differential equation.
  (a) Show that $b = -2am$ and $c = am^2$.
  (b) Verify by direct substitution that te^{mt} solves $ay'' + by' + cy = 0$.
  (c) Explain why $\{e^{mt}, te^{mt}\}$ is a fundamental set of solutions to the differential equation on $(-\infty, \infty)$.

16. Solve the equation $ay'' + 2y' + by = 0$ if a and b are real numbers and
  (a) $ab = 1$,
  (b) $ab > 1$.

17. In this problem you will use operator-factoring to solve $y'' + 6y' + 9y = 0$.
  (a) Write the operator $D^2 + 6D + 9$ in factored form as $(D+a)(D+b)$ for appropriate constants a and b. Verify that your factorization is correct by computing $(D^2 + 6D + 9)[y]$ and $(D+a)(D+b)[y]$.
  (b) Use the method of Example 4.5.4 to find the general solution to $y'' + 6y' + 9y = 0$.

18. Follow the same outline as in Exercise 17 to solve $4y'' + 4y' + y = 0$, beginning with factoring the operator $4D^2 + 4D + 1$.

19. Find the general solution on $(0, \infty)$ to each of the following Cauchy-Euler equations.
  (a)
  $$3t^2 y'' - 4ty' + 2y = 0.$$

(b)
$$t^2 y'' - 6y = 0.$$

(c)
$$2t^2 y'' + 7ty' + 2y = 0.$$

20. Solve the initial value problem
$$t^2 y'' - 2ty' + 2y = 0, \qquad y(1) = 2, \quad y'(1) = 3.$$

21. Any constant function $y = c$ is a solution to $ty'' - 2y' = 0$ on $(0, \infty)$. Find the general solution using reduction of order.

22. (a) Find all numbers m so that the function $y = t^m$ is a solution to
$$t^2 y'' - 5ty' + 9y = 0.$$
(b) Using reduction of order, find the general solution to the equation in (a).

23. (a) Using $y(t) = t^m$ as the form of a trial solution, find one solution to $t^2 y'' - 13ty' + 49y = 0$ on $(0, \infty)$.
(b) Find a second linearly independent solution by reduction of order.

24. (a) By direct substitution, show that the power function t^m will be a solution of the Cauchy-Euler equation
$$t^2 y'' + b_1 ty' + b_0 y = 0$$
on $(0, \infty)$ if

(4.93) $$m^2 + (b_1 - 1)m + b_0 = 0.$$

(b) If $(b_1 - 1)^2 - 4b_0 = 0$, then the work in (a) is not enough to produce a general solution. Show that in this case
$$y_1(t) = t^m,$$
where $m = (1 - b_1)/2$, is one solution, and use reduction of order to show that
$$y_2(t) = (\ln t)t^m = (\ln t)t^{(1-b_1)/2}$$
is a second linearly independent solution.

25. Any Cauchy-Euler equation,

(4.94) $$b_2 t^2 y'' + b_1 ty' + b_0 y = 0,$$

can be converted to an equation with constant coefficients by making the substitution
$$s = \ln t$$
(a change in the "time scale"). In this problem we see how this goes.
(a) Show that with this substitution we have
$$\frac{dy}{dt} = \frac{1}{t}\frac{dy}{ds}$$
and
$$\frac{d^2 y}{dt^2} = \frac{1}{t^2}\left[\frac{d^2 y}{ds^2} - \frac{dy}{ds}\right].$$
Hint: For the latter, recall that
$$\frac{d^2 y}{dt^2} = \frac{d}{dt}\left[\frac{dy}{dt}\right] = \left(\frac{d}{ds}\left[\frac{dy}{dt}\right]\right)\frac{ds}{dt}.$$
(b) Carry out this substitution to show that it converts equation (4.94) to
$$b_2 \frac{d^2 y}{ds^2} + (b_1 - b_2)\frac{dy}{ds} + b_0 y = 0.$$

(c) Use the result of (b) to solve
$$t^2 y'' - 2t y' + 2y = 0, \quad t > 0.$$
Give your final answer as a function of t, not s.

26. Find the general solution to
$$t y'' - (2t+1) y' + 2y = 0$$
on $(0, \infty)$, given that $y_1(t) = e^{2t}$ is one solution.

27. Find the general solution to
$$(\sin t) y'' - (2 \cos t) y' + \frac{1 + \cos^2 t}{\sin t} y = 0$$
on $(0, \pi)$, given that $y_1(t) = \sin t$ is one solution.

28. You are given the following two pieces of information about solutions to the linear homogeneous equation
$$y'' + p(t) y' + q(t) y = 0:$$
   (i) The function $y_1(t) = t^3$ is a solution.
   (ii) For any other solution $y_2(t)$, the Wronskian $W(y_1, y_2)(t)$ is a constant function.
Assuming p and q are continuous on $(0, \infty)$, find the general solution on $(0, \infty)$.

29. Suppose that a, b, and c are each nonnegative, with $a \neq 0$.
   (a) Show that if exactly one of the coefficients b and c is equal to 0, then any solution to $a y'' + b y' + c y = 0$ is bounded on the interval $(0, \infty)$; in other words, show that given a solution y to this equation, you can find a constant M so that $|y(t)| \leq M$ for all $t > 0$. Show that not every solution has the property
$$\lim_{t \to \infty} y(t) = 0.$$
   (b) Show that if both $b = 0$ and $c = 0$, then some solutions are not bounded on $(0, \infty)$.

30. Which of the following graphs shows a function that could be a solution to
$$y'' + a y' + b y = 0$$
if $a > 0$ and $b < 0$?

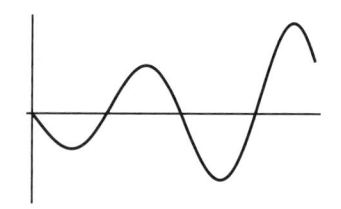

Figure 4.31. (A) 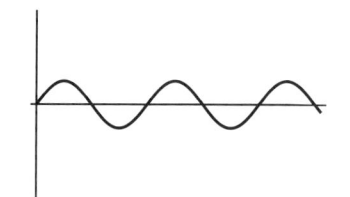   **Figure 4.32.** (B) 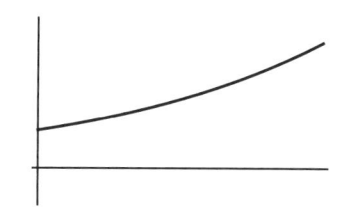   **Figure 4.33.** (C)

31. Find a general solution to
$$(1 - t^2) y'' + 2t y' - 2y = 0$$
given that $y_1(t) = t$ is one solution.

32. Use equation (4.84) to find the general solution of $t y'' - 2 y' + 9 t^5 y = 0$ if you are given that $y(t) = \cos(t^3)$ is one solution.

33. Equation (4.84) gives the reduction of order formula for a second (linearly independent) solution to
$$y'' + p(t) y' + q(t) y = 0$$

4.5. Repeated roots and reduction of order

if we know that $y_1(t)$ is one (nontrivial) solution. This problem gives a different way to obtain equation (4.84).
(a) Show that if y_2 is a second linearly independent solution, then
$$\left(\frac{y_2}{y_1}\right)' = \frac{W(y_1, y_2)}{y_1^2}$$
where W is the Wronskian.
(b) Using Abel's formula (equation (4.43)) for W and the result from (a), obtain the formula in (4.84).

34. The Hermite equation is the equation
$$y'' - 2ty' + 2\rho y = 0$$
where ρ is a constant. This differential equation arises in quantum mechanics. Show that with $\rho = 2$, one solution is $y_1(t) = 1 - 2t^2$. What does equation (4.84) give as an integral formula for a second linearly independent solution? We will return to the Hermite equation in Chapter 7.

35. (a) Compute $(D-3)(D+1)[y]$ and $(D+1)(D-3)[y]$ and show that both are equal to $(D^2 - 2D - 3)[y]$. Hint: Start with $(D-3)(D+1)[y] = (D-3)[(D+1)y] = (D-3)[y'+y] = \cdots$.
(b) Verify that the operator $(D+a)(D+b)$ is equal to $D^2 + (a+b)D + ab$ for any constants a and b, and check that $(D+a)(D+b) = (D+b)(D+a)$.

36. In this problem, you will use operator-factorization to solve the Cauchy-Euler equation $2t^2 y'' + 7ty' - 3y = 0$, which can also be written as $y'' + \frac{7}{2t}y' - \frac{3}{2t^2}y = 0$.
(a) Find two different factorizations of $D^2 + \frac{7}{2t}D - \frac{3}{2t^2}$ of the form $(D + \frac{a}{t})(D + \frac{b}{t})$ for constants a and b.
(b) Use the method in Example 4.5.7 and either of your answers from (a) to find the general solution of $2t^2 y'' + 7ty' - 3y = 0$ for $t > 0$.

37. (a) Is $(D+t)(D-t) = (D-t)(D+t)$?
(b) Is $(D-1)(D+t) = (D+t)(D-1)$?
(c) Give a factorization of $D^2 + (t-1)D + (1-t)$ using your work in (b).

38. The differential equation $y''(t) + 4ty'(t) + (4t^2 + 2)y(t) = 0$ is neither a constant coefficient nor Cauchy-Euler equation. However, we can solve it using the method of operator-factoring.
(a) By referencing the appropriate theorem, explain how you know that this differential equation will have general solution on $(-\infty, \infty)$ in the form $c_1 y_1 + c_2 y_2$, where y_1 and y_2 are linearly independent solutions on $(-\infty, \infty)$
(b) Verify that the operator $D^2 + 4tD + (4t^2 + 2)$ can be factored as $(D+2t)(D+2t)$ by computing $(D^2 + 4tD + (4t^2 + 2))[y]$ and $(D+2t)(D+2t)[y]$ and showing the results are the same.
(c) Suppose that y is any solution to $y'' + 4ty' + (4t^2 + 2)y = 0$, and set $h(t) = (D+2t)[y] = y' + 2ty$. Explain how you know that $(D+2t)[h] = 0$, and use this first-order equation to determine h.
(d) Use your answer to (c) and the definition of h as $h = y' + 2ty$ to find the general solution to $y'' + 4ty' + (4t^2 + 2)y = 0$.

39. Suppose that $s = s(t)$ and $v = v(t)$ are differentiable functions on the nonempty open interval I such that
(4.95) $$(D - s(t))(D - v(t))[y] = y'' + p(t)y' + q(t)y,$$
for all twice-differentiable functions y on I, where $D = \frac{d}{dt}$. Show that v must satisfy the following first-order equation:
$$\frac{dv}{dt} = -v(t)^2 - p(t)v(t) - q(t) \text{ on } I.$$

The preceding is a **Riccati equation**. (Definition: A Riccati equation with dependent variable y and independent variable t may be written in the form $\frac{dy}{dt} = q_2(t)y^2 + q_1(t)y + q_0(t)$; thus a Riccati equation is quadratic in the dependent variable.) It is known that you can't, in general, find solutions of Riccati equations by successive integration (which is the way we can solve linear equations and Bernoulli equations). Thus our work on this problem suggests that we cannot expect, in general, to be able to find formulas for functions s and v providing us with the operator-factorization (4.95). In order words, for many second-order equations $y'' + p(t)y' + q(t)y = 0$ with nonconstant coefficients, there is no nice factorization of the associated linear differential operator $D^2 + p(t)D + q(t)$.

4.6. Higher-order equations

4.6.1. General remarks. Much of the general theory of second-order linear homogeneous equations carries over in pleasant and predictable ways to higher-order linear homogeneous equations.

An nth-order linear differential equation has the form

(4.96) $$P_n(t)\frac{d^n y}{dt^n} + P_{n-1}(t)\frac{d^{n-1}y}{dt^{n-1}} + \cdots + P_1(t)\frac{dy}{dt} + P_0(t)y = G(t),$$

or more compactly

$$P_n(t)y^{(n)} + P_{n-1}(t)y^{(n-1)} + \cdots + P_1(t)y' + P_0(t)y = G(t).$$

When the right-hand side $G(t)$ is the zero function, the equation is said to be homogeneous. Throughout this section we assume that the coefficient functions $P_k(t), 0 \leq k \leq n$, and the function $G(t)$ are all continuous on some fixed open interval I and that the leading coefficient, $P_n(t)$, is not equal to zero anywhere in this interval. This means we can divide the equation by $P_n(t)$ and put it into the **standard form**

$$y^{(n)} + p_{n-1}(t)y^{(n-1)} + \cdots + p_1(t)y' + p_0(t)y = g(t),$$

where the functions $p_k(t)$ and $g(t)$ are continuous on I. An important special case is where all of the coefficients $p_k(t)$ are just constants.

What form should initial conditions for an nth-order equation take? For a second-order equation, initial conditions specify the value of the function and its derivative at a single point. By analogy, we might expect that for an nth-order equation we will specify the value of the function and its first $n-1$ derivatives at some point; that is, initial conditions for equation (4.96) look like

(4.97) $$y(t_0) = a_0,\ y'(t_0) = a_1, \ldots, y^{(n-1)}(t_0) = a_{n-1},$$

where t_0 is a point in some open interval on which the coefficient functions are continuous.

If we write L for the linear operator that takes an input function y to the output

(4.98) $$y^{(n)} + p_{n-1}(t)y^{(n-1)} + \cdots + p_1(t)y' + p_0(t)y,$$

we can immediately see that the Superposition Principle carries over to nth-order linear *homogeneous* equations:

Theorem 4.6.1 (Superposition Principle). *If* $y_1, y_2, \ldots, y_m$ *solve the homogeneous equation*

$$y^{(n)} + p_{n-1}(t)y^{(n-1)} + \cdots + p_1(t)y' + p_0(t)y = 0$$

on some open interval I, then so does $c_1 y_1 + c_2 y_2 + \cdots + c_m y_m$ *for any choice of constants* $c_1, c_2, \ldots, c_m$.

Just as for second-order linear equations, we have an Existence and Uniqueness Theorem which guarantees that initial value problems with nth-order linear equations (homogeneous or not) have one and only one solution.

4.6. Higher-order equations

Theorem 4.6.2 (Existence and Uniqueness Theorem). *Suppose that $p_0(t), p_1(t), \ldots, p_{n-1}(t)$ and $g(t)$ are continuous on some open interval I containing the point t_0. Let $a_0, a_1, \ldots, a_{n-1}$ be any specified numbers. There exists a unique solution to the differential equation*

$$y^{(n)} + p_{n-1}(t)y^{(n-1)} + \cdots + p_1(t)y' + p_0(t)y = g(t)$$

on the entire interval I that also satisfies the initial conditions

$$y(t_0) = a_0, \ y'(t_0) = a_1, \ldots, y^{(n-1)}(t_0) = a_{n-1}.$$

When we worked with second-order linear homogeneous equations, the notion of a fundamental set of solutions was basic to our study. We next want to extend this concept to higher-order linear homogeneous equations. A fundamental set of solutions is a collection of functions that solve the equation and serve as building blocks for all solutions in a particular way.

In Section 4.3 we discussed linear dependence/independence for a pair of functions and its role in solving a second-order linear homogeneous differential equation. The analogous idea of independence for sets of three or more functions is relevant to solving higher-order equations. We turn to this next.

4.6.2. Linear independence. The functions $f_1, f_2, \ldots, f_n$ are said to be **linearly dependent** on an open interval I if there are constants $c_1, c_2, \ldots, c_n$, **not all equal to** 0, so that

$$c_1 f_1 + c_2 f_2 + \cdots + c_n f_n = 0$$

on I. If the functions $f_1, f_2, \ldots, f_n$ are not linearly dependent on I, we say they are **linearly independent**. In other words, linear dependence means we can fill in the boxes in

$$\boxed{} f_1 + \boxed{} f_2 + \cdots + \boxed{} f_n = 0$$

with constants, at least one of which is not zero, to get a statement that is true at every point of the interval I. An expression of the form

$$c_1 f_1 + c_2 f_2 + \cdots + c_n f_n$$

for some constants $c_1, c_2, \ldots, c_n$ is called a **linear combination** of the functions $f_1, f_2, \ldots, f_n$, and linear independence requires that no **nontrivial** linear combination is equal to the zero function.

Example 4.6.3. Are the three functions $f_1(t) = \cos^2 t$, $f_2(t) = \sin^2 t$, and $f_3(t) = 3$ linearly independent or linearly dependent on $(-\infty, \infty)$? Since

$$\boxed{3} \cos^2 t + \boxed{3} \sin^2 t + \boxed{-1}(3) = 0$$

for every $-\infty < t < \infty$, we conclude these functions are linearly dependent.

Showing linear dependence requires finding **one** nontrivial linear combination that is equal to zero; showing linear independence requires showing that **no** nontrivial linear combination gives us 0; the latter appears to be a more difficult task.

Example 4.6.4. Let's show that the functions e^{2t}, e^{-t}, e^{4t} are linearly independent on any open interval I. Suppose that

(4.99) $$c_1 e^{2t} + c_2 e^{-t} + c_3 e^{4t} = 0$$

for all t in I. Since we are trying to show linear independence, our goal is to show that equation (4.99) forces all of the constants c_1, c_2, and c_3 to be zero. If we divide equation (4.99) by e^{2t}, we have

(4.100) $$c_1 + c_2 e^{-3t} + c_3 e^{2t} = 0.$$

Differentiating both sides of equation (4.100) with respect to t yields

(4.101) $$-3c_2 e^{-3t} + 2c_3 e^{2t} = 0.$$

Divide both sides of equation (4.101) by e^{-3t}, obtaining $-3c_2 + 2c_3 e^{5t} = 0$, and then differentiate on both sides of this equation to get

$$10 c_3 e^{5t} = 0.$$

Since the exponential function is never equal to zero, this forces $c_3 = 0$. Once we know this, equation (4.101) tells us that $c_2 = 0$, and finally equation (4.100) will imply $c_1 = 0$. The ideas in this example can be generalized to show that any collection of exponential functions $e^{r_1 t}, e^{r_2 t}, \ldots, e^{r_n t}$, where the numbers $r_1, r_2, \ldots, r_n$ are *distinct*, are linearly independent on any open interval I.

Another perspective on linear dependence. In Section 4.3, when we discussed pairs of functions, we had an easy intuitive way to think about linear dependence: The functions $\{f_1, f_2\}$ are linearly dependent if and only if one function is a constant multiple of the other. For sets of three or more functions, there is a related way to think about linear dependence, but it is a bit more complicated. Notice, for instance, that in Example 4.6.3, even though no function in the set $\{\cos^2 t, \sin^2 t, 3\}$ is a constant multiple of another, the functions were nonetheless dependent. For a set of three or more functions, linear dependence occurs exactly when one function is a *linear combination* of the others. For example,

$$3 = 3\cos^2 t + 3\sin^2 t,$$

so that the third function in Example 4.6.3 is a linear combination of the first two.

Example 4.6.5. We show that the functions $f_1(t) = t$, $f_2(t) = \ln t$, and $f_3(t) = \ln t^2 + 3t$ are linearly dependent on $(0, \infty)$. Properties of the logarithm say that $\ln t^2 = 2 \ln t$, so that $f_3 = 2f_2 + 3f_1$, and f_3 is expressed as a linear combination of f_2 and f_1. Said slightly differently (but equivalently), the relation

$$\boxed{3}\, f_1 + \boxed{2}\, f_2 + \boxed{-1}\, f_3 = 0$$

holds at all points of $(0, \infty)$, and hence the functions $\{f_1, f_2, f_3\}$ are linearly dependent there.

Fundamental set of solutions. We are primarily interested in linear independence when the functions in question are solutions to a homogeneous linear differential equation

(4.102) $$y^{(n)} + p_{n-1}(t) y^{(n-1)} + \cdots + p_1(t) y' + p_0(t) y = 0.$$

A set $\{y_1, y_2, \ldots, y_n\}$ of n *linearly independent* solutions of (4.102) on an open interval I is called a **fundamental set of solutions** for the equation. When $\{y_1, y_2, \ldots, y_n\}$ is a fundamental set of solutions, then any solution to equation (4.102) on I can be written as

(4.103) $$c_1 y_1 + c_2 y_2 + \cdots + c_n y_n,$$

for some choice of constants c_j. We call (4.103) the **general solution** of (4.102) on I. By the existence part of Theorem 4.6.2, if we specify any initial conditions at a point of I, where the coefficient functions $p_j(t)$ are continuous on I, then there is a choice of constants $c_1, c_2, \ldots, c_n$ so that the function in (4.103) solves the corresponding initial value problem.

Thus if we have managed, somehow, to produce n solutions $y_1, y_2, \ldots, y_n$ to equation (4.102), we then want to know if these n functions are linearly independent. As n increases from two on up, linear independence gets progressively more difficult to determine by inspection or from the definition. There is a mechanical procedure that will take n solutions to a nth-order linear

4.6. Higher-order equations

homogeneous differential equation and test them for linear independence. This procedure is a Wronskian determinant calculation. The Wronskian of the n solution functions is defined to be

$$W(y_1, y_2, \ldots, y_n) \equiv \det \begin{pmatrix} y_1 & y_2 & \cdots & y_n \\ y_1' & y_2' & \cdots & y_n' \\ \vdots & \vdots & & \vdots \\ y_1^{(n-1)} & y_2^{(n-1)} & \cdots & y_n^{(n-1)} \end{pmatrix}.$$

So long as the y_j's came to us as solutions to equation (4.102) on some open interval I (on which the coefficient functions p_j are continuous), a dichotomy exists: Either the Wronskian $W(y_1, y_2, \ldots, y_n)$ is **never** equal to 0 on I or it is **always** equal to 0 on I. The first case, W never 0, corresponds to linear independence of the functions $y_1, y_2, \ldots, y_n$, and the second case to linear dependence.

We haven't yet said how to compute the determinant of an $n \times n$ matrix for $n \geq 3$. The 3×3 case is reviewed in Exercise 16, and further information is presented in Section 8.2.

Theorem 4.3.8 for second-order linear homogeneous equations has an exact analogue for higher-order equations, which we can now state.

Theorem 4.6.6. *Suppose that $y_1, y_2, \ldots, y_n$ are solutions of the differential equation*

$$(4.104) \qquad y^{(n)} + p_{n-1}(t) y^{(n-1)} + \cdots + p_1(t) y' + p_0(t) y = 0$$

on an open interval I on which $p_0(t), p_1(t), \ldots, p_{n-1}(t)$ are continuous. Then the following conditions are equivalent:

(a) *The collection $\{y_1, y_2, \ldots, y_n\}$ is a fundamental set of solutions of equation (4.104) on I; that is, every solution on I has the form*

$$y = c_1 y_1 + c_2 y_2 + \cdots + c_n y_n$$

for some constants $c_1, c_2, \ldots, c_n$.

(b) *The Wronskian $W(y_1, y_2, \ldots, y_n)$ is nonzero at some point in I.*

(c) *The Wronskian $W(y_1, y_2, \ldots, y_n)$ is nonzero at every point in I.*

(d) *The functions $y_1, y_2, \ldots, y_n$ are linearly independent on I.*

4.6.3. Exercises.

1. (a) Show that the functions
   $$e^t, \quad e^{2t}, \quad e^{3t}$$
   are linearly independent on $(-\infty, \infty)$.
   (b) Show that the functions
   $$e^t, \quad e^{2t}, \quad 3e^{t+1} + e^{2t-4}$$
   are linearly dependent on $(-\infty, \infty)$.

2. This exercise outlines two ways to show that the set of functions $\{1, t, t^2, t^3, t^4\}$ is linearly independent on any open interval I.
   (a) Adapt the method of Example 4.6.4; that is, differentiate the identity
   $$(4.105) \qquad c_1 + c_2 t + c_3 t^2 + c_4 t^3 + c_5 t^4 = 0$$
   four times to conclude $c_5 = 0$. Then work backwards to conclude in turn that each of the other coefficients must also be 0.
   (b) How many roots can the polynomial
   $$c_1 + c_2 t + c_3 t^2 + c_4 t^3 + c_5 t^4$$
   have? Use this observation to conclude that if equation (4.105) holds for all t in some *open interval*, then $c_1 = c_2 = c_3 = c_4 = c_5 = 0$.

3. (a) Show that $f(t) = t$ is a linear combination of the functions $g(t) = t^2 + 5t$ and $h(t) = t^2 - 3t$.
   (b) Show that the set of functions $\{t, t^2 + 5t, t^2 - 3t\}$ is linearly dependent on $(-\infty, \infty)$.

In Exercises 4–12 decide whether the given functions are linearly independent or linearly dependent on $(-\infty, \infty)$. If dependent, express one function in the set as a linear combination of the others.

4. $\{1, t, t^2\}$.
5. $\{1, t^2 - 1, t^2 + 1\}$.
6. $\{t^2 + 4, t^2 + 4t, t - 1, t^3\}$.
7. $\{5, t + 3, t^2 + 2t, t^3 + 5t^2\}$.
8. $\{te^t, (2t+1)e^t, te^{t+2}\}$.
9. $\{1, \sin t, \cos t\}$.
10. $\{e^t, e^{-t}, e^{2t}\}$.
11. $\{e^t, e^{-t}, e^{2t}, e^{2t+2} + e^t\}$.
12. $\{e^t, e^{-t}, \sinh t\}$.

13. Identify each of the following statements as true or false, and explain your reasoning.
    (a) If the functions $\{f_1, f_2, f_3\}$ are linearly independent on an interval I, then so are $\{f_1, f_2\}$.
    (b) If the functions $\{f_1, f_2, f_3\}$ are linearly independent on an interval I, then so are $\{f_1, f_2, f_3, f_4\}$ for any choice of f_4.
    (c) If the functions $\{f_1, f_2, f_3\}$ are linearly dependent on an interval I, then so are $\{f_1, f_2, f_3, f_4\}$ for any choice of f_4.
    (d) No matter what f_1 and f_2 are, the functions $\{f_1, f_2, 0\}$ are linearly dependent on any interval I.

14. For each of the following, find the largest open interval on which the given initial value problem is guaranteed by Theorem 4.6.2 to have a unique solution.
    (a)
    $$\frac{d^3y}{dt^3} + \frac{1}{2-t}\frac{d^2y}{dt^2} + \frac{dy}{dt} + \ln(3+t)y = \cos t, \qquad y(1) = 1, \quad y'(1) = 2, \quad y''(1) = 3.$$
    (b)
    $$t^2\frac{d^4y}{dt^4} + \cos t\frac{d^2y}{dt^2} + 3y = \tan t, \qquad y(1) = 1, \quad y'(1) = 2, \quad y''(1) = 3, \quad y'''(1) = 4.$$
    (c)
    $$\sin t\frac{d^2y}{dt^2} + t^2\frac{dy}{dt} + y = \frac{1}{t^2 - 16}, \qquad y(5) = 1, \quad y'(5) = 2.$$

15. Sally claims that $\{1, t, t \ln t\}$ is a fundamental set of solutions for the differential equation
    $$t^3 y''' + t^2 y'' = 0$$
    on the interval $(0, \infty)$. Is she correct? Carefully justify your answer.

16. The determinant of a 3×3 matrix
    $$\begin{pmatrix} a_{11} & a_{12} & a_{13} \\ a_{21} & a_{22} & a_{23} \\ a_{31} & a_{32} & a_{33} \end{pmatrix}$$
    is defined to be
    $$a_{11}\det\begin{pmatrix} a_{22} & a_{23} \\ a_{32} & a_{33} \end{pmatrix} - a_{12}\det\begin{pmatrix} a_{21} & a_{23} \\ a_{31} & a_{33} \end{pmatrix} + a_{13}\det\begin{pmatrix} a_{21} & a_{22} \\ a_{31} & a_{32} \end{pmatrix}.$$
    (a) Compute the Wronskian of the functions $y_1(t) = e^t$, $y_2(t) = e^{-t}$, and $y_3(t) = e^{-2t}$ which solve $y''' + 2y'' - y' - 2y = 0$.
    (b) Solve the initial value problem $y''' + 2y'' - y' - 2y = 0$, $y(0) = 3, y'(0) = -4, y''(0) = 12$.
    (c) Compute the Wronskian of the functions $y_1(t) = 1, y_2(t) = \cos t$, and $y_3(t) = \sin t$ which solve $y''' + y' = 0$.
    (d) Solve the initial value problem $y''' + y' = 0$, $y(0) = -1, y'(0) = 2, y''(0) = 3$.

17. **Uniqueness of representation:** Let $\{y_1, y_2, \ldots, y_n\}$ be a fundamental set of solutions of

    (4.106) $$y^{(n)} + p_{n-1}(t)y^{(n-1)} + p_{n-2}(t)y^{(n-2)} + \cdots + p_1(t)y' + p_0(t)y = 0$$

    on a nonempty open interval I on which $p_0, \ldots, p_{n-1}$ are continuous. Show that every solution y of equation (4.106) on I may be written as a linear combination of $y_1, y_2, \ldots, y_n$ *in a unique way*.

18. Suppose that each function $g_1, g_2, \ldots, g_n$ is a linear combination of the functions $f_1, f_2, \ldots, f_m$. Show that any linear combination of the g_j's is also a linear combination of the f_k's.

4.7. Higher-order constant coefficient equations

From the work in Sections 4.4 and 4.5, you can now solve any second-order homogeneous linear equation with (real) constant coefficients. The basic principles we used there will carry over to higher-order constant coefficient equations.

Suppose we want to solve the linear homogeneous equation

(4.107) $$a_n y^{(n)} + a_{n-1} y^{(n-1)} + \cdots + a_1 y' + a_0 y = 0,$$

where the coefficients a_j are all real constants. Just as we did with second-order equations, we begin with a trial solution of the form

$$y(t) = e^{rt}.$$

Since the kth derivative of this trial solution is

$$y^{(k)}(t) = r^k e^{rt},$$

when we substitute our trial solution into equation (4.107) we get

(4.108) $$e^{rt}(a_n r^n + a_{n-1} r^{n-1} + \cdots + a_1 r + a_0) = 0.$$

By analogy with what we did with second-order equations, let's call the polynomial

(4.109) $$p(z) = a_n z^n + a_{n-1} z^{n-1} + \cdots + a_1 z + a_0$$

the **characteristic polynomial** of the differential equation (4.107); our trial solution e^{rt} therefore solves equation (4.107) exactly when r is a root of this polynomial.

Now the Fundamental Theorem of Algebra says that, as long as we are willing to consider complex roots, the polynomial in equation (4.109) factors as

(4.110) $$p(z) = a_n(z - r_1)(z - r_2) \cdots (z - r_n),$$

where $r_1, r_2, \ldots, r_n$ are the (possibly complex) roots of $p(z)$. It doesn't, however, give any clue as to how we are to *find* this factorization into linear factors; it just says that such a factorization *exists*. Some of the roots of the polynomial p may be repeated, meaning that some of the factors in the factorization (4.110) may be repeated.

To highlight the possibility of repeated roots, we might instead write our factorization of $p(z)$ into linear factors as

(4.111) $$p(z) = a_n(z - r_1)^{n_1}(z - r_2)^{n_2} \cdots (z - r_k)^{n_k},$$

where now the r_j's are distinct and the *multiplicities* $n_1, n_2, \ldots, n_k$ add up to n, the degree of p. Each of the roots r_j leads to a solution $e^{r_j t}$, but in the case that $n_j > 1$, the root r_j will produce other solutions to equation (4.107) as well; we describe these next. This will be a natural generalization of what you learned for second-order constant coefficient equations in Section 4.5.

First assume that $p(z)$ has exactly k different roots, $r_1, r_2, \ldots, r_k$, each of which is *real*. We will handle the complex case separately. Denoting the multiplicity of r_j by n_j, we describe n linearly independent solutions in the following table:

for r_1:	$e^{r_1 t}$,	$te^{r_1 t}$,	$t^2 e^{r_1 t}$, ...,	$t^{n_1 - 1} e^{r_1 t}$	(n_1 of these)
for r_2:	$e^{r_2 t}$,	$te^{r_2 t}$,	$t^2 e^{r_2 t}$, ...,	$t^{n_2 - 1} e^{r_2 t}$	(n_2 of these)
for r_3:	$e^{r_3 t}$,	$te^{r_3 t}$,	$t^2 e^{r_3 t}$, ...,	$t^{n_3 - 1} e^{r_3 t}$	(n_3 of these)
$\vdots$			$\vdots$		$\vdots$
for r_k:	$e^{r_k t}$,	$te^{r_k t}$,	$t^2 e^{r_k t}$, ...,	$t^{n_k - 1} e^{r_k t}$	(n_k of these)

There are several things to notice here. First, the row of solutions associated with the root r starts with e^{rt}, then proceeds with that exponential multiplied by increasing powers of t, stopping with the power that is one less than the multiplicity of r. For $n = 2$, you already know this from the case of *repeated roots* in Section 4.5. Next, the total number of solutions is $n_1 + n_2 + \cdots + n_k = n$. Call these solutions $y_1, y_2, \ldots, y_n$. They form a fundamental set of solutions, and thus the **general solution** is

$$y = c_1 y_1 + c_2 y_2 + c_3 y_3 + \cdots + c_n y_n,$$

where $c_1, c_2, \ldots, c_n$ are arbitrary constants.

Example 4.7.1. Consider the following differential equation:

$$y''' - 3y'' + 3y' - y = 0.$$

The characteristic polynomial is

$$p(z) = z^3 - 3z^2 + 3z - 1 = (z - 1)^3.$$

Our characteristic polynomial has one root, and its multiplicity is 3; i.e., $r_1 = 1$, $n_1 = 3$. Therefore, our fundamental set of solutions is

$$\{e^t, te^t, t^2 e^t\}$$

(one row only), and the general solution is $y = c_1 e^t + c_2 t e^t + c_3 t^2 e^t$.

Example 4.7.2. The differential equation $y''' - 2y'' = 0$ has characteristic polynomial $p(z) = z^3 - 2z^2 = z^2(z - 2)$. This has roots $r_1 = 0$ with multiplicity 2 and $r_2 = 2$ with multiplicity 1. This produces solution $y_1(t) = e^{0t} = 1$, $y_2(t) = te^{0t} = t$, and $y_3(t) = e^{2t}$. The general solution to the differential equation is $y(t) = c_1 + c_2 t + c_3 e^{2t}$.

Handling complex roots. Suppose that among the roots of the characteristic polynomial there are some complex roots. As long as the original differential equation has real coefficients, we know from Section 4.4 that any complex roots (with nonzero imaginary part) occur in conjugate pairs, so that if w is a complex root, so is $\overline{w}$. Let m be the multiplicity of both w and $\overline{w}$ (they always have the same multiplicity). The pair of roots w and $\overline{w}$ contribute *complex-valued* solutions

$$e^{wt}, \quad te^{wt}, \quad t^2 e^{wt}, \ldots, \quad t^{m-1} e^{wt},$$
$$e^{\overline{w}t}, \quad te^{\overline{w}t}, \quad t^2 e^{\overline{w}t}, \ldots, \quad t^{m-1} e^{\overline{w}t}.$$

As before, we prefer real-valued solutions. The real and imaginary parts of each of these complex-valued solutions is a real-valued solution. If $w = \lambda + i\mu$, the real part of both

$$t^j e^{wt} \quad \text{and} \quad t^j e^{\overline{w}t}$$

4.7. Higher-order constant coefficient equations

is $t^j e^{\lambda t} \cos(\mu t)$. The imaginary part of $t^j e^{wt}$ is $t^j e^{\lambda t} \sin(\mu t)$, while the imaginary part of $t^j e^{\overline{w}t}$ is $-t^j e^{\lambda t} \sin(\mu t)$, a scalar multiple of the imaginary part of $t^j e^{wt}$. From the complex roots w and $\overline{w}$, each with multiplicity m, we thus obtain $2m$ linearly independent real-valued solutions

$$e^{\lambda t}\cos(\mu t), \quad te^{\lambda t}\cos(\mu t), \ldots, \quad t^{m-1}e^{\lambda t}\cos(\mu t),$$

$$e^{\lambda t}\sin(\mu t), \quad te^{\lambda t}\sin(\mu t), \ldots, \quad t^{m-1}e^{\lambda t}\sin(\mu t).$$

We repeat this process for each pair of complex conjugate roots w and $\overline{w}$.

Summary: A mix of real and complex roots. Suppose the characteristic polynomial has q distinct real roots $r_1, r_2, \ldots, r_q$ with respective multiplicities n_i and $2s$ complex roots $w_1 = \lambda_1 + i\mu_1, \overline{w_1} = \lambda_1 - i\mu, w_2 = \lambda_2 + i\mu_2, \overline{w_2} = \lambda_2 - i\mu_2, \ldots, w_s = \lambda_s + i\mu_s, \overline{w_s} = \lambda_s - i\mu_s$ with multiplicities m_i. We then get the following **table of linearly independent real-valued solutions**:

$$e^{r_1 t}, \quad te^{r_1 t}, \quad t^2 e^{r_1 t}, \ldots, \quad t^{n_1-1}e^{r_1 t}$$
$$e^{r_2 t}, \quad te^{r_2 t}, \quad t^2 e^{r_2 t}, \ldots, \quad t^{n_2-1}e^{r_2 t}$$
$$\vdots$$
$$e^{r_q t}, \quad te^{r_q t}, \quad t^2 e^{r_q t}, \ldots, \quad t^{n_q-1}e^{r_q t}$$

$$\begin{bmatrix} e^{\lambda_1 t}\cos(\mu_1 t), & te^{\lambda_1 t}\cos(\mu_1 t), \ldots, & t^{m_1-1}e^{\lambda_1 t}\cos(\mu_1 t) \\ e^{\lambda_1 t}\sin(\mu_1 t), & te^{\lambda_1 t}\sin(\mu_1 t), \ldots, & t^{m_1-1}e^{\lambda_1 t}\sin(\mu_1 t) \end{bmatrix}$$

$$\begin{bmatrix} e^{\lambda_2 t}\cos(\mu_2 t), & te^{\lambda_2 t}\cos(\mu_2 t), \ldots, & t^{m_2-1}e^{\lambda_2 t}\cos(\mu_2 t) \\ e^{\lambda_2 t}\sin(\mu_2 t), & te^{\lambda_2 t}\sin(\mu_2 t), \ldots, & t^{m_2-1}e^{\lambda_2 t}\sin(\mu_2 t) \end{bmatrix}$$
$$\vdots$$
$$\begin{bmatrix} e^{\lambda_s t}\cos(\mu_s t), & te^{\lambda_s t}\cos(\mu_s t), \ldots, & t^{m_s-1}e^{\lambda_s t}\cos(\mu_s t) \\ e^{\lambda_s t}\sin(\mu_s t), & te^{\lambda_s t}\sin(\mu_s t), \ldots, & t^{m_s-1}e^{\lambda_s t}\sin(\mu_s t) \end{bmatrix}$$

While this may look a bit complicated, it's really not difficult to use in practice. We'll illustrate with several examples.

Example 4.7.3. Consider

$$y^{(4)} - y = 0.$$

The general solution has the form $y = c_1 y_1 + c_2 y_2 + c_3 y_3 + c_4 y_4$ where y_1, y_2, y_3, y_4 form a fundamental set of solutions. To find the roots of the characteristic polynomial $p(z) = z^4 - 1$, we factor

$$z^4 - 1 = (z^2 - 1)(z^2 + 1) = (z-1)(z+1)(z-i)(z+i).$$

We have $q = 2$ real roots and $s = 1$ pairs of complex conjugate roots, with multiplicities as shown:

$$r_1 = 1, \qquad n_1 = 1,$$
$$r_2 = -1, \qquad n_2 = 1,$$
$$w_1 = i, \qquad m_1 = 1,$$
$$\overline{w_1} = -i, \qquad m_1 = 1.$$

Therefore, our real solutions are

$$r_1 = 1: \quad e^t,$$
$$r_2 = -1: \quad e^{-t},$$
$$\left\{\begin{array}{rl} w_1 &= i \\ \overline{w_1} &= -i \end{array}\right\}: \quad \left\{\begin{array}{l} \cos t \\ \sin t \end{array}\right\}.$$

Thus the general solution is

$$y = c_1 e^t + c_2 e^{-t} + c_3 \cos t + c_4 \sin t.$$

Example 4.7.4. Let's solve $y^{(8)} - 2y^{(4)} + y = 0$. The characteristic polynomial is

$$p(z) = z^8 - 2z^4 + 1$$
$$= (z^4 - 1)^2$$
$$= (z^2 - 1)^2 (z^2 + 1)^2$$
$$= (z-1)^2 (z+1)^2 (z-i)^2 (z+i)^2,$$

so we have the following roots and multiplicities:

$$r_1 = 1, \qquad n_1 = 2,$$
$$r_2 = -1, \qquad n_2 = 2 \quad (q = 2),$$
$$w_1 = i, \qquad m_1 = 2,$$
$$\overline{w_1} = -i, \qquad m_1 = 2 \quad (s = 1).$$

Therefore, our real solutions are

$$r_1 = 1: \quad e^t, \; te^t,$$
$$r_2 = -1: \quad e^{-t}, \; te^{-t},$$
$$\left\{\begin{array}{rl} w_1 &= i \\ \overline{w_1} &= -i \end{array}\right\}: \quad \left\{\begin{array}{ll} \cos t, & t\cos t \\ \sin t, & t\sin t \end{array}\right\}.$$

The general solution is then

$$y = c_1 e^t + c_2 t e^t + c_3 e^{-t} + c_4 t e^{-t} + c_5 \cos t + c_6 t \cos t + c_7 \sin t + c_8 t \sin t.$$

In the examples so far, we were able to factor our characteristic polynomial into linear factors without too much difficulty. What about an equation like $y^{(3)} + 8y = 0$, which leads to the characteristic polynomial $z^3 + 8$? How do we find the roots of this polynomial? To answer this, we need to learn how to write complex numbers in *polar form* and then how to use the polar form to find the roots of a complex number.

4.7. Higher-order constant coefficient equations

4.7.1. Polar form for complex numbers. If we picture a complex number $x + iy$ as the point (x, y) in the plane, we can also describe this point by polar coordinates (r, θ), where $r \geq 0$, $x = r\cos\theta$, and $y = r\sin\theta$. In other words,

$$x + iy = r\cos\theta + ir\sin\theta.$$

Using Euler's identity $\cos\theta + i\sin\theta = e^{i\theta}$ from Section 4.4, we have

$$x + iy = re^{i\theta};$$

this is called the **polar form** of $x+iy$. Now polar form is not unique (equivalently, polar coordinates are not unique), since we can replace θ by $\theta + 2k\pi$ for any integer k. The value of r is the modulus of $x + iy$, and the allowable choices for θ are called the **arguments**, or polar angles, of $x + iy$. The following figure illustrates the polar coordinates r and θ for $x + iy$. We have $r = \sqrt{x^2 + y^2}$. If, as in the figure, $x > 0$, then we can choose $\theta = \tan^{-1}\left(\frac{y}{x}\right)$ (while for $x < 0$, we can choose $\theta = \tan^{-1}\left(\frac{y}{x}\right) + \pi$).

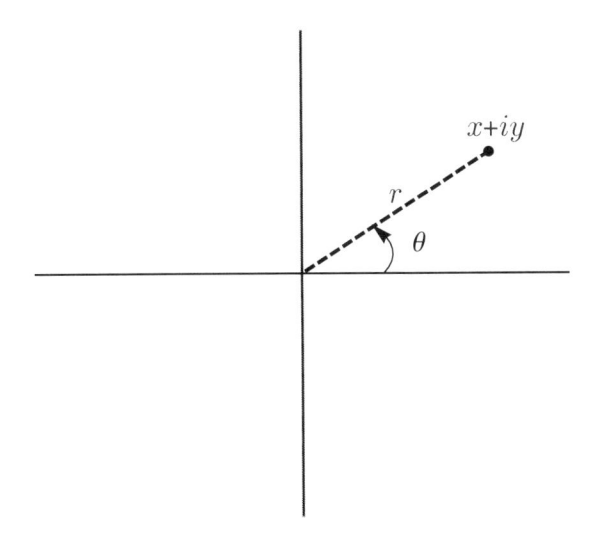

Figure 4.34. Polar coordinates for $z = x + iy$.

For example, since the polar coordinates of the point $x = 1$, $y = 1$ are $r = \sqrt{2}$, $\theta = \frac{\pi}{4} + 2k\pi$,

$$\begin{aligned} 1 + i &= \sqrt{2}e^{i\pi/4} = \sqrt{2}e^{i(9\pi/4)} = \sqrt{2}e^{i(17\pi/4)} = \cdots \\ &= \sqrt{2}e^{-i(7\pi/4)} = \sqrt{2}e^{-i(15\pi/4)} = \cdots \end{aligned}$$

are polar forms of $1 + i$, and

$$\begin{aligned} -8 &= 8e^{i\pi} = 8e^{i3\pi} = 8e^{i5\pi} = \cdots \\ &= 8e^{-i\pi} = e^{-i3\pi} = \cdots \end{aligned}$$

are polar forms of -8, since -8 has modulus 8 and arguments $(\pi + 2k\pi)$ for any integer k.

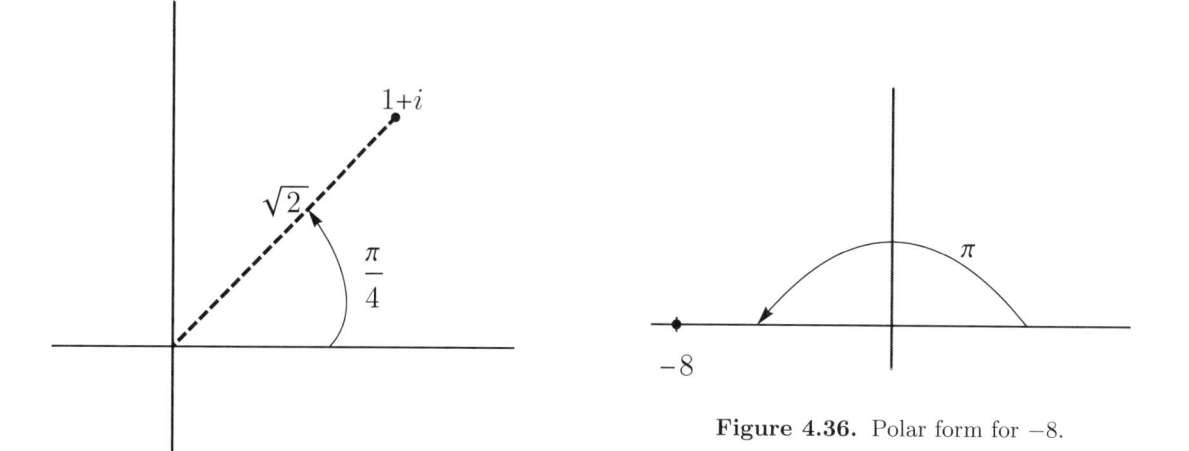

Figure 4.35. Polar form for $1+i$.

Figure 4.36. Polar form for -8.

Polar form is ideally suited for computing products and powers of complex numbers:

$$(re^{i\theta})(se^{i\phi}) = rse^{i(\theta+\phi)}$$

and for any integer n

(4.112) $$(re^{i\theta})^n = r^n e^{in\theta}.$$

These two formulas tell you how to find, in polar form, the product of two complex numbers written in polar form and how to compute any (integer) power of a complex number written in polar form. Exercise 13 outlines how to verify these formulas.

Finding nth roots. Polar form is also key to finding nth roots. The nth roots of a (real or complex) number w are all solutions to $z^n = w$. We illustrate how to find these in the next example, where we find the cube roots of -8.

Example 4.7.5. To solve $z^3 = -8$ we write our variable z in polar form as

$$z = re^{i\theta},$$

so that by equation (4.112)

$$z^3 = r^3 e^{i3\theta}.$$

We also write -8 in polar form, by noting that the modulus of -8 is merely 8, while the arguments of -8 are π, 3π, 5π,..., or in general $\pi + 2k\pi$ for any integer k. The equation we wish to solve, $z^3 = -8$, now looks like

$$(re^{i\theta})^3 = 8e^{i\pi} = 8e^{i(3\pi)} = 8e^{i(5\pi)} = \cdots = 8e^{i(\pi+2k\pi)},$$

or

(4.113) $$r^3 e^{i(3\theta)} = 8e^{i\pi} = 8e^{i(3\pi)} = 8e^{i(5\pi)} = \cdots.$$

If we match up the modulus r^3 on the left-hand side with the modulus 8 on the right-hand side, this tells us

$$r = 2,$$

since the modulus r must be positive and 2 is the unique positive cube root of 8. The polar angles are more interesting. If we match the argument 3θ on the left-hand side of (4.113) with the

4.7. Higher-order constant coefficient equations

arguments on the right-hand side, we obtain $3\theta = \pi + 2k\pi$ for any integer k. Thus

(4.114) $$\theta = \frac{\pi}{3} + \frac{2k\pi}{3}.$$

As we let $k = 0, 1, 2$, we get three different solutions to $z^3 = -8$, namely

$$z_1 = 2e^{i\pi/3} = 2\left(\cos\frac{\pi}{3} + i\sin\frac{\pi}{3}\right) = 2\left(\frac{1}{2} + i\frac{\sqrt{3}}{2}\right) = 1 + i\sqrt{3},$$

$$z_2 = 2e^{i\pi} = 2(\cos\pi + i\sin\pi) = 2(-1) = -2,$$

$$z_3 = 2e^{i(5\pi/3)} = 2\left(\cos\frac{5\pi}{3} + i\sin\frac{5\pi}{3}\right) = 2\left(\frac{1}{2} - i\frac{\sqrt{3}}{2}\right) = 1 - i\sqrt{3}.$$

Does this process continue indefinitely, with each choice of k in (4.114) yielding a different cube root? No, in fact any other integer value of k will duplicate one of the three cube roots we have already found. For example, with $k = 3$ we obtain the cube root

$$2e^{i(7\pi/3)} = 2\left(\cos\frac{7\pi}{3} + i\sin\frac{7\pi}{3}\right) = 2\left(\frac{1}{2} + i\frac{\sqrt{3}}{2}\right) = 1 + i\sqrt{3},$$

which is the same solution produced with $k = 0$. Similarly, the choice $k = -1$ yields the cube root $2e^{-i\pi/3} = 1 - i\sqrt{3}$. The three cube roots of -8 are shown pictorially in Fig. 4.37.

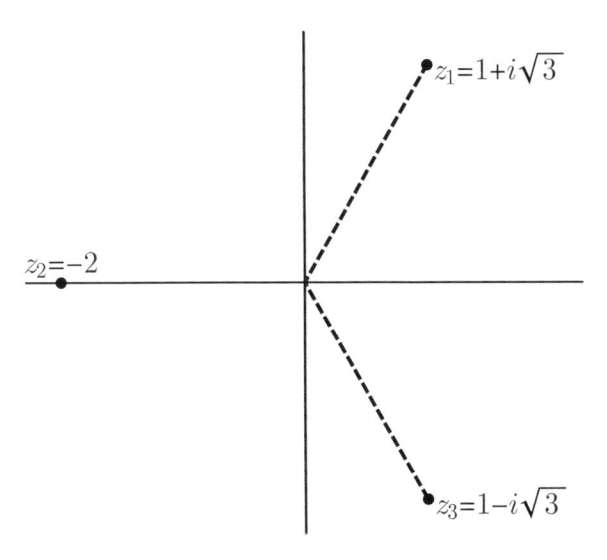

Figure 4.37. The cube roots of -8.

Summary: Solving $z^n = w$. To find the nth roots of any complex number w, that is, to solve $z^n = w$, let ρ be the modulus of w and let $\phi + 2k\pi$ be its arguments. As long as $w \neq 0$, there will be n distinct solutions to $z^n = w$. They are obtained, in polar form, as $z = re^{i\theta}$, where

$$r = \rho^{\frac{1}{n}}, \qquad \theta = \frac{\phi}{n} + \frac{2k\pi}{n} \quad \text{for } k = 0, 1, 2, \ldots, n-1.$$

Here $\rho^{(1/n)}$ denotes the positive nth root of the positive number ρ. If we show a picture of the nth roots of w, they will all lie on the circle centered at the origin with radius $\rho^{(1/n)}$, where $\rho = |w|$, and will be uniformly spaced around this circle, with the angle between any two consecutive roots being $2\pi/n$.

Example 4.7.6. We solve $z^4 = -1$. On the right-hand side, -1 has modulus 1 and arguments $\pi + 2k\pi$. Thus if $z = re^{i\theta}$ solves $z^4 = -1$, we have $r = 1$ and $\theta = (\pi + 2k\pi)/4$ for $k = 0, 1, 2, 3$. This gives four solutions for z,

$$e^{i\pi/4}, \quad e^{i3\pi/4}, \quad e^{i5\pi/4}, \quad e^{i7\pi/4},$$

or more transparently

(4.115) $$\frac{\sqrt{2}}{2} + i\frac{\sqrt{2}}{2}, \quad -\frac{\sqrt{2}}{2} + i\frac{\sqrt{2}}{2}, \quad -\frac{\sqrt{2}}{2} - i\frac{\sqrt{2}}{2}, \quad \frac{\sqrt{2}}{2} - i\frac{\sqrt{2}}{2}.$$

These are shown in Fig. 4.38.

In the next example we use our new skill of finding roots of complex numbers to solve a constant coefficient equation.

Example 4.7.7. We'll solve the equation $y''' + 8y = 0$. From the computations we did above, we know the characteristic polynomial

$$z^3 + 8$$

has roots $z = 1 + i\sqrt{3}, -2, 1 - i\sqrt{3}$. Notice that the nonreal solutions occur in conjugate pairs, as must happen whenever the original differential equation has only real coefficients. The roots $1 \pm i\sqrt{3}$ give rise to the real-valued solutions

$$e^t \cos(\sqrt{3}\,t), \quad e^t \sin(\sqrt{3}\,t)$$

and the root $z = -2$ gives a third solution

$$e^{-2t}.$$

These three solutions are linearly independent, and a general solution is thus

$$y = c_1 e^{-2t} + c_2 e^t \cos(\sqrt{3}\,t) + c_3 e^t \sin(\sqrt{3}\,t).$$

Example 4.7.8. Next we'll solve the differential equation $y^{(4)} + y = 0$ which has characteristic equation $z^4 + 1 = 0$, or $z^4 = -1$. We computed the four distinct solutions to this equation above, obtaining the roots in (4.115). Since there are four distinct roots for this polynomial of degree four, all of the multiplicities are one. They occur in the two conjugate pairs $w_1, \overline{w}_1$ and $w_2, \overline{w}_2$, where

$$w_1 = \frac{\sqrt{2}}{2} + i\frac{\sqrt{2}}{2}, \quad w_2 = -\frac{\sqrt{2}}{2} + i\frac{\sqrt{2}}{2},$$

as shown in Fig. 4.38. Our general solution is

$$y = c_1 e^{\frac{\sqrt{2}}{2}t} \cos\frac{\sqrt{2}\,t}{2} + c_2 e^{\frac{\sqrt{2}}{2}t} \sin\frac{\sqrt{2}\,t}{2} + c_3 e^{-\frac{\sqrt{2}}{2}t} \cos\frac{\sqrt{2}\,t}{2} + c_4 e^{-\frac{\sqrt{2}}{2}t} \sin\frac{\sqrt{2}\,t}{2}.$$

Write out the table of roots and real solutions to make sure you understand why.

4.7. Higher-order constant coefficient equations

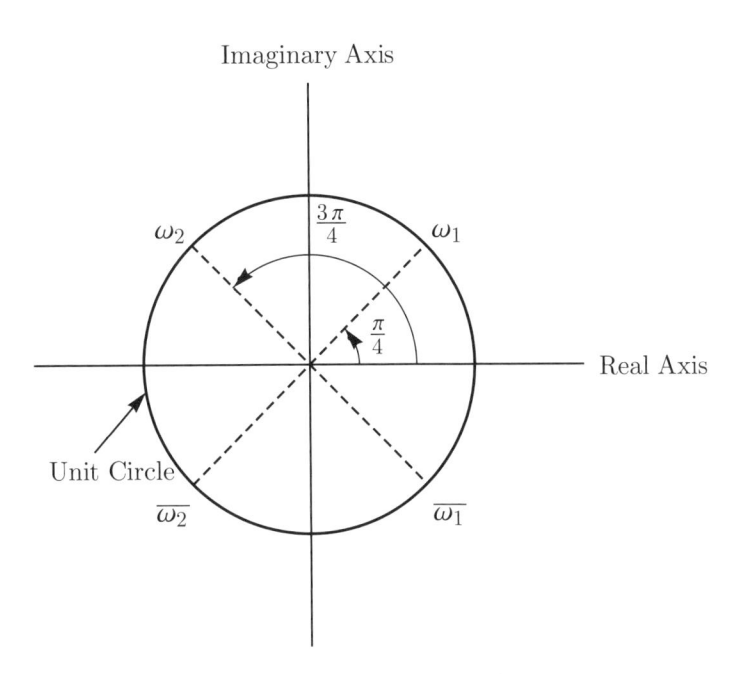

Figure 4.38. The solutions of $z^4 + 1 = 0$.

4.7.2. Exercises.

1. Write w in polar form and then find all solutions to the indicated equation.
   (a) $w = 16$; $z^4 = w$.
   (b) $w = 2 + 2i$; $z^3 = w$.
   (c) $w = 1 - \sqrt{3}i$; $z^4 = w$.
   (d) $w = -\sqrt{3} + i$; $z^4 = w$.

2. Find the three cube roots of 8, and show them graphically in the complex plane.

3. Find the four 4th roots of i, and show them graphically in the complex plane.

In Exercises 4–11 give the general solution.

4. $y''' + 2y'' + y' = 0$.
5. $y''' + 3y'' + 3y' + y = 0$.
6. $y''' - y = 0$.
7. $y''' - 2y'' + y' - 2y = 0$.
8. $y^{(4)} - 4y''' + 20y'' = 0$.
9. $y''' - 5y'' + 3y' + y = 0$.
10. $y^{(4)} + 2y^{(3)} + 10y^{(2)} = 0$.
11. $y^{(6)} + 64y = 0$.

12. Find the general solution of
$$y^{(12)} + 2y^{(6)} + y = 0.$$
The functions in your solution should be real valued.

13. (a) Verify the formula
$$(re^{i\theta})(se^{i\phi}) = rse^{i(\theta+\phi)}.$$
    Hint: Use the definition
$$re^{i\theta} = r\cos\theta + ir\sin\theta$$
    and trigonometric identities for $\cos(\theta + \psi)$ and $\sin(\theta + \psi)$.
    (b) Using (a), show by induction that
$$(re^{i\theta})^n = r^n e^{in\theta}$$
    for any positive integer n.

(c) Show that
$$\frac{re^{i\theta}}{se^{i\varphi}} = \frac{r}{s}e^{i(\theta-\varphi)}.$$

Hint:
$$\frac{re^{i\theta}}{se^{i\varphi}} = \frac{re^{i\theta}e^{-i\varphi}}{se^{i\varphi}e^{-i\varphi}}.$$

Now use (a).

(d) Show that
$$(re^{i\theta})^{-n} = r^{-n}e^{-in\theta}$$
for any positive integer n.

14. (a) Compute $(D-2)[te^{2t}]$ and $(D-2)[t^2 e^{2t}]$, simplifying your answer as much as possible. Here, D denotes differentiation, so the operator $D-a$ takes an input function y and produces $y' - ay$ as the output function.

(b) Compute $(D-2)^2[te^{2t}]$ and $(D-2)^2[t^2 e^{2t}]$. Note: $(D-2)^2[y] = (D-2)[(D-2)[y]]$.

(c) Show that te^{2t} and $t^2 e^{2t}$ are solutions to the differential equation
$$(D-2)^3[y] = 0.$$

15. Show that te^t, $t^2 e^t$, and $t^3 e^t$ are solutions to the differential equation
$$(D-1)^4[y] = 0.$$

16. (a) Give the general solution to $(D+1)(D-3)^2(D^2+4)[y] = 0$ if y is a function of t and D represents differentiation with respect to t.

(b) Similarly give the general solution to
$$(D-5)^3(D^2-4)(D^2 - 2D + 5)^2[y] = 0.$$

17. Consider the "differential-substitution" equation
(4.116)
$$y''(t) + y(-t) = 0.$$

Find all solutions $y(t)$. Hint: Differentiate equation (4.116) twice, and then combine the resulting equation with the original equation to obtain a differential equation you can solve. The solutions of (4.116) are among those you find.

18. The "mode shape function" $y = y(x)$ for the free vibration of a uniform beam satisfies
$$\frac{d^4 y}{dx^4} - \beta y = 0$$
where β is a positive constant depending on the frequency of vibration, the shape of the beam, and the material out of which the beam is made. Find the general solution of this differential equation.

4.8. Modeling with second-order equations

Linear differential equations in general and constant coefficient linear equations in particular appear in a number of problems that model basic physical systems. In this section, we focus on the modeling of a vertically oriented mass-spring system, both with and without damping. The same differential equation (4.119) that we will use to describe a damped mass-spring system appears in the analysis of "RLC" (resistor, inductor, capacitor) electric circuits (see Section 5.4, in particular equation (5.97)). This fact—that the same mathematical model may apply to a variety of physical problems—is part of what gives the mathematics its power. Another illustration of this fact is furnished by Exercises 23 and 24, in which a second-order linear equation of the form $y'' = \alpha(y+\beta)$, where $\alpha > 0$ and β are constants, models both a cable sliding off a peg and the steady-state temperature of a metal rod.

4.8. Modeling with second-order equations

4.8.1. Mass-spring systems: Free vibrations. A mass moving on the end of a spring gives a simple way to model the vibrations in a mechanical system. We'll think of the spring as suspended vertically from a fixed support (recall in Example 4.1.3 we looked at horizontal spring systems), and let ℓ denote the natural length of the spring. When a mass m is attached to the spring, it stretches the spring by some amount $\Delta\ell$. After the mass is attached, the system is allowed to come to rest. We call this the equilibrium position, and in this position the forces acting on the mass are in balance. These forces are

(a) the force due to gravity, i.e., the weight of the object, and

(b) the spring force.

Let's set up a vertical coordinate system, so that the origin corresponds to the rest position of the mass after it is attached to the spring and so that the downward direction is positive. The force due to gravity has magnitude mg, where m is the mass and g is the acceleration due to gravity, and acts in the downward, or positive, direction. Since the spring force acts in the direction to restore the spring to its natural length, it acts upwards at equilibrium. If we assume, as in Example 4.1.3, that our spring force obeys Hooke's law, then its magnitude at the equilibrium position is $k\Delta\ell$, where k is the positive spring force constant. These two forces are in balance, so

(4.117) $$mg = k\Delta\ell.$$

Typically we use (4.117) to calculate k, from the known values of m and g and the observed value of $\Delta\ell$.

Example 4.8.1. A mass weighing 2 lbs stretches a spring by 6 inches. Assuming the spring obeys Hooke's law, we calculate the spring constant by

$$2 = k\left(\frac{1}{2}\right).$$

Notice that we convert 6 inches to $\frac{1}{2}$ ft and that pounds are units of force in the British system of units. Thus $k = 4$ lb/ft, and the mass $m = \frac{2}{g} \approx \frac{2}{32}$ lb-sec^2/ft (or slugs).

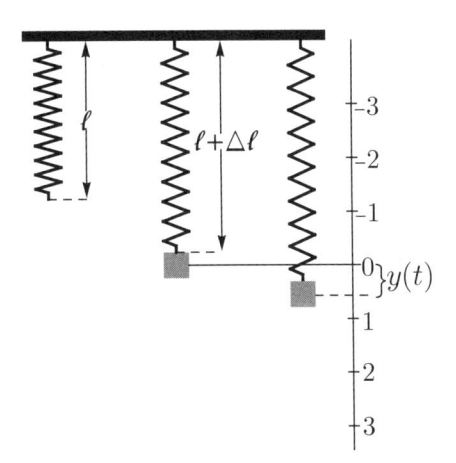

Figure 4.39. Measuring displacement from equilibrium.

Now imagine we put our mass into motion in some way; perhaps we pull it down and let it go, or give it a push either up or down. We want to describe the subsequent motion. We'll do this by means of the function $y(t)$ which gives its displacement *from the equilibrium position*. By the way we have chosen to orient our axis, a positive value of $y(t)$ means that the spring is further stretched from the equilibrium position; see Fig. 4.39.

To derive a differential equation for the function $y(t)$ we consider all the forces acting on the mass: the gravitational force, the spring force, and any drag or damping force. In this section we do not consider any externally applied force, except gravity itself, and the resulting motion is called "free vibration". The gravitational force is the weight mg; it acts in the positive direction. The spring force is proportional to the amount the spring is stretched or compressed from its *natural* length. An expression for this amount of compression or extension is

$$\Delta\ell + y(t).$$

When $\Delta\ell + y(t) > 0$, the spring is stretched from its natural length, so the spring force acts in the upward, or negative, direction and is given by

$$-k(\Delta\ell + y(t)).$$

When $\Delta\ell + y(t) < 0$, the spring is compressed from its natural length; the spring force acts in the positive direction and is still given by

$$-k(\Delta\ell + y(t))$$

(why?).

We also might want to consider a damping, or drag, force. This might simply be due to the medium in which the object moves, or it might be due to an attached dashpot damping device. In any case, we will assume that the drag force is proportional to the velocity of the mass and always acts opposite to the direction of motion. The derivative $y'(t)$ is the velocity; this will be positive if the mass is moving downward and negative if the mass is moving upward. When the mass moves downward, the drag force is upward, and we describe it by the expression

$$-cy'(t),$$

where c is a positive constant. What about the drag force when the object moves upward? The expression $-cy'(t)$ is still correct, since $y'(t) < 0$ and we want the drag force to have a positive sign.

Figure 4.40. Damping device.

By Newton's second law, the sum of the forces acting on our mass is equal to the product of the mass m and the object's acceleration:

(4.118) $$my''(t) = mg - k(\Delta\ell + y(t)) - cy'(t).$$

We can rewrite equation (4.118) as

$$my'' = mg - k\Delta\ell - ky - cy'$$

and use equation (4.117) to simplify this to

(4.119) $$my'' + cy' + ky = 0,$$

4.8. Modeling with second-order equations

a constant coefficient homogeneous equation. The constants $m, c,$ and k that appear in this equation are positive, unless we want to ignore damping, in which case we set $c = 0$. Let's compare the damped $(c > 0)$ and undamped $(c = 0)$ situations.

No damping. If we ignore damping, our differential equation for the displacement function $y(t)$ is
$$my'' + ky = 0.$$
This has general solution
$$y(t) = c_1 \cos\sqrt{\frac{k}{m}}t + c_2 \sin\sqrt{\frac{k}{m}}t.$$
If we set
$$\omega_0 = \sqrt{\frac{k}{m}},$$
we can write this a bit more cleanly as
$$y(t) = c_1 \cos(\omega_0 t) + c_2 \sin(\omega_0 t).$$
By our work in Section 4.4.2 we can rewrite this as
$$y(t) = R\cos(\omega_0 t - \delta),$$
where R and δ are determined from $R = \sqrt{c_1^2 + c_2^2}$ and
$$R\cos\delta = c_1, \quad R\sin\delta = c_2.$$
We conclude that in the undamped case, the motion of the mass is periodic, with period
$$T = \frac{2\pi}{\omega_0} = 2\pi\sqrt{\frac{m}{k}}$$
and amplitude R. The period is the time to finish one complete cycle of motion. The circular frequency of the system is $\omega_0 = \sqrt{\frac{k}{m}}$; its units are radians per unit of time. The number $\nu_0 = \frac{\omega_0}{2\pi}$ is the ordinary frequency, and its units are cycles per unit of time. One cycle per second is one hertz. The motion of the mass in this undamped case is called **simple harmonic motion**.

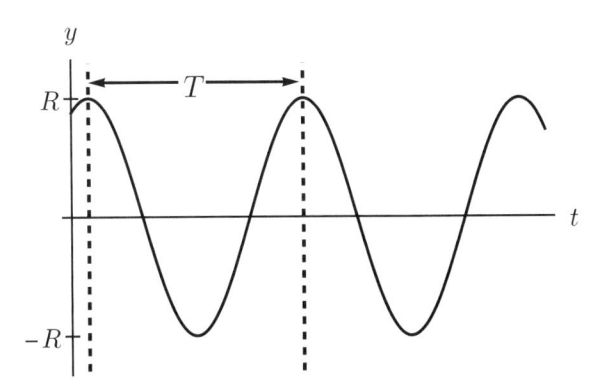

Figure 4.41. $y = R\cos(\omega_0 t - \delta)$ models simple harmonic motion.

Example 4.8.2. A 3 kg mass is attached to a vertical spring with spring constant $k = 24$ N/m. The damping force is negligible. Suppose the mass is pulled down 4 meters and then given an upward velocity of $8\sqrt{2}$ m/sec. The displacement function $y(t)$ solves the differential equation
$$3y'' + 24y = 0.$$

The associated characteristic equation has solutions $\pm 2\sqrt{2}\, i$, so the displacement function has the form
$$y(t) = c_1 \cos(2\sqrt{2}t) + c_2 \sin(2\sqrt{2}t).$$
From the initial conditions
$$y(0) = 4, \quad y'(0) = -8\sqrt{2}$$
we determine that $c_1 = 4$ and $c_2 = -4$. Thus the displacement of the mass at time t is given by
$$y(t) = 4\cos(2\sqrt{2}t) - 4\sin(2\sqrt{2}t),$$
which we can rewrite as
$$y(t) = 4\sqrt{2} \cos\left(2\sqrt{2}t - \frac{7\pi}{4}\right).$$
The amplitude of the motion is $4\sqrt{2}$m, and the period is $\pi/\sqrt{2}$ seconds.

Motion with damping. As soon as we include a damping term ($c \neq 0$) in our model, so that our equation
$$my'' + cy' + ky = 0$$
has positive coefficients, we know by Section 4.5.2 that, no matter the initial conditions, every solution $y(t)$ tends to 0 as $t \to \infty$. Let's give a bit more detailed analysis by considering three cases.

The underdamped case. If $c^2 - 4mk < 0$ (think of this as saying the damping constant is small), the system is said to be **underdamped**. The roots of the characteristic equation are a pair of complex conjugates $\lambda \pm i\mu$ where
$$\lambda = -\frac{c}{2m}, \quad \mu = \frac{\sqrt{|c^2 - 4mk|}}{2m} = \frac{\sqrt{4mk - c^2}}{2m}.$$
Notice that λ is negative. The general solution to $my'' + cy' + ky = 0$ in this case looks like
$$y(t) = c_1 e^{\lambda t} \cos(\mu t) + c_2 e^{\lambda t} \sin(\mu t) = Re^{\lambda t} \cos(\mu t - \delta)$$
where $R = \sqrt{c_1^2 + c_2^2}$, $R\cos\delta = c_1$, and $R\sin\delta = c_2$. From Section 4.4.2 we know that such a function shows oscillatory motion, with decreasing amplitude; a typical graph for such a function is shown in Fig. 4.42, along with the guidelines $y = \pm Re^{\lambda t}$.

The number $2\pi/\mu$ is called the **quasiperiod** of the motion, and it measures the time between consecutive maxima (or minima) of the displacement. The mass passes through the equilibrium $y = 0$ at regular intervals; the time between two consecutive such crossings is half the quasiperiod.

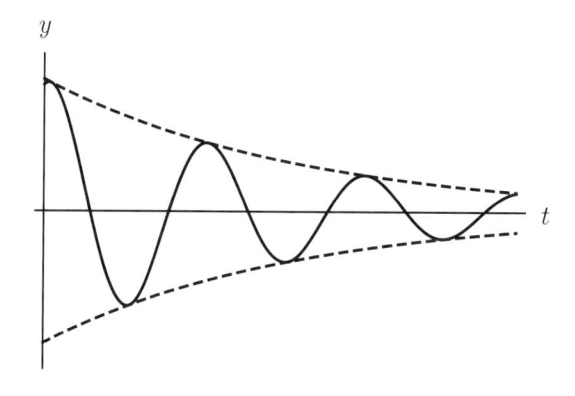

Figure 4.42. Typical graph of $y = Re^{\lambda t} \cos(\mu t - \delta)$.

4.8. Modeling with second-order equations

The overdamped case. By contrast, if the damping constant c is large enough that $c^2 - 4mk > 0$, the system is termed **overdamped**. In this case the characteristic equation will have two real unequal roots

$$r_1 = \frac{-c + \sqrt{c^2 - 4mk}}{2m}, \quad r_2 = \frac{-c - \sqrt{c^2 - 4mk}}{2m}.$$

Both of these roots are negative; this observation was made in Section 4.5.2. The general solution for the displacement function is now

$$y(t) = c_1 e^{r_1 t} + c_2 e^{r_2 t}.$$

There is no oscillatory behavior, and in fact, whatever the initial conditions, the mass can pass through the equilibrium position for at most one time $t > 0$; see Exercise 4. Typical graphs for the displacement function in the overdamped case are shown in Fig. 4.43.

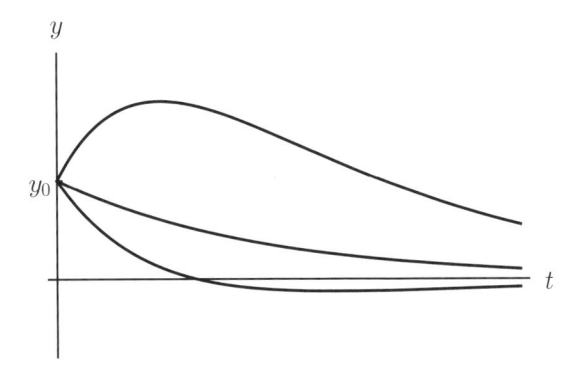

Figure 4.43. $y(0) = y_0$, different initial velocities.

The critically damped case. The transition from underdamped to overdamped occurs when $c^2 - 4mk = 0$; this is referred to as **critical damping**. Now the characteristic equation has a repeated root

$$r = -\frac{c}{2m}$$

and the general displacement function looks like

$$y(t) = c_1 e^{-\frac{c}{2m}t} + c_2 t e^{-\frac{c}{2m}t}.$$

When is the mass at equilibrium in the critically damped case? The only solution to

$$c_1 e^{-\frac{c}{2m}t} + c_2 t e^{-\frac{c}{2m}t} = 0$$

occurs when

$$c_1 + c_2 t = 0,$$

in other words only for

$$t = -\frac{c_1}{c_2}.$$

Whether this corresponds to a positive value of t will depend on the initial condition, which determines c_1 and c_2. Just as in the overdamped case, there is at most one time $t > 0$ at which the mass passes through the equilibrium position. Typical graphs for the displacement function in the critically damped case look similar to those in the overdamped case.

4.8.2. Exercises.

1. (a) The graph of the displacement function $y(t)$ for a particular (vertical) mass-spring system is shown below. At time $t = 2$ sec is the mass located above or below the equilibrium position? Is it moving up or down at this moment?

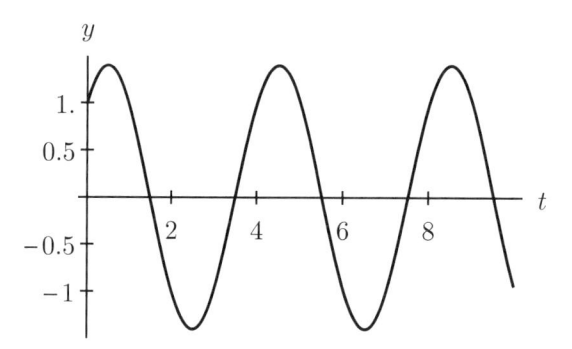

Figure 4.44. Displacement function.

   (b) If a 5-kg mass stretches a spring $\frac{1}{2}$ meter, how much does a 2-kg mass stretch the same spring?

2. When a 10-kg mass is attached to a vertically hung spring, the spring stretches $\frac{1}{2}$ meter. The mass is put into motion by pulling it down an additional $\frac{1}{4}$ meter and releasing it.
   (a) If we ignore any damping, what is the resulting displacement function? What is the period of the motion?
   (b) If the damping constant is $c = 12\sqrt{5}$ kg/sec, show that the system is underdamped, and find the quasiperiod.

3. A 2-lb weight stretches a spring 1.5 inches. The weight is pulled down 6 inches from equilibrium and released. Assume a damping force with damping constant $c = 2$ lb-sec/ft. Determine the position of the mass at any time $t > 0$. Does the mass ever pass through the equilibrium position for $t > 0$? If so, for what value(s) of t?

4. Show that for an overdamped mass-spring system, the mass can pass through the equilibrium position for at most one time $t > 0$.

5. The motion of a damped mass-spring system is modeled by the initial value problem
$$y'' + 4y' + 3y = 0, \qquad y(0) = 1, \quad y'(0) = -6.$$
   (a) Show that the system is overdamped.
   (b) Find any times $t > 0$ that the mass passes through the equilibrium position.

6. What value of c gives critical damping for the mass-spring system modeled by $y'' + cy' + 2y = 0$?

7. A 1-kg mass is attached to a spring with spring constant $k = 2$ N/m. The system is immersed in a viscous liquid that produces a damping force with damping constant $c = 3$ kg/sec. The mass is released with initial position $y(0) = -1$ and positive initial velocity v_0.
   (a) Suppose $v_0 = \frac{1}{2}$. Will the mass ever pass through the equilibrium position at a time $t > 0$?
   (b) Repeat (a) with $v_0 = 1, \frac{3}{2}, 3$.
   (c) Find the smallest number b with the following property: The mass will cross the equilibrium position at some positive time t, provided $v_0 > b$, but will never cross if $v_0 < b$. What happens if $v_0 = b$? Be careful about signs! A good picture of the graph of e^t may be helpful for reference.

4.8. Modeling with second-order equations

8. Suppose $c^2 - 4mk = 0$, so that the system $my'' + cy' + ky = 0$ is critically damped. Impose the initial conditions $y(0) = 0, y'(0) = v_0$ and solve the initial value problem, giving your answer in terms of m, c, and v_0. Assume $v_0 > 0$. At what time does the mass reach its lowest point?

9. A 1-kg mass is attached to a spring. It stretches the spring by 2 meters. What damping constant c gives critical damping?

10. In an underdamped system the mass makes 30 oscillations per minute, and the amplitude is halved after 2 minutes. Find a differential equation for the displacement function.

11. An 8-lb weight attached to a spring causes simple harmonic motion (neglecting damping) with period $\pi/2$. What is the spring constant? Give units on your answer.

12. A 1-lb weight stretches a spring 24 inches. If the weight is put into motion by pushing it up 2 in and releasing and we neglect damping, find the displacement function $y(t)$, where y is measured in inches. (Use $g = 384$ in/sec^2.) What is the displacement function if y is measured in feet?

13. Show that with critical damping, if the initial displacement and the initial velocity are both positive, then the mass never passes through equilibrium for $t > 0$.

14. Each of the following graphs shows the displacement function (in ft) for a vertical mass-spring system. Match the graph with the description of the initial condition.
    (a) The mass is pulled down 2 ft and released.
    (b) The mass is pulled down 2 ft and given an initial velocity of 1 ft/sec in the downward direction.
    (c) The mass is pulled down 2 ft and given an initial velocity of 1 ft/sec in the upward direction.

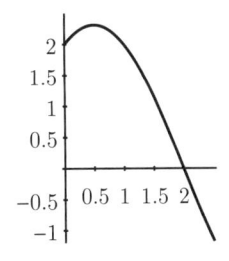

Figure 4.45. (A) 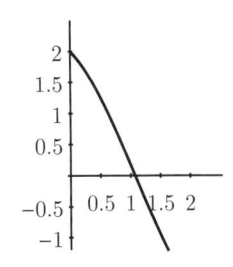 **Figure 4.46.** (B) 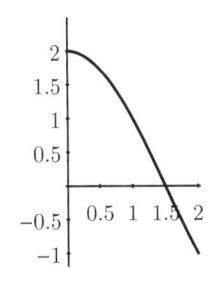 **Figure 4.47.** (C)

15. (a) Fig. 4.48 shows a typical graph of the displacement y as a function of t in the underdamped case. Reproduce this sketch, and then on the same graph show a possible graph of the velocity $v = y'$ as a function of t.

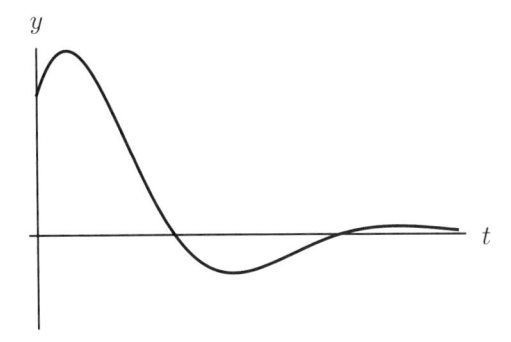

Figure 4.48. The displacement function $y(t)$.

(b) Next suppose that the functions $y = y(t)$, $v = v(t)$, $t \geq 0$, from (a) are parametric equations of a curve in the yv-plane. Which of the following could be a sketch of this curve? The arrows show the direction for increasing values of t, and the point P corresponding to $t = 0$ is indicated.

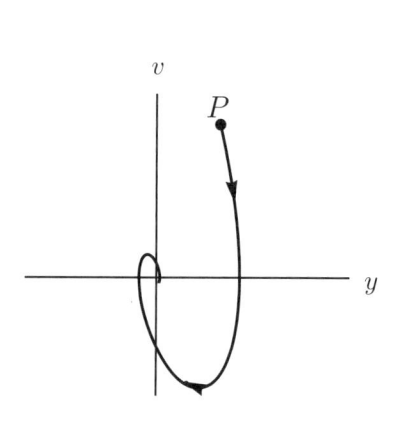

Figure 4.49. (A) $P = (y(0), v(0))$.

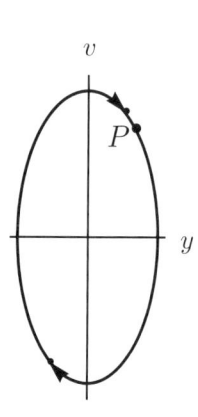

Figure 4.50. (B) $P = (y(0), v(0))$.

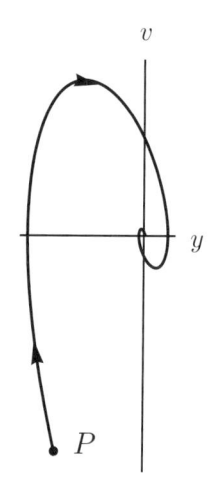

Figure 4.51. (C) $P = (y(0), v(0))$.

16. The point of this problem is to compare the undamped and underdamped solutions for a mass-spring system with the same initial condition.
    (a) Show that the initial value problem $my'' + ky = 0$, $y(0) = \alpha$, $y'(0) = \beta$, has solution
    $$\alpha \cos(\mu t) + \frac{\beta}{\mu} \sin(\mu t)$$
    where $\mu = \sqrt{\frac{k}{m}}$, so that the roots of the associated characteristic equation are $\pm i\mu$.
    (b) Show that if $c^2 - 4mk < 0$, the initial value problem
    $$my'' + cy' + ky = 0, \qquad y(0) = \alpha, \quad y'(0) = \beta$$
    (notice the initial condition is the same as in (a)) has solution
    $$\alpha e^{\lambda t} \cos(\nu t) + \frac{\beta - \lambda \alpha}{\nu} e^{\lambda t} \sin(\nu t)$$
    where the roots of the associated characteristic equation are $\lambda \pm i\nu$.

17. Suppose a vertical mass-spring system with no damping is put into motion by some initial displacement and/or some initial velocity and its maximum displacement from the equilibrium position is calculated. Will the maximum displacement be larger or smaller if the mass is replaced by a larger mass? Give an answer based on your intuition and then check it by examining the solution written in the form $R\cos(\mu t - \delta)$. (Hint: Solve $my'' + ky = 0$, $y(0) = \alpha, y'(0) = \beta$. If you write your answer as $R\cos(\mu t - \delta)$, what is an expression for R in terms of α, β, and m? If m gets larger (and α and β are unchanged), does R get larger or smaller?)

18. For an underdamped mass-spring system, does the quasiperiod increase or decrease as the damping constant c is decreased? What is the limit of the quasiperiod as $c \to 0$?

19. An initial value problem for a mass-spring system has solution
    $$y(t) = 1.3e^{-2t} - 0.5e^{-3t}.$$
    Can you tell by inspection whether the system is under, over, or critically damped? Explain.

4.8. Modeling with second-order equations

20. If the roots of the characteristic equation for a differential equation (with m and k positive and $c \geq 0$)

$$my'' + cy' + ky = 0$$

are r_1 and r_2, can you tell by looking at these roots whether the corresponding mass-spring system is damped or undamped? Can you distinguish under, over, and critical damping in the damped case? Explain.

21. Suppose a mass of m kg stretches a vertical spring d meters. If we put the mass into motion and ignore damping, give a formula for the frequency of the resulting simple harmonic motion in terms of d. What is the period of the motion in terms of d?

22. A greatly simplified model of a car's suspension system is obtained by assuming that it acts as a single mass-spring system and its shock absorbers act like a dashpot damping device on this system. Assume the car weighs 3,000 lbs and that without the shock absorbers it would exhibit undamped vibrations of 20 cycles/minute. Determine the spring constant k in lb/ft.

23. A perfectly flexible cable hangs over a frictionless peg as shown in Fig. 4.52, with 8 feet of cable on one side of the peg and 12 feet on the other. The goal of this problem is to determine how long it takes the cable to slide off the peg, starting from rest.
    (a) At time $t = 0$ what proportion of the whole cable is on the left side of the peg and what proportion is on the right side?
    (b) Let $y = y(t)$ be the amount of cable (in feet) that has slid from the left side to the right side after t seconds. Give expressions (in terms of y) for the proportion of the cable that is on the left side and the proportion that is on the right side at time $t > 0$
    (c) Using Newton's law $F = ma$, where m is mass and a is acceleration, find a differential equation for the function $y(t)$ (a "frictionless peg" means we are ignoring any drag force; the cable slides because of the excess weight on one side). Your equation should have the form $y'' = \alpha(y + \beta)$ for certain constants α and β. Find its general solution by means of the substitution $u = (y + \beta)$.
    (d) How long does it take for the cable, starting from rest, to slide completely off the peg?

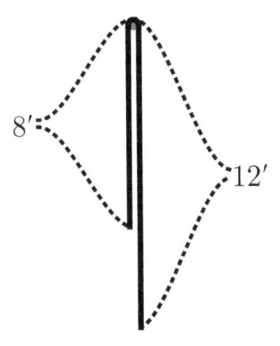

Figure 4.52. Cable sliding off peg.

24. A certain bare metal rod is in a room whose temperature is 25 degrees Celsius. The rod is 100 cm long; its left endpoint is held at 0 degrees while its right endpoint is held at 100 degrees. Assume the room's temperature as well as the endpoint temperatures have been held constant for many hours, so that the rod's temperature is not changing with respect to time. Let $T(x)$ be the temperature of the rod x cm from its left endpoint. We say T is a steady-state temperature function.

The heat equation, a partial differential equation discussed in Chapter 11, together with Newton's law of cooling provides the following model for T:

(4.120) $$\frac{d^2T}{dx^2} = k(T-25), \qquad T(0) = 0, \quad T(100) = 100,$$

where k is a positive constant. This is called a *boundary value problem*, and the requirement that $T(0) = 0$, $T(100) = 100$ is a *boundary condition*.

(a) Taking $k = 0.01$, solve the boundary value problem (4.120). (Hint: Note that the modeling equation has the same form as that arising in part (c) of Exercise 23.) Computations here can be simplified using the idea of Exercise 15 of Section 4.2.

(b) (CAS) Plot your solution to part (a) over the interval $0 \le x \le 100$. Check the plausibility of your plot: In particular, is the graph's concavity consistent with the sign of T'', which is specified by the differential equation of (4.120)?

Chapter 5

Higher-Order Linear Nonhomogeneous Equations

5.1. Introduction to nonhomogeneous equations

In this chapter we focus on linear nonhomogeneous differential equations of order two or more. To introduce a few basic principles, consider the nonhomogeneous equation

(5.1) $$y'' + p(t)y' + q(t)y = g(t)$$

and the associated homogeneous equation

(5.2) $$y'' + p(t)y' + q(t)y = 0,$$

which we will call the **complementary equation**. The coefficient functions p and q and the nonhomogeneous term g are assumed to be continuous on some (open) interval I on which we work. We make two simple, but crucial, observations about the relationships between solutions to (5.1) and to (5.2):

(a) If f_1 and f_2 solve (5.1), then $f_1 - f_2$ solves (5.2).

(b) If $\{y_1, y_2\}$ is a fundamental set of solutions to (5.2) and y_p is a particular, or specific, solution to (5.1), then for any constants c_1 and c_2,

$$c_1 y_1 + c_2 y_2 + y_p$$

solves (5.1).

These assertions are easily verified using the linearity of the differential operator L defined by

$$L[y] = y'' + p(t)y' + q(t)y.$$

If we know that $L[f_1] = g$ and $L[f_2] = g$ (this is the statement that f_1 and f_2 solve (5.1)), then

$$L[f_1 - f_2] = L[f_1] - L[f_2] = g - g = 0,$$

which says $f_1 - f_2$ is a solution to (5.2). Similarly, linearity tells us that

$$L[c_1 y_1 + c_2 y_2 + y_p] = c_1 L[y_1] + c_2 L[y_2] + L[y_p] = 0 + 0 + g = g,$$

given our assumptions in (b) about y_p, y_1, and y_2.

The general solution. The two-parameter solution identified in (b) is the general solution; that is, every solution to (5.1) may be written in the form

(5.3) $$c_1 y_1 + c_2 y_2 + y_p,$$

where y_p is any particular solution to (5.1), $\{y_1, y_2\}$ is a fundamental set of solutions to the associated homogeneous equation, and c_1 and c_2 are arbitrary constants. To see why this is so, suppose that y is any solution to the nonhomogeneous equation (5.1). By our observation in (a), $y - y_p$ solves the homogeneous equation (5.2), and so it can be written as

$$y - y_p = c_1 y_1 + c_2 y_2$$

for some choice of constants c_1 and c_2. This says that y is in the family (5.3).

The same principles hold for higher-order linear equations. The next result summarizes the situation.

Theorem 5.1.1. *Suppose that y_p is any particular solution to*

(5.4) $$y^{(n)} + p_{n-1}(t) y^{(n-1)} + \cdots + p_1(t) y' + p_0(t) y = g(t)$$

on an open interval I on which $p_0, p_1, \ldots, p_{n-1}$ and g are continuous. If $\{y_1, \ldots, y_n\}$ is a fundamental set of solutions to the complementary equation

$$y^{(n)} + p_{n-1}(t) y^{(n-1)} + \cdots + p_1(t) y' + p_0(t) y = 0,$$

then every solution to (5.4) on I has the form

(5.5) $$c_1 y_1 + \cdots + c_n y_n + y_p$$

for some choice of constants $c_1, c_2, \ldots, c_n$.

The general solution given in (5.5) is sometimes written as $y(t) = y_c(t) + y_p(t)$, where y_c, called the **complementary solution**, is given by

$$y_c = c_1 y_1 + c_2 y_2 + \cdots + c_n y_n.$$

What we haven't discussed yet is how to find a particular solution y_p to any of these nonhomogeneous second- or higher-order equations. We turn to this next.

5.1.1. Variation of parameters. In this section we discuss a method for solving a second-order linear nonhomogeneous equation, *written in standard form* as

(5.6) $$y'' + p(t) y' + q(t) y = g(t),$$

that applies when you *can* solve the associated homogeneous equation

(5.7) $$y'' + p(t) y' + q(t) y = 0.$$

By our previous discussion, the work is to find a particular solution $y_p(t)$ to (5.6). If y_1 and y_2 are two linearly independent solutions to (5.7), we'll show how to find a particular solution in the form

(5.8) $$y_p(t) = u_1(t) y_1(t) + u_2(t) y_2(t)$$

where u_1 and u_2 are two unknown but to be determined functions. Notice that if u_1 and u_2 are replaced by arbitrary constants, we have the general solution $c_1 y_1 + c_2 y_2$ to the homogeneous equation (5.7). The terminology "**variation of parameters**" refers to the process of replacing these *constant* parameters c_1 and c_2 by *functions* u_1 and u_2. At the moment the functions u_1 and u_2 are "unknowns"; we'll be happy to determine any choice for these two functions so that the resulting expression in (5.8) solves equation (5.6). Differentiating $y_p = u_1 y_1 + u_2 y_2$ using the product rule yields

(5.9) $$y_p' = u_1 y_1' + u_1' y_1 + u_2 y_2' + u_2' y_2.$$

5.1. Introduction to nonhomogeneous equations

Imposing an auxiliary condition. We'll also need an expression for y_p'', but the reader looking at equation (5.9) may feel discouraged; differentiating the expression for y_p' there (again using the product rule) will yield a formula for y_p'' with *eight* terms, which is a bit daunting. Instead, let's first simplify the right-hand side of (5.9). Because we have two unknowns and just one equation they must satisfy (namely (5.8)), we might expect to be able to impose an additional condition on the unknowns. We require that u_1 and u_2 satisfy the following:

(5.10) $$u_1' y_1 + u_2' y_2 = 0 \text{ (Auxiliary Condition)}.$$

With this auxiliary condition, y_p' simplifies to

(5.11) $$y_p' = u_1 y_1' + u_2 y_2'.$$

Differentiating this gives

(5.12) $$y_p'' = u_1 y_1'' + u_1' y_1' + u_2 y_2'' + u_2' y_2',$$

which has only four terms. So far, so good. However, the above simplification carries a danger: By imposing the auxiliary condition (5.10), could we be ruling out the very functions u_1 and u_2 we want to find? The only way to know is to try it.[1]

We want $y_p = u_1 y_1 + u_2 y_2$ to satisfy the nonhomogeneous equation (5.6). If we insert our expressions for y_p, y_p', and y_p'' into (5.6), we have

(5.13) $$u_1 y_1'' + u_1' y_1' + u_2 y_2'' + u_2' y_2' + p(t)(u_1 y_1' + u_2 y_2') + q(t)(u_1 y_1 + u_2 y_2) = g(t).$$

Let's simplify this by making use of what we know about y_1 and y_2, namely that

$$y_1'' + p(t) y_1' + q(t) y_1 = 0$$

and

$$y_2'' + p(t) y_2' + q(t) y_2 = 0$$

since y_1 and y_2 solve (5.7). To do this, first collect together the three terms in (5.13) containing u_1 as a factor and then the three terms that contain u_2 as a factor. When rewritten this way, (5.13) becomes

$$u_1(y_1'' + p(t) y_1' + q(t) y_1) + u_2(y_2'' + p(t) y_2' + q(t) y_2) + u_1' y_1' + u_2' y_2' = g(t)$$

where, as we have just noted, the two expressions in parentheses are zero, so that we are left with the simple equation

(5.14) $$u_1' y_1' + u_2' y_2' = g(t).$$

We pair this with our auxiliary condition from (5.10)

(5.15) $$u_1' y_1 + u_2' y_2 = 0$$

to get a system of two equations for the unknown functions u_1 and u_2. To finish, we need only solve this pair of equations (5.14) and (5.15) simultaneously for u_1 and u_2.

Solving (5.14) and (5.15) for u_1 and u_2. Assume we are working on an open interval I on which p, q, and g of (5.6) are continuous. Multiply equation (5.15) by y_1' and equation (5.14) by y_1, and subtract the resulting two equations. The result is

$$u_2'(y_1 y_2' - y_2 y_1') = g y_1,$$

which we also recognize as

(5.16) $$u_2' = \frac{g y_1}{W(y_1, y_2)}$$

---

[1] It has been said that mathematics is both a theoretical and experimental science. Pencil and paper (and sometimes a computer) provide the laboratory, and the experiments consist of a great deal of trial and error using these tools.

where the denominator $W(y_1, y_2) = y_1 y_2' - y_2 y_1'$ is the Wronskian of y_1 and y_2. The denominator is guaranteed to be nonzero since we have assumed that y_1 and y_2 constitute a fundamental set of solutions of (5.7). The right-hand side in (5.16) is a function of the independent variable t, and we can solve this first-order equation by integrating:

$$(5.17) \qquad u_2(t) = \int \frac{g(t) y_1(t)}{W(y_1, y_2)(t)} dt.$$

A specific choice for u_2 is obtained by choosing any specific antiderivative on the right-hand side. Alternately, we can fix any convenient point t_0 in the interval on which we are working and choose

$$u_2(t) = \int_{t_0}^{t} \frac{g(s) y_1(s)}{W(y_1, y_2)(s)} ds.$$

This gives a choice for u_2 that also satisfies $u_2(t_0) = 0$. Similarly, multiplying equation (5.15) by y_2' and equation (5.14) by y_2 and then subtracting the resulting equations gives

$$(5.18) \qquad u_1' = \frac{-g y_2}{W(y_1, y_2)}$$

so that

$$(5.19) \qquad u_1(t) = \int \frac{-g(t) y_2(t)}{W(y_1, y_2)(t)} dt,$$

or (if you want a choice of u_1 satisfying $u_1(t_0) = 0$)

$$u_1(t) = \int_{t_0}^{t} \frac{-g(s) y_2(s)}{W(y_1, y_2)(s)} ds.$$

Here's our particular solution to our nonhomogeneous equation (5.6):

$$y_p = u_1 y_1 + u_2 y_2 = y_1(t) \int \frac{-g(t) y_2(t)}{W(y_1, y_2)(t)} dt + y_2(t) \int \frac{g(t) y_1(t)}{W(y_1, y_2)(t)} dt.$$

From our earlier discussion the general solution to (5.6) is

$$c_1 y_1 + c_2 y_2 + y_p$$

for arbitrary constants c_1 and c_2. We emphasize that we are not making any particular assumptions about the form of the nonhomogeneous term g in (5.6), save that it is continuous on the interval in which we want to work. However, whether or not we can carry out the indicated integrations in (5.17) and (5.19) will depend on the particular functions y_1, y_2, and g at hand. The other limitation of this method is that we must be able to solve the associated homogeneous equation (5.7) to even get started. Of course, we know from the last chapter that if p and q are constants, this will always be possible.

Caution: To use formulas (5.17) and (5.19), it is crucial to put your differential equation in standard form to correctly identify $g(t)$. In particular, the coefficient of y'' must be 1. The next examples illustrate this.

5.1. Introduction to nonhomogeneous equations

Example 5.1.2. Find the general solution to

(5.20) $$2y'' - 4y' + 2y = \frac{e^t}{t}$$

for $t > 0$. The complementary equation

(5.21) $$2y'' - 4y' + 2y = 0$$

has characteristic polynomial $2z^2 - 4z + 2 = 0$ with root $z = 1$ of multiplicity 2. We may take $y_1(t) = e^t$ and $y_2(t) = te^t$ as two linearly independent solutions to (5.21). Their Wronskian is

$$W(e^t, te^t) = \det \begin{pmatrix} e^t & te^t \\ e^t & te^t + e^t \end{pmatrix} = e^{2t}.$$

To identify $g(t)$ we must put (5.20) in standard form:

$$y'' - 2y' + y = \frac{e^t}{2t},$$

so that $g(t) = e^t/(2t)$. From equations (5.16) and (5.18) we see that

$$u_1'(t) = \frac{-\left(\frac{e^t}{2t}\right)te^t}{e^{2t}} = -\frac{1}{2} \quad \text{and} \quad u_2'(t) = \frac{\left(\frac{e^t}{2t}\right)e^t}{e^{2t}} = \frac{1}{2t},$$

so that we may choose

(5.22) $$u_1(t) = -\frac{1}{2}t$$

and

(5.23) $$u_2(t) = \frac{1}{2}\ln t.$$

This gives a particular solution to (5.20):

(5.24) $$y_p(t) = -\frac{1}{2}te^t + \frac{1}{2}(\ln t)te^t.$$

Because we have already observed that te^t is a solution to the complementary equation (5.21), we can throw away the first term in (5.24) if we wish. The general solution to (5.20) is

$$y(t) = c_1 e^t + c_2 te^t + \frac{1}{2}(\ln t)te^t.$$

Example 5.1.3. We'll solve the initial value problem

$$t^2 y'' - 2y = t^3, \quad y(1) = 0, \quad y'(1) = 2.$$

The complementary equation

(5.25) $$t^2 y'' - 2y = 0$$

is a Cauchy-Euler equation (see Section 4.5.3). Using the trial solution $y = t^m$, we find that equation (5.25) has solutions $y_1(t) = t^2$ and $y_2(t) = \frac{1}{t}$. Using these, we'll find the general solution to $t^2 y'' - 2y = t^3$. With that in hand, we can deal with the initial conditions.

In standard form our nonhomogeneous equation is

$$y'' - \frac{2}{t^2}y = t$$

so we identify $g(t) = t$. The Wronskian $W(y_1, y_2)$ is easily computed to be -3. From equations (5.17) and (5.19) we have

$$u_1(t) = \int -\frac{gy_2}{W(y_1, y_2)} dt = \int -\frac{t(1/t)}{(-3)} dt = \frac{1}{3}t$$

and

$$u_2(t) = \int \frac{gy_1}{W(y_1, y_2)} dt = \int \frac{t(t^2)}{(-3)} dt = -\frac{t^4}{12}.$$

A particular solution is thus

$$y_p(t) = u_1 y_1 + u_2 y_2 = \frac{1}{3}t(t^2) - \frac{1}{12}t^4 \left(\frac{1}{t}\right) = \frac{1}{4}t^3.$$

The general solution to $t^2 y'' - 2y = t^2$ is

$$y(t) = c_1 t^2 + \frac{c_2}{t} + \frac{1}{4}t^3.$$

Meeting the initial conditions. To satisfy our initial conditions we must have

$$y(1) = c_1 + c_2 + \frac{1}{4} = 0 \quad \text{and} \quad y'(1) = 2c_1 - c_2 + \frac{3}{4} = 2.$$

Solving these equations for c_1 and c_2 gives

$$c_1 = \frac{1}{3}, \quad c_2 = -\frac{7}{12},$$

and the initial value problem has solution

$$y(t) = \frac{1}{3}t^2 - \frac{7}{12t} + \frac{1}{4}t^3, \quad 0 < t < \infty.$$

The method of variation of parameters can be used for higher-order linear nonhomogeneous equations. Exercise 18 outlines how this goes for a third-order equation.

Discovering variation of parameters by operator-factoring. The variation of parameters procedure may seem rather mysterious. Here we will show how an operator-factoring approach might lead to its discovery in the case of any equation

$$y'' + by' + cy = g(t),$$

where $g(t)$ is an arbitrary continuous function and the characteristic polynomial $z^2 + bz + c$ has two different real roots r_1 and r_2, so that the differential equation may be written as

(5.26) $$(D - r_1)(D - r_2)[y] = g(t).$$

Following the procedure in Section 4.2.1, we set

(5.27) $$h(t) = (D - r_2)[y],$$

so that by equation (5.26) we have $(D - r_1)[h] = g(t)$ or $h' - r_1 h = g(t)$. Solving this last equation for h using the integrating factor $e^{-r_1 t}$ gives

$$\frac{d}{dt}(e^{-r_1 t} h) = e^{-r_1 t} g(t)$$

and

$$e^{-r_1 t} h(t) = \int e^{-r_1 t} g(t) \, dt.$$

Let's temporarily denote any specific antiderivative of $e^{-r_1 t} g(t)$ by $H(t)$ so that we have

$$h(t) = e^{r_1 t}(H(t) + c),$$

5.1. Introduction to nonhomogeneous equations

or
$$h(t) = ce^{r_1 t} + e^{r_1 t} H(t).$$

By the definition of $h(t)$ in (5.27) this says
$$y' - r_2 y = ce^{r_1 t} + e^{r_1 t} H(t),$$

another linear equation which we solve with the integrating factor $e^{-r_2 t}$:
$$\frac{d}{dt}(e^{-r_2 t} y) = ce^{(r_1 - r_2)t} + e^{(r_1 - r_2)t} H(t)$$

so that
$$e^{-r_2 t} y = \frac{c}{r_1 - r_2} e^{(r_1 - r_2)t} + \int e^{(r_1 - r_2)t} H(t)\, dt + c_2.$$

This gives
$$y = c_1 e^{r_1 t} + c_2 e^{r_2 t} + e^{r_2 t} \int e^{(r_1 - r_2)t} H(t)\, dt$$

where $c_1 = c/(r_1 - r_2)$ and c_2 are arbitrary constants. We recognize the first two terms $c_1 e^{r_1 t} + c_2 e^{r_2 t}$ as the general solution of the complementary equation $y'' + by' + cy = 0$. Choosing a specific antiderivative for $e^{(r_1 - r_2)t} H(t)$, we see that the third term, $e^{r_2 t} \int e^{(r_1 - r_2)t} H(t)\, dt$, is a particular solution of the original nonhomogeneous equation. We want to write this particular solution in a more revealing form.

To do this, we use integration by parts with
$$u = H(t) \quad \text{and} \quad dv = e^{(r_1 - r_2)t}\, dt$$

so that
$$du = H'(t)\, dt = e^{-r_1 t} g(t)\, dt \quad \text{and} \quad v = \frac{e^{(r_1 - r_2)t}}{r_1 - r_2}.$$

The computation of du uses our definition of H as an antiderivative of $e^{-r_1 t} g(t)$. We have

$$\begin{aligned}
e^{r_2 t} \int e^{(r_1 - r_2)t} H(t)\, dt &= e^{r_2 t} \left(\frac{e^{(r_1 - r_2)t}}{r_1 - r_2} H(t) - \int \frac{e^{(r_1 - r_2)t}}{r_1 - r_2} e^{-r_1 t} g(t)\, dt \right) \\
&= \frac{e^{r_1 t}}{r_1 - r_2} H(t) - e^{r_2 t} \int \frac{e^{-r_2 t} g(t)}{r_1 - r_2}\, dt \\
&= e^{r_1 t} \int \frac{e^{-r_1 t} g(t)}{r_1 - r_2}\, dt - e^{r_2 t} \int \frac{e^{-r_2 t} g(t)}{r_1 - r_2}\, dt \\
&= e^{r_1 t} u_1 + e^{r_2 t} u_2 \\
&= y_1 u_1 + y_2 u_2,
\end{aligned}$$

where $y_1 = e^{r_1 t}$ and $y_2 = e^{r_2 t}$ form a fundamental set of solutions to the complementary equation $y'' + by' + cy = 0$ and
$$u_1 = \int \frac{e^{-r_1 t} g(t)}{r_1 - r_2}\, dt, \quad u_2 = \int \frac{-e^{-r_2 t} g(t)}{r_1 - r_2}\, dt.$$

Since the Wronskian $W[e^{r_1 t}, e^{r_2 t}] = (r_2 - r_1) e^{r_1 t} e^{r_2 t}$, some algebra lets us write
$$u_1 = \int \frac{e^{-r_1 t} g(t)}{r_1 - r_2}\, dt = \int \frac{-y_2(t) g(t)}{W[y_1, y_2](t)}\, dt,$$

and
$$u_2 = \int \frac{-e^{-r_2 t} g(t)}{r_1 - r_2}\, dt = \int \frac{y_1(t) g(t)}{W[y_1, y_2](t)}\, dt.$$

Thus we have obtained the variation of parameter solution for this class of nonhomogeneous equations.

5.1.2. Exercises.

In Exercises 1–7, use variation of parameters to find the general solution.

1. $y'' - y = 2e^t$.
2. $y'' + y = \sec t$, $0 < t < \pi/2$.
3. $y'' + y = \cos^2 t$.
4. $y'' = \ln t$, $t > 0$.
5. $t^2 y'' - 2ty' + 2y = t^4$, $t > 0$. Hint: The complementary equation is a Cauchy-Euler equation (see Section 4.5.3) and has trial solution $y = t^m$.
6. $t^2 y'' + (1 - 2\alpha)ty' + (\alpha^2)y = t^4$, $t > 0$, and α an arbitrary constant. Handle the case $\alpha = 4$ separately. Hint: The complementary equation is a Cauchy-Euler equation (see Section 4.5.3) and has trial solution $y = t^m$.
7. $y'' + 2y' + y = e^{-t}\sqrt{1+t}$, $t > -1$.
8. Sam solves the equation $y'' + y = \sin t$ and gets the general solution

$$c_1 \cos t + c_2 \sin t + \frac{1}{2}\sin t - \frac{1}{2}t \cos t.$$

   Sally solves the same equation and obtains

$$c_1 \cos t + c_2 \sin t - \frac{1}{2}t \cos t.$$

   Who is right? Hint: You may assume at least one of them is correct. Then, you can answer the question without doing any computations.

9. Solve the initial value problem

$$t^2 y'' + ty' - y = 4t \ln t, \qquad y(1) = 3, \quad y'(1) = 1.$$

   You'll need to start by solving the complementary equation

$$t^2 y'' + ty' - y = 0,$$

   which is a Cauchy-Euler equation (see Section 4.5.3); take $y = t^m$ as a trial solution to this homogeneous equation.

10. Suppose in Example 5.1.2 we make a different choice of antiderivative so that

$$u_1 = 5 - \frac{1}{2}t \quad \text{and} \quad u_2(t) = \frac{1}{2}\ln t + 2$$

   are our solutions for $u_1(t)$ and $u_2(t)$. How does this affect the general solution to (5.20)?

11. Suppose $L[y] = y'' + p(t)y' + q(t)y$. Show that if u solves the initial value problem

$$L[u] = 0, \qquad u(t_0) = \alpha, \quad u'(t_0) = \beta,$$

   and v solves the initial value problem

$$L[v] = g, \qquad v(t_0) = 0, \quad v'(t_0) = 0,$$

   then $y = u + v$ solves

$$L[y] = g, \qquad y(t_0) = \alpha, \quad y'(t_0) = \beta.$$

12. **A Generalized Superposition Principle.** Suppose that for each value of j, $j = 1, 2, \ldots, k$, we know that the function y_j solves the equation
$$y^{(n)} + p_{n-1}(t)y^{(n-1)} + \cdots + p_1(t)y' + p_0(t)y = g_j(t)$$
on some interval I. Show that the linear combination $c_1y_1 + c_2y_2 + \cdots + c_ky_k$ solves the equation
$$y^{(n)} + p_{n-1}(t)y^{(n-1)} + \cdots + p_1(t)y' + p_0(t)y = c_1g_1(t) + c_2g_2(t) + \cdots + c_kg_k(t)$$
on I.

13. Consider the initial value problem $y'' + p(t)y' + q(t)y = g(t)$, $y(t_0) = 0$, $y'(t_0) = 0$, where p, q, and g are continuous on an open interval I containing t_0. Suppose we know that y_1 and y_2 are linearly independent solutions on I to the complementary equation. Show that if we determine u_1 and u_2 by the definite integrals
$$u_1(t) = \int_{t_0}^{t} \frac{-g(s)y_2(s)}{W(y_1, y_2)(s)} ds$$
and
$$u_2(t) = \int_{t_0}^{t} \frac{g(s)y_1(s)}{W(y_1, y_2)(s)} ds,$$
then $u_1 y_1 + u_2 y_2$ solves our initial value problem on I. (You already know it solves the differential equation; the problem is to show it also satisfies the initial conditions.)

14. If f_1 and f_2 are solutions to $ay'' + by' + cy = g(t)$ on $(-\infty, \infty)$, where a, b, and c are *positive* constants, show that
$$\lim_{t \to \infty} (f_1(t) - f_2(t)) = 0.$$

15. The functions $f_1(t) = 2t$, $f_2(t) = 3 - 5t$, and $f_3(t) = 2t + e^{2t}$ all solve some second-order linear nonhomogeneous equation (5.1) on an open interval I on which p, q, and g are continuous. From this, can you determine the general solution to this equation on I? If your answer is yes, then do so.

16. Suppose that the equation
$$y'' + p(t)y' + q(t)y = g(t)$$
has solutions $f_1(t) = t$, $f_2(t) = t^2$, and $f_3(t) = t^3$ on an interval I on which p, q, and g are continuous.
    (a) Find a fundamental set of solutions for the associated homogeneous equation
    $$y'' + p(t)y' + q(t)y = 0.$$
    (b) Find the solution to the original equation
    $$y'' + p(t)y' + q(t)y = g(t)$$
    that satisfies the initial conditions
    $$y(2) = 2, \quad y'(2) = 3.$$

17. Suppose that the equation
$$y'' + p(t)y' + q(t)y = g(t)$$
has solutions $f_1(t) = t^2$, $f_2(t) = t$, and $f_3(t) = 1$ on an interval I on which p, q, and g are continuous. What is the general solution on I?

18. The idea of variation of parameters can be adapted to third- (or higher-) order linear nonhomogeneous equations as well. In this problem we outline the method for a third-order equation, written in standard form as
(5.28) $$y''' + p_2(t)y'' + p_1(t)y' + p_0(t)y = g(t).$$

Suppose $\{y_1, y_2, y_3\}$ is a fundamental set of solutions to the associated homogeneous equation

(5.29) $$y''' + p_2(t)y'' + p_1(t)y' + p_0(t)y = 0.$$

Our goal is to find a particular solution to (5.28) in the form

(5.30) $$y_p = u_1 y_1 + u_2 y_2 + u_3 y_3,$$

where the u_j's are functions to be determined.

(a) Impose two additional conditions on the u_j's:

(5.31) $$u_1' y_1 + u_2' y_2 + u_3' y_3 = 0 \text{ (Auxiliary Condition 1)}$$

and

(5.32) $$u_1' y_1' + u_2' y_2' + u_3' y_3' = 0 \text{ (Auxiliary Condition 2)}.$$

For y_p as in (5.30), compute an expression for y_p', where you use (5.31) to simplify your expression. Then compute y_p'', using (5.32) to simplify your answer. Finally compute y_p''' and substitute the expressions for $y_p, y_p', y_p'',$ and y_p''' into (5.28). Using the fact that y_1, y_2, y_3 solve (5.29), show that the result simplifies to

(5.33) $$u_1' y_1'' + u_2' y_2'' + u_3' y_3'' = g(t).$$

(b) Solve the system of three equations (5.31), (5.32), and (5.33) for the unknowns u_1', u_2', u_3'. Use Cramer's Rule to do this, which says that a system of equations in the form

$$a_{11}x + a_{12}y + a_{13}z = b_1,$$
$$a_{21}x + a_{22}y + a_{23}z = b_2,$$
$$a_{31}x + a_{32}y + a_{33}z = b_3$$

with unknowns $x, y,$ and z has solutions

$$x = \frac{\det\begin{pmatrix} b_1 & a_{12} & a_{13} \\ b_2 & a_{22} & a_{23} \\ b_3 & a_{32} & s_{33} \end{pmatrix}}{W}, \quad y = \frac{\det\begin{pmatrix} a_{11} & b_1 & a_{13} \\ a_{21} & b_2 & a_{23} \\ a_{31} & b_3 & a_{33} \end{pmatrix}}{W},$$

and

$$z = \frac{\det\begin{pmatrix} a_{11} & a_{12} & b_1 \\ a_{21} & a_{22} & b_2 \\ a_{31} & a_{32} & b_3 \end{pmatrix}}{W},$$

where

$$W = \det\begin{pmatrix} a_{11} & a_{12} & a_{13} \\ a_{21} & a_{22} & a_{23} \\ a_{31} & a_{32} & a_{33} \end{pmatrix},$$

provided W is not zero. Finally, solve for $u_1, u_2,$ and u_3. Calculation of 3×3 determinants was described in Exercise 16 of Section 4.6.

(c) As an application of (c), solve

$$y''' - 2y'' = 4e^t.$$

19. Suppose L is a linear differential operator of order m given by
$$L[y] = y^{(m)} + p_{m-1}(t)y^{(m-1)} + \cdots + p_2(t)y'' + p_1(t)y' + p_0(t)y,$$
where $p_0(t), p_1(t), \ldots, p_{m-1}(t)$ are continuous functions on an open interval I. Let
$$g(t) = c_1 g_1(t) + c_2 g_2(t) + \cdots + c_n g_n(t)$$
where $g_1(t), \ldots, g_n(t)$ are continuous functions on I and $c_1, \ldots, c_n$ are constants, at least one of which is nonzero.
   (a) Show that if $y_j(t)$ satisfies $L[y_j] = g_j(t)$ for each $j = 1, 2, \ldots, n$, then $y = c_1 y_1 + \cdots + c_n y_n$ satisfies $L[y] = g(t)$.
   (b) Show that if $y(t)$ satisfies $L[y] = g(t)$, then there exist functions $y_1(t), y_2(t), \ldots, y_n(t)$ such that $L[y_j] = g_j(t)$ for each j and
   $$y = c_1 y_1 + c_2 y_2 + \cdots + c_n y_n.$$

20. In this problem you will use operator-factoring, as an alternative to variation of parameters, to solve the nonhomogeneous equation

   (5.34) $$y'' - 2y' + y = \frac{e^t}{1 + t^2}.$$

   (a) Check that $D^2 - 2D + 1 = (D-1)(D-1)$.
   (b) Suppose that y is any solution of equation (5.34). Set $h(t) = (D-1)[y] = y' - y$ and observe that since $(D-1)[h] = \frac{e^t}{1+t^2}$, we have
   $$h' - h = \frac{e^t}{1 + t^2}.$$
   Solve this first-order equation for h to obtain
   $$h(t) = e^t \arctan t + c e^t$$
   for constant c.
   (c) By (b) we know
   $$y' - y = h(t) = e^t \arctan t + c e^t.$$
   Solve this for y, using an integrating factor and the integration formula
   $$\int \arctan t \, dt = t \arctan t - \frac{1}{2} \ln(1 + t^2)$$
   (this formula can be obtained by an integration by parts).

5.2. Annihilating operators

In this section we will look at an efficient method for solving certain kinds of *constant coefficient* nonhomogeneous linear differenindex tial equations
$$a_n y^{(n)} + a_{n-1} y^{(n-1)} + \cdots + a_1 y' + a_0 y = g(t),$$
where the a_j's are real numbers. This method will enable you to exploit your ability to solve constant coefficient *homogeneous* equations to find the form of particular solutions of certain *nonhomogeneous* equations.

In Section 5.1.1 we discussed variation of parameters as a general method for solving linear nonhomogeneous equations. However, even for constant coefficient equations this often leads to integrations that may be difficult or impossible to carry out, and even when they can be, variation of parameters can be increasingly tedious to implement as the order of the equation increases (e.g., compare equations (5.17) and (5.19) in the text for the case $n = 2$ with the analogous formulas in Exercise 18 of the previous section for the case $n = 3$). By contrast, the method we will discuss in this section will be easier to implement, but its disadvantage is that it can only be used for certain kinds of functions $g(t)$—those that have "annihilating operators".

Our first task is to explain what we mean by an annihilating operator. Write D for the linear operator that takes a function y to its derivative y'; recall that in Section 4.2 we visualized this operator as

$$y \longrightarrow \boxed{D} \longrightarrow y'$$

Similarly we have worked with operators like D^2, $D^2 + 3D$, $D^3 + 5$, and $aD^2 + bD + c$, which act as shown below. All of these operators have the form $p(D)$ for some polynomial p; for example $aD^2 + bD + c$ is $p(D)$ for the choice $p(z) = az^2 + bz + c$.

$$y \longrightarrow \boxed{D^2 + 3D} \longrightarrow y'' + 3y'$$

$$y \longrightarrow \boxed{D^3 + 5} \longrightarrow y''' + 5y$$

$$y \longrightarrow \boxed{aD^2 + bD + c} \longrightarrow ay'' + by' + cy$$

Keep in mind that the "+5" in the third example and the "+c" in the fourth example denote the operators of *multiplication* by 5 and c, respectively. These are all examples of constant coefficient linear differential operators. Suppressing the "black box" picture, we could describe, for example, the operator $aD^2 + bD + c$ by the rule

$$(aD^2 + bD + c)[y] = ay'' + by' + cy.$$

Consider the operator $(D - 3)$ and suppose we input the function $y = e^{3t}$ into this operator. The resulting output is

$$(D - 3)[e^{3t}] = (e^{3t})' - 3e^{3t} = 3e^{3t} - 3e^{3t} = 0.$$

We describe this by saying that the operator $D - 3$ **annihilates** the function e^{3t}.

Example 5.2.1. The operator $D^2 + 4$ annihilates the function $\sin(2t)$ since

$$(D^2 + 4)[\sin(2t)] = (\sin(2t))'' + 4\sin(2t) = -4\sin(2t) + 4\sin(2t) = 0.$$

Example 5.2.2. The function te^t is annihilated by the operator $D^2 - 2D + 1$ since

$$(D^2 - 2D + 1)[te^t] = (te^t)'' - 2(te^t)' + te^t = te^t + 2e^t - 2te^t - 2e^t + te^t = 0.$$

This last computation can be done a bit more simply if we use the observation that the operator $D^2 - 2D + 1$ is the same as the operator $(D-1)(D-1) = (D-1)^2$. This is verified by the computation

$$\begin{aligned}(D-1)(D-1)[y] = (D-1)(y' - y) &= y'' - y' - (y' - y) = y'' - 2y' + y \\ &= (D^2 - 2D + 1)y.\end{aligned}$$

5.2. Annihilating operators

With this observation, we recalculate

$$(D-1)^2[te^t] = (D-1)(D-1)[te^t] = (D-1)[te^t + e^t - te^t]$$
$$= (D-1)[e^t] = e^t - e^t = 0,$$

again verifying, this time more readily, that $D^2 - 2D + 1$ annihilates te^t.

Factoring operators. In the last example, we factored the constant coefficient operator $D^2 - 2D + 1$ as $(D-1)^2$. In general, when we factor a constant coefficient operator, we can write the order of the factors any way we want; that is, the factors *commute*. So for example, the operator $D^2 - 2D - 3$, which produces the output $y'' - 2y' - 3y$ when the input is y, can be written in factored form as either $(D-3)(D+1)$ or $(D+1)(D-3)$; see the discussion at the end of Section 4.5.3.

Finding annihilating operators. In order to use annihilating operators to solve certain constant coefficient nonhomogeneous equations, we will need to be able to

- decide what functions $g(t)$ have annihilating operators and
- find an annihilating operator for $g(t)$ when one exists.

We should point out that when $g(t)$ has an annihilating operator it will have many annihilating operators; for example, D^2 annihilates t, but so does D^3, D^4, and so forth. Not only does $D - 3$ annihilate e^{3t}, so does $D^2 - 3D = D(D-3)$ since

$$D(D-3)[e^{3t}] = D(0) = 0.$$

Is $D^3 - 3D^2 + D - 3 = (D^2 + 1)(D - 3)$ also an annihilating operator for e^{3t}? You should be able to check that the answer is yes.

The next example points the way to how to decide what functions have annihilating operators and how then to find an annihilating operator.

Example 5.2.3. Can we find an annihilating operator for the function te^{4t}? Using what we learned in Section 4.5, we know that the function te^{4t} is a solution to the constant coefficient homogeneous equation that has characteristic equation

(5.35) $$(z-4)(z-4) = 0;$$

the root $z = 4$ of this equation has multiplicity 2, and this gives te^{4t} as a solution to the corresponding differential equation. Since this polynomial can be written as $(z-4)(z-4) = z^2 - 8z + 16$, the homogeneous differential equation that has (5.35) as its characteristic equation is

$$y'' - 8y' + 16y = 0.$$

We can write this equation as

$$(D^2 - 8D + 16)[y] = 0.$$

The statement that te^{4t} solves this differential equation is exactly the statement that the operator $D^2 - 8D + 16$ annihilates te^{4t}. Thus we have found an annihilating operator for te^{4t}, namely $D^2 - 8D + 16$, or as we will prefer to write it, $(D-4)^2$. We can summarize this succinctly: Since te^{4t} is a solution to the homogeneous equation with characteristic equation $(z-4)^2 = 0$, the operator $(D-4)^2$ annihilates te^{4t}.

Which functions have annihilating operators? Precisely those functions that can appear as solutions of some constant coefficient linear homogeneous differential equation. Finding an annihilating operator becomes the problem of identifying this differential equation. This is done by working backwards from the characteristic equation. We always seek an annihilating operator of lowest possible order, so correspondingly we choose our characteristic equation to have lowest possible degree.

Example 5.2.4. We'll find annihilating operators for $\cos(3t)$ and $t\cos(3t)$. We know that the function $\cos(3t)$ is a solution to the constant coefficient equation that has characteristic equation $(z-3i)(z+3i) = 0$, or equivalently $z^2+9 = 0$. The corresponding differential equation is $y''+9y = 0$, or $(D^2 + 9)y = 0$. Thus $D^2 + 9$ is an annihilating operator for $\cos(3t)$. For the second function, $t\cos(3t)$, we want the characteristic equation to have the same roots $\pm 3i$, but this time with each having multiplicity 2. So we choose the characteristic equation

$$(z - 3i)^2(z + 3i)^2 = 0,$$

or $(z^2 + 9)^2 = 0$. An annihilating operator for $t\cos(3t)$ is then $(D^2 + 9)^2$. As we discussed above, there are other more complicated annihilators for this function, for example $(D+2)(D^2+9)^2$, but it will be to our advantage to choose the simplest, or most efficient, annihilator. We do this by choosing the lowest degree characteristic equation that gives rise to an annihilating operator.

Table 5.1 shows annihilating operators for some basic functions. In the table, m is a positive integer, and λ and μ are real numbers. In the exercises you are asked to show how some of these entries are determined. Keep in mind that nothing in this table needs to be memorized! For any of the functions listed there, you can always work out an annihilating operator as we did in Examples 5.2.3 and 5.2.4, by identifying a homogeneous constant coefficient differential equation with the function among its solutions.

Table 5.1. Annihilating Operators

Function	Annihilating Operator
constant	D
t^m	D^{m+1}
e^{rt}	$D - r$
$t^m e^{rt}$	$(D - r)^{m+1}$
$\cos(\mu t)$ or $\sin(\mu t)$	$D^2 + \mu^2$
$t^m \cos(\mu t)$ or $t^m \sin(\mu t)$	$\left(D^2 + \mu^2\right)^{m+1}$
$e^{\lambda t}\cos(\mu t)$ or $e^{\lambda t}\sin(\mu t)$	$D^2 - 2\lambda D + (\lambda^2 + \mu^2)$
$t^m e^{\lambda t}\cos(\mu t)$ or $t^m e^{\lambda t}\sin(\mu t)$	$\left(D^2 - 2\lambda D + (\lambda^2 + \mu^2)\right)^{m+1}$

Solving differential equations by annihilating operators. Here's how to use annihilating operators to solve any constant coefficient equation

(5.36) $$a_n y^{(n)} + a_{n-1} y^{(n-1)} + \cdots + a_1 y' + a_0 y = g(t),$$

where the function $g(t)$ has an annihilating operator. From Theorem 5.1.1, we know that the general solution to equation (5.36) has the form

$$y(t) = y_c(t) + y_p(t),$$

5.2. Annihilating operators

where the complementary solution y_c is the general solution to the associated homogeneous equation $a_n y^{(n)} + a_{n-1} y^{(n-1)} + \cdots + a_1 y' + a_0 y = 0$ and y_p is any particular solution to equation (5.36). Our work in Chapter 4 tells us how to find y_c, so the only remaining issue is to find a particular solution y_p. It's for this task we use annihilating operators. The procedure is best illustrated with a few examples.

Example 5.2.5. Let's solve the nonhomogeneous equation
$$y'' + 4y = t,$$
or
$$(5.37) \qquad L[y] = t,$$
where $L = (D^2 + 4)$. We will refer to L as the **principal operator** for the equation at hand.

Step 1: Solve the complementary equation. The associated homogeneous equation, $y'' + 4y = 0$, has characteristic equation $z^2 + 4 = 0$ with solutions $z = 2i$ and $z = -2i$. This tells us that
$$y_c(t) = c_1 \cos(2t) + c_2 \sin(2t).$$

Step 2: Find an annihilating operator A. The function $g(t) = t$, which is the nonhomogeneous term in our original equation, has annihilating operator $A = D^2$.

Step 3: Apply A and solve the resulting homogeneous equation. If we apply the annihilating operator $A = D^2$ from Step 2 to both sides of equation (5.37), we obtain
$$D^2 L[y] = D^2[t] = 0,$$
or
$$(5.38) \qquad D^2(D^2 + 4)[y] = 0.$$
The resulting equation, which is $y^{(4)} + 4y'' = 0$, is a fourth-order *homogeneous*, constant coefficient equation, and we can easily write down all solutions to it. The characteristic equation for equation (5.38) is
$$z^2(z^2 + 4) = 0,$$
which has solutions $z = 0$ (with multiplicity 2), $z = 2i$, and $z = -2i$. So the general solution to (5.38) is
$$(5.39) \qquad c_1 \cos(2t) + c_2 \sin(2t) + c_3 + c_4 t.$$

The crucial observation is that any solution to equation (5.37) (our ultimate goal) must be in this family of solutions to (5.38). (The converse statement is not true; there are solutions to (5.38) that are not solutions to (5.37).) Thus, at this point we know that equation (5.37) has a particular solution having the form (5.39).

Step 4: The demand step. Which functions in the family (5.39) solve equation (5.37)? One way to answer this is to simply plug the expression in (5.39) into equation (5.37) and see what it tells us about c_1, c_2, c_3, and c_4. However, we can save ourselves *much* computational effort if we first make the following key observation: The first two terms of (5.39) duplicate the complementary solution y_c identified in Step 1. Thus if L is our principal operator $L[y] = y'' + 4y$, then
$$L[c_1 \cos(2t) + c_2 \sin(2t)] = 0$$

no matter what the values of c_1 and c_2 are. This says that if we plug $c_1 \cos(2t) + c_2 \sin(2t) + c_3 + c_4 t$ into the linear operator L, the result will be

$$\begin{aligned} L[c_1 \cos(2t) + c_2 \sin(2t) + c_3 + c_4 t] &= L[c_1 \cos(2t) + c_2 \sin(2t)] + L[c_3 + c_4 t] \\ &= 0 + L[c_3 + c_4 t] \\ &= L[c_3 + c_4 t], \end{aligned}$$

where the linearity of L justifies the first equality. Hence, at this point, we know that equation (5.37) has a particular solution of the form $y_p = c_3 + c_4 t$. To find the undetermined coefficients c_3 and c_4 in y_p we *demand* that

$$L[c_3 + c_4 t] = t$$

where $L = D^2 + 4$ and determine the values for c_3 and c_4 so that this holds. It will help to first use the linearity of the operator L to "pull the constants out" and rewrite this equation as

$$c_3 L[1] + c_4 L[t] = t.$$

We can easily compute

$$L[1] = (D^2 + 4)[1] = 4$$

and

$$L[t] = (D^2 + 4)[t] = 4t,$$

so our equation becomes

(5.40) $$4c_3 + (4c_4 - 1)t = 0,$$

so that

(5.41) $$4c_3 = 0, \quad 4c_4 - 1 = 0.$$

What justifies this last step, going from (5.40) to (5.41)? Since $\{1, t\}$ are linearly independent (on any interval), if the linear combination in (5.40) is 0, then the coefficients $4c_3$ and $4c_4 - 1$ must both equal 0. Thus $c_3 = 0$ and $c_4 = \frac{1}{4}$. Therefore

$$y_p(t) = \frac{1}{4} t$$

is a particular solution to $y'' + 4y = t$. Putting this together with our work in Step 1, we conclude that the general solution to equation (5.37) is

$$c_1 \cos(2t) + c_2 \sin(2t) + \frac{1}{4} t,$$

where c_1 and c_2 are arbitrary constants.

Imposing initial conditions. If our ultimate goal is to solve an initial value problem, then there is a fifth step to our algorithm. With the general solution in hand (which contains arbitrary constants), we determine values for the arbitrary constants so as to meet the initial conditions. The next example shows how this goes. In this example we will be a bit more succinct in our presentation, but you should be able to see the five steps at work: (1) Find the complementary solution, (2) determine the annihilating operator, (3) apply the annihilating operator, (4) find a particular solution in the demand step and write the general solution, and then finally (5) satisfy the initial conditions.

Example 5.2.6. We'll solve the initial value problem

(5.42) $$y'' - 3y' - 4y = e^{-t}, \quad y(0) = \frac{3}{2}, \quad y'(0) = \frac{4}{5}.$$

Write the differential equation as $L[y] = e^{-t}$, for $L = D^2 - 3D - 4$.

5.2. Annihilating operators

Step 1. The complementary equation $y'' - 3y' - 4y = 0$ has characteristic equation $z^2 - 3z - 4 = 0$, with solutions $z = 4$ and $z = -1$. Thus $y_c(t) = c_1 e^{4t} + c_2 e^{-t}$, with c_1 and c_2 arbitrary constants.

Step 2. The nonhomogeneous term e^{-t} has annihilating operator $A = D + 1$.

Step 3. Applying this annihilating operator to both sides of the equation $L[y] = e^{-t}$, we obtain
$$(D+1)L[y] = (D+1)[e^{-t}] = 0,$$
or
$$(5.43) \qquad (D+1)(D^2 - 3D - 4)[y] = 0.$$

We seek a solution to the homogeneous equation (5.43) that also solves the original differential equation in (5.42); this will serve as y_p in building the general solution to the original equation. Since the polynomial
$$(z+1)(z^2 - 3z + 4) = (z+1)^2(z-4)$$
associated to equation (5.43) has roots $z = -1$ with multiplicity 2 and $z = 4$, the general solution to equation (5.43) is
$$(5.44) \qquad c_1 e^{4t} + c_2 e^{-t} + c_3 t e^{-t}.$$

Step 4. Upon comparing (5.44) with the complementary solution y_c that we already obtained, we see that $L[c_1 e^{4t} + c_2 e^{-t}] = 0$ and so to find a particular solution of the equation in (5.42) we demand that
$$L[c_3 t e^{-t}] = e^{-t},$$
that is (since L is linear),
$$(5.45) \qquad c_3 L[t e^{-t}] = e^{-t}.$$

It is easier to compute
$$L[t e^{-t}] = (D^2 - 3D - 4)[t e^{-t}]$$
if we write the operator $D^2 - 3D - 4$ in factored form as $(D-4)(D+1)$. We have
$$\begin{aligned} L[t e^{-t}] &= (D-4)(D+1)[t e^{-t}] \\ &= (D-4)[-t e^{-t} + e^{-t} + t e^{-t}] \\ &= (D-4)[e^{-t}] \\ &= -e^{-t} - 4 e^{-t} = -5 e^{-t}. \end{aligned}$$

Thus equation (5.45) becomes
$$c_3(-5 e^{-t}) = e^{-t},$$
so that $c_3 = -\frac{1}{5}$ and $y_p(t) = -\frac{1}{5} t e^{-t}$. Putting the pieces together we conclude that the general solution to $y'' - 3y' - 4y = e^{-t}$ is
$$y(t) = c_1 e^{4t} + c_2 e^{-t} - \frac{1}{5} t e^{-t}.$$

Step 5. The initial conditions require that $y(0) = \frac{3}{2}$ and $y'(0) = \frac{4}{5}$; thus
$$c_1 + c_2 = \frac{3}{2} \quad \text{and} \quad 4c_1 - c_2 - \frac{1}{5} = \frac{4}{5}.$$

Solving for c_1 and c_2 we obtain $c_1 = \frac{1}{2}, c_2 = 1$ and thus
$$y(t) = \frac{1}{2} e^{4t} + e^{-t} - \frac{1}{5} t e^{-t}$$
is the solution to the initial value problem.

Example 5.2.7. Find a particular solution to the equation

(5.46) $$y'' - 4y = e^{2t} + t.$$

Our approach is to separately find particular solutions y_{p_1} and y_{p_2} to

(5.47) $$y'' - 4y = e^{2t}$$

and

(5.48) $$y'' - 4y = t.$$

By linearity (see Exercise 12 in the last section), their sum $y_{p_1} + y_{p_2}$ will be a particular solution to (5.46).

Check that the complementary equation $y'' - 4y = 0$ has general solution $y_c(t) = c_1 e^{2t} + c_2 e^{-2t}$. An annihilating operator for e^{2t} is $D - 2$. Applying this in (5.47) gives

(5.49) $$(D-2)(D^2-4)[y] = 0, \quad \text{or} \quad (D-2)^2(D+2)[y] = 0,$$

which has characteristic equation $(z-2)^2(z+2) = 0$. Thus equation (5.49) has general solution $c_1 e^{2t} + c_2 e^{-2t} + c_3 t e^{2t}$, and to determine y_{p_1} we demand that $(D^2 - 4)[c_3 t e^{2t}] = e^{2t}$. Solving for c_3 we find $c_3 = \frac{1}{4}$ and thus $y_{p_1}(t) = \frac{1}{4} t e^{2t}$.

To find y_{p_2} we use D^2 as an annihilating operator for t. Applying this in (5.48) gives

$$D^2(D^2-4)[y] = 0$$

which has characteristic equation $z^2(z^2 - 4) = 0$. The demand step is

$$(D^2 - 4)[d_3 + d_4 t] = t.$$

A computation shows that $d_3 = 0$ and $d_4 = -\frac{1}{4}$, so that $y_{p_2}(t) = -\frac{t}{4}$. Putting the pieces together gives

$$y_p(t) = \frac{1}{4} t e^{2t} - \frac{t}{4}$$

as a particular solution to (5.46).

An alternate approach is to use $D^2(D-2)$ as an annihilating operator for the sum $e^{2t} + t$. You are asked to work out the details in Exercise 18.

In each of the last examples we solved a second-order nonhomogeneous equation. The method of annihilating operators can be used in exactly the same way to solve third- or higher-order constant coefficient nonhomogeneous equations, provided that the nonhomogeneous term has an annihilating operator. Here's one example; you will find others in the exercises.

Example 5.2.8. We solve

(5.50) $$L[y] = y''' - 3y'' + 3y' - y = e^t.$$

Note that $L = (D-1)^3$ (check this for yourself!). The complementary solution is

$$y_c(t) = c_1 e^t + c_2 t e^t + c_3 t^2 e^t.$$

The nonhomogeneous term e^t in (5.50) has annihilating operator

$$A = (D-1).$$

Applying this annihilator gives

(5.51) $$AL[y] = (D-1)(D-1)^3[y] = 0.$$

Thus

$$(D-1)^4[y] = 0.$$

5.2. Annihilating operators

The characteristic polynomial for this fourth-order homogeneous equation is $p(z) = (z-1)^4$. Therefore $z = 1$ is a root with multiplicity four. Equation (5.51) has general solution

$$y = c_1 e^t + c_2 t e^t + c_3 t^2 e^t + c_4 t^3 e^t.$$

We want $L[y] = e^t$. Since we know $L[c_1 e^t + c_2 t e^t + c_3 t^2 e^t] = 0$, the demand step is just

$$L[c_4 t^3 e^t] = e^t.$$

Check for yourself that this says

$$c_4 = \frac{1}{6},$$

so that we have a particular solution

$$y_p(t) = \frac{1}{6} t^3 e^t.$$

The general solution to (5.51) is

$$y = y_c + y_p = c_1 e^t + c_2 t e^t + c_3 t^2 e^t + \frac{1}{6} t^3 e^t.$$

5.2.1. Annihilating operators with complex coefficients. In the previous section we were mainly concerned with annihilating operators $p(D)$ where p is a polynomial with real coefficients. However, it is possible to allow p to have complex coefficients, and sometimes this expanded perspective will greatly simplify our computational work. To see how this will go, let's find a particular solution to

(5.52) $$y'' + 2y' = \cos(2t)$$

using annihilating operators. We'll do this in two ways, first using an annihilating operator with real coefficients, and then by "complexifying" equation (5.52) and using an annihilating operator with complex coefficients.

Method 1. The complementary solution to (5.52) is $y_c(t) = c_1 + c_2 e^{-2t}$. The operator $A = D^2 + 4$ annihilates $\cos(2t)$, and applying that operator to both sides of equation (5.52) yields

(5.53) $$(D^2 + 4)(D^2 + 2D)[y] = 0.$$

This has characteristic equation $(z^2 + 4)z(z + 2) = 0$ with solutions $z = \pm 2i, z = 0,$ and $z = -2$. The homogeneous equation (5.53) has general solution

$$c_1 + c_2 e^{-2t} + c_3 \cos(2t) + c_4 \sin(2t).$$

Upon comparing this with the complementary solution, we see that the demand step is

$$(D^2 + 2D)[c_3 \cos(2t) + c_4 \sin(2t)] = \cos(2t).$$

A calculation shows that $c_3 = -\frac{1}{8}$ and $c_4 = \frac{1}{8}$, and thus a particular solution is

$$y_p(t) = -\frac{1}{8} \cos(2t) + \frac{1}{8} \sin(2t).$$

For future reference we'll also write y_p (using the analysis in Section 4.4.2) as

$$R \cos(2t - \delta)$$

for

$$R = \sqrt{\left(-\frac{1}{8}\right)^2 + \left(\frac{1}{8}\right)^2} = \frac{\sqrt{2}}{8}$$

and δ satisfying

$$\cos \delta = -\frac{1}{\sqrt{2}}, \quad \sin \delta = \frac{1}{\sqrt{2}}.$$

Taking $\delta = 3\pi/4$ we have

(5.54) $$y_p(t) = \frac{\sqrt{2}}{8}\cos(2t - 3\pi/4).$$

Method 2. For a "complex" perspective on the equation $L[y] = \cos(2t)$, with $L = D^2 + 2D$, recall that $\cos(2t)$ is the real part of the function e^{2it}:

(5.55) $$e^{2it} = \cos(2t) + i\sin(2t).$$

The related differential equation. Our basic strategy will be to find a particular (complex-valued) solution Y_p to the related equation

(5.56) $$y'' + 2y' = e^{2it}.$$

Observe that the real part of Y_p will be a solution to our *original* equation

$$y'' + 2y' = \cos(2t).$$

The complex annihilator. The operator $A = (D - 2i)$ annihilates the function e^{2it}. You can check this by writing e^{2it} in terms of sines and cosines as in equation (5.55), but its much simpler to just recall that $(e^{2it})' = 2ie^{2it}$ and observe that then

$$(D - 2i)[e^{2it}] = (e^{2it})' - 2ie^{2it} = 0.$$

If we apply the annihilating operator $D - 2i$ to both sides of (5.56), we obtain

(5.57) $$(D - 2i)(D^2 + 2D)[y] = 0,$$

which has characteristic equation

$$(z - 2i)z(z + 2) = 0$$

with solutions $z = 0, z = -2, z = 2i$. (These roots no longer need to come in conjugate pairs, since the coefficients of the characteristic equation are not real.) Thus (5.57) has (complex-valued) solution $y(t) = c_1 + c_2 e^{-2t} + c_3 e^{2it}$. We want $L[y] = e^{2it}$, or

$$L[c_1 + c_2 e^{-2t} + c_3 e^{2it}] = e^{2it}.$$

Since $c_1 + c_2 e^{-2t}$ solves the complementary equation $y'' + 2y' = 0$, $L[c_1 + c_2 e^{-2t}] = 0$. Thus we demand

$$L[c_3 e^{2it}] = e^{2it}$$

or

$$c_3(D^2 + 2D)[e^{2it}] = e^{2it}.$$

Computing, we see that

$$c_3((2i)^2 e^{2it} + 2(2i)e^{2it}) = e^{2it}$$

so that

$$c_3(-4 + 4i) = 1.$$

This says $c_3 = 1/(-4 + 4i)$. It will help to write c_3 in polar form (as discussed in Section 4.4.1). Since $-4 + 4i = Re^{i\delta}$ for $R = \sqrt{(-4)^2 + 4^2} = 4\sqrt{2}$ and $\delta = 3\pi/4$, we have

$$c_3 = \frac{1}{-4 + 4i} = \frac{1}{4\sqrt{2}e^{i3\pi/4}}.$$

Thus a particular solution to (5.56) is

$$Y_p(t) = \frac{1}{4\sqrt{2}e^{i3\pi/4}}e^{2it} = \frac{\sqrt{2}}{8}\frac{e^{2it}}{e^{i3\pi/4}},$$

which is readily simplified since division is easy to carry out in polar form:

$$Y_p(t) = \frac{\sqrt{2}}{8} e^{2it - i3\pi/4} = \frac{\sqrt{2}}{8} e^{i(2t - 3\pi/4)}.$$

Extracting y_p for the original differential equation. Since Y_p is a particular solution to $y'' + 2y' = e^{2it}$ and the real part of e^{2it} is $\cos(2t)$, we can find a particular real-valued solution to $y'' + 2y' = \cos(2t)$ by taking the real part of Y_p:

$$(5.58) \qquad y_p(t) = \operatorname{Re}\left(\frac{\sqrt{2}}{8} e^{i(2t - 3\pi/4)}\right) = \frac{\sqrt{2}}{8} \cos(2t - 3\pi/4).$$

This is the same solution as that produced by the first method. It's worth noting that the "complex" method *automatically* produces a particular solution in the preferred form (5.58) in which its properties (in particular the amplitude and phase shift) are more clearly visible. In this form, we can easily give a quick sketch of the graph of y_p; see Fig. 5.1.

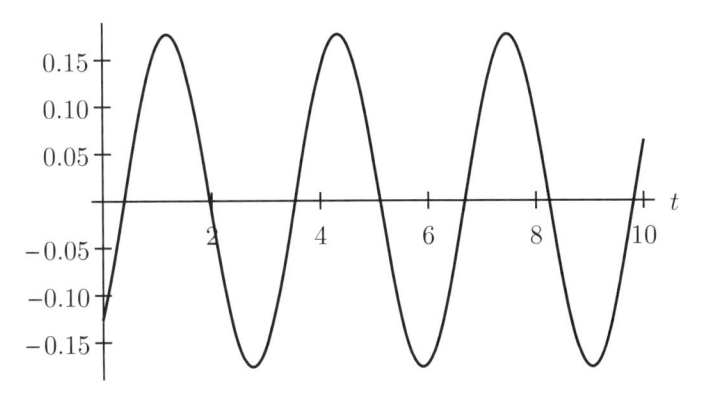

Figure 5.1. $y_p(t) = \frac{\sqrt{2}}{8} \cos(2t - 3\pi/4)$.

The advantages to using this sort of "complex" approach derive from being able to use an annihilating operator of lower-order (in this example we replaced the second-order operator $D^2 + 4$ with the first-order operator $(D - 2i)$) and the ease with which multiplication, division, and differentiation of exponentials can be carried out.

5.2.2. Exercises. Unless otherwise indicated, "annihilating operator" means a constant coefficient differential operator with real coefficients.

1. Find annihilating operators, of lowest possible order, for the functions $e^t, te^t, t^2 e^t$, and $t^m e^t$, where m is any positive integer.

2. Find an annihilating operator for $e^t \cos(2t)$.

3. Find an annihilating operator for $te^t \cos(2t)$.

4. If $D - 3$ annihilates $g(t)$, find an annihilating operator for $5g(t)$. What is the general principle here?

5. (a) Give three functions that are annihilated by $D + 2$.
   (b) Give an example of a function that is annihilated by $(D + 2)^2$ but not by $D + 2$.

6. Sam solves a (real number) constant coefficient homogeneous differential equation and gets $e^{2t} \cos t$ as one solution.

(a) What could the characteristic equation of his differential equation look like? Give an answer with real coefficients and the lowest possible degree.

(b) On the basis of your work in (a), give an annihilating operator for $e^{2t}\cos t$, with real coefficients and of lowest possible order.

7. (a) According to Table 5.1, an annihilating operator for $e^{\lambda t}\cos(\mu t)$ is
$$D^2 - 2\lambda D + (\lambda^2 + \mu^2).$$
Show how to obtain this formula by first writing down the characteristic equation for a constant coefficient homogeneous equation that has the function $e^{\lambda t}\cos(\mu t)$ among its solutions. (Your characteristic equation should have lowest possible degree.) It's helpful to first check by direct calculation that
$$(z - \lambda - i\mu)(z - \lambda + i\mu) = z^2 - 2\lambda z + \lambda^2 + \mu^2.$$

(b) Similarly, give the characteristic equation of lowest possible degree for a differential equation having the function $t^m e^{\lambda t}\cos(\mu t)$ (m is a positive integer) among its solutions. Use this to verify the entry in Table 5.1 giving an annihilating operator for $t^m e^{\lambda t}\cos(\mu t)$.

8. The "demand step" for solving the equation
$$y'' + 16y = 3\cos(4t)$$
reduces to the equation
$$-8c_3\sin(4t) + 8c_4\cos(4t) = 3\cos(4t).$$
Explain, using the phrase "linearly independent" in your answer, why this tells you that $c_3 = 0$ and $c_4 = \frac{3}{8}$.

In Exercises 9–16 give the general solution using the method of annihilating operators.

9. $y'' + y = t^2$.
10. $y'' - y = e^t$.
11. $y'' + 16y = 2\sin(4t)$.
12. $y'' - y' - 2y = 2\cos(2t)$.
13. $y'' - 4y = 8e^t$.
14. $y''' + 4y'' = 12$.
15. $y^{(4)} - 16y = e^{2t}$.
16. $y'' + 3y' + 2y = e^{-t} + e^{3t}$.

17. Find the general solution to the following equations using annihilating operators.
    (a) $y''' - 4y' = t$.
    (b) $y''' - 4y' = 3\cos t$.
    (c) $y''' - 4y' = t + 3\cos t$. (Use your work in the previous two parts; no new computations are needed!)

18. (a) Verify that $D^2(D-2)$ annihilates the function $e^{2t} + t$.
    (b) Rework Example 5.2.7 to find a particular solution to
$$y'' - 4y = e^{2t} + t$$
    by using the annihilating operator $D^2(D-2)$.

19. Find the general solution to the equation $y'' + y' = t^2 - 1$.
20. Find a particular solution to the equation $y'' + 5y' + 6y = e^{2t} + \sin t$.
21. Find a particular solution to the equation $y'' + 9y = 6e^{3t} + 9\tan(3t)$.
22. Find the general solution to the equation $y'' + 3y' + 2y = (2+t)e^t$.
23. Find the general solution to $y'' + 3y' + 2y = 5\cos^2 t$ using the identity $\cos(2t) = 2\cos^2 t - 1$.
24. Solve the initial value problem $y'' + 4y = -t$, $y(0) = 0$, $y'(0) = 1$.
25. Solve the initial value problem $y'' - y = \cos t$, $y(0) = 0$, $y'(0) = 0$.

26. Solve the initial value problem $y''' - 3y'' + 2y' = t$, $y(0) = 1$, $y'(0) = 23/4$, $y''(0) = 9$.

27. If y_{p_1} is a solution to $L[y] = g(t)$ and y_{p_2} is a solution to $L[y] = h(t)$, where L is a linear differential operator, find a solution to $L[y] = 3g(t) - 2h(t)$.

28. Solve the following equation in two ways: by annihilating operators and by variation of parameters (Section 5.1.1).
$$y'' + 3y' + 2y = 4e^t.$$

In Exercises 29–32 solve either by annihilating operators or by variation of parameters (Section 5.1.1). For some equations, either method can be used, but for others, only one is applicable.

29. $y'' - 4y' + 4y = e^{2t}/t$.

30. $y'' - 2y' + y = e^t \sin t$.

31. $y'' + 3y' + 2y = 3 \sin t$.

32. $t^2 y'' - 2ty' + 2y = t^2 \ln t$.

33. In this problem we use annihilating operators with complex coefficients.
   (a) Show that $(D - ib)$ annihilates the function e^{ibt} where b is any real number.
   (b) Suppose we wish to find a (necessarily complex-valued) particular solution to
   $$(5.59) \qquad y'' + 4y' + 3y = 6e^{3it}.$$
   By applying the appropriate annihilating operator from (a) to both sides of this equation, argue that a particular solution of the form ce^{3it} can be found. Now determine the value for c so that ce^{i3t} solves (5.59).
   (c) Use your work in (b) to find (real-valued) particular solutions to
   $$y'' + 4y' + 3y = 6\cos(3t) \qquad \text{and} \qquad v'' + 4v' + 3v = 6\sin(3t).$$

34. Solve $y'' + 3y' + 2y = \cos t$ by first solving $y'' + 3y' + 2y = e^{it}$ using an annihilating operator with complex coefficients and then extracting the desired solution from your result.

35. Follow the outline given below to solve
$$(5.60) \qquad y'' + 2y' + 2y = e^t \cos t$$
using a complex annihilating operator.
   (a) Check that $e^{(1+i)t} = e^t \cos t + ie^t \sin t$, and verify that $D - (1+i)$ is an annihilating operator for $e^{(1+i)t}$.
   (b) Find a particular (complex-valued) solution to
   $$y'' + 2y' + 2y = e^{(1+i)t}$$
   using the annihilating operator from (a).
   (c) Using the polar form of a complex number, compute the real part of your answer to (b) and explain why this must be a (real-valued) solution to (5.60). Give a quick sketch of the graph of this solution.
   (d) How would you use your answer to (b) to find a particular solution to
   $$y'' + 2y' + 2y = e^t \sin t?$$

36. Let $p(z) = a_n z^n + a_{n-1} z^{n-1} + \cdots + a_1 z + a_0$ be a polynomial with real coefficients and let L be the constant coefficient differential operator given by
$$L[y] = a_n y^{(n)} + a_{n-1} y^{(n-1)} + \cdots + a_1 y' + a_0 y.$$
Notice that
   (i) The characteristic equation for the homogeneous equation $L[y] = 0$ is $p(z) = 0$.
   (ii) We could describe the operator L as $p(D)$, where D is differentiation.
Make sure you understand both of these assertions before proceeding.

(a) Use the method of annihilating operators to show that a particular solution to the equation $L[y] = e^{bt}$ (b a real constant) is

$$y_p(t) = \frac{e^{bt}}{p(b)}$$

provided $p(b) \neq 0$.

(b) Suppose the $p(b)$ is equal to 0 and the multiplicity of the zero at b is 1. This means we can factor $p(z) = (z-b)q(z)$ where $q(b) \neq 0$. Show that a particular solution to $L[y] = e^{bt}$ is

$$y_p(t) = \frac{te^{bt}}{p'(b)}.$$

37. Use the preceding exercise to find a particular solution to each of the following equations.
    (a) $y^{(4)} + 2y^{(3)} + 3y' + 5y = e^{-2t}$.
    (b) $y^{(4)} + 2y^{(3)} + 3y' + 5y = 3e^{-2t}$.
    (c) $y^{(4)} + 2y^{(3)} + 3y' + 6y = e^{-2t}$.
    (d) $y^{(4)} + 2y^{(3)} + 3y' + 6y = 3e^{-2t}$.

5.3. Applications of nonhomogeneous equations

5.3.1. Oscillating systems with forcing.

In 1850 more than 200 soldiers died when the Angers Bridge, a suspension bridge spanning the Maine River in France, collapsed as a regiment of 478 soldiers marching in step attempted to cross it. In 1831 a similar fate befell soldiers crossing the Broughton Bridge near Manchester, England, as described in a newspaper at the time as follows:

> A very serious and alarming accident occurred on Tuesday last, in the fall of the Broughton suspension bridge, erected a few years ago by John Fitzgerald, Esq., whilst a company of the 60th rifles were passing over it; and, although fortunately no lives were lost, several of the soldiers received serious personal injuries, and damage was done to the structure, which will require a long time and very considerable expense to repair.
>
> It appears that, on the day when this accident happened, the 60th regiment had a field-day on Kersall Moor, and, about twelve o'clock, were on their way back to their quarters. The greater part of the regiment is stationed in the temporary barracks, in Dyche-street, St. George's Road, and took the route through Strangeways, but one company, commanded, as it happened singularly enough, by Lieut. P. S. Fitzgerald, the son of the proprietor of the bridge, being stationed at the Salford barracks, took the road over the suspension bridge, intending to go through Pendleton to the barracks. Shortly after they got upon the bridge, the men, who were marching four abreast, found that the structure vibrated in unison with the measured step with which they marched; and, as this vibration was by no means unpleasant, they were inclined to humour it by the manner in which they stepped. As they proceeded, and as a greater number of them got upon the bridge, the vibration went on increasing until the head of the column had nearly reached the Pendleton side of the river. They were then alarmed by a loud sound something resembling an irregular discharge of firearms; and immediately one of the iron pillars supporting the suspension chains...fell toward the bridge, carrying with it a large stone from the pier, to which it had been bolted. Of course that corner of the bridge, having lost the support of the pillar, immediately fell to the bottom on the river, a descent of about 16 or 18 feet; and from the great inclination thereby given to the roadway, nearly the whole of the soldiers who were upon it were precipitated into the river, where a scene of great confusion was exhibited.[2]

---

[2] *Fall of the Broughton suspension bridge*, The Manchester Guardian, April 16, 1831.

5.3. Applications of nonhomogeneous equations

The Millennium Bridge, nicknamed Wobbly Bridge, a pedestrian-only suspension bridge across the Thames River, was closed three days after opening in 2000 when large vibrations were caused by people walking over it. Spectacular oscillations in the Tacoma Narrows Bridge across Puget Sound, called "Galloping Gertie", resulted in the collapse of the bridge on the morning of November 7, 1940 (caught on film—amazing videos can be found on YouTube by searching on "Tacoma Narrows Bridge" or "Galloping Gertie"). In 1959 and 1960 two Electra turboprop planes, flown by Braniff Airlines and Northwest Orient Airlines, broke apart in mid-flight, killing everyone aboard. In all of these engineering disasters the phenomenon of resonance has been implicated as a causative factor.[3] In this section we use a modification of our basic mass-spring system from Section 4.8 to study resonance and other interesting phenomena arising from systems undergoing "forced vibrations".

Recall our mass-spring set-up from Section 4.8. A vertically suspended spring, with spring constant k, has a mass m attached to the end of the spring. The mass is put into motion and the displacement $y(t)$ of the mass from its equilibrium is measured at time t; $y = 0$ means the mass is in the equilibrium position. We derived a second-order constant coefficient homogeneous equation for the displacement function
$$my'' + cy' + ky = 0,$$
where $c \geq 0$ is the drag, or damping, force constant. We set $c = 0$ when we want to ignore any damping force.

Now we modify this set-up to allow an external force to be applied to the mass. For example, perhaps a motor is attached to the mass. We'll assume this external force is a function of time t so that the new situation is modeled by the equation

(5.61) $$my'' + cy' + ky = F(t)$$

for some function $F(t)$. When $F(t)$ is not 0, this is a constant coefficient nonhomogeneous equation.

Periodic forcing. We are particularly interested in the case that the external force is periodic, and we will begin by analyzing the solution of equation (5.61) when $c = 0$ (no damping) and $F(t) = F_0 \cos(\omega t)$:
$$my'' + ky = F_0 \cos(\omega t),$$
where F_0 is a constant. The associated homogeneous equation, $my'' + ky = 0$, has characteristic equation $mz^2 + k = 0$ with roots $\pm i\sqrt{k/m}$. As in Section 4.8, we let $\omega_0 = \sqrt{k/m}$, so that the complementary solution to (5.61) is $y_c(t) = c_1 \cos(\omega_0 t) + c_2 \sin(\omega_0 t)$. Recall also that ω_0 is the natural (circular) frequency of the system; with no damping and no external force our mass exhibits periodic motion with period $2\pi/\omega_0$. The forcing function $F_0 \cos(\omega t)$ has annihilating operator $A = D^2 + \omega^2$, and if we apply this annihilating operator to both sides of equation (5.61), we obtain

(5.62) $$(D^2 + \omega^2)(mD^2 + k)[y] = 0,$$

which, on dividing by m, we can equivalently write as

(5.63) $$(D^2 + \omega^2)(D^2 + \omega_0^2)[y] = 0.$$

To continue, we must distinguish two cases: $\omega = \omega_0$ **or** $\omega \neq \omega_0$. In words, we distinguish the case that the impressed force has the same frequency as the natural frequency of the system from the case that it does not. The solutions to equation (5.63) will be quite different in these two cases.

---

[3] Fifty years after the collapse of the Tacoma Narrows Bridge, mathematicians A. C. Lazar and P. J. McKenna carried out a critical look at the theory that resonance caused by high winds explained its collapse. They proposed instead a differential equation model for the vertical displacement of the center of the bridge that included a *nonlinear* term. You can read a simplified account of their model in [4].

Case 1: $c = 0, \omega = \omega_0$. When $\omega = \omega_0$, where the characteristic equation for equation (5.63) will have roots $\pm \omega_0 i$, each of multiplicity *two*. The general solution to equation (5.63) in this case is

$$c_1 \cos(\omega_0 t) + c_2 \sin(\omega_0 t) + c_3 t \cos(\omega_0 t) + c_4 t \sin(\omega_0 t).$$

In Exercise 9 you are asked to show that a particular solution to

(5.64) $$my'' + ky = F_0 \cos(\omega_0 t)$$

is

$$y_p(t) = \frac{F_0}{2m\omega_0} t \sin(\omega_0 t).$$

Thus the nonhomogeneous equation (5.64) has general solution

$$y(t) = c_1 \cos(\omega_0 t) + c_2 \sin(\omega_0 t) + \frac{F_0}{2m\omega_0} t \sin(\omega_0 t).$$

If initial conditions are specified, we can determine values for c_1 and c_2. For example, the initial conditions $y(0) = 0$, $y'(0) = 0$ gives $c_1 = c_2 = 0$ and this initial value problem has solution

$$y(t) = \frac{F_0}{2m\omega_0} t \sin(\omega_0 t).$$

Here's where things get interesting. The external force, $F_0 \cos(\omega_0 t)$, is certainly bounded (by $|F_0|$) but the displacement function $\frac{F_0}{2m\omega_0} t \sin(\omega_0 t)$ is *unbounded* as $t \to \infty$; see Fig. 5.2, where this function is graphed for specific values of the constants F_0, m, and ω_0. Moreover, this qualitative assertion doesn't change if other initial conditions are used, since no matter what c_1 and c_2 are, the terms $c_1 \cos(\omega_0 t) + c_2 \sin(\omega_0 t)$ stay bounded. This is the phenomenon of **pure resonance**. Our mathematical model for an undamped mass-spring system, with a simple harmonic externally applied force whose frequency matches the natural frequency of the system, will lead to an unbounded displacement function.

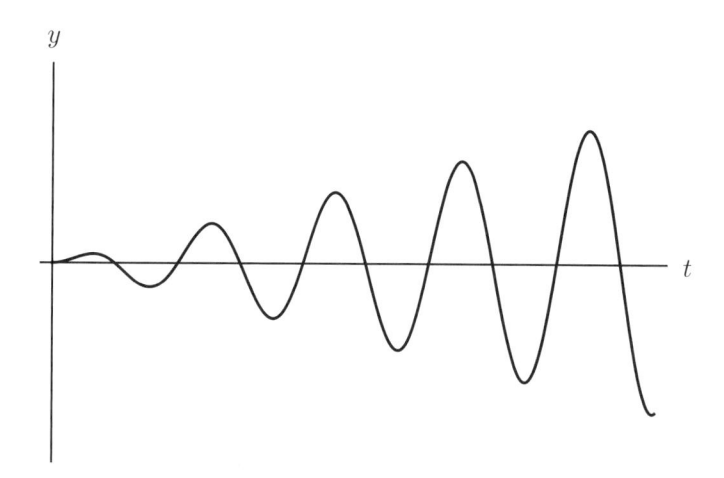

Figure 5.2. Typical graph of $y(t) = \frac{F_0}{2m\omega_0} t \sin(\omega_0 t)$.

Case 2: $c = 0, \omega \neq \omega_0$. Still assuming $c = 0$ and $F(t) = F_0 \cos(\omega t)$, how does the general solution to equation (5.61) change if $\omega \neq \omega_0$? The complementary solution is $y_c(t) = c_1 \cos(\omega_0 t) + c_2 \sin(\omega_0 t)$ just as in Case 1. You should check that we can look for a particular solution in the form $c_3 \cos(\omega t) + c_4 \sin(\omega t)$, and by substitution (the "demand step") we discover that

$$c_4 = 0 \quad \text{and} \quad c_3 = \frac{F_0}{m(\omega_0^2 - \omega^2)}.$$

5.3. Applications of nonhomogeneous equations

Thus the general solution is

(5.65) $$y(t) = c_1 \cos(\omega_0 t) + c_2 \sin(\omega_0 t) + \frac{F_0}{m(\omega_0^2 - \omega^2)} \cos(\omega t).$$

As we described in Section 4.4.2, we can rewrite

$$c_1 \cos(\omega_0 t) + c_2 \sin(\omega_0 t) = R \cos(\omega_0 t - \delta)$$

(where R and δ are determined from c_1 and c_2 by $R = \sqrt{c_1^2 + c_2^2}$, $\cos \delta = c_1/R$, and $\sin \delta = c_2/R$), so that any solution is a sum of two simple harmonic functions

(5.66) $$R \cos(\omega_0 t - \delta) + \frac{F_0}{m(\omega_0^2 - \omega^2)} \cos(\omega t).$$

This is a bounded function, no matter what the values of c_1 and c_2, but if ω_0 is close to ω, the amplitude of the second term in (5.66) may be large in comparison to F_0, the amplitude of the impressed force, so that a resonance-like phenomenon is possible.

Let's look at the solution (5.65) satisfying the initial conditions $y(0) = 0, y'(0) = 0$. From $y'(0) = 0$ we see that $c_2 = 0$, and then from $y(0) = 0$ we obtain

$$c_1 = -\frac{F_0}{m(\omega_0^2 - \omega^2)}$$

and

$$y(t) = \frac{F_0}{m(\omega_0^2 - \omega^2)} (\cos(\omega t) - \cos(\omega_0 t)).$$

To better understand this function, especially when ω is close to ω_0, we use the trigonometric formula

$$\cos \alpha - \cos \beta = 2 \sin \frac{\beta - \alpha}{2} \sin \frac{\beta + \alpha}{2}$$

to rewrite it as

(5.67) $$y(t) = \left[2 \frac{F_0}{m(\omega_0^2 - \omega^2)} \sin \left(\frac{\omega_0 - \omega}{2} t \right) \right] \sin \left(\frac{\omega_0 + \omega}{2} t \right).$$

The factor in square brackets in equation (5.67), representing a *slow sine wave* (long period), may be viewed as the time-varying amplitude of the faster sine wave (shorter period) $\sin((\omega_0 + \omega)t/2)$. A typical plot of the displacement function (5.67) is shown in Fig. 5.3, its amplitude being "modulated" by the sine curve that is shown dashed. As anyone who has played a stringed instrument knows, the larger the amplitude of the string displacement, the louder the sound produced. Thus the behavior of the solution (5.67) is easy to describe in terms of audible vibrations. For example, if you strike a piano key at the same time as a tuning fork of *almost* the same frequency, you will hear a tone of slowing varying intensity (loudness). This gives a way of hearing how far apart the two notes (frequencies) are, even if you are tone-deaf and can only distinguish the intensity of the sound, not the pitch. Thus if the vibrations depicted in Fig. 5.3 are audible and the time scale is seconds, you would hear a tone rise and fall in volume three times over the interval $[0, 1.5]$. These variations of the loudness of the tone are called **beats**. In general, the number of beats per second equals the frequency difference $|\omega_0/(2\pi) - \omega/(2\pi)|$, where frequencies are measured in hertz (cycles per second). This follows from the observation that the time varying amplitude given by the square-bracketed expression in (5.67) has two extrema over every cycle, and thus the number of beats per second is twice its frequency $2|\omega_0 - \omega|/(4\pi) = |\omega_0/(2\pi) - \omega/(2\pi)|$.

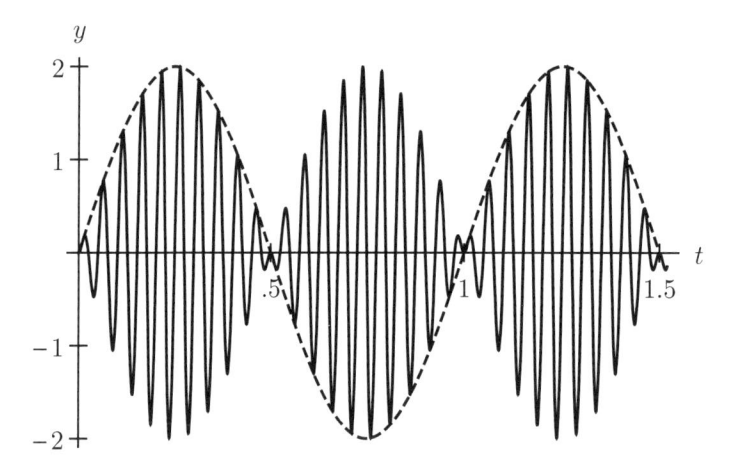

Figure 5.3. Three beats. Graph of equation (5.67) with $\omega_0 = 42\pi$, $\omega = 38\pi$, and $\frac{F_0}{m(\omega_0^2 - \omega^2)} = 1$. Dashed curve is graph of the square-bracketed factor.

Case 3: $c \neq 0$. More relevant to real life scenarios is what happens when the damping constant c is small (but nonzero). To analyze this situation, we'll solve

(5.68) $$my'' + cy' + ky = F_0 \cos(\omega t)$$

if $c \neq 0$. Our method will also apply if $c = 0$ and $\omega_0 \neq \omega$, giving an alternate approach to Case 2. We can avoid some fairly unpleasant algebraic computations if we convert (5.68) to a complex differential equation and solve it with an annihilating operator with complex coefficients, as discussed in Section 5.2.1. We think of this as a three-step process:

- Replace the real force $F_0 \cos(\omega t)$ by the complex force $F_0 e^{i\omega t}$, whose real part is the original forcing term: $\text{Re}(F_0 e^{i\omega t}) = F_0 \cos(\omega t)$.
- Find a particular (complex-valued) solution Y_p to the resulting complex differential equation by using the complex annihilating operator $D - i\omega$.
- Take the real part of this complex-valued particular solution to get a particular real-valued solution y_p of the original differential equation (5.68); $y_p = \text{Re}\, Y_p$.

Thus our complex differential equation is

(5.69) $$my'' + cy' + ky = F_0 e^{i\omega t},$$

and when we apply the annihilating operator $A = D - i\omega$ to both sides we get a third-order homogeneous equation with characteristic equation

$$(z - i\omega)(mz^2 + cz + k) = 0.$$

You should convince yourself that if either $c \neq 0$ or $\omega_0 \neq \omega$, then the root $z = i\omega$ cannot duplicate the roots of $mz^2 + cz + k = 0$. Thus, in the "demand step" of solving (5.69) we seek to determine the constant a so that

$$L[ae^{i\omega t}] = F_0 e^{i\omega t},$$

where $L[y] = my'' + cy' + ky$. Since $(e^{i\omega t})' = i\omega e^{i\omega t}$ and $(e^{i\omega t})'' = -\omega^2 e^{i\omega t}$ we must have

$$m(-\omega^2)a + ci\omega a + ka = F_0$$

or

$$a = \frac{F_0}{k - m\omega^2 + ic\omega}.$$

5.3. Applications of nonhomogeneous equations

We want to write a in polar form, but before doing so let's recall that

$$\sqrt{\frac{k}{m}} = \omega_0,$$

so that

$$a = \frac{F_0}{m(\omega_0^2 - \omega^2) + ic\omega} = \frac{F_0}{Re^{i\delta}}$$

where

$$R = \sqrt{m^2(\omega_0^2 - \omega^2)^2 + (c\omega)^2}$$

and

$$\sin\delta = c\omega/R, \quad \cos\delta = m(\omega_0^2 - \omega^2)/R.$$

This gives a particular complex-valued solution to (5.69)

$$Y_p(t) = \frac{F_0}{Re^{i\delta}} e^{i\omega t} = \frac{F_0}{R} e^{i(\omega t - \delta)}.$$

We immediately read off the real part of this complex solution to be

$$y_p(t) = \frac{F_0}{R} \cos(\omega t - \delta),$$

a particular solution to (5.68). Looking back at the expression for R you should observe that the amplitude of this cosine function can be large in relation to F_0 if c is small and $\omega \approx \omega_0$, since these conditions make the denominator R small. The general solution to (5.68) is obtained by adding the complementary solution:

$$y(t) = y_c(t) + y_p(t) = y_c(t) + \frac{F_0}{R} \cos(\omega t - \delta),$$

where y_c is the general solution to the homogeneous equation

(5.70) $$my'' + cy' + ky = 0.$$

Transient and steady-state solutions. When $m, k,$ and c are strictly positive, we know from Section 4.5.2 that any solution to (5.70) tends to 0 as $t \to \infty$. If we write the general solution to equation (5.68) (with $c \neq 0$) as

$$c_1 y_1(t) + c_2 y_2(t) + y_p(t),$$

the contribution from the complementary piece $c_1 y_1 + c_2 y_2$ tends to zero as t increases. Since the initial conditions determine the values of c_1 and c_2, another way to say this is that the effect of the initial conditions dissipates over time. Thus no matter what initial conditions we impose on (5.68), the solution behaves like

$$y_p = \frac{F_0}{R} \cos(\omega t - \delta)$$

as $t \to \infty$. The complementary solution $c_1 y_1 + c_2 y_2$ is called a **transient solution**, and y_p is called the **steady-state solution**. Note that y_p is periodic with the same period $T = 2\pi/\omega$ as the forcing function $F_0 \cos(\omega t)$, and in fact it is the only periodic solution of equation (5.68).

Example 5.3.1. We'll solve the initial value problem

$$y'' + 2y' + 10y = 2\cos(2t), \quad y(0) = 16/13, \quad y'(0) = -48/13.$$

The complementary equation $y'' + 2y' + 10y = 0$ has general solution

$$c_1 e^{-t} \cos(3t) + c_2 e^{-t} \sin(3t).$$

Complexifying, we solve the nonhomogeneous equation

$$y'' + 2y' + 10y = 2e^{2it},$$

using the annihilating operator $A = D - 2i$, and find that it has particular solution
$$Y_p(t) = ae^{2it}$$
for
$$a = \frac{2}{6+4i} = \frac{1}{3+2i} = \frac{1}{\sqrt{13}e^{i\delta}}$$
where $\cos\delta = 3/\sqrt{13}$ and $\sin\delta = 2/\sqrt{13}$, so that $\delta \approx 0.588$. The real part of Y_p is
$$y_p(t) = \frac{1}{\sqrt{13}}\cos(2t - \delta).$$
The general solution to $y'' + 2y' + 10y = 2\cos(2t)$ is thus
$$\frac{1}{\sqrt{13}}\cos(2t - \delta) + c_1 e^{-t}\cos(3t) + c_2 e^{-t}\sin(3t).$$
Using the initial conditions, we determine, after some calculation, that $c_1 = 1$ and $c_2 = -1$. Fig. 5.4 shows the graph of the solution, together with the graph of the steady-state solution
$$y_p(t) = \frac{1}{\sqrt{13}}\cos(2t - \delta) \approx \frac{1}{\sqrt{13}}\cos(2t - 0.588).$$
As you can see, the transient solution
$$y_c(t) = e^{-t}\cos(3t) - e^{-t}\sin(3t)$$
contributes little to the solution for all times exceeding 4.

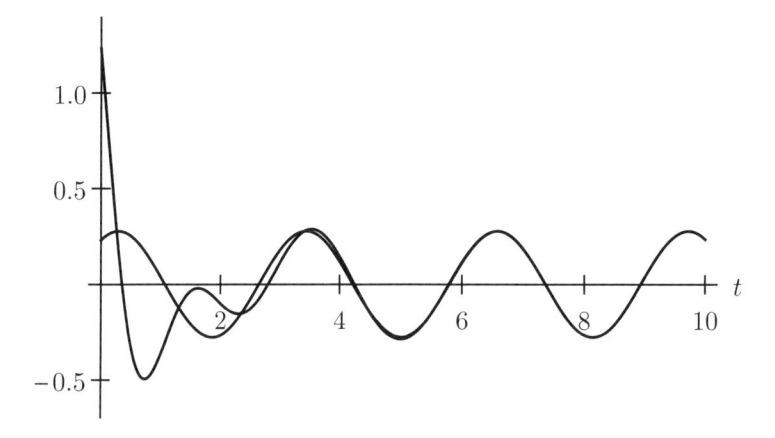

Figure 5.4. Solution and steady-state solution.

5.3.2. A compartment model for accidental drug overdose. Accidental poisoning by an overdose of aspirin has been a historically important problem. Children between the ages of one and three are especially at risk. In 1968 the maximum allowable number of baby aspirin tablets per package was reduced to thirty-six, and tablet strength was decreased to 81 mg. Shortly afterward, the Poison Prevention Packaging Act of 1970 mandated safety caps for certain medications, with aspirin being the first to require special packaging.[4] In this section we will develop two simple models to investigate the level of toxicity if a two-year-old child ingests thirty-six 81-mg aspirin tablets. While the models we present are standard ones from pharmacokinetics—the study of the absorption, distribution, metabolism, and elimination of drugs in the body—it is important to keep in mind that we are making many simplifications to what in reality is a complex problem.

---

[4]Since the 1980s concerns about an association between the sometimes fatal Reyes Syndrome and aspirin have resulted in warnings against aspirin use in children under age 19 with viral illnesses.

5.3. Applications of nonhomogeneous equations

For the purposes of modeling drug processing in the body it is often appropriate to view the body as being composed of several compartments, with the drug moving from one compartment to another over time. Here we focus on two compartments, the GI tract (the gastrointestinal tract including the stomach and intestines) and the bloodstream. Orally ingested drugs move from the GI tract to the bloodstream and then are eliminated from the blood, typically by the liver and or kidneys; this is illustrated below. We want to focus on the blood levels of the drug over time.

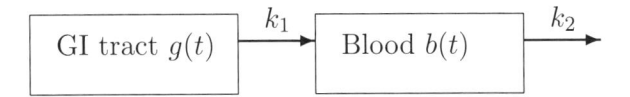

$$\boxed{\text{GI tract } g(t)} \xrightarrow{k_1} \boxed{\text{Blood } b(t)} \xrightarrow{k_2}$$

First model. One standard model is obtained by assuming "first-order kinetics" for the movement of the drug from one compartment to another. This means we assume that the aspirin leaves the GI tract and enters the bloodstream at a rate proportional to the amount present in the GI tract. Writing $g(t)$ for the amount of the drug (in mg) present in the GI tract at time t we have

(5.71) $$\frac{dg}{dt} = -k_1 g$$

for some positive constant k_1. In the bloodstream there are two rates to take into consideration, the rate at which the drug enters the blood from the GI tract and the rate at which the drug is eliminated from the blood by the liver and kidneys. Writing $b(t)$ for the amount of drug in the bloodstream at time t we have

(5.72) $$\frac{db}{dt} = k_1 g - k_2 b$$

for positive constants k_1 and k_2; again this represents first-order kinetics. Notice that the "rate in" term $k_1 g$ has the same constant of proportionality as the "rate out" term on the right-hand side of equation (5.71). We'll assume all of the drug is ingested at the same time (call this time $t = 0$) and is instantly dissolved in the stomach. This gives the initial condition $g(0) = g_0$, the amount ingested, in milligrams. Initially there is no drug in the bloodstream, so $b(0) = 0$. The two equations (5.71) and (5.72) are a *system* of first-order equations, but we will solve them by converting to a *single* constant coefficient second-order equation (see also Exercise 13 for a different approach).

Using primes for derivatives with respect to t and solving equation (5.72) for g gives

$$\frac{1}{k_1} b' + \frac{k_2}{k_1} b = g.$$

Substituting this into both sides of equation (5.71) yields

$$\frac{1}{k_1} b'' + \frac{k_2}{k_1} b' = -k_1 \left(\frac{1}{k_1} b' + \frac{k_2}{k_1} b \right)$$
$$= -b' - k_2 b.$$

Simplifying, we obtain the second-order constant coefficient homogeneous equation:

(5.73) $$b'' + (k_1 + k_2) b' + k_1 k_2 b = 0.$$

If we assume, as is the typical case, that $k_1 > k_2$ and use the quadratic formula to find the roots of the characteristic polynomial

$$z^2 + (k_1 + k_2) z + k_1 k_2,$$

the result is
$$z = \frac{-(k_1 + k_2) \pm \sqrt{(k_1 + k_2)^2 - 4k_1 k_2}}{2},$$
which simplifies nicely since
$$(k_1 + k_2)^2 - 4k_1 k_2 = k_1^2 - 2k_1 k_2 + k_2^2 = (k_1 - k_2)^2.$$
This gives
$$z = \frac{-(k_1 + k_2) + k_1 - k_2}{2} = -k_2 \quad \text{and} \quad \frac{-(k_1 + k_2) - (k_1 - k_2)}{2} = -k_1,$$
so that the general solution to (5.73) is

(5.74) $$b(t) = c_1 e^{-k_1 t} + c_2 e^{-k_2 t}.$$

We seek to determine c_1 and c_2 from our initial conditions. We know $b(0) = 0$. From equation (5.72) we see that
$$b'(0) = k_1 g(0) - k_2 b(0) = k_1 g_0.$$
Using the initial conditions $b(0) = 0, b'(0) = k_1 g_0$ in equation (5.74) gives the relations
$$c_1 + c_2 = 0 \quad \text{and} \quad -c_1 k_1 - c_2 k_2 = k_1 g(0).$$
Keep in mind that we regard $g(0)$, k_1, and k_2 as known quantities (depending on our particular scenario) and our goal is to determine values for c_1 and c_2 in terms of them. To this end, we must have $c_1 = -c_2$ and hence $c_2(k_1 - k_2) = k_1 g(0)$, or
$$c_2 = \frac{k_1 g(0)}{k_1 - k_2}, \quad c_1 = -\frac{k_1 g(0)}{k_1 - k_2},$$
and the amount of aspirin in the blood compartment at time t is

(5.75) $$b(t) = \frac{k_1 g(0)}{k_1 - k_2}(-e^{-k_1 t} + e^{-k_2 t}).$$

Let's analyze our aspirin poisoning scenario in light of this formula. Ingesting thirty-six 81-mg tablets all at once gives $g_0 = 2{,}916$ mg. The constants k_1 and k_2 are experimentally determined by the drug manufacturer to be approximately
$$k_1 = 1.38/\text{hr} \quad \text{and} \quad k_2 = 0.231/\text{hr}.$$
The maximum value of $b(t)$ occurs when $b'(t) = 0$ (why?). Since
$$b'(t) = \frac{k_1 g_0}{k_1 - k_2}(k_1 e^{-k_1 t} - k_2 e^{-k_2 t}),$$
solving $b'(t) = 0$ yields
$$t_{\max} = \frac{\ln k_1 - \ln k_2}{k_1 - k_2},$$
the time for peak blood values. Using the values of k_1 and k_2 from above gives $t_{\max} = 1.56$ hr. Substituting $t_{\max}$ into (5.75) and using $g(0) = 2{,}916$, we see that the maximum blood level is about 2,036 mg. The average two year old has a blood volume of approximately 1.5 l, so the maximum concentration according to this model is $2{,}035/1.5 = 1{,}356$ mg/l, well below the lethal concentration of 2,600 mg/l, but about 9 times the therapeutic level of 150 mg/l. In fact, serious toxicity side effects (hearing loss, respiratory failure, and cardiovascular collapse) can occur in children at blood levels above 800 mg/l and moderate toxic effects at concentrations of 400 mg/l. The graph of $b(t)$ is shown in Fig. 5.5.

5.3. Applications of nonhomogeneous equations

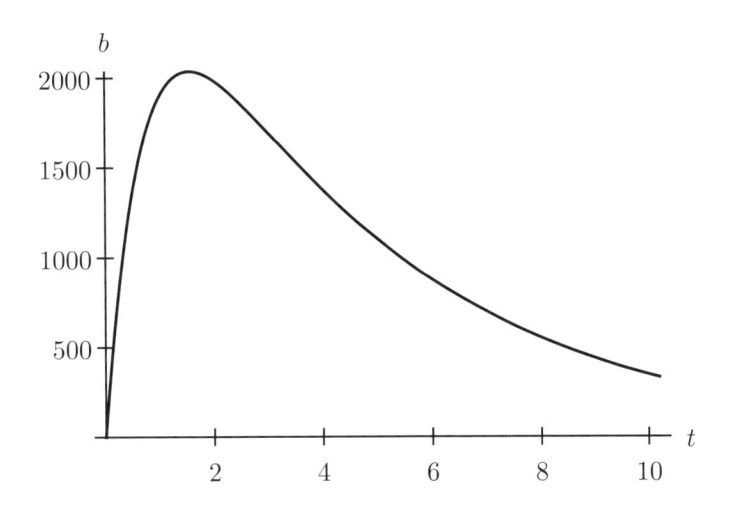

Figure 5.5. Graph of $b(t)$.

Improving the model. This model oversimplifies various important factors in analyzing aspirin overdose. The most significant of these is that the assumption that aspirin is eliminated from the bloodstream at a rate proportional to the amount present (this is the "$-k_2 b$" term in (5.72)) is not valid when the concentration in the bloodstream is high. This is because the metabolic pathways become saturated, and the net effect is that after $b(t)$ reaches a certain high level the differential equation for $\frac{db}{dt}$ should be replaced by

$$\frac{db}{dt} = k_1 g - k_3$$

where the "rate out" term has become a constant k_3. This is referred to as "zero-order kinetics", and we explore this next.

Second model. Now we assume that while equations (5.71) and (5.72) apply initially, after a certain period of time the relevant equations are

$$(5.76) \qquad \frac{dg}{dt} = -k_1 g$$

and

$$(5.77) \qquad \frac{db}{dt} = k_1 g - k_3$$

where k_3 is a constant, about which we will soon have more to say. To be a bit more explicit, we denote the amount of aspirin in the bloodstream for $0 \leq t \leq t_1$ by b_1 and the amount in the bloodstream for $t \geq t_1$ by b_2. The relevant differential equations are

$$(5.78) \qquad g' = -k_1 g, \quad b_1' = k_1 g - k_2 b_1, \qquad 0 \leq t \leq t_1,$$

exactly as in our first model, and

$$(5.79) \qquad g' = -k_1 g, \quad b_2' = k_1 g - k_3, \qquad t \geq t_1.$$

Solving the second equation in (5.79) for g gives

$$g = \frac{1}{k_1} b_2' + \frac{k_3}{k_1}.$$

Substitute this into both sides of the first equation in (5.79) to obtain

$$\frac{1}{k_1} b_2'' = -k_1 \left(\frac{1}{k_1} b_2' + \frac{k_3}{k_1} \right) = -b_2' - k_3.$$

Thus
(5.80) $$b_2'' = -k_1 b_2' - k_1 k_3.$$
This is a second-order nonhomogeneous equation, with nonhomogeneous term $-k_1 k_3$. Our goal is to solve this with the appropriate initial conditions.

Determining the appropriate choice for k_3. Before proceeding, we need to discuss the constant k_3. The shift from equations (5.78) to equations (5.79) occurs at time t_1 (i.e., at t_1 the metabolic pathways become saturated). So that the transition is smooth, we require that the value of $\frac{db}{dt}$ as given by each of these equations should agree at $t = t_1$:
$$k_1 g(t_1) - k_2 b_1(t_1) = k_1 g(t_1) - k_3,$$
which says
$$k_3 = k_2 b_1(t_1).$$
Equation (5.80) becomes
(5.81) $$b_2'' + k_1 b_2' = -k_1 k_2 b_1(t_1).$$

Solving equation (5.81). The nonhomogeneous term $-k_1 k_2 b_1(t_1)$ is a *constant*, so it has annihilating operator D. You should check that the complementary solution to (5.81) is
$$c_1 + c_2 e^{-k_1 t}$$
and that a particular solution has the form $b_p(t) = at$ for some constant a. We demand that
$$L[at] = -k_1 k_2 b_1(t_1)$$
for $L = D^2 + k_1 D$. It's easy to check that a must be $-k_2 b_1(t_1)$ (do it!), so that the general solution to (5.81) is
(5.82) $$b_2(t) = c_1 + c_2 e^{-k_1 t} - k_2 b_1(t_1) t.$$

Imposing the initial conditions. What are our initial conditions? Since our function b_2 applies for $t \geq t_1$, we'll set our initial conditions at the time $t = t_1$. We have
$$b_2(t_1) = b_1(t_1)$$
and
$$b_2'(t_1) = b_1'(t_1).$$
The first condition says exactly that
(5.83) $$c_1 + c_2 e^{-k_1 t_1} - k_2 b_1(t_1) t_1 = b_1(t_1).$$
The second, together with the second equation in (5.78), says that
(5.84) $$-k_1 c_2 e^{-k_1 t_1} - k_2 b_1(t_1) = k_1 g(t_1) - k_2 b_1(t_1).$$
From (5.84) we can solve for the constant c_2:
$$c_2 = -\frac{g(t_1)}{e^{-k_1 t_1}}.$$
We can simplify this a bit. Since
$$\frac{dg}{dt} = -k_1 g$$
we have
(5.85) $$g(t) = g(0) e^{-k_1 t}$$

5.3. Applications of nonhomogeneous equations

and $g(t_1) = g(0)e^{-k_1 t_1}$, so that

(5.86) $$c_2 = -\frac{g(t_1)}{e^{-k_1 t_1}} = -g(0).$$

Inserting this in (5.83) and solving for c_1 we get

(5.87) $$c_1 = g(0)e^{-k_1 t_1} + b_1(t_1)[1 + k_2 t_1].$$

We apply our work to the two year old who has ingested 2,916 mg of aspirin. Let's assume that the conversion from first-order kinetics to zero-order kinetics happens at time $t_1 = 1/2$ hr. The constants $k_1 = 1.38$ and $k_2 = 0.231$ are as before. From (5.75) we can compute

$$b_1(t_1) = b_1(1/2) = \frac{(1.38)(2,916)}{1.38 - 0.231}(e^{-(0.231)(1/2)} - e^{-(1.38)(1/2)}) \approx 1{,}363.$$

Straightforward computation using (5.86) and (5.87) now gives the (approximate) values of c_1 and c_2:

$$c_1 = 2{,}983 \quad \text{and} \quad c_2 = -2{,}916.$$

Using these numerical values in (5.82) gives

$$b_2(t) = 2{,}983 - 2{,}916 e^{-1.38 t} - 315 t.$$

What is the peak value of $b_2(t)$? It occurs when $b_2'(t) = 0$, that is, when

$$e^{-1.38t} = \frac{315}{(2{,}916)(1.38)}.$$

This gives $t = 1.85$ hr, and the corresponding maximum value of b_2 is then

$$b_2(1.85) = 2{,}983 - 2{,}916 e^{-(1.38)(1.85)} - 315(1.85) \approx 2{,}173,$$

for a somewhat higher concentration of $2{,}173/1.5 = 1{,}448$ mg/l than we obtained with our first model. In Fig. 5.6 we show the graph of $b_1, 0 \leq t \leq 1/2$, and $b_2(t), t \geq 1/2$, and compare this with our original solution for $b(t), t \geq 0$, obtained in the first model. By design, these graphs coincide for $0 \leq t \leq 1/2$ but diverge thereafter, with b_2 peaking later and at a larger value.

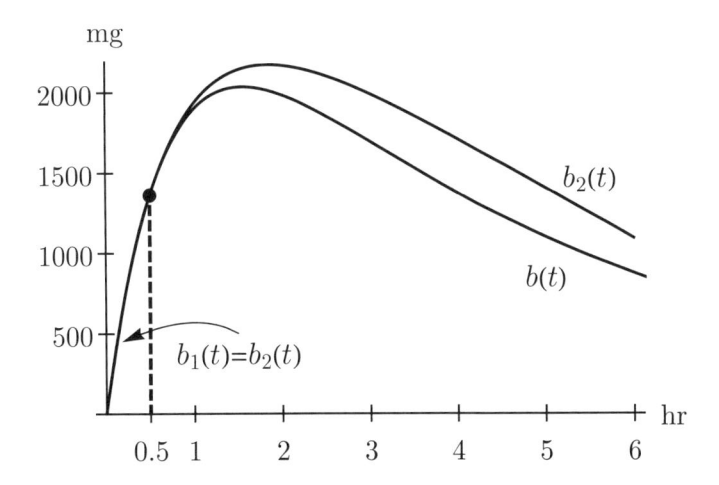

Figure 5.6. Blood levels in the two models.

No antidote exists for aspirin poisoning, but various intervention strategies are available when a toxic or lethal dose has been ingested. One of these, which has the effect of increasing the value of k_2, is discussed in Exercise 17. Cats and dogs react rather differently than humans to aspirin; Exercise 14 discusses the fact that even a single 325-mg aspirin ("adult" tablet size) can be fatal for a cat.

5.3.3. Exercises.

1. Do the solutions to $y'' + 4y = 5\cos(2t)$ exhibit resonance? Answer the same question for the equations $y'' + 4y = \sin(2t)$ and $y'' + 4y = -\cos\frac{t}{2}$.

2. If the solutions to $y'' + ay = g(t)$ exhibit resonance, will the solutions to $y'' + ay = -2g(t)$ do so also?

3. For which of the following equations would the solutions most clearly exhibit beats?
   (a) $y'' + 5y = \cos t$.
   (b) $y'' + y = \cos(4t)$.
   (c) $y'' + 6y = \cos(\frac{5}{2}t)$.

4. Do the solutions to $2y'' + 4y = \cos(\sqrt{2}\,t)$ display beats, resonance, or neither?

5. Do any of the solutions to $my'' + my = 5\sin t$ exhibit resonance? (Assume m is positive). Do all of them?

6. (a) A forced mass-spring system is modeled by the equation
$$y'' + 8y' + 16y = 2t.$$
Show that all solutions (regardless of the initial conditions) are unbounded as $t \to \infty$. This is not resonance, since the forcing term $2t$ is also unbounded as $t \to \infty$.
   (b) Repeat (a) for the equation
$$\frac{1}{2}y'' + y' + 5y = 2t.$$
   (c) True or false? If m, c, and k are positive, then all solutions to
$$my'' + cy' + ky = 2t$$
   are unbounded as $t \to \infty$.

7. A mass-spring system has an attached mass of 1 kg, a spring constant of 1 kg/sec^2, and a damping constant of 2 kg/sec. The system is initially hanging at rest and is then subjected to a constant force of 20 N. Determine the subsequent displacement function.

8. A mass-spring system has the property that the spring constant k is 4 times the mass m, and the damping constant c is 2 times the mass m. At $t = 0$ the mass, which is hanging at rest, is acted on by an external force $F(t) = 3\cos(2t)$. Find the displacement function.

9. In this problem you will show how to obtain
$$y_p(t) = \frac{F_0}{2m\omega_0} t \sin(\omega_0 t)$$
as a particular solution to $my'' + ky = F_0 \cos(\omega_0 t)$ when $\omega_0 = \sqrt{k/m}$.
   (a) Show that $F_0 \cos(\omega_0 t)$ has annihilating operator $D^2 + \omega_0^2$.
   (b) Keeping in mind that $my'' + ky = 0$ has solution $c_1 \cos(\omega_0 t) + c_2 \sin(\omega_0 t)$ for $\omega_0 = \sqrt{k/m}$, show that the "demand step" in solving $my'' + ky = F_0 \cos(\omega_0 t)$ using annihilating operators is
$$L[c_3 t \cos(\omega_0 t) + c_4 t \sin(\omega_0 t)] = F_0 \cos(\omega_0 t)$$
where $L[y] = my'' + ky$.
   (c) Determine c_3 and c_4 so that the demand step in (b) is satisfied.

10. In Chapter 2 we looked at population models based on first-order differential equations. For some scenarios, a second-order equation may be more appropriate.

(a) Suppose an explosion in population allows the population to temporarily exceed its maximal sustainable size. If the population is governed by the equation

$$\frac{d^2P}{dt^2} + \frac{dP}{dt} + \frac{5}{2}P = \frac{5}{2}P_{\max},$$

where $P_{\max}$ is a positive *constant*, $P(0) = \frac{2}{5}P_{\max}$, and $P'(0) = \frac{1}{10}P_{\max}$, solve for $P(t)$.
(b) Find the first time $t > 0$ for which $P(t) \geq P_{\max}$.
(c) What is $\lim_{t \to \infty} P(t)$?
(d) (CAS) Sketch the graph for t vs. $P(t)$ for $t > 0$.

11. A singer breaking a wine glass with a note sung loud enough at the right frequency is another example of the phenomenon of resonance. Actual, documented instances of this being done are rather hard to find. There was a Memorex commercial that ran in the 1970s in which Ella Fitzgerald was shown shattering a glass with her voice, but her voice was amplified through a speaker. More recently, the Discovery Channel, as part of the MythBusters program, enlisted the rock singer Jaime Vendera to try it without amplification. After a dozen tries he was successful—producing the right note at about 105 decibels. If the frequency of the wine glass was 556 cycles per second, what note should Vendera sing? (You may be able see Vendera doing this on a YouTube video.) You'll need to know some facts:
   - To compare the frequency of any two notes on the piano, use

   (frequency of note 1) = (frequency of note 2) $\times 2^{N/12}$

   where N is the number of half-steps between the two notes (N can be positive or negative).
   - The half-steps on the piano are C, C$^\sharp$, D, D$^\sharp$, E, F, F$^\sharp$, G, G$^\sharp$, A, A$^\sharp$, B, and back to C; these are the keys (white and black) in order, starting from C and moving up.
   - The frequency of middle C is 262 cycles per second.

12. In the piano/tuning fork example of beats discussed in Section 5.3.1, is the following statement true or false: The slower the beats, the more nearly the piano string is in tune. Explain.

13. In this problem you'll solve the initial value problem

(5.88) $$\frac{dg}{dt} = -k_1 g,$$

(5.89) $$\frac{db}{dt} = k_1 g - k_2 b,$$

$g(0) = g_0$, $b(0) = 0$ of Section 5.3.2 in a different way. Of course your answer should agree with that in (5.75) of the text.
   (a) Begin by solving (5.88) for $g(t)$ with initial condition $g(0) = g_0$.
   (b) Next, substitute your formula for $g(t)$ from (a) into (5.89), and then solve the resulting first-order equation for $b(t)$, using the initial condition $b(0) = 0$.

14. Cats metabolize aspirin much more slowly than humans (and dogs). The biological half-life of aspirin in a cat is about 38 hours; this translates into an estimated value of $k_2 = 0.018$/hr in our first-order kinetics model of Section 5.3.2. Assume $k_1 = 1.8$/hr. The Merck Veterinary Manual states that a single 325-mg aspirin tablet can prove fatal in a cat. Assuming that a cat on average has a blood volume of 140 ml and that the maximal concentration in the blood produced by ingestion of 325 mg of aspirin is a lethal level, determine this lethal concentration. How quickly is the maximum concentration reached?

15. Applying the first-order kinetics model of Section 5.3.2 to a dog ingesting one 325-mg adult aspirin, determine the maximum blood level of aspirin if $k_1 = 0.9$/hr and $k_2 = 0.09$/hr.

16. Suppose you take a standard dose of two 325-mg aspirin tablets. Use the first-order kinetics model of Section 5.3.2 and the values of k_1 and k_2 there. When does your peak blood level occur, and what is that level? If you have a blood volume of 5 liters, what is your peak concentration?

17. Activated charcoal is used as a treatment for many kinds of drug overdose. The charcoal acts to increase the elimination constant k_2 in our compartment model. Suppose that one hour after ingesting thirty-six 81-mg baby aspirin tablets a child is given a dose of activated charcoal, which acts to increase the value of k_2 from 0.231 to 0.4. What is the maximum concentration of aspirin in the blood for this child if his blood volume is 1.5 l? Use a first-order kinetics model (equations (5.71) and (5.72) in the text) and assume $k_1 = 1.38$.

18. In this problem we show how to convert the pair of first-order equations

$$(5.90) \qquad \frac{dx}{dt} = ax + by + f(t),$$

$$(5.91) \qquad \frac{dy}{dt} = cx + dy$$

to a second-order linear equation. Here, $a, b, c,$ and d are constants, and f is a differentiable function of t.

(a) Solve equation (5.91) for x to obtain $x = \frac{1}{c}y' - \frac{d}{c}y$, where primes denote derivatives with respect to t. Substitute this (on both sides) into equation (5.90) and show the result can be simplified to obtain

$$y'' - (a+d)y' + (ad - bc)y = cf(t).$$

Once this second-order equation is solved for y, x can then be determined by going back to the equation $x = \frac{1}{c}y' - \frac{d}{c}y$.

(b) Using your work in (a), solve

$$(5.92) \qquad \frac{dx}{dt} = x + y + e^{2t},$$

$$(5.93) \qquad \frac{dy}{dt} = 4x + y.$$

Solve for $y(t)$ first, as outlined in (a), and then obtain $x(t)$.

19. When a drug is given by intravenous injection, all of it is initially present in the bloodstream (though not uniformly distributed; we ignore this fact here). In this case, a two-compartment model is sometimes still used to study the pharmacokinetics of the drug, with the two compartments being the bloodstream and the "tissues" as shown below A bit more precisely, these compartments are sometimes called (1) the central compartment, composed of the blood and certain organs like the liver and kidneys, and (2) the peripheral compartment (other tissues). Unlike our GI/bloodstream scenario, the drug can pass in both directions between the two compartments, as indicated by the double arrows in the figure below. The purpose of this problem is to model the amount of the drug in each compartment under two different assumptions. We denote the amount of drug in the tissues at time t by $x(t)$, and the amount in the blood by $b(t)$.

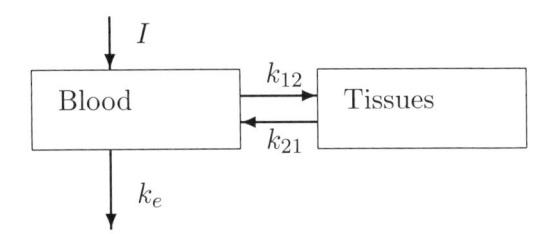

```
        │ I
        ▼
┌─────────┐   k₁₂   ┌─────────┐
│  Blood  │ ──────▶ │ Tissues │
│         │ ◀────── │         │
└─────────┘   k₂₁   └─────────┘
     │ kₑ
     ▼
```

(a) First suppose that the injection into the bloodstream occurs all at once ("an IV bolus injection"), so that at time $t = 0$ we have $b(0) = b_0$, the amount of drug injected, and $x(0) = 0$. Assume the drug passes from the blood to the tissues at a rate proportional to the amount in the blood with proportionality constant k_{12} and it passes from the tissues to the blood at a rate proportional to the amount in the tissues with proportionality constant k_{21}. It's eliminated from the body by passing from the blood at a rate proportional to the amount in the blood, with proportionality constant k_e. Set up a pair of two first-order equations for the amount $b(t)$ in the blood at time t, and the amount $x(t)$ in the tissues at time t. Convert this system to a single second-order equation for $b(t)$. What are the initial conditions for this second-order equation?

(b) Suppose instead that the drug is infused into the bloodstream at a constant rate I for the entire period of time under study. How does this change the equations you wrote in (a)? Again, convert your system of first-order equations to a second-order equation for $b(t)$. What are the initial conditions for this second-order equation?

20. Clonidine, a treatment for high blood pressure, is one of a number of drugs where ingestion of a single tablet by a toddler can cause a significant toxic effect (for clonidine, the effects include altered consciousness, low blood pressure, respiratory collapse, coma). Suppose a one year old, with a blood volume of 1 liter, ingests a single 0.3-mg tablet. Assume equations (5.71) and (5.72) apply to model the level of clonidine in the blood. To determine approximate values of k_1 and k_2, use the following facts: After 30 minutes, 75% of the drug has been absorbed from the GI tract. The half-life for elimination of the drug from the bloodstream is long, approximately 16 hours. Determine the peak blood concentration of clonidine in this child.

5.4. Electric circuits

Second-order differential equations provide exactly the right model for current flow in simple electric circuits.

5.4.1. Background.
An electric current is the rate of flow of electric charge past a given point in an electric circuit. Electric charges are of two kinds: positive and negative. In an electric circuit, the charge carriers are usually free electrons (which have a negative charge), positive ions (atoms that have lost one or more electrons and thus have a net positive charge), or negative ions (atoms that have acquired extra electrons). For example, current flow in a fluorescent lamp is a mixture of all three types.

What makes electric charge move to produce a current? The answer is electric potential, which is measured in volts. Imagine a voltage source such as a battery. It has two terminals, one labeled + and one labeled −. The voltage source pumps positive charges to its + terminal and negative charges to its − terminal. The positive terminal has a higher potential than the negative terminal. The potential difference between the terminals is the **voltage** of the source. If the terminals are connected by a conductor, positive charges within it will flow from the + terminal to the − terminal, and negative charges will flow backwards from the − terminal to the +. The opposite motions of positive and negative charges together make up the electric current. It is customary to think of this current as exclusively due to the flow of positive charges through the conductor. Thus an electron flowing backward is replaced in the mind's eye by a positive charge of the same magnitude flowing forward at the same rate.

Voltage sources other than batteries can produce voltages that vary with time. For example, the two terminals of a 220-volt household socket for an electric range or clothes dryer (connected to an alternating current generator in the power grid) "take turns" being positively and negatively charged at a rate of 60 times per second.

Charge is measured in **coulombs**, and the current I at a point in the circuit is the rate at which positive charge flows in the "forward direction" past that point in coulombs per second. A current of 1 coulomb per second is called 1 **ampere**. A coulomb is a rather large charge, about equal to the magnitude of 6.24×10^{18} electron charges. The forward direction in the circuit can be chosen arbitrarily, but let us agree to choose it to be the "clockwise" direction in any pictured one-loop circuit; see, e.g., Fig. 5.7. Thus if the charge carriers are electrons (which have negative charge) and at a given moment they are flowing counterclockwise (backwards), we think of this as positive charge flowing clockwise (forward), so at that moment $I > 0$. On the other hand, if the electrons are flowing clockwise at a given instant, we think of a corresponding counterclockwise (backwards) flow of positive change, and so at that instant $I < 0$.

We consider here electric circuits with three simple kinds of components. In order of increasing mysteriousness, the circuit components are resistors, capacitors, and inductors. These components will be connected with a metallic conductor like a copper wire. In such a conductor, all of the charge carriers are electrons, so by our convention mentioned above, the direction of the current is opposite to that of the electron flow.

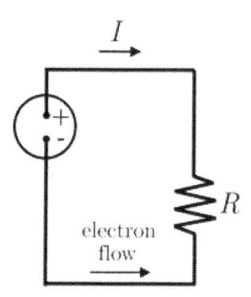

Figure 5.7. Here the current is positive.

Resistors. A resistor is a device inserted into a circuit that slows down the flow of charge. It is a conductor, but not as good a conductor as the wire itself. Its *resistance* R (in ohms) is a numerical measure of its ability to slow the flow of charge.

Fig. 5.7 shows a circuit containing a battery and a resistor with resistance R. The voltage V produced by the battery (which is the difference between the electric potential at the + terminal and the electric potential at the − terminal) is related to the resistance R and the current I by **Ohm's law:**

$$V = IR.$$

Thus, for example, if you double the voltage of the battery, the current through the resistor also doubles. If you double the value of the resistance R, the current will be cut in half if the battery voltage stays the same.

5.4. Electric circuits

We can make an analogy with water flowing through a pipe. Think of the battery as the analogue of two large reservoirs filled to different heights with water. The pressure at the outlet of a reservoir is proportional to the water depth. The pressure difference between the two reservoirs is analogous to the voltage of the battery. The water flows from the high pressure (deeper reservoir) to the low pressure (shallow reservoir) through the pipe. The wide part of the pipe corresponds to the circuit wire, and the narrower section of pipe to a resistor; see Fig. 5.8.

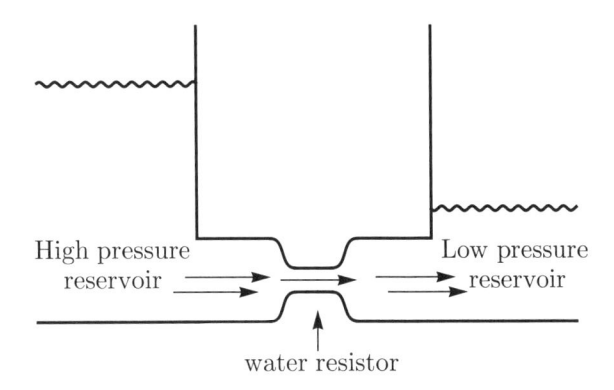

Figure 5.8. Water analogues of battery, wire, and resistor.

Capacitors. A capacitor is a device that will store charge when inserted in a circuit. We'll first describe the water-flow analogy. Imagine a pair of empty identical cylindrical water tanks as depicted in Fig. 5.9. We install each in a pipe to its respective reservoir as shown and let them fill with water. The water capacitor is now charged. If we disconnect it from the reservoirs and cap it off, it is still charged; see Fig. 5.11. The pressure difference between our two tanks is the same as the pressure difference between the two reservoirs. If the caps are removed and the tanks are connected by a pipe full of water, water will flow through the pipe from high pressure to low until the water depths have equalized; see Fig. 5.12. We then say the water capacitor has discharged through the pipe. The difference between the water capacitor and the pair of reservoirs is that the reservoirs are so large that their depths (and thus their pressures) don't noticeably change as water flows through a pipe from the deep to shallow reservoir.

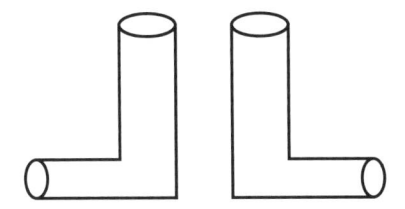

Figure 5.9. Water capacitor (a).

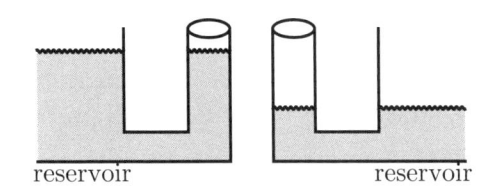

Figure 5.10. Water capacitor (b).

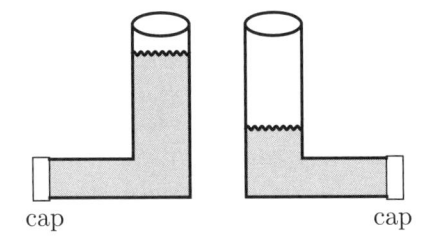

Figure 5.11. Water capacitor (c).

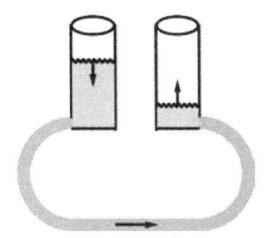

Figure 5.12. Water capacitor (d).

The electric version of a capacitor consists of two parallel metal plates set very close together. When the two plates are connected to separate terminals of a battery (see Fig. 5.13), positive charges accumulate on the "positive plate" (the plate connected to the positive terminal of the battery) and negative charges on the "negative plate". The capacitor is now charged and remains so if disconnected from the battery. (The positive and negative charges in close proximity attract each other and thus remain on the plates.) If we then connect the two terminals of the capacitor by a wire, the excess electrons on the negative plate flow through the wire to fill in the electron deficiency on the positive plate, discharging the capacitor. But remember that the current flow, by convention, is in the opposite direction.

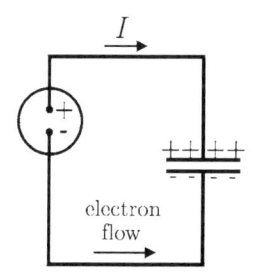

Figure 5.13. A capacitor connected to a battery.

Consider a (possibly charged) capacitor as pictured in Fig. 5.14. If the electric potentials at points a and b are V_a and V_b volts, respectively, the "voltage drop" across the capacitor is

$$V_{ab} = V_a - V_b.$$

If the left-hand plate is positively charged, and the right negatively charged, then $V_{ab} > 0$. If the positive charges are on the right, $V_{ab} < 0$. The charge Q on the left-hand plate is proportional to the voltage drop V_{ab}:

$$Q = CV_{ab}.$$

(By identifying the left-hand plate as where Q resides, we have implicitly decided that the "forward direction" of the circuit is from a to b.) The positive constant C is called the *capacitance*; its units are coulombs per volt or **farads**. One farad is a very large capacitance. Accordingly, in the real world capacitors are often rated in *microfarads*. One microfarad is 10^{-6} farads. In the water analogy, the "water capacitance" is proportional to the area of the base of either cylindrical tank.

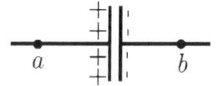

Figure 5.14. A charged capacitor.

Figure 5.15. An inductor.

Inductors. You can make an inductor simply by winding a wire into a coil. As current flows through the coil, it produces a magnetic field which opposes any *change* in the flow of current. The water analogy for an inductor is a frictionless paddle wheel partly immersed in the water flow and attached to a massive flywheel; see Fig. 5.16. The mass of the flywheel imparts inertia. It wants to keep turning freely at the same rate, so if the flow speeds up it resists, creating back-pressure until it is brought up to speed, while if the flow slows, it also resists (creating increased pressure on the front side of the wheel) until it has slowed enough to again match the flow rate in the pipe.

5.4. Electric circuits

Though an aid to understanding, the water analogy is not perfect. For example, even without the paddle wheel the water itself imparts inertia to the flow due to its considerable mass, a factor not present in electron flow.

An inductor produces a drop in voltage that is proportional to the rate of change of the current:

$$V_{ab} = L\frac{dI}{dt},$$

where I is the current and the positive constant L is called the inductance. Here the current I is the rate of flow of positive charge *from a to b*. The units on the constant L are volt-sec/ampere, or **henries**. In the electric circuit, the inductance plays the role of the inertia of the flywheel. Notice that if I is constant, there is no voltage drop across the inductor.

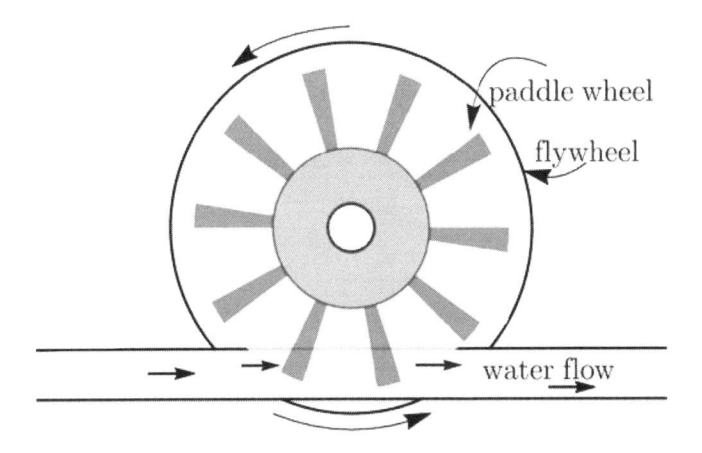

Figure 5.16. A water inductor.

5.4.2. One-loop circuits. Now consider a circuit containing all of these elements plus a battery or other voltage source as shown in Fig. 5.17. Kirchhoff's laws govern the current in, and distribution of potential around, the pictured circuit.

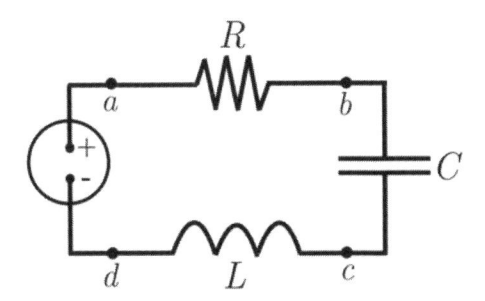

Figure 5.17. A one-loop RLC circuit.

Kirchhoff's voltage law: The sum of the voltage drops around a closed loop circuit is zero.

Thus if V_{ab}, V_{bc}, V_{cd}, and V_{da} denote the voltage drops from points a to b, b to c, and so on, then

(5.94) $$V_{ab} + V_{bc} + V_{cd} + V_{da} = 0.$$

Since $V_{da} = V_d - V_a = -(V_a - V_d) = -V$ where $V = V_a - V_d$ is the voltage of the battery, we may rearrange equation (5.94) to read:

(5.95) $$V_{ab} + V_{bc} + V_{cd} = V.$$

Note the preceding equation remains valid if the battery in Fig. 5.17 is replaced by some other, possibly variable, voltage source, for which V might be negative at times (or at all times). The water analogue for equation (5.95) would be "the sum of the pressure drops along a pipe (containing various 'water circuit elements') is equal to the pressure difference between the deep and shallow reservoirs". There is nothing mysterious about equation (5.94); it is an immediate consequence of the identities $V_{ab} = V_a - V_b, V_{bc} = V_b - V_c, V_{cd} = V_c - V_d$, and $V_{da} = V_d - V_a$. Copper wire itself has some resistance, but it is small and we are neglecting it here. This means that if b and b' are any two points on the wire between the resistor and the capacitor, then $V_b = V_b'$. In other words, V_b is independent of the placement of the point b on the wire between the resistor and capacitor, and analogously for V_a, V_c, and V_d.

> **Kirchhoff's current law for one-loop circuits.** The current is the same at any point in a closed loop circuit.

The water analogue of this fact is that the flow rate of water (say, in gallons per minute) through a single pipe is the same at every point in the pipe; that is, water cannot be compressed.

Let's see what these two laws and equation (5.95) have to say about the circuit. By Ohm's law the voltage drop across the resistor is $V_{ab} = RI$. The voltage drops across the capacitor and inductor are given by

$$V_{bc} = \frac{1}{C}Q \quad \text{and} \quad V_{cd} = L\frac{dI}{dt},$$

respectively. Thus equation (5.95) becomes

(5.96) $$RI + \frac{1}{C}Q + L\frac{dI}{dt} = V(t).$$

Here we have written $V(t)$ to indicate that the input voltage V may be a function of time.

The derivative $\frac{dQ}{dt}$ is the rate at which positive charge flows onto the left-hand capacitor plate, which, by Kirchoff's current law, is the same as the rate of flow of charge past any point in the loop. In other words,

$$\frac{dQ}{dt} = I.$$

If we use this fact and differentiate both sides of equation (5.96), we may rearrange the resulting

5.4. Electric circuits

equation to obtain

$$(5.97) \qquad L\frac{d^2I}{dt^2} + R\frac{dI}{dt} + \frac{1}{C}I = V'(t).$$

This is a second-order linear constant coefficient differential equation having the same form as the differential equation for the motion in a mass-spring system:

$$my'' + cy' + ky = F(t).$$

The coefficient L corresponds to the mass m, while R corresponds to the damping constant c and $\frac{1}{C}$ corresponds to the spring constant k.

Example 5.4.1. Let's consider a circuit with no inductor (an RC circuit). Suppose that at time $t = 0$ there is no charge on the capacitor and the voltage is a constant $V(t) = V_0$ volts. How much charge is on the capacitor at time $t > 0$?

In equation (5.96) we have $L = 0$ (no inductor), and since $I = \frac{dQ}{dt}$,

$$R\frac{dQ}{dt} + \frac{1}{C}Q = V_0.$$

This first-order linear equation is easily solved to obtain

$$Q(t) = CV_0 + ae^{-\frac{1}{RC}t}$$

where a is an arbitrary constant. Using the initial condition $Q(0) = 0$ we see that $a = -CV_0$ and

$$Q(t) = CV_0\left(1 - e^{-\frac{1}{RC}t}\right).$$

The graph of $Q(t)$ is shown in Fig. 5.18; it approaches the horizontal asymptote $Q = CV_0$ as $t \to \infty$ (corresponding, as we know, to the full charge CV_0).

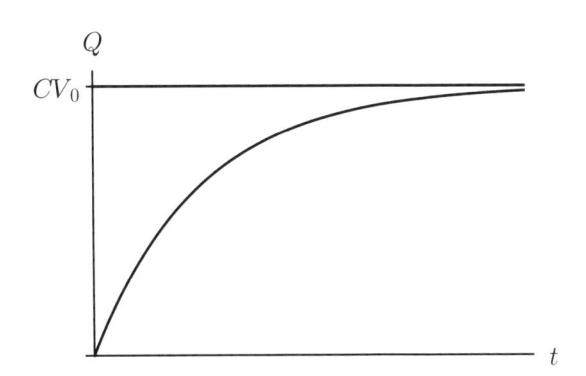

Figure 5.18. Charging the capacitor.

How long does it take for this capacitor to become 90% charged? To answer this we solve

$$Q(t) = 0.9CV_0,$$

that is

$$\left(1 - e^{-\frac{1}{RC}t}\right) = 0.9,$$

for time t. This shows that it is 90% charged after $t = (\ln 10)RC$ seconds. Note that if R is very small, so is t. In the case there is no resistance ($R = 0$), the capacitor reaches full charge instantly, but this doesn't happen in real life, since all actual circuit components (even wire) have some resistance.

Example 5.4.2. Now consider the full RLC circuit as in Fig. 5.17 above. Suppose that the voltage source supplies a periodically varying voltage given by

$$V(t) = V_0 \sin(\omega t).$$

We'll assume $\omega > 0$. Notice that $V(t)$ has frequency $\frac{\omega}{2\pi}$ cycles per second. Recall that a common synonym for "cycles per second" is "hertz", which is abbreviated Hz. Here our voltage oscillates between positive and negative values, so the two terminals of the voltage source take turns being positive and negative. The governing differential equation (5.97) becomes

(5.98) $$L\frac{d^2 I}{dt^2} + R\frac{dI}{dt} + \frac{1}{C}I = \omega V_0 \cos(\omega t).$$

The complementary equation

(5.99) $$L\frac{d^2 I}{dt^2} + R\frac{dI}{dt} + \frac{1}{C}I = 0$$

has characteristic equation

$$Lz^2 + Rz + \frac{1}{C} = 0$$

with roots

$$r_1 = -\frac{R}{2L} + \frac{\sqrt{R^2 - \frac{4L}{C}}}{2L}, \quad r_2 = -\frac{R}{2L} - \frac{\sqrt{R^2 - \frac{4L}{C}}}{2L}.$$

The case where $R^2 < \frac{4L}{C}$ is the most interesting. In this situation the roots of the characteristic equation are a pair of complex conjugates

$$r_1 = \lambda + i\mu, \quad r_2 = \lambda - i\mu$$

where

$$\lambda = -\frac{R}{2L} \quad \text{and} \quad \mu = \frac{\sqrt{\frac{4L}{C} - R^2}}{2L}.$$

This tells us that

$$I_1(t) = e^{\lambda t} \cos(\mu t) \quad \text{and} \quad I_2(t) = e^{\lambda t} \sin(\mu t)$$

form a fundamental set of solutions of the complementary equation. If we can find a particular solution $I_p(t)$ of (5.98), then the general solution to (5.98) will be

$$I(t) = c_1 I_1(t) + c_2 I_2(t) + I_p(t).$$

Since $\lambda < 0$, the complementary solution $c_1 I_1(t) + c_2 I_2(t)$ tends to 0 as $t \to \infty$ (the transient solution). Of the possible particular solutions, the one of greatest interest to us is the steady-state solution $I_s(t)$, which we can find using an appropriate annihilating operator. As in Section 5.2.1, the computations will be easiest if we replace the forcing term $\omega V_0 \cos(\omega t)$ in (5.98) by its complex analogue

$$\omega V_0 e^{i\omega t}$$

and use the complex annihilating operator $A = D - i\omega$. Write T for the linear differential operator defined by

$$T[I] = L\frac{d^2 I}{dt^2} + R\frac{dI}{dt} + \frac{1}{C}I,$$

so that the complexified version of (5.98) looks like

(5.100) $$T[I] = \omega V_0 e^{i\omega t}.$$

Applying our annihilating operator A to both sides of (5.100) gives

$$AT[I] = A[\omega V_0 e^{i\omega t}] = 0,$$

5.4. Electric circuits

or

(5.101) $$(D - i\omega)\left(LD^2 + RD + \frac{1}{C}\right)[I] = 0.$$

This homogeneous equation has characteristic equation

$$(z - i\omega)\left(Lz^2 + Rz + \frac{1}{C}\right) = 0,$$

which has three distinct roots, $i\omega$, r_1, and r_2. The corresponding complex-valued solutions to (5.101) are $e^{r_1 t}, e^{r_2 t}$, and $e^{i\omega t}$, so that our solution I to $AT[I] = 0$ looks like $I(t) = c_1 e^{r_1 t} + c_2 e^{r_2 t} + c_3 e^{i\omega t}$. Since the function $y = c_1 e^{r_1 t} + c_2 e^{r_2 t}$ solves $T[y] = 0$, we need only demand that

$$T[c_3 e^{i\omega t}] = \omega V_0 e^{i\omega t},$$

or

$$\left(LD^2 + RD + \frac{1}{C}\right)[c_3 e^{i\omega t}] = \omega V_0 e^{i\omega t}.$$

On computing on the left-hand side, we find

$$c_3\left(-L\omega^2 + R\omega i + \frac{1}{C}\right)e^{i\omega t} = \omega V_0 e^{i\omega t}$$

so

$$c_3 = \frac{\omega V_0}{\frac{1}{C} - \omega^2 L + R\omega i}.$$

We write the denominator in this expression for c_3 in polar form

$$\frac{1}{C} - L\omega^2 + R\omega i = re^{i\delta}$$

with

$$r = \sqrt{\left(\frac{1}{C} - L\omega^2\right)^2 + R^2\omega^2}$$

and δ satisfying

(5.102) $$\sin\delta = \frac{R\omega}{\sqrt{(\frac{1}{C} - L\omega^2)^2 + R^2\omega^2}}, \quad \cos\delta = \frac{\frac{1}{C} - L\omega^2}{\sqrt{(\frac{1}{C} - L\omega^2)^2 + R^2\omega^2}}.$$

This says

$$c_3 = \frac{\omega V_0}{re^{i\delta}} = \frac{\omega V_0}{r}e^{-i\delta}.$$

Thus equation (5.100) has the particular complex-valued solution

$$c_3 e^{i\omega t} = \frac{\omega V_0}{r}e^{-i\delta}e^{i\omega t} = \frac{\omega V_0}{\sqrt{\left(\frac{1}{C} - L\omega^2\right)^2 + R^2\omega^2}}e^{i(\omega t - \delta)}.$$

To get our real-valued steady-state solution $I_s(t)$ to (5.98) we take the real part of the complex-valued particular solution of (5.100):

(5.103) $$I_s(t) = \frac{\omega V_0}{\sqrt{\left(\frac{1}{C} - L\omega^2\right)^2 + R^2\omega^2}}\cos(\omega t - \delta).$$

Keep in mind that δ depends on ω. It's easy to see that the amplitude of $I_s(t)$ is a maximum when $\omega^2 = \frac{1}{LC}$. In this case you can check that $\delta = \frac{\pi}{2}$ (see Exercise 2) and

(5.104) $$I_s(t) = \frac{V_0}{R}\cos\left(\omega t - \frac{\pi}{2}\right).$$

Example 5.4.3. Tuning a circuit. Suppose that the input voltage $V(t)$ of our RLC circuit is a sum of several sinusoidal signals with different frequencies. We would like to "tune" the circuit to respond to one of the frequencies while in effect ignoring the others. One common way to do this is to choose fixed values for the resistance R and inductance L and vary the value of the capacitance C.

Suppose the circuit in Fig. 5.19 has inductance, resistance, and capacitance of 2 henries, 0.1 ohms, and C farads, respectively. Suppose that the input voltage is

$$V(t) = 3\sin(2t) + \sin(7t),$$

a linear combination of two sine functions with frequencies $\frac{2}{2\pi}$ and $\frac{7}{2\pi}$.

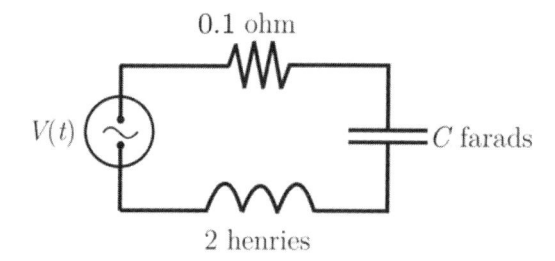

Figure 5.19. Circuit with a variable capacitor.

The circuit equation for current is

$$T[I] = 2\frac{d^2I}{dt^2} + \frac{1}{10}\frac{dI}{dt} + \frac{1}{C}I = V'(t) = 6\cos(2t) + 7\cos(7t).$$

We are interested in the steady-state solution I_s of this equation. If I_1 and I_2 are the steady-state solutions of

$$T[I] = 6\cos(2t)$$

and

$$T[I] = 7\cos(7t),$$

respectively, then

$$I_s = I_1 + I_2.$$

From equation (5.103) we have

$$I_1(t) = \frac{6}{\sqrt{(\frac{1}{C} - 8)^2 + (\frac{1}{5})^2}} \cos(2t - \delta_1)$$

and

$$I_2(t) = \frac{7}{\sqrt{(\frac{1}{C} - 98)^2 + (\frac{7}{10})^2}} \cos(7t - \delta_2),$$

where δ_1 and δ_2 are the appropriate phase shifts for I_1 and I_2, as given in equations (5.102). The phase shifts both depend on the choice of C.

To tune in the signal with frequency $\frac{2}{2\pi}$ Hz., choose C to maximize the amplitude of I_1. This means we choose $\frac{1}{C} = 8$, and with this choice,

$$I_1(t) = 30\cos\left(2t - \frac{\pi}{2}\right)$$

5.4. Electric circuits

and
$$I_2(t) = \frac{7}{\sqrt{(90)^2 + (\frac{7}{10})^2}} \cos(7t - \delta_2)$$

where $\delta_2 \approx 0.0077$ (computed from (5.102) with $R = 0.1, \omega = 7, V_0 = 1, L = 2$, and $C = \frac{1}{8}$). The amplitude of I_2 is less than $\frac{7}{90}$ while the amplitude of I_1 is 30, more than 385 times larger than that of I_2. Since $I_s = I_1 + I_2$ and I_1 is more than 385 times as strong as I_2, we have essentially eliminated I_2 and successfully tuned in the input $3\sin(2t)$. Figs. 5.20 and 5.21 show the voltage $V(t) = 3\sin(2t) + \sin(7t)$ and the steady-state solution I_s for the choice $\frac{1}{C} = 8$.

To tune in the input $\sin(7t)$ instead, we choose $\frac{1}{C} = 98$ to maximize the amplitude of I_2. With this choice of C we have
$$I_1(t) = \frac{6}{\sqrt{(90)^2 + (\frac{1}{5})^2}} \cos(2t - \delta_1)$$

with $\delta_1 \approx 0.0022$ and
$$I_2(t) = 10 \cos\left(7t - \frac{\pi}{2}\right).$$

Since the amplitude of I_1 is less than
$$\frac{6}{\sqrt{90^2}} = \frac{1}{15}$$

and the amplitude of I_2 is 10, I_2 is more than 150 times as strong as I_1, and we have successfully tuned in the effect of the input $\sin(7t)$ and tuned out the effect of $3\sin(2t)$. Fig. 5.22 shows the steady-state solution I_s for this choice of C.

We picked the numbers in this example to be illustrative rather than realistic. The reader familiar with circuits may note that the frequencies $\frac{2}{2\pi} = 0.3183$ Hz and $\frac{7}{2\pi} = 1.114$ Hz are extremely low and the capacitances $\frac{1}{8}$ farad and $\frac{1}{98}$ farad are impractically high. For a related example, with more realistic data, see Exercise 10.

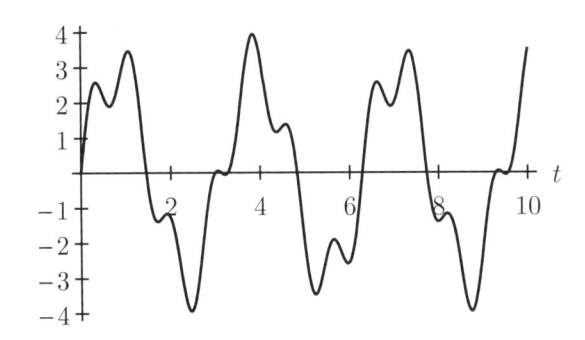

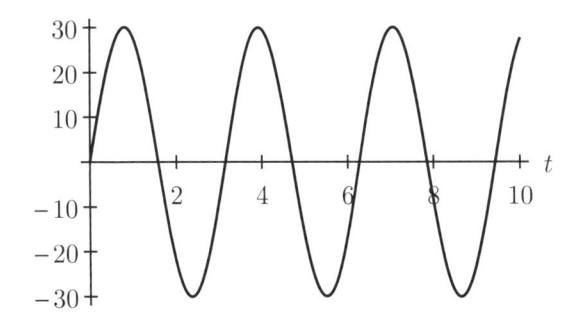

Figure 5.20. $V(t) = 3\sin(2t) + \sin(7t)$.

Figure 5.21. I_s for $1/C = 8$.

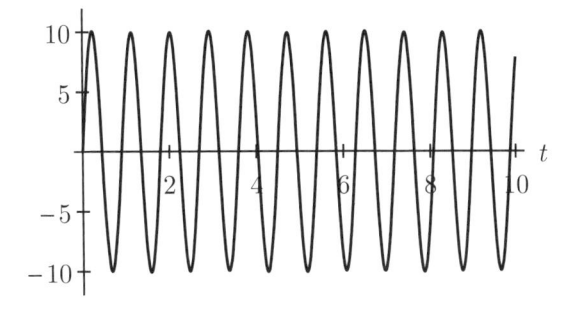

Figure 5.22. I_s for $1/C = 98$.

5.4.3. Exercises. The table below summarizes the units which are used in the exercises.

Table of Units

	current I	capacitance C	voltage V	inductance L	charge Q	resistance R
units	amperes	farads	volts	henries	coulombs	ohms

1. (a) Show that equation (5.96) can be rewritten as a second-order equation in which the unknown function is the charge $Q(t)$ on the capacitor, namely

$$a\frac{d^2Q}{dt^2} + b\frac{dQ}{dt} + cQ = V(t).$$

   What are a, b, and c?

   (b) Suppose that a one-loop circuit has no resistance and input voltage $V(t) = V_0 \sin(\omega t)$. Show that if $\omega^2 = \frac{1}{LC}$ where L and C are the inductance and capacitance, respectively, then oscillations of the charge $Q(t)$ on the capacitor grow increasingly large as $t \to \infty$. This is the phenomenon of resonance.

2. In Example 5.4.2 we claimed that the steady-state current I_s in an RLC circuit with input voltage $V_0 \sin(\omega t)$ has maximum amplitude when $\omega = (LC)^{-\frac{1}{2}}$.

   (a) Show that this is in fact true and that in this case

$$I(t) = \frac{V_0}{R}\cos(\omega t - \delta)$$

   with $\delta = \frac{\pi}{2}$, verifying equation (5.104).

   (b) Suppose you now apply the same input voltage to an "R-only" circuit as pictured in Fig. 5.23. It has the same value of R, but no inductor ($L = 0$) and no capacitor ($C = \infty$, or $\frac{1}{C} = 0$). Find the current as a function of time. How does it differ (if at all) from the steady-state current in the RLC circuit when $\omega = (LC)^{-\frac{1}{2}}$?

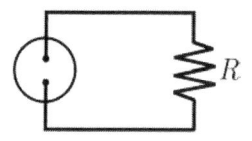

Figure 5.23. An R-only circuit with voltage source.

3. (a) Consider equation (5.97) for an RLC circuit with input voltage $V(t)$. Initial conditions at time $t = 0$ take the form

$$I(0) = I_0, \quad I'(0) = b$$

   for some constant b. Show that b can be expressed in terms of the initial current I_0, the initial charge Q_0 on the capacitor, and the initial voltage $V(0)$. Hint: Look at equation (5.96).

   (b) At time $t = 0$ the charge on the capacitor in an RLC circuit is 10^{-4} and the current is 0. If $R = 30, C = 10^{-4}, L = 0.2$, and $V(t) = 0$ for $t \geq 0$, find the current $I(t)$ for $t > 0$.

4. An industrial machine contains a capacitor with capacitance $C = \frac{1}{2000}$ farads. During routine operation a constant 20,000 volts is applied to the terminals of the capacitor by a voltage source, keeping the capacitor charged. At time $t = 0$, the capacitor is to be taken off-line for repairs. This is done by disconnecting the voltage source and at the same instant inserting the capacitor into a circuit loop with a resistor through which it can safely discharge (see Fig. 5.24).

5.4. Electric circuits

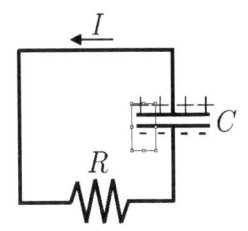

Figure 5.24. A discharge loop.

(a) The wire in the discharge loop can handle a maximum current of 50 amperes. What is the smallest value of the resistance R in the loop that will ensure the current does not exceed 50 amperes during discharge?

(b) The capacitor is considered safe to handle only after the magnitude of the voltage drop across its terminals has decreased to 20 volts. Assuming the resistance R in the discharge loop is the value you found in part (a), how long must the technician wait before starting to service the capacitor?

5. Consider an RLC circuit as in Fig. 5.17 having $R = 30$ Ohms, $L = \frac{1}{10}$ henries, $C = \frac{1}{2000}$ farads and containing a voltage source supplying $V(t) = 200\sin(100t)$ volts at time t. Suppose that at time $t = 0$ there is no current in the circuit and no charge on the capacitor. Find the current at time t for all $t > 0$.

6. In Example 5.4.2 we found the steady-state current $I_s(t)$ in an RLC circuit with input voltage $V_0 \sin(\omega t)$. Suppose we want an actual output voltage from this circuit to send to another circuit or device. There are three obvious choices: V_{ab}, V_{bc}, and V_{cd}. For example, to use V_{bc} as an output voltage, connect wires leading to another circuit to the terminals b and c as depicted in Fig. 5.25. Calculate the three possible output voltages V_{ab}, V_{bc}, and V_{cd}.

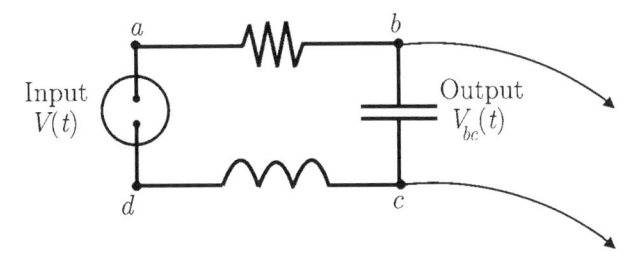

Figure 5.25. One possible output voltage.

7. (a) Convert the LC circuit equation

$$L\frac{dI}{dt} + \frac{Q}{C} = 0$$

to a pair of two first-order equations, with dependent variables Q and I, by recalling that

$$\frac{dQ}{dt} = I.$$

(b) Using your equations from (a), find a first-order differential equation for $\frac{dI}{dQ}$ and solve it. Graph some of your solutions in the QI-plane. These are called the **trajectories** of the pair of differential equations in (a). Identify these trajectories geometrically. In what direction are the trajectories traversed for increasing t?

(c) The magnetic energy stored in the inductor with inductance L is $\frac{1}{2}LI^2$ and the electric energy stored on the capacitor with capacitance C is $\frac{1}{2}\frac{Q^2}{C}$. Show that the equations of the trajectories in (b) have the "conservation of energy" form

$$\text{magnetic energy in inductor } + \text{ electric energy in capacitor } = \text{ constant.}$$

8. The pictures in Figs. 5.26–5.29 show the graphs of I vs. t and Q vs. t for two "battery-free" RLC loop circuits whose capacitors are initially charged. One corresponds to a large resistance, where $R^2 > \frac{4L}{C}$, and the other to a low resistance, where $R^2 < \frac{4L}{C}$.
   (a) Which pair of graphs corresponds to $R^2 < \frac{4L}{C}$?
   (b) For each pair of Q vs. t and I vs. t graphs, give a rough sketch of the corresponding phase plane trajectory in the QI-plane.

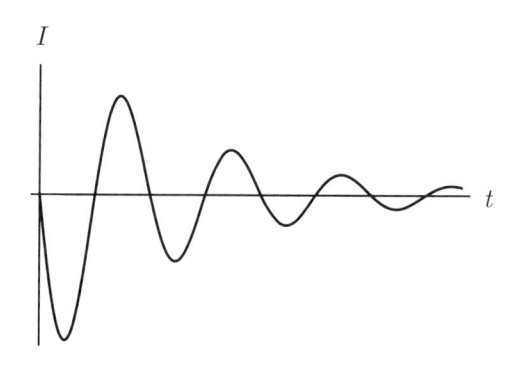

Figure 5.26.

Figure 5.27.

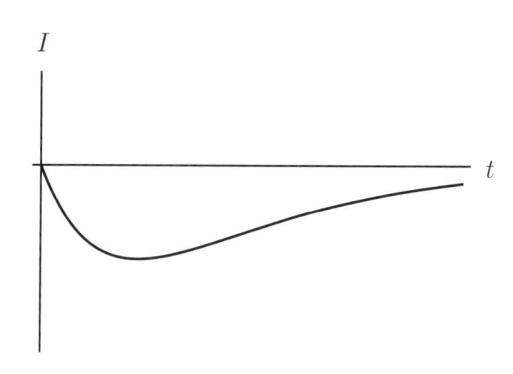

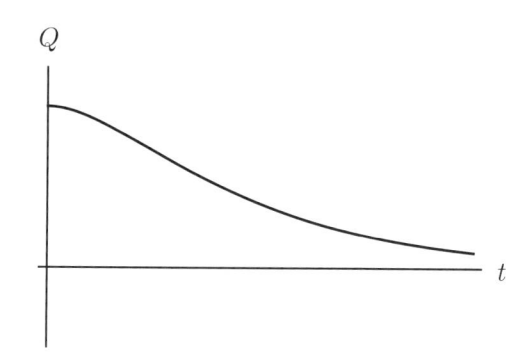

Figure 5.28.

Figure 5.29.

9. Suppose an RLC loop circuit has constant voltage source $V(t) = V_0$. Find $\lim_{t\to\infty} I(t)$ and $\lim_{t\to\infty} Q(t)$.

10. In Example 5.4.3 we saw how to tune in a selected frequency and tune out another by varying the capacitance C in an RLC circuit. If we are willing to vary both C and the inductance L, we can do better. Fix a positive number ω_0 and suppose we want to tune in the input voltage $V_0 \sin(\omega_0 t)$ and tune out all other input voltages $V_0 \sin(\omega t)$ (with $\omega > 0$) whose frequencies $\frac{\omega}{2\pi}$ are not too close to $\frac{\omega_0}{2\pi}$. For simplicity we assume all input voltages have the same amplitude V_0. Let

$$A(\omega) = \frac{\omega V_0}{\sqrt{(\frac{1}{C} - L\omega^2)^2 + R^2\omega^2}}$$

5.4. Electric circuits

be the amplitude of the steady-state current induced by any input voltage $V_0 \sin(\omega t)$. We have seen that if

(5.105) $$\frac{1}{C} = L\omega_0^2,$$

then $A(\omega_0) = \frac{V_0}{R}$, the maximum possible. Throughout this problem we assume equation (5.105) holds. We still have the freedom to simultaneously adjust C and L, so long as (5.105) continues to hold. In any such adjustment, C gets smaller if L is increased, and vice versa. For any other $\omega \neq \omega_0$,

$$A(\omega) = \frac{\omega V_0}{\sqrt{L^2(\omega_0^2 - \omega^2)^2 + R^2\omega^2}} < \frac{\omega V_0}{\sqrt{L^2(\omega_0^2 - \omega^2)^2}} = \frac{\omega V_0}{L|\omega_0^2 - \omega^2|}$$

since $\frac{1}{C} - L\omega^2 = L\omega_0^2 - L\omega^2 = L(\omega_0^2 - \omega^2)$.

(a) Using the facts that $\omega_0^2 - \omega^2 = (\omega_0 - \omega)(\omega_0 + \omega)$ and

$$\frac{\omega}{\omega_0 + \omega} < 1,$$

show that

$$A(\omega) < \frac{V_0}{L|\omega - \omega_0|}$$

if $\omega \neq \omega_0$.

(b) Let $R = 0.01$ ohms, and suppose $\omega_0 = 1000$ radians per second. Find values of L and C with

$$\frac{1}{C} = L\omega_0^2 = 10^6 L$$

such that $A(\omega) < (0.001)V_0/R$ whenever $\omega \leq 998$ or $\omega \geq 1002$. Of course we know that $A(1000) = V_0/R$, the maximum possible value.

(c) Copy Fig. 5.30 below and on it draw a reasonable sketch of the graph of $A(\omega)$ vs. ω for the numerical values in part (b), including $R = 0.01$, $\omega_0 = 1,000$, and the values that you found there for L and C.

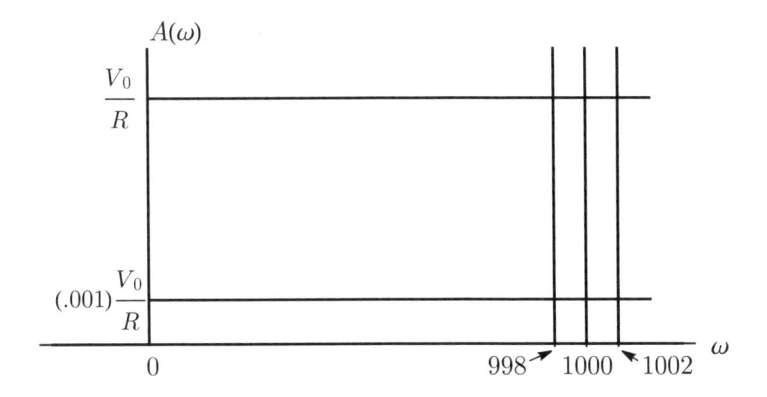

Figure 5.30. (Picture not to scale).

11. Consider an RLC circuit as pictured in Fig. 5.31, having input voltage $V(t)$ at time t. Show that the voltage V_{bc} across the capacitor satisfies the differential equation

$$LC\frac{d^2V_{bc}}{dt^2} + RC\frac{dV_{bc}}{dt} + V_{bc} = V(t).$$

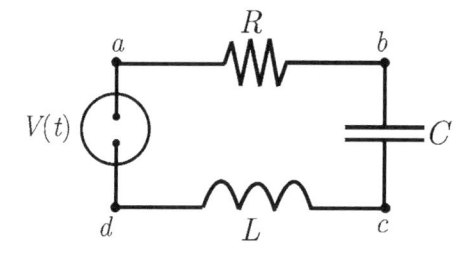

Figure 5.31. An RLC circuit.

Chapter 6

Laplace Transforms

6.1. Laplace transforms

In this chapter we will introduce a new tool, called the Laplace transform, which we'll use to solve linear constant coefficient initial value problems. It has several advantages over our previous methods: It changes the work of solving differential equations to *algebra*, it gives solutions satisfying initial conditions directly (bypassing the need to first find the general solution), and it is much better than our old methods for solving nonhomogeneous equations when the nonhomogeneous term has discontinuities. Because of the last property, the "Laplace transform method" is ideally suited for a variety of interesting applications. Finally, the method is the same regardless of the order of the equation (only the algebra gets more complicated as the order increases), and it can be easily adapted to solve *systems* of constant coefficient equations.

Starting with a function $f(t)$ defined for $0 \leq t < \infty$, we define a new function $F(s)$ by the relation

$$(6.1) \qquad F(s) = \int_0^\infty e^{-st} f(t)\, dt.$$

This new function $F(s)$ is called the **Laplace transform** of the original function f and is written

$$F(s) = \mathcal{L}[f(t)].$$

Notice that in the integral on the right-hand side of (6.1), the variable of integration is t, so that for the purposes of the integration s is treated as a constant. The integral used to define $F(s)$ is an *improper integral*; it is computed as a limit:

$$\int_0^\infty e^{-st} f(t)\, dt = \lim_{M \to \infty} \int_0^M e^{-st} f(t)\, dt.$$

The Laplace transform is defined only when this improper integral converges, i.e., when this limit exists. We'll say more about this below.

Think of the Laplace transform as an *operator* that takes a function $f(t)$ as the input and produces a function $F(s)$ as the output. Later we will see that this operator is linear.

$$f(t) \longrightarrow \boxed{\mathcal{L}} \longrightarrow F(s) = \mathcal{L}[f(t)]$$

Example 6.1.1. As a first example, let's find $\mathcal{L}[e^{2t}]$. By the definition in (6.1),

$$\mathcal{L}[e^{2t}] = \int_0^\infty e^{-st} e^{2t}\, dt = \int_0^\infty e^{(2-s)t}\, dt = \lim_{M\to\infty} \left.\frac{e^{(2-s)t}}{2-s}\right|_0^M = \lim_{M\to\infty} \frac{e^{(2-s)M}}{2-s} - \frac{1}{2-s}$$

so long as $s \ne 2$. When $s > 2$, this limit exists and is equal to

$$-\frac{1}{2-s} = \frac{1}{s-2}.$$

When $s < 2$, this limit is infinite. If $s = 2$, look back at the original integral defining $\mathcal{L}[e^{2t}]$ to see that it fails to converge when $s = 2$. Summarizing, we see that the Laplace transform of $f(t) = e^{2t}$ is the function $F(s) = \frac{1}{s-2}$ defined on the interval $2 < s < \infty$.

This example can readily be generalized to give the Laplace transform for any function $e^{\lambda t}$, for λ a constant:

$$(6.2) \qquad \mathcal{L}[e^{\lambda t}] = \frac{1}{s-\lambda} \quad \text{for } s > \lambda.$$

You are asked to verify this in Exercise 20. As a special case (setting $\lambda = 0$) we have

$$\mathcal{L}[1] = \frac{1}{s} \quad \text{for } s > 0.$$

For now λ is a *real* constant, but it is possible to consider λ to be complex, and doing so can be useful; see, e.g., Exercise 23.

Example 6.1.2. We find the Laplace transform of the function $\cos(\beta t)$. According to the definition,

$$\mathcal{L}[\cos(\beta t)] = \int_0^\infty e^{-st} \cos(\beta t)\, dt = \lim_{M\to\infty} \int_0^M e^{-st} \cos(\beta t)\, dt.$$

Integration by parts *twice* (see Exercise 22) gives

$$\int e^{-st} \cos(\beta t)\, dt = \frac{\beta}{\beta^2 + s^2}\left(e^{-st}\sin(\beta t) - \frac{s}{\beta} e^{-st}\cos(\beta t)\right).$$

Using this to evaluate the definite integral we see that

$$\int_0^M e^{-st}\cos(\beta t)\, dt = \frac{\beta}{\beta^2+s^2} e^{-sM}\sin(\beta M) - \frac{s}{\beta^2+s^2} e^{-sM}\cos(\beta M) + \frac{s}{\beta^2+s^2}.$$

If $s > 0$, both $e^{-sM}\sin(\beta M)$ and $e^{-sM}\cos(\beta M)$ approach 0 as M tends to infinity. Thus the Laplace transform of $\cos(\beta t)$ is

$$(6.3) \qquad \mathcal{L}[\cos(\beta t)] = \frac{s}{\beta^2+s^2} \quad \text{for } s > 0.$$

Similarly, you can show

$$(6.4) \qquad \mathcal{L}[\sin(\beta t)] = \frac{\beta}{\beta^2+s^2} \quad \text{for } s > 0.$$

6.1. Laplace transforms

Linearity. The Laplace transform is a linear operator. This means that if $\mathcal{L}[f(t)]$ and $\mathcal{L}[g(t)]$ exist, we have

- $\mathcal{L}[f(t) + g(t)] = \mathcal{L}[f(t)] + \mathcal{L}[g(t)]$ and
- $\mathcal{L}[cf(t)] = c\mathcal{L}[f(t)]$ for any constant c.

To verify these, we rely on the linearity of integration. Since

$$\int e^{-st}(f(t) + g(t))\, dt = \int e^{-st} f(t)\, dt + \int e^{-st} g(t)\, dt,$$

it follows that $\mathcal{L}[f(t) + g(t)] = \mathcal{L}[f(t)] + \mathcal{L}[g(t)]$. The property $\mathcal{L}[cf(t)] = c\mathcal{L}[f(t)]$ is verified similarly (see Exercise 25). Using linearity, we have, for example,

$$\mathcal{L}[3 - e^{-3t} + 5\cos(2t)] = \frac{3}{s} - \frac{1}{s+3} + 5\frac{s}{4+s^2}$$

for $s > 0$.

In the next example, we use linearity to compute $\mathcal{L}[\cosh(\beta t)]$ for any $\beta > 0$.

Example 6.1.3. Since

$$\cosh(\beta t) = \frac{e^{\beta t} + e^{-\beta t}}{2},$$

we have

$$\mathcal{L}[\cosh(\beta t)] = \frac{1}{2}\mathcal{L}[e^{\beta t}] + \frac{1}{2}\mathcal{L}[e^{-\beta t}] = \frac{1}{2}\frac{1}{s-\beta} + \frac{1}{2}\frac{1}{s+\beta} = \frac{s}{s^2 - \beta^2}$$

for $s > \beta$.

As a third application of linearity, we show an alternate method for computing the Laplace transforms of $\cos(\beta t)$ and $\sin(\beta t)$ that needs less computational effort than our first method. The idea is to write

$$\cos(\beta t) + i\sin(\beta t) = e^{i\beta t}$$

so that

$$\int e^{-st}(\cos(\beta t) + i\sin(\beta t))\, dt = \int e^{-st} e^{i\beta t}\, dt = \int e^{(i\beta - s)t}\, dt.$$

We have (see Exercise 23)

$$(6.5) \quad \int_0^M e^{(i\beta - s)t}\, dt = \frac{e^{(i\beta - s)M}}{i\beta - s} - \frac{1}{i\beta - s} = \frac{e^{-sM}(\cos(\beta M) + i\sin(\beta M)) - 1}{i\beta - s}.$$

As long as $s > 0$, we can evaluate the limit as $M \to \infty$ in equation (6.5) to see that

$$\lim_{M \to \infty} \int_0^M e^{(i\beta - s)t}\, dt = -\frac{1}{i\beta - s} = \left(\frac{1}{s - i\beta}\right)\left(\frac{s + i\beta}{s + i\beta}\right) = \frac{s + i\beta}{\beta^2 + s^2}$$

provided $s > 0$. Thus for $s > 0$,

$$\mathcal{L}[\cos(\beta t) + i\sin(\beta t)] = \frac{s + i\beta}{\beta^2 + s^2}.$$

Linearity of the transform $\mathcal{L}$ lets us rewrite this as

$$\mathcal{L}[\cos(\beta t)] + i\mathcal{L}[\sin(\beta t)] = \frac{s + i\beta}{\beta^2 + s^2} = \frac{s}{\beta^2 + s^2} + i\frac{\beta}{\beta^2 + s^2}.$$

Equating the real part of the left side with the real part of the right side and the imaginary part of the left side with the imaginary part of the right gives

$$\mathcal{L}[\cos(\beta t)] = \frac{s}{\beta^2 + s^2} \quad \text{and} \quad \mathcal{L}[\sin(\beta t)] = \frac{\beta}{\beta^2 + s^2}.$$

Not all functions have a Laplace transform. In deciding whether a particular function $f(t)$ has a Laplace transform, there are two issues at play.

(R_1) The definite integrals
$$\int_0^M e^{-st} f(t)\, dt$$
must exist for each $M > 0$.

(R_2) If the integrals $\int_0^M e^{-st} f(t)\, dt$ exist for each M, the *limit* of their values must be *finite* as $M \to \infty$.

The first requirement, (R_1), is rather mild. If f is continuous on $[0, \infty)$, then (R_1) will be satisfied. Some discontinuities are allowed as well:

Definition 6.1.4. Suppose that for each $M > 0$, f has at most a finite number of discontinuities in the interval $[0, M]$; suppose further that at each discontinuity $t = a$ in $(0, \infty)$ the one-sided limits
$$\lim_{t \to a^+} f(t) \quad \text{and} \quad \lim_{t \to a^-} f(t)$$
exist as finite values and that if the left endpoint $t = 0$ is a discontinuity, the one-sided limit $\lim_{t \to 0^+} f(t)$ is finite as well. Such a function is said to be **piecewise continuous** on $[0, \infty)$.

Condition (R_1) is satisfied for any piecewise continuous function. Note that since the value of a piecewise continuous function at a point $t = a$ of discontinuity has no bearing on the value of an integral of that function (see, for example, Exercise 33) we do not require that a piecewise continuous function even be defined at a point of discontinuity.

The second requirement, (R_2), is saying something about the rate of growth of $f(t)$. When $s > 0$ the factor e^{-st} tends to 0 rather rapidly as t gets large. This rapid decay of e^{-st} is helping the improper integral
$$\int_0^\infty e^{-st} f(t)\, dt$$
to converge, but if $f(t)$ grows too big too fast, e^{-st} may not be able to compensate, and
$$\lim_{M \to \infty} \int_0^M e^{-st} f(t)\, dt$$
may fail to exist. Also it may be the case that (R_2) is satisfied for $s > \alpha$ for some positive number α, but not for all $s > 0$. We saw an example of this in Example 6.1.1. This will not cause difficulties for us, and we will be content to have (R_2) hold for all sufficiently large s.

To ensure that (R_2) holds, at least for sufficiently large s, we will typically require that the growth of $f(t)$ is no worse than *some* exponential function. If for a given function f, you can find some constants A and b such that
$$|f(t)| \leq A e^{bt}$$
for all $0 \leq t < \infty$, we will say that f is **exponentially bounded**. If a function is both piecewise continuous and exponentially bounded, it will satisfy both (R_1) and (R_2) and its Laplace transform will exist (see Exercise 43).

Example 6.1.5. You can check (as a calculus problem) that the maximum value of $\dfrac{t}{e^t}$ on $[0, \infty)$ is $1/e$. This says $t \leq \frac{1}{e} e^t$ for $t \geq 0$. Thus $f(t) = t$ is exponentially bounded (in the definition, we can use $A = \frac{1}{e}, b = 1$). From this it follows that $t^2 \leq \frac{1}{e^2} e^{2t}$ for $t \geq 0$, so that t^2 is exponentially bounded too. The same conclusion holds for any positive power of t.

6.1. Laplace transforms

In Exercise 40 you'll show that the Laplace transform of e^{t^2} doesn't exist for *any* value of s. This function is not exponentially bounded.

Most of the functions we work with in this chapter are piecewise continuous and exponentially bounded. Moreover, when we use Laplace transforms to solve differential equations for an "unknown" function y, we will simply be assuming from the start that the Laplace transform $\mathcal{L}[y]$ exists, as well as the Laplace transforms of the various derivatives of y.

Piecewise-defined functions. One of the virtues of the Laplace transform method for solving differential equations is the ease with which it handles nonhomogeneous terms which are "piecewise-defined" functions. We will need to be able to compute Laplace transforms of such functions. As a first example, let's look at the **unit step function**, defined by

$$(6.6) \qquad u_a(t) = \begin{cases} 0 & \text{if } t < a, \\ 1 & \text{if } t \geq a. \end{cases}$$

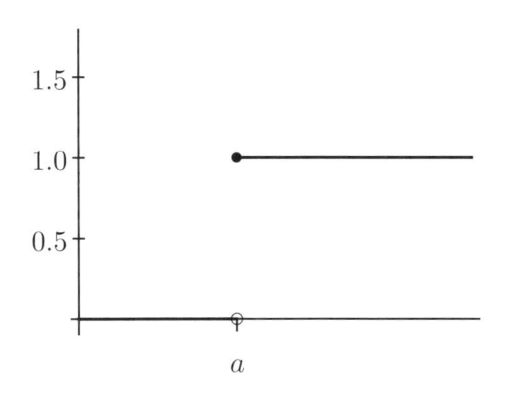

Figure 6.1. Graph of $y = u_a(t)$.

Notice that $u_a(t)$ is piecewise continuous and exponentially bounded. If $a \geq 0$, its Laplace transform is

$$\mathcal{L}[u_a(t)] = \int_0^\infty e^{-st} u_a(t)\, dt = \int_a^\infty e^{-st}\, dt = \lim_{M \to \infty} \frac{e^{-sM}}{(-s)} - \frac{e^{-sa}}{(-s)}.$$

This limit exists if $s > 0$ and is e^{-sa}/s. Thus

$$(6.7) \qquad \mathcal{L}[u_a(t)] = \frac{e^{-sa}}{s} \quad \text{for } s > 0.$$

In a physical system, the function $u_a(t)$ could be used to model a switch that is turned on at time $t = a$.

For $a < b$, the function $u_a(t) - u_b(t)$ is

$$(6.8) \qquad u_a(t) - u_b(t) = \begin{cases} 0 & \text{if } t < a, \\ 1 & \text{if } a \leq t < b, \\ 0 & \text{if } b \leq t, \end{cases}$$

with graph as shown in Fig. 6.2.

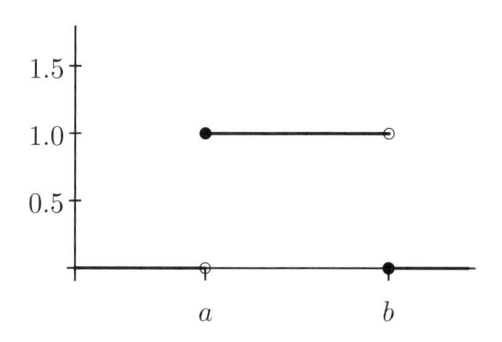

Figure 6.2. Graph of $u_a(t) - u_b(t)$, $0 < a < b$.

The function $f(t)[u_a(t) - u_b(t)]$ is

$$(6.9) \qquad f(t)[u_a(t) - u_b(t)] = \begin{cases} 0 & \text{if } 0 \leq t < a, \\ f(t) & \text{if } a \leq t < b, \\ 0 & \text{if } b \leq t. \end{cases}$$

Using this, we have a convenient way to describe other piecewise-defined functions, as the next example shows.

Example 6.1.6. The function

$$(6.10) \qquad g(t) = \begin{cases} t^2 & \text{if } 0 \leq t < 3, \\ 4 - 2t & \text{if } 3 \leq t < 4, \\ 2 & \text{if } 4 \leq t \end{cases}$$

can be written as

$$g(t) = t^2(u_0(t) - u_3(t)) + (4 - 2t)(u_3(t) - u_4(t)) + 2u_4(t).$$

A variation of $u_a(t)$. Suppose we modify the definition of the function $u_a(t)$, changing the value at the single point $t = a$, to obtain the function

$$(6.11) \qquad w_a(t) = \begin{cases} 0 & \text{if } t \leq a, \\ 1 & \text{if } t > a. \end{cases}$$

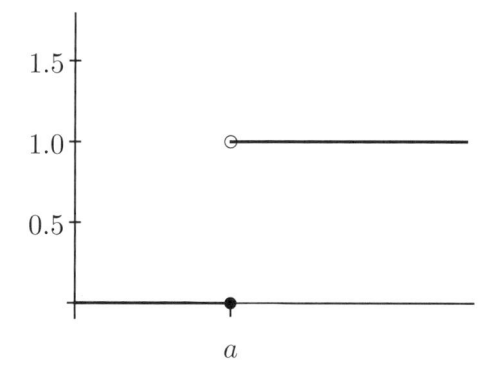

Figure 6.3. Graph of $y = w_a(t)$.

When the values of a function are changed at only finitely many points, it has no effect on the value of a definite integral of that function. Thus the Laplace transform of $w_a(t)$ is exactly the same as the Laplace transform of $u_a(t)$:

$$\mathcal{L}[u_a(t)] = \mathcal{L}[w_a(t)] = \frac{e^{-sa}}{s} \quad \text{for } s > 0.$$

6.1.1. Exercises.

In Exercises 1–9 use the definition in equation (6.1) to compute Laplace transforms of the given function. Indicate for what real numbers s the transform $\mathcal{L}[f(t)]$ is defined.

1. $f(t) = e^{3t}$.
2. $f(t) = t$.
3. $f(t) = t^2$.
4. $f(t) = 3$.
5. $f(t) = \sin t$.
6. $f(t) = \begin{cases} 0 & \text{if } t < 1, \\ e^{3t} & \text{if } t \geq 1. \end{cases}$

7. $f(t) = \begin{cases} e^{-t} & \text{if } t < 1, \\ 0 & \text{if } t \geq 1. \end{cases}$
8. $f(t) = \begin{cases} 0 & \text{if } t < 2, \\ t & \text{if } t \geq 2. \end{cases}$
9. $f(t) = \begin{cases} t & \text{if } t < 2, \\ 0 & \text{if } t \geq 2. \end{cases}$

In Exercises 10–19 use Table 6.3 in Section 6.3, which gives the Laplace transform for some basic functions of interest, and the linearity of the Laplace transform to find $\mathcal{L}[f(t)]$.

10. $f(t) = 3 + t$.
11. $f(t) = 2\cos(3t) - 5\sin(2t)$.
12. $f(t) = t^3 + 2e^{4t}\cos(3t)$.
13. $f(t) = (1+t)\cos(2t)$.
14. $f(t) = (t^3 + 1)^2$.
15. $f(t) = (t+1)^2 e^t$.
16. $f(t) = 2 + 3\sinh(5t)$.
17. $f(t) = \cos^2(2t)$. (Hint: Use a trigonometric identity.)
18. $f(t) = 1 + u_1(t)$.
19. $f(t) = e^{3t} - e^{-3t}$.

20. For any real number λ, show that the Laplace transform of $e^{\lambda t}$ is $\dfrac{1}{s - \lambda}$ for $s > \lambda$.

21. Verify the Laplace transform formula

$$\mathcal{L}[\sinh(\beta t)] = \frac{\beta}{s^2 - \beta^2}.$$

For what values of s is $\mathcal{L}[\sinh(\beta t)]$ defined? (You may assume $\beta > 0$.)

22. Use integration by parts twice to show that

$$\int e^{-st} \cos(\beta t)\, dt = \frac{\beta}{\beta^2 + s^2}\left(e^{-st}\sin(\beta t) - \frac{s}{\beta}e^{-st}\cos(\beta t)\right) + C.$$

23. Recall (see equation (4.65) in Section 4.4.1) that

$$\frac{d}{dt}\left(e^{\lambda t}\right) = \lambda e^{\lambda t}$$

holds even when $\lambda = \alpha + i\beta$ is complex. From this it follows that

$$\int e^{\lambda t}\, dt = \frac{e^{\lambda t}}{\lambda} + C$$

for any complex $\lambda \neq 0$.

(a) Show that if $\lambda = \alpha + i\beta$,
$$\lim_{M \to \infty} \int_0^M e^{-st} e^{\lambda t} \, dt = \frac{1}{s - \lambda}$$
if $s > \alpha$, the real part of λ.

(b) Show that
$$\frac{1}{s - \lambda} = \frac{s - \alpha}{(s - \alpha)^2 + \beta^2} + i \frac{\beta}{(s - \alpha)^2 + \beta^2}$$
if $\lambda = \alpha + i\beta$.

(c) Using your work in (a) and (b), conclude that
$$\mathcal{L}[e^{\alpha t} \cos(\beta t)] = \frac{s - \alpha}{(s - \alpha)^2 + \beta^2} \quad \text{and} \quad \mathcal{L}[e^{\alpha t} \sin(\beta t)] = \frac{\beta}{(s - \alpha)^2 + \beta^2}$$
for $s > \alpha$.

24. (a) Use integration by parts to compute
$$\int t e^t e^{-st} \, dt,$$
and then find $\mathcal{L}[t e^t]$ for $s > 1$.

(b) Using integration by parts, show that
$$\int t^n e^t e^{-st} \, dt = \frac{t^n e^{(1-s)t}}{1 - s} - \int \frac{n t^{n-1} e^{(1-s)t}}{1 - s} \, dt.$$

(c) Using (b), find $\mathcal{L}[t^2 e^t]$ and $\mathcal{L}[t^3 e^t]$, for $s > 1$.

(d) Verify by induction that
$$\mathcal{L}[t^n e^t] = \frac{n!}{(s - 1)^{n+1}}.$$

25. Show that if $\mathcal{L}[f(t)]$ exists, then $\mathcal{L}[cf(t)] = c\mathcal{L}[f(t)]$ for any constant c.

26. Find $\mathcal{L}[\cos(t + b)]$ and $\mathcal{L}[\sin(t + b)]$, where b is a constant, using the trigonometric identities for the cosine and sine of a sum.

27. For $0 < a < b$, find $\mathcal{L}[u_a(t) - u_b(t)]$.

28. Sketch the graph of each of the following functions for $t \geq 0$. The functions $u_a(t)$ are defined in equation (6.6).
   (a) $2u_1(t) - 3u_2(t) + 5u_4(t)$.
   (b) $t[u_1(t) - u_2(t)] + (2t - t^2)[u_2(t) - u_3(t)]$.
   (c) $tu_1(t) + (t^2 - t)[u_2(t) - u_1(t)]$.

29. Write each of the following piecewise-defined functions with a single formula, using the step functions $u_a(t)$ defined in equation (6.6).
   (a)
$$(6.12) \qquad g(t) = \begin{cases} t & \text{if } 0 \leq t < 3, \\ e^{-t} & \text{if } t \geq 3. \end{cases}$$

   (b)
$$(6.13) \qquad g(t) = \begin{cases} 2 & \text{if } 0 \leq t < 2, \\ 3t & \text{if } 2 \leq t < 3, \\ 5 & \text{if } t \geq 3. \end{cases}$$

30. **Laplace transforms of periodic and antiperiodic functions.**
    (a) A function $f(t)$ is periodic with period T if $f(t+T) = f(t)$ for all t in the domain of f. Show that if $f(t)$ has period T, then
    $$\mathcal{L}[f(t)] = \frac{\int_0^T e^{-st} f(t)\, dt}{1 - e^{-sT}}.$$

    Hint: Write
    $$\int_0^\infty e^{-st} f(t)\, dt = \int_0^T e^{-st} f(t)\, dt + \int_T^\infty e^{-st} f(t)\, dt.$$

    In the second integral, make the substitution $u = t - T$. You will then have an equation that you can solve for $\mathcal{L}[f(t)]$.

    (b) Suppose that for some positive constant T,
    $$f(t + T) = -f(t)$$
    for all t in the domain of f. Such a function is called "antiperiodic". Show that
    $$\mathcal{L}[f(t)] = \frac{\int_0^T e^{-st} f(t)\, dt}{1 + e^{-sT}}$$
    if f is antiperiodic.

31. Consider the function
    $$(6.14) \qquad f(t) = \begin{cases} \sin t & \text{if } 0 \leq t < \pi, \\ 0 & \text{if } \pi \leq t < 2\pi \end{cases}$$

    initially defined on the interval $[0, 2\pi]$. Extend it so that it is defined on $[0, \infty)$ by making it periodic with period 2π. The graph of the extended function is shown below.

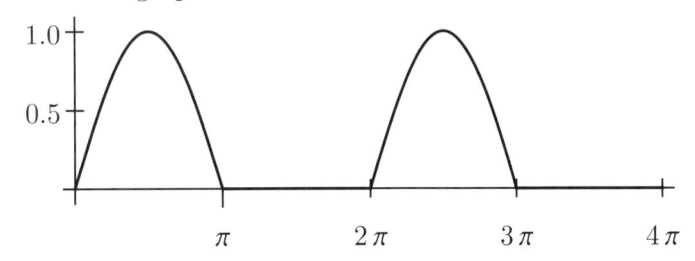

    Using Exercise 30(a), find the Laplace transform of this function.

32. Using the result of Exercise 30(a), show that the Laplace transform of the function whose graph is shown below is
    $$\frac{1}{s} \tanh \frac{s}{2}.$$

    (The hyperbolic tangent function is defined by
    $$\tanh x = \frac{\sinh x}{\cosh x} = \frac{e^x - e^{-x}}{e^x + e^{-x}}.)$$

33. Let f be given by
$$f(t) = \begin{cases} 2t & \text{if } 0 \leq t < 1, \\ 4 - t^2 & \text{if } 1 < t < 2, \\ 1 & \text{if } 2 \leq t < 3, \\ -1 & \text{if } t = 3, \\ 2 & \text{if } t > 3. \end{cases}$$

(a) Sketch the graph of f, verify that f is piecewise continuous on $[0, 4]$, and find $\int_0^4 f(t)\,dt$. Observe that the value of $\int_0^4 f(t)\,dt$ does not depend on how or whether f is defined at its discontinuities. Remark: Suppose that f is piecewise continuous on $[\alpha, \beta]$ and that $c_1, c_2, \ldots, c_k$ are the discontinuities of f in $[\alpha, \beta]$ in increasing order: $\alpha \leq c_1 < c_2 < \cdots < c_k \leq \beta$. Then,

$$\int_\alpha^\beta f(t)\,dt = \int_\alpha^{c_1} f(t)\,dt + \sum_{j=1}^{k-1} \int_{c_j}^{c_{j+1}} f(t)\,dt + \int_{c_k}^\beta f(t)\,dt.$$

(b) Is $f(t)e^{-st}$ piecewise continuous on $[0, \infty)$? At what value(s) of t does it fail to be continuous?

34. Using the definition of the Laplace transform in (6.1), show that
$$\mathcal{L}[e^{b(t-a)}u_a(t)] = \frac{e^{-as}}{s-b} \quad \text{for } s > b,$$
where a and b are real constants with $a > 0$.

35. For a small positive number ϵ, let
$$g_\epsilon(t) = \frac{1}{\epsilon}(u_0(t) - u_\epsilon(t)),$$
where $u_a(t)$ is the unit step function in equation (6.6).
(a) Sketch the graph of $g_\epsilon(t)$.
(b) Show that
$$\mathcal{L}[g_\epsilon(t)] = \frac{1 - e^{-\epsilon s}}{\epsilon s}.$$

(c) For fixed $s > 0$, what is the limit of
$$\frac{1 - e^{-\epsilon s}}{\epsilon s}$$
as ϵ decreases to 0? We'll return to this example in Section 6.7.

36. Decide which of the following functions are exponentially bounded. For those that are, give a choice of constants A and b so that $|f(t)| \leq Ae^{bt}$ holds for all $t \geq 0$.
(a) $f(t) = 1$.
(b) $f(t) = 10t$.
(c) $f(t) = t^2$.
(d) $f(t) = \cosh t$.
(e) $f(t) = t^2 e^{3t}$.
(f) $f(t) = \sin(e^{t^2})$.
(g) The derivative of the function in (f).

37. (a) Show that if $\mathcal{L}[f(t)] = F(s)$, then
$$\mathcal{L}[f(at)] = \frac{1}{a} F\left(\frac{s}{a}\right)$$

for a a positive constant. Hint:
$$\mathcal{L}[f(at)] = \int_0^\infty e^{-st} f(at) \, dt.$$
Make the substitution $u = at$ in this integral.

(b) If you know (see Exercise 24) that
$$\mathcal{L}[t^n e^t] = \frac{n!}{(s-1)^{n+1}},$$
what is $\mathcal{L}[t^n e^{at}]$ for a positive constant a?

38. The Laplace transform is but one example of an **integral transform**. In general, if you pick a function $K(s,t)$ and numbers a and b (∞ allowed), we can use these to define an operator whose rule is
$$T[f(t)] = \int_a^b K(s,t) f(t) \, dt.$$
(a) What choices of a, b, and $K(s,t)$ correspond to the Laplace transform?
(b) Show that any integral transform is linear.

39. Suppose that $f(t)$ is exponentially bounded on $[0, \infty)$.
(a) Show that there is some $s > 0$ so that
$$\lim_{t \to \infty} e^{-st} f(t) = 0.$$
Hint: If $|f(t)| \leq A e^{bt}$, consider $s = 2b$.
(b) If
$$\lim_{t \to \infty} e^{-s_0 t} f(t) = 0$$
for a particular $s_0 > 0$, explain why
$$\lim_{t \to \infty} e^{-st} f(t) = 0$$
for all $s > s_0$ too.

40. Show that the function e^{t^2} has no Laplace transform by showing that the integral
$$\int_0^\infty e^{t^2} e^{-st} \, dt$$
diverges for all $s > 0$.
Hint: Pick any $s > 0$. If $t > s + 1$, then $t^2 > (s+1)t$ so that
$$\int_{s+1}^\infty e^{t^2} e^{-st} \, dt \geq \int_{s+1}^\infty e^{(s+1)t} e^{-st} \, dt.$$

41. Show that the function $\frac{1}{t}$ has no Laplace transform, by showing that
$$\int_0^\infty \frac{1}{t} e^{-st} \, dt$$
diverges for all $s > 0$.
Hint: The problem is the rapid growth of $1/t$ near the discontinuity at $t = 0$. (In particular this function is not piecewise continuous on $[0, \infty)$.) Notice that for $0 \leq t \leq 1$ and any positive s, $e^{-st} \geq e^{-s}$, so that
$$\int_0^1 \frac{1}{t} e^{-st} \, dt \geq e^{-s} \int_0^1 \frac{1}{t} \, dt.$$

42. To show that the improper integral
$$\int_0^\infty e^{-st} f(t)\, dt$$
converges, it is enough to show that

(6.15)
$$\int_0^\infty e^{-st} |f(t)|\, dt$$

converges. Moreover, we can show that the integral in (6.15) converges if we can find a function $g(t)$ with
$$e^{-st}|f(t)| \leq g(t)$$
for all $t \geq 0$ such that
$$\int_0^\infty g(t)\, dt$$
converges. Assuming these two facts, show that if a piecewise continuous function $f(t)$ is exponentially bounded on $[0, \infty)$, then
$$\int_0^\infty e^{-st} f(t)\, dt$$
converges for s sufficiently large.

43. Suppose that f is piecewise continuous and exponentially bounded on $[0, \infty)$.

    (a) Show that there are positive constants C and s_0 so that
    $$|\mathcal{L}[f(t)]| \leq \frac{C}{s - s_0}$$
    for all $s > s_0$. Hint: Use the definition of $\mathcal{L}[f]$ and the fact that
    $$\left| \int_0^M e^{-st} f(t) dt \right| \leq \int_0^M e^{-st} |f(t)| dt$$
    for all $M > 0$.

    (b) Explain why (a) tells you that $\frac{s}{s+1}$ cannot be the Laplace transform of any piecewise continuous, exponentially bounded function. Hint: Use (a) to determine
    $$\lim_{s \to \infty} F(s)$$
    if f is piecewise continuous and exponentially bounded and $F(s) = \mathcal{L}[f(t)]$.

    (c) Can $F(s) = \sin s$ be the Laplace transform of any piecewise continuous and exponentially bounded function?

6.2. The inverse Laplace transform

If $F(s) = \mathcal{L}[f(t)]$, then we say that $f(t)$ is an **inverse Laplace transform** of $F(s)$ and write

(6.16) $$f(t) = \mathcal{L}^{-1}[F(s)].$$

For example, $\mathcal{L}^{-1}[\frac{1}{s-2}] = e^{2t}$, since $\mathcal{L}[e^{2t}] = \frac{1}{s-2}$. Technically, more than one choice of $f(t)$ could serve as an inverse Laplace transform for a given function $F(s)$. For example, both $u_a(t)$ and $w_a(t)$, as defined in equations (6.6) and (6.11) in the last section, are inverse Laplace transforms of the function $\frac{e^{-sa}}{s}$, $s > 0$. This nonuniqueness may look troublesome, but a couple of facts explain why it really will not be a problem.

- If $f(t)$ and $g(t)$ are *continuous* functions on $[0, \infty)$ with the same Laplace transform, then $f(t) = g(t)$ for all $t \geq 0$. This tells us that for a given $F(s)$, there is at most one continuous function $f(t)$ for which $f(t) = \mathcal{L}^{-1}[F(s)]$ holds.

6.2. The inverse Laplace transform

- If $f(t)$ and $g(t)$ are piecewise continuous functions that have the same Laplace transform, then $f = g$ at any point where f and g are continuous.

The functions $u_a(t)$ and $w_a(t)$ provide an illustration of the second of these facts. They have the same Laplace transform (namely e^{-sa}/s), and they are equal at every point except $t = a$, which is the only point of discontinuity of both $u_a(t)$ and $w_a(t)$.

Linearity of the inverse transform. The inverse Laplace transform is linear:
$$\mathcal{L}^{-1}[F_1(s) + F_2(s)] = \mathcal{L}^{-1}[F_1(s)] + \mathcal{L}^{-1}[F_2(s)]$$
(this says we can find an inverse Laplace transform for $F_1 + F_2$ by adding together an inverse transform for F_1 and an inverse transform for F_2), and
$$\mathcal{L}^{-1}[cF(s)] = c\mathcal{L}^{-1}[F(s)].$$

Example 6.2.1. From the formula
$$\mathcal{L}[\cos(\beta t)] = \frac{s}{\beta^2 + s^2}$$
we see that
$$\mathcal{L}^{-1}\left[\frac{3s}{4 + s^2}\right] = 3\mathcal{L}^{-1}\left[\frac{s}{4 + s^2}\right] = 3\cos(2t).$$

Example 6.2.2. Let's compute
$$\mathcal{L}^{-1}\left[\frac{5s^2 - 6s + 14}{(s+1)(s^2+4)}\right].$$
To get started, we use the partial fraction decomposition
$$\frac{5s^2 - 6s + 14}{(s+1)(s^2+4)} = \frac{5}{s+1} - \frac{6}{s^2+4}$$
so that
$$\mathcal{L}^{-1}\left[\frac{5s^2 - 6s + 14}{(s+1)(s^2+4)}\right] = \mathcal{L}^{-1}\left[\frac{5}{s+1} - \frac{6}{s^2+4}\right].$$
Using the linearity of $\mathcal{L}^{-1}$, we compute this as
$$5\mathcal{L}^{-1}\left[\frac{1}{s+1}\right] - 6\mathcal{L}^{-1}\left[\frac{1}{s^2+4}\right] = 5e^{-t} - 3\sin(2t).$$

6.2.1. Partial fraction review.
The last example made use of a partial fraction decomposition to compute an inverse transform of the rational function
$$\frac{5s^2 - 6s + 14}{(s+1)(s^2+4)}.$$
Because partial fraction decompositions will be used throughout this chapter, we summarize the basic ideas here. This is an *algebraic* procedure used to write a rational function
$$\frac{P(s)}{Q(s)},$$
where $P(s)$ and $Q(s)$ are polynomials, as a sum of simpler rational functions. We make two assumptions:

- The degree of $P(s)$ is less than the degree of $Q(s)$. If this isn't the case, then a preliminary long division will be needed. We'll also assume that the coefficient of the highest power of s in the denominator $Q(s)$ is one. We can always make this so by multiplying the numerator and denominator by a suitable constant.

- We can find a factorization of the denominator $Q(s)$ into linear factors $(s-b)$ (which may be repeated) and/or irreducible quadratic factors $(s^2 + as + b)$ (which also may be repeated). A factor $s^2 + as + b$ is irreducible if it cannot be factored further into linear factors using only real numbers. The quadratic formula tells us that $s^2 + as + b$ is irreducible if and only if $a^2 - 4b < 0$.

The form of the factorization of the denominator $Q(s)$ tells us the form of the partial fraction decomposition. Table 6.1 summarizes the most common situations that will arise in this chapter. In this table, "numerator" means a polynomial of degree less than that of the corresponding denominator.

Table 6.1. Basic Partial Fraction Forms.

rational function	form of partial fraction decomposition
$\dfrac{\text{numerator}}{(s+a)(s+b)},\ a \neq b$	$\dfrac{C_1}{s+a} + \dfrac{C_2}{s+b}$, C_1, C_2 constants
$\dfrac{\text{numerator}}{(s+a)^2}$	$\dfrac{C_1}{s+a} + \dfrac{C_2}{(s+a)^2}$, C_1, C_2 constants
$\dfrac{\text{numerator}}{(s+a)(s+b)^2},\ a \neq b$	$\dfrac{C_1}{s+a} + \dfrac{C_2}{s+b} + \dfrac{C_3}{(s+b)^2}$, C_1, C_2, C_3 constants
$\dfrac{\text{numerator}}{(s^2+a^2)(s+b)}$	$\dfrac{C_1 s + D_1}{s^2 + a^2} + \dfrac{C_2}{s+b}$, C_1, C_2, D_1 constants

The general rules for producing the form of a partial fraction decomposition are as follows:

- If linear factor $s + a$ appears a total of m times in the factorization of $Q(s)$, it contributes m terms
$$\frac{C_1}{s+a} + \frac{C_2}{(s+a)^2} + \cdots + \frac{C_m}{(s+a)^m} \quad \text{(for constants } C_j\text{)}$$
to the partial fraction decomposition.

- If an irreducible quadratic factor $s^2 + as + b$ appears a total of m times in the factorization of $Q(s)$, it contributes m terms
$$\frac{C_1 s + D_1}{s^2 + as + b} + \frac{C_2 s + D_2}{(s^2 + as + b)^2} + \cdots + \frac{C_m s + D_m}{(s^2 + as + b)^m} \quad \text{(for constants } C_j \text{ and } D_j\text{)}$$
to the partial fraction decomposition.

Once the form of the desired decomposition is determined, we must find the values of the constants appearing in the numerators of this decomposition. There are two methods for this: the **substitution method** and the **coefficient comparison method**. You can choose one or the other, or use both in combination. We'll illustrate with some examples.

Example 6.2.3. The rational function
$$\frac{8-s}{s^2-s-2} = \frac{8-s}{(s-2)(s+1)}$$

will have a decomposition in the form

(6.17) $$\frac{8-s}{(s-2)(s+1)} = \frac{C_1}{s-2} + \frac{C_2}{s+1}.$$

Multiply both sides of equation (6.17) by the common denominator of $(s-2)(s+1)$. The resulting equation is

(6.18) $$8 - s = C_1(s+1) + C_2(s-2).$$

To proceed by the substitution method, we choose some convenient values of s to substitute into equation (6.18). Setting $s = 2$ gives $8 - 2 = C_1(2+1)$, so that $C_1 = 2$. Another convenient substitution is $s = -1$ (convenient because it makes one of the terms on the right-hand side of (6.18) equal to zero), and with this choice we have $9 = C_2(-3)$, so that $C_2 = -3$. Thus

$$\frac{8-s}{(s-2)(s+1)} = \frac{2}{s-2} - \frac{3}{s+1}.$$

To use instead the coefficient comparison method, we compare the coefficients of s, and the constant terms, on both sides of equation (6.18). This gives the pair of equations

$$-1 = C_1 + C_2 \text{ (comparing the coefficient of } s\text{)}$$

and

$$8 = C_1 - 2C_2 \text{ (comparing the constant terms)}.$$

This pair of equations for the unknowns C_1 and C_2 can be solved to get $C_1 = 2$ and $C_2 = -3$, as before.

In the last example, the substitution method is probably the easier way to go. In the next example, we'll use a combination of the two methods.

Example 6.2.4. The rational function

$$\frac{5s^2 - 3s + 13}{s^3 - s^2 + 4s - 4} = \frac{5s^2 - 3s + 13}{(s^2 + 4)(s - 1)}$$

has a partial fraction decomposition in the form

(6.19) $$\frac{5s^2 - 3s + 13}{(s^2 + 4)(s - 1)} = \frac{C_1 s + D_1}{s^2 + 4} + \frac{C_2}{s - 1}.$$

Multiplying both sides of (6.19) by the common denominator $(s^2 + 4)(s - 1)$, we obtain

(6.20) $$5s^2 - 3s + 13 = (C_1 s + D_1)(s - 1) + C_2(s^2 + 4).$$

Making the substitution $s = 1$ in the last equation, we obtain $C_2 = 3$. Substituting $s = 0$ in (6.20) gives $D_1 = -1$. At this point, we switch to the coefficient comparison method to determine the remaining unknown C_1. Comparing the coefficient of s^2 on both sides of equation (6.20) (and remembering that $C_2 = 3$), we have $5 = C_1 + 3$, so $C_1 = 2$ and

$$\frac{5s^2 - 3s + 13}{(s^2 + 4)(s - 1)} = \frac{2s - 1}{s^2 + 4} + \frac{3}{s - 1}.$$

We could have also found C_1 by comparing the coefficient of s on both sides of (6.20). Comparing the constant terms, however, is the same as substituting $s = 0$.

Whether you use substitution, coefficient comparison, or a combination is up to you. If the factorization of $Q(s)$ includes a linear factor $s-a$, the substitution $s = a$ will be nice and will directly give one of the coefficients you seek. Other (different) linear factors will correspondingly give rise to other nice substitutions. Typically some coefficient comparison is used if the factorization of $Q(s)$ includes irreducible quadratics or if any linear factors are repeated.

The partial fraction decompositions of the functions featured in Examples 6.2.3 and 6.2.4 quickly yield inverse Laplace transforms of these functions.

Example 6.2.5. Find an inverse Laplace transform for
$$F(s) = \frac{5s^2 - 3s + 13}{s^3 - s^2 + 4s - 4}.$$

Since
$$\frac{5s^2 - 3s + 13}{(s^2 + 4)(s - 1)} = \frac{2s - 1}{s^2 + 4} + \frac{3}{s - 1},$$

we have
$$\mathcal{L}^{-1}[F(s)] = \mathcal{L}^{-1}\left[\frac{2s - 1}{s^2 + 4} + \frac{3}{s - 1}\right]$$
$$= 2\mathcal{L}^{-1}\left[\frac{s}{s^2 + 4}\right] - \frac{1}{2}\mathcal{L}^{-1}\left[\frac{2}{s^2 + 4}\right] + 3\mathcal{L}^{-1}\left[\frac{1}{s - 1}\right]$$
$$= 2\cos(2t) - \frac{1}{2}\sin(2t) + 3e^t.$$

6.2.2. Exercises.

In Exercises 1–3, give the partial fraction decomposition and then find the inverse Laplace transform.

1. $\dfrac{1}{s^2 - 5s + 6}$.

2. $\dfrac{2s^2 - 5s + 4}{(s - 1)(s^2 - 4s + 4)}$.

3. $\dfrac{s + 1}{(s + 2)(s^2 + 1)}$.

In Exercises 4–8 find $\mathcal{L}^{-1}[F(s)]$ for $F(s)$ as given. Begin with a partial fraction decomposition where necessary.

4. $F(s) = \dfrac{1}{s^2 + 4}$.

5. $F(s) = \dfrac{5s}{s^2 + 1}$.

6. $F(s) = \dfrac{s - 8}{(s + 2)(s - 3)}$.

7. $F(s) = \dfrac{1}{(s^2 + 4)(s + 2)}$.

8. $F(s) = \dfrac{3}{(s - 1)^2 + 4}$. (Use Exercise 23 of Section 6.1.)

9. For $n \geq 0$ an integer, the Laplace transform of t^n is $\dfrac{n!}{s^{n+1}}$. Using this, find $\mathcal{L}^{-1}[1/s^3]$ and $\mathcal{L}^{-1}[5/s^4]$.

10. (a) Complete the square on $s^2 + 6s + 13$, writing it in the form $(s - a)^2 + b^2$ for suitable choices of a and b. Then use the result of Exercise 23(c) in Section 6.1 to determine
$$\mathcal{L}^{-1}\left[\frac{1}{s^2 + 6s + 13}\right].$$

    (b) Find
$$\mathcal{L}^{-1}\left[\frac{s}{s^2 + 6s + 13}\right].$$

11. Suppose
$$F(s) = \frac{s + 2}{s^2 - 4s + 3}.$$

    (a) Give a partial fraction decomposition for $F(s)$ and then use it to find $\mathcal{L}^{-1}[F(s)]$.

(b) For a different method, complete the square on s^2-4s+3, writing it in the form $(s-a)^2-b^2$. Using the Laplace transform formulae

$$\mathcal{L}[e^{at}\cosh(bt)] = \frac{s-a}{(s-a)^2-b^2} \quad \text{and} \quad \mathcal{L}[e^{at}\sinh(bt)] = \frac{b}{(s-a)^2-b^2},$$

find $\mathcal{L}^{-1}[F(s)]$.

(c) Show that your answers in (a) and (b) are the same.

12. (a) Show that the partial fraction decomposition of the rational function $\frac{P(s)}{Q(s)}$ where

$$Q(s) = (s-a)(s-b)(s-c)$$

for distinct real numbers a, b, and c and where $P(s)$ is a polynomial of degree at most 2 is

$$\frac{P(s)}{Q(s)} = \frac{C_1}{s-a} + \frac{C_2}{s-b} + \frac{C_3}{s-c}$$

for

$$C_1 = \frac{P(a)}{Q'(a)}, \quad C_2 = \frac{P(b)}{Q'(b)}, \quad \text{and} \quad C_3 = \frac{P(c)}{Q'(c)}.$$

(b) What is an inverse Laplace transform of $\frac{P(s)}{Q(s)}$ if $P(s)$ and $Q(s)$ are as in (a)?

6.3. Solving initial value problems with Laplace transforms

Laplace transforms provide a new method for solving linear constant coefficient initial value problems. It's an especially useful method for nonhomogeneous equations where the nonhomogeneous term is discontinuous, a kind of problem that is awkward to do by our old methods (see Exercise 25). The Laplace transform method also builds the initial conditions into the solution directly, eliminating the need to first produce a general solution containing arbitrary constants.

The Laplace differentiation formulas. To solve initial value problems using Laplace transforms, we need to understand how Laplace transforms interact with differentiation. Specifically, we want to relate $\mathcal{L}[f'(t)]$ to $\mathcal{L}[f(t)]$. From the definition of the Laplace transform in equation (6.1) we have

$$\mathcal{L}[f'(t)] = \int_0^\infty e^{-st} f'(t)\, dt.$$

If we integrate by parts, using

$$u = e^{-st}, \quad dv = f'(t)dt,$$

so that

$$du = -se^{-st}\, dt \quad \text{and} \quad v = f(t),$$

we have

$$\int_0^M e^{-st} f'(t)\, dt = e^{-st} f(t) \Big|_0^M + \int_0^M s e^{-st} f(t)\, dt$$

(6.21)
$$= e^{-sM} f(M) - f(0) + s\int_0^M e^{-st} f(t)\, dt.$$

So long as f is exponentially bounded, we know that

$$e^{-sM} f(M) \to 0$$

as $M \to \infty$, for all sufficiently large s (see Exercise 39 in Section 6.1). Using this and letting $M \to \infty$ in (6.21), we obtain the formula

$$(6.22) \qquad \mathcal{L}[f'(t)] = s\mathcal{L}[f(t)] - f(0).$$

We can use this to compute the Laplace transform of the *second* derivative of $f(t)$:

$$\begin{aligned}\mathcal{L}[f''(t)] &= \mathcal{L}[(f'(t))'] = s\mathcal{L}[f'(t)] - f'(0) \\ &= s\left(s\mathcal{L}[f(t)] - f(0)\right) - f'(0) \\ &= s^2\mathcal{L}[f(t)] - sf(0) - f'(0).\end{aligned}$$

In the first line of this calculation we used (6.22) with f replaced by f' (we also tacitly assumed that f' is exponentially bounded). In the second line we used (6.22) again. Thus

$$(6.23) \qquad \mathcal{L}[f''(t)] = s^2\mathcal{L}[f(t)] - sf(0) - f'(0).$$

Repeating this procedure k times gives the **First Differentiation Formula**:

$$(6.24) \quad \mathcal{L}[y^{(k)}] = s^k\mathcal{L}[y] - s^{k-1}y(0) - s^{k-2}y'(0) - \cdots - sy^{(k-2)}(0) - y^{(k-1)}(0).$$

Here we have written y for $f(t)$ for convenience and have assumed that y and its first k derivatives are exponentially bounded. You are asked to verify equation (6.24) for $k = 3$ in Exercise 8.

The result in (6.24) becomes particularly simple if we have

$$(6.25) \qquad y(0) = y'(0) = \cdots = y^{(k-1)}(0) = 0,$$

which is called the **rest condition**. In this case equation (6.24) is just $\mathcal{L}[y^{(k)}] = s^k\mathcal{L}[y]$. Using this, you can show (see Exercise 27) that when y satisfies the rest condition (6.25), then

$$\mathcal{L}[a_n y^{(n)} + a_{n-1} y^{(n-1)} + \cdots + a_1 y' + a_0 y] = P(s)\mathcal{L}[y]$$

where $P(s)$ is the polynomial

$$P(s) = a_n s^n + a_{n-1} s^{n-1} + \cdots + a_1 s + a_0.$$

We're ready to see an example illustrating how Laplace transforms can be used to solve initial value problems.

Example 6.3.1. We'll solve the first-order initial value problem

$$(6.26) \qquad y' - y = 2\sin t, \quad y(0) = 0.$$

Of course, we could do this with methods from Chapter 2, but instead we take a new approach.

Step 1: Apply the Laplace transform, using the initial condition. If we apply the Laplace transform to both sides on the differential equation, the result is

$$\mathcal{L}[y' - y] = \mathcal{L}[2\sin t].$$

Using linearity and the First Differentiation Formula on the left-hand side and equation (6.4) on the right-hand side, this becomes

$$s\mathcal{L}[y] - y(0) - \mathcal{L}[y] = \frac{2}{s^2+1},$$

which simplifies to

$$(6.27) \qquad s\mathcal{L}[y] - \mathcal{L}[y] = \frac{2}{s^2+1},$$

6.3. Solving initial value problems with Laplace transforms

upon substituting the initial condition $y(0) = 0$.

Step 2: Solve for $\mathcal{L}[y]$. Simple algebra in equation (6.27) gives
$$\mathcal{L}[y] = \frac{2}{(s^2+1)(s-1)}.$$

Step 3: Compute the inverse transform. We have
$$y = \mathcal{L}^{-1}\left[\frac{2}{(s^2+1)(s-1)}\right].$$

To finish, we must compute this inverse Laplace transform. Working from the partial fraction decomposition
$$\frac{2}{(s^2+1)(s-1)} = \frac{1}{s-1} - \frac{s+1}{s^2+1}$$
and using the linearity of the inverse transform, we have
$$\begin{aligned} y &= \mathcal{L}^{-1}\left[\frac{1}{s-1}\right] - \mathcal{L}^{-1}\left[\frac{s+1}{s^2+1}\right] \\ &= \mathcal{L}^{-1}\left[\frac{1}{s-1}\right] - \mathcal{L}^{-1}\left[\frac{s}{s^2+1}\right] - \mathcal{L}^{-1}\left[\frac{1}{s^2+1}\right], \end{aligned}$$
giving the solution
$$y = e^t - \cos t - \sin t.$$

This strategy will apply to solve *any* constant coefficient equation
$$a_n y^{(n)} + a_{n-1} y^{(n-1)} + \cdots + a_1 y' + a_0 y = g(t)$$
with appropriate initial conditions specified at $t = 0$, provided that we can determine $\mathcal{L}[g(t)]$ and the needed inverse Laplace transform. The steps are the same regardless of the order of the equation:

Step 1. Apply the Laplace transform, inserting the initial condition along the way.
Step 2. Solve the resulting algebraic equation for $\mathcal{L}[y]$.
Step 3. Obtain y by the inverse Laplace transform.

The stickiest point is the last step, and before we give further examples illustrating the method, it's desirable to expand our list of known Laplace and inverse Laplace transforms. To help with this, we will develop some further "computational rules" for Laplace transforms.

The Second Differentiation Formula. In the First Differentiation Formula, we differentiate first and then compute the Laplace transform. There is a companion formula (called, not surprisingly, the Second Differentiation Formula) obtained when we reverse the order, computing the Laplace transform first and then differentiating. In other words, it's a formula for
$$\frac{d}{ds}\mathcal{L}[f(t)],$$
the derivative (with respect to s) of the Laplace transform $F(s)$ of f. To see how this goes, start with

(6.28) $$\frac{d}{ds}\mathcal{L}[f(t)] = \frac{d}{ds}\int_0^\infty e^{-st} f(t)\, dt.$$

To proceed, we need to know a fact from multivariable calculus, namely that under the right conditions we can move the differentiation on the right-hand side of (6.28) from *outside* the integral sign to *inside* the integral sign. It comes "inside" as $\frac{\partial}{\partial s}$ rather than as $\frac{d}{ds}$. In particular, this is legitimate whenever f is piecewise continuous and exponentially bounded for $t \geq 0$. Proceeding, we have

$$\frac{d}{ds}\mathcal{L}[f(t)] = \int_0^\infty \frac{\partial}{\partial s}(e^{-st}f(t))\,dt = \int_0^\infty -te^{-st}f(t)\,dt = -\mathcal{L}[tf(t)].$$

Inside the integral we computed the partial derivative, with respect to s, of $e^{-st}f(t)$, and then recognized the resulting integral, except for a minus sign, as the Laplace transform of the function $tf(t)$. We can state the end result compactly as:

$$\text{If } \mathcal{L}[f(t)] = F(s), \text{ then } \mathcal{L}[tf(t)] = -\tfrac{d}{ds}[F(s)].$$

A second differentiation gives

$$\mathcal{L}[t^2 f(t)] = -\frac{d}{ds}[\mathcal{L}[tf(t)]] = \frac{d^2}{ds^2}[\mathcal{L}[f(t)]].$$

This can be generalized to higher derivatives. The result is the **Second Differentiation Formula**:

$$(6.29) \qquad \mathcal{L}[t^n f(t)] = (-1)^n \frac{d^n}{ds^n}[\mathcal{L}[f(t)]].$$

This formula expands the list of functions for which we can compute Laplace transforms, as the next examples illustrate.

Example 6.3.2. We'll identify $\mathcal{L}[t^2 \sin(3t)]$, doing as little integral computation as possible. The Second Differentiation Formula (equation (6.29)), with $n = 2$, tells us that

$$\mathcal{L}[t^2 \sin(3t)] = \frac{d^2}{ds^2}\mathcal{L}[\sin(3t)].$$

From equation (6.4) we know that

$$\mathcal{L}[\sin(3t)] = \frac{3}{s^2 + 9},$$

so we have

$$\mathcal{L}[t^2 \sin(3t)] = \frac{d^2}{ds^2}\left(\frac{3}{s^2+9}\right) = 18\frac{s^2 - 3}{(s^2+9)^3}.$$

Example 6.3.3. If we use $f(t) = 1$ in equation (6.29), we see that

$$\mathcal{L}[t^n] = (-1)^n \frac{d^n}{ds^n}\mathcal{L}[1] = (-1)^n \frac{d^n}{ds^n}\left(\frac{1}{s}\right) = \frac{n!}{s^{n+1}}.$$

The shift formulas. The first shift formula tells us how the Laplace transform of $e^{\alpha t}f(t)$ can be obtained if we know the Laplace transform of $f(t)$. From the definition,

$$(6.30) \qquad \mathcal{L}[e^{\alpha t}f(t)] = \int_0^\infty e^{-st}e^{\alpha t}f(t)\,dt = \int_0^\infty e^{-(s-\alpha)t}f(t)\,dt.$$

If, as usual, we denote the Laplace transform of $f(t)$ by $F(s)$, so that

$$(6.31) \qquad F(s) = \int_0^\infty e^{-st}f(t)\,dt,$$

6.3. Solving initial value problems with Laplace transforms

then we have (by replacing s by $s - \alpha$ in (6.31))

$$F(s - \alpha) = \int_0^\infty e^{-(s-\alpha)t} f(t)\, dt.$$

Comparing this with (6.30), here is what we have learned; it is called the **First Shift Formula**:

$$\mathcal{L}[e^{\alpha t} f(t)] = F(s - \alpha), \text{ where } F(s) = \mathcal{L}[f(t)].$$

In words, this says "to find the Laplace transform of $e^{\alpha t} f(t)$, find the Laplace transform $F(s)$ of $f(t)$, and then replace s by $s - \alpha$". For example, since we know from Example 6.3.3 that

$$\mathcal{L}[t^3] = \frac{6}{s^4},$$

the First Shift Formula tells us immediately that

$$\mathcal{L}[e^{2t} t^3] = \frac{6}{(s-2)^4}$$

(replace s by $s - 2$). Similarly, since

$$\mathcal{L}[\cos(3t)] = \frac{s}{s^2 + 9},$$

we have

$$\mathcal{L}[e^{-t} \cos(3t)] = \frac{s+1}{(s+1)^2 + 9}$$

(replace s by $s + 1$). The function $F(s - \alpha)$ is a shift of the function $F(s)$; we get the graph of $F(s - \alpha)$ by translating the graph of $F(s)$ to the right by α units.

If we read the First Shift Formula backwards, we see that

$$(6.32) \qquad \mathcal{L}^{-1}[F(s - \alpha)] = e^{\alpha t} \mathcal{L}^{-1}[F(s)].$$

We'll illustrate this inverse transform property with several examples.

Example 6.3.4. We have

$$\mathcal{L}^{-1}\left[\frac{3}{(s-2)^5}\right] = e^{2t} \mathcal{L}^{-1}\left[\frac{3}{s^5}\right] = \frac{t^4 e^{2t}}{8}.$$

Here we've used (6.32) with $\alpha = 2$ and $F(s) = 3/s^5$, and the computation $\mathcal{L}^{-1}[3/s^5] = t^4/8$, which follows from Example 6.3.3.

Example 6.3.5. As a second example, we have

$$\mathcal{L}^{-1}\left[\frac{s}{s^2 - 2s + 5}\right] = \mathcal{L}^{-1}\left[\frac{s}{(s-1)^2 + 4}\right]$$

$$= \mathcal{L}^{-1}\left[\frac{s-1}{(s-1)^2 + 4} + \frac{1}{(s-1)^2 + 4}\right]$$

$$= \mathcal{L}^{-1}\left[\frac{s-1}{(s-1)^2 + 4}\right] + \mathcal{L}^{-1}\left[\frac{1}{(s-1)^2 + 4}\right]$$

(6.33) $$= e^t \cos(2t) + \frac{1}{2} e^t \sin(2t).$$

In the first step of these calculations, we completed the square on $s^2 - 2s + 5$. For the last step we recognized
$$\frac{s-1}{(s-1)^2+4}$$
as $F(s-1)$ for the choice
$$F(s) = \frac{s}{s^2+4}$$
and recognized
$$\frac{1}{(s-1)^2+4}$$
as $G(s-1)$ for the choice
$$G(s) = \frac{1}{s^2+4}.$$
Since $F(s)$ is the Laplace transform of $\cos(2t)$ and $G(s)$ is the Laplace transform of $\frac{1}{2}\sin(2t)$, (6.33) follows from (6.32).

The Second Shift Formula describes the Laplace transform of $u_a(t)f(t)$, where $u_a(t)$ is the unit step function defined in equation (6.6) of Section 6.1. Since $u_a(t)$ is equal to 1 for $t \geq a$ and 0 elsewhere,

$$\begin{aligned}
\mathcal{L}[u_a(t)f(t)] &= \int_0^\infty e^{-st} u_a(t) f(t) \, dt \\
&= \int_a^\infty e^{-st} f(t) \, dt \\
&= \int_0^\infty e^{-s(w+a)} f(w+a) \, dw \\
&= e^{-sa} \int_0^\infty e^{-sw} f(w+a) \, dw,
\end{aligned}$$

where we have made the substitution $w = t - a$ in the third line. We can rewrite the integral in the last line as
$$\int_0^\infty e^{-st} f(t+a) \, dt$$
by changing the dummy variable inside the integral from w to t. The end result is the **Second Shift Formula**:

$$\mathcal{L}[u_a(t)f(t)] = e^{-as} \mathcal{L}[f(t+a)].$$

In words, this says that to find the Laplace transform of $u_a(t)f(t)$, first determine the Laplace transform of the translated function $f(t+a)$, and then multiply the result by e^{-as}. The next examples illustrate this process.

Example 6.3.6. To compute $\mathcal{L}[u_2(t)e^{-t}]$, we use the Second Shift Formula with $a = 2$ and $f(t) = e^{-t}$. Since $f(t+2) = e^{-(t+2)} = e^{-2}e^{-t}$, linearity of $\mathcal{L}$ says
$$\mathcal{L}[e^{-(t+2)}] = e^{-2}\mathcal{L}[e^{-t}] = e^{-2} \frac{1}{s+1}.$$

By the Second Shift Formula,
$$\mathcal{L}[u_2(t)e^{-t}] = e^{-2s}\mathcal{L}[e^{-(t+2)}] = e^{-2s} e^{-2} \frac{1}{s+1} = \frac{e^{-2(s+1)}}{s+1}.$$

6.3. Solving initial value problems with Laplace transforms

Example 6.3.7. To compute the Laplace transform of the piecewise-defined function

(6.34)
$$g(t) = \begin{cases} 5 & \text{if } 0 \leq t < 1, \\ t & \text{if } 1 \leq t < 3, \\ e^{2t} & \text{if } 3 \leq t, \end{cases}$$

first write

$$g(t) = 5(u_0(t) - u_1(t)) + t(u_1(t) - u_3(t)) + e^{2t}u_3(t).$$

Using linearity and the Second Shift Formula we have

$$\begin{aligned}\mathcal{L}[g(t)] &= 5\,\mathcal{L}[u_0(t)] - 5\,\mathcal{L}[u_1(t)] + \mathcal{L}[tu_1(t)] - \mathcal{L}[tu_3(t)] + \mathcal{L}[e^{2t}u_3(t)] \\ &= \frac{5}{s} - \frac{5e^{-s}}{s} + e^{-s}\mathcal{L}[t+1] - e^{-3s}\mathcal{L}[t+3] + e^{-3s}\mathcal{L}[e^{2(t+3)}] \\ &= 5\frac{1-e^{-s}}{s} + e^{-s}\left(\frac{1}{s^2} + \frac{1}{s}\right) - e^{-3s}\left(\frac{1}{s^2} + \frac{3}{s}\right) + \frac{e^{-3(s-2)}}{s-2}.\end{aligned}$$

A convenient formulation of the Second Shift Formula for the inverse transform is

(6.35)
$$\mathcal{L}^{-1}[e^{-as}F(s)] = u_a(t)f(t-a),$$

where $f(t)$ is the inverse transform of $F(s)$. In words, "to find an inverse transform of $e^{-as}F(s)$, find an inverse transform $f(t)$ for $F(s)$, and then an inverse transform of $e^{-as}F(s)$ is $u_a(t)f(t-a)$". To verify this, notice that

$$\mathcal{L}[u_a(t)f(t-a)] = e^{-as}\mathcal{L}[f(t-a+a)] = e^{-as}F(s).$$

Example 6.3.8. We'll use (6.35) to determine

$$\mathcal{L}^{-1}\left[\frac{e^{-s}}{s^2}\right] \quad \text{and} \quad \mathcal{L}^{-1}\left[\frac{se^{-s}}{1+s^2}\right].$$

The presence of the factor e^{-s} is your clue that (6.35) is relevant. Since $f(t) = t$ has Laplace transform $F(s) = \frac{1}{s^2}$, equation (6.35) says that

$$\mathcal{L}^{-1}\left[\frac{e^{-s}}{s^2}\right] = u_1(t)(t-1).$$

Since $f(t) = \cos t$ has Laplace transform $\frac{s}{1+s^2}$,

$$\mathcal{L}^{-1}\left[\frac{se^{-s}}{1+s^2}\right] = u_1(t)\cos(t-1).$$

The next example shows the ease with which the Laplace transform method handles initial value problems with a piecewise-defined nonhomogeneous term.

Example 6.3.9. We'll solve the initial value problem

$$y'' + y = I(t), \quad y(0) = 0, \quad y'(0) = 0$$

where $I(t)$ is the function

(6.36)
$$I(t) = \begin{cases} 1 & \text{if } 0 \leq t < 1, \\ 0 & \text{if } t \geq 1. \end{cases}$$

We can write $I(t)$ in terms of unit step functions as $I(t) = u_0(t) - u_1(t)$. Applying the Laplace transform to both sides of our differential equation gives

$$s^2\mathcal{L}[y] - sy(0) - y'(0) + \mathcal{L}[y] = \mathcal{L}[u_0(t) - u_1(t)].$$

With our initial conditions, this simplifies to

$$(s^2 + 1)\mathcal{L}[y] = \frac{1}{s} - \frac{e^{-s}}{s}$$

so that

$$\mathcal{L}[y] = \frac{1}{s(1+s^2)} - \frac{e^{-s}}{s(1+s^2)}.$$

To finish, we need to compute an inverse transform of

$$\frac{1}{s(1+s^2)} - \frac{e^{-s}}{s(1+s^2)}.$$

Partial fractions are a help to get started. Write

$$\frac{1}{s(1+s^2)} = \frac{1}{s} - \frac{s}{1+s^2},$$

so that

$$\mathcal{L}^{-1}\left[\frac{1}{s(1+s^2)} - \frac{e^{-s}}{s(1+s^2)}\right] = \mathcal{L}^{-1}\left[\frac{1}{s} - \frac{s}{1+s^2}\right] - \mathcal{L}^{-1}\left[\frac{e^{-s}}{s} - \frac{se^{-s}}{1+s^2}\right].$$

The first term on the right-hand side is straightforward:

$$\mathcal{L}^{-1}\left[\frac{1}{s} - \frac{s}{1+s^2}\right] = 1 - \cos t.$$

We know that

$$\mathcal{L}^{-1}\left[\frac{e^{-s}}{s}\right] = u_1(t),$$

and in Example 6.3.8 we computed

$$\mathcal{L}^{-1}\left[\frac{se^{-s}}{1+s^2}\right] = u_1(t)\cos(t-1).$$

Putting the pieces together, our initial value problem has solution

$$y = 1 - \cos t - u_1(t) + u_1(t)\cos(t-1).$$

Translations in time. The unit step function

(6.37) $$u_a(t) = \begin{cases} 0 & \text{if } t < a, \\ 1 & \text{if } a \le t \end{cases}$$

can also be described at $u_0(t-a)$. The graph of $u_0(t-a)$ is obtained from the graph of $u_0(t)$ by translation to the right by a units. More generally, the graph of $f(t-a)$ is obtained from the graph of $f(t)$ by a translation a units to the right. If a is negative, a translation of a units to the right is the same as a translation of $|a|$ units to the left. If f is initially defined only for $t \ge 0$, we typically will extend f to be zero for $t < 0$ before doing any translation; see Figs. 6.4–6.5 for an example.

6.3. Solving initial value problems with Laplace transforms

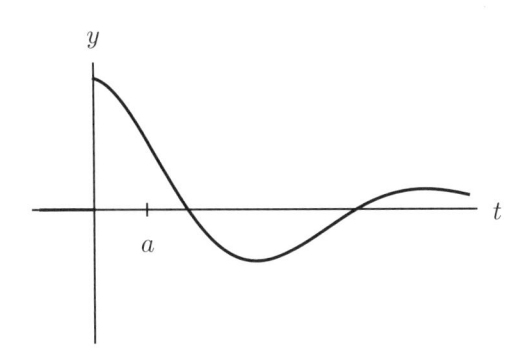

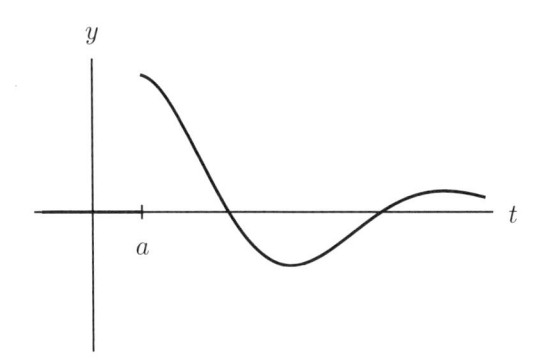

Figure 6.4. $y = f(t)$, $f(t) = 0$ for $t < 0$.

Figure 6.5. $y = f(t - a)$.

Thinking of the variable t as time, the function $f(t-a)$ is sometimes called a "delayed function". This language is most transparent when a is positive, so that the action of f starts at time $t = a$ rather than at time $t = 0$. If $f(t) = 0$ for $t < 0$, then $f(t - a) = 0$ for $t < a$, and we can write $f(t - a) = u_a(t)f(t - a) = u_0(t - a)f(t - a)$. With this notational change, we give another formulation of the Second Shift Formula as

$$\mathcal{L}[u_0(t - a)f(t - a)] = e^{-as}\mathcal{L}[f(t)].$$

Reading this backwards to give an inverse transform statement we have

$$\mathcal{L}^{-1}[e^{-as}F(s)] = u_0(t - a)f(t - a) = u_a(t)f(t - a)$$

where $F(s) = \mathcal{L}[f(t)]$.

Recap. Tables 6.2 and 6.3 summarize the computational rules for Laplace transforms and give a list of the Laplace transforms we have computed.

Table 6.2. Laplace Transform Computational Rules.	
First Differentiation Formula	$\mathcal{L}[y'] = s\mathcal{L}[y] - y(0)$, $\mathcal{L}[y''] = s^2\mathcal{L}[y] - sy(0) - y'(0)$, $\mathcal{L}[y^{(n)}] = s^n\mathcal{L}[y] - s^{n-1}y(0) - s^{n-2}y'(0) - \cdots - y^{(n-1)}(0)$.
Second Differentiation Formula	$\frac{d}{ds}\mathcal{L}[f(t)] = -\mathcal{L}[tf(t)]$, $\frac{d^n}{ds^n}\mathcal{L}[f(t)] = (-1)^n\mathcal{L}[t^n f(t)]$.
First Shift Formula	$\mathcal{L}[e^{\alpha t}f(t)] = F(s - \alpha)$, where $F(s) = \mathcal{L}[f(t)]$, $\mathcal{L}^{-1}[F(s - \alpha)] = e^{\alpha t}\mathcal{L}^{-1}[F(s)] = e^{\alpha t}f(t)$.
Second Shift Formula	$\mathcal{L}[u_a(t)f(t)] = e^{-as}\mathcal{L}[f(t + a)]$, $\mathcal{L}^{-1}[e^{-as}F(s)] = u_a(t)f(t - a)$, where $f(t) = \mathcal{L}^{-1}[F(s)]$.

The functions $f(t)$ and $F(s)$ in any row of Table 6.3 are called a **transform pair**. By reading this table from right to left, you have a table of inverse transforms.

Table 6.3. Some Laplace Transforms.			
$f(t)$	$F(s) = \mathcal{L}[f(t)]$	$f(t)$	$F(s) = \mathcal{L}[f(t)]$
1	$\dfrac{1}{s}$	$t\sin(\beta t)$	$\dfrac{2\beta s}{(s^2+\beta^2)^2}$
e^{at}	$\dfrac{1}{s-a}$	$t\cos(\beta t)$	$\dfrac{s^2-\beta^2}{(s^2+\beta^2)^2}$
t	$\dfrac{1}{s^2}$	$e^{at}\sin(\beta t)$	$\dfrac{\beta}{(s-a)^2+\beta^2}$
t^n $n=1,2,3,\ldots$	$\dfrac{n!}{s^{n+1}}$	$e^{at}\cos(\beta t)$	$\dfrac{s-a}{(s-a)^2+\beta^2}$
$t^n e^{at}$ $n=1,2,3,\ldots$	$\dfrac{n!}{(s-a)^{n+1}}$	$\sinh(\beta t)$	$\dfrac{\beta}{s^2-\beta^2}$
$\sin(\beta t)$	$\dfrac{\beta}{s^2+\beta^2}$	$\cosh(\beta t)$	$\dfrac{s}{s^2-\beta^2}$
$\cos(\beta t)$	$\dfrac{s}{s^2+\beta^2}$	$u_a(t)$	$\dfrac{e^{-as}}{s}$

6.3.1. Exercises.

In Exercises 1–6 use Table 6.3 to compute the Laplace transform of the given function.

1. $t\cos(\sqrt{3}t)$.
2. $e^{-2t}\sin(4t)$.
3. t/e^{2t}.
4. t^4.
5. $5\cos^2(2t)$. Hint: Use a trigonometric identity.
6. $\sin^2(3t)$.

7. Obtain the Laplace transform of $t^n e^{at}$, where n is a positive integer, using the Second Differentiation Formula. An alternate approach was used in Exercise 37 of Section 6.1.

8. Using (6.22) and (6.23), show that
$$\mathcal{L}[y'''(t)] = s^3 \mathcal{L}[y] - s^2 y(0) - s y'(0) - y''(0),$$
and check that this agrees with the $k=3$ statement in (6.24).

9. What is $\mathcal{L}[te^{2t}\cos(3t)]$?

10. Find the Laplace transform of the following piecewise-defined functions.
    (a)
    $$g(t) = \begin{cases} 1 & \text{if } 0 \leq t < 2\pi, \\ \sin t & \text{if } 2\pi \leq t. \end{cases}$$

(b)
$$g(t) = \begin{cases} t & \text{if } 0 \leq t < \pi/2, \\ \cos(2t) & \text{if } \pi/2 \leq t. \end{cases}$$

11. What is $\mathcal{L}[|2t-1|]$?

In Exercises 12–22 solve the given initial value problem using Laplace transforms.

12. $y'' + y = 0$, $y(0) = 3$, $y'(0) = 1$.

13. $y'' - y = 0$, $y(0) = 3$, $y'(0) = 1$.

14. $y'' - 2y' = 4$, $y(0) = -1$, $y'(0) = 2$.

15. $y'' + 4y = \sin(3t)$, $y(0) = y'(0) = 0$.

16. $y'' + 2y' - 3y = 3e^t$, $y(0) = 1$, $y'(0) = 0$.

17. $y'' + y' - 12y = e^{-2t}$, $y(0) = y'(0) = 0$.

18. $(D^2 + 2D + 2)y = 25te^t$, $y(0) = y'(0) = 0$.

19. $y'' + 4y' + 4y = t^2$, $y(0) = y'(0) = 0$.

20. $y' + 4y = g(t)$, $y(0) = 0$, where
$$g(t) = \begin{cases} t & \text{if } 0 \leq t < 2, \\ 1 & \text{if } 2 \leq t < \infty. \end{cases}$$

21. $(D^4 - 9D^2)y = t^2$, $y(0) = y'(0) = y''(0) = y'''(0) = 0$.

22. $(D^3 - D^2 - 4D + 4)y = \sin t$, $y(0) = y'(0) = y''(0) = 0$.

23. An inverse Laplace transform of
$$\frac{1}{(s^2 + b^2)^2}$$
has the form
$$c_1 \sin(bt) + c_2 t \cos(bt).$$
Determine the constants c_1 and c_2.

24. Use Laplace transforms to find the general solution of $y'' + y = 2t$ by using the generic initial conditions $y(0) = c_1$, $y'(0) = c_2$.

25. Sam and Sally are discussing how to solve

(6.38) $\qquad y''(t) + y(t) = u_{\pi/3}(t), \qquad y(0) = 1, \quad y'(0) = 0.$

Sally says, "I don't need to know anything about Laplace transforms to do this. I can just use the methods I learned in Sections 4.4 and 5.2. First I'll solve $y'' + y = 0, y(0) = 1, y'(0) = 0$; that's easy. Then I'll solve
$$y'' + y = 1$$
using annihilating operators, making sure to pick the solution that has the same value for $y(\frac{\pi}{3})$ and $y'(\frac{\pi}{3})$ as my first solution did. By piecing the two solutions together, I'll have the solution to (6.38)." Sam replies, "That sounds like a lot more work than just using Laplace transforms."

(a) Carry out the plan Sally has in mind, by solving (6.38) along the lines she proposes. Write your answer as a piecewise-defined function.

(b) Do you agree with Sam? Why or why not? To answer this, you should solve (6.38) using Laplace transforms and compare the amount of work to that in (a).

26. (a) Show that the solution to the initial value problem
$$ay'' + by' + cy = f(t), \quad y(0) = y'(0) = 0,$$
can be written as
$$y_1(t) = \mathcal{L}^{-1}\left[\frac{\mathcal{L}[f(t)]}{P(s)}\right]$$
where $P(s) = as^2 + bs + c$.

(b) Show that the solution to the initial value problem
$$ay'' + by' + cy = 0, \quad y(0) = k_1, \quad y'(0) = k_2,$$
for arbitrary constants k_1 and k_2 can be written as
$$y_2(t) = \mathcal{L}^{-1}\left[\frac{ak_1 s + ak_2 + bk_1}{P(s)}\right]$$
where $P(s)$ is as in (a).

(c) In (a) we solved a general nonhomogeneous second-order constant coefficient equation with a special initial condition, the rest condition, and called the solution y_1. In (b) we solved the corresponding homogeneous equation with arbitrary initial conditions and called the solution y_2. In terms of y_1 and y_2, what is the solution to
$$ay'' + by' + cy = f(t), \quad y(0) = k_1, \quad y'(0) = k_2?$$

27. In this problem we consider the constant coefficient equation
$$a_n y^{(n)} + a_{n-1} y^{(n-1)} + \cdots + a_1 y' + a_0 y = g(t)$$
with the rest condition $y(0) = y'(0) = \cdots = y^{(n-1)}(0) = 0$. Let $P(s)$ be the characteristic polynomial
$$P(s) = a_n s^n + a_{n-1} s^{n-1} + \cdots + a_1 s + a_0.$$

(a) Show that
$$\mathcal{L}[a_n y^{(n)} + a_{n-1} y^{(n-1)} + \cdots + a_1 y' + a_0 y] = P(s)\mathcal{L}[y].$$

(b) Show that the initial value problem has solution
$$y = \mathcal{L}^{-1}\left[\frac{\mathcal{L}[g(t)]}{P(s)}\right].$$

(c) Show that the function $\mathcal{L}^{-1}[1/P(s)]$ solves the differential equation
$$a_n y^{(n)} + a_{n-1} y^{(n-1)} + \cdots + a_1 y' + a_0 y = 0$$
with initial conditions
$$y(0) = y'(0) = \cdots = y^{(n-2)}(0) = 0, \quad y^{(n-1)}(0) = \frac{1}{a_n}.$$

28. The graph of $f(t)$ is as shown. Assume $f(t) = 0$ for $t < 0$.

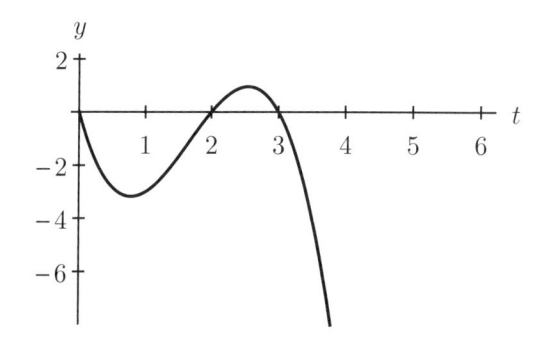

Figure 6.6. Graph of $y = f(t)$.

Match each of the following functions to its graph below.
(a) $u_2(t)f(t)$.
(b) $[u_2(t) - u_3(t)]f(t)$.
(c) $u_3(t)f(t-2)$.

(d) $u_2(t)f(t-2)$.
(e) $f(t) - u_3(t)f(t)$.

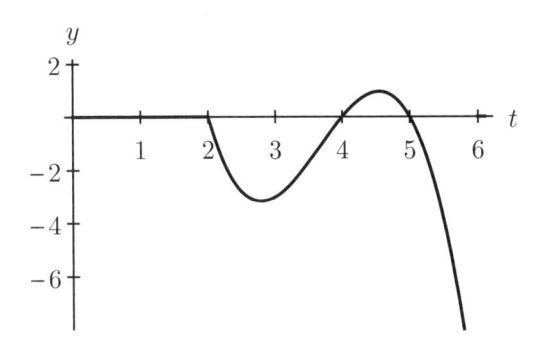

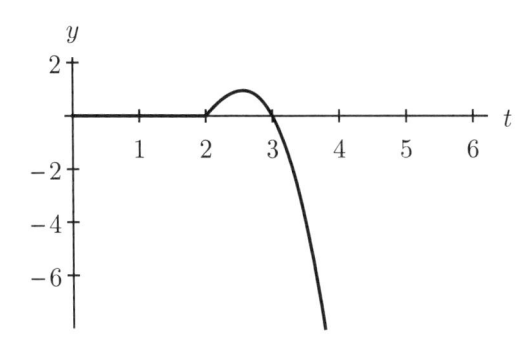

Figure 6.7. (I)

Figure 6.8. (II)

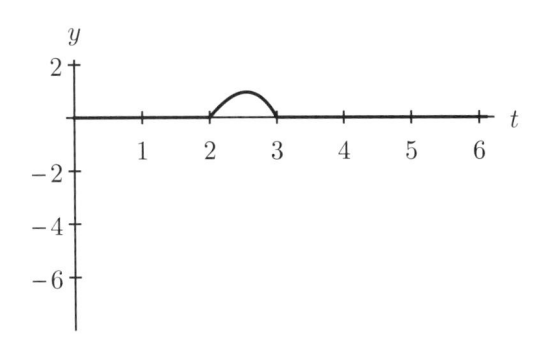

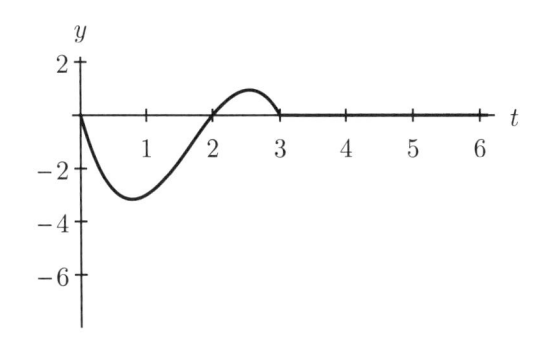

Figure 6.9. (III)

Figure 6.10. (IV)

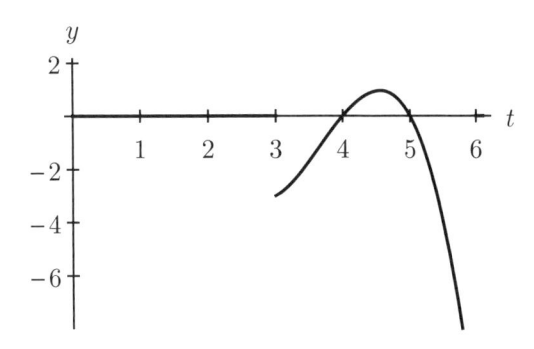

Figure 6.11. (V)

29. If $f(t) = t(t-2) + 3$, give the value of each of the following:
    (a) $u_2(1)f(1)$.
    (b) $u_2(3)f(3)$.
    (c) $f(t-4)$ evaluated at $t = 5$.
    (d) $[u_2(4) - u_5(4)]f(4)$.

30. Consider the equation $y'' - 2my' + m^2y = 0$ with initial conditions $y(0) = \alpha$, $y'(0) = \beta$.
    (a) Check that the characteristic equation for this differential equation has one repeated root.
    (b) Solve the initial value problem using Laplace transforms.

In Exercises 31–40 use Table 6.3 to find an inverse Laplace transform of the given function.

31. $F(s) = s^{-8}$.

32. $F(s) = \dfrac{1}{3s-2}$.

33. $F(s) = \dfrac{1}{s^2-4}$.

34. $F(s) = \dfrac{1}{(s-3)^2}$.

35. $F(s) = \dfrac{2}{(s-2)^3}$.

36. $F(s) = \dfrac{1}{s^2-6s+10}$.

37. $F(s) = \dfrac{3}{(s-2)(s^2-1)}$.

38. $F(s) = \dfrac{7}{2s^2+6}$.

39. $F(s) = \dfrac{s+3}{(s-3)(s+1)}$.

40. $F(s) = \dfrac{e^{-2s}}{s^2-2s-3}$.

41. We know that the solution to the initial value problem
$$y'' + y = 0, \qquad y(0) = 1, \quad y'(0) = 0,$$
is $y = \cos t$. Apply the Laplace transform, and use this known solution to determine $\mathcal{L}[\cos t]$. Similarly, determine $\mathcal{L}[\cos(kt)]$ by applying the Laplace transform to a constant coefficient initial value problem whose solution you know is $\cos(kt)$.

42. Suppose that $f(t)$ is periodic with period T, so that $f(t+T) = f(t)$ for all t in the domain of f. Using the Second Shift Formula, show that
$$\mathcal{L}[f(t)(1 - u_T(t))] = (1 - e^{-sT})\mathcal{L}[f(t)].$$
How does this result compare with that in Exercise 30 of Section 6.1?

43. The functions $u_a(t)$ and $w_a(t)$ (defined in equations (6.6) and (6.11)) are the same at every point except $t = a$. Explain why the formula
$$\mathcal{L}[w_a(t)f(t)] = e^{-as}\mathcal{L}[f(t+a)]$$
follows immediately from the Second Shift Formula.

44. In this problem you may use the fact that
$$\int_0^\infty e^{-u^2}\, du = \frac{\sqrt{\pi}}{2}.$$

(a) For any positive constant s, what is
$$\int_0^\infty e^{-su^2}\, du?$$

(b) Show that
$$\mathcal{L}[t^{-1/2}] = \frac{\sqrt{\pi}}{\sqrt{s}}.$$

Hint: In the integral defining the Laplace transform of $t^{-1/2}$, make the u-substitution $u = \sqrt{t}$. Notice that the function $t^{-1/2}$ is not piecewise continuous on $[0, \infty)$ since
$$\lim_{t \to 0^+} t^{-1/2} = \infty.$$

This example illustrates that while the conditions "piecewise continuous" and "exponentially bounded" are together *sufficient* to guarantee the existence of the Laplace transform, they are not *necessary* conditions.

6.3. Solving initial value problems with Laplace transforms

45. Using the result of Exercise 44, which showed that
$$\mathcal{L}[1/\sqrt{t}] = \frac{\sqrt{\pi}}{\sqrt{s}},$$
and the First Differentiation Formula, show that
$$\mathcal{L}[\sqrt{t}] = \frac{\sqrt{\pi}}{2s^{3/2}}.$$

46. Sam and Sally are debating the computation of $\mathcal{L}[t^{-\frac{3}{2}}]$. Sam says, "Since I know that
$$t^{-\frac{3}{2}}$$
is the derivative of
$$-2t^{-\frac{1}{2}},$$
I can use the First Differentiation Formula to compute
$$\mathcal{L}[t^{-\frac{3}{2}}] = -2\mathcal{L}[(t^{-\frac{1}{2}})']$$
and find $\mathcal{L}[t^{-3/2}]$ using the result of Exercise 44." Sally says, "Since
$$\int_0^1 \frac{1}{t^{3/2}}\, dt$$
diverges, the same kind of argument that was used in Exercise 41 of Section 6.1 will show that $\mathcal{L}[t^{-\frac{3}{2}}]$ does not exist." Who's correct?

47. Suppose we start with a piecewise continuous function $f(t)$ on $[0, \infty)$ and use it to define a new function $g(t)$ by the formula
$$g(t) = \int_0^t f(u)\, du.$$
Notice that $g(0) = 0$, and by the Fundamental Theorem of Calculus, $g'(t) = f(t)$ at points of continuity of f.
   (a) Using the First Differentiation Formula and the relationship $g'(t) = f(t)$, show that
$$\mathcal{L}[g(t)] = \frac{F(s)}{s}$$
   where $F(s) = \mathcal{L}[f(t)]$.
   (b) Use the result of (a) to determine
$$\mathcal{L}^{-1}\left[\frac{1}{s(s^2 + 4)}\right].$$
   You will NOT need partial fractions.

48. The method of Laplace transforms is designed to work when the initial conditions are specified at $t = 0$. But it can be adapted to deal with initial conditions at other values of t. This problem will lead you through a Laplace transform solution of the initial value problem
   (6.39) $\qquad y''(t) + 2y'(t) + y(t) = 4e^{-t}, \qquad y(1) = \frac{5}{e}, \quad y'(1) = 0.$

   The basic idea is to introduce a new timeline, so that time 0 in the new timeline corresponds to time 1 in the old timeline.
   (a) Let $v(t) = y(t+1)$ and check that $v(0) = y(1)$, $v'(t) = y'(t+1)$, and $v''(t) = y''(t+1)$.
   (b) In the differential equation of (6.39), replace t by $t+1$ on both sides of the equation, and then write the new equation in terms of the function $v(t)$ from (a). What is the initial condition $v(0)$?
   (c) Solve the initial value problem for $v(t)$ using Laplace transforms.
   (d) What is the corresponding solution for $y(t)$? Notice that by (a), $y(t) = v(t-1)$.

6.4. Applications

With a good table of Laplace transforms (and inverse transforms) at your disposal, solving a constant coefficient linear initial value problem becomes mainly the *algebraic* problem of solving the transformed equation for $\mathcal{L}[y]$. Thus the Laplace transform method is sometimes touted as being a "calculus-free" way to solve differential equations, especially in other disciplines that use differential equations in applications. In this section we will look at some models involving piecewise-defined nonhomogeneous terms.

Example 6.4.1. A fish population in a lake is growing according to the Malthusian model

$$\frac{dp}{dt} = kp$$

during the year, until fishing season begins (t measured in days). Starting on June 1, sport fishermen remove h fish per day from the lake. Fishing season ends on September 28. Suppose the fish population at the start of fishing season (we'll call this $t = 0$) is p_0. To model the fish population for a one-year period beginning June 1, we use the initial value problem

$$\frac{dp}{dt} = kp - f(t), \quad p(0) = p_0,$$

where

(6.40) $$f(t) = \begin{cases} h & \text{if } 0 \leq t < 120, \\ 0 & \text{if } 120 \leq t \leq 365, \end{cases}$$

since June 1 corresponds to $t = 0$ and September 28 corresponds to $t = 120$. We can write $f(t)$ as $h[u_0(t) - u_{120}(t)]$. Applying the Laplace transform gives

$$s\mathcal{L}[p(t)] - p_0 = k\mathcal{L}[p(t)] - h\mathcal{L}[u_0(t) - u_{120}(t)]$$
$$= k\mathcal{L}[p(t)] - h\left[\frac{1}{s} - \frac{e^{-120s}}{s}\right].$$

Solving this equation for $\mathcal{L}[p(t)]$ gives

$$\mathcal{L}[p(t)] = \frac{p_0}{s-k} - \frac{h}{s(s-k)} + h\frac{e^{-120s}}{s(s-k)}$$

where h and k are constants. We have the partial fraction decomposition

$$\frac{1}{s(s-k)} = -\frac{\frac{1}{k}}{s} + \frac{\frac{1}{k}}{s-k}$$

so that

$$\mathcal{L}[p(t)] = \frac{p_0}{s-k} + \frac{h}{k}\frac{1}{s} - \frac{h}{k}\frac{1}{s-k} - \frac{h}{k}\frac{e^{-120s}}{s} + \frac{h}{k}\frac{e^{-120s}}{s-k}.$$

Applying the inverse transform gives

$$p(t) = \left(p_0 - \frac{h}{k}\right)e^{kt} + \frac{h}{k}(1 - u_{120}(t)) + \frac{h}{k}u_{120}(t)e^{k(t-120)}.$$

We used the Second Shift Formula to compute

$$\mathcal{L}^{-1}\left[\frac{e^{-120s}}{s-k}\right] = u_{120}(t)e^{k(t-120)}.$$

We can write our solution piecewise as

(6.41) $$p(t) = \begin{cases} (p_0 - \frac{h}{k})e^{kt} + \frac{h}{k} & \text{if } 0 \leq t < 120, \\ (p_0 - \frac{h}{k})e^{kt} + \frac{h}{k}e^{k(t-120)} & \text{if } 120 \leq t \leq 365. \end{cases}$$

6.4. Applications

Fig. 6.12 shows the graph of this solution if $p_0 = 10{,}000$, $k = 0.005$, and the number of fish removed each day during the season is $h = 100$.

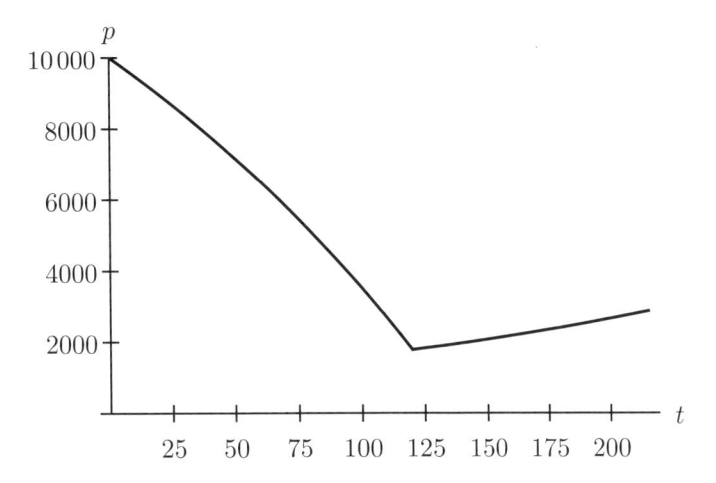

Figure 6.12. $p(t)$ for $0 \leq t \leq 200$, with $p_0 = 10{,}000$, $k = 0.005$, and $h = 100$.

If h is too large, then the population will become extinct before fishing season ends (at this point the model ceases to be meaningful). Fig. 6.13 shows the graph of $p(t)$ until the time of extinction for $p_0 = 10{,}000$, $k = 0.005$, and $h = 150$.

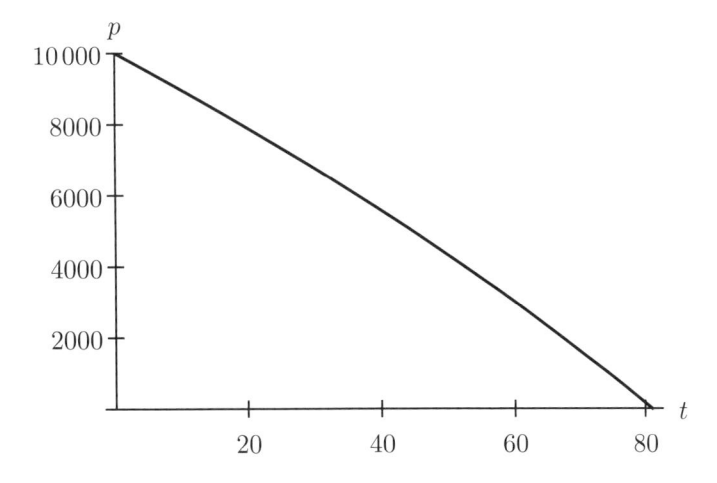

Figure 6.13. $p(t)$ with $p_0 = 10{,}000$, $k = 0.005$, and $h = 150$.

Example 6.4.2. Laplace transforms are often the preferred method for solving problems from pharmacology. In this example we're going to compare three dosing regimens for a medication that is given intravenously (directly into the bloodstream). We'll work with the simplest scenario, shown below, in which we think of the body as composed of a single compartment, labeled the blood compartment, and assume that once the drug is in the body, it is gradually removed from the blood at a rate proportional to the amount present. The amount of drug present in the blood at time t is denoted $y(t)$ (in mg), where t is measured in hours.

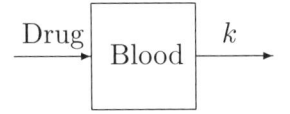

The same total amount of drug will be given to the patient under one of three dosing regimens:

(a) The drug is given by a gradual, continuous IV injection over a 12-hour period.
(b) The drug is given "all at once" by an IV bolus at the start of the 12-hour period.
(c) Half of the total amount of the drug is given by IV in each of two one-hour infusions: between $t = 0$ and $t = 1$ and then again between $t = 6$ and $t = 7$.

We want to compare how the drug levels in the blood vary for these different routes of administration.

In (a), where the drug is given by continuous infusion for 12 hours, the differential equation for $y(t)$ during the period $0 < t < 12$ is

$$\frac{dy}{dt} = I - ky,$$

where I is the constant rate of infusion, in mg/hr, and k is a constant. The "rate out" term $-ky$ reflects the assumption that the drug is eliminated from the body at a rate proportional to the amount present. For later times $t \geq 12$, the infusion is stopped and from then on the governing differential equation is

$$\frac{dy}{dt} = -ky.$$

Thus for all $t > 0$ we can write the differential equation to model scenario (a) as

(6.42) $$\frac{dy}{dt} = I(u_0(t) - u_{12}(t)) - ky.$$

We also know that $y(0) = 0$, since we assume that initially there is no drug present in the body. Applying the Laplace transform in (6.42), we have

$$s\mathcal{L}[y] = I\left(\frac{1}{s} - \frac{e^{-12s}}{s}\right) - k\mathcal{L}[y].$$

After some algebra we arrive at the equation

$$\mathcal{L}[y] = \frac{I}{k}\left(\frac{1}{s} - \frac{1}{s+k} - \frac{e^{-12s}}{s} + \frac{e^{-12s}}{s+k}\right),$$

where we have used the partial fraction decomposition

(6.43) $$\frac{1}{s(s+k)} = \frac{1}{k}\left(\frac{1}{s} - \frac{1}{s+k}\right).$$

Applying the inverse Laplace transform, we have

$$y(t) = \frac{I}{k}\left(1 - e^{-kt} - u_{12}(t) + u_{12}(t)e^{-k(t-12)}\right).$$

Fig. 6.14 shows the graph of $y(t)$, with $I = 10$ and $k = 1$, for 24 hours.

6.4. Applications

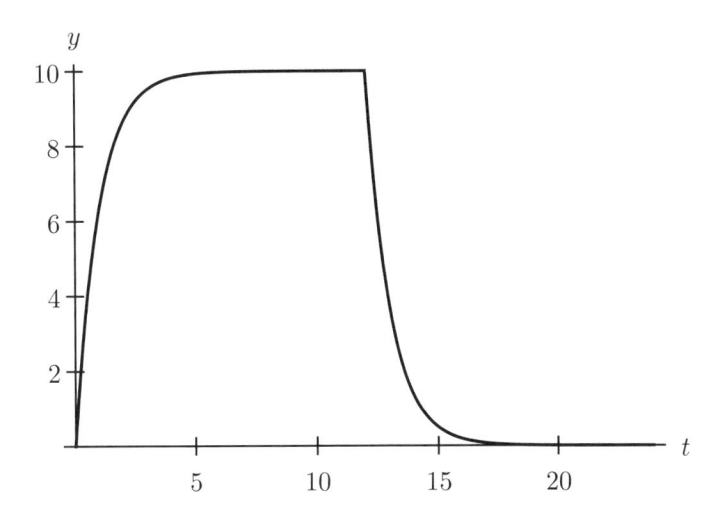

Figure 6.14. $y(t)$ under regimen (a), $0 < t < 24$.

In scenario (b) we have all of the drug given "instantaneously" at time $t = 0$. We'll model this by assuming that at time $t = 0$ the total amount of drug is present in the blood. Since we want the same total amount as given by a 12-hour continuous infusion at rate of I mg/hr, we have $y(0) = 12I$. For $t > 0$ the amount of drug in the body is governed by the equation

$$\frac{dy}{dt} = -ky.$$

Solving this equation with initial condition $y(0) = 12I$ we have

$$y(t) = 12Ie^{-kt}.$$

Setting $I = 10$ and $k = 1$ as before, we get the graph in Fig. 6.15. At some early times we have a much higher amount of drug present in the blood than we ever have under scenario (a). This is one reason why this dosing procedure, while more convenient to administer, may be undesirable if the high initial drug levels may not be well tolerated.

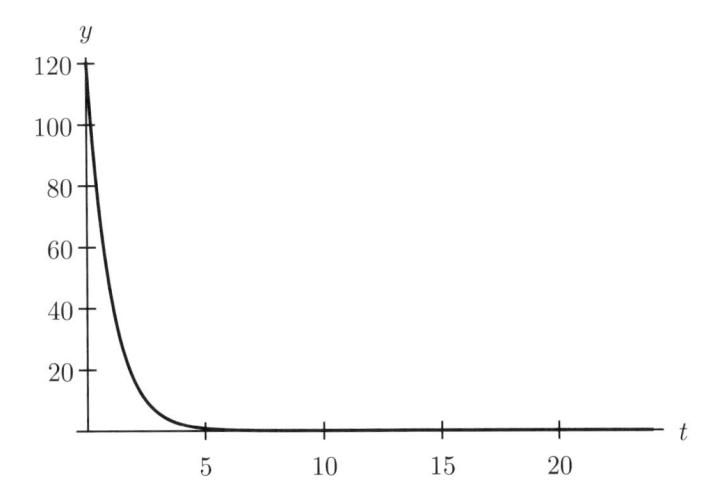

Figure 6.15. $y(t)$ under regimen (b), $0 < t < 24$.

In the final strategy we are going to infuse the total amount $12I$ mg of the drug in 2 one-hour sessions. This means the infusion rate will be $6I$ mg/hr between times $t = 0$ and $t = 1$, then again between $t = 6$ and $t = 7$. This leads us to the differential equation

$$(6.44) \qquad \frac{dy}{dt} = 6I(u_0(t) - u_1(t) + u_6(t) - u_7(t)) - ky$$

with initial condition $y(0) = 0$. Because of the form of the right-hand side of this equation, this is well suited to the method of Laplace transforms. Applying the Laplace transform to both sides of equation (6.44), we obtain

$$s\mathcal{L}[y] = 6I \left[\frac{1}{s} - \frac{e^{-s}}{s} + \frac{e^{-6s}}{s} - \frac{e^{-7s}}{s} \right] - k\mathcal{L}[y].$$

Solving this for $\mathcal{L}[y]$ we obtain

$$\mathcal{L}[y] = 6I \left[\frac{1}{s(s+k)} - \frac{e^{-s}}{s(s+k)} + \frac{e^{-6s}}{s(s+k)} - \frac{e^{-7s}}{s(s+k)} \right].$$

The partial fraction decomposition in (6.43) and repeated use of equation (6.35) gives $y(t)$

$$y(t) = \frac{6I}{k}[1 - e^{-kt} - u_1(t) + u_1(t)e^{-k(t-1)} + u_6(t) - u_6(t)e^{-k(t-6)}$$
$$- u_7(t) + u_7(t)e^{-k(t-7)}].$$

While this solution may look a bit intimidating, it's easy to think about it piecewise. For example, for $0 < t < 1$, it is just

$$y(t) = \frac{6I}{k}(1 - e^{-kt}),$$

since all the other terms are 0 for this range of t's. For $1 \leq t < 6$, we have

$$y(t) = \frac{6I}{k} \left[-e^{-kt} + e^{-k(t-1)} \right] = \frac{6I}{k} e^{-kt}(e^k - 1).$$

For $6 \leq t < 7$, we have

$$y(t) = \frac{6I}{k} \left(1 - e^{-kt} + e^k e^{-kt} - e^{6k} e^{-kt} \right).$$

Fig. 6.16 shows the graph of $y(t)$ for $0 < t < 24$, using $I = 10$ and $k = 1$. You can verify that $y(t)$ has local maxima at $t = 1$ and $t = 7$ and the maximum values are

$$y(1) = 60(1 - e^{-1}) \approx 37.93$$

and

$$y(7) = 60(1 - e^{-1} + e^{-6} - e^{-7}) \approx 38.02,$$

so the peaks are increasing slightly.

6.4. Applications

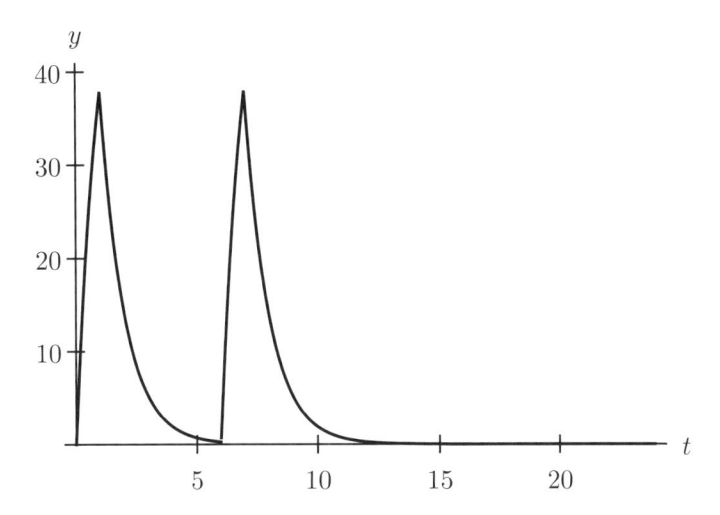

Figure 6.16. $y(t)$ under regimen (c), $0 < t < 24$.

In this example, the drug is eliminated at a rate proportional to the amount present (the $-ky$ term in the differential equation). This is called first-order kinetics. Some drugs are instead eliminated at a constant rate (zero-order kinetics), and this can have a profound effect on drug levels. Alcohol is an important example of a drug that is eliminated by zero-order kinetics. In Exercises 4 and 5 you can model some realistic problems about blood alcohol levels.

6.4.1. Exercises.

1. In Example 6.4.1, how large can the harvesting rate h be and still allow a positive fish population at the end of the first season (September 28)? Use $p_0 = 10{,}000$ and $k = 0.005$.

2. In Example 6.4.1, with $k = 0.005$ and $h = 100$, will the fish population return at least to its original level of 10,000 before the start of the next fishing season (the following June 1)? If the answer to this question is no, can the population still support a second year of fishing at the same level ($h = 100$)?

3. A dead body is found at midnight in a room with ambient temperature $A = 70°$. Murder is suspected. The body temperature is $79°$ at the time of discovery and $75°$ one hour later.
   (a) Estimate the time of death assuming the murder took place in the same room as where the body was found. Use Newton's law of cooling,
   $$\frac{dT}{dt} = -k(T - A),$$
   discussed in Section 2.3.1.
   (b) Suppose the murder actually took place outside, where the temperature is $50°$ and the body was moved to the room 1/2 hour after death. What is the estimated time of death for this scenario? Hint: In Newton's law, the value of k has not changed from (a), but there are two different values of A, corresponding to the two different time periods (the first half hour after the murder and the rest of the time). Thus the differential equation for $\frac{dT}{dt}$ will have a piecewise-defined term.

4. Sam is steadily drinking beer from 9:00 PM until 1:00 AM. He consumes eight 12-ounce beers (at a constant rate) during this four hour period. His blood alcohol level (BAC) is determined by

$$(6.45) \qquad \frac{dx}{dt} = \begin{cases} (\frac{0.025}{12})r - 0.018 & \text{if } 0 \leq t < 4, \\ -0.018 & \text{if } 4 \leq t < \infty, \end{cases}$$

where $x(t)$ is his BAC (expressed as a percentage), r is the (constant) rate at which he drinks (in ounces of beer per hour), and t is the time in hours after 9:00 PM. The particular numbers in this equation apply to Sam, a 150-lb male. Can he legally drive at 3:00 AM? At 4:00 AM? Assume Sam is of legal drinking age, that in his state legal driving requires a BAC under 0.08%, and he had not been drinking before 9:00 PM.

5. One ounce of 100-proof liquor is equivalent to 12 ounces of beer. How many ounces of liquor drunk steadily between 9:00 PM and 1:00 AM will result in a BAC of 0.15% at 2:00 AM? Use the differential equation

$$(6.46) \qquad \frac{dx}{dt} = \begin{cases} 0.025r - 0.018 & \text{if } 0 \leq t < 4, \\ -0.018 & \text{if } 4 \leq t < \infty, \end{cases}$$

where $x(t)$ is the BAC at time t hours after 9:00 PM and r is the constant drinking rate, in ounces of liquor per hour.

6. A newborn is treated with Ganciclovir for cytomegalovirus. The baby is initially given 15 mg of the drug infused by IV over a one-hour period. The elimination constant for this drug is $k = 0.288/\text{hr}$. The baby has an estimated blood volume of 0.3 liters. What is the expected concentration of Ganciclovir in the blood (in units of mg/l) at the end of the one-hour infusion period? Two hours after the start of the treatment? Eight hours after the start of the treatment? Use a one-compartment model as in Example 6.4.2, and assume none of the drug is present before the treatment is started.

7. The antibiotic Vancomycin is sometimes given for methicillin-resistant staphylococci infection. The drug is given by IV infusion. Suppose a total dose of 2,000 mg is to be given in a 24-hour period by either
   (I) 500 mg infused over a 1-hour period, every 6 hours, or
   (II) 1,000 mg infused over a 90-minute period, every 12 hours.
   (a) Give an initial value problem for the amount of drug $y(t)$ in the bloodstream at time t, $0 < t < 24$ hours, for each of these dosing regimens. Assume that the drug is removed at a rate proportional to the amount present, with proportionality constant $k = 0.34/\text{hr}$, and that initially none is present.
   (b) Determine the amount present, for each dosing regimen, at time $t = 2$ hours.
   (c) What is the maximum value of $y(t)$ for $0 < t < 24$ for dosing regimen II?
   (d) For a person with renal dysfunction, the value of k is greatly reduced. Answer (b) again, assuming $k = 0.03/\text{hr}$.

8. In Section 4.8 we used equations of the form $my'' + cy' + ky = 0$ to model a mass-spring system with damping, but having no externally applied force. In this problem we will solve this equation with $m = 1$ and initial conditions $y(0) = y_0$, $y'(0) = y_1$. For algebraic convenience, we will write $k = \omega^2$ (remembering that k, the spring constant, is positive).
   (a) Applying the Laplace transform to the equation, show that
   $$\mathcal{L}[y] = \frac{sy_0 + y_1 + cy_0}{(s + \tfrac{1}{2}c)^2 + (\omega^2 - \tfrac{c^2}{4})}.$$
   (b) Using the result of (a), show that if $c^2 - 4\omega^2 = 0$ (this is the critically damped case), then the solution is
   $$y(t) = y_0 e^{-\tfrac{1}{2}ct} + \left(y_1 + \frac{c}{2}y_0\right)\left(te^{-\tfrac{1}{2}ct}\right).$$
   Hint: Write $sy_0 + y_1 + cy_0 = y_0(s + \tfrac{1}{2}c) + (y_1 + \tfrac{1}{2}cy_0)$.
   (c) Using the result of (a), find $y(t)$ if the system is underdamped (so that $c^2 - 4\omega^2 < 0$) and $y_0 = 0$.
   (d) Find $y(t)$ in the overdamped case, assuming $y_0 = 0$.

9. The directions for cooking a beef tenderloin say to put the roast in a 450° oven for 15 minutes, and then turn the oven down to 350° and cook another hour. Using Newton's law of heating, set up an initial value problem for the temperature of the roast, assuming either
   (a) the initial temperature of the roast is 55° and the roast is instantaneously moved from a 450° oven to a 350° oven after 15 minutes or
   (b) the initial temperature of the roast is 55° and once the oven thermostat is adjusted, the oven temperature decreases linearly from 450° to 350° over a 10-minute period.

   "Set up" the initial value problem means to give the differential equation with initial condition, but not to solve it. Use the unit step functions $u_a(t)$ where appropriate in your equations. Your equation will involve a constant k from Newton's law of heating.

6.5. Laplace transforms, simple systems, and Iwo Jima

We will make a detailed study of *systems* of differential equations in Chapters 8–10. However in this section we will illustrate the use of Laplace transforms to solve simple systems of the form

$$\frac{dx}{dt} = ax + by + f(t),$$
$$\frac{dy}{dt} = cx + dy + g(t)$$

for unknown functions $x = x(t)$ and $y = y(t)$, when we have initial conditions specifying x and y at $t = 0$, say, $x(0) = x_0$ and $y(0) = y_0$. The coefficients a, b, c, and d are constants. If we apply the Laplace transform on both sides of both equations and use the First Differentiation Formula, we obtain the pair of equations

(6.47) $$s\mathcal{L}[x(t)] - x_0 = a\mathcal{L}[x(t)] + b\mathcal{L}[y(t)] + \mathcal{L}[f(t)]$$

and

(6.48) $$s\mathcal{L}[y(t)] - y_0 = c\mathcal{L}[x(t)] + d\mathcal{L}[y(t)] + \mathcal{L}[g(t)].$$

We can then solve this pair of equations *algebraically* for $\mathcal{L}[x(t)]$ and $\mathcal{L}[y(t)]$. For example, by equation (6.48) we have

(6.49) $$\mathcal{L}[y(t)] = \frac{y_0 + c\mathcal{L}[x(t)] + \mathcal{L}[g(t)]}{s - d}.$$

We can then substitute this expression for $\mathcal{L}[y(t)]$ into equation (6.47). The resulting equation can be solved for $\mathcal{L}[x(t)]$, and we can obtain $x(t)$ (provided we can compute the necessary inverse transform). Finally, with $x(t)$ in hand, we return to (6.49) to determine $y(t)$.

We'll illustrate this process with an analysis of the battle of Iwo Jima. Other "combat models" will be discussed in Section 8.1.

Example 6.5.1. A major World War II battle was fought on the island of Iwo Jima beginning February 19, 1945. Lasting a little over a month, the battle resulted in severe casualties on both sides. No Japanese reinforcements took place, and the Japanese fought "to the death"— there was no retreat. As a result, of the 22,000 Japanese soldiers initially in place on the island, only 216 were captured, and nearly all of the rest died in combat or by ritual suicide when defeat became clear.[1] The Americans landed 54,000 troops, mainly Marines, during the first day, another 6,000 on the third day, and a final 13,000 on the sixth day.

We'll use a "conventional combat" differential equation model developed by Frederick Lanchester in 1916. In this model, $A(t)$ and $J(t)$ are the number of active American and Japanese troops at time t, respectively. We don't distinguish between combat troops and support troops,

---

[1] A few Japanese hid in caves on the island long afterwards, with the last surrendering six years later, in 1951.

and a soldier is "active" until killed or wounded severely enough that he is removed from combat. Lanchester's conventional combat model assumes that $A(t)$ and $J(t)$ decrease at a rate proportional to the size of the *opposing* force. The proportionality constant in each case is a measure of the "effectiveness" of the opposing force. In the equation for $\frac{dA}{dt}$, we also build in a term that shows the reinforcements for the American forces. No corresponding term will appear in the equation for $\frac{dJ}{dt}$, since there were no Japanese reinforcements. We'll use "days" as the units for t. Our model is

$$(6.50) \qquad \frac{dA}{dt} = -bJ + R(t),$$

$$(6.51) \qquad \frac{dJ}{dt} = -cA,$$

where b and c are positive constants and

$$(6.52) \qquad R(t) = \begin{cases} 54{,}000 & \text{if } 0 \le t < 1, \\ 0 & \text{if } 1 \le t < 2, \\ 6{,}000 & \text{if } 2 \le t < 3, \\ 0 & \text{if } 3 \le t < 5, \\ 13{,}000 & \text{if } 5 \le t < 6, \\ 0 & \text{if } 6 \le t < 36. \end{cases}$$

The "combat-effectiveness coefficients" b and c represent the average effectiveness of a Japanese or American soldier, respectively.

The initial values of $A(t)$ and $J(t)$ are $A(0) = 0$ and $J(0) = 22{,}000$. Using these, we apply the Laplace transform to equations (6.50) and (6.51). The result is the pair of equations

$$(6.53) \qquad s\mathcal{L}[A(t)] = -b\mathcal{L}[J(t)] + \mathcal{L}[R(t)]$$

and

$$(6.54) \qquad s\mathcal{L}[J(t)] - 22{,}000 = -c\mathcal{L}[A(t)].$$

Solving the second of these for $\mathcal{L}[J(t)]$ gives

$$(6.55) \qquad \mathcal{L}[J(t)] = -\frac{c}{s}\mathcal{L}[A(t)] + \frac{22{,}000}{s}.$$

Substitute equation (6.55) into equation (6.53) to see that

$$s\mathcal{L}[A(t)] = \frac{bc}{s}\mathcal{L}[A(t)] - \frac{22{,}000}{s}b + \mathcal{L}[R(t)],$$

or

$$(6.56) \qquad \left(s - \frac{bc}{s}\right)\mathcal{L}[A(t)] = -\frac{22{,}000}{s}b + \mathcal{L}[R(t)].$$

To compute $\mathcal{L}[R(t)]$ we write $R(t)$ in terms of step functions as

$$R(t) = 54{,}000[u_0(t) - u_1(t)] + 6{,}000[u_2(t) - u_3(t)] + 13{,}000[u_5(t) - u_6(t)],$$

so that

$$\begin{aligned}\mathcal{L}[R(t)] &= 54{,}000\left(\frac{1}{s} - \frac{e^{-s}}{s}\right) + 6{,}000\left(\frac{e^{-2s}}{s} - \frac{e^{-3s}}{s}\right) \\ &\quad + 13{,}000\left(\frac{e^{-5s}}{s} - \frac{e^{-6s}}{s}\right).\end{aligned}$$

6.5. Laplace transforms, simple systems, and Iwo Jima

Substituting this into equation (6.56) and solving for $\mathcal{L}[A(t)]$, we obtain after some algebra

$$\begin{aligned}
\mathcal{L}[A(t)] &= -22{,}000 \frac{b}{s^2 - bc} + 54{,}000 \left(\frac{1}{s^2 - bc} - \frac{e^{-s}}{s^2 - bc} \right) \\
&\quad + 6{,}000 \left(\frac{e^{-2s}}{s^2 - bc} - \frac{e^{-3s}}{s^2 - bc} \right) \\
&\quad + 13{,}000 \left(\frac{e^{-5s}}{s^2 - bc} - \frac{e^{-6s}}{s^2 - bc} \right).
\end{aligned}$$
(6.57)

To solve for $A(t)$ by computing the inverse transform, it's helpful to write the positive constant bc as d^2 and recall that

$$\mathcal{L}^{-1}\left[\frac{1}{s^2 - d^2}\right] = \frac{\sinh(dt)}{d},$$

so that

$$\mathcal{L}^{-1}\left[\frac{e^{-ks}}{s^2 - d^2}\right] = u_k(t) \frac{\sinh(d(t-k))}{d}$$

by equation (6.35). Thus

$$\begin{aligned}
A(t) &= -22{,}000 \frac{b}{d} \sinh(dt) + 54{,}000 \left(\frac{\sinh(dt)}{d} - u_1(t) \frac{\sinh(d(t-1))}{d} \right) \\
&\quad + 6{,}000 \left(u_2(t) \frac{\sinh(d(t-2))}{d} - u_3(t) \frac{\sinh(d(t-3))}{d} \right) \\
&\quad + 13{,}000 \left(u_5(t) \frac{\sinh(d(t-5))}{d} - u_6(t) \frac{\sinh(d(t-6))}{d} \right).
\end{aligned}$$
(6.58)

This function is most easily understood by considering t in various ranges: $0 < t < 1$, $1 \leq t < 2$, $2 \leq t < 3$, $3 \leq t < 5$, and so on. For example, if $1 \leq t < 2$, we have $u_1(t) = 1$ but $u_2(t) = u_3(t) = u_5(t) = u_6(t) = 0$, so that for t in this range,

$$A(t) = -22{,}000 \frac{b}{d} \sinh(dt) + 54{,}000 \frac{\sinh(dt)}{d} - 54{,}000 \frac{\sinh(d(t-1))}{d}.$$

In Exercise 5 you are asked to similarly compute

$$\begin{aligned}
J(t) &= 22{,}000 \cosh(dt) + 54{,}000 \frac{c}{d^2} \left[1 - \cosh(dt) - u_1(t) + u_1(t) \cosh(d(t-1)) \right] \\
&\quad + 6{,}000 \frac{c}{d^2} \left[u_2(t) - u_2(t) \cosh(d(t-2)) - u_3(t) + u_3(t) \cosh(d(t-3)) \right] \\
&\quad + 13{,}000 \frac{c}{d^2} \left[u_5(t) - u_5(t) \cosh(d(t-5)) - u_6(t) + u_6(t) \cosh(d(t-6)) \right].
\end{aligned}$$

To proceed with our model, we need to know the constants b and c (and thus also $d = \sqrt{bc}$). Using detailed daily casualty figures from military records, estimates for b and c have been worked out,[2] giving $b \approx 0.0544$, $c \approx 0.0106$, and $d \approx 0.0240$. The method used to obtain these estimates is described in Exercise 8. With these values, Fig. 6.17 shows the graph of $A(t)$ for $0 < t < 36$. This compares very well with actual daily figures; see Fig. 6.18.

---
[2] J. H. Engel, *A Verification of Lanchester's Law*, J. Operations Research Society of America, Vol. 2, 1954, pp. 163–171.

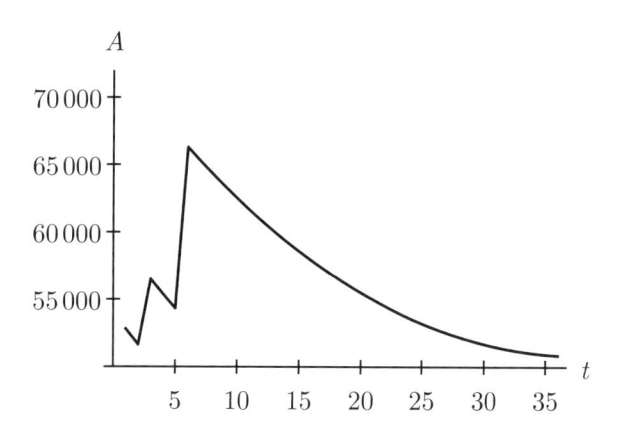

Figure 6.17. American troop strength from model.

Figure 6.18. American troop strength: actual (dotted) and model.

What does this actually show? The values of b and c have been computed from information available in military records, assuming the basic structure of our differential equation model is applicable. With the calculated values of b and c, we see that the solution function for $A(t)$ fits the known data very well. This is evidence that our differential equations (6.50)–(6.51) provide a good model for the Battle of Iwo Jima, but not necessarily that this is the only model which could be used.

6.5.1. Exercises. Note: Additional exercises involving the solution of systems by the Laplace transform method appear at the end of Section 8.8.3.

1. Solve
$$\frac{dx}{dt} = x + y + e^t,$$
$$\frac{dy}{dt} = 4x + y$$
with initial conditions $x(0) = 1$, $y(0) = 0$, using Laplace transforms.

2. Solve
$$\frac{dx}{dt} = y + 1,$$
$$\frac{dy}{dt} = 1 - x$$
with initial conditions $x(0) = 1$, $y(0) = 0$, using Laplace transforms.

3. Solve
$$\frac{dx}{dt} = -y,$$
$$\frac{dy}{dt} = -4x$$
with initial conditions $x(0) = -1$, $y(0) = 2$, using Laplace transforms.

4. Solve the equations
$$\frac{dx}{dt} = -ay,$$
$$\frac{dy}{dt} = -bx$$

6.5. Laplace transforms, simple systems, and Iwo Jima

for $x(t)$ and $y(t)$ using Laplace transforms, assuming $x(t) = x_0$ and $y(0) = y_0$. Here a and b are positive constants. The identity

(6.59) $$\frac{1}{s(s^2-d^2)} = \frac{1}{d^2}\left(\frac{s}{s^2-d^2} - \frac{1}{s}\right)$$

may be useful.

5. Solve the system in equations (6.50)–(6.51) for $J(t)$, using the expression for $\mathcal{L}[A(t)]$ from equation (6.57). Hint: Use the identity (6.59) in Exercise 4.

6. Use the formula (6.58) for $A(t)$ and equation (6.51) to find $J(t)$, by integrating both sides of (6.51) from 0 to t.

7. Solve the pair of equations

$$\frac{dg}{dt} = -k_1 g,$$

$$\frac{db}{dt} = k_1 g - k_2 b$$

for $g(t)$ and $b(t)$ using Laplace transforms and assuming $g(0) = g_0$ and $b(0) = 0$. We solved this system by another method in Section 5.3.2, where we used the equations to model aspirin levels in the blood and gastrointestinal track.

8. This exercise outlines a procedure for estimating the coefficients b and c in the Iwo Jima model.
    (a) From the equation
    $$\frac{dJ}{dt} = -cA$$
    we have
    $$\int_0^{36} \frac{dJ}{dt}\, dt = \int_0^{36} -cA(t)\, dt.$$
    By the Fundamental Theorem of Calculus, the left-hand side is $J(36) - J(0)$. On the right-hand side we have
    $$-c\int_0^{36} A(t)\, dt.$$
    Explain how this integral can be approximated if we have daily casualty figures for the American forces (so that we know $A(1), A(2), \ldots, A(36)$), and show how this leads to an approximation for c. Assume $J(36)$ and $J(0)$ are known.
    (b) Show that *if* we had daily casualty figures for the Japanese, we could estimate b by a similar procedure starting from
    $$\frac{dA}{dt} = -bJ + R(t).$$
    (c) However, figures on daily casualties for the Japanese are not directly available. To see how to *estimate* $J(i)$, the Japanese troop size on day i, start with
    $$\frac{dJ}{dt} = -cA$$
    and integrate on both sides from 0 to i. Show this gives
    $$J(i) = 22{,}000 - c\int_0^i A(t)\, dt.$$
    As in (a), show how to estimate
    $$\int_0^i A(t)\, dt$$

from data on daily American casualty figures. Since we know an estimate for c from (a), this allows us to estimate $J(i)$ for each i, and with these in hand, we can carry out the procedure in (b) to estimate b.

9. Consider the horizontal mass-spring system shown below. The first spring, with spring constant k_1, is attached to a wall on one end and has a mass m_1 on the other. This mass is connected by a second spring, with spring constant k_2, attached to a mass m_2. The horizontal surface along which the masses move is frictionless, and there is no externally acting force. When each spring is at its natural length (neither stretched nor compressed), each mass is at its equilibrium position. In the diagram below, 0_{m_1} and 0_{m_2} are, respectively, the equilibrium positions for the first and second masses. Denote by $U_1(t)$ and $U_2(t)$ the displacement of, respectively, the first and second mass from its equilibrium position (0_{m_1} and 0_{m_2}). A positive value for $U_1(t)$ (or $U_2(t)$) means the mass is to the right of its equilibrium position. The position functions satisfy the differential equations

$$m_1 U_1'' = -k_1 U_1 + k_2(U_2 - U_1),$$

$$m_2 U_2'' = -k_2(U_2 - U_1).$$

   (a) Determine the position functions U_1 and U_2 if the masses are $m_1 = 1$ kg and $m_2 = \frac{1}{2}$ kg, the spring constants are $k_1 = 2$ N/m and $k_2 = 1$ N/m, and we put the system into motion by holding the first mass at its equilibrium position, pulling the second mass 2 m to the right, and then releasing both masses simultaneously.
   (b) Show how to derive the differential equations for U_1 and U_2 using Hooke's law (see Section 4.1) and Newton's second law $F = ma$.

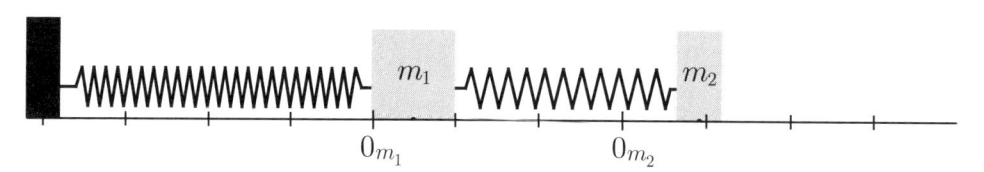

Figure 6.19. A double spring system.

10. (CAS) Use a computer algebra system that can compute inverse Laplace transforms to help with the computations in this problem. A patient is being treated with an intravenous medication. The amount of drug in the bloodstream at time t in minutes is $x(t)$ and the amount in the body tissues at time t is $y(t)$. If we initially have $y(0) = 0$ and $x(0) = 10$ mg and the patient continues to receive the drug intravenously (directly into the bloodstream) at a rate of 0.2 mg/min, the relevant differential equation model is

$$\frac{dx}{dt} = -0.07x + 0.04y + 0.2,$$

$$\frac{dy}{dt} = 0.07x - 0.04y$$

with $x(0) = 10$ and $y(0) = 0$.
   (a) Solve for $x(t)$ and $y(t)$ using Laplace transforms.
   (b) Does the concentration in the bloodstream stay within the therapeutic range of 1 mg/l and 5 mg/l for $0 \leq t \leq 180$? Assume the patient's blood volume is 5 liters.
   (c) If instead $x(0) = 0$, about how long does it take to reach the therapeutic concentration of 1 mg/l?

6.6. Convolutions

Because the inverse Laplace transform is linear, an inverse transform of the sum of two functions is the sum of inverse transforms of each function. What about the inverse transform of a *product* of two functions? The Second Shift Formula in equation (6.35) handled a very special kind of product

$$\mathcal{L}^{-1}[e^{-as}F(s)] = u_a(t)f(t-a),$$

where $\mathcal{L}[f(t)] = F(s)$, but we would like to be able to find inverse transforms of more general products. The key to doing this is a concept called the **convolution**. Here is how it is defined.

Definition 6.6.1. The convolution $f * g$ of two piecewise continuous functions $f(t)$ and $g(t)$ is defined as

(6.60) $$(f * g)(t) = \int_0^t f(t-u)g(u)\,du.$$

In the integral on the right-hand side of (6.60), the variable of integration is u, and t is treated as a constant in the integration. As an example, let's compute $e^{2t} * e^{3t}$:

$$e^{2t} * e^{3t} = \int_0^t e^{2(t-u)}e^{3u}\,du = \int_0^t e^{2t}e^u\,du = e^{2t}e^u\,\big|_{u=0}^{u=t} = e^{3t} - e^{2t}.$$

The convolution operation has certain "product-like" properties (see Exercise 12):

- Convolution is commutative: $f * g = g * f$.
- Convolution is associative: $f * (g * h) = (f * g) * h$.
- Convolution distributes over addition: $f * (g + h) = (f * g) + (f * h)$.

However, $1 * f \neq f$. Notice that according to the definition

$$1 * f(t) = \int_0^t f(u)\,du.$$

If we use this, for example, with $f(t) = t$, we see that

$$1 * t = \int_0^t u\,du = \frac{1}{2}u^2\,\big|_{u=0}^{u=t} = \frac{1}{2}t^2,$$

which is certainly different from $f(t) = t$.

The connection between convolution and Laplace transforms is described succinctly in the next theorem.

Theorem 6.6.2 (Convolution Theorem).

$$\mathcal{L}[f * g] = \mathcal{L}[f]\mathcal{L}[g].$$

On the right-hand side, you have the ordinary product of the two functions $\mathcal{L}[f]$ and $\mathcal{L}[g]$. The theorem asserts that this is equal to the Laplace transform of the convolution $f * g$. The proof of Theorem 6.6.2 is given below.

We can recast the statement of the Convolution Theorem as a result for inverse Laplace transforms. In this form, it says

(6.61) $$\mathcal{L}^{-1}[F(s)G(s)] = \mathcal{L}^{-1}[F(s)] * \mathcal{L}^{-1}[G(s)].$$

This gives us what we are seeking—a way of computing the inverse transform of a product of two functions if we know an inverse transform of each factor. Let's look at some examples.

Example 6.6.3. We'll determine
$$\mathcal{L}^{-1}\left[\frac{a}{s^2(s^2+a^2)}\right]$$
using Theorem 6.6.2. We can write
$$\frac{a}{s^2(s^2+a^2)} = \frac{1}{s^2} \cdot \frac{a}{s^2+a^2}$$
and then recognize that
$$\frac{1}{s^2}$$
is the Laplace transform of the function $f(t) = t$ while
$$\frac{a}{s^2+a^2}$$
is the Laplace transform of the function $g(t) = \sin(at)$. According to equation (6.61),
$$\mathcal{L}^{-1}\left[\frac{1}{s^2} \cdot \frac{a}{s^2+a^2}\right] = \mathcal{L}^{-1}\left[\frac{1}{s^2}\right] * \mathcal{L}^{-1}\left[\frac{a}{s^2+a^2}\right]$$
$$= t * \sin(at)$$
$$= \int_0^t (t-u)\sin(au)\,du.$$
This integral can be computed by integration by parts:
$$\int_0^t (t-u)\sin(au)\,du = -(t-u)\frac{\cos(au)}{a}\bigg|_{u=0}^{u=t} - \int_0^t \frac{\cos(au)}{a}\,du$$
$$= \frac{t}{a} - \frac{1}{a^2}\sin(at).$$

Example 6.6.4. Let's use the Convolution Theorem to find
$$\mathcal{L}^{-1}\left[\frac{1}{(s^2+1)^2}\right].$$
We know that
$$\mathcal{L}[\sin t] = \frac{1}{s^2+1}.$$
Equation (6.61) then tell us that
$$\mathcal{L}^{-1}\left[\frac{1}{(s^2+1)^2}\right] = \sin t * \sin t.$$
In Exercise 7 you'll show that
$$\sin t * \sin t = \frac{1}{2}\sin t - \frac{t}{2}\cos t.$$
Thus
$$\mathcal{L}^{-1}\left[\frac{1}{(s^2+1)^2}\right] = \frac{1}{2}\sin t - \frac{t}{2}\cos t.$$

As another application of the Convolution Theorem, we'll show how to solve the nonhomogeneous initial value problem
(6.62) $$y'' + a_1 y' + a_0 y = g(t), \quad y(0) = y'(0) = 0,$$
if we have solved the homogeneous problem
(6.63) $$y'' + a_1 y' + a_0 y = 0, \quad y(0) = 0, \quad y'(0) = 1.$$
Notice that the initial conditions for the second equation are *not* the same as for the first.

6.6. Convolutions

Call the solution to (6.62) $y_1(t)$ and the solution to (6.63) $y_2(t)$. Apply the Laplace transform to both sides of (6.63), using the First Differentiation Formula and the initial conditions to see that
$$s^2 \mathcal{L}[y_2(t)] - 1 + a_1 s \mathcal{L}[y_2(t)] + a_0 \mathcal{L}[y_2(t)] = 0.$$
Solve this for $\mathcal{L}[y_2(t)]$ to obtain

(6.64) $$\mathcal{L}[y_2(t)] = \frac{1}{s^2 + a_1 s + a_0}.$$

Applying the Laplace transform in equation (6.62) and using the initial conditions there gives
$$s^2 \mathcal{L}[y_1(t)] + a_1 s \mathcal{L}[y_1(t)] + a_0 \mathcal{L}[y_1(t)] = \mathcal{L}[g(t)].$$
A little algebra, together with equation (6.64), rewrites this as

(6.65) $$\mathcal{L}[y_1(t)] = \frac{\mathcal{L}[g(t)]}{s^2 + a_1 s + a_0} = \mathcal{L}[y_2(t)]\mathcal{L}[g(t)].$$

But the Convolution Theorem says that

(6.66) $$\mathcal{L}[y_2(t)]\mathcal{L}[g(t)] = \mathcal{L}[y_2(t) * g(t)].$$

Comparing (6.65) and (6.66), we conclude that
$$y_1(t) = y_2(t) * g(t).$$

Thus the nonhomogeneous initial value problem in (6.62) has solution $y_2 * g$ where y_2 solves homogeneous initial value problem in (6.63).

Proof of the Convolution Theorem. Our argument will use some ideas from multivariable calculus on double integrals. Using the definitions of the Laplace transform and of convolution, we have

(6.67) $$\mathcal{L}[f * g] = \int_0^\infty (f * g)(t) e^{-st} \, dt = \int_{t=0}^{t=\infty} e^{-st} \left(\int_{u=0}^{u=t} f(t-u) g(u) \, du \right) dt.$$

This describes $\mathcal{L}[f * g]$ as an *iterated integral*. The inequalities $0 \leq t < \infty$, $0 \leq u \leq t$, which correspond to the limits of integration on the "dt" and "du" integrals, respectively, describe the infinite triangular region R shown in Fig. 6.20 by "vertical cross sections"; each vertical line segment in R occurs at a t-value between 0 and ∞ and stretches from $u = 0$ to $u = t$.

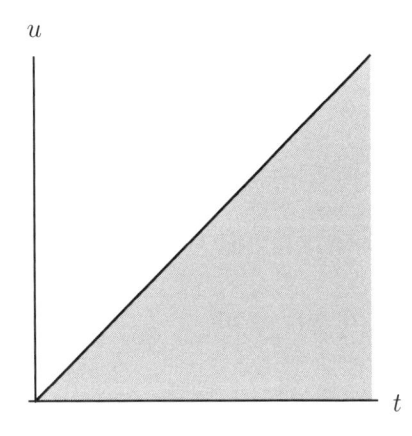

Figure 6.20. The region R.

The iterated integral in (6.67) can thus be written as a double integral over this region R:

(6.68) $$\mathcal{L}[f * g] = \int\int_R e^{-st} f(t-u) g(u) \, dA.$$

By viewing R as made up of *horizontal* cross sections, we will reverse the order of integration in the iterated integral in (6.67). Each horizontal cross section of R occurs for a fixed value of u between 0 and ∞ and stretches in the horizontal direction from $t = u$ to $t = \infty$. Thus the inequalities $0 \le u < \infty$, $u \le t < \infty$ describe the region by horizontal cross sections. The double integral in (6.68) is then given by the iterated integral (in reversed order from (6.67))

$$\int_{u=0}^{u=\infty} \int_{t=u}^{t=\infty} f(t-u) e^{-st} g(u) \, dt \, du.$$

We can compute the inner integral in this iterated integral:

$$\begin{aligned}
\int_{t=u}^{t=\infty} f(t-u) e^{-st} g(u) \, dt &= g(u) \int_{t=u}^{t=\infty} f(t-u) e^{-st} \, dt \\
&= g(u) \int_{w=0}^{w=\infty} f(w) e^{-s(u+w)} \, dw \\
&= g(u) e^{-su} \int_{w=0}^{w=\infty} f(w) e^{-sw} \, dw \\
&= g(u) e^{-su} \mathcal{L}[f](s).
\end{aligned}$$

In the second equality we made the substitution $w = t - u$, $dw = dt$. Thus

$$\mathcal{L}[f * g] = \int_{u=0}^{u=\infty} g(u) e^{-su} \mathcal{L}[f](s) \, du = \mathcal{L}[f](s) \int_0^\infty g(u) e^{-su} \, du = \mathcal{L}[f](s) \mathcal{L}[g](s).$$

The penultimate equality in the preceding line is justified since the variable of integration is u, so the function $\mathcal{L}[f](s)$ can be moved outside the integral sign.

6.6.1. Exercises.

In Exercises 1–4 evaluate the given convolution.

1. $t * t^3$.
2. $1 * 1$.
3. $t * e^{-4t}$.
4. $e^t * e^{2t}$.

5. (a) Show that
$$e^{at} * e^{bt} = \frac{e^{at} - e^{bt}}{a - b}$$
if $a \ne b$.

(b) What is $e^{bt} * e^{bt}$? Is your answer the same as what you get by computing
$$\lim_{a \to b} \frac{e^{at} - e^{bt}}{a - b}?$$
You can use l'Hôpital's Rule to compute this limit if you think of b and t as constant and a as your variable.

6. Show that $\cos t * \cos t$ is negative for some values of t. In particular, this shows that $f * f$ is not f^2.

7. Verify that
$$\sin(\alpha t) * \sin(\alpha t) = \frac{1}{2\alpha} \sin(\alpha t) - \frac{t}{2} \cos(\alpha t).$$

6.6. Convolutions

In Exercises 8–11 use equation (6.61) to determine the inverse Laplace transform. The result in Exercise 5 may also be helpful.

8. $\mathcal{L}^{-1}\left[\dfrac{1}{(s-1)(s+4)}\right].$

9. $\mathcal{L}^{-1}\left[\dfrac{1}{(s-a)(s-b)}\right], \; a \neq b.$

10. $\mathcal{L}^{-1}\left[\dfrac{1}{s(s^2+2s+2)}\right].$

11. $\mathcal{L}^{-1}\left[\dfrac{4}{(s-1)(s^2+4)}\right].$

12. (a) Show that $f * g = g * f$. Hint: In equation (6.60), make the substitution $w = t - u$ in the integral.
    (b) Show that $f * (g + h) = f * g + f * h$.

13. (a) Show that the initial value problem
    $$y'' + k^2 y = f(t), \quad y(0) = y'(0) = 0,$$
    has solution
    $$y = \dfrac{1}{k}\sin(kt) * f.$$

    (b) Show that the initial value problem
    $$y'' + k^2 y = f(t), \quad y(0) = \alpha, \quad y'(0) = \beta,$$
    has solution
    $$y = \dfrac{1}{k}\sin(kt) * f + \alpha \cos(kt) + \dfrac{\beta}{k}\sin(kt).$$

    (c) Consider any constant coefficient linear differential equation
    $$a_n y^{(n)} + a_{n-1} y^{(n-1)} + \cdots + a_1 y' + a_0 y = f(t)$$
    with initial conditions $y(0) = y'(0) = \cdots = y^{(n-1)}(0) = 0$. Show that this has solution
    $$y(t) = f * g$$
    if $g(t)$ is the inverse Laplace transform of $1/P(s)$, where $P(s)$ is the corresponding characteristic polynomial
    $$P(s) = a_n s^n + a_{n-1} s^{n-1} + \cdots + a_1 s + a_0.$$

    (d) Verify that the result of (a) is a special case of the result in (c).

14. The equation
    $$\int_0^t f(t-u) e^{-u}\, du = t^2 e^{-t}$$
    is called an integral equation for unknown function f. Find f.

15. (a) Show that $\mathcal{L}[f * g'] - \mathcal{L}[g * f'] = \mathcal{L}[f(0)g - g(0)f]$, and conclude that $f * g' - g * f' = f(0)g - g(0)f$.
    (b) Similarly show that
    $$(f * g)' = f * g' + g(0)f.$$

6.7. The delta function

In Example 6.4.2, we considered various ways that a dose of a medication could be given to a patient: all at once, at a constant rate over a period of time, or at a higher constant rate over shorter periods of time, with breaks between the periods of administration. Typically, the same total amount of drug is to be given no matter what the method of administration. To introduce the idea of the delta function, let's imagine we plan to give a total of A mg of a drug by IV infusion in a 24-hour period. If we infuse half of the total amount at a constant rate over a 1-hour period every 12 hours and the drug is removed from the body at a rate proportional to the amount present, we model the total amount $y(t)$ of drug in the body at time t (in hours) by the differential equation

(6.69) $$\frac{dy}{dt} = f_1(t) - ky,$$

where

$$f_1(t) = \frac{A}{2}[u_0(t) - u_1(t)] + \frac{A}{2}[u_{12}(t) - u_{13}(t)].$$

The graph of $f_1(t)$ is as shown in Fig. 6.21. The area under the graph of $f_1(t)$, which is the same as $\int_0^\infty f_1(t)\,dt$, is A, the total amount of drug given. Keep in mind that solving the differential equation (6.69) using Laplace transforms requires us to compute $\mathcal{L}[f_1(t)]$.

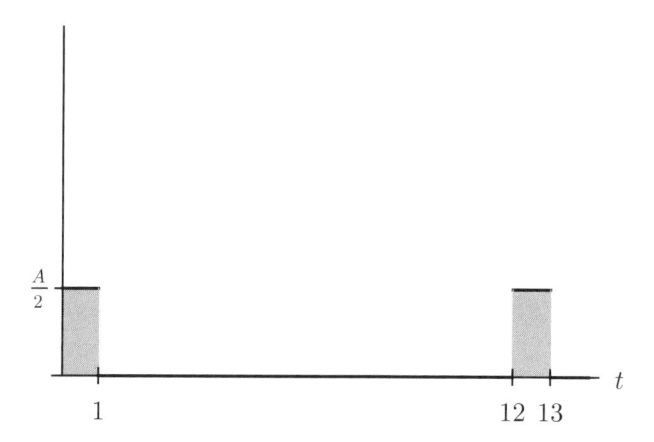

Figure 6.21. Graph of $f_1(t)$.

Now our patient would prefer to have the periods of infusion be shorter. If he convinces his doctor to let the medication run in at twice the rate for only 1/2-hour periods, the relevant differential equation is

$$\frac{dy}{dt} = f_2(t) - ky$$

where $f_2(t) = A[u_0(t) - u_{1/2}(t)] + A[u_{12}(t) - u_{25/2}(t)]$, with graph as shown. He's still receiving the same total amount A of the drug, and consequently we still have

$$\int_0^\infty f_2(t)\,dt = A.$$

6.7. The delta function

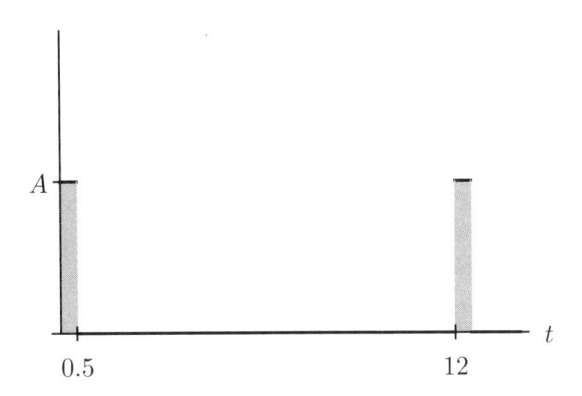

Figure 6.22. Graph of $f_2(t)$.

If we decrease the periods of infusion even more, from 1/2 hour to 15 minutes, we replace $f_2(t)$ by the function $f_3(t)$ whose graph is shown in Fig. 6.23. Compared to $f_2(t)$, $f_3(t)$ has thinner, but taller, spikes starting at $t = 0$ and $t = 12$, and the total area under the graph of f_3 is still A.

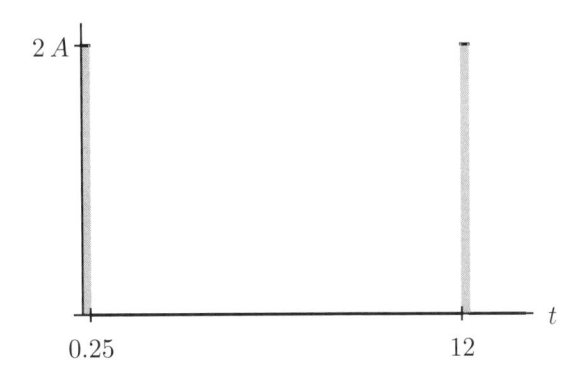

Figure 6.23. Graph of $f_3(t)$.

What happens in the limit as we decrease the infusion times "to 0"? It's not immediately clear what this means, since as we decrease the infusion time, the rate of infusion goes up (so that the patient gets the same total amount), and to decrease the time to 0 would require increasing the rate "to infinity". In other words, it looks like, in the limit, we want a "function" $f(t)$ which has value ∞ at $t = 0$ and $t = 12$ and is 0 everywhere else. The key to making sense of this is to recognize that what we're really interested in is not so much $f(t)$, but rather its Laplace transform. That's what's needed when we solve the associated differential equation. So even if by itself $f(t)$ doesn't appear to be a function in the usual sense, perhaps we can still find a way to make sense of $\mathcal{L}[f(t)]$.

To proceed further, let's introduce some notation and define the basic building blocks for our "function" $f(t)$. The definition is motivated by thinking about the functions $g_h(t)$ defined for $h > 0$ by

$$(6.70) \qquad g_h(t) = \begin{cases} \frac{1}{h} & \text{if } 0 \leq t < h, \\ 0 & \text{if } h \leq t < \infty \text{ or } t < 0. \end{cases}$$

No matter our choice of $h > 0$, the area under the graph of this function is 1. Expressed as an integral, this says

$$\int_{-\infty}^{\infty} g_h(t)\, dt = 1.$$

Notice that as h gets smaller, the height of the spike in the graph of g_h gets bigger, tending to ∞ as h decreases to 0.

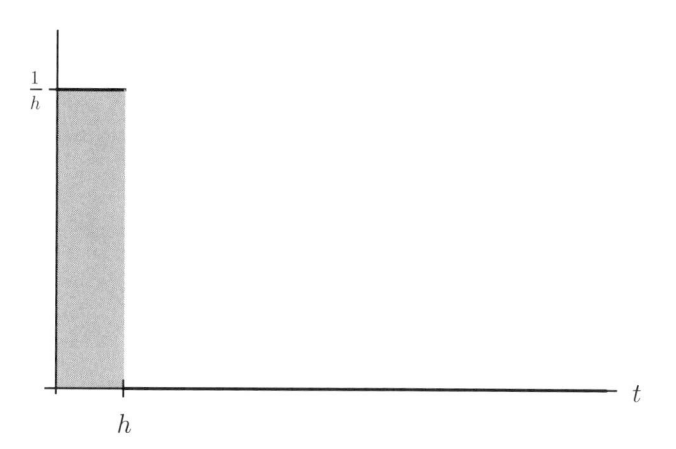

Figure 6.24. Graph of $g_h(t)$.

The delta function. The **unit impulse function** or **Dirac delta function**, denoted $\delta(t)$, is designed to capture the limiting behavior of the functions $g_h(t)$ as h decreases to 0. Thus we want $\delta(t)$ to have these two properties:

(i)
$$\int_{-\infty}^{\infty} \delta(t)\,dt = 1.$$

(ii) $\delta(t) = 0$ for any $t \neq 0$.

Because $\delta(t) = 0$ everywhere except at $t = 0$, condition (i) also says that the integral of $\delta(t)$ over *any* interval containing 0 is 1. We should really invent some alternative to the word "function" for $\delta(t)$, because it is not a function in the ordinary sense,[3] but we will follow convention and continue to refer to it as a function. When it comes to actually working with $\delta(t)$, we will use it within an integral. As an example, let's determine the Laplace transform of $\delta(t)$. Thinking of $\delta(t)$ as the limit of the functions $g_h(t)$ as h decreases to 0, we first calculate

$$\mathcal{L}[g_h(t)] = \int_0^{\infty} g_h(t) e^{-st}\,dt = \int_0^h \frac{1}{h} e^{-st}\,dt = \frac{1}{h} \int_0^h e^{-st}\,dt = \frac{1 - e^{-hs}}{hs}.$$

What happens to this as h decreases to 0? For fixed $s > 0$ we can use l'Hôpital's Rule to compute

$$\lim_{h \to 0} \frac{1 - e^{-hs}}{hs} = \lim_{x \to 0} \frac{1 - e^{-x}}{x} = \lim_{x \to 0} \frac{e^{-x}}{1} = 1.$$

L'Hôpital's Rule is used in the middle step, after we substituted x for hs. Since we think of the delta function as the limiting value of the functions $g_h(t)$ (as $h \to 0$), this suggests we should have

$$(6.71) \qquad \mathcal{L}[\delta(t)] = 1.$$

Keep in mind that $\delta(t)$ is not a function in the usual sense.

---
[3]The term *distribution* or *generalized function* is sometimes used.

6.7. The delta function

Formally, the definition of $\mathcal{L}[\delta(t)]$ is

$$\int_0^\infty e^{-st}\delta(t)\,dt,$$

and we computed this integral as

$$\lim_{h\to 0}\int_0^\infty e^{-st}g_h(t)\,dt.$$

More generally, we can compute an integral of the form

(6.72) $$\int_0^\infty f(t)\delta(t-a)dt,$$

for any $a \geq 0$ and function $f(t)$ that is continuous on some interval containing a, by thinking about $\delta(t-a)$ as the limiting value of the functions $g_h(t-a)$ in Fig. 6.25.

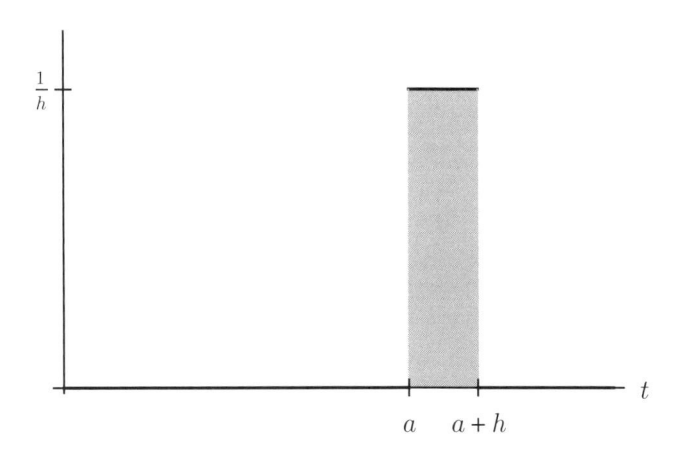

Figure 6.25. Graph of $g_h(t-a)$.

We have

$$\int_0^\infty g_h(t-a)f(t)\,dt = \int_a^{a+h} \frac{1}{h}f(t)\,dt = \frac{1}{h}\int_a^{a+h} f(t)\,dt$$
$$= \text{average value of } f \text{ over the interval } [a, a+h].$$

As h decreases to 0, the average value of f over $[a, a+h]$ tends to $f(a)$ (the continuity of f is being used to make this assertion). Thus we define

$$\int_0^\infty f(t)\delta(t-a)dt = f(a).$$

In particular, if $f(t) = e^{-st}$, we arrive at the Laplace transform of $\delta(t-a)$, where $a \geq 0$:

(6.73) $$\mathcal{L}[\delta(t-a)] = \int_0^\infty e^{-st}\delta(t-a)\,dt = e^{-sa}.$$

We also note that the "rectangular spike function" $g_h(t-a)$ can be expressed in terms of unit step functions by

$$g_h(t-a) = \frac{1}{h}[u_a(t) - u_{a+h}(t)].$$

Thus in the example of medication infusion at the beginning of this section, the infusion functions f_1, f_2, and f_3 can all be written as

$$\text{(6.74)} \qquad \frac{A}{2} g_h(t) + \frac{A}{2} g_h(t-12)$$

for appropriate values of h. The choice $h = 1$ gives $f_1(t)$, the choice $h = \frac{1}{2}$ gives $f_2(t)$, and $h = \frac{1}{4}$ yields $f_3(t)$. The units for h are hours, and h gives the length of each individual infusion period. By letting $h \to 0$, we should obtain the instantaneous infusion function

$$\text{(6.75)} \qquad f(t) = \frac{A}{2}\delta(t) + \frac{A}{2}\delta(t-12).$$

It follows from equation (6.73) that

$$\mathcal{L}[f(t)] = \frac{A}{2} + \frac{A}{2}e^{-12s}.$$

Units on the delta function are 1/time. You can easily check this in the context of our pharmacology example, where the units on $f(t)$ in (6.75) are mg/hour, while the units on A are mg, leaving units of "per hour", or 1/time, for the delta function.

Example 6.7.1. Suppose we give a patient an amount $A/2$ of a drug "instantaneously" at times $t = 0$ and $t = 12$, so that the total amount of drug administered is A. Assuming that the drug is removed from the body at a rate proportional to the amount present, we'll find the amount $y(t)$ in the body at time t hours after the first dose is administered. The differential equation for y is

$$\frac{dy}{dt} = -ky + \frac{A}{2}\delta(t) + \frac{A}{2}\delta(t-12).$$

Applying the Laplace transform and assuming $y(0) = 0$, we have

$$s\mathcal{L}[y] + k\mathcal{L}[y] = \frac{A}{2} + \frac{A}{2}e^{-12s}.$$

Solving for $\mathcal{L}[y(t)]$ gives

$$\mathcal{L}[y(t)] = \frac{A}{2}\left(\frac{1}{s+k} + \frac{e^{-12s}}{s+k}\right).$$

The inverse transform yields

$$y(t) = \frac{A}{2}\left(e^{-kt} + u_{12}(t)e^{-k(t-12)}\right).$$

The graph of $y(t)$, with $A = 2$ and $k = 0.2$, is shown in Fig. 6.26.

6.7. The delta function

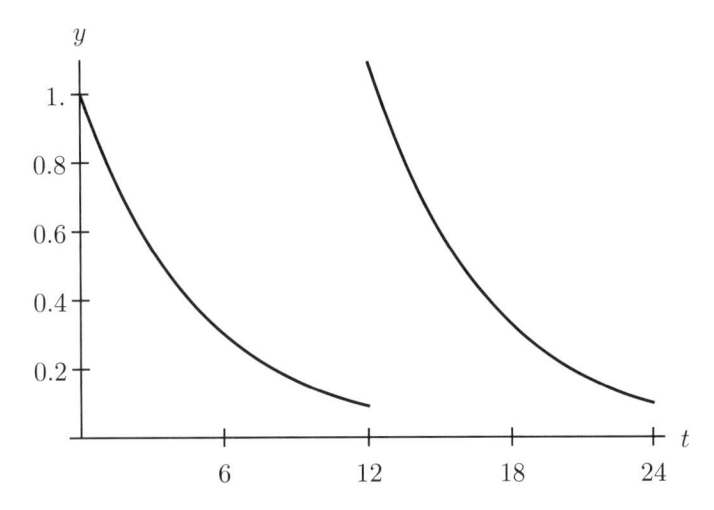

Figure 6.26. $y(t) = e^{-0.2t} + u_{12}(t)e^{-0.2(t-12)}$.

In a mass-spring problem, the delta function appears (as the limiting case) when we have an applied force which is large but only acts for a short time.

Example 6.7.2. A 1-kg mass hangs from a spring with spring constant $k = 10$ newtons/meter, and the mass is attached to a dashpot whose corresponding damping constant is $\frac{2}{5}$ kg/second. At time $t = 0$ the mass is hanging at rest in the equilibrium position when it is struck by a hammer, which imparts an impulse force of 20 kg-m/sec. The force delivered by the hammer strike is modeled by a delta function, and thus the displacement of the mass $y(t)$ is modeled by

$$y'' + \frac{2}{5}y' + 10y = 20\,\delta(t), \quad y(0) = y'(0) = 0.$$

The units on both sides of this equation are kg-m/sec^2, or newtons, and the delta function has units of 1/sec. Applying the Laplace transform, we have

$$s^2 \mathcal{L}[y] + \frac{2}{5}s\mathcal{L}[y] + 10\mathcal{L}[y] = 20.$$

This gives

$$\mathcal{L}[y] = 20\frac{1}{s^2 + \frac{2}{5}s + 10} = 20\frac{1}{(s+\frac{1}{5})^2 + \frac{249}{25}}.$$

Applying the inverse transform to solve for y we have

(6.76) $$y(t) = \frac{100}{\sqrt{249}}e^{-\frac{1}{5}t}\sin\frac{\sqrt{249}}{5}t, \quad t > 0,$$

or approximately

$$\frac{20}{\sqrt{10}}e^{-\frac{1}{5}t}\sin(\sqrt{10}\,t).$$

The graph of $y(t)$ is shown in Fig. 6.27.

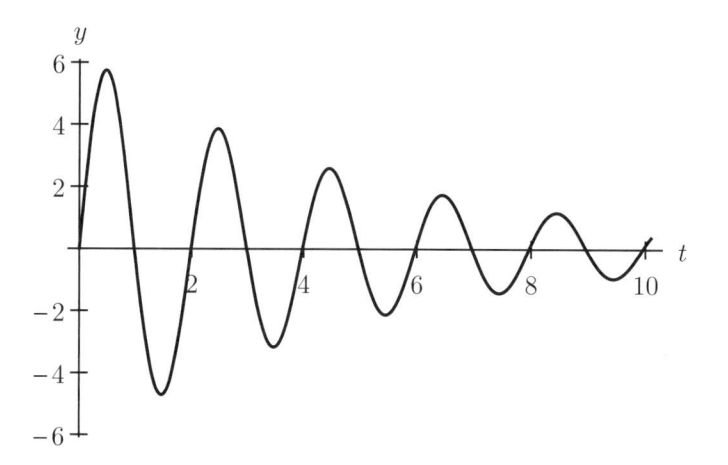

Figure 6.27. Graph of $y(t)$.

Notice that the solution in equation (6.76) doesn't actually satisfy the initial condition $y'(0) = 0$! The velocity $y'(t)$ has a jump discontinuity at $t = 0$. The mass is at rest for $t < 0$ so y' is 0 then, but the hammer strike gives it an instantaneous velocity that is close to 20 for t just bigger than 0. In terms of left- and right-hand limits, we have

$$0 = \lim_{t \to 0^-} y'(t) \neq \lim_{t \to 0^+} y'(t) = 20.$$

Moreover, these limit values correspond, respectively, to the left-hand and right-hand derivatives of y at 0, so that $y'(0)$ does not exist. We see this lack of differentiablity at $t = 0$ as a "corner" at $t = 0$ in the graph of the position function $y(t)$; see Figs. 6.28–6.29.

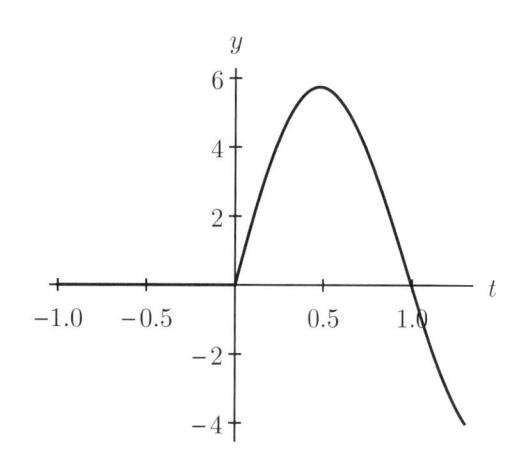

Figure 6.28. Corner at $t = 0$ in graph of $y(t)$.

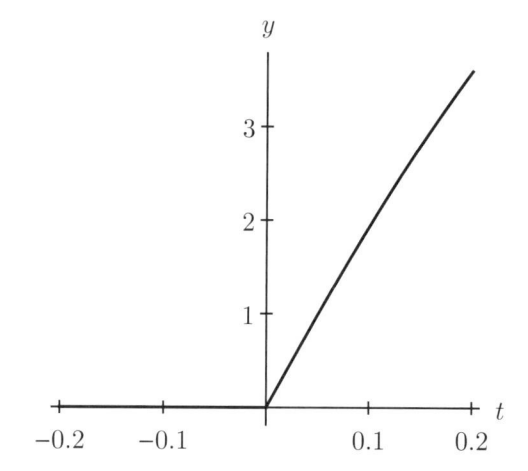

Figure 6.29. Expanded view near $t = 0$.

Generalized functions and generalized derivatives. We have noted that even though $\delta(t)$ and more generally $\delta(t-a)$ are not true functions, they can be thought of as "generalized functions" via how they integrate against actual functions. Recall that for a function f which is continuous on some interval containing $a > 0$ we should have

$$\int_0^\infty f(t)\delta(t-a)\, dt = f(a).$$

6.7. The delta function

Moreover, we saw that taking $f(t) = e^{-st}$ in this equation gives the formula

(6.77) $$\mathcal{L}[\delta(t-a)] = e^{-as}$$

for the Laplace transform of $\delta(t-a)$. This point of view allows us to think about finding "generalized derivatives" of actual functions that don't have ordinary derivatives at every point.

For example, if $a > 0$, consider the unit step function $u_a(t)$ defined by equation (6.6). The derivative $u_a'(t)$ exists and equals zero at every t *except* for $t = a$ where $u_a'(t)$ is undefined (this should be clear from the graph of $u_a(t)$ in Fig. 6.1). However, suppose that $u_a'(t)$ exists everywhere as a "generalized derivative" and has Laplace transform that obeys the First Differentiation Formula for $u_a(t)$:

(6.78) $$\mathcal{L}[u_a'(t)] = s\mathcal{L}[u_a(t)] - u_a(0).$$

Since $u_a(0) = 0$ and $\mathcal{L}[u_a(t)] = e^{-as}/s$, we have $\mathcal{L}[u_a'(t)] = e^{-as}$. Thus $u_a'(t)$ and $\delta(t-a)$ have the same Laplace transform (see equation (6.77)):

$$\mathcal{L}[u_a'(t)] = \mathcal{L}[\delta(t-a)].$$

Formally taking inverse Laplace transforms then gives

(6.79) $$u_a'(t) = \delta(t-a).$$

This can be made mathematically rigorous with the use of some more advanced ideas, but we should feel free to use equation (6.79) here as a possible tool in solving differential equations. You're invited to try this out in Exercise 20, which illustrates how the delta function appears naturally in applications of electric circuits. For a different (but still heuristic) approach to the formula in (6.79) when $a = 0$, see Exercise 19.

6.7.1. Exercises.

In Exercises 1–12 solve the initial value problem.

1. $y' = 4\delta(t), y(0) = 0.$
2. $y' = \delta(t-2), y(0) = 1.$
3. $y' = \delta(t) - \delta(t-2), y(0) = 0.$
4. $y'' + y = 2\delta(t-\pi), y(0) = y'(0) = 0.$
5. $y'' = 2\delta(t-\pi), y(0) = y'(0) = 0.$
6. $y'' + y = \delta(t) - \delta(t-1), y(0) = y'(0) = 0.$
7. $y'' + y = 3\delta(t - \frac{\pi}{2}), y(0) = 0, y'(0) = 1.$
8. $y'' + 2y' + 2y = \delta(t), y(0) = y'(0) = 0.$
9. $y'' + 2y' + 2y = \delta(t-4), y(0) = y'(0) = 0.$
10. $y'' + y = 2\sin t + \delta(t - 3\pi), y(0) = y'(0) = 0.$ Exercise 23 in Section 6.3 may be helpful.
11. $y'' + y = \delta(t) + \delta(t-\pi) + \delta(t-2\pi) + \delta(t-3\pi), y(0) = y'(0) = 0.$
12. $y'' + y = \delta(t) + \delta(t-2\pi) + \delta(t-4\pi) + \delta(t-6\pi), y(0) = y'(0) = 0.$

13. Solve the initial value problem
$$y'' + 3y' + 2y = \delta(t-5) + u_8(t), \qquad y(0) = 0, \quad y'(0) = 1/2.$$

14. In a (vertical) mass-spring system, a mass $m = 1$ g is attached to a spring with spring constant $k = 4$ gm/sec^2. At time $t = 0$ the mass is in equilibrium position and has velocity 1 cm/sec. At time $t = 4$ the mass is struck by a hammer which gives an impulse force of 2 gm-cm/sec to the mass. Find the displacement function and sketch its graph over $0 < t < 2\pi$, neglecting damping.

15. Sam and Sally are discussing the initial value problem
(6.80) $$y'' + 2y' + y = 4\delta(t-2), y(0) = y'(0) = 0.$$

Sam says, "Since $\delta(t-2)$ is 0 when $t \neq 2$ and ∞ when $t = 2$ and since $4 \cdot 0 = 0$ and $4 \cdot \infty = \infty$, the function $4\delta(t-2)$ is the same as $\delta(t-2)$. So the solution to (6.80) is the same as the solution to
$$y'' + 2y' + y = \delta(t-2), \quad y(0) = y'(0) = 0."$$
Sally says, "That can't be right. The factor of 4 must make *some* difference. Maybe there's something about $\delta(t-2)$ not being a genuine function...." Resolve this difficulty for Sam and Sally, and correctly solve (6.80).

16. In each part, two initial value problems are given. Call the solution to the first problem $y_1(t)$ and to the second $y_2(t)$. Show $y_1(t) = y_2(t)$ for $t > 0$.

    (a)
    $$y'(t) + by(t) = k\delta(t), \quad y(0) = 0,$$
    and
    $$y'(t) + by(t) = 0, \quad y(0) = k.$$

    (b)
    $$y'' + by' + cy = k\delta(t), \quad y(0) = y'(0) = 0,$$
    and
    $$y'' + by' + cy = 0, \quad y(0) = 0, \quad y'(0) = k.$$

    (c)
    $$y'' + by' + cy = f(t) + k\delta(t), \quad y(0) = y'(0) = 0,$$
    and
    $$y'' + by' + cy = f(t), \quad y(0) = 0, \quad y'(0) = k.$$

17. Answer the following question for each of the initial value problems in (a) and (b) of Exercise 16: Does the solution satisfy the initial conditions? For (b), you may assume $4c - b^2 > 0$.

18. (a) Fix $h > 0$ and solve
    $$y'' + y = g_h(t), \quad y(0) = y'(0) = 0,$$
    where

    (6.81) $$g_h(t) = \begin{cases} \frac{1}{h} & \text{if } 0 \leq t < h, \\ 0 & \text{if } h \leq t < \infty. \end{cases}$$

    Show that the solution $y(t)$ can be described as

    (6.82) $$y(t) = \begin{cases} \frac{1-\cos t}{h} & \text{if } 0 \leq t < h, \\ \frac{\cos(t-h) - \cos t}{h} & \text{if } h \leq t < \infty. \end{cases}$$

    (b) Show that
    $$\lim_{h \to 0} \frac{1 - \cos h}{h} = 0$$
    and that for $0 < h < \pi$ and $0 \leq t < h$ we have
    $$0 \leq \frac{1 - \cos t}{h} \leq \frac{1 - \cos h}{h}.$$
    Explain why it follows that
    $$\lim_{h \to 0} \frac{1 - \cos t}{h} = 0.$$
    Finally show that for t constant,
    $$\lim_{h \to 0} \frac{\cos(t-h) - \cos t}{h} = \sin t.$$
    Hint: If you use l'Hôpital's Rule for this last limit, keep in mind that the variable is h and t is a constant.

6.7. The delta function

(c) Solve
$$y'' + y = \delta(t), \quad y(0) = y'(0) = 0,$$
by applying the Laplace transform, using (6.71). Does your answer agree with the answer obtained by solving
$$y'' + y = g_h(t), \quad y(0) = y'(0) = 0,$$
and letting h tend to 0?

19. In this problem we give a "plausibility argument" that the derivative of the unit step function $u_0(t)$ is the delta function $\delta(t)$. Keep in mind that what we must mean by this is some kind of generalization of the usual meaning of derivative, since $\delta(t)$ is not a function in the strict sense of the word and $u_0(t)$ isn't even continuous at $t = 0$.

   (a) As an approximation to the step function $u_0(t)$, consider the function $v_h(t)$ whose graph is shown in Fig. 6.30. This function has the value 0 for $t \leq 0$ and the value 1 for $t \geq h$ and is linear on the interval $0 \leq t \leq h$, as shown. What is the slope of the portion of the graph corresponding to $0 \leq t \leq h$?

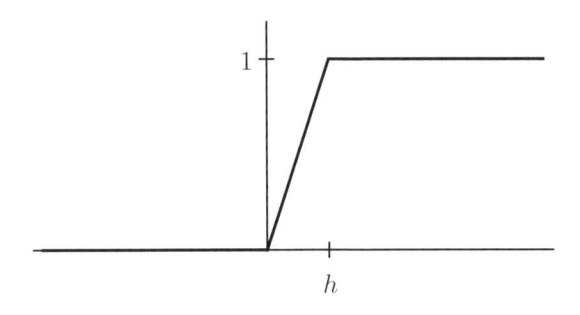

Figure 6.30. Graph of $v_h(t)$.

   (b) What is $v_h'(t)$ if $t > h$? What is $v_h'(t)$ if $t < 0$? What is $u_h'(t)$ if $0 < t < h$? Sketch the graph of $v_h'(t)$, using (approximately) the value of h shown in Fig. 6.30.
   (c) Sketch the graph of $v_h(t)$ using a smaller value of h than in Fig. 6.30. Also sketch the graph of $v_h'(t)$ for your smaller h.
   (d) What is the relationship between $v_h'(t)$ and the function $g_h(t)$ pictured in Fig. 6.24?
   (e) Explain why the above results intuitively say $v_0'(t) = \delta(t)$.

20. Consider the RLC circuit pictured in Fig. 6.31. Suppose the inductance L is 1 henry, the resistance R is 12 ohms, and the capacitance is $C = 0.01$ farad. Assume that circuit includes a battery producing 9 volts, as well as a switch initially (at time $t = 0$) in the open (off) position, as in the first picture. At time $t = 1$ the switch is closed as in Position 1, and at time $t = 2$ the switch is instantaneously moved to Position 2. The goal of this problem is to determine the current $I(t)$ flowing in the clockwise direction at the position labeled by the bold arrow for all $t \geq 0$.

   To get started, recall the set-up from Section 5.4. The current $I(t)$ at time t in such a one-loop circuit is the same at every point, and the charge on the initial plate of the capacitor (the top plate in the pictures) is $Q(t)$. If $V(t)$ is the voltage supplied to the circuit at time t, then Kirchhoff's voltage law implies
$$L\frac{dI}{dt} + RI + \frac{1}{C}Q = V(t);$$
see equation (5.96). Incorporating the values $L = 1$, $R = 12$, and $C = 0.01$, as well as the switching scenario described above, gives
$$\frac{dI}{dt} + 12I + 100Q = 9(u_1(t) - u_2(t)).$$

Since $I = Q'(t)$, we can write this as a differential equation for $Q(t)$:

(6.83) $$Q'' + 12Q' + 100Q = 9(u_1(t) - u_2(t)).$$

(a) Use the Laplace transform to solve (6.83) for $Q(t)$. Once you have $Q(t)$, differentiate to find the current $I(t)$. Note: For $t < 1$ the circuit is dead—there is no charge on the capacitor and no current in the circuit. This determines the initial conditions.

(b) Differentiating equation (6.83) and using equation (6.79) as well as the relation $Q'(t) = I(t)$ gives (at least formally) a differential equation for $I(t)$:

$$I'' + 12I' + 100I = 9(\delta(t-1) - \delta(t-2)).$$

Use Laplace transforms to solve this equation for $I(t)$.

(c) Are your answers for $I(t)$ in parts (a) and (b) the same? Which method is easier?

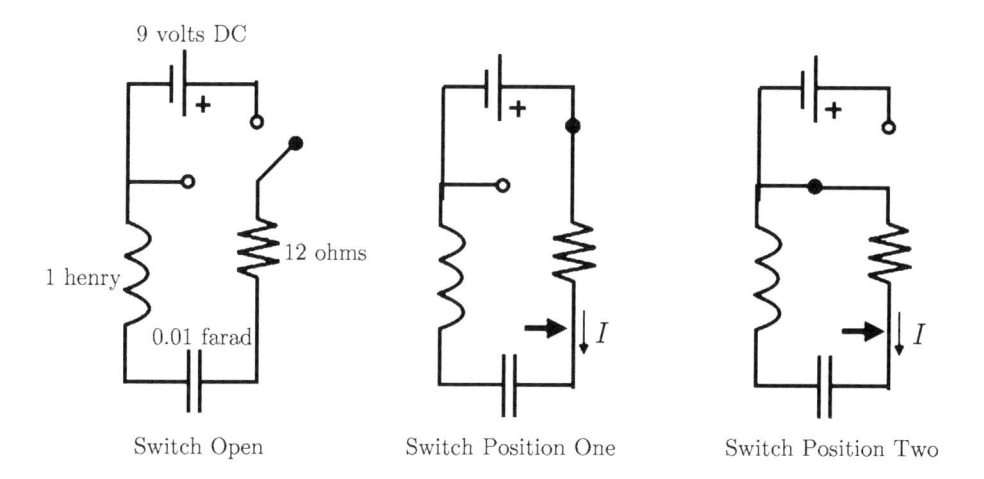

Figure 6.31. An RLC circuit.

21. (a) Suppose the differential equation $y'' + 4y' + 5y = 0$ models the displacement of a mass in a damped mass-spring system, with no externally applied force. With the initial condition $y(0) = y'(0) = 0$ show that the solution is (trivially) $y = 0$.

    (b) Now suppose that at time $t = 2$ the mass is struck with a hammer so that the displacement is given by
    $$y'' + 4y' + 5y = \delta(t-2), \quad y(0) = y'(0) = 0.$$
    Determine $y(t)$ and sketch its graph for $0 \le t \le 5$.

22. The equation $my'' + ky = A\delta(t)$ is used to model an undamped mass-spring system, with mass m in grams and spring constant k. The displacement $y(t)$ is measured in cm, and time in seconds. At $t = 0$ the mass is struck by a hammer, imparting an impulse force $A\delta(t)$.

    (a) What are the units on each term in the differential equation?
    (b) What are the units on A and $\delta(t)$?
    (c) What is the solution with the rest condition $y(0) = y'(0) = 0$?
    (d) If the solution from (c) is sketched for $-\infty < t < \infty$ (with $y \equiv 0$ for $t < 0$), is it continuous at $t = 0$? Is it differentiable at $t = 0$? If your answer to this second question is no, does the derivative have right-hand and left-hand limits at $t = 0$?

23. (a) Suppose we have a mass-spring system where the displacement $y(t)$ of the mass from equilibrium is governed by
    $$y'' + y = 0, \quad y(0) = a, \quad y'(0) = b.$$
    Check that the mass will oscillate forever, provided at least one of a and b is nonzero.

(b) Now suppose the mass is struck by a hammer at time $t = c$ so that the displacement is determined by
$$y'' + y = k\delta(t - c), \quad y(0) = a, \quad y'(0) = b,$$
for some constant k. Is there a value of k so that the oscillation can be canceled completely for $t > c$? That is, can you find k so that $y(t) = 0$ for $t > c$? Hint: Consider two cases: c chosen so that $y(c) = a\cos c + b\sin c = 0$ and c chosen so that $a\cos c + b\sin c \neq 0$.

24. In a certain (vertical) undamped mass-spring system, the displacement $y(t)$ of the mass from equilibrium is given by
$$y'' + y = 0, \quad y(0) = 0, \quad y'(0) = 1.$$

(a) Find the first time $t_1 > 0$ that the mass is in the equilibrium position and moving down. Sketch the displacement function from $t = 0$ to $t = t_1$.

(b) Suppose that as the mass passes through equilibrium at time t_1 from (a), it is struck by a hammer so that its position for $t > t_1$ is determined by
$$y'' + y = \delta(t - t_1), \quad y(0) = 0, \quad y'(0) = 1.$$
Sketch the displacement function from $t = 0$ until $t = t_2$, where t_2 is the next later time that the mass is at equilibrium and moving down.

(c) Suppose that the mass is struck by a hammer each time it passes through equilibrium moving down, so that
$$y'' + y = \delta(t - t_1) + \delta(t - t_2) + \delta(t - t_3) + \cdots, \quad y(0) = 0, \quad y'(0) = 1,$$
where the times $t_1 < t_2 < t_3 < \cdots$ are the times that the mass passes through equilibrium moving down. Describe the displacement function. In particular, is it bounded or unbounded as t becomes arbitrarily large?

25. Because of undesirable conditions, the population $y(t)$ of a fish species in a lake is decreasing according to the differential equation
$$\frac{dy}{dt} = -0.1y, \quad y(0) = y_0.$$

(a) Show that $y(t) \to 0$ as $t \to \infty$.

(b) If the lake is (instantaneously) restocked with fish at regular (yearly) intervals, the population is modeled by
$$\frac{dy}{dt} = -0.1y + A\delta(t - 1) + A\delta(t - 2) + A\delta(t - 3) + \cdots.$$
What is the smallest value of A that ensures the population at $t = 1$ is at least as big as y_0, the initial population? With this value of A, will $y(2), y(3)$, etc., be at least y_0?

26. In this problem we consider the two differential equations

(6.84) $$ay'' + by' + cy = \delta(t)$$

and

(6.85) $$ay'' + by' + cy = f(t),$$

each with the rest condition $y(0) = y'(0) = 0$. Denote the solution to (6.84) with the rest condition by $y_1(t)$. This is called the **impulse response function.** Show that the solution to (6.85) is $f * y_1$, the convolution of the nonhomogeneous term $f(t)$ with the impulse response function $y_1(t)$.

The philosophy of this result can be described as follows: If you know the response of a physical system (described by a constant coefficient linear differential equation) to the unit impulse function $\delta(t)$, then you "know" the response to an arbitrary function $f(t)$.

27. We want to find a function $y(t)$ that satisfies the *integral equation*

$$(6.86) \qquad y(t) + \int_0^t y(u)\, du = 1.$$

Identify the integral

$$\int_0^t y(u)\, du$$

as the convolution of y and some other function, and then solve equation (6.86) using Laplace transforms.

28. Solve the following integral equations using the method described in Exercise 27.
    (a) $y(t) + \int_0^t (t-u)y(u)\, du = 1$.
    (b) $y(t) + \int_0^t \sin(t-u)y(u)\, du = 2$.

29. Determine the convolutions
    (a) $\sin t * \delta(t)$.
    (b) $e^t * \delta(t)$.
    (c) $f(t) * \delta(t)$ for any continuous, exponentially bounded function f.

Chapter 7

Power Series Solutions

7.1. Motivation for the study of power series solutions

Power series will give us a tool to deal systematically with many second-order linear equations that do not have constant coefficients. There are two related issues here. First of all, our most complete methods in Chapter 4 for solving second-order linear equations apply to constant coefficient equations only. Secondly, once we move to variable coefficient equations, the (nontrivial) solutions to even some simple looking equations like

$$xy'' + y' + xy = 0$$

will be "new" types of functions for which we don't have familiar names. In other words, they are not made by adding, subtracting, multiplying, dividing, or composing finitely many polynomials, exponential, logarithmic, trigonometric, or inverse trigonometric functions. The terminology "special functions" or "higher transcendental functions" is sometimes used for these new types of functions. Second-order linear differential equations give a rich source of examples of how special functions arise.

To motivate the power series method of solution, we consider the second-order initial value problem

(7.1) $$y'' + (1+x)y' + xy = 0, \qquad y(0) = 1, \quad y'(0) = -1.$$

The Existence and Uniqueness Theorem, Theorem 4.1.4, tells us that the initial value problem in (7.1) has a unique solution valid on the entire real line. Because $y'' + (1+x)y' + xy = 0$ does not have constant coefficients and is not a Cauchy-Euler equation, the routine solution techniques discussed in Chapter 4 do not apply. We might approximate the solution of (7.1) numerically using, e.g., the ideas of Euler's method (see Exercise 29 in Section 4.1); however, there is another natural way to approximate the solution—by polynomials. Suppose we wish to construct a polynomial p of degree 3 or less that approximates the solution of (7.1) near $x = 0$. We'd want p to satisfy the initial conditions; thus, $p(0) = 1$ and $p'(0) = -1$. What value should be assigned to $p''(0)$? According to the differential equation of (7.1),

(7.2) $$y''(x) = -(1+x)y'(x) - xy(x);$$

substituting $x = 0$ into the preceding equation yields $y''(0) = -y'(0) = 1$. Thus, we set $p''(0) = 1$. What about $p^{(3)}(0)$? Because any solution of the differential equation of (7.1) must be twice differentiable, equation (7.2) reveals that the second derivative of the solution must also be differentiable and, inductively, that the solution is, in fact, infinitely differentiable. Differentiating both sides of

(7.2) and then substituting 0 for x yields $y'''(0) = -1$, so that we require $p'''(0) = -1$. Thus our approximating polynomial p satisfies

$$p(0) = 1, \quad p'(0) = -1, \quad p''(0) = 1, \quad \text{and} \quad p'''(0) = -1.$$

Using these values to determine the four coefficients of $p(x) = a_0 + a_1 x + a_2 x^2 + a_3 x^3$, we find $p(x) = 1 - x + x^2/2 - x^3/6$.

In general, an approximating polynomial near a point x_0 for a "sufficiently nice" function $f(x)$ (perhaps a solution to a differential equation) is chosen by matching the first n derivatives of f at x_0 with the first n derivatives at x_0 of a polynomial p of degree at most n. The larger the value of n, the better we expect the approximation to be near x_0. Thus our approximating polynomial for f of degree at most n should satisfy

(7.3) $\qquad p(x_0) = f(x_0) \quad \text{and} \quad p^{(j)}(x_0) = f^{(j)}(x_0) \quad \text{for } j = 1, \ldots, n.$

The polynomial p_n that does all this is given by

$$p_n(x) = f(x_0) + f'(x_0)(x - x_0) + \frac{f''(x_0)}{2!}(x - x_0)^2 + \cdots + \frac{f^{(n)}(x_0)}{n!}(x - x_0)^n.$$

Convince yourself that the preceding polynomial has the required "derivative matching" property (7.3). Notice that the coefficient of $(x - x_0)^j$, $j \leq n$, is the number

(7.4) $$\frac{f^{(j)}(x_0)}{j!},$$

where $f^{(0)}(x_0)$ is defined to be $f(x_0)$ and $0! = 0$. The polynomial p_n may be written more compactly as

$$p_n(x) = \sum_{j=0}^{n} \frac{f^{(j)}(x_0)}{j!}(x - x_0)^j.$$

We call p_n the nth **Taylor polynomial** for f at x_0. For example, the third Taylor polynomial at $x_0 = 0$ for the solution of (7.1) is $p_3(x) = 1 - x + x^2/2 - x^3/6$.

Returning to our work on the initial value problem (7.1), we assert that the nth Taylor polynomial at $x_0 = 0$ approximating its solution near 0 has the form

$$p_n(x) = \sum_{j=0}^{n} \frac{(-1)^j}{j!} x^j.$$

Thus, we are claiming that the jth derivative of the solution y of (7.1) satisfies $y^{(j)}(0) = (-1)^j$, a fact that can be verified by induction (see Exercise 20 in Section 7.2). Taking the limit of p_n as n approaches infinity, we obtain a **power series**

$$\sum_{j=0}^{\infty} \frac{(-1)^j}{j!} x^j,$$

whose partial sums are Taylor polynomials of the solution of (7.1). We have reason to hope that this power series is a solution to our initial value problem in (7.1). We verify this is the case in two different ways. The simplest is to recognize that

$$\sum_{j=0}^{\infty} \frac{(-1)^j}{j!} x^j = \sum_{j=0}^{\infty} \frac{(-x)^j}{j!} = e^{-x},$$

where we have used the familiar power series expansion

$$e^x = \sum_{j=0}^{\infty} \frac{x^j}{j!},$$

which is valid for every real number x. It's easy to check directly that $y(x) = e^{-x}$ solves (7.1).

As we discussed at the beginning of this section, we cannot, in general, expect our solutions to second-order initial value problems like (7.1) to have familiar forms, such as e^{-x}. In the next section, we review properties of power series, and there we will see that the Ratio Test and Theorem 7.2.3 show that the function

$$(7.5) \qquad y(x) = \sum_{j=0}^{\infty} \frac{(-1)^j}{j!} x^j$$

has domain $(-\infty, \infty)$, is infinitely differentiable, and may, in fact, be differentiated term by term, so that, for example, $y'(x) = \sum_{j=1}^{\infty} \frac{j(-1)^j}{j!} x^{j-1}$. Using these power series representations for y, y', and y'', you can verify that y given by (7.5) solves the initial value problem (7.1).

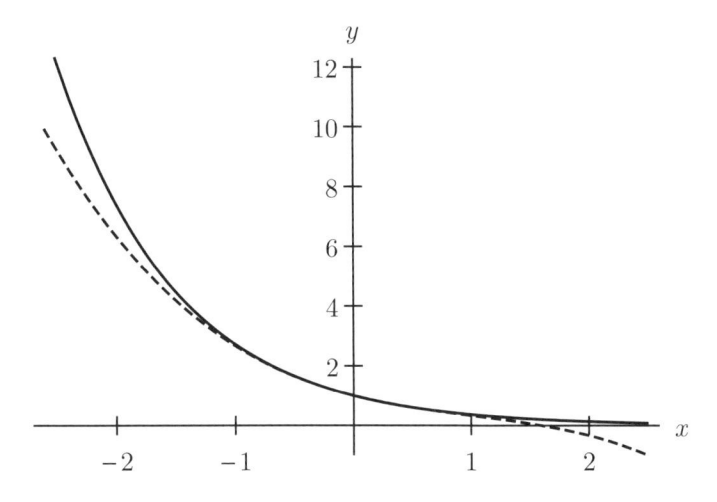

Figure 7.1. Solution to initial value problem (7.1) and its third Taylor polynomial (dashed).

7.2. Review of power series

Definition 7.2.1. A **power series centered at the point** x_0 is an infinite sum of the form

$$\sum_{n=0}^{\infty} a_n (x - x_0)^n.$$

The numbers a_n are called the coefficients of the series.

The power series $\sum_{n=0}^{\infty} a_n (x - x_0)^n$ **converges** at a particular point x if the **partial sums**

$$S_N(x) = a_0 + a_1(x - x_0) + a_2(x - x_0)^2 + \cdots + a_N(x - x_0)^N$$

have a (finite) limit as $N \to \infty$. We call the value of this limit the **sum** of the series at the point x and write

$$S(x) = \lim_{N \to \infty} S_N(x),$$

which is often just abbreviated as

$$S(x) = \sum_{n=0}^{\infty} a_n (x - x_0)^n.$$

We can think of this as defining a function $S(x)$ at each point where the series converges. At a point x where the partial sums don't have a limit we say the series **diverges**.

Example 7.2.2. The series
$$\sum_{n=0}^{\infty} x^n = 1 + x + x^2 + x^3 + \cdots$$
is called the **geometric series**. Notice it is centered at $x_0 = 0$. It converges for $-1 < x < 1$ and diverges for all other values of x. When $-1 < x < 1$, its sum is $1/(1-x)$:

$$\sum_{n=0}^{\infty} x^n = \frac{1}{1-x}, \quad -1 < x < 1.$$

To see why these statements are correct look at the Nth partial sum

(7.6) $$S_N(x) = 1 + x + x^2 + \cdots + x^N.$$

Multiplying equation (7.6) by x gives

(7.7) $$xS_N(x) = x + x^2 + \cdots + x^N + x^{N+1}.$$

Subtracting equation (7.7) from equation (7.6) gives
$$(1-x)S_N(x) = 1 - x^{N+1}.$$

When $x \neq 1$, this gives us a tidy formula for the Nth partial sum:
$$S_N(x) = \frac{1 - x^{N+1}}{1 - x}.$$

When $x = 1$ the Nth partial sum is $N + 1$. The question, "For what values of x does the geometric series converge?" is asking, "For what values of x does
$$\lim_{N \to \infty} \frac{1 - x^{N+1}}{1 - x}$$
exist?" When $-1 < x < 1$, the term $x^{N+1} \to 0$ as $N \to \infty$. Thus for these values of x,
$$\lim_{N \to \infty} \frac{1 - x^{N+1}}{1 - x} = \frac{1}{1 - x},$$
which is the desired conclusion. For all other values of x, the partial sums $S_N(x)$ don't have a finite limit as $N \to \infty$, and the geometric series diverges (see Exercise 3).

What we see in this example—convergence of the series at every point in an interval whose midpoint is the center of the series and divergence elsewhere— is typical of any power series. The formal statement is as follows.

Theorem 7.2.3. *For the power series $\sum_{n=0}^{\infty} a_n(x - x_0)^n$, exactly one of the following holds:*

(i) *The series converges only at $x = x_0$.*

(ii) *The series converges for all real numbers x.*

(iii) *There is a positive number R, called the **radius of convergence**, such that $\sum_{n=0}^{\infty} a_n(x - x_0)^n$ converges for all x in the interval $x_0 - R < x < x_0 + R$ and diverges for all x with $x > x_0 + R$ or $x < x_0 - R$. At the endpoints $x = x_0 + R$ and $x = x_0 - R$, we could have either convergence or divergence; these endpoints must be checked on a case-by-case basis.*

When (i) *holds, we say that the radius of convergence of the series is $R = 0$, and when* (ii) *holds, we have $R = \infty$.*

7.2. Review of power series

Moreover, when $R > 0$, the sum $\sum_{n=0}^{\infty} a_n(x - x_0)^n$ converges to a function $f(x)$ that is continuous and has derivatives of all orders on the interval $x_0 - R < x < x_0 + R$. The values of the derivatives of f at x_0 and the coefficients a_n are related by

$$a_n = \frac{f^{(n)}(x_0)}{n!}.$$

The formula for a_n in Theorem 7.2.3 appeared earlier, in (7.4), but our point of view here is slightly different. Now we are **starting** with the power series $\sum_{n=0}^{\infty} a_n(x-x_0)^n$, calling what it converges to $f(x)$, and then asserting that the nth derivative of f at x_0 exists and is equal to $n!a_n$. In (7.4), we **started** with a "sufficiently nice" function $f(x)$ (we have yet to explain what sufficiently nice means here) and used (7.4) to give approximating Taylor polynomials $p_n(x)$ for f. We hope to be able to assert that $f(x)$ is the limit of these Taylor polynomials, so that

$$f(x) = \sum_{n=0}^{\infty} \frac{f^{(n)}(x_0)}{n!}(x - x_0)^n;$$

we'll return to this matter shortly.

We can improve the conclusions of Theorem 7.2.3 using the notion of absolute convergence.

Definition 7.2.4. A series $\sum_{n=0}^{\infty} a_n(x - x_0)^n$ is said to **converge absolutely** at a point x if the series

$$\sum_{n=0}^{\infty} |a_n||x - x_0|^n$$

converges at x.

It is a fact that if a series converges absolutely, it converges, but the converse direction is not true—a series can converge but not converge absolutely. When the series $\sum_{n=0}^{\infty} a_n(x-x_0)^n$ has radius of convergence R as described by Theorem 7.2.3, then it converges **absolutely** at every point of the **open** interval $(x_0 - R, x_0 + R)$. Again, we can't make any blanket statement about what happens at the endpoints $x_0 - R$ and $x_0 + R$; these must be handled on a case-by-case basis.

Tests for convergence. We can use the Ratio Test and/or the Root Test to determine the radius of convergence of many series of interest to us.

Ratio Test. If

$$\lim_{n \to \infty} \frac{|a_{n+1}|}{|a_n|}$$

exists and has value L, where L can be a finite value or ∞, then the radius of convergence of $\sum_{n=0}^{\infty} a_n(x - x_0)^n$ is $1/L$. When $L = \infty$, we interpret $1/L$ to be 0 and when $L = 0$, we interpret $1/L$ to be ∞.

Root Test. If

$$\lim_{n \to \infty} |a_n|^{1/n}$$

exists and has value L, where L can be a finite value or ∞, then the radius of convergence of $\sum_{n=0}^{\infty} a_n(x - x_0)^n$ is $1/L$. When $L = \infty$, we interpret $1/L$ to be 0 and when $L = 0$, we interpret $1/L$ to be ∞.

However, not every power series yields to one of these tests, since the limits in question may fail to exist.

Manipulations with series. When two power series $\sum_{n=0}^{\infty} a_n(x-x_0)^n$ and $\sum_{n=0}^{\infty} b_n(x-x_0)^n$ both converge on some open interval $I = (x_0 - R, x_0 + R)$, (for some $R > 0$) the two series can be manipulated very much like two polynomials:

Sums: The two series can be added term by term to produce the series

$$\sum_{n=0}^{\infty}(a_n + b_n)(x-x_0)^n,$$

which will also converge at least on the interval I. If the sum of the first series is denoted $f(x)$ and the sum of the second series is $g(x)$ on I, then $\sum_{n=0}^{\infty}(a_n + b_n)(x-x_0)^n$ converges to $f(x) + g(x)$ on I. A similar property holds for **differences**, so that $\sum_{n=0}^{\infty}(a_n - b_n)(x-x_0)^n$ converges to $f(x) - g(x)$ on I.

Products: The two series can be multiplied, collecting common terms as in polynomial multiplication, to give a series which converges to $f(x)g(x)$ at least on I:

$$f(x)g(x) = \left(\sum_{n=0}^{\infty} a_n(x-x_0)^n\right)\left(\sum_{n=0}^{\infty} b_n(x-x_0)^n\right) = \sum_{n=0}^{\infty} c_n(x-x_0)^n$$

where

$$c_n = a_0 b_n + a_1 b_{n-1} + a_2 b_{n-2} + \cdots + a_{n-1} b_1 + a_n b_0.$$

Convince yourself that this formula for c_n is exactly what you would expect by treating the series for f and g like "big polynomials" and multiplying.

Differentiation: The series $f(x) = \sum_{n=0}^{\infty} a_n(x-x_0)^n$ can be differentiated term by term to give a series $\sum_{n=1}^{\infty} n a_n(x-x_0)^{n-1}$ that converges to $f'(x)$ at each point of I. The radius of convergence of the differentiated series is the same as the radius of convergence of the original series.

Integration: The series $f(x) = \sum_{n=0}^{\infty} a_n(x-x_0)^n$ may be integrated term by term to obtain a series $\sum_{n=0}^{\infty} a_n(x-x_0)^{n+1}/(n+1)$ converging on I to a function that is an antiderivative of f. The radius of convergence of the integrated series is the same as the radius of convergence of the original series.

Uniqueness: If $\sum_{n=0}^{\infty} a_n(x-x_0)^n$ has positive radius of convergence R and if

$$\sum_{n=0}^{\infty} a_n(x-x_0)^n = 0$$

for each x in $(x_0 - R, x_0 + R)$, then each coefficient $a_n = 0$.

This last property, the **Uniqueness Property**, tells you that if $\sum_{n=0}^{\infty} a_n(x-x_0)^n = \sum_{n=0}^{\infty} b_n(x-x_0)^n$ on $(x_0 - R, x_0 + R)$, where $R > 0$, then $a_n = b_n$ for all $n \geq 0$. This property will play a crucial role in using power series to solve differential equations.

Example 7.2.5. Using the geometric series

(7.8) $$\frac{1}{1-x} = 1 + x + x^2 + x^3 + \cdots,$$

which converges on the interval $(-1, 1)$, we will illustrate some of the above principles.

If we differentiate on both sides of equation (7.8), we see that

$$\frac{1}{(1-x)^2} = 1 + 2x + 3x^2 + 4x^3 + \cdots$$

7.2. Review of power series

for $-1 < x < 1$. If x is in the interval $(-1, 1)$, then $-x$ is also in this interval. Thus in (7.8) we may replace x by $-x$ to obtain

$$(7.9) \qquad \frac{1}{1+x} = 1 - x + x^2 - x^3 + x^4 + \cdots$$

for any x between -1 and 1. If we integrate term by term in equation (7.9) we see that

$$\ln(1+x) = C + x - \frac{x^2}{2} + \frac{x^3}{3} - \frac{x^4}{4} + \cdots$$

where the constant C arises as a constant of integration. Setting $x = 0$ on both sides we see that $C = 0$. Since the geometric series converges on $(-1, 1)$, the integrated series

$$(7.10) \qquad \ln(1+x) = x - \frac{x^2}{2} + \frac{x^3}{3} - \frac{x^4}{4} + \cdots$$

is guaranteed to converge *at least* on $(-1, 1)$. (In fact, it also converges at $x = 1$ to $\ln 2$.) The Uniqueness Property says that (7.10) is the *only* power series centered at 0 which converges to $\ln(1+x)$ on $(-1, 1)$.

Taylor series and analytic functions. Many of the functions of interest to us can be expanded in a power series that is written in the form

$$(7.11) \qquad f(x) = \sum_{n=0}^{\infty} a_n (x - x_0)^n$$

for some coefficients a_n and a point x_0 in the domain of the function. A function f having such a series expansion that is valid on an interval $(x_0 - R, x_0 + R)$ for some *positive* R is said to be **analytic** at x_0, and (7.11) is called the **Taylor series of** f **centered at** x_0. We've already observed that the coefficients in the Taylor series are connected to the values of the derivatives of f at x_0 by the formula

$$(7.12) \qquad a_n = \frac{f^{(n)}(x_0)}{n!}.$$

Polynomials and the functions e^x, $\sin x$, and $\cos x$ are analytic at every point x_0, and the function $\ln x$ is analytic at each point in its domain, i.e., at each $x_0 > 0$. Sums and products of analytic functions are analytic, and the quotient of two analytic functions is analytic at each point where the denominator is not zero. In particular, a rational function (the quotient of two polynomials) is analytic at each point where its denominator is nonzero. Analytic functions are the most nicely behaved functions. In particular, if f is analytic on an open interval I (meaning it is analytic at each point of I), then f has derivatives of all orders on I.

When a function is known to be analytic at x_0, equation (7.12) can in theory be used to find its Taylor series centered at x_0. However, except for certain simple functions, this may not be a practical method since it involves computing all derivatives of the function and evaluating them at x_0. More helpful, from a practical standpoint, is the fact that if we are able to obtain a power series representation for an analytic f by any legitimate method, it must be the Taylor series for f. Once we choose the center, an analytic function has a unique power series expansion with that center. For example, the function $\ln(1+x)$ is analytic for all $x > -1$, and thus it has, for example, a Taylor series centered at 0. We found that Taylor series in (7.10) by manipulating the geometric series.

As another example, let's also obtain the Taylor series expansion for the function $f(x) = \arctan x$ on the interval $(-1, 1)$ by manipulating the geometric series formula. Start with

$$\frac{1}{1-x} = 1 + x + x^2 + x^3 + \cdots$$

and replace x by $-x^2$ to obtain

(7.13) $$\frac{1}{1+x^2} = 1 - x^2 + x^4 - x^6 + \cdots,$$

which is guaranteed to be valid on the interval $(-1, 1)$. Integrating on both sides of (7.13) gives

$$\arctan x = C + x - \frac{x^3}{3} + \frac{x^5}{5} - \frac{x^7}{7} + \cdots.$$

Setting $x = 0$ tells us that $C = 0$. The Taylor series centered at 0 for $\arctan x$ must be

$$\arctan x = x - \frac{x^3}{3} + \frac{x^5}{5} - \frac{x^7}{7} + \cdots.$$

This series converges at least on the interval $(-1, 1)$. (In fact, it also converges at $x = 1$ to $\arctan 1 = \frac{\pi}{4}$ and at $x = -1$ to $\arctan(-1) = -\frac{\pi}{4}$.)

Shifting the index of summation. When we use power series to solve differential equations, we will need some facility with "shifting the index" in summation notation. The method is easy to explain through examples.

Consider the series

$$\sum_{n=0}^{\infty} \frac{2^n}{n!} x^n = 1 + 2x + \frac{4}{2}x^2 + \frac{8}{6}x^3 + \cdots$$

(recall that we define $0! = 1$). Suppose we want to write this series in the form

$$\sum_{k=?}^{\infty} a_k x^{k+1}$$

so that the index on the sum is now k and the generic term in the sum contains x^{k+1} instead of x^n. The connection between k and n is given by

$$n = k + 1.$$

This means that

$$\frac{2^n}{n!} = \frac{2^{k+1}}{(k+1)!}$$

and since the original sum started with $n = 0$, the shifted sum will start with $k = -1$. Thus

$$\sum_{n=0}^{\infty} \frac{2^n}{n!} x^n = \sum_{k=-1}^{\infty} \frac{2^{k+1}}{(k+1)!} x^{k+1}.$$

This could just as well be written as

$$\sum_{n=0}^{\infty} \frac{2^n}{n!} x^n = \sum_{n=-1}^{\infty} \frac{2^{n+1}}{(n+1)!} x^{n+1},$$

but you may initially find it easier to implement a shift of index by using two different letters (n and k) for the index.

Here's an example that hints at how we might use a shift of indices.

Example 7.2.6. We will write

$$\text{(7.14)} \qquad \sum_{n=1}^{\infty} nx^{n-1} + \sum_{n=2}^{\infty} n(n-1)x^{n-2}$$

as a single sum with generic term containing x^n. Let's shift the index on the first sum by setting $k = n - 1$ to see that

$$\text{(7.15)} \qquad \sum_{n=1}^{\infty} nx^{n-1} = \sum_{k=0}^{\infty} (k+1)x^k.$$

Similarly, shift the index on the second sum in (7.14) by setting $m = n - 2$ to obtain

$$\text{(7.16)} \qquad \sum_{n=2}^{\infty} n(n-1)x^{n-2} = \sum_{m=0}^{\infty} (m+2)(m+1)x^m.$$

The choice of letter name for the index doesn't matter so we can just as well write equations (7.15) and (7.16) as

$$\sum_{n=1}^{\infty} nx^{n-1} = \sum_{n=0}^{\infty} (n+1)x^n$$

and

$$\sum_{n=2}^{\infty} n(n-1)x^{n-2} = \sum_{n=0}^{\infty} (n+2)(n+1)x^n.$$

Now we can add the two series to obtain

$$\sum_{n=1}^{\infty} nx^{n-1} + \sum_{n=2}^{\infty} n(n-1)x^{n-2} = \sum_{n=0}^{\infty} (n+1)x^n + \sum_{n=0}^{\infty} (n+2)(n+1)x^n$$

$$= \sum_{n=0}^{\infty} [(n+1) + (n+2)(n+1)]x^n$$

$$= \sum_{n=0}^{\infty} (n+1)(n+3)x^n.$$

7.2.1. Exercises.

1. Shift indices on the series

$$\sum_{n=2}^{\infty} (n+1)a_n(x-2)^{n+1}$$

so that
   (a) the generic term in the series contains $(x-2)^n$,
   (b) the generic term in the series contains $(x-2)^{n-1}$,
   (c) the series starts with $n = 0$, rather than $n = 2$,
   (d) the series starts with $n = 1$, rather than $n = 2$.

2. Sam shifts indices on $\sum_{n=1}^{\infty} n(n-1)a_n x^{n-1}$ so as to write the series with generic term containing x^n and obtains

$$\sum_{n=1}^{\infty} (n+1)na_{n+1}x^n.$$

Sally similarly shifts indices on the same series but obtains the result
$$\sum_{n=0}^{\infty}(n+1)na_{n+1}x^n$$
(starting at $n=0$ instead of $n=1$). Who is correct?

3. Show that the geometric series $\sum_{n=0}^{\infty} x^n$ diverges for $x \geq 1$ or $x \leq -1$. (Hint: Recall that a necessary condition for a series to converge is that the terms of the series tend to 0.)

4. Write each of the following as a series in the form $\sum_{n=0}^{\infty} a_n x^n$ (say what the coefficients a_n are for each $n \geq 0$).

    (a) $(x+2) \sum_{n=0}^{\infty} \dfrac{2^n}{n!} x^n$.

    (b) $(x+2)^2 \sum_{n=0}^{\infty} \dfrac{2^n}{n!} x^n$.

5. Write the series
$$x \sum_{n=0}^{\infty} \frac{2^n}{n!} (x-1)^n$$
in the form $\sum_{n=0}^{\infty} a_n(x-1)^n$, with generic term containing $(x-1)^n$. Hint: Write $x=(x-1)+1$.

6. Write the series
$$(x-2) \sum_{n=0}^{\infty} \frac{2^n}{n!}(x-1)^n$$
in the form $\sum_{n=0}^{\infty} a_n(x-1)^n$, with generic term containing $(x-1)^n$. Hint: $x-2=(x-1)-1$.

7. Write the series
$$(x-2)^2 \sum_{n=0}^{\infty} \frac{2^n}{n!}(x-1)^n$$
in the form $\sum_{n=0}^{\infty} a_n(x-1)^n$, with generic term containing $(x-1)^n$.

8. If the series $\sum_{n=0}^{\infty} a_n x^n$ and $\sum_{n=0}^{\infty} b_n x^n$ are multiplied, what is the coefficient of x?

9. If the series $\sum_{n=0}^{\infty} a_n x^n$ and $\sum_{n=0}^{\infty} b_n x^n$ are multiplied, what is the coefficient of x^2?

10. If $e^x \sin x$ is written as a series $\sum_{n=0}^{\infty} a_n x^n$, what is the coefficient of x^2? Of x^3? Of x^4? You can do this in one of two ways: by multiplying the Taylor series for e^x and $\sin x$ (centered at 0) or by using equation (7.12).

11. Starting with the geometric series
$$(7.17) \qquad 1 + x + x^2 + x^3 + x^4 + \cdots = \frac{1}{1-x}$$
valid for $-1 < x < 1$, obtain similar formulas for the following series by manipulations of equation (7.17) such as differentiation, integration, or substitution. Also determine the largest open interval centered at 0 on which your formula is valid.

    (a) $1 + x^2 + x^4 + x^6 + \cdots$.
    (b) $1 + \frac{x}{2} + \frac{x^2}{4} + \frac{x^3}{8} + \cdots$.
    (c) $1 - x^2 + x^4 - x^6 + \cdots$.
    (d) $1 - 2x + 3x^2 - 4x^3 + \cdots$.
    (e) $x + \frac{x^2}{2} + \frac{x^3}{3} + \frac{x^4}{4} + \cdots$.

12. Use equation (7.10) and the result of part (e) in Exercise 11 to obtain a formula for
$$x + \frac{x^3}{3} + \frac{x^5}{5} + \frac{x^7}{7} + \cdots.$$

7.2. Review of power series

In Exercises 13–16 find the radius of convergence and give the largest open interval on which convergence is guaranteed.

13. $\sum_{n=0}^{\infty} \dfrac{n}{3^n} x^n.$

14. $\sum_{n=0}^{\infty} \dfrac{(-1)^n}{n+1} (x-2)^n.$

15. $\sum_{n=0}^{\infty} \dfrac{5^n}{(2n+1)!} x^n.$

16. $\sum_{n=0}^{\infty} \dfrac{n!}{2^n} (x+3)^n.$

17. In this problem we consider the radius of convergence of the power series

$$1 - \frac{1}{2}x^2 + \frac{1}{(4)(2)}x^4 - \frac{1}{(6)(4)(2)}x^6 + \frac{1}{(8)(6)(4)(2)}x^8 + \cdots$$

$$= 1 + \sum_{k=1}^{\infty} (-1)^k \frac{1}{(2k)(2k-2)\cdots(4)(2)} x^{2k}.$$

(a) Show that the Ratio Test cannot be applied with the series as written.

(b) With the substitution $w = x^2$ the series becomes

$$1 - \frac{1}{2}w + \frac{1}{(4)(2)}w^2 - \frac{1}{(6)(4)(2)}w^3 + \frac{1}{(8)(6)(4)(2)}w^4 + \cdots$$

$$= 1 + \sum_{k=1}^{\infty} (-1)^k \frac{1}{(2k)(2k-2)\cdots(4)(2)} w^k.$$

Use the Ratio Test to show that this series in w converges for all real w.

(c) What is the radius of convergence of

$$1 + \sum_{k=1}^{\infty} (-1)^k \frac{1}{(2k)(2k-2)\cdots(4)(2)} x^{2k}?$$

(d) Similarly, find the radius of convergence of

$$\sum_{k=0}^{\infty} \frac{1}{4^k} \frac{1 \cdot 3 \cdot 5 \cdots (2k-3)(2k-1)}{k!} x^{2k}.$$

18. Write the polynomial $p(x) = 1 + x - 2x^2 + 5x^4$ in the form

$$p(x) = \sum_{n=0}^{n=?} a_n (x-1)^n.$$

Hint: Use equation (7.12).

19. If $f(x) = e^x$ is written as a power series $\sum_{n=0}^{\infty} a_n (x-1)^n$, what are $a_0, a_1, a_2,$ and a_3? What is a_n?

20. Show by induction that if $y'' + (1+x)y' + xy = 0$, then for every $n \geq 3$,

$$y^{(n)} = -(1+x)y^{(n-1)} - (n-2)y^{(n-2)} - xy^{(n-2)} - (n-2)y^{(n-3)}.$$

Conclude that if $y(0) = 1$ and $y'(0) = -1$, then $y^{(n)}(0) = (-1)^n$ for every $n \geq 0$.

7.3. Series solutions

In this section we will describe a powerful method for solving a second-order linear equation

$$(7.18) \qquad P(x)\frac{d^2y}{dx^2} + Q(x)\frac{dy}{dx} + R(x)y = 0$$

on an open interval I where $P(x)$, $Q(x)$, and $R(x)$ are polynomials and $P(x) \neq 0$ in I. The hypothesis that $P(x) \neq 0$ on I plays an important role. Any point x_0 with $P(x_0) \neq 0$ is called an **ordinary point** for the equation (7.18). When $P(x) \neq 0$ on I we may divide by $P(x)$ in equation (7.18) to obtain

$$(7.19) \qquad \frac{d^2y}{dx^2} + p(x)\frac{dy}{dx} + q(x)y = 0,$$

where $p(x) = Q(x)/P(x)$ and $q(x) = R(x)/P(x)$ are rational functions whose denominators are nonzero on the interval I. Since $p(x)$ and $q(x)$ are continuous on I, Theorem 4.1.4 says that once we specify an initial condition $y(x_0) = \alpha, y'(x_0) = \beta$ at a point x_0 in I, there will be a unique solution valid on the entire interval I. We also know that the general solution to (7.19) on I is given as all linear combinations $c_1 y_1 + c_2 y_2$ of any two linearly independent solutions y_1 and y_2.

Our strategy is to look for solutions of equation (7.18) written as power series $\sum_{n=0}^{\infty} a_n(x-x_0)^n$. Of course, the coefficients a_n are initially unknown. Our aim is to determine them so that the differential equation is satisfied. Let's look at an example.

Example 7.3.1. We will find solutions to the equation

$$(7.20) \qquad y'' + xy' + y = 0$$

in terms of a power series centered at $x_0 = 0$. If

$$y(x) = \sum_{n=0}^{\infty} a_n x^n,$$

then term-by-term differentiation tells us that

$$y'(x) = \sum_{n=1}^{\infty} n a_n x^{n-1}$$

and

$$y''(x) = \sum_{n=2}^{\infty} n(n-1) a_n x^{n-2}.$$

Substituting into the differential equation gives

$$(7.21) \qquad \sum_{n=2}^{\infty} n(n-1) a_n x^{n-2} + x \sum_{n=1}^{\infty} n a_n x^{n-1} + \sum_{n=0}^{\infty} a_n x^n = 0.$$

We will combine these three series, shifting indices where needed so that the generic term in each contains x^n. We have

$$\sum_{n=2}^{\infty} n(n-1) a_n x^{n-2} = \sum_{n=0}^{\infty} (n+2)(n+1) a_{n+2} x^n$$

and

$$x \sum_{n=1}^{\infty} n a_n x^{n-1} = \sum_{n=1}^{\infty} n a_n x^n = \sum_{n=0}^{\infty} n a_n x^n.$$

7.3. Series solutions

The left-hand side of equation (7.21) is thus

(7.22) $$\sum_{n=0}^{\infty}(n+2)(n+1)a_{n+2}x^n + \sum_{n=0}^{\infty} na_n x^n + \sum_{n=0}^{\infty} a_n x^n,$$

and we seek values of the coefficients a_n so that

$$\sum_{n=0}^{\infty}[(n+2)(n+1)a_{n+2} + na_n + a_n]x^n = 0.$$

Invoking the Uniqueness Property we must have

(7.23) $$(n+2)(n+1)a_{n+2} + (n+1)a_n = 0 \text{ for } n \geq 0.$$

Rearranging (7.23), we see that

(7.24) $$a_{n+2} = -\frac{1}{n+2}a_n$$

for $n \geq 0$. This is an example of a **recursion relation**. It tells us how to determine a_2 from a_0, then how to determine a_3 from a_1, a_4 from a_2, and so on. We have the freedom initially to choose a_0 and a_1, but once those are decided, all of the remaining coefficients are determined for us by the recursion relation.

Let's use this to find two linearly independent solutions. The initial choice $a_0 = 1, a_1 = 0$ will give us one solution y_1 (we'll find this in a moment), while the initial choice $a_0 = 0, a_1 = 1$ gives us a second solution y_2. Will these two solutions be linearly independent? For the first, $y_1 = \sum_{n=0}^{\infty} a_n x^n$ with $a_0 = 1, a_1 = 0$, so that $y_1(0) = a_0 = 1$ and $y_1'(0) = a_1 = 0$. For the second, $y_2 = \sum_{n=0}^{\infty} a_n x^n$ with $a_0 = 0, a_1 = 1$, and we have $y_2(0) = 0$ and $y_2'(0) = 1$. At $x = 0$, the Wronskian

$$W(y_1, y_2) = \det\begin{pmatrix} y_1(0) & y_2(0) \\ y_1'(0) & y_2'(0) \end{pmatrix} = \det\begin{pmatrix} 1 & 0 \\ 0 & 1 \end{pmatrix} = 1.$$

By Theorem 4.3.8, this guarantees the linear independence we seek.

What are the functions y_1 and y_2? For y_1, $a_0 = 1, a_1 = 0$, and the recursion relation (7.24) allows us to determine all other values of a_n. Setting $n = 0$ in (7.24) we have

$$a_2 = -\frac{1}{2}a_0 = -\frac{1}{2}.$$

With $n = 1$ in (7.24), we determine that

$$a_3 = -\frac{1}{3}a_1 = 0.$$

In fact, since $a_1 = 0$, the recursion relation tells us that for every *odd* value of n, $a_n = 0$.

We obtain a_4 from the recursion relation with $n = 2$:

$$a_4 = -\frac{1}{4}a_2 = -\frac{1}{4}\left(-\frac{1}{2}\right) = \frac{1}{8}.$$

While we can continue in order, finding as many coefficients a_n as we want, if possible, we'd like to see a pattern for the coefficients a_n, for n even, that would allow us to describe their general form. Let's try a few more:

$$a_6 = -\frac{1}{6}a_4 = -\frac{1}{6 \cdot 4 \cdot 2} \quad \text{and} \quad a_8 = -\frac{1}{8}a_6 = \frac{1}{8 \cdot 6 \cdot 4 \cdot 2}.$$

Now it's clearer what is happening. Write an even integer as $n = 2k$. The sign of a_{2k} alternates and is equal to $(-1)^k$. We have

(7.25) $$a_n = a_{2k} = (-1)^k \frac{1}{(2k)(2k-2)\cdots(4)(2)}.$$

Putting this together we have our first solution

$$y_1(x) = 1 - \frac{1}{2}x^2 + \frac{1}{(4)(2)}x^4 - \frac{1}{(6)(4)(2)}x^6 + \frac{1}{(8)(6)(4)(2)}x^8 + \cdots$$

$$= 1 + \sum_{k=1}^{\infty}(-1)^k \frac{1}{(2k)(2k-2)\cdots(4)(2)}x^{2k}.$$

Our second (linearly independent) solution y_2 comes from making the initial choices

$$a_0 = 0, \quad a_1 = 1.$$

With these values, the recursion relation (7.24) gives

$$a_n = 0 \quad \text{for all even integers } n$$

and for $n = 2k + 1$ an odd integer,

(7.26) $$a_n = a_{2k+1} = (-1)^k \frac{1}{(2k+1)(2k-1)\cdots(5)(3)}.$$

You are asked to verify this in Exercise 8. This gives the second of our linearly independent solutions:

$$y_2(x) = \sum_{k=0}^{\infty}(-1)^k \frac{1}{(2k+1)(2k-1)\cdots(5)(3)}x^{2k+1}.$$

Factorial-like expressions such as the denominators in (7.26) and (7.25) can be put into "closed form" using factorials. To see how this goes, notice that the expression

$$(2k)(2k-2)\cdots(4)(2)$$

in (7.25) is a product of the first k even integers. If we factor out a 2 from each of these integers, we see that

(7.27) $$(2k)(2k-2)\cdots(4)(2) = 2^k(k)(k-1)\cdots(2)(1) = 2^k k!,$$

and our first solution may be written as

$$y_1(x) = \sum_{k=0}^{\infty}(-1)^k \frac{1}{2^k k!} x^{2k}.$$

The expression

$$(2k+1)(2k-1)\cdots(5)(3)$$

in (7.26) is a product of k odd integers. If we multiply and divide

$$\frac{1}{(2k+1)(2k-1)\cdots(5)(3)}$$

by the "missing" even integers

$$(2k)(2k-2)\cdots(4)(2),$$

we obtain

(7.28) $$\frac{1}{(2k+1)(2k-1)\cdots(5)(3)} = \frac{(2k)(2k-2)\cdots(4)(2)}{(2k+1)!}.$$

Using (7.27) the expression in (7.28) simplifies to

$$\frac{2^k k!}{(2k+1)!}.$$

7.3. Series solutions

Thus our second solution becomes

$$y_2(x) = \sum_{k=0}^{\infty} (-1)^k \frac{2^k k!}{(2k+1)!} x^{2k+1}.$$

We haven't yet discussed where the series expressions for y_1 and y_2 converge. While with a little cleverness (see Exercise 17(c) in Section 7.2) we could use the Ratio Test to show that the radius of convergence for both y_1 and y_2 is infinite, there is a theorem that quickly gets us to the same conclusion. This theorem applies to any equation

$$P(x)y'' + Q(x)y' + R(x)y = 0$$

where coefficient functions $P(x)$, $Q(x)$, and $R(x)$ are polynomials. It gives a lower bound estimate on the radius of convergence of any power series solution centered at an ordinary point x_0.

To state and use the theorem, we need the notion of the distance *in the complex plane* from x_0 to the nearest zero of $P(x)$. For example, if $P(x) = 1 + x^2$ and x_0 is chosen to be 0, then (since the solutions of $x^2 + 1 = 0$ are the complex numbers $x = i$ and $x = -i$), the distance from 0 to the nearest zero of $P(x)$ is 1 (see Fig. 7.2). If instead we choose our center to be at $x_0 = 2$, the distance from x_0 to the nearest zero of P is $\sqrt{5}$; see Fig. 7.3. On the other hand, if $P(x)$ is a nonzero constant (a polynomial of degree 0) such as in Example 7.3.1, then $P(x)$ has no zeros, and the "distance from any chosen center to the nearest zero of P" is infinity.

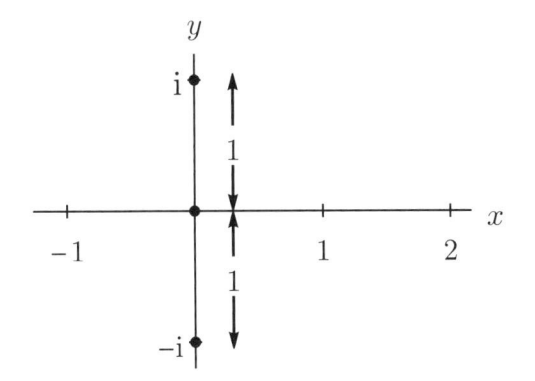

Figure 7.2. Distance from $x_0 = 0$ to the nearest zero of $1 + x^2$.

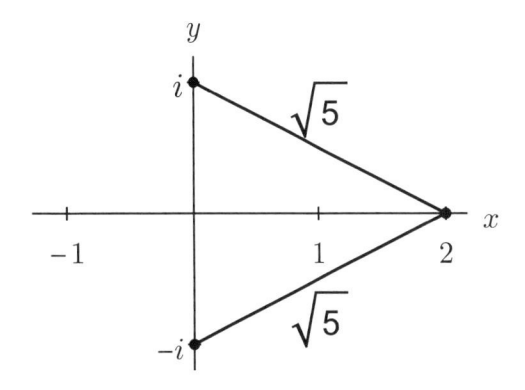

Figure 7.3. Distance from $x_0 = 2$ to the nearest zero of $1 + x^2$.

Theorem 7.3.2. *Consider the differential equation*

(7.29) $$P(x)y''(x) + Q(x)y'(x) + R(x)y = 0,$$

where $P(x)$, $Q(x)$, and $R(x)$ are polynomials with no factors common to all three. If x_0 is any point with $P(x_0) \neq 0$, then there are two linearly independent solutions to equation (7.29) of the form

(7.30) $$\sum_{n=0}^{\infty} a_n (x - x_0)^n.$$

Any solution of the form (7.30) converges at least on the interval $|x - x_0| < \rho$ where ρ is the distance in the complex plane from x_0 to the nearest zero of $P(x)$.

Applying this to our work in Example 7.3.1 where $P(x) = 1$, a polynomial with no zeros (and we've used $x_0 = 0$), tells us that our solutions y_1 and y_2 to equation (7.20) converge on the whole real line $(-\infty, \infty)$.

The general solution to $y'' + xy' + y = 0$. We have produced two linearly independent solutions y_1 and y_2 to this equation in Example 7.3.1. Linear combinations of these will give the general solution. Convince yourself that this is the same as the collection of power series solutions $\sum_{n=0}^{\infty} a_n x^n$ obtained by keeping a_0 and a_1 arbitrary and using the recursion relation (7.24) for all other coefficients.

Figs. 7.4 and 7.5 show the graphs of the partial sums of our series solutions y_1 and y_2 corresponding to their second, fourth, sixth and eighth Taylor polynomials centered at 0. Notice that over the interval $-1 \leq x \leq 1$ all of these Taylor polynomials have essentially the same graph. Thus it is reasonable to assume that Figs. 7.4 and 7.5 reveal what the graphs of y_1 and y_2, respectively, look like on $[-1, 1]$. The bell shape of the graph of y_1 could have been predicted; see Exercise 9.

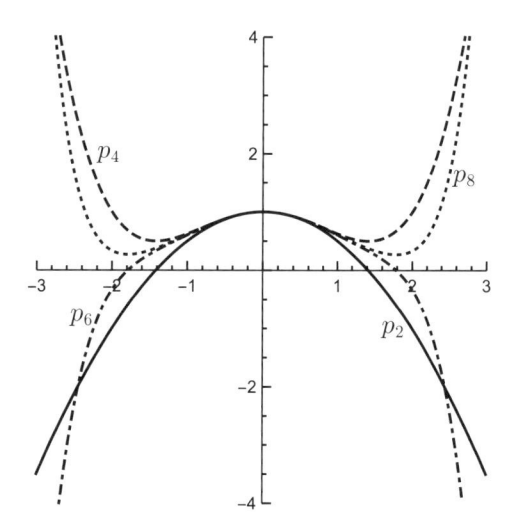

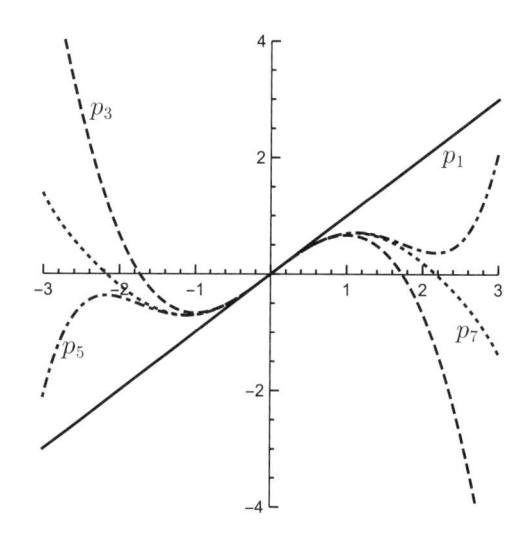

Figure 7.4. Some Taylor polynomials for y_1.

Figure 7.5. Some Taylor polynomials for y_2.

Example 7.3.3. Let's solve the initial value problem

$$(x^2 - 2)y'' + 3xy' + y = 0, \quad y(0) = 1, \quad y'(0) = 0.$$

Since our initial condition is given at $x_0 = 0$, we seek a power series solution centered at $x_0 = 0$. Let

$$y(x) = \sum_{n=0}^{\infty} a_n x^n$$

be our unknown solution. Then

$$y'(x) = \sum_{n=1}^{\infty} n a_n x^{n-1} \quad \text{and} \quad y''(x) = \sum_{n=2}^{\infty} n(n-1) a_n x^{n-2}.$$

Substituting these formulas for y, y', and y'' into the differential equation, we obtain

$$(x^2 - 2) \sum_{n=2}^{\infty} n(n-1) a_n x^{n-2} + 3x \sum_{n=1}^{\infty} n a_n x^{n-1} + \sum_{n=0}^{\infty} a_n x^n = 0.$$

Rewrite this as

(7.31) $$\sum_{n=2}^{\infty} n(n-1) a_n x^n - 2 \sum_{n=2}^{\infty} n(n-1) a_n x^{n-2} + 3 \sum_{n=1}^{\infty} n a_n x^n + \sum_{n=0}^{\infty} a_n x^n = 0.$$

7.3. Series solutions

Only the second series involves powers x^{n-2} instead of x^n, so we do an index shift on just this series, obtaining
$$\sum_{n=2}^{\infty} n(n-1)a_n x^{n-2} = \sum_{n=0}^{\infty} (n+2)(n+1)a_{n+2} x^n.$$
Substituting this into equation (7.31) gives
$$\sum_{n=2}^{\infty} n(n-1)a_n x^n - 2\sum_{n=0}^{\infty} (n+2)(n+1)a_{n+2} x^n + 3\sum_{n=1}^{\infty} na_n x^n + \sum_{n=0}^{\infty} a_n x^n = 0.$$

It's convenient to notice that we can start the summation in the first series here at $n = 0$ instead of $n = 2$. This is because $n(n-1)a_n x^n$ is zero if $n = 0$ or $n = 1$, so by starting the sum at zero, we are just adding two "zero" terms and thus changing nothing. Similarly, the summation in the third series can be started at $n = 0$ instead of $n = 1$ without changing the value of the sum (again, we are just adding a "zero" term).

If we make these changes and then write the whole left side as a *single* sum, collecting like powers of x within that single sum, we get
$$\sum_{n=0}^{\infty} [n(n-1)a_n - 2(n+2)(n+1)a_{n+2} + 3na_n + a_n] x^n = 0.$$

By the Uniqueness Property, the total coefficient of each x^n must be zero. That is, for $n \geq 0$,
$$n(n-1)a_n - 2(n+2)(n+1)a_{n+2} + 3na_n + a_n = 0.$$

Some algebra lets us rewrite this as
$$(n+1)^2 a_n - 2(n+2)(n+1)a_{n+2} = 0.$$

Solving for a_{n+2} gives our recursion relation:

(7.32) $$a_{n+2} = \frac{1}{2} \frac{n+1}{n+2} a_n.$$

for $n \geq 0$.

Imposing the initial condition. To find the solution y_1 satisfying $y_1(0) = 1$ and $y_1'(0) = 0$ we must have $a_0 = 1$ and $a_1 = 0$. The recursion relation (7.32) immediately shows that $a_3, a_5, a_7, \ldots$ are all zero (since $a_1 = 0$). On the other hand, setting $n = 0$ in (7.32) gives
$$a_2 = \frac{1}{2} \cdot \frac{1}{2} \cdot a_0 = \frac{1}{2} \cdot \frac{1}{2}.$$
Putting $n = 2$ in (7.32) gives
$$a_4 = \frac{1}{2} \cdot \frac{3}{4} \cdot a_2 = \frac{1}{2} \cdot \frac{3}{4} \cdot \frac{1}{2} \cdot \frac{1}{2}.$$
If there is a pattern, it is not clear yet. Keep going! If $n = 4$,
$$a_6 = \frac{1}{2} \cdot \frac{5}{6} \cdot a_4 = \frac{1}{2} \cdot \frac{5}{6} \cdot \frac{1}{2} \cdot \frac{3}{4} \cdot \frac{1}{2} \cdot \frac{1}{2}.$$
This looks promising. Let's try $n = 6$:
$$a_8 = \frac{1}{2} \cdot \frac{7}{8} \cdot a_6 = \frac{1}{2} \cdot \frac{7}{8} \cdot \frac{1}{2} \cdot \frac{5}{6} \cdot \frac{1}{2} \cdot \frac{3}{4} \cdot \frac{1}{2} \cdot \frac{1}{2}.$$
Regrouping, we see that
$$a_2 = \frac{1}{2} \cdot \frac{1}{2}, \quad a_4 = \frac{1}{2^2}\left(\frac{1 \cdot 3}{2 \cdot 4}\right), \quad a_6 = \frac{1}{2^3}\left(\frac{1 \cdot 3 \cdot 5}{2 \cdot 4 \cdot 6}\right), \quad \text{and} \quad a_8 = \frac{1}{2^4}\left(\frac{1 \cdot 3 \cdot 5 \cdot 7}{2 \cdot 4 \cdot 6 \cdot 8}\right).$$

This is enough to see the pattern
$$a_{2n} = \frac{1}{2^n}\left(\frac{1\cdot 3\cdot 5\cdots(2n-3)(2n-1)}{2\cdot 4\cdot 6\cdots(2n-2)(2n)}\right) = \frac{1\cdot 3\cdots(2n-3)(2n-1)}{2^n 2^n n!} = \frac{1\cdot 3\cdot 5\cdots(2n-3)(2n-1)}{4^n n!}.$$

We can put the numerator in closed form in the same way that we found a closed form expression for the denominator in (7.26). In Exercise 17 you are asked to show that
$$1\cdot 3\cdot 5\cdots(2n-3)(2n-1) = \frac{(2n-1)!}{2^{n-1}(n-1)!}.$$

Thus, (since $0 = a_1 = a_3 = a_5 = a_7 = \cdots$) our initial value problem has solution
$$y(x) = \sum_{n=0}^{\infty} a_n x^n = \sum_{n=0}^{\infty} a_{2n} x^{2n}$$

$$= \sum_{n=0}^{\infty} \frac{1}{4^n} \frac{1\cdot 3\cdot 5\cdots(2n-3)(2n-1)}{n!} x^{2n}$$

$$= \sum_{n=0}^{\infty} \frac{1}{4^n} \frac{(2n-1)!}{2^{n-1}(n-1)!n!} x^{2n}.$$

Since the zeros of $P(x) = x^2 - 2$ are $x = \sqrt{2}$ and $x = -\sqrt{2}$ and the distance from each of these to $x_0 = 0$ is $\sqrt{2}$, Theorem 7.3.2 guarantees that our power series solution y converges at least on the interval $(-\sqrt{2}, \sqrt{2})$. It's not difficult to show that the radius of convergence of the solution is exactly $\sqrt{2}$; see Exercise 17(d) in Section 7.2.

In this example, we were able to use the recursion relation to explicitly find the coefficients a_n. This will not always be convenient or even possible, and we may have to be content if computing by hand with finding only the first few of them.

Example 7.3.4. The differential equation
(7.33) $$y'' - xy = 0$$
is called **Airy's equation**, and some of its solutions are called Airy's functions. Besides having applications in modeling the diffraction of light, Airy's functions have some curious mathematical properties. Notice that Theorem 7.3.2 applies—here $P(x) = 1$ (a function with no zeros), $Q(x) = 0$, and $R(x) = -x$. We can choose x_0 as we wish and be guaranteed that a series solution centered at x_0 has infinite radius of convergence. Let's take $x_0 = 0$ and seek a solution
$$y(x) = \sum_{n=0}^{\infty} a_n x^n.$$

Computing $y'(x)$ and $y''(x)$ and substituting into equation (7.33), we obtain
(7.34) $$\sum_{n=2}^{\infty} n(n-1)a_n x^{n-2} - x\sum_{n=0}^{\infty} a_n x^n = 0.$$

To combine the series in the last line into a single sum, we shift indices so that the generic term contains x^n:
(7.35) $$\sum_{n=2}^{\infty} n(n-1)a_n x^{n-2} = \sum_{n=0}^{\infty} (n+2)(n+1)a_{n+2} x^n$$

7.3. Series solutions

and

(7.36) $$x\sum_{n=0}^{\infty} a_n x^n = \sum_{n=0}^{\infty} a_n x^{n+1} = \sum_{n=1}^{\infty} a_{n-1} x^n.$$

Rewrite the right-hand side of (7.35) as

$$\sum_{n=0}^{\infty}(n+2)(n+1)a_{n+2}x^n = 2a_2 + \sum_{n=1}^{\infty}(n+2)(n+1)a_{n+2}x^n,$$

by writing the $n = 0$ term explicitly, so that the summation begins with $n = 1$. This makes it easier to combine the two sums in equation (7.34):

$$\sum_{n=2}^{\infty} n(n-1)a_n x^{n-2} - x\sum_{n=0}^{\infty} a_n x^n = 2a_2 + \sum_{n=1}^{\infty}[(n+2)(n+1)a_{n+2} - a_{n-1}]x^n.$$

To guarantee that y solves the differential equation, we want the total coefficient of x^n on the right to be zero for all n. Thus

(7.37) $$2a_2 = 0,$$

and for $n \geq 1$,

$$(n+2)(n+1)a_{n+2} = a_{n-1}.$$

This gives the recursion relation

(7.38) $$a_{n+2} = \frac{1}{(n+2)(n+1)} a_{n-1}$$

for $n = 1, 2, 3, \ldots$. This is called a three-step recursion formula, since a_{n+2} is three steps away from a_{n-1}.

Using the recursion relation. The coefficients a_0 and a_1 may be chosen arbitrarily, but all other coefficients are then determined. From (7.37) we have $a_2 = 0$, and then the recursion relation tells us that a_5, a_8, a_{11}, and in general a_{3m+2} are all zero for $m = 0, 1, 2, 3, \ldots$.

Next we use (7.38) to see that

$$a_3 = \frac{1}{3 \cdot 2} a_0,$$

$$a_6 = \frac{1}{6 \cdot 5} a_3 = \frac{1}{6 \cdot 5 \cdot 3 \cdot 2} a_0,$$

$$a_9 = \frac{1}{9 \cdot 8} a_6 = \frac{1}{9 \cdot 8 \cdot 6 \cdot 5 \cdot 3 \cdot 2} a_0.$$

Can you see the pattern to describe a_{3m} for any positive integer m? We have

(7.39) $$a_{3m} = \frac{a_0}{(3m)(3m-1) \cdot (3m-3)(3m-4) \cdot (3m-6)(3m-7) \cdots (3)(2)}.$$

We can write this more compactly using product notation. The symbol $\prod$ is used for a product, analogously to the use of $\sum$ for a sum. With it, the denominator in equation (7.39) can be written as

$$\prod_{j=1}^{m} (3j)(3j-1)$$

so that

$$a_{3m} = \frac{a_0}{\prod_{j=1}^{m}(3j)(3j-1)}.$$

Alternately, we could write
$$a_{3m} = \frac{(3m-2)(3m-5)(3m-8)\cdots(7)(4)}{(3m)!} a_0.$$

You are asked to show in Exercise 20 that the remaining coefficients, which are a_{3m+1} for $m = 1, 2, 3, \ldots$, are given by
$$a_4 = \frac{1}{4\cdot 3} a_1, \quad a_7 = \frac{1}{7\cdot 6\cdot 4\cdot 3} a_1, \quad a_{10} = \frac{1}{10\cdot 9\cdot 7\cdot 6\cdot 4\cdot 3} a_1,$$

and in general

(7.40) $$a_{3m+1} = \frac{a_1}{(3m+1)(3m)\cdot(3m-2)(3m-3)\cdot(3m-5)(3m-6)\cdot(4)(3)},$$

or
$$a_{3m+1} = \frac{a_1}{\prod_{j=1}^{m}(3j+1)(3j)}.$$

Airy's equation with an initial condition. Suppose we want to solve Airy's equation with initial condition $y(0) = 1$, $y'(0) = 0$. This says $a_0 = 1$ and $a_1 = 0$. Using equations (7.39) and (7.40), we compute
$$a_2 = 0, \quad a_3 = \frac{1}{6}, \quad a_4 = a_5 = 0, \quad a_6 = \frac{1}{180}, \quad a_7 = a_8 = 0, \quad a_9 = \frac{1}{12{,}960}, \ldots.$$

Showing explicitly the first four nonzero terms in the power series solution to this initial value problem we get
$$y(x) = 1 + \frac{1}{6}x^3 + \frac{1}{180}x^6 + \frac{1}{12{,}960}x^9 + \cdots.$$

The ninth degree polynomial
$$1 + \frac{1}{6}x^3 + \frac{1}{180}x^6 + \frac{1}{12{,}960}x^9$$

is a partial sum approximation to this solution; we expect it to be a good approximation on some interval centered at 0, but not necessarily when we are far from 0. Better approximations, on larger intervals about 0, are obtained by taking Taylor polynomials of higher degree. Fig. 7.6 shows the graphs of the Taylor polynomials of degrees 3, 6, 9, and 12 for this initial value problem. Notice how they are indistinguishable near 0 but become quite different as we move away from 0.

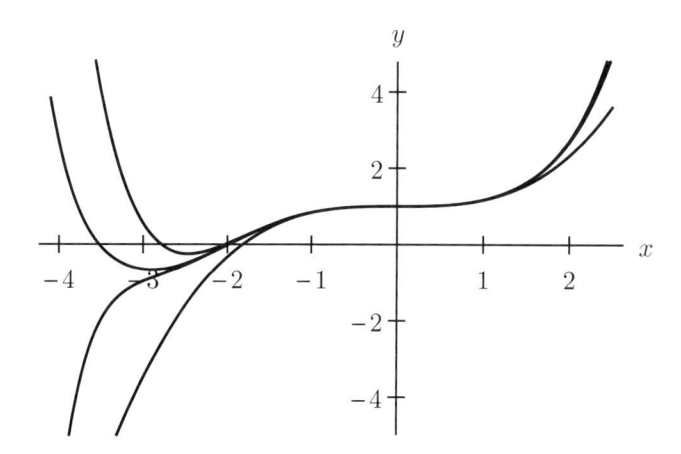

Figure 7.6. Approximate solutions to $y'' - xy = 0$, $y(0) = 1, y'(0) = 0$.

7.3. Series solutions

7.3.1. Exercises.

1. If $\sum_{n=0}^{\infty}[(n+1)a_{n+2} + (n-1)a_n]x^n = 0$ and $a_0 = 1$, $a_1 = 0$, find $a_2, a_3, a_4, a_5, a_6, a_8, a_{99}$, and a_{100}.

In Exercises 2–5 use the method of Example 7.3.1 to find the general solution in terms of power series centered at $x_0 = 0$. State the relevant recursion relation as well as the minimum radius of convergence guaranteed by Theorem 7.3.2 for your series solutions.

2. $(x^2+1)y'' + 4xy' + 2y = 0$.
3. $(x^2-4)y'' + 4xy' + 2y = 0$.
4. $(x^2-1)y'' - 2y = 0$.
5. $(x^2+9)y'' + 6xy' + 6y = 0$.

6. Consider the equation $y'' + 2xy' + 2y = 0$.
   (a) Find the general solution as a power series $\sum_{n=0}^{\infty} a_n x^n$.
   (b) Find the solution, as a power series, satisfying the initial condition $y(0) = 1$, $y'(0) = 0$. Show that the solution you get in this case is equal to e^{-x^2}.

7. In this problem, pretend you have never heard of the functions $\sin x$ and $\cos x$.
   (a) Find a power series solution of the form $\sum_{n=0}^{\infty} a_n x^n$ to the initial value problem
   $$y'' + y = 0, \quad y(0) = 0, \quad y'(0) = 1.$$
   (b) Use Theorem 7.3.2 to determine where the series in (a) converges.
   (c) Find a power series solution of the form $\sum_{n=0}^{\infty} a_n x^n$ to the initial value problem
   $$y'' + y = 0, \quad y(0) = 1, \quad y'(0) = 0.$$
   Where does this series converge?
   (d) Call the function given by the series in (a) $S(x)$, and the function given by the series in (c), $C(x)$. Using term-by-term differentiation, show that $S'(x) = C(x)$ and $C'(x) = -S(x)$.
   (e) Show that $S(x)$ is an odd function (meaning $S(-x) = -S(x)$) and $C(x)$ is an even function ($C(-x) = C(x)$).
   (f) Show that $[S(x)]^2 + [C(x)]^2 = 1$ for all x. (Hint: Use (d), and show that the derivative of $[S(x)]^2 + [C(x)]^2$ is identically zero.)

8. Verify that the coefficients given in Example 7.3.1 for the second solution y_2 are correct.

9. Write the solution y_1 of Example 7.3.1 in terms of the exponential function.

10. Find a power series solution $\sum_{n=0}^{\infty} a_n x^n$ to the first-order initial value problem
    $$y' - y = 0, \quad y(0) = 1.$$
    You should recognize your solution.

11. (a) With the differential equation in Example 7.3.3, solve the initial value problem
    $$(x^2 - 2)y'' + 3xy' + y = 0, \quad y(0) = 0, \quad y'(0) = 1.$$
    (b) What is the solution to the initial value problem
    $$(x^2 - 2)y'' + 3xy' + y = 0, \quad y(0) = 2, \quad y'(0) = 3?$$

12. Consider the differential equation
    $$(1 + x^2)y'' + 2xy' - 2y = 0.$$
    (a) Find a power series solution satisfying the initial condition $y(0) = 1, y'(0) = 0$.
    (b) Find a power series solution satisfying the initial condition $y(0) = 0, y'(0) = 1$.
    (c) Identify your series solution from (a) as a "known" function. Hint: The Taylor series for $\arctan x$ given in Section 7.2 is relevant.

13. Solve the initial value problem
$$y'' - 2xy' + y = 0, \qquad y(0) = 1, \quad y'(0) = -1,$$
showing the first six nonzero terms explicitly.

14. Find two linearly independent solutions in the form
$$\sum_{n=0}^{\infty} a_n x^n$$
to the equation $(x^2 - 1)y'' - 2y = 0$.

15. (a) If a power series solution in the form $\sum_{n=0}^{\infty} a_n x^n$ is found to the equation
$$(x^2 + 3)y'' + xy' + 2x^2 y = 0,$$
what does Theorem 7.3.2 guarantee as a minimum radius of convergence for this series?
(b) Answer the same question for a solution in the form $\sum_{n=0}^{\infty} b_n (x+1)^n$.

16. Consider the equation
$$(2x - 1)y'' - 4xy' + 4y = 0.$$
If a series solution centered at 0 is found for this equation, what does Theorem 7.3.2 guarantee as a minimum radius of convergence for this series? Verify that in fact $\{x, e^{2x}\}$ is a fundamental set of solutions to the equation on $(-\infty, \infty)$. This shows that you sometimes do better than the minimum radius of convergence guaranteed by the theorem.

17. Show that
$$1 \cdot 3 \cdot 5 \cdots (2n-3)(2n-1) = \frac{(2n-1)!}{2^{n-1}(n-1)!}.$$

18. (a) Show that if $y = \sum_{n=0}^{\infty} a_n (x-2)^n$ satisfies
$$xy'' + y' + y = 0,$$
then for $n \geq 1$,
$$a_{n+1} = -\frac{n}{2(n+1)} a_n - \frac{1}{2n(n+1)} a_{n-1}.$$
Hint: Use the idea of Exercise 5 of Section 7.2.
(b) Using the recursion relation for a_n obtained in part (a), find the first four nonzero terms of the power series solution $y = \sum_{n=0}^{\infty} a_n (x-2)^n$ to the initial value problem
(7.41) $$xy'' + y' + y = 0, \qquad y(2) = 1, \quad y'(2) = 0.$$

19. This exercise gives an alternate approach to finding the series solution of the initial value problem (7.41) above.
(a) Show that if y is a solution of (7.41), then $u(t) = y(t+2)$ is a solution of the initial value problem
(7.42) $$(t+2)u''(t) + u'(t) + u(t) = 0, \qquad u(0) = 1, \quad u'(0) = 0.$$
(b) Find the first four nonzero terms of the power series solution $u(t) = \sum_{n=0}^{\infty} a_n t^n$ of (7.42).
(c) Using $y(x) = u(x-2)$, find the first four nonzero terms of the power series solution $y = \sum_{n=0}^{\infty} a_n (x-2)^n$ to the initial value problem (7.41).

20. Verify that the coefficients a_{3m+1} in Example 7.3.4 have the form described by equation (7.40) for $m = 1, 2, 3, \ldots$. Also write a_{3m+1} in the form
$$\frac{???}{(3m+1)!} a_1,$$
identifying the numerator explicitly.

21. (a) Show that if f is a solution of Airy's equation (7.33), then $y = f(-x)$ is a solution of
    (7.43) $$y'' + xy = 0.$$
    (b) Using the observation of part (a) and the coefficient formula (7.39), show that the solution of the initial value problem
    (7.44) $$y'' + xy = 0, \quad y(0) = 1, \quad y'(0) = 0,$$
    may be written as
    $$y(x) = 1 + \sum_{m=1}^{\infty} \frac{(-1)^m}{(3m)(3m-1)\cdot(3m-3)(3m-4)\cdots(3)(2)} x^{3m}$$
    $$= 1 + \sum_{m=1}^{\infty} \frac{(-1)^m}{\prod_{j=1}^{m}(3j)(3j-1)} x^{3m}.$$
    (c) Fig. 7.7 shows the graph of the solution of the initial value problem (7.44) on the interval $[0, 12]$.

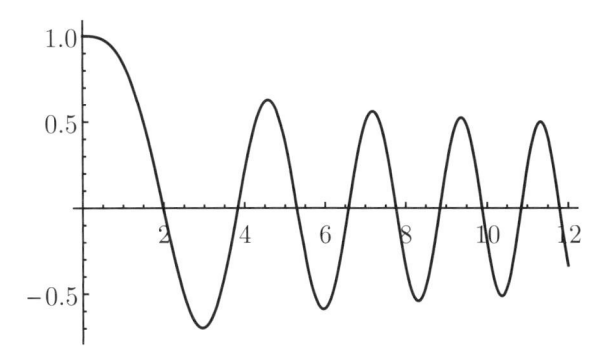

    Figure 7.7. Solution to the initial value problem.

    Changing the independent variable to time t, think of
    $$y''(t) + ty(t) = 0, \quad y(0) = 1, \quad y'(0) = 0,$$
    as modeling an undamped spring system whose spring stiffness increases with time. As the spring stiffens, intuitively what do you expect to happen to the amplitude and frequency of the oscillations? Is your answer consistent with what you can see in the graph of the solution?

22. This problem outlines a different approach to Airy's equation
    $$y'' - xy = 0.$$
    (a) Show that
    $$(xy)' = xy' + y,$$
    $$(xy)'' = xy'' + 2y',$$
    and
    $$(xy)''' = xy''' + 3y''$$
    where all differentiations are with respect to x.
    (b) Show by induction that
    $$(xy)^{(n)} = xy^{(n)} + ny^{(n-1)}$$
    where $^{(n)}$ denotes the nth derivative with respect to x.

(c) Differentiate both sides of Airy's equation n times to obtain
$$y^{(n+2)} = xy^{(n)} + ny^{(n-1)}$$
for $n \geq 1$. Evaluate this at $x = 0$ to obtain

(7.45) $$y^{(n+2)}(0) = ny^{(n-1)}(0) \text{ for } n \geq 1.$$

Also when $n = 0$ we have $y''(0) = 0$ directly from Airy's equation, so that (7.45) holds with $n = 0$ as well.

(d) Suppose that
$$y = \sum_{n=0}^{\infty} a_n x^n$$
is a solution to Airy's equation. Show that it follows from (c) that
$$a_{n+2} = \frac{1}{(n+2)(n+1)} a_{n-1},$$
the same three-step recursion relation we obtained in Example 7.3.4. Hint: Recall that
$$a_{n+2} = \frac{y^{(n+2)}(0)}{(n+2)!}.$$

23. A series solution centered at x_0 is found for the differential equation
$$(9 + x^2)y'' + (1 - 2x)y' + xy = 0.$$
On what interval, at least, must the series converge if
   (a) $x_0 = 0$,
   (b) $x_0 = 1$,
   (c) $x_0 = -2$.

24. Consider the equation $y'' - 2(x-1)y' + 2y = 0$.
   (a) Find the general solution as a power series centered at $x_0 = 1$.
   (b) Find the solution satisfying the initial condition $y(1) = 0$, $y'(1) = 1$, calling your answer y_1.
   (c) Using y_1 from (b) and the reduction of order formula (4.84), find an integral representation of a second linearly independent solution y_2. Evaluate the integral using series methods. If you give the general solution to the differential equation as $c_1 y_1 + c_2 y_2$, does this agree with your answer in (a)?

25. Suppose a series solution $\sum_{n=0}^{\infty} a_n x^n$ is sought for the equation
$$y'' + y' + xy = 0.$$
   (a) Show that for $n \geq 1$ the recursion relation is
$$a_{n+2} = -\frac{(n+1)a_{n+1} + a_{n-1}}{(n+2)(n+1)}.$$
   (b) Find a_2, a_3, a_4, and a_5 if we want $y(0) = 1$ and $y'(0) = 0$.

26. (a) Give a power series solution to the initial value problem
$$y'' - y = 0, \quad y(0) = 0, \quad y'(0) = 1.$$
Verify that your solution is equal to $\sinh x = \frac{1}{2}(e^x - e^{-x})$.
   (b) Give a power series solution to
$$y'' - y = 0, \quad y(0) = 1, \quad y'(0) = 0,$$
and verify that your solution is equal to $\cosh x = \frac{1}{2}(e^x + e^{-x})$.

27. (a) Show that if $y = \sum_{n=0}^{\infty} a_n x^n$ is a solution to

    (7.46) $\qquad (x-1)y'' + xy' - y = 0,$

    then for $n \geq 0$, a_n satisfies the recursion relation

    $$a_{n+2} = \frac{n}{n+2} a_{n+1} + \frac{n-1}{(n+1)(n+2)} a_n.$$

    (b) Using the recursion relation of part (a), show that the solution of the initial value problem

    $$(x-1)y'' + xy' - y = 0, \quad y(0) = 0, \quad y'(0) = 1,$$

    is $y(x) = x$, and check by substitution that $y(x) = x$ is indeed a solution of (7.46).

    (c) Using the recursion relation of part (a), find the first four nonzero terms of the series solution to

    $$(x-1)y'' + xy' - y = 0, \quad y(0) = 1, \quad y'(0) = 0.$$

28. (a) Find a power series $y(x) = \sum_{n=0}^{\infty} a_n x^n$ that solves $(1+x)y' = my$, $y(0) = 1$, where m is a real number.

    (b) Show that your series solution in (a) has radius of convergence at least 1.

    (c) Solve the initial value problem in (a) again, by separation of variables. By equating the separation of variable solution with the power series solution in (a), obtain the Binomial Theorem:

    $$(1+x)^m = 1 + mx + \frac{m(m-1)}{2!} x^2 + \frac{m(m-1)(m-2)}{3!} x^3 + \cdots$$

    for $|x| < 1$. Note that if m is a positive integer, the series has only finitely many nonzero terms, and thus it converges for all x.

    (d) Similarly, find the power series solution to

    $$(a+x)y' = my, \quad y(0) = a^m,$$

    for constant a, and use it to give the Binomial Theorem for $(a+x)^m$.

29. (a) A horizontal mass-spring system with mass 1 kg, damping constant 1 N-sec/m, and spring constant 19 N/m is set in motion so that the mass's initial displacement from equilibrium is 1/2 m and initial velocity is 0 m/sec. Assuming the mass's displacement from equilibrium at time t is $y(t)$ and there is no external forcing, the initial value problem modeling the system's motion is

    $$y'' + y' + 19y = 0, \quad y(0) = 1/2, \quad y'(0) = 0.$$

    Solve this initial value problem.

    (b) Suppose the system of part (a) resides in a tank that is gradually filled with a viscous fluid over the time interval $[0,5]$ resulting in its being modeled by the initial value problem

    (7.47) $\qquad y'' + (1+t)y' + 19y = 0, \quad y(0) = 1/2, \quad y'(0) = 0 \quad$ for $0 \leq t \leq 5.$

    Find a recursion relation for the coefficients of a power series solution $y = \sum_{n=0}^{\infty} a_n t^n$ of this new initial value problem, and determine the first four nonzero terms in the solution.

    (c) The graphs in Figs. 7.8–7.9 show the solutions to the initial value problems in (a) and (b) for $0 \leq t \leq 5$. Which is which? How do you know?

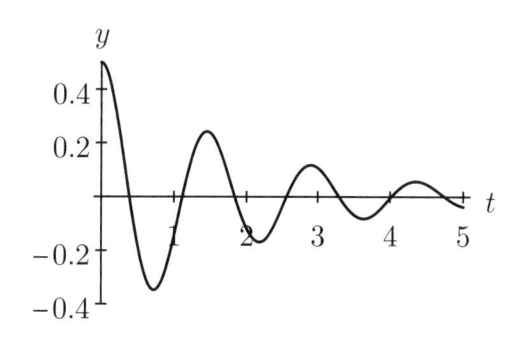

Figure 7.8.

Figure 7.9.

(d) (CAS) Use a computer algebra system to give a numerical solution to the initial value problem in (b) and plot this solution for $0 \leq t \leq 5$, giving a partial check on your answer in (c).

30. Imagine you are holding a short uniform pole made of spring-steel. If you hold the pole at its base so that the pole points vertically upward, then the entire pole assumes a vertical position. However, as you perform this experiment with longer and longer poles having the same composition and shape as the original pole, there will be a smallest length L such that pole readily bends over or perhaps even buckles.

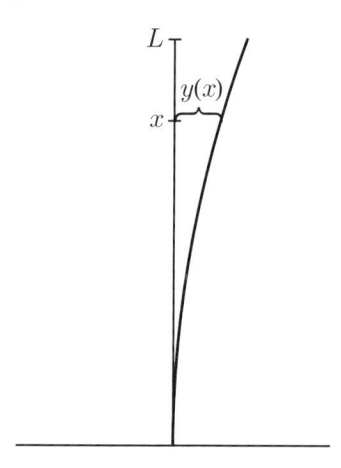

   In this problem we'll use a differential equation model to predict the height at which a uniformly constructed pole will readily bend under its own weight.

   Assume the pole has a base that is fixed so that the pole stands vertically there; i.e., the tangent line to the pole at its base is vertical. Also assume that the pole has cross sections that are disks of radius $r = 1/2$ inch. Let the x-axis be vertical, with origin at the base of the pole, and let the positive x-direction be upward. We assume that the pole has the smallest length L at which a slight horizontal push at its tip will cause the pole to bend over, eventually reaching an equilibrium, bent shape. The direction of the "horizontal push" determines the vertical plane containing the axis of the bent pole. Let $y(x)$ be the horizontal displacement, at height x, of the pole from the vertical axis through its base. To model this situation[1] we will use the equation

   (7.48) $$\rho y'''(x) + (L - x)y'(x) = 0,$$

---

[1] See, e.g., p. 258 of J. P. D. Hartog, *Advanced Strength of Materials*, McGraw-Hill, New York, 1952 or Section 276 of A. E. H. Love, *A Treatise on the Mathematical Theory of Elasticity*, 4th ed., Dover, New York, 1944. A simplifying assumption made in deriving the model is that during the bending process, particles of the pole are displaced only horizontally.

7.3. Series solutions

where ρ is a positive constant that depends on several physical characteristics of the pole, together with the condition

(7.49) $$y(0) = y'(0) = y''(L) = 0.$$

Notice this is *not* an *initial* condition, as y and its derivatives are being specified at two different values of x. The requirements in (7.49) are called **boundary conditions**. Two of the three boundary conditions obviously hold: We must have $y(0) = y'(0) = 0$ because the base of the pole lies at the origin and is in a fixed, upright position. The final boundary condition $y''(L) = 0$ may be understood intuitively as follows: If we take $y''(x)$ as a measure of the extent of bending of the pole at x and agree that bending at height $0 < x < L$ occurs owing to the weight of that part of the pole above level x, then there should be no bending at the upper tip of the pole, i.e., $y''(L) = 0$.

(a) Show that if y solves the boundary value problem (7.48)–(7.49), then

$$u(s) \equiv y'\left(L - \rho^{1/3}s\right)$$

solves

(7.50) $$u''(s) + su(s) = 0, \quad u\left(L\rho^{-\frac{1}{3}}\right) = 0, \quad u'(0) = 0.$$

Hint: Make the substitution $x = L - \rho^{1/3}s$.

(b) Suppose that u solves the boundary value problem (7.50) and set $u(0) = a_0$. From Exercise 21, we see that

$$\begin{aligned} u(s) &= a_0 \left(1 + \sum_{m=1}^{\infty} \frac{(-1)^m}{(3m)(3m-1) \cdot (3m-3)(3m-4) \cdots (3)(2)} s^{3m}\right) \\ &= a_0 \left(1 + \sum_{m=1}^{\infty} \frac{(-1)^m}{\prod_{j=1}^{m}[(3j)(3j-1)]} s^{3m}\right). \end{aligned}$$

Conclude that the boundary value problem (7.50) has a solution only if $a_0 = 0$ or if the function f defined by

$$f(s) \equiv 1 + \sum_{m=1}^{\infty} \frac{(-1)^m}{\prod_{j=1}^{m}[(3j)(3j-1)]} s^{3m}$$

vanishes at $s = L\rho^{-\frac{1}{3}}$. If $a_0 = 0$, then, applying the definition of u, we see that $y' = 0$ on $(0, L)$ so that y must be constant; however, the boundary condition $y(0) = 0$ of (7.48) then makes $y \equiv 0$. Thus, $a_0 = 0$ yields the unbent-pole solution of (7.48)–(7.49). A bent-pole solution then requires the length L of the pole to be such that

$$f(L\rho^{-\frac{1}{3}}) = 0.$$

Using the graph of $f(s)$ in Fig. 7.10, determine (approximately) the smallest value of L (in terms of the constant ρ) for a pole that bends readily according to our model.

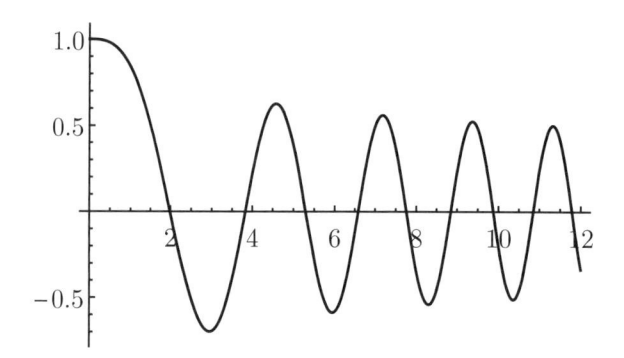

Figure 7.10. Graph of $f(s)$.

(c) The constant ρ in equation (7.48) is equal to
$$\frac{EI}{w}$$
where E is a constant called Young's modulus of elasticity (it depends on the material of the pole; for spring-steel, it is about 3×10^7 psi), I is a constant that depends on the shape of the cross sections of the pole (for circular cross sections of radius r, $I = \pi r^4/4$), and w is the linear weight density of the pole (about 0.22 lb/inch for our pole). The product EI is a measure of stiffness of the pole. Using your work in (b) and the given values of E, I, and w, what is the minimum length in feet at which our pole readily bends?

(d) Assume the pole still has cross sections that are 1/2 in radius but that the pole is hollow. That is, cross sections are annuli with outer radius 1/2. Assume that the inner radius is 1/4, that $I = \pi(r_0^2/4 - r_i^4/4)$ where $r_0 = 1/2$ and $r_i = 1/4$, and that $w = 0.165$ lb/in. Using these new values for I and w and your work in part (b), what is the minimum length at which bending readily occurs for this hollow pole?

31. The differential equation
$$y'' - 2xy' + 2\rho y = 0,$$
where ρ is a constant, is called Hermite's equation.

(a) Show that Hermite's equation has solutions
$$y_1(x) = 1 - \frac{2\rho}{2!}x^2 + \frac{2^2\rho(\rho-2)}{4!}x^4 - \frac{2^3\rho(\rho-2)(\rho-4)}{6!}x^6 + \cdots$$
and
$$y_2(x) = x - \frac{2(\rho-1)}{3!}x^3 + \frac{2^2(\rho-1)(\rho-3)}{5!}x^5 - \frac{2^3(\rho-1)(\rho-3)(\rho-5)}{7!}x^7 + \cdots.$$

(b) Show that if ρ is a positive integer or 0, one of the two solutions in (a) is just a polynomial.

(c) Find the polynomial solution from (b) corresponding to $\rho = 1, 2, 3, 4,$ and 5. When one of these polynomial solutions is multiplied by a constant chosen so that the highest degree term is $2^n x^n$, the resulting polynomial is called the Hermite polynomial $H_n(x)$. Find $H_2(x)$, $H_3(x)$, and $H_4(x)$.

7.4. Nonpolynomial coefficients

The power series method of the last section can be adapted to solve

(7.51) $$y'' + p(x)y' + q(x)y = 0$$

7.4. Nonpolynomial coefficients

when the coefficient functions $p(x)$ and $q(x)$ are analytic functions (but not necessarily polynomials or rational functions) at a point x_0. Recall that a function $f(x)$ is analytic at x_0 if the function can be written as a power series

$$f(x) = \sum_{n=0}^{\infty} a_n (x - x_0)^n$$

with a positive radius of convergence. The point $x = x_0$ is said to be an **ordinary point** for equation (7.51) if $p(x)$ and $q(x)$ are analytic at x_0. Notice that for an equation

$$P(x)y'' + Q(x)y' + R(x)y = 0$$

where $P(x)$, $Q(x)$, and $R(x)$ are polynomials (the kind of equation we considered in the last section), any point x_0 with $P(x_0) \neq 0$ is an ordinary point.

A version of Theorem 7.3.2 applies. In it we suppose that the coefficient functions $p(x)$ and $q(x)$ are given by power series

$$p(x) = \sum_{n=0}^{\infty} b_n (x - x_0)^n$$

and

$$q(x) = \sum_{n=0}^{\infty} c_n (x - x_0)^n$$

with radii of convergence $R_1 > 0$ and $R_2 > 0$, respectively.

Theorem 7.4.1. *Consider*

(7.52) $$y'' + p(x)y' + q(x)y = 0,$$

where $p(x)$ and $q(x)$ are analytic at x_0. Then equation (7.52) has two linearly independent solutions of the form

$$\sum_{n=0}^{\infty} a_n (x - x_0)^n.$$

Any solution in this form converges at least on the interval $(x_0 - R^*, x_0 + R^*)$, where R^* is the smaller of R_1 and R_2, the radii of convergence of the power series for $p(x)$ and $q(x)$ at $x = x_0$.

Example 7.4.2. We will find the general power series solution centered at $x_0 = 0$ for

(7.53) $$y'' - (\sin x)y = 0.$$

We know that

$$\sin x = x - \frac{x^3}{3!} + \frac{x^5}{5!} - \frac{x^7}{7!} + \cdots$$

for all $-\infty < x < \infty$. So $\sin x$ is analytic at $x = 0$, and its Taylor series centered at 0 has infinite radius of convergence. By Theorem 7.4.1 any series solution $\sum_{n=0}^{\infty} a_n x^n$ to (7.53) will converge everywhere. Substituting $y = \sum_{n=0}^{\infty} a_n x^n$, $y'' = \sum_{n=2}^{\infty} n(n-1) a_n x^{n-2}$, and our Taylor series for $\sin x$ into (7.53) gives

(7.54) $$\sum_{n=2}^{\infty} n(n-1) a_n x^{n-2} - \left(x - \frac{x^3}{3!} + \frac{x^5}{5!} - \frac{x^7}{7!} + \cdots \right) \left(\sum_{n=0}^{\infty} a_n x^n \right) = 0.$$

From this point on we obtain information about the coefficients a_n by determining the total coefficient of x^j on the left side of equation (7.54) for each $j = 0, 1, 2, 3, \ldots$ and setting this equal to 0.

To help carry this out, let's compute the first several terms in the product

$$\left(x - \frac{x^3}{3!} + \frac{x^5}{5!} - \frac{x^7}{7!} + \cdots\right)\left(\sum_{n=0}^{\infty} a_n x^n\right),$$

which we can also write as

(7.55) $$\left(x - \frac{x^3}{3!} + \frac{x^5}{5!} - \frac{x^7}{7!} + \cdots\right) \times (a_0 + a_1 x + a_2 x^2 + a_3 x^3 + \cdots).$$

Visualizing this multiplication as if we were multiplying two polynomials, we see that the product is equal to

$$a_0 x + a_1 x^2 + \left(a_2 - \frac{a_0}{3!}\right) x^3 + \left(a_3 - \frac{a_1}{3!}\right) x^4 + \left(a_4 - \frac{a_2}{3!} + \frac{a_0}{5!}\right) x^5 + \cdots.$$

Can you see how this was done? For example, the only way to get an x^2 term from the product is by multiplying the x term in the first factor with the $a_1 x$ term in the second factor, obtaining $a_1 x^2$. We can get an x^3 term in two ways: as the product of x from the first factor with $a_2 x^2$ from the second factor or as the product of $-\frac{x^3}{3!}$ from the first factor with a_0 from the second factor. Adding these together we see that the coefficient on x^3 in the product is $a_2 - a_0/3!$.

What is the constant term on the left-hand side of equation (7.54)? There is a constant term of $2a_2$ which comes from $n = 2$ in the first sum, $\sum_{n=2}^{\infty} n(n-1) a_n x^{n-2}$. By the calculation we just did, the product

$$\left(x - \frac{x^3}{3!} + \frac{x^5}{5!} - \frac{x^7}{7!} + \cdots\right)\left(\sum_{n=0}^{\infty} a_n x^n\right)$$

has **no** constant term. So the constant term on the left-hand side of (7.54) is just $2a_2$, and this must be 0:

$$2a_2 = 0$$

so that $a_2 = 0$.

Next, set the total coefficient of x on the left-hand side of (7.54) equal to 0:

$$6a_3 - a_0 = 0.$$

The "$6a_3$" comes from $n = 3$ in the first term, $\sum_{n=2}^{\infty} n(n-1) a_n x^{n-2}$.

The coefficient of x^2 on the left-hand side of (7.54) is $12a_4 - a_1$, and this too must be 0:

$$12a_4 - a_1 = 0.$$

Let's take this a few steps further to see that

$$20a_5 - a_2 + \frac{a_0}{3!} = 0 \text{ (coefficient of } x^3\text{)},$$

$$30a_6 - a_3 + \frac{a_1}{3!} = 0 \text{ (coefficient of } x^4\text{)},$$

$$42a_7 - a_4 + \frac{a_2}{3!} - \frac{a_0}{5!} = 0 \text{ (coefficient of } x^5\text{)}.$$

The values of a_0 and a_1 can be arbitrarily chosen, but once they are, these equations let us determine $a_2, a_3, a_4, \ldots$ in order.

For example, if $a_0 = 1$ and $a_1 = 0$, we have

$$a_2 = 0, \quad a_3 = \frac{1}{6}, \quad a_4 = 0, \quad a_5 = -\frac{1}{120}, \quad a_6 = \frac{1}{180}, \quad a_7 = \frac{1}{5{,}040}.$$

7.4. Nonpolynomial coefficients

Thus if we show the first five nonzero terms of the solution y_1 satisfying the initial condition $y_1(0) = 1, y_1'(0) = 0$ (so that $a_0 = 1$ and $a_1 = 0$) we have

$$y_1(x) = 1 + \frac{1}{6}x^3 - \frac{1}{120}x^5 + \frac{1}{180}x^6 + \frac{1}{5{,}040}x^7 + \cdots.$$

A second linearly independent solution y_2 comes from choosing $y_2(0) = 0$, $y_2'(0) = 1$ (so that $a_0 = 0$ and $a_1 = 1$). Showing the first four nonzero terms of this solution gives

$$y_2(x) = x + \frac{1}{12}x^4 - \frac{1}{180}x^6 + \frac{1}{504}x^7 + \cdots.$$

The general solution can be written as a linear combination of y_1 and y_2:

$$\begin{aligned} y(x) &= c_1 y_1 + c_2 y_2 \\ &= c_1 + c_2 x + \frac{c_1}{6}x^3 + \frac{c_2}{12}x^4 - \frac{c_1}{120}x^5 + \frac{c_1 - c_2}{180}x^6 + \frac{c_1 + 10c_2}{5{,}040}x^7 + \cdots. \end{aligned}$$

Nonhomogeneous equations. The techniques of this section can be adapted to solve some nonhomogeneous differential equations as well. If in the equation

$$y'' + p(x)y' + q(x)y = g(x),$$

the functions $p(x)$, $q(x)$, and $g(x)$ are analytic at $x = x_0$, we can find solutions of the form

$$y = \sum_{n=0}^{\infty} a_n (x - x_0)^n.$$

Each such solution will have radius of convergence at least as large as the smallest of the radii of convergence of the Taylor series for p, q, and g about x_0. Our next example illustrates the procedure.

Example 7.4.3. We'll solve the initial value problem

$$y'' - xy = e^x, \qquad y(0) = 1, \quad y'(0) = 0.$$

Set $y = \sum_{n=0}^{\infty} a_n x^n$, and substitute the Taylor series

$$e^x = \sum_{n=0}^{\infty} \frac{x^n}{n!}$$

for e^x. We have

$$\sum_{n=2}^{\infty} n(n-1) a_n x^{n-2} - \sum_{n=0}^{\infty} a_n x^{n+1} = \sum_{n=0}^{\infty} \frac{x^n}{n!}.$$

Shifting indices (so as to have generic term x^n in each series) and writing the constant term on the left-hand side separately, we have

$$2a_2 + \sum_{n=1}^{\infty} [(n+2)(n+1)a_{n+2} - a_{n-1}] x^n = \sum_{n=0}^{\infty} \frac{x^n}{n!}.$$

The Uniqueness Property tells us that for each j, the coefficient of x^j on the left-hand side must equal the coefficient of x^j on the right-hand side. This leads to the equations

$$2a_2 = 1,$$

and for $n \geq 1$, the recursion relation is

$$(n+2)(n+1)a_{n+2} - a_{n-1} = \frac{1}{n!}.$$

Thus $a_2 = \frac{1}{2}$, and for $n \geq 1$ we have the recursion relation
$$a_{n+2} = \frac{1}{(n+2)(n+1)}a_{n-1} + \frac{1}{(n+2)!}.$$
To meet our initial condition, we want $a_0 = 1$ and $a_1 = 0$. We can use the recursion relation to compute as many terms of the solution as we want:
$$a_3 = \frac{1}{6}a_0 + \frac{1}{3!} = \frac{1}{3},$$
$$a_4 = \frac{1}{12}a_1 + \frac{1}{4!} = \frac{1}{24},$$
$$a_5 = \frac{1}{20}a_2 + \frac{1}{5!} = \frac{1}{30},$$
and so on, giving the solution
$$y(x) = 1 + \frac{1}{2}x^2 + \frac{1}{3}x^3 + \frac{1}{24}x^4 + \frac{1}{30}x^5 + \cdots.$$
This series converges for all x because, e.g., it solves $y'' + p(x)y' + q(x)y = g(x)$, where $p(x) = 0, q(x) = -x$, and $g(x) = e^x$ all have Taylor series about 0 with radius of convergence ∞. (See Exercise 13 for a different justification.)

7.4.1. Exercises.

1. What is the coefficient of x^6 when the product in equation (7.55) is expanded?

2. Find the first four nonzero terms in the power series solution $\sum_{n=0}^{\infty} a_n x^n$ to the initial value problem
$$y'' - e^x y = 0, \quad y(0) = 1, \quad y'(0) = 0.$$

3. Give the first three nonzero terms in a power series solution $\sum_{n=0}^{\infty} a_n x^n$ to
$$y'' + (1 + 2\cos x)y = 0, \quad y(0) = 1, \quad y'(0) = 0.$$

4. Consider the differential equation
$$y'' + (\sin x)y = e^{x^2}.$$
If $y = \sum_{n=0}^{\infty} a_n x^n$ solves this equation and has $y(0) = 1$, $y'(0) = 0$, find the first five nonzero values of a_n.

5. Using Theorem 7.4.1, determine the minimum radius of convergence of a series solution $\sum_{n=0}^{\infty} a_n x^n$ to the differential equation
$$(1 + e^x)y'' - xy = 0.$$
Hint: The radius of convergence for the power series centered at $x_0 = 0$ of $q(x) = -x/(1+e^x)$ is equal to the distance from $x_0 = 0$ to the nearest point in the complex plane at which q is not analytic.

6. Explain why Theorem 7.4.1 guarantees that the equation
$$xy'' + (\sin x)y' + 2xy = 0$$
has a power series solution
$$\sum_{n=0}^{\infty} a_n x^n$$
which converges for all $-\infty < x < \infty$. Hint: In the notation of Theorem 7.4.1, what are $p(x)$ and $q(x)$? Why is $p(x)$ analytic at $x = 0$?

7. Find the first five nonzero terms in the power series solution $\sum_{n=0}^{\infty} a_n x^n$ to the initial value problem
$$xy'' + (\sin x)y' + 2xy = 0, \quad y(0) = 1, \quad y'(0) = 1.$$

8. Find the first five nonzero terms in the series solution to the initial value problem
$$(\cos x)y'' - x^2 y' + y = 0, \quad y(0) = 1, \quad y'(0) = -1.$$

9. Consider the initial value problem
$$my'' + k e^{\epsilon x} y = 0, \quad y(0) = 1, \quad y'(0) = 0,$$
where m, k, and ϵ are positive constants.

   (a) Find the first four nonzero terms in a power series solution
$$y = \sum_{n=0}^{\infty} a_n x^n.$$

   (b) What should your solution in (a) simplify to if $\epsilon = 0$? Does it?

10. Give a series solution to the nonhomogeneous initial value problem
$$(1 + x^2)y'' + 2xy = x^2, \quad y(0) = 0, \quad y'(0) = 0,$$
showing the first two nonzero terms.

11. A series solution centered at 0 to the initial value problem
$$y'' + 2xy' + 4y = e^x, \quad y(0) = 0, \quad y'(0) = 0,$$
has the form
$$y = \sum_{n=2}^{\infty} a_n x^n$$
(why?). Find the terms up to the x^5 term in this solution. Where must it converge?

12. Find the first three nonzero terms in the series solution $\sum_{n=0}^{\infty} a_n x^n$ to the initial value problem
$$y'' - x^2 y = \sin x, \quad y(0) = y'(0) = 0.$$

13. Let y be the power series solution of the initial value problem of Example 7.4.3. By applying an appropriate annihilating operator to both sides of the differential equation there, find a third-order *homogenous* linear equation that y must satisfy. Assuming the natural analog of Theorem 7.3.2 for third-order homogeneous equations is valid, explain why the power series representing y must have infinite radius of convergence.

7.5. Regular singular points

7.5.1. A warm-up: Cauchy-Euler equations. In Chapter 4 we learned how to solve any constant coefficient second-order linear homogeneous equation. Moving beyond the constant coefficient case, we also briefly looked at some Cauchy-Euler equations (see Section 4.5). Recall that a Cauchy-Euler equation has the form

(7.56) $$x^2 y'' + b_1 x y' + b_0 y = 0$$

for real constants b_1 and b_0. Writing the equation in standard form as
$$y'' + \frac{b_1}{x} y' + \frac{b_0}{x^2} y = 0$$
shows that in general we seek solutions on intervals not containing 0. Using $y = x^r$ as a trial solution and substituting $y = x^r$, $y' = r x^{r-1}$, and $y'' = r(r-1)x^{r-2}$ into (7.56), we obtain after

dividing by x^r

$$r(r-1) + b_1 r + b_0 = 0.$$

This is called the **indicial equation** for the Cauchy-Euler equation (7.56). There are three possible outcomes when we find the roots of the indicial equation:

Case I. There are two real different roots r_1 and r_2. The corresponding solutions x^{r_1} and x^{r_2} are a linearly independent pair on any open interval not containing 0, and the general solution is

$$y = c_1 x^{r_1} + c_2 x^{r_2}.$$

For example, the indicial equation for

(7.57) $$x^2 y'' + 6xy' + 4y = 0$$

is

$$r^2 + 5r + 4 = 0,$$

which has roots $r = -4$ and $r = -1$. It has linearly independent solutions $y_1 = x^{-1}$ and $y_2 = x^{-4}$ on any interval not containing 0. Thus the general solution to (7.57) on, e.g., $(0, \infty)$ is $y = c_1 x^{-1} + c_2 x^{-4}$.

Case II. The indicial equation has only one repeated (real) root; call it r_1. One solution to the differential equation is $y_1(x) = x^{r_1}$, and we can find a second linearly independent solution by reduction of order. On the interval $(0, \infty)$ this will give a solution

$$y_2(x) = x^{r_1} \ln x.$$

(See Exercise 24 in Section 4.5, or Exercise 9 in this section for a somewhat different approach.) As an example, you can check that the equation

(7.58) $$x^2 y'' + 5xy' + 4y = 0$$

has indicial equation $r^2 + 4r + 4 = 0$, which has the repeated root $r = -2$. The general solution to (7.58) on the interval $(0, \infty)$ is

$$y = c_1 x^{-2} + c_2 x^{-2} \ln x.$$

Case III. The third possibility is that the roots of the indicial equation are a pair of complex conjugates $r_1 = \alpha + i\beta$ and $r_2 = \alpha - i\beta$. To interpret the solution functions

$$x^{\alpha + i\beta} \quad \text{and} \quad x^{\alpha - i\beta} \quad (x > 0)$$

we define

$$x^{\alpha + i\beta} = x^\alpha x^{i\beta} = x^\alpha e^{i\beta \ln x} = x^\alpha [\cos(\beta \ln x) + i \sin(\beta \ln x)].$$

The second equality is suggested by the identity $x = e^{\ln x}$ and properties of exponents. and the third equality uses Euler's formula. Thus we have

$$y_1 = x^{\alpha + i\beta} = x^\alpha [\cos(\beta \ln x) + i \sin(\beta \ln x)]$$

as one complex-valued solution to the differential equation. The complex conjugate root $\alpha - i\beta$ similarly gives the complex-valued solution

$$y_2 = x^{\alpha - i\beta} = x^\alpha [\cos(\beta \ln x) - i \sin(\beta \ln x)].$$

The Superposition Principle says that

$$\frac{1}{2}y_1 + \frac{1}{2}y_2 = x^\alpha \cos(\beta \ln x) \quad \text{and} \quad \frac{1}{2i}y_1 - \frac{1}{2i}y_2 = x^\alpha \sin(\beta \ln x)$$

are also solutions. These real-valued solutions are the real and imaginary parts of y_1. The general (real-valued) solution on $(0, \infty)$ is

$$y = c_1 x^\alpha \cos(\beta \ln x) + c_2 x^\alpha \sin(\beta \ln x).$$

Example 7.5.1. We illustrate Case III with the equation

(7.59) $$x^2 y'' + 5xy' + 13y = 0.$$

Check that the indicial equation is $r^2 + 4r + 13 = 0$, which has roots $-2 + 3i$ and $-2 - 3i$. A complex-valued solution to (7.59) is

$$x^{-2+3i} = x^{-2}[\cos(3 \ln x) + i \sin(3 \ln x)].$$

Its real and imaginary parts give two linearly independent real-valued solutions, and the general real-valued solution on $(0, \infty)$ is

$$y = c_1 x^{-2} \cos(3 \ln x) + c_2 x^{-2} \sin(3 \ln x).$$

The behavior of a solution to a Cauchy-Euler equation *near* the point $x = 0$ can be quite varied. For example, the solutions $y_1(x) = x^{-1}$ and $y_2(x) = x^{-4}$ to $x^2 y'' + 6xy' + 4y = 0$ are both *unbounded* as x approaches 0. The equation $9x^2 y'' - 9xy' + 5y = 0$ has solutions $y_1(x) = x^{1/3}$ and $y_2(x) = x^{5/3}$ (see Exercise 7), both of which are continuous at $x = 0$. (Note also that the second is also differentiable at $x = 0$, while the first is not.) By contrast, all solutions to $x^2 y'' - 4xy' + 6y = 0$ are analytic on $(-\infty, \infty)$ (see Exercise 8).

Solutions on $(-\infty, 0)$. In general we seek to solve a Cauchy-Euler equation on an interval not containing 0. In our examples so far, we have chosen the interval to be $(0, \infty)$. Let's look a little closer at this issue. If the roots r_1 and r_2 are integers, then the functions $y_1 = x^{r_1}$ and $y_2 = x^{r_2}$ are defined for all $x \neq 0$, and they will solve the Cauchy-Euler equation on $(-\infty, 0)$ as well as $(0, \infty)$. (In fact, if r_1 and r_2 are *nonnegative* integers, the corresponding functions y_1 and y_2 solve the differential equation on $(-\infty, \infty)$.)

For negative values of x and (some) noninteger real values of r, the functions x^r and $x^r \ln x$ (which arises in Case II above) are not defined as real-valued functions. However, for any (real) value of r (integer or not) we claim that if x^r (or $x^r \ln x$) solves equation (7.56) on $(0, \infty)$, then $|x|^r$ (respectively, $|x|^r \ln |x|$) solves (7.56) on $(-\infty, 0)$. Exercise 12 outlines how to verify this. Since for positive values of x, $|x|^r = x^r$ and $|x|^r \ln |x| = x^r \ln x$, we can assert that $|x|^r$ (or $|x|^r \ln |x|$) solves the differential equation on any interval not containing 0.

Similarly, when $r = \alpha \pm i\beta$ are complex conjugate roots of the indicial equation, the functions $|x|^\alpha \cos(\beta \ln |x|)$ and $|x|^\alpha \sin(\beta \ln |x|)$ solve the Cauchy-Euler equation on any interval not containing 0.

7.5.2. Regular singular points and Frobenius equations. In this section we will generalize our analysis of Cauchy-Euler equations. Notice that if we rewrite the Cauchy-Euler equation

(7.60) $$x^2 y'' + b_1 xy' + b_0 y = 0$$

in standard form as

$$y'' + \frac{b_1}{x} y' + \frac{b_0}{x^2} y = 0,$$

the coefficient functions $p(x) = \frac{b_1}{x}$ and $q(x) = \frac{b_0}{x^2}$ fail to be analytic at $x = 0$, so $x = 0$ is not an ordinary point for equation (7.60). We'll say that $x = 0$ is a **singular point** for (7.60). Not all singular points are equally bad, and the next definition singles out those that are "better".

Definition 7.5.2. Consider the equation

(7.61) $$y'' + p(x)y' + q(x)y = 0.$$

The point $x = x_0$ is a **singular point** for (7.61) if at least one of the functions $p(x)$, $q(x)$ is not analytic at $x = x_0$. If $x = x_0$ is a singular point for (7.61), but both $(x-x_0)p(x)$ and $(x-x_0)^2 q(x)$ are analytic at x_0, then x_0 is said to be a **regular** singular point. A singular point that is not regular is termed **irregular**.

In particular, the point $x = 0$ is a regular singular point for any Cauchy-Euler equation.

Example 7.5.3. Let's find all singular points for the equation

(7.62) $$(x^2 - 1)^2 y'' + 5(x+1)y' + x(x+2)y = 0$$

and determine which are regular. Writing the equation in the form (7.61) we have

$$p(x) = \frac{5(x+1)}{(x^2-1)^2} = \frac{5}{(x-1)^2(x+1)}$$

and

$$q(x) = \frac{x(x+2)}{(x^2-1)^2} = \frac{x(x+2)}{(x-1)^2(x+1)^2}.$$

Thus (7.62) has singular points at $x = -1$ and $x = 1$. Both

$$(x+1)p(x) = \frac{5}{(x-1)^2} \quad \text{and} \quad (x+1)^2 q(x) = \frac{x(x+2)}{(x-1)^2}$$

are analytic at $x = -1$, so according to Definition 7.5.2 $x = -1$ is a regular singular point. But $x = 1$ is an irregular singular point since

$$(x-1)p(x) = \frac{5}{(x-1)(x+1)}$$

is not analytic at $x = 1$.

A class of equations with regular singular points. Suppose in the Cauchy-Euler equation (7.60) we replace the constants b_1 and b_0 by functions $P(x)$ and $Q(x)$ that are analytic at $x = 0$:

(7.63) $$x^2 y'' + P(x) x y' + Q(x) y = 0.$$

Any equation having $x = 0$ as a singular point that can be written in the form (7.63) is called a **Frobenius equation**. It can be solved *near* 0 by a method that combines the idea of power series solutions with the intuition we gained from solving Cauchy-Euler equations. Before describing this method, notice that if we rewrite equation (7.63) in standard form as

$$y'' + p(x)y' + q(x)y = 0,$$

we have

$$p(x) = \frac{P(x)}{x} \quad \text{and} \quad q(x) = \frac{Q(x)}{x^2}.$$

Moreover, since

$$x p(x) = P(x) \quad \text{and} \quad x^2 q(x) = Q(x),$$

7.5. Regular singular points

where we are assuming both $P(x)$ and $Q(x)$ are analytic at $x = 0$, the singular point $x = 0$ is a *regular* singular point for any Frobenius equation. Conversely, suppose the differential equation

(7.64) $$y'' + p(x)y' + q(x)y = 0$$

has a regular singular point at $x = 0$. Setting $P(x) = xp(x)$ and $Q(x) = x^2 q(x)$, regularity tells you that $P(x)$ and $Q(x)$ are analytic at $x = 0$, and equation (7.64) is equivalent to the Frobenius equation

$$x^2 y'' + P(x)xy' + Q(x)y = 0.$$

Thus a method for solving the Frobenius equation (7.63) is really a method for solving any equation (7.64) having $x = 0$ as a regular singular point. We turn to this next.

The method of Frobenius. We show how to produce solutions to (7.63) near 0 when $P(x)$ and $Q(x)$ are analytic at $x = 0$ (or, equivalently, to (7.64) when $x = 0$ is a regular singular point). (Exercise 33 will show how to adapt this method to solve (7.64) near a regular singular point other than 0.) The basic idea is to look for nontrivial solutions to (7.63) of the form

(7.65) $$y(x) = x^r \sum_{n=0}^{\infty} c_n x^n,$$

for some number r, where we assume that $c_0 \neq 0$ (notice we can always ensure this by adjusting r). A solution of the form (7.65) for some value of r is called a **Frobenius solution**. As with series solutions about ordinary points, we will be determining the coefficients c_n, but additionally we must find r as well.

Example 7.5.4. To see how this goes, let's solve

(7.66) $$x^2 y'' + (x^2 + \frac{x}{2})y' + xy = 0,$$

a Frobenius equation with $P(x) = x + \frac{1}{2}$ and $Q(x) = x$. Use

$$y = x^r \sum_{n=0}^{\infty} c_n x^n = c_0 x^r + c_1 x^{r+1} + c_2 x^{r+2} + c_3 x^{r+3} + \cdots$$

(with an unknown, but to be determined, value of r) as a trial solution. Computing, we have

$$y' = c_0 r x^{r-1} + c_1(r+1)x^r + c_2(r+2)x^{r+1} + c_3(r+3)x^{r+2} + \cdots$$

and

$$y'' = c_0 r(r-1)x^{r-2} + c_1(r+1)r x^{r-1} + c_2(r+2)(r+1)x^r + c_3(r+3)(r+2)x^{r+1} + \cdots.$$

Substitute these expressions into (7.66) to obtain

$$x^2[c_0 r(r-1)x^{r-2} + c_1(r+1)r x^{r-1} + c_2(r+2)(r+1)x^r + c_3(r+3)(r+2)x^{r+1} + \cdots]$$
$$+ \left(x^2 + \frac{1}{2}x\right)[c_0 r x^{r-1} + c_1(r+1)x^r + c_2(r+2)x^{r+1} + c_3(r+3)x^{r+2} + \cdots]$$
$$+ x[c_0 x^r + c_1 x^{r+1} + c_2 x^{r+2} + c_3 x^{r+3} + c_4 x^{r+4} + \cdots] = 0.$$

When we collect like terms in the left-hand side, the lowest power of x that appears is x^r, and its coefficient is

$$c_0 r(r-1) + \frac{1}{2} c_0 r.$$

The coefficients for the next several higher powers of x can be determined in a similar way. The results are summarized below:

coefficient of x^r	$r(r-1)c_0 + \frac{r}{2}c_0$
coefficient of x^{r+1}	$(r+1)rc_1 + rc_0 + \frac{(r+1)}{2}c_1 + c_0$
coefficient of x^{r+2}	$(r+2)(r+1)c_2 + (r+1)c_1 + \frac{(r+2)}{2}c_2 + c_1$
coefficient of x^{r+3}	$(r+3)(r+2)c_3 + (r+2)c_2 + \frac{(r+3)}{2}c_3 + c_2$
coefficient of x^{r+4}	$(r+4)(r+3)c_4 + (r+3)c_3 + \frac{(r+4)}{2}c_4 + c_3$

The coefficient expressions are simple enough in this example that we can predict that the coefficient of x^{r+j}, for any $j \geq 1$, will be

$$(r+j)(r+j-1)c_j + (r+j-1)c_{j-1} + \frac{(r+j)}{2}c_j + c_{j-1}.$$

Thus the equation that results from substituting the Frobenius solution from (7.65) into our differential equation (7.66) is

$$x^r \left(r(r-1)c_0 + \frac{c_0 r}{2} + \sum_{j=1}^{\infty} \left[(r+j)(r+j-1)c_j + (r+j-1)c_{j-1} + \frac{(r+j)}{2}c_j + c_{j-1} \right] x^j \right) = 0.$$

This equation holds on an interval of positive length if and only if the power series within parentheses vanishes on the interval. Hence, by the Uniqueness Property, we have

(7.67) $$r(r-1)c_0 + \frac{r}{2}c_0 = 0,$$

and for $j \geq 1$,

(7.68) $$(r+j)(r+j-1)c_j + (r+j-1)c_{j-1} + \frac{(r+j)}{2}c_j + c_{j-1} = 0.$$

We are assuming that $c_0 \neq 0$, so by equation (7.67) we must have

$$r(r-1) + \frac{r}{2} = 0.$$

This equation, which determines r, is called the **indicial equation**. The solutions to the indicial equation are $r = \frac{1}{2}$ and $r = 0$. We'll use each of these values of r, starting with the larger one, $r = \frac{1}{2}$.

Solutions with $r = \frac{1}{2}$. With $r = \frac{1}{2}$ in equation (7.68) we see that

$$\left(j+\frac{1}{2}\right)\left(j-\frac{1}{2}\right)c_j + \frac{(j+\frac{1}{2})}{2}c_j = -c_{j-1} - \left(j-\frac{1}{2}\right)c_{j-1}.$$

After some algebra, we can rewrite this as

$$c_j = -\frac{1}{j}c_{j-1}.$$

Setting $j = 1, 2, 3, \ldots$, this gives

$$c_1 = -c_0, \quad c_2 = -\frac{1}{2}c_1 = \frac{1}{2}c_0, \quad c_3 = -\frac{1}{3}c_2 = -\frac{1}{6}c_0, \quad c_4 = -\frac{1}{4}c_3 = \frac{1}{24}c_0,$$

and in general,
$$c_j = (-1)^j \frac{1}{j!} c_0.$$
Once c_0 is specified, the remaining coefficients c_j, $j \geq 1$, are determined. Setting $c_0 = 1$, we assemble the pieces to produce our first solution to (7.65):

(7.69) $$y_1(x) = x^{1/2}\left(1 - x + \frac{1}{2!}x^2 - \frac{1}{3!}x^3 + \frac{1}{4!}x^4 + \cdots\right).$$

We can recognize this solution as $y_1(x) = x^{1/2} e^{-x}$, valid on $(0, \infty)$. Fig. 7.11 shows the graph of $y_1(x)$, and Fig. 7.12 shows this solution together with several approximations to $y_1(x)$, obtained by taking only the first few terms in the infinite series factor in (7.69).

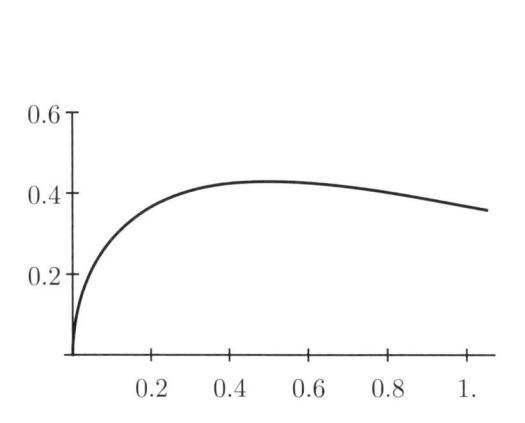

Figure 7.11. $y_1(x) = x^{1/2} e^{-x}$.

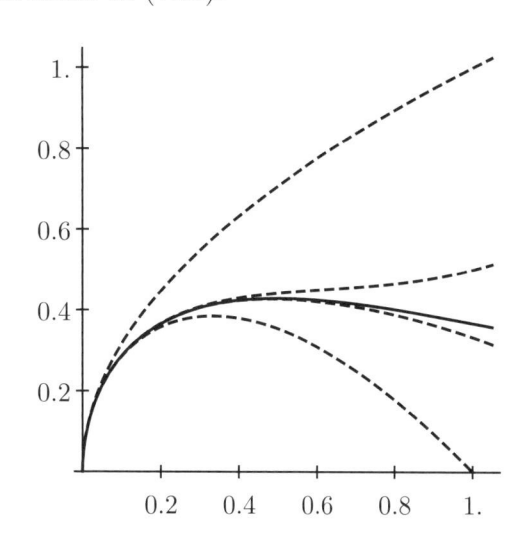

Figure 7.12. $y_1(x)$ and several approximations (dashed).

Solutions with $r = 0$. Next, set $r = 0$ in equation (7.68) to obtain
$$j(j-1)c_j + (j-1)c_{j-1} + \frac{j}{2}c_j + c_{j-1} = 0, \quad j \geq 1,$$
which simplifies to
$$c_j = -\frac{1}{j - \frac{1}{2}} c_{j-1}.$$
From this we learn that
$$c_1 = -2c_0, \quad c_2 = \frac{2^2}{(3)(1)} c_0, \quad c_3 = -\frac{2^3}{(5)(3)(1)} c_0,$$
and in general
$$c_j = (-1)^j \frac{2^j}{(2j-1)(2j-3)\cdots(5)(3)(1)} c_0.$$
Again, setting $c_0 = 1$ and remembering $r = 0$, we obtain a second solution
$$y_2(x) = 1 - 2x + \frac{4}{3}x^2 - \frac{8}{15}x^3 + \frac{16}{105}x^4 - \cdots.$$
The Ratio Test shows this series converges for all $-\infty < x < \infty$. Our solutions y_1 and y_2, arising from two different roots of the indicial equation, are linearly independent on the interval $(0, \infty)$ (where both are defined); see Exercise 32.

The algorithm for solving a Frobenius equation. Motivated by our work in the last example, we can give an algorithm for solving any Frobenius equation

(7.70) $$x^2 y'' + P(x) x y' + Q(x) y = 0,$$

with $P(x)$ and $Q(x)$ analytic at $x = 0$. We seek a solution in the form

$$y(x) = x^r \sum_{n=0}^{\infty} c_n x^n$$

for $c_0 \neq 0$. In Step 1 of our solution process, we will work with the **indicial equation** for (7.70). This, a generalization of the indicial equation for a Cauchy-Euler equation, is the equation

(7.71) $$r(r-1) + rP(0) + Q(0) = 0.$$

To understand where this comes from, you are asked to show in Exercise 19 that when $y(x) = x^r \sum_{n=0}^{\infty} c_n x^n$, with $c_0 \neq 0$, is substituted into equation (7.70) and the resulting coefficient of x^r is set equal to 0, equation (7.71) is obtained.

Step 1. Find the roots of the indicial equation for (7.70)

$$r(r-1) + rP(0) + Q(0) = 0.$$

We will only consider the case that the roots of this indicial equation are real numbers.

Step 2. Assuming the roots of the indicial equation are real, they will either be two different real roots $r_1 > r_2$ or a single repeated real root r_1. Begin with the larger (or only) root, r_1, and substitute the trial solution

$$y(x) = x^{r_1} \sum_{n=0}^{\infty} c_n x^n$$

into the differential equation (7.70). By finding an expression for the total coefficient A_1 of x^{r_1+1}, A_2 of x^{r_1+2}, A_3 of x^{r_1+3}, ..., obtain

(7.72) $$A_1 x^{r_1+1} + A_2 x^{r_1+2} + A_3 x^{r_1+3} + \cdots = 0.$$

There should be **no** x^{r_1} term, since r_1 is a root of the indicial equation and the roots of the indicial equation are exactly the values that make the coefficient of the x^{r_1} term equal to 0.

Step 3. Set each coefficient $A_1, A_2, A_3, \ldots$ from Step 2 equal to 0. In general, if the coefficient of x^{r_1+n} is A_n, the Uniqueness Property for power series and equation (7.72) say

$$A_n = 0, \quad n = 1, 2, 3, \ldots,$$

and this allows us to describe c_n in terms of the values c_k, $k < n$. Sometimes (if it is simple enough) the recursion relation will allow us to determine a general formula for c_n in terms of c_0, but often we must be content if computing by hand to simply determine the first "several" coefficients $c_1, c_2, \ldots$ from c_0. We may set c_0 to be any nonzero value; for simplicity we often choose $c_0 = 1$. The resulting solution

$$y_1(x) = x^{r_1} \sum_{n=0}^{\infty} c_n x^n$$

is valid on some interval $(0, \rho)$ for some positive ρ. Theorem 7.5.6 below will give information about ρ.

7.5. Regular singular points

Step 4. If there is a second real root r_2 to the indicial equation, repeat Steps 2 and 3 using r_2 instead of r_1. In general, this may or may not produce a second linearly independent solution, but if $r_1 - r_2$ is not an integer, it will give a second linearly independent solution—see Theorem 7.5.6 below.

If there is only one root r_1 to the indicial equation, there will be only one (linearly independent) Frobenius solution. In theory one could use reduction of order to find a second linearly independent solution, but in practice this is often too cumbersome to carry out explicitly. In fact, there will always be a second solution of the form

$$y_2(x) = y_1 \ln x + x^{r_1} \sum_{n=0}^{\infty} d_n x^n$$

but finding the d_n's by substitution may not be practical.

A few remarks on the implementation of the algorithm: If equation (7.70) contains fractions or if the terms on the left-hand side have a common factor, Step 2 is often made computationally simpler if we clear the denominators and cancel any common factors to replace (7.70) by an equivalent equation before we substitute the trial solution. For example, in implementing Step 2 for the Frobenius equation

$$x^2 y'' + \frac{3x}{1-x} y' + \frac{2x}{1-x} y = 0$$

we first rewrite the equation as

$$(x - x^2) y'' + 3y' + 2y = 0$$

and then substitute $y = x^{r_1} \sum_{n=0}^{\infty} c_n x^n$; see Exercise 28.

Once we complete Step 3 to obtain a solution on an interval to the right of 0, if we replace x^{r_1} by $|x|^{r_1}$, we will produce a solution valid on an interval $|x| < \rho$, except possibly at $x = 0$, for some positive ρ.

Example 7.5.5. To illustrate this algorithm and focus some attention on Step 4, we'll solve the equation

(7.73) $$x^2 y'' - (2 + x^2) y = 0.$$

Step 1: Roots of the indicial equation. Here, $P(x) = 0$ and $Q(x) = -(2 + x^2)$ (both analytic everywhere), so that $P(0) = 0$ and $Q(0) = -2$. The indicial equation is

$$r(r-1) - 2 = 0.$$

It has roots $r_1 = 2$ and $r_2 = -1$.

Step 2: Setting up the first Frobenius solution. We work with the larger root $r_1 = 2$ first. We seek a solution in the form

$$y = x^2 \sum_{n=0}^{\infty} c_n x^n = \sum_{n=0}^{\infty} c_n x^{n+2}$$

where $c_0 \neq 0$. Computing, we have

$$y' = \sum_{n=0}^{\infty} (n+2) c_n x^{n+1}, \quad y'' = \sum_{n=0}^{\infty} (n+2)(n+1) c_n x^n.$$

Substituting these values into (7.73) we have

$$x^2 \sum_{n=0}^{\infty}(n+2)(n+1)c_n x^n - (2+x^2)\sum_{n=0}^{\infty} c_n x^{n+2} = 0,$$

or

(7.74) $$\sum_{n=0}^{\infty}(n+2)(n+1)c_n x^{n+2} - 2\sum_{n=0}^{\infty} c_n x^{n+2} - \sum_{n=0}^{\infty} c_n x^{n+4} = 0.$$

Rewrite equation (7.74) by a shift of index in the last sum to obtain

(7.75) $$\sum_{n=0}^{\infty}(n+2)(n+1)c_n x^{n+2} - 2\sum_{n=0}^{\infty} c_n x^{n+2} - \sum_{n=2}^{\infty} c_{n-2} x^{n+2} = 0.$$

If we separate out the $n=0$ and $n=1$ terms, we can combine the three series to write this as

(7.76) $$2c_0 x^2 + 6c_1 x^3 - 2c_0 x^2 - 2c_1 x^3 + \sum_{n=2}^{\infty}[(n+2)(n+1)c_n - 2c_n - c_{n-2}]x^{n+2} = 0.$$

Next we identify the total coefficient A_n in front of x^{n+2} for each value of $n = 0, 1, 2, 3, \ldots$. For $n = 0$, this is just a check on our computation of the indicial equation—the coefficient should be 0 for this term. The table below shows the coefficients for $n = 0, 1, 2, 3, 4$, and 5. We can keep going in this table as long as we wish.

$n = 0$	coefficient of x^2	$2c_0 - 2c_0 = 0$
$n = 1$	coefficient of x^3	$A_1 = 6c_1 - 2c_1 = 4c_1$
$n = 2$	coefficient of x^4	$A_2 = (4)(3)c_2 - 2c_2 - c_0$
$n = 3$	coefficient of x^5	$A_3 = (5)(4)c_3 - 2c_3 - c_1$
$n = 4$	coefficient of x^6	$A_4 = (6)(5)c_4 - 2c_4 - c_2$
$n = 5$	coefficient of x^7	$A_5 = (7)(6)c_5 - 2c_5 - c_3$

Step 3: Computing the coefficients. Each of the coefficients A_n must be 0, so that from the second row of the table we learn that $c_1 = 0$. From this, in turn, we learn that $c_3 = 0$ (look at the coefficient of x^5), and $c_5 = 0$ (look at the coefficient of x^7). The expression for the coefficient of x^4 tells us that $c_2 = \frac{1}{10}c_0$ and the coefficient of x^6 tells us that $c_4 = \frac{1}{28}c_2 = \frac{1}{280}c_0$.

Let's see if we can find a general formula for c_n in terms of c_0. Looking at equation (7.76), we see that for $n \geq 2$, the total coefficient in front of x^{n+2} is

$$A_n = c_n(n+2)(n+1) - 2c_n - c_{n-2},$$

which is therefore zero. Thus for $n \geq 2$ we must have

$$c_n = \frac{1}{(n+2)(n+1)-2} c_{n-2} = \frac{1}{(n+3)n} c_{n-2}.$$

Since we've already determined that $c_1 = 0$, this tells us that for every *odd* value of n, $c_n = 0$. In Exercise 21, you are asked to show that when n is even, say, $n = 2k$, then

$$c_n = c_{2k} = \frac{3(2k+2)}{(2k+3)!} c_0.$$

Choosing $c_0 = 1$ and putting the pieces together, we have the first solution

$$y_1(x) = x^2 \left(1 + \sum_{k=1}^{\infty} \frac{3(2k+2)}{(2k+3)!} x^{2k}\right).$$

Step 4: Trying for a second Frobenius solution. We work now with the second root $r_2 = -1$. Since $r_1 - r_2$ is an integer, we are not guaranteed to be able to produce a second solution from r_2, but we will at least try. Our trial solution is

$$y_2(x) = x^{-1} \sum_{n=0}^{\infty} d_n x^n = \sum_{n=0}^{\infty} d_n x^{n-1}.$$

Computing y_2' and y_2'' and substituting into equation (7.73) gives

$$\sum_{n=0}^{\infty} d_n(n-1)(n-2)x^{n-1} - 2\sum_{n=0}^{\infty} d_n x^{n-1} - \sum_{n=0}^{\infty} d_n x^{n+1} = 0.$$

Rewrite the third series on the left side by a shift of index as

$$\sum_{n=0}^{\infty} d_n x^{n+1} = \sum_{n=2}^{\infty} d_{n-2} x^{n-1},$$

so that we have

$$\sum_{n=0}^{\infty} d_n(n-1)(n-2)x^{n-1} - 2\sum_{n=0}^{\infty} d_n x^{n-1} - \sum_{n=2}^{\infty} d_{n-2} x^{n-1} = 0.$$

Writing the $n = 0$ and $n = 1$ terms explicitly and combining the series, this becomes

$$-2d_1 + \sum_{n=2}^{\infty} [d_n(n^2 - 3n) - d_{n-2}]x^{n-1} = 0.$$

Setting the coefficient of each power of x equal to 0 gives

(7.77) $\qquad d_1 = 0 \quad \text{and} \quad d_n(n^2 - 3n) = d_{n-2} \quad \text{for } n \geq 2.$

For every value of $n \geq 2$, except $n = 3$, the last equation can be rewritten as

(7.78) $\qquad d_n = \dfrac{1}{n(n-3)} d_{n-2},$

our recursion relation to determine d_n from d_{n-2}. When $n = 3$, however, equation (7.77) says

(7.79) $\qquad d_3(0) = d_1.$

Since we already know $d_1 = 0$, this puts **no** restriction on d_3; in other words, d_3 is arbitrary. (On the other hand, if d_1 was not 0, then no value of d_3 would satisfy (7.79)!)

We can choose d_0 to be any nonzero value, and we can choose d_3 to be any value whatsoever, so let's make the simple choices

$$d_0 = 1, \quad d_3 = 0$$

and use the recursion formula (7.78) to determine d_n for $n = 2$ and $n \geq 4$. For n any odd integer greater than or equal to 5, $d_n = 0$; this follows from (7.78) and our choice $d_3 = 0$. For even values of n, we have

$$d_2 = -\frac{1}{2}d_0 = -\frac{1}{2}, \quad d_4 = \frac{1}{4}d_2 = -\frac{1}{8}, \quad d_6 = \frac{1}{18}d_4 = -\frac{1}{144}, \quad d_8 = \frac{1}{40}d_6 = -\frac{1}{5{,}760},$$

giving the solution

$$y_2(x) = x^{-1}\left(1 - \frac{1}{2}x^2 - \frac{1}{8}x^4 - \frac{1}{144}x^6 - \frac{1}{5{,}760}x^8 - \cdots\right).$$

This solution is linearly independent from the first solution y_1 on any interval on which both converge (see Exercise 32).

Given that we now have two linearly independent solutions, what are we to make of the fact that we can still go back and choose d_3 arbitrarily? You can show that if you choose $d_3 = 1$ and $d_0 = 0$ (instead of $d_3 = 0$ and $d_0 = 1$ as we did above), then y_2 coincides with y_1.

The next result summarizes our work in this section and provides some information on where a Frobenius solution converges.

Theorem 7.5.6. *A Frobenius equation*
$$x^2 y'' + P(x) x y' + Q(x) y = 0,$$
where $P(x)$ and $Q(x)$ are analytic at $x = 0$, has indicial equation
$$r(r-1) + rP(0) + Q(0) = 0.$$
If the indicial equation has real roots $r_1 \geq r_2$, the differential equation has a solution of the form
$$y_1 = x^{r_1} \sum_{n=0}^{\infty} c_n x^n$$
with $c_0 = 1$. Moreover, if $r_1 - r_2$ is not an integer, then the differential equation has a second solution of the form
$$y_2 = x^{r_2} \sum_{n=0}^{\infty} d_n x^n,$$
with $d_0 = 1$, such that y_1 and y_2 are linearly independent on any interval where they both exist.

The power series factor in the formula for y_1 or for y_2 has radius of convergence at least as large as the distance from 0 to the nearest point in the complex plane at which either $P(x)$ or $Q(x)$ fails to be analytic.

As a practical matter, the description of the radius of convergence in Theorem 7.5.6 is most useful to us when $P(x)$ and $Q(x)$ are rational functions. In this case, the theorem says that the radius of convergence is at least as large as the distance from 0 to the closest zero (in the *complex plane*) of the denominators of P and Q. If P and Q are actually polynomials, we can take this to be ∞. So, for example, looking back at equation (7.66), with $P(x) = 0$ and $Q(x) = -(2 + x^2)$, we see that the first solution we produced,
$$y_1(x) = x^2 \left(1 + \sum_{k=1}^{\infty} \frac{3(2k+2)}{(2k+3)!} x^{2k}\right),$$
is valid on $(-\infty, \infty)$, while the second solution,
$$y_2(x) = x^{-1} \left(1 - \frac{1}{2}x^2 - \frac{1}{8}x^4 - \frac{1}{144}x^6 - \frac{1}{5{,}760}x^8 - \cdots\right),$$
is valid on both of the intervals $(-\infty, 0)$ and $(0, \infty)$. The point $x = 0$ cannot be included because of the factor x^{-1}.

7.5.3. Exercises.

In Exercises 1–5 solve the Cauchy-Euler equation on $(0, \infty)$.

1. $x^2 y'' - 3xy' + 4y = 0$.
2. $x^2 y'' + 4xy' + 2y = 0$.
3. $x^2 y'' + xy' + y = 0$.
4. $x^2 y'' + 9xy' + 17y = 0$.
5. $y'' + \dfrac{3}{x} y' + \dfrac{5}{x^2} y = 0$.

6. How does the solution in Exercise 1 change if we want the solution on $(-\infty, 0)$?

7.5. Regular singular points

7. Solve the Cauchy-Euler equation $9x^2y'' - 9xy' + 5y = 0$ on $(0, \infty)$, and show that all solutions are continuous at $x = 0$.

8. Solve the Cauchy-Euler equation $x^2y'' - 4xy' + 6y = 0$ and show that all solutions are polynomials and hence analytic on $(-\infty, \infty)$.

9. (a) If the indicial equation $r(r-1) + b_1 r + b_0 = 0$ has only one (repeated) root, show that $(b_1 - 1)^2 = 4b_0$ and that the one root is $r = (1 - b_1)/2$.
   (b) Show that when $y = x^r \ln x$ is substituted into the Cauchy-Euler expression
   $$x^2 y'' + b_1 x y' + b_0 y$$
   the result is
   $$x^r(2r - 1 + b_1) + x^r \ln x (r^2 - r + b_1 r + b_0).$$
   (c) Using the result in (b), show that if $(b_1 - 1)^2 = 4b_0$ and $r = (1 - b_1)/2$, then $y = x^r \ln x$ is a solution to
   $$x^2 y'' + b_1 x y' + b_0 y = 0$$
   on $(0, \infty)$.

10. (a) Find the general solution to the Cauchy-Euler equation $x^2 y'' - 6y = 0$ on $(0, \infty)$.
    (b) The equation $x^2 y'' - 6y = 4x^2$ has a particular solution that is a polynomial. Find it, and then give the general solution to $x^2 y'' - 6y = 4x^2$ on $(0, \infty)$.

11. Suppose the roots of the indicial equation for a Cauchy-Euler equation are $\alpha \pm i\beta$ for real numbers α and β with $\alpha > 0$. Which of the following best describes the behavior of the solution $y(x) = x^\alpha \cos(\beta \ln x)$?
    (a) It tends to 0 as $x \to 0^+$.
    (b) It tends to ∞ as $x \to 0^+$.
    (c) It tends to $-\infty$ as $x \to 0^+$.
    (d) You can't say anything without more explicit information about α and β.

12. If x^r solves the Cauchy-Euler equation
    $$x^2 y'' + b_1 x y' + b_0 y = 0$$
    on $(0, \infty)$ for some real number r, show that $|x|^r$ solves the equation on $(-\infty, 0)$. Similarly, show that if $x^r \ln x$ solves the equation on $(0, \infty)$, then $|x|^r \ln |x|$ is a solution on $(-\infty, 0)$. Hint: Substitute into the differential equation, remembering that when $x < 0$, $|x| = -x$.

13. Find the singular points for each of the following equations and classify them as regular or irregular.
    (a) $x^2 y'' + 3xy' + xy = 0$.
    (b) $x^2 y'' + 3y' + xy = 0$.
    (c) $(1 - x^2) y'' - 2xy' + 2y = 0$.
    (d) $x(x - 2) y'' + (\sin x) y' + xy = 0$.
    (e) $(x^2 + 3x + 2)^2 y'' + (x + 1) y' + y = 0$.

14. Solve the following third-order analogue of a Cauchy-Euler equation by seeking solutions in the form x^r:
    $$x^3 y''' - x^2 y'' - 2xy' + 6y = 0.$$

15. Give the general solution to the "Cauchy-Euler type" equation
    $$(x - 4)^2 y'' - (x - 4) y' + 5y = 0$$
    on $(4, \infty)$ by
    (a) first making the substitution $t = x - 4$ or
    (b) using $y = (x - 4)^r$ as a trial solution.

16. Find the general solution to the equation
$$(x-1)^2 y'' + 4(x-1)y' + 2y = 0$$
on the interval $(1, \infty)$ by making the change of independent variable $s = x - 1$ and solving the resulting equation on $(0, \infty)$.

17. Find the general solution to the equation
$$y'' + \frac{1}{x+1}y' - \frac{1}{(x+1)^2}y = 0$$
on $(-1, \infty)$ by making a change of variable in the independent variable, as in Exercise 16.

18. Solve the initial value problem
$$(x-3)^2 y'' - 2y = 0, \quad y(4) = 2, \quad y'(4) = 1,$$
by first making a change of variable in the independent variable, as in Exercise 16.

19. This problem shows how the indicial equation

(7.80) $$r(r-1) + rP(0) + Q(0) = 0$$

for the differential equation

(7.81) $$x^2 y'' + P(x) xy' + Q(x) y = 0,$$

where $P(x)$ and $Q(x)$ are analytic at $x = 0$, arises.

(a) Explain why $P(x)$ has a power series expansion
$$P(x) = P(0) + P'(0)x + \frac{P''(0)}{2!}x^2 + \cdots$$
in an interval containing 0, and obtain a similar power series expansion of $Q(x)$ near 0.

(b) Show that if we make the substitution
$$y = x^r \sum_{n=0}^{\infty} c_n x^n,$$
with $c_0 \neq 0$ in equation (7.81), the lowest power of x that appears on the left-hand side is x^r and that it has coefficient
$$c_0 [r(r-1) + rP(0) + Q(0)].$$
Thus if $c_0 \neq 0$, r must satisfy (7.80) if y is a solution to (7.81).

20. Suppose $x = 0$ is a regular singular point for the equation $y'' + p(x)y' + q(x)y = 0$. Write this as a Frobenius equation

(7.82) $$x^2 y'' + P(x) xy' + Q(x) y = 0$$

by identifying $P(x)$ and $Q(x)$ in terms of $p(x)$ and $q(x)$. Show that the indicial equation for (7.82) is the same as the indicial equation for the Cauchy-Euler equation
$$x^2 y'' + P(0) xy' + Q(0) y = 0.$$

21. (a) Verify that the recursion relation
$$c_n = \frac{1}{(n+3)n} c_{n-2},$$
from Example 7.5.5, tells you that
$$c_2 = \frac{1}{(5)(2)} c_0 = \frac{3(4)}{5!} c_0, \quad c_4 = \frac{1}{(7)(4)(5)(2)} c_0 = \frac{3(6)}{7!} c_0,$$

and
$$c_6 = \frac{1}{(9)(6)(7)(4)(5)(2)}c_0 = \frac{3(8)}{9!}c_0.$$
(b) Obtain expressions in a similar form for c_8 and c_{10} in terms of c_0.
(c) In general, we have
$$c_{2k} = \frac{3(2k+2)}{(2k+3)!}c_0,$$
for $k = 1, 2, 3, \ldots$. Verify this by induction. To do this, assume
$$c_{2k} = \frac{3(2k+2)}{(2k+3)!}c_0$$
for some k and prove, using the recursion relation from (a), that
$$c_{2(k+1)} = \frac{3(2(k+1)+2)}{(2(k+1)+3)!}c_0.$$

22. Follow the outline below to solve the Frobenius equation
$$3x^2 y'' + 2x(1-x)y' - 4y = 0.$$
In parts (b) and (c), if finding the general term is not feasible, find the first four nonzero terms.
   (a) Find the indicial equation and show that its roots are $r_1 = 4/3$ and $r_2 = -1$.
   (b) Find the Frobenius solution
$$y_1(x) = x^{4/3} \sum_{n=0}^{\infty} c_n x^n$$
with $c_0 = 1$.
   (c) Find the Frobenius solution
$$y_2(x) = x^{-1} \sum_{n=0}^{\infty} c_n x^n$$
with $c_0 = 1$.
   (d) Explain why both solutions from (b) and (c) must exist on $(0, \infty)$. Are they linearly independent there?

23. (a) Find two linearly independent Frobenius solutions to
$$(7.83) \qquad x^2 y'' + \frac{x + x^2}{2} y' = 0$$
on $(0, \infty)$.
   (b) As a completely different approach to the same equation, solve (7.83) by making the substitution $v = y'$ and solving the resulting first-order equation. As in (a), obtain two linearly independent solutions for y. Hint: You should get $y' = Cx^{-\frac{1}{2}} e^{-\frac{1}{2}x}$. Solve for y (as a power series) by using the Taylor series for $e^{-\frac{1}{2}x}$.
   (c) Reconcile your solutions in (a) and (b) by showing each solution you obtained in (b) is a linear combination of the solutions you gave in (a), and each solution in (a) is a linear combination of your solutions in (b).

24. Consider the equation $x^2 y'' + x^3 y' - 2y = 0$ on $(0, \infty)$.
   (a) Show that the indicial equation has roots $r_1 = 2$ and $r_2 = -1$.
   (b) Find a solution $y_1 = x^2 \sum_{n=0}^{\infty} a_n x^n$ where $a_0 = 1$. Either give a general formula for a_n or find the first four nonzero values.
   (c) If possible, find a second linearly independent solution of the form $y_2 = x^{-1} \sum_{n=0}^{\infty} c_n x^n$ with $c_0 = 1$.

25. Consider the Frobenius equation $x^2 y'' + xy' + (x^2 - 1)y = 0$.
    (a) Determine the indicial equation and find its roots $r_1 \geq r_2$. Observe that $r_1 - r_2 = 2$, an integer.
    (b) Find a Frobenius solution of the form
    $$x^{r_1} \sum_{n=0}^{\infty} c_n x^n$$
    with $c_0 = 1$. If finding the general term of this Frobenius solution is not feasible, find the first four nonzero terms.
    (c) Show that there is no linearly independent second Frobenius solution of the form
    $$x^{r_2} \sum_{n=0}^{\infty} d_n x^n$$
    with $d_0 \neq 0$.

26. Consider the Frobenius equation $x^2 y'' + 4xy' + (2 - x)y = 0$.
    (a) Show that the indicial equation has roots $r_1 = -1$ and $r_2 = -2$.
    (b) Find the Frobenius solution
    $$y_1(x) = x^{-1} \sum_{n=0}^{\infty} c_n x^n$$
    with $c_0 = 1$, showing $c_1, c_2, c_3,$ and c_4 explicitly.
    (c) Show that if we use
    $$y_2(x) = x^{-2} \sum_{n=0}^{\infty} d_n x^n$$
    as a trial solution, we obtain the recursion relation
    $$(n^2 - n)d_n = d_{n-1}$$
    for $n \geq 1$. Show that with $d_0 \neq 0$, there is no choice of d_1 that will satisfy this.
    (d) Show that if in (c) we set $d_0 = 0$, then there are allowable values of d_1, but with any of them, the solution y_2 is just a scalar multiple of the solution y_1 obtained in (a).

27. Consider the equation $x^2 y'' + x(x + 1)y' - y = 0$.
    (a) Determine the indicial equation and find its roots r_1 and r_2. Check that the difference of the roots is an integer.
    (b) Find solutions of the form
    $$y_1(x) = x^{r_1} \sum_{n=0}^{\infty} c_n x^n$$
    where r_1 is the larger of the two roots from (a) and $c_0 = 1$.
    (c) Find a second linearly independent solution
    $$y_2(x) = x^{r_2} \sum_{n=0}^{\infty} d_n x^n$$
    with $d_0 = 1$.
    (d) Show $y_1(x) = 2x^{-1} e^{-x} - 2x^{-1} + 2$ and $y_2(x) = x^{-1} e^{-x}$.

28. Obtain two linearly independent Frobenius solutions to
$$x^2 y'' + x \frac{3}{1-x} y' + \frac{2x}{1-x} y = 0.$$
For what values of x are your solutions valid? Hint: The equation
$$(x - x^2) y'' + 3y' + 2y = 0$$
is equivalent to the given one and simpler to work with.

29. (a) Show that the hyperbolic cosine function has power series
$$\cosh x = 1 + \frac{x^2}{2!} + \frac{x^4}{4!} + \frac{x^6}{6!} + \cdots.$$
(b) Find two linearly independent Frobenius solutions to
$$4xy'' + 2y' - y = 0 \tag{7.84}$$
on $(0, \infty)$.
(c) Show directly that $y = \cosh \sqrt{x}$ is a solution to equation (7.84) on $(0, \infty)$. Is this solution some linear combination of the solutions you found in (b)? Explain.

30. The equation $xy'' - y' + 4x^3 y = 0$ has a regular singular point at $x = 0$.
(a) Show that the indicial equation has roots $r_1 = 2$ and $r_2 = 0$.
(b) Using the root $r_1 = 2$, obtain a Frobenius solution and show that it is equal to $c \sin(x^2)$.
(c) Show that there is a second linearly independent Frobenius solution corresponding to the root $r_2 = 0$.
(d) Show that the equation has general solution
$$y = c_1 \sin(x^2) + c_2 \cos(x^2).$$

31. In this problem we consider the equation
$$xy'' - (x + m)y' + my = 0 \tag{7.85}$$
where m is a positive integer.
(a) Show that $y = e^x$ is a solution.
(b) Show that if y satisfies equation (7.85), then y' solves the same equation with m replaced by $m - 1$:
$$xy'' - (x + m - 1)y' + (m - 1)y = 0.$$
(c) Show that
$$y = 1 + x + \frac{x^2}{2!} + \cdots + \frac{x^m}{m!}$$
(a finite sum) is a solution to (7.85).

32. Suppose the indicial equation for $y'' + P(x)xy' + Q(x)y = 0$ has two distinct real roots, r_1 and r_2 with $r_1 > r_2$. Suppose further that on an interval $(0, \rho)$ we have found solutions
$$y_1(x) = x^{r_1} \sum_{n=0}^{\infty} c_n x^n \tag{7.86}$$
and
$$y_2(x) = x^{r_2} \sum_{n=0}^{\infty} d_n x^n \tag{7.87}$$
with $c_0 = d_0 = 1$. In this problem we show y_1 and y_2 must be linearly independent on $(0, \rho)$.

(a) Explain why
$$y_1(x) = x^{r_1}(1+f(x)) \quad \text{and} \quad y_2(x) = x^{r_2}(1+g(x))$$
on $(0, \rho)$, for some continuous functions $f(x)$ and $g(x)$ having $f(0) = g(0) = 0$.

(b) Show that if $y_1 = cy_2$ for some constant $c \neq 0$, then
$$x^{r_1 - r_2} = c\frac{1+g(x)}{1+f(x)}$$
for small positive values of x.

(c) Using your work in (b), show that y_1 and y_2 must be linearly independent on $(0, \rho)$. Hint: Let x tend to zero through positive values in the expression for $x^{r_1-r_2}$ in (b).

(d) Suppose y_1 and y_2 are as in equations (7.86)–(7.87) with $c_0 \neq 0$ and $d_0 \neq 0$ (but not necessarily $c_0 = 1$, $d_0 = 1$). Explain why y_1 and y_2 must be linearly independent on $(0, \rho)$.

33. Suppose that the point $x = x_0$ is a regular singular point for the equation
$$y'' + p(x)y' + q(x)y = 0,$$
so that at least one of $p(x)$, $q(x)$ fails to be analytic at $x = x_0$. A Frobenius solution to such an equation on an interval $(x_0, x_0 + \rho)$ has the form
$$y(x) = (x - x_0)^r \sum_{n=0}^{\infty} c_n(x - x_0)^n$$
where $c_0 \neq 0$. The point of this problem is to convince you that no new ideas are needed for this situation.

(a) Show that the equation
$$(x^2 + 2x + 1)y'' - (x^2 + 2x + 3)y = 0$$
has a regular singular point at $x = -1$.

(b) Show that the substitution $v = x + 1$ converts the differential equation in (a) into
$$v^2 \frac{d^2 y}{dv^2} - (2 + v^2)y = 0,$$
that is, into the equation in (a) written with independent variable v instead of x.

(c) In Example 7.5.5 we found two linearly independent solutions to
$$x^2 y'' - (2 + x^2)y = 0$$
on $(0, \infty)$. Using these solutions, give two linearly independent solutions to the differential equation in (a) on the interval $(-1, \infty)$. You should not have to do much computation.

7.6. Bessel's equation

Bessel's equation is

(7.88) $$x^2 y'' + xy' + (x^2 - p^2)y = 0,$$

where p is a nonnegative constant. Solutions to equation (7.88) give the **Bessel functions**, which have diverse applications in physics and engineering, especially in problems involving circular symmetry. Regardless of the value of p, equation (7.88) is a Frobenius equation and Theorem 7.5.6 applies. In the notation of this theorem, $P(x) = 1$ and $Q(x) = x^2 - p^2$, and so Bessel's equation has indicial equation
$$r(r-1) + r - p^2 = 0.$$
This has roots $r_1 = p$ and $r_2 = -p$.

7.6. Bessel's equation

Solutions from $r_1 = p$. Working with the root $r_1 = p$, we know by Theorem 7.5.6 that there will be a solution
$$y = x^p \sum_{n=0}^{\infty} c_n x^n$$
with $c_0 \neq 0$, valid at least on $(0, \infty)$. We'll find it with the Frobenius Algorithm. Computing y' and y'' and substituting into equation (7.88) we obtain

(7.89) $$\sum_{n=0}^{\infty} c_n(n+p)(n+p-1)x^{n+p} + \sum_{n=0}^{\infty} c_n(n+p)x^{n+p} + (x^2 - p^2)\sum_{n=0}^{\infty} c_n x^{n+p} = 0.$$

Since
$$(x^2 - p^2)\sum_{n=0}^{\infty} c_n x^{n+p} = \sum_{n=0}^{\infty} c_n x^{n+p+2} - \sum_{n=0}^{\infty} p^2 c_n x^{n+p} = \sum_{n=2}^{\infty} c_{n-2} x^{n+p} - \sum_{n=0}^{\infty} p^2 c_n x^{n+p},$$
we may rewrite equation (7.89) (after some algebraic simplification) as
$$c_1(1+2p)x^{1+p} + \sum_{n=2}^{\infty} [c_n(n+p)(n+p-1) + c_n(n+p) + c_{n-2} - p^2 c_n]x^{n+p} = 0.$$

The total coefficient of each power of x must be zero, so that

(7.90) $$c_1(1+2p) = 0$$

and
$$c_n[(n+p)(n+p-1) + (n+p) - p^2] = -c_{n-2}$$
for $n \geq 2$. A little algebra lets us rewrite this recursion relation as
$$n(n+2p)c_n = -c_{n-2} \quad (n \geq 2).$$
Since $p \geq 0$, we certainly have $n(n+2p) \neq 0$, and so we can write

(7.91) $$c_n = -\frac{c_{n-2}}{n(n+2p)}$$

for $n \geq 2$. Equation (7.90) tells us that $c_1 = 0$ (since $p \geq 0$ and hence $1 + 2p \neq 0$). With $c_1 = 0$, the recursion relation in (7.91) then says
$$0 = c_1 = c_3 = c_5 = c_7 = \cdots.$$

We next use (7.91) to determine c_2, c_4, c_6, and in general c_{2k}, in terms of c_0. We have
$$c_2 = -\frac{c_0}{2(2+2p)} = -\frac{c_0}{2^2(1+p)},$$
$$c_4 = -\frac{c_2}{4(4+2p)} = \frac{c_0}{(2\cdot 4)(2+2p)(4+2p)} = \frac{c_0}{2^4 2!(1+p)(2+p)},$$
and
$$c_6 = -\frac{c_4}{6(6+2p)} = -\frac{c_0}{(2\cdot 4\cdot 6)(2+2p)(4+2p)(6+2p)} = -\frac{c_0}{2^6 3!(1+p)(2+p)(3+p)}.$$

At this point we can see the pattern, and we have
$$c_{2k} = (-1)^k \frac{c_0}{2^{2k} k!(1+p)(2+p)\cdots(k+p)}$$
for $k = 1, 2, 3, \ldots$.

Thus Bessel's equation has the Frobenius solution
$$c_0 x^p \left(1 - \frac{1}{2^2(1+p)}x^2 + \frac{1}{2^4 2!(1+p)(2+p)}x^4 - \frac{c_0}{2^6 3!(1+p)(2+p)(3+p)}x^6 + \cdots\right),$$

or

$$(7.92) \quad y(x) = c_0 x^p \left(1 + \sum_{k=1}^{\infty} (-1)^k \frac{1}{2^{2k} k!(1+p)(2+p)\cdots(k+p)} x^{2k}\right)$$

for arbitrary constant c_0.

So far, we have no restriction on our nonnegative constant p. However, when p is a nonnegative *integer* we can write our solution more compactly. Since

$$\frac{1}{(1+p)(2+p)\cdots(k+p)} = \frac{p!}{(p+k)!},$$

we can write our first Frobenius solution, for nonnegative integer p, as

$$y = c_0 x^p \left(1 + \sum_{k=1}^{\infty} (-1)^k \frac{p!}{2^{2k} k!(p+k)!} x^{2k}\right) = c_0 x^p \sum_{k=0}^{\infty} (-1)^k \frac{p!}{2^{2k} k!(p+k)!} x^{2k}.$$

In equation (7.92) it is customary to single out a particular choice for c_0. This choice for c_0 depends on p, and when p is a nonnegative integer it is

$$(7.93) \quad c_0 = \frac{1}{2^p p!}.$$

To interpret (7.93) when $p = 0$, recall that $0!$ is defined to be 1. Thus one solution to (7.88), for a nonnegative integer value of p, is

$$y = \sum_{k=0}^{\infty} \frac{(-1)^k}{k!(k+p)!} \left(\frac{x}{2}\right)^{2k+p}.$$

This is called the **Bessel function of order** p (this is a different use of the term "order" than our usual one) and is denoted $J_p(x)$. For p a nonnegative integer we have

$$(7.94) \quad J_p(x) = \sum_{k=0}^{\infty} \frac{(-1)^k}{k!(k+p)!} \left(\frac{x}{2}\right)^{2k+p}.$$

We are particularly interested in the Bessel functions of order $p = 0$ and $p = 1$,

$$(7.95) \quad J_0(x) = \sum_{k=0}^{\infty} \frac{(-1)^k}{(k!)^2} \left(\frac{x}{2}\right)^{2k}$$

and

$$(7.96) \quad J_1(x) = \sum_{k=0}^{\infty} \frac{(-1)^k}{k!(1+k)!} \left(\frac{x}{2}\right)^{2k+1}.$$

The graphs of J_0 and J_1 are shown in Fig. 7.13. Each exhibits damped oscillations. Each has infinitely many zeros, and between any two consecutive zeros of one function is a zero of the other function. This is reminiscent of the relationship between the functions $\sin x$ and $\cos x$ (see Exercise 1).

7.6. Bessel's equation

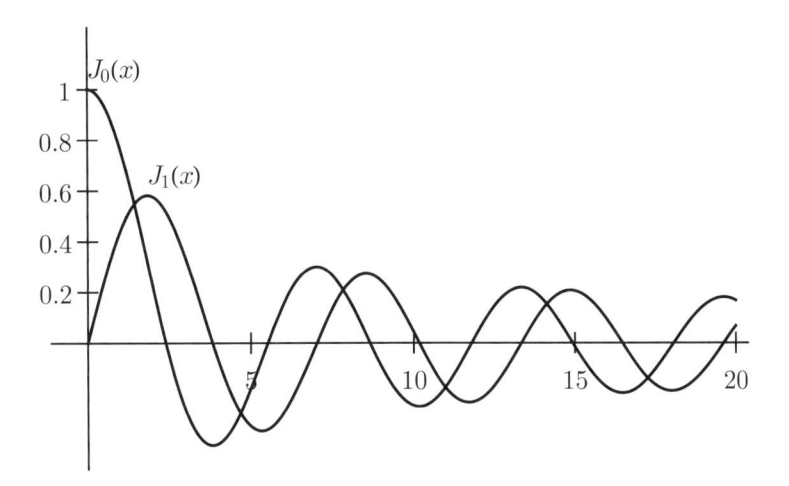

Figure 7.13. The functions J_0 and J_1.

Finding a second linearly independent solution. When $p \neq 0$, the indicial equation for (7.88) has a second root, $r_2 = -p$. We can try to find a second linearly independent solution to (7.88) of the form
$$y = x^{-p} \sum_{n=0}^{\infty} d_n x^n.$$

We know, however, from our work in Section 7.5, that we may not be successful if $p - (-p) = 2p$ is an integer. Thus the values of p that are potentially problematic (in addition to $p = 0$, where the indicial equation has only one root) are

(7.97) $$p = \frac{1}{2}, 1, \frac{3}{2}, 2, \frac{5}{2}, 3, \ldots.$$

Substituting
$$y = x^{-p} \sum_{n=0}^{\infty} d_n x^n$$

into
$$x^2 y'' + xy' + (x^2 - p^2)y = 0$$

we obtain after some computations
$$(1 - 2p)d_1 x^{1-p} + \sum_{n=2}^{\infty} [n(n - 2p)d_n + d_{n-2}] x^{n-p} = 0.$$

This tells us that

(7.98) $$d_1(1 - 2p) = 0,$$

and for $n \geq 2$,

(7.99) $$n(n - 2p)d_n = -d_{n-2}.$$

We haven't yet solved (7.99) for d_n in terms of d_{n-2}, since $n - 2p$ is zero for some value of n when $2p$ is an integer. However, when $2p$ is *not* an integer, we can write

(7.100) $$d_n = -\frac{d_{n-2}}{n(n - 2p)} \quad (n \geq 2).$$

This will lead to a second Frobenius solution

$$(7.101) \qquad y = d_0 x^{-p} \left(1 + \sum_{k=1}^{\infty} \frac{(-1)^k x^{2k}}{2^{2k} k! (1-p)(2-p) \cdots (k-p)}\right).$$

You are asked to verify this in Exercise 4. Notice that equation (7.101) is the same as what would be obtained by replacing p by $-p$ in equation (7.92). With a suitable choice for d_0, the function in (7.101) is denoted $J_{-p}(x)$. We're still assuming that $2p$ is not an integer, so that p is not in the list of problematic values in (7.97).

Let's look briefly at the problematic values from the standpoint of obtaining a second linearly independent Frobenius solution. It turns out that the half-integer values

$$p = \frac{1}{2}, \frac{3}{2}, \frac{5}{2}, \cdots$$

are not in fact a problem. We'll illustrate this with $p = \frac{1}{2}$.

The Bessel functions $J_{1/2}(x)$ and $J_{-1/2}(x)$. Setting $p = \frac{1}{2}$ in equations (7.90) and (7.91) we see that our first Frobenius solution

$$x^{1/2} \sum_{n=0}^{n} c_n x^n$$

has $0 = c_1 = c_3 = c_5 = \cdots$, and for even values $n = 2k$,

$$c_{2k} = (-1)^k \frac{1}{(2k+1)!} c_0.$$

This gives solutions

$$y = c_0 x^{\frac{1}{2}} \sum_{k=0}^{\infty} (-1)^k \frac{x^{2k}}{(2k+1)!}.$$

Multiplying and dividing by x we can rewrite this as

$$y = c_0 x^{-\frac{1}{2}} \sum_{k=0}^{\infty} (-1)^k \frac{x^{2k+1}}{(2k+1)!}.$$

In this form, we recognize the solution as $y = c_0 x^{-\frac{1}{2}} \sin x$. The customary choice for c_0 is $c_0 = \sqrt{2/\pi}$, giving the solution

$$(7.102) \qquad J_{1/2}(x) = \sqrt{\frac{2}{\pi}} \, x^{-\frac{1}{2}} \sin x.$$

The graph of $J_{1/2}(x)$ is shown in Fig. 7.14.

7.6. Bessel's equation

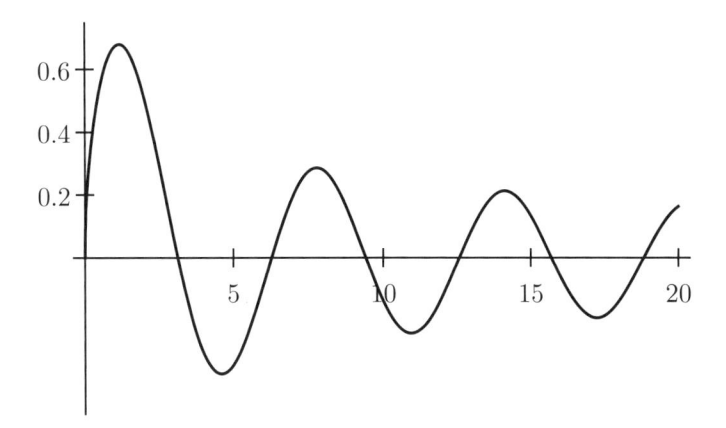

Figure 7.14. $y(x) = J_{1/2}(x)$.

In working for a second linearly independent solution $x^{-1/2} \sum_{n=0}^{\infty} d_n x^n$, keep in mind that the computations leading up to equations (7.98) and (7.99) are still valid. With $p = \frac{1}{2}$, equation (7.98) no longer *forces* $d_1 = 0$, but we are still permitted to *choose* $d_1 = 0$, which we do. With $p = \frac{1}{2}$ we can write the recursion relation as in (7.100) since $n - 2p = n - 1$ is nonzero for $n \geq 2$. Thus we have

$$d_1 = 0, \quad d_n = -\frac{d_{n-2}}{n(n-1)} \quad \text{for } n \geq 2.$$

This tells us that

$$0 = d_3 = d_5 = d_7 = \cdots$$

and

$$d_2 = -\frac{d_0}{2}, \quad d_4 = \frac{d_0}{4!}, \quad d_6 = -\frac{d_0}{6!}, \cdots$$

or in general

$$d_{2k} = (-1)^k \frac{d_0}{(2k)!}.$$

This gives solutions

$$y = d_0 x^{-1/2} \left(1 - \frac{1}{2!}x^2 + \frac{1}{4!}x^4 - \frac{1}{6!}x^6 + \cdots\right)$$

which we recognize as constant multiplies of $x^{-\frac{1}{2}} \cos x$. So long as $d_0 \neq 0$, any one of these will provide a solution which is linearly independent from our first solution, $J_{1/2}(x)$. Again, there is a customary choice of d_0 to give the solution that is denoted $J_{-\frac{1}{2}}(x)$. This choice is $\sqrt{2/\pi}$, so that

$$J_{-1/2}(x) = \sqrt{\frac{2}{\pi}} x^{-\frac{1}{2}} \cos x.$$

Its graph is shown in Fig. 7.15. The general solution to Bessel's equation with $p = \frac{1}{2}$ on $(0, \infty)$ is

$$a_1 x^{-\frac{1}{2}} \sin x + a_2 x^{-\frac{1}{2}} \cos x$$

for arbitrary constants a_1 and a_2.

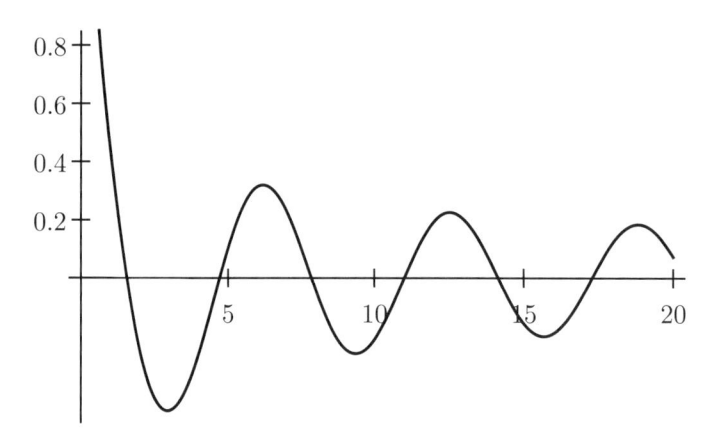

Figure 7.15. $y(x) = J_{-1/2}(x)$.

In Exercises 8 and 9 you are asked to similarly work out two linearly independent solutions for the Bessel equation with $p = 3/2$. Other half-integer values of p work similarly. The integer values of p are another story. In Exercise 3 you are asked to show that, with $p = 1$, you cannot find a solution of the form

$$x^{-1} \sum_{n=0}^{\infty} d_n x^n$$

that is linearly independent from the solution $J_1(x)$ already obtained. In fact, whenever p is a positive integer, there will fail to exist a second linearly independent Frobenius solution. Of course there must be *some* linearly independent second solution. When $p = 1$, it can be found as

$$(\ln x) J_1(x) + x^{-1} \sum_{n=0}^{\infty} b_n x^n$$

for certain coefficients b_n.

In general, using the method of reduction of order, one can obtain a second solution of Bessel's equation (7.88) where p is nonnegative integer such that it and J_p are linearly independent on $(0, \infty)$. Appropriately normalized, this second solution is denoted Y_p and is called the Bessel function of order p of the second kind (while J_p is called the Bessel function of order p of the first kind). For applications, such as those discussed in Exercises 13, the following property of Y_p is of critical importance:

(7.103) $$\lim_{x \to 0^+} Y_p(x) = -\infty.$$

The Gamma function. The choice of c_0 given in equation (7.93) for defining $J_p(x)$ when p is a nonnegative integer involved $p!$. For noninteger values of p, it turns out there is a useful generalization of $p!$ related to the **Gamma function**. This function, which is ubiquitous in mathematics, is denoted $\Gamma(p)$ and is defined for positive values of p by

(7.104) $$\Gamma(p) = \int_0^{\infty} t^{p-1} e^{-t} dt.$$

This is an improper integral, but it converges for all $p > 0$. For example,

$$\Gamma(1) = \int_0^{\infty} e^{-t} dt = \lim_{M \to \infty} \int_0^M e^{-t} dt = \lim_{M \to \infty} (-e^{-M} + e^0) = 1.$$

7.6. Bessel's equation

A first hint of how the Gamma function behaves like a factorial is the property

$$\Gamma(p+1) = p\Gamma(p).$$

You are asked to verify this property in Exercise 7. It holds for all $p > 0$, and if we apply this with $p = 1, 2, 3, \ldots$, we see that

$$\Gamma(2) = 1\Gamma(1) = 1, \quad \Gamma(3) = 2\Gamma(2) = 2, \quad \Gamma(4) = 3\Gamma(3) = 3!, \quad \Gamma(5) = 4\Gamma(4) = 4!,$$

and so on. In general, for a positive integer n we have

$$\Gamma(n+1) = n!.$$

Our reason for defining the Gamma function is to show the role it plays in defining the Bessel functions $J_p(x)$ for $p \geq 0$. We have already defined $J_p(x)$ when p is a nonnegative integer in equation (7.94). This is just our first Frobenius solution to Bessel's equation with the specific choice

$$c_0 = \frac{1}{2^p p!}.$$

For positive, but not necessarily integer, values of p, we'll take our first Frobenius solution as given by equation (7.92) and set

(7.105) $$c_0 = \frac{1}{2^p \Gamma(p+1)}$$

to obtain the Bessel function $J_p(x)$. In Exercise 10 you are asked to show this gives

$$J_p(x) = \sum_{k=0}^{\infty} (-1)^k \frac{1}{k!\Gamma(k+p+1)} \left(\frac{x}{2}\right)^{2k+p}.$$

7.6.1. Exercises.

1. For $J_0(x)$ and $J_1(x)$ as defined in equations (7.95)–(7.96), show that $J_0'(x) = -J_1(x)$. Using this and the Mean Value Theorem from calculus, show that if $J_0(x)$ has consecutive (positive) zeros at $x = r_1$ and $x = r_2$, then $J_1(x)$ has a zero between r_1 and r_2.

2. Here's another approach to solving the Bessel equation with $p = \frac{1}{2}$, which avoids all series computations. We look for a solution in the form

$$y(x) = \frac{u(x)}{\sqrt{x}}.$$

(This is suggested by the fact that $r = -\frac{1}{2}$ is a root of the indicial equation.) By substitution, show that y solves the Bessel equation with $p = 1/2$ on $(0, \infty)$ if u satisfies

$$u'' + u = 0.$$

From this, give the general solution to

$$x^2 y'' + xy' + \left(x^2 - \frac{1}{4}\right) y = 0$$

on $(0, \infty)$.

3. Suppose we try to look for a Frobenius solution

$$y = x^{-1} \sum_{n=0}^{\infty} d_n x^n$$

for the Bessel equation of order 1 corresponding to the second root $r_2 = -1$ of the indicial equation.
   (a) Use equations (7.98)–(7.99) with $p = 1$ to show that both d_1 and d_0 must be 0.
   (b) Show that $0 = d_3 = d_5 = d_7 = \cdots$.
   (c) Show that d_2 is arbitrary, and show that for $k \geq 2$,

$$d_{2k} = (-1)^{k+1} \frac{d_2}{k!(k-1)!2^{2k-2}}.$$

   (d) Show that regardless of the choice of d_2, the resulting solution is a constant multiple of $J_1(x)$ as given by equation (7.96). Hence, no linearly independent second Frobenius solution can be obtained.

4. Show that when $2p$ is not an integer, Bessel's equation has the solution given in equation (7.101).

5. There are many formulas that relate the Bessel functions of various orders to each other. One of these is the recursion formula

$$xJ_{p-1}(x) + xJ_{p+1}(x) = 2pJ_p(x).$$

Use this and the determination of $J_{1/2}(x)$ and $J_{-1/2}(x)$ in the text to show that

$$J_{3/2}(x) = \sqrt{\frac{2}{\pi}} \, [x^{-\frac{3}{2}} \sin x - x^{-\frac{1}{2}} \cos x].$$

6. Show that if the change of variable $t = s^2$ is made in the integral

$$\int_0^\infty t^{-\frac{1}{2}} e^{-t} dt,$$

the result gives

$$\Gamma\left(\frac{1}{2}\right) = 2\int_0^\infty e^{-s^2} ds.$$

Since

$$\int_0^\infty e^{-s^2} ds$$

is known to be $\frac{\sqrt{\pi}}{2}$, this gives

$$\Gamma\left(\frac{1}{2}\right) = \sqrt{\pi}.$$

7. (a) According to the definition of the Gamma function,

$$\Gamma(p+1) = \int_0^\infty t^p e^{-t} dt = \lim_{M \to \infty} \int_0^M t^p e^{-t} dt.$$

   Using an integration by parts, show that

$$\Gamma(p+1) = p\Gamma(p).$$

   (b) Using (a) and the fact that $\Gamma(\frac{1}{2}) = \sqrt{\pi}$, compute $\Gamma(\frac{3}{2})$, $\Gamma(\frac{5}{2})$, and $\Gamma(\frac{7}{2})$.
   (c) We've defined the Gamma function $\Gamma(p)$ for positive values of p by the integral in equation (7.104). However, we can use the property $\Gamma(p+1) = p\Gamma(p)$ from (a) to extend the definition to negative values of p that are not integers. Show that $\Gamma(-\frac{1}{2}) = -2\sqrt{\pi}$, and then compute $\Gamma(-\frac{3}{2})$ and $\Gamma(-\frac{7}{2})$.

7.6. Bessel's equation

8. Verify that the expansion of the function $J_{3/2}(x)$ begins
$$J_{3/2}(x) = \frac{2}{3\sqrt{2\pi}}x^{3/2} - \frac{1}{15\sqrt{2\pi}}x^{7/2} + \frac{1}{420\sqrt{2\pi}}x^{11/2} - \cdots.$$
Hint: You may use the result of Exercise 5.

9. Suppose we seek a solution to the Bessel equation with $p = \frac{3}{2}$ of the form
$$x^{-\frac{3}{2}}\sum_{n=0}^{\infty}d_n x^n$$
(a Frobenius solution corresponding to the root $r_2 = -\frac{3}{2}$).
   (a) Show that d_1 must be 0, $d_2 = \frac{d_0}{2}$, and that we *may* choose $d_3 = 0$.
   (b) With $d_3 = 0$, determine d_5, d_7, and in general d_{2k+1} for $k \geq 2$.
   (c) Find d_4, d_6, and d_8, each in terms of d_0.
   (d) Obtain a formula for d_{2k} in terms of d_0 for $k \geq 2$.
   (e) The function $J_{-3/2}(x)$ is obtained by choosing
$$d_0 = \frac{2^{\frac{3}{2}}}{\Gamma(-\frac{1}{2})}.$$
   Show that
$$J_{-3/2}(x) = -\sqrt{\frac{2}{\pi}}\, x^{-3/2} - \frac{1}{\sqrt{2\pi}}\, x^{1/2} + \frac{1}{4\sqrt{2\pi}}\, x^{5/2} - \frac{1}{72\sqrt{2\pi}}\, x^{9/2} + \cdots.$$
   (Hint: By Exercise 7(c), $\Gamma(-\frac{1}{2}) = -2\sqrt{\pi}$.)

10. Using equation (7.92) with c_0 as given in (7.105), show that
$$J_p(x) = \sum_{k=0}^{\infty}(-1)^k \frac{1}{k!\Gamma(k+p+1)}\left(\frac{x}{2}\right)^{2k+p}.$$

11. Consider the equation $y'' + e^{-2t}y = 0$ which is used to model an "aging undamped spring". Make the change of variable
$$x = e^{-t}$$
in the independent variable, and show that the resulting equation is a Bessel equation with $p = 0$.

12. The parametric Bessel equation of order p is
    (7.106) $$x^2 y'' + xy' + (\lambda^2 x^2 - p^2)y = 0$$
where λ is a positive parameter. Assuming p is a nonnegative integer, show that the general solution of (7.106) on $(0, \infty)$ is
$$y(x) = c_1 J_p(\lambda x) + c_2 Y_p(\lambda x),$$
where $Y_p(x)$ is the Bessel function of order p of the second kind; see the discussion preceding equation (7.103) in the text.

13. In this problem we investigate the connection between the Bessel function J_0 and the vibrations of a circular drumhead. We assume the edge (circumference) on the drumhead is held fixed and place the origin at the center of the drumhead. We also assume that the vibrations are *radially symmetric*, so that to describe the motion of the drumhead we give the displacement $u(r,t)$ of a point at distance r from the center at time t. Our drumhead has radius b cm and the units on $u(r,t)$ are cm. Since the edge is clamped, $u(b,t) = 0$ for all times t. The motion of the drumhead may be modeled as a superposition of "natural modes" of vibration.

If the drumhead is vibrating in a natural mode, its displacement function varies periodically and has the form $u(r,t) = y(r)s(t)$ where $y(r)$ and $s(t)$ are one-variable functions satisfying the differential equations (see Exercise 23 in Section 11.10):

(7.107) $$ry''(r) + y'(r) + kry(r) = 0$$

and

(7.108) $$s''(t) + \alpha^2 k s(t) = 0,$$

where k is a constant and α is a positive constant (whose value depends on the tension in the drumhead and the material from which it is made).

(a) Show that if $u(r,t) = y(r)s(t)$ and $u(b,t) = 0$ for all t, then either $s(t) = 0$ for all t or $y(b) = 0$. Explain why the first option, $s(t) = 0$ for all t, doesn't lead to an interesting displacement function $u(r,t)$. So we will assume that $y(b) = 0$.

(b) Using equation (7.108), show that if we want $u(r,t) = y(r)s(t)$ to be periodic in time, the constant k must be positive. For a positive choice of k, find the period and frequency of vibration (in Hz) for the corresponding solution to (7.108).

(c) Since k is positive (by (b)), equation (7.107) is a parametric Bessel equation of order 0. According to Exercise 12, its solutions are

$$y(r) = c_1 J_0(\sqrt{k}\, r) + c_2 Y_0(\sqrt{k}\, r)$$

for constants c_1 and c_2. Since we want $y(r)$ to be continuous on $[0, b]$, explain why $c_2 = 0$. Thus $y(r) = c_1 J_0(\sqrt{k}r)$, where, to avoid the trivial case $u(r,t) = 0$ for all t and r, we have $c_1 \neq 0$.

(d) The function J_0 has an infinite number of positive zeros, which we list in increasing order as $\gamma_1, \gamma_2, \gamma_3, \ldots$. From (a) we know that $y(b) = 0$ and from (c) we have $y(r) = c_1 J_0(\sqrt{k}\, r)$. What are the possible values of k in terms of the zeros γ_j? Substituting these values of k into your frequency formula from (b), obtain an expression (in terms of α, γ_j, and b) for the frequencies of the natural modes of vibration.

(e) The human ear is sensitive to ratios of frequencies (pitches), perceiving frequencies f_1 and f_2 to be harmonious when f_2/f_1 reduces to a rational number whose numerator and denominator are small integers (e.g., 2/1 corresponds to an octave while 3/2 corresponds to a "fifth"). Let f_1, f_2, f_3 be the sequence of natural-mode frequencies obtained in (c), in increasing order. Simplify the ratio f_j/f_k for positive integers j and k, and show that our ears' perception of sound produced by these radially symmetric vibrations of a circular drumhead depends on the ratios of the positive zeros of J_0.

14. Suppose we have a uniform flexible rope or chain of length L feet suspended from a point located L feet above the ground. Assume the rope swings freely, no friction at the pivot, e.g., and that the only force influencing its motion (after it is released from its initial displacement configuration) is that due to tension in the rope. When the rope is in its equilibrium position, it is vertical with upper tip L feet above the ground and lower tip 0 feet above the ground. Assume that the rope's initial displacement from equilibrium is small; for example, it is pulled to one side so that it remains straight, making an angle of 10 degrees with respect to the vertical at its pivot.

Let $u(x,t)$ be the rope's displacement from equilibrium at time t sec at height x ft; see Fig. 7.16. We have $u(L,t) = 0$ for all t. The motion of the rope may be modeled as a superposition of "natural modes" of oscillation. If the rope is oscillating in a natural mode, its motion is periodic and its displacement function takes the form

$$u(x,t) = y(x)s(t),$$

7.6. Bessel's equation

where y and s are nonconstant functions that satisfy the following ordinary differential equations:

(7.109) $$xy''(x) + y'(x) + ky(x) = 0$$

and

(7.110) $$s''(t) + kg\, s(t) = 0,$$

where g is the acceleration due to gravity and k is a positive constant. (See Exercise 15 in Section 11.3 for more on the derivation of these equations.) Because the displacement of the rope from equilibrium is zero at height L at all times, we have $y(L) = 0$; we also assume that y is continuous on $[0, L]$.

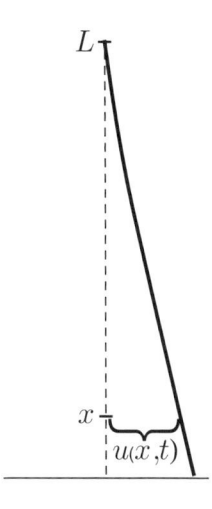

Figure 7.16. A hanging rope.

(a) Show that $x = 0$ is a regular singular point of equation (7.109) and that $r = 0$ is a (repeated) root of its indicial equation. Thus, Theorem 7.5.6 assures us that (7.109) has a power series solution $y_1 = \sum_{n=0}^{\infty} c_n x^n$.

(b) Substitute $y_1 = \sum_{n=0}^{\infty} c_n x^n$ for y in equation (7.109), finding a recursion relation satisfied by the series coefficients c_n.

(c) Use your recursion relation from (b) to find a formula for each coefficient c_n in terms of c_0. Show that if we choose $c_0 = 1$, we then have $y_1 = J_0(2\sqrt{kx})$ for $x \geq 0$.

(d) Show by substitution that $y_2 = Y_0(2\sqrt{kx})$ is also a solution of equation (7.109) on $(0, \infty)$, where Y_0 is the Bessel function of order 0 of the second kind. Because y_1 and y_2 are linearly independent on $(0, \infty)$, we conclude that $y = c_1 J_0(2\sqrt{kx}) + c_2 Y_0(2\sqrt{kx})$ for some choice of constants c_1 and c_2.

(e) Using the form of y given in part (d) and the continuity of y at 0, argue that $y(x) = c_1 J_0(2\sqrt{kx})$ for some constant c_1 (nonzero, because we are assuming y is nonconstant).

(f) Let $\gamma_1, \gamma_2, \gamma_3, \ldots$ be the sequence of positive zeros of J_0, in increasing order. Using $y(x) = c_1 J_0(2\sqrt{kx})$ and the requirement that $y(L) = 0$, determine the possible values of the parameter k.

(g) Give the frequency, in hertz, of the vibration modeled by equation (7.110) in terms of $\sqrt{k}$ and g. By substituting the possible values of k from part (f) into the frequency formula, obtain the natural-mode frequencies of the oscillating rope, in terms of L, g, and γ_j. The smallest frequency, which corresponds to γ_1, is called the fundamental frequency.

(h) Using the graph in Fig. 7.13, give the approximate value of γ_1, the smallest positive zero of J_0. Using this approximation for γ_1, $g = 32$ ft/sec, and $L = 6$ ft, and your work in (g), what is the fundamental frequency of oscillation of a 6-ft rope?

(i) Assume the rope is pulled to the right side so that it remains straight, making an angle of about 10 degrees with respect to the vertical at its pivot, and then released. In this situation, the reciprocal of the fundamental frequency represents the model's prediction of how much time will pass from the moment of release to the time the rope returns to its rightmost position, which we will call its period. Hang a flexible, uniform rope (or chain) from a hook and measure its length L. Compute the fundamental frequency as in part (h). What is the predicted period of the motion? Displace the rope from equilibrium as described and start a timer upon its release. Count the number of returns to the rightmost position over a time interval T of at least 30 seconds. Compare your experimental period, T/number-of-returns, with the predicted period.

Chapter 8

Linear Systems I

8.1. Nelson at Trafalgar and phase portraits

We'll begin our study of systems of first-order differential equations with a discussion of the military strategy of Lord Nelson in the Battle of Trafalgar.

On October 21, 1805, a massive sea battle, the Battle of Trafalgar, was waged by the British against the combined French and Spanish fleets of Napoleon's navy. The decisive British victory, despite the numerical superiority of the combined French and Spanish fleets with respect to the number of ships, guns, and men, began a long era of British naval supremacy and effectively prevented an invasion of England by Napoleon. Nelson's victory at Trafalgar was due in significant part to his brilliant strategy for this engagement, which came to be known as "the Nelson touch". Nelson summarized his tactical ideas in a now famous "Memorandum" written on October 8, 1805. We're going to analyze Nelson's plan as detailed in this Memorandum from a differential equations perspective to understand its success and its significance in military strategy.

Mathematical analysis of combat by a system of differential equations goes back at least as far as the First World War, when F. Lanchester proposed several simple differential equation models for different kinds of military conflict.[1] For a naval battle of the sort fought at the beginning of the nineteenth century, Lanchester's model looks like

(8.1)
$$\begin{aligned} \frac{dx}{dt} &= -ay, \\ \frac{dy}{dt} &= -bx, \end{aligned}$$

where $x(t)$ and $y(t)$ are the strengths of the two opposing forces as a function of time t. We'll measure "strength" by the number of ships each side has; more generally these would be the number of combatants on each side. Without reinforcements (as is appropriate for a naval battle that takes place in the course of perhaps just a single day), the strength of each force is decreasing, and in this model we are assuming that it decreases at a rate proportional to the size of the opposing force. The positive proportionality constants a and b in these equations would reflect the "effectiveness" of the opposing force. For example, a is much larger than b would indicate that the y-force is more effective than the x-force. We assume that the initial strengths of both sides are known; say, $x(0) = x_0$ and $y(0) = y_0$; these are **initial conditions** for the system.

---

[1] We used one such model in Section 6.5 to analyze the WWII battle at Iwo Jima.

The system (8.1) is simple enough that we can make a preliminary analysis of it with ad hoc methods. A solution to (8.1) is a pair of functions

(8.2) $$x = x(t), \quad y = y(t)$$

that give the strength of each force as a function of time. We can think of the pair of equations in (8.2) as giving parametric equations of a curve in the xy-plane; that is, for each t we imagine plotting the point $(x(t), y(t))$, and as t varies this point sketches out a curve in the xy-plane. This curve is called the **trajectory** of the solution. As we will see, the trajectory encodes information about how the battle progresses (and in particular who wins and by how much). At the moment, we are more interested in identifying this curve in the xy-plane than in giving an explicit formula for x and y in terms of t. From calculus you know that if $\frac{dx}{dt} = -ay$ and $\frac{dy}{dt} = -bx$, then

$$\frac{dy}{dx} = \frac{\frac{dy}{dt}}{\frac{dx}{dt}} = \frac{bx}{ay}.$$

This is a first-order differential equation for the curves we are trying to identify. We can solve it as a separable equation, obtaining

$$\int ay \, dy = \int bx \, dx,$$

so that

$$\frac{ay^2}{2} = \frac{bx^2}{2} + c,$$

or equivalently

(8.3) $$ay^2 - bx^2 = C.$$

If we use the initial conditions $x(0) = x_0$, $y(0) = y_0$, we see that

$$ay_0^2 - bx_0^2 = C,$$

so that C is determined by the initial force strengths and the constants a and b. For $C \neq 0$, the curves with equation (8.3) are hyperbolas centered at $(0,0)$ in the xy-plane. Whether the hyperbolas cross the x-axis or the y-axis depends on whether C is positive or negative; see Figs. 8.1–8.2.

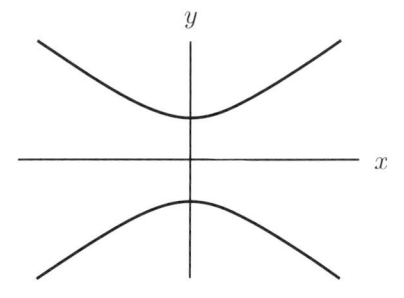

Figure 8.1. $ay^2 - bx^2 = C, C > 0$.

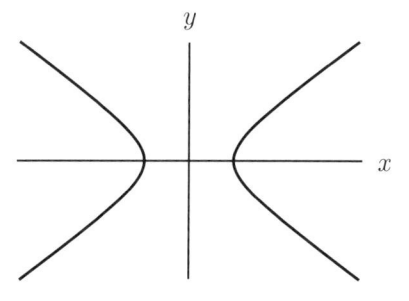

Figure 8.2. $ay^2 - bx^2 = C, C < 0$.

8.1. Nelson at Trafalgar and phase portraits

The choice $C = 0$ corresponds to a pair of lines:

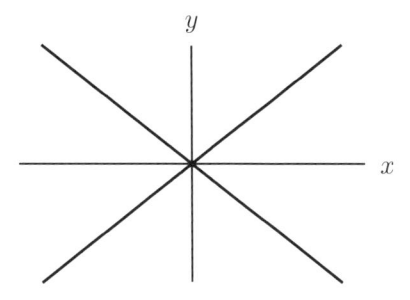

Figure 8.3. $ay^2 - bx^2 = 0$.

Keep in mind that C is determined by the initial conditions. The condition $C > 0$ corresponds to $ay_0^2 > bx_0^2$, and $C < 0$ to the reverse inequality. Imagine now that we are going to sketch these curves for a variety of initial conditions. We'll only show this sketch in the first quadrant of the xy-plane, since the physical meaning of x and y makes negative values of either meaningless. The arrows on the curves in Fig. 8.4 show the direction each curve is traversed as t increases. Since $\frac{dx}{dt}$ and $\frac{dy}{dt}$ are negative, both x and y decrease with time. This picture is called a **phase portrait** of the system (8.1).

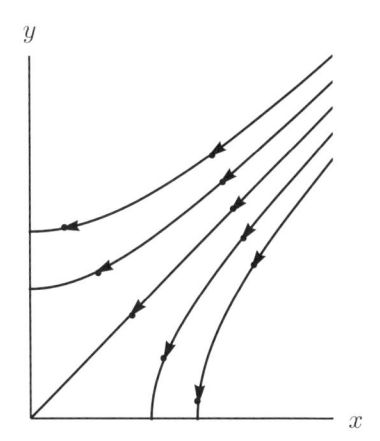

Figure 8.4. Phase portrait for the system (8.1).

Let's return to Nelson's battle plan for Trafalgar. Standard tactics for naval battles at this time involved the ships of opposing fleets arranging themselves into two parallel lines at close quarters; each would engage the nearest ship in the opposing line; this is why the battleships were referred to as "ships-of-the-line". Nelson assumed that the initial strength for the French-Spanish fleet would be 46 ships-of-the-line and that the British would have 40 ships-of-the-line. If we assume they had equal effectiveness[2] so that $a = b$ and if there is a single battle between the two forces, the equation of the relevant curve in the phase plane is

$$ay^2 - ax^2 = a(46)^2 - a(40)^2,$$

where $x = x(t)$ is the strength of the British fleet and $y = y(t)$ is the strength of the combined French-Spanish fleet. This curve is shown in Fig. 8.5. The result of this scenario is defeat for the British, since $x = 0$ when $y = \sqrt{46^2 - 40^2} \approx 23$. In other words, by the time all the British ships are destroyed or captured, the French-Spanish fleet still has about 23 ships available.

---

[2]In reality, the British had several distinct advantages, including better training, more experience, better morale, and the technical advantage of flint-lock firing mechanisms on their guns rather than the slow-burning fuses used by the French and Spanish, which made it easier to aim accurately in heavy sea swell.

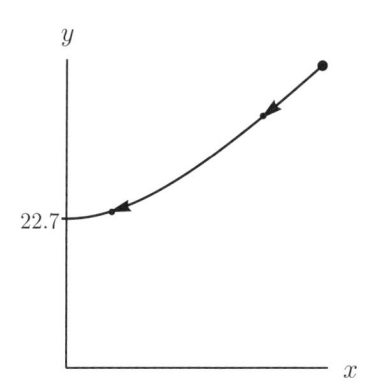

Figure 8.5. Without Nelson's strategy.

However, Admiral Nelson's brilliant plan called for a different strategy. He described his plan to the captains of the ships-of-the-line under his command over dinner onboard his ship, the *Victory*, and followed up with a memorandum written a few days later and sent to all. The reaction to his plan among his commanders was described in a letter written by Nelson[3] as follows: "...when I came to explain to them the Nelson touch, it was like an electric shock. Some shed tears all approved, it was new, it was singular, it was simple and, from Admirals downwards, it was repeated, it must succeed, if ever they will allow us to get at them." The plan called for the British forces to arrange themselves into three groups, two with 16 ships each and one group of 8, with the expectation that the ships of the combined French-Spanish fleet would have arranged themselves into a long single line, as was the custom. One group of 16, led by Vice-Admiral Collingwood on board the *Royal Sovereign* would cut off and engage 12 of the French-Spanish fleet, the other group, led by Nelson aboard the *Victory*, would cut off and engage another 12 of the combined fleet, and the squadron of 8 smaller, faster ships would prevent the remaining ships of the combined fleet from effectively participating in either of the first two battles.

Figure 8.6. Lord Nelson, Britain's greatest admiral, in a poster celebrating the victory at Trafalgar. Source: https://en.wikipedia.org/wiki/File:Battle_of_Trafalgar_Poster_1805.jpg.

If we apply the Lanchester model to the two individual battles envisioned by Nelson, the result of each will be victory for the British, with

$$\sqrt{16^2 - 12^2} \approx 10.6$$

British ships surviving each battle, assuming equal effectiveness of the two sides. This would assure Nelson an almost even battle should the other 22 ships of the combined fleet be eventually able to maneuver into battle, even assuming all 8 ships of the third group to be destroyed by this point. Thus, by arranging a succession of sequential battles, Nelson could manipulate a victory with a smaller initial force, even assuming equal effectiveness of each ship in the two forces. In

---

[3]Letter from Nelson to Emma Hamilton, his mistress, October 1, 1805, the British Library.

8.1. Nelson at Trafalgar and phase portraits 447

reality, the number of ships present on both sides on October 21 was somewhat smaller than Nelson anticipated (28 British and 33 French-Spanish), and the details of how the British divided up their fleet differed somewhat from Nelson's memorandum, but the basic idea was still implemented. In the end, Britain lost no ships, and the French-Spanish lost 18 (destroyed or captured).[4]

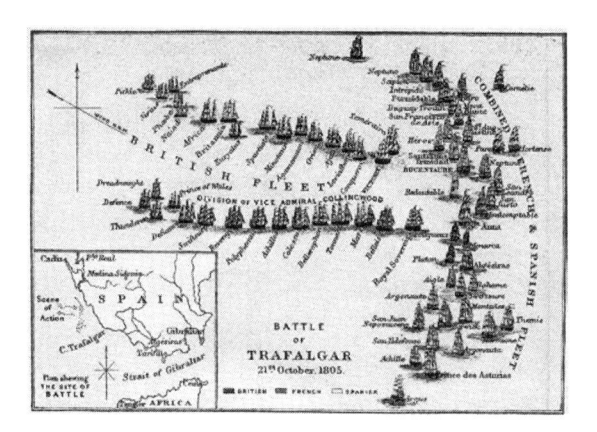

Figure 8.7. Advancing British fleet with Nelson's group on the left and Collingwood's on the right. Source: Wikimedia Commons. Public domain.

8.1.1. Basic notions. The system of differential equations in (8.1) is an example of a two-dimensional, or planar, **linear system**. In general, a linear system of two first-order differential equations has the form

(8.4)
$$\begin{aligned} \frac{dx}{dt} &= p_{11}(t)x + p_{12}(t)y + g_1(t), \\ \frac{dy}{dt} &= p_{21}(t)x + p_{22}(t)y + g_2(t). \end{aligned}$$

In these equations, t is the independent variable, and there are two dependent variables, x and y. In applications, t is frequently time. We use double subscripts on the **coefficient functions** $p_{ij}(t)$; notice, for example, that $p_{12}(t)$ is the coefficient in the *first* equation of the *second* dependent variable y. If $g_1(t) = g_2(t) = 0$, this system is said to be **homogeneous**. More generally, a system (linear or not) of two first-order equations looks like

$$\frac{dx}{dt} = F(x, y, t),$$

$$\frac{dy}{dt} = G(x, y, t)$$

for some functions F and G. Initial conditions for a planar system specify x and y at a particular point t_0. If the independent variable t doesn't explicitly appear on the right-hand side (so that F and G are functions of x and y only), the system is said to be **autonomous**. The Lanchester model in (8.1) is an autonomous system.

Phase portraits. For an *autonomous* system of two first-order equations,

(8.5)
$$\frac{dy}{dx} = f(x, y), \quad \frac{dy}{dx} = g(x, y),$$

---

[4]Nelson himself was killed, however, a victim of a musket ball fired by an unknown Frenchman aboard the *Redoubtable* as the *Victory* and *Redoubtable* were side by side in battle.

we often give graphical information about the solutions by sketching a phase portrait (as in Fig. 8.4). A solution
$$x = x(t), \quad y = y(t)$$
to the system gives parametric equations for a curve in the xy-plane. Other solutions of the system similarly give rise to other curves. A phase portrait shows a representative collection of these parametric solution curves (called trajectories or orbits) in the xy-plane, and the xy-plane is then called the **phase plane**. In the exercises you will be asked to sketch phase portraits for several simple autonomous systems. In this section these will be done as we did for the Trafalgar model (8.1), by solving a single first-order equation for $\frac{dy}{dx}$. In later sections, we will learn a variety of other methods for producing phase portraits of planar autonomous systems. For example, in Section 8.2.1 we show how a "vector field sketch" for an autonomous linear system of two first-order equations facilitates construction of a phase portrait in the same way that a direction field for a single first-order equation helps us sketch its solution curves.

Trajectories. Assume that $x = x(t), y = y(t)$ is a solution of the system (8.5) valid on the interval I. You can think of the ordered pair $(x(t), y(t))$ as the state of the system at time t and the corresponding trajectory $\{(x(t), y(t)) : t \text{ in } I\}$ plotted in the xy-plane as a way to visualize one of the ways that the system's states can evolve. Often we will put arrows on the trajectories to show the direction in which they are traversed as t increases. Note that a trajectory is a geometric object—a directed curve in the plane—and that two different solutions of an autonomous system can have the same trajectory. For instance, the unit circle, directed counterclockwise, is the trajectory of the solutions $x = \cos t, y = \sin t$ and $x = \sin t, y = -\cos t$ of the system $dx/dt = -y$, $dy/dt = x$. We use the terms "trajectory" and "orbit" interchangeably.

Linear systems more generally. Equation (8.4) is a two-dimensional linear system. An n-dimensional linear system has the form

(8.6)
$$\begin{aligned}
\frac{dy_1}{dt} &= p_{11}(t)y_1 + \cdots + p_{1n}(t)y_n + g_1(t), \\
\frac{dy_2}{dt} &= p_{21}(t)y_1 + \cdots + p_{2n}(t)y_n + g_2(t), \\
&\vdots \\
\frac{dy_n}{dt} &= p_{n1}(t)y_1 + \cdots + p_{nn}(t)y_n + g_n(t).
\end{aligned}$$

The "unknowns" are the functions $y_j(t)$, and a solution is a set of n differentiable functions
$$y_1 = \varphi_1(t), \ y_2 = \varphi_2(t), \ldots, y_n = \varphi_n(t)$$
which satisfy (8.6) over some interval of values of t. The y_j's are dependent variables, and t is the independent variable; in applications t is frequently time. There are the same number n of equations in (8.6) as unknowns. When each $g_i(t)$ is the zero function, the system (8.6) is homogeneous. If the functions $p_{ij}(t)$ are constant functions, the system is called a **constant coefficient** system.

Initial conditions for the n-dimensional system (8.6) take the form

(8.7)
$$y_1(t_0) = \alpha_1, \ y_2(t_0) = \alpha_2, \ldots, y_n(t_0) = \alpha_n,$$

so that the values of each y_j are prescribed at a common point. Notice that initial conditions for an n-dimensional system have n data points. Generally we seek to determine the solution functions on some open interval I on which the functions p_{ij} and g_j are all continuous, with the initial condition prescribing the values of the y_j's at a point t_0 in I. An existence and uniqueness theorem, which is the analogue of Theorem 2.4.4, applies in this setting:

8.1. Nelson at Trafalgar and phase portraits

Theorem 8.1.1 (Existence and Uniqueness for Linear Systems). *In the system (8.6), suppose p_{ij} and g_j are continuous on an open interval I, for $1 \leq i, j \leq n$. Given any initial conditions of the form (8.7) with $t_0 \in I$, there is a unique solution*

$$y_1 = \varphi_1(t), \; y_2 = \varphi_2(t), \ldots, \; y_n = \varphi_n(t)$$

to (8.6) valid on the entire interval I and satisfying (8.7).

In Section 9.2 we'll give a proof of Theorem 8.1.1 for a constant coefficient homogeneous system.

Can trajectories touch or cross? The uniqueness conclusion of Theorem 8.1.1 states that there cannot be two different solutions to the same initial value problem (as long as the hypotheses of Theorem 8.1.1 hold). This leads to the question of whether the *paths* followed by two different solutions can touch if the solutions arrive at the "touch point" at different times. We've seen that the answer to this question is yes—the paths followed by the solutions $y = \cos t, y = \sin t$ and $y = \sin t, y = -\cos t$ of the system $dx/dt = -y, dy/dt = x$ are identical (touching, e.g., the point $(1, 0)$ at times $t = 0$ and $t = \pi/2$, respectively). For an autonomous linear system this can happen only if one solution is merely a time translate of the other. This means that both solutions travel the same path and pass each given point on that path at the same speed; the only difference is that one solution follows this path ahead of the other in time.[5] This also means (for autonomous linear systems) that if the trajectories of two solutions have a point in common, then the trajectories must be identical (assuming the domains of the solutions are not restricted). Thus, trajectories cannot cross; moreover, they cannot touch unless they coincide. In particular, phase portraits for such systems (of two equations with two dependent variables) will never contain trajectories forming shapes as shown in Fig. 8.8 .

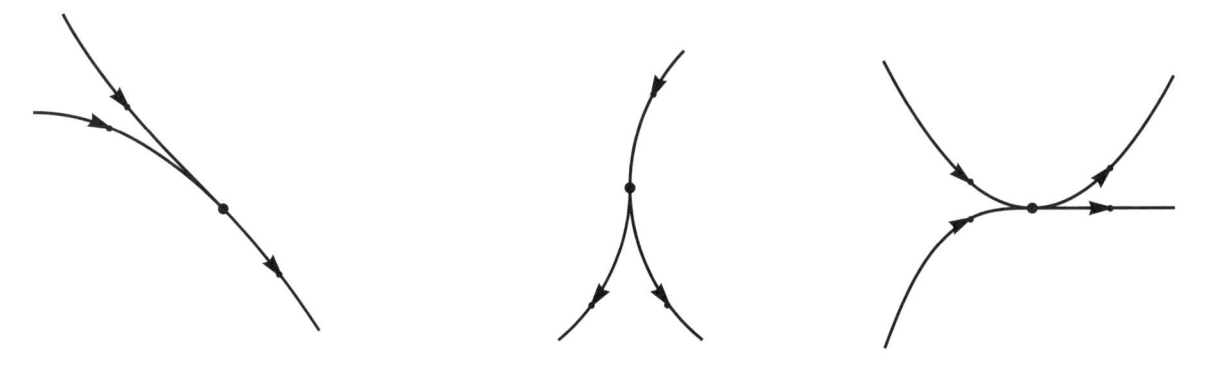

Figure 8.8. Forbidden behavior for trajectories of an autonomous linear system.

Note that a linear system (8.6) is autonomous precisely when all the coefficient functions p_{ij} as well as the functions g_i are constant, in which case the interval I of continuity described in Theorem 8.1.1 is $(-\infty, \infty)$. An autonomous linear system with two dependent variables x and y and independent variable t takes the form

(8.8) $$\frac{dx}{dt} = a_{11}x + a_{12}y + c_1, \quad \frac{dy}{dt} = a_{21}x + a_{22}y + c_2,$$

where $a_{11}, a_{12}, a_{21}, a_{22}$ as well as c_1 and c_2 are constants. We have the following corollary of Theorem 8.1.1:

Corollary 8.1.2. *Let (x_0, y_0) be any point in the plane. There is a unique solution $x = \varphi_1(t), y = \varphi_2(t)$ of the system (8.8) valid on the interval $(-\infty, \infty)$ and satisfying the initial conditions $x(0) = x_0 \; y(0) = y_0$.*

---
[5]We'll see a proof of this in Section 10.1.

Equilibrium points. A constant solution of a differential equation system is called an **equilibrium solution**. For example, the system in equation (8.1) has equilibrium solution $x(t) \equiv 0$, $y(t) \equiv 0$. In the phase portrait for (8.1), this equilibrium solution appears as a point at $(0,0)$. The value of a constant solution is called an **equilibrium point**; the term **critical point** is also used.

Example 8.1.3. To find the equilibrium points of the (nonlinear) system
$$\frac{dx}{dt} = x - y, \quad \frac{dy}{dt} = 4 - x^2 - y^2,$$
we find all values of x and y that make
$$x - y = 0 \quad \text{and} \quad 4 - x^2 - y^2 = 0.$$
Thus $(\sqrt{2}, \sqrt{2})$ and $(-\sqrt{2}, -\sqrt{2})$ are the equilibrium points.

If the matrix of coefficients a_{ij} of the autonomous linear system (8.8) has nonzero determinant, then the system has exactly one equilibrium point. In this case, observe that because trajectories that touch must be identical,[6] no trajectory of the system can include the equilibrium point, unless the trajectory consists of *only* the equilibrium point.

Systems from higher-order equations. A differential equation of order two or higher can always be converted into a system of first-order equations. The next example illustrates how this goes.

Example 8.1.4. We convert the second-order linear equation
(8.9) $$u'' + (\sin t)\, u' + (t^2 - t)u = t^3$$
into a linear system of two first-order equations using the substitution
$$x = u, \quad y = u'.$$
(The primes denote differentiation with respect to t.) Since $y' = u''$ we may rewrite equation (8.9) as
$$y' + (\sin t)y + (t^2 - t)x = t^3,$$
so that equation (8.9) is equivalent to the nonhomogeneous linear system
$$\frac{dx}{dt} = y,$$
$$\frac{dy}{dt} = (t - t^2)x - (\sin t)\, y + t^3.$$

As far as initial conditions go, notice that specifying x and y at a point t_0 is the same as specifying u and u' at t_0.

Higher-order *systems* of differential equations can be converted to first-order systems via substitutions similar to those used in the preceding example; see Exercise 14. This permits the application of the theory and techniques for first-order systems to higher-order systems, as is illustrated by Exercise 24. In addition, converting a higher-order equation to a system of first-order equations facilitates numerical solution of the equation—see Exercise 29 in Section 4.1.

Higher-order equations from systems. In the next example we will convert a simple first-order linear system to a single higher-order differential equation.

---
[6]We are assuming the domains of the corresponding solutions have not been restricted.

8.1. Nelson at Trafalgar and phase portraits

Example 8.1.5. (**Method of Elimination.**) We will solve the system

(8.10)
$$\frac{dx}{dt} = 3x + 2y,$$
$$\frac{dy}{dt} = x + 2y$$

by first converting it to a single second-order equation with dependent variable y and independent variable t.

Solving the second equation of (8.10) for x and substituting the result into the first equation of the system, we obtain $(y' - 2y)' = 3(y' - 2y) + 2y$, which simplifies to

$$y'' - 5y' + 4y = 0.$$

This second-order equation has characteristic equation $r^2 - 5r + 4 = 0$, which has roots 4 and 1. Thus, the equation has general solution $y = c_1 e^{4t} + c_2 e^t$. Using $x = y' - 2y$, we find $x = 2c_1 e^{4t} - c_2 e^t$.

If we think of the system in (8.10) as describing a particle's velocity $(\frac{dx}{dt}, \frac{dy}{dt})$ at time t, then we see that the general form for the particle's position at time t is

$$(x, y) = (2c_1 e^{4t} - c_2 e^t, c_1 e^{4t} + c_2 e^t).$$

Taking advantage of the standard vector operators of addition and scalar multiplication (vectors and operations on vectors are reviewed in the next section), we can write this as

$$(x, y) = c_1(2e^{4t}, e^{4t}) + c_2(-e^t, e^t).$$

If we write our vectors as columns and set

$$\mathbf{X} = \begin{pmatrix} x(t) \\ y(t) \end{pmatrix}, \quad \mathbf{X}_1(t) = \begin{pmatrix} 2e^{4t} \\ e^{4t} \end{pmatrix}, \quad \text{and} \quad \mathbf{X}_2(t) = \begin{pmatrix} -e^t \\ e^t \end{pmatrix},$$

then our solution to the system (8.10) takes the form

$$\mathbf{X} = c_1 \mathbf{X}_1(t) + c_2 \mathbf{X}_2(t)$$

for arbitrary constants c_1 and c_2. This way of writing the solution allows us to anticipate that the theory of linear systems has much in common with the theory of higher-order linear equations, an idea that will be explored fully in later sections.

8.1.2. Exercises.

1. The Lanchester combat model for a conventional force fighting against a guerrilla force, with no reinforcements and negligible operational losses (operational losses are losses due to noncombat issues) is given by

   $$\frac{dx}{dt} = -axy, \quad \frac{dy}{dt} = -bx$$

   where $x(t)$ is the number of guerrilla force combatants and $y(t)$ is the number of conventional force combatants.

   (a) Find a single differential equation for $\frac{dy}{dx}$ and solve it. Describe the solution curves.

   (b) Using your work from (a), sketch some phase portrait trajectories for the system in the first quadrant of the xy-plane; show the direction of motion along each trajectory with increasing t.

   (c) If the initial size of the guerrilla force is x_0 and the initial size of the conventional force is y_0, who wins? Your answer should depend on x_0, y_0, a, and b.

2. The Lanchester model for two guerrilla forces, with no reinforcements and negligible operation losses is

   $$\frac{dx}{dt} = -axy, \quad \frac{dy}{dt} = -bxy.$$

By finding and solving a single differential equation for $\frac{dy}{dx}$, sketch some trajectories in the phase plane. If the initial force sizes are $x(0) = x_0, y(0) = y_0$, under what conditions on a, b, x_0, and y_0 does the x-force win?

3. (a) Sketch some orbits in the phase plane for the system
$$\frac{dx}{dt} = -y, \quad \frac{dy}{dt} = x.$$
(b) Check by substitution that
$$x(t) = \cos t, \quad y(t) = \sin t$$
is a solution to this system. Is this fact consistent with your answer to (a)?

4. Check that
(8.11) $$x(t) = e^{-t}\cos t, \quad y(t) = e^{-t}\sin t$$
is a solution to the system
$$\frac{dx}{dt} = -x - y, \quad \frac{dy}{dt} = x - y.$$
Using (8.11), give a rough sketch of the corresponding solution curve in the xy-plane, noting the direction your curve is traversed as t increases. (Notice that $x^2 + y^2 = e^{-2t}$ and $e^{-2t} \to 0$ as $t \to \infty$.)

5. (a) Show that the system
$$\frac{dx}{dt} = x + y, \quad \frac{dy}{dt} = x - y$$
has only one equilibrium point, and find it.
(b) Show that the system
$$\frac{dx}{dt} = x + y, \quad \frac{dy}{dt} = -x - y$$
has infinitely many equilibrium points.
(c) Show that the system
$$\frac{dx}{dt} = x + 4y - 6, \quad \frac{dy}{dt} = 2x - y - 3$$
has only one equilibrium point, and find it.

6. Consider the system
$$\frac{dx}{dt} = -y, \quad \frac{dy}{dt} = -y - 2xy.$$
(a) Find a single differential equation for $\frac{dy}{dx}$ and solve it.
(b) Use your answer to (a) to show that the trajectories for the system lie in certain parabolas, and sketch these parabolas.
(c) Find all the equilibrium points of this system.

7. Consider a battle between and x- and y-force described by the equations
$$\frac{dx}{dt} = -ay, \quad \frac{dy}{dt} = -bx$$
with initial conditions $x(0) = x_0$, $y(0) = y_0$. Suppose you are the commander of the x-force and the x-force is the losing side in this initial value problem.
(a) Is $ay_0^2 - bx_0^2$ positive or negative? How do you know?
(b) If you have the ability to either double your initial force or double your effectiveness (by this we mean doubling the value of the appropriate coefficient a or b), which would you choose to do? Explain.

8. **Differential equations and love affairs.** This problem is based on an article of Steven Strogatz in [48]. A related article by Strogatz appeared in the New York Times Olivia Judson Blog on May 26, 2009. Suppose two lovers, let's call them Romeo and Juliet, have romantic styles that are summarized as follows:
   (I) The more Juliet loves Romeo, the less he loves her, but when she appears less interested in him, his romantic feelings for her increase.
   (II) Juliet's feelings tend to mirror Romeo's; the more he loves her, the more she loves him, and conversely, the more he dislikes her, the more she dislikes him.

   We propose to model their relationship with a system of differential equations and use it to predict the course of their relationship. Let $R(t)$ denote Romeo's feelings for Juliet and $J(t)$ denote Juliet's feelings for Romeo. A positive value of either $R(t)$ or $J(t)$ denotes love, and a negative value dislike. Consider the equations
   $$\frac{dR}{dt} = -aJ, \quad \frac{dJ}{dt} = bR$$
   where a, b are positive constants.
   (a) Explain how the overall sign on the right-hand side of these equations fits with principles (I) and (II) above.
   (b) Sketch a phase portrait for this system in the RJ-plane. What does it tell you about the course of their romance? Are Romeo and Juliet ever in a state of simultaneous love? Do they ever simultaneously dislike each other?

9. The system
   $$\frac{dp}{dt} = a_1 p - b_1 p^2 - c_1 pq, \quad \frac{dq}{dt} = a_2 q - b_2 q^2 + c_2 pq$$
   is proposed to model the populations of two interacting species that are in a predator-prey relationship. The constants a_1, b_1, a_2, b_2, c_1, and c_2 are positive. Which is the predator and which is the prey?

10. Identify each of the systems in Exercises 1–9 as linear or nonlinear, and autonomous or not autonomous. Of the linear equations, which are homogeneous?

11. Consider the system of equations
    $$\frac{dx}{dt} = -ay, \quad \frac{dy}{dt} = -bx$$
    discussed in the text.
    (a) Verify that
    $$x(t) = -\sqrt{\frac{a}{b}} e^{\sqrt{ab}\,t}, \quad y(t) = e^{\sqrt{ab}\,t}$$
    is a solution to this system, and show that these parametric equations describe a portion of the pair of lines $ay^2 - bx^2 = 0$. Which portion?
    (b) Verify that
    $$x(t) = \sqrt{\frac{a}{b}} e^{-\sqrt{ab}\,t}, \quad y(t) = e^{-\sqrt{ab}\,t}$$
    is also a solution to this system, and show that these parametric equations describe a portion of the pair of lines $ay^2 - bx^2 = 0$. Which portion?
    (c) Finally verify that for any constants c_1 and c_2
    $$x(t) = -c_1 \sqrt{\frac{a}{b}} e^{\sqrt{ab}\,t} + c_2 \sqrt{\frac{a}{b}} e^{-\sqrt{ab}\,t}$$
    and
    $$y(t) = c_1 e^{\sqrt{ab}\,t} + c_2 e^{-\sqrt{ab}\,t}$$

is a solution to this system. If c_1 and c_2 are nonzero, show that these are parametric equations for a portion of the hyperbola $ay^2 - bx^2 = C$ for some nonzero C.

12. Convert the differential equation $u''' + e^t u'' + tu' - 2u = 0$ to a system of three first-order equations by means of the substitutions $y_1 = u$, $y_2 = u'$, $y_3 = u''$.

13. (a) Convert the initial value problem
$$u''' + 2u'' + 3u' - 2u = 0, \quad u(0) = 1, \quad u'(0) = -2, \quad u''(0) = 5$$
to an initial value problem for three first-order equations.
(b) Is the system you gave in (a) linear? Is it homogeneous?

14. The pair of equations
$$u'' = tu' - 2v' - e^t v + \sin t,$$
$$v'' = u - 3v'$$
is a system of two second-order equations. Convert this to a system of four first-order equations.

15. Verify by substitution that
$$x(t) = -t^2 + c_1 + c_2(2t+1), \quad y(t) = 2t - 2t^2 + 2c_1 + 4tc_2$$
is a solution of the linear system
$$\frac{dx}{dt} = 2x - y, \quad \frac{dy}{dt} = 4x - 2y + 2$$
for any constants c_1 and c_2.

16. (a) Verify by substitution that
$$x(t) = 2e^{-t}, \quad y(t) = e^{-t}$$
is a solution of the linear system

(8.12) $$\frac{dx}{dt} = -4x + 6y,$$

(8.13) $$\frac{dy}{dt} = -3x + 5y.$$

Then verify that
$$x(t) = e^{2t}, \quad y(t) = e^{2t}$$
is also a solution of the same system.

(b) Show that for arbitrary constants c_1 and c_2
$$x(t) = 2c_1 e^{-t} + c_2 e^{2t}, \quad y(t) = c_1 e^{-t} + c_2 e^{2t}$$
also solves the system in (a).

(c) Show that (b) gives *every* solution to the system via the following process:
   (i) Solve equation (8.12) for y and substitute the result into (8.13) to show that x must satisfy
   $$x'' - x' - 2x = 0.$$
   (ii) Show that the general solution of $x'' - x' - 2x = 0$ may be written in the form

   (8.14) $$x(t) = 2c_1 e^{-t} + c_2 e^{2t},$$

   where c_1 and c_2 are arbitrary constants.
   (iii) Use (8.14) and the expression for y in terms of x and x' from part (i) to see that y must take the form $y(t) = c_1 e^{-t} + c_2 e^{2t}$.

17. Consider the system
$$\frac{dx}{dt} = 2x + y, \quad \frac{dy}{dt} = x + 2y.$$
Use the method of elimination to obtain a second-order linear equation for y. Solve this equation for y, and then solve the original system.

18. Use the method of elimination to solve the following initial value problems, where primes denote derivatives with respect to time t.
    (a) $x' = 3x + 4y, y' = 2x + y, \ x(0) = 3, \ y(0) = 0$.
    (b) $x' = x + 2y, y' = -2x - 3y, \ x(0) = 1, \ y(0) = 2$.
    (c) $x' = -x - y, y' = x - y, \ x(0) = 1, \ y(0) = 0$.
    (d) $x' = x + y + t, y' = 3x - y, \ x(0) = 0, \ y(0) = 0$.
    (e) $x' = y, y' = \frac{3}{t^2}x + \frac{1}{t}y, \ x(1) = 4, \ y(1) = 0$.
    In part (c) you should obtain the solution (8.11) in Exercise 4.

19. (a) Use the method of elimination to convert the system
$$\frac{dx}{dt} = x + 4y - 6, \quad \frac{dy}{dt} = 2x - y - 3$$
to a single nonhomogeneous second-order differential equation.
    (b) Solve the second-order equation you obtained in (a) (for example, by the method of annihilating operators).
    (c) Solve the initial value problem consisting of the system in (a) with the initial conditions $x(0) = 11, y(0) = 0$.

20. (a) Verify that Theorem 8.1.1 guarantees that the following initial value problem has a unique solution valid on $(0, \infty)$:
$$\frac{dy_1}{dt} = y_2,$$
$$\frac{dy_2}{dt} = \frac{2}{t^2}y_1 - \frac{2}{t}y_2, \quad y_1(1) = 0, \ y_2(1) = 3.$$
    (b) Find the solution of the initial value problem of part (a) using the method of elimination.

21. Saline solution is exchanged between the two tanks pictured below, where Tank 1 contains 50 gallons of liquid and Tank 2 contains 25. The exchange rate is 10 gallons per minute, the initial amount of salt in Tank 1 is 15 pounds, and the initial amount of salt in Tank 2 is 0 pounds. Let $x(t)$ be the amount of salt in t minutes later in Tank 1 and let $y(t)$ be the amount of salt t minutes later in Tank 2.

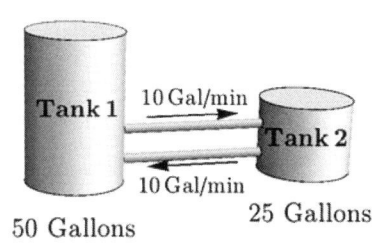

    (a) Set up a system of two first-order differential equations with dependent variables x and y modeling this mixing problem. What are the initial conditions?
    (b) Solve the initial value problem in part (a).
    (c) Compute the long-term amount of salt in each tank; that is, compute $\lim_{t\to\infty} x(t)$ and $\lim_{t\to\infty} y(t)$.

22. To model the spread of swine flu in a closed population (a population of fixed size, with no births or nondisease related deaths), we use a compartment model as shown below. The compartments are
    - the susceptibles: people who haven't had swine flu and are therefore at risk,
    - the infectives: people who are sick and thus infectious,
    - the removed group: this includes those who have recovered and thus have immunity from reinfection, as well as people who have died from the disease.

    At any moment each person in the population is in one of these three compartments. People move from one compartment to another only in the direction indicated by the arrows below. Let $S(t)$ be the number of susceptibles at time t, $I(t)$ the number of infectives at time t, and $R(t)$ the number in the removed compartment at time t. Assume that the rate at which people become infected is jointly proportional to both I and S; that is, the rate is proportional to the product of I and S. Also assume that people move from the I compartment to the R compartment at a rate proportional to the number of infectives.

    Susceptibles → Infectives → Removed

    (a) Set up a system of 3 differential equations (for $\frac{dS}{dt}$, $\frac{dI}{dt}$, and $\frac{dR}{dt}$) to model the spread of this disease. For any constants that appear in your equations, clearly identify if they are positive or negative.

    (b) Your equations in (a) should have the property that
    $$\frac{dS}{dt} + \frac{dI}{dt} + \frac{dR}{dt} = 0.$$
    Explain why you would know a priori that this is the case.

    (c) Use your equations for $\frac{dS}{dt}$ and $\frac{dI}{dt}$ to find a single first-order equation for $\frac{dI}{dS}$ and solve it using the initial condition $S = S_0, I = I_0$ when $t = 0$.

    (d) For a certain choice of constants of proportionality in the system of part (a), we have graphed below some of the solution functions from (c) (in the SI-plane). The arrows on the graphs show how the corresponding trajectories $(S(t), I(t))$ are traced out as time increases. On each graph, the initial pair (S_0, I_0) determining the graph is its rightmost point.

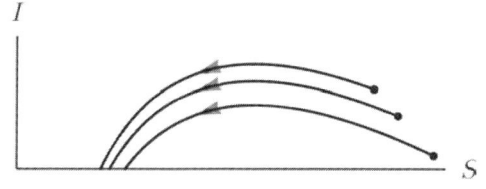

    Based on this phase portrait, which of the following statements is correct?

    The spread of the disease eventually stops because there is no one left in the susceptible compartment.

    The spread of the disease eventually stops because there is no one left in the infective compartment.

    (e) The phase portrait in (d) suggests that if a trajectory starts with a sufficiently large value of S_0 (the initial size of the susceptible population), the number of people in the infective class will increase until it reaches a peak, then decrease. Show that the value of S corresponding to this peak is the same for different trajectories, and determine this value in terms of the parameters in your original system of equations.

8.2. Vectors, vector fields, and matrices

23. Let $x = x(t)$ and $y = y(t)$ be parametric equations of a curve in the plane, and let $r = r(t)$ be the distance from $(x(t), y(t))$ to $(0,0)$, so that

(8.15) $$r(t)^2 = x(t)^2 + y(t)^2.$$

   (a) Differentiate equation (8.15) with respect to t to obtain
   $$r\frac{dr}{dt} = x\frac{dx}{dt} + y\frac{dy}{dt}.$$
   Use this to show that any solution $x = x(t), y = y(t)$ of the system
   $$\frac{dx}{dt} = y, \quad \frac{dy}{dt} = -x$$
   other than the constant solution $x(t) = 0, y(t) = 0$ lies on a circle centered at the origin.

   (b) Next consider the system
   $$\frac{dx}{dt} = y + x(x^2 + y^2), \quad \frac{dy}{dt} = -x + y(x^2 + y^2)$$
   with initial conditions $x(0) = x_0, y(0) = y_0$. Using the equation for $\frac{dr}{dt}$ from (a), show that
   $$\frac{dr}{dt} = r^3.$$
   Show that for any solution of the system other than $(0,0)$, $r(t) \to \infty$ as $t \to (2r_0^2)^{-1}$, where $r_0^2 = x_0^2 + y_0^2$.

24. The following system of two second-order equations models the oscillatory motion of two masses subject to certain spring forces, where $u_1(t)$ and $u_2(t)$ are the masses' displacements from equilibrium at time t. See Exercise 9 of Section 6.5 for details.

(8.16) $$\begin{aligned} m_1 u_1'' &= -k_1 u_1 + k_2(u_2 - u_1), \\ m_2 u_2'' &= -k_2(u_2 - u_1). \end{aligned}$$

   (a) Convert (8.16) into a system of four first-order equations with dependent variables y_1, y_2, y_3, and y_4.
   (b) Using your work from (a) and Theorem 8.1.1 show that the following initial value problem has a unique solution on $I = (-\infty, \infty)$:
   $$\begin{aligned} u_1'' &= -3u_1 + u_2, \\ u_2'' &= 2u_1 - 2u_2, \end{aligned}$$
   $u_1(0) = 0, u_1'(0) = 0, u_2(0) = 2, u_2'(0) = 0.$
   (c) Use the method of elimination to solve the initial value problem of part (b) (so that, after elimination, you'll be solving a single fourth-order equation).

8.2. Vectors, vector fields, and matrices

In the rest of this chapter we will focus on linear systems. Both computationally and conceptually, the simplest linear system is a homogeneous, constant coefficient system of two equations, which we may write as

$$\frac{dx}{dt} = ax + by,$$

$$\frac{dy}{dt} = cx + dy.$$

For a good part of this chapter we will be particularly interested in this type of system. The tools we will need for this, as well as other more general first-order linear systems, involve some basic notions from matrix theory, so we'll begin our discussion there.

Vectors in the plane. The xy-plane $\mathbb{R}^2$ is called two-dimensional Euclidean space. Its elements are ordered pairs $\mathbf{X} = (x, y)$ of real numbers. We can think of the element (x, y) of $\mathbb{R}^2$ as the location of a point in the plane or as a **vector**. *Visually*, one way to represent the vector (x, y) is by an arrow with tail at the origin $\mathbf{0} = (0, 0)$ and tip at the point (x, y); however *mathematically*, it is just the ordered pair (x, y). Sometimes we write $\mathbf{X} = (x_1, x_2)$, rather than $\mathbf{X} = (x, y)$.

Vector addition: We **add** two vectors $\mathbf{X} = (x_1, x_2)$ and $\mathbf{Y} = (y_1, y_2)$ in $\mathbb{R}^2$ by the rule

$$\mathbf{X} + \mathbf{Y} = (x_1 + y_1, x_2 + y_2).$$

Vector addition is visualized by the **parallelogram law**, which says that the arrowhead for $\mathbf{X} + \mathbf{Y}$ is at the far corner from the origin of a parallelogram, two of whose sides are arrows for $\mathbf{X}$ and $\mathbf{Y}$.

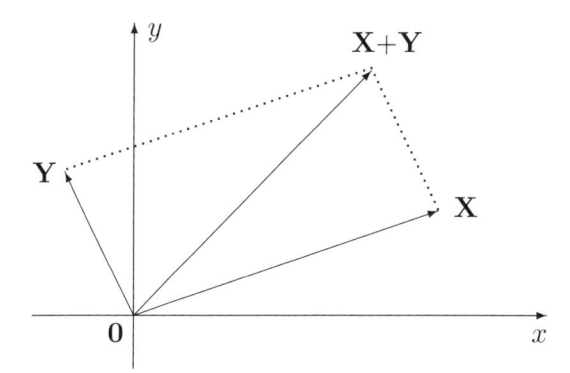

Visualizing vector addition.

Example 8.2.1. If $\mathbf{X} = (3, -2)$ and $\mathbf{Y} = (4, 4)$, then $\mathbf{X} + \mathbf{Y} = (7, 2)$. For your own amusement, draw the points $(0, 0), (3, -2), (4, 4)$, and $(7, 2)$ well enough that the parallelogram is apparent.

Scalar multiplication: You can also multiply a vector by a real number c: If $\mathbf{X} = (x_1, x_2)$,

$$c\mathbf{X} = (cx_1, cx_2).$$

If $c > 0$, the vector $c\mathbf{X}$ points in the same direction as $\mathbf{X}$ and is proportionally longer or shorter, depending on whether $c > 1$ or $0 < c < 1$. If $c < 0$, then $c\mathbf{X}$ points in the direction *opposite* to that of $\mathbf{X}$.

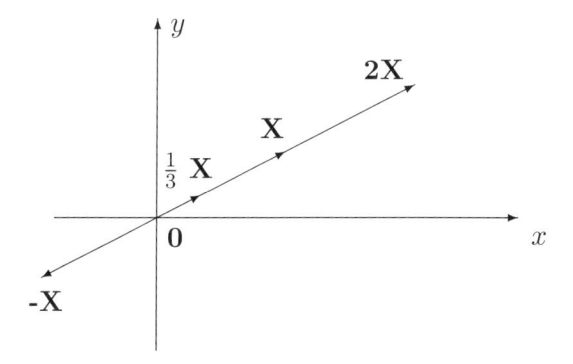

Visualizing scalar multiplication.

Vectors in higher dimensions. Three-dimensional Euclidean space $\mathbb{R}^3$ (sometimes called xyz-space) consists of all ordered triples of real numbers

$$\mathbf{X} = (x_1, x_2, x_3), \quad \text{or} \quad \mathbf{X} = (x, y, z).$$

As with $\mathbb{R}^2$, we visualize $\mathbf{X}$ as a directed line segment from the origin $(0,0,0)$ to (x,y,z). Addition is again performed by adding like coordinates: If

$$\mathbf{X} = (x_1, x_2, x_3) \quad \text{and} \quad \mathbf{Y} = (y_1, y_2, y_3),$$

then

$$\mathbf{X} + \mathbf{Y} = (x_1 + y_1, x_2 + y_2, x_3 + y_3).$$

The parallelogram law still holds: $\mathbf{0} = (0,0,0)$, $\mathbf{X}, \mathbf{Y}$, and $\mathbf{X} + \mathbf{Y}$ are the four corners of a planar parallelogram sitting in $\mathbb{R}^3$. As you would expect, we define scalar multiplication by

$$c\mathbf{X} = (cx_1, cx_2, cx_3),$$

and the geometric interpretation is the same as in $\mathbb{R}^2$.

Once we have $\mathbb{R}^2$ and $\mathbb{R}^3$, n-dimensional Euclidean space $\mathbb{R}^n$ comes easily. Its elements are points $\mathbf{X} = (x_1, x_2, \ldots, x_n)$ with n coordinates. Addition and scalar multiplication are defined coordinatewise as in $\mathbb{R}^2$ and $\mathbb{R}^3$.

The **length**, or **norm**, of a vector $\mathbf{X}$ is

$$\|\mathbf{X}\| = \sqrt{x_1^2 + x_2^2} \quad \text{(in } \mathbb{R}^2\text{)},$$

$$\|\mathbf{X}\| = \sqrt{x_1^2 + x_2^2 + x_3^2} \quad \text{(in } \mathbb{R}^3\text{)},$$

$$\|\mathbf{X}\| = \sqrt{x_1^2 + x_2^2 + \cdots + x_n^2} \quad \text{(in } \mathbb{R}^n\text{)},$$

the distance from its arrowhead to the origin.

Keep in mind that mathematically the *vector* (x, y) is merely an ordered pair with coordinates (or components) x and y. However, the vector (x, y) is also characterized by its length and direction, which, as we have indicated, are represented by an arrow with tail at the origin and tip at the point (x, y). In drawing pictures we may translate this arrow in the plane, so that its tail is at (a, b) and tip at $(a + x, b + y)$, to obtain another representation (or visualization) of the same vector. The convenience of having these different representations is illustrated, for example, when we think of a parametric curve $\mathbf{X}(t) = (x(t), y(t))$ as giving the position of a moving particle. The **velocity vector** at time t is defined by

$$\mathbf{X}'(t) = \frac{d\mathbf{X}}{dt} = \left(\frac{dx}{dt}, \frac{dy}{dt}\right) = (x'(t), y'(t)).$$

This vector points in the direction of motion of the particle along its path $\mathbf{X}(t)$ at time t, while the *length* of $\mathbf{X}'(t)$,

$$\|\mathbf{X}'(t)\| = \sqrt{\left(\frac{dx}{dt}\right)^2 + \left(\frac{dy}{dt}\right)^2},$$

is the **speed** of the particle at time t. We often draw $\mathbf{X}'(t)$ as an arrow with its tail at $\mathbf{X}(t)$, rather than at the origin, so that this translated arrow is tangent to the path of motion at the point $\mathbf{X}(t)$; see Fig. 8.9. Keep in mind that this is just a *picture*, however illuminating. The coordinates of $\mathbf{X}'(t)$ are the coordinates of the arrowhead when the tail is placed at the origin.

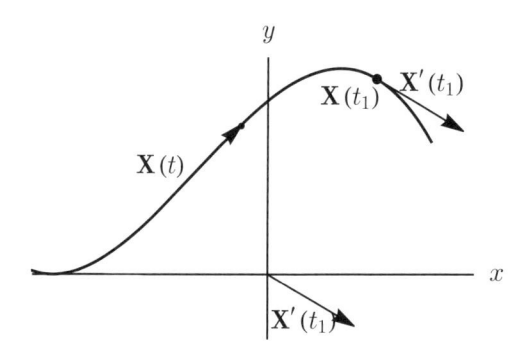

Figure 8.9. The path of motion $\mathbf{X}(t)$, the velocity vector $\mathbf{X}'(t_1)$ at time t_1, as well as one of its alternate representations (with its tail at $\mathbf{X}(t_1)$).

In the next section, we show how this way of visualizing velocity vectors can help us construct phase portraits for planar first-order systems that are autonomous.

8.2.1. Vector fields and phase portrait sketching. Consider the general autonomous system

(8.17)
$$\frac{dx}{dt} = f(x, y),$$
$$\frac{dy}{dt} = g(x, y).$$

Think of a solution of the preceding system $(x(t), y(t))$ as giving the position of a particle at time t, so that the particle traces out one of the trajectories of the system as time increases. Then the velocity of the particle at time t satisfies $(\frac{dx}{dt}, \frac{dy}{dt}) = (f(x(t), y(t)), g(x(t), y(t)))$; thus, we see that $(f(x, y), g(x, y))$ specifies the velocity vector for the particle as it passes through the point (x, y). If we plot the vector $\mathbf{F}(x, y) = (f(x, y), g(x, y))$ with its tail at (x, y), then it will be tangent at (x, y) to the trajectory traced out by our particle. Moreover, the vector $(f(x, y), g(x, y))$ will point in the direction of motion of the particle (which is the direction of the trajectory) and the length of $(f(x, y), g(x, y))$ will be the speed at which the particle is traveling when its location is (x, y).

Suppose we choose an array of points (x, y) in some region R of the phase plane and at each point we draw the vector $(f(x, y), g(x, y))$ with its tail at (x, y). Fig. 8.10 shows this for a particular choice of functions f and g. Any trajectory for the corresponding system (8.17) that passes through R must be tangent to the vectors $(f(x, y), g(x, y))$ at the chosen points (x, y).

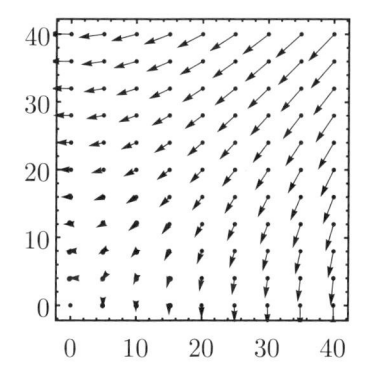

Figure 8.10. A vector field for the system (8.17), with $f(x, y) = -0.1y$, $g(x, y) = -0.1x$, and $R = \{(x, y) : 0 \leq x \leq 40, 0 \leq y \leq 40\}$.

8.2. Vectors, vector fields, and matrices

Here's a sample computation: With $f(x,y) = -0.1y$ and $g(x,y) = -0.1x$, the vector in Fig. 8.10 with tail at $(10, 20)$ is $(f(10, 20), g(10, 20)) = (-2, -1)$, consistent with the figure.

From a purely mathematical perspective, Fig. 8.10 constitutes a way of plotting the function $\mathbf{F}(x,y) = (-0.1y, -0.1x)$, which is a vector-valued function of vector variable—you plug a vector, say, $(10, 20)$, into the function $\mathbf{F}$ and get a vector, $\mathbf{F}(10, 20) = (-2, -1)$, as an output. Such vector-valued functions $\mathbf{F}$ of vector variables are called **vector fields**. Because Fig. 8.10 provides a way of visualizing the vector field $\mathbf{F}(x,y) = (-0.1y, -0.1x)$ as well as trajectories of the corresponding system, we call Fig. 8.10 a vector field for the system

(8.18) $$\frac{dx}{dt} = -0.1y, \quad \frac{dy}{dt} = -0.1x.$$

In the next figure, we have added to Fig. 8.10 a representative collection of trajectories of (8.18), guided by the vector field arrows, thereby obtaining a phase portrait for the system (8.18) with a vector field background.

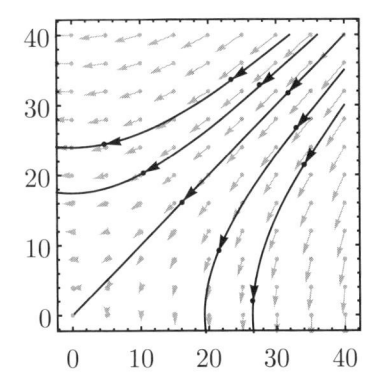

Figure 8.11. A phase portrait for the system (8.18) with a vector field for the system in the background.

Readers should recognize the system (8.18) as a possible Lanchester model for the Trafalgar battle with both fleets having equal effectiveness. Compare the phase portrait plot in Fig. 8.11 to that in Fig. 8.4.

Sketching a trajectory on a vector field plot for the system (8.17) is reminiscent of threading a solution curve through a direction field as in Chapter 1; see Fig. 1.5 in Section 1.2. However, note that the vectors in the vector field contain more information than the slope lines in a direction field: They indicate the direction (as shown by the arrowhead), as well as the speed of the motion (the length of the arrow). Likewise the trajectory $\mathbf{X}(t) = (x(t), y(t))$ tells you both x and y as functions of time t, whereas the solution curves of a single first-order equation $\frac{dy}{dx} = f(x,y)$ merely give y as a function of x.

Example 8.2.2. Consider the process of constructing a vector field for the system

(8.19) $$\frac{dx}{dt} = -4x + 3y, \quad \frac{dy}{dt} = -3x - 4y.$$

If we choose to feature a vector with tail at $(2, 2)$, then that vector would be $(-2, -14)$, while one with tail at $(-1, 1)$ would be $(7, -1)$. These vectors' lengths are so large that a representative collection of field vectors for the system (8.19) plotted over, say, the region $\{(x, y) : -2 \leq x \leq 2, -2 \leq y \leq 2\}$ would be hard to read. Thus, typically, vector fields are displayed with vectors scaled so that they do not intersect. Pictured below left is a vector field for the system (8.19) with vectors scaled to be 0.035 of their actual lengths; below right is a vector field for the system in which all nonzero vectors are scaled to have the same length.

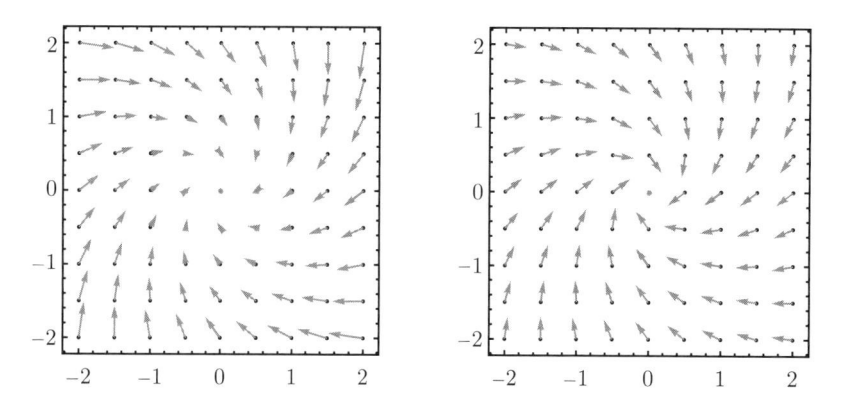

Figure 8.12. Scaled versions of the vector field for the system (8.19).

Either of the preceding two figures allows phase portrait visualization. Here's one possible such portrait:

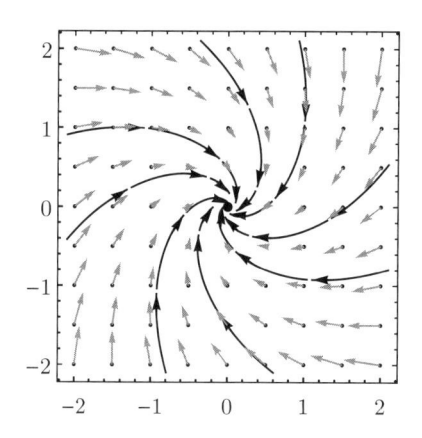

Figure 8.13. Phase portrait for (8.19) with vector field in background.

In Fig. 8.12, the point at the origin should be viewed as the tail of the zero vector, since evaluation of the vector field $\mathbf{F}(x,y) = (-4x + 3y, -3x - 4y)$ for the system at $(0,0)$ yields the vector $(0,0)$. In Fig. 8.13, we can view the point at the origin as the trajectory of the equilibrium solution $x(t) = 0, y(t) = 0$. Generalizing our terminology from Section 2.5 in the natural way, we say that the equilibrium trajectory $x(t) = 0, y(t) = 0$ is (asymptotically) stable or that the equilibrium point $(0,0)$ is (asymptotically) stable. Beginning in Section 8.4, analysis of equilibrium points will play a large role in our discussion of linear systems, and in Chapter 10, equilibrium point analysis will be a principal focus of our discussion of nonlinear systems. For now, note that **0** is always an equilibrium point for any homogeneous linear system.

8.2.2. Matrices. A matrix is a rectangular array of numbers (real or complex). We describe the size of the matrix by stating its number of rows and its number of columns; for example, the matrix

$$\begin{pmatrix} 3 & 1 & 2 & -3 \\ 4 & 0 & -1 & 0 \end{pmatrix}$$

is a 2×4 matrix. The entry in the ith row and jth column of a matrix $\mathbf{A}$ is often denoted a_{ij}. A matrix with the same number of rows as columns is called a **square matrix**.

We'll have the notions of addition for two matrices (of the same size), scalar multiplication, and matrix multiplication for two matrices of compatible sizes, as described below. In general,

8.2. Vectors, vector fields, and matrices

when **A** and **B** are both $n \times m$ matrices, their sum is the $n \times m$ matrix with ij-entry $a_{ij} + b_{ij}$. For any number s, the scalar product $s\mathbf{A}$ is the $n \times m$ matrix with ij-entry sa_{ij}.

For example, the 2×2 matrices

$$\mathbf{A} = \begin{pmatrix} a_{11} & a_{12} \\ a_{21} & a_{22} \end{pmatrix} \quad \text{and} \quad \mathbf{B} = \begin{pmatrix} b_{11} & b_{12} \\ b_{21} & b_{22} \end{pmatrix}$$

have sum

$$\mathbf{A} + \mathbf{B} = \begin{pmatrix} a_{11} + b_{11} & a_{12} + b_{12} \\ a_{21} + b_{21} & a_{22} + b_{22} \end{pmatrix},$$

and for any scalar s,

$$s\mathbf{A} = \begin{pmatrix} sa_{11} & sa_{12} \\ sa_{21} & sa_{22} \end{pmatrix}.$$

Matrix multiplication. To compute a matrix product $\mathbf{AB}$ the matrices $\mathbf{A}$ and $\mathbf{B}$ must have *compatible sizes*. If $\mathbf{A}$ is size $n \times m$, then $\mathbf{B}$ must be size $m \times p$, so that the number of columns of $\mathbf{A}$ is equal to the number of rows of $\mathbf{B}$. The product $\mathbf{AB}$ has size $n \times p$.

As a first example, we describe the product of a 2×2 matrix $\mathbf{A}$ with a 2×1 matrix $\mathbf{X}$. Here is the definition:

$$\begin{pmatrix} a & b \\ c & d \end{pmatrix} \begin{pmatrix} x \\ y \end{pmatrix} = \begin{pmatrix} ax + by \\ cx + dy \end{pmatrix}.$$

This is also called the product of the matrix $\mathbf{A}$ and the (column) vector $\mathbf{X}$. A nice visual trick for remembering this is the following. Write

$$\begin{pmatrix} a & b \\ c & d \end{pmatrix} \begin{pmatrix} x \\ y \end{pmatrix} = \begin{pmatrix} ? \\ ? \end{pmatrix}.$$

The 1-1 entry is obtained by using the first row of $\mathbf{A}$ and the column $\mathbf{X}$ by multiplying corresponding entries and adding the results:

$$\begin{pmatrix} \boxed{a \quad b} \\ c \quad d \end{pmatrix} \begin{pmatrix} \boxed{x} \\ y \end{pmatrix} = \begin{pmatrix} ax + by \\ ? \end{pmatrix}.$$

We compute the 2-1 entry similarly, using the second row of $\mathbf{A}$ and the column $\mathbf{X}$:

$$\begin{pmatrix} a \quad b \\ \boxed{c \quad d} \end{pmatrix} \begin{pmatrix} \boxed{x} \\ y \end{pmatrix} = \begin{pmatrix} ax + by \\ cx + dy \end{pmatrix}.$$

To describe the product of a 2×2 matrix $\mathbf{A}$ and a 2×2 matrix $\mathbf{B}$, set

$$\mathbf{A} = \begin{pmatrix} a_{11} & a_{12} \\ a_{21} & a_{22} \end{pmatrix} \quad \text{and} \quad \mathbf{B} = \begin{pmatrix} b_{11} & b_{12} \\ b_{21} & b_{22} \end{pmatrix}.$$

The product $\mathbf{AB}$ is defined by

$$\mathbf{AB} = \begin{pmatrix} a_{11}b_{11} + a_{12}b_{21} & a_{11}b_{12} + a_{12}b_{22} \\ a_{21}b_{11} + a_{22}b_{21} & a_{21}b_{12} + a_{22}b_{22} \end{pmatrix}$$

$$= \begin{pmatrix} c_{11} & c_{12} \\ c_{21} & c_{22} \end{pmatrix}.$$

So, for example, the 1-1 entry c_{11} of **AB** is obtained from the first row of **A** and the first column of **B**

$$\begin{pmatrix} \boxed{a_{11} \ a_{12}} \\ a_{21} \ a_{22} \end{pmatrix} \begin{pmatrix} \boxed{b_{11}} \ b_{12} \\ \boxed{b_{21}} \ b_{22} \end{pmatrix}$$

by

$$c_{11} = a_{11}b_{11} + a_{12}b_{21}.$$

Similarly, for the 1-2 entry c_{12} of the product, we use the first row of **A** and the second column of **B**, multiply corresponding elements of each, and add the results to get $c_{12} = a_{11}b_{12} + a_{12}b_{22}$:

$$\begin{pmatrix} \boxed{a_{11} \ a_{12}} \\ a_{21} \ a_{22} \end{pmatrix} \begin{pmatrix} b_{11} \ \boxed{b_{12}} \\ b_{21} \ \boxed{b_{22}} \end{pmatrix}.$$

In general, when **A** is size $n \times m$ and **B** is size $m \times p$, the ij-entry of the product **AB** is computed using the ith row of **A** and the jth column of **B** as

$$a_{i1}b_{1j} + a_{i2}b_{2j} + \cdots + a_{im}b_{mj}.$$

For example, the product

$$\begin{pmatrix} 3 & 1 & 2 \\ \boxed{4 & 0 & -1} \end{pmatrix} \begin{pmatrix} -1 & 1 & \boxed{0} \\ 2 & 3 & \boxed{-1} \\ 4 & -2 & \boxed{2} \end{pmatrix}$$

is a matrix with two rows and three columns. Its 2-3 entry, obtained from the second row of the first factor and the third column of the second factor, is equal to $(4)(0) + (0)(-1) + (-1)(2) = -2$.

Matrix multiplication is not in general commutative. For example, if

$$\mathbf{A} = \begin{pmatrix} 2 & 3 \\ 1 & -5 \end{pmatrix}, \quad \mathbf{B} = \begin{pmatrix} 4 & 0 \\ -3 & -1 \end{pmatrix},$$

then

$$\mathbf{AB} = \begin{pmatrix} -1 & -3 \\ 19 & 5 \end{pmatrix}$$

but

$$\mathbf{BA} = \begin{pmatrix} 8 & 12 \\ -7 & -4 \end{pmatrix}.$$

However, matrix multiplication is associative, so in a product of three matrices we have

$$(\mathbf{AB})\mathbf{C} = \mathbf{A}(\mathbf{BC}).$$

Associativity means we may omit the parentheses in this three factor product without ambiguity.

Some elementary properties of matrix multiplication are described in the next theorem. You are asked to prove some special cases of these in Exercise 10.

Theorem 8.2.3. *If* **A**, **B**, *and* **C** *are matrices and c is a scalar, then*

(a) $\mathbf{A}(\mathbf{B} + \mathbf{C}) = \mathbf{AB} + \mathbf{AC}$,

(b) $(\mathbf{A} + \mathbf{B})\mathbf{C} = \mathbf{AC} + \mathbf{BC}$,

(c) $\mathbf{A}(c\mathbf{B}) = c(\mathbf{AB}) = (c\mathbf{A})\mathbf{B}$

provided **A**, **B**, *and* **C** *are of compatible sizes for the indicated sums and products.*

Writing a linear system of differential equations in matrix form. The definition of matrix multiplication makes it natural to write linear differential equation systems compactly using matrix notation. The next example illustrates this.

Example 8.2.4. Consider the system
$$x'(t) = x + 12y,$$
$$y'(t) = 3x + y.$$
Set $\mathbf{A}$ to be the 2×2 matrix
$$\begin{pmatrix} 1 & 12 \\ 3 & 1 \end{pmatrix},$$
and let $\mathbf{X}(t)$ be the 2×1 matrix whose entries are the functions $x(t)$ and $y(t)$:
$$\begin{pmatrix} x(t) \\ y(t) \end{pmatrix}.$$
We have
$$\mathbf{X}'(t) = \begin{pmatrix} x'(t) \\ y'(t) \end{pmatrix},$$
and our differential equation system can be written as
$$\mathbf{X}'(t) = \mathbf{A}\mathbf{X}(t), \text{ or more briefly, } \mathbf{X}' = \mathbf{A}\mathbf{X}.$$
As we will see later, matrix tools will help us to solve such systems.

Some special matrices. A matrix (or vector) of any size that has all of its entries equal to 0 is called the **zero matrix** and is denoted $\mathbf{0}$. It should be clear from the context what the size is. The **identity matrices**
$$\mathbf{I} = \begin{pmatrix} 1 & 0 \\ 0 & 1 \end{pmatrix}, \quad \mathbf{I} = \begin{pmatrix} 1 & 0 & 0 \\ 0 & 1 & 0 \\ 0 & 0 & 1 \end{pmatrix},$$
or in general the $n \times n$ matrix with 1's on the main diagonal[7] and 0's elsewhere, have the property that $\mathbf{IA} = \mathbf{AI} = \mathbf{A}$ for any square matrix $\mathbf{A}$ of appropriate size. The **inverse** of a square matrix $\mathbf{A}$ is the unique matrix $\mathbf{A}^{-1}$ such that $\mathbf{A}^{-1}\mathbf{A} = \mathbf{A}\mathbf{A}^{-1} = \mathbf{I}$. Not every square matrix has an inverse, and we will see in Theorem 8.2.6 below that an $n \times n$ matrix $\mathbf{A}$ is invertible if and only if the determinant of $\mathbf{A}$ is not 0.

Observe that if $\mathbf{A}$ is an invertible $n \times n$ matrix and $\mathbf{B}$ is any matrix having n rows, then the matrix equation

(8.20) $$\mathbf{A}\mathbf{X} = \mathbf{B} \text{ has the unique solution } \mathbf{X} = \mathbf{A}^{-1}\mathbf{B}.$$

Determinants, invertibility, and the nullspace. Recall that the determinant of the 2×2 matrix $\mathbf{A} = \begin{pmatrix} a & b \\ c & d \end{pmatrix}$ is defined by $\det \mathbf{A} = ad - bc$. When this determinant is nonzero, the inverse of $\mathbf{A}$ is given by

(8.21) $$\mathbf{A}^{-1} = \frac{1}{\det \mathbf{A}} \begin{pmatrix} d & -b \\ -c & a \end{pmatrix}.$$

Check for yourself that $\mathbf{A}\mathbf{A}^{-1} = \mathbf{A}^{-1}\mathbf{A} = \mathbf{I}$.

The determinant of the 3×3 matrix
$$\mathbf{A} = \begin{pmatrix} a_{11} & a_{12} & a_{13} \\ a_{21} & a_{22} & a_{23} \\ a_{31} & a_{32} & a_{33} \end{pmatrix},$$

---

[7] The main diagonal of an $n \times n$ matrix consists of the entries $a_{11}, a_{22}, \ldots, a_{nn}$.

which is denoted det **A**, is defined in terms of three 2×2 determinants by a process called "expansion about the first row":

$$\det \mathbf{A} = a_{11}\det\begin{pmatrix} a_{22} & a_{23} \\ a_{32} & a_{33} \end{pmatrix} - a_{12}\det\begin{pmatrix} a_{21} & a_{23} \\ a_{31} & a_{33} \end{pmatrix} + a_{13}\det\begin{pmatrix} a_{21} & a_{22} \\ a_{31} & a_{32} \end{pmatrix}.$$

In each term on the right-hand side, an entry from the first row is multiplied by the determinant of the 2×2 matrix obtained by deleting the row and column to which the entry belongs. Note there is also a minus sign that appears with the second summand. Computing the 2×2 determinants, we see that det **A** is equal to

(8.22) $\quad a_{11}a_{22}a_{33} + a_{12}a_{23}a_{31} + a_{13}a_{21}a_{32} - a_{31}a_{22}a_{13} - a_{32}a_{23}a_{11} - a_{33}a_{21}a_{12}.$

There is a shortcut to easily remember this: Recopy the first two columns of **A** as below:

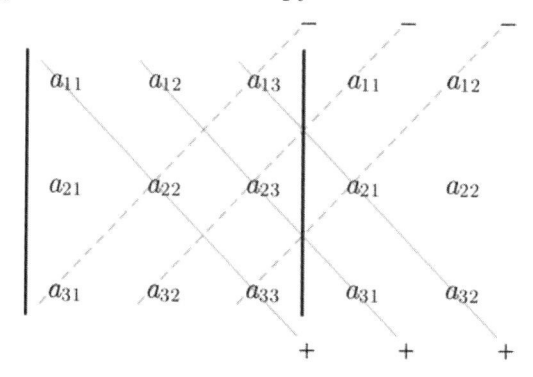

$$\begin{vmatrix} a_{11} & a_{12} & a_{13} \\ a_{21} & a_{22} & a_{23} \\ a_{31} & a_{32} & a_{33} \end{vmatrix} \begin{matrix} a_{11} & a_{12} \\ a_{21} & a_{22} \\ a_{31} & a_{32} \end{matrix}$$

The first three terms in (8.22) for the determinant are found by multiplying diagonally down from left to right and the last three subtracted terms are found by multiplying diagonally up from left to right as shown. This shortcut comes with an important *warning*: It can only be used for determinants of 3×3 matrices.

Example 8.2.5. Suppose that

$$\mathbf{A} = \begin{pmatrix} 3 & 1 & 2 \\ 4 & 0 & -1 \\ 5 & 2 & 1 \end{pmatrix}.$$

Expanding about the first row, we see that

$$\det \mathbf{A} = 3(0+2) - 1(4+5) + 2(8-0) = 13.$$

You can check that the shortcut method gives the same result.

Our definition for the determinant of a 3×3 matrix involves expanding about the first row. Determinants can also be computed by expanding about a different row or about a column. This is discussed below, in Section 8.3.1. There the method is generalized to show how to compute the determinant of a matrix of size $n \times n$, $n \geq 3$, by expanding about a row or column. While the computational work becomes greater as the size increases, the following general principle holds regardless of the size of n.

Theorem 8.2.6. *For any $n \times n$ matrix **A**, the following conditions are equivalent (this means that if any one of them holds, so do the others):*

(a) $\det \mathbf{A} \neq 0$;

(b) $\mathbf{A}^{-1}$ *exists*;

(c) *the only vector* **X** *with* $\mathbf{AX} = \mathbf{0}$ *is* $\mathbf{X} = \mathbf{0}$.

8.2. Vectors, vector fields, and matrices

Proof. We'll give the proof for the case of a 2×2 matrix $\mathbf{A} = \begin{pmatrix} a & b \\ c & d \end{pmatrix}$. Let's use the notation $P \Rightarrow Q$ for "P implies Q", that is, "if P, then Q".

First we show (a) $\Rightarrow$ (b). If $\det \mathbf{A} = ad - bc \neq 0$, then we have already observed (see (8.21)) that the matrix defined by

$$\frac{1}{ad - bc} \begin{pmatrix} d & -b \\ -c & a \end{pmatrix}$$

is $\mathbf{A}^{-1}$. Thus (a) $\Rightarrow$ (b).

For the implication (b) $\Rightarrow$ (c) we argue as follows. Assume $\mathbf{A}^{-1}$ exists. If $\mathbf{AX} = \mathbf{0}$, then

$$\begin{aligned} \mathbf{X} &= \mathbf{IX} \\ &= (\mathbf{A}^{-1} \mathbf{A}) \mathbf{X} \\ &= \mathbf{A}^{-1} (\mathbf{AX}) \\ &= \mathbf{A}^{-1} \mathbf{0} \\ &= \mathbf{0}, \end{aligned}$$

and we have shown that (b) $\Rightarrow$ (c).

To prove that (c) $\Rightarrow$ (a) we may show the equivalent implication **not** (a) $\Rightarrow$ **not** (c): Here we are using the fact that "if P, then Q" means the same thing as "if not Q, then not P".[8] If **not** (a) holds, that is, if $ad - bc = 0$, we consider the vectors $\begin{pmatrix} d \\ -c \end{pmatrix}$ and $\begin{pmatrix} b \\ -a \end{pmatrix}$. Check that $\mathbf{A}$ multiplies both of these vectors to $\mathbf{0}$:

$$\begin{pmatrix} a & b \\ c & d \end{pmatrix} \begin{pmatrix} d \\ -c \end{pmatrix} = \begin{pmatrix} ad - bc \\ cd - dc \end{pmatrix} = \begin{pmatrix} 0 \\ 0 \end{pmatrix}; \quad \begin{pmatrix} a & b \\ c & d \end{pmatrix} \begin{pmatrix} b \\ -a \end{pmatrix} = \begin{pmatrix} ab - ba \\ bc - ad \end{pmatrix} = \begin{pmatrix} 0 \\ 0 \end{pmatrix}.$$

If one of these two vectors is nonzero, **not** (c) holds and we're done. If both are $\mathbf{0}$, then $a = b = c = d = 0$, so $\mathbf{A} = \mathbf{0}$, and *any* vector $\mathbf{X}$ satisfies $\mathbf{AX} = \mathbf{0}$. Therefore **not** (c) holds in this case, too.

To summarize, we have proved that $(a) \Rightarrow (b) \Rightarrow (c) \Rightarrow (a)$, so each condition implies the others. $\square$

The proof of Theorem 8.2.6 in the case of $n \times n$ matrices with $n > 2$ is more complicated. This is something you will see in a linear algebra course.

Condition (c) of the theorem leads us to the next definition.

Definition 8.2.7. Given a matrix $\mathbf{A}$, the set of vectors $\mathbf{X}$ with $\mathbf{AX} = \mathbf{0}$ is called the **nullspace** of $\mathbf{A}$ and is denoted Null $\mathbf{A}$.

The nullspace always contains the zero vector. Notice that condition (c) in Theorem 8.2.6 can be rephrased as Null $\mathbf{A} = \{\mathbf{0}\}$. The meaning of "null" in nullspace should be clear: Null $\mathbf{A}$ consists of those vectors that are "nullified" when multiplied on the left by $\mathbf{A}$.

A shortcut for 2×2 matrices. For a 2×2 matrix $\mathbf{A}$ with determinant zero, it is especially easy to write down some nonzero vectors in Null $\mathbf{A}$. This will be helpful in Section 8.3 when we discuss computing the eigenvectors of $\mathbf{A}$. The idea is contained in the proof of the theorem above. Specifically, if

$$\mathbf{A} = \begin{pmatrix} a & b \\ c & d \end{pmatrix}$$

---

[8]For example, "If it is raining, then it is cloudy," means the same thing as "If it is not cloudy, then it is not raining."

and det $\mathbf{A} = 0$, we saw in the proof of "not (a) $\Rightarrow$ not (c)" that

$$\begin{pmatrix} b \\ -a \end{pmatrix} \quad \text{and} \quad \begin{pmatrix} d \\ -c \end{pmatrix} \tag{8.23}$$

lie in Null $\mathbf{A}$. This is worth remembering! As long as not every entry of $\mathbf{A}$ is 0, at least one of these two vectors will be a nonzero vector. If every entry of $\mathbf{A}$ is zero, then *every* vector $\begin{pmatrix} x \\ y \end{pmatrix}$ lies in Null $\mathbf{A}$.

Finding nonzero vectors in Null $\mathbf{A}$ when $\mathbf{A}$ is a 3×3 or larger matrix takes more work. We will see some examples later. Always remember that Null $\mathbf{A}$ contains only the zero vector if det $\mathbf{A} \neq 0$.

The next definition describes linear independence for a collection of vectors.

Definition 8.2.8. A collection of k vectors of size $n \times 1$, $\mathbf{V}_1, \mathbf{V}_2, \ldots, \mathbf{V}_k$, is said to be **linearly independent** if the only linear combination

$$c_1 \mathbf{V}_1 + c_2 \mathbf{V}_2 + \cdots + c_k \mathbf{V}_k$$

which gives the zero vector $\mathbf{0}$ is the trivial one $c_1 = c_2 = \cdots = c_k = 0$.

When there are only two vectors $\mathbf{V}_1$ and $\mathbf{V}_2$ in question, linear independence of $\mathbf{V}_1$ and $\mathbf{V}_2$ is equivalent to neither vector being a scalar multiple of the other.

Readers should recognize that for collections of vectors in $\mathbb{R}^n$, both the definition of linear independence and the notion of linear combination are the natural analogues of those given earlier for collections of functions defined on an interval I (see Section 4.6.2).

A new perspective on matrix-vector multiplication. If we let $\mathbf{V}_j$ be the jth column of the $n \times n$ matrix $\mathbf{A}$,

$$\mathbf{V}_j = \begin{pmatrix} a_{1j} \\ a_{2j} \\ \vdots \\ a_{nj} \end{pmatrix},$$

then the product

$$\mathbf{A} \begin{pmatrix} c_1 \\ c_2 \\ \vdots \\ c_n \end{pmatrix} = \begin{pmatrix} \uparrow & \uparrow & & \uparrow \\ \mathbf{V}_1 & \mathbf{V}_2 & \cdots & \mathbf{V}_n \\ \downarrow & \downarrow & & \downarrow \end{pmatrix} \begin{pmatrix} c_1 \\ c_2 \\ \vdots \\ c_n \end{pmatrix}$$

is equal to

$$c_1 \mathbf{V}_1 + c_2 \mathbf{V}_2 + \cdots + c_n \mathbf{V}_n \tag{8.24}$$

(see Exercise 7). In words, this says that the product

$$\mathbf{A} \begin{pmatrix} c_1 \\ c_2 \\ \vdots \\ c_n \end{pmatrix}$$

gives a linear combination of the columns of $\mathbf{A}$.

8.2. Vectors, vector fields, and matrices

As a special case of (8.24), note that if $\mathbf{E}_j$ is the $n \times 1$ matrix with a "1" in the jth position and 0's elsewhere, then

(8.25) $\qquad\qquad \mathbf{A}\mathbf{E}_j = \mathbf{V}_j$, the jth column of $\mathbf{A}$.

We will use this fact frequently.

Rephrasing condition (c) of Theorem 8.2.6. This new perspective on matrix-vector multiplication will be quite useful. For example, it shows that the linear combination in (8.24) is equal to the zero vector, i.e., that

$$c_1 \mathbf{V}_1 + c_2 \mathbf{V}_2 + \cdots + c_n \mathbf{V}_n = \mathbf{0}$$

if and only if

$$\mathbf{A} \begin{pmatrix} c_1 \\ c_2 \\ \vdots \\ c_n \end{pmatrix} = \begin{pmatrix} 0 \\ 0 \\ \vdots \\ 0 \end{pmatrix},$$

where $\mathbf{A}$ is the matrix with columns $\mathbf{V}_1, \mathbf{V}_2, \ldots, \mathbf{V}_n$. Thus we can rephrase condition (c) of Theorem 8.2.6 as

The vectors formed from the columns of $\mathbf{A}$ are linearly independent.

For future reference, we provide the following expanded version of Theorem 8.2.6, which incorporates the preceding observation as well as our earlier observation (8.20) concerning the role of inverse matrices in solving equations. Note that observation (8.20) shows that (b) implies (c), while (d) follows from (c) by choosing $\mathbf{V} = \mathbf{0}$.

Theorem 8.2.9. *For any $n \times n$ matrix $\mathbf{A}$, the following conditions are equivalent:*

(a) $\det \mathbf{A} \neq 0$;

(b) $\mathbf{A}^{-1}$ *exists*;

(c) *for every $n \times 1$ vector $\mathbf{V}$, the matrix equation $\mathbf{A}\mathbf{X} = \mathbf{V}$ has a unique solution;*

(d) *the only vector $\mathbf{X}$ with $\mathbf{A}\mathbf{X} = \mathbf{0}$ is $\mathbf{X} = \mathbf{0}$;*

(e) *the columns of $\mathbf{A}$ are linearly independent.*

The following corollary is a useful consequence.

Corollary 8.2.10. *Suppose $\mathbf{V}_1, \mathbf{V}_2, \ldots, \mathbf{V}_n$ are n linearly independent vectors in $\mathbb{R}^n$. Then every vector $\mathbf{X}$ in $\mathbb{R}^n$ can be written as a linear combination of $\mathbf{V}_1, \mathbf{V}_2, \ldots, \mathbf{V}_n$:*

$$\mathbf{X} = c_1 \mathbf{V}_1 + c_2 \mathbf{V}_2 + \cdots + c_n \mathbf{V}_n.$$

Moreover, the scalars $c_1, c_2, \ldots, c_n$ are unique.

Proof. Let $\mathbf{A}$ be the $n \times n$ matrix whose columns are $\mathbf{V}_1, \mathbf{V}_2, \ldots, \mathbf{V}_n$. Since the vectors $\mathbf{V}_1, \mathbf{V}_2, \ldots, \mathbf{V}_n$ are linearly independent, Theorem 8.2.9 tells us that $\mathbf{A}^{-1}$ exists. Given any vector $\mathbf{X}$ in $\mathbb{R}^n$, we have

$$\mathbf{X} = \mathbf{I}\mathbf{X} = (\mathbf{A}\mathbf{A}^{-1})\mathbf{X} = \mathbf{A}(\mathbf{A}^{-1}\mathbf{X}).$$

If $c_1, c_2, \ldots, c_n$ are the coordinates of $\mathbf{A}^{-1}\mathbf{X}$, so that

$$\mathbf{A}^{-1}\mathbf{X} = \begin{pmatrix} c_1 \\ c_2 \\ \vdots \\ c_n \end{pmatrix},$$

then our expression for $\mathbf{X}$ becomes

$$\mathbf{X} = \mathbf{A} \begin{pmatrix} c_1 \\ c_2 \\ \vdots \\ c_n \end{pmatrix} = c_1 \mathbf{V}_1 + c_2 \mathbf{V}_2 + \cdots + c_n \mathbf{V}_n,$$

where the last equality follows from the discussion preceding (8.24). Thus $\mathbf{X}$ can be written as a linear combination of the vectors $\mathbf{V}_j$.

For the uniqueness statement, note that if $\mathbf{X} = b_1 \mathbf{V}_1 + b_2 \mathbf{V}_2 + \cdots + b_n \mathbf{V}_n$ is another representation of $\mathbf{X}$, then

$$(c_1 - b_1)\mathbf{V}_1 + (c_2 - b_2)\mathbf{V}_2 + \cdots + (c_n - b_n)\mathbf{V}_n = \mathbf{X} - \mathbf{X} = \mathbf{0},$$

and $c_1 - b_1, c_2 - b_2, \ldots, c_n - b_n$ are all zero by the linear independence of the $\mathbf{V}_j$'s. $\square$

8.2.3. Exercises.

1. For
$$\mathbf{A} = \begin{pmatrix} 2 & 1 & -3 \\ 0 & -2 & 4 \\ -2 & 3 & 0 \end{pmatrix} \quad \text{and} \quad \mathbf{B} = \begin{pmatrix} 1 & -1 & 4 \\ 5 & 0 & 1 \\ -2 & -3 & -1 \end{pmatrix}$$
compute
   (a) $\mathbf{A} + 2\mathbf{B}$,
   (b) $\mathbf{AB}$,
   (c) $\mathbf{BA}$,
   (d) $\mathbf{A}^2 = \mathbf{AA}$.

2. If
$$\mathbf{A} = \begin{pmatrix} 3 & 2 & 1 \\ 0 & 2 & 4 \\ 4 & 2 & 3 \end{pmatrix} \quad \text{and} \quad \mathbf{B} = \begin{pmatrix} 1 & 1 \\ -1 & 2 \\ 0 & -3 \end{pmatrix},$$
only one of the products $\mathbf{AB}$ and $\mathbf{BA}$ is defined. Decide which one, and then compute it.

3. Compute the inverse, if it exists, of each of the following matrices:
   (a)
   $$\mathbf{A} = \begin{pmatrix} 1 & 4 \\ 3 & -2 \end{pmatrix},$$
   (b)
   $$\mathbf{A} = \begin{pmatrix} 2 & 6 \\ 1 & 3 \end{pmatrix}.$$

4. Find examples of 2×2 matrices $\mathbf{A}$ and $\mathbf{B}$ with $\mathbf{AB} = \mathbf{0}$ but neither $\mathbf{A} = \mathbf{0}$ nor $\mathbf{B} = \mathbf{0}$.

5. Suppose that $\mathbf{A}$ is a 2×2 matrix with the products
$$\mathbf{A}\begin{pmatrix} 1 \\ 1 \end{pmatrix} = \begin{pmatrix} -2 \\ 3 \end{pmatrix} \quad \text{and} \quad \mathbf{A}\begin{pmatrix} 2 \\ 0 \end{pmatrix} = \begin{pmatrix} 3 \\ 4 \end{pmatrix}.$$
   (a) Using the fact that
   $$\begin{pmatrix} 1 \\ 3 \end{pmatrix} = 3\begin{pmatrix} 1 \\ 1 \end{pmatrix} - \begin{pmatrix} 2 \\ 0 \end{pmatrix}$$
   find
   $$\mathbf{A}\begin{pmatrix} 1 \\ 3 \end{pmatrix}.$$

8.2. Vectors, vector fields, and matrices

(b) Find constants c_1, c_2 so that
$$\begin{pmatrix} 5 \\ 2 \end{pmatrix} = c_1 \begin{pmatrix} 1 \\ 1 \end{pmatrix} + c_2 \begin{pmatrix} 2 \\ 0 \end{pmatrix},$$
and then compute
$$\mathbf{A} \begin{pmatrix} 5 \\ 2 \end{pmatrix}.$$

6. Let θ be a real number. There is a 2×2 matrix $\mathbf{A}$ having the property that for each 2×1 vector $\mathbf{V}$, the vector $\mathbf{AV}$ is the 2×1 vector obtained by rotating $\mathbf{V}$ by θ radians about the origin (in the counterclockwise direction).

   (a) Find $\mathbf{A} \begin{pmatrix} 1 \\ 0 \end{pmatrix}$ and $\mathbf{A} \begin{pmatrix} 0 \\ 1 \end{pmatrix}$.

   (b) Using $\begin{pmatrix} x \\ y \end{pmatrix} = x \begin{pmatrix} 1 \\ 0 \end{pmatrix} + y \begin{pmatrix} 0 \\ 1 \end{pmatrix}$, find $\mathbf{A} \begin{pmatrix} x \\ y \end{pmatrix}$.

   (c) What is the matrix $\mathbf{A}$?

7. Show that if the n-dimensional vectors $\mathbf{V}_1, \mathbf{V}_2, \ldots, \mathbf{V}_n$ are used as the columns of a matrix $\mathbf{A}$, then the matrix product
$$\mathbf{A} \begin{pmatrix} c_1 \\ c_2 \\ \vdots \\ c_n \end{pmatrix}$$
is equal to $c_1 \mathbf{V}_1 + c_2 \mathbf{V}_2 + \cdots + c_n \mathbf{V}_n$.

8. Verify that for any 2×2 invertible matrix $\mathbf{A}$,
$$\det(\mathbf{A}^{-1}) = \frac{1}{\det \mathbf{A}}.$$

9. For what values of t is the matrix-valued function (that is, a matrix whose entries are functions)
$$\mathbf{A}(t) = \begin{pmatrix} t & t^2 \\ -2t & 3t^3 \end{pmatrix}$$
invertible? Determine the inverse for such t.

10. (a) If $\mathbf{A}, \mathbf{B},$ and $\mathbf{C}$ are $n \times n$ matrices, then Theorem 8.2.3 says that
$$\mathbf{A}(\mathbf{B} + \mathbf{C}) = \mathbf{AB} + \mathbf{AC}.$$
Prove this in the case $n = 2$.

    (b) If s is a scalar, $\mathbf{A}$ is an $n \times n$ matrix, and $\mathbf{X}$ is an $n \times 1$ matrix, then
$$s(\mathbf{AX}) = \mathbf{A}(s\mathbf{X}) = (s\mathbf{A})\mathbf{X}.$$
Prove this when $n = 2$.

11. **Cramer's Rule.** In this exercise you will use equation (8.21) for the inverse of a 2×2 invertible matrix to verify Cramer's Rule for solving a pair of linear equations
$$ax + by = k_1,$$
$$cx + dy = k_2$$
for the unknowns x and y.

    (a) Write the equations in matrix form as
$$(8.26) \qquad \begin{pmatrix} a & b \\ c & d \end{pmatrix} \begin{pmatrix} x \\ y \end{pmatrix} = \begin{pmatrix} k_1 \\ k_2 \end{pmatrix}.$$

Assuming that
$$\det \mathbf{A} = \det \begin{pmatrix} a & b \\ c & d \end{pmatrix} \neq 0$$
multiply both sides of equation (8.26) (on the left!) by $\mathbf{A}^{-1}$ to obtain
$$\begin{pmatrix} x \\ y \end{pmatrix} = \frac{1}{ad - bc} \begin{pmatrix} dk_1 - bk_2 \\ -ck_1 + ak_2 \end{pmatrix}.$$

(b) Using the result of (a) verify that
$$x = \frac{dk_1 - bk_2}{ad - bc} = \frac{\det \begin{pmatrix} k_1 & b \\ k_2 & d \end{pmatrix}}{\det \mathbf{A}}$$
and
$$y = \frac{ak_2 - ck_1}{ad - bc} = \frac{\det \begin{pmatrix} a & k_1 \\ c & k_2 \end{pmatrix}}{\det \mathbf{A}}.$$
This is the formula of Cramer's Rule.

12. (CAS)
    (a) Sketch a vector field for the system
    $$\frac{dx}{dt} = -y, \quad \frac{dy}{dt} = x$$
    featuring vectors with tails at (i, j), where i and j are integers satisfying $-1 \leq i, j \leq 1$ and where each vector is scaled to have length $1/2$.
    (b) Use your answer in (a) to sketch the trajectory of the solution to the system in (a) that satisfies the initial conditions $x(0) = 1$, $y(0) = 0$.
    (c) Using the method of elimination (see Example 8.1.5 in Section 8.1), find a formula for the trajectory of (b) and plot it on your vector field sketch.

13. (CAS) Using any technology at your disposal, produce a vector field plot for
    $$\frac{dx}{dt} = -x - y, \quad \frac{dy}{dt} = x - y$$
    over the region $\{(x, y) : -1 \leq x \leq 1, -1 \leq y \leq 1\}$. Is your field consistent with your work for Exercise 4 of Section 8.1?

14. Consider the first-order system of equations
    $$x' - y' - x + y = 0,$$
    $$x' + y' - 3x - 3y = 0.$$
    Using matrix techniques, rewrite the preceding system in matrix form
    $$\begin{pmatrix} x' \\ y' \end{pmatrix} = \mathbf{A} \begin{pmatrix} x \\ y \end{pmatrix},$$
    where $\mathbf{A}$ is a 2×2 matrix.

8.3. Eigenvalues and eigenvectors

In Example 8.1.5, we used the method of elimination to show that the system $\mathbf{X}'(t) = \mathbf{A}\mathbf{X}$, where
$$\mathbf{A} = \begin{pmatrix} 3 & 2 \\ 1 & 2 \end{pmatrix},$$

8.3. Eigenvalues and eigenvectors

has solutions
$$\mathbf{X}(t) = c_1 e^{4t} \begin{pmatrix} 2 \\ 1 \end{pmatrix} + c_2 e^{t} \begin{pmatrix} -1 \\ 1 \end{pmatrix}$$
for arbitrary constants c_1 and c_2. It is natural to wonder if there is a connection between the matrix $\mathbf{A}$ and the ingredients in our solution, namely the vectors $\begin{pmatrix} 2 \\ 1 \end{pmatrix}$ and $\begin{pmatrix} -1 \\ 1 \end{pmatrix}$ and the scalars 4 and 1. Eigenvalues and eigenvectors will provide this connection.

The concept of eigenvalues and eigenvectors is ubiquitous in mathematics and has applications in such diverse fields as finance, engineering, and quantum mechanics. For us, it will be the key tool in solving certain linear systems of differential equations.

The idea of eigenvalues and eigenvectors is simple enough. Fix an $n \times n$ matrix $\mathbf{A}$, and think of "multiplication by $\mathbf{A}$" as an operator on the set of $n \times 1$ vectors (column matrices) that transforms an input vector $\mathbf{V}$ into the output vector $\mathbf{AV}$:
$$\mathbf{V} \to \mathbf{AV},$$
or pictorially

$$\mathbf{V} \longrightarrow \boxed{\mathbf{A} \times} \longrightarrow \mathbf{AV}$$

The input and output are *vectors*.

When is there a vector $\mathbf{V}$ and a scalar r (real or complex) so that $\mathbf{AV} = r\mathbf{V}$ (so that the output vector is just a scalar multiple of the input vector)? Of course if $\mathbf{V} = \mathbf{0}$, the vector of all 0's, then $\mathbf{AV} = r\mathbf{V}$ for *any* choice of r. So we should modify our question to ask, "When is there a **nonzero** vector $\mathbf{V}$ and scalar r so that $\mathbf{AV} = r\mathbf{V}$?" This leads to the next definition.

Definition 8.3.1. A real or complex number r is an **eigenvalue** for the matrix $\mathbf{A}$ if there exists a nonzero vector $\mathbf{V}$ (meaning that not all entries of $\mathbf{V}$ are 0) with
$$\mathbf{AV} = r\mathbf{V},$$
or equivalently,
$$(\mathbf{A} - r\mathbf{I})\mathbf{V} = \mathbf{0}.$$
A nonzero vector $\mathbf{V}$ satisfying $\mathbf{AV} = r\mathbf{V}$ is called an **eigenvector** corresponding to the eigenvalue r.

The "or equivalently" part of the definition may need some explanation. Write $\mathbf{AV} = r\mathbf{V}$ as $\mathbf{AV} - r\mathbf{V} = \mathbf{0}$. Now we can't quite factor $\mathbf{V}$ out of this equation, since $(\mathbf{A} - r)\mathbf{V} = \mathbf{0}$ doesn't make sense (because $\mathbf{A}$ is a matrix and r is a scalar, so the difference $\mathbf{A} - r$ isn't defined). Instead, we write $\mathbf{AV} - r\mathbf{V} = \mathbf{0}$ as $\mathbf{AV} - r\mathbf{IV} = \mathbf{0}$, where $\mathbf{I}$ is the appropriately sized identity matrix, and *then* we can factor to obtain $(\mathbf{A} - r\mathbf{I})\mathbf{V} = \mathbf{0}$ as desired.

Example 8.3.2. The vector
$$\mathbf{V}_1 = \begin{pmatrix} 2 \\ 1 \end{pmatrix}$$
is an eigenvector of
$$\mathbf{A} = \begin{pmatrix} 3 & 2 \\ 1 & 2 \end{pmatrix}.$$

What is the corresponding eigenvalue? Since

$$\mathbf{A}\mathbf{V}_1 = \begin{pmatrix} 3 & 2 \\ 1 & 2 \end{pmatrix} \begin{pmatrix} 2 \\ 1 \end{pmatrix} = \begin{pmatrix} 8 \\ 4 \end{pmatrix} = 4\mathbf{V}_1,$$

the eigenvalue is 4. You can check that

$$\mathbf{V}_2 = \begin{pmatrix} -1 \\ 1 \end{pmatrix}$$

is also an eigenvector of $\mathbf{A}$ with corresponding eigenvalue 1.

The matrix $\mathbf{A}$ in the preceding example is the matrix of coefficients of the system of Example 8.1.5, which we solved by the method of elimination. We now see that the solution of that system takes the form

$$\mathbf{X}(t) = c_1 e^{4t} \mathbf{V}_1 + c_2 e^{t} \mathbf{V}_2,$$

where $\mathbf{V}_1$ is an eigenvector for $\mathbf{A}$ with corresponding eigenvalue 4 and $\mathbf{V}_2$ is an eigenvector for $\mathbf{A}$ with corresponding eigenvalue 1. This observation suggests that we should be able to solve at least some constant coefficient linear systems using eigenvalues of and eigenvectors of coefficient matrices. We now discuss how to find eigenvalues and eigenvectors.

Finding eigenvalues and eigenvectors. On combining the definition of eigenvalue and eigenvector with Theorem 8.2.9, we arrive at the following important fact:

Theorem 8.3.3. *For any $n \times n$ matrix $\mathbf{A}$ and any real or complex number r, the following conditions are equivalent.*

(a) *r is an eigenvalue of $\mathbf{A}$.*

(b) *$\mathbf{A} - r\mathbf{I}$ is not invertible.*

(c) *$\det(\mathbf{A} - r\mathbf{I}) = 0$.*

Moreover, the eigenvectors corresponding to the eigenvalue r are the nonzero vectors in the nullspace of $\mathbf{A} - r\mathbf{I}$.

Notice that eigenvectors for a given eigenvalue are *not unique*! If $\mathbf{V}$ is an eigenvector, with eigenvalue r, and $c \neq 0$, then $c\mathbf{V}$ is also an eigenvector with the same eigenvalue r; see Exercise 10.

One way to *find* the eigenvalues of a matrix $\mathbf{A}$ is to use condition (c) of Theorem 8.3.3. For example, for a 2×2 matrix,

$$\mathbf{A} = \begin{pmatrix} a & b \\ c & d \end{pmatrix},$$

we compute

$$\begin{aligned} \det(\mathbf{A} - z\mathbf{I}) &= \det \begin{pmatrix} a - z & b \\ c & d - z \end{pmatrix} \\ &= (a-z)(d-z) - bc \\ &= z^2 - (a+d)z + (ad - bc). \end{aligned}$$

The eigenvalues of $\mathbf{A}$ are the solutions (real or complex) to the quadratic equation

$$z^2 - (a+d)z + (ad - bc) = 0.$$

Definition 8.3.4. For an $n \times n$ matrix $\mathbf{A}$, we define the **characteristic polynomial** of $\mathbf{A}$ to be

$$p(z) = \det(\mathbf{A} - z\mathbf{I}).$$

8.3. Eigenvalues and eigenvectors

By Theorem 8.3.3, we find the eigenvalues by solving the **characteristic equation**

$$\det(\mathbf{A} - z\mathbf{I}) = 0$$

for z. That is, the eigenvalues are exactly the roots of the characteristic polynomial.

Example 8.3.5. Find the eigenvalues and eigenvectors of

$$\mathbf{A} = \begin{pmatrix} 1 & 12 \\ 3 & 1 \end{pmatrix}.$$

We have

$$\begin{aligned} p(z) &= \det(\mathbf{A} - z\mathbf{I}) = \det\begin{pmatrix} 1 - z & 12 \\ 3 & 1 - z \end{pmatrix} = (1-z)^2 - 36 \\ &= z^2 - 2z - 35 = (z-7)(z+5). \end{aligned}$$

Thus $\mathbf{A}$ has two eigenvalues $r_1 = 7$ and $r_2 = -5$.

Next, let's find the eigenvectors for the eigenvalue $r_1 = 7$. We'll show two slightly different methods for this, each having its own advantage.

Method 1. By the definition of "eigenvector", we are looking for a (nonzero) vector $\mathbf{V} = \begin{pmatrix} v_1 \\ v_2 \end{pmatrix}$ so that

$$\mathbf{AV} = 7\mathbf{V},$$

or

$$\begin{pmatrix} 1 & 12 \\ 3 & 1 \end{pmatrix}\begin{pmatrix} v_1 \\ v_2 \end{pmatrix} = 7\begin{pmatrix} v_1 \\ v_2 \end{pmatrix}.$$

This says that we want

$$v_1 + 12v_2 = 7v_1 \quad \text{and} \quad 3v_1 + v_2 = 7v_2.$$

Rewrite this pair of equations as

$$-6v_1 + 12v_2 = 0 \tag{8.27}$$

and

$$3v_1 - 6v_2 = 0. \tag{8.28}$$

Since equation (8.27) is a multiple of equation (8.28), these equations tell us exactly the same thing, precisely that $v_1 = 2v_2$. Choosing $v_2 = 1$, we see that

$$\mathbf{V}_1 = \begin{pmatrix} 2 \\ 1 \end{pmatrix}$$

is an eigenvector for the eigenvalue $r_1 = 7$, as is any nonzero scalar multiple.

Method 2. Using $r_1 = 7$, we have

$$\begin{aligned} \mathbf{A} - r_1\mathbf{I} &= \mathbf{A} - 7\mathbf{I} \\ &= \begin{pmatrix} 1 & 12 \\ 3 & 1 \end{pmatrix} - \begin{pmatrix} 7 & 0 \\ 0 & 7 \end{pmatrix} \\ &= \begin{pmatrix} -6 & 12 \\ 3 & -6 \end{pmatrix}. \end{aligned}$$

By Theorem 8.3.3, the eigenvectors for $r_1 = 7$ are the nonzero vectors in the nullspace of $\mathbf{A} - 7\mathbf{I}$. Since this is a 2×2 matrix, the "shortcut" (8.23) tells us that corresponding to the eigenvalue

$r_1 = 7$ we have eigenvectors $\begin{pmatrix} b \\ -a \end{pmatrix} = \begin{pmatrix} 12 \\ 6 \end{pmatrix}$ and $\begin{pmatrix} d \\ -c \end{pmatrix} = \begin{pmatrix} -6 \\ -3 \end{pmatrix}$. These are multiples of each other and of $\begin{pmatrix} 2 \\ 1 \end{pmatrix}$. Choosing the "simplest" one, we identify $\mathbf{V}_1 = \begin{pmatrix} 2 \\ 1 \end{pmatrix}$ as an eigenvector for the eigenvalue $r_1 = 7$.

Comparing these two methods, perhaps the second one is a bit quicker. However the first method has a built-in check on our work: If the pair of equations (8.27)–(8.28) were **not** multiples of each other, the only solution would be $v_1 = v_2 = 0$, and we would know we had made an error, either in the computation of the eigenvalues or in setting up to find the eigenvectors. If you use the second method, you can similarly check your work by making sure that the determinant of $\mathbf{A} - r_1\mathbf{I}$ really is 0. Of course, the final answer is easy to check directly:

$$\mathbf{A}\begin{pmatrix} 2 \\ 1 \end{pmatrix} = \begin{pmatrix} 14 \\ 7 \end{pmatrix} = 7\begin{pmatrix} 2 \\ 1 \end{pmatrix}.$$

Finally, we find the eigenvectors for $r_2 = -5$. Now

$$\mathbf{A} - r_2\mathbf{I} = \mathbf{A} + 5\mathbf{I}$$

$$= \begin{pmatrix} 1 & 12 \\ 3 & 1 \end{pmatrix} + \begin{pmatrix} 5 & 0 \\ 0 & 5 \end{pmatrix} = \begin{pmatrix} 6 & 12 \\ 3 & 6 \end{pmatrix}.$$

Notice that $\det(\mathbf{A} - r_2\mathbf{I}) = 0$, as must be the case. By (8.23) we see that $\begin{pmatrix} d \\ -c \end{pmatrix} = \begin{pmatrix} 6 \\ -3 \end{pmatrix}$ and $\begin{pmatrix} b \\ -a \end{pmatrix} = \begin{pmatrix} 12 \\ -6 \end{pmatrix}$ are eigenvectors. These are multiples of each other and of $\begin{pmatrix} 2 \\ -1 \end{pmatrix}$. We may choose $\mathbf{V}_2 = \begin{pmatrix} 2 \\ -1 \end{pmatrix}$ as an eigenvector corresponding to the eigenvalue $r_2 = -5$.

In the next example, we find the eigenvalues and eigenvectors of a 3×3 matrix.

Example 8.3.6. Set

$$\mathbf{A} = \begin{pmatrix} 3 & 1 & 1 \\ 2 & 4 & 2 \\ -1 & -1 & 1 \end{pmatrix}.$$

After some computation (see Exercise 7), the characteristic equation $\det(\mathbf{A} - z\mathbf{I}) = 0$ can be written as

$$-(z-4)(z-2)^2 = 0.$$

Thus $\mathbf{A}$ has eigenvalues $r_1 = 4$ and $r_2 = 2$. The matrix $\mathbf{A} - 4\mathbf{I}$ is

$$\mathbf{A} = \begin{pmatrix} -1 & 1 & 1 \\ 2 & 0 & 2 \\ -1 & -1 & -3 \end{pmatrix}.$$

An eigenvector for the eigenvalue 4 is any nonzero vector in the nullspace of $\mathbf{A} - 4\mathbf{I}$, that is, any nonzero vector satisfying

$$\begin{pmatrix} -1 & 1 & 1 \\ 2 & 0 & 2 \\ -1 & -1 & -3 \end{pmatrix} \begin{pmatrix} v_1 \\ v_2 \\ v_3 \end{pmatrix} = \begin{pmatrix} 0 \\ 0 \\ 0 \end{pmatrix}.$$

Thus we want a solution to the set of equations

$$-v_1 + v_2 + v_3 = 0,$$
$$2v_1 + 2v_3 = 0,$$
$$-v_1 - v_2 - 3v_3 = 0$$

that is different from $v_1 = v_2 = v_3 = 0$.

8.3. Eigenvalues and eigenvectors

Substituting $v_1 = -v_3$ from the second equation into the third equation we see that $v_2 = -2v_3$. With $v_1 = -v_3$ and $v_2 = -2v_3$, the first equation $-v_1 + v_2 + v_3$ is automatically satisfied. Choosing $v_3 = 1$ we see that

$$\mathbf{V} = \begin{pmatrix} -1 \\ -2 \\ 1 \end{pmatrix}$$

is a solution to $(\mathbf{A} - 4\mathbf{I})\mathbf{V} = \mathbf{0}$, and hence this vector, or any nonzero multiple of it, can serve as an eigenvector for the eigenvalue 4. For the eigenvalue 2 we compute $\mathbf{A} - 2\mathbf{I}$ to be

$$\begin{pmatrix} 1 & 1 & 1 \\ 2 & 2 & 2 \\ -1 & -1 & -1 \end{pmatrix},$$

so that $(\mathbf{A} - 2\mathbf{I})\mathbf{V} = \mathbf{0}$ exactly when $v_1 + v_2 + v_3 = 0$. Any choice of v_1, v_2, v_3, not all zero, that satisfies this equation will provide an eigenvector for the eigenvalue 2. For example, both

$$\begin{pmatrix} 1 \\ 0 \\ -1 \end{pmatrix} \quad \text{and} \quad \begin{pmatrix} 0 \\ 1 \\ -1 \end{pmatrix}$$

are eigenvectors, with neither being a scalar multiple of the other, so that these two vectors are linearly independent.

In the last example, the eigenvalue $r_2 = 2$ came from the factor $z - 2$ which appeared to the power 2 in the characteristic equation. To keep track of this information, we will say that $r_2 = 2$ is an *eigenvalue of multiplicity* 2 for the matrix $\mathbf{A}$. In this example, we were able to find two linearly independent eigenvectors for the eigenvalue 2, but this is not always possible (see Exercise 9).

Diagonal matrices and similarity. A **diagonal** matrix $\mathbf{D}$ is a square matrix all of whose off-diagonal entries are 0; that is, $\mathbf{D}$ is diagonal if $d_{ij} = 0$ whenever $i \neq j$. In Exercise 19 you will show that the eigenvalues of a diagonal matrix are precisely the entries along the main diagonal (the numbers $d_{11}, d_{22}, \ldots$) and that an eigenvector for d_{jj} is the vector

(8.29)
$$\mathbf{E}_j = \begin{pmatrix} 0 \\ \vdots \\ 1 \\ \vdots \\ 0 \end{pmatrix},$$

where the 1 is the jth entry. Thus the eigenvalues and eigenvectors of a diagonal matrix can be found "by inspection".

Definition 8.3.7. Two $n \times n$ matrices $\mathbf{A}$ and $\mathbf{B}$ are said to be **similar** if there is an invertible $n \times n$ matrix $\mathbf{T}$ with $\mathbf{A} = \mathbf{T}^{-1}\mathbf{BT}$.

Theorem 8.3.8. *Similar matrices have the same eigenvalues.*

Proof. Suppose $\mathbf{A}$ and $\mathbf{B}$ are similar so that $\mathbf{A} = \mathbf{T}^{-1}\mathbf{BT}$. We want to show that $\mathbf{A}$ and $\mathbf{B}$ have the same eigenvalues. First suppose that r is an eigenvalue of $\mathbf{B}$, so that there is a nonzero vector $\mathbf{V}$ with $\mathbf{BV} = r\mathbf{V}$. Theorem 8.2.9 says that $\mathbf{T}^{-1}\mathbf{V} \neq \mathbf{0}$. We have

$$\mathbf{A}(\mathbf{T}^{-1}\mathbf{V}) = \mathbf{T}^{-1}\mathbf{BT}(\mathbf{T}^{-1}\mathbf{V}) = \mathbf{T}^{-1}\mathbf{BV} = \mathbf{T}^{-1}(r\mathbf{V}) = r(\mathbf{T}^{-1}\mathbf{V}).$$

This shows that r is an eigenvalue of $\mathbf{A}$, with eigenvector $\mathbf{T}^{-1}\mathbf{V}$.

Conversely, suppose that s is an eigenvector of $\mathbf{A}$ with eigenvector $\mathbf{W}$, so that $\mathbf{AW} = s\mathbf{W}$. Since $\mathbf{A} = \mathbf{T}^{-1}\mathbf{BT}$, this means
$$\mathbf{T}^{-1}\mathbf{BT}(\mathbf{W}) = s\mathbf{W},$$
and upon multiplying both sides of this equation on the left by $\mathbf{T}$ we obtain

(8.30) $$\mathbf{B}(\mathbf{TW}) = \mathbf{T}(s\mathbf{W}) = s(\mathbf{TW}).$$

Since $\mathbf{W}$ is an eigenvector of $\mathbf{A}$, it is by definition not the vector $\mathbf{0}$. Theorem 8.2.9 assures us then that $\mathbf{TW}$ is not the zero vector either, and (8.30) shows that $\mathbf{TW}$ is an eigenvector for $\mathbf{B}$ with eigenvalue s, as desired. □

A different approach to the proof, working directly with the characteristic equation, is outlined in Exercise 25.

In particular, our discussion above shows that if $\mathbf{A}$ is similar to a diagonal matrix $\mathbf{D}$, then the eigenvalues of $\mathbf{A}$ are just the diagonal entries of $\mathbf{D}$.

8.3.1. More on determinants. Determinants of square matrices can be computed by a process called "expansion about a row" (or column). We illustrated this in the preceding section in a 3×3 example, expanding about the first row. Before describing the procedure in general, we compute the determinant of a 4×4 matrix. Set

$$\mathbf{A} = \begin{pmatrix} a_{11} & a_{12} & a_{13} & a_{14} \\ a_{21} & a_{22} & a_{23} & a_{24} \\ a_{31} & a_{32} & a_{33} & a_{34} \\ a_{41} & a_{42} & a_{43} & a_{44} \end{pmatrix}.$$

The determinant of $\mathbf{A}$ is computed as

(8.31) $$a_{11}\det \mathbf{A}_{11} - a_{12}\det \mathbf{A}_{12} + a_{13}\det \mathbf{A}_{13} - a_{14}\det \mathbf{A}_{14},$$

where $\mathbf{A}_{ij}$ is the 3×3 matrix obtained from $\mathbf{A}$ by deleting the ith row and jth column of $\mathbf{A}$. For example,

$$\mathbf{A}_{11} = \begin{pmatrix} a_{22} & a_{23} & a_{24} \\ a_{32} & a_{33} & a_{34} \\ a_{42} & a_{43} & a_{44} \end{pmatrix}.$$

In (8.31) we have "expanded about the first row" and written the determinant of our 4×4 matrix in terms of determinants of certain 3×3 matrices. Notice the alternating plus and minus signs in front of the coefficients a_{ij} in (8.31); the sign in front of a_{ij} can be described by $(-1)^{i+j}$. Analogous formulas give the determinant expanded about a different row or about a column. For example, expanding about the second column leads to the formula

$$(-1)^{1+2}a_{12}\det \mathbf{A}_{12} + (-1)^{2+2}a_{22}\det \mathbf{A}_{22} + (-1)^{3+2}a_{32}\det \mathbf{A}_{32} + (-1)^{4+2}a_{42}\det \mathbf{A}_{42}$$

for the determinant of $\mathbf{A}$. In a linear algebra course you would verify that no matter which row or column you expand by, the same value results. Having the choice of any row or column allows you to exploit computational simplifications; for example a row or column with 0's in it leads to a sum with fewer terms to compute.

Example 8.3.9. We'll evaluate the determinant of the matrix

$$\mathbf{A} = \begin{pmatrix} 1 & 2 & -1 & 3 \\ 2 & 0 & 0 & 5 \\ -1 & -2 & -1 & 4 \\ 2 & 2 & 1 & 4 \end{pmatrix}$$

8.3. Eigenvalues and eigenvectors

by expanding about the second row. We have

$$\det \mathbf{A} = -2 \det \begin{pmatrix} 2 & -1 & 3 \\ -2 & -1 & 4 \\ 2 & 1 & 4 \end{pmatrix} + 5 \det \begin{pmatrix} 1 & 2 & -1 \\ -1 & -2 & -1 \\ 2 & 2 & 1 \end{pmatrix} = 44.$$

Determinants of larger matrices are defined recursively. If $\mathbf{A}$ is size $n \times n$, its determinant is defined in terms of n determinants of size $(n-1) \times (n-1)$ by its ith row expansion formula

$$\det \mathbf{A} = (-1)^{i+1} a_{i1} \det \mathbf{A}_{i1} + (-1)^{i+2} a_{i2} \det \mathbf{A}_{i2} + \cdots + (-1)^{i+n} a_{in} \det \mathbf{A}_{in}$$

where, as before, $\mathbf{A}_{ij}$ is the $(n-1) \times (n-1)$ matrix obtained from $\mathbf{A}$ by deleting the ith row and jth column. Similarly, expanding about the jth column yields the formula

$$\det \mathbf{A} = (-1)^{1+j} a_{1j} \det \mathbf{A}_{1j} + (-1)^{2+j} a_{2j} \det \mathbf{A}_{2j} + \cdots + (-1)^{n+j} a_{nj} \det \mathbf{A}_{nj}.$$

Determinants may seem like rather mysterious objects even though we have seen, at least in the 2×2 case, that they appear naturally in some questions of interest to us. Determinants are nicely behaved with respect to multiplication; this is the content of the next result.

Theorem 8.3.10. *If $\mathbf{A}$ and $\mathbf{B}$ are $n \times n$ matrices and s is a scalar, then:*

(a) $\det(\mathbf{AB}) = (\det \mathbf{A})(\det \mathbf{B})$.

(b) $\det(s\mathbf{A}) = s^n \det \mathbf{A}$.

(c) *If $\mathbf{A}$ is invertible, then* $\det(\mathbf{A}^{-1}) = \dfrac{1}{\det \mathbf{A}}$.

We won't prove this result, although in Exercise 17 you are asked to show how (b) and (c) follow from (a).

8.3.2. Exercises.

In Exercises 1–6 find all eigenvalues and corresponding eigenvectors for the given matrix.

1. $\mathbf{A} = \begin{pmatrix} 1 & -1 \\ 1 & 3 \end{pmatrix}$.

2. $\mathbf{A} = \begin{pmatrix} 1 & -3 \\ 1 & 5 \end{pmatrix}$.

3. $\mathbf{A} = \begin{pmatrix} 3 & -2 \\ 4 & -1 \end{pmatrix}$.

4. $\mathbf{A} = \begin{pmatrix} 4 & -4 \\ 3 & -3 \end{pmatrix}$.

5. $\mathbf{A} = \begin{pmatrix} 1 & i \\ -i & 1 \end{pmatrix}$.

6. $\mathbf{A} = \begin{pmatrix} 3 & 2 & 2 \\ 1 & 4 & 1 \\ -2 & -4 & -1 \end{pmatrix}$.

7. Show that if

$$\mathbf{A} = \begin{pmatrix} 3 & 1 & 1 \\ 2 & 4 & 2 \\ -1 & -1 & 1 \end{pmatrix},$$

then $\det(\mathbf{A} - z\mathbf{I}) = -(z-4)(z-2)^2$.

8. For the matrices below, one eigenvalue is given. Find the other eigenvalues, and for each eigenvalue, find an eigenvector.

(a)
$$\mathbf{A} = \begin{pmatrix} 5 & 4 & 4 \\ -7 & -3 & -1 \\ 7 & 4 & 2 \end{pmatrix}, \text{ with eigenvalue } r = 1.$$

(b)
$$\mathbf{A} = \begin{pmatrix} 4 & 0 & 1 \\ 2 & 3 & 2 \\ 1 & 0 & 4 \end{pmatrix}, \text{ with eigenvalue } r = 5.$$

9. Show that
$$\mathbf{A} = \begin{pmatrix} 2 & 0 & 0 \\ 1 & 2 & 0 \\ 1 & 1 & 2 \end{pmatrix}$$
has only one eigenvalue, $r = 2$, and that the only eigenvectors corresponding to this eigenvalue are nonzero scalar multiples of
$$\begin{pmatrix} 0 \\ 0 \\ 1 \end{pmatrix}.$$

10. Show that if r is an eigenvalue of $\mathbf{A}$ with eigenvector $\mathbf{V}$, then $c\mathbf{V}$ is also an eigenvector corresponding to the eigenvalue r, for any scalar $c \neq 0$. Why is $c = 0$ excluded?

11. From Exercise 6 of Section 8.2, you know that if
$$\mathbf{A} = \begin{pmatrix} \cos\theta & -\sin\theta \\ \sin\theta & \cos\theta \end{pmatrix} \quad \text{and} \quad \mathbf{V} = \begin{pmatrix} v_1 \\ v_2 \end{pmatrix},$$
then $\mathbf{AV}$ gives the vector that results when $\mathbf{V}$ is rotated about the origin by θ radians in the counterclockwise direction.
   (a) Thinking about rotation geometrically, for what values of θ should the matrix $\mathbf{A}$ have real eigenvalues? What should these eigenvalues be? No computations are needed to answer this question.
   (b) Now compute the eigenvalues of $\mathbf{A}$ (for any real value of θ), and verify your answer to (a).

12. Suppose that all the entries of an $n \times n$ matrix $\mathbf{A}$ are real. Show that if $\mathbf{V}$ is an eigenvector of $\mathbf{A}$ corresponding to the complex eigenvalue $\alpha + i\beta$, then $\overline{\mathbf{V}}$ is also an eigenvector of $\mathbf{A}$, with complex conjugate eigenvalue $\alpha - i\beta$. In short, this shows that the nonreal eigenvalues of a real matrix come in complex conjugate pairs and have complex conjugate eigenvectors.

   Hints: For any matrix $\mathbf{C}$, write $\overline{\mathbf{C}}$ for the matrix all of whose entries are the conjugates of the corresponding entries of $\mathbf{C}$. Use the easily verified properties $\overline{\mathbf{AB}} = \overline{\mathbf{A}}\,\overline{\mathbf{B}}$ and $\overline{r\mathbf{A}} = \overline{r}\overline{\mathbf{A}}$ for matrices $\mathbf{A}, \mathbf{B}$ and scalar r.

13. Show that 0 is an eigenvalue of $\mathbf{A}$ if and only if $\mathbf{A}$ is not invertible.

14. Suppose that $\mathbf{A}$ is an invertible matrix and that r is an eigenvalue of $\mathbf{A}$ (by the last problem, $r \neq 0$), with corresponding eigenvector $\mathbf{V}$. Show that $\frac{1}{r}$ is an eigenvalue of $\mathbf{A}^{-1}$, with eigenvector $\mathbf{V}$.

15. Suppose that $\mathbf{A}$ is a matrix and that r is an eigenvalue of $\mathbf{A}$ with corresponding eigenvector $\mathbf{V}$.
   (a) Show that r^2 is an eigenvalue of $\mathbf{A}^2$, with eigenvector $\mathbf{V}$. Similarly, show that r^3 is an eigenvalue of $\mathbf{A}^3$, with eigenvector $\mathbf{V}$.
   (b) Show that for each positive integer n, r^n is an eigenvalue of $\mathbf{A}^n$ with eigenvector $\mathbf{V}$.

16. Given that the 3×3 matrix $\mathbf{A}$ has eigenvalue λ, Sally proposes the following method for finding an eigenvector $\mathbf{V}$ for $\mathbf{A}$ corresponding to λ.

   Step 1. Try to solve $(\mathbf{A} - \lambda \mathbf{I}) \begin{pmatrix} v_1 \\ v_2 \\ 1 \end{pmatrix} = \mathbf{0}$. If a solution v_1 and v_2 can be found, then $\mathbf{V} = \begin{pmatrix} v_1 \\ v_2 \\ 1 \end{pmatrix}$ will be an eigenvector.

Step 2. If Step 1 fails, then $(\mathbf{A} - \lambda \mathbf{I}) \begin{pmatrix} v_1 \\ v_2 \\ 0 \end{pmatrix} = \mathbf{0}$ must have a solution, which provides an eigenvector

$\mathbf{V}$ of the form $\mathbf{V} = \begin{pmatrix} v_1 \\ v_2 \\ 0 \end{pmatrix}$.

Is Sally's method correct?

17. (a) Find the determinant of the $n \times n$ identity matrix $\mathbf{I}$ and of $s\mathbf{I}$.
    (b) Show that (b) and (c) of Theorem 8.3.10 follow from (a) of that theorem. Hint: Consider $(s\mathbf{I})(\mathbf{A})$ and $\mathbf{A}\mathbf{A}^{-1}$.

18. **Invertibility and matrix products.**
    (a) Let $\mathbf{A}$ and $\mathbf{B}$ be invertible $n \times n$ matrices. Show that $\mathbf{AB}$ is invertible and in fact that $(\mathbf{AB})^{-1} = \mathbf{B}^{-1}\mathbf{A}^{-1}$.
    (b) Let $\mathbf{A}$ and $\mathbf{B}$ be $n \times n$ matrices. Show that if $\mathbf{AB}$ is invertible, so are $\mathbf{A}$ and $\mathbf{B}$. Hint: See Theorem 8.2.9 and part (a) of Theorem 8.3.10.
    (c) More generally, a finite product of invertible $n \times n$ matrices will be invertible. State and prove versions of parts (a) and (b) for products of three matrices.

19. **Eigenvalues of diagonal and triangular matrices.**
    (a) Show that the eigenvalues of a diagonal matrix are precisely the entries on the main diagonal and that an eigenvector corresponding to the jth diagonal entry is the vector $\mathbf{E}_j$ having a 1 in the jth position and 0's elsewhere (see equation (8.29)).
    (b) A square matrix is **upper triangular** if all of the entries below the main diagonal are 0 (so that $a_{ij} = 0$ if $i > j$). Show that the eigenvalues of an upper triangular matrix are also the entries on the main diagonal. Does this result also hold for **lower triangular** matrices, defined analogously?

20. Show that if an $n \times n$ matrix $\mathbf{A}$ has eigenvectors $\mathbf{E}_j$, $1 \leq j \leq n$ (where $\mathbf{E}_j$ is as defined in Exercise 19) and corresponding eigenvalues r_j, then $\mathbf{A}$ is diagonal with diagonal entires r_j. Hint: Use the equation (8.25) in Section 8.2.

21. **Similarity to a diagonal matrix.** Suppose that an $n \times n$ matrix $\mathbf{A}$ is similar to a diagonal matrix $\mathbf{D}$, so that $\mathbf{T}^{-1}\mathbf{AT} = \mathbf{D}$ for some invertible $\mathbf{T}$. Show that the columns of $\mathbf{T}$ are eigenvectors of $\mathbf{A}$, with eigenvalues being the corresponding diagonal entries of $\mathbf{D}$.

22. **Powers of a diagonal matrix.** If $\mathbf{D}$ is a diagonal matrix, describe the matrices $\mathbf{D}^2$, $\mathbf{D}^3$, and in general $\mathbf{D}^k$, where $\mathbf{D}^k$ is the product of $\mathbf{D}$ with itself k times.

23. Find the determinant of
$$\mathbf{A} = \begin{pmatrix} 0 & 2 & 1 & 4 \\ 2 & 0 & -1 & 5 \\ 0 & 2 & -3 & 1 \\ 3 & 2 & 0 & 4 \end{pmatrix}.$$

24. Show that the nonzero eigenvalues of the matrix
$$\mathbf{A} = \begin{pmatrix} a_{11} & 0 & a_{13} & 0 \\ a_{21} & 0 & a_{23} & 0 \\ a_{31} & 0 & a_{33} & 0 \\ a_{41} & 0 & a_{43} & 0 \end{pmatrix}$$
are the nonzero eigenvalues of the matrix
$$\begin{pmatrix} a_{11} & a_{13} \\ a_{31} & a_{33} \end{pmatrix}.$$

Hint: Start by expanding $\det(\mathbf{A} - z\mathbf{I})$ about the fourth column.

25. Here is another approach to obtaining the proof of Theorem 8.3.8: If $\mathbf{B}$ is any $n \times n$ matrix and $\mathbf{T}$ is any $n \times n$ invertible matrix, show that for any scalar λ, $\det(\mathbf{B} - \lambda\mathbf{I}) = \det(\mathbf{T}^{-1}\mathbf{B}\mathbf{T} - \lambda\mathbf{I})$, by first showing that $\mathbf{T}^{-1}\mathbf{B}\mathbf{T} - \lambda\mathbf{I} = \mathbf{T}^{-1}(\mathbf{B} - \lambda\mathbf{I})\mathbf{T}$ and then using Theorem 8.3.10. Conclude that $\mathbf{B}$ and $\mathbf{T}^{-1}\mathbf{B}\mathbf{T}$ have the same eigenvalues.

8.4. Solving linear systems

Let's now apply what we learned in the last two sections to solve some first-order linear systems of differential equations. We begin by writing our system in matrix notation. Consider the linear system

$$(8.32) \quad \begin{aligned} y_1' &= p_{11}(t)y_1 + \cdots + p_{1n}(t)y_n + g_1(t), \\ y_2' &= p_{21}(t)y_1 + \cdots + p_{2n}(t)y_n + g_2(t), \\ &\vdots \\ y_n' &= p_{n1}(t)y_1 + \cdots + p_{nn}(t)y_n + g_n(t), \end{aligned}$$

with independent variable t and dependent variables $y_1, y_2, \ldots, y_n$. The primes denote differentiation with respect to t. Set

$$\mathbf{X}(t) = \begin{pmatrix} y_1(t) \\ y_2(t) \\ \vdots \\ y_n(t) \end{pmatrix}.$$

We call $\mathbf{X}(t)$ a **vector-valued function** (or just a "vector function"); its entries are functions of t. When those entries are differentiable functions, we say $\mathbf{X}(t)$ is a differentiable vector function and write

$$\mathbf{X}'(t) = \frac{d\mathbf{X}}{dt} = \begin{pmatrix} y_1'(t) \\ y_2'(t) \\ \vdots \\ y_n'(t) \end{pmatrix}.$$

Let $\mathbf{P}(t)$ denote the $n \times n$ matrix whose jkth entry is the function $p_{jk}(t)$:

$$\mathbf{P}(t) = \begin{pmatrix} p_{11}(t) & \cdots & p_{1n}(t) \\ p_{21}(t) & \cdots & p_{2n}(t) \\ \vdots & & \vdots \\ p_{n1}(t) & \cdots & p_{nn}(t) \end{pmatrix}.$$

This is called a **matrix-valued function** or a matrix function. Finally, let $\mathbf{G}(t)$ be the vector function

$$\mathbf{G}(t) = \begin{pmatrix} g_1(t) \\ g_2(t) \\ \vdots \\ g_n(t) \end{pmatrix}.$$

Because of the way matrix multiplication is defined, the system in equation (8.32) can be written in matrix form as $\mathbf{X}'(t) = \mathbf{P}(t)\mathbf{X}(t) + \mathbf{G}(t)$. Initial conditions for (8.32), say, $y_1(t_0) = \alpha_1$,

8.4. Solving linear systems

$y_2(t_0) = \alpha_2, \ldots, y_n(t_0) = \alpha_n$, can be written as

$$\mathbf{X}(t_0) = \begin{pmatrix} \alpha_1 \\ \alpha_2 \\ \vdots \\ \alpha_n \end{pmatrix}.$$

In this section we will be exclusively interested in homogeneous systems, so that $\mathbf{G}(t) = \mathbf{0}$, the $n \times 1$ column matrix of zeros. For any homogeneous linear system, the Superposition Principle (the analogue of Theorem 4.6.1 in Section 4.6) applies:

Theorem 8.4.1 (Superposition Principle). *If $\mathbf{X}_1(t), \mathbf{X}_2(t), \ldots, \mathbf{X}_m(t)$ are solutions to the linear homogeneous system*

(8.33) $$\mathbf{X}' = \mathbf{P}(t)\mathbf{X},$$

then so is any linear combination $c_1\mathbf{X}_1(t) + c_2\mathbf{X}_2(t) + \cdots + c_m\mathbf{X}_m(t)$ for arbitrary constants $c_1, c_2, \ldots, c_m$.

You are asked to verify this in Exercise 13. One of the solutions of *any* homogeneous linear system $\mathbf{X}' = \mathbf{P}(t)\mathbf{X}$, corresponding to the choice $c_1 = c_2 = \cdots = c_m = 0$, is $\mathbf{X}(t) \equiv \mathbf{0}$; this is the equilibrium solution

$$y_1(t) \equiv 0, \ y_2(t) \equiv 0, \ldots, y_n(t) \equiv 0.$$

Sometimes a homogeneous system will have other equilibrium points in addition to $\mathbf{0} = (0, 0, \ldots, 0)$; for an example, see Exercise 16. However, the equivalence of (a) and (d) of Theorem 8.2.9 yields the following:

> If $\mathbf{A}$ is an $n \times n$ matrix with nonzero determinant, then $\mathbf{0}$ is the only equilibrium point of $\mathbf{X}' = \mathbf{A}\mathbf{X}$.

Two natural questions may occur to you at this point:

(Q_1) To make use of the Superposition Principle, we must first be able to produce *some* solutions of equation (8.33). How is this best done?

(Q_2) How do we produce the family of *all* solutions of (8.33)?

Borrowing language we used for single differential equations, let's call a description of the family of all solutions of (8.33) the **general solution**. Our experience with second- or higher-order equations in Chapter 4 suggests that if we have n *properly chosen* solutions $\mathbf{X}_1, \mathbf{X}_2, \ldots, \mathbf{X}_n$ to a linear homogeneous system of n first-order equations, we should be able to produce the general solution as

$$c_1\mathbf{X}_1 + c_2\mathbf{X}_2 + \cdots + c_n\mathbf{X}_n$$

for arbitrary constants c_j. Furthermore, the notion of "properly chosen" should be a linear independence requirement. This will be made precise in Theorems 8.4.3 and 8.4.4 below and the remarks that follow the theorems.

We can make some progress on (Q_1) in the case of a *constant coefficient* homogeneous system. In this case the matrix function $\mathbf{P}(t)$ in equation (8.33) is just a matrix $\mathbf{A}$ of constants, and we will write the system as $\mathbf{X}'(t) = \mathbf{A}\mathbf{X}(t)$, or more briefly as $\mathbf{X}' = \mathbf{A}\mathbf{X}$.

Producing a solution from an eigenvalue/eigenvector pair. We have seen through Examples 8.1.5 and 8.3.2 that the linear system $\mathbf{X}' = \mathbf{A}\mathbf{X}$, where

$$A = \begin{pmatrix} 3 & 2 \\ 1 & 2 \end{pmatrix},$$

has solutions of the form $\mathbf{X}(t) = e^{4t}\mathbf{V}_1$ and $\mathbf{X}(t) = e^{t}\mathbf{V}_2$, where $\mathbf{V}_1$ is an eigenvector for $\mathbf{A}$ with corresponding eigenvalue 4 and $\mathbf{V}_2$ is an eigenvector for $\mathbf{A}$ with corresponding eigenvalue 1. This suggests that if $\mathbf{A}$ is any $n \times n$ matrix of constants such that r is an eigenvalue of $\mathbf{A}$ with eigenvector $\mathbf{V}$, then

$$\mathbf{X}(t) = e^{rt}\mathbf{V}$$

should be solution of the system $\mathbf{X}' = \mathbf{A}\mathbf{X}$. This is easy to verify. Since the entries of $\mathbf{V}$ are just constants, we have

$$\begin{aligned} \mathbf{X}'(t) &= re^{rt}\mathbf{V} = e^{rt}(r\mathbf{V}) \\ &= e^{rt}(\mathbf{A}\mathbf{V}) = \mathbf{A}\left(e^{rt}\mathbf{V}\right) \\ &= \mathbf{A}\mathbf{X}(t). \end{aligned}$$

Thus, we have established the following:

> If $\mathbf{A}$ is any $n \times n$ matrix with eigenvalue r and corresponding eigenvector $\mathbf{V}$, then $\mathbf{X}(t) = e^{rt}\mathbf{V}$ is a solution to $\mathbf{X}' = \mathbf{A}\mathbf{X}$.

Such a solution is called an **exponential solution** or a **ray solution**. To see where this terminology comes from, let's specialize to the case that $\mathbf{A}$ is a 2×2 matrix with *real* eigenvalue r and eigenvector

$$\mathbf{V} = \begin{pmatrix} v_1 \\ v_2 \end{pmatrix}$$

having real coordinates; then the ray solution

$$\mathbf{X}(t) = e^{rt}\mathbf{V} = \begin{pmatrix} v_1 e^{rt} \\ v_2 e^{rt} \end{pmatrix}$$

has components

$$x(t) = v_1 e^{rt}, \quad y(t) = v_2 e^{rt}.$$

These are parametric equations of a *ray* in the xy-plane passing through the point (v_1, v_2). If $r > 0$, the vector $e^{rt}\mathbf{V}$ gets longer as t increases, and we call this solution an **outgoing ray**. If $r < 0$, the vector $e^{rt}\mathbf{V}$ shrinks as t increases, and we call the solution an **incoming ray**. An incoming ray approaches the origin (but never reaches it) as t goes forward into the distant future ($t \to +\infty$). An outgoing ray approaches the origin (but never reaches it) as t goes backwards into the distant past ($t \to -\infty$). We call the entire line through the origin and containing the point (v_1, v_2) the **eigenvector line** for $\mathbf{V}$. It has equation $y = (v_2/v_1)x$ if $v_1 \neq 0$, while if $v_1 = 0$, it coincides with the y-axis ($x = 0$). The trajectory of a ray solution $e^{rt}\mathbf{V}$, or more generally $ce^{rt}\mathbf{V}$ where c is a nonzero real constant, occupies the half of this line lying on one side or the other of the origin. For any nonzero constant c, $ce^{rt}\mathbf{V}$ is itself a ray solution, since $c\mathbf{V}$ is also an eigenvector for the eigenvalue r.

8.4. Solving linear systems

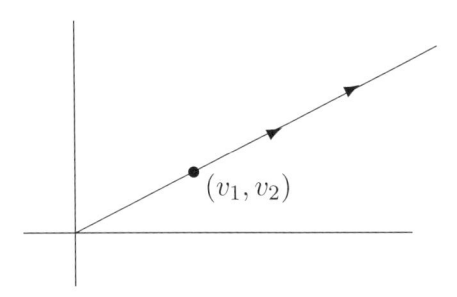

Trajectory of the outgoing ray solution $\mathbf{X}(t) = e^{rt}\mathbf{V}$, $r > 0$.

We will exploit what we have just learned to describe all solutions of a planar constant coefficient homogeneous system $\mathbf{X}' = \mathbf{A}\mathbf{X}$ where $\mathbf{A}$ is a 2×2 matrix of real numbers. Our analysis, which occupies the remainder of this section as well as Section 8.7, will be organized by the type of eigenvalues that $\mathbf{A}$ possesses.

Distinct real eigenvalues. Suppose the 2×2 matrix $\mathbf{A}$ has two distinct real eigenvalues, r_1 and r_2. Let $\mathbf{V}_1$ and $\mathbf{V}_2$ be eigenvectors for r_1 and r_2, respectively. The system

$$\text{(8.34)} \qquad \mathbf{X}' = \mathbf{A}\mathbf{X}$$

has ray solutions $\mathbf{X}_1(t) = e^{r_1 t}\mathbf{V}_1$ and $\mathbf{X}_2(t) = e^{r_2 t}\mathbf{V}_2$.

By the Superposition Principle, any linear combination

$$\text{(8.35)} \qquad c_1 e^{r_1 t}\mathbf{V}_1 + c_2 e^{r_2 t}\mathbf{V}_2$$

will also be a solution. We will show that as c_1 and c_2 vary over all constants, every solution to $\mathbf{X}' = \mathbf{A}\mathbf{X}$ is produced, so that (8.35) is the general solution. To do this we will use the Existence and Uniqueness Theorem, Theorem 8.1.1, in a way that is reminiscent of our discussion of general solutions for single higher-order equations in Chapter 4. In addition to Theorem 8.1.1, the other key fact we will need is that because $r_1 \neq r_2$, the eigenvectors $\mathbf{V}_1$ and $\mathbf{V}_2$ are *linearly independent*. To see why this is so, suppose that

$$\text{(8.36)} \qquad k_1 \mathbf{V}_1 + k_2 \mathbf{V}_2 = \mathbf{0}.$$

Multiplying equation (8.36) by r_1 gives

$$\text{(8.37)} \qquad k_1 r_1 \mathbf{V}_1 + k_2 r_1 \mathbf{V}_2 = \mathbf{0},$$

while multiplying (8.36) by $\mathbf{A}$ gives

$$k_1 \mathbf{A}\mathbf{V}_1 + k_2 \mathbf{A}\mathbf{V}_2 = \mathbf{0}.$$

Since $\mathbf{V}_1$ and $\mathbf{V}_2$ are eigenvectors, we can rewrite the last equation as

$$\text{(8.38)} \qquad k_1 r_1 \mathbf{V}_1 + k_2 r_2 \mathbf{V}_2 = \mathbf{0}.$$

Subtracting equation (8.38) from equation (8.37) gives $k_2(r_1 - r_2)\mathbf{V}_2 = \mathbf{0}$. Since $r_1 \neq r_2$ and $\mathbf{V}_2 \neq \mathbf{0}$, this tells us that $k_2 = 0$, and then by equation (8.36) we must also have $k_1 = 0$. Thus $\mathbf{V}_1$ and $\mathbf{V}_2$ are linearly independent. Notice that the same argument shows two eigenvectors belonging to two distinct eigenvalues of an $n \times n$ matrix must be linearly independent.

Why is (8.35) the general solution? Suppose $\mathbf{X}(t)$ is an arbitrary solution of (8.34). Can we find it in the family (8.35)? Is there a choice of c_1 and c_2 so that

$$\text{(8.39)} \qquad \mathbf{X}(t) = c_1 e^{r_1 t}\mathbf{V}_1 + c_2 e^{r_2 t}\mathbf{V}_2$$

for all t, $-\infty < t < \infty$? Notice we're asking for one choice of constants c_1, c_2 that work simultaneously for all t. Let's start by asking if we can find c_1 and c_2 that work for $t = 0$. Take our solution vector function $\mathbf{X}(t)$, plug in $t = 0$, and call the resulting value $\begin{pmatrix} \alpha \\ \beta \end{pmatrix}$. After plugging 0 into the right-hand side of (8.39), we are asking for a choice of c_1 and c_2 so that

$$\begin{pmatrix} \alpha \\ \beta \end{pmatrix} = c_1 \mathbf{V}_1 + c_2 \mathbf{V}_2.$$

Setting

$$\mathbf{V}_1 = \begin{pmatrix} v_{11} \\ v_{21} \end{pmatrix} \quad \text{and} \quad \mathbf{V}_2 = \begin{pmatrix} v_{12} \\ v_{22} \end{pmatrix}$$

this becomes the pair of equations

(8.40)
$$\begin{aligned} \alpha &= v_{11}c_1 + v_{12}c_2, \\ \beta &= v_{21}c_1 + v_{22}c_2 \end{aligned}$$

with "unknowns" c_1 and c_2. Because the eigenvectors $\mathbf{V}_1$ and $\mathbf{V}_2$ are linearly independent, Theorem 8.2.9 tells us that

$$\det \begin{pmatrix} v_{11} & v_{12} \\ v_{21} & v_{22} \end{pmatrix} \neq 0.$$

Thus, by Theorem 8.2.9, there is one (and only one) solution for c_1 and c_2 in equation (8.40). We could describe the solution by Cramer's Rule (see Exercise 11 in Section 8.2) or as

$$\begin{pmatrix} c_1 \\ c_2 \end{pmatrix} = \begin{pmatrix} v_{11} & v_{12} \\ v_{21} & v_{22} \end{pmatrix}^{-1} \begin{pmatrix} \alpha \\ \beta \end{pmatrix}.$$

Now for the coup de grace: If c_1 and c_2 are the solutions to (8.40), then both $c_1 e^{r_1 t} \mathbf{V}_1 + c_2 e^{r_2 t} \mathbf{V}_2$ and $\mathbf{X}(t)$ solve the initial value problem

$$\mathbf{X}' = \mathbf{A}\mathbf{X}, \quad \mathbf{X}(0) = \begin{pmatrix} \alpha \\ \beta \end{pmatrix}.$$

The uniqueness part of Theorem 8.1.1 says this initial value problem has only one solution. Thus we conclude

$$\mathbf{X}(t) = c_1 e^{r_1 t} \mathbf{V}_1 + c_2 e^{r_2 t} \mathbf{V}_2$$

for all $-\infty < t < \infty$, exactly as desired.

We summarize our discussion so far as follows:

> If $\mathbf{A}$ is a 2×2 matrix with two distinct (real) eigenvalues r_1, r_2, then the general solution to $\mathbf{X}' = \mathbf{A}\mathbf{X}$ is
>
> $$\mathbf{X}(t) = c_1 e^{r_1 t} \mathbf{V}_1 + c_2 e^{r_2 t} \mathbf{V}_2,$$
>
> where $\mathbf{V}_1, \mathbf{V}_2$ are eigenvectors for $\mathbf{A}$ corresponding, respectively, to the eigenvalues r_1 and r_2.

Indeed, this result holds whenever $\mathbf{A}$ has distinct eigenvalues, real **or** complex valued. However, when the eigenvalues are complex, our final goal will be to produce *real-valued* solutions, so there is more work to do. This will be discussed in Section 8.7.

Example 8.4.2. Find the general solution to the system $\mathbf{X}' = \mathbf{A}\mathbf{X}$, where

$$\mathbf{A} = \begin{pmatrix} 2 & 1 \\ 2 & 3 \end{pmatrix},$$

and then find the solution which satisfies

$$\mathbf{X}(0) = \begin{pmatrix} -1 \\ 7 \end{pmatrix}.$$

8.4. Solving linear systems

The eigenvalues of $\mathbf{A}$, which are found by solving $\det(\mathbf{A} - z\mathbf{I}) = (2-z)(3-z) - 2 = 0$, are $z = 4$ and $z = 1$. Corresponding to the eigenvalue $z = 4$ we have an eigenvector $\binom{1}{2}$, and an eigenvector for the eigenvalue $z = 1$ is $\binom{1}{-1}$. This tells us that the general solution is

$$\mathbf{X}(t) = c_1 e^{4t} \binom{1}{2} + c_2 e^{t} \binom{1}{-1}.$$

For the initial value problem, we must choose c_1 and c_2 so that

$$\mathbf{X}(0) = c_1 \binom{1}{2} + c_2 \binom{1}{-1} = \binom{-1}{7}.$$

This gives the equations $c_1 + c_2 = -1$, $2c_1 - c_2 = 7$, which have solution $c_1 = 2, c_2 = -3$. Thus the solution satisfying the desired initial condition is

$$\mathbf{X}(t) = 2 e^{4t} \binom{1}{2} - 3 e^{t} \binom{1}{-1}.$$

Linear homogeneous systems in general. We expected that the general solution to any two-dimensional homogeneous system $\mathbf{X}' = \mathbf{P}(t)\mathbf{X}$ could be built from *two* properly chosen solutions $\mathbf{X}_1(t)$ and $\mathbf{X}_2(t)$. Adapting the above arguments we can now make this precise for any linear homogeneous planar system (constant coefficient or not).

Theorem 8.4.3. *Consider a two-dimensional linear homogeneous system $\mathbf{X}' = \mathbf{P}(t)\mathbf{X}$, where the entries of $\mathbf{P}(t)$ are continuous on an open interval I. Suppose $\mathbf{X}_1(t)$ and $\mathbf{X}_2(t)$ are solutions of $\mathbf{X}' = \mathbf{P}(t)\mathbf{X}$ on I. Pick any point t_0 in I. The following are equivalent:*

(a) $\mathbf{X} = c_1 \mathbf{X}_1(t) + c_2 \mathbf{X}_2(t)$ *is the general solution of $\mathbf{X}' = \mathbf{P}(t)\mathbf{X}$ on I.*

(b) *Given any initial condition $\mathbf{X}(t_0) = \binom{\alpha}{\beta}$, there is a choice of c_1, c_2 so that*

$$c_1 \mathbf{X}_1(t_0) + c_2 \mathbf{X}_2(t_0) = \binom{\alpha}{\beta}.$$

(c) *The vectors $\mathbf{X}_1(t_0)$ and $\mathbf{X}_2(t_0)$ are linearly independent.*

(d) *The determinant of the matrix whose columns are $\mathbf{X}_1(t_0)$ and $\mathbf{X}_2(t_0)$ is nonzero.*

You are invited to write out a proof of this in Exercise 21. Notice how the theorem makes clear what we mean by saying that the general solution is all linear combinations of two "properly chosen" solutions—the solutions have to yield linearly independent vectors when evaluated at a point t_0 of I. Also note that if $\mathbf{X}_1(t)$ and $\mathbf{X}_2(t)$ are solutions to a planar system $\mathbf{X}' = \mathbf{P}(t)\mathbf{X}$ on an interval I on which the entries of $\mathbf{P}(t)$ are continuous, then $\mathbf{X}_1(t_0)$ and $\mathbf{X}_2(t_0)$ are linearly independent at *some* $t_0 \in I$ if and only if they are linearly independent at *all* $t \in I$. This follows from the equivalence of (a) and (c) in Theorem 8.4.3 and the fact that t_0 can be chosen to be any point of I.

Theorem 8.4.3 generalizes in the expected way to linear homogeneous systems with arbitrary dimension:

Theorem 8.4.4. *Consider an n-dimensional system $\mathbf{X}' = \mathbf{P}(t)\mathbf{X}$, where the entries of $\mathbf{P}$ are continuous functions on some open interval I. Suppose we have n solutions $\mathbf{X}_1(t), \mathbf{X}_2(t), \ldots, \mathbf{X}_n(t)$ to this system on I, and let t_0 be in I. The following are equivalent:*

(a) *The general solution of $\mathbf{X}' = \mathbf{P}(t)\mathbf{X}$ on I is $\mathbf{X} = c_1 \mathbf{X}_1(t) + c_2 \mathbf{X}_2(t) + \cdots + c_n \mathbf{X}_n(t)$.*

(b) *Given any constants $\alpha_1, \alpha_2, \ldots, \alpha_n$ there are constants $c_1, c_2, \ldots, c_n$ so that*

$$c_1 \mathbf{X}_1(t_0) + c_2 \mathbf{X}_2(t_0) + \cdots + c_n \mathbf{X}_n(t_0) = \begin{pmatrix} \alpha_1 \\ \alpha_2 \\ \vdots \\ \alpha_n \end{pmatrix}.$$

(c) *The vectors $\mathbf{X}_1(t_0), \mathbf{X}_2(t_0), \ldots, \mathbf{X}_n(t_0)$ are linearly independent.*

(d) *The determinant of the $n \times n$ matrix*

$$\begin{pmatrix} \uparrow & \uparrow & & \uparrow \\ \mathbf{X}_1(t_0) & \mathbf{X}_2(t_0) & \cdots & \mathbf{X}_n(t_0) \\ \downarrow & \downarrow & & \downarrow \end{pmatrix},$$

whose columns are the vectors $\mathbf{X}_1(t_0), \mathbf{X}_2(t_0), \ldots, \mathbf{X}_n(t_0)$, is nonzero.

Exercise 22 outlines a proof of Theorem 8.4.4.

The Wronskian. Because of part (d) in Theorems 8.4.3 and 8.4.4, the determinant of the matrix whose columns are solutions of $\mathbf{X}' = \mathbf{P}(t)\mathbf{X}$ is of special interest. If $\mathbf{X}_1(t), \mathbf{X}_2(t), \ldots, \mathbf{X}_n(t)$ are any n such solutions on the interval I, the determinant of the $n \times n$ matrix function whose columns are $\mathbf{X}_j(t)$ is called the **Wronskian** of these solutions and is denoted $\mathbf{W}(t)$:

$$(8.41) \qquad \mathbf{W}(t) = \det \begin{pmatrix} \uparrow & \uparrow & & \uparrow \\ \mathbf{X}_1(t) & \mathbf{X}_2(t) & \cdots & \mathbf{X}_n(t) \\ \downarrow & \downarrow & & \downarrow \end{pmatrix}.$$

According to Theorem 8.4.4, the linear combinations $c_1 \mathbf{X}_1(t) + c_2 \mathbf{X}_2(t) + \cdots + c_n \mathbf{X}_n(t)$ give the general solution precisely when this Wronskian is nonzero at any point of I. As a consequence of Theorem 8.4.4, we have the following dichotomy: Either the Wronskian is nonzero for all $t \in I$ or the Wronskian is identically equal to 0 on I.

Remark: It may seem that the definition of "Wronskian" for solutions to a system uses different ingredients than the notion of Wronskian for a single higher-order differential equation (as in Section 4.2). Exercise 17 reconciles this difference.

Linear independence of vector-valued functions. One of the equivalent conditions in Theorem 8.4.4 is the linear independence of the *constant* vectors $\mathbf{X}_1(t_0), \mathbf{X}_2(t_0), \ldots, \mathbf{X}_n(t_0)$ where t_0 is any point of I. It is natural to extend the definition of linear independence from a set of constant vectors to a set of *vector-valued* functions.

Definition 8.4.5. (i) A set of vector-valued functions $\mathbf{X}_1(t), \mathbf{X}_2(t), \ldots, \mathbf{X}_n(t)$ is **linearly dependent** on an interval I if there are constants $c_1, c_2, \ldots, c_n$, not all of which are 0, so that

$$(8.42) \qquad c_1 \mathbf{X}_1(t) + c_2 \mathbf{X}_2(t) + \cdots + c_n \mathbf{X}_n(t) = \mathbf{0}$$

for all $t \in I$.

(ii) If the set of vector-valued functions is not linearly dependent on I, so that the only choice of constants $c_1, c_2, \ldots, c_n$ for which equation (8.42) holds is $c_1 = c_2 = \cdots = c_n = 0$, we say the vector-valued functions are **linearly independent** on I.

Remark. With this definition in hand we can add a fifth equivalent condition (e) to Theorem 8.4.4:

(e) The vector-valued functions $\mathbf{X}_1(t), \mathbf{X}_2(t), \ldots, \mathbf{X}_n(t)$ are linearly independent on I.

8.4. Solving linear systems

Verification of the equivalence of (e) with statement (a) in Theorem 8.4.4 is left to you as Exercise 20. A similar addition applies to the two-dimensional result in Theorem 8.4.3, namely that a fifth equivalent condition is that $\mathbf{X}_1(t)$ and $\mathbf{X}_2(t)$ are linearly independent vector-valued functions on I.

Fundamental set of solutions. Our expanded statement of Theorem 8.4.4 says that if the system $\mathbf{X}' = \mathbf{P}(t)\mathbf{X}$ has solutions $\mathbf{X}_1(t), \mathbf{X}_2(t), \ldots, \mathbf{X}_n(t)$ on I, then $c_1\mathbf{X}_1(t) + c_2\mathbf{X}_2(t) + \cdots + c_n\mathbf{X}_n(t)$ is the general solution on I if and only if the vector-valued functions $\mathbf{X}_1(t), \mathbf{X}_2(t), \ldots, \mathbf{X}_n(t)$ are linearly independent on I. Borrowing language from Section 4.2, we will say that a collection of n linearly independent vector-valued solutions $\mathbf{X}_1(t), \mathbf{X}_2(t), \ldots, \mathbf{X}_n(t)$ to an n-dimensional linear system $\mathbf{X}' = \mathbf{P}(t)\mathbf{X}$ as in Theorem 8.4.4 is a **fundamental set of solutions** to the system. Linear combinations of the functions in a fundamental set of solutions produces the general solution. The Wronskian test can be used to check if our solutions form a fundamental set: It suffices to pick any point t_0 in I and verify that $\mathbf{W}(t_0) \neq 0$.

Constant coefficient systems. Returning to constant coefficient systems, suppose an $n \times n$ matrix $\mathbf{A}$ has n **distinct** eigenvalues. We claim that the general solution of $\mathbf{X}' = \mathbf{A}\mathbf{X}$ is the set of linear combinations of the ray solutions

$$\mathbf{X}_j(t) = e^{r_j t}\mathbf{V}_j, \quad 1 \leq j \leq n,$$

which come from the eigenvalue/eigenvector pairs for $\mathbf{A}$. This follows from Theorem 8.4.4, provided we can show that the vectors

$$\mathbf{X}_1(0) = \mathbf{V}_1, \ \mathbf{X}_2(0) = \mathbf{V}_2, \ldots, \ \mathbf{X}_n(0) = \mathbf{V}_n$$

are linearly independent. The general principle is that **eigenvectors corresponding to distinct eigenvalues are linearly independent**:

Theorem 8.4.6. *If $\mathbf{A}$ is an $n \times n$ matrix having distinct eigenvalues $r_1, r_2, \ldots, r_j$, where $j \leq n$, then the corresponding eigenvectors $\mathbf{V}_1, \mathbf{V}_2, \ldots, \mathbf{V}_j$ are linearly independent.*

Proof. Our goal is to show that a collection of j eigenvectors corresponding to j distinct eigenvalues must be linearly independent. We will argue by induction. The case of $j = 2$ eigenvectors has already been done; see the discussion following equation (8.35). For some $j \geq 2$, suppose that

(8.43) $$c_1\mathbf{V}_1 + c_2\mathbf{V}_2 + \cdots + c_j\mathbf{V}_j = \mathbf{0},$$

where $\mathbf{V}_k$ is an eigenvector corresponding to the eigenvalue r_k and the r_k's are distinct. To verify linear independence of these vectors, we must show that $c_1 = c_2 = \cdots = c_j = 0$. Multiply equation (8.43) by r_1 to obtain

(8.44) $$c_1 r_1\mathbf{V}_1 + c_2 r_1\mathbf{V}_2 + \cdots + c_j r_1\mathbf{V}_j = \mathbf{0}.$$

Next, multiply equation (8.43) by $\mathbf{A}$ to obtain

$$c_1\mathbf{A}\mathbf{V}_1 + c_2\mathbf{A}\mathbf{V}_2 + \cdots + c_j\mathbf{A}\mathbf{V}_j = \mathbf{0}.$$

Since the $\mathbf{V}$'s are eigenvectors, we may rewrite the last line as

(8.45) $$c_1 r_1\mathbf{V}_1 + c_2 r_2\mathbf{V}_2 + \cdots + c_j r_j\mathbf{V}_j = \mathbf{0}.$$

Subtracting equation (8.44) from equation (8.45) gives

$$c_2(r_2 - r_1)\mathbf{V}_2 + c_3(r_3 - r_1)\mathbf{V}_3 + \cdots + c_j(r_j - r_1)\mathbf{V}_j = \mathbf{0}.$$

Now the induction hypothesis comes into play. In the last line we have a linear combination of $j-1$ eigenvectors belonging to distinct eigenvalues which sums to $\mathbf{0}$. By the induction hypothesis, these eigenvectors are linearly independent. Hence the coefficients $c_2(r_2 - r_1)$, $c_3(r_3 - r_1)$, $\ldots$, $c_j(r_j - r_1)$

must all be zero. Since the r_k's are distinct, this forces $c_2 = c_3 = \cdots = c_j = 0$. Once we know this, equation (8.43) also tells us that $c_1 = 0$. □

Earlier we described the general solution of a planar system $\mathbf{X} = \mathbf{AX}$ if the 2×2 matrix $\mathbf{A}$ has two distinct real eigenvalues. Theorem 8.4.6 allows us to generalize this higher dimensions:

> If an $n \times n$ matrix $\mathbf{A}$ has n distinct real eigenvalues $r_1, r_2, \ldots, r_n$, with corresponding eigenvectors $\mathbf{V}_1, \mathbf{V}_2, \ldots, \mathbf{V}_n$, then the general solution of $\mathbf{X}' = \mathbf{AX}$ on $(-\infty, \infty)$ is
> $$\mathbf{X} = c_1 e^{r_1 t} \mathbf{V}_1 + c_2 e^{r_2 t} \mathbf{V}_2 + \cdots + c_n r^{r_n t} \mathbf{V}_n.$$

Example 8.4.7. Find the general solution to the system $\mathbf{X}' = \mathbf{AX}$, where
$$\mathbf{A} = \begin{pmatrix} 3 & 0 & 4 \\ 0 & 2 & 0 \\ -2 & 0 & -3 \end{pmatrix}.$$

Some calculations show that $\mathbf{A}$ has characteristic equation $(3-z)(2-z)(-3-z) + 8(2-z) = (2-z)(z^2-1) = 0$ and that $\mathbf{A}$ has eigenvalue/eigenvector pairs

$$r_1 = 2, \ \mathbf{V}_1 = \begin{pmatrix} 0 \\ 1 \\ 0 \end{pmatrix}, \quad r_2 = 1, \ \mathbf{V}_2 = \begin{pmatrix} -2 \\ 0 \\ 1 \end{pmatrix}, \quad \text{and} \quad r_3 = -1, \ \mathbf{V}_3 = \begin{pmatrix} 1 \\ 0 \\ -1 \end{pmatrix}.$$

Since the eigenvalues are distinct, these eigenvectors are linearly independent and the general solution is
$$\mathbf{X}(t) = c_1 e^{2t} \begin{pmatrix} 0 \\ 1 \\ 0 \end{pmatrix} + c_2 e^t \begin{pmatrix} -2 \\ 0 \\ 1 \end{pmatrix} + c_3 e^{-t} \begin{pmatrix} 1 \\ 0 \\ -1 \end{pmatrix}.$$

Similarity and solving systems. When $\mathbf{D}$ is a diagonal matrix, the corresponding differential equation system $\mathbf{X}' = \mathbf{DX}$ is particularly simple, since it is "uncoupled": Each equation in the system involves only one unknown function. For example, if
$$\mathbf{D} = \begin{pmatrix} 3 & 0 & 0 \\ 0 & 4 & 0 \\ 0 & 0 & 1 \end{pmatrix},$$
then the system $\mathbf{X}' = \mathbf{DX}$ is just
$$\frac{dx}{dt} = 3x, \quad \frac{dy}{dt} = 4y, \quad \text{and} \quad \frac{dz}{dt} = z.$$

We may solve for each variable *separately* by the methods of Chapter 2 to obtain
$$x(t) = c_1 e^{3t}, \quad y(t) = c_2 e^{4t}, \quad z(t) = c_3 e^t,$$
for arbitrary constants c_1, c_2, and c_3. Convince yourself that the eigenvalue/eigenvector approach leads to the same solutions.

Now suppose that $\mathbf{A}$ is similar to some diagonal matrix $\mathbf{D}$, so that there is an invertible matrix $\mathbf{T}$ with $\mathbf{T}^{-1}\mathbf{AT} = \mathbf{D}$. In Exercise 21 of Section 8.3 you were asked to show that the columns of $\mathbf{T}$ are eigenvectors of $\mathbf{A}$, with eigenvalues being the corresponding diagonal entries of $\mathbf{D}$. Since $\mathbf{T}$ is invertible, Theorem 8.2.9 tells us that these eigenvectors are linearly independent. This gives us

8.4. Solving linear systems

the ingredients for building the general solution to $\mathbf{X}' = \mathbf{AX}$. If $\mathbf{V}_j$ is the jth column of $\mathbf{T}$ and d_{jj} is the jth diagonal entry of $\mathbf{D}$ for $1 \leq j \leq n$, then the general solution to $\mathbf{X}' = \mathbf{AX}$ is

$$c_1 e^{d_{11} t} \mathbf{V}_1 + c_2 e^{d_{22} t} \mathbf{V}_2 + \cdots + c_n e^{d_{nn} t} \mathbf{V}_n.$$

This does *not* require that the diagonal entries d_{jj} (which are the eigenvalues of $\mathbf{A}$) be distinct.

On the other hand, suppose $\mathbf{A}$ is an $n \times n$ matrix having n distinct real eigenvalues $r_1, r_2, \ldots, r_n$ and corresponding eigenvectors $\mathbf{V}_j$. By Theorem 8.4.6, these eigenvectors are linearly independent. Build a matrix $\mathbf{T}$ by using $\mathbf{V}_j$ for the jth column of $\mathbf{T}$. Theorem 8.2.9 says that $\mathbf{T}$ is invertible. In Exercise 23 you will show that $\mathbf{T}^{-1} \mathbf{A T}$ is diagonal, with diagonal entries r_j. In short, an $n \times n$ matrix with n distinct real eigenvalues is similar to a diagonal matrix.

Recap. We have seen in this section how to produce the general solution of a planar linear constant coefficient homogeneous system $\mathbf{X}' = \mathbf{AX}$ when $\mathbf{A}$ has two distinct real eigenvalues, and we have indicated how the method generalizes to n-dimensional systems. A solution satisfying any desired initial condition can be found from the general solution. In Section 8.7 we will use the eigenvalue/eigenvector approach to solve $\mathbf{X}' = \mathbf{AX}$ when $\mathbf{A}$ is a 2×2 matrix of real constants whose eigenvalues are **not** two distinct real numbers.

8.4.1. Exercises.

In Exercises 1–9 write the system in the form $\mathbf{X}' = \mathbf{AX}$ for appropriate choice of $\mathbf{A}$ and $\mathbf{X}$ and then find the general solution.

1. $\dfrac{dx}{dt} = x + 2y,$
   $\dfrac{dy}{dt} = 2x + y.$

2. $\dfrac{dx}{dt} = x + 3y,$
   $\dfrac{dy}{dt} = 3x + y.$

3. $\dfrac{dx}{dt} = -x - 3y,$
   $\dfrac{dy}{dt} = y.$

4. $\dfrac{dx}{dt} = x,$
   $\dfrac{dy}{dt} = 3x + 2y.$

5. $\dfrac{dx}{dt} = x + y,$
   $\dfrac{dy}{dt} = 9x + y.$

6. $\dfrac{dx}{dt} = -2x + y,$
   $\dfrac{dy}{dt} = x - 2y.$

7. $\dfrac{dx}{dt} = x,$
   $\dfrac{dy}{dt} = 3x - 2y.$

8. $\dfrac{dx}{dt} = -x,$
   $\dfrac{dy}{dt} = x + 3y,$
   $\dfrac{dz}{dt} = x - y + 2z.$

9. $\dfrac{dx}{dt} = -x + y,$
   $\dfrac{dy}{dt} = 3y - 12z,$
   $\dfrac{dz}{dt} = x - y.$

10. Find the solution to the system in Exercise 1 satisfying $x(0) = -1, y(0) = 5$.
11. Find the solution to the system in Exercise 4 satisfying $x(0) = -1, y(0) = 6$.
12. Find the solution to the system in Exercise 4 satisfying $x(0) = 0, y(0) = 0$.

13. (a) Use Theorem 8.2.3 to show that if $\mathbf{X}_1$ and $\mathbf{X}_2$ solve $\mathbf{X}' = \mathbf{P}(t)\mathbf{X}$, then so do $\mathbf{X}_1 + \mathbf{X}_2$ and $c\mathbf{X}_1$, for any constant c.
    (b) Explain how the Superposition Principle (Theorem 8.4.1) follows from (a).

14. (a) Confirm that $x(t) = 2$, $y(t) = t$ and $x(t) = t^2$, $y(t) = t^3$ are both solutions to the system
$$\frac{dx}{dt} = -\frac{2}{t}x + \frac{4}{t^2}y, \quad \frac{dy}{dt} = -2x + \frac{5}{t}y$$
on the interval $(0, \infty)$.
    (b) Is
$$x(t) = 2c_1 + c_2 t^2, \quad y(t) = c_1 t + c_2 t^3$$
the general solution to this system on $(0, \infty)$? Why or why not?

15. (a) Write the system
$$\frac{dx}{dt} = -(\tan t)x - y, \quad \frac{dy}{dt} = x - (\tan t)y$$
in the form $\mathbf{X}' = \mathbf{P}(t)\mathbf{X}$, identifying $\mathbf{P}(t)$.
    (b) Confirm that $x(t) = \cos^2 t$, $y(t) = \cos t \sin t$ is a solution to this system on $(-\frac{\pi}{2}, \frac{\pi}{2})$.
    (c) Confirm that $x(t) = 2 - 2\sin^2 t$, $y(t) = \sin(2t)$ is also a solution to this system on $(-\frac{\pi}{2}, \frac{\pi}{2})$.
    (d) Is $x(t) = c_1 \cos^2 t + c_2(2 - 2\sin^2 t)$, $y(t) = c_1 \cos t \sin t + c_2 \sin(2t)$ the general solution to the system on $(-\frac{\pi}{2}, \frac{\pi}{2})$? Why or why not?

16. For $\mathbf{A}$ as given below, describe all equilibrium points of the system $\mathbf{X}' = \mathbf{A}\mathbf{X}$.
    (a)
$$\mathbf{A} = \begin{pmatrix} 2 & -1 \\ 4 & -2 \end{pmatrix}.$$
    (b)
$$\mathbf{A} = \begin{pmatrix} 1 & 0 & 0 \\ -2 & 0 & 0 \\ 3 & 1 & 5 \end{pmatrix}.$$

17. Recall that the substitution $x = u$, $y = u'$ converts the second-order differential equation
    (8.46) $$u'' + p(t)u' + q(t)u = 0$$
    into the system
    (8.47) $$\frac{dx}{dt} = y, \quad \frac{dy}{dt} = -q(t)x - p(t)y.$$
    (a) Suppose $u = f_1$ and $u = f_2$ are solutions to equation (8.46). What is the Wronskian $W(f_1, f_2)$?
    (b) What are the solutions $\mathbf{X}_1(t)$ and $\mathbf{X}_2(t)$ to equation (8.47) corresponding to $u = f_1$ and $u = f_2$? What is the Wronskian of these solutions?

18. Suppose that $\mathbf{X}_1(t), \mathbf{X}_2(t), \ldots, \mathbf{X}_n(t)$ are solutions to $\mathbf{X}' = \mathbf{P}(t)\mathbf{X}$ on an interval I, where the entries of $\mathbf{P}(t)$ are continuous on I, and let $\mathbf{W}(t)$ be the Wronskian of the $\mathbf{X}_j(t)$'s as defined in equation (8.41). Identify each of the following statements as TRUE or FALSE, with justification:
    (a) If $\mathbf{W}(t) \neq 0$ for *all* $t \in I$, then $c_1 \mathbf{X}_1(t) + c_2 \mathbf{X}_2(t) + \cdots + c_n \mathbf{X}_n(t)$ is the general solution of $\mathbf{X}' = \mathbf{P}(t)\mathbf{X}$ on I, where the c_j's are arbitrary constants.
    (b) If $\mathbf{W}(t) \neq 0$ for *some* $t_0 \in I$, then $c_1 \mathbf{X}_1(t) + c_2 \mathbf{X}_2(t) + \cdots + c_n \mathbf{X}_n(t)$ is the general solution of $\mathbf{X}' = \mathbf{P}(t)\mathbf{X}$ on I, where the c_j's are arbitrary constants.
    (c) Either $\mathbf{W}(t) = 0$ for all $t \in I$ or $\mathbf{W}(t) = 0$ at no point of I.
    (d) If the vectors $\mathbf{X}_1(t_0), \mathbf{X}_2(t_0), \ldots, \mathbf{X}_n(t_0)$ are linearly independent for *some* $t_0 \in I$, then the vectors $\mathbf{X}_1(t), \mathbf{X}_2(t), \ldots, \mathbf{X}_n(t)$ are linearly independent for *any* $t \in I$.

(e) If $\mathbf{W}(t_0) = 0$ for some t_0 in I, then the vector-valued functions $\mathbf{X}_1(t), \mathbf{X}_2(t), \ldots, \mathbf{X}_n(t)$ are linearly dependent on I.

19. Consider the vector-valued functions
$$\mathbf{X}_1(t) = \begin{pmatrix} t \\ 2t \end{pmatrix} \quad \text{and} \quad \mathbf{X}_2(t) = \begin{pmatrix} t \\ t^2 \end{pmatrix}.$$

   (a) Are $\mathbf{X}_1(t)$ and $\mathbf{X}_2(t)$ linearly independent on $(-\infty, \infty)$?
   (b) Is there any point t_0 at which the constant vectors $\mathbf{X}_1(t_0)$ and $\mathbf{X}_2(t_0)$ are linearly dependent?
   (c) Can $\mathbf{X}_1(t)$ and $\mathbf{X}_2(t)$ be solutions to any system $\mathbf{X}' = \mathbf{P}(t)\mathbf{X}$ on $(-\infty, \infty)$ where the entries of $\mathbf{P}(t)$ are continuous on $(-\infty, \infty)$?

20. Suppose that $\mathbf{X}_1(t), \mathbf{X}_2(t), \ldots, \mathbf{X}_n(t)$ are solutions to the system $\mathbf{X}' = \mathbf{P}(t)\mathbf{X}$ on an interval I, where the entries of $\mathbf{P}(t)$ are continuous on I.
   (a) Show that if the vector-valued functions $\mathbf{X}_1(t), \mathbf{X}_2(t), \ldots, \mathbf{X}_n(t)$ are linearly dependent on I, then $c_1\mathbf{X}_1(t) + c_2\mathbf{X}_2(t) + \cdots + c_n\mathbf{X}_n(t)$ is *not* the general solution to the system on I, where the c_j's are arbitrary constants. You may use the equivalence of (a)–(d) in Theorem 8.4.4.
   (b) Show that if the vector-valued functions $\mathbf{X}_1(t), \mathbf{X}_2(t), \ldots, \mathbf{X}_n(t)$ are linearly independent on I, then $c_1\mathbf{X}_1(t) + c_2\mathbf{X}_2(t) + \cdots + c_n\mathbf{X}_n(t)$, for arbitrary constants c_j, is the general solution to the system on I.

   Remark: Since (a) is logically equivalent to "If $c_1\mathbf{X}_1(t) + c_2\mathbf{X}_2(t) + \cdots + c_n\mathbf{X}_n(t)$ is the general solution, then $\mathbf{X}_1(t), \mathbf{X}_2(t), \ldots, \mathbf{X}_n(t)$ are linearly independent," this exercise shows that we may add the equivalent condition "(e) The vector-valued functions $\mathbf{X}_1(t), \mathbf{X}_2(t), \ldots, \mathbf{X}_n(t)$ are linearly independent on I" to Theorem 8.4.4.

21. Following the outline below, provide a proof of Theorem 8.4.3, using Theorems 8.1.1 and 8.2.9 as needed.
   (i) Show that properties (c) and (d) are equivalent.
   (ii) Show that properties (a) and (b) are equivalent.
   (iii) Show that property (d) implies property (b).
   (iv) Suppose that $\mathbf{X}_1(t_0)$ and $\mathbf{X}_2(t_0)$ are linearly *dependent* vectors. Show that there is a vector $\begin{pmatrix} \alpha \\ \beta \end{pmatrix}$ so that the equation
   $$c_1\mathbf{X}_1(t_0) + c_2\mathbf{X}_2(t_0) = \begin{pmatrix} \alpha \\ \beta \end{pmatrix}$$
   has no solution for c_1 and c_2. This establishes the implication **not**(c) $\Longrightarrow$ **not**(b), which is logically the same as the implication (b) $\Longrightarrow$ (c).
   (v) Check that parts (i)–(iv) provide a complete proof of Theorem 8.4.3.

22. This problem outlines a proof of Theorem 8.4.4. Suppose that
$$\mathbf{X}_1(t), \mathbf{X}_2(t), \ldots, \mathbf{X}_n(t)$$
are solutions to $\mathbf{X}' = \mathbf{P}(t)\mathbf{X}$ on an open interval I, where the entries of $\mathbf{P}(t)$ are continuous on I. Using Theorems 8.1.1 and 8.2.9 and Exercise 7 in Section 8.2, show the following:
   (i) For any $t_0 \in I$, the vectors $\mathbf{X}_1(t_0), \mathbf{X}_2(t_0), \ldots, \mathbf{X}_n(t_0)$ are linearly independent if and only if
   $$\det \begin{pmatrix} \uparrow & \uparrow & & \uparrow \\ \mathbf{X}_1(t_0) & \mathbf{X}_2(t_0) & \cdots & \mathbf{X}_n(t_0) \\ \downarrow & \downarrow & & \downarrow \end{pmatrix} \neq 0.$$
   Thus in the statement of Theorem 8.4.4, this shows (c) and (d) are equivalent.
   (ii) The general solution of $\mathbf{X}' = \mathbf{P}(t)\mathbf{X}$ on I is
   $$c_1\mathbf{X}_1(t) + c_2\mathbf{X}_2(t) + \cdots + c_n\mathbf{X}_n(t)$$

if and only if given any $t_0 \in I$ and constants $\alpha_1, \alpha_2, \ldots, \alpha_n$, there are constants $c_1, c_2, \ldots, c_n$ so that
$$c_1 \mathbf{X}_1(t_0) + c_2 \mathbf{X}_2(t_0) + \cdots + c_n \mathbf{X}_n(t_0) = \begin{pmatrix} \alpha_1 \\ \alpha_2 \\ \vdots \\ \alpha_n \end{pmatrix}.$$

This shows (a) and (b) in Theorem 8.4.4 are equivalent.

(iii) Let t_0 be in I and form the $n \times n$ matrix whose columns are the vectors $\mathbf{X}_j(t_0)$. If
$$\det \begin{pmatrix} \uparrow & \uparrow & & \uparrow \\ \mathbf{X}_1(t_0) & \mathbf{X}_2(t_0) & \cdots & \mathbf{X}_n(t_0) \\ \downarrow & \downarrow & & \downarrow \end{pmatrix} \neq 0,$$
then given any constants $\alpha_1, \alpha_2, \ldots, \alpha_n$ show that there are constants $c_1, c_2, \ldots, c_n$ so that
$$c_1 \mathbf{X}_1(t_0) + c_2 \mathbf{X}_2(t_0) + \cdots + c_n \mathbf{X}_n(t_0) = \begin{pmatrix} \alpha_1 \\ \alpha_2 \\ \vdots \\ \alpha_n \end{pmatrix}.$$

This verifies (d)$\Rightarrow$(b) in Theorem 8.4.4.

(iv) In a linear algebra course you will learn the following fact: If you have k vectors $\mathbf{V}_1, \mathbf{V}_2, \ldots, \mathbf{V}_k$ in $\mathbb{R}^n$, where $k < n$, then there is a vector $\mathbf{W}$ in $\mathbb{R}^n$ that cannot be written as a linear combination of the vectors $\mathbf{V}_1, \mathbf{V}_2, \ldots, \mathbf{V}_k$. Use this fact to show that if for some $t_0 \in I$ the vectors $\mathbf{X}_1(t_0), \mathbf{X}_2(t_0), \ldots, \mathbf{X}_n(t_0)$ are linearly *dependent*, then there are constants $\alpha_1, \alpha_2, \ldots, \alpha_n$ so that the equation
$$c_1 \mathbf{X}_1(t_0) + c_2 \mathbf{X}_2(t_0) + \cdots + c_n \mathbf{X}_n(t_0) = \begin{pmatrix} \alpha_1 \\ \alpha_2 \\ \vdots \\ \alpha_n \end{pmatrix}$$
has *no* solution for $c_1, c_2, \ldots, c_n$. This shows the implication **not**(c)$\Rightarrow$**not**(b), or equivalently (b)$\Rightarrow$(c) in the theorem.

23. Suppose that the $n \times n$ matrix $\mathbf{A}$ has n distinct real eigenvalues $r_1, r_2, \ldots, r_n$, and suppose an eigenvector corresponding to r_j is $\mathbf{V}_j$. Let $\mathbf{T}$ be the matrix whose jth column is $\mathbf{V}_j$. Explain why $\mathbf{T}$ is invertible, and show that $\mathbf{T}^{-1}\mathbf{A}\mathbf{T}$ is diagonal, with diagonal entries r_j.

Exercises 24–28 concern solutions to **autonomous linear systems**.

24. Consider the (nonhomogeneous) system
$$\frac{dx}{dt} = -3x + 4y + 10 = -3(x-2) + 4(y+1),$$
$$\frac{dy}{dt} = -2x + 3y + 7 = -2(x-2) + 3(y+1).$$
Make the substitution $u = x - 2, v = y + 1$ to rewrite the equations with dependent variables u and v instead of x and y. Solve the resulting linear homogeneous system for u and v and then use your answer to give the solution to the original system.

25. Consider the system
$$\frac{dx}{dt} = x + 4y - 6, \quad \frac{dy}{dt} = 2x - y - 3.$$
    (a) Find the equilibrium point.

(b) If (x_0, y_0) is the equilibrium point from (a), make the substitution $u = x - x_0$, $v = y - y_0$ to convert the original system to a homogeneous system for u and v. Solve the resulting system for u and v and then give the solution to the original system.

26. This exercise codifies the procedure in Exercise 25. Consider any (planar) autonomous linear system
$$\frac{dx}{dt} = ax + by + k_1, \quad \frac{dy}{dt} = cx + dy + k_2$$
where a, b, c, d, k_1, and k_2 are constants. Write this in matrix form as $\mathbf{X}' = \mathbf{AX} + \mathbf{C}$ for
$$\mathbf{A} = \begin{pmatrix} a & b \\ c & d \end{pmatrix} \quad \text{and} \quad \mathbf{C} = \begin{pmatrix} k_1 \\ k_2 \end{pmatrix}.$$

(a) Show that if $\det \mathbf{A} \neq 0$, then there is exactly one equilibrium point (x_0, y_0) and
$$\mathbf{X}_p(t) = \begin{pmatrix} x_0 \\ y_0 \end{pmatrix}$$
is a particular solution of the system $\mathbf{X}' = \mathbf{AX} + \mathbf{C}$.

(b) Show that if $\mathbf{X}_c(t)$ is the general solution of the homogeneous system $\mathbf{X}' = \mathbf{AX}$, then $\mathbf{X}(t) = \mathbf{X}_p(t) + \mathbf{X}_c(t)$ solves the original system $\mathbf{X}' = \mathbf{AX} + \mathbf{C}$. (Theorem 9.4.3 in the next chapter extends this principle to nonhomogeneous linear systems that are not necessarily autonomous.)

(c) Show that $\mathbf{X}_c(t) + \mathbf{X}_p(t)$ is the general solution to $\mathbf{X}' = \mathbf{AX} + \mathbf{C}$ by showing that if $\mathbf{Y}$ is any solution of $\mathbf{X}' = \mathbf{AX} + \mathbf{C}$, then $\mathbf{Y} - \mathbf{X}_p$ solves $\mathbf{X}' = \mathbf{AX}$.

In Exercises 27–28 find the equilibrium point of the given system and then the general solution of the system using the method outlined in Exercise 26.

27.
$$\frac{dx}{dt} = x - 6y + 3, \quad \frac{dy}{dt} = -x + 2y + 1.$$

28.
$$\frac{dx}{dt} = -2x + 3y + 8, \quad \frac{dy}{dt} = 3x - 2y - 7.$$

29. Consider the Cauchy-Euler type system
$$t\frac{dx}{dt} = -2y, \quad t\frac{dy}{dt} = -3x + y.$$

(a) Show that there are two values of r so that $x(t) = t^r$, $y(t) = bt^r$ solves the system on $(0, \infty)$, and find the corresponding values of b.
(b) Give the general solution to the system on $(0, \infty)$.

8.5. Phase portraits via ray solutions

In this section we will develop some quick tools for understanding phase portraits for certain planar linear systems, using their ray solutions and the parallelogram law for vector addition. We consider a system $\mathbf{X}' = \mathbf{AX}$, where $\mathbf{A}$ is a 2×2 constant matrix with real entries. In this section we assume that $\mathbf{A}$ has two distinct, nonzero real eigenvalues r_1 and r_2.

Example 8.5.1. Consider the linear system
$$\frac{dx}{dt} = x + 12y, \quad \frac{dy}{dt} = 3x + y,$$

which we write in matrix notation as $\mathbf{X}' = \mathbf{A}\mathbf{X}$, for $\mathbf{A} = \begin{pmatrix} 1 & 12 \\ 3 & 1 \end{pmatrix}$. The matrix $\mathbf{A}$ has eigenvalue/eigenvector pairs

$$r_1 = 7,\ \mathbf{V}_1 = \begin{pmatrix} 2 \\ 1 \end{pmatrix} \quad \text{and} \quad r_2 = -5,\ \mathbf{V}_2 = \begin{pmatrix} 2 \\ -1 \end{pmatrix}$$

(see Example 8.3.5). We can write the general solution to $\mathbf{X}' = \mathbf{A}\mathbf{X}$ as

$$\tag{8.48} \mathbf{X}(t) = c_1 e^{r_1 t} \mathbf{V}_1 + c_2 e^{r_2 t} \mathbf{V}_2 = c_1 e^{7t} \begin{pmatrix} 2 \\ 1 \end{pmatrix} + c_2 e^{-5t} \begin{pmatrix} 2 \\ -1 \end{pmatrix},$$

or

$$\tag{8.49} x(t) = 2c_1 e^{7t} + 2c_2 e^{-5t}, \quad y(t) = c_1 e^{7t} - c_2 e^{-5t}.$$

Understanding shapes of trajectories. For each choice of c_1 and c_2, the equations in (8.49) are parametric equations for a curve in the xy-plane, called a trajectory or orbit. Recall that a sketch of a representative collection of trajectories is called a phase portrait for the system. A phase portrait is one way to show the solutions graphically. Ray solutions, which are obtained by choosing one of c_1, c_2 to be 0, can be quickly drawn or visualized.

Let's see how to understand the behavior of the trajectory of (8.48) when both c_1 and c_2 are nonzero. We consider a concrete example by choosing $c_1 = \frac{1}{2}$ and $c_2 = -1$ in our general solution:

$$\tag{8.50} \mathbf{X}(t) = \frac{1}{2} e^{7t} \mathbf{V}_1 - e^{-5t} \mathbf{V}_2,$$

and we show how to quickly obtain a rough sketch of the trajectory of this solution. Start with a sketch showing the vectors $\mathbf{V}_1$ and $\mathbf{V}_2$ with their tails at the origin, and then scale these vectors by $c_1 = \frac{1}{2}$ and $c_2 = -1$, respectively (remembering that a negative scalar reverses the direction):

Figure 8.14. The vectors $\mathbf{V}_1$ and $\mathbf{V}_2$.

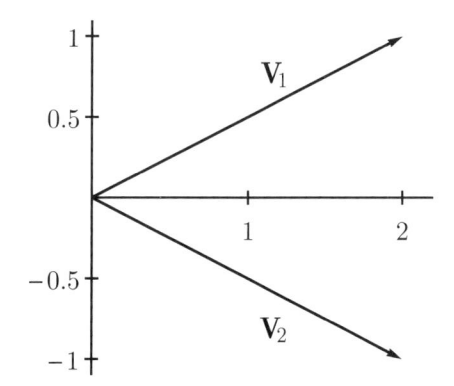

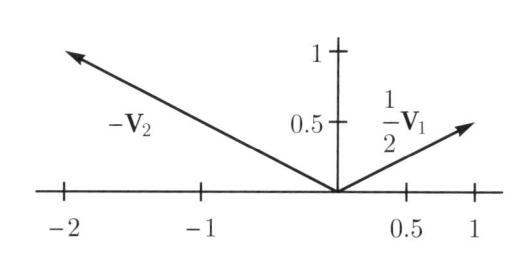

Figure 8.15. The vectors $\frac{1}{2}\mathbf{V}_1$ and $-\mathbf{V}_2$.

To find the point on the trajectory of (8.50) corresponding to $t = 0$ we compute $\mathbf{X}(0) = \frac{1}{2}\mathbf{V}_1 - \mathbf{V}_2$. Visually, we do this computation with the parallelogram law:

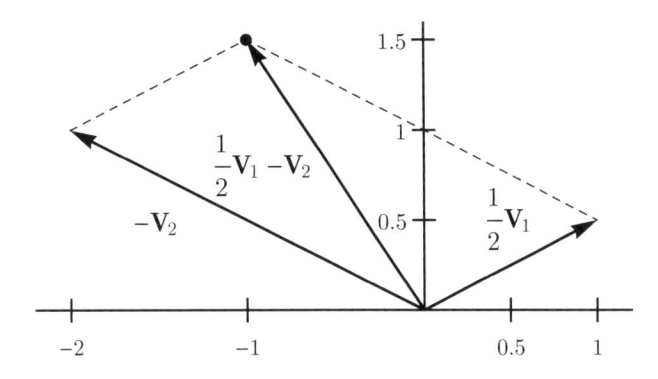

8.5. Phase portraits via ray solutions

Figure 8.16. $\mathbf{X}(0) = \frac{1}{2}\mathbf{V}_1 - \mathbf{V}_2$.

For a point on the trajectory corresponding to a time $t > 0$, we scale the vectors $\mathbf{V}_1$ and $\mathbf{V}_2$ by $\frac{1}{2}e^{7t}$ and $-e^{-5t}$, respectively, and then add. The scalar $\frac{1}{2}e^{7t}$ gets large (quickly!) as t increases beyond 0, while the scalar $-e^{-5t}$ gets small quickly as t increases from 0. The picture for a particular $t > 0$ might look something like that shown in Fig. 8.17.

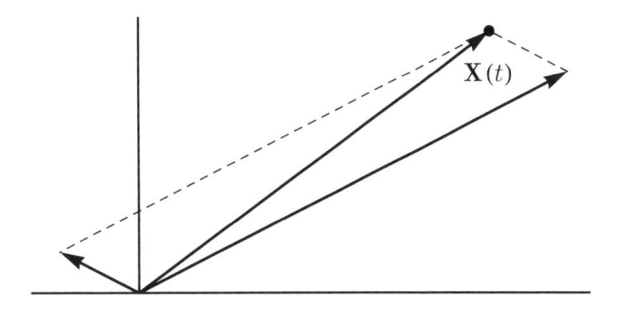

Figure 8.17. $\mathbf{X}(t) = \frac{1}{2}e^{7t}\mathbf{V}_1 - e^{-5t}\mathbf{V}_2$ for some $t > 0$.

Notice that the term $\frac{1}{2}e^{7t}\mathbf{V}_1$ is more important (or dominant) in determining the location of the sum $\mathbf{X}(t)$ when $t > 0$ is "large".

A similar analysis applies to determine points on the trajectory corresponding to negative values of t. For large negative values of t, the term $-e^{-5t}\mathbf{V}_2$ will dominate the term $\frac{1}{2}e^{7t}\mathbf{V}_1$, since e^{-5t} will be much larger than e^{7t}, as shown in Fig. 8.18.

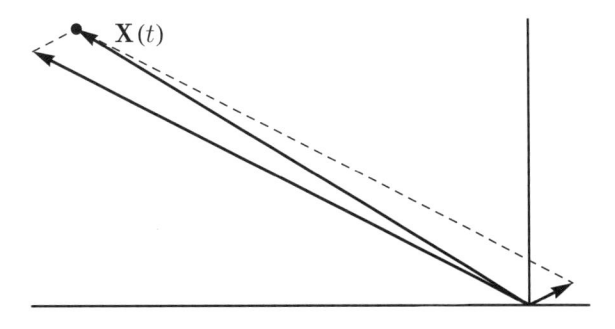

Figure 8.18. $\mathbf{X}(t) = \frac{1}{2}e^{7t}\mathbf{V}_1 - e^{-5t}\mathbf{V}_2$ for some $t < 0$.

If you carry out this analysis for a number of different values of t, positive and negative, you will see that the corresponding (hyperbola-like) trajectory looks something like that shown in Fig. 8.19. This is one trajectory which might be featured in a phase portrait. The dashed lines in Fig. 8.19

are along the trajectories of the ray solutions $e^{7t}\mathbf{V}_1$ and $-e^{-5t}\mathbf{V}_2$. Other trajectories come from making different choices for c_1 and c_2. A sketch of a phase portrait is shown in Fig. 8.20. Notice how this phase portrait includes four ray trajectories (two outgoing and two incoming). Keep in mind that the point $(0,0)$ is itself an (equilibrium) solution and that no other trajectory passes through this point (though the ray solutions approach it either as $t \to \infty$ or $t \to -\infty$).

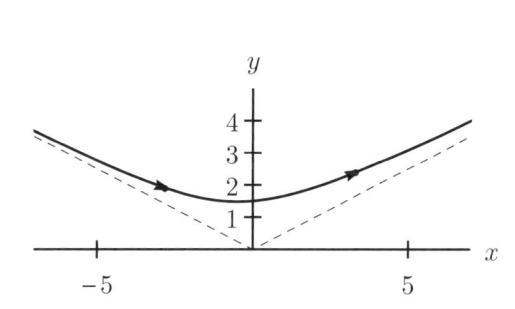

Figure 8.19. The trajectory $\mathbf{X}(t) = \frac{1}{2}e^{7t}\mathbf{V}_1 - e^{-5t}\mathbf{V}_2$.

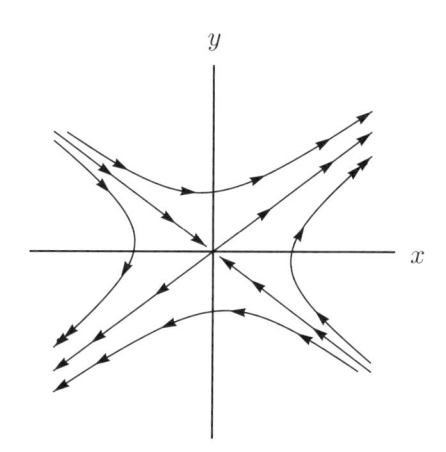

Figure 8.20. Phase portrait.

If, as in the preceding example, the eigenvalues of $\mathbf{A}$ are two real numbers with opposite signs, the equilibrium point $\mathbf{0}$ is called a **saddle point**, and it is said to be **unstable**, reflecting the fact that most solution curves that start "near" $\mathbf{0}$ eventually move far away from $\mathbf{0}$, as we saw in Fig. 8.20. Phase portraits for two other systems $\mathbf{X}' = \mathbf{A}\mathbf{X}$ where $\mathbf{A}$ has two real eigenvalues of opposite signs are shown in Figs. 8.21 and 8.22.

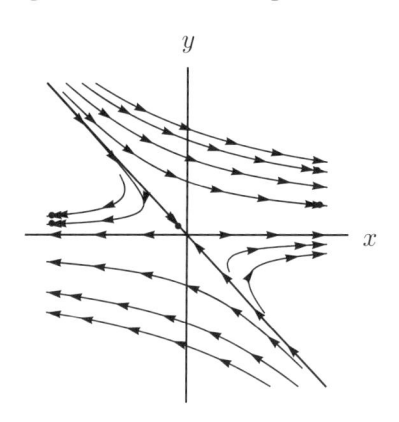

Figure 8.21. Real eigenvalues, opposite signs.

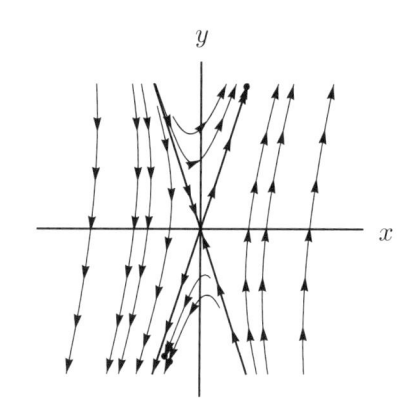

Figure 8.22. Real eigenvalues, opposite signs.

In the next example, there are two distinct real eigenvalues with the *same* sign.

Example 8.5.2. Let $\mathbf{A} = \begin{pmatrix} 3 & 1 \\ 4 & 6 \end{pmatrix}$ be the matrix associated with the linear system

$$\frac{dx}{dt} = 3x + y, \quad \frac{dy}{dt} = 4x + 6y.$$

The characteristic polynomial is $p(z) = \det(\mathbf{A} - z\mathbf{I}) = z^2 - 9z + 14 = (z-2)(z-7)$, and $\mathbf{A}$ has eigenvalues $r_1 = 2$ and $r_2 = 7$. Corresponding eigenvectors are

$$\mathbf{V}_1 = \begin{pmatrix} 1 \\ -1 \end{pmatrix} \text{ for } r_1 = 2 \quad \text{and} \quad \mathbf{V}_2 = \begin{pmatrix} 1 \\ 4 \end{pmatrix} \text{ for } r_2 = 7.$$

8.5. Phase portraits via ray solutions

There are outgoing ray solutions
$$\mathbf{X}_1(t) = e^{2t}\mathbf{V}_1 = e^{2t}\begin{pmatrix} 1 \\ -1 \end{pmatrix} \quad \text{and} \quad \mathbf{X}_2(t) = e^{7t}\mathbf{V}_2 = e^{7t}\begin{pmatrix} 1 \\ 4 \end{pmatrix}.$$

The general solution is
$$\mathbf{X}(t) = c_1 e^{2t}\begin{pmatrix} 1 \\ -1 \end{pmatrix} + c_2 e^{7t}\begin{pmatrix} 1 \\ 4 \end{pmatrix}.$$

The vectors $c_1 e^{2t}\mathbf{V}_1$ and $c_2 e^{7t}\mathbf{V}_2$ both get longer as t increases. Since $c_2 e^{7t}\mathbf{V}_2$ has the higher exponential power, it gets longer faster, which causes the solution curve to bend more towards the ray $c_2 e^{7t}\mathbf{V}_2$ than towards the ray $c_1 e^{2t}\mathbf{V}_1$. This yields the parabola-like shape in Fig. 8.23.

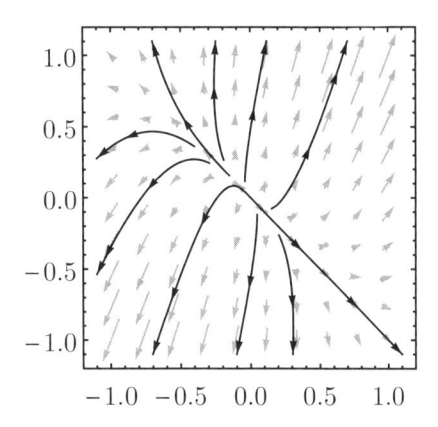

Figure 8.23. Phase portrait for Example 8.5.2 with vector field in background. The origin is called an unstable node.

This picture is typical of the case where $\mathbf{A}$ has two different real eigenvalues, both positive. In this situation, the equilibrium point $\mathbf{0}$ is called an **unstable node**. If instead $\mathbf{A}$ had two different negative eigenvalues, the solution curves would be shaped similarly to those in Fig. 8.23 but would all approach (but never reach) $\mathbf{0}$ as $t \to \infty$. In this case, $\mathbf{0}$ is called a **stable node**.

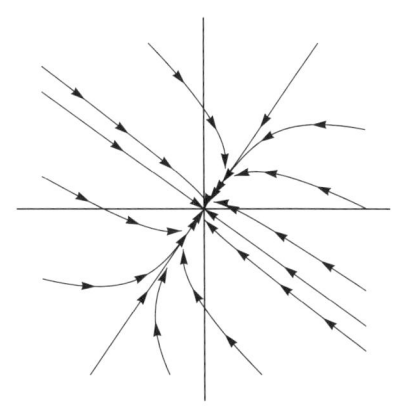

Figure 8.24. Stable node.

A simple trick for visualizing trajectories: Relating ray solutions. We continue to assume that $\mathbf{A}$ is a 2×2 matrix with two distinct real eigenvalues r_1 and r_2, neither of which is 0. The general solution to the system $\mathbf{X}' = \mathbf{A}\mathbf{X}$ is then

(8.51) $$\mathbf{X}(t) = c_1 e^{r_1 t}\mathbf{V}_1 + c_2 e^{r_2 t}\mathbf{V}_2,$$

where $\mathbf{V}_1$ and $\mathbf{V}_2$ are eigenvectors, respectively, for r_1 and r_2, and c_1, c_2 are arbitrary real constants. Keep in mind that $c_1 e^{r_1 t} \mathbf{V}_1$ and $c_2 e^{r_2 t} \mathbf{V}_2$ are ray solutions. For any value of t, the parallelogram law tells us how to draw a picture of the vector equation in (8.51); see Fig. 8.25.

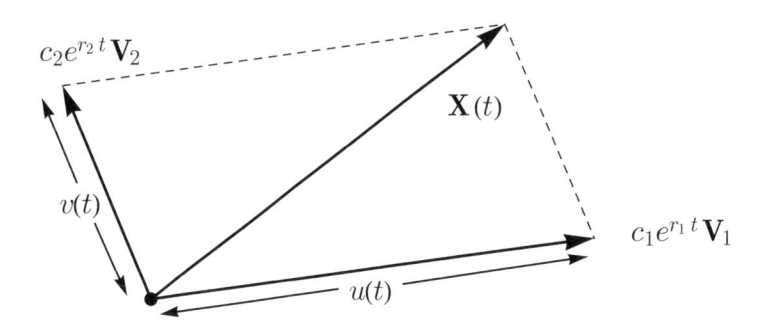

Figure 8.25. Picture of the vector equation (8.51).

For each t, let $u(t)$ and $v(t)$ be the *lengths* of the vectors $c_1 e^{r_1 t} \mathbf{V}_1$ and $c_2 e^{r_2 t} \mathbf{V}_2$, respectively, as pictured. We have

$$u(t) = \|c_1 e^{r_1 t} \mathbf{V}_1\| = |c_1| \|\mathbf{V}_1\| e^{r_1 t} \quad \text{and} \quad v(t) = \|c_2 e^{r_2 t} \mathbf{V}_2\| = |c_2| \|\mathbf{V}_2\| e^{r_2 t}.$$

Assuming neither c_1 nor c_2 is 0, we may divide by $|c_1|$ or $|c_2|$ to obtain

$$\frac{u(t)}{|c_1| \|\mathbf{V}_1\|} = e^{r_1 t} \quad \text{and} \quad \frac{v(t)}{|c_2| \|\mathbf{V}_2\|} = e^{r_2 t}.$$

Since

$$e^{r_2 t} = (e^{r_1 t})^{r_2/r_1},$$

we have

$$\frac{v(t)}{|c_2| \|\mathbf{V}_2\|} = \left(\frac{u(t)}{|c_1| \|\mathbf{V}_1\|} \right)^{r_2/r_1},$$

or, writing u for $u(t)$ and v for $v(t)$,

(8.52) $$v = a u^{r_2/r_1}$$

where a is the positive constant

$$a = |c_2| \|\mathbf{V}_2\| (|c_1| \|\mathbf{V}_1\|)^{-r_2/r_1}.$$

equation (8.52) relates the lengths of the vectors $c_1 e^{r_1 t} \mathbf{V}_1$ and $c_2 e^{r_2 t} \mathbf{V}_2$ (which give ray solutions) to each other. We can plot this relationship $v = a u^{r_2/r_1}$ in the first quadrant of the uv-plane. In the case that r_1 and r_1 have opposite signs, so that r_2/r_1 is negative, this is done in Fig. 8.26, yielding a hyperbola-like curve. Possible trajectories in a phase portrait are shown in Fig. 8.27, indicating a saddle point at $(0,0)$. The boldface trajectory containing $\mathbf{X}(t)$ should be considered as a skewed version of the curve shown in the uv-plot. If the vectors $\mathbf{V}_1$ and $\mathbf{V}_2$ happen to be perpendicular, this trajectory will exactly duplicate the curve in Fig. 8.26, up to a rotation of the axes about the origin. The remaining three (nonboldface) trajectories in Fig. 8.27 are possible trajectories in the remaining three "skewed quadrants".

8.5. Phase portraits via ray solutions

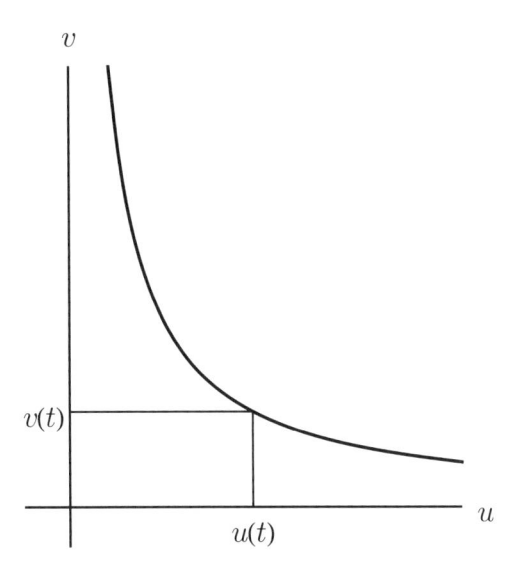

Figure 8.26. Typical graph of $v = au^{r_2/r_1}$ if r_1 and r_2 have different signs.

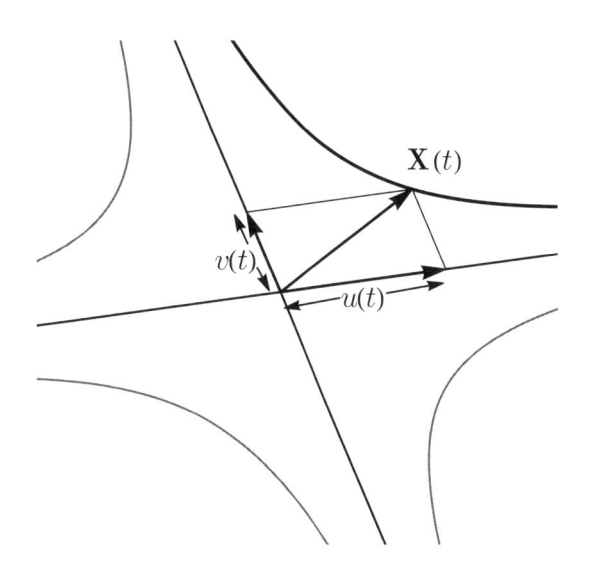

Figure 8.27. Possible trajectories if r_1 and r_2 have different signs.

When $r_2/r_1 < 0$, the curve $v = au^{r_2/r_1}$ has the u- and v-axes as asymptotes. From this observation, we correctly predict that the nonray trajectories in Fig. 8.27 are asymptotic to the eigenvector lines (which contain the ray trajectories), either as $t \to \infty$ or as $t \to -\infty$. This feature will be explored further in Section 8.6.

If $0 < r_1 < r_2$ or $r_2 < r_1 < 0$, then $r_2/r_1 > 1$ and the graph of $v = au^{r_2/r_1}$ in the uv-plane is parabola-like and tangent to the u-axis at the origin, as pictured in Fig 8.28. In Fig. 8.29 we see a skewed version of this curve giving the corresponding trajectory in the phase portrait, indicating a node at $(0,0)$. Note the trajectory is tangent to the eigenvector line for $\mathbf{V}_1$, the eigenvector corresponding to the eigenvalue r_1 having smaller absolute value.

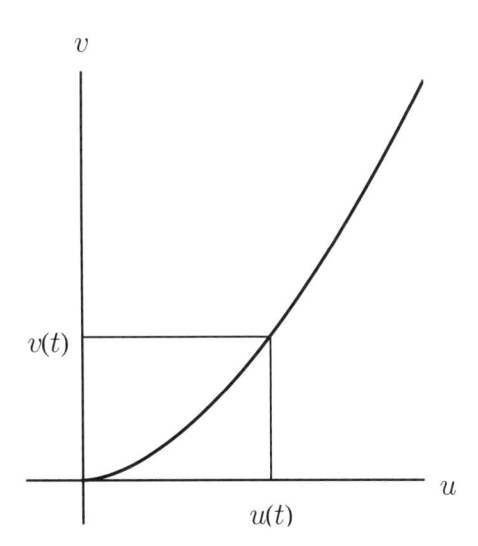

Figure 8.28. Typical graph of $v = au^{r_2/r_1}$ when $r_2/r_1 > 1$.

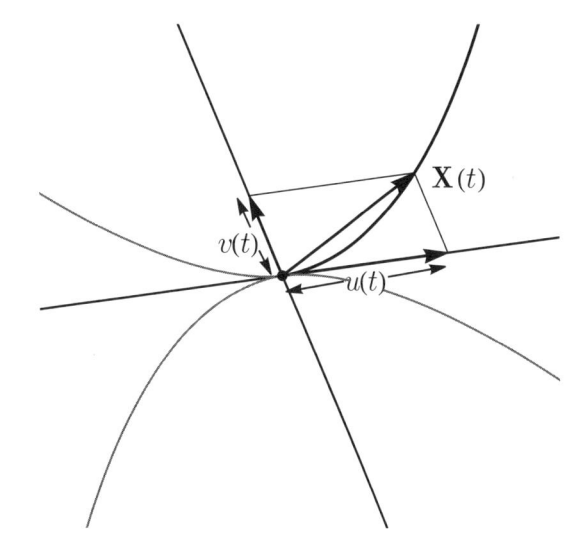

Figure 8.29. Possible trajectories when $r_2/r_1 > 1$.

If $0 < r_2 < r_1$ or $r_1 < r_2 < 0$, then $0 < r_2/r_1 < 1$, and the graph of $v = au^{r_2/r_1}$ is still parabola-like, but now tangent to the v-axis at the origin; see Figs. 8.30. Fig. 8.31 shows a possible phase portrait; note the nonray trajectories are tangent to the eigenvector line for $\mathbf{V}_2$, which is the eigenvector corresponding to the eigenvalue r_2 having smaller absolute value. The equilibrium point $\mathbf{0}$ is again a node.

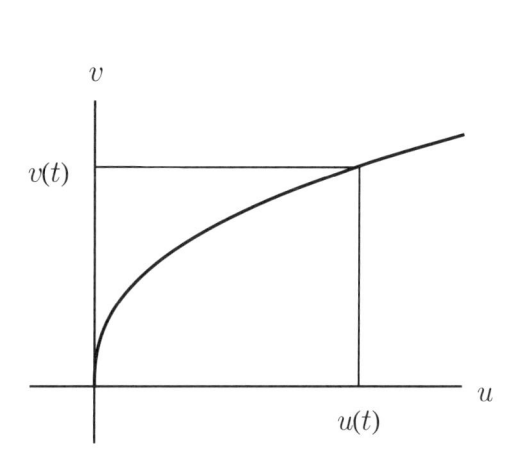

Figure 8.30. Typical graph of $v = au^{r_2/r_1}$ if $0 < r_2/r_1 < 1$.

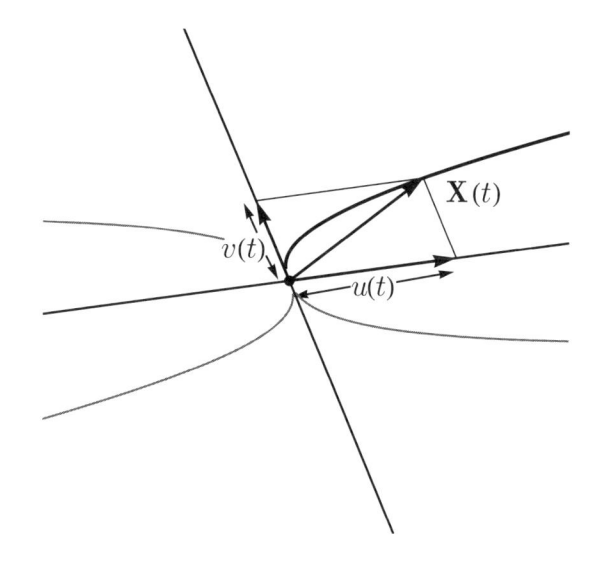

Figure 8.31. Possible trajectories in the case $0 < r_2/r_1 < 1$.

It is worth remembering that in all cases the lengths $u(t)$ and $v(t)$ of the ray solutions $c_1 e^{r_1 t} \mathbf{V}_1$ and $c_2 e^{r_2 t} \mathbf{V}_2$ whose sum is the solution $\mathbf{X}(t)$ are related by a simple power law: $v(t) = au(t)^{r_2/r_1}$. The pictures in Figs. 8.27, 8.29, and 8.31 do not include arrows showing the directions the trajectories are traversed since we have not specified the signs of the individual eigenvalues r_1 and r_2.

When $\mathbf{0}$ is a node, all trajectories will approach $\mathbf{0}$ either as $t \to \infty$ (if both eigenvalues are negative) or as $t \to -\infty$ (if both eigenvalues are positive). The uv-plane pictures correctly suggest that the nonray trajectories approach the origin tangent to one particular line—the eigenvector line for the eigenvalue with smaller absolute value. In fact, the only trajectories that don't approach tangent to this line are rays along the eigenvector line for the eigenvalue with larger absolute value. We call the eigenvector lines the **nodal tangents** for the system, and we call the one that corresponds to the eigenvalue with smaller absolute value the **preferred nodal tangent**. The next result summarizes these observations.

Theorem 8.5.3 (Preferred Nodal Tangent Theorem). *Suppose that $\mathbf{A}$ is a 2×2 matrix having distinct real eigenvalues of the same sign so that $(0,0)$ is a node for the system*

$$\mathbf{X}' = \mathbf{A}\mathbf{X}.$$

The preferred nodal tangent for this system is the eigenvector line L corresponding to the eigenvalue of $\mathbf{A}$ that has smaller absolute value. The trajectory of any solution, except for the ray solutions corresponding to the eigenvalue of larger absolute value, will approach the origin (either as $t \to \infty$ or as $t \to -\infty$) tangent to L.

In Exercise 10 you are invited to provide a proof of Theorem 8.5.3 using substitution.

8.5. Phase portraits via ray solutions

Recap. We have developed some simple tools for understanding the shapes of the trajectories of solutions to $\mathbf{X}' = \mathbf{AX}$ when the 2×2 matrix has two distinct real eigenvalues. When the eigenvalues have opposite signs, the trajectories are hyperbola-like, while when the eigenvalues have the same sign, the trajectories are parabola-like. In the latter case, the nonray trajectories approach the origin (either as $t \to \infty$ or as $t \to -\infty$) tangent to the eigenvector line corresponding to the eigenvalue with smaller absolute value. In the next section we will give additional methods for enhancing geometric detail in phase portrait production. While in that section we still focus on the "distinct real eigenvalues" cases, the tools we develop have much broader applicability; in particular in Chapter 10 they will be the primary means for producing a phase portrait for *nonlinear* autonomous planar systems.

8.5.1. Exercises.

1. (a) Sketch, in the first quadrant and on the same set of axes, the curves $v = u^2$, $v = u^3$, $v = u^{1/2}$, and $v = u^{1/3}$, for $0 \leq u \leq 3/2$. Your curves should show the correct tangency at $(0,0)$ and the correct relationships to each other.
   (b) All four curves intersect at $(0,0)$ and in one other point. Which one?

2. (a) The vector function
$$\mathbf{X}(t) = e^t \begin{pmatrix} 1 \\ 0 \end{pmatrix} + e^{2t} \begin{pmatrix} 1 \\ 1 \end{pmatrix}$$
is a solution of some linear system $\mathbf{X}' = \mathbf{AX}$. One of the three pictures below shows the trajectory of this solution. Which one?

   (b) The vector function
$$\mathbf{X}(t) = e^{-t} \begin{pmatrix} 1 \\ 0 \end{pmatrix} + e^t \begin{pmatrix} 1 \\ 1 \end{pmatrix}$$
is a solution of another linear system. One of the pictures below shows the trajectory of this solution. Which one?

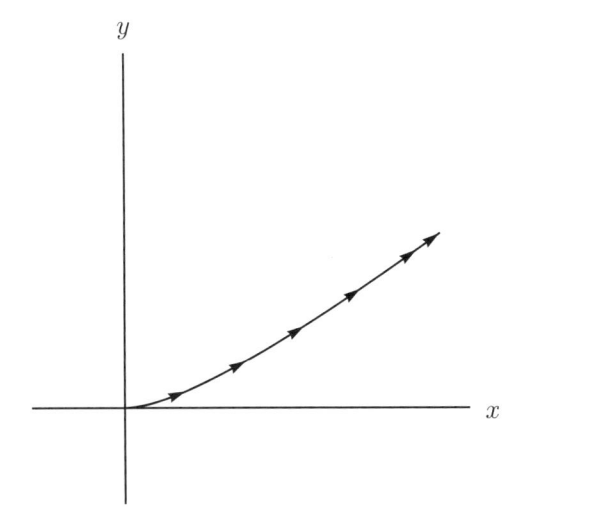

Figure 8.32. (A)

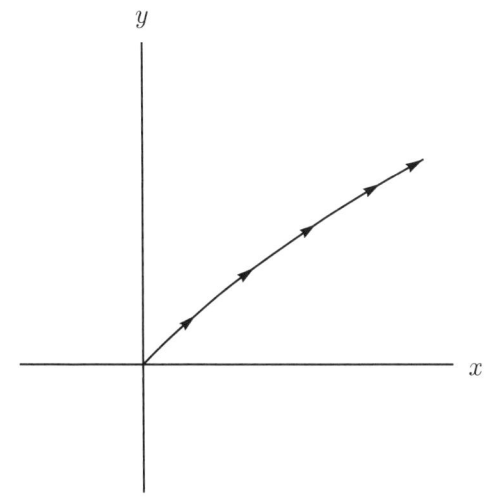

Figure 8.33. (B)

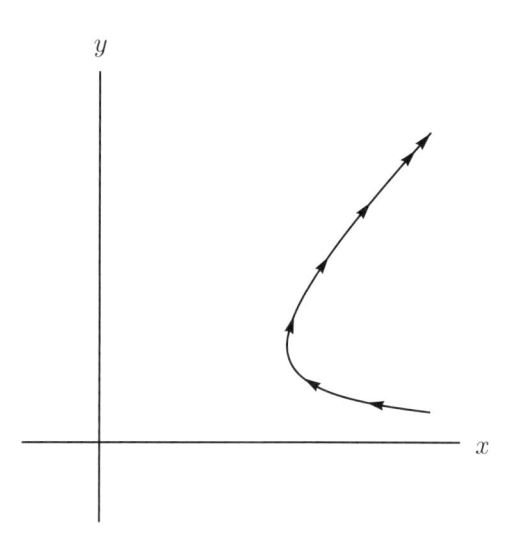

Figure 8.34. (C)

3. (a) The vector function

$$\mathbf{X}(t) = e^{2t} \begin{pmatrix} 1 \\ 0 \end{pmatrix} + e^t \begin{pmatrix} -1 \\ 1 \end{pmatrix}$$

is a solution to some linear system $\mathbf{X}' = \mathbf{A}\mathbf{X}$. One of the pictures below shows the trajectory of this solution. Which one?

(b) For the trajectory in your answer to (a), this trajectory approaches $\mathbf{0}$ (as $t \to -\infty$) tangent to what line?

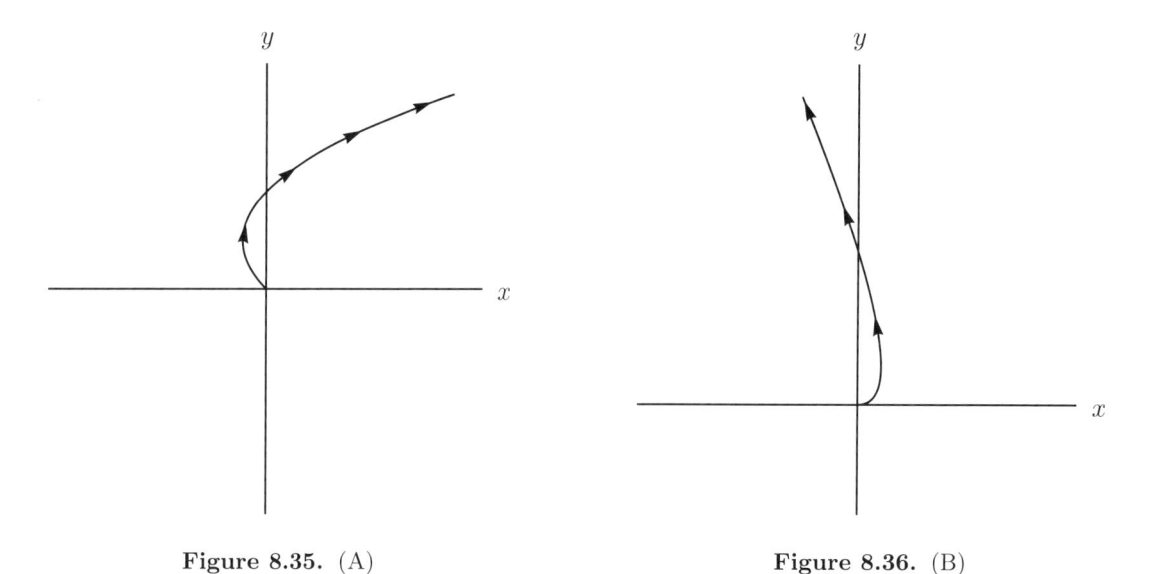

Figure 8.35. (A)   **Figure 8.36.** (B)

4. The term *parabola-like*, used in the text in reference to trajectories near a node, is intended to be flexible enough to encompass some variation. Suppose for instance that $\mathbf{A}$ and $\mathbf{B}$ are two particular 2×2 matrices with real entries and you know that $\mathbf{A}$ has eigenvalues $r_1 = 2$ and $r_2 = 16$, while $\mathbf{B}$ has eigenvalues $r_1 = 4$ and $r_2 = 6$.
   (a) (CAS) Use a computer to produce a uv-plot of the relation $v = u^{r_2/r_1}$, for $0 \le u \le 1$ and $0 \le v \le 1$, for the system $\mathbf{X}' = \mathbf{A}\mathbf{X}$ (for definiteness we take $a = 1$ in equation (8.52)).
   (b) (CAS) Repeat (a) for the system $\mathbf{X}' = \mathbf{B}\mathbf{X}$.

8.5. Phase portraits via ray solutions

(c) Figs. 8.37–8.38 show phase portraits for the two systems $\mathbf{X}' = \mathbf{AX}$ and $\mathbf{X}' = \mathbf{BX}$. Which phase portrait goes with which system? Your pictures in (a) and (b) should offer a clue.

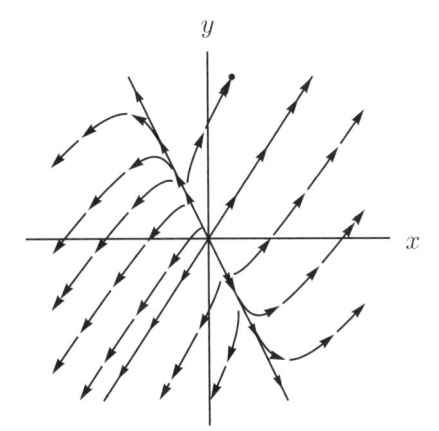

Figure 8.37. Phase portrait I.

Figure 8.38. Phase portrait II.

5. Phase portraits for several planar systems $\mathbf{X}' = \mathbf{AX}$ are shown in Figs. 8.39–8.41. For which one(s) does $\mathbf{A}$ have one positive and one negative eigenvalue? For which one(s) does $\mathbf{A}$ have two positive eigenvalues? Two negative eigenvalues?

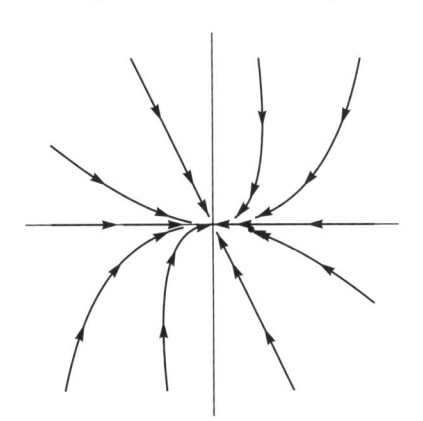

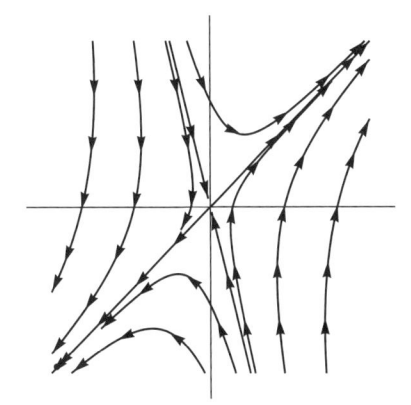

Figure 8.39. (A)

Figure 8.40. (B)

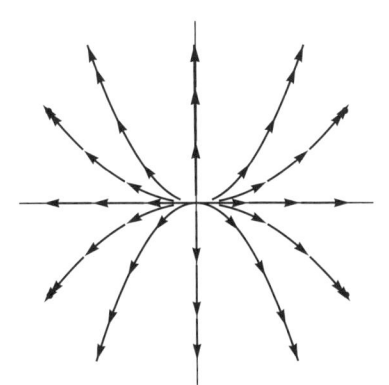

Figure 8.41. (C)

6. The system $\mathbf{X}' = \mathbf{AX}$ with
$$\mathbf{A} = \begin{pmatrix} 1 & 0 \\ 3 & -2 \end{pmatrix}$$
has general solution
$$\mathbf{X}(t) = c_1 e^t \begin{pmatrix} 1 \\ 1 \end{pmatrix} + c_2 e^{-2t} \begin{pmatrix} 0 \\ 1 \end{pmatrix}.$$

   (a) Sketch the ray solutions, including arrows to show if they are incoming or outgoing rays.
   (b) Do you expect the nonray solutions to have trajectories that are hyperbola-like or parabola-like? Explain.
   (c) Give a rough sketch by hand of the trajectory of the solution obtained by setting $c_1 = 1$ and $c_2 = -1$. Include arrows on the trajectory to show the direction it is traversed.

7. A linear system $\mathbf{X}' = \mathbf{AX}$ has general solution
$$\mathbf{X}(t) = c_1 e^{-2t} \begin{pmatrix} 1 \\ 2 \end{pmatrix} + c_2 e^{-3t} \begin{pmatrix} -1 \\ 3 \end{pmatrix}.$$

   (a) Explain how you can tell that the nonray solutions are parabola-like.
   (b) Identify the nodal tangent lines and specify which is the preferred nodal tangent.
   (c) Sketch the ray solutions, including arrows to show if they are incoming or outgoing.
   (d) Sketch the solution obtained by setting $c_1 = 1$ and $c_2 = -1$.

8. Consider the linear system $\mathbf{X}' = \mathbf{AX}$ where $\mathbf{A}$ is a 2×2 matrix with real entries having eigenvalues $r_1 = 1$ and $r_2 = -1$ with corresponding eigenvectors
$$\mathbf{V}_1 = \begin{pmatrix} 2 \\ 1 \end{pmatrix} \quad \text{and} \quad \mathbf{V}_2 = \begin{pmatrix} 1 \\ 2 \end{pmatrix},$$
respectively.

   (a) Carefully sketch the trajectories of the ray solutions, indicating the directions in which they are traversed by arrows.
   (b) Add to your picture from (a) the point $(3, 7)$ and a sketch, for $t \geq 0$, of your prediction for the trajectory of the solution $(x(t), y(t))$ with initial value $(x(0), y(0)) = (3, 7)$. Include a direction arrow on this trajectory.
   (c) Also add to your picture from (a) the point $(4, 7)$ and a sketch, for $t \geq 0$, of your prediction for the trajectory of the solution $(z(t), w(t))$ with initial value $(z(0), w(0)) = (4, 7)$. Include a direction arrow on this trajectory.

9. A certain system of homogeneous linear differential equations with constant coefficients has general solution
$$\mathbf{X}(t) = c_1 e^{2t} \begin{pmatrix} 1 \\ 3 \end{pmatrix} + c_2 e^{3t} \begin{pmatrix} -2 \\ 2 \end{pmatrix}.$$
Sketch the trajectory in the phase plane corresponding to $c_1 = -1, c_2 = 0$ and label it (A). Sketch the trajectory corresponding to $c_1 = -1, c_2 = 1$ and label it (B).

10. Suppose that $\mathbf{A}$ has distinct real eigenvalues r_1 and r_2 having the same sign, with $|r_1| < |r_2|$, and suppose that
    (8.53) $$\mathbf{X}(t) = c_1 e^{r_1 t} \mathbf{V}_1 + c_2 e^{r_2 t} \mathbf{V}_2$$
    is the general solution to $\mathbf{X}' = \mathbf{AX}$.
    (a) Make the change of variable $s = e^{r_1 t}$. If the eigenvalues are negative, what happens to s as $t \to \infty$? If the eigenvalues are positive, what happens to s as $t \to -\infty$?
    (b) Show that if c_1 and c_2 are nonzero, then the trajectory of (8.53) will be tangent at $(0,0)$ to the eigenvector line L for the eigenvalue r_1. Hint: Rewrite $\mathbf{X}(t)$ as a parametric curve with new independent variable $s = e^{r_1 t}$ (so that $\mathbf{Y}(s) = \mathbf{X}(t)$) and calculate $\mathbf{Y}(0)$ and $\mathbf{Y}'(0)$.

8.6. More on phase portraits: Saddle points and nodes

(c) What are the only trajectories of the system $\mathbf{X}' = \mathbf{A}\mathbf{X}$ that are *not* tangent to L at $(0,0)$?

8.6. More on phase portraits: Saddle points and nodes

Fig. 8.42 shows a computer generated phase portrait for the autonomous linear system
$$\frac{dx}{dt} = 4x + 2y, \qquad \frac{dy}{dt} = 3x - y.$$
Let's begin by recalling what the previous section would tell us about this picture. The system can be written as $\mathbf{X}' = \mathbf{A}\mathbf{X}$ for
$$\mathbf{A} = \begin{pmatrix} 4 & 2 \\ 3 & -1 \end{pmatrix}.$$
The eigenvalues of $\mathbf{A}$ are $r_1 = 5$ and $r_2 = -2$. Since they have opposite signs, $(0,0)$ is a saddle point for this system. We can check that
$$\mathbf{V}_1 = \begin{pmatrix} 2 \\ 1 \end{pmatrix} \quad \text{and} \quad \mathbf{V}_2 = \begin{pmatrix} -1 \\ 3 \end{pmatrix}$$
are eigenvectors of $\mathbf{A}$ corresponding to $r_1 = 5$ and $r_2 = -2$, respectively, and the corresponding eigenvector lines have equations $y = \frac{1}{2}x$ and $y = -3x$. These lines contain the outgoing and incoming ray trajectories, which are clearly visible in Fig. 8.42. According to Section 8.5, Fig. 8.42 certainly looks like a picture of trajectories near a saddle point. But without a computer to do it for us, can we generate and understand the fine detail of Fig. 8.42 ourselves? We will discuss more tools for doing so in this section. The first of these tools is called a **nullcline-and-arrow diagram**[9] which identifies the regions of the plane where solution curves—trajectories—of an autonomous system

(8.54) $$\frac{dx}{dt} = f(x,y), \qquad \frac{dy}{dt} = g(x,y)$$

will have one of four basic directional headings (as t increases): rising rightward, rising leftward, falling rightward, and falling leftward.

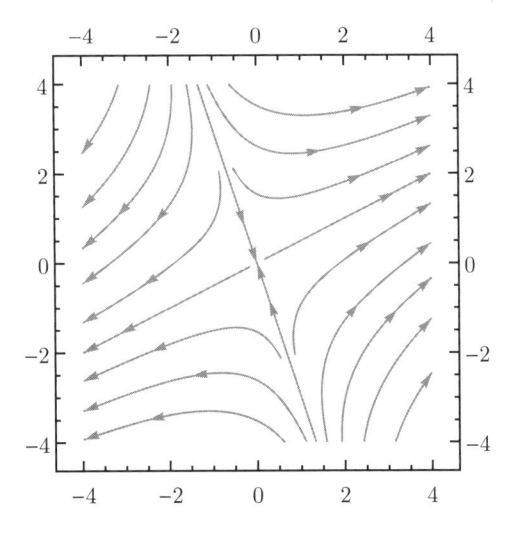

Figure 8.42. A phase portrait.

---

[9] You can think of this as an analogue of the phase line, which we introduced in Chapter 2, to quickly identify horizontal strips of the ty-plane in which solution curves of an autonomous equation $\frac{dy}{dt} = f(y)$ are increasing or decreasing.

Nullclines. An x-nullcline for the autonomous (not necessarily linear) system (8.54) is a curve C in the xy-plane such that each point (x,y) on C satisfies $f(x,y) = 0$. Thus, an x-nullcline of (8.54) is a curve along which field vectors for (8.54) will have x-component 0, so that they point up or down (assuming they are nonzero). Similarly, a y-nullcline for (8.54) is a curve along which field vectors for (8.54) have y-component 0, so that they point left or right (if they are nonzero). Note that a point where an x-nullcline meets a y-nullcline will be an equilibrium point. To produce x-nullclines (if any) for (8.54), we graph the equation $f(x,y) = 0$ and to produce y-nullclines, if any, we graph $g(x,y) = 0$.

Example 8.6.1. The parabola $y = x^2$ is the x-nullcline for

$$\frac{dx}{dt} = y - x^2, \quad \frac{dy}{dt} = xy + x$$

while the lines $x = 0$ and $y = -1$ are y-nullclines.

Nullcline-and-arrow diagrams. In Chapter 10, we will use nullcline-and-arrow diagrams to analyze the behavior of trajectories of nonlinear systems such as the one in the preceding example. Here, we will focus on developing and applying this tool in the context of *linear* systems, as illustrated in the next example.

Example 8.6.2. For the linear system

(8.55) $$\frac{dx}{dt} = 4x + 2y, \quad \frac{dy}{dt} = 3x - y,$$

whose phase portrait appears in Fig. 8.42 above, we see that the line $y = -2x$ is the x-nullcline while the line $y = 3x$ is the y-nullcline.

To construct a nullcline-and-arrow diagram for system (8.55), we begin by plotting the nullclines, which separate the plane into four sectors, I–IV:

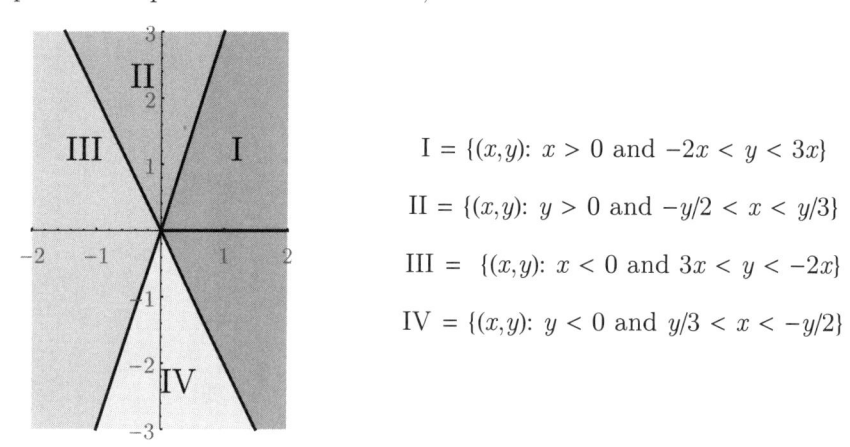

I = {(x,y): $x > 0$ and $-2x < y < 3x$}

II = {(x,y): $y > 0$ and $-y/2 < x < y/3$}

III = {(x,y): $x < 0$ and $3x < y < -2x$}

IV = {(x,y): $y < 0$ and $y/3 < x < -y/2$}

Figure 8.43. The x-nullcline $y = -2x$ and the y-nullcline $y = 3x$ create four sectors.

Note that because the inequalities in the definitions of sectors I–IV are strict, the half-line portions of the nullclines that form their boundaries are not included in the sectors. Thus, within these sectors, both $dx/dt = 4x + 2y$ and $dy/dt = 3x - y$ are nonzero; moreover, dx/dt and dy/dt do not change sign within a given sector—this follows quickly from the Intermediate Value Theorem (see Exercise 15). Hence, to determine the signs of dx/dt and dy/dt within a sector, that is, the signs of $f(x,y) = 4x + 2y$ and $g(x,y) = 3x - y$, we can choose a single test point (x_0, y_0) in the sector, The sign of $f(x_0, y_0)$ will be the sign of dx/dt throughout the sector and the sign of $g(x_0, y_0)$ will be the sign of dy/dt throughout the sector. For instance, in sector I, we might use

8.6. More on phase portraits: Saddle points and nodes

the test value $(1,0)$. Since $f(1,0) = 4 > 0$ and $g(1,0) = 3 > 0$, we know that dx/dt and dy/dt are positive throughout sector I. In a nullcline-and-arrow diagram, we convey the sign of dx/dt within a sector by placing in the sector a horizontal arrow—pointing right if dx/dt is positive and left if it is negative. To convey the sign of dy/dt within the sector, we place in the sector a vertical arrow—pointing up if dy/dt is positive and pointing down if it is negative. These **direction-indicator arrows** are drawn with common tail. The result is Fig. 8.44.

In Fig. 8.44, direction-indicator arrows tell us the basic behavior of trajectories of the system (8.55). For example, the combination of the rightward- and upward-pointing vectors in sector I tell us that trajectories in this sector move to the right and rise as time increases. Note this is consistent with the computer generated phase portrait of (8.55) in Fig. 8.42 above. In fact, the four direction-indicator pairs of arrows give us a global qualitative picture of the rightward/leftward and upward/downward behavior of trajectories consistent with Fig. 8.42. Because all trajectories display the same basic direction of motion in a given sector, we call these sectors **basic regions** for the system (8.55).

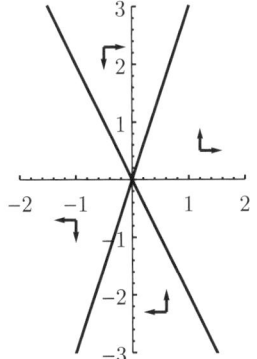

Figure 8.44. Nullclines $y = -2x$ and $y = 3x$ together with direction-indicator arrows.

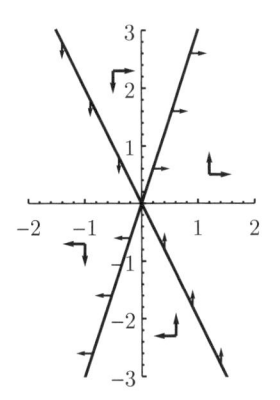

Figure 8.45. Complete nullcline-and-arrow diagram for the system (8.55).

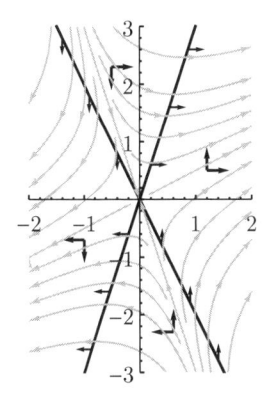

Figure 8.46. Nullcline-and-arrow diagram with phase portrait.

Crossing a nullcline. What about behavior of trajectories as they exit one basic region—crossing a nullcline—and enter another? If they cross an x-nullcline, they must have a vertical tangent line at the point crossed; if they cross a y-nullcline, they must have a horizontal tangent at the point crossed. To complete the nullcline-and-arrow diagram, we add some field vectors for the system (8.55) with tails along the nullclines. As we have already discussed, unless the tail lies at the equilibrium point $(0,0)$, those along the x-nullcline will be vertical and those along the y-nullcline will be horizontal. We scale these field vectors along the nullclines so that they have the same length. It remains to decide in which of the two possible directions these field vectors point. For example, consider the two half-lines that make up the x-nullcline $y = -2x$. The upper half-line (above the x-axis) serves as the boundary between sectors II and III, in which the vertical direction-indicator arrows point downward, and hence the field vectors along the top half of the x-nullcline $y = -2x$ must point downward (as shown in Fig. 8.45). Similarly, the lower half-line of the nullcline $y = -2x$ serves as the boundary between sectors I and IV, in which the vertical direction-indicator arrows point up, leading to upward pointing field vectors along this half-line. Similar reasoning gives the right- and left-pointing field vectors on the two halves of the y-nullcline $y = 3x$. Our complete nullcline-and-arrow diagram is Fig. 8.45. The field vectors along the nullclines (which we call nullcline arrows) remind us that any trajectory for the system that crosses an x-nullcline must have a vertical tangent at the crossing point (and be traveling in the direction of the nullcline arrow at the crossing point) and that any trajectory for the system that crosses a y-nullcline must have a horizontal tangent at the crossing point (and be traveling in the direction of the nullcline arrow at the crossing point).

We may view the nullcline-and-arrow diagram Fig. 8.45 as a "shorthand" version of the vector field of the system (8.55). Scaled field arrows appear along the nullclines, and the direction-indicator arrows in the basic regions tell us the basic orientation of field vectors in those regions.

Summary. Assume that $f = f(x,y)$ and $g = g(x,y)$ are continuous functions. To produce a nullcline-and-arrow diagram for the system

$$(8.56) \qquad \frac{dx}{dt} = f(x,y), \quad \frac{dy}{dt} = g(x,y)$$

do the following:

(i) Plot the x- and y-nullclines of (8.56) by graphing $f(x,y) = 0$ and $g(x,y) = 0$, respectively.

(ii) In each basic region of the plane free of nullclines, draw a pair of direction-indicator arrows with common tail and equal length—one vertical and one horizontal, indicating the signs of x' and y' in the region and thus the right/left and up/down direction of trajectories there.

(iii) Add some nullcline arrows—short field vectors of uniform length with tails along the x- and y-nullclines.

Beyond the information provided by the nullcline-and-arrow diagram, there are other details that will help improve accuracy for the trajectories we want to sketch. In particular, we will want to be able to understand the long-term behavior of solutions, and the next result (which generalizes the result of Exercise 23 in Section 2.6) is a key ingredient.

Theorem 8.6.3. *If $x = x(t)$, $y = y(t)$ is a solution of the autonomous system*

$$\frac{dx}{dt} = f(x,y), \quad \frac{dy}{dt} = g(x,y),$$

where f and g are continuous, and if $(x(t), y(t)) \to (x_*, y_*)$ as $t \to \infty$, then (x_*, y_*) must be an equilibrium point of the system.

Sketching the solution to an initial value problem: An example. We will use the nullcline and arrow diagram in Fig. 8.45, plus additional information about concavity and long-term behavior of trajectories, to sketch the solution of the initial value problem

$$(8.57) \qquad \frac{dx}{dt} = 4x + 2y, \quad \frac{dy}{dt} = 3x - y, \qquad x(0) = 0, \quad y(0) = 2.$$

Using the nullcline-and-arrow diagram Fig. 8.45, we can begin a sketch in the phase plane of the solution of the initial value problem in equation (8.57). Starting at $(0,2)$, we use the nearby direction-indictor arrows to draw a curve moving down and to the right. Realizing that our curve is heading toward the y-nullcline $y = 3x$, we wonder if the curve must cross the nullcline. The answer is yes. If it did not cross the nullcline, the curve moving downward and rightward from $(0,2)$ would have to approach a limit (x_*, y_*), with (x_*, y_*) lying in the triangle with vertices $(0,0), (0,2)$, and $(2/3, 2)$, and satisfying $x_* > 0$ and $y_* < 2$. By Theorem 8.6.3, (x_*, y_*) would be an equilibrium point of the system in (8.57), but $(0,0)$ is the system's only equilibrium point. Thus, the solution curve does cross the nullcline, as claimed. Because the solution curve will cross the y-nullcline $y = 3x$ at a point where it will have horizontal tangent, we guess that the decreasing curve we're sketching should be concave up; and, after entering the basic region to the right of the nullcline $y = 3x$, we guess it will continue to be concave up but will start increasing—moving upward and

8.6. More on phase portraits: Saddle points and nodes

rightward, consistent with the direction-indicator arrows in this region. *We do not have to guess about the concavity of our trajectory.* We can get information about its concavity, in fact global concavity information, via a **concavity chart** for the system in (8.57), which separates the phase plane into regions where trajectories must be concave up and regions where they must be concave down. The definition of concavity to apply here is the following: A curve C in the plane is concave up in a region R provided lines tangent to C at points in R will lie below C throughout R; similarly, C is concave down in R if lines tangent to C at points in R will lie above C throughout R; see Fig. 8.47.

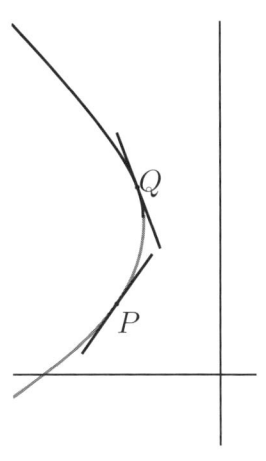

Figure 8.47. Curve is concave up at P (tangent below curve) and concave down at Q (tangent above curve).

Producing our concavity chart. As expected, we obtain concavity information via the second derivative. Eliminating the parameter from the system in (8.57), we obtain

$$\frac{dy}{dx} = \frac{\frac{dy}{dt}}{\frac{dx}{dt}} = \frac{3x - y}{4x + 2y}.$$

Thus,

$$\frac{d^2y}{dx^2} = \frac{(4x + 2y)\left(3 - \frac{dy}{dx}\right) - (3x - y)\left(4 + 2\frac{dy}{dx}\right)}{(4x + 2y)^2} = \frac{5\left(y - x\frac{dy}{dx}\right)}{2(2x + y)^2}.$$

Upon making the substitution $\frac{dy}{dx} = \frac{3x - y}{4x + 2y}$ in the preceding and simplifying, we obtain

(8.58) $$\frac{d^2y}{dx^2} = \frac{-5(3x + y)(x - 2y)}{4(2x + y)^3}.$$

We see that $\frac{d^2y}{dx^2}$ is zero along the lines $y = -3x$ and $y = \frac{1}{2}x$ and undefined along $y = -2x$. These lines separate the plane into six sectors where $\frac{d^2y}{dx^2}$ is defined and nonzero—see Fig. 8.48. The continuity of the function $\frac{-5(3x+y)(x-2y)}{4(2x+y)^3}$ on these regions together with the Intermediate Value Theorem (see Exercise 15) shows that $\frac{d^2y}{dx^2}$ does not change signs within these sectors. Thus a single test value from each sector determines the signs of the second derivative. For instance, substituting the point $(1,0)$ into the right-hand side of equation (8.58) yields $-15/32$, so that $\frac{d^2y}{dx^2}$ is negative in the sector containing $(1,0)$, and thus trajectories are concave down in this region. Choosing test values from the remaining regions and converting plus signs to "CU" (concave up) and minus signs to "CD" (concave down), we obtain the concavity chart in Fig. 8.49. In Fig. 8.50, we have combined the concavity chart and a phase portrait to see how the chart does indeed provide global concavity information for trajectories of the system in (8.57). In particular, the chart confirms that our guess about the concavity of the trajectory passing through $(0, 2)$ is correct.

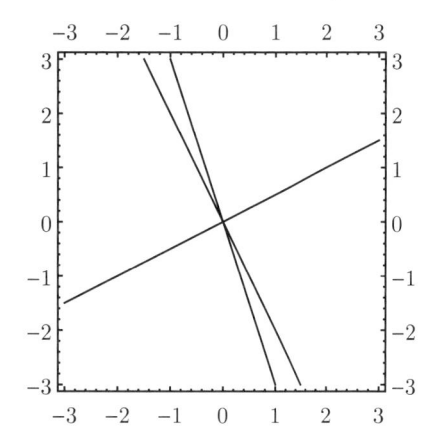

Figure 8.48. Lines along which $\frac{d^2y}{dx^2}$ is zero or undefined create six sectors within which $\frac{d^2y}{dx^2}$ doesn't change sign.

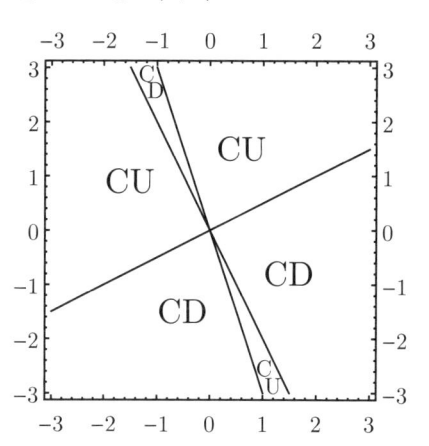

Figure 8.49. Concavity chart for the system in (8.57), with concavity regions bordered by the lines $y = -2x$, $y = -3x$, and $y = x/2$.

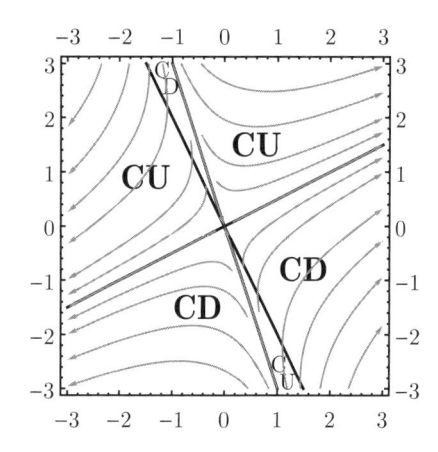

Figure 8.50. Concavity chart with phase portrait

Connections. The computation of the second derivative $\frac{d^2y}{dx^2}$ may seem somewhat tedious, but by making a connection with our work in Section 8.4 we can see how the lines dividing the phase plane into its concavity sectors can identified without the second derivative computation. The system in (8.57) has matrix of coefficients $\mathbf{A} = \begin{pmatrix} 4 & 2 \\ 3 & -1 \end{pmatrix}$. The matrix $\mathbf{A}$ has eigenvalue 5 with corresponding eigenvector $\begin{pmatrix} 2 \\ 1 \end{pmatrix}$ and eigenvalue -2 with eigenvector $\begin{pmatrix} -1 \\ 3 \end{pmatrix}$, so that the system $\mathbf{X}' = \mathbf{A}\mathbf{X}$ has ray solutions

$$\mathbf{X}_1(t) = e^{5t} \begin{pmatrix} 2 \\ 1 \end{pmatrix} \quad \text{and} \quad \mathbf{X}_2(t) = e^{-2t} \begin{pmatrix} -1 \\ 3 \end{pmatrix}.$$

Linear combinations of these two ray solutions provide the general solution to the system:

$$(8.59) \qquad \mathbf{X}(t) = c_1 e^{5t} \begin{pmatrix} 2 \\ 1 \end{pmatrix} + c_2 e^{-2t} \begin{pmatrix} -1 \\ 3 \end{pmatrix}.$$

8.6. More on phase portraits: Saddle points and nodes

The trajectory of the ray solution $\mathbf{X}_1$ follows the eigenvector line $y = \frac{1}{2}x$. Similarly, the trajectory for the ray solution $\mathbf{X}_2$ lies along the eigenvector line $y = -3x$. These eigenvector lines are precisely the "second-derivative zero" lines in our concavity chart in Fig. 8.49. We can see that this should be the case at a glance: A trajectory following a straight line should have no concavity. The third line $y = -2x$ in the concavity chart in Fig. 8.49 is the x-nullcline of the system (8.55). Because any trajectory of the system (8.55) will have a vertical tangent at a point where it crosses the x-nullcline, we anticipate its concavity will change at that point. Summarizing, we can understand, without using a second derivative computation, the lines appearing in our cavity chart:

- $\frac{d^2y}{dx^2} = 0$ along the eigenvector lines $y = \frac{1}{2}x$ and $y = -3x$, and
- $\frac{d^2y}{dx^2}$ does not exist along the x-nullcline $y = -2x$.

We return now to the task of sketching or visualizing, for $t \geq 0$, the solution of the initial value problem in equation (8.57). We have concavity information for the trajectory of this solution, and the direction-indicator arrows in our nullcline-and-arrow diagram Fig. 8.45 provide "increasing-decreasing" information. Thus, our task is similar to that encountered in a Calculus I course when curves are sketched by hand using increasing-decreasing information as well as concavity information. To improve the accuracy of our sketch, we note that our trajectory is initially tangent to the field vector $(4, -2)$, obtained by substituting $t = 0$ into our system's equations and using the initial condition $x(0) = 0, y(0) = 2$. Moving with increasing t from the point $(0, 2)$, we sketch a curve that is concave up and moving to the right and down, until it crosses the y-nullcline $y = 3x$ (crossing with a horizontal tangent). After this crossing it will move right and up but remain concave up.

To add more detail to our sketch, we can determine *where* it crosses the y-nullcline. To do this, we use the solution to our initial value problem,

$$(8.60) \qquad \mathbf{X}(t) = \frac{2}{7}e^{5t}\begin{pmatrix}2\\1\end{pmatrix} + \frac{4}{7}e^{-2t}\begin{pmatrix}-1\\3\end{pmatrix},$$

which is obtained from the general solution in equation (8.59) and the initial conditions $x(0) = 0, y(0) = 2$. Using this solution, in which $x(t) = \frac{4}{7}e^{5t} - \frac{4}{7}e^{-2t}$ and $y(t) = \frac{2}{7}e^{5t} + \frac{12}{7}e^{-2t}$, we identify the point where its trajectory will cross the y-nullcline $y = 3x$ by solving

$$\frac{2}{7}e^{5t} + \frac{12}{7}e^{-2t} = 3\left(\frac{4}{7}e^{5t} - \frac{4}{7}e^{-2t}\right).$$

Simplifying, we obtain $e^{7t} = 12/5$, so that $t = \frac{1}{7}\ln\left(\frac{12}{5}\right)$. Substituting this value of t into our solution (8.60), we find that the solution crosses the y-nullcline at the point $\left(\frac{2^{10/7}}{3^{2/7}5^{5/7}}, \frac{2^{10/7}3^{5/7}}{5^{5/7}}\right) \approx (0.623, 1.869)$.

Long-term behavior of trajectories: Slant asymptotes. Based on our informal discussion in Section 8.5, we expect that the trajectory of the solution (8.60) will be asymptotic to the eigenvector line $y = \frac{1}{2}x$ as $t \to \infty$. Here we will make this precise. In most calculus texts, you'll find the following definition: The line $y = mx + b$ is a slant asymptote of the graph of $h(x)$ provided $\lim_{x \to \infty}\left(h(x) - (mx + b)\right) = 0$ or $\lim_{x \to -\infty}\left(h(x) - (mx + b)\right) = 0$. Thus, a line is a slant asymptote of a graph provided the vertical distance from the graph to the line approaches 0 as we move rightward to ∞ or leftward to $-\infty$. To apply this definition in the context of the trajectory under discussion we take a vectorial perpective.

A vectorial perspective. We visualize the point $\mathbf{X}(t) = (x(t), y(t))$ on the trajectory of (8.60) as the point with position vector equal to the sum of $\mathbf{V}_1 = \frac{2}{7}e^{5t}\binom{2}{1}$ (represented as an arrow with tail at the origin) and $\mathbf{V}_2 = \frac{4}{7}e^{-2t}\binom{-1}{3}$ (represented as an arrow with tail at the tip of $\mathbf{V}_1$), as shown in Fig. 8.51. Note that $\mathbf{V}_1$ lies along the line $y = \frac{1}{2}x$ and thus the point on that line directly below $(x(t), y(t))$ has coordinates $(x(t), x(t)/2)$. Hence, the vertical distance from the point $(x(t), y(t))$ to the line $y = \frac{1}{2}x$ is

$$y(t) - \frac{1}{2}x(t) = \frac{2}{7}e^{5t} + \frac{12}{7}e^{-2t} - \frac{1}{2}\left(\frac{4}{7}e^{5t} - \frac{4}{7}e^{-2t}\right) = 2e^{-2t}.$$

Because this vertical distance goes to 0 as $t \to \infty$, this verifies that the line $y = \frac{1}{2}x$ is a slant asymptote for the trajectory of (8.60).

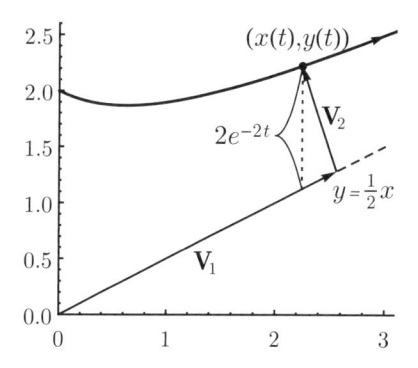

Figure 8.51. The eigenvector line $y = \frac{1}{2}x$ for $\mathbf{V}_1$ is a slant asymptote of the trajectory of (8.60).

Because the slant asymptote $y = \frac{1}{2}x$ is itself traced as a trajectory of a ray solution and because trajectories of autonomous linear systems that touch must be identical, the trajectory of (8.60) must remain *above* the line $y = \frac{1}{2}x$ as it approaches this asymptote. Note that our vectorial point of view also shows us that the trajectory of (8.60) remains above the slant asymptote $y = \frac{1}{2}x$ as $t \to \infty$. Understanding now the long-term behavior of the trajectory of (8.60), as well as knowing approximately where it will cross the y-nullcline $y = 3x$, we can produce an accurate sketch for $t \geq 0$.

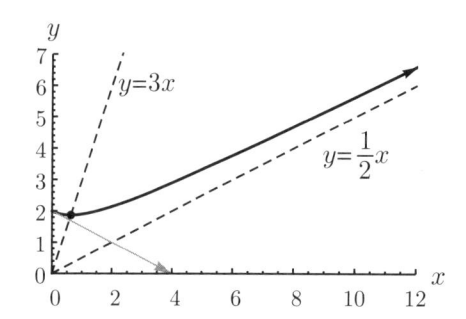

Figure 8.52. Trajectory of the solution of the initial value problem (8.57), which starts at $(0, 2)$ traveling in the direction of the vector $(4, -2)$, crosses the y-nullcline $y = 3x$ at approximately $(0.623, 1.869)$ and has $y = \frac{1}{2}x$ as a slant asymptote.

Extending the trajectory for $t < 0$ and sketching a phase portrait. In the next figure, we have extended the trajectory of Fig. 8.52, adding points corresponding to negative t values. Note that $y = -3x$, the eigenvector line containing the trajectory of the ray solution $\mathbf{X}_2 = e^{-2t}\binom{-1}{3}$, becomes a second slant asymptote (approached as $t \to -\infty$). You are asked to verify this in

8.6. More on phase portraits: Saddle points and nodes

Exercise 9. We have added a few other trajectories, including the four ray trajectories (two outgoing and two incoming) to produce a phase portrait for the system (8.55) near $(0,0)$.

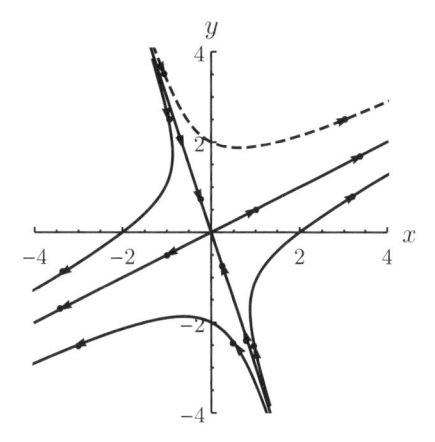

Figure 8.53. Phase portrait for the system (8.55): $\frac{dx}{dt} = 4x + 2y$, $\frac{dy}{dt} = 3x - y$, with the trajectory of (8.60) dashed.

Classifying the equilibrium point. Keep in mind that the point $(0,0)$ is itself an (equilibrium) solution of (8.55) and that no other trajectory passes through this point (though the ray solutions approach it either as $t \to \infty$ or $t \to -\infty$). Recall that if, as in this example, the eigenvalues of the matrix of coefficients $\mathbf{A}$ of a planar linear system $\mathbf{X}' = \mathbf{AX}$ are two real numbers with opposite signs, then the equilibrium point $\mathbf{0}$ is called a saddle point, and it is said to be unstable, reflecting the fact that most solution curves that start "near" $\mathbf{0}$ eventually move far away from $\mathbf{0}$, as we see in Fig. 8.53. Phase portraits for two other systems $\mathbf{X}' = \mathbf{AX}$ where $\mathbf{A}$ has two real eigenvalues of opposite signs are shown in Figs. 8.54–8.55.

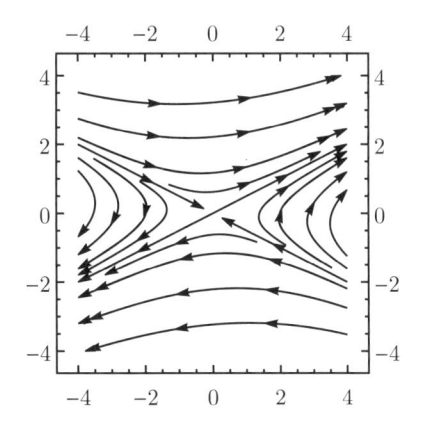

Figure 8.54. Real eigenvalues, opposite signs.

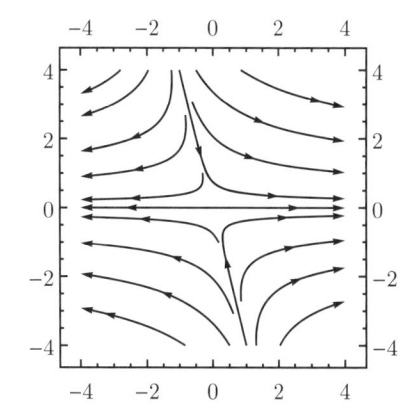

Figure 8.55. Real eigenvalues, opposite signs.

In each of the phase portraits in Figs. 8.53–8.55, we can see that the ray solutions traverse lines that form slant asymptotes for trajectories of the system. This is the case for any planar system $\mathbf{X}' = \mathbf{AX}$ in which the matrix $\mathbf{A}$ has one positive and one negative eigenvalue: All trajectories, except those corresponding to ray solutions, will have two slant asymptotes, $y = m_1 x$ and $y = m_2 x$, which are the eigenvector lines. Moreover, because these slant asymptotes correspond to trajectories of ray solutions and because trajectories of autonomous linear systems cannot cross, the slant asymptotes confine any trajectory to one of the four sectors created by the asymptotes (as illustrated in Figs. 8.53–8.55).

Other kinds of graphs. The trajectories in a phase portrait do not give detailed information on how a point moves along the trajectory in terms of the time t. This information is provided graphically with the graph of x vs. t and the graph of y vs. t. For the trajectory of the solution (8.60)

$$\mathbf{X}(t) = \frac{2}{7}e^{5t}\begin{pmatrix} 2 \\ 1 \end{pmatrix} + \frac{4}{7}e^{-2t}\begin{pmatrix} -1 \\ 3 \end{pmatrix}$$

of the initial value problem (8.57), which appears dashed in Fig. 8.53, we sketch below the x vs. t graph in Fig. 8.56 and the y vs. t graph in Fig. 8.57. These pictures are quite different from the trajectory sketch in the phase plane since they show precisely how x and y depend on t, but you should be able to reconcile qualitative information from these with what you can see in the corresponding trajectory in the phase plane. For example, looking at the dashed trajectory in Fig. 8.53, we can see that x is increasing for all t, while y decreases initially and then increases. Figs. 8.56 and 8.57 make this precise.

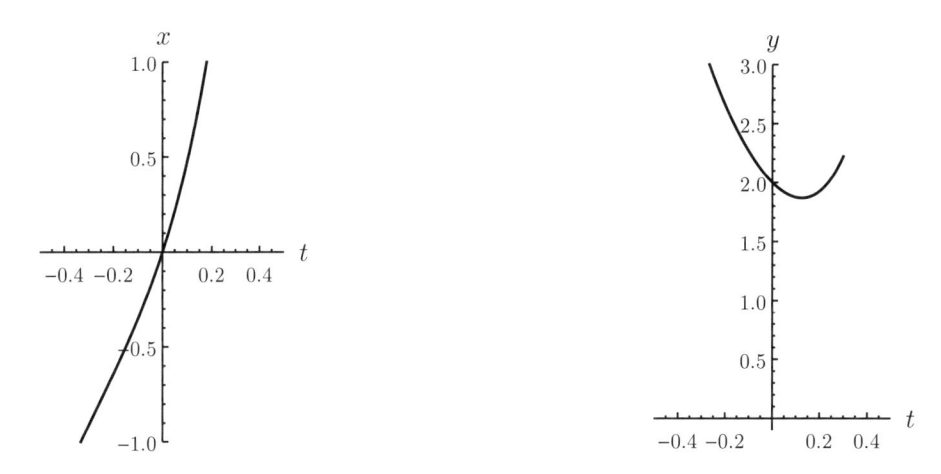

Figure 8.56. Graph of $x = \frac{4}{7}e^{5t} - \frac{4}{7}e^{-2t}$.

Figure 8.57. Graph of $y = \frac{2}{7}e^{5t} + \frac{12}{7}e^{-2t}$.

We conclude this section by applying the tools we have developed—nullcline-and-arrow diagrams and concavity charts—together with an analysis of asymptotic behavior of trajectories, to understand, qualitatively, the behavior of trajectories of a system $\mathbf{X}' = \mathbf{A}\mathbf{X}$ in an example where $\mathbf{A}$ has two distinct real eigenvalues with the *same* sign.

Example 8.6.4. We follow the steps (i)–(iii) appearing below equation (8.56) to produce a nullcline-and-arrow diagram for the system

(8.61) $$\frac{dx}{dt} = -2x + 2y, \quad \frac{dy}{dt} = x - 3y.$$

The system has x-nullcline $-2x + 2y = 0$, i.e., $y = x$, and y-nullcline $x - 3y = 0$, i.e. $y = \frac{1}{3}x$. Plotting these nullclines, adding direction-indicator arrows to each of the four sectors created by the nullclines, and adding the appropriate scaled field vectors to the nullclines, we obtain the nullcline-and-arrow diagram in Fig. 8.58. To the right of the nullcline-and-arrow diagram, we have placed a concavity chart for the system (8.61).

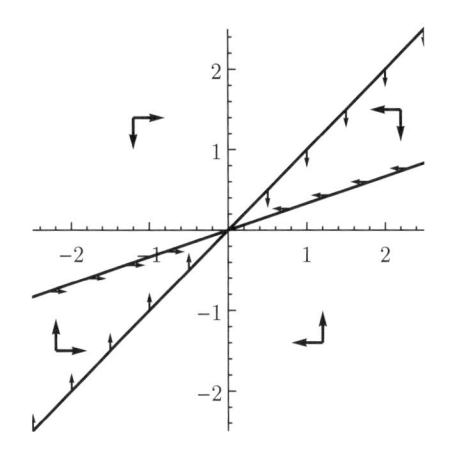

8.6. More on phase portraits: Saddle points and nodes

Figure 8.58. Nullcline-and-arrow diagram for system (8.61). The nullclines are $y = x$ and $y = x/3$.

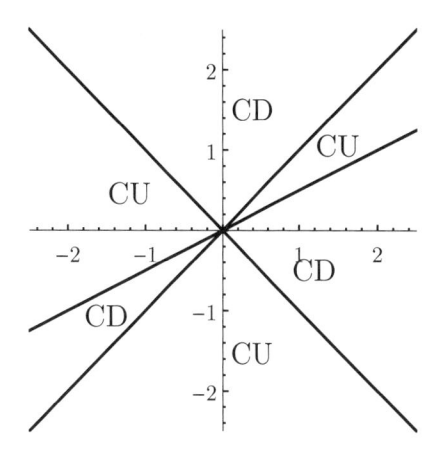

Figure 8.59. Concavity chart for (8.61). The concavity regions are bordered by the lines $y = x$, $y = x/2$, and $y = -x$.

To obtain the concavity chart, we compute

(8.62) $$\frac{dy}{dx} = \frac{x - 3y}{-2x + 2y} \quad \text{and} \quad \frac{d^2y}{dx^2} = \frac{4x\frac{dy}{dx} - 4y}{4(y-x)^2} = \frac{(x - 2y)(x + y)}{2(y - x)^3}.$$

As expected, the x-nullcline $y = x$ becomes one of the three lines separating regions of different concavity. The other two lines $y = \frac{1}{2}x$ and $y = -x$ should be lines traced by ray solutions (the eigenvector lines). We check this: The first line is parallel to the vector $\mathbf{V}_1 = \binom{2}{1}$ while the second is parallel to $\mathbf{V}_2 = \binom{-1}{1}$. Letting $\mathbf{A}$ be the matrix of coefficients of (8.61), the vectors $\mathbf{V}_1$ and $\mathbf{V}_2$ should be eigenvectors of $\mathbf{A}$. Are they? We compute

$$\mathbf{AV}_1 = \begin{pmatrix} -2 & 2 \\ 1 & -3 \end{pmatrix} \begin{pmatrix} 2 \\ 1 \end{pmatrix} = \begin{pmatrix} -2 \\ -1 \end{pmatrix} = -1 \cdot \mathbf{V}_1 \quad \text{and} \quad \mathbf{AV}_2 = \begin{pmatrix} 4 \\ -4 \end{pmatrix} = -4 \cdot \mathbf{V}_2.$$

Thus, we see that $\mathbf{A}$ does indeed have $\mathbf{V}_1$ and $\mathbf{V}_2$ as eigenvectors, corresponding, respectively, to eigenvalues -1 and -4. The general solution of the system (8.61) becomes

(8.63) $$\mathbf{X} = c_1 e^{-t} \begin{pmatrix} 2 \\ 1 \end{pmatrix} + c_2 e^{-4t} \begin{pmatrix} -1 \\ 1 \end{pmatrix},$$

and we see that our lines $y = \frac{1}{2}x$ and $y = -x$ appearing in our concavity chart are indeed those followed by ray solutions of the system. We also see from this general solution that all trajectories of the system (8.61) must approach $(0, 0)$ as $t \to \infty$ since both e^{-t} and e^{-4t} approach 0 as $t \to \infty$. This is consistent with Theorem 8.6.3, since $(0, 0)$ is an equilibrium point.

Let's use the information provided in Figs. 8.58 and 8.59 above to sketch the solution, for $t \geq 0$, of the initial value problem

(8.64) $$\frac{dx}{dt} = -2x + 2y, \quad \frac{dy}{dt} = x - 3y, \quad x(0) = 1, \quad y(0) = 2.$$

Let T denote the trajectory of the solution of (8.64). For $t \geq 0$, the first portion of our sketch of T appears in Fig. 8.60. We know that T is initially tangent to the vector $(-2x(0) + 2y(0), x(0) - 3y(0)) = (2, -5)$, which yields a tangent line to T at $(1, 2)$ having slope $-5/2$. This tangent line appears dotted in Fig. 8.60. Following the direction-indicator arrows for the basic region in which our trajectory starts and glancing at the concavity chart Fig. 8.59 above, we see that T moves down and to the right (initially in the direction of the vector $(2, -5)$) and is concave down until it crosses the x-nullcline $y = x$ at a point where it has a vertical tangent. Because T is initially concave down, the dotted tangent must remain above T in the region above $y = x$; this forces T to cross $y = x$ to the left of where the dotted tangent crosses. Once T enters the sector between the nullclines $y = x$ and $y = \frac{1}{3}x$, it must travel down and to the left and become concave up. Furthermore, it cannot cross the line $y = \frac{1}{2}x$ because this line is traced by a ray solution. As we have already observed, every trajectory approaches $(0,0)$ as $t \to \infty$. To complete an accurate sketch of T, we need to understand its asymptotic behavior: How does it approach $(0,0)$ as $t \to \infty$?

Theorem 8.5.3 on nodal tangents provides the answer. Since the eigenvalue -1 has smaller absolute value than the eigenvalue -4, every nonray trajectory, and in particular the trajectory T, approaches $(0,0)$ tangent to the eigenvector line $y = \frac{1}{2}x$ (the preferred nodal tangent) as $t \to \infty$. Understanding its asymptotic behavior, we complete our sketch of the trajectory T for $t \geq 0$ in Fig. 8.61.

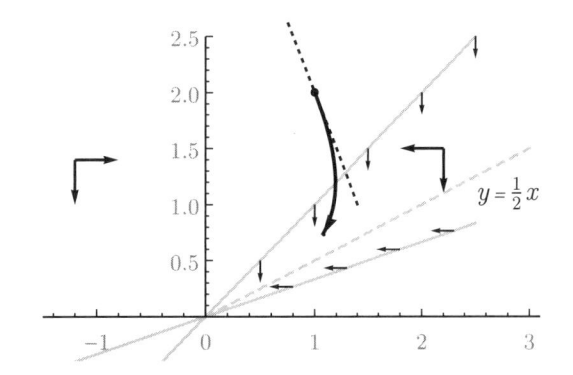

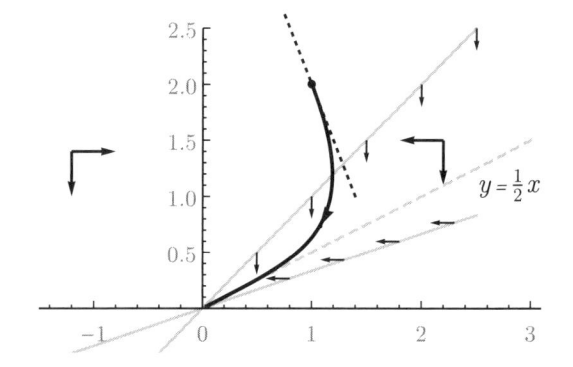

Figure 8.60. The trajectory T of (8.65), $0 \leq t \leq 0.5$.

Figure 8.61. The trajectory T of (8.65), $0 \leq t < \infty$.

In Fig. 8.62 we have extended T, adding points corresponding to negative t values—see the dashed trajectory. Referring to the nullcline-and-arrow diagram in Fig. 8.58, we note the direction-indicator arrows in the sector containing $(1, 2)$ tell us that we should extend T leftward and upward (as $t \to -\infty$), and the concavity chart in Fig. 8.59, coupled with the fact that T cannot cross that line $y = -x$ in the chart (because the line is traced by a ray solution), tell us that this extension of T must be a concave down curve. Using the general solution (8.63), we find the initial value problem (8.64) has solution

$$(8.65) \qquad \mathbf{X}(t) = e^{-t}\begin{pmatrix}2\\1\end{pmatrix} + e^{-4t}\begin{pmatrix}-1\\1\end{pmatrix} = e^{-t}\left(\begin{pmatrix}2\\1\end{pmatrix} + e^{-3t}\begin{pmatrix}-1\\1\end{pmatrix}\right).$$

From this it follows that the solution whose trajectory is T has derivative

$$\mathbf{X}'(t) = -e^{-t}\begin{pmatrix}2\\1\end{pmatrix} - 4e^{-4t}\begin{pmatrix}-1\\1\end{pmatrix} = -e^{-4t}\left(e^{3t}\begin{pmatrix}2\\1\end{pmatrix} + 4\begin{pmatrix}-1\\1\end{pmatrix}\right).$$

Because e^{3t} goes to 0 (quickly) as $t \to -\infty$, we see that tangent vectors to T become nearly parallel to the vector $(-1, 1)$ (or equivalently to the eigenvector line $y = -x$) at points of T corresponding to t's that are negative and have large absolute value. An examination of the general solution (8.63), shows that all trajectories (other than the two ray trajectories along the line $y = \frac{1}{2}x$) have this property, as is illustrated by Fig. 8.62.

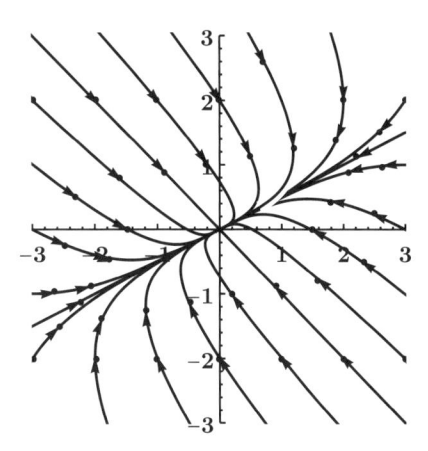

8.6. More on phase portraits: Saddle points and nodes

Figure 8.62. Phase portrait for the system (8.61): $\frac{dx}{dt} = -2x + 2y$, $\frac{dy}{dt} = x - 3y$, with trajectory T (extended) dashed.

We know from Section 8.5 that the picture in Fig. 8.62 is typical of any system $\mathbf{X}' = \mathbf{AX}$ where $\mathbf{A}$ is a 2×2 matrix having two different real eigenvalues, both negative, and that in this situation, the equilibrium point $\mathbf{0}$ is called a stable node. Moreover, if instead $\mathbf{A}$ had two different *positive* eigenvalues, then the trajectories would be shaped qualitatively similarly but would be directed outward—away from $(0,0)$, reflecting the fact that all solutions $x = x(t)$, $y = y(t)$ of such a system (except the equilibrium solution $x = 0$, $y = 0$) would have the property that the distance from $(x(t), y(t))$ to $(0,0)$ has limit ∞ as $t \to \infty$. In this case, the equilibrium point $\mathbf{0}$ is an unstable node.

Knowing (from the eigenvalues of $\mathbf{A}$) that $(0,0)$ is a saddle point, a stable node, or an unstable node for the system $\mathbf{X}' = \mathbf{AX}$ tells us the characteristic features of the trajectories in the phase portrait for the system. If we add to this knowledge a hand sketch of the ray solutions and a nullcline-and-arrow diagram, we can produce a reasonable phase portrait sketch with minimal computations.

Further information about the classification of the equilibrium point $(0,0)$ of $\mathbf{X}' = \mathbf{AX}$, based on the eigenvalues of the 2×2 matrix $\mathbf{A}$, is provided in the next section as well as in Section 8.9.

8.6.1. Exercises.

1. (a) Use the general solution
$$\mathbf{X}(t) = c_1 e^{5t} \begin{pmatrix} 2 \\ 1 \end{pmatrix} + c_2 e^{-2t} \begin{pmatrix} -1 \\ 3 \end{pmatrix}$$
   in Example 8.6.2 to solve the initial value problem
$$\frac{dx}{dt} = 4x + 2y, \quad \frac{dy}{dt} = 3x - y, \quad x(0) = -1, \quad y(0) = 1.$$
   (b) Sketch the trajectory of the solution you found in (a) for $-\infty < t < \infty$, using the nullcline-and-arrow sketch in Fig. 8.46 and the concavity chart in Fig. 8.49.
   (c) Does the trajectory in (b) cross the x-nullcline for the system? Does it cross the y-nullcline? If yes, say where.

2. Consider the linear system
$$\frac{dx}{dt} = -4x + 3y, \quad \frac{dy}{dt} = -4x + 4y.$$
   (a) Identify the x- and y-nullclines and give a nullcline-and-arrow diagram for the system.

(b) Using the fact that
$$\frac{d^2y}{dx^2} = \frac{4(y-2x)(3y-2x)}{(-4x+3y)^3}$$
produce the concavity chart for the system.

(c) From your work in (b) you should (without further computation) be able to identify two eigenvectors for **A**, the matrix of coefficients for the system. Do so, and determine the associated eigenvalue for each eigenvector. Using your answer give the general solution to the system.

(d) Using your work in (c), find the solution satisfying the initial condition $x(0) = -2$, $y(0) = 0$, and then sketch the trajectory of this solution for $-\infty < t < \infty$, using the information from (a) and (b). Identify any slant asymptotes (as $t \to \infty$ or as $t \to -\infty$). Does this trajectory cross any x- or y-nullcline? If so, find approximately the crossing point.

(e) Repeat (d) with the initial condition $x(0) = 2, y(0) = 1.5$.

(f) To your sketches from (d) and (e), add several more trajectories, including any ray solutions, to obtain a phase portrait.

3. Consider the linear system
$$\frac{dx}{dt} = \frac{11}{4}x - \frac{1}{4}y, \quad \frac{dy}{dt} = -\frac{3}{4}x + \frac{9}{4}y.$$

(a) Identify the x- and y-nullclines and give a nullcline-and-arrow diagram for the system.

(b) Using the fact that
$$\frac{d^2y}{dx^2} = \frac{96(y-3x)(y+x)}{(11x-y)^3}$$
produce the concavity chart for the system.

(c) From your work in (b) you should (without further computation) be able to identify two eigenvectors for **A**, the matrix of coefficients for the system. Do so, and determine the associated eigenvalue for each eigenvector, and give the general solution to the system.

(d) Using your work in (c), find the solution satisfying the initial condition $x(0) = 3$, $y(0) = 1$, and then sketch the trajectory of this solution for $-\infty < t < \infty$, using the information from (a) and (b). As $t \to \infty$, tangent vectors to corresponding points on this trajectory are nearly parallel to what vector? What line is the trajectory tangent to at $(0,0)$?

(e) To your sketch from (d), add several more trajectories, including any ray solutions, to obtain a phase portrait.

4. Consider the linear system
$$\frac{dx}{dt} = x, \quad \frac{dy}{dt} = 3x + 2y.$$

(a) Identify the x- and y-nullclines and give a nullcline-and-arrow diagram for the system.

(b) Compute $\frac{dy}{dx}$ and $\frac{d^2y}{dx^2}$, and produce the concavity chart for the system.

(c) Determine the eigenvalues and eigenvectors for **A**, the matrix of coefficients for the system, and give the general solution to $\mathbf{X}' = \mathbf{AX}$.

(d) Using the general solution in (c), find the solution satisfying the initial condition $x(0) = 2, y(0) = -2$, and then sketch the trajectory of this solution for $-\infty < t < \infty$, using the information from (a) and (b). As $t \to \infty$, tangent vectors to corresponding points on this trajectory are nearly parallel to what vector? What line is the trajectory tangent to at $(0,0)$?

(e) To your sketch from (d), add several more trajectories, including any ray solutions, to obtain a phase portrait.

8.6. More on phase portraits: Saddle points and nodes 521

5. (CAS) The first-order equation

(8.66) $$\frac{dy}{dx} = \frac{3x - y}{4x + 2y}$$

results from elimination of the parameter t from the system $\frac{dx}{dt} = 4x + 2y$, $\frac{dy}{dt} = 3x - y$, whose phase portraits appear in Figs. 8.42 and 8.53.

(a) Note that (8.66) may be written $\frac{dy}{dx} = \frac{3 - y/x}{4 + 2y/x}$. By making the homogeneous-equation substitution $v = y/x$ (see Section 2.7 of Chapter 2) show that

$$(2y - x)^5 (y + 3x)^2 = C$$

is a one-parameter family of implicit solutions to (8.66), with parameter C. You may use a computer algebra system to help with the integrations.

(b) Find explicit solutions to (8.66) corresponding to the choice of $C = 0$ in the preceding implicit solution. Indicate how these solutions relate to the general solution

$$\mathbf{X}(t) = c_1 e^{5t} \begin{pmatrix} 2 \\ 1 \end{pmatrix} + c_2 e^{-2t} \begin{pmatrix} -1 \\ 3 \end{pmatrix}$$

of the system $\frac{dx}{dt} = 4x + 2y$, $\frac{dy}{dt} = 3x - y$.

(c) Use the solution of part (a) above to find an implicit solution of the initial value problem

$$\frac{dy}{dx} = \frac{3x - y}{4x + 2y}, \quad y = 2 \text{ when } x = 0;$$

then plot the solution and compare it to the dashed trajectory appearing in Fig. 8.53.

6. The phase portrait in Fig. 8.20 shows four ray trajectories. Give values of c_1 and c_2 so that the solution

$$c_1 e^{7t} \begin{pmatrix} 2 \\ 1 \end{pmatrix} + c_2 e^{-5t} \begin{pmatrix} 2 \\ -1 \end{pmatrix}$$

gives each of these rays. Are there one or many choices of c_1 and c_2 for each ray?

7. The following figure shows vector field plots for three different systems of the form $\mathbf{X}' = \mathbf{AX}$, where $\mathbf{A}$ is a 2×2 matrix having distinct, real, nonzero eigenvalues.
  (a) For each of the fields (A), (B), and (C), determine the signs of the eigenvalues of the corresponding matrix of coefficients $\mathbf{A}$.
  (b) In each vector field picture, draw any ray solutions you see, and draw at least two other trajectories in the phase portrait.

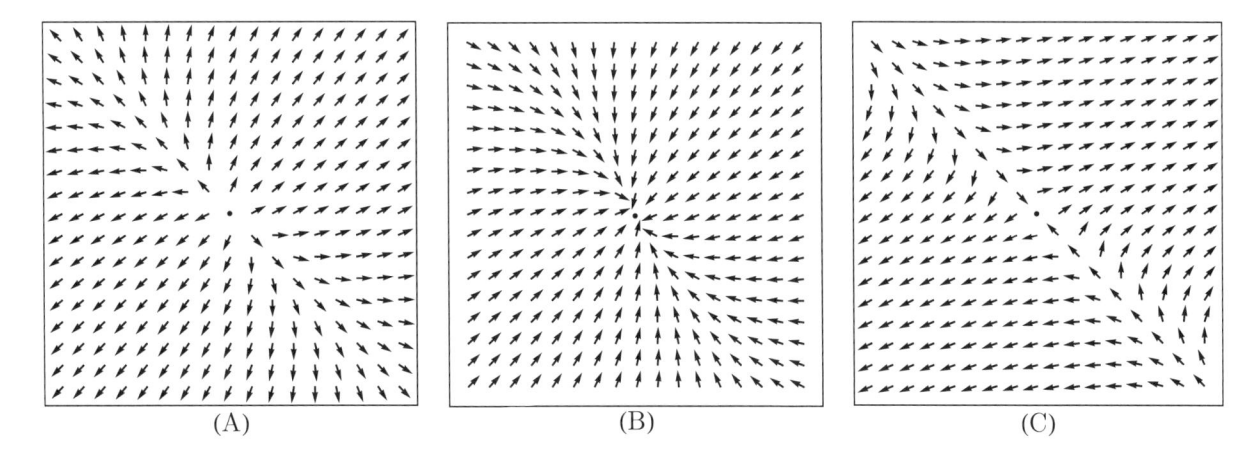

(A) (B) (C)

8. Figs. 8.63 and 8.64 show x vs. t and y vs. t for one of the trajectories in Fig. 8.53. Which trajectory?

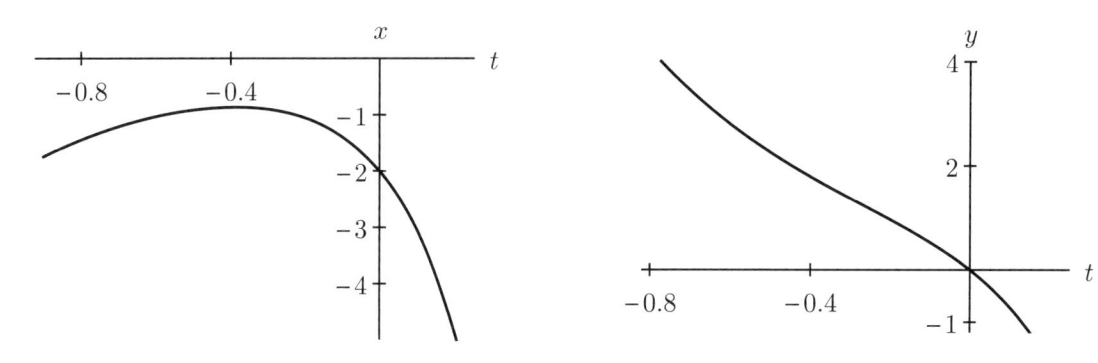

Figure 8.63. $x = x(t)$.

Figure 8.64. $y = y(t)$.

9. Show that the line $y = -3x$ is a slant asymptote of the solution (8.60) to the initial value problem (8.57) by computing the vertical distance from the point $(x(t), y(t))$ on the trajectory of the solution to the point on $y = -3x$ directly below $(x(t), y(t))$ and showing the distance goes to 0 as $t \to -\infty$.

10. Consider the system
$$\frac{dx}{dt} = x + y, \quad \frac{dy}{dt} = 3x - y.$$

(a) Verify that $(0,0)$ is a saddle point of the system by showing that its matrix of coefficients has real eigenvalues having opposite signs.
(b) Find the solution $x = x(t)$, $y = y(t)$ of the system satisfying $x(0) = 1$, $y(0) = 0$.
(c) What are the slant asymptotes of the trajectory of the solution of (b)?
(d) What is the vertical distance from the point on the trajectory $(x(\ln 2), y(\ln 2))$ of the solution of (b) to the slant asymptote nearest the point?

11. Consider the system
$$\frac{dx}{dt} = 2x + 6y, \quad \frac{dy}{dt} = x - 3y.$$

(a) Verify that $(0,0)$ is a saddle point of the system by showing that its matrix of coefficients has real eigenvalues having opposite signs.
(b) Find the solution $x = x(t)$, $y = y(t)$ of the system satisfying $x(0) = 1$, $y(0) = 0$.
(c) There are two points on the trajectory of the solution of (b) having x-coordinate 48. Quickly estimate the values of the corresponding y-coordinates of each of these points (without using a calculator or computer). Explain your reasoning.
(d) (CAS) Use a computer algebra system to find, approximately, the times when the trajectory is located at each of its points with x-coordinate 48. Substitute the times into the solution of (b) to find approximations for the y-coordinates of the points on the trajectory of the solution having x-coordinate 48. Compare your answers to those you found for part (c).

12. Consider the system
$$\frac{dx}{dt} = -x + 2y, \quad \frac{dy}{dt} = x - 2y.$$

(a) Write this in matrix form $\mathbf{X}' = \mathbf{A}\mathbf{X}$ and show that one of the eigenvalues of $\mathbf{A}$ is zero, and find a corresponding eigenvector.
(b) Find the nonzero eigenvalue and a corresponding eigenvector.

(c) Show that the general solution to the system can be written as
$$\mathbf{X}(t) = c_1 e^{-3t} \begin{pmatrix} -1 \\ 1 \end{pmatrix} + c_2 \begin{pmatrix} 2 \\ 1 \end{pmatrix}.$$

(d) Sketch the ray solutions
$$e^{-3t} \begin{pmatrix} -1 \\ 1 \end{pmatrix} \quad \text{and} \quad -e^{-3t} \begin{pmatrix} -1 \\ 1 \end{pmatrix}.$$
Are they incoming or outgoing rays?

(e) For each choice of c_2, the equilibrium solution
$$c_2 \begin{pmatrix} 2 \\ 1 \end{pmatrix}$$
gives a single-point trajectory in the phase portrait. In your sketch from (c), show these equilibrium points for $c_2 = -3, -2, -1, 0, 1, 2,$ and 3.

(f) Also in your sketch from (c), carefully show the solution
$$e^{-3t} \begin{pmatrix} -1 \\ 1 \end{pmatrix} + 2 \begin{pmatrix} 2 \\ 1 \end{pmatrix}.$$
What special kind of curve is this?

13. Suppose $\mathbf{A}$ is a 2×2 matrix with determinant equal to 0. To avoid a trivial situation, assume that not every entry of $\mathbf{A}$ is zero.
  (a) Show that the linear system $\mathbf{X}' = \mathbf{A}\mathbf{X}$ has infinitely many equilibrium points.
  (b) Show that these equilibrium points fill out a line in the phase plane.
  (c) Suppose
  $$\mathbf{A} = \begin{pmatrix} 0 & 1 \\ 0 & 3 \end{pmatrix}.$$
  Find all equilibrium points of $\mathbf{X}' = \mathbf{A}\mathbf{X}$. Solve this system, and sketch a phase portrait. Your sketch should show the equilibrium points. The equilibrium points are said to be "nonisolated". Why does this terminology make sense?

14. Prove Theorem 8.6.3 by an argument similar to that used in Exercise 23 of Section 2.6.

15. (a) Suppose $\mathcal{R}$ is a region in the plane with the property that given any two points in $\mathcal{R}$, the line segment joining these points lies in $\mathcal{R}$. Observe that any of the sectors I–IV in Fig. 8.43 has this property. Suppose that $h = h(x, y)$ is a continuous real-valued function on $\mathcal{R}$ that never takes the value 0 at points in $\mathcal{R}$. Use the Intermediate Value Theorem (discussed in a typical Calculus I class) to show that h is either always positive on $\mathcal{R}$ or always negative on $\mathcal{R}$. Hint: For any two points (x_0, y_0) and (x_1, y_1) the points on the line segment joining (x_0, y_0) and (x_1, y_1) can be described by $((1-t)x_0 + tx_1, (1-t)y_0 + ty_1)$ for $0 \le t \le 1$. Consider the one-variable function $g(t) = h((1-t)x_0 + tx_1, (1-t)y_0 + ty_1)$ on the interval $[0, 1]$.

  (b) The result in (a) can be generalized to regions $\mathcal{R}$ that are path-connected. A region in the plane is said to be path-connected provided that every pair of points in the region may be joined by a continuous curve lying entirely in the region (not necessarily a line segment); that is, $\mathcal{R}$ is path-connected provided that whenever (x_0, y_0) and (x_1, y_1) are points in $\mathcal{R}$, there are continuous functions $x = \varphi(t)$ and $y = \psi(t)$ defined on some closed interval $[a, b]$, such that $\varphi(a) = x_0$, $\psi(a) = y_0$, $\varphi(b) = x_1$, $\psi(b) = y_1$, and $\{(\varphi(t), \psi(t)) : a \le t \le b\}$ is contained in $\mathcal{R}$. Show that if $\mathcal{R}$ is a path-connected region in the plane and that $h = h(x, y)$ is a continuous real-valued function on $\mathcal{R}$ that never takes the value 0 at points in $\mathcal{R}$, then h is either always positive on $\mathcal{R}$ or always negative on $\mathcal{R}$.

8.7. Complex and repeated eigenvalues

In this section we use the eigenvalue/eigenvector approach to solve planar systems $\mathbf{X}' = \mathbf{AX}$, where $\mathbf{A}$, a matrix with real number entries, either has complex eigenvalues $\alpha \pm i\beta, \beta \neq 0$, or a single eigenvalue of multiplicity two. We begin with an example having complex conjugate eigenvalues.

Example 8.7.1. Let
$$\mathbf{A} = \begin{pmatrix} 0 & 4 \\ -1 & 0 \end{pmatrix}$$
be the matrix associated with the system of equations
$$\frac{dx}{dt} = 4y,$$
$$\frac{dy}{dt} = -x.$$

The characteristic polynomial is
$$p(z) = \det(\mathbf{A} - z\mathbf{I}) = \det\begin{pmatrix} -z & 4 \\ -1 & -z \end{pmatrix} = z^2 + 4,$$
so that $\mathbf{A}$ has purely imaginary eigenvalues $r_1 = 2i$ and $r_2 = -2i$. We know that when a matrix $\mathbf{A}$ has all of its entries real, any complex eigenvalues of $\mathbf{A}$ occur in conjugate pairs (Exercise 12 of Section 8.3). To find an eigenvector for $r_1 = 2i$, we compute
$$\mathbf{A} - r_1\mathbf{I} = \begin{pmatrix} 0 & 4 \\ -1 & 0 \end{pmatrix} - \begin{pmatrix} 2i & 0 \\ 0 & 2i \end{pmatrix} = \begin{pmatrix} -2i & 4 \\ -1 & -2i \end{pmatrix}.$$

Using (8.23), both $\binom{4}{2i}$ and $\binom{-2i}{1}$ are eigenvectors for the eigenvalue $2i$. These vectors should be scalar multiples of each other. Are they? Yes, since
$$2i\begin{pmatrix} -2i \\ 1 \end{pmatrix} = \begin{pmatrix} 4 \\ 2i \end{pmatrix}.$$

We can choose either one of these as our eigenvector for the eigenvalue $2i$, or even choose the simpler scalar multiple
$$\mathbf{V}_1 = \begin{pmatrix} 2 \\ i \end{pmatrix}.$$

This gives us the **exponential**, or **complex ray**, solution
$$e^{r_1 t}\mathbf{V}_1 = e^{2it}\begin{pmatrix} 2 \\ i \end{pmatrix} = \begin{pmatrix} 2e^{2it} \\ ie^{2it} \end{pmatrix}.$$

The conjugate of $\mathbf{V}_1$,
$$\mathbf{V}_2 = \begin{pmatrix} 2 \\ -i \end{pmatrix},$$
will be an eigenvector corresponding to the conjugate eigenvalue, $-2i$. Again, this is a feature of any matrix with real entries; see Exercise 12 in Section 8.3. The corresponding solution is
$$e^{r_2 t}\mathbf{V}_2 = e^{-2it}\begin{pmatrix} 2 \\ -i \end{pmatrix} = \begin{pmatrix} 2e^{-2it} \\ -ie^{-2it} \end{pmatrix}.$$

Extracting real-valued solutions. Our two exponential solutions have the form of a scalar multiple of a vector. Both the scalar and the vector are complex valued, and this is not something we can draw in the xy-plane. Using Euler's formula we can write
$$e^{2it} = \cos(2t) + i\sin(2t) \quad \text{and} \quad e^{-2it} = \cos(2t) - i\sin(2t).$$

8.7. Complex and repeated eigenvalues

We find

$$e^{r_1 t}\mathbf{V}_1 = \begin{pmatrix} 2e^{2it} \\ ie^{2it} \end{pmatrix} = \begin{pmatrix} 2(\cos(2t) + i\sin(2t)) \\ i(\cos(2t) + i\sin(2t)) \end{pmatrix}$$

$$= \begin{pmatrix} 2\cos(2t) + 2i\sin(2t) \\ -\sin(2t) + i\cos(2t) \end{pmatrix}$$

$$= \begin{pmatrix} 2\cos(2t) \\ -\sin(2t) \end{pmatrix} + i \begin{pmatrix} 2\sin(2t) \\ \cos(2t) \end{pmatrix}$$

and

$$e^{r_2 t}\mathbf{V}_2 = \begin{pmatrix} 2e^{-2it} \\ -ie^{-2it} \end{pmatrix} = \begin{pmatrix} 2\cos(2t) \\ -\sin(2t) \end{pmatrix} + i \begin{pmatrix} -2\sin(2t) \\ -\cos(2t) \end{pmatrix}.$$

By the Superposition Principle, both

$$\mathbf{X}_1(t) = \frac{1}{2}e^{r_1 t}\mathbf{V}_1 + \frac{1}{2}e^{r_2 t}\mathbf{V}_2 = \begin{pmatrix} 2\cos(2t) \\ -\sin(2t) \end{pmatrix}$$

and

$$\mathbf{X}_2(t) = \frac{1}{2i}e^{r_1 t}\mathbf{V}_1 - \frac{1}{2i}e^{r_2 t}\mathbf{V}_2 = \begin{pmatrix} 2\sin(2t) \\ \cos(2t) \end{pmatrix}$$

are real-valued solutions to our system. Notice that $\mathbf{X}_1(t)$ is the real part of our complex-valued solution $e^{r_1 t}\mathbf{V}_1$ and $\mathbf{X}_2(t)$ is the imaginary part of $e^{r_1 t}\mathbf{V}_1$. We don't get anything substantially new by looking at the real and imaginary parts of our second complex exponential solution $e^{r_2 t}\mathbf{V}_2$; the real part of this solution is just $\mathbf{X}_1(t)$ again, and the imaginary part is the negative of $\mathbf{X}_2(t)$. By the Superposition Principle,

$$\mathbf{X}(t) = c_1 \mathbf{X}_1(t) + c_2 \mathbf{X}_2(t) = \begin{pmatrix} 2c_1 \cos(2t) + 2c_2 \sin(2t) \\ -c_1 \sin(2t) + c_2 \cos(2t) \end{pmatrix}$$

is a (real-valued) solution for any (real) constants c_1 and c_2. Exercise 9 asks you to verify this is the general solution. Note that we can get all of the information we need to build the general solution from one eigenvalue/eigenvector pair.

To find the solution with $\mathbf{X}(0) = \binom{3}{4}$, for example, set $t = 0$ in the general solution and solve for c_1 and c_2. We obtain $c_1 = \frac{3}{2}$ and $c_2 = 4$, and this initial value problem has solution

$$\mathbf{X}(t) = \begin{pmatrix} 3\cos(2t) + 8\sin(2t) \\ -3/2 \sin(2t) + 4\cos(2t) \end{pmatrix}.$$

Sketching this orbit in a phase portrait. Since the functions $x(t) = 3\cos(2t) + 8\sin(2t)$ and $y(t) = -\frac{3}{2}\sin(2t) + 4\cos(2t)$ are both periodic with period π, the orbit for this solution is a closed loop. In Exercise 10 you are asked to sketch this orbit and show that it is in fact an ellipse.

When, as in the last example, the eigenvalues are a purely imaginary conjugate pair, the equilibrium point $(0,0)$ is called a **center**, and a phase portrait consists of ellipses (see Exercise 13 in Section 8.9) encircling the origin. All of these orbits have the same orientation—either all are clockwise or all are counterclockwise.

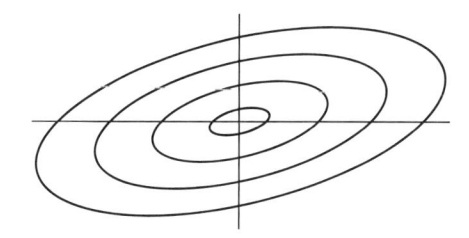

Figure 8.65. Elliptical orbits arise from purely imaginary eigenvalues.

Summary. Our analysis in Example 8.7.1 relied on a number of general principles, which we summarize as follows:

- Suppose $\mathbf{A}$ is a 2×2 matrix with real entries. Any complex eigenvalues of $\mathbf{A}$ occur in conjugate pairs $\alpha \pm i\beta$. Moreover, if $\mathbf{V}$ is an eigenvector for $\alpha + i\beta$, then $\overline{\mathbf{V}}$ is an eigenvector for $\alpha - i\beta$ (Exercise 12 in Section 8.3).

- Furthermore, the equation $\mathbf{X}' = \mathbf{AX}$ has complex-valued solutions

$$e^{(\alpha+i\beta)t}\mathbf{V} \quad \text{and} \quad e^{(\alpha-i\beta)t}\overline{\mathbf{V}}.$$

- The real parts of these two complex-valued solutions are the same and are solutions to $\mathbf{X}' = \mathbf{AX}$. In Exercise 16 you will show that the real part $e^{(\alpha+i\beta)t}\mathbf{V}$ is

$$\mathbf{X}_1(t) = (e^{\alpha t}\cos(\beta t))\mathbf{U} - (e^{\alpha t}\sin(\beta t))\mathbf{W}$$

where $\mathbf{U}$ is the real part of the eigenvector $\mathbf{V}$ and $\mathbf{W}$ is the imaginary part of $\mathbf{V}$.

- The imaginary parts of these complex-valued solutions are negatives of each other and also solve $\mathbf{X}' = \mathbf{AX}$. The imaginary part of $e^{(\alpha+i\beta)t}\mathbf{V}$ is

$$\mathbf{X}_2(t) = (e^{\alpha t}\sin(\beta t))\mathbf{U} + (e^{\alpha t}\cos(\beta t))\mathbf{W},$$

where $\mathbf{U}$ and $\mathbf{W}$ are as just described.

- The general real-valued solution to the system is

$$\mathbf{X}(t) = c_1\mathbf{X}_1(t) + c_2\mathbf{X}_2(t)$$

where $\mathbf{X}_1(t)$ is the real part of $e^{(\alpha+i\beta)t}\mathbf{V}$ and $\mathbf{X}_2(t)$ is the imaginary part of $e^{(\alpha+i\beta)t}\mathbf{V}$. A proof of this assertion is outlined in Exercise 16.

Example 8.7.2. We'll solve the initial value problem $\mathbf{X}' = \mathbf{AX}$, $\mathbf{X}(0) = \mathbf{X}_0$, where

$$\mathbf{A} = \begin{pmatrix} -3 & 4 \\ -2 & 1 \end{pmatrix} \quad \text{and} \quad \mathbf{X}_0 = \begin{pmatrix} 2 \\ 1 \end{pmatrix}$$

and sketch the corresponding orbit in the phase plane. The eigenvalues of $\mathbf{A}$ are $-1 \pm 2i$, and an eigenvector for $-1 + 2i$ is

$$\mathbf{V} = \begin{pmatrix} 1+i \\ i \end{pmatrix}.$$

We have the complex-valued solution

$$\begin{aligned} e^{(-1+2i)t}\mathbf{V} &= e^{-t}(\cos(2t) + i\sin(2t))\begin{pmatrix} 1+i \\ i \end{pmatrix} \\ &= \begin{pmatrix} e^{-t}\cos(2t) - e^{-t}\sin(2t) \\ -e^{-t}\sin(2t) \end{pmatrix} + i\begin{pmatrix} e^{-t}\cos(2t) + e^{-t}\sin(2t) \\ e^{-t}\cos(2t) \end{pmatrix}.\end{aligned}$$

8.7. Complex and repeated eigenvalues

The general real-valued solution $\mathbf{X}(t)$ is built as linear combinations of the real and imaginary parts of $e^{(-1+2i)t}\mathbf{V}$:

$$\mathbf{X}(t) = c_1 \begin{pmatrix} e^{-t}\cos(2t) - e^{-t}\sin(2t) \\ -e^{-t}\sin(2t) \end{pmatrix} + c_2 \begin{pmatrix} e^{-t}\cos(2t) + e^{-t}\sin(2t) \\ e^{-t}\cos(2t) \end{pmatrix}.$$

To satisfy the given initial condition, we must choose $c_1 = 1$ and $c_2 = 1$.

To sketch the orbit of this solution, write it as

$$e^{-t} \begin{pmatrix} 2\cos(2t) \\ \cos(2t) - \sin(2t) \end{pmatrix}.$$

Focus first on the vector function

(8.67)
$$\begin{pmatrix} 2\cos(2t) \\ \cos(2t) - \sin(2t) \end{pmatrix}.$$

It describes a closed loop (ellipse), traversed clockwise as t increases, as shown in Fig. 8.66. The points corresponding to $t = 0$, $t = \pi/4$, $t = \pi/2$, and $t = 3\pi/4$ on the ellipse are shown. Notice that $t = 0$ gives the point $(2, 1)$, and as t increases we move clockwise on the ellipse. The scalar factor e^{-t} decreases to 0 as $t \to \infty$ and increases to ∞ as $t \to -\infty$. To get points on the orbit of our solution, we take the scalar e^{-t} for the appropriate value of t and multiply the corresponding point on the ellipse by this scalar. We can do this visually. For example, when $t = \frac{\pi}{2}$, the corresponding point on the ellipse is $(-2, -1)$ and the scalar factor is $e^{-\pi/2} \approx 0.207$. Similarly, when $t = -\frac{\pi}{4}$, the point on the ellipse is $(0, 1)$ and the scalar factor is $e^{\pi/4} \approx 2.19$. The resulting orbit is a clockwise spiral, tending towards the origin as $t \to 0$ (and moving away from the origin as $t \to -\infty$).

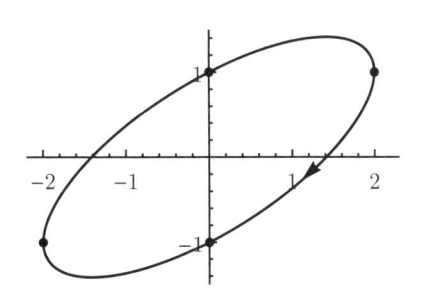

Figure 8.66. Ellipse with equation (8.67).

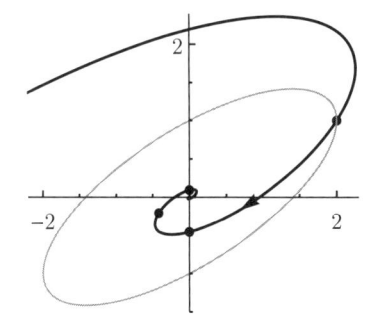

Figure 8.67. Clockwise spiral orbit.

Whenever the 2×2 matrix $\mathbf{A}$ has complex conjugate eigenvalues $\alpha \pm i\beta$ with α and β nonzero, the orbits for the system $\mathbf{X}' = \mathbf{AX}$ are spirals. If $\alpha < 0$, the orbits spiral towards the equilibrium point $(0,0)$ as $t \to \infty$, and $(0,0)$ is called a **stable spiral point**. If $\alpha > 0$, the orbits spiral away from the origin as $t \to \infty$, and $(0,0)$ is called an **unstable spiral point**.

The final scenario is that of a 2×2 matrix $\mathbf{A}$ with real entries where $\mathbf{A}$ has one repeated eigenvalue. Let's look at an example.

Example 8.7.3. One repeated eigenvalue. We solve the equation $\mathbf{X}' = \mathbf{AX}$ where

$$\mathbf{A} = \begin{pmatrix} 3 & 4 \\ -1 & -1 \end{pmatrix}.$$

The eigenvalues of $\mathbf{A}$ are the roots of the characteristic equation $(3-z)(-1-z)+4=0$, which simplifies to $(z-1)^2 = 0$. Thus $\mathbf{A}$ has only one eigenvalue, $z = 1$, with multiplicity two. It is

easy to check that the only eigenvectors corresponding to this eigenvalue are the nonzero scalar multiples of
$$\mathbf{V} = \begin{pmatrix} -2 \\ 1 \end{pmatrix}.$$

Thus $\mathbf{X}_1(t) = e^t \mathbf{V}$ is one solution to $\mathbf{X}' = \mathbf{A}\mathbf{X}$. By analogy with our work on constant coefficient homogeneous equations in Chapter 5, we might guess that a second solution could be $\mathbf{X}(t) = te^t \mathbf{V}$. Is this correct? If $\mathbf{X}(t) = te^t\mathbf{V}$, then $\mathbf{X}'(t) = te^t\mathbf{V} + e^t\mathbf{V}$. To verify this computation, note that since $\mathbf{V}$ is a vector of constants,

$$\frac{d}{dt}(te^t\mathbf{V}) = \frac{d}{dt}(te^t)\mathbf{V} = (te^t + e^t)\mathbf{V} = te^t\mathbf{V} + e^t\mathbf{V},$$

where we have used the product rule. Also, $\mathbf{A}\mathbf{X} = \mathbf{A}(te^t\mathbf{V}) = te^t(\mathbf{A}\mathbf{V}) = te^t\mathbf{V}$, and thus in order for $\mathbf{X}(t) = te^t\mathbf{V}$ to be a solution we must have $te^t\mathbf{V} + e^t\mathbf{V} = te^t\mathbf{V}$. Because $\mathbf{V} \neq \mathbf{0}$, this is clearly **not** true; so our trial solution $\mathbf{X}(t) = te^t\mathbf{V}$ **does not work**. Based on this calculation though, let's modify our guess for a second solution and try

$$\mathbf{X}_2(t) = e^t[t\mathbf{V} + \mathbf{W}] = te^t\mathbf{V} + e^t\mathbf{W}$$

where $\mathbf{W}$ is a "to be determined" vector of constants. Now

$$\mathbf{X}_2'(t) = te^t\mathbf{V} + e^t\mathbf{V} + e^t\mathbf{W}$$

and $\mathbf{X}_2'$ will equal $\mathbf{A}\mathbf{X}_2$ provided

(8.68) $$te^t\mathbf{V} + e^t\mathbf{V} + e^t\mathbf{W} = \mathbf{A}(te^t\mathbf{V} + e^t\mathbf{W}).$$

Since in this example we have $\mathbf{A}\mathbf{V} = 1\mathbf{V}$, equation (8.68) is equivalent to

$$te^t\mathbf{V} + e^t\mathbf{V} + e^t\mathbf{W} = te^t\mathbf{V} + e^t(\mathbf{A}\mathbf{W}),$$

which can be rearranged as $e^t\mathbf{V} = e^t(\mathbf{A}\mathbf{W}) - e^t\mathbf{W}$, or just $\mathbf{V} = (\mathbf{A} - \mathbf{I})\mathbf{W}$. This says that we want $\mathbf{W} = \begin{pmatrix} w_1 \\ w_2 \end{pmatrix}$ to satisfy

$$\begin{pmatrix} -2 \\ 1 \end{pmatrix} = \begin{pmatrix} 2 & 4 \\ -1 & -2 \end{pmatrix} \begin{pmatrix} w_1 \\ w_2 \end{pmatrix}$$

or
$$2w_1 + 4w_2 = -2 \quad \text{and} \quad -w_1 - 2w_2 = 1.$$

Notice these two equations are multiples of each other. Can we solve this for w_1 and w_2? Yes, since any choice of w_1 and w_2 satisfying $2w_1 + 4w_2 = -2$ will give a solution. For example we may choose $w_1 = -1$ and $w_2 = 0$ and obtain a second solution

$$\mathbf{X}_2(t) = te^t\mathbf{V} + e^t\mathbf{W} \quad \text{where} \quad \mathbf{V} = \begin{pmatrix} -2 \\ 1 \end{pmatrix} \text{ and } \mathbf{W} = \begin{pmatrix} -1 \\ 0 \end{pmatrix}.$$

Taking linear combinations of $\mathbf{X}_1(t)$ and $\mathbf{X}_2(t)$ we have

(8.69) $$\mathbf{X}(t) = c_1\left[e^t\mathbf{V}\right] + c_2\left[te^t\mathbf{V} + e^t\mathbf{W}\right].$$

In Exercise 17 you are asked to verify this is the general solution.

Sketching some trajectories in the phase plane. In the general solution

$$\mathbf{X}(t) = c_1 e^t \begin{pmatrix} -2 \\ 1 \end{pmatrix} + c_2 \left[te^t \begin{pmatrix} -2 \\ 1 \end{pmatrix} + e^t \begin{pmatrix} -1 \\ 0 \end{pmatrix} \right]$$

the choice $c_1 = 1$, $c_2 = 0$ gives the solution

$$e^t \begin{pmatrix} -2 \\ 1 \end{pmatrix}.$$

8.7. Complex and repeated eigenvalues

This is an outgoing ray, in the direction of the vector $\begin{pmatrix} -2 \\ 1 \end{pmatrix}$. The choice $c_1 = -1$, $c_2 = 0$ gives an outgoing ray in the opposite direction. Both of these rays are shown in Fig. 8.69. The choice $c_1 = 0$, $c_2 = 1$ gives

$$te^t \begin{pmatrix} -2 \\ 1 \end{pmatrix} + e^t \begin{pmatrix} -1 \\ 0 \end{pmatrix}.$$

This is not a ray, but something more complicated. To sketch the trajectory, write the solution as

$$e^t \left[\begin{pmatrix} -1 \\ 0 \end{pmatrix} + t \begin{pmatrix} -2 \\ 1 \end{pmatrix} \right].$$

The expression in the square brackets is the equation of a line that passes through the point $(-1, 0)$ and is in the direction of the vector $(-2, 1)$; we call this the guideline. To obtain points on the trajectory, we multiply points on the line by the scalar factor e^t for the appropriate value of t. For example, when $t = 0$ the point on the line is $(-1, 0)$ and the scalar factor is $e^0 = 1$, while when $t = -\frac{1}{2}$, the point on the line is $(0, -\frac{1}{2})$ and the scalar factor is $e^{-1/2} \approx 0.61$. We can get a rough sketch of the trajectory by doing these computations visually, remembering that e^t is always positive, increasing to ∞ as $t \to \infty$ and decreasing to 0 as $t \to -\infty$; see Fig. 8.68. As t becomes increasingly negative, we are moving down and to the right on the line and scaling by a positive factor that is (rapidly) approaching 0. For $t > 0$ and increasing, we are moving up and to the left on the line and scaling by a positive factor that is rapidly growing larger. Fig. 8.68 shows our guideline and how points on this line are scaled to give the trajectory of the solution.

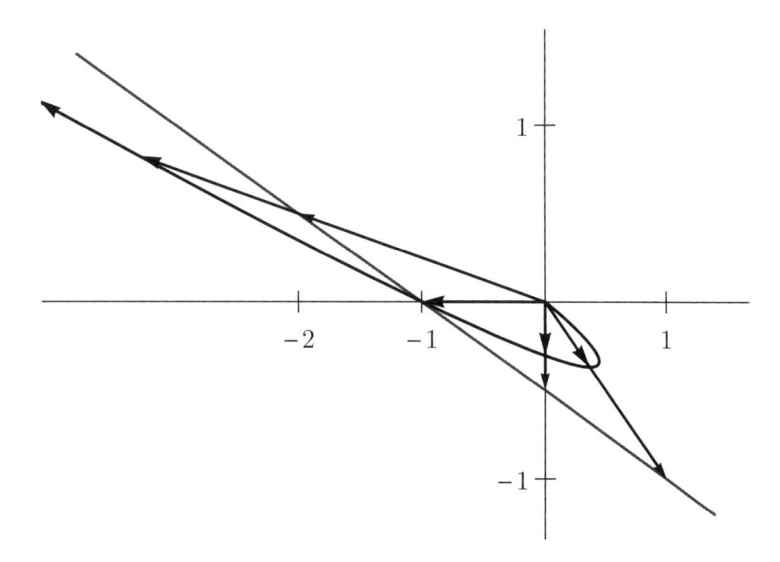

Figure 8.68. Sketching the trajectory from the guideline.

Fig. 8.69 shows a phase portrait for this system, with trajectories obtained from several choices of c_1 and c_2. The equilibrium point $(0, 0)$ is an **unstable node** in this case. For reasons to be discussed shortly, additional terms describing this node are "degenerate" and "improper".

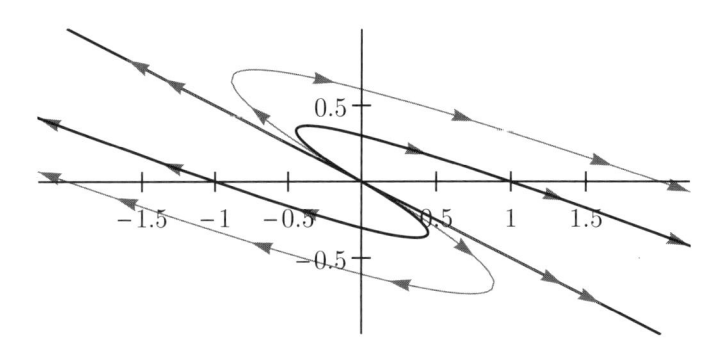

Figure 8.69. Phase portrait for system in Example 8.7.3.

One repeated eigenvalue: the general principle. Suppose that $\mathbf{A}$ is a 2×2 matrix with a single repeated eigenvalue r. When we turn to find an eigenvector for r, two rather different things can happen.

Case (1). The typical situation is that, as in Example 8.7.3, all eigenvectors corresponding to r are (nonzero) scalar multiples of a single vector $\mathbf{V}$. We will be able to find $\mathbf{W}$ so that

$$(\mathbf{A} - r\mathbf{I})\mathbf{W} = \mathbf{V}$$

(see Exercise 18). Using $r, \mathbf{V},$ and $\mathbf{W}$ we can build the general solution:

> In Case (1), the system $\mathbf{X}' = \mathbf{A}\mathbf{X}$ has general solution
> $$\mathbf{X} = c_1 \left[e^{rt}\mathbf{V} \right] + c_2 \left[te^{rt}\mathbf{V} + e^{rt}\mathbf{W} \right],$$
> where $\mathbf{W}$ is any solution to
> $$(\mathbf{A} - r\mathbf{I})\mathbf{W} = \mathbf{V}.$$
> The equilibrium point is classified as a **degenerate improper node**; it is unstable if $r > 0$ and stable if $r < 0$.

A typical phase portrait in Case (1) is shown in Fig. 8.70. All trajectories are directed away from the origin if $r > 0$ and towards the origin if $r < 0$.

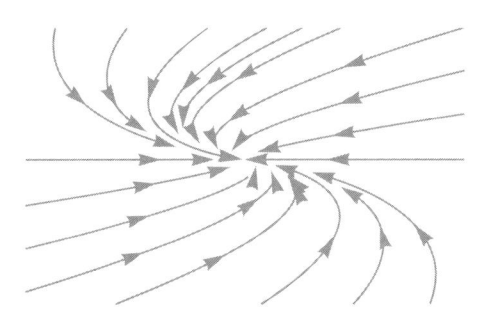

Figure 8.70. One repeated eigenvalue, Case (1).

Case (2). In special circumstances it may turn out that $\mathbf{A}$ has two *linearly independent* eigenvectors corresponding to the single eigenvalue r. This case is explored in Exercise 19, where you will show that for this to happen, the matrix $\mathbf{A}$ must be

$$\mathbf{A} = \begin{pmatrix} r & 0 \\ 0 & r \end{pmatrix}$$

8.7. Complex and repeated eigenvalues

and *every* nonzero vector is an eigenvector corresponding to the eigenvalue r. From this, it is easy to see that the general solution is

$$\mathbf{X} = c_1 e^{rt} \begin{pmatrix} 1 \\ 0 \end{pmatrix} + c_2 e^{rt} \begin{pmatrix} 0 \\ 1 \end{pmatrix} = e^{rt} \begin{pmatrix} c_1 \\ c_2 \end{pmatrix}.$$

Every solution is a ray solution, outgoing if $r > 0$ and incoming if $r < 0$.

In Case (2), the equilibrium point $(0,0)$ is called a **degenerate proper node**.[10] In this case, the phase portrait consists of infinitely many rays as shown below. Moreover, every ray to/from the origin occurs.

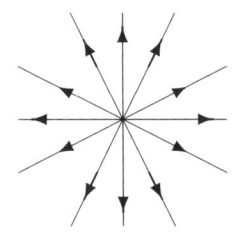

Proper node at $(0,0), r > 0$.

In everyday English usage, "junction" is a synonym for "node". Note that in the two preceding phase portraits (as well as that of Fig. 8.69), if we extend trajectories so that they include the origin, then the origin is indeed a junction point for trajectories of the system. In general, the equilibrium point $(0,0)$ is a **nondegenerate node** for the planar linear system $\mathbf{X}' = \mathbf{AX}$ provided $\mathbf{A}$ has two distinct eigenvalues having the same sign. If $\mathbf{A}$ has a single, repeated nonzero eigenvalue, then the origin is a **degenerate node**. We have indicated above that degenerate nodes come in two varieties: "proper" and "improper". The origin is a **proper node** for a planar linear system provided every ray approaching the origin (as $t \to \infty$ or $t \to -\infty$) is a trajectory of the system and it is **improper** otherwise. In fact, if the origin is an improper degenerate node, then there is a single line L through the origin such that every nonzero trajectory approaches the origin[11] tangent to L. This line L is the eigenvector line. If the origin is a nondegenerate node, then it is necessarily improper and there is a line through the origin (the preferred nodal tangent) such that all but two trajectories approach the origin tangent to this line (either as $t \to \infty$ or as $t \to -\infty$).

Example 8.7.4. We will analyze the system

$$\frac{dx}{dt} = y, \quad \frac{dy}{dt} = -9x + 6y.$$

The matrix

$$\mathbf{A} = \begin{pmatrix} 0 & 1 \\ -9 & 6 \end{pmatrix}$$

has only one eigenvalue, 3, with corresponding eigenvector

$$\mathbf{V} = \begin{pmatrix} 1 \\ 3 \end{pmatrix}.$$

All other eigenvectors are scalar multiples of this one. Next solve $(\mathbf{A} - 3\mathbf{I})\mathbf{W} = \mathbf{V}$ for $\mathbf{W}$. The equation

$$\begin{pmatrix} -3 & 1 \\ -9 & 3 \end{pmatrix} \begin{pmatrix} w_1 \\ w_2 \end{pmatrix} = \begin{pmatrix} 1 \\ 3 \end{pmatrix}$$

says

$$-3w_1 + w_2 = 1 \quad \text{and} \quad -9w_1 + 3w_2 = 3.$$

---
[10]The term **star node** is sometimes used instead of proper node.
[11]As $t \to \infty$ in the stable case and as $t \to -\infty$ in the unstable case.

Since the second equation is just a multiple of the first, there are infinitely many choices for w_1 and w_2. Any one will do, and we choose

$$\mathbf{W} = \begin{pmatrix} 1 \\ 4 \end{pmatrix}.$$

The general solution to $\mathbf{X}' = \mathbf{A}\mathbf{X}$ is

$$\begin{aligned} \mathbf{X}(t) &= c_1 e^{3t} \mathbf{V} + c_2 (t e^{3t} \mathbf{V} + e^{3t} \mathbf{W}) \\ &= c_1 e^{3t} \begin{pmatrix} 1 \\ 3 \end{pmatrix} + c_2 \left(t e^{3t} \begin{pmatrix} 1 \\ 3 \end{pmatrix} + e^{3t} \begin{pmatrix} 1 \\ 4 \end{pmatrix} \right). \end{aligned}$$

To aid in understanding what a phase portrait for the system should look like, we'll do a nullcline-and-arrow diagram for the system. The x-nullcline is the line $y = 0$; above this line trajectories move to the right, while below the line trajectories move to the left. The y-nullcline is the line $-9x + 6y = 0$. Check that above this line trajectories move up and below this line trajectories move down. Fig. 8.71 shows this information. In Fig. 8.72 we sketch some trajectories in the phase portrait, using the nullcline-and-arrow diagram as well as our qualitative understanding of the appearance of trajectories near a unstable degenerate improper node. Trajectories of the ray solutions $e^{3t} \mathbf{V}$ and $-e^{3t} \mathbf{V}$ are explicitly shown. All trajectories are directed outward, and, as $t \to -\infty$, all trajectories approach the origin tangent to the line containing the trajectories of the ray solutions (which is the only eigenvector line).

Figure 8.71. Nullcline-and-arrow diagram.

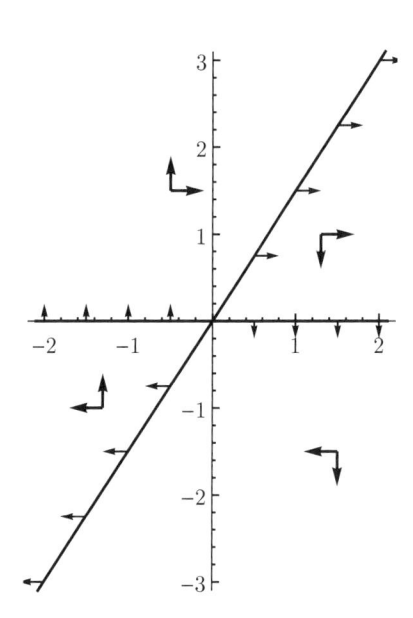

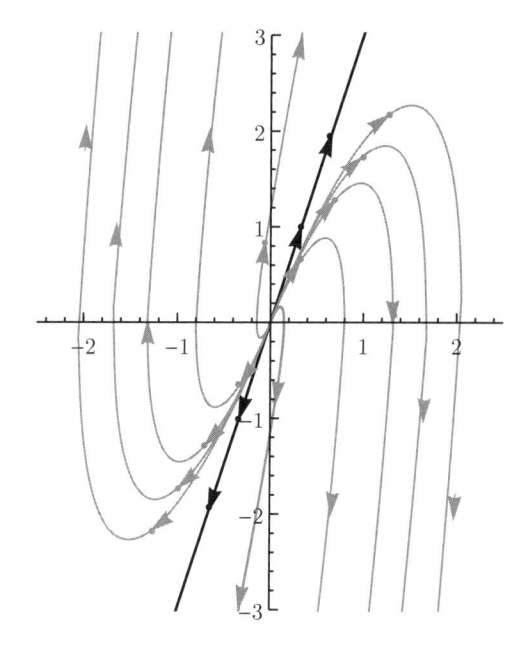

Figure 8.72. Phase portrait.

8.7.1. Exercises.

In Exercises 1–4 write the system in the form $\mathbf{X}' = \mathbf{A}\mathbf{X}$ for appropriate choice of $\mathbf{A}$ and $\mathbf{X}$ and then find the general solution.

1. $\dfrac{dx}{dt} = x - y,$
   $\dfrac{dy}{dt} = 5x - 3y.$

2. $\dfrac{dx}{dt} = x - 2y,$
   $\dfrac{dy}{dt} = 2x - y.$

3. $\dfrac{dx}{dt} = x - 2y,$
   $\dfrac{dy}{dt} = 2x + y.$

4. $\dfrac{dx}{dt} = 3x + y,$
   $\dfrac{dy}{dt} = -4x - y.$

5. Find the general (real-valued) solution to $\mathbf{X}' = \mathbf{A}\mathbf{X}$ for
$$\mathbf{A} = \begin{pmatrix} 1 & 1 \\ -1 & 1 \end{pmatrix}.$$

6. Suppose the 2×2 matrix $\mathbf{A}$ has eigenvalues $1+i$ and $1-i$, with eigenvectors
$$\begin{pmatrix} 2+i \\ 3 \end{pmatrix} \quad \text{and} \quad \begin{pmatrix} 2-i \\ 3 \end{pmatrix},$$
respectively. What is the general *real-valued* solution to the system $\mathbf{X}' = \mathbf{A}\mathbf{X}$?

7. Solve the initial value problem
$$\mathbf{X}' = \begin{pmatrix} 1 & -4 \\ 4 & -7 \end{pmatrix} \mathbf{X}, \quad \mathbf{X}(0) = \begin{pmatrix} 3 \\ 2 \end{pmatrix}.$$

8. Solve the initial value problem
$$\dfrac{dx}{dt} = x - y, \quad \dfrac{dy}{dt} = x + y, \quad x(0) = 1, \quad y(0) = 0,$$
and sketch the trajectory of your solution.

9. For the system in Example 8.7.1 we found the real-valued solutions
$$\mathbf{X}_1(t) = \begin{pmatrix} 2\cos(2t) \\ -\sin(2t) \end{pmatrix} \quad \text{and} \quad \mathbf{X}_2(t) = \begin{pmatrix} 2\sin(2t) \\ \cos(2t) \end{pmatrix}.$$

   (a) Show that the vectors $\mathbf{X}_1(0)$ and $\mathbf{X}_2(0)$ are linearly independent.
   (b) Explain why it follows that $c_1\mathbf{X}_1(t) + c_2\mathbf{X}_2(t)$ is the general solution to this system. Hint: Invoke a theorem from Section 8.4.

10. In Example 8.7.1 we found that the solution to the initial value problem
$$\mathbf{X}' = \mathbf{A}\mathbf{X}, \quad \mathbf{X}(0) = \begin{pmatrix} 3 \\ 4 \end{pmatrix},$$
is
$$\mathbf{X}(t) = \begin{pmatrix} 3\cos(2t) + 8\sin(2t) \\ -3/2\sin(2t) + 4\cos(2t) \end{pmatrix}.$$
Here we sketch the orbit of this solution in the phase plane.

   (a) Write $3\cos(2t) + 8\sin(2t)$ in the form $R\cos(2t - \delta)$ for appropriate choice of R and δ. Show that $-3/2\sin(2t) + 4\cos(2t) = R_2\sin(2t - \delta)$ for a new choice of (not necessarily positive) R_2 but the same δ. Hint: Use the identity $\sin(A - B) = \sin A \cos B - \cos A \sin B$.
   (b) Show that the orbit is an ellipse, and sketch it.

11. The solution to
$$\mathbf{X}'(t) = \begin{pmatrix} 1 & 1 \\ -1 & 3 \end{pmatrix} \mathbf{X}, \quad \mathbf{X}(0) = \begin{pmatrix} 1 \\ 0 \end{pmatrix},$$
is
$$e^{2t}\left(\begin{pmatrix} 1 \\ 0 \end{pmatrix} + t\begin{pmatrix} -1 \\ -1 \end{pmatrix}\right).$$
In this problem you will sketch the corresponding trajectory in the phase plane.

(a) Sketch the line
$$\begin{pmatrix} 1 \\ 0 \end{pmatrix} + t \begin{pmatrix} -1 \\ -1 \end{pmatrix}.$$
Show the points on this line corresponding to $t = 0, \frac{1}{2}, -\frac{1}{2}$, and -1.

(b) Compute the scaling factor e^{2t} for $t = 0, \frac{1}{2}, -\frac{1}{2}$, and -1.

(c) Using your picture from (a) and computations in (b), show the points on the trajectory corresponding to $t = 0, \frac{1}{2}, -\frac{1}{2}$, and -1. Then sketch the trajectory.

12. In Example 8.7.3, the general solution was
$$\mathbf{X}(t) = c_1 e^t \begin{pmatrix} -2 \\ 1 \end{pmatrix} + c_2 \left(t e^t \begin{pmatrix} -2 \\ 1 \end{pmatrix} + e^t \begin{pmatrix} -1 \\ 0 \end{pmatrix} \right).$$
Sketch the trajectories corresponding to the choices $c_1 = 1, c_2 = 1$. Hint: Write this solution as
$$e^t \left(\begin{pmatrix} -3 \\ 1 \end{pmatrix} + t \begin{pmatrix} -2 \\ 1 \end{pmatrix} \right)$$
and begin by sketching the line through the point $(-3, 1)$ and in the direction of the vector $(-2, 1)$. Points on this line are scaled by the factor e^t.

13. Consider the system
$$\frac{dx}{dt} = 2x + 5y, \quad \frac{dy}{dt} = -x - 2y.$$

(a) At every point in the first quadrant, $\frac{dx}{dt}$ has the same sign. Is it positive or negative? (There is no need to solve the system to answer this!)

(b) Similarly, at every point in the first quadrant, $\frac{dy}{dt}$ has the same sign. Is it positive or negative?

(c) The matrix
$$\mathbf{A} = \begin{pmatrix} 2 & 5 \\ -1 & -2 \end{pmatrix}$$
has eigenvalues $\pm i$, so the phase portrait trajectories are closed loops around the origin. Are they clockwise or counterclockwise? Use your answers to (a) and (b) to decide, with no additional computation needed!

14. The eigenvalues of
$$\mathbf{A} = \begin{pmatrix} 1 & 2 \\ -2 & 0 \end{pmatrix}$$
are $\frac{1}{2} \pm i\frac{\sqrt{15}}{2}$.

(a) In quadrant 1 of the phase plane for the system $\mathbf{X}' = \mathbf{AX}$, is $\frac{dx}{dt}$ positive or negative? Is $\frac{dy}{dt}$ positive or negative? No computations are needed to answer this!

(b) The orbits in the phase plane are spirals. Are they clockwise or counterclockwise spirals?

15. Suppose that $\mathbf{A}$ is a 2×2 matrix with real entries and eigenvalues $\alpha \pm i\beta$, with $\beta \neq 0$. Let $\mathbf{V}$ be an eigenvector for the eigenvalue $\alpha + i\beta$, and write
$$\mathbf{V} = \mathbf{U} + i\mathbf{W}$$
where $\mathbf{U}$ and $\mathbf{W}$ are vectors with real entries. The goal of this problem is to show that the vectors $\mathbf{U}$ and $\mathbf{W}$ are linearly independent.

(a) Suppose that $c_1 \mathbf{U} + c_2 \mathbf{W} = \mathbf{0}$ for some scalars c_1 and c_2. Show that
$$(c_1 - ic_2)\mathbf{V} + (c_1 + ic_2)\overline{\mathbf{V}} = \mathbf{0}.$$

(b) Keeping in mind the result of Theorem 8.4.6, which tells you that $\mathbf{V}$ and $\overline{\mathbf{V}}$ are linearly independent, show that $c_1 = c_2 = 0$ as desired.

8.7. Complex and repeated eigenvalues

16. Suppose that $\mathbf{A}$ is a 2×2 matrix with real entries and eigenvalues $\alpha \pm i\beta$, with $\beta \neq 0$. Let $\mathbf{V}$ be an eigenvector for the eigenvalue $\alpha + i\beta$, and write

$$\mathbf{V} = \mathbf{U} + i\mathbf{W} = \begin{pmatrix} u_1 \\ u_2 \end{pmatrix} + i \begin{pmatrix} w_1 \\ w_2 \end{pmatrix},$$

where $\mathbf{U}$ and $\mathbf{W}$ are the real and imaginary parts of $\mathbf{V}$. Consider the complex exponential solution

$$e^{\alpha t}(\cos(\beta t) + i \sin(\beta t)) \begin{pmatrix} u_1 + iw_1 \\ u_2 + iw_2 \end{pmatrix}.$$

(a) Show that the real and imaginary parts of this complex exponential solution are

$$\mathbf{X}_1(t) = (e^{\alpha t} \cos(\beta t))\mathbf{U} - (e^{\alpha t} \sin(\beta t))\mathbf{W}$$

and

$$\mathbf{X}_2(t) = (e^{\alpha t} \sin(\beta t))\mathbf{U} + (e^{\alpha t} \cos(\beta t))\mathbf{W},$$

respectively.

(b) Using Exercise 15, show that $\mathbf{X}_1(0)$ and $\mathbf{X}_2(0)$ are linearly independent vectors.

(c) Explain why it follows that $c_1 \mathbf{X}_1(t) + c_2 \mathbf{X}_2(t)$ is the general solution to $\mathbf{X}' = \mathbf{A}\mathbf{X}$.

17. Verify that equation (8.69) gives the general solution of the differential equation in Example 8.7.3.

18. Suppose $\mathbf{A}$ is a 2×2 matrix with one repeated (real) eigenvalue r, and suppose that the only eigenvectors corresponding to r are scalar multiples of a single vector $\mathbf{V}$.

    (a) Writing

    $$\mathbf{A} = \begin{pmatrix} a & b \\ c & d \end{pmatrix},$$

    show using the quadratic formula that $(a - d)^2 + 4bc = 0$ and $r = \frac{a+d}{2}$.

    (b) Using the result of (a), show that

    $$(\mathbf{A} - r\mathbf{I})^2 = \begin{pmatrix} a - r & b \\ c & d - r \end{pmatrix}^2 = \begin{pmatrix} 0 & 0 \\ 0 & 0 \end{pmatrix}$$

    (this is a special case of the Cayley-Hamilton Theorem discussed in the next chapter).

    (c) Show that there are infinitely many vectors

    $$\mathbf{W} = \begin{pmatrix} w_1 \\ w_2 \end{pmatrix}$$

    satisfying $(\mathbf{A} - r\mathbf{I})\mathbf{W} = \mathbf{V}$. Hint: Suppose $\mathbf{U}$ is any nonzero vector that is not an eigenvector of $\mathbf{A}$. Then $(\mathbf{A} - r\mathbf{I})[(\mathbf{A} - r\mathbf{I})\mathbf{U}] = \mathbf{0}$ by (b).

19. Suppose $\mathbf{A}$ is a 2×2 matrix with a single (repeated) real eigenvalue r. Suppose further that we are in the special case that we can find two *linearly independent* eigenvectors $\mathbf{V}_1$ and $\mathbf{V}_2$ corresponding to r.

    (a) Show that any vector in $\mathbb{R}^2$ can be written as a linear combination of $\mathbf{V}_1$ and $\mathbf{V}_2$.

    (b) Using the result of (a), show that every nonzero vector in $\mathbb{R}^2$ is an eigenvector for $\mathbf{A}$.

    (c) Show that

    $$\mathbf{A} = r\mathbf{I} = \begin{pmatrix} r & 0 \\ 0 & r \end{pmatrix}.$$

    Hint: By (b), both

    $$\begin{pmatrix} 1 \\ 0 \end{pmatrix} \text{ and } \begin{pmatrix} 0 \\ 1 \end{pmatrix}$$

    are eigenvectors of $\mathbf{A}$.

(d) Find the general solution of $\mathbf{X}' = \mathbf{AX}$ if $\mathbf{A} = r\mathbf{I}$, and sketch a phase portrait for this system.

20. A charged particle moves in the xy-plane subject to the force it experiences from a uniform magnetic field perpendicular to the plane. Suppose that $\mathbf{V}(t) = \begin{pmatrix} v_1(t) \\ v_2(t) \end{pmatrix}$ is the particle's velocity vector; then

$$\frac{dv_1}{dt} = \alpha v_2(t),$$
$$\frac{dv_2}{dt} = -\alpha v_1(t),$$

where $\alpha = qB/m$ is a positive constant determined by the mass m and charge q of the particle as well as the magnetic field strength B (all in appropriate units).

(a) Find the general solution of the preceding system.
(b) Via integration of each component of the velocity, find the general form for the components of the particle's position and show that the particle travels a circular path.

8.8. Applications: Compartment models

Systems of first-order equations often arise in applications that are based on compartment models. We have used such models in several earlier examples; for instance in Section 5.3.2 we used a pharmacokinetic compartment model to study a drug overdose, and in Exercise 22 of Section 8.1 we modeled the spread of swine flu by dividing the population into three compartments. Now we will be a bit more systematic in discussing how compartment models arise in applications.

A basic compartment model allows for the continuous transfer of a "substance" from one compartment to another. For example, the substance could be a drug, and the compartments could correspond to various parts of the body—the bloodstream, the gastrointestinal tract, the liver, the kidneys, etc., as appropriate for the particular drug under study. We might also consider an "external" compartment, to capture the losses of the drug from the body. In another example, the substance might be a pollutant, and the compartments various components of an ecological system, e.g., soil, plants, and animals. In a compartment model, the "substance" need not be a physical substance. For example, in Exercise 22 of Section 8.1 we divided a population into the three compartments of "infective", "susceptible", and "removed". Similarly, we could use a compartment model to study a population organized into various age classes; e.g., we might have "child","adolescent", and "adult" compartments. A common feature of any compartment system is that each compartment should be a well-mixed homogeneous unit.

Basic assumption. In the models of this section, we assume that the rate of transfer of the substance *into* the kth compartment *from* the jth compartment is proportional to the amount x_j present in the jth compartment. The proportionality constant (called the transfer coefficient or rate constant) is denoted a_{kj}. This is sometimes called *linear, donor-controlled* transfer. As we will see, it leads to a system of first-order constant coefficient equations.

$$\boxed{\text{Compartment } j} \xrightarrow{a_{kj}} \boxed{\text{Compartment } k}$$

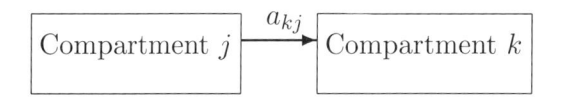

Example 8.8.1. A small amount of radioactive iodine ^{131}I is introduced into the bloodstream. From there it's taken up by the thyroid gland, where it is gradually removed. The compartment

8.8. Applications: Compartment models

diagram for this situation is shown below. We'll set up and solve the differential equation model for this system, assuming the transfer coefficients a_{21} and a_{32} are not equal.

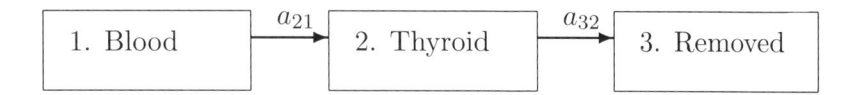

$$\boxed{\text{1. Blood}} \xrightarrow{a_{21}} \boxed{\text{2. Thyroid}} \xrightarrow{a_{32}} \boxed{\text{3. Removed}}$$

We'll let $x(t)$ denote the amount of ^{131}I (in mg) present in the blood at time t (in hr) and let $y(t)$ be the amount in the thyroid gland. We might imagine several ways in which the ^{131}I is introduced into the bloodstream:

Case 1. All at once, so that at time $t = 0$, $x(0) = d_0$ for some constant d_0, but there is no further input of ^{131}I into the system.

Case 2. Through a constant intravenous infusion, so that $x(0) = 0$, but ^{131}I is added to the blood compartment at a constant rate throughout the period of time under study.

Case 3. By a constant intravenous infusion over the first T_0 hours, but with no additional ^{131}I added after T_0 hours.

Let's set up a model for Case 1, focusing on the amounts $x(t)$ and $y(t)$ in the blood and thyroid compartments at time t, using our assumption that the rate of transfer of iodine ^{131}I to compartment k from compartment j is proportional to the amount in compartment j, with proportionality constant a_{kj}. This leads to the equations

(8.70)
$$\begin{aligned} x' &= -a_{21}x, \\ y' &= a_{21}x - a_{32}y. \end{aligned}$$

There is also ongoing radioactive decay of the ^{131}I in the system. Losses from the thyroid gland due to this decay are included in the constant a_{32}. However, we have ignored losses due to decay from the blood. This is justified by the fact that the corresponding rate constant is very small in comparison to the standard value of a_{21}; see Exercise 12.

We could add a third equation to our system,

(8.71)
$$z' = a_{32}\, y,$$

for the amount $z(t)$ in the "removed" compartment at time t. As a three-dimensional system, equations (8.70)–(8.71) take the form $\mathbf{X}' = \mathbf{A}\mathbf{X}$ for

$$\mathbf{A} = \begin{pmatrix} -a_{21} & 0 & 0 \\ a_{21} & -a_{32} & 0 \\ 0 & a_{32} & 0 \end{pmatrix} \quad \text{and} \quad \mathbf{X} = \begin{pmatrix} x(t) \\ y(t) \\ z(t) \end{pmatrix}.$$

But since we are mainly interested in x and y and since z does not appear in the equation for either x' or y', we can concentrate on the planar system (8.70), writing it as $\mathbf{X}' = \mathbf{A}\mathbf{X}$ for

$$\mathbf{X}(t) = \begin{pmatrix} x(t) \\ y(t) \end{pmatrix} \quad \text{and} \quad \mathbf{A} = \begin{pmatrix} -a_{21} & 0 \\ a_{21} & -a_{32} \end{pmatrix}.$$

In Case 1 there is an instantaneous injection of ^{131}I into compartment 1 at time $t = 0$, and thus we have the initial condition $x(0) = d_0$, $y(0) = 0$, or

$$\mathbf{X}(0) = \begin{pmatrix} d_0 \\ 0 \end{pmatrix}.$$

Does this system looks familiar? It is the same set of equations as we used in Section 5.3.2 to model the pharmacokinetics of an aspirin overdose (where the compartments were the gastrointestinal track and bloodstream). There we solved the system by converting it to a single second-order differential equation, but now we will solve (8.70) using linear system methods.

The eigenvalues of $\mathbf{A}$ are $-a_{21}$ and $-a_{32}$ (these are different numbers). An eigenvector for $-a_{32}$ is
$$\mathbf{V}_1 = \begin{pmatrix} 0 \\ 1 \end{pmatrix}$$
and for $-a_{21}$ is
$$\mathbf{V}_2 = \begin{pmatrix} 1 \\ \frac{a_{21}}{a_{32}-a_{21}} \end{pmatrix}.$$

The general solution to $\mathbf{X}' = \mathbf{A}\mathbf{X}$ is

(8.72) $$c_1 e^{-a_{32}t} \begin{pmatrix} 0 \\ 1 \end{pmatrix} + c_2 e^{-a_{21}t} \begin{pmatrix} 1 \\ \frac{a_{21}}{a_{32}-a_{21}} \end{pmatrix}.$$

Using the initial condition we see that
$$c_1 = \frac{d_0 a_{21}}{a_{21} - a_{32}}, \quad c_2 = d_0.$$

Focusing on the amount y in the thyroid gland at time t we have

(8.73) $$y(t) = \frac{d_0 a_{21}}{a_{21} - a_{32}} e^{-a_{32}t} + \frac{d_0 a_{21}}{a_{32} - a_{21}} e^{-a_{21}t}.$$

Ensuring safe ^{131}I levels. Typically the value of a_{21} will be considerably larger than the value of a_{32}. Reasonable approximations are $a_{21} = 0.2$/hr and $a_{32} = 0.05$/hr. Suppose that we want to ensure that the amount of ^{131}I in the thyroid never exceeds 25 mg. What is the maximum loading dose d_0 that can be safely used? The maximum value of
$$y(t) = \frac{0.2 d_0}{0.2 - 0.05} e^{-0.05t} + \frac{0.2 d_0}{0.05 - 0.2} e^{-0.2t} = \frac{4}{3} d_0 e^{-0.05t} - \frac{4}{3} d_0 e^{-0.2t}$$
occurs when $y'(t) = 0$. A computation shows that $y'(t) = 0$ when $0.2 e^{-0.2t} = 0.05 e^{-0.05t}$ or $\ln \frac{0.2}{0.05} = 0.15t$. This gives a t-value of about 9.2 hours. The value of y at $t = 9.2$ is
$$\frac{4}{3} d_0 (e^{-0.46} - e^{-1.84}).$$
To ensure this stays below 25 mg, we solve
$$\frac{4}{3} d_0 (e^{-0.46} - e^{-1.84}) < 25$$
to see that we must have $d_0 < 39.6$ mg.

Other infusion methods. In Case 2 (constant intravenous infusion), we modify our system of equations to be

(8.74) $$\begin{aligned} x' &= -a_{21}x + I, \\ y' &= a_{21}x - a_{32}y \end{aligned}$$

for some constant I. For Case 3 (constant infusion over a fixed period of time), we have the equations

(8.75) $$\begin{aligned} x' &= -a_{21}x + I(t), \\ y' &= a_{21}x - a_{32}y \end{aligned}$$

8.8. Applications: Compartment models

where

$$I(t) = \begin{cases} I & \text{if } 0 \le t < T_0, \\ 0 & \text{if } T_0 \le t. \end{cases}$$

In both of these cases, the system is no longer homogeneous, and the initial conditions are $x(0) = 0$ and $y(0) = 0$.

Exercise 19 outlines a way to solve the system in (8.74) by a change of variable which converts it to a homogeneous system. Laplace transforms give a method to solve both (8.74) and (8.75); see Section 8.8.1 below. Yet another solution method will be discussed later (see Section 9.4).

In the next example, we have a five-compartment system, each compartment representing one of the Great Lakes.

Example 8.8.2. Pollution in the Great Lakes

In Exercise 22 of Section 2.3.1, we used a single differential equation to model pollution in Lake Superior (or Lake Michigan). To study the pollution levels in the lower Great Lakes (Huron, Erie, and Ontario), we must take into account how water flows into these lakes from other lakes, since even if all new pollution into the Great Lakes is stopped, contaminates will continue to flow into the lower lakes from the upper lakes. Below we show a compartment diagram to indicate the transfer of pollutants between the various Great Lakes.

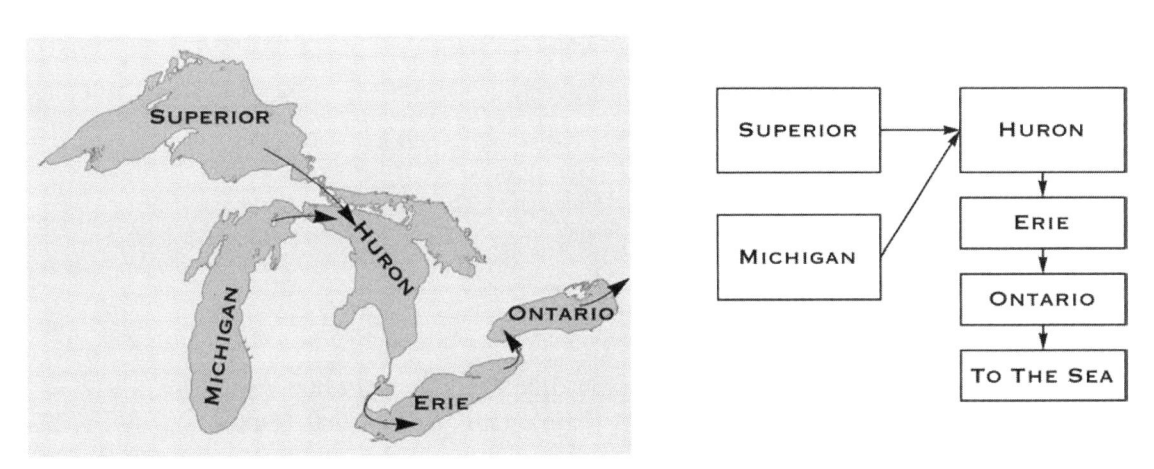

Figure 8.73. The Great Lakes and corresponding compartment diagram. Derived from: Wikimedia Commons, author: Phizzy, licensed under the GNU Free Documentation license and the Creative Commons Attribution-ShareAlike 3.0 Unported (https://creativecommons.org/licenses/by-sa/3.0/deed.en) license.

To get started on a model, consider x_S, x_M, x_H, x_E, and x_O, the amounts of pollutants (in kg) in Lakes Superior, Michigan, Huron, Erie, and Ontario, respectively, at time t (in years).[12]

Our basic assumptions are:

- the total water inflow (in km^3/year) into each lake (including precipitation) is equal to the total water outflow from the lake (including evaporation), so that the volumes V_S, V_M, V_H, V_E, and V_O (in km^3) of Lakes Superior, Michigan, Huron, and so on, remain constant;
- pollutants are uniformly distributed throughout each lake;
- at time $t = 0$ all entry of pollutants into the Great Lakes as a whole is stopped.

---

[12] Rather than talking about pollution as a whole, we might choose to model a specific pollutant, such as phosphorus or chloride.

The rates of flow between the lakes, as indicated by the arrows in Fig. 8.73, are as shown:

Direction of flow	Flow rate in km^3/year
From Superior to Huron	60
From Michigan to Huron	160
From Huron to Erie	280
From Erie to Ontario	350
Outward from Ontario	400

Keeping in mind that the concentration of pollutants, in kg/km^3, in any of the lakes is obtained as the quotient of the amount x of pollutant divided by the lake volume V, we have the differential equation system

$$\frac{dx_S}{dt} = -\frac{60}{V_S}x_S,$$
$$\frac{dx_M}{dt} = -\frac{160}{V_M}x_M,$$
$$\frac{dx_H}{dt} = \frac{60}{V_S}x_S + \frac{160}{V_M}x_M - \frac{280}{V_H}x_H,$$
$$\frac{dx_E}{dt} = \frac{280}{V_H}x_H - \frac{350}{V_E}x_E,$$
$$\frac{dx_O}{dt} = \frac{350}{V_E}x_E - \frac{400}{V_O}x_O.$$

The volume of each lake, in km^3, is approximately $V_S = 12,000$ (Superior), $V_M = 5,000$ (Michigan), $V_H = 3,500$ (Huron), $V_E = 470$ (Erie), and $V_O = 1,600$ (Ontario). With this, our system becomes $\mathbf{X}'(t) = \mathbf{A}\mathbf{X}(t)$, where

$$\mathbf{A} = \begin{pmatrix} -0.005 & 0 & 0 & 0 & 0 \\ 0 & -0.032 & 0 & 0 & 0 \\ 0.005 & 0.032 & -0.08 & 0 & 0 \\ 0 & 0 & 0.08 & -0.75 & 0 \\ 0 & 0 & 0 & 0.75 & -0.25 \end{pmatrix} \quad \text{and} \quad \mathbf{X} = \begin{pmatrix} x_S \\ x_M \\ x_H \\ x_E \\ x_O \end{pmatrix}.$$

Since $\mathbf{A}$ is a lower triangular matrix (all entries above the main diagonal are 0), the eigenvalues of $\mathbf{A}$ are the entries on the main diagonal (see Exercise 19). While corresponding eigenvectors could be found by hand, the larger size of $\mathbf{A}$ makes the use of some technology preferable in finding these eigenvectors. Since there are 5 distinct real eigenvalues, the general solution can be built from linear combinations of the ray solutions coming from the eigenvalue/eigenvector pairs:

$$\mathbf{X}(t) = c_1 e^{-0.005t}\mathbf{W}_1 + c_2 e^{-0.032t}\mathbf{W}_2 + c_3 e^{-0.08t}\mathbf{W}_3 + c_4 e^{-0.75t}\mathbf{W}_4 + c_5 e^{-0.25t}\mathbf{W}_5,$$

8.8. Applications: Compartment models

where $\mathbf{W}_j$ is an eigenvector corresponding to the jth diagonal entry of $\mathbf{A}$. We used *Mathematica* to compute these eigenvectors, obtaining for the general solution

$$(8.76) \quad \mathbf{X}(t) = c_1 e^{-0.005t} \begin{pmatrix} 1 \\ 0 \\ 0.067 \\ 0.0071 \\ 0.022 \end{pmatrix} + c_2 e^{-0.032t} \begin{pmatrix} 0 \\ 0.81 \\ 0.54 \\ 0.06 \\ 0.21 \end{pmatrix}$$

$$+ c_3 e^{-0.08t} \begin{pmatrix} 0 \\ 0 \\ 0.88 \\ 0.11 \\ 0.46 \end{pmatrix} + c_4 e^{-0.75t} \begin{pmatrix} 0 \\ 0 \\ 0 \\ 0.55 \\ -0.83 \end{pmatrix} + c_5 e^{-0.25t} \begin{pmatrix} 0 \\ 0 \\ 0 \\ 0 \\ 1 \end{pmatrix}.$$

If we know the values of x_S, x_M, x_H, x_E, and x_O at a particular time (call this time $t = 0$), this allows us to determine values for the constants c_j to solve that particular initial value problem. In Exercise 15 you are asked to do this when the pollutant being studied is phosphorus.

8.8.1. Solving a system by Laplace transforms. In Section 6.5 we used Laplace transforms to solve some simple systems. Here we'll review the method by solving the pharmacology model in equation (8.74).

Example 8.8.3. We'll solve the system

$$x' = -0.2x + I, \quad y' = 0.2x - 0.05y,$$

where I is a constant, with the initial conditions $x(0) = 0$, $y(0) = 0$. Applying the Laplace transform on both sides in each equation and remembering the basic formula

$$\mathcal{L}[f'(t)] = s\mathcal{L}[f(t)] - f(0),$$

we have

$$(8.77) \quad s\mathcal{L}[x] = -0.2\mathcal{L}[x] + \frac{I}{s}$$

and

$$(8.78) \quad s\mathcal{L}[y] = 0.2\mathcal{L}[x] - 0.05\mathcal{L}[y].$$

We've used the fact that the Laplace transform of the constant I is the function I/s. Now equation (8.77) can be solved for x as follows: We have

$$(s + 0.2)\mathcal{L}[x] = \frac{I}{s},$$

so that, using a partial fraction decomposition,

$$\mathcal{L}[x] = \frac{I}{s(s+0.2)} = \frac{I}{0.2}\left(\frac{1}{s} - \frac{1}{s+0.2}\right).$$

Taking inverse Laplace transforms gives

$$x(t) = \frac{I}{0.2}(1 - e^{-0.2t}).$$

Similarly, from equation (8.78) we have

$$(s + 0.05)\mathcal{L}[y] = 0.2\mathcal{L}[x] = 0.2\frac{I}{s(s+0.2)}.$$

Thus
$$\mathcal{L}[y] = 0.2I \frac{1}{s(s+0.2)(s+0.05)}.$$
Using the partial fraction decomposition
$$\frac{1}{s(s+0.2)(s+0.05)} = \frac{100}{s} + \frac{100/3}{s+0.2} - \frac{400/3}{s+0.05}$$
we have
$$\mathcal{L}[y] = 0.2I \left(\frac{100}{s} + \frac{100/3}{s+0.2} - \frac{400/3}{s+0.05} \right).$$
The inverse Laplace transform gives
$$y(t) = I \left(20 + \frac{20}{3} e^{-0.2t} - \frac{80}{3} e^{-0.05t} \right).$$

In Exercise 22 you are asked to use Laplace transforms to solve equation (8.75) with initial conditions $x(0) = 0$, $y(0) = 0$.

8.8.2. Compartment models with two-way transfer. In the compartment models considered in Example 8.8.1 and Example 8.8.2, the substance being modeled moved unidirectionally; e.g., ^{131}I moves from compartment 1 (blood) to compartment 2 (thyroid) to the "removed" compartment and not backwards between any of these compartments. In other situations, a compartment model that allows two-way transfer between some compartments may be appropriate. Specific examples will arise in the exercises. For example, we may have a compartment model which looks like this:

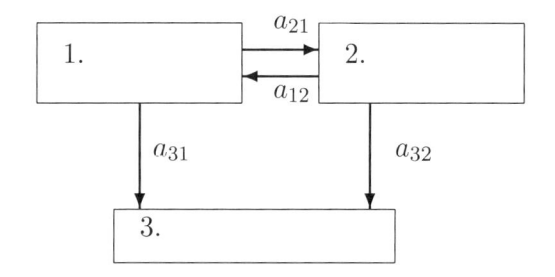

with transfer coefficients a_{ij} as shown. If $y_j(t)$ denotes the quantity in the jth compartment at time t, the corresponding differential equations are
$$\frac{dy_1}{dt} = -a_{21} y_1 + a_{12} y_2 - a_{31} y_1 = -(a_{21} + a_{31}) y_1 + a_{12} y_2,$$
$$\frac{dy_2}{dt} = a_{21} y_1 - a_{12} y_2 - a_{32} y_2 = a_{21} y_1 - (a_{12} + a_{32}) y_2,$$
$$\frac{dy_3}{dt} = a_{31} y_1 + a_{32} y_2.$$
Writing this in matrix form, we have $\mathbf{X}' = \mathbf{A}\mathbf{X}$ for
$$\mathbf{A} = \begin{pmatrix} -(a_{21} + a_{31}) & a_{12} & 0 \\ a_{21} & -(a_{12} + a_{32}) & 0 \\ a_{31} & a_{32} & 0 \end{pmatrix}.$$

Example 8.8.4. It's possible to work backwards and reconstruct the compartment diagram and transfer coefficients simply by knowing the matrix $\mathbf{A}$. For example, suppose in a three-compartment model we have
$$\mathbf{A} = \begin{pmatrix} -8 & 4 & 0 \\ 3 & -6 & 2 \\ 5 & 2 & -2 \end{pmatrix}.$$

Initially we'll show arrows going both ways between any pair of compartments:

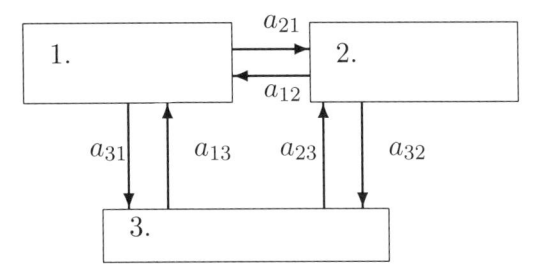

Our goal is to determine all a_{ij} values; a value that is 0 indicates an arrow that can be removed. From the matrix we reconstruct the equations as

$$\frac{dy_1}{dt} = -8y_1 + 4y_2,$$

$$\frac{dy_2}{dt} = 3y_1 - 6y_2 + 2y_3,$$

$$\frac{dy_3}{dt} = 5y_1 + 2y_2 - 2y_3.$$

Focus on the positive coefficient terms in these equations. The "$4y_2$" term in the first equation, which comes from transfer from compartment 2 to compartment 1, tells us that $a_{12} = 4$. Similarly, the "$3y_1$" term in the second equation tells us that $a_{21} = 3$ and the "$2y_3$" term in the same equation says $a_{23} = 2$. From the terms "$5y_1$" and "$2y_2$" in the third equation we learn that $a_{31} = 5$ and $a_{32} = 2$. Since there is no "y_3" term in the first equation, we must have $a_{13} = 0$, and we remove the arrow from compartment 3 to compartment 1. We determined all a_{ij}, and as a check on our work, you can verify that the terms with negative coefficients in the equations are correct, since $8 = a_{21} + a_{31}$ and $6 = a_{12} + a_{32}$. Here's the diagram again, showing the numerical values of the transfer coefficients:

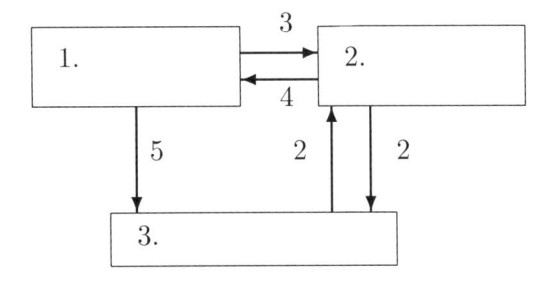

8.8.3. Exercises.

1. Two tanks are connected by pipes as shown. Tank 1 initially contains 10 lbs of salt dissolved in 100 gallons of water, and Tank 2 initially contains 100 gallons of pure water. At time $t = 0$, the tanks are connected and the mixture in each tank flows at a rate of 10 gal/min into the other tank. How much salt is in each tank at time t?

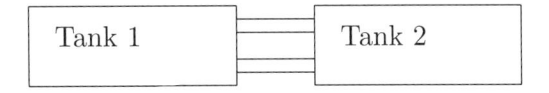

2. Two 100-gallon tanks are connected as shown below. Tank A is initially full of salt water with a concentration of 0.5 lb/gal, while tank B is initially full of pure water. At time $t = 0$ pure water begins flowing into tank A at a rate of 2 gal/min, and the well-mixed solution exits tank A and flows into tank B at a rate of 2 gal/min. The well-mixed solution in tank B exits that

tank at the same rate of 2 gal/min. Find the amount $x(t)$ of salt in tank A at time t and the amount $y(t)$ of salt in tank B at time t.

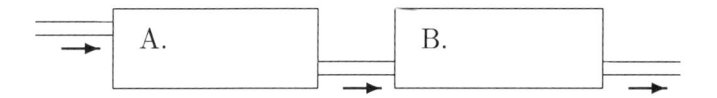

3. A three-compartment model gives rise to the differential equation system $\mathbf{X}' = \mathbf{A}\mathbf{X}$ with
$$\mathbf{A} = \begin{pmatrix} -\frac{3}{2} & 2 & 0 \\ 1 & -2 & 0 \\ \frac{1}{2} & 0 & 0 \end{pmatrix}.$$
Recreate the compartment diagram, showing the numerical values of the transfer coefficients.

4. (a) The matrix $\mathbf{A}$ in Example 8.8.4 has eigenvalues 0, -10, and -6. Find the general solution of the system $\mathbf{X}' = \mathbf{A}\mathbf{X}$ given there.
   (b) Find the equilibrium solution having first component $y_1(t) = 1$.

5. A four-compartment model gives the differential equation system $\mathbf{X}' = \mathbf{A}\mathbf{X}$ with
$$\mathbf{A} = \begin{pmatrix} -2 & 0 & 0 & 0 \\ 1 & -3 & \frac{1}{2} & 1 \\ 0 & 2 & -1 & 0 \\ 1 & 1 & \frac{1}{2} & -1 \end{pmatrix}.$$
Recreate the compartment diagram, showing the numerical values of the transfer coefficients.

6. A general two-compartment model with an external "losses" compartment is shown below. A planar homogeneous system $\mathbf{X}' = \mathbf{A}\mathbf{X}$ is written for the amounts $x(t)$ and $y(t)$ in compartments 1 and 2 at time t. If
$$\mathbf{A} = \begin{pmatrix} -3 & 1 \\ 2 & -4 \end{pmatrix},$$
find all of the transfer coefficients a_{ij}.

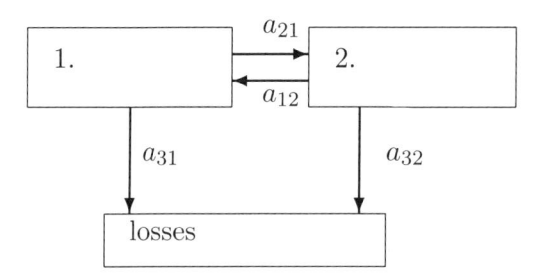

7. Albumin is a protein substance found in many animal tissues and fluids. In humans, it is the main protein found in the blood plasma. To study the synthesis, transfer, and breakdown of albumin in rabbits, a compartment model is proposed to follow the course of albumin to which a radioactive tracer ^{131}I is applied.[13] This model has four compartments:
   (1) the vascular compartment (blood plasma),
   (2) the extravascular compartment (lymph and tissue fluids),
   (3) the breakdown compartment,
   (4) the excretion compartment,
with diagram as shown below. The rate of transfer from compartment j to compartment k is proportional to the amount of albumin in compartment j, with proportionality constant a_{kj}.

---

[13] This problem is adapted from *The kinetics of the distribution and breakdown of ^{131}I-albumin in the rabbit*, by E. Reeve and J. Roberts, J. Gen. Physiol. 43 (1959), 415–444.

8.8. Applications: Compartment models

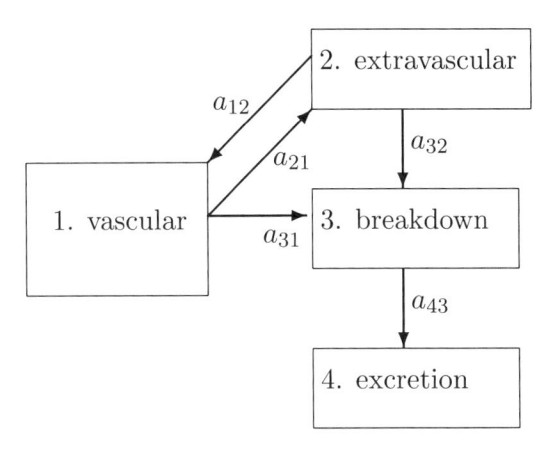

(a) Write the system of differential equations for the amount $y_j(t)$ of ^{131}I-albumin in compartment j at time t. Assume that after $t = 0$ there is no intake of albumin to the system.

(b) What is
$$\frac{dy_1}{dt} + \frac{dy_2}{dt} + \frac{dy_3}{dt} + \frac{dy_4}{dt}?$$
Why should this be so?

(c) Suppose that transfer to the breakdown compartment can only occur from the extravascular compartment. How would this change your system of equations in (a)? Answer the same question if transfer to the breakdown compartment only occurs from the vascular compartment.

8. **Set up** the system of differential equations for the compartment model shown below[14] for the transfer of Cesium^{137} in an arctic food chain, where the compartments are: 1. atmosphere, 2. lichens, 3. caribou, 4. man, and 5. removed.

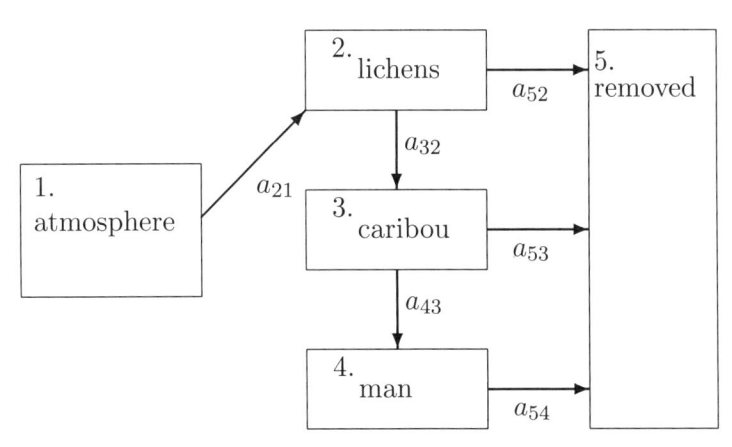

9. A general two-compartment model with an external "losses" compartment is shown below. A planar homogeneous system $\mathbf{X}' = \mathbf{AX}$ is written for the amounts $x(t)$ and $y(t)$ in compartments 1 and 2 at time t.

(a) Show that if a_{31} and a_{32} are both 0, then $\mathbf{A}$ has an eigenvalue equal to 0.

(b) Show that as long as all transfer coefficients a_{ij} are nonnegative, then the eigenvalues of $\mathbf{A}$ are real. Hint: $(b_1 + b_2)^2 - 4b_1b_2 = (b_1 - b_2)^2$; use this with $b_1 = a_{31} + a_{21}$ and $b_2 = a_{12} + a_{32}$.

---

[14]This compartment model comes from L. Eberhardt and W. Hanson, *A simulation model for an arctic food chain*, Health Physics 17 (1969), 793–806.

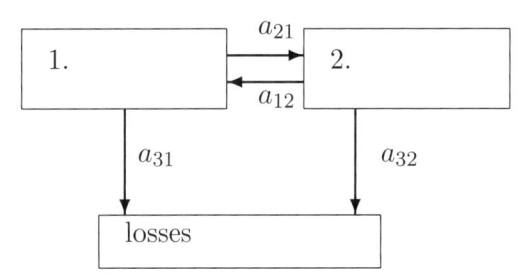

10. A four-compartment "chain" model is as shown. Let $y_j(t)$ be the amount of substance in the jth compartment. **Set up** a system of differential equations for this model, writing it in matrix form $\mathbf{X}' = \mathbf{A}\mathbf{X}$. Your matrix $\mathbf{A}$ should be 4×4.

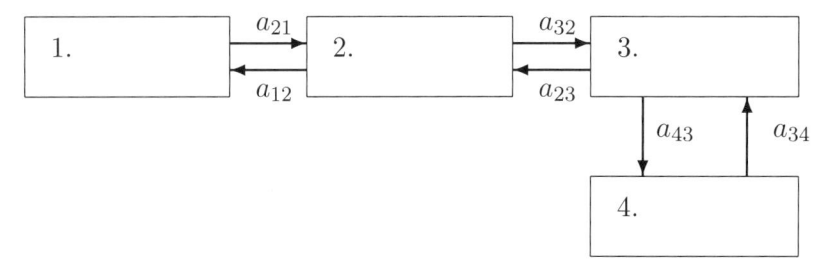

11. In Example 8.8.1 we considered the three-dimensional system $\mathbf{X}' = \mathbf{A}\mathbf{X}$ where

$$\mathbf{A} = \begin{pmatrix} -a_{21} & 0 & 0 \\ a_{21} & -a_{32} & 0 \\ 0 & a_{32} & 0 \end{pmatrix},$$

assuming $a_{21} \neq a_{32}$.
   (a) Find the eigenvalues of $\mathbf{A}$. How do these compare with the eigenvalues of the related 2×2 matrix
   $$\begin{pmatrix} -a_{21} & 0 \\ a_{21} & -a_{32} \end{pmatrix}?$$
   (b) What are all of the equilibrium solutions of $\mathbf{X}' = \mathbf{A}\mathbf{X}$? What is the physical meaning of these equilibrium solutions?
   (c) Find the general solution to the three-dimensional system $\mathbf{X}' = \mathbf{A}\mathbf{X}$.

12. Suppose in Example 8.8.1 we wish to include the loss of ^{131}I from the blood compartment due to the radioactive decay of the iodine. Schematically this adds an arrow from the blood compartment directly to the removed compartment, with corresponding transfer coefficient a_{31}. If the half-life of ^{131}I is about 192.5 hours, what is the constant a_{31}?

13. A three-compartment chain model is depicted below, with numerical values for the transfer coefficients as shown. It is a *closed* system, since once material is introduced into the system, it never leaves.
   (a) Write the three-dimensional system of differential equations for this model in matrix form $\mathbf{X}' = \mathbf{A}\mathbf{X}$.
   (b) Show that $\mathbf{A}$ has eigenvalue 0. This is a feature of any closed system.
   (c) What are the equilibrium points of this system? Does it make physical sense that there would be equilibrium points different from $(0, 0, 0)$?

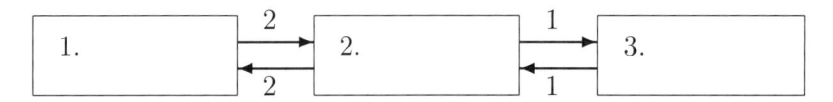

14. A five-compartment model with one central compartment exchanging with four peripheral compartments is as shown. **Set up** a five-dimensional homogeneous system of differential equations for this model, writing it in matrix form.

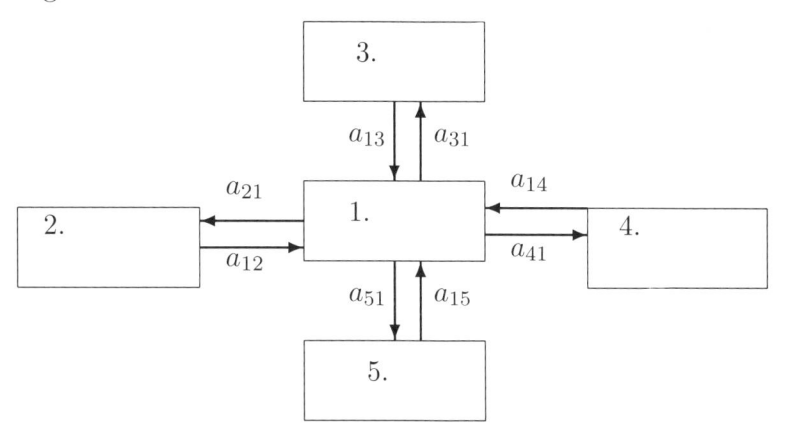

15. (CAS) City water in Toledo, Ohio, was unsafe to drink for several days in 2014 when phosphorus runoff in Lake Erie caused thick green algae blooms, which produced a toxin called microcystin that contaminated the city's water supply. Nearly a half million people were left without usable water. In this problem you will use the model in Example 8.8.2 to study phosphorus in the Great Lakes.

    In 2008 average concentration levels of phosphorus in each of the Great Lakes were approximately:

Superior	2,500 kg/km^3
Michigan	3,500 kg/km^3
Huron	2,000 kg/km^3
Erie	13,500 kg/km^3
Ontario	4,000 kg/km^3

    (a) Using the volume values from Example 8.8.2, give estimates for the amounts of phosphorus (in kg) $x_S(0)$, $x_M(0)$, $x_H(0)$, $x_E(0)$, and $x_O(0)$ in each of the lakes, where $t=0$ corresponds to 2008.

    (b) Using the initial conditions in (a), find (either by hand or using a computer algebra system) the the corresponding values for the constants c_1, c_2, c_3, c_4, and c_5 in equation (8.76) of Example 8.8.2.

    (c) Using technology, graph the functions $x_S(t)$, $x_M(t)$, $x_H(t)$, $x_E(t)$, and $x_O(t)$ that solve the system in Example 8.8.2 with the initial conditions from (a), for $0 \le t \le 20$ years.

    (d) Using your graphs from (c), answer the following question: Assuming all phosphorus pollution into the Great Lakes as a whole is stopped in 2008, for which of the Great Lakes will the phosphorus levels drop to less that 20% of their starting values within a decade?

16. (CAS) Lead poisoning is an important public health problem, especially in children. Lead is inhaled or ingested from various sources—from the air (leaded gas exhaust used to be a significant source), from contaminated water (as recently happened in Flint, Michigan), or from food or consumer products. Homes built prior to the mid-1970s usually have paint containing lead, and peeling paint (or paint-chip contaminated soil or dust) is sometimes eaten by young children. The simplest useful model for studying lead and lead poisoning in the human body uses two compartments as shown: Compartment 1 represents the blood, together with certain organs like liver and kidneys. Compartment 2 is bone, in which lead becomes bound, so that the transfer coefficient from bone to blood is much smaller than from blood to bone. A third external component collects the losses, chiefly in the urine.

(a) Set up an initial value problem for the amount of lead in compartments 1 and 2 (in micrograms) at time t (in days), if the initial amount x_0 in the blood is 100 micrograms and the initial amount y_0 in the bone is 650 micrograms and there is no further lead introduced into the system.

(b) Solve your initial value problem from (a), and determine how much lead is in the bloodstream and bone after 100 days. Some heavy calculations are required; use any technology at your disposal.

(c) Compute the ratio of the amount of lead in the bone to the amount in the body (the sum of the amounts in the blood and the bone) at 100 days, 500 days, 1,000 days, and 10,000 days. In the long term, what appears to be the approximate value of this ratio?

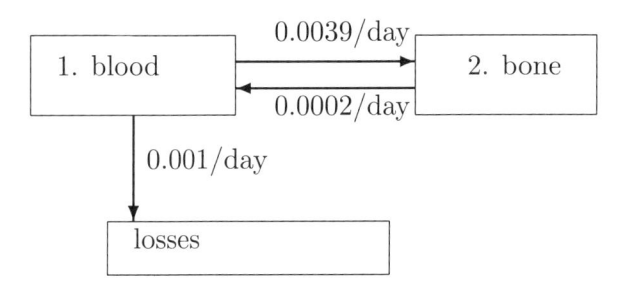

17. A population is divided into two groups, adults and children. In each of these groups, there is loss due to death, at a rate proportional to the size of the group, with proportionality constants δ_1 (adults) and δ_2 (children). Children "graduate" to the adult group at a rate proportional to the size of the group of children, with proportionality constant g. The children have no offspring, but the adult group gives rise to offspring at a rate proportional to the adult group size, with proportionality constant b. Derive a system of first-order equations for the number of children $c(t)$ and number of adults $a(t)$ at time t.

18. (CAS) This problem continues an analysis of the population model in Exercise 17.
    (a) Use any technology at your disposal to solve the system of Exercise 17 if $\delta_1 = 0.02, \delta_2 = 0.01, g = 0.15$, and $b = 0.03$. Assume initial population sizes of $c_0 = 20$ for children and $a_0 = 100$ for adults, where c and a are in millions of people. Assume t is in years. What are the sizes of the child and adult populations after 20 years?
    (b) Plot the adult and child populations (on the same coordinate plane), using the data in part (a), over the time interval $[0, 100]$.
    (c) For this part, you will need to use software that allows you to animate graphical information. In your computation of the solution to (a), replace the adult death rate δ_1 with a parameter s. Then animate the plot from (b) to show how the adult and child population plots vary as s changes from 0.02 to 0.1. Repeat with s replacing δ_2 and have s vary from 0.01 to 0.1.

19. In this problem we show an alternate approach to solving

    (8.79) $$x' = -0.2x + I, \quad y' = 0.2x - 0.05y$$

    from Example 8.8.3, where I is a constant.
    (a) Show that this system has equilibrium solution $x = 5I$, $y = 20I$.
    (b) Make the substitutions $u = x - 5I$, $v = y - 20I$ to convert the nonhomogeneous system (8.79) to a homogeneous system with dependent variables u and v. What initial conditions on u and v correspond to the initial conditions $x(0) = 0, y(0) = 0$?
    (c) Solve the initial value problem from (b) for u and v. (The work in Example 8.8.1 leading up to equation (8.72) is helpful here.)
    (d) Using your answer to (c), give the solution to (8.79) satisfying $x(0) = 0, y(0) = 0$.

20. Use Laplace transforms to solve
$$x'(t) = -x + e^t, \quad y'(t) = 2x + 2y$$
if $x(0) = 0$ and $y(0) = 0$.

21. Use Laplace transforms to solve the system
$$x'(t) = -x - y, \quad y'(t) = x + y + 1$$
if $x(0) = 0$ and $y(0) = 0$.

22. Use Laplace transforms to solve the system
$$x'(t) = -0.2x + I(t), \quad y'(t) = 0.2x - 0.05y$$
where
$$I(t) = \begin{cases} I & \text{if } 0 \leq t < 2, \\ 0 & \text{if } 2 \leq t \end{cases}$$
for some constant I, with initial conditions $x(0) = 0$, $y(0) = 0$. Notice that $I(t) = I - Iu_2(t)$, where $u_2(t)$ is a unit step function (see Section 6.1).

8.9. Classifying equilibrium points

Sometimes we are interested in knowing what *kind* of eigenvalues a 2×2 matrix **A** has (e.g., real eigenvalues with opposite sign or nonreal complex conjugate eigenvalues) without going through the calculational work of explicitly finding them. There is an efficient way to do this, which we'll describe next. This in turn allows us to easily classify the equilibrium point $(0,0)$ for the linear system $\mathbf{X}' = \mathbf{AX}$ (saddle point, spiral point, center, node) and thus predict the general features of corresponding phase portraits. This analysis can also be helpful if we don't have the actual numerical values for the entries of the matrix, but we want to make qualitative predictions about the behavior of solutions to $\mathbf{X}' = \mathbf{AX}$. We'll generally be assuming that the determinant of **A** is not zero; this tells us that $(0,0)$ is the only equilibrium point for the system $\mathbf{X}' = \mathbf{AX}$.

8.9.1. The trace-determinant plane.
Starting with a 2×2 matrix
$$\mathbf{A} = \begin{pmatrix} a & b \\ c & d \end{pmatrix},$$
there are two numbers we can compute "by inspection": the determinant
$$\det \mathbf{A} = ad - bc$$
(for short, we'll denote $\det \mathbf{A}$ by Δ) and the **trace** of **A**, denoted τ and defined by
$$\tau = a + d.$$
The trace is just the sum of the entries on the main diagonal. Encoded in these two numbers, Δ and τ, is information about the eigenvalues of **A**:

- The product of the two eigenvalues of **A** is Δ, the determinant of **A**.
- The sum of the two eigenvalues of **A** is τ, the trace of **A**.

When we say "the two eigenvalues of **A**" we use the convention that if **A** has only one eigenvalue, we list it twice. For example, if
$$\mathbf{A} = \begin{pmatrix} 3 & 0 \\ 1 & 3 \end{pmatrix},$$

then the two eigenvalues of $\mathbf{A}$ are 3 and 3, $\tau = 3 + 3 = 6$, and $\Delta = (3)(3) = 9$. To see why these two properties are true in general, consider the characteristic equation $\det(\mathbf{A} - z\mathbf{I}) = 0$, or $(a - z)(d - z) - bc = 0$, which is the same as

(8.80) $$z^2 - (a + d)z + ad - bc = 0.$$

Suppose we've factored this on the left-hand side to get

$$(z - r_1)(z - r_2) = 0.$$

This tells us that the eigenvalues of $\mathbf{A}$ are r_1 and r_2; they could be real or complex. This last equation is the same as

(8.81) $$z^2 - (r_1 + r_2)z + r_1 r_2 = 0.$$

Comparing equations (8.80) and (8.81) we see that

$$r_1 + r_2 = a + d = \text{ trace } \mathbf{A}$$

and

$$r_1 r_2 = ad - bc = \det \mathbf{A}.$$

How is this perspective useful? Let's focus on a 2×2 matrix $\mathbf{A}$ with *real* entries. Then both the determinant and trace are real numbers, and we know from Section 8.3 that any complex eigenvalues occur in conjugate pairs. Suppose first that the determinant of $\mathbf{A}$ is negative. This means the product of the eigenvalues is negative. The only way this can happen is if the matrix has one positive and one negative eigenvalue. Notice that if $\mathbf{A}$ had complex conjugate eigenvalues $\alpha \pm i\beta$, then their product would be $(\alpha + i\beta)(\alpha - i\beta) = \alpha^2 + \beta^2 > 0$. Thus a negative value for Δ means $\mathbf{A}$ has two real eigenvalues with opposite signs, and

$$\boxed{\Delta < 0 \implies (0, 0) \text{ is a saddle point for } \mathbf{X}' = \mathbf{AX}.}$$

To analyze the case that $\Delta > 0$ is a bit more work. Equation (8.80) says that the characteristic equation is

$$z^2 - \tau z + \Delta = 0.$$

The solutions to this equation—which are the eigenvalues of $\mathbf{A}$—are

(8.82) $$r_1 = \frac{\tau + \sqrt{\tau^2 - 4\Delta}}{2} \quad \text{and} \quad r_2 = \frac{\tau - \sqrt{\tau^2 - 4\Delta}}{2}.$$

What kind of numbers are r_1 and r_2? It depends on what's underneath the square root. Remember, we're assuming right now that Δ is positive, so $\tau^2 - 4\Delta$ could be positive, negative, or 0. Let's make a small chart to tabulate the possibilities, taking into account the sign of τ as well as that of $\tau^2 - 4\Delta$.

8.9. Classifying equilibrium points

Table 8.1. If $\Delta > 0$, what kind of eigenvalues does A have?

	$\tau^2 - 4\Delta > 0$	$\tau^2 - 4\Delta < 0$	$\tau^2 - 4\Delta = 0$
$\tau > 0$	two + eigenvalues	complex conjugate eigenvalues with positive real part	one + eigenvalue
$\tau < 0$	two − eigenvalues	complex conjugate eigenvalues with negative real part	one − eigenvalue
$\tau = 0$	case does not occur	purely imaginary conjugate eigenvalues	case does not occur

Verifying the entries in this table is a familiar exercise. For example, if $\Delta > 0, \tau > 0$, and $\tau^2 - 4\Delta > 0$, then $\sqrt{\tau^2 - 4\Delta}$ is positive but smaller than τ. So $\tau + \sqrt{\tau^2 - 4\Delta}$ and $\tau - \sqrt{\tau^2 - 4\Delta}$ are both positive. By equation (8.82), this says that the eigenvalues r_1 and r_2 are both positive. On the other hand, if $\Delta > 0$ and $\tau^2 - 4\Delta < 0$, then $\sqrt{\tau^2 - 4\Delta}$ is imaginary and the eigenvalues r_1 and r_2 are complex conjugates. They will have positive real part precisely when $\tau > 0$. Make sure you can verify all of the entries in Table 8.1 (see Exercise 9).

The information on the eigenvalues immediately leads to a classification of the equilibrium point $(0,0)$:

Table 8.2. If $\Delta > 0$, what is the classification of $(0,0)$?

	$\tau^2 - 4\Delta > 0$	$\tau^2 - 4\Delta < 0$	$\tau^2 - 4\Delta = 0$
$\tau > 0$	unstable node	unstable spiral	degenerate node (unstable)
$\tau < 0$	stable node	stable spiral	degenerate node (stable)
$\tau = 0$	case does not occur	center	case does not occur

Notice how in Table 8.2 the sign of τ determines whether $(0,0)$ is stable or unstable. The terminology "**spiral source**" and "**spiral sink**" are sometimes used for "unstable spiral point" and "stable spiral point", respectively. Similarly, unstable nodes are sometimes called **nodal sources**, while stable nodes are called **nodal sinks**.

When the determinant Δ is equal to 0, at least one eigenvalue of **A** must be 0 (why?) and the differential equation system has nonisolated equilibrium points (see Exercise 13 in Section 8.6)—either a line of equilibrium points when **A** is not the zero matrix or when **A** *is* the zero matrix; every point in the phase plane is an equilibrium point!

The trace-determinant plane. We can summarize all of this information in a picture. Draw a plane with coordinate axes τ and Δ (we call this the "trace-determinant" plane), and sketch the graph of the parabola $4\Delta = \tau^2$ in this plane; see Fig. 8.74.

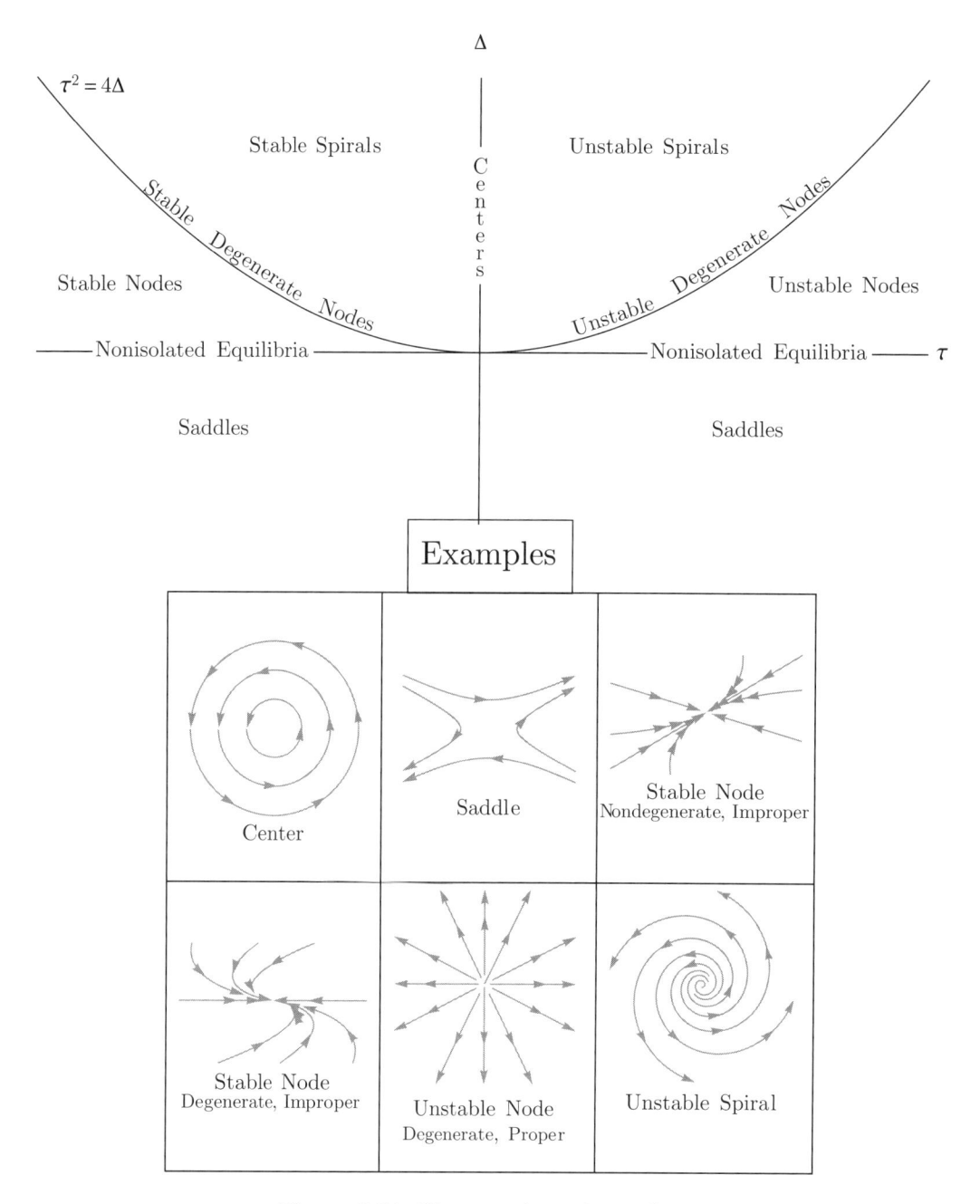

Figure 8.74. The trace-determinant plane.

The axes, together with this parabola, divide the plane into various regions. Given a planar linear system $\mathbf{X}' = \mathbf{AX}$, compute $\Delta = \det \mathbf{A}$ and $\tau = \text{trace } \mathbf{A}$ and locate the point (Δ, τ) on the plane. The region, or bordering edge, in which you find yourself tells you the classification of $(0,0)$. The borderline condition $\tau = 0, \Delta > 0$ corresponds to the classification of a "center", while the borderline condition $\tau^2 - 4\Delta = 0$ corresponds to the classification of a "degenerate node". The condition $\Delta = 0$ corresponds to "nonisolated equilibrium points". Degenerate nodes are sometimes called "borderline" nodes, and for emphasis, nondegenerate nodes will also be called "major nodes". Of the various borderline cases, centers are the most important, because they appear in key applications. Check to see that Fig. 8.74 encapsulates all of the information from

8.9. Classifying equilibrium points

Tables 8.1 and 8.2 and our discussion above. We'll return to this picture in Section 10.2, where it will help us analyze nonlinear systems.

8.9.2. Classification type and trajectory behavior. In this section we'll summarize the behavior of the trajectories of a constant coefficient system $\mathbf{X}' = \mathbf{A}\mathbf{X}$ near the equilibrium point $(0,0)$. We assume that $\det \mathbf{A} \neq 0$, so that $(0,0)$ is the only equilibrium point for the system.

Nodes. A characteristic feature of nodes is that all trajectories approach the origin, either as $t \to \infty$ or as $t \to -\infty$, from a definite direction; that is, each approaches tangent to a particular line through the origin. For **nondegenerate nodes** (which arise when $\mathbf{A}$ has two distinct real eigenvalues with the same sign), two such directions are possible, corresponding to the two eigenvector lines. By Theorem 8.5.3, one of these two directions is preferred, and the preferred nodal tangent is along the eigenvector line corresponding to the eigenvalue with smaller absolute value. All trajectories, except those for the ray solutions corresponding to the *other* eigenvalue, approach $\mathbf{0}$ tangent to this eigenvector line. This approach occurs as $t \to \infty$ if the eigenvalues are negative and as $t \to -\infty$ if the eigenvalues are positive. In the first case, the equilibrium point $\mathbf{0}$ is stable, and in the second case, unstable. The "only two approach directions" property gives all nondegenerate nodes the further classification of **improper nodes**. These features are illustrated in Figs. 8.75-8.76.

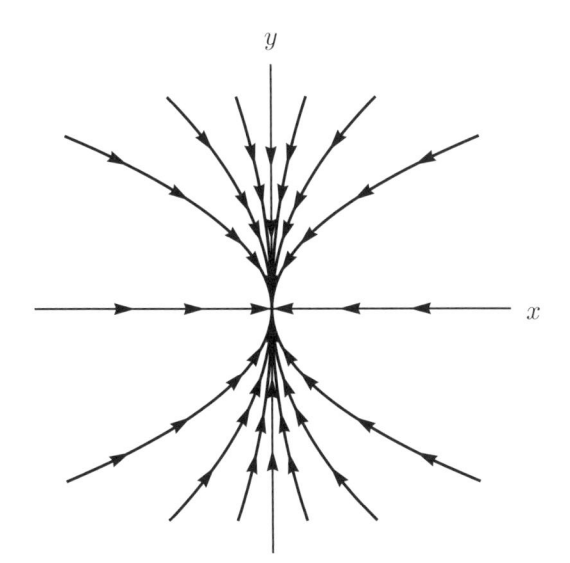

Figure 8.75. Phase portrait with stable nondegenerate node at $\mathbf{0}$. All trajectories approach $\mathbf{0}$ as $t \to \infty$ from one of two directions, with one direction preferred.

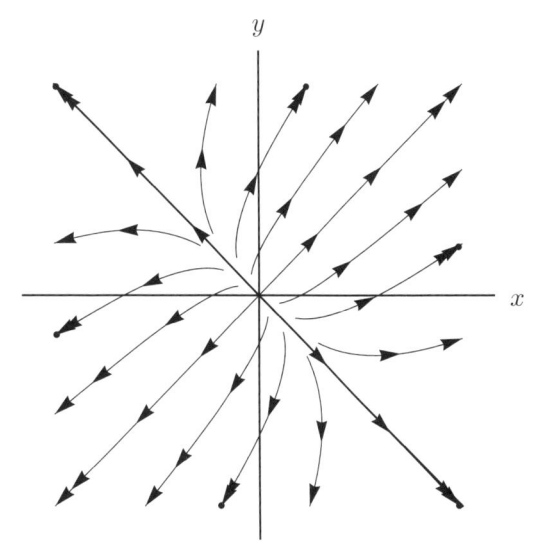

Figure 8.76. Phase portrait with unstable nondegenerate node at $\mathbf{0}$. All trajectories approach $\mathbf{0}$ as $t \to -\infty$ from one of two directions, with one direction preferred.

Degenerate nodes occur when $\mathbf{A}$ has a single repeated eigenvalue r. As with nondegenerate nodes, every trajectory approaches $\mathbf{0}$, from a definite direction, either as $t \to \infty$ (if $r < 0$) or as $t \to -\infty$ (if $r > 0$). **Proper** nodes arise when there are two linearly independent eigenvectors for the single eigenvalue r. This can only happen when $\mathbf{A} = r\mathbf{I}$. In this case, all nonzero trajectories are rays, either all directed towards the origin or all directed away from the origin. All directions of approach occur; see Fig. 8.77.

Improper degenerate nodes arise when $\mathbf{A}$ has a single eigenvalue r having only one eigenvector, up to constant multiples. Every trajectory will approach $\mathbf{0}$ tangent to the eigenvector line, either as $t \to \infty$ if $r < 0$ or as $t \to -\infty$ if $r > 0$, so there is only one nodal tangent, which we still call the preferred nodal tangent. Fig. 8.78 shows a phase portrait where $\mathbf{0}$ is an improper degenerate node.

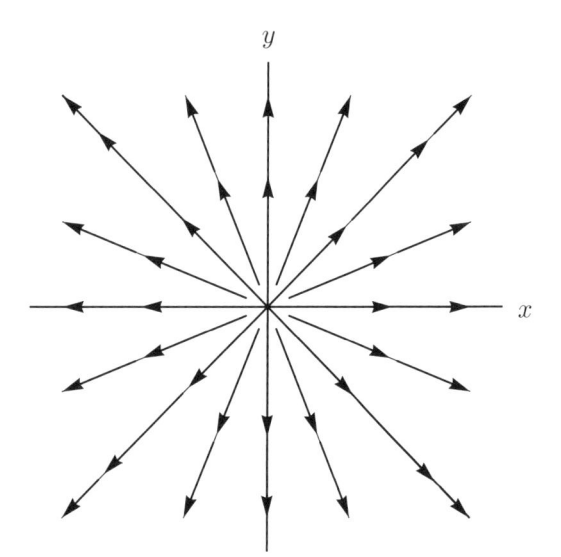

Figure 8.77. Phase portrait with a proper (degenerate) node at $(0,0)$. Approach from any direction is possible.

Figure 8.78. Phase portrait with an improper degenerate node at $(0,0)$. All trajectories approach from one direction.

Spiral points. The behavior of trajectories near a node contrasts with the behavior of trajectories near a spiral point. When $(0,0)$ is a spiral point, then all trajectories approach the origin (either as $t \to \infty$ if the real part of the eigenvalues of $\mathbf{A}$ is negative or as $t \to -\infty$ if the eigenvalues have positive real part), but not with a specific direction. It makes a certain aesthetic sense that the degenerate nodes are on the border between nondegenerate nodes and spirals. The trajectories in the improper degenerate node case are trying to spiral around $(0,0)$ but they cannot cross the two half-line trajectories (why not?) and thus are prevented from spiraling. The half-line trajectories referred to correspond to ray solutions of the form $\mathbf{X}(t) = \pm e^{rt}\mathbf{V}$, where $\mathbf{V}$ is an eigenvector for the matrix $\mathbf{A}$, with corresponding eigenvalue r.

Saddle points. If $(0,0)$ is a saddle point for $\mathbf{X}'(t) = \mathbf{A}\mathbf{X}$, all trajectories that approach $(0,0)$ as $t \to \infty$ lie on a single line through the origin, called the **stable line** (containing the incoming ray solutions), while all trajectories that approach $(0,0)$ as $t \to -\infty$ lie on another line through the origin, called the **unstable line** (containing the outgoing ray trajectories). The stable line is the eigenvector line for the negative eigenvalue, and the unstable line is the eigenvector line for the positive eigenvalue. No other nonzero trajectories will approach the equilibrium point as $t \to \infty$ or as $t \to -\infty$, and $(0,0)$ is an unstable equilibrium point. Notice that in any saddle point phase portrait, such as that shown in Fig. 8.80, solutions that lie on different sides of the stable line have very different fates as $t \to \infty$. For this reason, the stable line is called a **separatrix** of the system: It separates two different types of long-term behavior.

8.9. Classifying equilibrium points

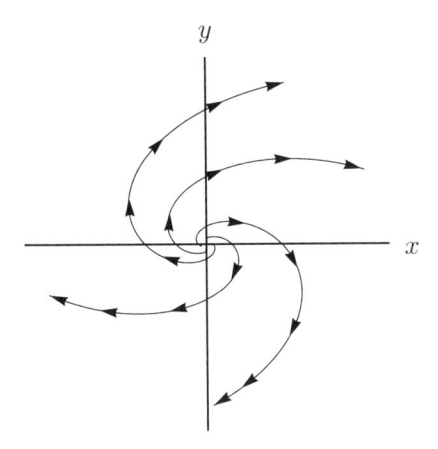

Figure 8.79. Phase portrait with spiral point at $(0,0)$.

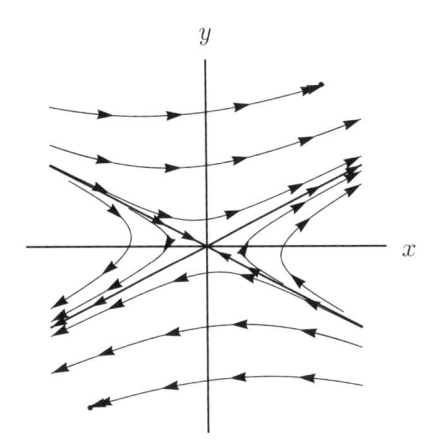

Figure 8.80. Phase portrait with saddle point at $(0,0)$.

Centers. If $(0,0)$ is a center for $\mathbf{X}'(t) = \mathbf{AX}$, the eigenvalues of $\mathbf{A}$ are a pair of purely imaginary complex conjugates ($\pm i\beta$) and all nonzero solutions are periodic functions. The closed-loop orbits in the phase plane are actually ellipses, although their axes may be neither horizontal nor vertical; see Exercise 13 for an outline of how to verify this.

8.9.3. Stability. We have informally used the terminology "stable" and "unstable" to classify the isolated equilibrium point $(0,0)$ of a 2×2 linear system $\mathbf{X}' = \mathbf{AX}$ with $\det \mathbf{A} \neq 0$. Now we'll be a bit more precise with this terminology, so that it may also be applied in Chapter 10 to equilibrium points in more general autonomous (but not necessarily linear) planar systems; such systems can have multiple isolated equilibrium points. It's intended to capture the following intuitive idea: If a trajectory starts "near" an equilibrium point (x_e, y_e), must the trajectory

(i) approach (x_e, y_e) as $t \to \infty$ or

(ii) at least stay within a quantifiable distance of (x_e, y_e) for all $t \geq 0$?

If (ii) holds, we'll call the equilibrium point (x_e, y_e) **stable**. If both (i) and (ii) hold for every trajectory that starts near (x_e, y_e), then we will say that this equilibrium point is **asymptotically stable**. For a linear system $\mathbf{X}' = \mathbf{AX}$, $(0,0)$ is asymptotically stable if the eigenvalues of $\mathbf{A}$ are either (strictly) negative or have (strictly) negative real parts. If the eigenvalues are purely imaginary, then $(0,0)$ is stable but not asymptotically stable.

If (ii) does not hold, the point (x_e, y_e) is termed **unstable**. An equilibrium point (x_e, y_e) is unstable if there is some "quantifiable distance" d so that we can find trajectories starting as close to (x_e, y_e) as desired whose distance to (x_e, y_e) will eventually exceed d. All of this discussion so far relies on rather imprecise terminology; for example, what exactly do we mean by saying a trajectory starts "near" the equilibrium point (x_e, y_e)? Our formal definitions will make this precise. They will use the notation $\|(x,y) - (u,v)\|$ for the distance between the points (x,y) and (u,v) in the plane, so $\|(x,y) - (u,v)\| = \sqrt{(x-u)^2 + (y-v)^2}$. The next two definitions apply not just to linear systems, but to any autonomous planar system, as will be discussed in Chapter 10.

Definition 8.9.1. An equilibrium point (x_e, y_e) is called **stable** if, given any $\epsilon > 0$, there is a $\delta > 0$ such that every solution $x = x(t)$, $y = y(t)$ of the system which satisfies

$$\|(x(0), y(0)) - (x_e, y_e)\| < \delta$$

exists for all positive t and satisfies
$$\|(x(t), y(t)) - (x_e, y_e)\| < \epsilon,$$
for all $t > 0$.

This means that if a solution starts sufficiently close (within δ) to the equilibrium solution (x_e, y_e), then it stays close (within ϵ) to (x_e, y_e) as time goes on.

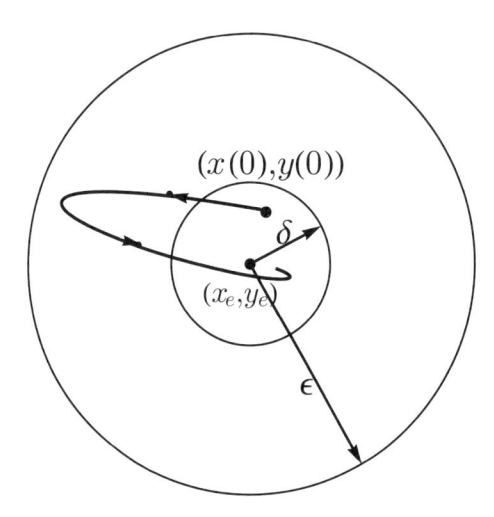

Figure 8.81. The idea of stability.

Asymptotic stability is a particularly strong form of stability:

Definition 8.9.2. An equilibrium point (x_e, y_e) is called **asymptotically stable** if it is stable and there is a number $\rho > 0$ such that every solution $x = x(t), y = y(t)$ of the system which satisfies
$$\|(x(0), y(0)) - (x_e, y_e)\| < \rho$$
has the property that
$$\lim_{t \to \infty} (x(t), y(t)) = (x_e, y_e).$$

This means that if a solution starts sufficiently close to the equilibrium point (x_e, y_e), it not only stays close, but it actually approaches this equilibrium point as $t \to \infty$.

When an equilibrium point (x_e, y_e) is unstable, you're guaranteed to be able to find a disk centered at (x_e, y_e), so that there are trajectories starting arbitrarily close to (x_e, y_e) that will leave this disk for some $t > 0$. For a linear system $\mathbf{X}' = \mathbf{AX}$, $(0, 0)$ is unstable if $\mathbf{A}$ has one positive and one negative eigenvalue (so $(0,0)$ is a saddle point) or two positive eigenvalues or complex conjugate eigenvalues with positive real part or a single repeated positive eigenvalue.

8.9.4. Recap. We finish with a table, on the next page, summarizing the classification of the equilibrium point $(0, 0)$ for the planar system $\mathbf{X}' = \mathbf{AX}$.

8.9. Classifying equilibrium points

Table 8.3. Summary for the planar system $\mathbf{X}' = \mathbf{AX}$ if det $\mathbf{A} \neq 0$.		
Classification of $(0,0)$ and eigenvalues of $\mathbf{A}$	Stability	Characteristic Features
Saddle point: $\mathbf{A}$ has two real eigenvalues with opposite signs.	Unstable	• Incoming rays along eigenvector line for negative eigenvalue. • Outgoing rays along eigenvector line for positive eigenvalue. • Nonray trajectories asymptotic to eigenvector lines.
Nondegenerate node; improper: $\mathbf{A}$ has two real eigenvalues of the same sign, $0 < r_1 < r_2$ or $r_2 < r_1 < 0$.	Asymptotically stable if the eigenvalues are negative. Unstable if the eigenvalues are positive.	• Eigenvector lines are nodal tangents. • Preferred nodal tangent is the eigenvector line corresponding to r_1.
Degenerate node; proper: $\mathbf{A}$ has one eigenvalue with two linearly independent eigenvectors.	Asymptotically stable if the eigenvalue is negative. Unstable if the eigenvalue is positive.	• All trajectories are rays. • All rays to/from $\mathbf{0}$ occur. • This case occurs only if $\mathbf{A} = r\mathbf{I}$.
Degenerate node; improper: $\mathbf{A}$ has only one eigenvalue, with one eigenvector, up to constant multiples.	Asymptotically stable if the eigenvalue is negative. Unstable if the eigenvalue is positive.	• One nodal tangent, which is the eigenvector line. • Every trajectory is tangent to this nodal line, either as $t \to \infty$ or as $t \to -\infty$.
Spiral: $\mathbf{A}$ has complex conjugate eigenvalues $\alpha \pm i\beta$, $\alpha \neq 0, \beta \neq 0$.	Asymptotically stable if $\alpha < 0$. Unstable if $\alpha > 0$.	• All trajectories approach $\mathbf{0}$ as $t \to \infty$ ($\alpha < 0$) or as $t \to -\infty$ ($\alpha > 0$), but not from a specific direction.
Center: $\mathbf{A}$ has purely imaginary eigenvalues $\pm i\beta$, $\beta \neq 0$.	Stable, not asymptotically stable.	• Trajectories are ellipses; all either directed clockwise or counterclockwise.

8.9.5. Exercises.

In Exercises 1–6, compute Δ and τ, locate the point (τ, Δ) on the trace-determinant plane, and then give the classification of the equilibrium point $(0,0)$ for the system $\mathbf{X}' = \mathbf{A}\mathbf{X}$.

1. $\mathbf{A} = \begin{pmatrix} -2 & 1 \\ 1 & -2 \end{pmatrix}$.

2. $\mathbf{A} = \begin{pmatrix} 4 & -2 \\ 7 & -5 \end{pmatrix}$.

3. $\mathbf{A} = \begin{pmatrix} 3 & -1 \\ 1 & 1 \end{pmatrix}$.

4. $\mathbf{A} = \begin{pmatrix} -\frac{1}{2} & 1 \\ -1 & -\frac{1}{2} \end{pmatrix}$.

5. $\mathbf{A} = \begin{pmatrix} 3 & 5 \\ -1 & 9 \end{pmatrix}$.

6. $\mathbf{A} = \begin{pmatrix} 3 & -4 \\ 1 & -1 \end{pmatrix}$.

7. Find and classify the equilibrium point of the following nonhomogeneous linear autonomous systems. Note that Exercises 25 and 26 of Section 8.4 show that for a nonhomogenous autonomous planar system $\mathbf{X}' = \mathbf{A}\mathbf{X} + \mathbf{C}$, for which $\det \mathbf{A} \neq 0$, $\mathbf{C}$ is a constant vector, and the equilibrium point is (a, b), the phase portrait will be the same as that of $\mathbf{X}' = \mathbf{A}\mathbf{X}$ but shifted a in the horizontal direction and b in the vertical direction. Thus, the classification of the nonzero equilibrium point for $\mathbf{X}' = \mathbf{A}\mathbf{X} + \mathbf{C}$ is the same as the classification of the equilibrium point $(0,0)$ of $\mathbf{X}' = \mathbf{A}\mathbf{X}$.
   (a) $\dfrac{dx}{dt} = 2x - y + 1$, $\dfrac{dy}{dt} = x + 2y + 3$.
   (b) $\dfrac{dx}{dt} = x + 4y + 6$, $\dfrac{dy}{dt} = x + 2y + 2$.

8. For what values of c does the system

$$\frac{dx}{dt} = -x + (1-c)y, \qquad \frac{dy}{dt} = x - y$$

have a stable node at $(0,0)$? A saddle point? A stable spiral point?

9. Verify the entries of Table 8.1 corresponding to
   (a) $\tau > 0, \tau^2 - 4\Delta = 0$.
   (b) $\tau < 0$.
   (c) $\tau = 0$.

10. In Section 4.8 we studied the second-order differential equation $mx'' + cx' + kx = 0$ for the position x of a mass m on the end of a damped spring. Converting this differential equation into a system of two first-order equations by the substitution $y = \frac{dx}{dt}$, we obtain

$$\frac{dx}{dt} = y, \qquad \frac{dy}{dt} = -\frac{k}{m}x - \frac{c}{m}y.$$

The phase portrait for this system depends on whether it is overdamped ($c^2 - 4mk > 0$), underdamped ($c^2 - 4mk < 0$), or critically damped ($c^2 - 4mk = 0$).
   (a) Classify according to type the equilibrium point $(0,0)$ in each of these three cases.
   (b) Typical phase portraits for these three cases are shown in Figs. 8.82–8.84; all trajectories approach the origin as $t \to \infty$. Which is which?
Use the methods of this section to answer these with minimal computation.

8.9. Classifying equilibrium points

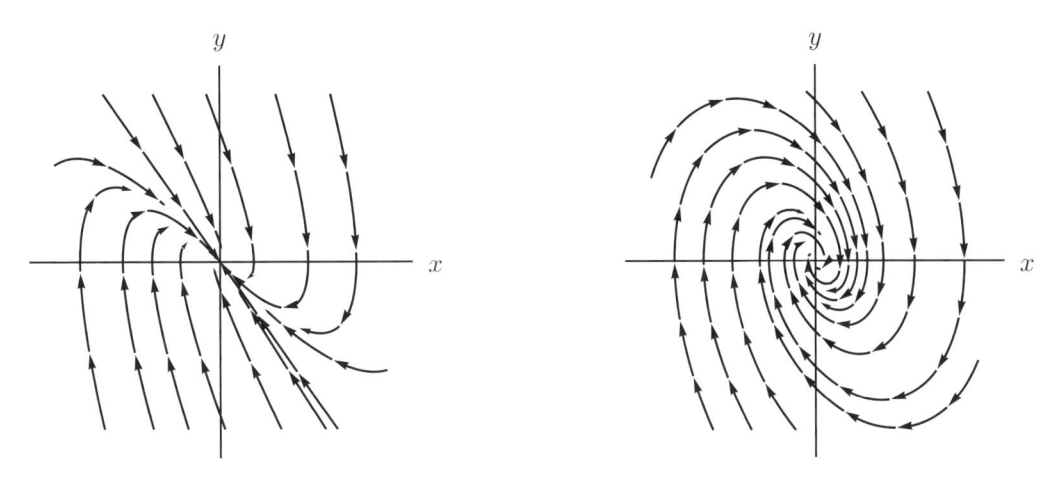

Figure 8.82. (A.)

Figure 8.83. (B.)

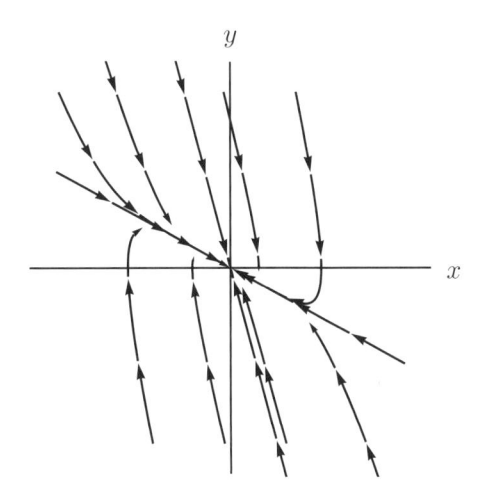

Figure 8.84. (C.)

11. Consider a variation on the Romeo and Juliet model of Exercise 8 in Section 8.1, given by

$$\frac{dR}{dt} = aR + bJ, \quad \frac{dJ}{dt} = bR + aJ$$

where a, b are constants but could be either positive or negative. Romeo's feelings for Juliet at time t are represented by $R(t)$; a positive value indicated affection and a negative value, dislike. Similarly, $J(t)$ gives Juliet's feelings for Romeo at time t.

(a) Based on trace/determinant calculations, predict the classification of the equilibrium point $(0,0)$ in each of the following scenarios.
  (i) $a < 0$ and $a^2 - b^2 > 0$.
  (ii) $a < 0$ and $a^2 - b^2 < 0$.
  (iii) $a > 0$ and $a^2 - b^2 < 0$.
  (iv) $a > 0$ and $a^2 - b^2 > 0$.

(b) In which of the scenario(s) in (a) do Romeo and Juliet always end up in a state of "mutual indifference"?

(c) Verify that

$$\begin{pmatrix} 1 \\ 1 \end{pmatrix} \quad \text{and} \quad \begin{pmatrix} -1 \\ 1 \end{pmatrix}$$

are eigenvectors of
$$\mathbf{A} = \begin{pmatrix} a & b \\ b & a \end{pmatrix},$$
and give the corresponding eigenvalues.

(d) If $a + b > 0$ and $a - b < 0$, sketch a phase portrait and verify that in this case they nearly always end up either in mutual love or mutual hate. If $R(0) = 1$ and $J(0) = 0$, does their relationship end in love or war?

12. Find, if possible, a linear system $\mathbf{X}' = \mathbf{A}\mathbf{X}$ with the given properties:
    (a) The positive and negative halves of the x-axis are trajectories, both directed towards the origin. All other nonzero trajectories approach $(0,0)$ as $t \to \infty$.
    (b) Same as in (a), except the trajectories along the x-axis are directed away from the origin.
    (c) There are infinitely many equilibrium points.
    (d) All nonconstant trajectories move infinitely far away from the origin as $t \to \infty$.
    (e) As $t \to \infty$, all nonconstant trajectories tend to $(0,0)$ tangent to the x-axis.

    Hint: When an example exists, you can find a diagonal matrix
    $$\mathbf{A} = \begin{pmatrix} a & 0 \\ 0 & d \end{pmatrix}$$
    or an upper triangular matrix
    $$\mathbf{A} = \begin{pmatrix} a & b \\ 0 & d \end{pmatrix}$$
    that will work.

13. Suppose that the system
    $$\frac{dx}{dt} = a_1 x + b_1 y, \quad \frac{dy}{dt} = a_2 x + b_2 y$$
    has a center at $(0,0)$. By Table 8.2, this says the trace τ must be 0 and the determinant Δ is positive.

    (a) By writing the first-order equation
    $$\frac{dy}{dx} = \frac{a_2 x + b_2 y}{a_1 x + b_1 y}$$
    in the form
    $$-a_2 x - b_2 y + (a_1 x + b_1 y)\frac{dy}{dx} = 0,$$
    show that this equation is exact and find its implicit solutions.

    (b) Show that the orbits in the phase portrait are ellipses. You may use the fact that
    $$Ax^2 + Bxy + Cy^2 = 1$$
    is the equation of an ellipse if $B^2 - 4AC < 0$.

    (c) Suppose instead that the system has a saddle point at $(0,0)$, and suppose further that the trace τ is 0. Show that trajectories in the phase plane are hyperbolas; it may be helpful to know that
    $$Ax^2 + Bxy + Cy^2 = 1$$
    is the equation of a hyperbola if $B^2 - 4AC > 0$.

14. You and I play the following game: I choose a 2×2 matrix $\mathbf{A}$ and tell you what it is. You then choose an allowable error "ϵ" (this is just a small but positive number) and challenge me to change one or more of the entries of the original matrix by an amount no more than ϵ to get a new matrix $\mathbf{A}_\epsilon$. I win if I can do this in such a way that the classification of the equilibrium point $(0,0)$ (saddle, spiral, etc.) for $\mathbf{X}' = \mathbf{A}_\epsilon \mathbf{X}$ is different from the classification of $(0,0)$ for

$\mathbf{X}' = \mathbf{A}\mathbf{X}$. What kinds of matrices $\mathbf{A}$ will always allow me to win? Hint: Changing one or several entries of $\mathbf{A}$ by a small amount causes the trace and determinant to change by a small amount. Think about the trace-determinant plane.

15. Under the set-up of the previous problem, suppose I win if I am able to change the stability of $(0,0)$. Under what condition(s) am I guaranteed to be able to do this? Assume that initially $(0,0)$ is an *isolated* equilibrium point.

16. For each of the following systems, (i) confirm that $(0,0)$ is an improper node, (ii) characterize the node as unstable or asymptotically stable, (iii) find an equation of the preferred nodal tangent, and (iv) (CAS) use a computer algebra system to generate a phase portrait of the system to check your answer to (iii) for plausibility.
    (a) $\dfrac{dx}{dt} = -5x + 2y$, $\dfrac{dy}{dt} = x - 4y$.
    (b) $\dfrac{dx}{dt} = 3x + 2y$, $\dfrac{dy}{dt} = x + 4y$.
    (c) $\dfrac{dx}{dt} = 2x + y$, $\dfrac{dy}{dt} = 3x + 4y$.
    (d) $\dfrac{dx}{dt} = -5x + y$, $\dfrac{dy}{dt} = 3x - 3y$.

17. Two doomed strains of bacteria compete for space and food in a hostile culture environment. Let $x(t)$ be the number (in thousands) of strain-one bacteria present at time t and let $y(t)$ be the number (in thousands) of strain-two bacteria present at time t. Assume that t is measured in hours and that the following system models the demise of these unfortunate bacteria:

$$(8.83) \qquad \frac{dx}{dt} = -2x - y, \quad \frac{dy}{dt} = -x - 2y.$$

   (a) Confirm that $(0,0)$ is an asymptotically stable improper node of the system (8.83).
   (b) Draw a nullcline-and-arrow diagram for the system (8.83).
   (c) Add to your nullcline-and-arrow diagram the nodal tangents of the system (8.83).
   (d) Add to your diagram a quick sketch of the trajectory $(t \geq 0)$ of the solution satisfying (8.83) and the initial conditions $x(0) = 1$, $y(0) = 2$.
   (e) Based on you sketch of (d), if the initial population levels are $x = 1$ thousand and $y = 2$ thousand, which strain dies off first—the x strain (strain one) or the y strain (strain two)?

18. Suppose that $\mathbf{A}$ is a 3×3 matrix with ij-entry a_{ij} and eigenvalues λ_1, λ_2, and λ_3 (which may not be distinct).
    (a) Show that the determinant of $\mathbf{A}$ is the product $\lambda_1 \lambda_2 \lambda_3$.
    (b) Define the trace of $\mathbf{A}$ to be the sum of the entries along the main diagonal:
    $$\text{trace } \mathbf{A} = a_{11} + a_{22} + a_{33}.$$
    Show that trace $\mathbf{A} = \lambda_1 + \lambda_2 + \lambda_3$.
    (c) If $\mathbf{A}$ has real entries, explain why (b) tells you that if the trace of $\mathbf{A}$ is positive, $\mathbf{A}$ must have an eigenvalue with strictly positive real part (this means either a real eigenvalue $r > 0$ or a complex eigenvalue $\alpha + i\beta$ with α, β real and $\alpha > 0$).
    (d) Conjecture the extension of (a)–(c) to $n \times n$ matrices.

19. For each of the following systems, (i) verify that $(0,0)$ is a saddle point, (ii) find the stable and unstable lines, (iii) give the general solution and based on its form describe the "different fates" as $t \to \infty$ of trajectories of the system lying on different sides of the stable line, and (iv) (CAS) create a phase portrait for the system and use it to verify your answer in (iii) about different fates.

(a) $\dfrac{dx}{dt} = 6x + 4y, \quad \dfrac{dy}{dt} = 2x - y.$

(b) $\dfrac{dx}{dt} = x + 2y, \quad \dfrac{dy}{dt} = -y.$

(c) $\dfrac{dx}{dt} = -3x + 8y, \quad \dfrac{dy}{dt} = 3x + 2y.$

20. The currents I_1, I_2, and I_3 flowing in the three branches of the circuit pictured below right are modeled by the algebraic equation $I_1 = I_2 + I_3$ (obtained through considering current entering and leaving node a) and the differential equations

    (8.84) $\quad \begin{aligned} \dfrac{dI_1}{dt} &= -I_1 + I_2, \\ \dfrac{dI_2}{dt} &= -4I_1 - 6I_2 + 9. \end{aligned}$

    (a) Find and classify the equilibrium point of the system (8.84).
    (b) Find the general solution of the system (8.84). (Use the result of Exercise 26 of Section 8.4.)
    (c) Solve the system (8.84) subject to the initial conditions $I_1(0) = 0$, $I_2(0) = 0$.
    (d) Use your solution of part (c) to find $\lim_{t \to \infty} I_1(t)$ and $\lim_{t \to \infty} I_2(t)$.
    (e) (CAS) Graph the currents I_1 and I_2, for $0 \le t \le 2$, on the same coordinate grid.

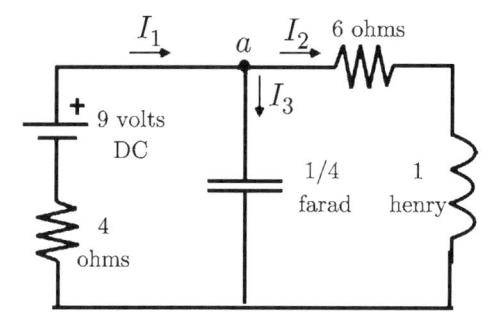

Chapter 9

Linear Systems II

9.1. The matrix exponential, Part I

For a 2×2 matrix $\mathbf{A}$ with real entries, we saw in Sections 8.4 and 8.7 how to obtain the general solution to the planar system $\mathbf{X}' = \mathbf{AX}$ using the eigenvalues and eigenvectors of $\mathbf{A}$. We could then use the general solution to find the solution that satisfies any given initial conditions; this required finding appropriate values for the arbitrary constants in the general solution. In this section we introduce the **matrix exponential** $e^{t\mathbf{A}}$, which serves as a black box that automatically hands you the solution when you plug in any desired initial conditions. This will also lead to a procedure for solving initial value problems with equation $\mathbf{X}' = \mathbf{AX}$ when $\mathbf{A}$ is an $n \times n$ matrix, even if $\mathbf{A}$ does not have n distinct eigenvalues.

To motivate the concept of the matrix $e^{t\mathbf{A}}$, consider the initial value problem

$$\frac{dx}{dt} = ax, \quad x(0) = x_0.$$

From Chapter 2 we know this has solution $x(t) = x_0 e^{at} = e^{at} x_0$. By analogy, can we make sense of $\mathbf{X} = e^{t\mathbf{A}} \mathbf{X}_0$ as the solution to the *system* $\mathbf{X}' = \mathbf{AX}$ with initial condition $\mathbf{X}(0) = \mathbf{X}_0$? Our first task is simply to propose a reasonable definition for the "exponential of a matrix". To do this, we'll use the familiar power series representation for e^z, $e^z = 1 + z + z^2/2! + z^3/3! + \cdots$, and replace z by a (square) matrix.

Definition 9.1.1. Let $\mathbf{A}$ be an $n \times n$ matrix. Define the **matrix exponential** $e^{\mathbf{A}}$ by

$$\begin{aligned} e^{\mathbf{A}} &= \mathbf{I} + \mathbf{A} + \frac{1}{2!} \mathbf{A}^2 + \frac{1}{3!} \mathbf{A}^3 + \frac{1}{4!} \mathbf{A}^4 + \cdots \\ &= \sum_{k=0}^{\infty} \frac{1}{k!} \mathbf{A}^k. \end{aligned}$$

We use the convention that $\mathbf{A}^0 = \mathbf{I}$ (where $\mathbf{I}$ is the appropriate size identity matrix) and $0! = 1$ for the $k = 0$ term in this sum.

Using methods of real analysis one can show that this series always converges to an $n \times n$ matrix (which we are calling $e^{\mathbf{A}}$). Let's accept that on faith and see whether this definition behaves as we hope with respect to our differential equation $\mathbf{X}' = \mathbf{AX}$. For t a real number, the definition above of the matrix exponential gives

$$e^{t\mathbf{A}} = \mathbf{I} + t\mathbf{A} + \frac{1}{2!}t^2\mathbf{A}^2 + \frac{1}{3!}t^3\mathbf{A}^3 + \frac{1}{4!}t^4\mathbf{A}^4 + \cdots = \sum_{k=0}^{\infty} \frac{t^k}{k!}\mathbf{A}^k,$$

which converges for all $-\infty < t < \infty$. Since the entries of $\mathbf{A}$ are all constants, the same is true for $\mathbf{A}^k$, and the variable t only appears in the scalar factors in this sum. For example, if

(9.1) $$\mathbf{A} = \begin{pmatrix} 0 & 0 & 0 \\ 1 & 0 & 0 \\ 2 & 3 & 0 \end{pmatrix},$$

then

$$\mathbf{A}^2 = \begin{pmatrix} 0 & 0 & 0 \\ 0 & 0 & 0 \\ 3 & 0 & 0 \end{pmatrix} \quad \text{and} \quad \mathbf{A}^3 = \mathbf{A}^4 = \cdots = \begin{pmatrix} 0 & 0 & 0 \\ 0 & 0 & 0 \\ 0 & 0 & 0 \end{pmatrix}.$$

We see that

(9.2) $$e^{t\mathbf{A}} = \mathbf{I} + t\mathbf{A} + \frac{t^2}{2}\mathbf{A}^2 = \begin{pmatrix} 1 & 0 & 0 \\ t & 1 & 0 \\ 2t + \frac{3}{2}t^2 & 3t & 1 \end{pmatrix}.$$

The preceding equation shows that $e^{t\mathbf{A}}$ is a **matrix-valued function**; it is a function whose value at each t is an ordinary matrix. For example, a 2×2 matrix function $\mathbf{F}(t)$ looks like

$$\mathbf{F}(t) = \begin{pmatrix} f(t) & g(t) \\ u(t) & v(t) \end{pmatrix},$$

where f, g, u, and v are ordinary scalar-valued functions. Alternatively we can think of a matrix-valued function as a matrix whose entries are functions of t. The matrix-valued function $\mathbf{F}$ is differentiable exactly when all of its entries are differentiable in the usual sense, and its derivative $\mathbf{F}'$ is obtained by differentiating every entry. For the 2×2 example above we have

$$\mathbf{F}'(t) = \begin{pmatrix} f'(t) & g'(t) \\ u'(t) & v'(t) \end{pmatrix}.$$

Note that the derivative $\mathbf{F}'$ is again a matrix-valued function.

Example 9.1.2. Consider the matrix $\mathbf{A}$ in (9.1) and let $\mathbf{F}(t) = e^{t\mathbf{A}}$, which is written out explicitly in (9.2). Clearly $\mathbf{F}$ is differentiable and

$$\mathbf{F}'(t) = \begin{pmatrix} 0 & 0 & 0 \\ 1 & 0 & 0 \\ 2 + 3t & 3 & 0 \end{pmatrix}.$$

Differentiation rules for matrix-valued functions. The following familiar looking sum and product differentiation rules apply to matrix-valued functions.

Theorem 9.1.3. *If $\mathbf{B}(t)$ and $\mathbf{C}(t)$ are differentiable $n \times n$ matrix functions and $\mathbf{X}(t)$ is a differentiable $n \times 1$ matrix function (i.e., a vector function), then $\mathbf{B}(t) + \mathbf{C}(t)$, $\mathbf{B}(t)\mathbf{C}(t)$, and $\mathbf{B}(t)\mathbf{X}(t)$ are differentiable with*

(a)
$$\frac{d}{dt}[\mathbf{B}(t) + \mathbf{C}(t)] = \mathbf{B}'(t) + \mathbf{C}'(t),$$

(b)
$$\frac{d}{dt}[\mathbf{B}(t)\mathbf{C}(t)] = \mathbf{B}'(t)\mathbf{C}(t) + \mathbf{B}(t)\mathbf{C}'(t), \quad \text{and}$$

9.1. The matrix exponential, Part I

(c)
$$\frac{d}{dt}[\mathbf{B}(t)\mathbf{X}(t)] = \mathbf{B}'(t)\mathbf{X}(t) + \mathbf{B}(t)\mathbf{X}'(t).$$

In Exercise 16 you are asked to prove this when $n = 2$. Keep in mind that since matrix multiplication is not in general commutative, the order of the factors in the product rules (b) and (c) is important.

For any $n \times n$ matrix $\mathbf{A}$, the matrix-valued function $\mathbf{F}(t) = e^{t\mathbf{A}}$ is differentiable, and it has this nice additional property: Its derivative $\mathbf{F}'(t)$ can also be obtained from the infinite sum definition of $e^{t\mathbf{A}}$ by differentiating "term by term" with respect to t, giving

$$\mathbf{F}'(t) = \frac{d}{dt}\left(e^{t\mathbf{A}}\right) = \mathbf{0} + \mathbf{A} + t\mathbf{A}^2 + \frac{1}{2!}t^2\mathbf{A}^3 + \frac{1}{3!}t^3\mathbf{A}^4 + \cdots$$

$$= \mathbf{A}\left(\mathbf{I} + t\mathbf{A} + \frac{1}{2!}t^2\mathbf{A}^2 + \frac{1}{3!}t^3\mathbf{A}^3 + \frac{1}{4!}t^4\mathbf{A}^4 + \cdots\right)$$

$$= \mathbf{A}e^{t\mathbf{A}}.$$

Repeating these calculations, but now factoring out the $\mathbf{A}$ on the right-hand side in the second line gives

$$\frac{d}{dt}\left(e^{t\mathbf{A}}\right) = e^{t\mathbf{A}}\mathbf{A}$$

as well. Keep in mind that since matrix multiplication is not in general commutative, it is not *automatic* that $\mathbf{A}e^{t\mathbf{A}} = e^{t\mathbf{A}}\mathbf{A}$, but our calculations above show that it is true:

$$(9.3) \qquad \frac{d}{dt}\left(e^{t\mathbf{A}}\right) = e^{t\mathbf{A}}\mathbf{A} = \mathbf{A}e^{t\mathbf{A}}.$$

As an example, we'll use equation (9.3) to compute again the derivative of $e^{t\mathbf{A}}$, where $\mathbf{A}$ is given in equation (9.1):

$$\frac{d}{dt}\left(e^{t\mathbf{A}}\right) = \mathbf{A}e^{t\mathbf{A}} = \begin{pmatrix} 0 & 0 & 0 \\ 1 & 0 & 0 \\ 2 & 3 & 0 \end{pmatrix}\begin{pmatrix} 1 & 0 & 0 \\ t & 1 & 0 \\ 2t + \frac{3}{2}t^2 & 3t & 1 \end{pmatrix} = \begin{pmatrix} 0 & 0 & 0 \\ 1 & 0 & 0 \\ 2 + 3t & 3 & 0 \end{pmatrix},$$

which agrees with the result obtained in Example 9.1.2.

Now given an initial condition vector $\mathbf{X}_0$ (a vector of constants), set $\mathbf{X}(t) = e^{t\mathbf{A}}\mathbf{X}_0$. When $t = 0$, the definition of $e^{t\mathbf{A}}$ gives $e^{\mathbf{0}} = \mathbf{I}$, so that

$$\mathbf{X}(0) = \mathbf{I}\mathbf{X}_0 = \mathbf{X}_0.$$

Using equation (9.3) and property (c) of Theorem 9.1.3, we have

$$\mathbf{X}'(t) = \frac{d}{dt}(e^{t\mathbf{A}}\mathbf{X}_0) = \mathbf{A}e^{t\mathbf{A}}\mathbf{X}_0 = \mathbf{A}\mathbf{X}(t).$$

Thus we have a solution to $\mathbf{X}'(t) = \mathbf{A}\mathbf{X}(t)$, with the initial condition built in!

The initial value problem
$$\mathbf{X}'(t) = \mathbf{A}\mathbf{X}(t), \quad \mathbf{X}(0) = \mathbf{X}_0,$$
has solution $\mathbf{X}(t) = e^{t\mathbf{A}}\mathbf{X}_0$.

This result is easily adjusted when the initial condition is specified at $t = t_0$ rather than $t = 0$; see Exercise 8 in Section 9.2.

The eigenvalues and eigenvectors of $e^{t\mathbf{A}}$. To make practical use of the matrix exponential, we need a way of calculating $e^{t\mathbf{A}}$ that doesn't just rely on the "infinite sum" definition. If r is an eigenvalue of $\mathbf{A}$ with eigenvector $\mathbf{V}$, we have $\mathbf{A}\mathbf{V} = r\mathbf{V}$. Therefore,

$$\mathbf{A}^2 \mathbf{V} = \mathbf{A}\mathbf{A}\mathbf{V} = \mathbf{A}(r\mathbf{V}) = r\mathbf{A}\mathbf{V} = r(r\mathbf{V}) = r^2 \mathbf{V}.$$

Similarly, we can compute $\mathbf{A}^3 \mathbf{V} = r^3 \mathbf{V}$, and, in general, $\mathbf{A}^n \mathbf{V} = r^n \mathbf{V}$. With these facts in hand, we compute for the eigenvector $\mathbf{V}$,

$$e^{t\mathbf{A}}\mathbf{V} = \left(\mathbf{I} + t\mathbf{A} + \frac{t^2}{2!}\mathbf{A}^2 + \frac{t^3}{3!}\mathbf{A}^3 + \frac{t^4}{4!}\mathbf{A}^4 + \cdots\right)\mathbf{V}$$

$$= \mathbf{I}\mathbf{V} + t\mathbf{A}\mathbf{V} + \frac{t^2}{2!}\mathbf{A}^2\mathbf{V} + \frac{t^3}{3!}\mathbf{A}^3\mathbf{V} + \cdots$$

$$= \mathbf{V} + tr\mathbf{V} + \frac{t^2}{2!}r^2\mathbf{V} + \frac{t^3}{3!}r^3\mathbf{V} + \cdots$$

$$= \left(1 + tr + \frac{t^2}{2!}r^2 + \frac{t^3}{3!}r^3 + \cdots\right)\mathbf{V}$$

$$= e^{rt}\mathbf{V}.$$

Thus we have shown

(9.4) $$e^{t\mathbf{A}}\mathbf{V} = e^{rt}\mathbf{V},$$

so that $\mathbf{V}$ *is also an eigenvector for* $e^{t\mathbf{A}}$ *with associated eigenvalue* e^{rt}.

The Cayley-Hamilton Theorem. To proceed further, we need another fact from matrix theory, the Cayley-Hamilton Theorem. Given an $n \times n$ matrix $\mathbf{A}$, let $p(z) = \det(\mathbf{A} - z\mathbf{I})$ be its characteristic polynomial. It has degree n. If $\mathbf{B}$ is any $n \times n$ matrix, we can *define* $p(\mathbf{B})$ by substituting $\mathbf{B}$ for z in each term of $p(z)$; as before $\mathbf{B}^0 = \mathbf{I}$. For example, if $\mathbf{A}$ is 2×2, say,

$$\mathbf{A} = \begin{pmatrix} a & b \\ c & d \end{pmatrix},$$

then we know

(9.5) $$p(z) = z^2 - (a+d)z + (ad - bc),$$

and for any 2×2 matrix $\mathbf{B}$,

$$p(\mathbf{B}) = \mathbf{B}^2 - (a+d)\mathbf{B} + (ad - bc)\mathbf{I},$$

another 2×2 matrix. Taking $\mathbf{B}$ to be $\mathbf{A}$ itself yields something surprising.

Theorem 9.1.4. Cayley-Hamilton Theorem. *If $\mathbf{A}$ is an $n \times n$ matrix with characteristic polynomial $p(z) = \det(\mathbf{A} - z\mathbf{I})$, then*

$$p(\mathbf{A}) = \mathbf{0},$$

the zero matrix (of size $n \times n$).

Example 9.1.5. Specializing the Cayley-Hamilton Theorem to the case $n = 2$ we see that if
$$\mathbf{A} = \begin{pmatrix} a & b \\ c & d \end{pmatrix},$$
then, using equation (9.5),

(9.6) $$p(\mathbf{A}) = \mathbf{A}^2 - (a+d)\mathbf{A} + (ad - bc)\mathbf{I} = \begin{pmatrix} 0 & 0 \\ 0 & 0 \end{pmatrix} = \mathbf{0}.$$

You are invited to verify this in Exercise 20.

Our goal for the rest of the section is to give formulas for $e^{t\mathbf{A}}$ when $\mathbf{A}$ is a 2×2 matrix with real entries, using the Cayley-Hamilton Theorem. These formulas will depend on the kind of eigenvalues $\mathbf{A}$ possesses.

Computing $e^{t\mathbf{A}}$ when the 2×2 matrix A has one repeated eigenvalue. Consider a 2×2 matrix $\mathbf{A}$ with eigenvalues $r_1 = r_2 = r$, which must be a real number if the entries of $\mathbf{A}$ are real. The characteristic polynomial is
$$p(z) = z^2 - 2rz + r^2 = (z - r)^2,$$
which has a single root of multiplicity two. By the Cayley-Hamilton Theorem, we get
$$p(\mathbf{A}) = \mathbf{A}^2 - 2r\mathbf{A} + r^2\mathbf{I} = (\mathbf{A} - r\mathbf{I})^2 = \mathbf{0}.$$
Now, we need another fact about exponentials of matrices: If $\mathbf{S}$ and $\mathbf{T}$ are matrices which commute (that is, $\mathbf{ST} = \mathbf{TS}$), then

(9.7) $$e^{\mathbf{S}+\mathbf{T}} = e^{\mathbf{S}} e^{\mathbf{T}}.$$

To see why this is true, recall our discussion in Section 4.4.1 showing that $e^{z+w} = e^z e^w$ for any complex numbers z and w. Since $\mathbf{S}$ and $\mathbf{T}$ commute, you can do the same calculation, replacing z by $\mathbf{S}$ and w by $\mathbf{T}$, to get $e^{\mathbf{S}+\mathbf{T}} = e^{\mathbf{S}}e^{\mathbf{T}}$. (Another approach is outlined in Section 9.2.1). We stress, though, that without commutativity of $\mathbf{S}$ and $\mathbf{T}$, equation (9.7) need not hold; see Exercise 15.

Let's apply equation (9.7) to compute $e^{t\mathbf{A}}$ when $\mathbf{A}$ has the repeated eigenvalue r. Adding and subtracting $t(r\mathbf{I})$ in the exponent gives us
$$e^{t\mathbf{A}} = e^{t(\mathbf{A}-r\mathbf{I})+t(r\mathbf{I})} = e^{t(\mathbf{A}-r\mathbf{I})}e^{t(r\mathbf{I})}.$$
The second equality holds because $t(\mathbf{A} - r\mathbf{I})$ and $t(r\mathbf{I})$ commute (because $\mathbf{I}$ commutes with *any* matrix). Now you can show from the definition that $e^{t(r\mathbf{I})} = e^{rt}\mathbf{I}$ (see Exercise 12), and therefore
$$e^{t\mathbf{A}} = e^{t(\mathbf{A}-r\mathbf{I})}e^{t(r\mathbf{I})}$$
$$= e^{t(\mathbf{A}-r\mathbf{I})}e^{rt}\mathbf{I}$$
$$= e^{rt}\mathbf{I}e^{t(\mathbf{A}-r\mathbf{I})}$$
$$= e^{rt}\left(\mathbf{I} + t(\mathbf{A}-r\mathbf{I}) + \frac{t^2}{2!}(\mathbf{A}-r\mathbf{I})^2 + \frac{t^3}{3!}(\mathbf{A}-r\mathbf{I})^3 + \cdots\right).$$

In the last line we've just used Definition 9.1.1 for $e^{t(\mathbf{A}-r\mathbf{I})}$. But we've observed that the Cayley-Hamilton Theorem tells us that $(\mathbf{A}-r\mathbf{I})^2 = \mathbf{0}$, and thus the higher powers of $\mathbf{A}-r\mathbf{I}$ are $\mathbf{0}$ as well. Hence in this situation we have the simple formula

(9.8) $$e^{t\mathbf{A}} = e^{rt}\left(\mathbf{I} + t(\mathbf{A}-r\mathbf{I})\right).$$

Examining equation (9.8) a bit further, we see that there are really two cases here: Either $\mathbf{A} - r\mathbf{I} = \mathbf{0}$ or $\mathbf{A} - r\mathbf{I} \neq \mathbf{0}$. In the first case,

$$\mathbf{A} = r\mathbf{I} = \begin{pmatrix} r & 0 \\ 0 & r \end{pmatrix},$$

and equation (9.8) merely says

$$e^{t\mathbf{A}} = e^{t(r\mathbf{I})} = e^{rt}\mathbf{I} = \begin{pmatrix} e^{rt} & 0 \\ 0 & e^{rt} \end{pmatrix},$$

so that

$$\mathbf{X}(t) = e^{t\mathbf{A}}\mathbf{X}_0 = e^{rt}\mathbf{I}\mathbf{X}_0 = e^{rt}\mathbf{X}_0$$

is the solution to $\mathbf{X}' = \mathbf{A}\mathbf{X}$ with $\mathbf{X}(0) = \mathbf{X}_0$. Note that this is a *ray solution* passing through $\mathbf{X}_0$ at time $t = 0$. Thus, for any initial condition, the solution is a ray (*outgoing* or *incoming* according as $r > 0$ or $r < 0$), and the equilibrium point $(0,0)$ is a proper degenerate node.

In the second case, $\mathbf{A} \neq r\mathbf{I}$, so $\mathbf{A} - r\mathbf{I}$ is not the zero matrix. Moreover, the only eigenvectors of $\mathbf{A}$ corresponding to r are (nonzero) constant multiples of a single eigenvector $\mathbf{V}_0$ (see, e.g., Exercise 19 in Section 8.7). From equation (9.8), the solution satisfying the initial condition $\mathbf{X}(0) = \mathbf{X}_0$ is

$$\begin{aligned} \mathbf{X}(t) &= e^{rt}[\mathbf{I} + t(\mathbf{A} - r\mathbf{I})]\mathbf{X}_0 \\ &= e^{rt}(\mathbf{X}_0 + t\mathbf{V}), \end{aligned}$$

where $\mathbf{V} = (\mathbf{A} - r\mathbf{I})\mathbf{X}_0$.

If $\mathbf{V} \neq \mathbf{0}$, $\mathbf{X}_0 + t\mathbf{V}$ describes parametrically a line passing through $\mathbf{X}_0$ at time $t = 0$ and parallel to $\mathbf{V}$; see the figure below. Note that $\mathbf{V}$ must be an eigenvector of $\mathbf{A}$ corresponding to r: $(\mathbf{A} - r\mathbf{I})\mathbf{V} = (\mathbf{A} - r\mathbf{I})^2\mathbf{X}_0 = \mathbf{0}$; hence $\mathbf{V}$ is a constant multiple of the eigenvector $\mathbf{V}_0$.

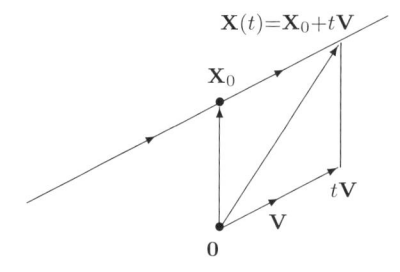

The line $\mathbf{X}_0 + t\mathbf{V}$ through $\mathbf{X}_0$ in the direction of $\mathbf{V}$.

We can sketch the trajectory of the solution $e^{rt}(\mathbf{X}_0 + t\mathbf{V})$ by first sketching the line $\mathbf{X}_0 + t\mathbf{V}$ and then scaling points on this line by e^{rt} for appropriate values of t, similarly to what we did in Section 8.7. The trajectory approaches $(0,0)$ (as $t \to \infty$ if $r < 0$ or as $t \to -\infty$ if $r > 0$) tangent to the line L through the origin parallel to $\mathbf{V}$. Since $\mathbf{V}$ is a multiple of the eigenvector $\mathbf{V}_0$, L can also be described as the unique line through the origin in the direction of any eigenvector for $\mathbf{A}$, i.e., the eigenvector line. For typical trajectory pictures, see Fig. 9.1 (where $r > 0$) and Fig. 9.2 (where $r < 0$). Compare to Fig. 8.68, which shows a trajectory of the same type, but with a differently situated guideline $\mathbf{X}_0 + t\mathbf{V}$.

9.1. The matrix exponential, Part I

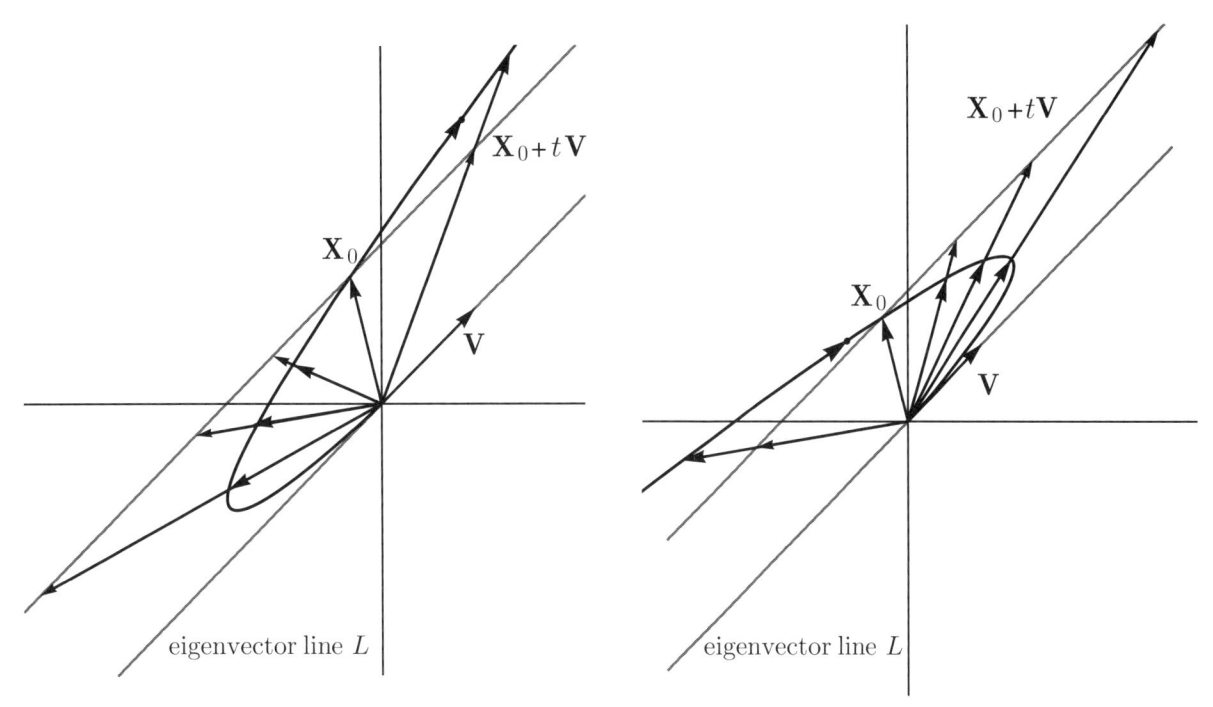

Figure 9.1. Trajectory of a solution $\mathbf{X}(t) = e^{rt}(\mathbf{X}_0 + t\mathbf{V})$ where $r > 0$, with guideline $\mathbf{X}_0 + t\mathbf{V}$.

Figure 9.2. Trajectory of a solution $\mathbf{X}(t) = e^{rt}(\mathbf{X}_0 + t\mathbf{V})$ where $r < 0$, with guideline $\mathbf{X}_0 + t\mathbf{V}$.

Example 9.1.6. Let's solve the initial value problem $\mathbf{X}' = \mathbf{A}\mathbf{X}$, $\mathbf{X}(0) = \mathbf{X}_0$, where

$$\mathbf{A} = \begin{pmatrix} 3 & 4 \\ -1 & -1 \end{pmatrix} \quad \text{and} \quad \mathbf{X}_0 = \begin{pmatrix} 0 \\ 1 \end{pmatrix}.$$

We have $\det(\mathbf{A} - z\mathbf{I}) = z^2 - 2z + 1 = (z-1)^2$. Thus there is one eigenvalue, $r = 1$, which has multiplicity two. Using equation (9.8) with $r = 1$ we have

$$e^{t\mathbf{A}} = e^t\left(\mathbf{I} + t(\mathbf{A} - \mathbf{I})\right) = e^t\left(\begin{pmatrix} 1 & 0 \\ 0 & 1 \end{pmatrix} + t\begin{pmatrix} 2 & 4 \\ -1 & -2 \end{pmatrix}\right).$$

The solution satisfying the desired initial condition is

$$\begin{aligned}
\mathbf{X}(t) &= e^{t\mathbf{A}}\begin{pmatrix} 0 \\ 1 \end{pmatrix} = e^t\left[\begin{pmatrix} 1 & 0 \\ 0 & 1 \end{pmatrix} + t\begin{pmatrix} 2 & 4 \\ -1 & -2 \end{pmatrix}\right]\begin{pmatrix} 0 \\ 1 \end{pmatrix} \\
&= e^t\left[\begin{pmatrix} 1 & 0 \\ 0 & 1 \end{pmatrix}\begin{pmatrix} 0 \\ 1 \end{pmatrix} + t\begin{pmatrix} 2 & 4 \\ -1 & -2 \end{pmatrix}\begin{pmatrix} 0 \\ 1 \end{pmatrix}\right] \\
&= e^t\left[\begin{pmatrix} 0 \\ 1 \end{pmatrix} + t\begin{pmatrix} 4 \\ -2 \end{pmatrix}\right] = \begin{pmatrix} 4te^t \\ (1-2t)e^t \end{pmatrix}.
\end{aligned}$$

Fig. 9.3 shows this solution in the phase plane.

To obtain this picture by hand, start by sketching the line

$$\begin{pmatrix} 0 \\ 1 \end{pmatrix} + t\begin{pmatrix} 4 \\ -2 \end{pmatrix}$$

and scale points on this line by the scalar factor e^t for the appropriate value of t; see Exercise 10.

Next we turn to the case that $\mathbf{A}$ has two distinct eigenvalues. As before, the Cayley-Hamilton Theorem will provide the key tool for providing a formula for $e^{t\mathbf{A}}$.

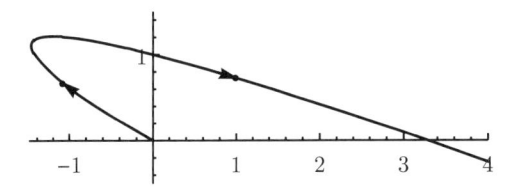

Figure 9.3. Trajectory in the phase plane.

Computing $e^{t\mathbf{A}}$ for a 2×2 matrix with two distinct eigenvalues. Suppose $\mathbf{A}$ is a 2×2 matrix with *distinct* eigenvalues r_1 and r_2 (which could be real or complex). We define two special matrices
$$\mathbf{B} = \frac{1}{r_1 - r_2}(\mathbf{A} - r_2 \mathbf{I}),$$
which we abbreviate as

(9.9) $$\mathbf{B} = \frac{\mathbf{A} - r_2 \mathbf{I}}{r_1 - r_2}$$

and

(9.10) $$\mathbf{C} = \frac{1}{r_2 - r_1}(\mathbf{A} - r_1 \mathbf{I}) = \frac{\mathbf{A} - r_1 \mathbf{I}}{r_2 - r_1}.$$

The matrices $\mathbf{B}$ and $\mathbf{C}$ have some remarkable properties, and we will be able to give a simple formula for $e^{t\mathbf{A}}$ in terms of $\mathbf{B}$ and $\mathbf{C}$.

Properties of the special matrices B and C. Since the characteristic polynomial $p(z)$ for $\mathbf{A}$ can be factored as
$$p(z) = (z - r_1)(z - r_2),$$
the Cayley-Hamilton Theorem tells us that

(9.11) $$(\mathbf{A} - r_1 \mathbf{I})(\mathbf{A} - r_2 \mathbf{I}) = p(\mathbf{A}) = \mathbf{0}.$$

Comparing this with the definition of $\mathbf{B}$ in equation (9.9), we see that

(9.12) $$(\mathbf{A} - r_1 \mathbf{I})\mathbf{B} = \mathbf{0}.$$

Suppose now that $\mathbf{X}_0$ is any vector in $\mathbb{R}^2$. By equation (9.12) we have $(\mathbf{A} - r_1 \mathbf{I})\mathbf{B}\mathbf{X}_0 = \mathbf{0}$, or what is the same thing
$$\mathbf{A}(\mathbf{B}\mathbf{X}_0) = r_1(\mathbf{B}\mathbf{X}_0).$$

Thus if $\mathbf{B}\mathbf{X}_0$ is not the zero vector, it is an eigenvector for $\mathbf{A}$ with associated eigenvalue r_1. Then by equation (9.4) applied to the eigenvector $\mathbf{V} = \mathbf{B}\mathbf{X}_0$,

(9.13) $$e^{t\mathbf{A}}(\mathbf{B}\mathbf{X}_0) = e^{r_1 t}(\mathbf{B}\mathbf{X}_0).$$

Even if $\mathbf{B}\mathbf{X}_0$ is equal to $\mathbf{0}$, this equation remains true, for then it just says $\mathbf{0} = \mathbf{0}$. Therefore, equation (9.13) holds for all vectors $\mathbf{X}_0$ in $\mathbb{R}^2$.

Similarly, the definition of $\mathbf{C}$ and equation (9.11), which is valid with the order of the factors on the left reversed, tell us that $(\mathbf{A} - r_2 \mathbf{I})\mathbf{C} = \mathbf{0}$, and thus, for any vector $\mathbf{X}_0$ in $\mathbb{R}^2$, $(\mathbf{A} - r_2 \mathbf{I})\mathbf{C}\mathbf{X}_0 = \mathbf{0}$, or
$$\mathbf{A}(\mathbf{C}\mathbf{X}_0) = r_2(\mathbf{C}\mathbf{X}_0).$$

Arguing just as above, with $\mathbf{C}$ replacing $\mathbf{B}$ and r_2 replacing r_1 we find that

(9.14) $$e^{t\mathbf{A}}(\mathbf{C}\mathbf{X}_0) = e^{r_2 t}(\mathbf{C}\mathbf{X}_0)$$

9.1. The matrix exponential, Part I

for any vector $\mathbf{X}_0$ in $\mathbb{R}^2$. Finally, we add the matrices $\mathbf{B}$ and $\mathbf{C}$ to get

$$\mathbf{B} + \mathbf{C} = \frac{\mathbf{A} - r_2\mathbf{I}}{r_1 - r_2} + \frac{\mathbf{A} - r_1\mathbf{I}}{r_2 - r_1} = \frac{\mathbf{A} - r_2\mathbf{I}}{r_1 - r_2} - \frac{\mathbf{A} - r_1\mathbf{I}}{r_1 - r_2}$$

$$= \frac{1}{r_1 - r_2}(r_1 - r_2)\mathbf{I} = \mathbf{I},$$

so that

(9.15) $$\mathbf{B} + \mathbf{C} = \mathbf{I}.$$

Further interesting properties of $\mathbf{B}$ and $\mathbf{C}$ are explored in Exercises 28–31.

A formula for $e^{t\mathbf{A}}$ using B and C. Now let's put this together: For any vector $\mathbf{X}_0$ in $\mathbb{R}^2$,

$$\begin{aligned} e^{t\mathbf{A}}\mathbf{X}_0 &= e^{t\mathbf{A}}\mathbf{I}\mathbf{X}_0 \\ &= e^{t\mathbf{A}}(\mathbf{B} + \mathbf{C})\mathbf{X}_0 \\ &= e^{t\mathbf{A}}(\mathbf{B}\mathbf{X}_0) + e^{t\mathbf{A}}(\mathbf{C}\mathbf{X}_0) \\ &= e^{r_1 t}(\mathbf{B}\mathbf{X}_0) + e^{r_2 t}(\mathbf{C}\mathbf{X}_0) \\ &= (e^{r_1 t}\mathbf{B} + e^{r_2 t}\mathbf{C})\mathbf{X}_0, \end{aligned}$$

where we have used equations (9.15), (9.13), and (9.14). Since this is true for *every* vector $\mathbf{X}_0$, it must be the case (see Exercise 27) that the corresponding matrices are equal:

(9.16) $$e^{t\mathbf{A}} = e^{r_1 t}\mathbf{B} + e^{r_2 t}\mathbf{C} = e^{r_1 t}\frac{\mathbf{A} - r_2\mathbf{I}}{r_1 - r_2} + e^{r_2 t}\frac{\mathbf{A} - r_1\mathbf{I}}{r_2 - r_1}.$$

Note that by collecting the terms on the right-hand side in equation (9.16) involving the matrix $\mathbf{I}$ and the matrix $\mathbf{A}$, we can restate the formula given there as

(9.17) $$e^{t\mathbf{A}} = \left(\frac{r_1 e^{r_2 t} - r_2 e^{r_1 t}}{r_1 - r_2}\right)\mathbf{I} + \left(\frac{e^{r_1 t} - e^{r_2 t}}{r_1 - r_2}\right)\mathbf{A},$$

which is sometimes (but not always) easier to use than (9.16).

Example 9.1.7. Let's check equation (9.16) by redoing Example 8.4.2 in Section 8.4. There we solved the initial value problem $\mathbf{X}' = \mathbf{A}\mathbf{X}$, $\mathbf{X}(0) = \mathbf{X}_0$, where

$$\mathbf{A} = \begin{pmatrix} 2 & 1 \\ 2 & 3 \end{pmatrix} \quad \text{and} \quad \mathbf{X}_0 = \begin{pmatrix} -1 \\ 7 \end{pmatrix}.$$

Let's now find the solution using $e^{t\mathbf{A}}$. Recall that $\mathbf{A}$ has distinct eigenvalues $r_1 = 4$ and $r_2 = 1$. On calculating the special matrices $\mathbf{B}$ and $\mathbf{C}$, we find that

$$\mathbf{B} = \frac{\mathbf{A} - \mathbf{I}}{4 - 1} = \frac{1}{3}(\mathbf{A} - \mathbf{I}) = \begin{pmatrix} 1/3 & 1/3 \\ 2/3 & 2/3 \end{pmatrix}$$

and

$$\mathbf{C} = \frac{\mathbf{A} - 4\mathbf{I}}{1 - 4} = -\frac{1}{3}(\mathbf{A} - 4\mathbf{I}) = \begin{pmatrix} 2/3 & -1/3 \\ -2/3 & 1/3 \end{pmatrix}.$$

Then by equation (9.16),

$$e^{t\mathbf{A}} = e^{4t}\mathbf{B} + e^t\mathbf{C} = e^{4t}\begin{pmatrix} 1/3 & 1/3 \\ 2/3 & 2/3 \end{pmatrix} + e^t\begin{pmatrix} 2/3 & -1/3 \\ -2/3 & 1/3 \end{pmatrix}.$$

The solution satisfying the desired initial condition is

$$\begin{aligned}
\mathbf{X}(t) &= e^{t\mathbf{A}}\mathbf{X}_0 = e^{t\mathbf{A}}\begin{pmatrix} -1 \\ 7 \end{pmatrix} \\
&= e^{4t}\begin{pmatrix} 1/3 & 1/3 \\ 2/3 & 2/3 \end{pmatrix}\begin{pmatrix} -1 \\ 7 \end{pmatrix} + e^t\begin{pmatrix} 2/3 & -1/3 \\ -2/3 & 1/3 \end{pmatrix}\begin{pmatrix} -1 \\ 7 \end{pmatrix} \\
&= e^{4t}\begin{pmatrix} 2 \\ 4 \end{pmatrix} + e^t\begin{pmatrix} -3 \\ 3 \end{pmatrix} \\
&= \begin{pmatrix} 2e^{4t} - 3e^t \\ 4e^{4t} + 3e^t \end{pmatrix},
\end{aligned}$$

the same (fortunately) as found earlier!

Purely imaginary eigenvalues. Equations (9.16) and (9.17) can be used whenever the 2×2 matrix $\mathbf{A}$ has two distinct eigenvalues r_1 and r_2, even if they are complex. For example, suppose $\mathbf{A}$ is a matrix whose eigenvalues are a conjugate pair of purely imaginary numbers, $r_1 = i\mu$, $r_2 = -i\mu$. By (9.17) we have

$$\begin{aligned}
e^{t\mathbf{A}} &= \left(\frac{r_1 e^{r_2 t} - r_2 e^{r_1 t}}{r_1 - r_2}\right)\mathbf{I} + \left(\frac{e^{r_1 t} - e^{r_2 t}}{r_1 - r_2}\right)\mathbf{A} \\
&= \left(\frac{i\mu e^{-i\mu t} - (-i\mu)e^{i\mu t}}{i\mu - (-i\mu)}\right)\mathbf{I} + \left(\frac{e^{i\mu t} - e^{-i\mu t}}{i\mu - (-i\mu)}\right)\mathbf{A} \\
&= \left(\frac{e^{-i\mu t} + e^{i\mu t}}{2}\right)\mathbf{I} + \left(\frac{e^{i\mu t} - e^{-i\mu t}}{2i\mu}\right)\mathbf{A}
\end{aligned}$$

(9.18)
$$= \cos(\mu t)\mathbf{I} + \sin(\mu t)\frac{1}{\mu}\mathbf{A}$$

where the last equality uses Euler's formulas $e^{i\mu t} = \cos(\mu t) + i\sin(\mu t)$ and $e^{-i\mu t} = \cos(-\mu t) + i\sin(-\mu t) = \cos(\mu t) - i\sin(\mu t)$. The solution to the initial value problem $\mathbf{X}' = \mathbf{A}\mathbf{X}$, $\mathbf{X}(0) = \mathbf{X}_0$, is then

(9.19)
$$\mathbf{X}(t) = e^{t\mathbf{A}}\mathbf{X}_0 = \cos(\mu t)\mathbf{X}_0 + \sin(\mu t)\frac{1}{\mu}\mathbf{A}\mathbf{X}_0.$$

Understanding the solution (9.19) graphically. The solution given by (9.19) can be analyzed graphically. Note that the vectors $\mathbf{X}_0$ and $\mathbf{A}\mathbf{X}_0$ point in different directions. If not, $\mathbf{A}\mathbf{X}_0$ would be a multiple of $\mathbf{X}_0$, say, $\mathbf{A}\mathbf{X}_0 = r\mathbf{X}_0$ for some real number r. But then r would be an eigenvalue for $\mathbf{A}$, with eigenvector $\mathbf{X}_0$. Since $\mathbf{A}$ has only the imaginary eigenvalues $\pm i\mu$, this is impossible.

9.1. The matrix exponential, Part I

Thus typical pictures for the vectors $\mathbf{X}_0$ and $\frac{1}{\mu}\mathbf{A}\mathbf{X}_0$ look like this:

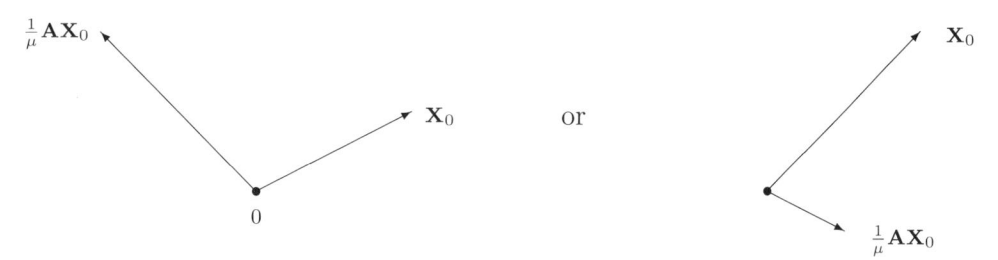

As t increases from 0 to $\frac{\pi}{2\mu}$, then to $\frac{\pi}{\mu}$, then to $\frac{3\pi}{2\mu}$, and finally to $\frac{2\pi}{\mu}$, $\cos(\mu t)$ goes from 1 to 0 to -1 to 0 to 1, while $\sin(\mu t)$ goes from 0 to 1 to 0 to -1 to 0. At the same time $\mathbf{X}(t)$ goes from $\mathbf{X}_0$ to $\frac{1}{\mu}\mathbf{A}\mathbf{X}_0$ to $-\mathbf{X}_0$ to $-\frac{1}{\mu}\mathbf{A}\mathbf{X}_0$ and back to $\mathbf{X}_0$. Thus the solution is periodic with period $\frac{2\pi}{\mu}$ and looks, for example, like in the pictures below.

t	$\cos(\mu t)$	$\sin(\mu t)$	$\mathbf{X}(t) = \cos(\mu t)\mathbf{X}_0 + \sin(\mu t)\frac{1}{\mu}\mathbf{A}\mathbf{X}_0$
0	1	0	$\mathbf{X}_0$
$\frac{\pi}{2\mu}$	0	1	$\frac{1}{\mu}\mathbf{A}\mathbf{X}_0$
$\frac{\pi}{\mu}$	-1	0	$-\mathbf{X}_0$
$\frac{3\pi}{2\mu}$	0	-1	$-\frac{1}{\mu}\mathbf{A}\mathbf{X}_0$
$\frac{2\pi}{\mu}$	1	0	$\mathbf{X}_0$

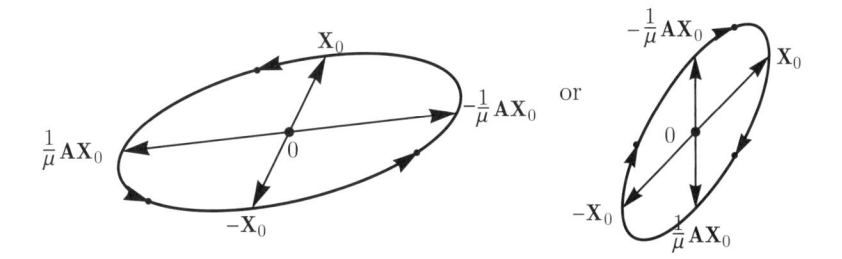

Whether the oval-shaped curve (which is actually an ellipse by Exercise 13 in Section 8.9) is traced clockwise or counterclockwise as t increases depends on the relative position of $\mathbf{X}_0$ and $\frac{1}{\mu}\mathbf{A}\mathbf{X}_0$ (as in the pictures above). A phase portrait for the system $\mathbf{X}' = \mathbf{A}\mathbf{X}$ where the matrix $\mathbf{A}$ has eigenvalues $\pm i\mu$ (so that $(0,0)$ is a center) consists of a family of ellipses encircling the origin. All of these elliptical orbits have the same orientation—either all are clockwise or all are counterclockwise. This orientation can actually be determined simply by looking at the signs of two of the entries in the matrix $\mathbf{A}$; see Exercise 11.

Example 9.1.8. We'll solve the initial value problem $\mathbf{X}' = \mathbf{A}\mathbf{X}$, $\mathbf{X}(0) = \mathbf{X}_0$, where

$$\mathbf{A} = \begin{pmatrix} 2 & -1 \\ 8 & -2 \end{pmatrix} \quad \text{and} \quad \mathbf{X}_0 = \begin{pmatrix} 1 \\ 0 \end{pmatrix}.$$

The eigenvalues of $\mathbf{A}$, which are found using the characteristic polynomial $p(z) = z^2 + 4 = (z - 2i)(z + 2i)$, are $r_1 = 2i$, $r_2 = -2i$. Hence in equation (9.19) we use $\mu = 2$, and our initial value problem has solution

$$\mathbf{X}(t) = e^{t\mathbf{A}}\begin{pmatrix} 1 \\ 0 \end{pmatrix} = \cos(2t)\begin{pmatrix} 1 \\ 0 \end{pmatrix} + \sin(2t)\frac{1}{2}\begin{pmatrix} 2 \\ 8 \end{pmatrix} = \begin{pmatrix} \cos(2t) + \sin(2t) \\ 4\sin(2t) \end{pmatrix}.$$

The trajectory of this solution in the phase plane is pictured below. We began this picture by sketching the vectors $\mathbf{X}_0$ and

$$\frac{1}{\mu}\mathbf{A}\mathbf{X}_0 = \frac{1}{2}\begin{pmatrix} 2 & -1 \\ 8 & -2 \end{pmatrix}\begin{pmatrix} 1 \\ 0 \end{pmatrix} = \begin{pmatrix} 1 \\ 4 \end{pmatrix}.$$

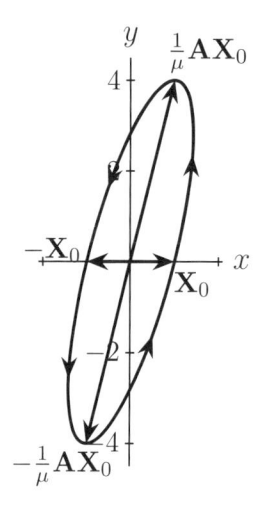

Computing $e^{t\mathbf{A}}$ when $\mathbf{A}$ has complex conjugate eigenvalues. In general, if $\mathbf{A}$ has complex eigenvalues, $r_1 = \lambda + i\mu$ and $r_2 = \lambda - i\mu$ with $\mu \neq 0$, equation (9.17) applies (since r_1 and r_2 are distinct) and

$$e^{t\mathbf{A}} = \left(\frac{r_1 e^{r_2 t} - r_2 e^{r_1 t}}{r_1 - r_2}\right)\mathbf{I} + \left(\frac{e^{r_1 t} - e^{r_2 t}}{r_1 - r_2}\right)\mathbf{A}.$$

Substituting $r_1 = \lambda + i\mu$, $r_2 = \lambda - i\mu$ and computing as in the imaginary case, we find

$$e^{t\mathbf{A}} = e^{\lambda t}\left(\cos(\mu t)\mathbf{I} + \sin(\mu t)\frac{1}{\mu}(\mathbf{A} - \lambda\mathbf{I})\right).$$

You are asked to verify this computation in Exercise 24. The solution $\mathbf{X}(t)$ to $\mathbf{X}' = \mathbf{A}\mathbf{X}$ with initial condition $\mathbf{X}(0) = \mathbf{X}_0$ is

$$\mathbf{X}(t) = e^{t\mathbf{A}}\mathbf{X}_0 = e^{\lambda t}\left(\cos(\mu t)\mathbf{X}_0 + \sin(\mu t)\frac{1}{\mu}(\mathbf{A} - \lambda\mathbf{I})\mathbf{X}_0\right).$$

The factor in the big parentheses,

$$\mathbf{Z}(t) = \cos(\mu t)\mathbf{X}_0 + \sin(\mu t)\frac{1}{\mu}(\mathbf{A} - \lambda\mathbf{I})\mathbf{X}_0,$$

is an elliptical periodic parametric curve which moves clockwise or counterclockwise, depending on the relative positions of $\mathbf{X}_0$ and $\frac{1}{\mu}(\mathbf{A} - \lambda\mathbf{I})\mathbf{X}_0$ (these vectors cannot be parallel; why not?). If $\lambda \neq 0$, the solution to the system, $\mathbf{X}(t) = e^{\lambda t}\mathbf{Z}(t)$, traverses a *spiral* outgoing or incoming according as $\lambda > 0$ or $\lambda < 0$; see Fig. 9.4. To sketch this spiral orbit, we first sketch the graph of $\mathbf{Z}(t)$; points on this loop are then scaled by the factor $e^{\lambda t}$ for appropriate choice of t. Fig. 9.4, in which the graph of $\mathbf{Z}$ appears dashed, indicates this scaling process.

Recap. The solution to $\mathbf{X}'(t) = \mathbf{A}\mathbf{X}(t)$ satisfying $\mathbf{X}(0) = \mathbf{X}_0$ is $\mathbf{X}(t) = e^{t\mathbf{A}}\mathbf{X}_0$. To use this when $\mathbf{A}$ is a 2×2 matrix with real entries and eigenvalues r_1 and r_2, we have formulas for $e^{t\mathbf{A}}$, which are summarized in Table 9.1.

9.1. The matrix exponential, Part I

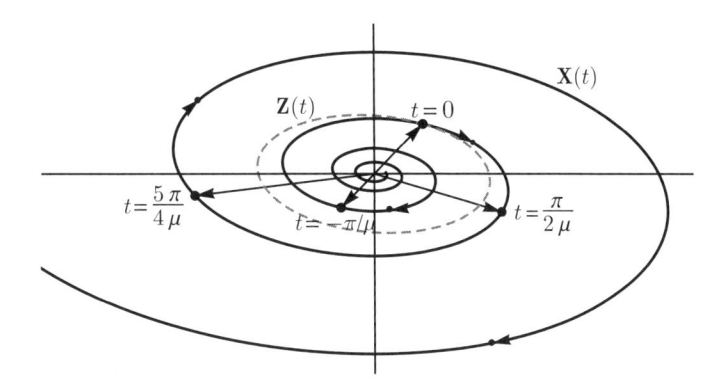

Figure 9.4. Outgoing spiral $\mathbf{X}(t) = e^{\lambda t} \mathbf{Z}(t)$ ($\lambda > 0$), with $\mathbf{Z}(t)$ shown dashed.

Table 9.1. Matrix exponential formulas.	
Eigenvalues of $\mathbf{A}$	Formula for $e^{t\mathbf{A}}$
r_1, r_2 not equal	$e^{t\mathbf{A}} = e^{r_1 t} \frac{\mathbf{A} - r_2 \mathbf{I}}{r_1 - r_2} + e^{r_2 t} \frac{\mathbf{A} - r_1 \mathbf{I}}{r_2 - r_1}$ $= \left(\frac{r_1 e^{r_2 t} - r_2 e^{r_1 t}}{r_1 - r_2}\right) \mathbf{I} + \left(\frac{e^{r_1 t} - e^{r_2 t}}{r_1 - r_2}\right) \mathbf{A}$
complex roots $r_1 = \lambda + i\mu$, $r_2 = \lambda - i\mu$ (including the purely imaginary case $\lambda = 0$)	$e^{t\mathbf{A}} = e^{\lambda t}\left[\cos(\mu t)\mathbf{I} + \sin(\mu t)\frac{1}{\mu}(\mathbf{A} - \lambda\mathbf{I})\right]$
$r_1 = r_2 = r$; equal real roots	$e^{t\mathbf{A}} = e^{rt}\left(\mathbf{I} + t(\mathbf{A} - r\mathbf{I})\right)$

We have seen that the expression for $e^{t\mathbf{A}}$ that appears in the first entry of this table actually applies whenever the eigenvalues r_1 and r_2 are distinct (real or not). However, if the entries of $\mathbf{A}$ are real and complex eigenvalues occur, they occur as a conjugate pair, and in this case the formula in the second row of the table is generally more useful.

9.1.1. Exercises.

In Exercises 1–8, calculate the matrix $e^{t\mathbf{A}}$ and use it to solve the system $\mathbf{X}'(t) = \mathbf{A}\mathbf{X}$ with the given initial condition. Sketch the trajectory of this solution in the phase plane, showing the direction the trajectory is traversed with increasing t.

1. $\mathbf{A} = \begin{pmatrix} 0 & 1 \\ -1 & 0 \end{pmatrix}$, $\mathbf{X}(0) = \begin{pmatrix} 1 \\ 1 \end{pmatrix}$.

2. $\mathbf{A} = \begin{pmatrix} 1 & -2 \\ 2 & -3 \end{pmatrix}$, $\mathbf{X}(0) = \begin{pmatrix} -1 \\ 1 \end{pmatrix}$.

3. $\mathbf{A} = \begin{pmatrix} 0 & -2 \\ 5 & 2 \end{pmatrix}$, $\mathbf{X}(0) = \begin{pmatrix} 0 \\ 1 \end{pmatrix}$.

4. $\mathbf{A} = \begin{pmatrix} -1 & 1 \\ -1 & -1 \end{pmatrix}$, $\mathbf{X}(0) = \begin{pmatrix} 0 \\ 1 \end{pmatrix}$.

5. $\mathbf{A} = \begin{pmatrix} -1 & 1 \\ -2 & -2 \end{pmatrix}$, $\mathbf{X}(0) = \begin{pmatrix} 0 \\ 1 \end{pmatrix}$.

6. $\mathbf{A} = \begin{pmatrix} 1 & 1 \\ -1 & 3 \end{pmatrix}$, $\mathbf{X}(0) = \begin{pmatrix} 0 \\ 1 \end{pmatrix}$.

7. $\mathbf{A} = \begin{pmatrix} 2 & -1 \\ 8 & -2 \end{pmatrix}$, $\mathbf{X}(0) = \begin{pmatrix} 1 \\ 1 \end{pmatrix}$.

8. $\mathbf{A} = \begin{pmatrix} -1 & -1 \\ 1 & -1 \end{pmatrix}$, $\mathbf{X}(0) = \begin{pmatrix} 1 \\ 0 \end{pmatrix}$.

9. Show that the orbit for the solution in Exercise 1 lies on a circle. What is the radius of this circle?

10. Sketch the line
$$\begin{pmatrix} 0 \\ 1 \end{pmatrix} + t \begin{pmatrix} 4 \\ -2 \end{pmatrix}.$$
With this as a guide, show how to obtain the trajectory of the solution in Example 9.1.6.

11. Consider the system $\mathbf{X}' = \mathbf{A}\mathbf{X}$ where
$$\mathbf{A} = \begin{pmatrix} a & b \\ c & d \end{pmatrix},$$
for real numbers $a, b, c,$ and d.
    (a) If the eigenvalues of $\mathbf{A}$ are $\alpha \pm i\beta$ where $\beta \neq 0$, then explain why b and c must have opposite signs.
    (b) When $\mathbf{A}$ has eigenvalues $\alpha \pm i\beta$ with $\beta \neq 0$, the orbits in a phase portrait for $\mathbf{X}' = \mathbf{A}\mathbf{X}$ are either spirals or closed loops (ellipses). Whether these orbits are traversed clockwise or counterclockwise is determined by the signs of b and c. By (a) the two possible cases are (i) $b > 0, c < 0$ or (ii) $b < 0, c > 0$. Predict, with explanation, which of these two possibilities corresponds to clockwise traversal and which to counterclockwise.[1] Hint: What is $\frac{dy}{dt}$ at $(1, 0)$, and what does this say geometrically (where $y(t)$ is the second component of $\mathbf{X}(t)$)?

12. If $\mathbf{I}$ is the $n \times n$ identity matrix and r is any constant, show that $e^{r\mathbf{I}} = e^r \mathbf{I}$ and $e^{rt\mathbf{I}} = e^{rt}\mathbf{I}$ using the infinite sum definition of the matrix exponential.

13. Using Definition 9.1.1, calculate $e^{t\mathbf{A}}$ if
    (a) $\mathbf{A} = \begin{pmatrix} 0 & 1 & 0 \\ 0 & 0 & 1 \\ 0 & 0 & 0 \end{pmatrix}$.
    (b) $\mathbf{A} = \begin{pmatrix} 0 & 0 & 0 \\ 1 & 0 & 0 \\ 2 & 1 & 0 \end{pmatrix}$.

14. (a) If
$$\mathbf{A} = \begin{pmatrix} 0 & -1 \\ 1 & 0 \end{pmatrix},$$
then compute $\mathbf{A}^n$ for $n = 2, 3, 4, 5, \ldots$.
    (b) Use Definition 9.1.1 to compute $e^{t\mathbf{A}}$ and $e^{t\mathbf{B}}$ where
$$\mathbf{B} = \begin{pmatrix} 0 & -b \\ b & 0 \end{pmatrix}.$$
    (c) Write
$$\mathbf{C} = \begin{pmatrix} a & -b \\ b & a \end{pmatrix}$$
as $\mathbf{C} = a\mathbf{I} + \mathbf{B}$ and compute $e^{t\mathbf{C}}$, using your answer from (b).

15. Suppose
$$\mathbf{A} = \begin{pmatrix} 2 & 2 \\ 0 & 0 \end{pmatrix} \text{ and } \mathbf{B} = \begin{pmatrix} 1 & -1 \\ 0 & 0 \end{pmatrix}.$$
Compute $\mathbf{AB}$, $\mathbf{BA}$, $e^{\mathbf{A}}e^{\mathbf{B}}$, $e^{\mathbf{B}}e^{\mathbf{A}}$, and $e^{\mathbf{A}+\mathbf{B}}$. This example shows that $e^{\mathbf{A}}e^{\mathbf{B}}$ need not be equal to $e^{\mathbf{A}+\mathbf{B}}$ if $\mathbf{A}$ and $\mathbf{B}$ do not commute.

---

[1] For a proof of the results given in this problem, see John Sanders, *An easy way to determine the sense of rotation of phase plane trajectories*, Pi Mu Epsilon Journal, vol. 13, Spring 2013, 491–494. The author wrote this article while a student in an introductory differential equations course.

16. Consider the matrix and vector functions
$$\mathbf{B} = \begin{pmatrix} f_1(t) & f_2(t) \\ f_3(t) & f_4(t) \end{pmatrix}, \quad \mathbf{C} = \begin{pmatrix} g_1(t) & g_2(t) \\ g_3(t) & g_4(t) \end{pmatrix}, \quad \mathbf{X} = \begin{pmatrix} h_1(t) \\ h_2(t) \end{pmatrix}.$$
Verify the differentiation rules:

(a)
$$\frac{d}{dt}[\mathbf{B}(t) + \mathbf{C}(t)] = \mathbf{B}'(t) + \mathbf{C}'(t).$$

(b)
$$\frac{d}{dt}[\mathbf{B}(t)\mathbf{C}(t)] = \mathbf{B}'(t)\mathbf{C}(t) + \mathbf{B}(t)\mathbf{C}'(t).$$

(c)
$$\frac{d}{dt}[\mathbf{B}(t)\mathbf{X}(t)] = \mathbf{B}'(t)\mathbf{X}(t) + \mathbf{B}(t)\mathbf{X}'(t).$$

17. Suppose that
$$\mathbf{A}(t) = \begin{pmatrix} \cos t & \sin t \\ -\sin t & t - \cos t \end{pmatrix}.$$
Is
$$\frac{d}{dt}(\mathbf{A}(t)^2) = 2\mathbf{A}(t)\frac{d}{dt}(\mathbf{A}(t))?$$
If not, find a correct formula for $\frac{d}{dt}(\mathbf{A}(t)^2)$.

18. (a) Suppose
$$\mathbf{A}(t) = \begin{pmatrix} t & 0 \\ 2t & t^2 \end{pmatrix}.$$
Compute $\mathbf{A}(t)^{-1}$ for any $t \neq 0$, and then compute $\frac{d}{dt}(\mathbf{A}(t)^{-1})$.

(b) When $\mathbf{A}$ is any $n \times n$ differentiable and invertible matrix-valued function, then $\mathbf{A}(t)^{-1}$ is also differentiable. Find an expression for $\frac{d}{dt}\mathbf{A}(t)^{-1}$ by differentiating the identity $\mathbf{A}(t)\mathbf{A}(t)^{-1} = \mathbf{I}$, using the product rule from Theorem 9.1.3.

(c) Verify your formula from (b) for the matrix in (a).

19. If
$$e^{t\mathbf{A}} = \begin{pmatrix} e^t & 0 \\ -e^{-t} + e^t & e^{-t} \end{pmatrix},$$
compute $\frac{d}{dt}(e^{t\mathbf{A}})$ and then use it to find $\mathbf{A}$. Hint: What is $\frac{d}{dt}(e^{t\mathbf{A}})$ evaluated at $t = 0$?

20. Prove the Cayley-Hamilton Theorem for 2×2 matrices: If
$$\mathbf{A} = \begin{pmatrix} a & b \\ c & d \end{pmatrix}$$
and $p(z) = z^2 - (a+d)z + (ad - bc)$, show $p(\mathbf{A}) = \begin{pmatrix} 0 & 0 \\ 0 & 0 \end{pmatrix}$.

21. Verify that the Cayley-Hamilton Theorem for the matrix
$$\mathbf{A} = \begin{pmatrix} 1 & 0 & 0 \\ 2 & 3 & 0 \\ 0 & 1 & 1 \end{pmatrix}.$$

22. Suppose that $\mathbf{A}$ is any 3×3 upper triangular matrix
$$\mathbf{A} = \begin{pmatrix} a & b & c \\ 0 & d & e \\ 0 & 0 & f \end{pmatrix}.$$

Verify that the Cayley-Hamilton Theorem holds for $\mathbf{A}$. Hint: It is enough to show that $(\mathbf{A} - a\mathbf{I})(\mathbf{A} - d\mathbf{I})(\mathbf{A} - f\mathbf{I}) = \mathbf{0}$. Why?

23. Suppose that $\mathbf{A}$ is a 3×3 matrix with eigenvalues r_1, r_2, and r_3.
    (a) Use the Cayley-Hamilton Theorem to show that any nonzero column of the matrix $(\mathbf{A} - r_2\mathbf{I})(\mathbf{A} - r_3\mathbf{I})$ is an eigenvector for $\mathbf{A}$ with corresponding eigenvalue r_1. Hint: What is $(\mathbf{A} - r_1\mathbf{I})(\mathbf{A} - r_2\mathbf{I})(\mathbf{A} - r_3\mathbf{I})$?
    (b) Suppose that $\mathbf{W}$ is any vector for which $(\mathbf{A} - r_2\mathbf{I})(\mathbf{A} - r_3\mathbf{I})\mathbf{W}$ is nonzero. Show that $(\mathbf{A} - r_2\mathbf{I})(\mathbf{A} - r_3\mathbf{I})\mathbf{W}$ is an eigenvector of $\mathbf{A}$ with eigenvalue r_1.

24. Suppose the 2×2 matrix $\mathbf{A}$ has complex eigenvalues, $r_1 = \lambda + i\mu$ and $r_2 = \lambda - i\mu$ with $\lambda \neq 0$ and $\mu \neq 0$, so that by equation (9.17)
$$e^{t\mathbf{A}} = \left(\frac{r_1 e^{r_2 t} - r_2 e^{r_1 t}}{r_1 - r_2}\right)\mathbf{I} + \left(\frac{e^{r_1 t} - e^{r_2 t}}{r_1 - r_2}\right)\mathbf{A}.$$
Verify that this can be written as
$$e^{t\mathbf{A}} = e^{\lambda t}\left(\cos(\mu t)\mathbf{I} + \sin(\mu t)\frac{1}{\mu}(\mathbf{A} - \lambda\mathbf{I})\right).$$

25. Suppose $\mathbf{D}$ is a 2×2 diagonal matrix, with diagonal entries λ_1 and λ_2.
    (a) Show that $\mathbf{D}^k$ is a diagonal matrix for each positive integer k and determine the diagonal entries.
    (b) Using (a) and the series definition of the matrix exponential, determine $e^{t\mathbf{D}}$.
    (c) Extend the results of (a) and (b) to diagonal matrices of size $n \times n$.
    (d) Suppose $\mathbf{D}$ is a diagonal matrix and $\mathbf{A}$ is any invertible matrix of the same size. Set $\mathbf{B} = \mathbf{A}\mathbf{D}\mathbf{A}^{-1}$. Show that
$$e^{t\mathbf{B}} = \mathbf{A}(e^{t\mathbf{D}})\mathbf{A}^{-1}.$$
    Hint: Check that $\mathbf{B}^n = \mathbf{A}\mathbf{D}^n\mathbf{A}^{-1}$. Use the infinite sum definition of $e^{t\mathbf{B}}$ and factor out an $\mathbf{A}$ on the left and an $\mathbf{A}^{-1}$ on the right.

26. The currents I_1, I_2, and I_3 flowing in the three branches of the circuit pictured below right are modeled by the algebraic equation $I_1 = I_2 + I_3$ (obtained through considering current flow entering and leaving node a) and the differential equations

$$(9.20) \quad \begin{array}{l} \frac{dI_1}{dt} = -12I_1 + 12I_2, \\ \frac{dI_2}{dt} = -3I_1. \end{array}$$

   (a) Write the system (9.20) in matrix form $\mathbf{I}'(t) = \mathbf{A}\mathbf{I}$, where $\mathbf{I} = \begin{pmatrix} I_1 \\ I_2 \end{pmatrix}$, identifying $\mathbf{A}$.
   (b) Find $e^{t\mathbf{A}}$.
   (c) Solve the initial value problem consisting of the system (9.20) with initial conditions $I_1(0) = -4$, $I_2(0) = 0$.
   (d) (CAS) Graph the three currents I_1, I_2, I_3, for $0 \leq t \leq 2$, on the same coordinate grid.
   (e) What is the maximum value of the I_2 over $(0, \infty)$?

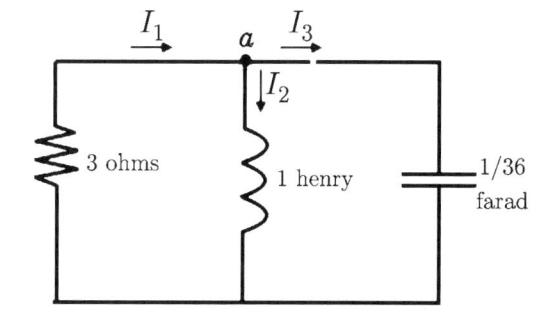

27. Let $\mathbf{A}, \mathbf{B}$, and $\mathbf{C}$ be $n \times n$ matrices.
    (a) Show that if $\mathbf{A}\mathbf{X} = \mathbf{0}$ for all n-dimensional vectors $\mathbf{X}$, then $\mathbf{A}$ is the zero matrix.
    (b) Show that if $\mathbf{B}\mathbf{X} = \mathbf{C}\mathbf{X}$ for all n-dimensional vectors $\mathbf{X}$, then $\mathbf{B} = \mathbf{C}$.

28. In this problem, let $\mathbf{B}$ and $\mathbf{C}$ be the particular matrices computed in Example 9.1.7.

9.1. The matrix exponential, Part I

(a) Check that $\mathbf{BC} = \mathbf{CB} = \mathbf{0}$.
(b) Check that $\mathbf{B} + \mathbf{C} = \mathbf{I}$.
(c) Check that $\mathbf{B}^2 = \mathbf{B}$ and $\mathbf{C}^2 = \mathbf{C}$.
(d) Consider the vectors
$$\mathbf{X}_0 = \begin{pmatrix} 6 \\ 4 \end{pmatrix}, \quad \mathbf{X}_1 = \begin{pmatrix} 8 \\ 6 \end{pmatrix}.$$
Draw a picture of the 4-sided figure with corners at $\mathbf{0}$, $\mathbf{BX}_0$, $\mathbf{CX}_0$, and $\mathbf{X}_0$ and a picture of the figure with corners at $\mathbf{0}$, $\mathbf{BX}_1$, $\mathbf{CX}_1$, and $\mathbf{X}_1$. What kind of figures are these?
(e) Add to your pictures for part (d) the eigenvector lines of the matrix $\mathbf{A}$ discussed in Example 9.1.7. Describe in geometric terms what the operations "multiply a vector by $\mathbf{B}$" and "multiply a vector by $\mathbf{C}$" do to the vector.

29. Let $\mathbf{A}$ be a 2×2 matrix with two distinct eigenvalues r_1 and r_2, and let $\mathbf{B}$ and $\mathbf{C}$ be the matrices defined in equations (9.9) and (9.10).
    (a) Show that $\mathbf{BC} = \mathbf{0} = \mathbf{CB}$.
    (b) Show that $\mathbf{B}^2 = \mathbf{B}$ and $\mathbf{C}^2 = \mathbf{C}$. Hint: Multiply equation (9.15) by $\mathbf{B}$. Then multiply it by $\mathbf{C}$.

30. As in Exercise 29, suppose that $\mathbf{A}$ is a 2×2 matrix with distinct eigenvalues r_1 and r_2, and let $\mathbf{B}$ and $\mathbf{C}$ be the matrices defined in equations (9.9) and (9.10).
    (a) Show that $\mathbf{A} = r_1\mathbf{B} + r_2\mathbf{C}$. Hint: $\mathbf{A} = \mathbf{AI}$ and $\mathbf{I} = \mathbf{B} + \mathbf{C}$.
    (b) Show that for any positive integer n, $\mathbf{A}^n = r_1^n\mathbf{B} + r_2^n\mathbf{C}$.
    (c) For any polynomial $p(z) = c_m z^m + c_{m-1} z^{m-1} + \cdots + c_2 z^2 + c_1 z + c_0$ and any square matrix $\mathbf{S}$, recall that $p(\mathbf{S})$ is defined by
    $$p(\mathbf{S}) = c_m \mathbf{S}^m + c_{m-1} \mathbf{S}^{m-1} + \cdots + c_2 \mathbf{S}^2 + c_1 \mathbf{S} + c_0 \mathbf{I}.$$
    Show that for the matrix $\mathbf{A}$, $p(\mathbf{A}) = p(r_1)\mathbf{B} + p(r_2)\mathbf{C}$.

31. As in Exercises 29 and 30, let $\mathbf{A}$ be a 2×2 matrix with distinct eigenvalues r_1 and r_2, and let $\mathbf{B}$ and $\mathbf{C}$ be the matrices defined by equations (9.9) and (9.10).
    (a) Given numbers b_1, c_1, b_2, and c_2, define the matrices $\mathbf{S} = b_1 \mathbf{B} + c_1 \mathbf{C}$ and $\mathbf{T} = b_2 \mathbf{B} + c_2 \mathbf{C}$. Show that
    $$\mathbf{ST} = b_1 b_2 \mathbf{B} + c_1 c_2 \mathbf{C} = \mathbf{TS}.$$
    Hint: Use Exercise 29.
    (b) Suppose that neither b_1 nor c_1 is equal to 0. Show that the matrix $\mathbf{S}$ in part (a) is invertible and
    $$\mathbf{S}^{-1} = \frac{1}{b_1}\mathbf{B} + \frac{1}{c_1}\mathbf{C}.$$
    (c) Show that b_1 and c_1 are eigenvalues of $\mathbf{S}$.
    (d) If λ is a real or complex number different from both b_1 and c_1, show that $\mathbf{S} - \lambda\mathbf{I}$ is invertible and
    $$(\mathbf{S} - \lambda\mathbf{I})^{-1} = \frac{1}{b_1 - \lambda}\mathbf{B} + \frac{1}{c_1 - \lambda}\mathbf{C}.$$
    Hint: In writing $\mathbf{S} - \lambda\mathbf{I}$, recall that $\mathbf{I} = \mathbf{B} + \mathbf{C}$.

32. Suppose that $\mathbf{A}$ is an $n \times n$ matrix and $p(z) = c_m z^m + c_{m-1} z^{m-1} + \cdots + c_1 z + c_0$ is a polynomial, so that
$$p(\mathbf{A}) = c_m \mathbf{A}^m + c_{m-1} \mathbf{A}^{m-1} + \cdots + c_2 \mathbf{A}^2 + c_1 \mathbf{A} + c_0 \mathbf{I}.$$
We know that if r is an eigenvalue of $\mathbf{A}$ with corresponding eigenvector $\mathbf{V}$, then for every positive integer k
$$\mathbf{A}^k \mathbf{V} = r^k \mathbf{V},$$

so that r^k is an eigenvalue of $\mathbf{A}^k$, also with eigenvalue $\mathbf{V}$. Show that $p(r)$ is an eigenvalue of $p(\mathbf{A})$, with eigenvector $\mathbf{V}$.

9.2. A return to the Existence and Uniqueness Theorem

In Section 8.1 we stated an "existence and uniqueness" theorem (Theorem 8.1.1) for a general linear system. Armed with what we have since learned, we can now prove this theorem in the special case of a constant coefficient homogeneous linear initial value problem

(9.21) $$\mathbf{X}'(t) = \mathbf{A}\mathbf{X}(t), \quad \mathbf{X}(0) = \mathbf{X}_0,$$

where $\mathbf{A}$ is any $n \times n$ matrix of constants. The goal is to show that (9.21) has one and only one solution. Assembling the relevant facts, we recall the following:

(a) The $n \times n$ matrix function $e^{t\mathbf{A}}$ satisfies $\frac{d}{dt}\left(e^{t\mathbf{A}}\right) = \mathbf{A}e^{t\mathbf{A}} = e^{t\mathbf{A}}\mathbf{A}$.

(b) The $n \times 1$ vector function $\mathbf{X}_1(t) = e^{t\mathbf{A}}\mathbf{X}_0$ is a solution to the initial value problem (9.21), which is valid on $-\infty < t < \infty$.

By virtue of (b) the *existence* of a solution to (9.21) is already known. What about uniqueness? Could there be more than one solution? Suppose $\mathbf{X}_2(t)$ is a second solution to (9.21), so that $\mathbf{X}_2' = \mathbf{A}\mathbf{X}_2$ and $\mathbf{X}_2(0) = \mathbf{X}_0$. Consider the vector function

(9.22) $$\mathbf{Y}(t) = e^{-t\mathbf{A}}\mathbf{X}_2(t).$$

By the product rule in part (c) of Theorem 9.1.3,

$$\mathbf{Y}'(t) = -\mathbf{A}e^{-t\mathbf{A}}\mathbf{X}_2(t) + e^{-t\mathbf{A}}\mathbf{X}_2'(t) = -\mathbf{A}e^{-t\mathbf{A}}\mathbf{X}_2(t) + e^{-t\mathbf{A}}\mathbf{A}\mathbf{X}_2(t) = \mathbf{0},$$

since $\mathbf{A}$ and $e^{-t\mathbf{A}}$ commute. This tells us that each entry of the vector function $\mathbf{Y}(t)$ must be a constant. To identify the values of these constant entries, we see from equation (9.22) that $\mathbf{Y}(0) = e^{\mathbf{0}}\mathbf{X}_2(0) = \mathbf{I}\mathbf{X}_2(0) = \mathbf{X}_0$. Thus, for all t,

$$\mathbf{Y}(t) = \mathbf{X}_0.$$

Substituting this in equation (9.22) we see that

$$\mathbf{X}_0 = e^{-t\mathbf{A}}\mathbf{X}_2(t).$$

Multiplying both sides of this equation, on the left, by the matrix $e^{t\mathbf{A}}$ we conclude that

$$e^{t\mathbf{A}}\mathbf{X}_0 = e^{t\mathbf{A}}e^{-t\mathbf{A}}\mathbf{X}_2(t) = \mathbf{X}_2(t),$$

since $e^{t\mathbf{A}}e^{-t\mathbf{A}} = \mathbf{I}$ (see Exercise 7). Thus our presumed second solution, $\mathbf{X}_2$, is actually the same as the first solution, $\mathbf{X}_1(t) = e^{t\mathbf{A}}\mathbf{X}_0$. We summarize our conclusions in the next result.

Theorem 9.2.1. *If $\mathbf{A}$ is an $n \times n$ matrix of constants, the linear homogeneous initial value problem*

$$\mathbf{X}'(t) = \mathbf{A}\mathbf{X}(t), \quad \mathbf{X}(0) = \mathbf{X}_0,$$

has the unique solution $\mathbf{X}(t) = e^{t\mathbf{A}}\mathbf{X}_0$.

As a first application of the preceding theorem, observe that if $\mathbf{X} = \mathbf{X}_*$ is any solution to $\mathbf{X}' = \mathbf{A}\mathbf{X}$ and we set $\mathbf{X}_0 = \mathbf{X}_*(0)$, then the uniqueness part of the theorem assures us that $\mathbf{X}_* = e^{t\mathbf{A}}\mathbf{X}_0$. We conclude that if $\mathbf{A}$ is an $n \times n$ matrix of constants, then the general solution of $\mathbf{X}' = \mathbf{A}\mathbf{X}$ may be expressed as $\mathbf{X}(t) = e^{t\mathbf{A}}\mathbf{C}$, where $\mathbf{C}$ is an $n \times 1$ matrix whose entries $c_1, c_2, \ldots, c_n$ are arbitrary constants. The uniqueness part of Theorem 9.2.1 also can be used to establish uniqueness of the solution to a nonhomogeneous initial value problem $\mathbf{X}'(t) = \mathbf{A}\mathbf{X}(t) + \mathbf{G}(t)$, $\mathbf{X}(0) = \mathbf{X}_0$; see Exercise 9.

9.2. A return to the Existence and Uniqueness Theorem

Fundamental matrices. Suppose we have found n solutions $\mathbf{X}_1(t), \mathbf{X}_2(t), \ldots, \mathbf{X}_n(t)$ to the system $\mathbf{X}' = \mathbf{A}\mathbf{X}$, where $\mathbf{A}$ is an $n \times n$ matrix. We know that if the matrix

$$(9.23) \qquad \begin{pmatrix} \uparrow & \uparrow & & \uparrow \\ \mathbf{X}_1(t) & \mathbf{X}_2(t) & \cdots & \mathbf{X}_n(t) \\ \downarrow & \downarrow & & \downarrow \end{pmatrix}$$

has nonzero determinant at *one* value of t, then it has nonzero determinant at *all* values of t (see the discussion of the Wronskian following Theorem 8.4.4). In this case the matrix in equation (9.23) is called a **fundamental matrix** for the system $\mathbf{X}' = \mathbf{A}\mathbf{X}$; its columns form a fundamental set of solutions to $\mathbf{X}' = \mathbf{A}\mathbf{X}$ on $(-\infty, \infty)$. From a fundamental matrix we can write the general solution to $\mathbf{X}' = \mathbf{A}\mathbf{X}$ as $c_1\mathbf{X}_1(t) + c_2\mathbf{X}_2(t) + \cdots + c_n\mathbf{X}_n(t)$.

Example 9.2.2. The matrix $\mathbf{Y}(t) = e^{t\mathbf{A}}$ is one example of a fundamental matrix for the system $\mathbf{X}' = \mathbf{A}\mathbf{X}$. To verify this, first note that each column of $e^{t\mathbf{A}}$ solves this system; specifically the solution to the initial value problem

$$\mathbf{X}' = \mathbf{A}\mathbf{X}, \quad \mathbf{X}(0) = \mathbf{E}_j,$$

where $\mathbf{E}_j$ is the column vector with a "1" in the jth row and 0's elsewhere, is $e^{t\mathbf{A}}\mathbf{E}_j$, which is the jth column of $e^{t\mathbf{A}}$. Moreover, at $t = 0$ we have $\mathbf{Y}(0) = e^{\mathbf{0}} = \mathbf{I}$ and $\det \mathbf{I} = 1$.

Building a fundamental matrix from eigenvalues/eigenvectors. Can we produce a fundamental matrix from an eigenvalue/eigenvector computation? The easiest situation is when $\mathbf{A}$ has n distinct eigenvalues $r_1, r_2, \ldots, r_n$, with corresponding eigenvectors $\mathbf{V}_1, \mathbf{V}_2, \ldots, \mathbf{V}_n$. For each j, $e^{r_j t}\mathbf{V}_j$ solves $\mathbf{X}' = \mathbf{A}\mathbf{X}$, and the matrix $\mathbf{Y}$ whose columns are

$$e^{r_1 t}\mathbf{V}_1, e^{r_2 t}\mathbf{V}_2, \ldots, e^{r_n t}\mathbf{V}_n$$

satisfies $\det \mathbf{Y}(0) \neq 0$. This follows because the vectors $\mathbf{V}_1, \mathbf{V}_2, \ldots, \mathbf{V}_n$ belong to distinct eigenvalues and are thus linearly independent, so that

$$\det \mathbf{Y}(0) = \det \begin{pmatrix} \uparrow & \uparrow & & \uparrow \\ \mathbf{V}_1 & \mathbf{V}_2 & \cdots & \mathbf{V}_n \\ \downarrow & \downarrow & & \downarrow \end{pmatrix} \neq 0$$

by Theorem 8.2.9. Thus $\mathbf{Y}$ is a fundamental matrix. When $\mathbf{A}$ has fewer than n distinct eigenvalues, the situation is more complicated, and $\mathbf{A}$ may not have a full set of n linearly independent eigenvectors.

Using a fundamental matrix to compute $e^{t\mathbf{A}}$. Remarkably, if we can produce a fundamental matrix, we can use it to compute the matrix $e^{t\mathbf{A}}$ directly. The next result describes how this is done. Note that if $\mathbf{Z}(t)$ is a fundamental matrix, then $\det \mathbf{Z}(0) \neq 0$ so that $\mathbf{Z}(0)^{-1}$ exists.

Theorem 9.2.3. *If $\mathbf{Z}(t)$ is any fundamental matrix for the system $\mathbf{X}' = \mathbf{A}\mathbf{X}$, then*

$$e^{t\mathbf{A}} = \mathbf{Z}(t)\mathbf{Z}(0)^{-1},$$

the product of $\mathbf{Z}(t)$ with its inverse at $t = 0$.

Proof. For $j = 1, 2, \ldots, n$, let $\mathbf{E}_j$ be the column vector with a 1 in the jth position and 0's elsewhere. By Theorem 9.2.1 we know that $e^{t\mathbf{A}}\mathbf{E}_j$ is the *unique* solution to the initial value problem $\mathbf{X}'(t) = \mathbf{A}\mathbf{X}(t), \mathbf{X}(0) = \mathbf{E}_j$. We claim that $\mathbf{Y}(t) = \mathbf{Z}(t)\mathbf{Z}(0)^{-1}\mathbf{E}_j$ also solves this initial value problem. Once the claim is verified, uniqueness tells us that

$$e^{t\mathbf{A}}\mathbf{E}_j = \mathbf{Z}(t)\mathbf{Z}(0)^{-1}\mathbf{E}_j$$

for each j, $1 \leq j \leq n$. By equation (8.25) in Section 8.2.2, this tells us that for each j, the jth column of $e^{t\mathbf{A}}$ is the same as the jth column of $\mathbf{Z}(t)\mathbf{Z}(0)^{-1}$, giving the desired conclusion that $e^{t\mathbf{A}} = \mathbf{Z}(t)\mathbf{Z}(0)^{-1}$.

Verifying the claim. Let j be an arbitrary integer between 1 and n. To verify the claim, observe first that $\mathbf{Y}(0) = \mathbf{Z}(0)\mathbf{Z}(0)^{-1}\mathbf{E}_j = \mathbf{E}_j$ and $\mathbf{Y}(t)$ satisfies the desired initial condition at $t = 0$. Thus we need only show that $\mathbf{Y}'(t) = \mathbf{A}\mathbf{Y}(t)$. Viewing $\mathbf{Y}(t)$ as the product of the $n \times n$ matrix function $\mathbf{Z}(t)$ and the $n \times 1$ constant vector $\mathbf{Z}^{-1}(0)\mathbf{E}_j$, we compute $\mathbf{Y}'(t) = \mathbf{Z}'(t)\mathbf{Z}(0)^{-1}\mathbf{E}_j$ using the differentiation rule in Theorem 9.1.3(c). If $\mathbf{Z}_j$ denotes the jth column of $\mathbf{Z}$, then

$$\begin{aligned}\mathbf{A}\mathbf{Y}(t) &= \mathbf{A}\mathbf{Z}(t)\mathbf{Z}(0)^{-1}\mathbf{E}_j = \mathbf{A}\begin{pmatrix} \uparrow & \uparrow & & \uparrow \\ \mathbf{Z}_1 & \mathbf{Z}_2 & \cdots & \mathbf{Z}_n \\ \downarrow & \downarrow & & \downarrow \end{pmatrix}\mathbf{Z}(0)^{-1}\mathbf{E}_j \\ &= \begin{pmatrix} \uparrow & \uparrow & & \uparrow \\ \mathbf{A}\mathbf{Z}_1 & \mathbf{A}\mathbf{Z}_2 & \cdots & \mathbf{A}\mathbf{Z}_n \\ \downarrow & \downarrow & & \downarrow \end{pmatrix}\mathbf{Z}(0)^{-1}\mathbf{E}_j \\ &= \begin{pmatrix} \uparrow & \uparrow & & \uparrow \\ \mathbf{Z}'_1 & \mathbf{Z}'_2 & \cdots & \mathbf{Z}'_n \\ \downarrow & \downarrow & & \downarrow \end{pmatrix}\mathbf{Z}(0)^{-1}\mathbf{E}_j \\ &= \mathbf{Z}'(t)\mathbf{Z}(0)^{-1}\mathbf{E}_j = \mathbf{Y}'(t),\end{aligned}$$

where the penultimate equality uses the fact that, by the definition of "fundamental matrix", each column of $\mathbf{Z}$ satisfies $\mathbf{X}' = \mathbf{A}\mathbf{X}$. The claim is verified. $\square$

9.2.1. Commuting properties of $e^{t\mathbf{A}}$ and $e^{t\mathbf{B}}$ when A and B commute. If $\mathbf{A}$ and $\mathbf{B}$ are $n \times n$ matrices satisfying $\mathbf{A}\mathbf{B} = \mathbf{B}\mathbf{A}$, then we have the following nice properties of $e^{t\mathbf{A}}$ and $e^{t\mathbf{B}}$:

(9.24) $$e^{t\mathbf{A}+t\mathbf{B}} = e^{t\mathbf{A}}e^{t\mathbf{B}},$$

(9.25) $$e^{t\mathbf{A}}e^{t\mathbf{B}} = e^{t\mathbf{B}}e^{t\mathbf{A}}.$$

In particular, setting $t = 1$ says $e^{\mathbf{A}+\mathbf{B}} = e^{\mathbf{A}}e^{\mathbf{B}}$ and $e^{\mathbf{A}}e^{\mathbf{B}} = e^{\mathbf{B}}e^{\mathbf{A}}$ when $\mathbf{A}$ and $\mathbf{B}$ commute. Recall that in Exercise 15 of Section 9.1 you showed that these properties need not hold when $\mathbf{A}\mathbf{B} \neq \mathbf{B}\mathbf{A}$.

Equation (9.25) follows immediately from equation (9.24) since $t\mathbf{A} + t\mathbf{B} = t\mathbf{B} + t\mathbf{A}$, so that

$$e^{t\mathbf{A}}e^{t\mathbf{B}} = e^{t\mathbf{A}+t\mathbf{B}} = e^{t\mathbf{B}+t\mathbf{A}} = e^{t\mathbf{B}}e^{t\mathbf{A}},$$

where the first and last equality use equation (9.24).

To prove equation (9.24), observe that if $\mathbf{E}_j$ is the $n \times 1$ column vector with a "1" in the jth position and 0's elsewhere, then by Theorem 9.2.1, $e^{t(\mathbf{A}+\mathbf{B})}\mathbf{E}_j$, which is the same as $e^{t\mathbf{A}+t\mathbf{B}}\mathbf{E}_j$, is the unique solution to the initial value problem

$$\mathbf{X}'(t) = (\mathbf{A}+\mathbf{B})\mathbf{X}(t), \quad \mathbf{X}(0) = \mathbf{E}_j.$$

This holds for each j, $1 \leq j \leq n$. Using Theorem 9.1.3(b) we compute

$$\left(e^{t\mathbf{A}}e^{t\mathbf{B}}\right)' = (e^{t\mathbf{A}})'e^{t\mathbf{B}} + e^{t\mathbf{A}}(e^{t\mathbf{B}})' = \left(\mathbf{A}e^{t\mathbf{A}}\right)e^{t\mathbf{B}} + e^{t\mathbf{A}}\left(\mathbf{B}e^{t\mathbf{B}}\right).$$

In Exercise 11 you are asked to show that if $\mathbf{B}$ commutes with $\mathbf{A}$, then $\mathbf{B}$ commutes with $e^{t\mathbf{A}}$. Thus

$$\left(e^{t\mathbf{A}}e^{t\mathbf{B}}\right)' = \mathbf{A}e^{t\mathbf{A}}e^{t\mathbf{B}} + \mathbf{B}e^{t\mathbf{A}}e^{t\mathbf{B}} = (\mathbf{A}+\mathbf{B})e^{t\mathbf{A}}e^{t\mathbf{B}}.$$

Setting $\mathbf{Y}(t) = e^{t\mathbf{A}}e^{t\mathbf{B}}\mathbf{E}_j$, we see that by Theorem 9.1.3(c) that $\mathbf{Y}'(t) = (\mathbf{A}+\mathbf{B})e^{t\mathbf{A}}e^{t\mathbf{B}}\mathbf{E}_j = (\mathbf{A}+\mathbf{B})\mathbf{Y}$ and $\mathbf{Y}(0) = \mathbf{E}_j$. By the uniqueness part of Theorem 9.2.1, we must have

$$e^{t\mathbf{A}+t\mathbf{B}}\mathbf{E}_j = e^{t\mathbf{A}}e^{t\mathbf{B}}\mathbf{E}_j$$

9.2. A return to the Existence and Uniqueness Theorem

for each j, $1 \leq j \leq n$. This says (see equation (8.25) in Section 8.2.2) that for each j, the jth column of $e^{t\mathbf{A}+t\mathbf{B}}$ is the same as the jth column of $e^{t\mathbf{A}}e^{t\mathbf{B}}$, and hence we have equation (9.24).

9.2.2. Exercises.

1. (a) Find the eigenvalues and eigenvectors of
$$\mathbf{A} = \begin{pmatrix} 0 & 1 \\ 1 & 0 \end{pmatrix}$$
and then give a fundamental matrix for the system $\mathbf{X}' = \mathbf{AX}$.
(b) Use Theorem 9.2.3 to show that
$$e^{t\mathbf{A}} = \begin{pmatrix} \cosh t & \sinh t \\ \sinh t & \cosh t \end{pmatrix}.$$

2. Suppose that
$$\mathbf{A} = \begin{pmatrix} 3 & -2 & 0 \\ -1 & 3 & -2 \\ 0 & -1 & 3 \end{pmatrix}.$$
Given that $\mathbf{A}$ has the eigenvalue/eigenvector pairs
$$r_1 = 1,\ \mathbf{V}_1 = \begin{pmatrix} 2 \\ 2 \\ 1 \end{pmatrix},\quad r_2 = 5,\ \mathbf{V}_2 = \begin{pmatrix} -2 \\ 2 \\ -1 \end{pmatrix},\quad r_3 = 3,\ \mathbf{V}_3 = \begin{pmatrix} 2 \\ 0 \\ -1 \end{pmatrix}$$
and that
$$\begin{pmatrix} 2 & -2 & 2 \\ 2 & 2 & 0 \\ 1 & -1 & -1 \end{pmatrix}^{-1} = \begin{pmatrix} \frac{1}{8} & \frac{1}{4} & \frac{1}{4} \\ -\frac{1}{8} & \frac{1}{4} & -\frac{1}{4} \\ \frac{1}{4} & 0 & -\frac{1}{2} \end{pmatrix},$$
find $e^{t\mathbf{A}}$ using Theorem 9.2.3.

3. Suppose that
$$\mathbf{A} = \begin{pmatrix} 2 & 0 & 9 \\ 0 & 3 & 0 \\ 1 & 0 & 2 \end{pmatrix}.$$
Given that $\mathbf{A}$ has the eigenvalue/eigenvector pairs
$$r_1 = 3,\ \mathbf{V}_1 = \begin{pmatrix} 0 \\ 1 \\ 0 \end{pmatrix},\quad r_2 = -1,\ \mathbf{V}_2 = \begin{pmatrix} 3 \\ 0 \\ -1 \end{pmatrix},\quad r_3 = 5,\ \mathbf{V}_3 = \begin{pmatrix} 3 \\ 0 \\ 1 \end{pmatrix}$$
and that
$$\begin{pmatrix} 0 & 3 & 3 \\ 1 & 0 & 0 \\ 0 & -1 & 1 \end{pmatrix}^{-1} = \begin{pmatrix} 0 & 1 & 0 \\ \frac{1}{6} & 0 & -\frac{1}{2} \\ \frac{1}{6} & 0 & \frac{1}{2} \end{pmatrix},$$
find $e^{t\mathbf{A}}$ using Theorem 9.2.3.

4. Suppose $\mathbf{A}$ is a 2×2 matrix with eigenvalue $r_1 = 2$ and corresponding eigenvector $\binom{1}{1}$, and eigenvalue $r_2 = -3$ with eigenvector $\binom{3}{0}$. Which of the following is a fundamental matrix solution for $\mathbf{X}' = \mathbf{AX}$?

(a) $\begin{pmatrix} e^{2t} & 3e^{-3t} \\ e^{2t} & 0 \end{pmatrix}$. (b) $\begin{pmatrix} e^{2t} & e^{-3t} \\ e^{2t} & 0 \end{pmatrix}$. (c) $\begin{pmatrix} e^{2t} & 3e^{-3t}+e^{2t} \\ e^{2t} & 0 \end{pmatrix}$.
(d) Both (a) and (b). (e) None of the above.

5. Explain why

$$\begin{pmatrix} e^t & e^t + 3e^{-t} & e^{-t} \\ 2e^t & 2e^t - 3e^{-t} & -e^{-t} \\ e^t & e^t + 6e^{-t} & 2e^{-t} \end{pmatrix}$$

cannot be a fundamental matrix for any constant coefficient system $\mathbf{X}' = \mathbf{AX}$.

6. Let $\mathbf{A} = \begin{pmatrix} -4 & -1 \\ 9 & 2 \end{pmatrix}$. Find $e^{t\mathbf{A}}$, and then give the general solution of $\mathbf{X}' = \mathbf{AX}$.

7. (a) Suppose that $\mathbf{A}$ is an $n \times n$ matrix of constants. Use Theorem 9.1.3 to show that

$$\frac{d}{dt}(e^{t\mathbf{A}} e^{-t\mathbf{A}}) = \mathbf{0}.$$

   (b) Using (a), show that $e^{t\mathbf{A}} e^{-t\mathbf{A}} = \mathbf{I}$ and $e^{-t\mathbf{A}} e^{t\mathbf{A}} = \mathbf{I}$, so that $e^{t\mathbf{A}}$ is invertible, with inverse $e^{-t\mathbf{A}}$. In particular, $(e^{\mathbf{A}})^{-1} = e^{-\mathbf{A}}$.

8. Show that

$$\mathbf{X}(t) = e^{(t-t_0)\mathbf{A}} \mathbf{X}_0$$

solves the initial value problem $\mathbf{X}' = \mathbf{AX}$, $\mathbf{X}(t_0) = \mathbf{X}_0$.

9. Let $\mathbf{A}$ be an $n \times n$ matrix of constants and suppose I is an open interval containing 0. Show that if the nonhomogeneous initial value problem

$$\mathbf{X}'(t) - \mathbf{AX}(t) = \mathbf{G}(t), \quad \mathbf{X}(0) = \mathbf{X}_0,$$

has a solution on I, then the solution is unique. Hint: Use Theorem 9.2.1.

10. Suppose we have solution vectors $\mathbf{X}_j(t)$ to the initial value problems

$$\mathbf{X}' = \mathbf{AX}, \quad \mathbf{X}(0) = \mathbf{E}_j, \quad 1 \le j \le n,$$

where $\mathbf{E}_j$ is the $n \times 1$ column vector with a "1" in the jth row and 0's elsewhere. Form the $n \times n$ matrix $\mathbf{Y}$ whose jth column is $\mathbf{X}_j(t)$. What is the relationship between $\mathbf{Y}$ and $e^{t\mathbf{A}}$, and why?

11. Suppose $\mathbf{A}$ and $\mathbf{B}$ are $n \times n$ constant matrices.
   (a) Suppose $\mathbf{B}$ commutes with $\mathbf{A}$. Show that $\mathbf{B}$ commutes with $\mathbf{A}^2$, and then use mathematical induction to show that $\mathbf{B}$ commutes with every power $\mathbf{A}^k$ of $\mathbf{A}$ (where k is any positive integer).
   (b) Using (a) and the definition of the matrix exponential, show that if $\mathbf{B}$ commutes with $\mathbf{A}$, then $\mathbf{B}$ commutes with $e^{t\mathbf{A}}$.

9.3. The matrix exponential, Part II

In Section 9.1 we saw that for an $n \times n$ matrix $\mathbf{A}$, the initial value problem

$$\mathbf{X}' = \mathbf{AX}, \quad \mathbf{X}(0) = \mathbf{X}_0,$$

has solution $\mathbf{X}(t) = e^{t\mathbf{A}} \mathbf{X}_0$. We also saw how to determine $e^{t\mathbf{A}}$ in the case that $\mathbf{A}$ is any 2×2 matrix with real entries. Here we look at methods of finding $e^{t\mathbf{A}}$ for 3×3 and larger $\mathbf{A}$. Our "all-purpose" method will be Putzer's Algorithm, described in Section 9.3.3 below. Before discussing that, we look at methods that apply under special conditions on the eigenvalues of $\mathbf{A}$.

9.3. The matrix exponential, Part II

9.3.1. A single eigenvalue of multiplicity n. For a 2×2 matrix $\mathbf{A}$ with only one eigenvalue, we saw in Section 9.1 how to compute $e^{t\mathbf{A}}$ using the Cayley-Hamilton theorem. Our method generalizes to compute $e^{t\mathbf{A}}$ for an $n \times n$ matrix $\mathbf{A}$ with only one eigenvalue r. The characteristic equation is $(z - r)^n = 0$, and by the Cayley-Hamilton Theorem $(\mathbf{A} - r\mathbf{I})^n = \mathbf{0}$. We can write

$$\begin{aligned} e^{t\mathbf{A}} &= e^{t(\mathbf{A} - r\mathbf{I} + r\mathbf{I})} \\ &= e^{t(r\mathbf{I})} e^{t(\mathbf{A} - r\mathbf{I})}. \end{aligned}$$

Recall that for the second equality, we need to know that $t(r\mathbf{I})$ and $t(\mathbf{A} - r\mathbf{I})$ commute, but this is clear, since a scalar multiple of the identity commutes with any matrix. We also know (see Exercise 12 in Section 9.1) that $e^{t(r\mathbf{I})} = e^{rt}\mathbf{I}$. Moreover,

$$e^{t(\mathbf{A} - r\mathbf{I})} = \mathbf{I} + t(\mathbf{A} - r\mathbf{I}) + \frac{t^2}{2!}(\mathbf{A} - r\mathbf{I})^2 + \cdots + \frac{t^{n-1}}{(n-1)!}(\mathbf{A} - r\mathbf{I})^{n-1},$$

where the sum on the right-hand side is a *finite* sum, since we have already observed that $(\mathbf{A} - r\mathbf{I})^n = \mathbf{0}$ (and hence all higher powers of $\mathbf{A} - r\mathbf{I}$ are $\mathbf{0}$ too). Thus we can write $e^{t\mathbf{A}}$ as the finite sum

$$e^{t\mathbf{A}} = e^{rt} \sum_{k=0}^{n-1} \frac{t^k}{k!} (\mathbf{A} - r\mathbf{I})^k.$$

Here we use our usual conventions that $(\mathbf{A} - r\mathbf{I})^0 = \mathbf{I}$ and $0! = 1$. To use this formula in a particular example, the only computations that need to be done are the first $n - 1$ powers of $\mathbf{A} - r\mathbf{I}$.

Example 9.3.1. The matrix

$$\mathbf{A} = \begin{pmatrix} 4 & 0 & -1 \\ -2 & 2 & 3 \\ -2 & -2 & 6 \end{pmatrix}$$

has only one eigenvalue, $r = 4$, so that $(\mathbf{A} - 4\mathbf{I})^3 = \mathbf{0}$. We have

$$e^{t\mathbf{A}} = e^{4t}\left(\mathbf{I} + t(\mathbf{A} - 4\mathbf{I}) + \frac{t^2}{2}(\mathbf{A} - 4\mathbf{I})^2\right).$$

Since

$$\mathbf{A} - 4\mathbf{I} = \begin{pmatrix} 0 & 0 & -1 \\ -2 & -2 & 3 \\ -2 & -2 & 2 \end{pmatrix} \quad \text{and} \quad (\mathbf{A} - 4\mathbf{I})^2 = \begin{pmatrix} 2 & 2 & -2 \\ -2 & -2 & 2 \\ 0 & 0 & 0 \end{pmatrix},$$

this gives

$$e^{t\mathbf{A}} = e^{4t} \begin{pmatrix} 1 + t^2 & t^2 & -t - t^2 \\ -2t - t^2 & 1 - 2t - t^2 & 3t + t^2 \\ -2t & -2t & 1 + 2t \end{pmatrix}.$$

9.3.2. Distinct eigenvalues. At the other extreme, suppose that the $n \times n$ matrix has n distinct eigenvalues. In Section 9.1 we saw that for a 2×2 matrix $\mathbf{A}$ with two distinct eigenvalues r_1 and r_2,

$$e^{t\mathbf{A}} = e^{r_1 t}\mathbf{B} + e^{r_2 t}\mathbf{C},$$

where

$$\mathbf{B} = \frac{\mathbf{A} - r_2\mathbf{I}}{r_1 - r_2} \quad \text{and} \quad \mathbf{C} = \frac{\mathbf{A} - r_1\mathbf{I}}{r_2 - r_1}.$$

This generalizes to $n \times n$ matrices $\mathbf{A}$ with n distinct eigenvalues $r_1, r_2, \ldots, r_n$.

Here we'll concentrate on the 3×3 case, so that $\mathbf{A}$ now has distinct eigenvalues r_1, r_2, and r_3. The 3×3 analogues of the 2×2 matrices $\mathbf{B}$ and $\mathbf{C}$ above are

$$\mathbf{B} = \frac{(\mathbf{A} - r_2\mathbf{I})(\mathbf{A} - r_3\mathbf{I})}{(r_1 - r_2)(r_1 - r_3)}, \quad \mathbf{C} = \frac{(\mathbf{A} - r_1\mathbf{I})(\mathbf{A} - r_3\mathbf{I})}{(r_2 - r_1)(r_2 - r_3)},$$

and (we need a third one!)
$$\mathbf{D} = \frac{(\mathbf{A} - r_1\mathbf{I})(\mathbf{A} - r_2\mathbf{I})}{(r_3 - r_1)(r_3 - r_2)}.$$

We will see that

(9.26) $$e^{t\mathbf{A}} = e^{r_1 t}\mathbf{B} + e^{r_2 t}\mathbf{C} + e^{r_3 t}\mathbf{D}$$

and that $\mathbf{B}$, $\mathbf{C}$, and $\mathbf{D}$, like their 2×2 counterparts, have interesting properties. It is helpful to think of $\mathbf{B}$, $\mathbf{C}$, and $\mathbf{D}$ as associated with the eigenvalues r_1, r_2, and r_3, respectively, as suggested by equation (9.26). From this point of view, you can probably discern a pattern in the definitions of $\mathbf{B}$, $\mathbf{C}$, and $\mathbf{D}$.

The characteristic polynomial $p(z) = \det(\mathbf{A} - z\mathbf{I})$ can be factored as
$$p(z) = -(z - r_1)(z - r_2)(z - r_3).$$

(It is a fact that the leading coefficient of the characteristic polynomial is equal to 1 in even dimensions and -1 in odd dimensions.) By the Cayley-Hamilton Theorem,

(9.27) $$(\mathbf{A} - r_1\mathbf{I})(\mathbf{A} - r_2\mathbf{I})(\mathbf{A} - r_3\mathbf{I}) = -p(\mathbf{A}) = \mathbf{0}.$$

The factors $\mathbf{A} - r_1\mathbf{I}$, $\mathbf{A} - r_2\mathbf{I}$, and $\mathbf{A} - r_3\mathbf{I}$ commute, so the left-hand side of equation (9.27) can be written in any order. Thus from the definitions of $\mathbf{B}$, $\mathbf{C}$, and $\mathbf{D}$, we see that

$$(\mathbf{A} - r_1\mathbf{I})\mathbf{B} = \mathbf{0}, \quad (\mathbf{A} - r_2\mathbf{I})\mathbf{C} = \mathbf{0}, \quad \text{and} \quad (\mathbf{A} - r_3\mathbf{I})\mathbf{D} = \mathbf{0}.$$

Therefore (arguing exactly as in the 2×2 case; see equations (9.13) and (9.14)), for any vector $\mathbf{X}_0$ in $\mathbb{R}^3$

$$e^{t\mathbf{A}}(\mathbf{B}\mathbf{X}_0) = e^{r_1 t}(\mathbf{B}\mathbf{X}_0), \quad e^{t\mathbf{A}}(\mathbf{C}\mathbf{X}_0) = e^{r_2 t}(\mathbf{C}\mathbf{X}_0), \quad \text{and} \quad e^{t\mathbf{A}}(\mathbf{D}\mathbf{X}_0) = e^{r_3 t}(\mathbf{D}\mathbf{X}_0).$$

In the 2×2 case we saw that $\mathbf{B} + \mathbf{C} = \mathbf{I}$. Here the 3×3 analogue is

(9.28) $$\mathbf{B} + \mathbf{C} + \mathbf{D} = \mathbf{I}.$$

You can verify this by substituting the above definitions of $\mathbf{B}$, $\mathbf{C}$, and $\mathbf{D}$, multiplying out the products like $(\mathbf{A} - r_2\mathbf{I})(\mathbf{A} - r_3\mathbf{I})$ therein, getting a common denominator, and adding. But here is a slicker argument that also works in higher dimensions: Consider the polynomial

$$q(z) = \frac{(z - r_2)(z - r_3)}{(r_1 - r_2)(r_1 - r_3)} + \frac{(z - r_1)(z - r_3)}{(r_2 - r_1)(r_2 - r_3)} + \frac{(z - r_1)(z - r_2)}{(r_3 - r_1)(r_3 - r_2)}.$$

Notice that z is the only variable here and $q(z)$ has degree at most two. The polynomial $q(z) - 1$ also has degree at most two and takes the value zero at r_1, r_2, and r_3 (three distinct points!). A polynomial of exact degree two has at most two distinct zeros, and a polynomial of degree one has only one zero. Therefore $q(z) - 1$ must have degree zero; that is, it is *constant*. Since $q(z) - 1$ is zero at r_1, r_2, and r_3, the constant value must be zero, so $q(z) = 1$ for all z. It follows that $q(\mathbf{A}) = \mathbf{I}$, which is exactly equation (9.28).

Let's assemble the pieces as we did in the 2×2 case. For any three-dimensional vector $\mathbf{X}_0$,

$$\begin{aligned}
e^{t\mathbf{A}}\mathbf{X}_0 = e^{t\mathbf{A}}\mathbf{I}\mathbf{X}_0 &= e^{t\mathbf{A}}(\mathbf{B} + \mathbf{C} + \mathbf{D})\mathbf{X}_0 \\
&= e^{t\mathbf{A}}(\mathbf{B}\mathbf{X}_0) + e^{t\mathbf{A}}(\mathbf{C}\mathbf{X}_0) + e^{t\mathbf{A}}(\mathbf{D}\mathbf{X}_0) \\
&= e^{r_1 t}(\mathbf{B}\mathbf{X}_0) + e^{r_2 t}(\mathbf{C}\mathbf{X}_0) + e^{r_3 t}(\mathbf{D}\mathbf{X}_0) \\
&= (e^{r_1 t}\mathbf{B} + e^{r_2 t}\mathbf{C} + e^{r_3 t}\mathbf{D})\mathbf{X}_0.
\end{aligned}$$

9.3. The matrix exponential, Part II

(Note that the second equality uses equation (9.28).) Since this is true for any choice of $\mathbf{X}_0$, the corresponding matrices multiplying $\mathbf{X}_0$ must be equal (see Exercise 27 in Section 9.1); that is, our desired equation (9.26) is true.

Example 9.3.2. Let's see formula (9.26) in action. Consider the matrix

$$\mathbf{A} = \begin{pmatrix} -2 & 3 & 1 \\ -2 & 4 & 0 \\ -1 & 3 & 0 \end{pmatrix}.$$

Its characteristic polynomial, det $(\mathbf{A} - z\mathbf{I})$, is $-z^3 + 2z^2 + z - 2 = (-z^2+1)(z-2)$, so that $\mathbf{A}$ has eigenvalues $r_1 = 2$, $r_2 = 1$, and $r_3 = -1$. Since these are distinct, we can use equation (9.26) to compute $e^{t\mathbf{A}}$ as follows:

$$e^{t\mathbf{A}} = e^{2t}\mathbf{B} + e^{t}\mathbf{C} + e^{-t}\mathbf{D}, \tag{9.29}$$

where

$$\mathbf{B} = \frac{(\mathbf{A}-\mathbf{I})(\mathbf{A}+\mathbf{I})}{3} = \begin{pmatrix} -4/3 & 3 & -2/3 \\ -4/3 & 3 & -2/3 \\ -4/3 & 3 & -2/3 \end{pmatrix}, \quad \mathbf{C} = \frac{(\mathbf{A}-2\mathbf{I})(\mathbf{A}+\mathbf{I})}{(-2)} = \begin{pmatrix} 3/2 & -3 & 3/2 \\ 1 & -2 & 1 \\ 3/2 & -3 & 3/2 \end{pmatrix},$$

and

$$\mathbf{D} = \frac{(\mathbf{A}-2\mathbf{I})(\mathbf{A}-\mathbf{I})}{6} = \begin{pmatrix} 5/6 & 0 & -5/6 \\ 1/3 & 0 & -1/3 \\ -1/6 & 0 & 1/6 \end{pmatrix}.$$

Substituting into (9.29) gives

$$e^{t\mathbf{A}} = e^{2t}\begin{pmatrix} -4/3 & 3 & -2/3 \\ -4/3 & 3 & -2/3 \\ -4/3 & 3 & -2/3 \end{pmatrix} + e^{t}\begin{pmatrix} 3/2 & -3 & 3/2 \\ 1 & -2 & 1 \\ 3/2 & -3 & 3/2 \end{pmatrix} + e^{-t}\begin{pmatrix} 5/6 & 0 & -5/6 \\ 1/3 & 0 & -1/3 \\ -1/6 & 0 & 1/6 \end{pmatrix}.$$

9.3.3. Putzer's Algorithm. Here we will give a different approach for determining $e^{t\mathbf{A}}$ which can be used whenever the eigenvalues of the $n \times n$ matrix $\mathbf{A}$ can be found, regardless of whether or not some of the eigenvalues are repeated. This method, called **Putzer's Algorithm**, will give an expression for $e^{t\mathbf{A}}$ as a sum of n terms

$$e^{t\mathbf{A}} = p_1(t)\mathbf{I} + p_2(t)\mathbf{M}_1 + p_3(t)\mathbf{M}_2 + \cdots + p_n(t)\mathbf{M}_{n-1}$$

where each $\mathbf{M}_j$ is an $n \times n$ matrix of constants and each $p_j(t)$ is a function of t. Putzer's Algorithm, described below, will tell us how to get the $\mathbf{M}_j$'s and $p_j(t)$'s. To use this algorithm, we have to be able to determine the eigenvalues of $\mathbf{A}$ and identify the multiplicity of each eigenvalue. For example, if $\mathbf{A}$ has characteristic equation $-(z-1)^2(z+4)^3 = 0$, then the eigenvalues of $\mathbf{A}$ are 1, with multiplicity 2, and -4, with multiplicity 3. To show the multiplicities, we list the eigenvalues as $1, 1, -4, -4, -4$. Notice that in this case the matrix $\mathbf{A}$ must be of size 5×5. Why? It doesn't matter in what order you list the eigenvalues, but you must list each eigenvalue the same number of times as its multiplicity.

After describing Putzer's Algorithm, we'll show some examples of its application and then discuss why it works. Here is *how* it works.

Putzer's Algorithm: How it works. Starting with an $n \times n$ matrix $\mathbf{A}$, our goal is to write

$$e^{t\mathbf{A}} = p_1(t)\mathbf{I} + p_2(t)\mathbf{M}_1 + p_3(t)\mathbf{M}_2 + \cdots + p_n(t)\mathbf{M}_{n-1} \tag{9.30}$$

for some matrices $\mathbf{M}_j$ and functions $p_j(t)$. Recall that the characteristic equation $\det(\mathbf{A} - z\mathbf{I}) = 0$ may be written in the form $(-1)^n(z-r_1)(z-r_2)\cdots(z-r_n) = 0$, where the r_j's are the eigenvalues of **A** (and which could be real or complex, with repeats allowed), so that the Cayley-Hamilton Theorem assures us that

(9.31) $$(\mathbf{A} - r_1\mathbf{I})(\mathbf{A} - r_2\mathbf{I}) \cdots (\mathbf{A} - r_n\mathbf{I}) = \mathbf{0}.$$

Step 1. Find the eigenvalues of A. Determine the eigenvalues of **A** and list them (in any order) according to their multiplicities: $r_1, r_2, \ldots, r_n$.

Step 2. Find the matrices $\mathbf{M}_j$. For $1 \leq j \leq n-1$, define matrices $\mathbf{M}_j$ by

$$\begin{aligned} \mathbf{M}_1 &= (\mathbf{A} - r_1\mathbf{I}), \\ \mathbf{M}_2 &= (\mathbf{A} - r_1\mathbf{I})(\mathbf{A} - r_2\mathbf{I}), \\ &\vdots \\ \mathbf{M}_{n-1} &= (\mathbf{A} - r_1\mathbf{I})(\mathbf{A} - r_2\mathbf{I}) \cdots (\mathbf{A} - r_{n-1}\mathbf{I}), \end{aligned}$$

so that in general

$$\mathbf{M}_j = (\mathbf{A} - r_1\mathbf{I})(\mathbf{A} - r_2\mathbf{I}) \cdots (\mathbf{A} - r_j\mathbf{I})$$

for $1 \leq j \leq n-1$. Notice that if we continued and used the formula for $\mathbf{M}_j$ with $j = n$, we would get $\mathbf{M}_n = \mathbf{0}$ by equation (9.31).

Step 3. Find the functions $p_j(t)$. Each function $p_1(t), p_2(t), \ldots, p_n(t)$ will be the solution of a certain first-order linear initial value problem. The function $p_1(t)$ satisfies

$$p_1'(t) = r_1 p_1(t), \quad p_1(0) = 1,$$

so that

$$p_1(t) = e^{r_1 t}.$$

The function $p_2(t)$ is the solution to the initial value problem

$$p_2'(t) = r_2 p_2(t) + p_1(t), \quad p_2(0) = 0.$$

Since we have $p_1(t) = e^{r_1 t}$, this is the same as

$$p_2'(t) = r_2 p_2(t) + e^{r_1 t}, \quad p_2(0) = 0,$$

and we determine p_2 by solving this first-order linear equation. It has integrating factor $e^{-r_2 t}$, as you should check. The remaining functions $p_3(t), \ldots, p_n(t)$ are similarly determined by solving the first-order linear initial value problems

$$\begin{aligned} p_3'(t) &= r_3 p_3(t) + p_2(t), \quad p_3(0) = 0, \\ p_4'(t) &= r_4 p_4(t) + p_3(t), \quad p_4(0) = 0, \\ &\vdots \\ p_n'(t) &= r_n p_n(t) + p_{n-1}(t), \quad p_n(0) = 0, \end{aligned}$$

one after the other. To solve the equation

$$p_k'(t) = r_k p_k(t) + p_{k-1}(t)$$

for $p_k(t)$, first plug in your answer for $p_{k-1}(t)$ from the previous equation.

Step 4. Assemble the pieces to give $e^{t\mathbf{A}}$. Using the computed values of $\mathbf{M}_j$ and $p_j(t)$, the matrix exponential is obtained from equation (9.30).

9.3. The matrix exponential, Part II

This completes the description of the Putzer Algorithm. Let's illustrate it with some examples.

Example 9.3.3. Find $e^{t\mathbf{A}}$ by Putzer's Algorithm, and then use it to solve
$$\mathbf{X}' = \mathbf{A}\mathbf{X}, \quad \mathbf{X}(0) = \begin{pmatrix} 1 \\ 0 \\ 2 \end{pmatrix},$$
where
$$\mathbf{A} = \begin{pmatrix} 3 & 1 & -1 \\ -1 & 1 & 2 \\ 0 & 0 & 1 \end{pmatrix}.$$

The characteristic equation, $\det(\mathbf{A} - z\mathbf{I}) = 0$, is $(1-z)(2-z)(2-z) = 0$. The eigenvalues of $\mathbf{A}$, listed according to their multiplicities, are $r_1 = 1, r_2 = 2$, and $r_3 = 2$. (Any ordering can be used, for example, $r_1 = 2, r_2 = 1, r_3 = 2$; this will lead to different $\mathbf{M}_j$ and $p_j(t)$, but the same final answer for $e^{t\mathbf{A}}$.)

By Step 2 of the Putzer Algorithm we have
$$\mathbf{M}_1 = \mathbf{A} - \mathbf{I} = \begin{pmatrix} 2 & 1 & -1 \\ -1 & 0 & 2 \\ 0 & 0 & 0 \end{pmatrix}$$
and
$$\mathbf{M}_2 = (\mathbf{A} - \mathbf{I})(\mathbf{A} - 2\mathbf{I}) = \begin{pmatrix} 1 & 1 & 1 \\ -1 & -1 & -1 \\ 0 & 0 & 0 \end{pmatrix}.$$

Implementing Step 3, we have
$$p_1(t) = e^t.$$
Next we solve
$$p_2'(t) = 2p_2(t) + e^t, \quad p_2(0) = 0,$$
to get
$$p_2(t) = -e^t + e^{2t}.$$
Finally, for p_3 we solve
$$p_3'(t) = 2p_3(t) + p_2(t), \quad p_3(0) = 0,$$
or, inserting our solution for $p_2(t)$,
$$p_3'(t) = 2p_3(t) - e^t + e^{2t}, \quad p_3(0) = 0,$$
obtaining
$$p_3(t) = te^{2t} + e^t - e^{2t}.$$

Substituting the matrices $\mathbf{M}_j$ and functions $p_j(t)$ into equation (9.30) we obtain, after a few simplifying computations,
$$\begin{aligned} e^{t\mathbf{A}} &= p_1(t)\mathbf{I} + p_2(t)\mathbf{M}_1 + p_3(t)\mathbf{M}_2 \\ &= \begin{pmatrix} te^{2t} + e^{2t} & te^{2t} & te^{2t} + 2e^t - 2e^{2t} \\ -te^{2t} & -te^{2t} + e^{2t} & -te^{2t} + 3e^{2t} - 3e^t \\ 0 & 0 & e^t \end{pmatrix}. \end{aligned}$$

The solution to the initial value problem is then
$$e^{t\mathbf{A}} \begin{pmatrix} 1 \\ 0 \\ 2 \end{pmatrix} = \begin{pmatrix} 3te^{2t} + 4e^t - 3e^{2t} \\ -3te^{2t} - 6e^t + 6e^{2t} \\ 2e^t \end{pmatrix}.$$

Example 9.3.4. Suppose
$$\mathbf{A} = \begin{pmatrix} 4 & 0 & 1 \\ 2 & 3 & 2 \\ 1 & 0 & 4 \end{pmatrix},$$
which has characteristic equation $(5-z)(3-z)^2 = 0$. The eigenvalues are $r_1 = 5, r_2 = 3$, and $r_3 = 3$. In Step 2 of Putzer's Algorithm we get
$$\mathbf{M}_1 = \mathbf{A} - 5\mathbf{I} = \begin{pmatrix} -1 & 0 & 1 \\ 2 & -2 & 2 \\ 1 & 0 & -1 \end{pmatrix}$$
and
$$\mathbf{M}_2 = (\mathbf{A} - 5\mathbf{I})(\mathbf{A} - 3\mathbf{I}) = \begin{pmatrix} 0 & 0 & 0 \\ 0 & 0 & 0 \\ 0 & 0 & 0 \end{pmatrix}.$$

We have $p_1(t) = e^{5t}$ and we solve
$$p_2'(t) = 3p_2(t) + e^{5t}, \quad p_2(0) = 0,$$
to obtain $p_2(t) = \frac{1}{2}e^{5t} - \frac{1}{2}e^{3t}$. Since $\mathbf{M}_2 = \mathbf{0}$, we do not need to compute $p_3(t)$. Why? Assembling the pieces using equation (9.30), we have
$$e^{t\mathbf{A}} = e^{5t}\mathbf{I} + \left(\frac{1}{2}e^{5t} - \frac{1}{2}e^{3t}\right)\begin{pmatrix} -1 & 0 & 1 \\ 2 & -2 & 2 \\ 1 & 0 & -1 \end{pmatrix},$$
giving
$$e^{t\mathbf{A}} = \begin{pmatrix} \frac{1}{2}e^{5t} + \frac{1}{2}e^{3t} & 0 & \frac{1}{2}e^{5t} - \frac{1}{2}e^{3t} \\ e^{5t} - e^{3t} & e^{3t} & e^{5t} - e^{3t} \\ \frac{1}{2}e^{5t} - \frac{1}{2}e^{3t} & 0 & \frac{1}{2}e^{5t} + \frac{1}{2}e^{3t} \end{pmatrix}.$$

Putzer's Algorithm: Why it works. We'll show why the Putzer Algorithm indeed gives the matrix $e^{t\mathbf{A}}$ in the case that $\mathbf{A}$ is a 3×3 matrix. The general $n \times n$ case follows from exactly the same ideas. With $p_1(t), p_2(t), p_3(t), \mathbf{M}_1$, and $\mathbf{M}_2$ as defined in the Putzer Algorithm, set

(9.32) $$\mathbf{M}(t) = p_1(t)\mathbf{I} + p_2(t)\mathbf{M}_1 + p_3(t)\mathbf{M}_2.$$

Its entries are functions of t, so that $\mathbf{M}(t)$ is a 3×3 matrix-valued function. Our goal is to show $\mathbf{M}(t) = e^{t\mathbf{A}}$, and the main work in establishing this will be to show that

(9.33) $$\mathbf{M}'(t) = \mathbf{AM}(t).$$

Once we have verified equation (9.33) it follows that for each j, $1 \leq j \leq 3$, $\mathbf{M}(t)\mathbf{E}_j$ solves the initial value problem $\mathbf{X}' = \mathbf{AX}$, $\mathbf{X}(0) = \mathbf{E}_j$, where $\mathbf{E}_j$ is the column vector with a 1 in the jth place and 0's elsewhere, since $\mathbf{M}(0) = \mathbf{I}$ by equation (9.32). But $e^{t\mathbf{A}}\mathbf{E}_j$ solves the same initial value problem, so by the uniqueness part of Theorem 9.2.1 we see that $\mathbf{M}(t)\mathbf{E}_j = e^{t\mathbf{A}}\mathbf{E}_j$ for each j. Since $\mathbf{M}(t)\mathbf{E}_j$ is the jth column of $\mathbf{M}(t)$ and $e^{t\mathbf{A}}\mathbf{E}_j$ is the jth column of $e^{t\mathbf{A}}$, our desired conclusion will follow.

To verify equation (9.33) suppose that the eigenvalues of $\mathbf{A}$ are r_1, r_2, r_3 (with repeats allowed). We have

(9.34) $$\mathbf{A} - r_1\mathbf{I} = \mathbf{M}_1$$

and

(9.35) $$(\mathbf{A} - r_2\mathbf{I})\mathbf{M}_1 = \mathbf{M}_2,$$

by Step 2 of the algorithm. The second equation can be rewritten as

(9.36) $$\mathbf{AM}_1 = \mathbf{M}_2 + r_2\mathbf{M}_1.$$

Since the characteristic polynomial for $\mathbf{A}$ is $-(z-r_3)(z-r_2)(z-r_1)$, the Cayley-Hamilton Theorem assures us that
$$(\mathbf{A} - r_3\mathbf{I})(\mathbf{A} - r_2\mathbf{I})(\mathbf{A} - r_1\mathbf{I}) = \mathbf{0}.$$
From this and equations (9.34)–(9.35) it follows immediately that
$$(\mathbf{A} - r_3\mathbf{I})\mathbf{M}_2 = \mathbf{0},$$
or

(9.37) $$\mathbf{AM}_2 = r_3\mathbf{M}_2.$$

On one hand we have
$$\begin{aligned}\mathbf{AM}(t) &= \mathbf{A}[p_1(t)\mathbf{I} + p_2(t)\mathbf{M}_1 + p_3(t)\mathbf{M}_2] \\ &= p_1(t)\mathbf{A} + p_2(t)\mathbf{AM}_1 + p_3(t)\mathbf{AM}_2 \\ &= p_1(t)\mathbf{A} + p_2(t)(\mathbf{M}_2 + r_2\mathbf{M}_1) + r_3 p_3(t)\mathbf{M}_2,\end{aligned}$$
where we have used equations (9.36) and (9.37). On the other hand, differentiating in the definition of $\mathbf{M}(t)$ in equation (9.32) gives
$$\mathbf{M}'(t) = p_1'(t)\mathbf{I} + p_2'(t)\mathbf{M}_1 + p_3'(t)\mathbf{M}_2.$$
Using the definitions of $p_j(t)$ from Step 3 of the algorithm gives
$$\mathbf{M}'(t) = r_1 p_1(t)\mathbf{I} + [r_2 p_2(t) + p_1(t)]\mathbf{M}_1 + [r_3 p_3(t) + p_2(t)]\mathbf{M}_2.$$
Subtracting this expression for $\mathbf{M}'(t)$ from our formula for $\mathbf{AM}(t)$ gives
$$\begin{aligned}\mathbf{AM}(t) - \mathbf{M}'(t) &= p_1(t)\mathbf{A} + p_2(t)(\mathbf{M}_2 + r_2\mathbf{M}_1) + r_3 p_3(t)\mathbf{M}_2 \\ &\quad - r_1 p_1(t)\mathbf{I} - [r_2 p_2(t) + p_1(t)]\mathbf{M}_1 - [r_3 p_3(t) + p_2(t)]\mathbf{M}_2 \\ &= p_1(t)\mathbf{A} - r_1 p_1(t)\mathbf{I} - p_1(t)\mathbf{M}_1 \\ &= p_1(t)(\mathbf{A} - r_1\mathbf{I} - \mathbf{M}_1).\end{aligned}$$
Since $\mathbf{M}_1 = \mathbf{A} - r_1\mathbf{I}$, this verifies that
$$\mathbf{AM}(t) = \mathbf{M}'(t),$$
as desired.

9.3.4. An application: Dissipative systems. Most mechanical and electrical systems operate with some dissipation of energy, through damping, friction, or (in the case of an electric circuit) the presence of electrical resistance. The mathematical expression of this in a linear system $\mathbf{X}' = \mathbf{AX}$, or a corresponding nonhomogeneous system $\mathbf{X}' = \mathbf{AX} + \mathbf{G}(t)$ (see Section 9.4), is simply that all eigenvalues of the matrix $\mathbf{A}$ are negative or have negative real part. An intuitively reasonable consequence is that in such a "dissipative" system without a forcing function $\mathbf{G}(t)$, the activity of the system "winds down" over time. The next result verifies this intuition.

Theorem 9.3.5. *Suppose that $\mathbf{A}$ is an $n \times n$ matrix all of whose eigenvalues are negative or have negative real part. Then any solution $\mathbf{X}(t)$ of the system $\mathbf{X}' = \mathbf{AX}$ tends to zero as $t \to \infty$.*

We'll illustrate why this theorem is true, by giving a proof using Putzer's Algorithm when $\mathbf{A}$ is 3×3. It will be convenient to distinguish three cases:

(a) all three eigenvalues of $\mathbf{A}$ are different or

(b) all eigenvalues of $\mathbf{A}$ are the same or

(c) **A** has two different eigenvalues, one of which is repeated.

In case (a) we'll denote the eigenvalues by r_1, r_2, and r_3, in case (b) by r_1, r_1, r_1, and in case (c) by r_1, r_2, r_2. In every case, we are assuming that each eigenvalue r_j is either a negative real number or a complex number with negative real part.

A solution to our system $\mathbf{X}' = \mathbf{A}\mathbf{X}$ looks like $\mathbf{X}(t) = e^{t\mathbf{A}}\mathbf{X}_0$ where $\mathbf{X}_0 = \mathbf{X}(0)$. Putzer's Algorithm tells us that
$$e^{t\mathbf{A}} = p_1(t)\mathbf{I} + p_2(t)\mathbf{M}_1 + p_3(t)\mathbf{M}_2$$
for certain *constant* matrices $\mathbf{M}_1$ and $\mathbf{M}_2$. The exact form of the $\mathbf{M}_j$'s is not important for us right now. What is important is the form of the coefficient functions $p_1(t)$, $p_2(t)$, and $p_3(t)$. By Putzer's Algorithm, these functions are found by solving certain first-order initial value problems (see Step 3 in the description of Putzer's Algorithm). The resulting solutions are:

- In case (a): $p_1(t) = e^{r_1 t}$, $p_2(t) = (e^{r_1 t} - e^{r_2 t})/(r_1 - r_2)$, and
$$p_3(t) = \frac{e^{r_1 t}}{(r_1 - r_2)(r_1 - r_3)} - \frac{e^{r_2 t}}{(r_1 - r_2)(r_2 - r_3)} + \frac{e^{r_3 t}}{(r_2 - r_3)(r_1 - r_3)}.$$
- In case (b): $p_1(t) = e^{r_1 t}$, $p_2(t) = te^{r_1 t}$, and $p_3(t) = t^2 e^{r_1 t}/2$.
- In case (c): $p_1(t) = e^{r_1 t}$, $p_2(t) = (e^{r_1 t} - e^{r_2 t})/(r_1 - r_2)$, and
$$p_3(t) = \frac{e^{r_1 t} - e^{r_2 t}}{(r_1 - r_2)^2} - \frac{te^{r_2 t}}{r_1 - r_2}.$$

In every case, the assumption that the eigenvalues are negative or have negative real part tells us that the functions $p_1(t)$, $p_2(t)$, and $p_3(t)$ tend to 0 as $t \to \infty$. To verify this if some $r = \alpha + i\beta$ is complex with $\alpha < 0$, recall that $e^{rt} = e^{\alpha t}(\cos \beta t + i \sin \beta t)$ and $\lim_{t \to \infty} e^{\alpha t} = 0$. Moreover, $te^{\alpha t}$ and $t^2 e^{\alpha t}$ also tend to 0 as $t \to \infty$ by l'Hôpital's Rule. Because the entries of $e^{t\mathbf{A}}$ are simply linear combinations of the functions p_1, p_2, and p_3, every entry of $e^{t\mathbf{A}}$ tends to 0 as $t \to \infty$, and the same is true for $\mathbf{X}(t) = e^{t\mathbf{A}}\mathbf{X}_0$, as desired.

9.3.5. Exercises.

In Exercises 1–4, compute the matrix $e^{t\mathbf{A}}$, and then use it to solve the initial value problem
$$\mathbf{X}' = \mathbf{A}\mathbf{X}, \quad \mathbf{X}(0) = \begin{pmatrix} 1 \\ 1 \\ 0 \end{pmatrix}.$$

1. $\mathbf{A} = \begin{pmatrix} 1 & 0 & 0 \\ 0 & 2 & 0 \\ 0 & 2 & 2 \end{pmatrix}$.

2. $\mathbf{A} = \begin{pmatrix} 1 & 0 & 0 \\ 1 & 3 & 0 \\ 1 & 2 & 5 \end{pmatrix}$.

3. $\mathbf{A} = \begin{pmatrix} 1 & 0 & 0 \\ 1 & 0 & 1 \\ 0 & 1 & 0 \end{pmatrix}$.

4. $\mathbf{A} = \begin{pmatrix} 1 & 1 & 0 \\ 0 & 0 & -1 \\ 0 & 1 & 0 \end{pmatrix}$.

5. In Exercise 9 of Section 8.2, you showed that the matrix
$$\mathbf{A} = \begin{pmatrix} 2 & 0 & 0 \\ 1 & 2 & 0 \\ 1 & 1 & 2 \end{pmatrix}$$
has only one eigenvalue $r = 2$ and only one linearly independent eigenvector. Find $e^{t\mathbf{A}}$, and then give the general solution of $\mathbf{X}' = \mathbf{A}\mathbf{X}$.

9.3. The matrix exponential, Part II

6. Use the Putzer Algorithm to derive the formula for $e^{t\mathbf{A}}$ given in Table 9.1 (Section 9.1) for the case of unequal eigenvalues r_1 and r_2 (for $\mathbf{A}$ a 2×2 matrix).

7. Using the work in Example 9.3.2, solve the system
$$\frac{dx}{dt} = -2x + 3y + z, \quad \frac{dy}{dt} = -2x + 4y, \quad \frac{dz}{dt} = -x + 3y,$$
with initial conditions $x(0) = 1, y(0) = 2, z(0) = -1$.

8. (a) Use Section 9.3.1 to show that a 3×3 matrix $\mathbf{A}$ with only one eigenvalue r has
$$e^{t\mathbf{A}} = e^{rt}\left(\left(\frac{r^2 t^2}{2} - rt + 1\right)\mathbf{I} + (t - rt^2)\mathbf{A} + \frac{t^2}{2}\mathbf{A}^2\right).$$

   (b) Use (a) to compute $e^{t\mathbf{A}}$ for
$$\mathbf{A} = \begin{pmatrix} 3 & 1 & 0 \\ 0 & 3 & 1 \\ 0 & 0 & 3 \end{pmatrix}.$$

   (c) Write $\mathbf{A} = 3\mathbf{I} + \mathbf{N}$, where
$$\mathbf{N} = \begin{pmatrix} 0 & 1 & 0 \\ 0 & 0 & 1 \\ 0 & 0 & 0 \end{pmatrix}.$$

   Explain why
$$e^{t\mathbf{A}} = e^{3t} e^{t\mathbf{N}}$$
   and recompute $e^{t\mathbf{A}}$ from this.

9. Suppose that $\mathbf{D}$ is a 3×3 diagonal matrix, with two distinct eigenvalues r_1 and r_2, with r_2 having multiplicity two.
   (a) Show by direct computation that $(\mathbf{D} - r_1\mathbf{I})(\mathbf{D} - r_2\mathbf{I}) = \mathbf{0}$.
   (b) Suppose that $\mathbf{A} = \mathbf{B}\mathbf{D}\mathbf{B}^{-1}$ for $\mathbf{D}$ as above and $\mathbf{B}$ is any 3×3 invertible matrix. Show $(\mathbf{A} - r_1\mathbf{I})(\mathbf{A} - r_2\mathbf{I}) = \mathbf{0}$. Hint: Check that $\mathbf{B}\mathbf{D}\mathbf{B}^{-1} - r\mathbf{I} = \mathbf{B}(\mathbf{D} - r\mathbf{I})\mathbf{B}^{-1}$.
   (c) If $\mathbf{A}$ is as in (b), show that
$$e^{t\mathbf{A}} = e^{r_1 t}\mathbf{I} + \left(\frac{1}{r_1 - r_2}e^{r_1 t} - \frac{1}{r_1 - r_2}e^{r_2 t}\right)(\mathbf{A} - r_1\mathbf{I}).$$

10. Suppose $\mathbf{A}$ is an $n \times n$ matrix which satisfies the relation $\mathbf{A}^2 = 2\mathbf{A}$.
    (a) Show that $\mathbf{A}^3 = 4\mathbf{A}$, $\mathbf{A}^4 = 8\mathbf{A}$, and in general $\mathbf{A}^m = 2^{m-1}\mathbf{A}$.
    (b) Using (a), simplify
$$\mathbf{I} + \mathbf{A} + \frac{1}{2!}\mathbf{A}^2 + \frac{1}{3!}\mathbf{A}^3 + \frac{1}{4!}\mathbf{A}^4 + \cdots$$
    and then show that
$$e^{t\mathbf{A}} = \mathbf{I} + \frac{1}{2}(e^{2t} - 1)\mathbf{A}.$$

    (c) What is the analogous formula for $e^{t\mathbf{A}}$ if instead $\mathbf{A}^2 = \beta\mathbf{A}$ for some constant β?
    (d) Find $e^{t\mathbf{A}}$ for
$$\mathbf{A} = \begin{pmatrix} 1 & 1 & 1 & 1 \\ 1 & 1 & 1 & 1 \\ 1 & 1 & 1 & 1 \\ 1 & 1 & 1 & 1 \end{pmatrix}.$$

11. (a) The matrix
$$\mathbf{A} = \begin{pmatrix} 0 & 1 & 2 \\ 0 & 0 & 1 \\ 0 & 0 & 0 \end{pmatrix}$$
satisfies $\mathbf{A}^3 = \mathbf{0}$. Find $e^{t\mathbf{A}}$.
(b) Find all powers of the matrix
$$\mathbf{A} = \begin{pmatrix} 0 & a & b \\ 0 & 0 & c \\ 0 & 0 & 0 \end{pmatrix}$$
and then determine $e^{t\mathbf{A}}$.
(c) Suppose that $\mathbf{B}$ is an $n \times n$ matrix whose only eigenvalue is 0. Show that $\mathbf{B}^n = \mathbf{0}$.

12. Let $\mathbf{A}$ be a 3×3 matrix with three distinct eigenvalues r_1, r_2, and r_3. Let $\mathbf{B}, \mathbf{C}$, and $\mathbf{D}$ be the matrices associated to $\mathbf{A}$ as in Section 9.3.2.
    (a) Show that $\mathbf{BC}, \mathbf{CB}, \mathbf{BD}, \mathbf{DB}, \mathbf{CD}$, and $\mathbf{DC}$ are all equal to the zero matrix.
    (b) Show that $\mathbf{B}^2 = \mathbf{B}, \mathbf{C}^2 = \mathbf{C}$, and $\mathbf{D}^2 = \mathbf{D}$. Hint: Look at Exercise 29 in Section 9.1.

13. Let $\mathbf{A}$ be a 3×3 matrix with distinct eigenvalues r_1, r_2, and r_3. Suppose that $\mathbf{B}, \mathbf{C}$, and $\mathbf{D}$ are the matrices associated with $\mathbf{A}$ as in Section 9.3.2. Show that if $p(z)$ is any polynomial, then
$$p(\mathbf{A}) = p(r_1)\mathbf{B} + p(r_2)\mathbf{C} + p(r_3)\mathbf{D}.$$

Hint: See Exercise 30 in Section 9.1 for the 2×2 case and Exercise 12 above for useful properties of $\mathbf{B}, \mathbf{C}$, and $\mathbf{D}$.

14. This problem is the 3×3 analogue of Exercise 31 in Section 9.1. Let $\mathbf{A}, \mathbf{B}, \mathbf{C}$, and $\mathbf{D}$ be as in Exercise 12. Given numbers b_1, b_2, c_1, c_2, d_1, and d_2 define the matrices
$$\mathbf{S} = b_1\mathbf{B} + c_1\mathbf{C} + d_1\mathbf{D}.$$
and
$$\mathbf{T} = b_2\mathbf{B} + c_2\mathbf{C} + d_2\mathbf{D}.$$

(a) Show that
$$\mathbf{ST} = b_1 b_2 \mathbf{B} + c_1 c_2 \mathbf{C} + d_1 d_1 \mathbf{D} = \mathbf{TS}.$$

(b) Suppose that none of b_1, c_1, or d_1 is equal to 0. Show that the matrix $\mathbf{S}$ defined above is invertible and
$$\mathbf{S}^{-1} = \frac{1}{b_1}\mathbf{B} + \frac{1}{c_1}\mathbf{C} + \frac{1}{d_1}\mathbf{D}.$$

(c) Show that b_1, c_1, and d_1 are eigenvalues of $\mathbf{S}$. (You may use the fact that none of $\mathbf{B}, \mathbf{C}$, and $\mathbf{D}$ are the zero matrix.)

(d) If λ is a real or complex number different from each of b_1, c_1, and d_1, show that $\mathbf{S} - \lambda\mathbf{I}$ is invertible and
$$(\mathbf{S} - \lambda\mathbf{I})^{-1} = \frac{1}{b_1 - \lambda}\mathbf{B} + \frac{1}{c_1 - \lambda}\mathbf{C} + \frac{1}{d_1 - \lambda}\mathbf{D}.$$

Hint: In writing $\mathbf{S} - \lambda\mathbf{I}$, recall that $\mathbf{I} = \mathbf{B} + \mathbf{C} + \mathbf{D}$.

9.4. Nonhomogeneous constant coefficient systems

Consider a nonhomogeneous system
$$\frac{dx}{dt} = ax + by + g_1(t), \quad \frac{dy}{dt} = cx + dy + g_2(t),$$
which we write in matrix form as

(9.38) $$\mathbf{X}' = \mathbf{A}\mathbf{X} + \mathbf{G}(t),$$

where
$$\mathbf{A} = \begin{pmatrix} a & b \\ c & d \end{pmatrix}, \quad \mathbf{X}(t) = \begin{pmatrix} x(t) \\ y(t) \end{pmatrix}, \quad \text{and} \quad \mathbf{G}(t) = \begin{pmatrix} g_1(t) \\ g_2(t) \end{pmatrix}.$$

Single first-order linear equations provide a guide for solving this system. Recall the procedure used in Section 2.1 to solve the linear equation
$$\frac{dx}{dt} = ax + g(t).$$
We multiplied by the integrating factor e^{-at} to obtain
$$\frac{d}{dt}(e^{-at}x) = e^{-at}g(t)$$
and then integrated both sides of this last equation to get
$$e^{-at}x = \int e^{-at}g(t)dt + c$$
where c is a constant of integration. Multiplying by e^{at} gives the solution

(9.39) $$x(t) = e^{at}\int e^{-at}g(t)dt + ce^{at}.$$

This is equation (2.12) of Section 2.1 in the special case $p(t) = -a$. If we want $x(t)$ to satisfy the initial condition $x(0) = x_0$, we can alter this by replacing the indefinite integral in equation (9.39) with the definite integral
$$\int_0^t e^{-as}g(s)ds.$$
Since this definite integral is 0 when $t = 0$, this forces c to be chosen to be x_0, and the initial value problem has solution

(9.40) $$x(t) = x_0 e^{at} + e^{at}\int_0^t e^{-as}g(s)ds.$$

To solve our system (9.38) with an initial condition $\mathbf{X}(0) = \mathbf{X}_0$ we can duplicate this argument using matrix and vector functions. To carry this out, we will need to be able to *integrate* vector functions $\mathbf{F}(t)$.

Integration of vector functions. For concreteness, let's consider the two-dimensional case, so
$$\mathbf{F}(t) = \begin{pmatrix} f_1(t) \\ f_2(t) \end{pmatrix}.$$
The n-dimensional case is a straightforward generalization. We define $\int_a^b \mathbf{F}(t)dt$ to be the vector
$$\int_a^b \mathbf{F}(t)dt = \begin{pmatrix} \int_a^b f_1(t)dt \\ \int_a^b f_2(t)dt \end{pmatrix}.$$

Notice that if $\mathbf{F}$ is continuously differentiable and we integrate its derivative, we have

$$\int_a^b \mathbf{F}'(t)dt = \begin{pmatrix} \int_a^b f_1'(t)dt \\ \int_a^b f_2'(t)dt \end{pmatrix} = \begin{pmatrix} f_1(b) - f_1(a) \\ f_2(b) - f_2(a) \end{pmatrix}$$
$$= \mathbf{F}(b) - \mathbf{F}(a),$$

a vector version of the Fundamental Theorem of Calculus.

With these facts in hand, let's solve the nonhomogeneous equation

(9.41) $$\mathbf{X}' - \mathbf{A}\mathbf{X} = \mathbf{G}(t)$$

where $\mathbf{A}$ is a constant matrix. The matrix function $e^{-t\mathbf{A}}$ will serve as a (matrix-valued) integrating factor. Since $\frac{d}{dt}e^{-t\mathbf{A}} = -\mathbf{A}e^{-t\mathbf{A}} = -e^{-t\mathbf{A}}\mathbf{A}$, the product rule in Theorem 9.1.3(b) says

$$\frac{d}{dt}\left(e^{-t\mathbf{A}}\mathbf{X}(t)\right) = -e^{-t\mathbf{A}}\mathbf{A}\mathbf{X}(t) + e^{-t\mathbf{A}}\mathbf{X}'(t) = e^{-t\mathbf{A}}[\mathbf{X}'(t) - \mathbf{A}\mathbf{X}(t)].$$

Thus if we multiply equation (9.41) on the left by $e^{-t\mathbf{A}}$, we obtain the equation

$$\frac{d}{dt}(e^{-t\mathbf{A}}\mathbf{X}(t)) = e^{-t\mathbf{A}}\mathbf{G}(t).$$

Integrate on both sides from 0 to t to get

(9.42) $$\int_0^t \frac{d}{ds}(e^{-s\mathbf{A}}\mathbf{X}(s))ds = \int_0^t e^{-s\mathbf{A}}\mathbf{G}(s)ds$$

where we have used s as the variable of integration. By the vector form of the Fundamental Theorem of Calculus just discussed, the left side of (9.42) equals

$$e^{-t\mathbf{A}}\mathbf{X}(t) - e^{-0\mathbf{A}}\mathbf{X}(0) = e^{-t\mathbf{A}}\mathbf{X}(t) - \mathbf{X}(0).$$

Setting $\mathbf{X}(0) = \mathbf{X}_0$, we have (by equation (9.42))

(9.43) $$e^{-t\mathbf{A}}\mathbf{X}(t) - \mathbf{X}_0 = \int_0^t e^{-s\mathbf{A}}\mathbf{G}(s)ds.$$

Multiplying in equation (9.43) (on the left!) by $e^{t\mathbf{A}}$ we obtain our solution to the initial value problem $\mathbf{X}' = \mathbf{A}\mathbf{X} + \mathbf{G}(t)$, $\mathbf{X}(0) = \mathbf{X}_0$:

(9.44) $$\mathbf{X}(t) = e^{t\mathbf{A}}\mathbf{X}_0 + e^{t\mathbf{A}}\int_0^t e^{-s\mathbf{A}}\mathbf{G}(s)ds.$$

This is the system analogue of equation (9.40). We can adjust equation (9.44) if the initial condition is specified at a point different from $t = 0$; see Exercise 17.

The next two examples illustrate using equation (9.44) to solve a nonhomogeneous system.

Example 9.4.1. Use equation (9.44) to solve $\mathbf{X}' = \mathbf{A}\mathbf{X} + \mathbf{G}(t)$, $\mathbf{X}(0) = \mathbf{X}_0$, where

$$\mathbf{A} = \begin{pmatrix} 2 & 1 \\ 2 & 3 \end{pmatrix}, \quad \mathbf{G}(t) = \begin{pmatrix} 2e^t \\ 0 \end{pmatrix}, \quad \text{and} \quad \mathbf{X}_0 = \begin{pmatrix} 1 \\ -1 \end{pmatrix}.$$

For this choice of $\mathbf{A}$, we computed $e^{t\mathbf{A}}$ in Example 9.1.7 and obtained

$$e^{t\mathbf{A}} = \frac{1}{3}\begin{pmatrix} e^{4t} + 2e^t & e^{4t} - e^t \\ 2e^{4t} - 2e^t & 2e^{4t} + e^t \end{pmatrix},$$

9.4. Nonhomogeneous constant coefficient systems

so that
$$e^{t\mathbf{A}}\mathbf{X}_0 = \begin{pmatrix} e^t \\ -e^t \end{pmatrix}$$

and
$$e^{-s\mathbf{A}}\mathbf{G}(s) = \frac{1}{3}\begin{pmatrix} e^{-4s} + 2e^{-s} & e^{-4s} - e^{-s} \\ 2e^{-4s} - 2e^{-s} & 2e^{-4s} + e^{-s} \end{pmatrix}\begin{pmatrix} 2e^s \\ 0 \end{pmatrix} = \begin{pmatrix} 2e^{-3s}/3 + 4/3 \\ 4e^{-3s}/3 - 4/3 \end{pmatrix}.$$

Integration gives
$$\int \left(\frac{2}{3}e^{-3s} + \frac{4}{3}\right) ds = -\frac{2}{9}e^{-3s} + \frac{4}{3}s \quad \text{and} \quad \int \left(\frac{4}{3}e^{-3s} - \frac{4}{3}\right) ds = -\frac{4}{9}e^{-3s} - \frac{4}{3}s,$$

so that
$$\int_0^t e^{-s\mathbf{A}}\mathbf{G}(s)\,ds = \begin{pmatrix} \int_0^t (2e^{-3s}/3 + 4/3)\,ds \\ \int_0^t (4e^{-3s}/3 - 4/3)\,ds \end{pmatrix} = \begin{pmatrix} -2e^{-3t}/9 + 4t/3 + 2/9 \\ -4e^{-3t}/9 - 4t/3 + 4/9 \end{pmatrix}.$$

By equation (9.44), the solution to our initial value problem is

$$\begin{aligned}
\mathbf{X}(t) &= e^{t\mathbf{A}}\mathbf{X}_0 + e^{t\mathbf{A}}\int_0^t e^{-s\mathbf{A}}\mathbf{G}(s)\,ds \\
&= \begin{pmatrix} e^t \\ -e^t \end{pmatrix} + \frac{1}{3}\begin{pmatrix} e^{4t} + 2e^t & e^{4t} - e^t \\ 2e^{4t} - 2e^t & 2e^{4t} + e^t \end{pmatrix}\begin{pmatrix} -2e^{-3t}/9 + 4t/3 + 2/9 \\ -4e^{-3t}/9 - 4t/3 + 4/9 \end{pmatrix} \\
&= \begin{pmatrix} e^t \\ -e^t \end{pmatrix} + \begin{pmatrix} -2e^t/9 + 2e^{4t}/9 + 4te^t/3 \\ -4e^t/9 + 4e^{4t}/9 - 4te^t/3 \end{pmatrix} \\
&= \begin{pmatrix} 7e^t/9 + 2e^{4t}/9 + 4te^t/3 \\ -13e^t/9 + 4e^{4t}/9 - 4te^t/3 \end{pmatrix}.
\end{aligned}$$

Example 9.4.2. Solve the system
$$\frac{dx}{dt} = y + t, \quad \frac{dy}{dt} = -x + t,$$

with the initial condition $x(0) = 1$, $y(0) = 1$; that is,

(9.45) $$\mathbf{X}(0) = \mathbf{X}_0 = \begin{pmatrix} 1 \\ 1 \end{pmatrix}.$$

Writing the system in matrix form we have $\mathbf{X}' = \mathbf{A}\mathbf{X} + \mathbf{G}(t)$, where
$$\mathbf{A} = \begin{pmatrix} 0 & 1 \\ -1 & 0 \end{pmatrix} \quad \text{and} \quad \mathbf{G}(t) = \begin{pmatrix} t \\ t \end{pmatrix}.$$

The characteristic polynomial of $\mathbf{A}$ is $z^2 + 1 = (z-i)(z+i)$, and we have two purely imaginary eigenvalues $r_1 = i$ and $r_2 = -i$. From equation (9.18) in Section 9.1, we see that
$$e^{t\mathbf{A}} = (\cos t)\,\mathbf{I} + (\sin t)\,\mathbf{A} = \cos t\begin{pmatrix} 1 & 0 \\ 0 & 1 \end{pmatrix} + \sin t\begin{pmatrix} 0 & 1 \\ -1 & 0 \end{pmatrix} = \begin{pmatrix} \cos t & \sin t \\ -\sin t & \cos t \end{pmatrix}.$$

Thus
$$e^{-s\mathbf{A}}\mathbf{G}(s) = \begin{pmatrix} \cos(-s) & \sin(-s) \\ -\sin(-s) & \cos(-s) \end{pmatrix}\begin{pmatrix} s \\ s \end{pmatrix} = \begin{pmatrix} s\cos s - s\sin s \\ s\sin s + s\cos s \end{pmatrix}.$$

Integrating on both sides yields
$$\int_0^t e^{-s\mathbf{A}}\mathbf{G}(s) = \begin{pmatrix} \int_0^t (s\cos s - s\sin s)\,ds \\ \int_0^t (s\sin s + s\cos s)\,ds \end{pmatrix}.$$

Integration by parts gives the formulas

$$\int_0^t s\cos s\, ds = t\sin t + \cos t - 1 \quad \text{and} \quad \int_0^t s\sin s\, ds = -t\cos t + \sin t,$$

so that

$$\int_0^t e^{-s\mathbf{A}}\mathbf{G}(s) = \begin{pmatrix} t\sin t + \cos t - 1 + t\cos t - \sin t \\ -t\cos t + \sin t + t\sin t + \cos t - 1 \end{pmatrix}.$$

Multiplying the last line by $e^{t\mathbf{A}}$ gives

$$e^{t\mathbf{A}}\int_0^t e^{-s\mathbf{A}}\mathbf{G}(s) = \begin{pmatrix} \cos t & \sin t \\ -\sin t & \cos t \end{pmatrix}\begin{pmatrix} t\sin t + \cos t - 1 + t\cos t - \sin t \\ -t\cos t + \sin t + t\sin t + \cos t - 1 \end{pmatrix}.$$

After some computation (making use of the identity $\cos^2 t + \sin^2 t = 1$), the matrix product on the right simplifies to

$$\begin{pmatrix} 1 + t - \cos t - \sin t \\ 1 - t - \cos t + \sin t \end{pmatrix}.$$

If we impose our initial condition in (9.45), equation (9.44) gives the solution

$$\begin{aligned}
\mathbf{X}(t) &= e^{t\mathbf{A}}\mathbf{X}_0 + e^{t\mathbf{A}}\int_0^t e^{-s\mathbf{A}}\mathbf{G}(s)ds \\
&= \begin{pmatrix} \cos t & \sin t \\ -\sin t & \cos t \end{pmatrix}\begin{pmatrix} 1 \\ 1 \end{pmatrix} + \begin{pmatrix} 1 + t - \cos t - \sin t \\ 1 - t - \cos t + \sin t \end{pmatrix} \\
&= \begin{pmatrix} 1 + t \\ 1 - t \end{pmatrix}.
\end{aligned}$$

In Exercise 29 you are invited to find what happens when the initial value $(1, 1)$ in Example 9.4.2 is replaced by any other point in the plane.

Some general principles. Several basic principles relate the solutions of

(9.46) $$\mathbf{X}' = \mathbf{P}(t)\mathbf{X} + \mathbf{G}(t)$$

and the associated homogeneous system

(9.47) $$\mathbf{X}' = \mathbf{P}(t)\mathbf{X}.$$

These are the system analogues of familiar facts from Section 5.1. We assume that the entries of $\mathbf{P}(t)$ and $\mathbf{G}(t)$ are continuous functions on some interval $a < t < b$ on which we work.

- If $\mathbf{X}_1(t)$ and $\mathbf{X}_2(t)$ are solutions to the nonhomogeneous system in (9.46), then $\mathbf{X}_1(t) - \mathbf{X}_2(t)$ solves the homogeneous system in (9.47).
- If $\mathbf{Z}(t)$ is a solution of the homogeneous system (9.47) and $\mathbf{Y}(t)$ is a solution of the nonhomogeneous system (9.46), then $\mathbf{Z}(t) + \mathbf{Y}(t)$ is also a solution of the nonhomogeneous system (9.46).

You are asked to verify these assertions in Exercises 11–12. As a consequence of them, we can describe the form of the general solution to (9.46).

Theorem 9.4.3 (General solution for a nonhomogeneous linear system). *If $\mathbf{X}_p(t)$ is a particular solution of $\mathbf{X}' = \mathbf{P}(t)\mathbf{X} + \mathbf{G}(t)$ on an open interval (a, b) on which $\mathbf{P}(t)$ and $\mathbf{G}(t)$ are continuous and if $\mathbf{X}_c(t)$ is the general solution of $\mathbf{X}' = \mathbf{P}(t)\mathbf{X}$, then $\mathbf{X}_c(t) + \mathbf{X}_p(t)$ is the general solution of $\mathbf{X}' = \mathbf{P}(t)\mathbf{X} + \mathbf{G}(t)$ on (a, b).*

9.4. Nonhomogeneous constant coefficient systems

Let's revisit equation (9.44) in light of this theorem. If we set

$$\mathbf{X}_0 = \begin{pmatrix} c_1 \\ c_2 \\ \vdots \\ c_n \end{pmatrix},$$

a vector of arbitrary constants, then $\mathbf{X}_c(t) = e^{t\mathbf{A}}\mathbf{X}_0$ is the general solution to the homogeneous equation $\mathbf{X}' = \mathbf{A}\mathbf{X}$. You can check that

$$\mathbf{X}_p(t) = e^{t\mathbf{A}} \int_0^t e^{-s\mathbf{A}} \mathbf{G}(s)\, ds$$

is a particular solution to the nonhomogeneous equation $\mathbf{X}' = \mathbf{A}\mathbf{X} + \mathbf{G}(t)$. The general solution to $\mathbf{X}' = \mathbf{A}\mathbf{X} + \mathbf{G}(t)$ is

$$\mathbf{X}_c(t) + \mathbf{X}_p(t) = e^{t\mathbf{A}}\mathbf{X}_0 + e^{t\mathbf{A}} \int_0^t e^{-s\mathbf{A}} \mathbf{G}(s)\, ds,$$

as in equation (9.44), with $\mathbf{X}_0$ a vector of arbitrary constants.

Example 9.4.4. We'll use Theorem 9.4.3 to find a simple expression for the general solution to

(9.48) $$\mathbf{X}'(t) = \mathbf{A}\mathbf{X}(t) + \mathbf{D},$$

where $\mathbf{A}$ is an *invertible* $n \times n$ matrix (of constants) and the nonhomogeneous term $\mathbf{D}$ is an $n \times 1$ *constant* vector. First note that (9.48) has equilibrium solution $\mathbf{X}_e(t) = -\mathbf{A}^{-1}\mathbf{D}$. As with all equilibrium solutions, this is obtained by setting $\mathbf{X}'(t) = \mathbf{A}\mathbf{X}(t) + \mathbf{D}$ equal to $\mathbf{0}$ and solving for $\mathbf{X}(t)$. Now observe that $\mathbf{X}_e$ is a particular solution of (9.48), and thus the general solution of (9.48) takes the form

$$\mathbf{X}(t) = e^{t\mathbf{A}}\mathbf{C} + \mathbf{X}_e,$$

where $\mathbf{C}$ is an n-dimensional vector of arbitrary constants $c_1, \ldots, c_n$. If we wish to solve equation (9.48) with the initial condition $\mathbf{X}(0) = \mathbf{X}_0$, substitute $t = 0$ into the general solution to find that $\mathbf{C} = \mathbf{X}_0 - \mathbf{X}_e$, and so

(9.49) $$\mathbf{X}(t) = e^{t\mathbf{A}}(\mathbf{X}_0 - \mathbf{X}_e) + \mathbf{X}_e$$

solves the initial value problem consisting of (9.48) with initial condition $\mathbf{X}(0) = \mathbf{X}_0$. Exercise 21 outlines how to obtain (9.49) directly from formula (9.44).

9.4.1. Chernobyl and Fukushima: Nuclear accidents at a distance. On the morning of April 26, 1986, in the course of a poorly designed "safety test", Unit 4 of the Chernobyl Power Complex exploded, releasing into the atmosphere radioactive substances such as xenon ^{133}Xe, iodine ^{131}I, and cesium ^{137}Cs. According to a report by the Nuclear Energy Agency,[2] the total amount of radioactivity eventually released by the explosion is estimated to be approximately 11×10^{18} Bq.[3] Beginning on March 11, 2011, as a result of a magnitude 9.0 earthquake and subsequent tsunami, multiple reactors at the Fukushima Daiichi power plant in Japan lost their cooling capabilities and underwent at least a partial meltdown. By March 31, 2011, an estimated 5.2×10^{17} Bq of radioactivity had been released.

After such accidents, even people at some distance removed from the site are exposed to radionuclide hazard in various ways: through inhalation of airborne particles, through contaminated water, and through food chain contamination. In this example, we focus on one aspect of the

---

[2] *Chernobyl, Ten Years On: Assessment of Radiological and Health Impact*, OECD-NEA. 1995.
[3] A becquerel, or Bq, is a unit of radioactivity, with 1 Bq describing radioactive decay occurring at 1 nuclear disintegration per second. The total-release estimate 11×10^{18} Bq represents a sum of the radioactivity levels, at the time of release, of all the radioactive material that eventually escaped from the Chernobyl reactor as a result of the April 26, 1986, accident.

latter, the quick and significant appearance of radionuclides in cow's milk that occurs when the animal ingests contaminated pasture. In this example, we explore via a compartment-type model how ^{131}I moves from the atmosphere into the food chain through cow's milk. Numerical values for parameters in this model come from data collected in northwestern Greece in the month after the Chernobyl accident; see [**2**].

For ^{131}I, a two-compartment model (with an additional "external losses" compartment) gives a reasonably accurate picture of the situation. The two compartments are labeled "grass" and "milk", as shown below. We study the *concentration* $G(t)$ and $C(t)$ of ^{131}I in the grass and milk compartments, using units of Bq/kg for $G(t)$, Bq/L for $C(t)$, and days for t.

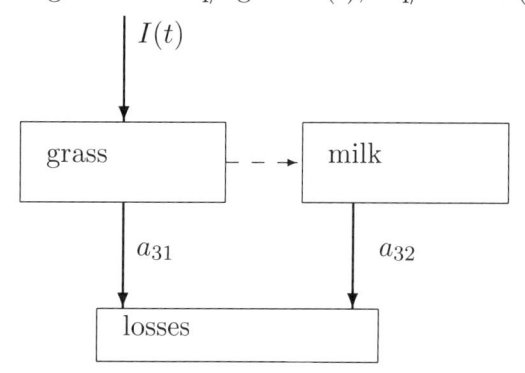

The concentration of $G(t)$ is determined by several factors: airborne "input" $I(t)$ (atmospheric deposition), measured in Bq/kg per day, as well as losses caused by the passage of ^{131}I from the grass to the soil, the radioactive decay of ^{131}I, and "weathering", which is loss due to wind and water removal. We'll lump these losses together with the single arrow from the "grass" compartment to the "losses" compartment, with transfer coefficient a_{31} having units 1/day. Similarly, a portion of $C(t)$ is lost due to radioactive decay and to elimination from the cow's milk by excretion; these losses are reflected in the arrow from "milk" to "losses" with transfer coefficient a_{32}/day.

As a cow eats the contaminated grass, some portion of the ingested contaminant is converted into radioactivity in its milk. The rate of change of $C(t)$ due to this ingestion depends on how much the cow eats and a constant which is called the "grass to milk transfer rate". For the dairy farm studied in Greece in May, 1986, it was estimated that a cow ate 50 kg/day of contaminated pasture and the grass to milk transfer rate was 4×10^{-3}/L.

We're ready to write the differential equations for $\frac{dG}{dt}$ and $\frac{dC}{dt}$. For the latter, our discussion above says we have
$$\frac{dC}{dt} = (50)(4 \times 10^{-3})G(t) - a_{32}C(t),$$
or
$$\frac{dC}{dt} = 0.2G(t) - a_{32}C(t).$$
Check that the units on the left-hand side of this equation and on both terms on the right-hand side are Bq/L per day. To write a differential equation for $\frac{dG}{dt}$, notice first that since $G(t)$ measures the *concentration* of ^{131}I in the grass, it is not changed by the cow's grazing. It is changed by the losses represented by the arrow from "grass" to "losses" compartments in our diagram and by the input $I(t)$ of radioactivity on the grass from the atmosphere. Thus
$$\frac{dG}{dt} = I(t) - a_{31}G(t),$$
with units of Bq/kg per day on both sides of this equation. Estimates for the value of a_{31} and a_{32} are
$$a_{31} = 0.1/\text{day}, \quad a_{32} = 0.8/\text{day}.$$

9.4. Nonhomogeneous constant coefficient systems

We have the nonhomogeneous linear system

(9.50) $$\frac{dG}{dt} = I(t) - 0.1G(t),$$

(9.51) $$\frac{dC}{dt} = 0.2G(t) - 0.8C(t).$$

To go further with this, we need some information on the atmospheric input $I(t)$. As a radioactive plume arrives at a location from the site of a distant accident, the concentration of radionuclides in the atmosphere increases sharply to a maximum and then slowly decreases as the plume moves away and contaminants fall to the ground. Thus we would want the basic shape of the graph of $I(t)$ to look like this:

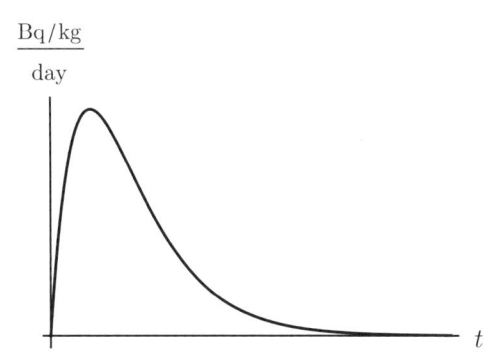

Figure 9.5. Desired shape for the atmospheric input function $I(t)$.

A function which fits this form is $I(t) = Mte^{-\lambda t}$ for positive constants M and λ. Atmospheric contamination data from different locations in northwestern Greece in May 1986 fit this form for $I(t)$ well and allow an estimate of $\lambda = 0.7$/day. The value of M varied considerably with the location, so we'll leave this unspecified for now. (See also Exercise 27, in which this form for $I(t)$ is justified via another compartment model.)

In matrix form equations (9.50) and (9.51) are written as $\mathbf{X}'(t) = \mathbf{AX} + \mathbf{F}(t)$ for

$$\mathbf{A} = \begin{pmatrix} -0.1 & 0 \\ 0.2 & -0.8 \end{pmatrix}, \quad \mathbf{X}(t) = \begin{pmatrix} G(t) \\ C(t) \end{pmatrix}, \quad \text{and} \quad \mathbf{F}(t) = \begin{pmatrix} Mte^{-0.7t} \\ 0 \end{pmatrix}.$$

We'll solve this nonhomogeneous system with initial conditions $G(0) = 0, C(0) = 0$. By equation (9.44) we have solution

$$\mathbf{X}(t) = e^{t\mathbf{A}} \int_0^t e^{-s\mathbf{A}} \mathbf{F}(s) ds.$$

The lower triangular matrix $\mathbf{A}$ has eigenvalues -0.1 and -0.8 by inspection. The first entry of Table 9.1 in Section 9.1 tells us that

$$e^{t\mathbf{A}} = \begin{pmatrix} e^{-0.1t} & 0 \\ \frac{2}{7}(e^{-0.1t} - e^{-0.8t}) & e^{-0.8t} \end{pmatrix}$$

so that

$$e^{-s\mathbf{A}} \mathbf{F}(s) = \begin{pmatrix} e^{0.1s} & 0 \\ \frac{2}{7}(e^{0.1s} - e^{0.8s}) & e^{0.8s} \end{pmatrix} \begin{pmatrix} Mse^{-0.7s} \\ 0 \end{pmatrix} = \begin{pmatrix} Mse^{-0.6s} \\ \frac{2}{7}Ms(e^{-0.6s} - e^{0.1s}) \end{pmatrix}.$$

Integrating this last vector function from 0 to t gives

$$\begin{pmatrix} M(-\frac{10}{6}te^{-0.6t} - \frac{100}{36}e^{-0.6t} + \frac{100}{36}) \\ M(-\frac{250}{9} + (-\frac{50}{63} - \frac{10}{21}t)e^{-0.6t} + (\frac{200}{7} - \frac{20}{7}t)e^{0.1t}) \end{pmatrix}.$$

Upon multiplying the last vector by $e^{t\mathbf{A}}$ (on the left) we obtain our functions $G(t)$ and $C(t)$:

$$G(t) = M\left(-\frac{5}{3}te^{-0.7t} - \frac{25}{9}e^{-0.7t} + \frac{25}{9}e^{-0.1t}\right),$$

$$C(t) = M\left(-\frac{200}{7}e^{-0.8t} + \frac{50}{63}e^{-0.1t} + \frac{250}{9}e^{-0.7t} - \frac{10}{3}te^{-0.7t}\right).$$

This solution can also be obtained using Laplace transforms; see Exercise 26. Fig. 9.6 shows the graph of $C(t)$ for $0 \leq t \leq 31$, using the value $M = 1100$, chosen so that the concentrations $G(t)$ and $C(t)$ are consistent with those observed in northwestern Greece t days after May 1, 1986.

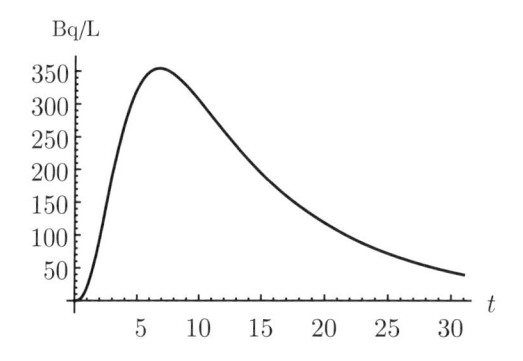

Figure 9.6. Milk compartment concentration $C(t)$ for $0 \leq t \leq 30$ days.

9.4.2. Exercises.

1. For the system in Example 9.4.2, find the solution satisfying the initial condition

$$\mathbf{X}(0) = \begin{pmatrix} 0 \\ 1 \end{pmatrix}.$$

2. The matrix

$$\mathbf{A} = \begin{pmatrix} 0 & -2 \\ 2 & 0 \end{pmatrix}$$

has

$$e^{t\mathbf{A}} = \begin{pmatrix} \cos(2t) & -\sin(2t) \\ \sin(2t) & \cos(2t) \end{pmatrix}.$$

Using this, solve the initial value problem $\mathbf{X}' = \mathbf{A}\mathbf{X} + \mathbf{G}(t)$, $\mathbf{X}(0) = \mathbf{X}_0$, for the given choices of $\mathbf{G}(t)$ and $\mathbf{X}_0$.

(a) $\mathbf{G}(t) = \begin{pmatrix} t\cos(2t) \\ t\sin(2t) \end{pmatrix}$, $\mathbf{X}_0 = \begin{pmatrix} 0 \\ 0 \end{pmatrix}$. (b) $\mathbf{G}(t) = \begin{pmatrix} t\cos(2t) \\ t\sin(2t) \end{pmatrix}$, $\mathbf{X}_0 = \begin{pmatrix} 3 \\ 1 \end{pmatrix}$.

In Exercises 3–5, find the first component function $x(t)$ in the solution of the initial value problem $\mathbf{X}' = \mathbf{A}\mathbf{X} + \mathbf{G}(t)$, $\mathbf{X}(0) = \mathbf{X}_0$, for the given choices of $\mathbf{G}(t)$ and $\mathbf{X}_0$. The matrix $e^{t\mathbf{A}}$ is provided.

3.
$$\mathbf{A} = \begin{pmatrix} 4 & 3 \\ 2 & -1 \end{pmatrix}, \quad \mathbf{X}(t) = \begin{pmatrix} x(t) \\ y(t) \end{pmatrix}, \quad \mathbf{G}(t) = \begin{pmatrix} 3 \\ 1 \end{pmatrix}, \quad \mathbf{X}_0 = \begin{pmatrix} 0 \\ 1 \end{pmatrix};$$

$$e^{t\mathbf{A}} = \frac{1}{7}\begin{pmatrix} e^{-2t} + 6e^{5t} & -3e^{-2t} + 3e^{5t} \\ -2e^{-2t} + 2e^{5t} & 6e^{-2t} + e^{5t} \end{pmatrix}.$$

9.4. Nonhomogeneous constant coefficient systems

4.
$$\mathbf{A} = \begin{pmatrix} -2 & -1 & 1 \\ -1 & 2 & -1 \\ -2 & 6 & -3 \end{pmatrix}, \quad \mathbf{X}(t) = \begin{pmatrix} x(t) \\ y(t) \\ z(t) \end{pmatrix}, \quad \mathbf{G}(t) = \begin{pmatrix} e^{-t} \\ 0 \\ 0 \end{pmatrix}, \quad \mathbf{X}_0 = \begin{pmatrix} 0 \\ 0 \\ 0 \end{pmatrix};$$

$$e^{t\mathbf{A}} = e^{-t} \begin{pmatrix} 1-t & 2t^2-t & t-t^2 \\ -t & 1+3t+2t^2 & -t-t^2 \\ -2t & 6t+4t^2 & 1-2t-2t^2 \end{pmatrix}.$$

5.
$$\mathbf{A} = \begin{pmatrix} 1 & -3 \\ -2 & 2 \end{pmatrix}, \quad \mathbf{X}(t) = \begin{pmatrix} x(t) \\ y(t) \end{pmatrix}, \quad \mathbf{G}(t) = \begin{pmatrix} e^{-2t} \\ 0 \end{pmatrix}, \quad \mathbf{X}_0 = \begin{pmatrix} 0 \\ 0 \end{pmatrix};$$

$$e^{t\mathbf{A}} = \begin{pmatrix} \frac{1}{5}e^{-t}(3+2e^{5t}) & -\frac{3}{5}e^{-t}(-1+e^{5t}) \\ -\frac{2}{5}e^{-t}(-1+e^{5t}) & \frac{1}{5}e^{-t}(2+3e^{5t}) \end{pmatrix}.$$

6. Solve $\mathbf{X}' = \mathbf{A}\mathbf{X} + \mathbf{G}(t)$ with initial condition $\mathbf{X}(0) = \mathbf{X}_0$, where

$$\mathbf{A} = \begin{pmatrix} 0 & 1 \\ -1 & 0 \end{pmatrix}, \quad \mathbf{G}(t) = \begin{pmatrix} 0 \\ \cos(2t) \end{pmatrix}, \quad \text{and} \quad \mathbf{X}_0 = \begin{pmatrix} 0 \\ 0 \end{pmatrix}.$$

For this choice of $\mathbf{A}$, the matrix

$$e^{t\mathbf{A}} = \begin{pmatrix} \cos t & \sin t \\ -\sin t & \cos t \end{pmatrix}$$

was computed in Example 9.4.2. The identities

$$\cos\alpha\cos\beta = \frac{1}{2}[\cos(\alpha+\beta) + \cos(\alpha-\beta)]$$

and

$$\sin\alpha\cos\beta = \frac{1}{2}[\sin(\alpha+\beta) + \sin(\alpha-\beta)]$$

may be helpful.

In Exercises 7–9, solve the nonhomogeneous system $\mathbf{X}' = \mathbf{A}\mathbf{X} + \mathbf{G}(t)$ with initial condition $\mathbf{X}(0) = \mathbf{X}_0$, where $\mathbf{A}$, $\mathbf{G}(t)$, and $\mathbf{X}_0$ are as given.

7. $\mathbf{A} = \begin{pmatrix} 0 & 1 \\ 1 & 0 \end{pmatrix}$, $\mathbf{G}(t) = \begin{pmatrix} -e^{-t} \\ e^{-t} \end{pmatrix}$, and $\mathbf{X}_0 = \begin{pmatrix} 0 \\ 0 \end{pmatrix}$.

8. $\mathbf{A} = \begin{pmatrix} 2 & 3 \\ 3 & 2 \end{pmatrix}$, $\mathbf{G}(t) = \begin{pmatrix} 1 \\ 1 \end{pmatrix}$, and $\mathbf{X}_0 = \begin{pmatrix} 1 \\ -1 \end{pmatrix}$.

9. $\mathbf{A} = \begin{pmatrix} 1 & -2 \\ 1 & 4 \end{pmatrix}$, $\mathbf{G}(t) = \begin{pmatrix} 4e^{-t} \\ -4e^{-t} \end{pmatrix}$, and $\mathbf{X}_0 = \begin{pmatrix} -1 \\ 0 \end{pmatrix}$.

10. For the nonhomogeneous system $\mathbf{X}' = \mathbf{A}\mathbf{X} + \mathbf{G}$ with $\mathbf{A}$ and $\mathbf{G}$ as given, find any equilibrium solutions, that is, any solutions $\mathbf{X}(t)$ where each component of $\mathbf{X}$ is constant.

   (a) $\mathbf{A} = \begin{pmatrix} 1 & 2 \\ -1 & 0 \end{pmatrix}$, $\mathbf{G} = \begin{pmatrix} 3 \\ 2 \end{pmatrix}$.
   (b) $\mathbf{A} = \begin{pmatrix} 1 & 2 \\ 2 & 4 \end{pmatrix}$, $\mathbf{G} = \begin{pmatrix} 3 \\ 6 \end{pmatrix}$.

   (c) $\mathbf{A} = \begin{pmatrix} 1 & 2 \\ 2 & 4 \end{pmatrix}$, $\mathbf{G} = \begin{pmatrix} 3 \\ 2 \end{pmatrix}$.

In Exercises 11–13, assume the entries of $\mathbf{P}(t)$ and $\mathbf{G}(t)$ are continuous functions of t.

11. Suppose that $\mathbf{Z}$ is a solution of the homogeneous system $\mathbf{X}' = \mathbf{P}(t)\mathbf{X}$ and $\mathbf{Y}$ is a solution of the nonhomogeneous system $\mathbf{X}' = \mathbf{P}(t)\mathbf{X} + \mathbf{G}(t)$. Show that $\mathbf{Z} + \mathbf{Y}$ is also a solution to $\mathbf{X}' = \mathbf{P}(t)\mathbf{X} + \mathbf{G}(t)$.

12. Suppose we have found two solutions $\mathbf{X}_1$ and $\mathbf{X}_2$ to $\mathbf{X}' = \mathbf{P}(t)\mathbf{X} + \mathbf{G}(t)$. Show that $\mathbf{X}_1 - \mathbf{X}_2$ solves the associated homogeneous system $\mathbf{X}' = \mathbf{P}(t)\mathbf{X}$.

13. Suppose that $\mathbf{Y}$ solves $\mathbf{X}' = \mathbf{P}(t)\mathbf{X} + \mathbf{G}(t)$. If c is any constant, show that $c\mathbf{Y}$ solves $\mathbf{X}'(t) = \mathbf{P}(t)\mathbf{X}(t) + c\mathbf{G}(t)$.

14. (a) Suppose we have found that $\mathbf{Y}_1$ solves the system $\mathbf{X}' = \mathbf{P}(t)\mathbf{X} + \mathbf{G}_1(t)$ and $\mathbf{Y}_2$ solves the system $\mathbf{X}' = \mathbf{P}(t)\mathbf{X} + \mathbf{G}_2(t)$. Show that $\mathbf{Y}_1 + \mathbf{Y}_2$ solves the system $\mathbf{X}' = \mathbf{P}(t)\mathbf{X} + \mathbf{G}_1(t) + \mathbf{G}_2(t)$.

    (b) Extend (a) to obtain the **Generalized Superposition Principle** for systems: Show that if $\mathbf{Y}_j$ solves $\mathbf{X}' = \mathbf{P}(t)\mathbf{X} + \mathbf{G}_j(t)$ for each j, $1 \leq j \leq k$, then $c_1\mathbf{Y}_1 + c_2\mathbf{Y}_2 + \cdots + c_k\mathbf{Y}_k$ solves $\mathbf{X}' = \mathbf{P}(t)\mathbf{X} + \mathbf{G}$ for $\mathbf{G} = c_1\mathbf{G}_1 + c_2\mathbf{G}_2 + \cdots + c_k\mathbf{G}_k$.

15. Given an n-dimensional vector function $\mathbf{F}(t)$ with the special form $\mathbf{F}(t) = h(t)\mathbf{C}$ where $h(t)$ is a scalar function and

$$\mathbf{C} = \begin{pmatrix} c_1 \\ c_2 \\ \vdots \\ c_n \end{pmatrix}$$

is a constant vector, show that

$$\int_a^b \mathbf{F}(t)dt = \left(\int_a^b h(t)dt\right)\mathbf{C}$$

where $\int_a^b h(t)dt$ is the ordinary scalar integral of h from a to b.

16. (a) If $\mathbf{B}(t)$ is an $n \times n$ matrix function and $\mathbf{C}$ is an $n \times 1$ vector of constants, then

$$\int_a^b \mathbf{B}(t)\mathbf{C}\,dt = \left(\int_a^b \mathbf{B}(t)\,dt\right)\mathbf{C},$$

so that the constant vector $\mathbf{C}$ can be moved outside the integration (on the right). Verify this when $n = 2$, so that

$$\mathbf{B}(t) = \begin{pmatrix} b_{11}(t) & b_{12}(t) \\ b_{21}(t) & b_{22}(t) \end{pmatrix} \quad \text{and} \quad \mathbf{C} = \begin{pmatrix} c_1 \\ c_2 \end{pmatrix}.$$

    (b) If $\mathbf{C}$ is a constant $n \times n$ matrix and $\mathbf{B}(t)$ is an $n \times 1$ vector-valued function, then

$$\int_a^b \mathbf{C}\mathbf{B}(t)\,dt = \mathbf{C}\int_a^b \mathbf{B}(t)\,dt,$$

so that the constant matrix $\mathbf{C}$ can be moved outside the integration (on the left). Verify this formula when $n = 2$, so that

$$\mathbf{C} = \begin{pmatrix} c_{11} & c_{12} \\ c_{21} & c_{22} \end{pmatrix} \quad \text{and} \quad \mathbf{B}(t) = \begin{pmatrix} b_1(t) \\ b_2(t) \end{pmatrix}.$$

17. Check that the solution to the initial value problem $\mathbf{X}' = \mathbf{A}\mathbf{X} + \mathbf{G}(t)$, $\mathbf{X}(t_0) = \mathbf{X}_0$, can be written

$$\mathbf{X}(t) = e^{\mathbf{A}(t-t_0)}\mathbf{X}_0 + e^{t\mathbf{A}}\int_{t_0}^t e^{-s\mathbf{A}}\mathbf{G}(s)ds.$$

18. (CAS) Lidocaine is a drug used to treat irregular heartbeats. The compartment model shown below is appropriate to model the pharmacokinetics of this drug, with the approximate values of the transfer coefficients (in units of $\min^{-1}$) shown.

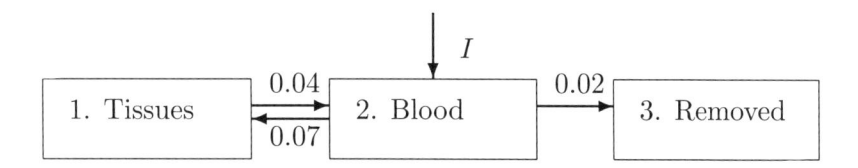

9.4. Nonhomogeneous constant coefficient systems

```
              ↓ I
  ┌─────────┐ 0.04  ┌─────────┐ 0.02  ┌─────────┐
  │1. Tissues│ ────→ │ 2. Blood │ ────→ │3. Removed│
  │         │ ←──── │         │       │         │
  └─────────┘ 0.07  └─────────┘       └─────────┘
```

Suppose lidocaine is to be infused into the bloodstream at a constant rate of I mg/min starting at time $t = 0$.

(a) *Set up* a system of differential equations for the amount of lidocaine in the tissue ($x(t)$) and blood ($y(t)$) compartments at time t. Write your system in matrix form.

(b) Find any equilibrium solutions of your system in (a). Your answer will be in terms of the constant I.

(c) If $I = 2.5$ mg/min, will the equilibrium solution be above the minimum therapeutic concentration of 1.5 mg/liter in the blood? Assume the blood compartment volume is 30 liters (the blood compartment contains, in addition to blood, other body fluids in equilibrium with the blood).

(d) (CAS) Use a computer algebra system to find $y(t)$, the amount of lidocaine in the blood compartment, assuming $I = 2.5$ and that $x(0) = 0, y(0) = 0$.

(e) If $I = 2.5$ mg/min, will the concentration in the blood ever exceed the maximum safe concentration of 6 mg/liter? Again, assume the blood compartment contains 30 liters.[4]

19. (CAS) This exercise gives a refinement of the compartment model for lead poisoning from Exercise 16 in Section 8.8.[5] Now we consider three compartments in the body: 1. blood, 2. soft tissues, and 3. bones, plus an external compartment, as shown below. Lead can move directly between compartments 1 and 2 and between 1 and 3, as indicated. It leaves the body from 1 and 2 and enters the body first through compartment 1. An exchange from compartment i to compartment j takes place at a rate proportional to the amount of lead in compartment i, with proportionality constant a_{ji}.

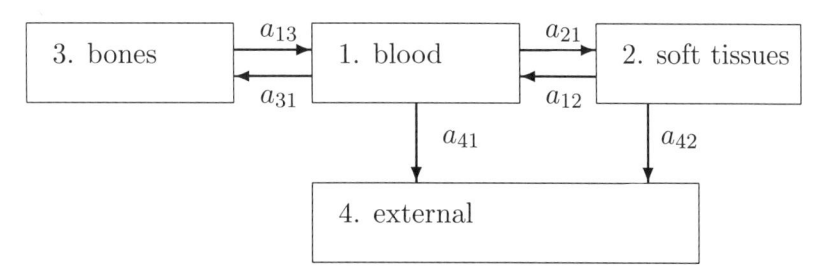

```
  ┌─────────┐ a₁₃  ┌─────────┐ a₂₁  ┌──────────────┐
  │3. bones │ ───→ │1. blood │ ───→ │2. soft tissues│
  │         │ ←─── │         │ ←─── │              │
  └─────────┘ a₃₁  └─────────┘ a₁₂  └──────────────┘
                     │ a₄₁            │ a₄₂
                     ↓                ↓
                   ┌──────────────────┐
                   │  4. external     │
                   └──────────────────┘
```

The constants a_{ji} have been estimated in studies, and using their values, we have the following equations for the amount of lead x_i, in micrograms, at time t in days, in compartment i:

$$\frac{dx_1}{dt} = -.036x_1 + .0124x_2 + .000035x_3, \quad \frac{dx_2}{dt} = .0111x_1 - .0286x_2, \quad \frac{dx_3}{dt} = .0039x_1 - .000035x_3,$$

assuming no additional lead enters the body.

(a) What are the values of the constants $a_{12}, a_{13}, a_{21}, a_{31}, a_{41}$, and a_{42}?

(b) How is this system of equations modified if there is a constant intake of lead into the body (all entering into the blood compartment) of I micrograms/day?

(c) Find the equilibrium solutions of the system in part (b) in terms of I.

---

[4] Some data in this problem is taken from M. R. Cullen, *Linear Models in Biology: Linear Systems Analysis with Biological Applications*, Ellis Horwood, 1985.

[5] The model in this problem comes from *Lead Metabolism in the Normal Human: Stable Isotope Studies*, by M. Rabinowitz, G. Wetherill, and J. Kopple, Science 182 (1973), 725–727.

(d) Writing the system in (b) in matrix notation as $\mathbf{X}' = \mathbf{A}\mathbf{X} + \mathbf{B}$, identify the 3×3 matrix $\mathbf{A}$ and the column matrix $\mathbf{B}$.

(e) (CAS) Using any technology at your disposal, show that $\mathbf{A}$ has three negative eigenvalues.

(f) Suppose that $I = 50$ mcg/day. What are the limiting values of $x_1(t)$, $x_2(t)$, and $x_3(t)$, as $t \to \infty$? Hint: Essentially no new computations are needed if you use your work in the earlier parts of this problem and equation (9.49). Your answer will not depend on the initial values of x_1, x_2, and x_3.

20. Two 1,000-gallon tanks are initially both full of pure water. Starting at time $t = 0$, water containing salt in a concentration of 1 lb/gal is added to the first tank at a rate of 1 gal/hour. The well-mixed solution flows into the second tank at a rate of 3 gal/hour and back from the second tank to the first at a rate of 2 gal/hour. Pure water evaporates from the second tank at a rate of 1 gal/hour.

   (a) Set up an initial value problem for the amount of salt in each tank at time t.

   (b) Does the nonhomogeneous system in (a) have any equilibrium solutions?

   (c) Let $x(t)$ and $y(t)$ be the amounts of salt in, respectively, the first and second tanks at time t. What should be the values of $\lim_{t \to \infty} x(t)$ and $\lim_{t \to \infty} y(t)$? How would you assess the long-term validity of the mathematical model of (a)?

   (d) Solve the initial value problem of part (a) and confirm your answer to (c) is correct. Remark: A hand computation of the solution is facilitated by rewriting equation (9.44) as

   $$\mathbf{X}(t) = e^{t\mathbf{A}}\mathbf{X}_0 + \int_0^t e^{(t-s)\mathbf{A}}\mathbf{G}(s)\,ds,$$

   which follows from the identity $e^{t\mathbf{A}}e^{-s\mathbf{A}} = e^{(t-s)\mathbf{A}}$ (see Section 9.2.1).

21. In this problem you will use equation (9.44) directly to give the general solution to

    (9.52) $$\mathbf{X}'(t) = \mathbf{A}\mathbf{X}(t) + \mathbf{D},$$

    where $\mathbf{D}$ is a *constant* vector and $\mathbf{A}$ is an invertible $n \times n$ matrix of constants.

    (a) Verify that equation (9.52) has equilibrium solution $\mathbf{X}_e = -\mathbf{A}^{-1}\mathbf{D}$.

    (b) Verify that $-e^{-s\mathbf{A}}\mathbf{A}^{-1}$ is an antiderivative for $e^{-s\mathbf{A}}$. Using this antiderivative, show that

    $$\int_0^t e^{-s\mathbf{A}}\,ds = \mathbf{A}^{-1} - e^{-t\mathbf{A}}\mathbf{A}^{-1}.$$

    (c) According to equation (9.44), the solution to $\mathbf{X}'(t) = \mathbf{A}\mathbf{X}(t) + \mathbf{D}$ satisfying the initial condition $\mathbf{X}(0) = \mathbf{X}_0$ is

    $$\mathbf{X}(t) = e^{t\mathbf{A}}\mathbf{X}_0 + e^{t\mathbf{A}}\int_0^t e^{-s\mathbf{A}}\mathbf{D}\,ds.$$

    Since $\mathbf{D}$ is a constant vector, Exercise 16(a) allows us to write

    $$\int_0^t e^{-s\mathbf{A}}\mathbf{D}\,ds = \left(\int_0^t e^{-s\mathbf{A}}\,ds\right)\mathbf{D}.$$

    Show that $\mathbf{X}(t) = e^{t\mathbf{A}}(\mathbf{X}_0 - \mathbf{X}_e) + \mathbf{X}_e$ where $\mathbf{X}_e$ is the equilibrium solution $-\mathbf{A}^{-1}\mathbf{D}$ from (a).

22. In this problem we consider the nonhomogeneous equation

    (9.53) $$\mathbf{X}'(t) = \mathbf{A}\mathbf{X}(t) + \mathbf{G}(t),$$

9.4. Nonhomogeneous constant coefficient systems

where $\mathbf{A}$ is an $n \times n$ constant matrix and

$$\mathbf{G}(t) = \begin{pmatrix} c_1 e^{rt} \\ c_2 e^{rt} \\ \vdots \\ c_n e^{rt} \end{pmatrix} = e^{rt} \begin{pmatrix} c_1 \\ c_2 \\ \vdots \\ c_n \end{pmatrix} = e^{rt} \mathbf{C}$$

for constants $c_1, c_2, \ldots, c_n$. If r is *not* an eigenvalue of $\mathbf{A}$, then $\mathbf{A} - r\mathbf{I}$ is invertible. Show that

$$\mathbf{X}(t) = -e^{rt}(\mathbf{A} - r\mathbf{I})^{-1}\mathbf{C}$$

is a particular solution to (9.53). Hint: Substitute the proposed solution into (9.53) and verify that it works. In computing $\mathbf{X}'(t)$, notice that t only appears in the factor e^{rt}. In computing $\mathbf{AX}$, write $\mathbf{A}(-e^{rt}(\mathbf{A} - r\mathbf{I})^{-1}\mathbf{C})$ as $((\mathbf{A} - r\mathbf{I}) + r\mathbf{I})(-e^{rt}(\mathbf{A} - r\mathbf{I})^{-1}\mathbf{C})$. The scalar factor e^{rt} in this product can be moved around.

23. Verify the computations of $G(t)$ and $C(t)$ in Section 9.4.1.

24. In terms of M and λ, what is the maximum value of $Mte^{-\lambda t}$ and for what value of t does it occur? If $\lambda = 0.7$ as in Section 9.4.1 and the maximum value of $Mte^{-0.7t}$ is known (approximately), can M be determined?

25. The same post-Chernobyl study of ^{131}I contamination in northwestern Greece as discussed in Section 9.4.1 also looked at contamination of sheep's milk with the same basic model. The parameter a_{32} is now approximately 0.7/day. A sheep eats about 4 kg/day of grass, and the grass to milk transfer rate for sheep's milk is about 0.7/L. The parameter a_{31} is unchanged. What is the differential equation system analogous to equations (9.50) and (9.51) for the concentrations $G(t)$ and $S(t)$ of ^{131}I in grass and sheep milk, respectively?

26. Use Laplace transforms to solve the system

$$\frac{dG}{dt} = -0.1G + Mte^{-0.7t}, \quad \frac{dC}{dt} = 0.2G - 0.8C,$$

with initial conditions $G(0) = C(0) = 0$.

27. To give further justification for the choice $I(t) = Mte^{-\lambda t}$ for the input of ^{131}I from the atmosphere to the "grass" compartment, the authors of the article [2], on which Section 9.4.1 is based, model the relevant portion of the atmosphere as composed of three compartments of equal volume:

    - Compartment 2, consisting of a box of atmosphere above the location of interest (the dairy farm).
    - Compartment 1, a box of atmosphere just upstream of compartment 2, which feeds into compartment 2.
    - Compartment 3, a box of atmosphere just downstream from compartment 2, which is fed by compartment 2.

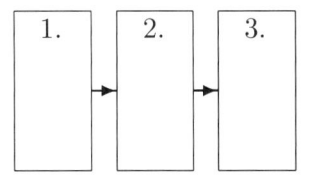

Suppose the concentration of ^{131}I in compartment 1 is given by $C_1(T)$ and in compartment 2 by $C_2(t)$ (say, with units of Bq/m^3). We would also have a third concentration function $C_3(t)$ for the third compartment, but as we will see, we will not need this. This leads to the differential equation system

$$\frac{dC_1}{dt} = -\beta C_1, \quad \frac{dC_2}{dt} = \gamma C_1 - \beta C_2.$$

The parameter β, with units 1/day, reflects both the flow of the contaminant to compartment 2, as well as radioactive decay and deposition on the ground.

Show that the solution to this system of equations, with initial conditions $C_1(0) = C$, $C_2(0) = 0$, leads to a formula for $C_2(t)$ of the form $C_2(t) = Mte^{-\lambda t}$. How are M and λ related to β, γ, and the initial concentration $C_1(0)$?

28. Suppose that in equations (9.50)–(9.51) we assume the atmospheric input function $I(t)$ is a constant I for $t > 0$. Solve the system with the initial conditions $G(0) = 0, C(0) = 0$.

29. In Example 9.4.2, $\mathbf{X}(t) = (1+t, 1-t)$ was shown to be the (necessarily unique) solution to the system
$$\frac{dx}{dt} = x + t, \quad \frac{dy}{dt} = -x + t$$
satisfying the initial conditions $x(0) = 1$, $y(0) = 1$.

(a) Show that for any point (x_0, y_0), the solution of this system satisfying $x(0) = x_0, y(0) = y_0$ is
$$\mathbf{X}(t) = \begin{pmatrix} 1 + t + R\cos(\delta - t) \\ 1 - t + R\sin(\delta - t) \end{pmatrix}$$
where R and δ are the polar coordinates of the point $(x_0 - 1, y_0 - 1)$: $x_0 - 1 = R\cos\delta$, $y_0 - 1 = R\sin\delta$, and $R^2 = (x_0 - 1)^2 + (y_0 - 1)^2 = \|(x_0, y_0) - (1,1)\|^2$. Hint: Use the calculations in Example 9.4.2, at the end substituting (x_0, y_0) for $(1, 1)$. The trigonometric identities $\cos(a - b) = \cos a \cos b + \sin a \sin b$ and $\sin(a - b) = \sin a \cos b - \cos a \sin b$ should be useful.

(b) We can write the formula for $\mathbf{X}(t)$ as

(9.54)
$$\mathbf{X}(t) = \begin{pmatrix} 1 \\ 1 \end{pmatrix} + t\begin{pmatrix} 1 \\ -1 \end{pmatrix} + \begin{pmatrix} R\cos(\delta - t) \\ R\sin(\delta - t) \end{pmatrix}.$$

What curve is described parametrically by the third term? Draw this curve and the parametric line represented by the sum of the first two terms in separate pictures. Include arrows showing the direction they are traversed with increasing t.

(c) (CAS) Give pictures of the trajectories, with $0 \le t \le 4\pi$, for the solution in (9.54) when $(x_0, y_0) = (3/2, 1)$ and when $(x_0, y_0) = (3, 1)$. A noticeable qualitative difference in the two trajectories should be apparent in your pictures; describe it.

9.5. Periodic forcing and the steady-state solution

Sometimes the forcing term $\mathbf{G}(t)$ in a nonhomogeneous system is a periodic function of t, so that there is a positive number T (called a period of $\mathbf{G}$) such that $\mathbf{G}(t + T) = \mathbf{G}(t)$ for all t. If T is a period of $\mathbf{G}(t)$, will the system

(9.55)
$$\mathbf{X}' = \mathbf{A}\mathbf{X} + \mathbf{G}(t)$$

have a periodic solution also having period T? By equation (9.44), we know the solutions to (9.55) are
$$\mathbf{X}(t) = e^{t\mathbf{A}}\mathbf{X}_0 + e^{t\mathbf{A}}\int_0^t e^{-s\mathbf{A}}\mathbf{G}(s)ds$$
where $\mathbf{X}_0 = \mathbf{X}(0)$. If $\mathbf{X}(t)$ has period T, then $\mathbf{X}(T) = \mathbf{X}(0)$, so that

(9.56)
$$\mathbf{X}_0 = \mathbf{X}(T) = e^{T\mathbf{A}}\mathbf{X}_0 + e^{T\mathbf{A}}\int_0^T e^{-s\mathbf{A}}\mathbf{G}(s)ds.$$

9.5. Periodic forcing and the steady-state solution

Suppose we call the second vector on the right-hand side of (9.56) $\mathbf{W}$: $\mathbf{W} = e^{T\mathbf{A}} \int_0^T e^{-s\mathbf{A}} \mathbf{G}(s) ds$. Then in terms of $\mathbf{W}$, equation (9.56) says exactly that

(9.57) $$(\mathbf{I} - e^{T\mathbf{A}})\mathbf{X}_0 = \mathbf{W}.$$

If the matrix $\mathbf{I} - e^{T\mathbf{A}}$ is invertible, there is a unique choice of the initial condition value $\mathbf{X}_0$ satisfying (9.57), namely

(9.58) $$\mathbf{X}_0 = (\mathbf{I} - e^{T\mathbf{A}})^{-1}\mathbf{W}.$$

But recall that $\mathbf{I} - e^{T\mathbf{A}}$ *fails* to be invertible exactly when there is a nonzero vector $\mathbf{V}$ in its nullspace (see Theorem 8.2.9):

$$(\mathbf{I} - e^{T\mathbf{A}})\mathbf{V} = \mathbf{0},$$

or what is the same thing,

$$e^{T\mathbf{A}}\mathbf{V} = \mathbf{V}.$$

This last equation says that 1 is an eigenvalue for $e^{T\mathbf{A}}$. To interpret this condition recall that in Section 9.1 we saw that if r is an eigenvalue of $\mathbf{A}$ and $t > 0$, then e^{tr} is an eigenvalue of $e^{t\mathbf{A}}$. In fact, every eigenvalue of $e^{t\mathbf{A}}$ arises in this way:

> The set of eigenvalues of $e^{t\mathbf{A}}$ is exactly equal to $\{e^{tr} : r \text{ is an eigenvalue of } \mathbf{A}\}$.

We will accept this on faith. It follows that the only way 1 can be an eigenvalue for $e^{T\mathbf{A}}$ is for $\mathbf{A}$ to have an eigenvalue r with $e^{rT} = 1$. The only way this can be true is if $rT = 2\pi i k$ for some integer k. Let's turn this around and summarize:

Theorem 9.5.1. *If $\mathbf{G}(t)$ is a continuous function having period T and no eigenvalue of $\mathbf{A}$ is an integer multiple of $\frac{2\pi i}{T}$, then the system (9.55) has a unique solution $\mathbf{X}(t)$ with $\mathbf{X}(t+T) = \mathbf{X}(t)$ for all t.*

In particular, note that if an invertible matrix $\mathbf{A}$ has no *imaginary* eigenvalues, there is a unique periodic solution to equation (9.55).

The reader may object that we have only shown that this eigenvalue condition on $\mathbf{A}$ guarantees a unique solution $\mathbf{X}(t)$ with $\mathbf{X}(T) = \mathbf{X}(0)$. But in Exercise 1 you can verify that such a solution is actually periodic with period T, so that $\mathbf{X}(t+T) = \mathbf{X}(t)$ for all t, not just for $t = 0$. This justifies the full statement above.

Now let's consider a *dissipative* system

(9.59) $$\mathbf{X}' = \mathbf{AX} + \mathbf{G}(t),$$

where $\mathbf{G}(t)$ is periodic with period T. Recall that being dissipative means that all eigenvalues of $\mathbf{A}$ have negative real part. In particular, there are no purely imaginary eigenvalues. Therefore, by the above discussion, there is a unique periodic solution of (9.59), which is often called the **steady-state solution**. Also, we saw in Theorem 9.3.5 that every solution of the homogeneous (that is, the unforced) system $\mathbf{X}' = \mathbf{AX}$ tends to zero as time tends to infinity, reflecting, in real-world situations, the presence of friction, damping, electrical resistance, or other forms of drag. For this reason, all solutions of the homogeneous system can reasonably be called **transient**.

We know that every solution of equation (9.59) has the form

(9.60) $$\mathbf{X}(t) = e^{t\mathbf{A}}\mathbf{X}_0 + e^{t\mathbf{A}} \int_0^t e^{-s\mathbf{A}}\mathbf{G}(s)\, ds,$$

where $\mathbf{X}(0) = \mathbf{X}_0$. The first term (a transient solution) solves the unforced system with exactly this initial condition, while the second solves the nonhomogeneous system and has initial value *zero* at $t = 0$. For this reason, the second term

$$(9.61) \qquad e^{t\mathbf{A}} \int_0^t e^{-s\mathbf{A}} \mathbf{G}(s)\, ds$$

is sometimes called the **zero initial state solution** of equation (9.59).

Here we are more interested in another way of decomposing $\mathbf{X}(t)$ into a sum. Letting $\mathbf{P}_0$ be the initial value giving rise to the unique periodic solution, we have

$$\mathbf{X}(t) = e^{t\mathbf{A}}(\mathbf{X}_0 - \mathbf{P}_0) + \left(e^{t\mathbf{A}}\mathbf{P}_0 + e^{t\mathbf{A}} \int_0^t e^{-s\mathbf{A}} \mathbf{G}(s)\, ds \right)$$
$$= \text{another transient solution } + \text{ the steady-state solution.}$$

The transient solution $e^{t\mathbf{A}}(\mathbf{X}_0 - \mathbf{P}_0)$ solves the unforced system and so tends to zero as $t \to \infty$ (and hence is "transient") and can usually in practice be ignored. The other summand (the steady-state solution) is just the unique periodic solution. It repeats over and over with period T and represents, in practical terms, the essential long-term behavior of the system. It is usually what we most want to find.

Let's consider how to find the steady-state solution $\mathbf{S}(t)$ when $\mathbf{G}(t)$ is the simplest possible periodic function, a "complex sinusoid"

$$\mathbf{G}(t) = e^{i\omega t} \mathbf{U}.$$

Here $\mathbf{U}$ is a constant vector and $\omega > 0$ is the circular frequency. Note that $\mathbf{G}(t)$ is periodic with period $T = \frac{2\pi}{\omega}$.

We would expect the steady-state solution $\mathbf{S}(t)$ to also involve $e^{i\omega t}$. Could it be as simple as $\mathbf{S}(t) = e^{i\omega t}\mathbf{B}$, where $\mathbf{B}$ is another constant vector? Let's check by plugging this $\mathbf{S}(t)$ into our system: $\mathbf{S}'(t) = i\omega e^{i\omega t}\mathbf{B}$, so

$$\mathbf{S}'(t) - \mathbf{A}\mathbf{S}(t) = (i\omega \mathbf{I} - \mathbf{A})e^{i\omega t}\mathbf{B}.$$

We want the right side here to equal $\mathbf{G}(t) = e^{i\omega t}\mathbf{U}$; that is, we want

$$(9.62) \qquad (i\omega \mathbf{I} - \mathbf{A})\mathbf{B} = \mathbf{U}.$$

Since $i\omega$ is not an eigenvalue of $\mathbf{A}$, $(i\omega \mathbf{I} - \mathbf{A})$ is invertible, and we can multiply equation (9.62) on the left by $(i\omega \mathbf{I} - \mathbf{A})^{-1}$ to obtain

$$\mathbf{B} = (i\omega \mathbf{I} - \mathbf{A})^{-1} \mathbf{U}.$$

Thus:

> (9.63) $\qquad \mathbf{S}(t) = e^{i\omega t}(i\omega \mathbf{I} - \mathbf{A})^{-1}\mathbf{U}$
>
> solves our dissipative system
>
> $$\mathbf{X}' = \mathbf{A}\mathbf{X} + e^{i\omega t}\mathbf{U}$$
>
> and is certainly periodic with period T, so it must be the unique steady-state solution.

Even if the system is not dissipative, the above procedure produces a periodic solution with period $T = \frac{2\pi}{\omega}$ as long as $i\omega$ is not an eigenvalue of $\mathbf{A}$.

9.5. Periodic forcing and the steady-state solution

In applications one is more likely to encounter forcing functions $\mathbf{G}(t)$ involving the trigonometric functions $\cos(\omega t)$ or $\sin(\omega t)$ than $e^{i\omega t}$. However, it is easy to use our result above and the formulas

$$\cos(\omega t) = \frac{1}{2} e^{i\omega t} + \frac{1}{2} e^{-i\omega t}, \quad \sin(\omega t) = \frac{1}{2i} e^{i\omega t} - \frac{1}{2i} e^{-i\omega t} \tag{9.64}$$

to handle this situation.

For example, suppose we want to find the steady-state solution $\mathbf{S}(t)$ of the system

$$\mathbf{X}' = \mathbf{A}\mathbf{X} + \cos(\omega t)\mathbf{U}, \tag{9.65}$$

where $\mathbf{U}$ is a constant vector and $i\omega$ is not an eigenvalue of $\mathbf{A}$. On writing the forcing function $\mathbf{G}(t) = \cos(\omega t)\mathbf{U}$ as a sum corresponding to the first equation in (9.64), we have

$$\mathbf{G}(t) = \cos(\omega t)\mathbf{U} = \frac{1}{2} e^{i\omega t}\mathbf{U} + \frac{1}{2} e^{-i\omega t}\mathbf{U}. \tag{9.66}$$

By our discussion above,

$$\mathbf{S}_1(t) = \frac{1}{2} e^{i\omega t}(i\omega \mathbf{I} - \mathbf{A})^{-1}\mathbf{U} \quad \text{and}$$

$$\mathbf{S}_2(t) = \frac{1}{2} e^{-i\omega t}(-i\omega \mathbf{I} - \mathbf{A})^{-1}\mathbf{U}$$

$$= -\frac{1}{2} e^{-i\omega t}(i\omega \mathbf{I} + \mathbf{A})^{-1}\mathbf{U}$$

are the respective steady-state solutions of the systems

$$\mathbf{X}' = \mathbf{A}\mathbf{X} + \frac{1}{2} e^{i\omega t}\mathbf{U} \quad \text{and} \quad \mathbf{X}' = \mathbf{A}\mathbf{X} + \frac{1}{2} e^{-i\omega t}\mathbf{U}.$$

Here we have used the fact that for a real matrix, complex eigenvalues occur in conjugate pairs, and so $-i\omega$, as well as $i\omega$, fails to be an eigenvalue of $\mathbf{A}$. It follows from the Generalized Superposition Principle for systems (see Exercise 14 of Section 9.4) and equation (9.66) that the steady-state solution $\mathbf{S}(t)$ of the system (9.65) is

$$\mathbf{S}(t) = \mathbf{S}_1(t) + \mathbf{S}_2(t)$$

$$= \frac{1}{2} e^{i\omega t}(i\omega \mathbf{I} - \mathbf{A})^{-1}\mathbf{U} - \frac{1}{2} e^{-i\omega t}(i\omega \mathbf{I} + \mathbf{A})^{-1}\mathbf{U}$$

$$= -\cos(\omega t)(\omega^2 \mathbf{I} + \mathbf{A}^2)^{-1}\mathbf{A}\mathbf{U} + \omega \sin(\omega t)(\omega^2 \mathbf{I} + \mathbf{A}^2)^{-1}\mathbf{U}. \tag{9.67}$$

The last equality follows from Euler's formulas for $e^{i\omega t}$ and $e^{-i\omega t}$ and a few lines of algebra. The reader is invited to fill in the details in Exercise 4.

A similar argument will show that the steady-state solution of the system

$$\mathbf{X}' = \mathbf{A}\mathbf{X} + \sin(\omega t)\mathbf{U} \tag{9.68}$$

is

$$\mathbf{S}(t) = -\omega \cos(\omega t)(\omega^2 \mathbf{I} + \mathbf{A}^2)^{-1}\mathbf{U} - \sin(\omega t)(\omega^2 \mathbf{I} + \mathbf{A}^2)^{-1}\mathbf{A}\mathbf{U}; \tag{9.69}$$

again see Exercise 4.

Example 9.5.2. Consider the system
$$\frac{dx}{dt} = -x - y + 2\cos t, \quad \frac{dy}{dt} = 2x - y.$$

This has the form (9.65) where
$$\mathbf{A} = \begin{pmatrix} -1 & -1 \\ 2 & -1 \end{pmatrix}, \quad \omega = 1, \quad \mathbf{U} = \begin{pmatrix} 2 \\ 0 \end{pmatrix}.$$

Note that the eigenvalues of $\mathbf{A}$ are $-1 \pm i\sqrt{2}$, so the system is dissipative and there is a unique periodic solution $\mathbf{S}(t)$ with period $T = \frac{2\pi}{\omega} = 2\pi$. To find it, we compute
$$\mathbf{A}^2 = \begin{pmatrix} -1 & 2 \\ -4 & -1 \end{pmatrix}, \quad \omega^2\mathbf{I} + \mathbf{A}^2 = \mathbf{I} + \mathbf{A}^2 = \begin{pmatrix} 0 & 2 \\ -4 & 0 \end{pmatrix}$$

and
$$(\omega^2\mathbf{I} + \mathbf{A}^2)^{-1} = \begin{pmatrix} 0 & 2 \\ -4 & 0 \end{pmatrix}^{-1} = \begin{pmatrix} 0 & -\frac{1}{4} \\ \frac{1}{2} & 0 \end{pmatrix}.$$

It follows from equation (9.67) that
$$\mathbf{S}(t) = -\cos t \begin{pmatrix} 0 & -\frac{1}{4} \\ \frac{1}{2} & 0 \end{pmatrix} \begin{pmatrix} -1 & -1 \\ 2 & -1 \end{pmatrix} \begin{pmatrix} 2 \\ 0 \end{pmatrix} + \sin t \begin{pmatrix} 0 & -\frac{1}{4} \\ \frac{1}{2} & 0 \end{pmatrix} \begin{pmatrix} 2 \\ 0 \end{pmatrix}$$
$$= \begin{pmatrix} \cos t \\ \cos t + \sin t \end{pmatrix}.$$

9.5.1. Exercises.

1. Suppose that $\mathbf{G}(t)$ has period $T > 0$. Suppose that $\mathbf{X}(t)$ is a solution of $\mathbf{X}' = \mathbf{A}\mathbf{X} + \mathbf{G}(t)$ having $\mathbf{X}(0) = \mathbf{X}(T)$. Show that $\mathbf{X}(t) = \mathbf{X}(t + T)$ for all t, so that $\mathbf{X}(t)$ is periodic with period T. Hint: Show that the vector function $\mathbf{Y}(t) = \mathbf{X}(t) - \mathbf{X}(t + T)$ solves some simple homogeneous system of differential equations.

2. Let $\mathbf{A}$ be an $n \times n$ matrix with no imaginary eigenvalues. Show that $\mathbf{A}^2$ has no negative eigenvalues and therefore $(\omega^2\mathbf{I} + \mathbf{A}^2)^{-1}$ exists for all nonzero real numbers ω. Suggestion: Establish the contrapositive.

3. Let $\mathbf{A}$ and $\mathbf{B}$ be $n \times n$ matrices such that $\mathbf{A}\mathbf{B} = \mathbf{B}\mathbf{A}$ and $\mathbf{B}$ is invertible. Show that $\mathbf{A}\mathbf{B}^{-1} = \mathbf{B}^{-1}\mathbf{A}$.

4. (a) Let $\mathbf{A}$ be an $n \times n$ matrix with no imaginary eigenvalues. Show that for any nonzero real number ω,
$$(i\omega\mathbf{I} + \mathbf{A})^{-1} = -(\omega^2\mathbf{I} + \mathbf{A}^2)^{-1}(i\omega\mathbf{I} - \mathbf{A})$$
and
$$(i\omega\mathbf{I} - \mathbf{A})^{-1} = -(\omega^2\mathbf{I} + \mathbf{A}^2)^{-1}(i\omega\mathbf{I} + \mathbf{A}).$$

Hint: We have $-(\omega^2\mathbf{I} + \mathbf{A}^2) = (i\omega\mathbf{I} - \mathbf{A})(i\omega\mathbf{I} + \mathbf{A})$, where all factors are invertible under our hypothesis on ω, by Exercise 2 above and Exercise 18 in Section 8.3.

(b) Use your result in (a) and the formulas in equation (9.64) to derive the formula (9.67) for the unique periodic solution of the system (9.65).

(c) Derive the formula (9.69) for the unique periodic solution of the system (9.68).

9.5. Periodic forcing and the steady-state solution

5. Find the steady-state solution for each of the following systems.
   (a) $\dfrac{dx}{dt} = -x - y + 3\sin(2t)$, $\dfrac{dy}{dt} = x - y$.
   (b) $\dfrac{dx}{dt} = -x - y$, $\dfrac{dy}{dt} = x - y - \cos(2t)$.

6. (a) Note that the forcing function in the nonhomogeneous system
   $$\frac{dx}{dt} = -y + \cos(2t), \quad \frac{dy}{dt} = 16x$$
   is periodic with period $T = \pi$. Show that **every** solution of this system is periodic with period $T = \pi$.
   (b) Explain why your result in (a) does not contradict Theorem 9.5.1.
   (c) The system
   $$\frac{dx}{dt} = -y + \cos(2t), \quad \frac{dy}{dt} = 9x$$
   has the same forcing function as the system in part (a). How many periodic solutions with period $T = \pi$ does this new system have? Justify your answer.
   (d) Does the system
   $$\frac{dx}{dy} = -y + \cos(2t), \quad \frac{dy}{dt} = 4x$$
   have **any** periodic solutions? Justify your answer.

7. The currents I_1, I_2, and I_3 flowing in the three branches of the pictured circuit are modeled by the algebraic equation $I_1 = I_2 + I_3$ (obtained through considering current flow entering and leaving node a) and the differential equations

   (9.70) $\begin{aligned}\dfrac{dI_1}{dt} &= -12I_1 + 12I_2 + 2\cos(6t), \\ \dfrac{dI_2}{dt} &= -3I_1 + \sin(6t).\end{aligned}$

   (a) Find the steady-state solution—yielding steady-state currents—for the system (9.70).
   (b) Find the amplitudes of the steady-state currents I_1 and I_2
   (c) Using your steady-state solution vector function as a particular solution $\mathbf{I}_p$, apply Theorem 9.4.3 on the general solution for nonhomogenous systems to solve the system (9.70) subject to the initial conditions $I_1(0) = 0$, $I_2(0) = 0$.
   (d) (CAS) Graph $I_1(t)$ and $I_2(t)$ for $0 \le t \le 4$ and check your answers to part (b).

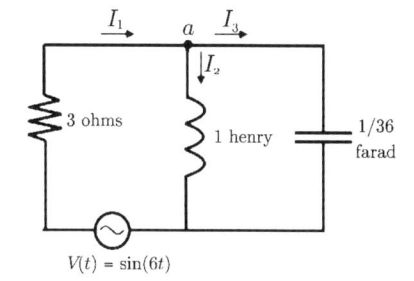

Chapter 10

Nonlinear Systems

10.1. Introduction: Darwin's finches

In 1835, Charles Darwin, as part of the second voyage of the *Beagle*, visited the Galapagos Islands where he collected 13 species of finch-like birds. These bird species were all quite similar in size and color, except for showing a remarkable variation in beak size. Different sized beaks translated into birds which had adapted to using different food sources; one species fed on seeds, another on insects, another on grubs, and so on. Darwin's finches, as they came to be known, provide one of the clearest examples of the *Principle of Competitive Exclusion*, which says that when two highly similar species living in the same environment compete for the same resources (that is, the species occupy the same ecological niche), the competition between the species for these resources will be so fierce that either

(1) one species will be driven to extinction or

(2) an evolutionary adaption will occur to move the species into different niches.

Here we will propose a mathematical model that predicts this principle and use this to introduce nonlinear systems and some of the tools we will apply to analyze them.

Consider two species, with populations $x(t)$ and $y(t)$ at time t, that separately (in isolation) would grow according to logistic models

$$(10.1) \qquad \frac{dx}{dt} = ax - bx^2, \qquad \frac{dy}{dt} = cy - dy^2$$

for positive constants a, b, c, and d. Recall from Section 2.5 that the terms $-bx^2$ and $-dy^2$ in these equations take into account the competition for resources *within* the x and y populations. We want to modify the equations to also reflect competition *between* the two species. To see how we might do this under the assumption that the species occupy the same niche, we rewrite the equations in (10.1) as

$$(10.2) \qquad \frac{dx}{dt} = ax\left(\frac{K_1 - x}{K_1}\right), \qquad \frac{dy}{dt} = cy\left(\frac{K_2 - y}{K_2}\right),$$

where $K_1 = a/b$ and $K_2 = c/d$ are the carrying capacities for the logistically growing x and y populations, respectively (recall from Section 2.5 that the carrying capacity is the maximum sustainable population). Think of the factors $(K_1 - x)$ and $(K_2 - y)$ in the equations in (10.2) as the "unused capacity" in the two individual populations. Since we're assuming the species occupy the same niche, meaning that they use available resources in the same way, once we put the species in direct competition with each other, the available capacity for the x-population becomes $K_1 - x - y$ and for the y-population is $K_2 - y - x$. This says each member of the y-population takes up the space of one member of the x-population and vice versa. So to model our two competing species we propose the system

(10.3) $$\frac{dx}{dt} = ax\left(\frac{K_1 - x - y}{K_1}\right), \quad \frac{dy}{dt} = cy\left(\frac{K_2 - y - x}{K_2}\right).$$

The interaction between the two species is detrimental to both; if you multiply out the right-hand sides of the equations in (10.3), each will contain a term "xy" with a *negative* constant in front of it. The presence of the xy term, as well as the x^2 and y^2 terms, makes (10.3) a nonlinear system.

Because the independent variable t does not appear on the right-hand sides of the equations in (10.3), the system is **autonomous**. We'll deal nearly exclusively with *planar (two-dimensional) autonomous systems* in this chapter, and generally speaking our goal will be not to solve the system analytically, but rather to give a qualitative solution. This will typically mean that we want to sketch a phase portrait for the system, showing the trajectories of various solutions and making qualitative predictions about the behavior of the solutions from this picture.

The tools we will use to sketch our phase portrait fall into two broad categories:

(1) a global analysis using a nullcline-and-arrow diagram and

(2) a local analysis, focusing on the phase portrait near each equilibrium point. Recall that an equilibrium solution is a constant solution $(x(t), y(t)) = (c_1, c_2)$ to the differential equation system, and the constant value of the equilibrium solution—in this case (c_1, c_2)—is called an equilibrium point.

In this section, we focus on the global analysis, using the system in equation (10.3) to illustrate the procedure. Because of the biological meaning attached to this system, our goal is to sketch a phase portrait in the first quadrant of the xy-plane. To get started, consider the carrying capacities K_1 and K_2. We will assume that $K_1 > K_2$; that is, in isolation the maximum sustainable population for the x species is greater than that of the y species. The opposite assumption, $K_2 > K_1$, leads to a completely parallel mathematical analysis (but a different biological outcome, as we will see). The case of equality, $K_1 = K_2$, is less important from a biological perspective; it is discussed in Exercise 8. We'll also assume that the coefficients a and c, while possibly similar, are not exactly equal.

Nullclines. Recall that a **nullcline** for an autonomous system

(10.4) $$\frac{dx}{dt} = f(x, y), \quad \frac{dy}{dt} = g(x, y),$$

is a curve in the xy-plane along which either $\frac{dx}{dt} = 0$ or $\frac{dy}{dt} = 0$; these are the curves with equations $f(x, y) = 0$ and $g(x, y) = 0$, respectively. For emphasis, we will sometimes call these the x-nullclines and y-nullclines, respectively.

What are the x-nullclines for the system (10.3)? By the first equation in (10.3) we have

$$\frac{dx}{dt} = 0 \iff x = 0 \quad \text{or} \quad K_1 - x - y = 0,$$

and so the derivative $\frac{dx}{dt}$ is 0 along the vertical line $x = 0$ and along the line $x + y = K_1$ as shown in Fig. 10.1.

The x-nullcline $x + y = K_1$ divides the first quadrant into two regions, as shown in Fig. 10.1. At any point in the small triangular region, $\frac{dx}{dt} = x(K_1 - x - y)$ is positive, and we indicate this by putting a right-pointing arrow in that region of the first quadrant. If we move out of this region,

10.1. Introduction: Darwin's finches

staying in the first quadrant but crossing over the line $x + y = K_1$, the derivative $\frac{dx}{dt}$ becomes negative. To indicate this, we mark the other region in the first quadrant with a left-pointing arrow. These arrows tell us about the right/left direction of any trajectory in the first quadrant.

Turning to the y-nullclines, we use the second equation in (10.3) to see that

$$\frac{dy}{dt} = 0 \iff y = 0 \quad \text{or} \quad K_2 - x - y = 0,$$

and the y-nullclines are the horizontal line $y = 0$ and the line $x + y = K_2$. This pair of lines is shown in Fig. 10.2. The nullcline $x + y = K_2$ divides the first quadrant into two regions. In the small triangular piece, $\frac{dy}{dt}$ is positive, and that region gets marked with an upward-pointing arrow. From here, as we cross over the line $x + y = K_2$ but stay in the first quadrant, $\frac{dy}{dt}$ becomes negative. This is indicated with the downward-pointing arrow in Fig. 10.2. The up/down arrows tell us where in the first quadrant trajectories are moving up or moving down.

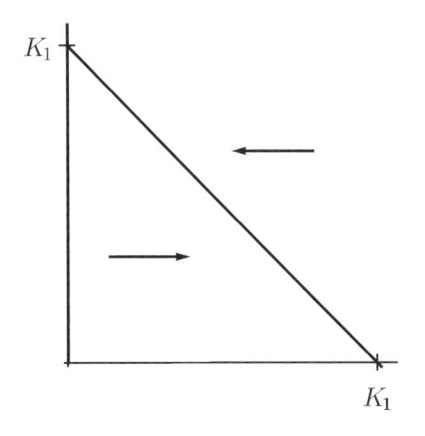

Figure 10.1. The x-nullclines.

Figure 10.2. The y-nullclines.

To put the information in Figs. 10.1 and 10.2 together into a single picture, recall our assumption that $K_1 > K_2$. This tells us how the parallel lines $x + y = K_1$ and $x + y = K_2$ are placed in relation to each other; see Fig. 10.3.

The combined collection of nullclines divides the first quadrant into three basic regions, labeled I, II, and III in Fig. 10.3. The right/left and up/down direction-indicator arrows from the Figs. 10.1 and 10.2 are transferred to our combined sketch. They indicate that in region I, both $\frac{dx}{dt}$ and $\frac{dy}{dt}$ are positive, while both are negative in region III. Region II is marked with right and down arrows; there $\frac{dx}{dt}$ is positive and $\frac{dy}{dt}$ is negative. We have also added scaled field vectors at points along the nullclines $x + y = K_1$ and $x + y = K_2$. These nullcline arrows are horizontal along the y-nullcline $x + y = K_2$ and vertical along the x-nullcline $x + y = K_1$.

The direction-indicator arrows in Fig. 10.3 tell us qualitatively how to move (with increasing time) through each region along a trajectory in a phase portrait. If $\frac{dx}{dt} > 0$, then x increases with t, and similarly if $\frac{dy}{dt} > 0$, then y increases with t. Thus any portion of a trajectory that is in region I must be moving *to the right* and *up* in that region. Because region I is bounded above and to the right by the y-nullcline $y = K_2 - x$, a trajectory in region I must cross this nullcline (horizontally) passing into region II[1] and then begin moving down as it continues to move right. A trajectory in region III moves down and to the left. It may pass into region II, crossing the x-nullcline vertically (as indicated by the field vectors along $x + y = K_1$), and then move down and to the right. Using

---
[1] The trajectory cannot "stall" before or at the nullcline because the portion of the nullcline the orbit can reach by moving up and to the right, as well as the portion of region I it can reach, is free of equilibrium points; see Theorem 8.6.3.

this information, we obtain a reasonable sketch of a phase portrait in the first quadrant, as shown in Fig. 10.4. Moreover, the x-axis (a y-nullcline) and the y-axis (an x-nullcline) consist entirely of equilibrium points and trajectories that traverse some portion of the axis, as shown in Fig. 10.3. Explicit formulas for the solutions whose trajectories lie on the y-axis can be found by substituting $x = 0$ in the equation for $\frac{dy}{dt}$ and solving the resulting logistic equation; see Exercise 9. Similarly, the solutions whose trajectories lie on the x-axis are found by substituting $y = 0$ into the equation for $\frac{dx}{dt}$.

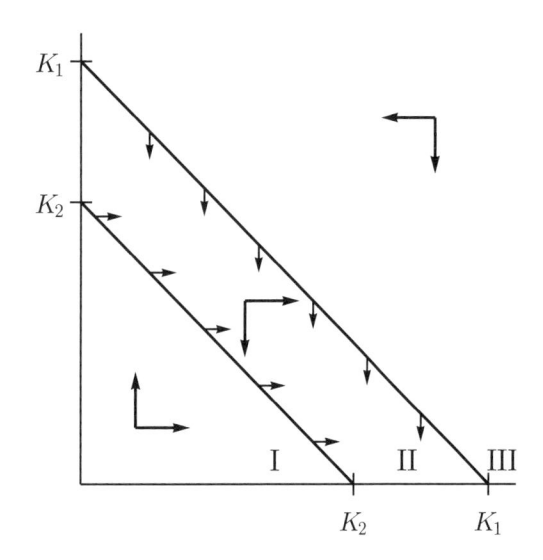

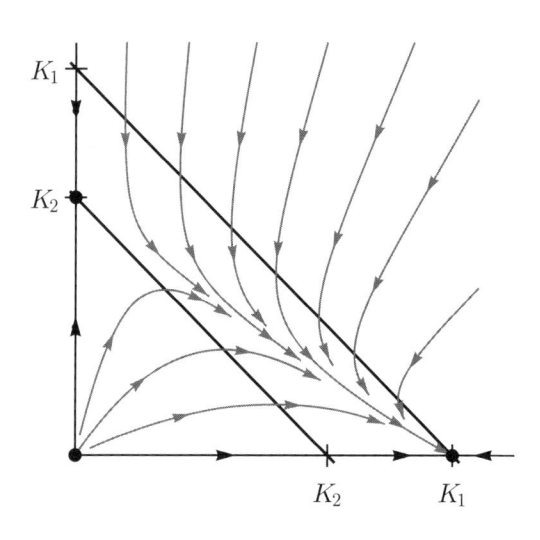

Figure 10.3. Nullcline-and-arrow diagram.

Figure 10.4. Some trajectories in the first quadrant.

Is what we see in our sketch consistent with Darwin's principle of competitive exclusion? Our information suggests that no matter where a trajectory starts in the first quadrant (in other words no matter what the initial populations of the two species), as time goes on, the x-population approaches K_1 and the y-population tends to 0. Recall we assumed at the beginning that $K_1 > K_2$. So it appears that the the species with the larger carrying capacity thrives, while the other species is driven to extinction. Exercise 10 discusses how to *verify* this prediction.

The system in equation (10.3) is an example of a **quadratic competing species model**. In general these models take the form

(10.5) $$\frac{dx}{dt} = a_1 x - b_1 x^2 - c_1 xy, \quad \frac{dy}{dt} = a_2 y - b_2 y^2 - c_2 xy.$$

You can easily verify that the two species occupy the "same niche" (meaning that the system can be rewritten in the form (10.3)) exactly when $c_1 = b_1$ and $c_2 = b_2$. Depending on the values of the parameters in these equations, peaceful coexistence of the two species may or may not be possible. Exercises 1–4 and 17–18 illustrate some possibilities for the behavior of such systems.

Example 10.1.1. Nullclines need not be lines. For example, consider the system

(10.6) $$\begin{aligned} \frac{dx}{dt} &= 1 - x^2 - y^2, \\ \frac{dy}{dt} &= x - y. \end{aligned}$$

The x-nullcline has equation $x^2 + y^2 = 1$, the equation of a circle of radius one centered at $(0, 0)$. The y-nullcline is the line $y = x$. In Fig. 10.5 we show the x-nullcline and the corresponding right/left arrows, while Fig. 10.6 shows the y-nullcline and corresponding up/down arrows.

10.1. Introduction: Darwin's finches

Figure 10.5. The x-nullcline.

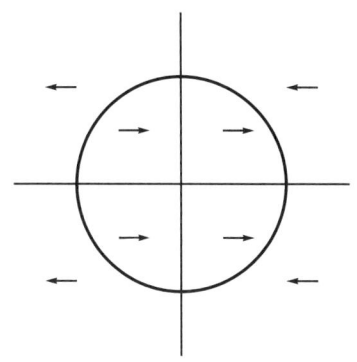

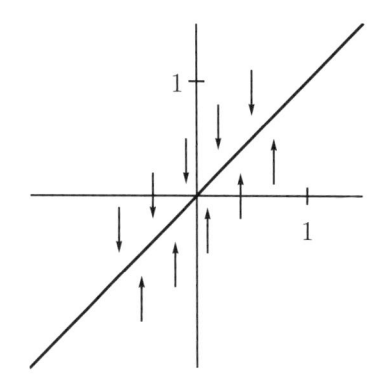

Figure 10.6. The y-nullcline.

In Fig. 10.7 we combine Figs. 10.5 and (10.6), specializing to the first quadrant, and show the direction-indicator arrows in the four basic regions together with some scaled field vectors along the nullclines. Using Fig. 10.7, we then show a few trajectories in Fig. 10.8.

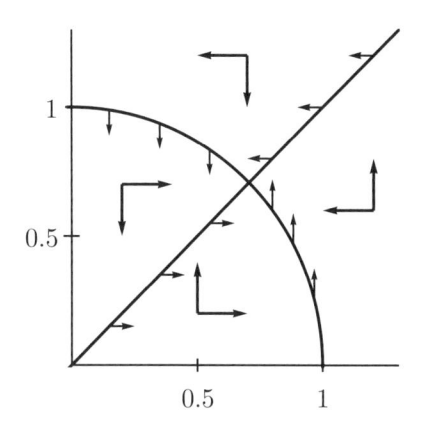

Figure 10.7. Nullcline-and-arrow diagram in first quadrant.

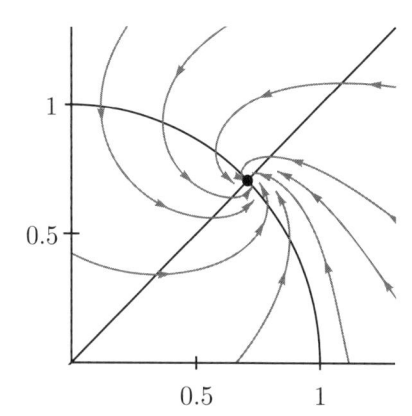

Figure 10.8. Some trajectories in first quadrant.

Vector fields and an existence and uniqueness theorem. A nullcline-and-arrow diagram is a quick substitute for a more detailed **vector field** sketch (see Section 8.2.1). Compare the vector field sketch for equation (10.6) shown in Fig. 10.9 to Fig. 10.7.

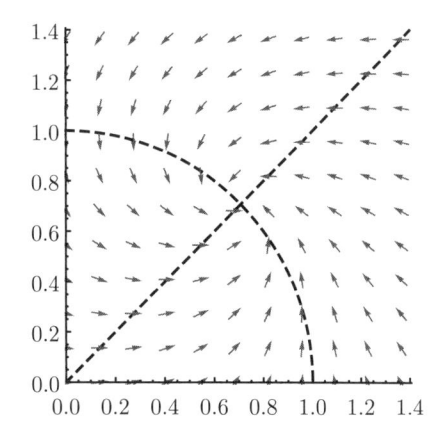

Figure 10.9. A (scaled) vector field for equation (10.6), with the nullclines dashed.

Keep in mind that because the system of Example 10.1.1 is autonomous, the vector field plot in Fig. 10.9 remains stationary over time—the arrows do not "wiggle" as time varies.

Here is one interpretation for a vector field $\mathbf{F}(x,y) = (f(x,y), g(x,y))$ (corresponding to our general autonomous system (10.4)): Think of the surface of a stream seen from above. Each point $\mathbf{X} = (x,y)$ represents an infinitesimal patch of water centered at $\mathbf{X}$ and $\mathbf{F}(\mathbf{X})$ is its *velocity vector*. The direction of the vector is the direction that patch at $\mathbf{X}$ is moving, and the length of the velocity vector, $\|\mathbf{F}(\mathbf{X})\|$, is the speed of the water in that patch. The system of differential equations (10.4) can be abbreviated in vector form as the single vector equation

$$\frac{d\mathbf{X}}{dt} = \mathbf{F}(\mathbf{X}(t)), \quad \text{or} \quad \frac{d\mathbf{X}}{dt} = \mathbf{F}(\mathbf{X}), \quad \text{or just} \quad \mathbf{X}' = \mathbf{F}(\mathbf{X}).$$

The solution $\mathbf{X}(t)$ is the position at time t of a leaf floating on the water surface. At every instant t, its velocity vector $\frac{d\mathbf{X}}{dt}$ matches the velocity vector $\mathbf{F}(\mathbf{X}(t))$ of the patch of water on which it's floating at time t. Put another way, the differential equation is specified by the velocity field $\mathbf{F}$ of the water surface. When you find a solution of the differential equation, you're finding a parametric curve followed by a floating object on that surface. As long as the vector field is sufficiently nice—for example, $\mathbf{F}$ is continuously differentiable, meaning that its coordinate functions f and g have continuous partial derivatives—then given an initial time t_0 and initial position $\mathbf{X}_0$, there is one and only one solution $\mathbf{X}(t)$ on some open interval I containing t_0, of the initial value problem

$$\frac{d\mathbf{X}}{dt} = \mathbf{F}(\mathbf{X}), \quad \mathbf{X}(t_0) = \mathbf{X}_0.$$

This is the **Existence and Uniqueness Theorem** for autonomous systems. The notion of continuously differentiable vector fields $\mathbf{F}(x,y) = (f(x,y), g(x,y))$ is discussed further in the next section. For now we note that whenever the coordinate functions f and g are polynomials[2], then $\mathbf{F}$ will be continuously differentiable. In particular, the Existence and Uniqueness Theorem for autonomous systems applies to the system of Example 10.1.1 as well as to the competing species model given in equation (10.3).

In the following plot, we have added to the vector field sketch of Fig. 10.9 a trajectory, over $0 \leq t < \infty$, of the unique solution to the initial value problem consisting of the system of Example 10.1.1 subject to the initial condition $x(0) = 0.1, y(0) = 0.2$.

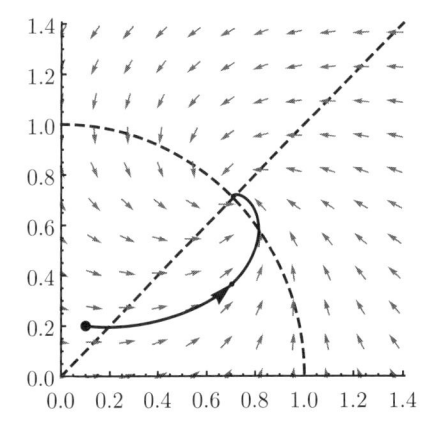

Figure 10.10. Vector field with solution curve to an initial value problem.

---

[2] A polynomial in the two variables x and y is a finite sum of terms of the form $cx^n y^m$ where c is a constant and n and m are nonnegative integers; for example $f(x,y) = 2x^2 + 5xy + 3y^2$ or $g(x,y) = 1 + x^3 y$.

10.1. Introduction: Darwin's finches

Even though two different solutions of an autonomous system cannot take the same value at a given time t_0, can the trajectories followed by two different solutions of

$$\frac{d\mathbf{X}}{dt} = \mathbf{F}(\mathbf{X})$$

touch (by passing through the same point at different times)? Yes, but only if one solution is merely a time translate of the other. This means that both solutions travel the same path and pass each point on that path at the same speed; the only difference is that one solution follows this path ahead of the other in time. The next theorem makes this precise.

Theorem 10.1.2. *Let $\mathbf{F} = (f(x,y), g(x,y))$ be a continuously differentiable vector field and let $\mathbf{X}(t)$ and $\mathbf{Y}(t)$ be solutions of the autonomous system*

(10.7) $$\frac{d\mathbf{X}}{dt} = \mathbf{F}(\mathbf{X})$$

satisfying $\mathbf{X}(t_0) = \mathbf{Y}(t_1)$. Let $T = t_0 - t_1$. Then $\mathbf{Y}(t) = \mathbf{X}(t+T)$ for all t for which both sides are defined.

Proof. Let $\mathbf{Z}(t) = \mathbf{X}(t+T)$. Notice that $\mathbf{Z}'(t) = \mathbf{X}'(t+T) = \mathbf{F}(\mathbf{X}(t+T)) = \mathbf{F}(\mathbf{Z}(t))$, so $\mathbf{Z}(t)$ solves the system (10.7). Also, $\mathbf{Z}(t_1) = \mathbf{X}(t_1+T) = \mathbf{X}(t_0) = \mathbf{Y}(t_1)$. By the uniqueness part of the Existence and Uniqueness Theorem for autonomous systems just discussed, $\mathbf{Z}(t) = \mathbf{Y}(t)$ for all t for which both sides are defined. This is the desired conclusion. □

A consequence of the preceding theorem is that if the trajectories of two solutions $\mathbf{X} = \mathbf{X}(t)$ and $\mathbf{Y} = \mathbf{Y}(t)$ of the system (10.7) (with $\mathbf{F}$ continuously differentiable) have a point in common, then the trajectories must be identical (assuming the domains of the solutions are not restricted). In terms of phase portraits, the preceding theorem tells us that given any point $\mathbf{X}_0 = (x_0, y_0)$ in the phase plane of (10.7), there is one and only one phase portrait curve passing through $\mathbf{X}_0$; every trajectory passing through $\mathbf{X}_0$ follows this curve.

Equilibrium points. An equilibrium point of the system $\frac{d\mathbf{X}}{dt} = \mathbf{F}(\mathbf{X})$ is a point $\mathbf{X}_e = (x_e, y_e)$, where $\mathbf{F}(\mathbf{X}_e) = \mathbf{0}$; that is, $f(x_e, y_e) = 0$ and $g(x_e, y_e) = 0$. Then the constant curve $\mathbf{X}(t) = \mathbf{X}_e$ solves the system because

$$\frac{d\mathbf{X}}{dt} = \frac{d\mathbf{X}_e}{dt} = \frac{d(\mathbf{constant})}{dt} = \mathbf{0},$$

while $\mathbf{F}(\mathbf{X}_e) = \mathbf{0}$, too. In the analogy of the stream, a leaf at $\mathbf{X}_e$ stays at $\mathbf{X}_e$ forever; it is becalmed at a still point in the flow. Since the equilibrium solutions occur at points (x_e, y_e) where both $\frac{dx}{dt}$ and $\frac{dy}{dt}$ are simultaneously 0 they occur at any point where an x-nullcline intersects a y-nullcline. In the next section, we will focus on tools producing more accurate sketches of phase portraits near equilibrium points. This will help us refine our phase portrait sketches with local information near these key points.

Example 10.1.3. Consider the system

(10.8) $$\frac{dx}{dt} = x\left(5 - \frac{1}{20}x - \frac{1}{20}y\right), \quad \frac{dy}{dt} = y\left(4 - \frac{1}{50}y - \frac{1}{50}x\right),$$

which is obtained from equation (10.3) by setting $a = 5, K_1 = 100, c = 4,$ and $K_2 = 200$. The equilibrium solutions of (10.8) are $(0,0), (100,0),$ and $(0,200)$, found by solving the pair of equations

$$0 = x\left(5 - \frac{1}{20}x - \frac{1}{20}y\right) \quad \text{and} \quad 0 = y\left(4 - \frac{1}{50}y - \frac{1}{50}x\right).$$

The solutions occur when one of the factors in *each* product is equal to 0:

$$x = 0, \ y = 0 \text{ gives the equilibrium point } (0,0),$$

$$x = 0, \ 4 - \frac{1}{50}y - \frac{1}{50}x = 0 \text{ gives the equilibrium point } (0, 200),$$

$$5 - \frac{1}{20}x - \frac{1}{20}y = 0, \ y = 0 \text{ gives the equilibrium point } (100, 0).$$

Since the fourth possible pair of equations,

$$5 - \frac{1}{20}x - \frac{1}{20}y = 0, \quad 4 - \frac{1}{50}y - \frac{1}{50}x = 0,$$

has no solution (the two lines are parallel), these are all of the equilibrium points.

Fig. 10.11 shows the solution of (10.8) satisfying the initial condition $x(0) = 8, y(0) = 15$, plotted over a (scaled) vector-field background. Observe that the solution appears to approach the equilibrium point $(0, 200)$ as $t \to \infty$. This is consistent with Theorem 8.6.3, which says that if a solution tends to a finite limit, it must be an equilibrium point.

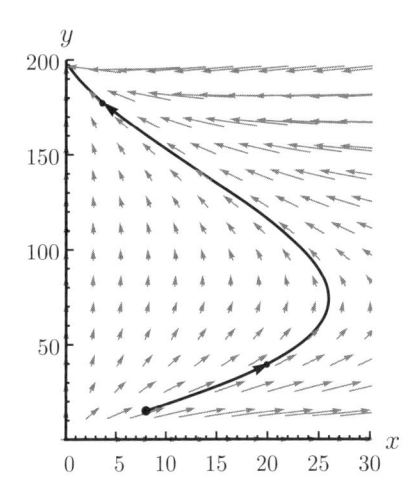

Figure 10.11. Solution to (10.8) with initial condition $x(0) = 8, y(0) = 15$.

10.1.1. Exercises.

In Exercises 1–4 two competing (dissimilar) species whose populations (in millions of individuals) are modeled by the given equations. For each, find the equilibrium points for the system, and do a nullcline-and-arrow diagram for a phase portrait in the first quadrant. Based on your sketch, can you make a prediction about the long-term fate of the two populations?

1. $\begin{aligned} x'(t) &= x(5 - x - 4y), \\ y'(t) &= y(2 - y - x). \end{aligned}$

2. $\begin{aligned} x'(t) &= x(40 - x - y), \\ y'(t) &= y(90 - x - 2y). \end{aligned}$

3. $\begin{aligned} x'(t) &= x(5 - 2x - 4y), \\ y'(t) &= y(7 - 3y - 4x). \end{aligned}$

4. $\begin{aligned} x'(t) &= x(6 - x - 2y), \\ y'(t) &= y(3 - x - y). \end{aligned}$

5. The following differential equation system is proposed for two competing species whose populations would grow exponentially (rather than logistically) in isolation:

$$\frac{dx}{dt} = x - xy, \quad \frac{dy}{dt} = 2y - 2xy.$$

   (a) What are the equilibrium points for this system?

10.1. Introduction: Darwin's finches

(b) Do a nullcline-and-arrow diagram for the phase portrait in the first quadrant, showing a few trajectories.

6. The following model[3] has been proposed for competition between Neanderthal man (with population $N(t)$ at time t in years) and early modern man (with population $E(t)$ at time t):

$$\frac{dN}{dt} = N[a - b - d(N + E)], \quad \frac{dE}{dt} = E[a - sb - d(N + E)]$$

where a, b, and d are positive constants with $a > b$ and s is a positive constant with $0 < s < 1$. The constant s is sometimes called the "parameter of similarity" and it reflects the different average life spans of Neanderthal man and early man. It is estimated to be about 0.995.

(a) Show that these equations have the same form as other quadratic competing species models; that is, they reflect logistic growth for each species in the absence of the other and a (mutually detrimental) interaction term that is proportional to the product of the two species.

(b) Show that
$$\frac{d}{dt}\left(\frac{N}{E}\right) = \frac{Nb(s-1)}{E}.$$

(c) Set $R = N/E$, so that the differential equation in (b) is
$$\frac{dR}{dt} = Rb(s-1).$$

Solve this equation for R to show that
$$\frac{N}{E} = \frac{N(0)}{E(0)} e^{-b(1-s)t}.$$

(d) Explain how your work in (c) and the assumption that $b > 0$ and $0 < s < 1$ show that Neanderthal man is driven to extinction by early modern man.

(e) Use $s = 0.995$ and $b = 0.033 = 1/30$ (this reflects an average life span of 30 years for Neanderthal man) to determine how long it takes until the ratio $N(t)/E(t)$ is 10 percent of its initial ratio $N(0)/E(0)$. Compare this to the estimated extinction time of Neanderthal man of 10,000 years.

7. The time from infection with HIV to the clinical diagnosis of AIDS is only partially known. To study this, patients for whom an estimate of date of infection with HIV can be given are important.[4] Suppose a group of K hemophiliacs are known to have been given an HIV-infected blood product at time $t = 0$. The number $x(t)$ of people who are HIV-positive but do not yet have AIDS and the number $y(t)$ who have converted from being HIV-positive to having AIDS are modeled by the differential equations

$$\frac{dx}{dt} = -atx, \quad \frac{dy}{dt} = atx$$

(a is a positive constant, and t is in years), with initial condition $x(0) = K, y(0) = 0$. Notice that this is **not** an autonomous system. However, it can be solved explicitly because the first equation, which doesn't involve y, is a first-order separable equation.

(a) Solve the equation
$$\frac{dx}{dt} = -atx$$
explicitly for x as a function of t, using the initial condition $x(0) = K$.

---
[3] This exercise is based on the article *A mathematical model for Neanderthal extinction* by J. C. Flores, J. Theor. Biol. 191 (1998), 295–298. See also [31, p. 115].
[4] This exercise is a simplification of a problem that appears in [31, p. 394].

(b) Substitute your answer for (a) into the second equation
$$\frac{dy}{dt} = atx$$
and then solve the resulting separable equation explicitly for y in terms of t, using the initial condition $y(0) = 0$.

(c) If 1/3 of the initial population of K has converted from HIV-positive status to AIDS after 6 years, how long does it take for 90% of the initial population to convert?

8. What does the nullcline-and-arrow diagram for the system in equation (10.3) look like if $K_1 = K_2$? Show that there are infinitely many equilibrium points in this case.

9. (a) Show that in the competing species system (10.3), the x-axis and the y-axis consist entirely of equilibrium points and trajectories that traverse some portion of the axis (as shown in Fig. 10.4). Explicit formulas for the solutions whose trajectories lie on the y-axis can be found by substituting $x = 0$ in the equation for $\frac{dy}{dt}$ and solving the resulting logistic equation. A similar comment applies to find explicit formulas for the solutions whose trajectories lie on the x-axis.

(b) Give a careful argument to show that the trajectory of any solution of the competing species system (10.3) satisfying $x(0) > 0$, $y(0) > 0$ must lie entirely in the first quadrant.

10. Some of the observations "suggested" by a nullcline-and-arrow diagram can be verified using the following properties, which apply to an autonomous system
$$\frac{dx}{dt} = f(x, y), \quad \frac{dy}{dt} = g(x, y)$$
when f and g are continuous.

Fact (i). If a solution $x(t)$ is bounded and monotone increasing or monotone decreasing for all $t \geq t_0$, then $\lim_{t \to \infty} x(t)$ exists as a finite number. The same applies to $y(t)$. To say that an increasing function $x(t)$ is bounded means $x(t) \leq c$ for some constant c. A decreasing function is bounded if it is greater than or equal to some constant.

Fact (ii). According to Theorem 8.6.3, if a trajectory $(x(t), y(t))$ has a limit (x_*, y_*) as $t \to \infty$, then (x_*, y_*) must be an equilibrium point.

Assuming $K_1 > K_2$, explain how Fact (i) and Fact (ii) help justify the assertion that in the system in (10.3), one species is driven to extinction while the other tends to the carrying capacity for that population. You may also use the fact that any trajectory that starts in region II of Fig. 10.3 must stay in that region for all later times.

11. In this problem we consider the competing species system (10.8) of Example 10.1.3. Recall $(0, 0)$ is an equilibrium solution and that on the positive half of the x-axis the system has equilibrium point $x = 100$, $y = 0$, while on the positive half of the y-axis it has equilibrium point $x = 0$, $y = 200$.

(a) Give an exact solution to the system with initial condition $x(0) = 2$, $y(0) = 0$ using the solution in equation (2.87) of the logistic equation (2.80) in Section 2.5. Describe the behavior of your solution as $t \to \infty$ and as $t \to -\infty$, and verify that the trajectory of this solution is the interval $0 < x < 100$ on the x-axis, traversed from left to right as t increases from $-\infty$ to ∞.

(b) Again using equation (2.87), give an exact solution of the system (10.8) with initial condition $x(0) = 200$, $y(0) = 0$. Show that this solution tends to 100 as $t \to \infty$ and traces out the segment $100 < x < \infty$, $y = 0$ on the x-axis as t ranges from $-\frac{\ln 2}{5}$ to ∞.

(c) By parts (a) and (b) the portion of the positive half of the x-axis to the left of the equilibrium point $(100, 0)$ and the portion to the right of this point are each traversed by trajectories of solutions. A similar result holds on the positive half of the y-axis: The portion of the positive half of the y-axis below the equilibrium point $(0, 200)$ and the portion above this point are traversed by solutions of the system satisfying the initial condition $x(0) = 0$, $y(0) = y_0$ for $0 < y_0 < 200$ and $200 < y_0$, respectively (you are not required to verify this). Explain why a trajectory for the system (10.8) whose initial point (x_0, y_0) lies in the open first quadrant (so that $x_0 > 0$ and $y_0 > 0$) can never leave the open first quadrant. Hint: Can trajectories cross?

12. Sam and Sally have each obtained a numerical solution of the initial value problem

$$\frac{dx}{dt} = (x+y^2)/10, \quad \frac{dy}{dt} = xy - x, \quad x(0) = 1, \quad y(0) = 2.$$

According to Sam, $x(1) \approx 2.11$ and $y(1) \approx 5.05$, while Sally says $x(1) \approx 2.11$ and $y(1) \approx 1.22$. Which of the two solutions **must** be incorrect? Explain.

13. Suppose we know that $\mathbf{X}(t) = (2e^{-3t}, -e^{-3t})$ is a solution of some autonomous system $\frac{d\mathbf{X}}{dt} = \mathbf{F}(\mathbf{X})$ where $\mathbf{F}$ is a continuously differentiable vector field.
    (a) Sketch the trajectory of this solution for $-\infty < t < \infty$, and show that $\mathbf{Y}_1(t) = (6e^{-3t}, -3e^{-3t})$, $\mathbf{Y}_2(t) = (2e^{-5t}, -e^{-5t})$, and $\mathbf{Y}_3(t) = (2e^{-3(t-1)}, -e^{-3(t-1)})$ all have the same trajectory as $\mathbf{X}(t)$.
    (b) Which of the functions $\mathbf{Y}_1(t)$, $\mathbf{Y}_2(t)$, or $\mathbf{Y}_3(t)$ are merely "time translates" of $\mathbf{X}(t)$; that is, which can be written as $\mathbf{X}(t+T)$ for some constant T (and all t)?
    (c) Which of $\mathbf{Y}_1(t)$, $\mathbf{Y}_2(t)$, or $\mathbf{Y}_3(t)$ *cannot* be a solution of the same system $\frac{d\mathbf{X}}{dt} = \mathbf{F}(\mathbf{X})$? Justify your answer.
    (d) *Could* $(0,0)$ be an equilibrium point of the system $\frac{d\mathbf{X}}{dt} = \mathbf{F}(\mathbf{X})$? *Must* $(0,0)$ be an equilibrium point of the system $\frac{d\mathbf{X}}{dt} = \mathbf{F}(\mathbf{X})$? Justify your answer.

14. Suppose we know that $\mathbf{X}(t) = (2\cos t, 3\sin t)$ is a solution of some autonomous system $\frac{d\mathbf{X}}{dt} = \mathbf{F}(\mathbf{X})$ where $\mathbf{F}$ is a continuously differentiable vector field.
    (a) Sketch the trajectory of this solution for $-\infty < t < \infty$, and show that $\mathbf{Y}_1(t) = (-2\sin t, 3\cos t)$, $\mathbf{Y}_2(t) = (2\cos(2t), 3\sin(2t))$, and $\mathbf{Y}_3(t) = (2\cos(t/2), 3\sin(t/2))$ all have the same trajectory as $\mathbf{X}(t)$.
    (b) Which of the functions $\mathbf{Y}_1(t)$, $\mathbf{Y}_2(t)$, or $\mathbf{Y}_3(t)$ are merely "time translates" of $\mathbf{X}(t)$; that is, which can be written as $\mathbf{X}(t+T)$ for some constant T (and all t)?
    (c) Which of $\mathbf{Y}_1(t)$, $\mathbf{Y}_2(t)$, or $\mathbf{Y}_3(t)$ *cannot* be a solution of the system $\frac{d\mathbf{X}}{dt} = \mathbf{F}(\mathbf{X})$? Justify your answer.

15. In this problem you will work with the system

$$\frac{dx}{dt} = y - \frac{x^2}{2}, \quad \frac{dy}{dt} = y\left(y - \frac{1}{2}x - 1\right).$$

    (a) Sketch the nullcline-and-arrow diagram for this system in the first quadrant, and identify any equilibrium points in your sketch. Include the y-nullcline $y = 0$ forming the lower boundary of your diagram.
    (b) Suppose that $x = x(t), y = y(t)$ is a solution of the system such that $x(0) = 2$ and $y(0) = 1$. Use your diagram of (a) and Theorem 8.6.3 to predict $\lim_{t \to \infty} x(t)$ and $\lim_{t \to \infty} y(t)$.
    (c) (CAS) Use a computer algebra system to produce a phase portrait of the system over the rectangle $0 < x < 3, 0 < y < 3$. Include nullclines in your portrait. Check that your nullcline-and-arrow diagram of (a) and your solution of (b) are consistent with the portrait.

16. Consider the system

$$\frac{dx}{dt} = 4 - x^2 - y, \quad \frac{dy}{dt} = x(1 - y^2).$$

    (a) Sketch the nullcline-and-arrow diagram for this system in the first quadrant and find any equilibrium points there.

(b) Suppose that $x = x(t), y = y(t)$ is a solution of the system such that $x(0) = 1$ and $y(0) = 3$. Use your diagram of (a) and Theorem 8.6.3 to predict $\lim_{t\to\infty} x(t)$ and $\lim_{t\to\infty} y(t)$.

(c) (CAS) Use a computer algebra system to produce a phase portrait of the system over the rectangle $0 < x < 3, 0 < y < 3$. Include nullclines in your portrait. Check that your nullcline-and-arrow diagram of (a) and your solution of (b) are consistent with the portrait.

17. (CAS) The figure below shows a phase portrait, in the first quadrant, for the system

(10.9) $$\frac{dx}{dt} = x(2 - x - y), \quad \frac{dy}{dt} = y(1 - x - y).$$

Notice this is the competing species system (10.3) with $a = 2, c = 1, K_1 = 2$, and $K_2 = 1$.

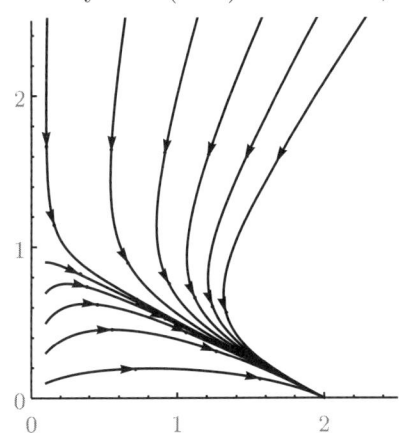

From (10.9) we have

(10.10) $$\frac{dy}{dx} = \frac{y(1 - x - y)}{x(2 - x - y)}.$$

After some computation and simplification we can also obtain

(10.11) $$\frac{d^2y}{dx^2} = \frac{y(2 - x - 2y)}{x^2(-2 + x + y)^3}.$$

(a) Use the second derivative (10.11) to create a concavity chart for the system (10.9) in the first quadrant, and check that your chart is consistent with phase portrait given above.

(b) Verify by substitution that $y = 1 - \frac{x}{2}$ is a solution of the first-order equation (10.10). The line with equation $y = 1 - \frac{x}{2}$ is also a "zero-concavity line" that you should have obtained in (a). Draw some field vectors *for the system* at points on this line (in the first quadrant). Finally, find an explicit formula for a solution to the system (10.9), for $-\infty < t < \infty$, whose trajectory traverses the portion of this line in the first quadrant by solving the initial value problem

$$\frac{dx}{dt} = x(2 - x - y), \quad \frac{dy}{dt} = y(1 - x - y), \quad x(0) = 1, \quad y(0) = 1/2.$$

To do this, substitute $1 - x/2$ for y in the first equation and $2 - 2y$ for x in the second equation to obtain

$$\frac{dx}{dt} = x\left(1 - \frac{x}{2}\right), \quad \frac{dy}{dt} = y(y - 1).$$

Solve both of these first-order equations, with the respective initial conditions $x(0) = 1$ and $y(0) = 1/2$, by separation of variables or as Bernoulli equations or using a CAS.

(c) Suppose we have any solution to the system (10.9) satisfying $x(0) > 0$ and $y(0) > 0$. Show that the trajectory of the solution for $t \geq 0$ has at most one inflection point.

18. (CAS) Consider the system

(10.12) $$\frac{dx}{dt} = x(2 - x - y), \quad \frac{dy}{dt} = 4y(1 - x - y).$$

(a) Using
$$\frac{dy}{dx} = \frac{4y(1 - x - y)}{x(2 - x - y)}$$

and
$$\frac{d^2y}{dx^2} = \frac{4y(-4 + 14x - 12x^2 + 3x^3 + (16 - 27x + 9x^2)y + 3(-5 + 3x)y^2 + 3y^3)}{x^2(-2 + x + y)^3},$$

produce a concavity chart in the first quadrant for the system (10.12). Hint: Use a CAS to determine the curves in the first quadrant where $-4 + 14x - 12x^2 + 3x^3 + (16 - 27x + 9x^2)y + 3(-5 + 3x)y^2 + 3y^3 = 0$ and also sketch the line $y = 2 - x$ (where the second derivative is undefined) in this quadrant. These divide the first quadrant into regions where the concavity can be determined using test values in the formula for $\frac{d^2y}{dx^2}$.

(b) Using a CAS, add to your concavity chart from (a) a trajectory having two points of inflection (in the first quadrant).

10.2. Linear approximation: The major cases

The basic idea of this section is both simple and appealing: We can often predict what the phase portrait of a nonlinear system looks like *near an isolated equilibrium point* by relating it to an associated linear system whose phase portrait we know. Our first task is to explain how to find this associated linear system.

We begin with a brief review of linear approximation from calculus. Suppose we have a differentiable function $f(x)$ and we pick a point $(x_0, f(x_0))$ on the graph of f. To get the best straight line approximation to the graph of f near the point $(x_0, f(x_0))$, we use the tangent line passing through $(x_0, f(x_0))$. This line is the graph of the function

(10.13) $$y = f(x_0) + f'(x_0)(x - x_0),$$

and (10.13) is the best linear approximation to f near x_0. We have

(10.14) $$f(x) = f(x_0) + f'(x_0)(x - x_0) + R_f(x),$$

where the "error term" $R_f(x)$ (the difference between $f(x)$ and the linear approximation (10.13)) is not just small when x is near x_0, but small in comparison to the distance $|x - x_0|$ between x and x_0. By this we mean

(10.15) $$\frac{|R_f(x)|}{|x - x_0|} \to 0 \quad \text{as } x \to x_0.$$

The condition in (10.15) is just a restatement of the familiar definition of the derivative of f at x_0 from calculus.

For a differentiable function of two variables, say, $f(x, y)$, the graph of $z = f(x, y)$ is a surface in $\mathbb{R}^3$. The best linear approximation to this surface near a point $P = (x_0, y_0, f(x_0, y_0))$ is obtained by using the *tangent plane* to the surface passing through P. This tangent plane has equation

$$z = f(x_0, y_0) + \frac{\partial f}{\partial x}(x_0, y_0)(x - x_0) + \frac{\partial f}{\partial y}(x_0, y_0)(y - y_0).$$

Keep in mind that for a given point P, $\frac{\partial f}{\partial x}(x_0, y_0)$ and $\frac{\partial f}{\partial y}(x_0, y_0)$ are just numbers. Thus, the best linear approximation to f near (x_0, y_0) should be given by the function that takes (x, y) to

$$f(x_0, y_0) + \frac{\partial f}{\partial x}(x_0, y_0)(x - x_0) + \frac{\partial f}{\partial y}(x_0, y_0)(y - y_0).$$

If the partial derivatives $\frac{\partial f}{\partial x}$ and $\frac{\partial f}{\partial y}$ exist and are continuous at and near (x_0, y_0), this is a good approximation in the following sense: If we move away from (x_0, y_0) to the point (x, y), then

(10.16) $\quad f(x, y) = f(x_0, y_0) + \frac{\partial f}{\partial x}(x_0, y_0)(x - x_0) + \frac{\partial f}{\partial y}(x_0, y_0)(y - y_0) + R_f(x, y),$

where the error term $R_f(x, y)$, now a function of two variables, is not only small when (x, y) is near (x_0, y_0), but small compared to the distance

$$\|\mathbf{X} - \mathbf{X}_0\| = \|(x, y) - (x_0, y_0)\| = \sqrt{(x - x_0)^2 + (y - y_0)^2}$$

between the points $\mathbf{X} = (x, y)$ and $\mathbf{X}_0 = (x_0, y_0)$. This means that

(10.17) $\quad \dfrac{|R_f(\mathbf{X})|}{\|\mathbf{X} - \mathbf{X}_0\|} \to 0 \quad \text{as } \mathbf{X} \to \mathbf{X}_0,$

which is the two-variable analogue of (10.15).

Verifying (10.17) requires more work than the one-variable case and really does use the assumption that the partial derivatives $\frac{\partial f}{\partial x}$ and $\frac{\partial f}{\partial y}$ exist and are themselves continuous functions of (x, y). This assumption on a function $f(x, y)$ is important enough that we give it an official name.

Definition 10.2.1. A function $f(x, y)$ is *continuously differentiable* if the partial derivatives $\frac{\partial f}{\partial x}$ and $\frac{\partial f}{\partial y}$ exist at all points of the domain of f and are themselves continuous functions of (x, y).

Now, let's consider a vector field $\mathbf{F}(x, y) = (f(x, y), g(x, y))$. We say that $\mathbf{F}$ is continuously differentiable if both coordinate functions f and g are continuously differentiable. In this case the remarks above apply to the second coordinate function g, as well as to the function f. Specifically,

$$z = g(x_0, y_0) + \frac{\partial g}{\partial x}(x_0, y_0)(x - x_0) + \frac{\partial g}{\partial y}(x_0, y_0)(y - y_0)$$

is the best linear approximation to g near $\mathbf{X}_0 = (x_0, y_0)$, and its graph is the tangent plane to the graph of g at the point $(x_0, y_0, g(x_0, y_0))$. Moreover,

(10.18) $\quad g(x, y) = g(x_0, y_0) + \frac{\partial g}{\partial x}(x_0, y_0)(x - x_0) + \frac{\partial g}{\partial y}(x_0, y_0)(y - y_0) + R_g(x, y)$

where the error term R_g satisfies

(10.19) $\quad \dfrac{|R_g(\mathbf{X})|}{\|\mathbf{X} - \mathbf{X}_0\|} \to 0 \quad \text{as } \mathbf{X} \to \mathbf{X}_0.$

10.2. Linear approximation: The major cases

Let's put equations (10.16) and (10.18) together in vector form. Writing vectors as columns, we have

$$\mathbf{F}(\mathbf{X}) = \begin{pmatrix} f(x,y) \\ g(x,y) \end{pmatrix} = \begin{pmatrix} f(x_0, y_0) \\ g(x_0, y_0) \end{pmatrix}$$

$$+ \begin{pmatrix} \frac{\partial f}{\partial x}(x_0, y_0)(x - x_0) + \frac{\partial f}{\partial y}(x_0, y_0)(y - y_0) \\ \frac{\partial g}{\partial x}(x_0, y_0)(x - x_0) + \frac{\partial g}{\partial y}(x_0, y_0)(y - y_0) \end{pmatrix}$$

(10.20)
$$+ \begin{pmatrix} R_f(x,y) \\ R_g(x,y) \end{pmatrix}.$$

The first and third terms on the right-hand side of equation (10.20) can be written concisely in vector form as

$$\begin{pmatrix} f(x_0, y_0) \\ g(x_0, y_0) \end{pmatrix} = \mathbf{F}(x_0, y_0) = \mathbf{F}(\mathbf{X}_0)$$

(where $\mathbf{X}_0 = (x_0, y_0)$) and

$$\begin{pmatrix} R_f(x,y) \\ R_g(x,y) \end{pmatrix} = \mathbf{R}(x,y) = \mathbf{R}(\mathbf{X}),$$

where the "error" vector field $\mathbf{R}$ is defined by this last equation.

Since $\|\mathbf{R}(x,y)\| = \sqrt{R_f(x,y)^2 + R_g(x,y)^2}$, the limit requirements (10.17) and (10.19) together say that

(10.21)
$$\frac{\|\mathbf{R}(\mathbf{X})\|}{\|\mathbf{X} - \mathbf{X}_0\|} \to 0 \quad \text{as } \mathbf{X} \to \mathbf{X}_0.$$

On the other hand, the middle term on the right-hand side of equation (10.20) is the product of a 2×2 matrix $\mathbf{J}$ and a 2×1 vector $\mathbf{X} - \mathbf{X}_0$, where

(10.22)
$$\mathbf{J} = \begin{pmatrix} \frac{\partial f}{\partial x}(x_0, y_0) & \frac{\partial f}{\partial y}(x_0, y_0) \\ \frac{\partial g}{\partial x}(x_0, y_0) & \frac{\partial g}{\partial y}(x_0, y_0) \end{pmatrix}$$

and

$$\mathbf{X} - \mathbf{X}_0 = \begin{pmatrix} x - x_0 \\ y - y_0 \end{pmatrix}.$$

The matrix $\mathbf{J}$ is called the **Jacobian matrix** of $\mathbf{F}$ at $\mathbf{X}_0 = (x_0, y_0)$. To emphasize that the partial derivatives in equation (10.22) are evaluated at $\mathbf{X}_0 = (x_0, y_0)$ we sometimes write $\mathbf{J}(\mathbf{X}_0)$ instead of just $\mathbf{J}$.

On putting this together, we can rewrite equation (10.20) succinctly as

(10.23)
$$\mathbf{F}(\mathbf{X}) = \mathbf{F}(\mathbf{X}_0) + \mathbf{J}(\mathbf{X}_0)(\mathbf{X} - \mathbf{X}_0) + \mathbf{R}(\mathbf{X}),$$

where the error term $\mathbf{R}(\mathbf{X})$ satisfies (10.21). This is the analogue for a continuously differentiable vector field of equation (10.14) for a one-variable differentiable function. The Jacobian matrix $\mathbf{J}(\mathbf{X}_0)$ for $\mathbf{F}$ at (x_0, y_0) plays the same role as the derivative $f'(x_0)$ in equation (10.14).

Now let's return to our nonlinear system

$$\frac{dx}{dt} = f(x,y), \quad \frac{dy}{dt} = g(x,y),$$

where f and g are continuously differentiable. In terms of the vector field $\mathbf{F} = (f, g)$, we can write this system in vector form as

(10.24) $$\frac{d\mathbf{X}}{dt} = \mathbf{F}(\mathbf{X}).$$

Suppose $\mathbf{X}_e = (x_e, y_e)$ is an equilibrium point of the system (10.24), so that $\mathbf{F}(\mathbf{X}_e) = \mathbf{0} = (0,0)$. Since $\mathbf{F}$ is continuously differentiable, we can invoke equation (10.23) with $\mathbf{X}_0$ chosen to be this equilibrium point $\mathbf{X}_e$. Since $\mathbf{F}(\mathbf{X}_e) = \mathbf{0}$, (10.23) reduces to

$$\mathbf{F}(\mathbf{X}) = \mathbf{J}(\mathbf{X}_e)(\mathbf{X} - \mathbf{X}_e) + \mathbf{R}(\mathbf{X}).$$

If the determinant of $\mathbf{J}(\mathbf{X}_e)$ is not zero, the Jacobian matrix $\mathbf{J}(\mathbf{X}_e)$ is invertible and $\|\mathbf{J}(\mathbf{X}_e)(\mathbf{X} - \mathbf{X}_e)\|$ tends to zero as $\mathbf{X} \to \mathbf{X}_e$ at the same rate as $\|\mathbf{X} - \mathbf{X}_e\|$. However, the limit condition (10.21) says that the length of $\mathbf{R}(\mathbf{X})$ tends to zero even faster than $\|\mathbf{X} - \mathbf{X}_e\|$, leaving $\mathbf{J}(\mathbf{X}_e)(\mathbf{X} - \mathbf{X}_e)$ as the dominant part of $\mathbf{F}(\mathbf{X})$ when $\mathbf{X}$ is near $\mathbf{X}_e$. Thus it is reasonable to expect the trajectories of the system (10.24) to resemble those of the system

(10.25) $$\frac{d\mathbf{X}}{dt} = \mathbf{J}(\mathbf{X}_e)(\mathbf{X} - \mathbf{X}_e)$$

for $\mathbf{X}$ near $\mathbf{X}_e$.

The system (10.25) is "translated linear" with an equilibrium point at $\mathbf{X}_e$ and is easily solved: If $\mathbf{Y}$ is a solution of the truly linear system

(10.26) $$\frac{d\mathbf{Y}}{dt} = \mathbf{J}(\mathbf{X}_e)\mathbf{Y},$$

then it's easy to check that $\mathbf{X}(t) = \mathbf{Y}(t) + \mathbf{X}_e$ satisfies (10.25). Conversely, if $\mathbf{X}$ satisfies (10.25), then $\mathbf{Y}(t) = \mathbf{X}(t) - \mathbf{X}_e$ satisfies (10.26). Note that we can quickly solve the linear system (10.26) using techniques discussed in Chapter 8 or Chapter 9.

Recall that when $\mathbf{J}(\mathbf{X}_e)$ is invertible, $\mathbf{0} = (0,0)$ is the only equilibrium point of the linear system (10.26); in addition, not only is $\mathbf{X}_e$ the only equilibrium point of the translated linear system (10.25), but also it is possible to show that $\mathbf{X}_e$ must be an *isolated* equilibrium point of the *original nonlinear system*. The equation $\mathbf{X}(t) = \mathbf{Y}(t) + \mathbf{X}_e$ relating the solutions of (10.25) and (10.26) shows the equilibrium solution $\mathbf{Y}(t) \equiv \mathbf{0}$ for (10.26) is translated to the equilibrium solutions $\mathbf{X}(t) = \mathbf{X}_e$ of (10.25). In fact, all solutions of (10.26) become solutions of (10.25) when translated (shifted) by $\mathbf{X}_e$, and all solutions of (10.25) become solutions of (10.26) when translated by $-\mathbf{X}_e$. Thus the phase portrait of (10.25) is obtained simply by translating the phase portrait of (10.26) by $\mathbf{X}_e$. For example, if $(0,0)$ is a saddle point for (10.26), $\mathbf{X}_e$ will be a saddle point for the translated linear system (10.25); see Figs. 10.12 and 10.13. Except for being translated, the phase portrait of (10.25) is identical to that of (10.26). Moreover, because the translated linear system (10.25) approximates the original nonlinear system (10.24) near $\mathbf{X}_e$, we expect the phase portrait of (10.25) to *resemble*, near $\mathbf{X}_e$, the phase portrait of the nonlinear system (10.24). We now explore to what extent this expectation is valid through examples and theorems.

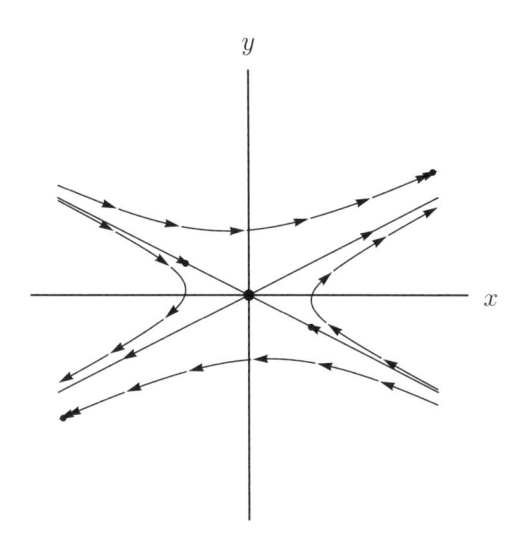

10.2. Linear approximation: The major cases

Figure 10.12. A possible phase portrait for the linear system in equation (10.26).

Figure 10.13. A phase portrait for the corresponding translated linear system in equation (10.25).

Example 10.2.2. Consider the nonlinear system
$$\frac{dx}{dt} = 1 - y, \quad \frac{dy}{dt} = x^2 - y^2.$$
The equilibrium points, which occur where $1-y$ and x^2-y^2 are simultaneously zero, are $(1,1)$ and $(-1,1)$. In the notation from our discussion above we have $f(x,y) = 1-y$ and $g(x,y) = x^2 - y^2$. Computing the first-order partial derivatives of f and g gives
$$\frac{\partial f}{\partial x} = 0, \quad \frac{\partial f}{\partial y} = -1, \quad \frac{\partial g}{\partial x} = 2x, \quad \frac{\partial g}{\partial y} = -2y.$$
Evaluating these derivatives at $(1,1)$ and $(-1,1)$, we see that the Jacobian matrix $\mathbf{J}_1$ at $(1,1)$ and the Jacobian matrix $\mathbf{J}_2$ at $(-1,1)$ are
$$\mathbf{J}_1 = \begin{pmatrix} 0 & -1 \\ 2 & -2 \end{pmatrix} \quad \text{and} \quad \mathbf{J}_2 = \begin{pmatrix} 0 & -1 \\ -2 & -2 \end{pmatrix}.$$
Let's focus on the equilibrium point $(1,1)$. The eigenvalues of $\mathbf{J}_1$ are the roots of the characteristic polynomial
$$\det(\mathbf{J}_1 - z\mathbf{I}) = \det\begin{pmatrix} -z & -1 \\ 2 & -2-z \end{pmatrix} = z^2 + 2z + 2,$$
which are easily calculated to be the complex conjugate pair $-1 \pm i$. The point $(0,0)$ is thus a stable spiral point of the linear system $\mathbf{X}' = \mathbf{J}_1 \mathbf{X}$. We can reach the same conclusion with less computational work by applying the methods of Section 8.9. The matrix $\mathbf{J}_1$ has determinant Δ equal to 2 and trace τ equal to -2. Since $\tau^2 - 4\Delta < 0$, Table 8.2 (or Fig. 8.74) in Section 8.9 tells us that the linear system $\mathbf{X}' = \mathbf{J}_1\mathbf{X}$ has a stable spiral point at the origin. In a similar way you can either compute that $\mathbf{J}_2$ has eigenvalues $-1 \pm \sqrt{3}$ (one positive and one negative value) or simply observe that the determinant of $\mathbf{J}_2$ is negative, to conclude that the linear system $\mathbf{X}' = \mathbf{J}_2\mathbf{X}$ has a saddle point at the origin. We'll see next how this information carries over to the original nonlinear system.

Major types. Roughly speaking, the next result says that under most conditions the phase portrait for a nonlinear system $\frac{d\mathbf{X}}{dt} = \mathbf{F}(\mathbf{X})$ near one of its equilibrium points $\mathbf{X}_e$ will resemble that of the related linear system $\frac{d\mathbf{Y}}{dt} = \mathbf{J}(\mathbf{X}_e)\mathbf{Y}$ near its equilibrium point $(0,0)$, where $\mathbf{J}(\mathbf{X}_e)$ is the Jacobian matrix of $\mathbf{F}$ evaluated at $\mathbf{X}_e$. We use the familiar classification terminology from Chapter 8 of "saddle point", "node", and "spiral point", and so on to capture this idea. Recall that nondegenerate nodes correspond to two different real eigenvalues with the same sign, while a planar linear system has a degenerate node at $(0,0)$ if the associated matrix has a repeated eigenvalue. We will refer to saddle points, spiral points, and nondegenerate nodes as **major types** and get the best information in these cases:

Theorem 10.2.3. *Suppose we have an autonomous nonlinear system*

$$(10.27) \qquad \frac{dx}{dt} = f(x,y), \qquad \frac{dy}{dt} = g(x,y),$$

where $f(x,y)$ and $g(x,y)$ are continuously differentiable. Suppose (x_e, y_e) is an isolated equilibrium point for (10.27). Consider the related linear system

$$(10.28) \qquad \mathbf{X}' = \mathbf{J}\mathbf{X}$$

where $\mathbf{J}$ is the Jacobian matrix

$$\begin{pmatrix} \frac{\partial f}{\partial x}(x_e, y_e) & \frac{\partial f}{\partial y}(x_e, y_e) \\ \frac{\partial g}{\partial x}(x_e, y_e) & \frac{\partial g}{\partial y}(x_e, y_e) \end{pmatrix},$$

and assume that $(0,0)$ is an isolated equilibrium point for this linear system (this is the same as saying $\det \mathbf{J} \neq 0$).

If $(0,0)$ is one of the major types (saddle point, spiral point, or nondegenerate node) for the linear system (10.28), then (x_e, y_e) is the same type for the nonlinear system. Moreover, for these major types, the stability of the equilibrium point is the same for the nonlinear system as it is for the linear system.

Borderline types. Note that Theorem 10.2.3 does not apply if $(0,0)$ is a center or degenerate node for the linear system $\mathbf{X}' = \mathbf{J}\mathbf{X}$. These cases are discussed in the next theorem, which will be further explored in Section 10.3.

Theorem 10.2.4. *Under the same hypotheses as Theorem 10.2.3, we have the following:*

If the linear system $\mathbf{X}' = \mathbf{J}\mathbf{X}$ has a degenerate node at $(0,0)$, then the nonlinear system has either a node or spiral at (x_e, y_e), and the stability is the same for both systems.

If the linear system $\mathbf{X}' = \mathbf{J}\mathbf{X}$ has a center at $(0,0)$, then the nonlinear system has either a center, a spiral, or a hybrid center/spiral at (x_e, y_e). In this case, we cannot predict the stability of (x_e, y_e) for (10.27).

It is helpful to keep the picture of the trace-determinant plane from Section 8.9 in mind here. Degenerate nodes correspond there to points on the parabola $4\Delta = \tau^2$; to either side of this parabola are regions corresponding to spirals and nodes. Centers correspond to the positive half of the Δ-axis; the bordering regions on either side of this ray correspond to asymptotically stable spirals or unstable spirals.

The terminology "hybrid center/spiral" refers to a situation where in each disk centered at the equilibrium point there are closed curve orbits surrounding the equilibrium point, but between two such orbits there may be spiral trajectories. We'll see an example in Exercise 8 of Section 10.3. This situation does not occur if the component functions $f(x,y)$ and $g(x,y)$ are sufficiently nice (as will be the case for nearly every example we consider)[5].

---
[5]In particular, it cannot occur if f and g are polynomials in x and y.

10.2. Linear approximation: The major cases

Tables 10.1 and 10.2 summarize the conclusions of Theorem 10.2.3 and 10.2.4. In these tables, we have a nonlinear system

$$\frac{dy}{dx} = f(x,y), \quad \frac{dy}{dt} = g(x,y),$$

where f and g are continuously differentiable, and we have an isolated equilibrium point at (x_e, y_e). We denote the Jacobian matrix at (x_e, y_e) by $\mathbf{J}$.

Table 10.1. Major Cases

Eigenvalues of $\mathbf{J}$	Classification for $\mathbf{X}' = \mathbf{JX}$	Classification for nonlinear
Real; $r_1 < 0$, $r_2 > 0$	saddle point (unstable)	saddle point (unstable)
Real; $r_1, r_2 > 0$, $r_1 \neq r_2$	unstable node	unstable node
Real; $r_1, r_2 < 0$, $r_1 \neq r_2$	asymptotically stable node	asymptotically stable node
Complex $\alpha \pm i\beta$; $\alpha > 0$, $\beta \neq 0$	unstable spiral	unstable spiral
Complex $\alpha \pm i\beta$; $\alpha < 0$, $\beta \neq 0$	asymptotically stable spiral	asymptotically stable spiral

Table 10.2. Borderline Cases

Eigenvalues of $\mathbf{J}$	Classification for $\mathbf{X}' = \mathbf{JX}$	Classification for nonlinear
Purely imaginary $\pm i\beta$	center (stable)	center, spiral or hybrid center/spiral stability not determined
Real; $r_1 = r_2 > 0$	degenerate node; unstable	node or spiral point unstable
Real; $r_1 = r_2 < 0$	degenerate node asymptotically stable	node or spiral point; asymptotically stable

Theorem 10.2.3 asserts that when the equilibrium point $(0,0)$ of the linear system $\mathbf{X}' = \mathbf{JX}$, where $\mathbf{J}$ is the Jacobian matrix at (x_e, y_e), is one of the major types, the corresponding nonlinear equilibrium (x_e, y_e) inherits the same type. This is very useful but a bit ambiguous since we have not given a mathematical definition of a *nonlinear* node, saddle, or spiral point. An accurate (even though heuristic) way to think about this is that trajectories near a nonlinear node, saddle, or spiral point (x_e, y_e) look like slightly distorted versions of the trajectories of the corresponding linear system near its equilibrium $(0,0)$. The resemblance is strongest near (x_e, y_e) but fades (or

even disappears) far from (x_e, y_e).[6] However, it is reasonable to seek information more concrete and mathematically detailed. Such results do exist, depending on the type of the equilibrium. For instance, here's a general definition of a **node**.

Nonlinear nodes. Let

(10.29) $$\frac{dx}{dt} = f(x,y), \quad \frac{dy}{dt} = g(x,y)$$

be an autonomous system such that f and g are continuously differentiable. Roughly speaking, an equilibrium point (x_e, y_e) for (10.29) is a **stable node** provided that the trajectory of any solution that starts sufficiently close to (x_e, y_e) will approach (x_e, y_e), as $t \to \infty$, tangent to some line through (x_e, y_e) (so that it approaches in a specific direction). More precisely, (x_e, y_e) is a stable node if there is a $\rho > 0$ such that each solution $x = x(t)$, $y = y(t)$ of the system satisfying

(10.30) $$\|(x(0), y(0)) - (x_e, y_e)\| < \rho$$

exists for all $t \geq 0$, has limit (x_e, y_e) as $t \to \infty$, and the trajectory of the solution approaches (x_e, y_e) tangent to some line through (x_e, y_e). *Observe that a stable node is necessarily asymptotically stable.*

The analogous notion of an unstable node involves approach to (x_e, y_e) as $t \to -\infty$: An equilibrium point (x_e, y_e) for the system (10.29) is an **unstable node** provided that there is some $\rho > 0$ such that each solution $x = x(t)$, $y = y(t)$ of the system satisfying inequality (10.30) exists for all $t \leq 0$, has limit (x_e, y_e) as $t \to -\infty$, and the trajectory of the solution approaches (x_e, y_e) (as $t \to -\infty$) tangent to some line through (x_e, y_e).

In keeping with the terminology we have used to describe nodes in the linear setting, we say a node (x_e, y_e) of the system (10.29) is **proper node** provided that for *every* line L through (x_e, y_e), there is a trajectory of the system that approaches (x_e, y_e) tangent to L. Nodes that are not proper are **improper**. If the Jacobian matrix $\mathbf{J}(x_e, y_e)$ of the system (10.29) has two distinct eigenvalues having the same sign, then, just as in the linear case, there will be exactly two lines through (x_e, y_e) to which trajectories will be tangent. Thus, (x_e, y_e) will be an improper node. Moreover, just as in the linear case, one of the lines will be tangent to many distinct trajectories while the other will be tangent to only two trajectories, approaching (x_e, y_e) from opposite directions. Just as in the linear case, we will call these two tangent lines the **nodal tangents** for (10.29) at (x_e, y_e). It can be shown that these nodal tangents are precisely those of the linear system $\mathbf{X}' = \mathbf{J}(x_e, y_e)\mathbf{X}$, but shifted by x_e in the x-direction and by y_e in the y-direction.

Example 10.2.5. Consider the system

(10.31) $$\frac{dx}{dt} = -5xy + 4y + x, \quad \frac{dy}{dt} = -x^2 - y^5 + 2,$$

which has $(1,1)$ as an equilibrium point. The system's Jacobian matrix evaluated at $(1,1)$ is

$$\mathbf{J} = \begin{pmatrix} -4 & -1 \\ -2 & -5 \end{pmatrix},$$

which has eigenvalues -3 and -6, with corresponding eigenvectors

$$\mathbf{V}_1 = \begin{pmatrix} -1 \\ 1 \end{pmatrix} \text{ for the eigenvalue } -3 \quad \text{and} \quad \mathbf{V}_2 = \begin{pmatrix} 1 \\ 2 \end{pmatrix} \text{ for the eigenvalue } -6.$$

The eigenvectors yield the nodal tangents $y = -x$ and $y = 2x$ for the linear system $\mathbf{X}' = \mathbf{J}\mathbf{X}$. Thus, those for the nonlinear system (10.31) are $y = -(x-1) + 1$ and $y = 2(x-1) + 1$; see Figs. 10.14 and 10.15.

---

[6] You can see an illustration of this remark in Figs. 10.14 and 10.15 by looking at the trajectory in the lower right corner of Fig. 10.15.

10.2. Linear approximation: The major cases

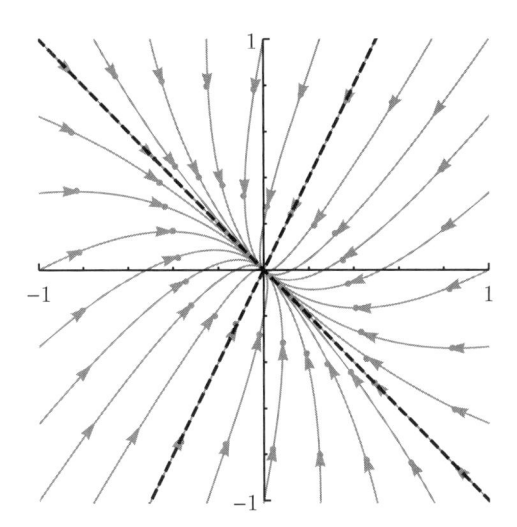

Figure 10.14. The lines $y = -x$ and $y = 2x$, dashed, are nodal tangents of $\mathbf{X}' = \mathbf{J}\mathbf{X}$, where $\mathbf{J} = \mathbf{J}(1,1)$ for the system (10.31).

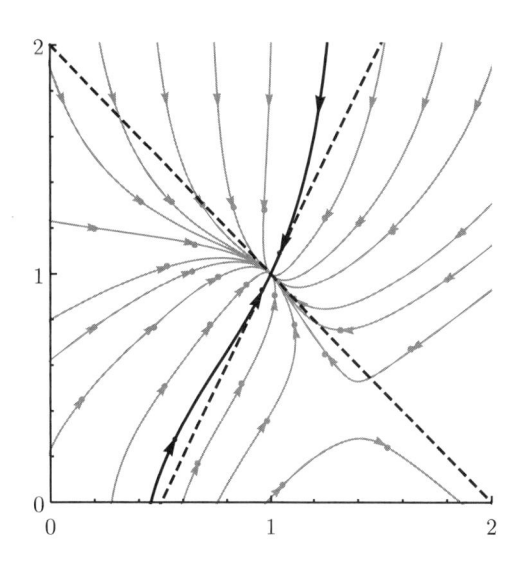

Figure 10.15. The lines $y = -(x-1) + 1$ and $y = 2(x-1) + 1$, dashed, are nodal tangents for the system (10.31), with the first being the preferred nodal tangent.

Note that for the linear system $\mathbf{X}' = \mathbf{J}\mathbf{X}$ of Fig. 10.14, the line $y = -x$ is the "preferred nodal tangent" in that all trajectories of the system, except those along the ray solutions $\mathbf{X} = e^{-6t}\mathbf{V}_2$ and $\mathbf{X} = -e^{-6t}\mathbf{V}_2$, are tangent to $y = -x$ at $(0,0)$. Looking to Fig. 10.15, we see the translate $y = -(x-1) + 1$ of $y = -x$ is the preferred nodal tangent for the nonlinear system's node at $(1,1)$. This reflects a general result. Recall that the preferred nodal tangent for a 2×2 linear system $\mathbf{X}' = \mathbf{A}\mathbf{X}$, where $\mathbf{A}$ has distinct real eigenvalues of the same sign, is the eigenvector line for the eigenvector of $\mathbf{A}$ corresponding to the eigenvalue of smaller absolute value. Thus, we have the following general result:

Theorem 10.2.6. *Assume that the vector field* $\mathbf{F} = (f, g)$ *is continuously differentiable,* (x_e, y_e) *is an equilibrium point of the system*

$$\frac{dx}{dt} = f(x,y), \quad \frac{dy}{dt} = g(x,y),$$

and $\mathbf{J}$ *is the Jacobian matrix of* $\mathbf{F}$ *at* (x_e, y_e). *Suppose that* $\mathbf{J}$ *has distinct real eigenvalues* λ_1 *and* λ_2 *of the same sign, having, respectively, corresponding eigenvectors* $\mathbf{V}_1$ *and* $\mathbf{V}_2$. *Then* (x_e, y_e) *is a node for this system whose nodal tangents will be the lines* L_1 *and* L_2, *passing through* (x_e, y_e), *and parallel to* $\mathbf{V}_1$ *and* $\mathbf{V}_2$, *respectively. Moreover, the preferred nodal tangent will be* L_j *if* λ_j *is the eigenvalue having smaller absolute value.*

We can similarly give a nice description of saddle points for nonlinear systems; see below. For a mathematical description of nonlinear spiral points, see Exercise 31.

Nonlinear saddle points. Suppose that $\mathbf{F} = (f, g)$ is a continuously differentiable vector field and that (x_e, y_e) is an equilibrium point of the system

(10.32) $$\frac{dx}{dt} = f(x,y), \quad \frac{dy}{dt} = g(x,y).$$

Suppose that the Jacobian matrix $\mathbf{J}$ at (x_e, y_e) has eigenvalues r_1 and r_2 with $r_1 < 0 < r_2$ and associated eigenvectors $\mathbf{V}_1$ for r_1 and $\mathbf{V}_2$ for r_2. The equilibrium $\mathbf{0} = (0,0)$ for the linear system $\mathbf{X}' = \mathbf{J}\mathbf{X}$ is a saddle point. We know from Section 8.4 that the incoming ray solutions for this system all lie on the line through **0** and in the direction $\mathbf{V}_1$. Since these ray solutions tend to the equilibrium **0** as $t \to \infty$, we call this eigenvector line the **stable line**. Similarly, the outgoing ray solutions all lie on the line containing **0** and in the direction $\mathbf{V}_2$. We call this the **unstable line** since these solutions all leave the vicinity of **0** as $t \to \infty$; see Fig. 10.16.

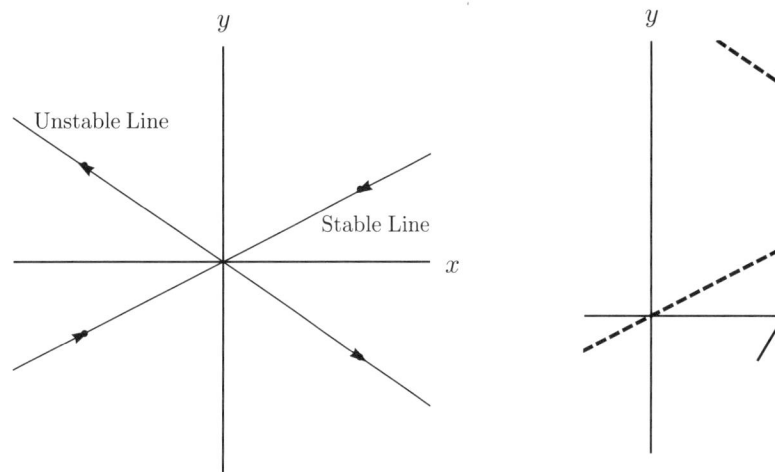

Figure 10.16. Stable and unstable lines for $\mathbf{X}' = \mathbf{JX}$.

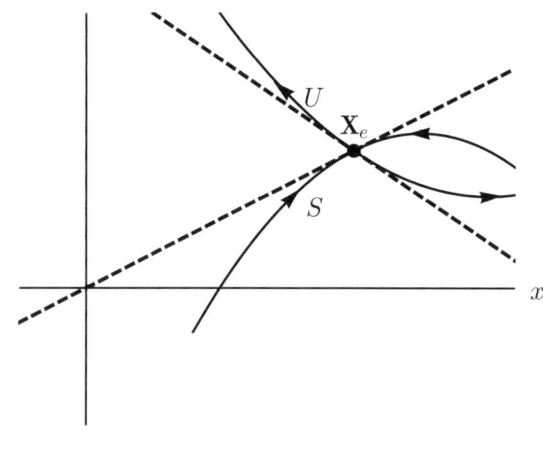

Figure 10.17. Stable and unstable curves S and U for the nonlinear system $\mathbf{X}' = \mathbf{F}(\mathbf{X})$.

The new ingredient here is that our nonlinear system possesses analogues of the stable and unstable lines, namely the **stable curve** S and **unstable curve** U associated to the equilibrium point $\mathbf{X}_e$. The stable curve S and the unstable curve U both pass through $\mathbf{X}_e$ and have continuously turning tangent lines. The stable curve S has the following properties:

(S_1) Any solution $\mathbf{X}(t)$ of the system (10.32) having initial value $\mathbf{X}(0)$ that lies on S must exist and remain on S for all times $t \geq 0$.

(S_2) $\mathbf{X}(t) \to \mathbf{X}_e$ as $t \to \infty$.

The unstable curve U has the following companion properties:

(U_1) Any solution $\mathbf{X}(t)$ of the system (10.32) with $\mathbf{X}(0)$ lying on U must exist and remain on U for all time $t \leq 0$.

(U_2) $\mathbf{X}(t) \to \mathbf{X}_e$ as $t \to -\infty$.

Note that a solution $\mathbf{X}(t)$ lying on S tends to $\mathbf{X}_e$ as time t goes forward towards ∞, while a solution $\mathbf{X}(t)$ lying on U approaches $\mathbf{X}_e$ as time t goes backwards towards $-\infty$. We have the following useful result:

Theorem 10.2.7 (Stable Curve Theorem). *Assume that the vector field $\mathbf{F} = (f, g)$ is continuously differentiable, (x_e, y_e) is an equilibrium point of the system (10.32), and $\mathbf{J}$ is the Jacobian matrix of $\mathbf{F}$ at (x_e, y_e). Suppose that $(0,0)$ is a saddle point for the linear system $\mathbf{X}' = \mathbf{JX}$. The stable curve S and unstable curve U exist with the properties $(S_1), (S_2), (U_1), (U_2)$ described above, and the tangent lines to S and U at (x_e, y_e) are parallel to the stable line and unstable line, respectively, of the associated linear system $\mathbf{X}' = \mathbf{JX}$.*

Fig. 10.17 above shows possible curves S and U and their tangent lines (dashed) at $\mathbf{X}_e$ for a nonlinear system whose approximating linear system has stable and unstable lines as pictured in Fig. 10.16.

10.2. Linear approximation: The major cases

Example 10.2.8. The nonlinear system

$$\frac{dx}{dt} = -x, \quad \frac{dy}{dt} = x^2 + y \tag{10.33}$$

has $(0,0)$ as its only equilibrium point. The Jacobian matrix there is

$$\mathbf{J} = \begin{pmatrix} -1 & 0 \\ 0 & 1 \end{pmatrix},$$

which has eigenvalues 1 and -1. By Theorem 10.2.3, the nonlinear system has a saddle point at $(0,0)$, and we expect its phase portrait nearby to resemble the phase portrait of the linear system $\mathbf{X}' = \mathbf{JX}$. Here we can verify this directly because the nonlinear system is simple enough to permit computation of the form of all solutions. Since y does not appear in the first equation of (10.33) we can solve for x in terms of t to obtain

$$x(t) = c_1 e^{-t}.$$

Substitute this into the second equation of (10.33) to get

$$\frac{dy}{dt} = c_1^2 e^{-2t} + y,$$

a first-order linear equation which we can solve for y:

$$y(t) = -\frac{c_1^2}{3} e^{-2t} + c_2 e^t.$$

Choosing $c_1 = 0, c_2 = 1$ gives the solution $x(t) = 0, y(t) = e^t$, which describes the positive half of the y-axis, directed away from the origin with increasing time. The trajectory corresponding to the choice $c_1 = 0, c_2 = -1$ is along the negative half of the y-axis, still directed away from the origin. This trajectory lies on the unstable curve U, which consists of the y-axis (note that as t goes backward in time, that is, as $t \to -\infty$, the point $(x(t), y(t)) = (0, -e^t)$ tends to the equilibrium point $(0,0)$, in agreement with the definition of U). When $c_1 = 1$ and $c_2 = 0$ we have the solution

$$x(t) = e^{-t}, \quad y(t) = -\frac{1}{3} e^{-2t}.$$

These are parametric equations for the part of the parabola $y = -\frac{1}{3}x^2$ in the fourth quadrant (why?), and as $t \to \infty$ we approach $(0,0)$ along this parabolic curve; see Fig. 10.19. Similarly, the choice $c_1 = -1, c_2 = 0$ gives a trajectory along the other half of the same parabola, directed in to $(0,0)$. The entire parabola is the stable curve S.

A sketch of the phase portrait is shown in Fig. 10.19; notice how the overall picture near $(0,0)$ looks like a distorted version of the phase portrait in Fig. 10.18 for the approximating linear system

$$\mathbf{X}' = \begin{pmatrix} -1 & 0 \\ 0 & 1 \end{pmatrix} \mathbf{X}.$$

The trajectories for the nonlinear system (Fig. 10.19) have the following properties, which are characteristic of the trajectories of a linear system with a saddle point at $\mathbf{0}$:

- All trajectories that approach the equilibrium point $\mathbf{0}$ as $t \to \infty$ lie on a single curve through $\mathbf{0}$—the stable curve $y = -x^2/3$.
- All trajectories that approach $\mathbf{0}$ as $t \to -\infty$ lie on another curve through the origin—the unstable curve $x = 0$.
- No other trajectory (aside from the equilibrium solution $\mathbf{X}(t) = \mathbf{0}$) will approach $\mathbf{0}$ as $t \to \infty$ or as $t \to -\infty$ and $\mathbf{0}$ is an unstable equilibrium point.

Finally, just as in the linear case, solutions that lie on different sides of the stable curve have very different behavior (here, tending upward without bound if above the stable curve and downward without bound if below the stable curve). For this reason, we call the stable curve a **separatrix** of the nonlinear system: It separates two different types of behavior for trajectories.

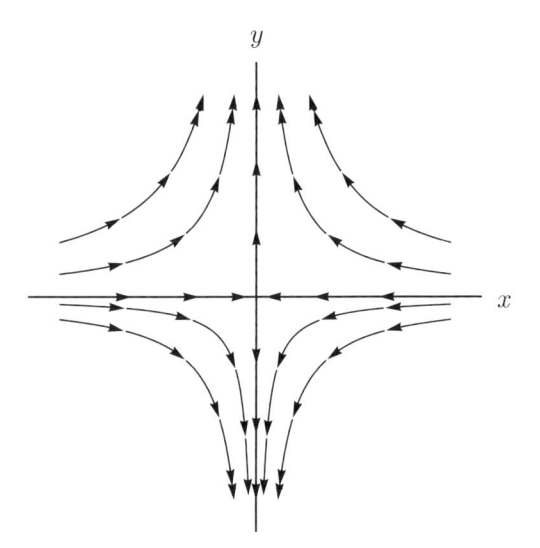

Figure 10.18. Phase portrait for approximating linear system.

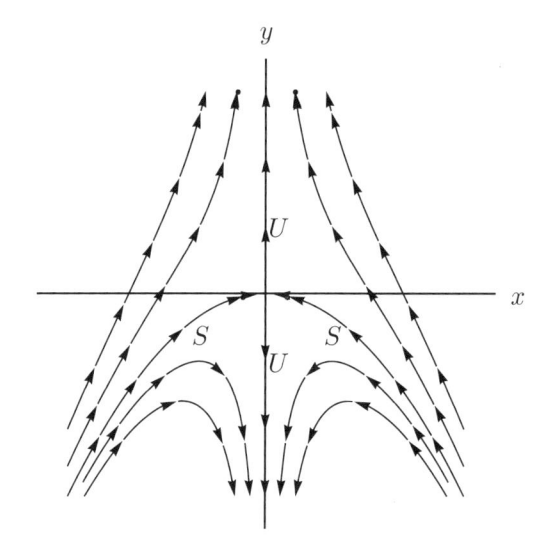

Figure 10.19. Phase portrait with stable curve S and unstable curve U.

Example 10.2.9. The nonlinear system

(10.34) $$\frac{dx}{dt} = y, \quad \frac{dy}{dt} = -y - \sin x$$

has equilibrium points at $(\pm n\pi, 0)$ for $n = 0, 1, 2, 3 \ldots$. The Jacobian matrix is

$$\begin{pmatrix} 0 & 1 \\ -\cos x & -1 \end{pmatrix}.$$

When n is an even integer, $n = 0, \pm 2, \pm 4, \ldots$, we have $\cos(n\pi) = 1$, and when n is an odd integer, $n = \pm 1, \pm 3, \ldots$, we have $\cos(n\pi) = -1$. So when n is even, the Jacobian matrix at $(\pm n\pi, 0)$ is

$$\mathbf{J}_e = \begin{pmatrix} 0 & 1 \\ -1 & -1 \end{pmatrix},$$

and when n is odd, it is

$$\mathbf{J}_o = \begin{pmatrix} 0 & 1 \\ 1 & -1 \end{pmatrix}.$$

The matrix $\mathbf{J}_e$ has eigenvalues $-\frac{1}{2} \pm i\frac{\sqrt{3}}{2}$ while $\mathbf{J}_o$ has eigenvalues $-\frac{1}{2} \pm \frac{\sqrt{5}}{2}$. By Theorem 10.2.3 we expect the phase portrait of the nonlinear system to look like a saddle point at $(\pm \pi, 0)$, $(\pm 3\pi, 0)$, $(\pm 5\pi, 0), \ldots$ and like a stable spiral at $(0, 0), (\pm 2\pi, 0), (\pm 4\pi, 0), (\pm 6\pi, 0), \ldots$; see Fig 10.20. You should "see" the stable and unstable curves at each saddle point, as well as the different behavior of trajectories on opposite sides of the pictured stable curves (separatrices) $S_{-\pi}$ and S_{π}.

10.2. Linear approximation: The major cases

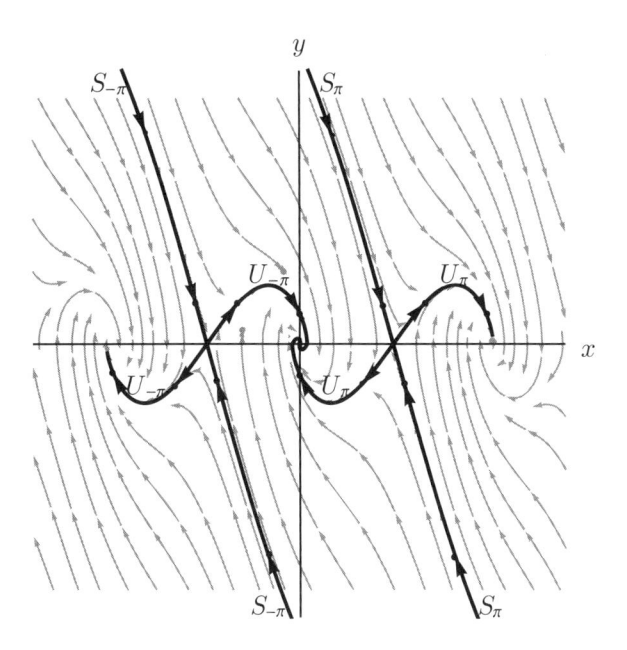

Figure 10.20. The phase portrait.

Example 10.2.10. We will revisit the competing species model introduced in the last section, now including an analysis of the equilibrium points. Our model is described by the equations

$$\text{(10.35)} \qquad \frac{dx}{dt} = ax\left(\frac{K_1 - x - y}{K_1}\right), \qquad \frac{dy}{dt} = cy\left(\frac{K_2 - y - x}{K_2}\right)$$

for the populations of two similar species competing for the same resources. As in Section 10.1, we will assume $K_1 > K_2$. The equilibrium points are $(0,0)$, $(K_1, 0)$, and $(0, K_2)$. The Jacobian matrix is

$$\begin{pmatrix} a - \frac{2a}{K_1}x - \frac{a}{K_1}y & -\frac{a}{K_1}x \\ -\frac{c}{K_2}y & c - \frac{2c}{K_2}y - \frac{c}{K_2}x \end{pmatrix},$$

which is obtained by computing the first-order partial derivatives of

$$f(x,y) = ax - \frac{a}{K_1}x^2 - \frac{a}{K_1}xy \quad \text{and} \quad g(x,y) = cy - \frac{c}{K_2}y^2 - \frac{c}{K_2}xy.$$

To classify the equilibrium point $(0,0)$, we evaluate the Jacobian matrix at this point, obtaining the matrix

$$\mathbf{J}_1 = \mathbf{J}(0,0) = \begin{pmatrix} a & 0 \\ 0 & c \end{pmatrix},$$

which has two positive eigenvalues, a and c (recall, as in Section 10.1, we assume $a \neq c$). The equilibrium point $(0,0)$ is an unstable node for our nonlinear system, and we expect the phase portrait near $(0,0)$ to look like a mildly distorted version of the phase portrait for the linear system $\mathbf{X}' = \mathbf{J}_1 \mathbf{X}$.

Evaluated at the equilibrium point $(K_1, 0)$, the Jacobian matrix is

$$\mathbf{J}_2 = \mathbf{J}(K_1, 0) = \begin{pmatrix} -a & -a \\ 0 & c - c\frac{K_1}{K_2} \end{pmatrix}.$$

The eigenvalues of this matrix are $-a$ and $c - c\frac{K_1}{K_2}$. These are both negative since we are assuming $K_1 > K_2$ and thus $\mathbf{X}' = \mathbf{J}_2 \mathbf{X}$ has stable node at $(K_1, 0)$ (nondegenerate if $a \neq c(\frac{K_1}{K_2} - 1)$).

At the equilibrium point $(0, K_2)$, the Jacobian matrix is

$$\mathbf{J}_3 = \begin{pmatrix} a - a\frac{K_2}{K_1} & 0 \\ -c & -c \end{pmatrix}$$

which has eigenvalues $-c$ and $a - a\frac{K_2}{K_1}$. Using our assumption $K_1 > K_2$ we see that one eigenvalue is positive and one is negative, so that the point $(0, K_2)$ is classified as a saddle point for the original nonlinear system.

The classification of the equilibrium points helps us fill out the sketch of the phase portrait as begun in the previous section; see Fig. 10.21. With the phase portrait in hand, we predict that for any solution $(x(t), y(t))$ with $x(0) > 0$ and $y(0) > 0$ we have $x(t) \to K_1$ and $y(t) \to 0$ as $t \to \infty$. Thus our model predicts that one species is driven to extinction, while the other approaches its carrying capacity. Moreover, the surviving species is the one corresponding to the larger initial carrying capacity.

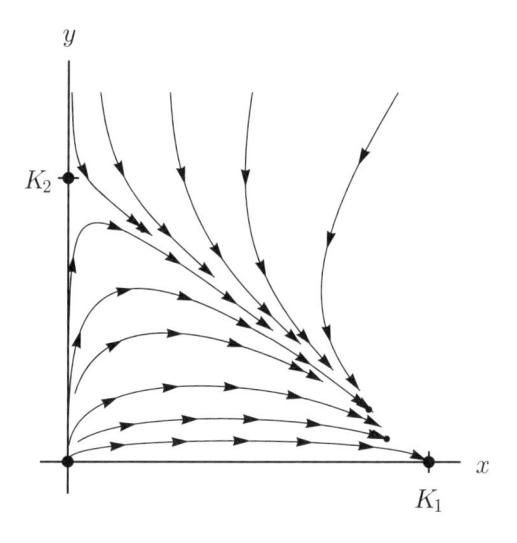

Figure 10.21. Typical phase portrait for (10.35); $K_1 > K_2$.

Example 10.2.11. The system

$$\frac{dx}{dt} = x(1 - x - 2y), \quad \frac{dy}{dt} = y(1 - y - 3x)$$

could be given the interpretation of a competing species model—two interacting populations that in isolation would grow logistically but together compete with each other for resources. Observe that this system fits the description of a quadratic competing species model (see equation (10.5)). The units for x and y could be tens of thousands (or millions) of individuals. Note that these two populations do not occupy the same niche. Thus, although the two species compete for at least some of the same resources, they don't use them (or fight for them) in exactly the same way.

The x-nullclines are the line $x + 2y = 1$ and the vertical line $x = 0$. The y-nullclines are the line $3x + y = 1$ and the horizontal line $y = 0$. Fig. 10.22 shows the first quadrant of the phase plane, with these nullclines and the corresponding direction-indicator arrows in the four basic regions partitioned by the nullclines. The equilibrium points are $(0,0), (1,0), (0,1)$, and $(1/5, 2/5)$; they occur when an x-nullcline intersects a y-nullcline.

The Jacobian matrix is

$$\mathbf{J} = \begin{pmatrix} 1 - 2x - 2y & -2x \\ -3y & 1 - 2y - 3x \end{pmatrix}.$$

10.2. Linear approximation: The major cases

Figure 10.22. Nullcline-and-arrow diagram.

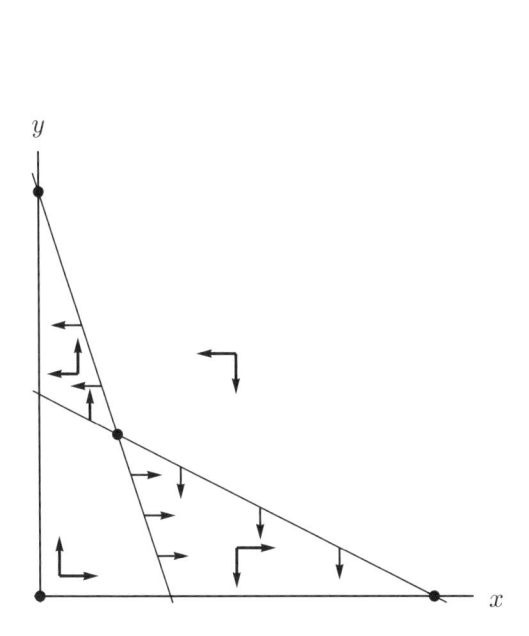

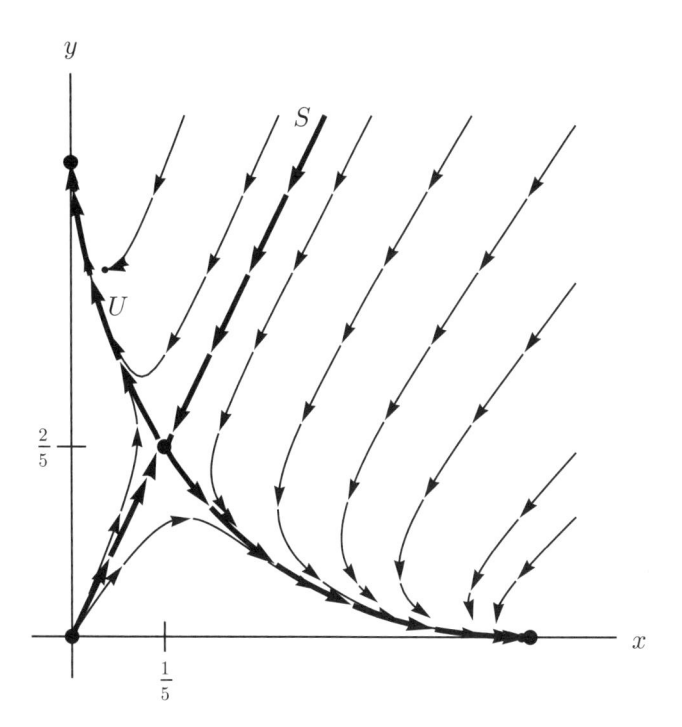

Figure 10.23. Phase portrait showing stable curve S and unstable curve U in bold.

The table below shows the eigenvalues of the Jacobian matrix **J** at each equilibrium point and the corresponding conclusion for the equilibrium point classification.

Equilibrium point classification for Example 10.2.11		
Equilibrium point	Eigenvalues of **J**	Classification
$(1,0)$	$-1, -2$	asymptotically stable node
$(0,1)$	$-1, -1$	asymptotically stable node or spiral point
$(1/5, 2/5)$	$-1, \frac{2}{5}$	saddle
$(0,0)$	$1, 1$	unstable node or spiral point

The Jacobian matrix at $(0,0)$ puts us into a borderline case, and by Theorem 10.2.4 $(0,0)$ is either an unstable spiral or an unstable node. Because the y-axis is an x-nullcline, it consists of trajectories and equilibrium points (the idea is similar to that in Exercise 9 of Section 10.1). Thus, no trajectory can cross the y-axis, preventing $(0,0)$ from being a spiral point; hence, it is an unstable node. The other borderline case—that of the equilibrium point $(0,1)$—is settled by the same argument: $(0,1)$ is an asymptotically stable node (not a spiral point).

Notice that we can determine the saddle point classification at $(1/5, 2/5)$ without the computational fuss of actually determining the eigenvalues; the Jacobian matrix has determinant $-\frac{2}{5} < 0$, and by Fig. 8.74 (and Theorem 10.2.3) this tells us that we have a saddle at $(1/5, 2/5)$. There are two trajectories that approach $(1/5, 2/5)$ as $t \to \infty$ (these follow the stable curve), but unless we are so lucky to have an initial condition that puts us exactly on one of these curves, or at the equilibrium point $(1/5, 2/5)$ itself, first-quadrant trajectories of our system will be such that either $x(t) \to 1$ and $y(t) \to 0$ as $t \to \infty$ or $x(t) \to 0$ and $y(t) \to 1$ as $t \to \infty$. If these represent populations, then extinction of one species occurs. Which of these two cases occurs depends on which on which side of the separatrix the initial condition has put us. Again, we see that the separatrix separates qualitatively different behavior.

Basin of attraction. The preceding example gives a nice way to introduce the following notion: The **basin of attraction** of an asymptotically stable equilibrium point is the set of all initial points of trajectories that approach the equilibrium point as $t \to \infty$. More precisely, a point (x_0, y_0) belongs to the basin of attraction of the asymptotically stable equilibrium point (x_e, y_e) of the system $dx/dt = f(x, y), dy/dt = g(x, y)$ provided that if $x = x(t), y = y(t)$ solves the system and satisfies $x(0) = x_0$ and $y(0) = y_0$, then $\lim_{t\to\infty}(x(t), y(t)) = (x_e, y_e)$.

In Example 10.2.11, what points in the first quadrant $x > 0$, $y > 0$ lie in the basin of attraction of the equilibrium point $(0, 1)$? Fig. 10.23 provides the answer: Any point in the first quadrant that lies strictly to the left of the stable curve S is in the basin of attraction of $(0, 1)$. In Exercise 21 you are asked to describe the points in the first quadrant that lie in the basin of attraction of the equilibrium point $(1, 0)$.

Remark: When considered in the entire plane, the basin of attraction of an asymptotically stable equilibrium point must include an open disk of positive radius centered at the equilibrium point; this is a consequence of the definition of an asymptotically stable equilibrium. Notice that for a planar *linear system* with a stable node or stable spiral point at $(0, 0)$, the basin of attraction is the whole plane.

10.2.1. Exercises.

In Exercises 1–15, (i) find all equilibrium points, (ii) find the Jacobian matrix for the linear approximation at each equilibrium point, and (iii) for each equilibrium point where Theorem 10.2.3 applies, classify the point as to type (spiral, saddle, etc.) and stability. If Theorem 10.2.3 does not apply, say so, and then say what information Theorem 10.2.4 gives.

1. $\dfrac{dx}{dt} = x - y^2$, $\dfrac{dy}{dt} = x - y$.

2. $\dfrac{dx}{dt} = x(2x + 3y - 7)$, $\dfrac{dy}{dt} = y(3x - 4y - 2)$.

3. $\dfrac{dx}{dt} = x(x^2 + y^2 - 10)$, $\dfrac{dy}{dt} = y(xy - 3)$.

4. $\dfrac{dx}{dt} = x - xy$, $\dfrac{dy}{dt} = xy - y$.

5. $\dfrac{dx}{dt} = y$, $\dfrac{dy}{dt} = -x + (1 - x^2)y$.

6. $\dfrac{dx}{dt} = y$, $\dfrac{dy}{dt} = -x - (1 - x^2)y$.

7. $\dfrac{dx}{dt} = y$, $\dfrac{dy}{dt} = -x + 3(1 - x^2)y$.

8. $\dfrac{dx}{dt} = x(1 - x - y)$, $\dfrac{dy}{dt} = y(2 - x - 4y)$.

9. $\dfrac{dx}{dt} = \sin(x + y)$, $\dfrac{dy}{dt} = e^x - 1$.

10. $\dfrac{dx}{dt} = y + x^2 - 3$, $\dfrac{dy}{dx} = x(1 + y)$.

11. $\dfrac{dx}{dt} = y - x^3$, $\dfrac{dy}{dt} = -x$.

12. $\dfrac{dx}{dt} = x + xy$, $\dfrac{dy}{dt} = x - y$.

13. $\dfrac{dx}{dt} = y^3 - x^3$, $\dfrac{dy}{dt} = 1 - y^3$.

14. $\dfrac{dx}{dt} = -3x + x^2 - xy$, $\dfrac{dy}{dt} = -5y + xy$.

15. $\dfrac{dx}{dt} = x + y$, $\dfrac{dy}{dt} = y + x^2 y$.

16. In Exercise 1 of Section 10.1, you did a nullcline-and-arrow diagram for the system
$$\frac{dx}{dt} = x(5 - x - 4y), \quad \frac{dy}{dt} = y(2 - y - x).$$

10.2. Linear approximation: The major cases

Show that there are four equilibrium points, and find the linear approximation at each. Use Theorem 10.2.3 to classify each equilibrium point, if possible. Put together this classification information with the nullcline-and-arrow diagram from Exercise 1 in Section 10.1 to give a sketch of a phase portrait in the first quadrant.

17. Two competing species are modeled by the system
$$\frac{dx}{dt} = x(120 - 3x - y), \quad \frac{dy}{dt} = y(80 - x - y).$$
What is the outcome of this competition? Extinction of one species or peaceful coexistence?

18. The linear approximation for an autonomous system at an equilibrium point (x_e, y_e) has matrix $\mathbf{A}$. Under what conditions on the trace τ of $\mathbf{A}$ and the determinant Δ of $\mathbf{A}$ can you **not** apply Theorem 10.2.3 to classify (x_e, y_e)?

19. Consider the system
$$\frac{dx}{dt} = y - x, \quad \frac{dy}{dt} = \frac{5x^2}{x^2 + 4} - y.$$
   (a) Find all equilibrium points in the first quadrant and the Jacobian matrix at each. Classify the equilibrium points if possible.
   (b) Do a nullcline-and-arrow diagram for the phase portrait in the first quadrant. Sketch some trajectories, keeping in mind your answers to (a).

20. Show that $(1, 1)$ is a nonisolated equilibrium point of
$$\frac{dx}{dt} = x^2 - y^2, \quad \frac{dy}{dt} = xy - x^2,$$
and compute the determinant of the associated Jacobian matrix $\mathbf{J}(1, 1)$.

21. In Example 10.2.11, what points in the first quadrant $x > 0$, $y > 0$ are in the basin of attraction of the equilibrium point $(1, 0)$?

22. Consider the system
$$(10.36) \qquad \frac{dx}{dt} = -x + y, \quad \frac{dy}{dt} = x - xy^2.$$
   (a) Confirm that $(1, 1)$ is an asymptotically stable equilibrium point of (10.36).
   (b) Give a nullcline-and-arrow sketch in the first quadrant for the system (10.36).
   (c) Is $(3, 2)$ in the basin of attraction of $(1, 1)$? Justify your answer. You should be able to construct a convincing argument, not involving any computer-based work.

23. Consider the system
$$(10.37) \qquad \frac{dx}{dt} = y - 4x, \quad \frac{dy}{dt} = -xy - 3x.$$
   (a) Confirm that $(0, 0)$ is an asymptotically stable equilibrium point of (10.37).
   (b) Identify a point that is not in the basin of attraction of $(0, 0)$. Justify your answer.
   (c) Identify an entire region of points that are not in the basin of attraction of $(0, 0)$. Justify your answer.

24. Consider the system
$$\frac{dx}{dt} = -x + y, \quad \frac{dy}{dt} = x - xy^2.$$
   (a) In Exercise 22 above you were asked to verify that $(1, 1)$ is an asymptotically stable equilibrium point. Confirm it is an improper node and find its nodal tangents. Which is the preferred nodal tangent?
   (b) (CAS) Use a computer algebra system to create a phase portrait for this system and check that your nodal tangent information of (a) is consistent with your portrait.

25. Consider the competing species system

$$\frac{dx}{dt} = x(1 - x - y), \quad \frac{dy}{dt} = y(2 - x - y)$$

in which x and y are measured in thousands and t in years.
  (a) Show that $(0, 2)$ is an improper node for the system and find its nodal tangents. Which is the preferred nodal tangent?
  (b) Suppose that $x(0) = 2$ and $y(0) = 0.2$. Suppose the population level of y has risen to 1.9 (i.e., 1,900); use your preferred nodal tangent information to estimate corresponding x-population level.
  (c) (CAS) Use a computer algebra system to create a phase portrait for this system and check that your nodal tangent information of (a) as well as your answer to (b) is consistent with your portrait.
  (d) (CAS) Solve numerically the initial value problem consisting of the given competing species system with $x(0) = 2$ and $y(0) = 0.2$. According to your numerical solution when does the y reach 1.9? What is x at that time?

26. Consider the system

$$\frac{dx}{dt} = 5 - x^2 - y, \quad \frac{dy}{dt} = x(1 - y^2).$$

  (a) Find the only equilibrium point of the system with $x > 0$ and $y > 0$.
  (b) Use Theorems 10.2.3 and 10.2.4 to say as much as you can about the equilibrium point you found in (a).
  (c) Sketch a nullcline-and-arrow diagram for the system in the first quadrant and use it to argue that the equilibrium point you found in (a) cannot be a spiral point.
  (d) (CAS) Use a computer algebra system to create, near the equilibrium point from (a), a phase portrait for both the original nonlinear system as well as for the approximating linear system there.
  (e) Comment on the extent to which your phase portraits are consistent with the information provided in response to (b) and (c).

27. Consider the system

$$\frac{dx}{dt} = x - xy + (x - 2)^3, \quad \frac{dy}{dt} = -2y + xy + (y - 1)^3.$$

  (a) Confirm that $(2, 1)$ is an equilibrium point of the system.
  (b) Use Theorems 10.2.3 and 10.2.4 to say as much as you can about the equilibrium point $(2, 1)$.
  (c) (CAS) Use a computer algebra system to create, near $(2, 1)$, a phase portrait for both the original nonlinear system as well as for the approximating linear system there.
  (d) Comment on the extent to which your phase portraits are consistent with the information provided in response to (b).

28. Consider the system

$$\frac{dx}{dt} = y + x^2 - 3, \quad \frac{dy}{dt} = x(1 + y).$$

  (a) Verify that $(2, -1)$ is an unstable improper node of this system and find its nodal tangents. Which is the preferred nodal tangent?
  (b) (CAS) Use a computer algebra system to create a phase portrait for this system and confirm that your nodal tangent information from (a) is consistent with the portrait.

10.2. Linear approximation: The major cases

29. The population levels (in thousands) of two competing species on a small island are modeled by the system

    (10.38) $$\frac{dx}{dt} = x(1 - 2x - 2y), \quad \frac{dy}{dt} = y(1 - x - y),$$

    where time t is measured in years.
    (a) What is the outcome of this competition? Extinction of one species or peaceful coexistence?
    (b) (CAS) Solve numerically the initial value problem consisting of (10.38) with $x(0) = 0.1, y(0) = 0.02$, so that the initial x-population is 100 and the initial y-population is 20. After approximately how many years will x reach its maximum value? What is the approximate maximum value?

30. Pictured below is a phase portrait for the system $\frac{dx}{dt} = \frac{1}{2}x^2 + y^4 - \frac{3}{2}, \quad \frac{dy}{dt} = xy - 1$ near its equilibrium point $(1, 1)$.

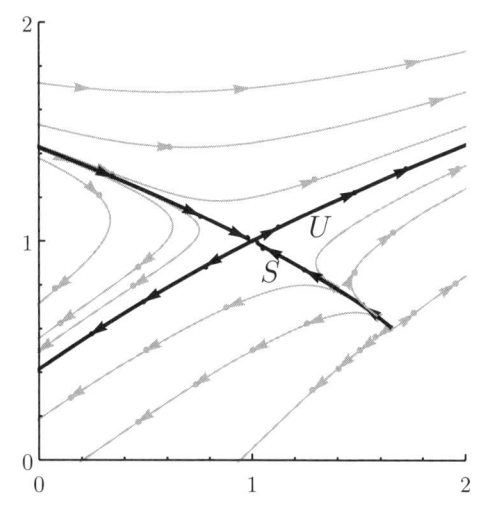

    (a) Verify that $(1, 1)$ is a saddle point of the system using Theorem 10.2.3.
    (b) The stable curve S and unstable curve U for $(1, 1)$ are shown in the figure above. Find equations of the lines through $(1, 1)$ that are tangent at $(1, 1)$ to S and to U, respectively.

31. **Nonlinear spiral points.** Consider a nonlinear system

    (10.39) $$\frac{dx}{dt} = f(x, y), \quad \frac{dy}{dt} = g(x, y)$$

    with continuously differentiable vector field $\mathbf{F} = (f, g)$, for which the origin $(0, 0)$ is an isolated equilibrium point. Suppose this equilibrium point is also an asymptotically stable spiral point for the corresponding linear system

    $$\frac{d\mathbf{X}}{dt} = \mathbf{J}\mathbf{X},$$

    where

    $$\mathbf{J} = \mathbf{J}(0, 0) = \begin{pmatrix} a & b \\ c & d \end{pmatrix}$$

    is the Jacobian matrix of $\mathbf{F}$ at $(0, 0)$. Thus $\mathbf{J}$ is assumed to have complex eigenvalues $\alpha \pm i\beta$ where $\alpha < 0$ and $\beta \neq 0$. By Theorem 10.2.3, $(0, 0)$ is also an asymptotically stable spiral point for the nonlinear system (10.39). A reasonable mathematical description of what this means is that the following two conditions should hold:
    (I) $(0, 0)$ is asymptotically stable for the system (10.39). In particular, if $(x(t), y(t))$ is a solution with $(x(0), y(0))$ sufficiently close to $(0, 0)$, then $(x(t), y(t)) \to (0, 0)$ as $t \to \infty$.

(II) Any solution $(x(t), y(t))$ with $(x(0), y(0))$ sufficiently close to $(0,0)$ will "spiral around" $(0,0)$ an infinite number of times as $t \to \infty$. More specifically, if $\theta(t)$ is a continuous choice of polar angle for $(x(t), y(t))$, then $\theta(t)$ tends either to $+\infty$ or $-\infty$ as $t \to \infty$.

In this exercise you will prove (II). In doing so, you may assume that (I) is true.

(a) Show that b and c have opposite signs.

(b) For definiteness, assume $c > 0$ so that $-b > 0$ by (a). Also assume throughout that $(x(t), y(t))$ is a solution to (10.39) with $(x(0), y(0))$ sufficiently close to $(0, 0)$ so that $(x(t), y(t)) \to (0, 0)$ as $t \to \infty$. A continuous choice of polar angle $\theta(t)$ for the point $(x(t), y(t))$ satisfies
$$\tan(\theta(t)) = \frac{y(t)}{x(t)}.$$
Writing x, y, and θ for $x(t)$, $y(t)$, and $\theta(t)$, show that
$$\theta' = \frac{xy' - yx'}{r^2},$$
where primes denote derivatives with respect to t and $r^2 = x^2 + y^2$.

(c) We have
$$x' = f(x, y) = ax + by + R_f(x, y)$$
and
$$y' = g(x, y) = cx + dy + R_g(x, y)$$
where R_f and R_g are the respective error terms. Show that
$$\theta' = \frac{cx^2 + (d-a)xy - by^2}{r^2} + \frac{xR_g(x,y) - yR_f(x,y)}{r^2}.$$

(d) Show that
$$\frac{|xR_g(x,y) - yR_f(x,y)|}{r^2} \to 0$$
as $(x, y) \to (0, 0)$, that is, as $r \to 0$. Hint: $xR_g(x,y) - yR_f(x,y) = (x, -y) \cdot (R_g(x, y), R_f(x, y))$, a dot product in $\mathbb{R}^2$. Recall that for any two vectors $\mathbf{A}$ and $\mathbf{B}$ in $\mathbb{R}^2$, $|\mathbf{A} \cdot \mathbf{B}| \leq \|\mathbf{A}\|\|\mathbf{B}\|$. Don't forget the crucial property (10.21) for the error term.

(e) Consider two regions in $\mathbb{R}^2$, the first defined by $|y| \leq |x|$ and the second by $|y| \geq |x|$. (You may find it helpful to draw a picture.) Show that in the first region
$$\frac{cx^2 + (d-a)xy - by^2}{r^2} \geq \frac{1}{2} p\left(\frac{y}{x}\right),$$
where $p(u) = -bu^2 + (d-a)u + c$, while in the second region
$$\frac{cx^2 + (d-a)xy - by^2}{r^2} \geq \frac{1}{2} q\left(\frac{x}{y}\right),$$
where $q(u) = cu^2 + (d-a)u - b$.

(f) Since $-b$ and c are positive, the graphs of the two quadratic polynomials $p(u)$ and $q(u)$ are concave up parabolas. Show that the minimum values of $p(u)$ and $q(u)$ are both positive. Hint: Can $p(u)$ or $q(u)$ have a real zero?

(g) Combine the above results to show that there exist positive numbers B and t_0 such that $\theta'(t) \geq B$ for all $t > t_0$.

(h) Deduce from (g) that there is a real number C such that $\theta(t) \geq Bt + C$ for all $t > t_0$. Thus $\theta(t) \to \infty$ as $t \to \infty$ and $(x(t), y(t))$ spirals counterclockwise. Remark: If at the start you had instead assumed $b > 0$ and $c < 0$, then a similar argument would lead to $\theta(t) \to -\infty$ as $t \to \infty$, giving a clockwise spiral.

10.3. Linear approximation: The borderline cases

We'll look at some examples to illustrate what can happen when the classification of the equilibrium point for the approximating linear system falls into one of the borderline cases. Of these various borderline cases, centers are the most important, because they appear in key applications, and so we concentrate on these.

Example 10.3.1. Consider the nonlinear system

$$(10.40) \qquad \frac{dx}{dt} = -y - 2xy, \quad \frac{dy}{dt} = x + 2x^2,$$

which has an equilibrium point at $(0,0)$. The Jacobian matrix for the linear approximation near $(0,0)$ is

$$\mathbf{J} = \begin{pmatrix} 0 & -1 \\ 1 & 0 \end{pmatrix}.$$

Since $\mathbf{J}$ has eigenvalues $\pm i$, $(0,0)$ is a center for the approximating linear system. According to Theorem 10.2.4 the nonlinear system could have either a center or a spiral at $(0,0)$, and if a spiral, it could be either asymptotically stable or unstable.[7] Can we determine which? Our original equations are simple enough that we can write a separable first-order equation for $\frac{dy}{dx}$:

$$\frac{dy}{dx} = \frac{dy/dt}{dx/dt} = \frac{x + 2x^2}{-y - 2xy} = -\frac{x}{y},$$

so that

$$\int y\, dy = \int -x\, dx,$$

or $x^2 + y^2 = C$. The orbits of (10.40) are therefore circles, and this nonlinear system has a center at $(0,0)$. Centers are stable, but not asymptotically stable.

It's the rare equation that is simple enough to be able to analyze as we did in the last example. A tool that is useful for some other borderline cases is a **change of variables to polar coordinates** (r, θ), where we assume $r \geq 0$. The next example illustrates this.

Example 10.3.2. We start with the nonlinear system

$$(10.41) \qquad \frac{dx}{dt} = -y - x(x^2 + y^2), \quad \frac{dy}{dt} = x - y(x^2 + y^2),$$

which has the same linear approximation at $(0,0)$ as in the preceding example. So again the question is whether this nonlinear system has a center or spiral at $(0,0)$. We will make a change of variables to polar coordinates to decide which.

Differentiation with respect to t on both sides of the polar coordinate change of variable equation $r^2 = x^2 + y^2$ gives $2rr' = 2xx' + 2yy'$. Thus, denoting the derivative with respect to t with primes,

$$(10.42) \qquad r' = \frac{xx' + yy'}{r}.$$

---

[7] According to Theorem 10.2.4, there is a third possibility: a hybrid center/spiral. An example of a hybrid center/spiral (which we have not formally defined) is found in Exercise 8. This hybrid case occurs only rarely and never when, as in Examples 10.3.1 and 10.3.2, the component functions f and g are polynomials (in x and y).

Making the substitutions $x' = -y - x(x^2 + y^2)$ and $y' = x - y(x^2 + y^2)$ from (10.41) into equation (10.42) we obtain after a short calculation

$$r' = \frac{-(x^2+y^2)^2}{r} = -r^3.$$

This is a separable first-order equation for r with solution $r(t) = (2t+c_1)^{-1/2}$ for arbitrary constant c_1. Similarly, differentiating the relation $\tan\theta = y/x$ with respect to t gives

$$(\sec^2\theta)\,\theta' = \frac{xy' - yx'}{x^2}, \quad \text{or equivalently,} \quad \theta' = \frac{xy' - yx'}{x^2}\cos^2\theta.$$

Using the relation $x = r\cos\theta$ from the polar coordinate change of variables, we obtain

$$\boxed{\theta' = \frac{xy' - yx'}{r^2}.}$$

Again making the substitutions for x' and y' from the original equations in (10.41), we have

$$\theta'(t) = \frac{x^2 + y^2}{r^2} = 1,$$

so that $\theta(t) = t + c_2$, for arbitrary constant c_2.

What does a solution to (10.41) that starts out near the origin look like? Let's interpret the statement "starts near the origin" to mean $r(0) < 1$. This says $c_1 > 1$, and as t increases from 0, $r(t) = (2t + c_1)^{-1/2}$ decreases, while $\theta(t) = t + c_2$ increases. Thus a trajectory with $r(0) < 1$ spirals into the origin in a counterclockwise fashion. Now observe that in fact all orbits have this behavior, including those for which $r(0) \geq 1$. The equilibrium point $(0,0)$ is a stable spiral point for this nonlinear system. A sketch of some orbits is shown in Fig. 10.24.

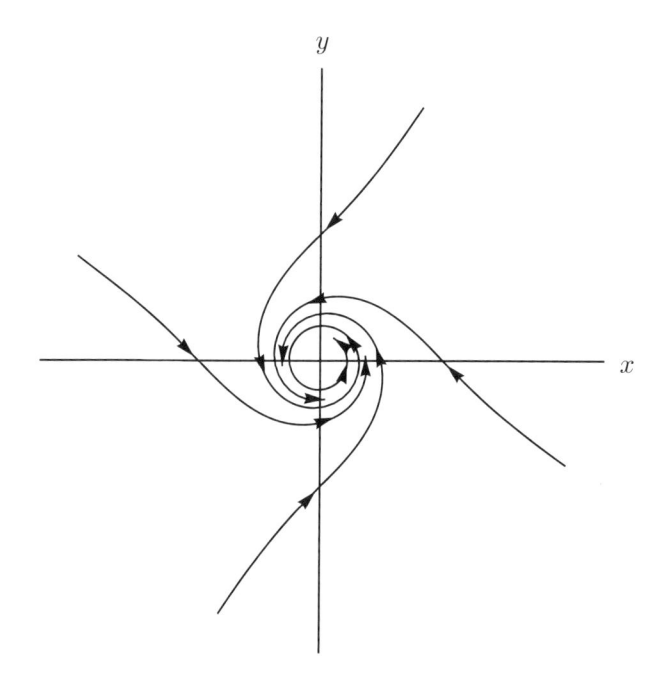

Figure 10.24. Phase portrait for equation (10.41).

10.3. Linear approximation: The borderline cases

Our final example, while not a borderline case, continues the theme of polar coordinate change of variable and also serves to emphasize the fact that the linear approximation gives *local information* only.

Example 10.3.3. Consider the nonlinear system

$$(10.43) \qquad \frac{dx}{dt} = -y + x(1 - x^2 - y^2), \quad \frac{dy}{dt} = x + y(1 - x^2 - y^2),$$

which has an equilibrium point at $(0,0)$. The Jacobian matrix for the linear approximation there is

$$\mathbf{J} = \begin{pmatrix} 1 & -1 \\ 1 & 1 \end{pmatrix}.$$

Since the eigenvalues of $\mathbf{J}$ are $1 \pm i$, $(0,0)$ is an unstable spiral for the approximating linear system. By Theorem 10.2.3, the same is true for the original nonlinear system. We will verify this independently by a polar coordinate change of variable. By equations (10.42) and (10.43) we have

$$r' = \frac{xx' + yy'}{r} = \frac{x(-y + x(1 - x^2 - y^2)) + y(x + y(1 - x^2 - y^2))}{r} = r - r^3.$$

Also

$$\theta'(t) = \frac{xy' - yx'}{r^2} = 1$$

so $\theta(t) = t + c$, for c an arbitrary constant. The first-order differential equation $\frac{dr}{dt} = r - r^3$ can be solved as a Bernoulli equation (or by separation of variables) to obtain

$$r(t) = \frac{1}{\sqrt{1 + ke^{-2t}}}$$

for some constant k. The constant k is determined by the value of r at $t = 0$:

$$\frac{1}{1+k} = r(0)^2, \quad \text{or} \quad k = \frac{1}{r(0)^2} - 1.$$

If we start close to the origin (say, $r(0) < 1$), then k is positive and $r(t) \to 0$ as $t \to -\infty$. Since $\theta(t) = t + c$, we see that the trajectories that start near the origin spiral away from the origin in the counterclockwise direction with increasing t. This is what we expect from Theorem 10.2.3. However, as $t \to \infty$, $r(t) \to 1$. Notice that the nonlinear system has a periodic solution $x(t) = \cos t, y(t) = \sin t$, whose orbit is the circle $r(t) = 1$, or $x^2 + y^2 = 1$. As $t \to \infty$, our spirals approach this circle but cannot cross it; see Fig. 10.25. This is a phenomena we never see in a linear system. This emphasizes the fact that the method of linear approximation gives only *local* information *near* an equilibrium point.

In Example 10.3.3 we call the unit circle a **limit cycle** for the system (10.43). It is a closed-curve orbit that is approached by other orbits as $t \to \infty$. We will return to the notion of limit cycles in Section 10.8.

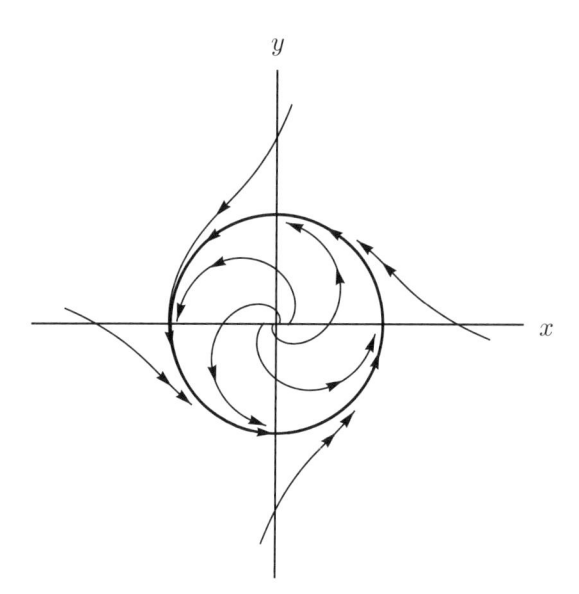

Figure 10.25. Phase portrait for the system (10.43).

10.3.1. Exercises.

1. Does the nonlinear system
$$\frac{dx}{dt} = y - xy^2, \quad \frac{dy}{dt} = -x + x^2 y$$
have a center or spiral at $(0,0)$?

2. This problem generalizes Example 10.3.2. Consider the system
$$\frac{dx}{dt} = -y + kx(x^2 + y^2), \quad \frac{dy}{dt} = x + ky(x^2 + y^2),$$
where k is a constant. Show that $(0,0)$ is a stable spiral if $k < 0$, an unstable spiral if $k > 0$, and a center if $k = 0$.

3. In Example 10.3.3, suppose that $r(0) > 1$, so that
$$r(t) = \frac{1}{\sqrt{1 + ke^{-2t}}},$$
where k is now negative. What happens to the corresponding orbit as $t \to \infty$?

4. Consider the solution $x(t) = (1 - \frac{3}{4}e^{-2t})^{-1/2} \cos t$, $y(t) = (1 - \frac{3}{4}e^{-2t})^{-1/2} \sin t$ of the system of Example 10.3.3. Note the solution satisfies the initial conditions $x(0) = 2, y(0) = 0$.
   (a) Show that $(-\frac{1}{2}\ln(4/3), \infty)$ is the interval of validity of this solution.
   (b) Let $T = -\frac{1}{2}\ln(4/3)$. Write the solution $x(t) = (1 - \frac{3}{4}e^{-2t})^{-1/2} \cos t$, $y(t) = (1 - \frac{3}{4}e^{-2t})^{-1/2} \sin t$ in polar coordinates $(r(t), \theta(t))$, and show that $r(t) \to \infty$ and $\theta(t) \to T$ as t approaches T from the right.
   (c) Part (b) suggests that the line $y = (\tan T)x$ is an asymptote for the given solution in the sense that the distance from the point
$$(10.44) \quad \left(\left(1 - \frac{3}{4}e^{-2t}\right)^{-1/2} \cos t, \left(1 - \frac{3}{4}e^{-2t}\right)^{-1/2} \sin t\right) \text{ on the trajectory}$$
   to the point
$$(10.45) \quad \left(\left(1 - \frac{3}{4}e^{-2t}\right)^{-1/2} \cos(T), \left(1 - \frac{3}{4}e^{-2t}\right)^{-1/2} \sin(T)\right) \text{ on the line } y = \tan(T)x$$

10.3. Linear approximation: The borderline cases 651

approaches 0 as $t \to T^+$. Verify this. Note that each of the two points displayed above has the same distance to the origin, namely, $(1 - \frac{3}{4}e^{-2t})^{-1/2}$.

(d) (CAS) Illustrate the result in (c) by creating a parametric plot, over the interval $T+0.01 \leq t \leq 5$, of the point (10.44) on the trajectory and the point (10.45) on the line.

5. Consider the nonlinear system

(10.46) $$\frac{dx}{dt} = x - x^2, \quad \frac{dy}{dt} = y.$$

(a) Find the equilibrium points.
(b) Find the Jacobian matrix $\mathbf{J}$ at $(0,0)$ and classify the equilibrium point for the linear system $\mathbf{X}' = \mathbf{J}\mathbf{X}$.
(c) Solve the separable equation
$$\frac{dy}{dx} = \frac{y}{x - x^2}$$
to find equations for the trajectories of (10.46). Sketch these trajectories near $(0,0)$.
(d) True or false: For every ray through the origin, there is a trajectory tangent to that ray.
(e) How would you classify the equilibrium point $(0,0)$ for (10.46)?

6. Consider the nonlinear system
$$\frac{dx}{dt} = y, \quad \frac{dy}{dt} = x - x^3.$$

(a) Show that $(0,0), (1,0)$, and $(-1,0)$ are the equilibrium points for this system, and compute the Jacobian matrix at each of these points.
(b) Show that $(0,0)$ is a saddle point for the approximating linear system and the other equilibrium points are centers for the approximating linear systems there. On the basis of this alone, what (if anything) can you conclude about the classification of the three equilibrium points for the nonlinear system?
(c) Show that the trajectories for the nonlinear system have equations $y^2 - x^2 + \frac{x^4}{2} = C$ for some constant C. (Hint: Find a first-order separable differential equation for $\frac{dy}{dx}$ and solve it.)
(d) The graph of the surface $z = y^2 - x^2 + \frac{x^4}{2}$ is shown in Fig. 10.26. How does this help you determine whether the equilibrium points $(1,0)$ and $(-1,0)$ are centers or spiral points?

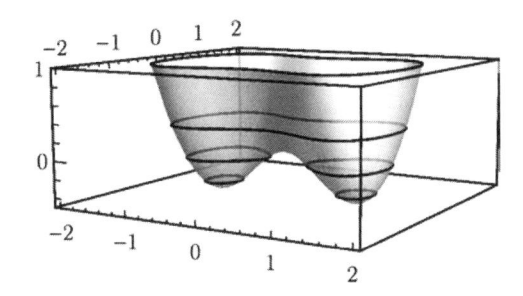

Figure 10.26. The surface $z = y^2 - x^2 + x^4/2$.

(e) Using your work in (a)–(d), sketch the phase portrait for the nonlinear system. Your picture should include the trajectories that correspond to each of the cross sections shown in Fig. 10.26.

7. Consider the nonlinear system
$$\frac{dx}{dt} = cx + y - x(x^2 + y^2), \quad \frac{dy}{dt} = -x + cy - y(x^2 + y^2),$$
where c is a nonzero constant.

(a) Show that $(0,0)$ is the only equilibrium point of this system.
(b) Show that if $c > 0$, then $x = \sqrt{c}\sin t, y = \sqrt{c}\cos t$ is a solution.
(c) Show that if we change to polar coordinates, we have

$$\frac{dr}{dt} = r(c - r^2).$$

This is a Bernoulli equation, but instead of solving it explicitly, give a qualitative solution of this first-order equation by sketching some solutions in the tr-plane for $r > 0$. Begin your sketch by showing any equilibrium solutions $r = $ constant. Your sketch will depend on whether c is positive or negative, so you'll need two separate graphs.
(d) Show that $\theta' = -1$, so $\theta(t) = -t + k$.
(e) Make a prediction about what the phase portrait of the nonlinear system looks like if $c > 0$ and if $c < 0$.
(f) What is the stability of $(0,0)$ if $c > 0$? If $c < 0$?

8. Consider the system

$$\frac{dx}{dt} = -y + xr^3 \sin\left(\frac{\pi}{r}\right), \quad \frac{dy}{dt} = x + yr^3 \sin\left(\frac{\pi}{r}\right),$$

where $r = \sqrt{x^2 + y^2}$.
(a) Show that $(0,0)$ is an equilibrium point by showing that $-y + xr^3 \sin(\frac{\pi}{r})$ and $x + yr^3 \sin(\frac{\pi}{3})$ both tend to 0 as $(x, y) \to (0, 0)$.
(b) Show that in polar coordinates the system is

$$\frac{dr}{dt} = r^4 \sin\left(\frac{\pi}{r}\right), \quad \frac{d\theta}{dt} = 1.$$

(c) Using the result of (b), show that each circle $r = \frac{1}{n}$, n a nonzero integer, is an orbit in the phase portrait for this system.
(d) A trajectory which starts between two of the circular orbits $r = \frac{1}{n}$ and $r = \frac{1}{n+1}$ must remain between these circles for all later times. What can it look like? Hint: Notice $\theta(t) = t + c$ and $\frac{dr}{dt}$ is either always positive or always negative in the region between the circles $r = \frac{1}{n}$ and $r = \frac{1}{n+1}$. The origin is called a hybrid center/spiral for the nonlinear system.

9. This problem illustrates the possibly "exotic" nature of the phase portrait when $(0,0)$ is not an isolated equilibrium point of the associated linear system.
(a) Show that the nonlinear system

$$\frac{dx}{dt} = x(x^3 - 2y^3), \quad \frac{dy}{dt} = y(2x^3 - y^3)$$

has an isolated equilibrium point at $(0,0)$.
(b) Find the Jacobian matrix at $(0,0)$ for the system in (a), and show that the linear approximation system has a nonisolated equilibrium point at $(0,0)$.
(c) Show that the equations for the orbits of the system in (a) are given by $x^3 + y^3 = cxy$ for arbitrary constant c. To do this, solve the first-order equation

$$\frac{dy}{dx} = \frac{y(2x^3 - y^3)}{x(x^3 - 2y^3)}$$

for the orbits (substitute $v = y/x$ as in Section 2.7). Some of the orbits are shown in Fig. 10.27.

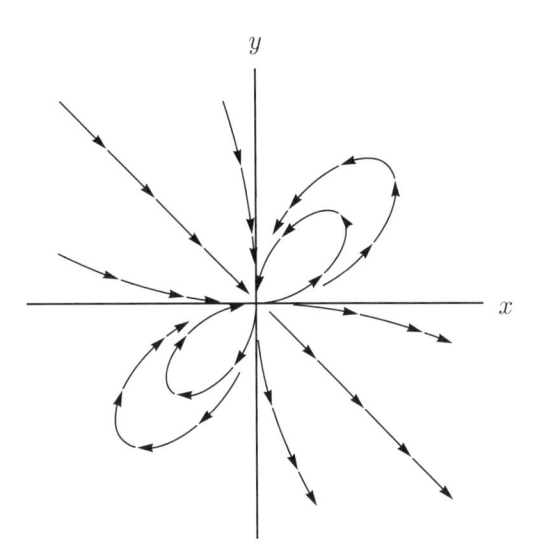

Figure 10.27. Some orbits for Exercise 9.

10.4. More on interacting populations

In Section 10.1 we proposed a model to analyze Darwin's principle of competitive exclusion. In this section we'll expand our study of interacting species, still concentrating on two-species systems.

In the 1920s the mathematician Vito Volterra was asked by his son-in-law, the Italian biologist Umberto D'Ancona, if a mathematical explanation could be found for some puzzling data he had from fish markets in three Italian cities along the Adriatic. D'Ancona had noticed that during World War I, when commercial fishing had greatly decreased in the Adriatic, there was a dramatic increase in the percentage of sharks and other predator fish among the relatively few catches that were being made. He asked Volterra if he could explain mathematically why a decrease in the overall level of fishing would benefit predator fish to a greater extent than it benefited their prey. Volterra relished the opportunity to apply differential equations to this interesting biological problem and initiated work which now appears in essentially every text on theoretical ecology. The differential equation model Volterra proposed, which goes by the name of the Lotka-Volterra equations (due to related work of a physical chemist/statistician named Alfred Lotka), investigates two interacting populations that are in a predator-prey relationship. It gives an explanation of D'Ancona's biological phenomenon, which otherwise seemed inexplicable. This model contains a warning to all gardeners contemplating control of a garden pest by a pesticide which also attacks beneficial insects; see Exercise 3.

Volterra proposed the following system of differential equations for the population $x(t)$ (the prey fish) and $y(t)$ (sharks and other predator fish) in the Adriatic Sea:

$$(10.47) \qquad \frac{dx}{dt} = ax - bxy, \qquad \frac{dy}{dt} = -dy + cxy,$$

where $a, b, c,$ and d are positive constants. In formulating these equations, Volterra made the following assumptions:

(i) Plankton, the food source for the prey fish, is abundant in the Adriatic, so there is negligible competition within the prey population for food. In the absence of predators, the prey population would thus grow, at least in the short term, according to a Malthusian model.

(ii) The predator fish are dependent upon the prey fish for their food supply; in the absence of prey, predator population would decline at an exponential rate.

These two assumptions explain the "ax" and "$-dy$" terms in our equations. The interaction terms "$-bxy$" and "cxy" occur with a positive sign for the predator (which is benefited by the interaction) and a negative sign for the prey. Think of the number of "encounters" between predator and prey to be proportional to the product of the two populations; this explains the form of the interaction terms. Another perspective on how these equations arise is to start with the basic exponential growth and decay equations

$$\frac{dx}{dt} = ax, \quad \frac{dy}{dt} = -dy$$

and replace the constants a and d by linear functions which measure "how good life is". For the predator, a good life means abundant prey, so $-d$ is replaced by $-d + cx$, a linear function which increases as the prey population x increases. For the prey, a good life means a scarcity of predators, so a is replaced by $a - by$, a linear function which decreases as the predator population increases.

We carry out a nullcline-and-arrow analysis of the system (10.47) (in the first quadrant only because of the physical meaning of x and y as populations) and a local analysis near the equilibrium points. The x-nullclines consist of the y-axis ($x = 0$) and the horizontal line $y = a/b$. The y-nullclines are the x-axis ($y = 0$) and the vertical line $x = d/c$. Fig. 10.28 shows these nullclines and the corresponding pairs of direction-indicator arrows in the four basic regions in the first quadrant. You should verify that the various right/left and up/down configurations follow from (10.47).

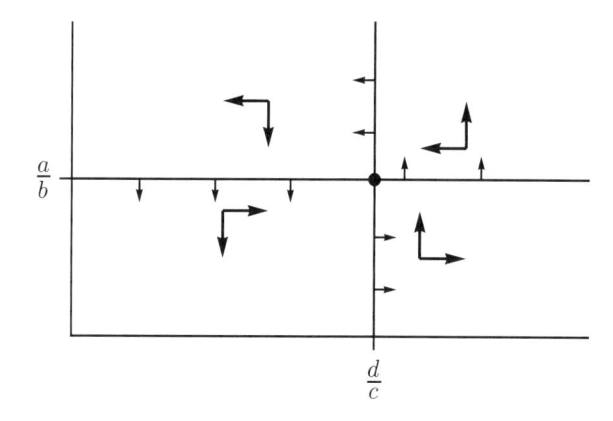

Figure 10.28. Nullcline-and-arrow diagram for (10.47).

We also see that the equilibrium points for the Lokta-Volterra system are $(0,0)$ and $(d/c, a/b)$. The Jacobian matrix for (10.47) is

$$\begin{pmatrix} a - by & -bx \\ cy & -d + cx \end{pmatrix}.$$

Evaluated at the origin, this gives

$$\begin{pmatrix} a & 0 \\ 0 & -d \end{pmatrix},$$

a matrix with eigenvalues $a > 0$ and $-d < 0$. This puts us in a major case and we may apply Theorem 10.2.3 to see that the nonlinear Lotka-Volterra system has a saddle point at the origin. Things don't go as smoothly at the other equilibrium point, $(d/c, a/b)$. Now the Jacobian matrix is

$$\begin{pmatrix} 0 & -bd/c \\ ac/b & 0 \end{pmatrix},$$

10.4. More on interacting populations

which has purely imaginary eigenvalues $\pm i\sqrt{ad}$, and we are in a borderline case. The equilibrium point $(d/c, a/b)$ for (10.47) could be a center or a spiral, both of which are consistent with the nullcline-and-arrow diagram in Fig. 10.28. Can we decide which is correct?

We can find equations for the trajectories in the phase portrait by a familiar technique: From (10.47) we have

$$\frac{dy}{dx} = \frac{dy/dt}{dx/dt} = \frac{y(-d+cx)}{x(a-by)}$$

and we separate variables obtaining

$$\int \frac{a-by}{y}\,dy = \int \frac{-d+cx}{x}\,dx$$

or $a\ln|y| - by = -d\ln|x| + cx + k$. These are equations for the trajectories we want to sketch, but we cannot solve explicitly for y in terms of x and hope to identify the orbits that way. However, we will see below that they are the equations of closed curves encircling the equilibrium point, so that $(d/c, a/b)$ is a center for the system (10.47). A sketch of the phase portrait is shown in Fig. 10.29.

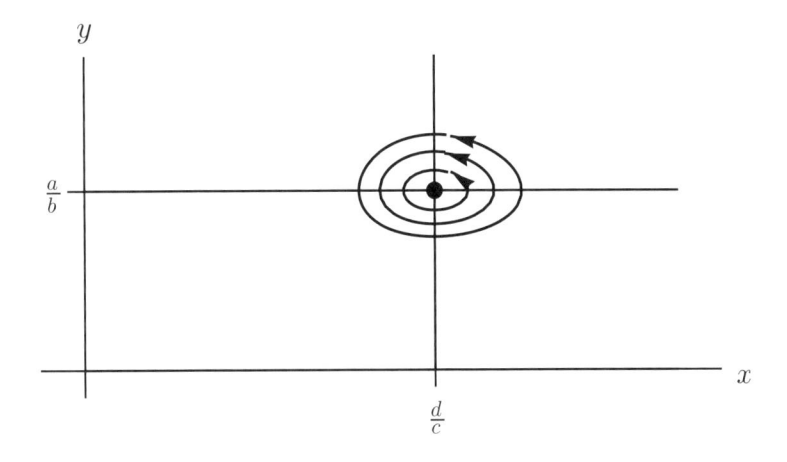

Figure 10.29. Phase portrait for equation (10.47).

Assuming this for the moment, let's return to Volterra's original puzzle: Why does a decrease in fishing benefit the predator population more than the prey population? We need to incorporate fishing into our predator/prey model, and we'll do this in the simplest natural way. Let's assume that when fishing occurs, members of both the predator and prey populations are harvested at a rate proportional to the population sizes. This leads to a modification of equation (10.47):

$$\frac{dx}{dt} = ax - bxy - hx, \qquad \frac{dy}{dt} = -dy + cxy - hy.$$

Here h is a positive constant representing fishing effort—how many are fishing, how much time they spend fishing, equipment used, and so on.[8] The larger h is, the larger the harvests, per unit of time, of prey (hx) and predator (hy). We assume that $h < a$, since otherwise the prey population will simply decrease to 0 because $\frac{dx}{dt}$ is negative. As long as $h < a$ this new system of equations has exactly the same form as the original Lotka-Volterra equations, and the only change has been in the value of the constants:

$$\frac{dx}{dt} = (a-h)x - bxy, \qquad \frac{dy}{dt} = -(d+h)y + cxy.$$

---

[8] Here we are using the same coefficient h in both equations, so that we are harvesting the same proportion of each population. A more general model would allow for different "harvesting coefficients" h_1 and h_2.

All of our previous analysis still applies, and a new phase portrait is sketched in Fig. 10.30. Notice that the equilibrium points are $(0,0)$ and $(\frac{d+h}{c}, \frac{a-h}{b})$.

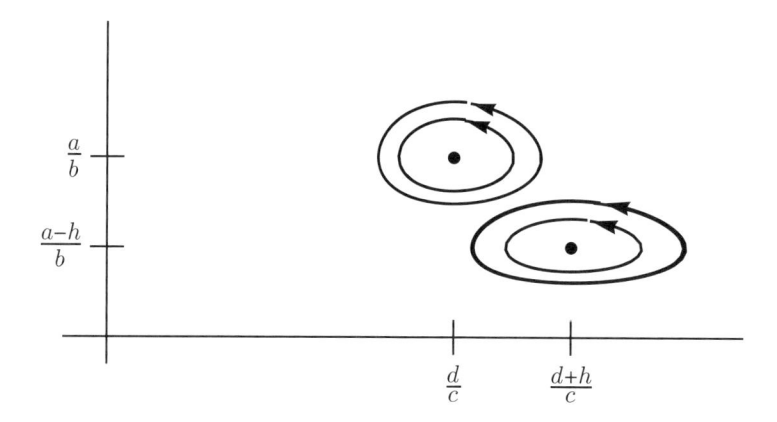

Figure 10.30. The effect of fishing intensity.

The effect of including fishing in the model has been to move the equilibrium point from $(d/c, a/b)$ to $((d+h)/c, (a-h)/b)$. This means that as h decreases, the equilibrium point for the fishing-included model moves leftward and upward. Thus when fishing intensity was lowered during World War I, the orbits circled around a point with a smaller x-coordinate and a larger y-coordinate. It seems plausible to expect that the data on numbers of predator and prey fish caught, which reflects the total populations, is related to these equilibrium values. This is indeed the case. The closed loop orbits in the phase portrait come from *periodic* functions $x(t)$ and $y(t)$. The average value of $x(t)$ over one period is exactly the x-coordinate of the equilibrium point, and the average of $y(t)$ over one period is the y-coordinate of the equilibrium point; see Exercise 2. Thus Volterra's model predicts that when fishing intensity is decreased we will see an increase in the percentage of predator fish, and a decrease in the percentage of prey fish, among the total catch.

We finish our analysis of the Lotka-Volterra equations by discussing why the orbits are closed curves.

Why the orbits in the Lotka-Volterra system must be closed curves. Here we will give an ad hoc argument that will show the curves with equation $a \ln y - by + d \ln x - cx = k$ are either closed curves or a single point for the relevant values of the constant k. (In Section 10.6 we will develop some machinery to analyze this system in a different way.) We begin with some elementary analysis of the two functions

$$f(x) = d \ln x - cx \quad \text{and} \quad g(y) = a \ln y - by.$$

A calculus argument shows that on their domains $x > 0$ and $y > 0$, the graphs of f and g are as depicted below, with f strictly increasing from $-\infty$ to its maximum at $x = \frac{d}{c}$ and then strictly decreasing to $-\infty$ as $x \to \infty$. We'll denote the maximum value of f by M_1. Similarly, g strictly increases from $-\infty$ to its maximum at $y = \frac{a}{b}$ and then strictly decreases to $-\infty$ as $y \to \infty$. Denote the maximum value of g by M_2. Whether the graphs of f and g look like those in Figs. 10.31–10.32 or Figs. 10.33–10.34 depend on the relative sizes of a and b and of c and d. Our analysis will be the same in either case.

10.4. More on interacting populations

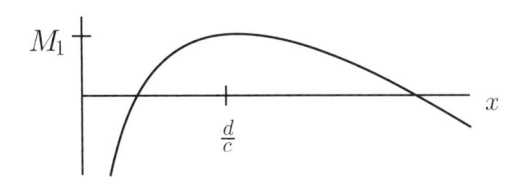

Figure 10.31. $f(x) = d \ln x - cx$.

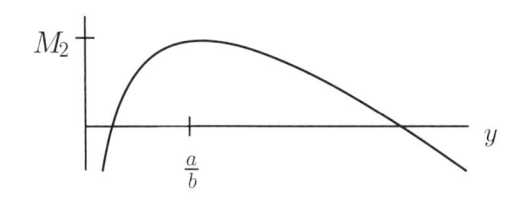

Figure 10.32. $g(y) = a \ln y - by$.

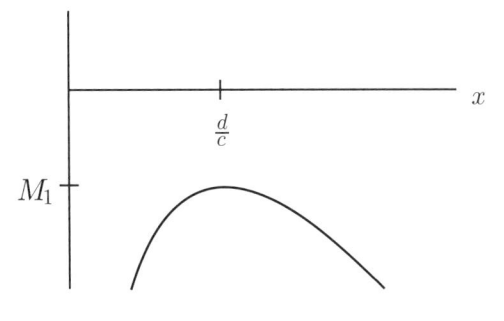

Figure 10.33. $f(x) = d \ln x - cx$.

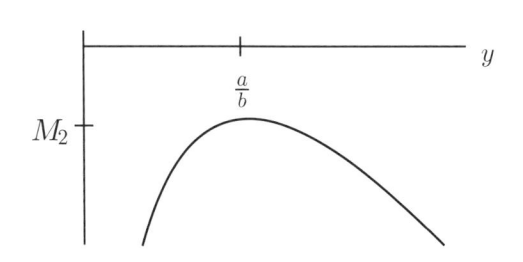

Figure 10.34. $g(y) = a \ln y - by$.

Consider an orbit with equation $f(x) + g(y) = k$. If $k > M_1 + M_2$, then there are no values of x and y for which $f(x) + g(y) = k$. The only solution to $f(x) + g(y) = M_1 + M_2$ is $x = \frac{d}{c}, y = \frac{a}{b}$, so the orbit given by $f(x) + g(y) = M_1 + M_2$ is just the equilibrium point $(\frac{d}{c}, \frac{a}{b})$. Thus we are mainly interested in the solutions to

$$f(x) + g(y) = k \quad \text{where } k < M_1 + M_2.$$

Note that a value of k less than $M_1 + M_2$ can be described as $M_1 + M_2 - \alpha$ for $\alpha > 0$.

Pick a positive constant α and notice that the equation $f(x) = M_1 - \alpha$ has exactly two solutions. Call the smaller solution x_m and the larger solution x_M; we have $x_m < \frac{d}{c} < x_M$. We make three observations (see Fig. 10.35):

(i) If $y = \frac{a}{b}$, so that $g(y) = M_2$, then the equation $f(x) + g(y) = M_1 + M_2 - \alpha$ has exactly two solutions for x, namely x_m and x_M (with, as already noted, $x_m < \frac{d}{c} < x_M$). This says that the orbit $f(x) + g(y) = M_1 + M_2 - \alpha$ intersects the horizontal line $y = \frac{a}{b}$ in exactly two points, $(x_m, \frac{a}{b})$ and $(x_M, \frac{a}{b})$.

(ii) If $x^* < x_m$ or $x^* > x_M$, then $f(x^*) < M_1 - \alpha$ and the equation $f(x^*) + g(y) = M_1 + M_2 - \alpha$ has no solution for y, since $g(y)$ is never greater than M_2. This says that the orbit $f(x) + g(y) = M_1 + M_2 - \alpha$ lies within the vertical strip $x_m \leq x \leq x_M$.

(iii) If $x_m < x^* < x_M$, then $f(x^*) > M_1 - \alpha$, and the equation $f(x^*) + g(y) = M_1 + M_2 - \alpha$ has exactly two solutions for y, one less than $\frac{a}{b}$ and one greater than $\frac{a}{b}$. This says that the vertical line $x = x^*$ intersects the orbit $f(x) + g(y) = M_1 + M_2 - \alpha$ in exactly two points.

Thus in the first quadrant, the orbits of the Lotka-Volterra system (10.47) are either closed curves encircling the equilibrium point $(d/c, a/b)$ or the equilibrium solution $x(t) = d/c, y(t) = a/b$ itself. We conclude that $(d/c, a/b)$ is a center for this nonlinear system.

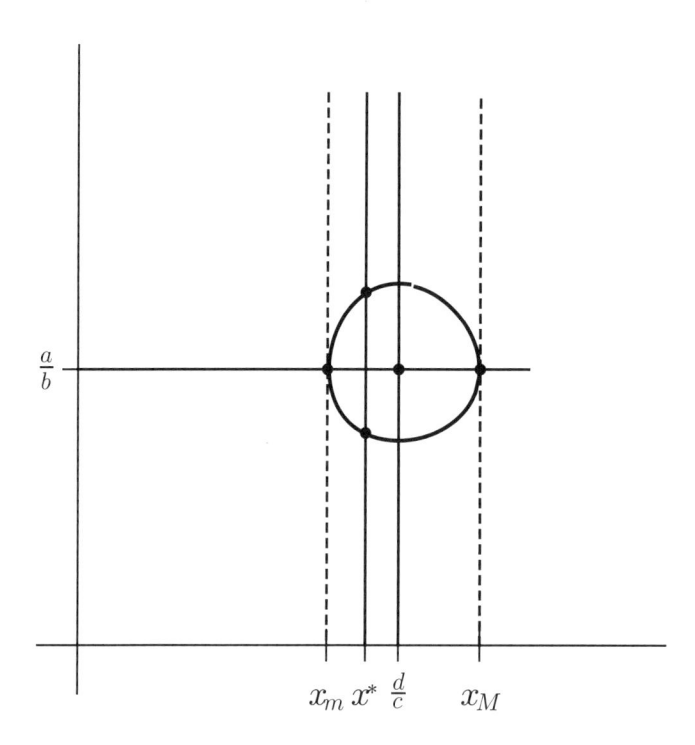

[Figure: phase portrait diagram showing vertical lines at x_m, x^*, $\frac{d}{c}$, x_M, horizontal line at $\frac{a}{b}$, with a closed orbit]

Figure 10.35. Determining an orbit for the Lotka-Volterra system.

10.4.1. Competition for space and habitat destruction. Interacting-species models are sometimes refined by specifying the nature of the competition between the species. We'll finish this section by analyzing a model for species that compete for *space*. This model has particular interest when studying the effects of habitat destruction.

Example 10.4.1. Corals in an ocean reef, barnacles growing in a rocky tide pool, plants in a tropical rain forest, different grass species in a prairie field—these are all examples of organisms that ecologists might view as "competing for space". In this example we model two species in such spatial competition. Our basic assumption is that the first species outcompetes the second, in the sense that it can colonize any space occupied by the second species just as easily as it can colonize empty space. We write $p_1(t)$ for the *proportion* of total available space occupied at time t by the first species, and $p_2(t)$ for the proportion occupied by the second species, so that $0 \leq p_1(t)$, $0 \leq p_2(t)$, and $p_1(t) + p_2(t) \leq 1$ for all times t.

The differential equation for p_1 does not depend at all on p_2, since as far as the first species is concerned, space occupied by the second species is as good as empty space. The first species colonizes at a rate jointly proportional to p_1 and the fraction of space not currently occupied by the first species, namely $1 - p_1$. There is also ordinary mortality, or local extinction, of sites occupied by the first species, which occurs at a rate proportional to p_1. The resulting equation for $\frac{dp_1}{dt}$ is

$$(10.48) \qquad \frac{dp_1}{dt} = a_1 p_1 (1 - p_1) - m p_1.$$

The constant a_1 is called the **colonization-rate constant** for the first species, and m is called the **mortality rate**.

For the second species, the only space available for colonization is the currently empty space, so the portion of space available for its colonization is $1 - p_1 - p_2$. It colonizes at a rate jointly proportional to p_2 and the proportion of available space, $1 - p_1 - p_2$. Some of the space currently occupied by the second species is lost due to colonization by the first species, this occurs at a rate

10.4. More on interacting populations

$a_1 p_1 p_2$. Putting this all together we have

$$\frac{dp_2}{dt} = a_2 p_2 (1 - p_1 - p_2) - a_1 p_1 p_2 - m p_2, \tag{10.49}$$

where the last term represents the ordinary mortality of the second species. Notice we use the same mortality rate m as in the equation for $\frac{dp_1}{dt}$; a more general model would allow for two different mortality constants.

The p_1-nullclines are the pair of lines

$$p_1 = 0 \quad \text{and} \quad p_1 = 1 - \frac{m}{a_1}.$$

We will make the assumption that $m < a_1$, so that the p_1-nullclines are situated as shown in Fig. 10.36. The left and right arrows indicate the regions in the first quadrant in which $\frac{dp_1}{dt}$ is positive or negative.

The p_2-nullclines consist of the line $p_2 = 0$ and the line

$$p_2 = 1 - \frac{m}{a_2} - \left(1 + \frac{a_1}{a_2}\right) p_1.$$

We assume that $m < a_2$. The p_2-nullclines are shown in Fig. 10.37, along with the up and down arrows, indicating the regions in the first quadrant in which $\frac{dp_2}{dt}$ is positive or negative.

Figure 10.36. The p_1-nullcline; $m < a_1$.

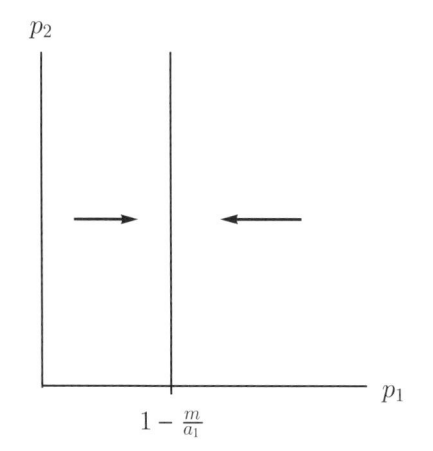

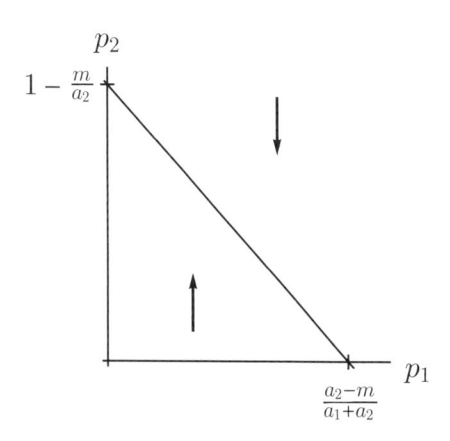

Figure 10.37. The p_2-nullcline; $m < a_2$.

How Fig. 10.36 and Fig. 10.37 fit together depends on the size of the parameters a_1, a_2, and m. The two possibilities are shown in Fig. 10.38 and Fig. 10.39. We can distinguish them by thinking about equilibrium points.

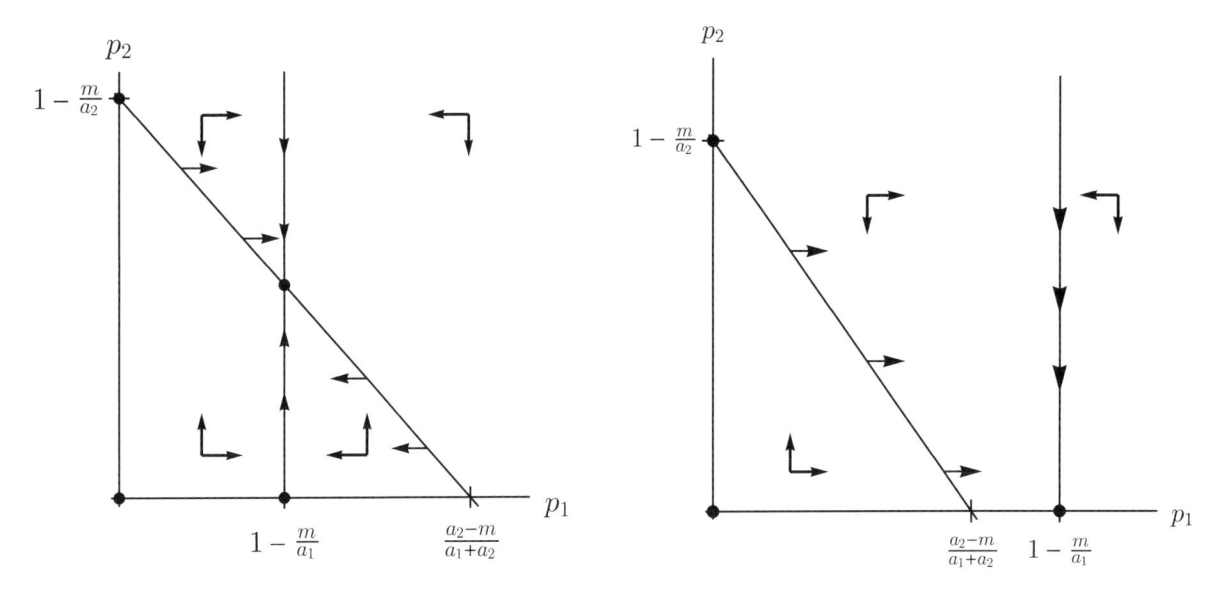

Figure 10.38. Case 1.

Figure 10.39. Case 2.

One equilibrium point is clearly $(0,0)$. There are equilibrium points at

$$\left(0, 1 - \frac{m}{a_2}\right) \quad \text{and} \quad \left(1 - \frac{m}{a_1}, 0\right),$$

and these are meaningful if $m < a_2$ and $m < a_1$. Under what conditions do we have an equilibrium point with positive values for p_1 and p_2? Some algebra shows that the lines

$$p_1 = 1 - \frac{m}{a_1} \quad \text{and} \quad p_2 = 1 - \left(1 + \frac{a_1}{a_2}\right) p_1 - \frac{m}{a_2}$$

intersect in the point

(10.50) $$P = \left(\frac{a_1 - m}{a_1}, \frac{ma_2 - a_1^2}{a_1 a_2}\right).$$

This will be a biologically meaningful equilibrium point if

(10.51) $$a_1 > m \quad \text{and} \quad ma_2 > a_1^2.$$

For these two conditions to hold, we must have $a_2 > a_1$ (see Exercise 15), so that the weaker competitor (species 2) must be able to more effectively colonize empty space. This is sometimes called the "competition-colonization trade off". For the rest of this example we will assume that conditions (10.51) hold. This also means that the nullcline-and-arrow diagram is as in Fig. 10.38.

The Jacobian matrix for equations (10.48)–(10.49) is

$$\begin{pmatrix} a_1 - 2a_1 p_1 - m & 0 \\ -a_2 p_2 - a_1 p_2 & a_2 - a_2 p_1 - 2a_2 p_2 - a_1 p_1 - m \end{pmatrix}.$$

At the equilibrium point $(0,0)$, this is

$$\mathbf{J}_1 = \begin{pmatrix} a_1 - m & 0 \\ 0 & a_2 - m \end{pmatrix}$$

and our assumption on the parameters tells us that the eigenvalues of $\mathbf{J}_1$ are both positive and $(0, 0)$ is an unstable node.

10.4. More on interacting populations

Some computation shows that the Jacobian matrix at the equilibrium point P is

$$\mathbf{J}_2 = \begin{pmatrix} m - a_1 & 0 \\ (-a_1 - a_2)(\frac{m}{a_1} - \frac{a_1}{a_2}) & \frac{a_1^2 - a_2 m}{a_1} \end{pmatrix}.$$

This lower triangular matrix has eigenvalues

$$m - a_1 \quad \text{and} \quad \frac{a_1^2 - a_2 m}{a_1}.$$

Since we are assuming the conditions in equation (10.51), both of these eigenvalues are negative and this equilibrium point is a stable node. In Exercise 15 you are asked to show that the equilibrium points $(0, 1 - \frac{m}{a_2})$ and $(1 - \frac{m}{a_1}, 0)$ are saddle points. Using all of this information and the nullcline-and-arrow analysis above, we sketch the phase portrait in Fig. 10.40.

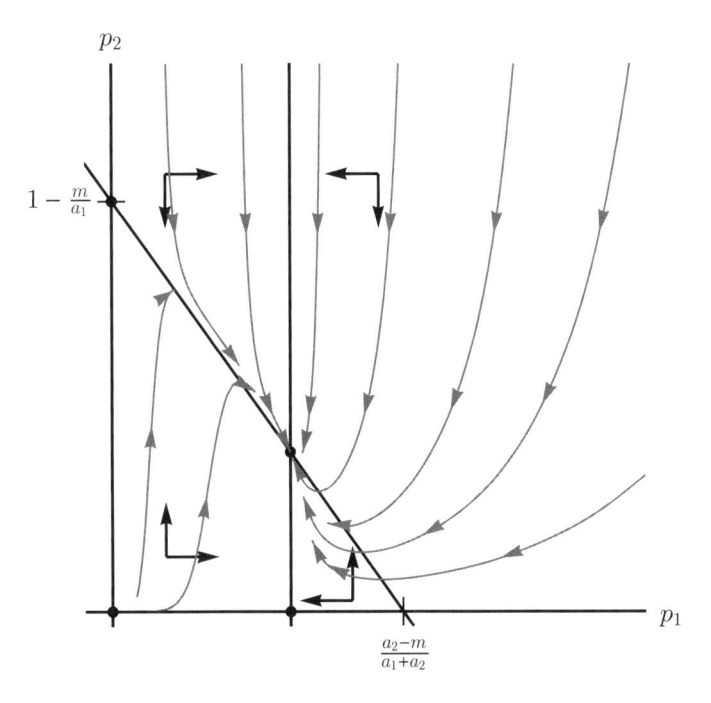

Figure 10.40. Phase portrait when (10.51) holds.

The model predicts coexistence of the two species under our assumptions on the parameters in (10.51).

Habitat destruction. The preceding example gives the simplest model for the utilization of space by species that interact in a hierarchical web (meaning, in general, that species 1 outcompetes species 2, which outcompetes species 3, etc.). There has been recent interest in incorporating a "habitat destruction" feature into such a model to see how loss of habitat affects the various species in the community. The results are unexpected and warn that a surprising "selective extinction" may occur. This is explored in Exercise 19.

10.4.2. Exercises.

1. Consider an orbit in the phase portrait for the predator-prey system in (10.47), as shown below. Suppose the point P corresponds to $t = 0$ and the orbit is traversed counterclockwise as t increases. Which of the following two graphs could show x and y as functions of t for this orbit?

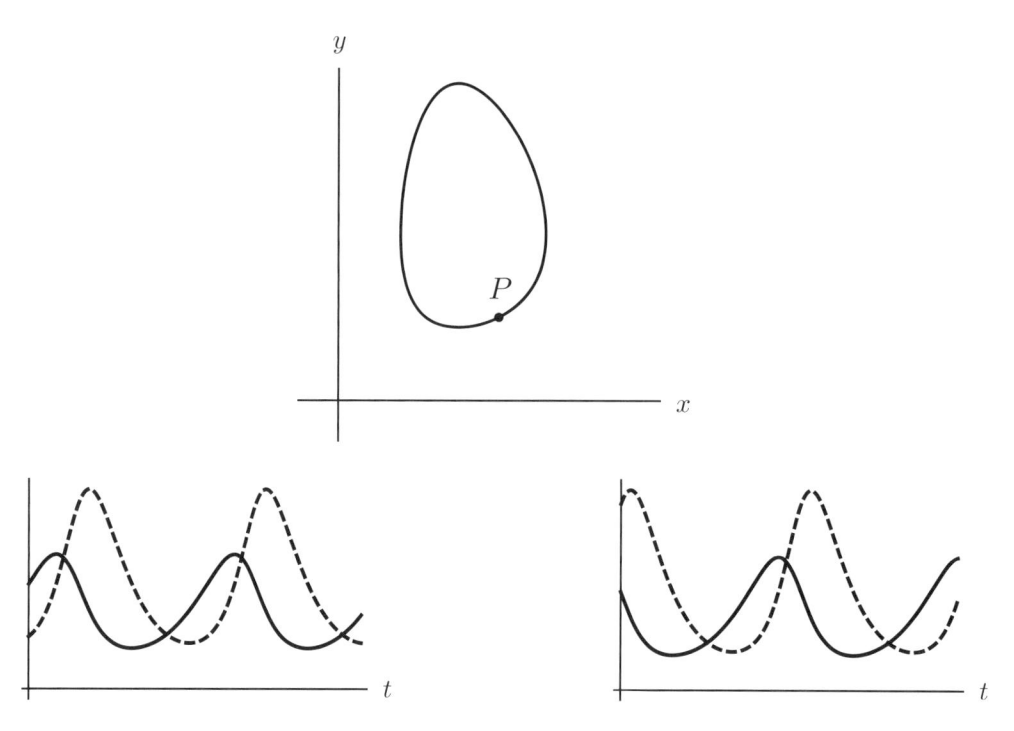

Figure 10.41. Predator population curve shown dashed.

Figure 10.42. Predator population curve shown dashed.

2. Since the first-quadrant orbits in the phase portrait for the Lotka-Volterra equations (10.47) are closed loops, the solution functions $x(t)$ and $y(t)$ are periodic. The purpose of this problem is to show: *The average value of $x(t)$ over one period is d/c, the x-coordinate of the equilibrium point, and the average value of $y(t)$ is a/b, the y-coordinate of the equilibrium point.*

   Suppose T denotes the period of $y(t)$; then the average value of $y(t)$ is defined to be
   $$\frac{1}{T}\int_0^T y(t)\,dt.$$
   Our goal is to show this is a/b.

   (a) By (10.47) we have $x'(t) = x(a - by)$, so that
   $$(10.52) \qquad \int_0^T \frac{x'(t)}{x(t)}\,dt = \int_0^T (a - by)\,dt.$$
   Using the fact that $x(0) = x(T)$, show that the *left-hand* side of equation (10.52) is 0.

   (b) From (a) you now know that
   $$\int_0^T a\,dt = \int_0^T by\,dt.$$
   From this, show
   $$\frac{1}{T}\int_0^T y(t)\,dt = \frac{a}{b}.$$

   (c) With analogous computations show
   $$\frac{1}{T}\int_0^T x(t)\,dt = \frac{d}{c}.$$

3. Aphids are eating the tomato plants in your garden. The aphid population (prey) is kept in check by ladybugs (predator). You decide to use a pesticide to try to reduce the aphid population.

The pesticide also kills ladybugs, at the same rate that it kills the aphids. Modify the Lotka-Volterra predator-prey equations for this scenario, and then comment on the wisdom of your decision.

4. The Lotka-Volterra predator-prey equations are modified by assuming logistic growth for the prey in the absence of the predator:

$$x'(t) = x(a - bx - cy), \quad y'(t) = y(-d + fx).$$

   (a) Which is the prey population, x or y?
   (b) Sketch the phase portrait for the system

$$x'(t) = x(1 - x - y), \quad y'(t) = y(-1 + 2x).$$

   Identify any equilibrium points and give their stability.

5. Here's another modification of the Lotka-Volterra predator-prey equations:

$$x'(t) = ax - by\sqrt{x}, \quad y'(t) = -cy + dy\sqrt{x}.$$

   (a) Which is the prey population, x or y?
   (b) Sketch the phase portrait for the system

$$x'(t) = x - 2y\sqrt{x}, \quad y'(t) = -2y + y\sqrt{x}.$$

   Identify any equilibrium points in the open first quadrant $x > 0$, $y > 0$, and give their stability.

6. Two competing fish species are modeled by the equations

$$x'(t) = x(60 - 2x - y), \quad y'(t) = y(100 - x - 4y).$$

   (a) Show there is an asymptotically stable equilibrium at $(20, 20)$.
   (b) The y-population begins to be harvested with "constant effort"—this means that fishing removes members of the y-population at a rate proportional to the y-population; the proportionality constant is called the effort coefficient. If the effort coefficient is E, what is the new system of equations for the two populations?
   (c) Find the equilibrium points for your answer to (b) if $E = 10$ and if $E = 80$.
   (d) For what values of $E < 100$ are there no equilibrium points in the open first quadrant ($x > 0$ and $y > 0$)?

7. Lionfish, a favorite for salt water aquariums due to its exotic and beautiful appearance, is also a voracious predator. Speculation is that lionfish were accidentally introduced into the Atlantic at Biscayne Bay, Florida, when a small number of fish escaped from an aquarium during Hurricane Andrew in 1992. Since then, lionfish have spread rapidly into the Caribbean and along the Eastern Seaboard, at great detriment to a variety of coral reef fishes, upon which lionfish prey. Losses of nearly 80% of native reef fishes in just 5 weeks have been documented in some locations. Lionfish have been observed eating both large quantities of smaller fish (in one case, of 20 small wrasses in 30 minutes) as well as fish up to 2/3 of their own length. They have few natural predators in the Atlantic and Caribbean.

Figure 10.43. A common variety of lionfish in the Atlantic and Caribbean. Photo by the authors.

A recent article in the Food section of the Washington Post[9] highlighted a campaign to promote lionfish as a tasty human food option, with the hope that by aggressively encouraging commercial fishing of lionfish, they might be removed from these locations in which they are not native. A follow-up article in the Washington Post indicates that this strategy has had considerable success.[10] Indeed, some Whole Foods stores in Florida and elsewhere now sell lionfish for food.

(a) Suppose that a model for lionfish population $y(t)$ in an Atlantic coral reef and their prey $x(t)$ is given by
$$x'(t) = x(r - ax - by), \quad y'(t) = y(s + cx - dy)$$
for positive constants $a, b, c, d, r,$ and s. Do a nullcline-and-arrow diagram in the first quadrant of the phase portrait for this system under the assumption that $s/d > r/b$. Show all equilibrium points.

(b) Evaluate the Jacobian matrix at each equilibrium point from (a), and classify them if possible. You may assume $r \neq s$.

(c) Using your work in (a) and (b), what do you predict for the long-term behavior of the two populations?

(d) Suppose now we selectively fish for just the lionfish, in such a way that they are harvested at a rate proportional to their population. This gives rise to a new system with equations
$$x'(t) = x(r - ax - by), \quad y'(t) = y(s + cx - dy) - fy.$$
Suppose that f is less than s but is large enough that $(s - f)/d < r/b$. What happens to the populations now? To answer this, sketch the phase portrait for the new system.

8. A predator-prey model is proposed with the equations
$$x'(t) = ax\left(1 - \frac{x}{K}\right) - bxy \text{ (prey)}, \quad y'(t) = cy\left(1 - \frac{y}{dx}\right) \text{ (predator)}.$$

These equations are based on the following principles:

The prey population grows logistically, with carrying capacity K, in the absence of predators; encounters with predators are detrimental to the prey. The predator population grows according to a modified logistic equation with *variable* carrying capacity that is proportional to the amount of available prey.

---

[9] Juliet Eilperin, The Washington Post, July 7, 2010.
[10] Ramit Masti, *Can U.S. crush invasive species by eating them?*, The Washington Post, May 26, 2014.

(a) Explain how these lead to the form of the differential equations in the model. If y is the population of whales and x is the population of krill (small shrimp-like creatures), do you expect the constant d to be small or large? Hint: Think of d as a measure of how many whales can be sustained by one krill.

(b) Give a nullcline-and-arrow analysis in the first quadrant for the system

$$x'(t) = x - 2x^2 - xy, \quad y'(t) = y - 3\frac{y^2}{x},$$

which is obtained from particular values of $a, b, c, d,$ and K.

(c) What are the equilibrium points with $x > 0$ and $y > 0$ of the system in (b)? Classify them according to type and stability.

(d) On the basis of your work in (b) and (c), sketch the phase portrait and predict the long-term fate of the two populations modeled by the equations in (b).

9. Rabbits ($x(t)$) and foxes ($y(t)$) are in a predator-prey relationship governed by the Lotka-Volterra equations

$$x'(t) = ax - bxy, \quad y'(t) = -dy + cxy.$$

Assume that a superpredator (for example, human hunters) with a fixed population size begins to prey on both rabbits and foxes, so that the new differential equation system is

$$x'(t) = ax - bxy - h_1 x,$$
$$y'(t) = -dy + cxy - h_2 y,$$

where h_1 and h_2 are positive constants, possibly different.

(a) If $a - h_1 > 0$, show that the presence of the superpredator benefits the rabbits.

(b) What happens to the rabbit and fox populations if $a - h_1 < 0$?

10. *Reptilian dominance in Australia and New Guinea.*[11] On the continent of Australia, neighboring New Guinea, and adjacent islands there are large reptiles, such as monitor lizards and the Komodo dragon, that are predators, but few large mammalian predators. This is true not just in present times, but going back millions of years. The only carnivore mammals weighing more than 10 pounds that can be found today in Australia are the Tasmanian devil (a marsupial hyena) and the spotted-tailed quoll, which resembles a weasel. Now extinct, there was previously also a dog-like marsupial and a marsupial lion. By contrast, in the United States even today one can find dozens of kinds of mammalian carnivores and more many that existed previously but are now extinct.

What can account for this difference? The land in Australia is, by comparison with the Americas, Europe, and Asia, relatively infertile, as measured by the quality of the soil, the potential for prolonged droughts, and other climatic factors. This impedes the growth of high quality vegetation which in turn means that fewer large herbivores will be available as prey for carnivores. Carnivorous reptiles, being cold-blooded, generally require less food and energy than their warm-blooded counterparts. Thus the low productivity of the land may favor the development of meat-eating reptiles over meat-eating mammals. In this problem, we explore this theory.

Begin with the modified Lotka-Volterra model

$$x'(t) = x(r - ax - by), \quad y'(t) = -sy + cxy$$

with positive constants $a, b, c, r,$ and s for interacting predator (y) and prey (x) populations in Australia.

(a) Show that if $r/a < s/c$, there is no equilibrium point (x_e, y_e) with x_e and y_e both positive.

---

[11] The ideas in this problem are adapted from Clifford Taubes, *Modeling Differential Equations in Biology*, 2nd ed., Cambridge University Press, New York, 2008, Chapter 12.

(b) Show $(r/a, 0)$ is an equilibrium point and that under the same condition $r/a < s/c$, this equilibrium point is stable.

(c) Show by contrast that if $r/a > s/c$, then there is an equilibrium point (x_e, y_e) with both x_e and y_e positive, and moreover, this equilibrium point is stable.

(d) Assume that the constants r, a which appear in the differential equation for the prey do not change when the specific type of predator changes. If there is indeed a greater chance for stable, positive equilibrium values of both prey and predator with a reptilian predator than with a mammalian predator, would you expect the ratio s/c to be larger or smaller for a reptilian predator, as compared with a mammalian one? Do you expect that a species with lower food energy needs would have a larger or smaller value of c?

11. Sperm whales in the Southern Ocean eat squid, and the squid in turn eat krill. The following equations are proposed to model the three populations:

$$x'(t) = r_1 x(1 - b_1 x - cz), \quad y'(t) = r_2 y(1 - b_2 y/z), \quad z'(t) = r_3 z(1 - b_3 z/x - dy),$$

where all constants are positive. These equations reflect the assumption that the "carrying capacity" of the two predator populations (sperm whale and squid) is proportional to the population of their respective prey. Which variables go with which species?

12. Start with the predator-prey equations

$$x'(t) = x(-d - kx + my), \quad y'(t) = y(a - bx - cy),$$

where the parameters a, b, c, d, k, and m are positive. Suppose a pesticide is going to be used to try to control the prey population. Assume the pesticide targets both predator and prey in an equal fashion, so that a term hx is subtracted from the first equation and a term hy from the second.

(a) Which is the predator population, x or y?

(b) Is there any condition (on the various coefficients in the differential equations) under which the prey will be driven to extinction?

(c) Show that if

$$h > \frac{ma - dc}{m + c},$$

then the predator will be driven to extinction.

13. (CAS) A small island supports a population of rabbits and foxes. The rabbit and fox populations, in thousands, are given at time t, in years, by $x(t)$ and $y(t)$, respectively. Suppose that these populations are modeled by the predator-prey system

$$x'(t) = \frac{1}{4}x - xy, \quad y'(t) = -3y + xy$$

and that $x(0) = 2.4$ and $y(0) = 0.2$.

(a) Solve this initial value problem numerically, and use your solution to find the approximate maximum and minimum values of the rabbit and fox populations. Also, approximate the period of this predator-prey system—the number of years that pass before the rabbit and fox populations first return to the levels $x = 2.4$ and $y = 0.2$.

(b) Plot your solution of (a) over one period.

(c) Find an implicit solution of the initial value problem

$$\frac{dy}{dx} = \frac{-3y + xy}{\frac{1}{4}x - xy}, \quad y(2.4) = 0.2.$$

Then plot your solution and compare it to your plot for (b).

10.4. More on interacting populations

14. (CAS)
    (a) Construct a concavity chart, over the square $-4 < x < 4, -4 < y < 4$ for the Lokta-Volterra system
    $$\frac{dx}{dt} = x - xy, \quad \frac{dy}{dt} = -2y + xy.$$
    Hint: Differentiation with respect to x of $\frac{dy}{dx} = \frac{-2y+xy}{x-xy}$ gives
    $$\frac{d^2y}{dx^2} = -\frac{y(6 - 4x + x^2 - 4y + 2y^2)}{x^2(-1+y)^3}.$$
    Complete the square on the x and on the y terms to analyze $\frac{d^2y}{dx^2}$.
    (b) Produce a phase portrait over the same square. Include in your portrait any curves on which $\frac{d^2y}{dx^2}$ is 0 or undefined.

15. (a) Show why the conditions in equation (10.51) of Example 10.4.1 are exactly what is needed for there to be an equilibrium point with positive coordinates for the differential equation system of Example 10.4.1, and verify that this equilibrium point lies in the biologically meaningful triangle $p_1 \geq 0$, $p_2 \geq 0$, $p_1 + p_2 \leq 1$.
    (b) Explain why $a_2 > a_1$ is a consequence of the conditions in equation (10.51).
    (c) At these equilibrium values of p_1 and p_2, is there any empty (uncolonized) space in the environment?
    (d) Still assuming the conditions of equation (10.51), compute the Jacobian matrix at the equilibrium points
    $$\left(0, 1 - \frac{m}{a_2}\right) \quad \text{and} \quad \left(1 - \frac{m}{a_1}, 0\right),$$
    and show they are saddle points.

16. (CAS) Using the values $a_1 = 1$, $a_2 = 2$, and $m = 0.25$ in equations (10.48)–(10.49), verify that Fig. 10.39 applies and find and classify the equilibrium points having $p_1 \geq 0$ and $p_2 \geq 0$. Then use a computer algebra system to produce a phase portrait.

17. Suppose in the model of Example 10.4.1 we have $a_1 = 0.2/\text{year}$, $a_2 = 0.8/\text{year}$, and $m = 0.1/\text{year}$. If both species start at some positive value, what do you expect to happen to p_1 and p_2 over a long period of time? Is there any empty space?

18. (CAS) Consider the initial value problem consisting of the model of Example 10.4.1 with constants a_1, a_2, and m as in the preceding exercise and the initial condition $p_1(0) = 0.8$, $p_2(0) = 0.05$.
    (a) Using a computer algebra system, solve this initial value problem numerically and plot the solutions for $p_1(t)$ and $p_2(t)$ over the time interval $[0, 100]$ (time measured in years).
    (b) Using your numerical solution, estimate the values of p_1 and p_2 at $t = 5, 10, 30, 50,$ and 100 years.

19. This problem continues the ideas of Example 10.4.1, which modeled the competition for space by species that interact in a hierarchical web (in a hierarchical web, species 1 outcompetes species 2, which outcompetes species 3, etc.). In this problem we incorporate a "habitat destruction" feature into the model. We assume that some proportion D of the sites available for colonization is randomly and permanently destroyed. Modifying equations (10.48)–(10.49) to reflect this, the new system is
    $$\frac{dp_1}{dt} = a_1 p_1 (1 - p_1 - D) - m p_1 = p_1(a_1 - a_1 p_1 - a_1 D - m),$$
    $$\frac{dp_2}{dt} = a_2 p_2(1 - p_1 - p_2 - D) - a_1 p_1 p_2 - m p_2 = p_2(a_2 - a_2 p_1 - a_2 p_2 - a_2 D - a_1 p_1 - m).$$

We will see that an unanticipated consequence of habitat destruction is the different ways in which the dominant competitor (species 1) and the weaker competitor (species 2) are affected by the loss of habitat.

(a) Find the equilibrium point P_D obtained by solving the pair of equations

$$a_1 - a_1 p_1 - a_1 D - m = 0 \quad \text{and} \quad a_2 - a_2 p_1 - a_2 p_2 - a_2 D - a_1 p_1 - m = 0.$$

(When $D = 0$ your answer should agree with equation (10.50).)

(b) We know from Example 10.4.1 that when $D = 0$, under certain conditions on the parameters a_1, a_2, and m, this equilibrium point will lie in the open first quadrant $p_1 > 0$, $p_2 > 0$. Assuming these conditions hold, how large can D be so that the equilibrium point P_D from (a) has both coordinates positive? You may find it helpful to write the coordinates of P_D in the form

$$\text{first coordinate} = \frac{a_1 - m}{a_1} + \text{an expression in terms of } D$$

and

$$\text{second coordinate} = \frac{ma_2 - a_1^2}{a_1 a_2} + \text{an expression in terms of } D$$

so that the relationship between the coordinates of P_D and the equilibrium point in equation (10.50) (corresponding to $D = 0$) is clear.

(c) True or false: The dominant competitor is predicted to become extinct at a lower level of habitat destruction than the weaker competitor.

(d) Does the phrase "the enemy of my enemy is my friend" have any relevance to this model?

10.5. Modeling the spread of disease

Communicable diseases have had profound impacts on the course of history. The Antonine plague (probably smallpox or measles or both) in AD 165–180 contributed to the demise of the Roman Empire. Bubonic plague caused the death of up to one third of the population of Europe in the fourteenth century. It reappeared again in London in 1665, killing one fifth of the population there. It forced the closure of Cambridge University for a while, where coincidently Issac Newton was a student. Returning home, Newton had a spectacularly prolific period of scientific work during the university's closure. The 1918–1919 pandemic influenza affected one third of the world's population and killed more (up to 50 million) than died in World War I (16 million); healthy young adults were particularly hard hit. HIV, SARS, swine or bird flu, the possibility of weaponized anthrax, or a terrorist release of the smallpox virus are just a few of the current concerns of public health officials.

The use of mathematical models to analyze the spread of diseases goes back centuries. In 1760, the mathematician Daniel Bernoulli used a mathematical argument to show that cowpox vaccination against smallpox would significantly increase the average life span—his model showed that if deaths from smallpox could be eliminated, the average life span would increase by about 3 years (from the then-current value of 26.5 years). His work (see Exercise 4) represents the first important use of a mathematical model to address a practical vaccination proposal. In the early twentieth century, differential equation compartment models for disease propagation started to appear and win acceptance. As with other types of mathematical models, there is a trade-off between simple models which incorporate only a few broad details and more refined models whose solutions may be more difficult to obtain and to analyze. In this section we will look at several compartment models and see how even simple models give interesting predictions which can be tested against observed phenomena.

10.5. Modeling the spread of disease

Example 10.5.1. Diseases with permanent immunity.

We'll model the spread of a disease—like measles—which confers immunity. This means, once infected and recovered, a person can never get the disease again. We will also modify this basic model, called an SIR model, to include the possibility of vaccination against the disease, and look at the idea of "herd immunity".

Setting up the SIR model. Imagine our population divided into three nonoverlapping compartments: the susceptibles, with population $S(t)$ at time t; the infectives, with population $I(t)$; and the recovered, with population $R(t)$, consisting of those people who have had the disease and recovered (and are forevermore immune from it). There is also a fourth compartment, consisting of people who have died (either from the disease or from other causes). The diagram below shows how people move from one compartment to another. Notice the model allows for births, and all newborns begin life in the susceptible compartment. We will set up our model so that total size of the population is a fixed value K. This means we assume that deaths balance out births. Here are the assumptions we make to describe the movement between the compartments:

- Infectives recover from the disease at a rate proportional to the number of infectives.
- Susceptibles become infective at a rate jointly proportional to the number of infectives and the number of susceptibles.
- The birth rate is constant and proportional to the total size K of the population.
- The total death rate is equal to the birth rate, and the death rate from each compartment is proportional to the size of the compartment, with the same proportionality constant μ for each of the three compartments, S, I, and R.

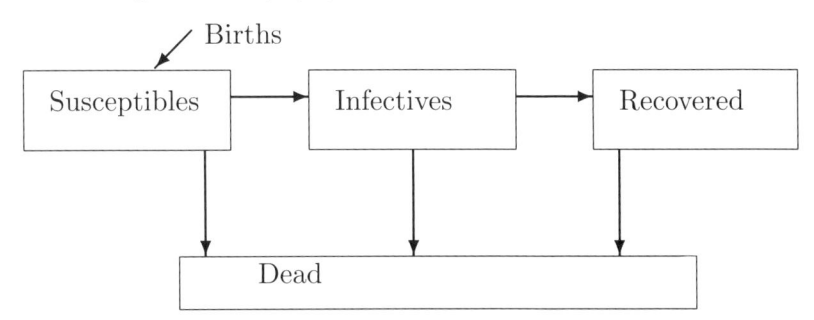

[Diagram: Births → Susceptibles → Infectives → Recovered, with arrows from each to Dead]

These assumptions give us the following equations:

$$(10.53) \qquad \frac{dS}{dt} = \mu K - \beta SI - \mu S, \quad \frac{dI}{dt} = \beta SI - \mu I - \gamma I, \quad \frac{dR}{dt} = \gamma I - \mu R,$$

where $\mu, \beta,$ and γ are positive constants. Check that $\frac{dS}{dt} + \frac{dI}{dt} + \frac{dR}{dt} = 0$, reflecting the assumption that $S + I + R = K$ for all times. Since the equations for $\frac{dS}{dt}$ and $\frac{dI}{dt}$ involve only S and I, we can focus on those so as to have a planar system to analyze.

Nullcline-and-arrow diagrams. From the equation for $\frac{dI}{dt}$ in (10.53) we see that the I-nullclines are the lines with equations $I = 0$ and $S = (\gamma + \mu)/\beta$, shown in Fig. 10.44, along with the relevant up/down arrows. Using the equation for $\frac{dS}{dt}$ in (10.53) we see that the S-nullcline has equation

$$I = \frac{\mu K}{\beta} \frac{1}{S} - \frac{\mu}{\beta},$$

which is shown in Fig. 10.45 along with the relevant right/left arrows. How Figs. 10.45 and 10.44 fit together depend on whether $K \leq (\gamma + \mu)/\beta$ or $K > (\gamma + \mu)/\beta$.

Figure 10.44. The I-nullclines $I = 0$ and $S = \frac{\gamma+\mu}{\beta}$.

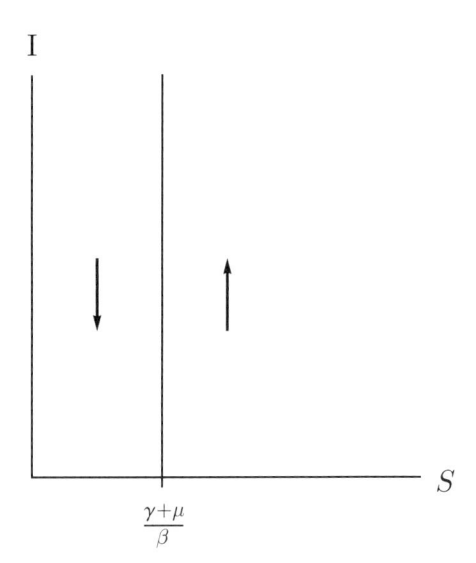

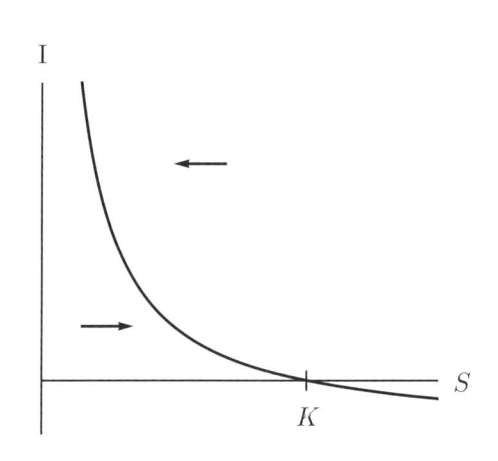

Figure 10.45. The S-nullcline $I = \frac{\mu K}{\beta}\frac{1}{S} - \frac{\mu}{\beta}$.

For the rest of this example, we make the following assumption:

(10.54) $$K > \frac{\gamma+\mu}{\beta}.$$

With this assumption, the S- and I-nullclines intersect in the first quadrant as shown in Fig. 10.46.

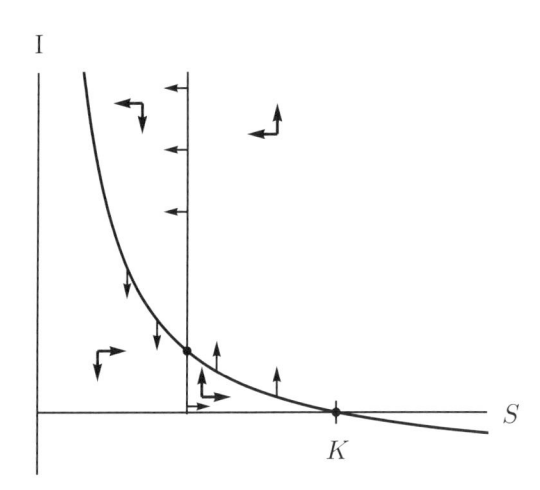

Figure 10.46. Nullcline-and-arrow diagram for first two equations in (10.53) if $\beta K > \gamma + \mu$.

Equilibrium points. Still assuming $\beta K > \gamma + \mu$, we see that there are two biologically relevant equilibrium points, obtained as the intersection of the S-nullcline with an I-nullcline. One is the "disease-free" equilibrium point $(K, 0)$. This should make perfect sense: If there are no infectives and $S = K$, then S should remain K, the total population size, for all times. The second equilibrium point is

(10.55) $$\left(\frac{\gamma+\mu}{\beta}, \frac{\mu}{\beta(\gamma+\mu)}(\beta K - \gamma - \mu)\right).$$

10.5. Modeling the spread of disease

To analyze these equilibrium points we compute the Jacobian matrix for the first two equations in (10.53), obtaining

$$\mathbf{J} = \begin{pmatrix} -\beta I - \mu & -\beta S \\ \beta I & \beta S - \gamma - \mu \end{pmatrix}.$$

At the equilibrium point $(K, 0)$ this is

$$\begin{pmatrix} -\mu & -\beta K \\ 0 & \beta K - \gamma - \mu \end{pmatrix}.$$

This matrix has eigenvalues $-\mu$ and $\beta K - \gamma - \mu$. Under assumption equation (10.54), one eigenvalue is positive and one is negative. By Theorem 10.2.3, we have an (unstable) saddle point at $(K, 0)$.

Computing the Jacobian at the second equilibrium point in equation (10.55) looks tedious, but we can extract what we need to know by observing that the 2-2 entry of the Jacobian at this point is 0: $\beta S - \gamma - \mu = 0$ when $S = (\gamma + \mu)/\beta$. From this it is immediate that the trace of the Jacobian is negative and the determinant, which is $\beta^2 SI$, is positive. This tells us that the second equilibrium point is stable (either a node or spiral).

Because we know that the equilibrium point (10.55) is stable, we can predict that the disease will become *endemic* in the population, with $I(t)$ tending to a constant, positive value as $t \to \infty$. An endemic disease is one that is consistently present in the population. Exercise 1 asks you to predict the long-term behavior if (10.54) does not hold.

The condition of (10.54) can be written as $\beta K/(\gamma + \mu) > 1$, and the qualitative long-term behavior depends on whether the quantity

$$(10.56) \qquad R_0 = \frac{\beta K}{\gamma + \mu}$$

is greater than or less than 1. This is an example of a **threshold** quantity; R_0 is sometimes called the **basic reproduction number**.

The effect of vaccination. How can vaccination change the course of the disease? Assume that a fixed fraction ρ of all newborns is vaccinated against the disease, so they are born into the recovered compartment. Since the birth rate is μ, this adds a "rate in" term of $\rho\mu K$ to the equation for $\frac{dR}{dt}$ and modifies the "rate in" term for $\frac{dS}{dt}$ to be $(1-\rho)\mu K$. Our equations are now

$$(10.57) \qquad \frac{dS}{dt} = (1-\rho)\mu K - \beta SI - \mu S, \quad \frac{dI}{dt} = \beta SI - \mu I - \gamma I, \quad \frac{dR}{dt} = \gamma I - \mu R + \rho\mu K.$$

Again we can focus on just the first two equations in (10.57). Our previous analysis still applies, except where we had "K" before, we now have $(1-\rho)K$. What determines whether we have one equilibrium point or two (in the first quadrant $I \geq 0, S \geq 0$) is whether $\beta(1-\rho)K > \gamma + \mu$ or not; this is (10.54) with K replaced by $(1-\rho)K$.

What fraction of newborns needs to be vaccinated so that instead of having a stable endemic equilibrium value, we have only a (stable) disease-free equilibrium? By our analysis and Exercise 1 we know this will happen if $\beta(1-\rho)K \leq \gamma + \mu$, or equivalently, if the new basic reproduction number

$$\widetilde{R_0} = \frac{\beta(1-\rho)K}{\gamma + \mu}$$

is less than 1. (The case of equality is ignored, as exact equality would not be expected to occur in practice.) Some algebra shows $\widetilde{R_0} < 1$ precisely if

$$(10.58) \qquad \rho > 1 - \frac{\gamma + \mu}{\beta K}, \quad \text{or equivalently, if } \rho > 1 - \frac{1}{R_0}$$

where R_0 is the original basic reproduction number in (10.56). For different diseases, the value of R_0 can be estimated from epidemiological data. For example, measles has a high value, estimated at about 12–18 in urban areas. By equation (10.58) this translates into a need of vaccinating about 91.6–94.4 percent of the newborn population to ensure that measles cannot become endemic. Exercise 2 asks you to verify this and to similarly determine what fraction of newborns needs to be vaccinated to control several other diseases with different basic reproduction numbers. If enough of the population has been vaccinated to prevent the disease from becoming endemic, the population is said to have "herd immunity".

Example 10.5.2. Modeling the spread of a sexually transmitted disease. We start with a fixed population of N at-risk (sexually active) people and assume that men can only be infected by women and women can only be infected by men. Let M be the total number of men, and W the total number of women, so that $M + W = N$; M, W, and N are constants. As is typical of many bacterial STDs, we assume that an infection does not confer immunity, so after a person is infected and recovers, he or she is immediately susceptible to reinfection. At any point in time there will be $x(t)$ infected men and $M - x(t)$ susceptible men, and $y(t)$ infected women and $W - y(t)$ susceptible women. Our model will give differential equations for $\frac{dx}{dt}$ and $\frac{dy}{dt}$. To derive the form of these equations we make several basic assumptions:

- Men are infected at a rate jointly proportional to the number of susceptible men and the number of infected women.
- Women are infected at a rate jointly proportional to the number of susceptible women and the number of infected men.
- Infected men recover at a rate proportional to the number $x(t)$ of infected men.
- Infected women recover at a rate proportional to the number $y(t)$ of infected women.

These assumptions lead to the equations

$$(10.59) \qquad \frac{dx}{dt} = a_1(M-x)y - b_1 x, \qquad \frac{dy}{dt} = a_2(W-y)x - b_2 y$$

for some positive constants a_1, a_2, b_1, and b_2. You should be able to identify which terms in these equations correspond to which "in" and "out" arrows in the schematic diagram below.

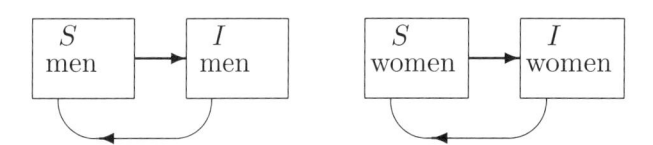

For a number of important STDs, like gonorrhea and chlamydia, women typically show no or few symptoms and thus do not seek (antibiotic) treatment. This fact is reflected in the value of b_2 being relatively small. If men do typically show symptoms and seek treatment (for example, this is the case with gonorrhea), then we would expect b_1 to be significantly larger than b_2. The reciprocals $1/b_1$ and $1/b_2$ can be interpreted as the average time a man or woman remains infective.

Equilibrium points. The system (10.59) has an equilibrium point at $(0, 0)$. There is a possibility of a second biologically meaningful equilibrium point. Solving the equations

$$a_1(M-x)y - b_1 x = 0 \quad \text{and} \quad a_2(W-y)x - b_2 y = 0,$$

10.5. Modeling the spread of disease

we get a solution point

(10.60) $$\left(\frac{MWa_1a_2 - b_1b_2}{b_1a_2 + Wa_1a_2}, \frac{MWa_1a_2 - b_1b_2}{b_2a_1 + Ma_1a_2}\right).$$

This will be of interest if it lies in the first quadrant of the xy-plane; this happens if

(10.61) $$MWa_1a_2 - b_1b_2 > 0.$$

Before going further, let's rewrite the coordinates of the point in (10.60) by dividing each numerator and denominator by a_1a_2 to obtain

$$\frac{MWa_1a_2 - b_1b_2}{b_1a_2 + Wa_1a_2} = \frac{MW - \rho_1\rho_2}{W + \rho_1}, \quad \frac{MWa_1a_2 - b_1b_2}{b_2a_1 + Ma_1a_2} = \frac{MW - \rho_1\rho_2}{M + \rho_2}$$

where $\rho_1 = b_1/a_1$ and $\rho_2 = b_2/a_2$. The equilibrium point

$$\left(\frac{MW - \rho_1\rho_2}{W + \rho_1}, \frac{MW - \rho_1\rho_2}{M + \rho_2}\right)$$

lies in the first quadrant if

(10.62) $$MW - \rho_1\rho_2 > 0.$$

From this point on we assume that (10.62), or equivalently (10.61), holds. The Jacobian matrix for the linear approximation of (10.59) near $(0,0)$ is

$$\mathbf{J}_1 = \mathbf{J}(0,0) = \begin{pmatrix} -b_1 & Ma_1 \\ Wa_2 & -b_2 \end{pmatrix}.$$

Under our assumption (10.61), this is a matrix with negative determinant, so according to Section 8.9 and Theorem 10.2.3, $(0,0)$ is a saddle point for the system (10.59).

The Jacobian matrix at the second equilibrium point is

$$\mathbf{J}_2 = \begin{pmatrix} -a_1y_0 - b_1 & a_1M - a_1x_0 \\ a_2W - a_2y_0 & -a_2x_0 - b_2 \end{pmatrix},$$

where $x_0 = (MW - \rho_1\rho_2)/(W + \rho_1)$ and $y_0 = (MW - \rho_1\rho_2)(M + \rho_2)$ are the coordinates of the nonzero equilibrium point. This is a little unpleasant to deal with, but we can make two observations with minimal calculation:

- The trace of $\mathbf{J}_2$ is negative, since both entries on the main diagonal are negative.
- The determinant of $\mathbf{J}_2$ is positive. To see this, compute

$$a_1M - a_1x_0 = a_1\left(M - \frac{MW - \rho_1\rho_2}{W + \rho_1}\right) = a_1\rho_1\frac{M + \rho_2}{W + \rho_1}$$

and

$$a_2W - a_2y_0 = a_2\left(W - \frac{MW - \rho_1\rho_2}{M + \rho_2}\right) = a_2\rho_2\frac{W + \rho_1}{M + \rho_2}.$$

Using the definitions of ρ_1 and ρ_2 we see that the determinant of $\mathbf{J}_2$ is

$$(a_1y_0 + b_1)(a_2x_0 + b_2) - a_1a_2\rho_1\rho_2 = a_1a_2y_0x_0 + b_1a_2x_0 + b_2a_1y_0$$

which is clearly positive.

A negative trace and positive determinant tell us that the equilibrium point is stable, either a node or a spiral. We'll return to this shortly.

Nullcline-and-arrow diagrams. What are the nullclines for our system? We are only interested in sketching the phase portrait in the rectangle $0 \leq x \leq M, 0 \leq y \leq W$, since this is the only portion of the xy-plane that is biologically relevant. The restrictions $x \geq 0, y \geq 0$ are clear, and the restrictions $x \leq M, y \leq W$ follow since the number of infected men (women) cannot exceed the total number of men (women) in the population. The x-nullcline is the curve with equation

$$y = \frac{b_1 x}{a_1(M-x)};\tag{10.63}$$

let's call the right-hand side of (10.63) $F(x)$. To sketch the graph, it's helpful to use a little calculus. From (10.63) we compute

$$F'(x) = \frac{Ma_1 b_1}{[a_1(M-x)]^2} \quad \text{and} \quad F''(x) = \frac{2Mb_1}{a_1(M-x)^3}.$$

Since F' is always positive, the graph of F is increasing. We've observed that $x \leq M$, so the graph of F is concave up for $0 < x < M$. Finally, $F(0) = 0$ and $\lim_{x \to M^-} F(x) = \infty$. The x-nullcline is sketched in Fig. 10.47, where we have also shown the correct right/left arrows on either side of this nullcline. We've only shown the portion of the nullcline in the biologically relevant rectangle.

A similar analysis is possible for the y-nullcline, which has equation

$$y = \frac{Wa_2 x}{b_2 + a_2 x}.$$

The graph of this is an increasing, concave down curve, passing through $(0,0)$ and which has limit W as $x \to \infty$. You are asked to verify these facts in Exercise 5. Fig. 10.48 shows the y-nullcline and the corresponding up/down arrows in the biologically relevant rectangle.

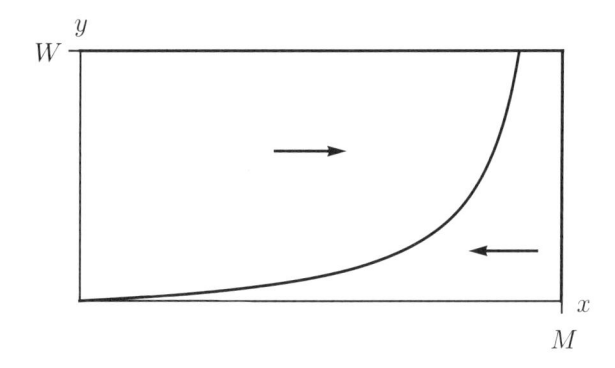

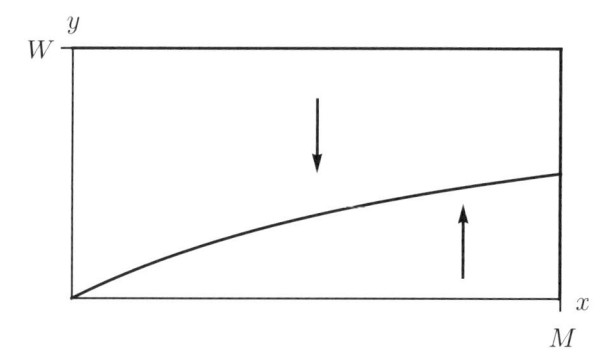

Figure 10.47. The x-nullcline.

Figure 10.48. The y-nullcline.

We want to put Figs. 10.47 and 10.48 together. There are two possibilities here: Either the x- and y-nullclines intersect in two points, $(0,0)$ and a second equilibrium point, or they only intersect in $(0,0)$. From our earlier discussion of the equilibrium points, we know the first case occurs exactly when $MWa_1 a_2 - b_1 b_2 > 0$. (Exercise 5(c) gives another perspective on joining Fig. 10.47 and Fig. 10.48.) We focus on this case, leaving the case $MWa_1 a_2 - b_1 b_2 \leq 0$ for Exercise 5.

The nullcline-and-arrow picture looks like Fig. 10.49. Remembering that $(0,0)$ is a saddle point, we sketch some trajectories. We hadn't classified the second equilibrium point, since the algebra involved in doing so was a bit off-putting, but we did note it was stable, and we can see from the nullcline-and-arrow diagram that it must be a node: A trajectory cannot spiral about the first quadrant equilibrium point because once it enters the region between the nullclines and to the left of the equilibrium point, it cannot leave. In that basic region trajectories rise, but they cannot rise above the y-nullcline because in so doing they would violate the downward direction-indicator arrow in the region above the y-nullcline.

10.5. Modeling the spread of disease

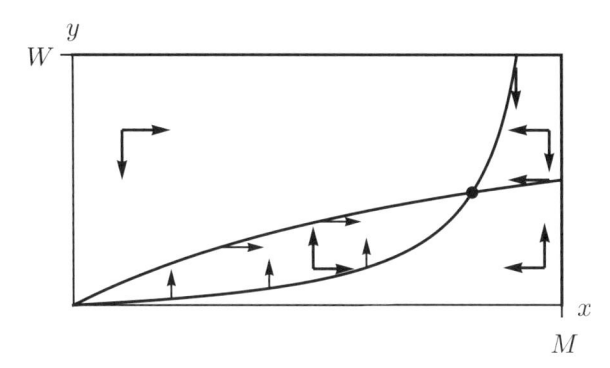

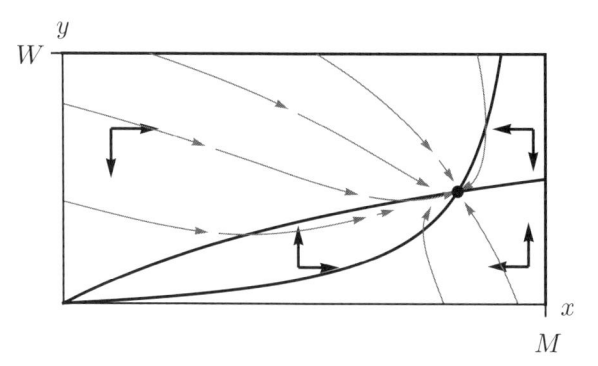

Figure 10.49. Nullcline-and-arrow diagram; $MWa_1a_2 - b_1b_2 > 0$.

Figure 10.50. Phase portrait; $MWa_1a_2 - b_1b_2 > 0$.

What does this predict for the long-term behavior of $x(t)$ and $y(t)$? As $t \to \infty$,

$$x(t) \to \frac{MW - \rho_1\rho_2}{W + \rho_1} \quad \text{and} \quad y(t) \to \frac{MW - \rho_1\rho_2}{M + \rho_2}.$$

Since these limiting values are nonzero, the disease will continue at an endemic level in the population.

Example 10.5.3. Modeling smoking. In this example, we model cigarette smoking among high school students. We assume that peer pressure plays some role in recruiting nonsmokers to become smokers. A similar system of equations could be used to model drug or alcohol use. We consider the student population to be divided into three compartments: nonsmokers $x(t)$ (whom we think of as "potential smokers"), smokers $y(t)$, and students who have quit smoking $z(t)$. We will assume that the total population size is fixed. Thus while students graduate, an equal number of new students matriculate, so that $x(t) + y(t) + z(t) = N$, where N is the constant total size of the school. The nonsmoker compartment plays the role of the "susceptible" compartment in the disease models, and we assume that the rate at which nonsmokers become smokers is jointly proportional to the number of smokers and the number of nonsmokers. This is the quantification of the role of peer pressure, and it will give rise to a term $-(\text{constant})xy$ in the differential equation for $\frac{dx}{dt}$ and a corresponding term, with a positive sign, in the differential equation for $\frac{dy}{dt}$. Remembering that N is constant, we will actually write this term in the form

$$-bx\frac{y}{N},$$

where the factor $\frac{y}{N}$ is the *proportion* of smokers in the student body and b is a positive constant that reflects, for example, the overall amount of social interaction and the likelihood of a nonsmoker being influenced to start smoking by the presence of smokers. We will also assume that the graduation rates are the same from each of the three compartments and that all newly matriculated students are initially nonsmokers. Finally we assume that smokers "recover"—that is, move into the "quitters" compartment at a rate proportional to y, the number of smokers. Our first model doesn't consider the possibility of "relapse" for someone who has quit smoking. With these assumptions, we have the differential equation system

(10.64) $$\frac{dx}{dt} = gN - bx\frac{y}{N} - gx \qquad \text{(nonsmokers)},$$

(10.65) $$\frac{dy}{dt} = bx\frac{y}{N} - cy - gy \qquad \text{(smokers)},$$

(10.66) $$\frac{dz}{dt} = cy - gz \qquad \text{(quitters)}$$

for some positive constants g, b, and c. The terms $-gx$, $-gy$, and $-gz$ represent the graduation rates from the three compartments; these are exactly balanced by the term $gN = g(x + y + z)$ in the first equation, representing the addition of new students to the school, all initially in the nonsmoker category. In the equation for $\frac{dy}{dt}$ the term $-cy$ represents smokers quitting. The units on g and c would be, for example, $\frac{1}{\text{years}}$, and their reciprocals $\frac{1}{g}$ and $\frac{1}{c}$ have the meaning of "average time in school" and "average time as a smoker". Thus if the average time that an student spends in high school is 4 years, we have $g = \frac{1}{4}$/year. If we take 12 years as the average number of years that a smoker continues to smoke, then $c = \frac{1}{12}$/year.

You may notice that the *form* of equations (10.64)–(10.66) is exactly the same as in the SIR disease model in Example 10.5.1. So the analysis we did there applies here as well. However, we are going to proceed a little differently and begin by using the relation $x(t) + y(t) + z(t) = N$, to reduce the system to a planar system, with dependent variables y and z. Making the substitution $x = N - y - z$ in equation (10.65), we obtain the equations

$$\frac{dy}{dt} = b\frac{y(N - y - z)}{N} - (g + c)y,$$

$$\frac{dz}{dt} = cy - gz.$$

Next we will make the substitutions $S = \frac{y}{N}$ and $Q = \frac{z}{N}$, so that S and Q are the *proportions* of smokers and of smokers who have quit, respectively. These are "dimensionless" variables; they have no units. Since $y = NS$ and $z = NQ$, where N is constant, this gives

$$N\frac{dS}{dt} = b\frac{NS(N - NS - NQ)}{N} - (g + c)NS,$$

$$N\frac{dQ}{dt} = cNS - gNQ,$$

or simply

(10.67) $$\frac{dS}{dt} = bS(1 - S - Q) - (g + c)S, \qquad \frac{dQ}{dt} = cS - gQ.$$

Because $\frac{x(t)}{N} + \frac{y(t)}{N} + \frac{z(t)}{N} = 1$ for all t, the only relevant part of the SQ-plane is the triangular region $S \geq 0, Q \geq 0, S + Q \leq 1$ shown in Fig. 10.51.

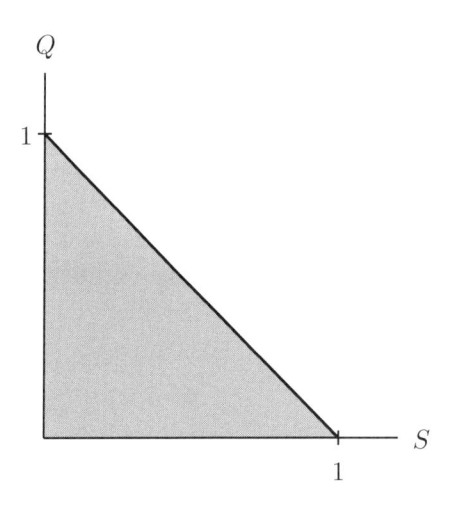

Figure 10.51. The biorelevant region $S \geq 0, Q \geq 0, S + Q \leq 1$.

10.5. Modeling the spread of disease

Note that the Jacobian matrix for the system in equation (10.67) is

$$\begin{pmatrix} b - 2bS - bQ - (g+c) & -bS \\ c & -g \end{pmatrix}.$$

To proceed with our analysis of this system, we will assign values to the constants b, g, and c. As previously discussed, we will use $g = \frac{1}{4}$ and $c = \frac{1}{12}$, and we will set $b = \frac{1}{2}$. With these values our system becomes

$$(10.68) \qquad \frac{dS}{dt} = \frac{1}{2}S(1 - S - Q) - \frac{1}{3}S, \quad \frac{dQ}{dt} = \frac{1}{12}S - \frac{1}{4}Q.$$

Nullclines and arrows. Since $\frac{dS}{dt} = S(\frac{1}{6} - \frac{1}{2}S - \frac{1}{2}Q)$, the S-nullclines are the pair of lines $S = 0$ and $S + Q = \frac{1}{3}$. The Q-nullcline is the line $Q = \frac{1}{3}S$. Fig. 10.52 shows the nullcline-and-arrow picture in the biologically relevant triangle.

Equilibrium points. Equilibrium points for the system (10.68) appear as the intersection of an S-nullcline with a Q-nullcline. One equilibrium point is $S = 0, Q = 0$, which corresponds in the original system (10.64)–(10.66) to $x = N, y = 0, z = 0$. A second equilibrium point is $S = \frac{1}{4}, Q = \frac{1}{12}$. To classify these equilibrium points we use the Jacobian matrix

$$\begin{pmatrix} \frac{1}{6} - S - \frac{1}{2}Q & -\frac{1}{2}S \\ \frac{1}{12} & -\frac{1}{4} \end{pmatrix}.$$

Evaluating the Jacobian matrix at the equilibrium point $(0,0)$ gives

$$\mathbf{J}_1 = \begin{pmatrix} \frac{1}{6} & 0 \\ \frac{1}{12} & -\frac{1}{4} \end{pmatrix},$$

and the Jacobian matrix at the equilibrium point $(\frac{1}{4}, \frac{1}{12})$ is

$$\mathbf{J}_2 = \begin{pmatrix} -\frac{1}{8} & -\frac{1}{8} \\ \frac{1}{12} & -\frac{1}{4} \end{pmatrix}.$$

Since $\mathbf{J}_1$ has one positive eigenvalue ($\frac{1}{6}$) and one negative eigenvalue ($-\frac{1}{4}$), we conclude from Theorem 10.2.3 that $(0,0)$ is a saddle point and thus unstable. Since $\mathbf{J}_2$ has trace $\tau = -\frac{3}{8} < 0$ and determinant $\Delta = \frac{1}{24} > 0$ with $\tau^2 - 4\Delta = \frac{9}{64} - \frac{1}{6} < 0$, Table 8.2 from Section 8.9 together with Theorem 10.2.3 tell us that $(\frac{1}{4}, \frac{1}{12})$ is a stable spiral point. Figs. 10.52 and 10.53 show the nullcline-and-arrow diagram and a phase portrait. Trajectories approach the equilibrium point $(\frac{1}{4}, \frac{1}{12})$, and long term we expect $S(t) \to \frac{1}{4}$ and $Q(t) \to \frac{1}{12}$.

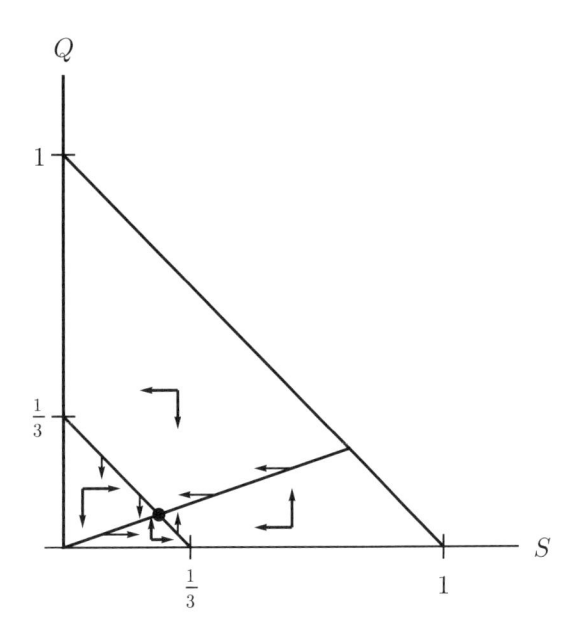

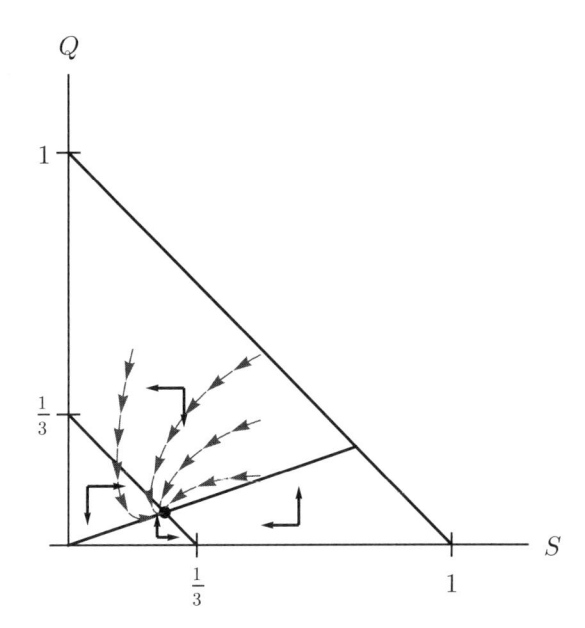

Figure 10.52. Nullcline-and-arrow diagram.

Figure 10.53. A phase portrait.

Allowing for relapse. Let's make this model more realistic. Smokers who have quit may very well start smoking again; this adds the arrow from compartment 3 back to compartment 2 in the diagram below.

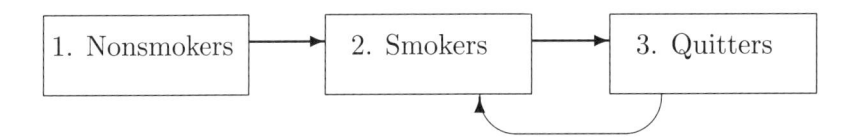

```
1. Nonsmokers  →  2. Smokers  →  3. Quitters
```

We assume that the rate at which quitters relapse is jointly proportional to z, the number of students in compartment 3, and y/N, the proportion of smokers in the school population. So we modify equations (10.64)–(10.66) to

$$(10.69) \qquad \frac{dx}{dt} = gN - bx\frac{y}{N} - gx, \quad \frac{dy}{dt} = bx\frac{y}{N} - (g+c)y + rz\frac{y}{N}, \quad \frac{dz}{dt} = cy - gz - rz\frac{y}{N}.$$

As before, we focus on the second two equations (substituting $x = N - y - z$) and make the change of variable $S = y/N$, $Q = z/N$ to obtain (see Exercise 8)

$$(10.70) \qquad \frac{dS}{dt} = bS(1 - S - Q) - (g+c)S + rQS, \quad \frac{dQ}{dt} = cS - gQ - rQS.$$

The Jacobian matrix for this new system is

$$\begin{pmatrix} b - 2bS - bQ - (g+c) + rQ & -bS + rS \\ c - rQ & -g - rS \end{pmatrix}.$$

Analyzing this model with values for the parameters. As before, we will continue our analysis using specific values for the parameters. Now we will use the values $b = 0.62$, $g = 0.25$,

10.5. Modeling the spread of disease

$c = 0.06$, and $r = 1.4$.[12] With these values the S-nullclines are the pair of lines

$$S = 0 \quad \text{and} \quad Q = \frac{0.62}{0.78}S - \frac{0.31}{0.78}$$

and the Q-nullcline is the curve with equation

$$Q = \frac{0.06S}{0.25 + 1.4S}.$$

This is a curve passing through $(0,0)$ which is increasing and concave down for S in the relevant range $0 \leq S \leq 1$. These nullclines are shown separately in Figs. 10.54–10.55. (Note the scale on the Q-axis in Fig. 10.55.) In Exercise 11 you are asked to show the appropriate left/right and up/down arrows to complete this nullcline-and-arrow diagram and to finish the analysis of the phase portrait for these particular parameters.

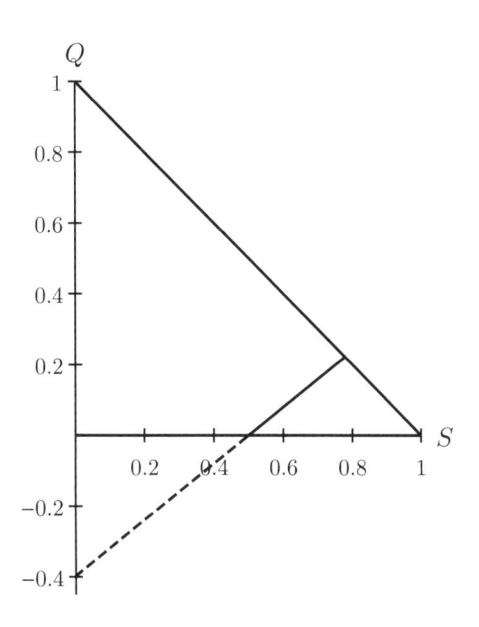

Figure 10.54. The S-nullcline in the biorelevant triangle.

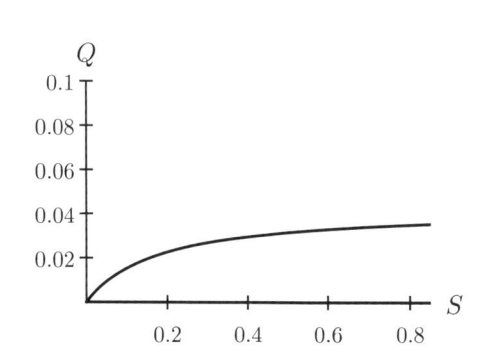

Figure 10.55. The Q-nullcline in a portion of the biorelevant triangle.

10.5.1. Exercises.

1. Suppose that in Example 10.5.1 we assume that $\beta K < \gamma + \mu$, so that the only biologically relevant equilibrium point is $(K, 0)$.
   (a) Show that this equilibrium point is a stable node.
   (b) Give the nullcline-and-arrow sketch for this case, showing several trajectories.
   (c) What happens to $S(t)$ and $I(t)$ as $t \to \infty$?

2. Use the following estimates for the basic reproduction number for the listed diseases (all are diseases where recovery confers immunity) to estimate what percentage of newborns need to be successfully immunized to prevent the disease from becoming endemic.

---

[12]The values of b and r are taken from *Mathematical models for the dynamics of tobacco use, recovery, and relapse* by C. Castillo-Garsow, G. Jordan-Salivia, and A. Rodriguez-Herrera, Biometrics Unit Technical Reports, Number BU-1505-M, Cornell University, Ithaca, 1997.

Table 10.3. Comparing R_0.	
Disease	Basic reproduction number R_0
measles	12–18
mumps	5–7
diphtheria	6–7
pertussis	12–17
smallpox	5–7

Only smallpox has been eliminated on a worldwide basis, by intensive vaccination efforts. The last known case was in 1977 in Somalia. Now it exists only in laboratories, and routine smallpox vaccination is no longer done. The deliberate release of smallpox as a terrorist act has been a recent concern, especially since 9/11/2001. Mathematical models are an important tool in planning for a response to such a bioterrorism attack. One recent such model assumes the release of smallpox in the New York City subway.[13] While the model is more complicated than that considered in Example 10.5.1 (in particular it separates "subway users" from "nonsubway users", and it considers behavioral changes people might make after such an attack), many of the basic features of our model are still present.

3. For endemic diseases that confer immunity, there is a rule of thumb for estimating the basic reproduction number R_0:

$$R_0 \approx 1 + \frac{L}{A}$$

where L is the average life span and A is the average age of contracting the disease. Suppose that in 1955 the average life span in the US was 70 and the average age of contracting polio was 17.9. What percentage of the population would have to be vaccinated for herd immunity?

4. This problem outlines Daniel Bernoulli's work on smallpox in 1760. We start with a group of people all born at the same time $t = 0$.
   - Let $x(t)$ denote the number of this original group who are still alive t years later.
   - Let $y(t)$ be the number who are still alive at time t and have not yet had smallpox.

The y-population is thus the susceptibles who are alive at time t. There are two "rate out" terms for the y-population, corresponding to the fact that some members contract smallpox and others die from other causes. Bernoulli assumes that susceptibles contract smallpox at a rate proportional to the y-population and deaths occur from nonsmallpox causes at a rate which depends on time t but is the same for both the y- and x-populations. These assumptions give the equation

(10.71) $$\frac{dy}{dt} = -\beta y - d(t) y$$

where β is a positive constant. The x-population changes for two reasons: Some people die from smallpox, and some die from other causes. The "die from other causes" factor gives rise to a term $-d(t)x$ in the differential equation for $\frac{dx}{dt}$. The people who die from smallpox are some fraction δ of those who contract the disease. Thus the differential equation for the x-population is

(10.72) $$\frac{dx}{dt} = -\delta \beta y - d(t) x.$$

---

[13]G. Chowell, A. Cintron-Arias, S. Del Valle, F. Sanchez, B. Song, J. Hyman, H. Hethcote, and C. Castillo-Chavez, *Mathematical applications associated with the deliberate release of infectious agents*, in Mathematical Studies on Human Disease Dynamics, Contemporary Mathematics, Vol. 410, A. Gumel, Editor-in-Chief, American Mathematical Society, 2006, pp. 51–71.

10.5. Modeling the spread of disease

Since t appears explicitly in the equations for $\frac{dx}{dt}$ and $\frac{dy}{dt}$ and we don't know what the function $d(t)$ is, we will need to do something clever to solve the system (10.71)–(10.72).

(a) Set $z = y/x$ and show that $\frac{dz}{dt} = -\beta z + \delta\beta z^2$.

(b) Solve the equation in (a) for z, with the initial condition $z(0) = 1$. The rationale for this initial condition is that at a very young age, no survivors have smallpox, since it is almost always fatal to infants. So $z(t) = \frac{y(t)}{x(t)} = 1$ as $t \to 0^+$.

(c) In your answer to (b), use the values $\beta = \delta = \frac{1}{8}$, estimated by Bernoulli, and compute $z(10)$ and $z(20)$. What percentage of 10-year-olds have not had smallpox? What percentage of 20-year-olds have not?

5. (a) In Example 10.5.2, suppose t is measured in days. What are the units on $a_1, b_1, a_2, b_2, \rho_1$, and ρ_2?

(b) Sketch the graph of
$$y = \frac{Wa_2 x}{b_2 + a_2 x}$$
showing only the portion that lies in the rectangle $0 \le x \le M, 0 \le y \le W$.

(c) For the functions
$$F_1(x) = \frac{b_1 x}{a_1(M-x)} \quad \text{and} \quad F_2(x) = \frac{Wa_2 x}{b_2 + a_2 x}$$
compute $F_1'(0)$ and $F_2'(0)$. Under what condition is $F_1'(0) > F_2'(0)$? How does this help explain how to fit together the two pictures in Figs 10.47 and 10.48?

(d) Give the nullcline-and-arrow diagram for equation (10.59) under the assumption $MWa_1 a_2 - b_1 b_2 < 0$. Predict the course of the disease in this case.

6. Suppose we model the spread of gonorrhea in a population of 1,000 sexually active women and 1,000 sexually active men by the system in equation (10.59), with $b_1 = 1/50$ days and $b_2 = 1/10$ days (this corresponds to an average infective period of 50 days for women and 10 days for men). Also suppose that $a_1 = 1/5,000$ and $a_2 = 1/25,000$.

(a) Find the equilibrium point that has both coordinates positive. What are the total number of infected persons (men + women) corresponding to this equilibrium solution?

(b) Suppose there is a public health policy that would halve the number of at-risk women. What is the new first quadrant equilibrium point, and is the total number of infected persons corresponding to this equilibrium?

(c) Next suppose there is a different public health policy that would instead halve the number of at-risk men. Find the equilibrium point in the first quadrant and the total number of infected persons corresponding to this equilibrium.

(d) Of the two public health policies just discussed, which is more effective in reducing the number of infected individuals at the equilibrium solution?

7. In equation (10.59), show that the substitutions $X = \frac{x}{M}$ and $Y = \frac{y}{W}$ lead to the system

(10.73) $$\frac{dX}{dt} = a_1(1-X)WY - b_1 X \quad \text{and} \quad \frac{dY}{dt} = a_2(1-Y)MX - b_2 Y.$$

Notice that X represents the proportion of the male population that is infected and Y represents the proportion of the female population that is infected.

8. Show the details of obtaining the equations (10.70) from the second two equations in (10.69).

9. Show that for the system in equation (10.67), the equilibrium point $(0,0)$ is a saddle point if $b > g + c$ and a stable node if $b < g + c$.

10. In the model for "smoking with relapse" do you think it is more likely for a "quitter" to start smoking again or for a nonsmoker to start smoking? Based on your answer, would you have $b > r$ or $b < r$?

11. In this problem you will further analyze the smoking with relapse model in equation (10.70) with the values $b = 0.62, g = 0.25, c = 0.06$, and $r = 1.4$.
    (a) Using Figs. 10.54–10.55 give the nullcline-and-arrow diagram in the biologically relevant region $S \geq 0$, $Q \geq 0$, $S + Q \leq 1$.
    (b) Find (approximately) all biorelevant equilibrium points and classify them according to type and stability.
    (c) Sketch a phase portrait using (a) and (b). What do you predict for the long-term behavior of $S(t)$ and $Q(t)$?

12. Consider the initial value problem consisting of the system in equation (10.70), with the values for b, g, c, and r as given in the preceding exercise and initial condition $S(0) = 0.2$, $Q(0) = 0.01$.
    (a) Sketch the solution curve of this initial value problem based on your nullcline-and-arrow diagram from Exercise 11.
    (b) (CAS) Using a computer algebra system, solve this initial value problem numerically, and plot the corresponding trajectory over the time interval $[0, 18]$.
    (c) Using your numerical solution, compute $S(6)$, $S(12)$, and $S(18)$, as well as $Q(6)$, $Q(12)$, and $Q(18)$.

13. Suppose in Example 10.5.3 we model relapse in a different way, by assuming that quitters become smokers again at a rate proportional to the number of people who have quit.
    (a) What system of equations does this assumption give?
    (b) Convert your model in (a) to a planar system for $S = y/N$ and $Q = z/N$. When is the equilibrium point $S = 0$, $Q = 0$ stable? Unstable?

14. Norovirus ("stomach flu") is a common and unpleasant illness caused by a virus.[14] College students, especially those living in dorms, are frequent victims, as are cruise ship passengers. In this problem we will look at a model for the spread of this disease which divides the population into 5 compartments: (S) susceptibles, (E) exposed (these are people who do not yet show symptoms but soon will), (I) infected individuals showing symptoms, (A) asymptotic infected people (these are people who do not have symptoms but are still shedding the virus), and (R), the recovered class of people with temporary immunity to norovirus. Movement between these compartments follows these rules:

(1) All births go into the susceptible category with constant births of B per day.
(2) Susceptibles move into the exposed compartment at a rate proportional to the product of the number of susceptibles ($S(t)$) and the number of symptomatic infectives ($I(t)$), with proportionality constant β. Notice this means we are assuming only symptomatic infected people are infectious, not the "presymptomatic" exposed or "postsymptomatic" asymptomatic infected people.
(3) Exposed individuals move into the symptomatic infected compartment at a rate proportional to the number of exposed ($E(t)$), with proportionality constant $1/\mu_s$, where μ_s is the average length of the incubation period in days.
(4) Symptomatic individuals move into the asymptomatic infected compartment at a rate proportional to the number of (symptomatic) infecteds ($I(t)$), with proportionality constant $1/\mu_a$, where μ_a is the average length of symptoms in days.
(5) Asymptomatic infected individuals move into the temporarily immune category (R compartment) at a rate proportional to the number of asymptomatic infected people ($A(t)$),

---

[14] The model discussed in this exercise comes from the article *Duration of Immunity to Norovirus Gastroenteritis* by K. Simmons, M. Gambhir, J. Leon, and B. Lopman, Emerging Infectious Diseases, Aug. 2013, 19(8), 1260–1267.

with proportionality constant $1/\rho$, where ρ is average number of days a person sheds the virus without showing symptoms.

(6) People leave the R compartment in one of two ways: Either they lose immunity entirely, moving into the S (susceptible) compartment, or they "lose immunity to infection without losing immunity to disease"; this means they move back into the A compartment, shedding the virus but not showing any symptoms. The first case occurs at a rate proportional to R with proportionality constant $1/\theta$ where θ is the average duration of temporary immunity. The second case occurs at a rate jointly proportional to $R(t)$ and $I(t)$, with proportionality constant β as in (2). Moving from R to A has the effect of temporarily boosting immunity again, and these individuals eventually move back into the R compartment.

(7) Deaths occur from all compartments, at a rate proportional to the compartment size with proportionality constant δ, which we assume is the same for all compartments.

(a) Draw a compartment diagram, labeled S, E, I, A, and R, with arrows showing what transfers we have between compartments. You may include a sixth compartment, D, for dead, or you may omit this.

(b) Using your answer to (a), give a system of differential equations for $\frac{dS}{dt}$, $\frac{dE}{dt}$, $\frac{dI}{dt}$, $\frac{dA}{dt}$, and $\frac{dR}{dt}$, using the following facts: The average incubation period for norovirus is 1 day, the average duration of symptoms is 2 days, the average duration of asymptotic virus shedding is 10 days, and the average duration of temporary immunity is 5 *years*. Your equations will include the parameter β from (2) above; we do not give a numerical value for this. Whether or not you included a "dead" compartment in (a), your equations should include births of B people per day and deaths as described in (7) above. We do not give a numerical value for δ.

(c) Another model allows for susceptibles to become infected by people in the exposed and asymptomatic compartments, but at a lower rate than by symptomatic infected persons. This changes the "rate in" terms for $\frac{dE}{dt}$ to be $\beta_1 S(t)E(t) + \beta_2 S(t)I(t) + \beta_3 S(t)A(t)$ with β_1 and β_3 to be smaller than β_2. What are the corresponding changes in $\frac{dS}{dt}$? In $\frac{dR}{dt}$? In $\frac{dA}{dt}$?

10.6. Hamiltonians, gradient systems, and Lyapunov functions

Conserved quantities. In Section 4.8 we used the second-order equation $my'' + ky = 0$ to model an undamped mass-spring system, where m is the mass, k is the spring constant, and $y(t)$ is the displacement of the mass from equilibrium at time t. We can convert this to a first-order system, with dependent variables y and v, using the substitution $y' = v$:

$$(10.74) \qquad y' = v, \quad v' = -\frac{k}{m}y.$$

This linear system has a center at $(0,0)$. The orbits are ellipses in the yv-plane with equations determined by

$$\frac{dv}{dy} = \frac{dv/dt}{dy/dt} = -\frac{ky}{mv}.$$

Separating variables in this first-order equation gives $\frac{1}{2}mv^2 + \frac{1}{2}ky^2 = C$; for $C > 0$ these describe ellipses as shown in Fig. 10.56.

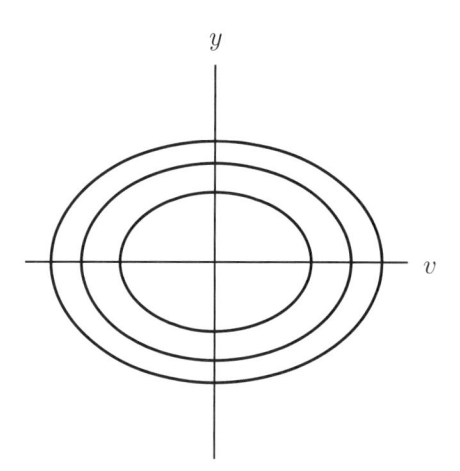

Figure 10.56. Ellipses $\frac{1}{2}mv^2 + \frac{1}{2}ky^2 = C$, $C > 0$.

The first term, $\frac{1}{2}mv^2$, is the kinetic energy of the system, while the second term, $\frac{1}{2}ky^2$, is the potential energy stored in the stretched or compressed spring. Their sum is the total energy, and we have just shown that the total energy is constant along any solution curve. We say that the total energy $E(y,v) = \frac{1}{2}mv^2 + \frac{1}{2}ky^2$ is a **conserved quantity**, and we call (10.74) a **conservative system**. In general, we have the following definition.

Definition 10.6.1. A conserved quantity for an autonomous system

$$\text{(10.75)} \qquad \frac{dx}{dt} = f(x,y), \quad \frac{dy}{dt} = g(x,y)$$

on an open rectangle $\mathcal{R}$ is a differentiable function $E = E(x,y)$ such that E is constant along any trajectory of the system that lies in $\mathcal{R}$.

The Chain Rule shows that if $E(x,y) = K$ is an (implicit) solution of

$$\frac{dy}{dx} = \frac{g(x,y)}{f(x,y)}$$

on a open rectangle $\mathcal{R}$, then E is a conserved quantity for the system (10.75) on $\mathcal{R}$. We will be interested in conserved quantities that are not constant on any nonempty open disks (or rectangles) in their domains.

Let's also think about a conserved quantity with reference to a three-dimensional picture. In the mass-spring system in (10.74) notice that the graph of the total energy $z = E(y,v) = \frac{1}{2}mv^2 + \frac{1}{2}ky^2$ is a paraboloid surface in 3-space as shown in Fig. 10.57. A **level curve** for the function $E(y,v)$ is the set of points in the yv-plane along which $E(y,v)$ is constant: $\frac{1}{2}mv^2 + \frac{1}{2}ky^2 = C$. These are our elliptical orbits from Fig. 10.56. Every point on the paraboloid which lies directly above the level curve $\frac{1}{2}mv^2 + \frac{1}{2}ky^2 = C$ is at height C, reflecting the total energy of the system at any point (v,y) on the curve.

By contrast, the *damped* mass-spring system with equation $my'' + cy' + ky = 0$ converts to the linear system

$$y' = v, \quad v' = -\frac{c}{m}v - \frac{k}{m}y,$$

which has a spiral point at the origin. If you revisit Exercise 21 in Section 4.1, you can see that what you showed in part (b) of that problem is that the total energy $E(y,v) = \frac{1}{2}mv^2 + \frac{1}{2}ky^2$ is not conserved for this system. The curve that lies on the surface of the paraboloid $z = \frac{1}{2}mv^2 + \frac{1}{2}ky^2$ and directly above a trajectory in the phase plane spirals down to $(0,0,0)$; see Fig. 10.58.

10.6. Hamiltonians, gradient systems, and Lyapunov functions

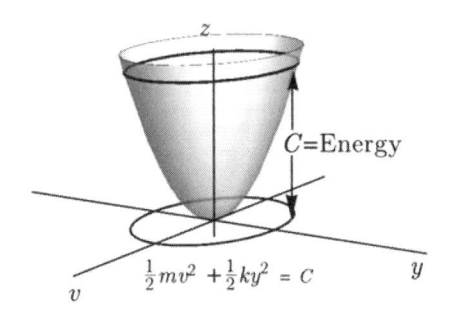

Figure 10.57. Graph of $z = \frac{1}{2}mv^2 + \frac{1}{2}ky^2$ and the energy C level curve.

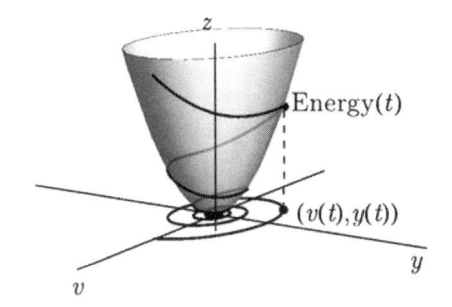

Figure 10.58. Energy is not conserved in the damped mass-spring system.

Example 10.6.2. We revisit the Lotka-Volterra system
$$\frac{dx}{dt} = ax - bxy, \quad \frac{dy}{dt} = -dy + cxy$$
from Section 10.4. This has an equilibrium point $P = (\frac{d}{c}, \frac{a}{b})$. The linear approximation at P has a center; this gives no information on the stability of P. We saw in (10.4) that the orbits are described by the equations $a \ln y - by + d \ln x - cx = K$, for constant K, and we gave an ad hoc argument that these are closed curves encircling P.

In our new terminology, we say that the function $E(x, y) = a \ln y - by + d \ln x - cx$ is a conserved quantity for the Lotka-Volterra system on the open first quadrant $\{(x, y) : 0 < x < \infty, 0 < y < \infty\}$. We claim that the equilibrium point P cannot be asymptotically stable. If it were, then every orbit $(x(t), y(t))$ starting close enough to P would approach P as $t \to \infty$. By continuity of $E(x, y)$ (in the first quadrant), this would say that

$$(10.76) \qquad E(x(t), y(t)) \to E\left(\frac{d}{c}, \frac{a}{b}\right) \quad \text{as } t \to \infty.$$

But $E(x(t), y(t))$ is constant; this is what it means for E to be a conserved quantity, and equation (10.76) says the constant value must be $E(\frac{d}{c}, \frac{a}{b})$. This applies to any orbit that starts sufficiently close to P and thus gives the conclusion that $E(x, y)$ is constant in some entire disk centered at P, which is not the case (note the only point at which both $\frac{\partial E}{\partial x}$ and $\frac{\partial E}{\partial y}$ are zero is $(d/c, a/b)$). So P cannot be asymptotically stable.

This argument says something quite general about any autonomous system

$$(10.77) \qquad \frac{dx}{dt} = f(x, y), \quad \frac{dy}{dt} = g(x, y):$$

If there is a conserved quantity $E = E(x, y)$ for (10.77) on some open rectangle $\mathcal{R}$ of the plane such that E is not constant in $\mathcal{R}$, then no equilibrium point of (10.77) in $\mathcal{R}$ is asymptotically stable.

Hamiltonian systems. Suppose that we start with a function $H(x, y)$ all of whose first- and second-order partial derivatives exist and are continuous functions; we say that $H(x, y)$ is **twice continuously differentiable**. Form the autonomous system

$$(10.78) \qquad \frac{dx}{dt} = \frac{\partial H}{\partial y}, \quad \frac{dy}{dt} = -\frac{\partial H}{\partial x}.$$

Such a system is called a **Hamiltonian system**, and the function $H(x, y)$ is called a **Hamiltonian function**. The trajectories in the phase portrait for this system satisfy the first-order differential

equation
$$\frac{dy}{dx} = -\frac{\frac{\partial H}{\partial x}}{\frac{\partial H}{\partial y}}.$$

Rewrite this as

(10.79)
$$\frac{\partial H}{\partial x} + \frac{\partial H}{\partial y}\frac{dy}{dx} = 0.$$

This an exact equation, whose solutions are given implicitly by $H(x,y) = C$ for arbitrary constants C (see Section 2.8). In other words:

> Each trajectory for the system (10.78) is contained in a level curve $H(x,y) = C$ of the Hamiltonian function H, and $H(x,y)$ is a conserved quantity for the system.

Can we recognize when an autonomous system

(10.80)
$$\frac{dx}{dt} = f(x,y), \quad \frac{dy}{dt} = g(x,y)$$

is a Hamiltonian system? Is there a twice continuously differentiable function $H(x,y)$ with

$$\frac{\partial H}{\partial y} = f(x,y) \quad \text{and} \quad \frac{\partial H}{\partial x} = -g(x,y)?$$

This is a variant of a question we have considered before, in Section 2.8. If such a function $H(x,y)$ exists, then

$$\frac{\partial^2 H}{\partial x \partial y} = \frac{\partial f}{\partial x} \quad \text{and} \quad \frac{\partial^2 H}{\partial y \partial x} = -\frac{\partial g}{\partial y}.$$

Since the mixed second-order partials of H must be the same, this forces

(10.81)
$$\frac{\partial f}{\partial x} = -\frac{\partial g}{\partial y}$$

if (10.80) is Hamiltonian. In other words, the relationship (10.81) is a *necessary* condition for the system to be Hamiltonian. The next theorem says that the converse holds as well, when (10.81) holds in the plane $\mathbb{R}^2$ or in some open rectangle $\mathcal{R}$ in $\mathbb{R}^2$.

Theorem 10.6.3. *Given an autonomous system*

$$\frac{dx}{dt} = f(x,y), \quad \frac{dy}{dt} = g(x,y),$$

where $f(x,y)$ and $g(x,y)$ are continuously differentiable functions in an open rectangle $\mathcal{R}$, the system is Hamiltonian if and only if

$$\frac{\partial f}{\partial x} = -\frac{\partial g}{\partial y}$$

for all (x,y) in $\mathcal{R}$.

Example 10.6.4. Show that the system

$$\frac{dx}{dt} = 2y + 3 + x\sin y, \quad \frac{dy}{dt} = x^2 + \cos y$$

is Hamiltonian, and find a Hamiltonian function. Since

$$\frac{\partial}{\partial x}(2y + 3 + x\sin y) = \sin y \quad \text{and} \quad \frac{\partial}{\partial y}(x^2 + \cos y) = -\sin y,$$

10.6. Hamiltonians, gradient systems, and Lyapunov functions

Theorem 10.6.3 applies. A Hamiltonian function $H(x,y)$ must satisfy

(10.82) $$\frac{\partial H}{\partial y} = 2y + 3 + x\sin y$$

and

(10.83) $$\frac{\partial H}{\partial x} = -x^2 - \cos y.$$

From (10.82) we see that
$$H(x,y) = y^2 + 3y - x\cos y + \varphi(x)$$
for some function $\varphi(x)$. Then
$$\frac{\partial H}{\partial x} = -\cos y + \varphi'(x)$$
and we want this to equal $-x^2 - \cos y$. Choosing $\varphi(x) = -\frac{1}{3}x^3$ we see that
$$H(x,y) = y^2 + 3y - x\cos y - \frac{1}{3}x^3$$
is a Hamiltonian function.

10.6.1. Lyapunov functions. For a conservative system, there is a nonconstant function which is constant on trajectories. In this section, we will focus on systems which are not necessarily conservative, but for which we can find a function which *decreases* along the trajectories. This will provide a method for determining the stability of some equilibrium points whose stability cannot be determined from linear approximation (i.e., Theorem 10.2.3 does not apply).

Consider an autonomous system

(10.84) $$\frac{dx}{dt} = f(x,y), \quad \frac{dy}{dt} = g(x,y)$$

with an isolated equilibrium point at (x_e, y_e). Suppose we have a continuously differentiable function $E(x,y)$ in a neighborhood U of (x_e, y_e) with $E(x_e, y_e) = 0$. By a neighborhood of (x_e, y_e) we mean some disk centered at (x_e, y_e) with positive radius; that is, all points whose distance to (x_e, y_e) is less than this radius. If $x = x(t), y = y(t)$ is a trajectory of the system (10.84), then $E(x(t), y(t))$ gives the value of E along this trajectory at time t. So we are thinking now of $E(x(t), y(t))$ as a function of the single variable t and we write
$$\frac{dE}{dt} = \frac{d}{dt}[E(x(t), y(t))]$$
for the time rate of change of E along the trajectory at the point $(x(t), y(t))$. By the Multivariable Chain Rule

(10.85) $$\begin{aligned} \frac{dE}{dt} &= \frac{d}{dt}[E(x(t), y(t))] \\ &= \frac{\partial E}{\partial x}(x(t), y(t))\frac{dx}{dt} + \frac{\partial E}{\partial y}(x(t), y(t))\frac{dy}{dt} \\ &= \frac{\partial E}{\partial x}(x(t), y(t))f(x(t), y(t)) + \frac{\partial E}{\partial y}(x(t), y(t))g(x(t), y(t)), \end{aligned}$$

where we have used (10.84) to obtain the last equality. This leads us to *define*:

(10.86) $$\dot{E}(x,y) = \frac{\partial E}{\partial x}(x,y)f(x,y) + \frac{\partial E}{\partial y}(x,y)g(x,y),$$

which is purely a function of x and y, with no reference to t. Even so, $\dot{E}$ is sometimes called the **time derivative** of E, or the **derivative of E along trajectories**, because by equation (10.85)

$$\frac{dE}{dt} = \dot{E}(x(t), y(t)).$$

It tells you how the function $E(x, y)$ is changing along the solution curves of (10.84):

- If $\dot{E}(x, y) > 0$, then E is increasing along the trajectory that passes through the point (x, y).
- If $\dot{E}(x, y) < 0$, then E is decreasing along the trajectory that passes through the point (x, y).

Notice that we can compute $\dot{E}(x, y)$ (using equation (10.86)) without actually knowing the solution curve passing through (x, y).

These observations about the time derivative $\dot{E}(x, y)$ lead to the important notion of a Lyapunov function, which we define next.

Definition 10.6.5. A function $E = E(x, y)$ is a **Lyapunov function** of the system (10.84) corresponding to its equilibrium point (x_e, y_e) provided E is continuously differentiable on a neighborhood U of (x_e, y_e) and satisfies

(i) $E(x_e, y_e) = 0$,

(ii) $E(x, y) > 0$ for all (x, y) in U for which $(x, y) \neq (x_e, y_e)$, and

(iii) $\dot{E}(x, y) \leq 0$ for all (x, y) in U, where $\dot{E}$ is given by (10.86).

Example 10.6.6. For the system

$$y' = f(y, v) = v, \quad v' = g(y, v) = -\frac{c}{m}v - \frac{k}{m}y,$$

arising from the damped mass-spring equation $my'' + cy' + ky = 0$, set

$$E(y, v) = \frac{1}{2}mv^2 + \frac{1}{2}ky^2.$$

At the equilibrium point $(0, 0)$ we have $E(0, 0) = 0$ and $E(x, y) > 0$ for all $(x, y) \neq (0, 0)$. Computing,

$$\dot{E}(y, v) = \frac{\partial E}{\partial y}f(y, v) + \frac{\partial E}{\partial v}g(y, v) = kyv + mv\left(-\frac{c}{m}v - \frac{k}{m}y\right) = -cv^2 \leq 0.$$

Thus $E(y, v)$ is a Lyapunov function for this system (we may choose $U = \mathbb{R}^2$ for our neighborhood of $(0, 0)$).

The next result shows that finding a Lyapunov function can help determine the stability of the equilibrium point. It is helpful when the eigenvalues of the linear approximation Jacobian matrix are purely imaginary or one is 0.

Theorem 10.6.7. *Suppose (x_e, y_e) is an isolated equilibrium point of the autonomous system (10.84), and suppose further that we can find a corresponding Lyapunov function, that is, a continuously differentiable function $E(x, y)$ with the following properties:*

(a) $E(x_e, y_e) = 0$ *and* $E(x, y) > 0$ *for all* $(x, y) \neq (x_e, y_e)$ *in some neighborhood U of (x_e, y_e).*

(b) $\dot{E}(x, y) \leq 0$ *for all* (x, y) *in* U.

Then (x_e, y_e) is a stable equilibrium. If we have the strict inequality $\dot{E}(x, y) < 0$ for all $(x, y) \neq (x_e, y_e)$ in U, then (x_e, y_e) is asymptotically stable.

10.6. Hamiltonians, gradient systems, and Lyapunov functions

When E is as in (a) and $\dot{E}$ is strictly less than 0 at all points $(x,y) \neq (x_e, y_e)$ in U, we say E is a **strong Lyapunov function**.

In Example 10.6.6 we have $\dot{E} = -cv^2 \leq 0$ in $\mathbb{R}^2$. According to Theorem 10.6.7, $(0,0)$ is stable, but we cannot use this theorem to make the stronger conclusion (which we already know!) that $(0,0)$ is asymptotically stable, since $\dot{E} = 0$ whenever $v = 0$. However, perhaps a different Lyapunov function can be found which will give the stronger conclusion; see Exercise 19.

Theorem 10.6.7 doesn't tell us anything about how to *find* a Lyapunov function. For systems that arise from mechanical (or electrical) models with a notion of energy, total energy can often serve as a Lyapunov function. In the case of the Lotka-Volterra model, we can modify the conserved quantity E identified in Example 10.6.2 to obtain a Lyapunov function for the model corresponding to its first-quadrant equilibrium point. In other cases, we are forced to rely on trial and error. For an equilibrium point at $(0,0)$, functions of the form $ax^2 + bxy + cy^2$ are often considered as possible Lyapunov functions; such functions are said to be of *quadratic form*[15].

Example 10.6.8. The system
$$\frac{dx}{dt} = -y - x^3 - xy^2, \qquad \frac{dy}{dt} = x - y^3 - x^2 y$$
has $(0,0)$ as its only equilibrium point. The function $E(x,y) = x^2 + y^2$ is a strong Lyapunov function, since $E(0,0) = 0$, $E(x,y) > 0$ when $(x,y) \neq (0,0)$ and
$$\begin{aligned}
\dot{E}(x,y) &= \frac{\partial E}{\partial x} f(x,y) + \frac{\partial E}{\partial y} g(x,y) \\
&= 2x(-y - x^3 - xy^2) + 2y(x - y^3 - x^2 y) \\
&= -2x^4 - 4x^2 y^2 - 2y^4,
\end{aligned}$$
which is 0 at $(0,0)$ and is strictly less than 0 for all $(x,y) \neq 0$. By Theorem 10.6.7, $(0,0)$ is an asymptotically stable equilibrium point. We looked at the same system in Example 10.3.2 of Section 10.3, arriving at the same conclusion by a rather different argument.

A Lyapunov function for the Lotka-Volterra model For the Lotka-Volterra system
$$\frac{dx}{dt} = ax - bxy, \qquad \frac{dy}{dt} = -dy + cxy,$$
set $(x_e, y_e) = (d/c, a/b)$, so that (x_e, y_e) is the (only) first-quadrant equilibrium point of the system. Let
$$E(x,y) = c\left(x - x_e - x_e \ln\left(\frac{x}{x_e}\right)\right) + b\left(y - y_e - y_e \ln\left(\frac{y}{y_e}\right)\right).$$
It's clear that $E(x_e, y_e) = 0$. That $E(x,y) > 0$ for all x, y in the first quadrant for which $(x,y) \neq (x_e, y_e)$ follows from the observation that for every positive number a, the function $f(x) = x - a - a \ln(x/a)$ is strictly decreasing on $(0, a)$, strictly increasing on (a, ∞), and $f(a) = 0$. Finally, the reader should check that $\dot{E}(x,y) = 0$ for all (x,y) in the first quadrant. Thus, E is a Lyapunov function of the Lotka-Volterra system corresponding to its equilibrium point (x_e, y_e). Hence, by Theorem 10.6.7, (x_e, y_e) is stable. From Example 10.6.2 we know that (x_e, y_e) is not asymptotically stable. These observations together with Theorem 10.2.4 tell us that (x_e, y_e) is a center or hybrid center/spiral. The hybrid center/spiral case cannot occur since $f(x,y) = ax - bxy$ and $g(x,y) = -dy + cxy$ are polynomials in the variables x and y (the hybrid center/spiral case is also ruled out by the ad hoc argument in Section 10.4) and (x_e, y_e) must be a center.

---

[15] More generally, for an equilibrium point (x_e, y_e) one can look for a Lyapunov function of the quadratic form $a(x - x_e)^2 + b(x - x_e)(y - y_e) + c(y - y_e)^2$.

Visualizing Theorem 10.6.7 geometrically. For simplicity we assume that the curves $E(x,y) = k$ encircle the isolated equilibrium point and if $k_1 < k_2$, then the curve $E(x,y) = k_1$ lies inside the curve $E(x,y) = k_2$ (see, for example, Fig. 10.59). Notice that we can write the time derivative $\dot{E}(x,y)$ as the dot product of the vectors

$$\left(\frac{\partial E}{\partial x}, \frac{\partial E}{\partial y}\right) \quad \text{and} \quad (f(x,y), g(x,y)),$$

since

$$\left(\frac{\partial E}{\partial x}, \frac{\partial E}{\partial y}\right) \cdot (f(x,y), g(x,y)) = \frac{\partial E}{\partial x} f(x,y) + \frac{\partial E}{\partial y} g(x,y) = \dot{E}(x,y).$$

The vector

$$\operatorname{grad} E = \left(\frac{\partial E}{\partial x}, \frac{\partial E}{\partial y}\right)$$

is called the gradient vector of the function $E(x,y)$. The gradient is also written as ∇E.

The gradient vector has an important geometric meaning: At any (x,y) where it is nonzero, grad $E(x,y)$ is a vector which is perpendicular to the level curve of E passing through that point; in other words, grad $E(x,y)$ is a **normal vector** to this level curve. It points in the direction in which E is increasing most rapidly. Fig. 10.59 illustrates this for the function $E(x,y) = x^2 + 2y^2$, which has gradient grad $E(x,y) = (2x, 4y)$. The level curve of E passing through the point $(a,b) \neq (0,0)$ is an ellipse with equation $x^2 + 2y^2 = k$, where $k = a^2 + 2b^2$. At the point (a,b), the gradient vector $(2a, 4b)$ is perpendicular to this ellipse and points "outward", in the direction in which E is increasing most rapidly.

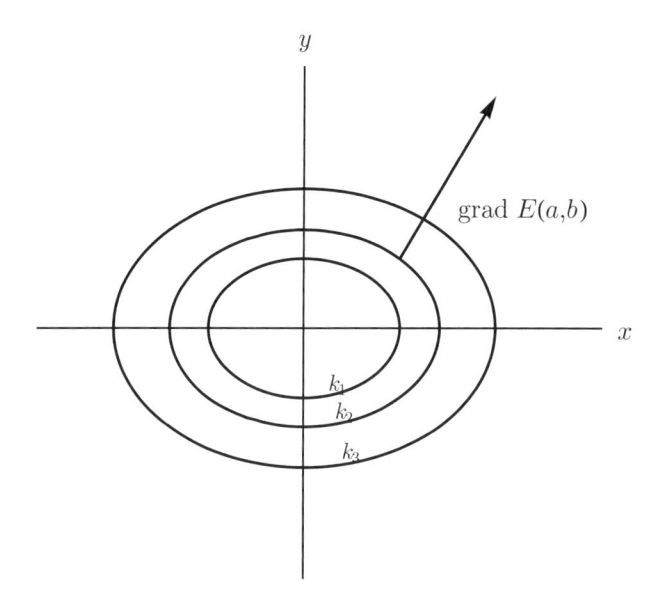

Figure 10.59. $x^2 + 2y^2 = k_j, k_1 < k_2 < k_3$.

Returning to $\dot{E}(x,y)$ as the dot product of grad $E(x,y)$ and $(f(x,y), g(x,y))$, recall that the dot product of two vectors is negative if and only if the angle between the vectors is greater than $\frac{\pi}{2}$, i.e., the angle between them is obtuse.

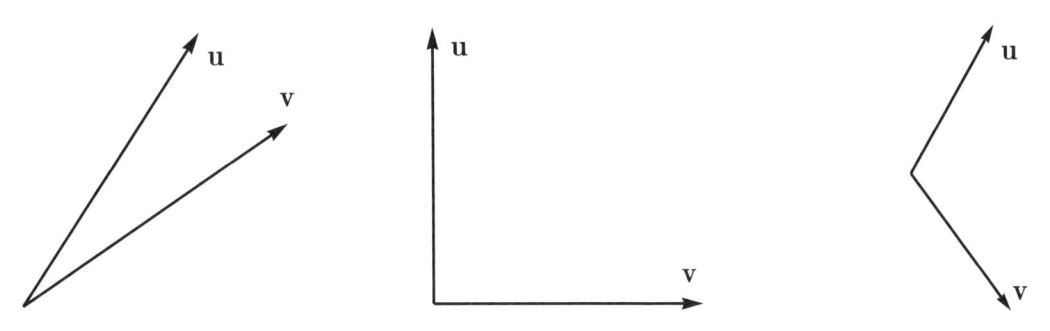

10.6. Hamiltonians, gradient systems, and Lyapunov functions

Figure 10.60. $\mathbf{u} \cdot \mathbf{v} > 0$.

Figure 10.61. $\mathbf{u} \cdot \mathbf{v} = 0$.

Figure 10.62. $\mathbf{u} \cdot \mathbf{v} < 0$.

Now $(f(x, y), g(x, y))$ is a vector in the vector field for our system. At each nonequilibrium point, it gives the tangent vector to the trajectory passing through that point, and it points along the direction of motion for the trajectory. Suppose we are in a neighborhood of an equilibrium point (x_e, y_e) with $\dot{E}$ negative at points different from (x_e, y_e) in this neighborhood. Pick any point $(x, y) \neq (x_e, y_e)$ and find the trajectory passing through that point and also the level curve $E(x, y) = k$ of E passing through the point. The hypothesis that $\dot{E}(x, y) < 0$ says that the angle between the vectors grad $E(x, y)$ and $(f(x, y), g(x, y))$ is obtuse, i.e., in the range $(\frac{\pi}{2}, \frac{3\pi}{2})$. This tells us the picture looks like Fig. 10.63, rather than like Fig. 10.64. The trajectory is driven *inside* the level curve $E(x, y) = k$ surrounding the equilibrium point by the requirement that $\dot{E}(x, y) < 0$ for (x, y) near (x_e, y_e), as asserted by Theorem 10.6.7.

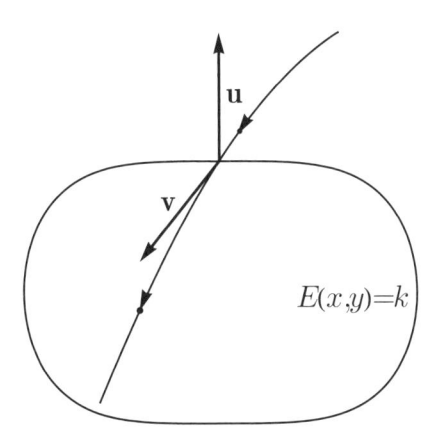

Figure 10.63. Level curve of $E(x, y)$, $\mathbf{u} = \operatorname{grad} E(x, y)$, and $\mathbf{v} = (\frac{dx}{dt}, \frac{dy}{dt})$.

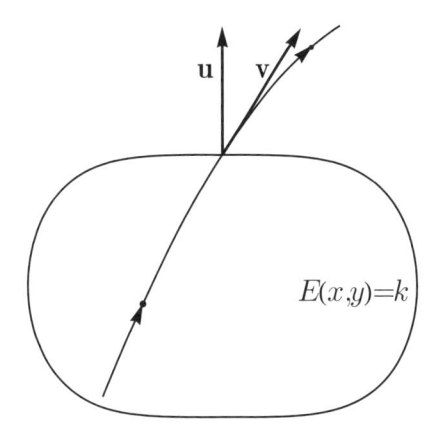

Figure 10.64. Configuration not possible if $\mathbf{u} \cdot \mathbf{v} < 0$.

The values of a Lyapunov function decrease (or at least can't increase) as you move along a trajectory in the phase portrait in the direction of increasing time. So if a trajectory crosses several level curves $E(x, y) = k$ for different values of k as t increases, it must do so in order, from the largest value of k to the smallest. These ideas are illustrated in the next example.

Example 10.6.9. The system

$$(10.87) \qquad \frac{dx}{dt} = -y - x^3, \quad \frac{dy}{dt} = x - 2y^3$$

has equilibrium point $(0, 0)$. The linear approximation near $(0, 0)$ has a center. If $E(x, y) = x^2 + y^2$, then

$$\dot{E}(x, y) = 2x(-y - x^3) + 2y(x - 2y^3) = -2x^4 - 4y^4,$$

and $E(x,y)$ is a strong Lyapanov function. Theorem 10.6.7 tells us that $(0,0)$ is asymptotically stable. The level curves of $E(x,y)$ are circles $x^2 + y^2 = k$ centered at the origin. The larger the value of k, the larger the radius of a circle. Fig. 10.65 shows a trajectory in the phase portrait (satisfying the initial condition $x(0) = 1, y(0) = 1$). Note that it crosses these circles from larger to smaller with increasing time, and, as the trajectory crosses one of the circles, the angle between the tangent vector to the trajectory and the "outward pointing" normal vector to the circle at the point of intersection is more than $90°$ (as shown at the point P).

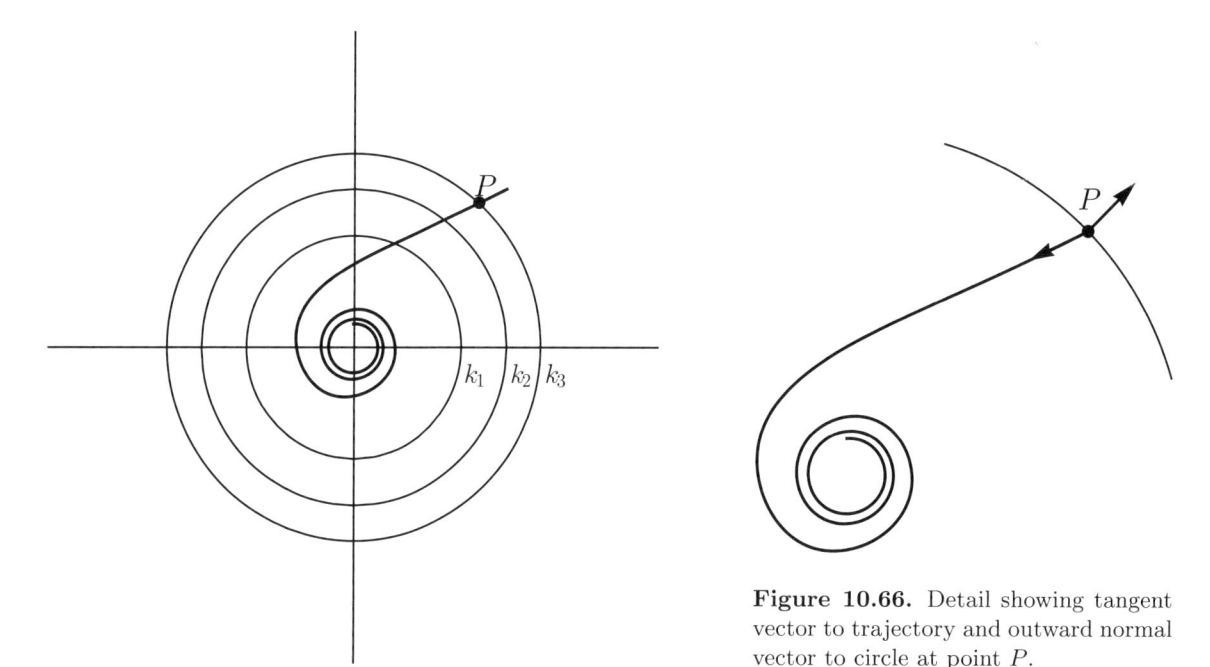

Figure 10.65. A trajectory for (10.87) and level curves $x^2 + y^2 = k_j$, $k_1 < k_2 < k_3$.

Figure 10.66. Detail showing tangent vector to trajectory and outward normal vector to circle at point P.

There is some terminology that goes along with the hypotheses in Theorem 10.6.7.

- If $E(x_e, y_e) = 0$ and $E(x, y) > 0$ on a neighborhood U of (x_e, y_e), except at (x_e, y_e), then E is said to be **positive definite** on U. Notice that these conditions say that the graph of the surface $z = E(x, y)$ over U will be roughly like a bowl turned right-side up, with its lowest point at $(x_e, y_e, 0)$; see Fig. 10.67

- If $\dot{E}(x, y) \leq 0$ on U, we say that $\dot{E}$ is **negative semidefinite**, and if we have strict inequality $\dot{E}(x, y) < 0$, for all $(x, y) \neq (x_e, y_e)$ in U, we say that $\dot{E}$ is **negative definite**. In the latter case, we see that the curve on the surface of Fig. 10.67 that lies directly above a trajectory in the phase portrait for (10.84) is moving down, towards $(x_e, y_e, 0)$.

In this language, a Lyapunov function for (10.84) at an equilibrium point (x_e, y_e) is a function $E(x, y)$ that is positive definite on a neighborhood U of (x_e, y_e) and for which the time derivative $\dot{E}$ is negative semidefinite. It is a strong Lyapunov function if $\dot{E}$ is negative definite in U.

10.6. Hamiltonians, gradient systems, and Lyapunov functions

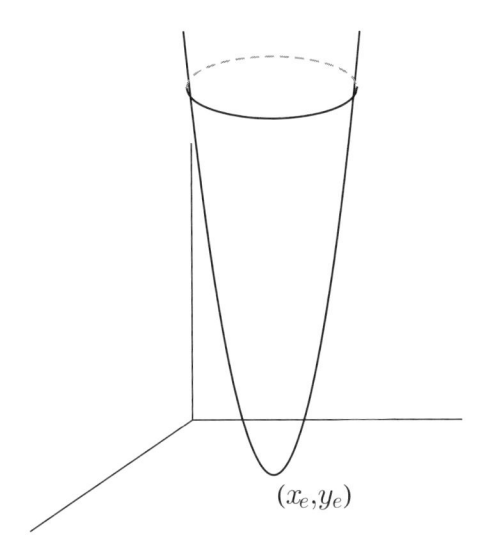

Figure 10.67. The surface $z = E(x, y)$.

To look for a Lyapunov function in the class $ax^2 + bxy + cy^2$ it is helpful to know, as a first step, when a quadratic form is positive definite.

Theorem 10.6.10. *The function $E(x, y) = ax^2 + bxy + cy^2$, having $E(0, 0) = 0$, is positive definite on $\mathbb{R}^2$ if and only if $a > 0$ and $b^2 < 4ac$.*

A proof is outlined in Exercise 15.

10.6.2. Gradient systems. Recall that the gradient of a differentiable function $F(x, y)$ is

$$\operatorname{grad} F(x, y) = \left(\frac{\partial F}{\partial x}, \frac{\partial F}{\partial y} \right).$$

A point (x_0, y_0) at which $\operatorname{grad} F(x_0, y_0) = (0, 0)$ is called a **critical point** of F, and other points at which $\operatorname{grad} F(x, y) \neq (0, 0)$ are called **regular points**. In $\mathbb{R}^2$, a **gradient system** has the form

$$(10.88) \qquad \frac{dx}{dt} = -\frac{\partial F}{\partial x}, \quad \frac{dy}{dt} = -\frac{\partial F}{\partial y}$$

for some twice continuously differentiable function $F(x, y)$. An equilibrium point for this system is a point (x_e, y_e) such that $\operatorname{grad} F(x_e, y_e) = (0, 0)$; in other words the equilibrium points are exactly the critical points of the $F(x, y)$.

If we draw the vector field for the gradient system (10.88), at every regular point of F we are drawing the vector

$$\left(-\frac{\partial F}{\partial x}, -\frac{\partial F}{\partial y} \right) = -\operatorname{grad} F(x, y).$$

Since $\operatorname{grad} F(x, y)$, and thus also $-\operatorname{grad} F(x, y)$, is perpendicular to the level curves of F, this tells us that:

> The trajectories of the gradient system (10.88) are perpendicular to the level curves of $F(x, y)$.

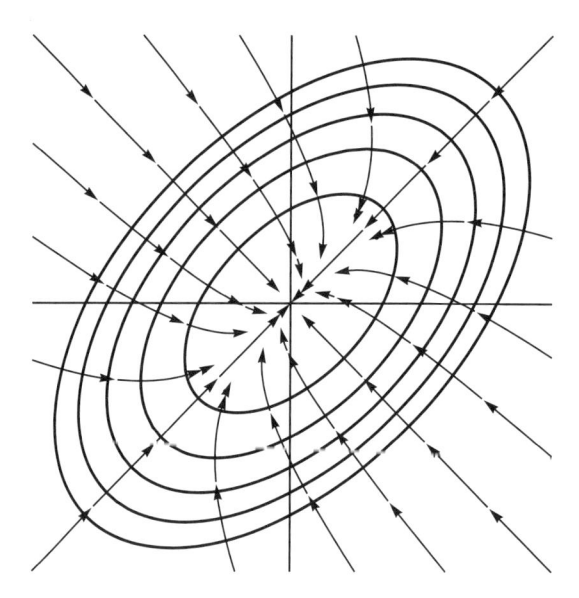

Figure 10.68. Trajectories of the gradient system for $F(x, y) = x^2 - xy + y^2$ are perpendicular to the level curves of F.

By comparison, the trajectories of a Hamiltonian system lie **in** the level curves of the Hamiltonian function.

For the gradient system (10.88), we have

$$\begin{aligned}
\dot{F}(x, y) &= \frac{\partial F}{\partial x} \frac{dx}{dt} + \frac{\partial F}{\partial y} \frac{dy}{dt} \\
&= \frac{\partial F}{\partial x} \left(-\frac{\partial F}{\partial x} \right) + \frac{\partial F}{\partial y} \left(-\frac{\partial F}{\partial y} \right) \\
&= -\left(\frac{\partial F}{\partial x} \right)^2 - \left(\frac{\partial F}{\partial y} \right)^2
\end{aligned}$$

which is 0 at the equilibrium points of (10.88) and strictly less than 0 elsewhere. This means that if (x_e, y_e) is an isolated equilibrium point for (10.88) at which F has a strict minimum, then $F(x, y) - F(x_e, y_e)$ is a strong Lyapunov function, and Theorem 10.6.7 tells us that (x_e, y_e) is asymptotically stable.

The Jacobian matrix for the linear approximation at (x_e, y_e) is

$$\begin{pmatrix} -\frac{\partial^2 F}{\partial x^2} & -\frac{\partial^2 F}{\partial y \partial x} \\ -\frac{\partial^2 F}{\partial x \partial y} & -\frac{\partial^2 F}{\partial y^2} \end{pmatrix}$$

where each derivative is evaluated at (x_e, y_e). Because the mixed second-order partial derivatives of F are equal, this matrix has the form

$$\begin{pmatrix} a & b \\ b & c \end{pmatrix},$$

and we say it is a **symmetric** matrix (the first column and first row are equal, and the second column and second row are equal). In Exercise 13 you are asked to show that eigenvalues of any 2×2 symmetric matrix with real entries are real numbers. Thus (x_e, y_e) could be a node or a saddle, but not a spiral point or a center.

10.6.3. Exercises.

1. Show that a linear system
$$x'(t) = ax + by, \quad y'(t) = cx + dy$$
is a Hamiltonian system if and only if $d = -a$. Find a Hamiltonian function if $d = -a$.

2. Write the second-order equation $x'' - x + 2x^3 = 0$ as a system of two first-order equations by means of the substitution $x' = y$. Is your system Hamiltonian? If so, describe a Hamiltonian function for it.

3. Write the second-order equation $x'' + f(x) = 0$ as a system of two first-order equations by means of the substitution $x' = y$. Is your system Hamiltonian? If so, describe a Hamiltonian function for it.

4. Show that any system of the form
$$\frac{dx}{dt} = f(y), \quad \frac{dy}{dt} = g(x),$$
where f and g are continuous, is a Hamiltonian system. What is a Hamiltonian function?

5. This problem continues the differential equation models of Romeo and Juliet, first introduced in Exercise 8 of Section 8.1. The functions $R(t)$ and $J(t)$ measure Romeo's feelings for Juliet and Juliet's feelings for Romeo, respectively. A positive value denotes love, and a negative value, dislike. Romeo's emotions are complicated. When Juliet is somewhat enthusiastic about Romeo, his feelings for her increase. But if Juliet gets too enamored of Romeo, he gets cold feet and his feelings for her start to wane. However, if Juliet dislikes Romeo, then Romeo's feelings for Juliet decrease sharply. To quantify this, suppose the rate of change of Romeo's feelings are given by

(10.89) $$\frac{dR}{dt} = J - J^2.$$

Notice that $\frac{dR}{dt} > 0$ when $0 < J < 1$ but $\frac{dR}{dt} < 0$ if $J > 1$ or $J < 0$. Juliet's feelings are more straightforward. The more Romeo loves her, the more her feelings for him grow, and the more he dislikes her, the more her feelings for him decrease. We quantify this by the differential equation

(10.90) $$\frac{dJ}{dt} = R.$$

(a) What are the equilibrium points of the system given by equations (10.89) and (10.90)?
(b) Show that this system is Hamiltonian and that a Hamiltonian function is
$$H(R, J) = \frac{J^2}{2} - \frac{J^3}{3} - \frac{R^2}{2}.$$

(c) Some level curves for the function $\frac{J^2}{2} - \frac{J^3}{3} - \frac{R^2}{2}$ are shown below. We know that the trajectories of the solutions to our system lie along level curves of the Hamiltonian function. The level curve shown in boldface, which passes through the point $R = 1, J = 0$ is indeed a (portion of) a trajectory in the phase portrait. Indicate the direction on this trajectory. If at time $t = 0$, $R = 1$ and $J = 0$, what is the fate of the relationship as $t \to \infty$? Mutual dislike? Romeo in love but Juliet not? Something else?

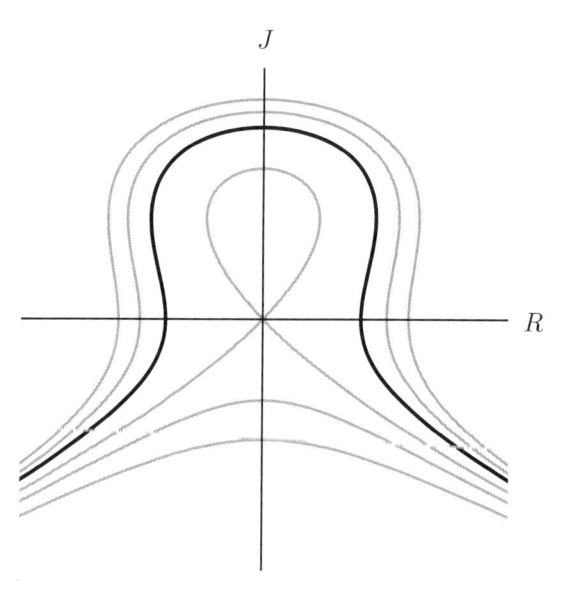

6. **Equilibrium points of a Hamiltonian system.** If
$$\frac{dx}{dt} = \frac{\partial H}{\partial y}, \quad \frac{dy}{dt} = -\frac{\partial H}{\partial x},$$
where $H(x,y)$ is twice continuously differentiable, describe the Jacobian matrix at an equilibrium point. What possibilities exist for the eigenvalues of the Jacobian? What does this mean about the possible types of equilibrium points for a Hamiltonian system? You may assume the eigenvalues are nonzero.

7. In this problem we compare two systems:
$$\frac{dx}{dt} = yx^3, \quad \frac{dy}{dt} = x + 2$$
and
$$\frac{dx}{dt} = y, \quad \frac{dy}{dt} = \frac{1}{x^2} + \frac{2}{x^3}.$$

(a) Show that the second system is Hamiltonian, but the first is not.
(b) Find a conserved quantity for the second system. Is it also conserved for the first? Hint: A function $E(x,y)$ is conserved for a planar differential equation system if $E(x,y)$ is constant along any solution $x = x(t), y = y(t)$ of the system. The Multivariable Chain Rule says
$$\frac{d}{dt}[E(x(t), y(t))] = \frac{\partial E}{\partial x}\frac{dx}{dt} + \frac{\partial E}{\partial y}\frac{dy}{dt}.$$
(c) How does the direction field for the first system compare to that for the second system? What does this imply about the phase portraits for the two systems?

8. Show that $E(x,y) = x^2 + \frac{1}{2}y^2$ is a Lyapunov function for
$$\frac{dx}{dt} = -x^3, \quad \frac{dy}{dt} = -2x^2 y - \frac{1}{2}y^3,$$
corresponding to its equilibrium point $(0,0)$. Determine the stability of this equilibrium point.

9. Find a Lyapunov function for
$$\frac{dx}{dt} = -y - x^3, \quad \frac{dy}{dt} = x - y^3,$$
corresponding to its equilibrium point $(0,0)$, and use it to determine the stability of $(0,0)$.

10. Find values of a and b so that $E(x,y) = ax^2 + by^4$ is a Lyapunov function for
$$\frac{dx}{dt} = -2y^3, \quad \frac{dy}{dt} = x - 3y^3,$$
corresponding to its equilibrium point $(0,0)$.

11. Find values of a and b so that $E(x,y) = ax^2 + by^2$ is a Lyapunov function for
$$\frac{dx}{dt} = y - xf(x,y), \quad \frac{dy}{dt} = -x - yf(x,y),$$
corresponding to its equilibrium point $(0,0)$, where $f(x,y)$ is continuously differentiable with $f(x,y) > 0$ in $\mathbb{R}^2$. Determine the stability of $(0,0)$.

12. Consider the system
$$\frac{dx}{dt} = y - f(x,y), \quad \frac{dy}{dt} = -x - g(x,y),$$
where we have the following information about f and g:
- f and g are continuously differentiable, with $f(0,0) = 0$, $g(0,0) = 0$,
- when $x > 0$, $f(x,y) > 0$ and when $x < 0$, $f(x,y) < 0$, and
- when $y > 0$, $g(x,y) > 0$ and when $y < 0$, $g(x,y) < 0$.

Show that $E(x,y) = x^2 + y^2$ is a Lyapunov function and that $(0,0)$ is asymptotically stable.

13. Show that a 2×2 symmetric matrix with real entries must have real eigenvalues.

14. (a) Show that
$$\frac{dx}{dt} = -2x + 2y, \quad \frac{dy}{dt} = 2x - 4y^3 - 2y$$
is a gradient system.
   (b) Determine the stability of the equilibrium point $(0,0)$ by finding a Lyapunov function.
   (c) Can you use Theorems 10.2.3 or 10.2.4 to determine the stability of $(0,0)$?

15. Follow the outline to show that $E(x,y) = ax^2 + bxy + cy^2$ is positive definite on $\mathbb{R}^2$ if and only if $a > 0$ and $b^2 < 4ac$.
   (a) Suppose $E(x,y) > 0$ for all $(x,y) \neq (0,0)$. Show that $a > 0$.
   (b) Show that if $E(x,y) > 0$ for all $(x,y) \neq (0,0)$, then $b^2 < 4ac$. Hint: If $y \neq 0$, write $ax^2 + bxy + cy^2 = y^2(av^2 + bv + c)$ where $v = x/y$. By (a) you know $a > 0$.
   (c) Conversely, assume $a > 0$ and $b^2 < 4ac$. Show that $E(x,y) > 0$ for all $(x,y) \neq (0,0)$. Hint: As in (b), write $ax^2 + bxy + cy^2 = y^2(av^2 + bv + c)$ where $v = x/y$.

16. Give statements, analogous to that in Theorem 10.6.10, that characterize when the quadratic form $E(x,y) = ax^2 + bxy + cy^2$ (which has $E(0,0) = 0$) is
   (i) positive semidefinite (meaning $E(x,y) \geq 0$ for all (x,y)),
   (ii) negative definite (meaning $E(x,y) < 0$ for all $(x,y \neq (0,0))$),
   (iii) negative semidefinite (meaning $E(x,y) \leq 0$ for all (x,y)).

17. Show that the function $E(x,y) = x^3 + x^2y + xy^2 + y^3$ is neither positive definite nor negative definite.

18. Consider the system
$$\frac{dx}{dt} = -x - 2y, \quad \frac{dy}{dt} = 2x.$$
   (a) Show that with the choice $E(x,y) = x^2 + y^2$, $\dot{E}$ is negative semidefinite, but not negative definite.
   (b) Show that with the choice $E(x,y) = 9x^2 + 4xy + 10y^2$, $E(x,y)$ is positive definite and $\dot{E}$ is negative definite.

(c) Let's see how the function $E(x,y)$ of (b) was found. Suppose we seek E in the form $E(x,y) = ax^2 + bxy + cy^2$. From Theorem 10.6.10 we know $E(x,y)$ is positive definite if $a > 0$ and $b^2 < 4ac$. A computation shows that
$$\dot{E} = (2ax + by)(-x - 2y) + (bx + 2cy)(2x).$$
Show that one way to ensure that $\dot{E}$ is negative definite is to choose a, b, and c so that $b > 0$, $b - a < 0$, and $4c - 4a - b = 0$. The choices $a = 9$, $b = 4$, and $c = 10$ meet all five of the desired conditions. Find a different choice of a, b, and c which also meets the desired conditions, and hence provides a different Lyapunov function.

19. Consider the damped mass-spring system
$$\frac{dy}{dt} = v, \quad \frac{dv}{dt} = -2v - y.$$
   (a) Show that $E(y,v) = y^2 + v^2$ is a Lyapunov function for this system, but not a strong Lyapunov function.
   (b) Find a strong Lyapunov function of the form $E(y,v) = ay^2 + byv + cv^2$. Hint: Use the idea of part (c) of Exercise 18.

20. (a) Show that the function $E(x,y) = x^2 + y^2$ is a Lyapunov function for the system
$$\frac{dx}{dt} = -xy - y^2, \quad \frac{dy}{dt} = -y + x^2$$
   on the open disk of radius 1 centered at $(0,0)$, but not on the whole plane $\mathbb{R}^2$.
   (b) Can you use Theorem 10.6.7 to classify the stability of the equilibrium point $(0,0)$?
   (c) Why can't Theorems 10.2.3 or 10.2.4 be applied to classify the stability of $(0,0)$?

21. Suppose $H(x,y) = 2x^2 + y^2$.
   (a) Sketch the level curves of $H(x,y) = c$ for $c = 1, 2, 4$, and 8.
   (b) In the same picture as (a), sketch some trajectories of the gradient system
$$\frac{dx}{dt} = -4x, \quad \frac{dy}{dt} = -2y.$$
   What geometric relationship should you see in your picture?

In Exercises 22–25 show that the given system is either Hamiltonian or a gradient system (or possibly both). Sketch a phase portrait for each, near the specified equilibrium point, showing the particular features the trajectories have by virtue of the system being a Hamiltonian or gradient system.

22.
$$\frac{dx}{dt} = -2y, \quad \frac{dy}{dt} = -2x, \text{ near}(0,0).$$

23.
$$\frac{dx}{dt} = \sin x, \quad \frac{dy}{dt} = -y \cos x, \text{ near}(0,0).$$

24.
$$\frac{dx}{dt} = -y \cos x, \quad \frac{dy}{dt} = -\sin x, \text{ near}(0,0).$$

25.
$$\frac{dx}{dt} = 2y - 4, \quad \frac{dy}{dt} = 2x - 2, \text{ near}(1,2).$$

26. Show that if

   (10.91)
$$\frac{dx}{dt} = f(x,y), \quad \frac{dy}{dt} = g(x,y)$$

10.7. Pendulums

is a Hamiltonian system, then

(10.92) $$\frac{dx}{dt} = g(x,y), \quad \frac{dy}{dt} = -f(x,y)$$

is a gradient system. Show that both systems have the same equilibrium points. How are the trajectories of (10.91) related to the trajectories of (10.92)?

27. The purpose of this problem is to prove the "stable equilibrium" conclusion of Theorem 10.6.7. To simplify notation, we assume the isolated equilibrium point is $(0,0)$. By the hypothesis we have $E(0,0) = 0$, $E(x,y) > 0$ for all $(x,y) \neq (0,0)$ in a neighborhood U of $(0,0)$ and $\dot{E}(x,y) \leq 0$ for all (x,y) in U. Our goal is to show that $(0,0)$ is a stable equilibrium point.

    Pick any positive number r such that the closed disk of radius r and centered at $(0,0)$ is contained in U. Set $f(t) = E(r\cos t, r\sin t)$. As t ranges over $[0, 2\pi]$, $f(t)$ gives the values of E at points on the circle of radius r centered at $(0,0)$.

    (a) By citing the appropriate Calculus I result, explain why f attains a minimum value m at some point t_0 in the interval $[0, 2\pi]$, so that $m = f(t_0) \leq f(t)$ for all $0 \leq t \leq 2\pi$. Explain why $m > 0$.
    (b) Explain why we can find a positive number δ so that $E(x,y) < m$ on the closed disk D of radius δ centered at $(0,0)$. Show that $\delta < r$.
    (c) Suppose that $x = x(t), y = y(t)$ is a trajectory of the system with $(x(0), y(0))$ in the disk D from (b). Explain why, for $t \geq 0$, this trajectory cannot intersect with the circle of radius r centered at $(0,0)$.
    (d) By appealing to the definition of "stable equilibrium" explain why $(0,0)$ must be a stable equilibrium.

10.7. Pendulums

A pendulum consists of a rigid rod (of negligible weight) with length L, fixed to a pivot point at one end and with mass m (called a bob) at the other end. The rod is free to move in a plane, so that the mass moves along a circular path. We describe the motion by giving the angle θ as shown in Fig. 10.69, so that θ is 0 when the pendulum is in its rest position. The force due to gravity of the bob is mg, where g is the acceleration due to gravity. We think of this as a vector pointing down, and we resolve this vector into two components as shown in Fig. 10.70, with one component tangent to the circle on which the bob moves.

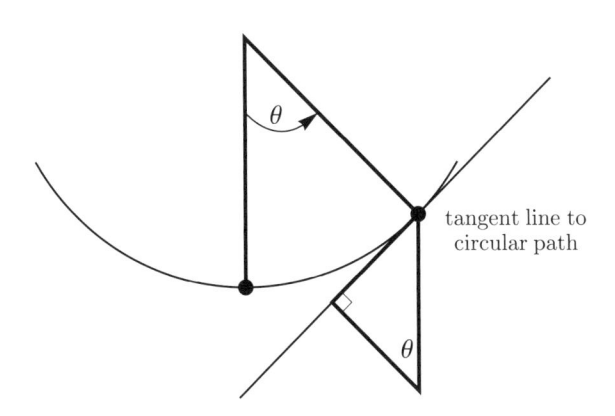

Figure 10.69. Pendulum.

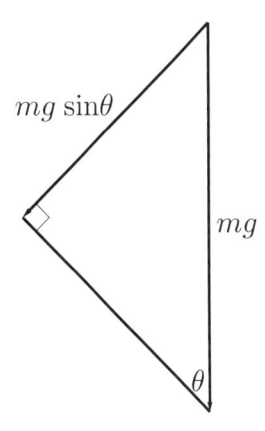

Figure 10.70. Resolving the force vector into components.

Only the tangential force affects the motion of the pendulum. We also allow for a damping force and make the familiar assumption that it is proportional to the velocity and acts opposite to the direction of motion. When the bob moves through an angle θ, it travels a distance $L\theta$ and has velocity $L\frac{d\theta}{dt}$ and acceleration $L\frac{d^2\theta}{dt^2}$. From Newton's second law we have

$$mL\frac{d^2\theta}{dt^2} = -cL\frac{d\theta}{dt} - mg\sin\theta$$

for some positive constant c. The first term on the right-hand side represents damping, and the minus sign reflects our assumption that the damping force acts opposite to the direction of motion. The minus sign with the second term indicates that the tangential component of the gravitational force always pushes θ towards zero when $-\pi < \theta < \pi$. Dividing by mL gives

$$\frac{d^2\theta}{dt^2} + \frac{c}{m}\frac{d\theta}{dt} + \frac{g}{L}\sin\theta = 0.$$

Convert this to an autonomous system by the substitution $v = \frac{d\theta}{dt}$:

(10.93) $$\frac{d\theta}{dt} = v, \quad \frac{dv}{dt} = -\frac{c}{m}v - \frac{g}{L}\sin\theta.$$

The undamped pendulum. To ignore damping, we set $c = 0$ and obtain the equations

(10.94) $$\frac{d\theta}{dt} = v, \quad \frac{dv}{dt} = -\frac{g}{L}\sin\theta.$$

This system has equilibrium points $(n\pi, 0)$ for any integer n. For even integers $n = 0, \pm 2, \pm 4, \ldots$ these correspond to the pendulum hanging straight down. Odd values of n correspond to the pendulum being in the straight up position. You can probably guess at the stability of these equilibrium points based on your physical intuition.

At the equilibrium point $(n\pi, 0)$ the linear approximation to this system is $\mathbf{X}' = \mathbf{J}\mathbf{X}$ for

$$\mathbf{X} = \begin{pmatrix} \theta \\ v \end{pmatrix} \quad \text{and} \quad \mathbf{J} = \begin{pmatrix} 0 & 1 \\ -\frac{g}{L}\cos(n\pi) & 0 \end{pmatrix}.$$

For odd values of n, the eigenvalues of $\mathbf{J}$ are $\pm\sqrt{g/L}$ and thus the equilibrium points $(n\pi, 0)$ for n odd are saddles. When n is an even integer, $\mathbf{J}$ has eigenvalues $\pm i\sqrt{g/L}$ and the linear approximation has a center. Since this is a borderline case, linear approximation by itself doesn't give a classification of the equilibrium points $(n\pi, 0)$, n even, or a determination of their stability, for the nonlinear pendulum. We proceed further by recognizing that (10.94) is a Hamiltonian system with Hamiltonian function

$$H(\theta, v) = \frac{1}{2}v^2 + \frac{g}{L}(1 - \cos\theta).$$

In fact, the function

$$E(\theta, v) = mL^2 H(\theta, v) = \frac{1}{2}mL^2 v^2 + mgL(1 - \cos\theta)$$

is the total energy function: The term $\frac{1}{2}mL^2v^2$ is the bob's kinetic energy and the second term $mgh = mgL(1 - \cos\theta)$ is its potential energy due to the gravitational force (see Exercise 10; we assume the bob has zero potential energy when $\theta = 0$). The orbits lie in the level curves of $H(\theta, v)$, or what is the same thing, in the level curves of the energy function $E(\theta, v)$.

For a sufficiently small value of K, the level curve

(10.95) $$\frac{1}{2}v^2 + \frac{g}{L}(1 - \cos\theta) = K$$

10.7. Pendulums

is a collection of closed curves encircling the equilibrium points $(n\pi, 0)$ for n even, as illustrated in Fig. 10.71; see Exercise 5. The corresponding motion of the pendulum is the familiar back and forth oscillation. When θ is in the range $-\pi < \theta < \pi$, the value of θ oscillates back and forth between a maximum $M < \pi$ and a minimum $-M$.

The nullcline-and-arrow diagram for (10.94), with θ-nullcline $v = 0$ and v-nullclines $\theta = n\pi$ for n any integer, is shown in Fig. 10.72. This gives the clockwise direction on the closed loop orbits in Fig. 10.71.

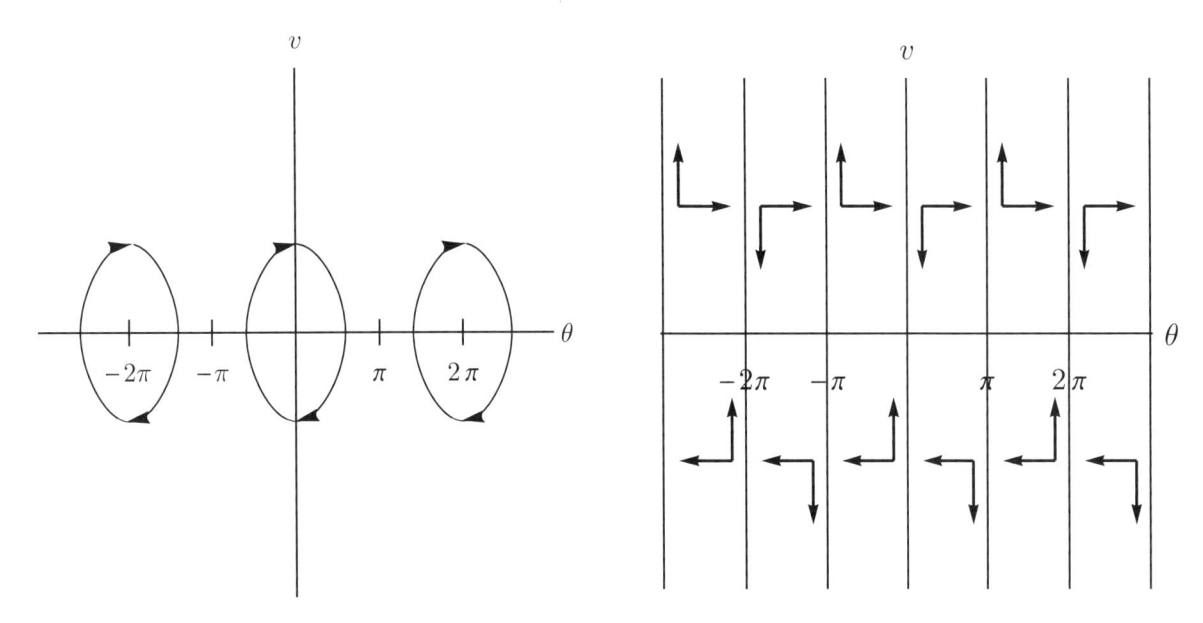

Figure 10.71. $\frac{1}{2}v^2 + \frac{g}{L}(1 - \cos\theta) = K$, $K < \frac{2g}{L}$.

Figure 10.72. Nullcline-and-arrow diagram for undamped pendulum.

For large values of K, the level curves look like those shown in Fig. 10.73. Again, the nullcline-and-arrow diagram in Fig. 10.72 gives the direction on these "high energy" orbits, which correspond to the pendulum whirling around and around in complete counterclockwise circles (top) or clockwise circles (bottom).

The value of K that separates the two very different types of motion shown in Figs. 10.71 and 10.73 is $K = \frac{2g}{L}$. With this choice of K we have the level curve

$$\frac{1}{2}v^2 = \frac{g}{L}(1 + \cos\theta),$$

or

$$v = \pm\sqrt{\frac{2g}{L}(1 + \cos\theta)}.$$

Included in this level curve are the unstable equilibrium points $(n\pi, 0)$ (n an odd integer) and the separatrices which join these saddle points; see Fig. 10.74 and Exercise 3. This corresponds to motion we would never observe in real life—for example, the pendulum swinging up towards its straight up position, never actually reaching that position but approaching it as $t \to \infty$.

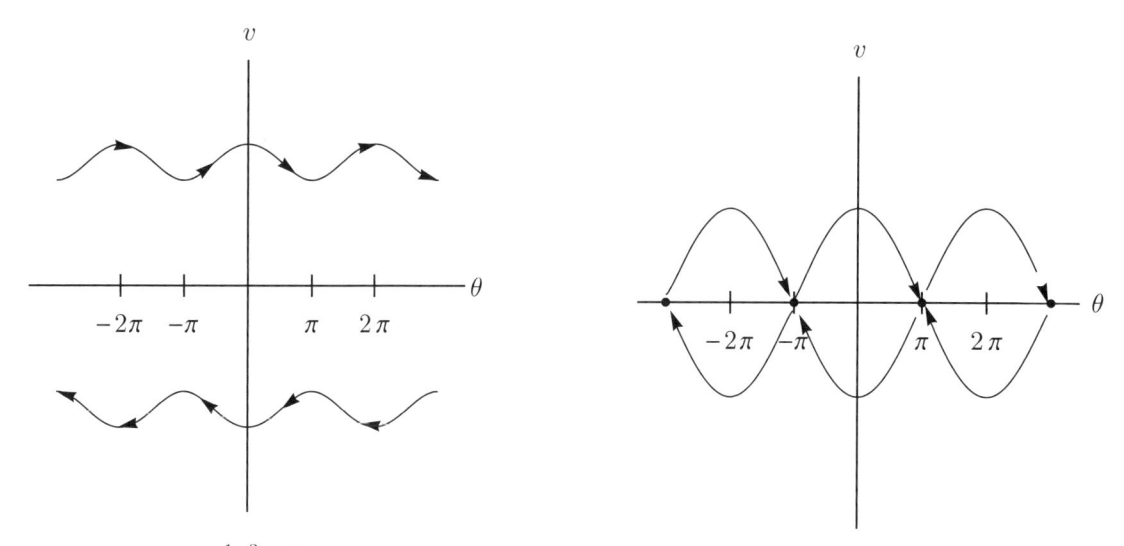

Figure 10.73. $\frac{1}{2}v^2 + \frac{g}{L}(1-\cos\theta) = K$, $K > \frac{2g}{L}$.

Figure 10.74. $\frac{1}{2}v^2 = \frac{g}{L}(1+\cos\theta)$.

Putting together Figs. 10.71, 10.73, and 10.74 we get a sketch of the phase portrait in Fig. 10.75.

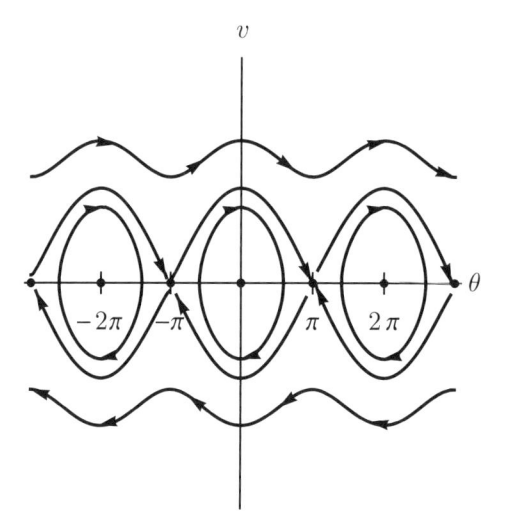

Figure 10.75. Phase portrait for undamped pendulum.

The damped pendulum. Suppose that $c \neq 0$ in the system (10.93). The equilibrium points are still $(n\pi, 0)$ for any integer n. Exercise 9 asks you to determine the linear approximation at each of these points and show that for odd values of n, the equilibrium point is a saddle, while for even values of n, the point $(n\pi, 0)$ is a stable spiral for small positive values of c but becomes a stable node for larger values of c. We'll focus here on the small values of $c > 0$, assembling information we get from various perspectives about the phase portrait and pendulum motion.

When $c \neq 0$ the energy function

$$E(\theta, v) = \frac{1}{2}mL^2v^2 + mgL(1-\cos\theta)$$

is no longer a conserved quantity, and in fact

$$\dot{E}(\theta, v) = -cL^2v^2$$

10.7. Pendulums

(see Exercise 1). Moreover, $E(\theta, v) \geq 0$ and $E(\theta, v) = 0$ if and only if $v = 0$ and $\theta = n\pi$ for n an even integer. Thus the function $E(\theta, v)$ is a Lyapunov function for (10.93) in a neighborhood of each of the equilibrium points $(n\pi, 0)$ for n even. Since $\dot{E}$ is negative semidefinite, but not negative definite ($\dot{E} = 0$ whenever $v = 0$), Theorem 10.6.7 only lets us conclude that we have stability at each equilibrium point $(n\pi, 0)$, n even. Moreover, the orbits at points with $v \neq 0$ must cross the level curves $E = k$ from larger to smaller values. So we can still use the level curves of E as guidelines for sketching the phase portrait, but instead of the orbits staying in a level curve, they will cross over the level curves from higher energy to lower energy. You should be able to see this in Fig. 10.77.

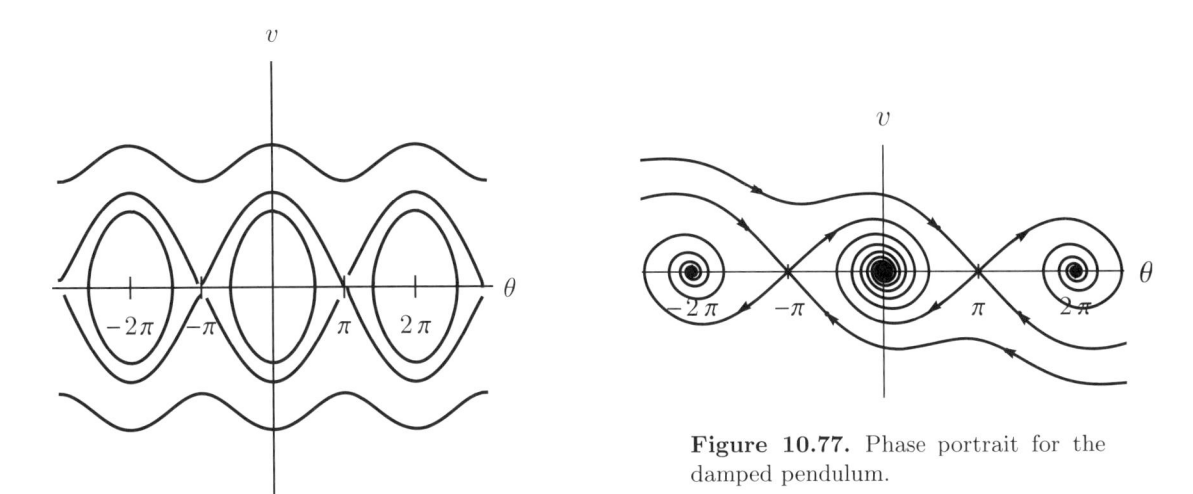

Figure 10.76. Level curves of $E(\theta, v)$.

Figure 10.77. Phase portrait for the damped pendulum.

The nullclines of (10.93) are the horizontal line $v = 0$ (the θ-nullcline) and the (scaled) sine curve $v = -\frac{gm}{cL} \sin \theta$ (the v-nullcline). These are sketched in Fig. 10.78, along with the direction-indicator arrows. A sketch of the phase portrait, when c is small enough that we have spiral points at the stable equilibria, is shown in Fig. 10.77.

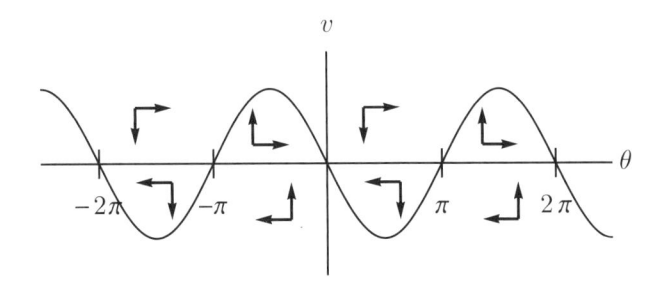

Figure 10.78. Nullcline-and-arrow diagram for damped pendulum.

Basin of attraction. We revisit the concept of "basin of attraction" (introduced at the end of Section 10.2) in the context of the damped pendulum. Recall, for example, that the point (a, b) is in the basin of attraction of the asymptotically stable equilibrium point $(0, 0)$ if the solution $(x(t), y(t))$ of (10.93) which satisfies $x(0) = a$, $y(0) = b$ will have $\lim_{t \to \infty} x(t) = 0$ and $\lim_{t \to \infty} y(t) = 0$. Each of the asymptotically stable equilibrium points $(n\pi, 0)$, n even, for (10.93) has its own basin of attraction and the separatrices that enter each unstable equilibria ($(n\pi, 0)$ for n odd) serve to separate the basins of attraction of the stable equilibria. In Fig. 10.79, the basin of attraction for $(0,0)$ is shown shaded.

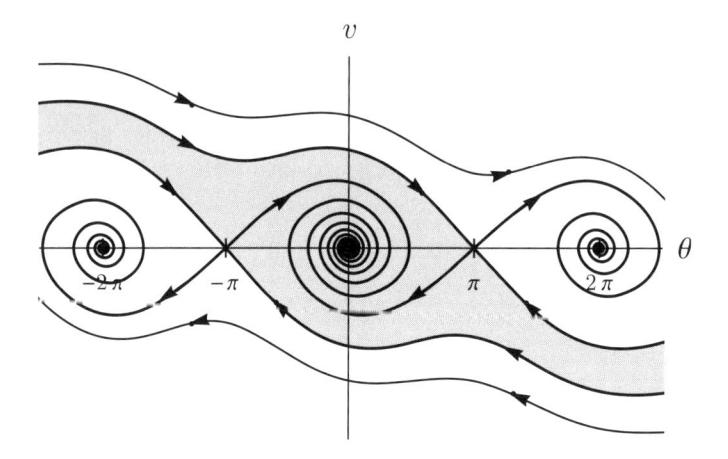

Figure 10.79. Basin of attraction for $(0,0)$ in the damped pendulum model.

10.7.1. Exercises.

1. Compute $\dot{E}(\theta, v)$ for the system (10.93) if $E(\theta, v) = \frac{1}{2}mL^2v^2 + mgL(1 - \cos\theta)$.

2. The figure below shows a particular orbit for a damped pendulum. The point P on this orbit corresponds to $t = 0$, so that at time $t = 0$ the pendulum is in the straight down position and moving counterclockwise. Which of the following describes the motion for $t > 0$?
   (a) The pendulum never reaches the straight up position but just oscillates back and forth, with decreasing amplitude and tending to the straight down position as t goes to ∞.
   (b) The pendulum makes one complete revolution, before settling down to back and forth oscillations with decreasing amplitude and tending to the straight down position as t goes to ∞.
   (c) The pendulum makes two complete revolutions, before settling down to back and forth oscillations with decreasing amplitude and tending to the straight down position as t goes to ∞.
   (d) None of the above are correct.

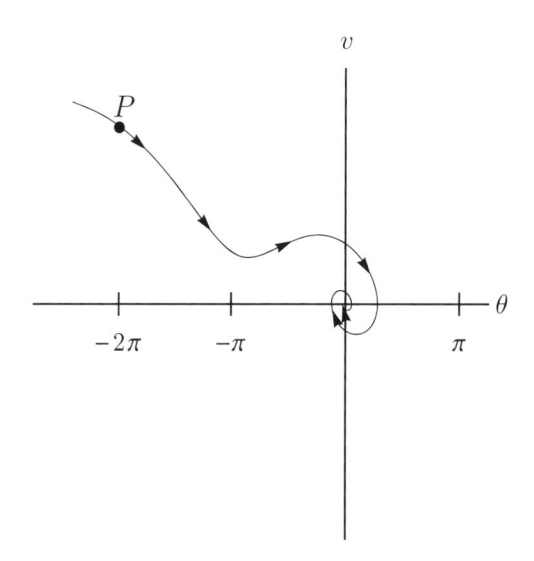

3. Working with Fig. 10.74, identify the stable curve S and unstable curve U corresponding to the saddle point $(\pi, 0)$.

4. The level curve $\frac{1}{2}v^2 + (1 - \cos\theta) = 4$ for the undamped pendulum

$$\frac{d\theta}{dt} = v, \quad \frac{dv}{dt} = -\sin\theta$$

is shown below. Notice that if the point P corresponds to $t = 0$, the motion of the pendulum has the following description: At time 0 it is in the straight down position and moving counterclockwise. It continues to revolve counterclockwise around the pivot over and over. Similarly, give a description of the pendulum if the point Q corresponds to the time $t = 0$.

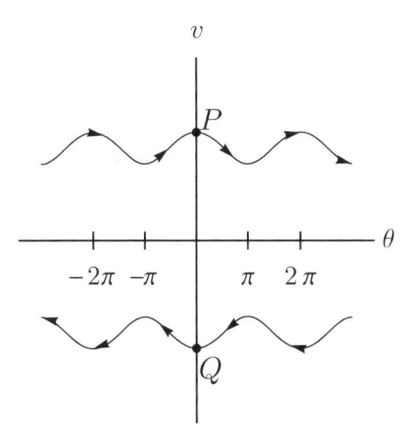

5. Show that the level curve

$$\frac{1}{2}v^2 + (1 - \cos\theta) = \frac{1}{2}$$

(a special case of equation (10.95), with $L = g$ and $K = \frac{1}{2}$) consists of closed curves, each encircling one point $(n\pi, 0)$ with n even. For the orbit encircling $(0, 0)$, how high does the pendulum go? What are the maximum and minimum values of θ along this orbit?

6. Why isn't $E(\theta, v) = \frac{1}{2}mL^2v^2 + mgL(1 - \cos\theta)$ a Lyapunov function for the damped pendulum near an equilibrium point $(n\pi, 0)$, n odd?

7. The figure below shows four orbits in the phase portrait of the undamped pendulum (equation (10.94)). Match these four orbits with the corresponding graph of θ vs. t for $t \geq 0$, assuming $\theta = 0$ at $t = 0$. The pendulum equations are

$$\frac{d\theta}{dt} = v, \quad \frac{dv}{dt} = -\sin\theta$$

and the four orbits lie, respectively, in the level curves

$$\frac{1}{2}v^2 + (1 - \cos\theta) = k$$

for $k = 4$ (with $v > 0$), $k = 2$, $k = 1.5$, and $k = 4$ (with $v < 0$).

Figure 10.80. (A).

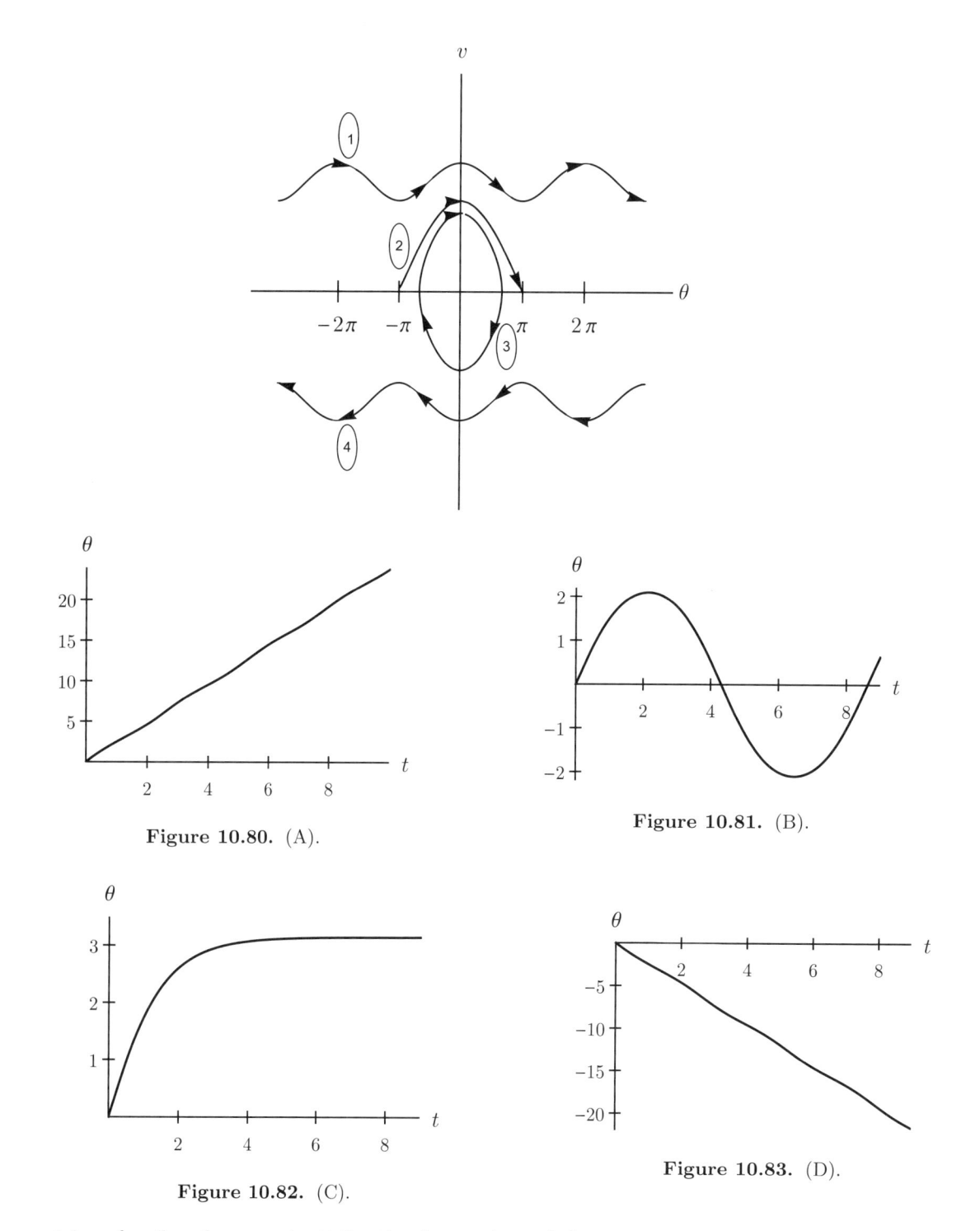

Figure 10.81. (B).

Figure 10.82. (C).

Figure 10.83. (D).

8. Separatrices for the phase portrait for the damped pendulum system

$$\frac{d\theta}{dt} = v, \quad \frac{dv}{dt} = -0.2v - \sin\theta$$

are shown in the figure below.

10.7. Pendulums

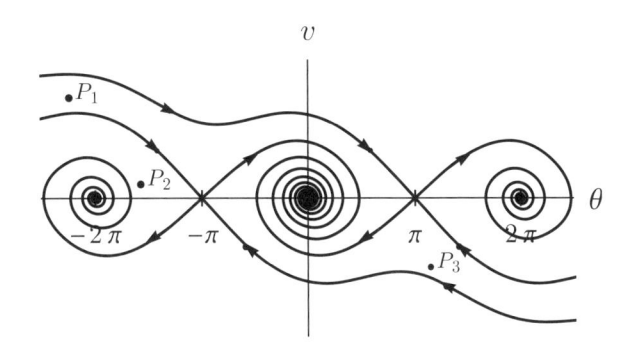

(a) If the point P_1 corresponds to time $t = 0$, how many complete revolutions will the pendulum make (for $t > 0$) before settling into a back and forth oscillation?

(b) The point labeled P_1 lies in the basin of attraction for which equilibrium point? Answer the same question for the points P_2 and P_3.

9. Using linear approximation, show that the equilibrium points $(n\pi, 0)$ with n odd are saddle points of the damped pendulum described by equation (10.93) with $c \neq 0$. For n even, show that $(n\pi, 0)$ could be a spiral point or a node, but in either case it is stable. What condition on c, m, and L distinguishes nodes from spiral points?

10. The undamped pendulum equation

$$\text{(10.96)} \qquad \frac{d^2\theta}{dt^2} + \frac{g}{L}\sin\theta = 0$$

can also be derived by starting with the conservation of energy principle

$$\text{kinetic energy} + \text{potential energy} = C$$

applied to the pendulum's bob. The kinetic energy is $\frac{1}{2}mV^2$ where m is the mass and V is the velocity. The potential energy is the product mgh where h is the height above the reference position which we take to be the rest position.

(a) Show that the kinetic energy is given by

$$\frac{1}{2}mL^2\left(\frac{d\theta}{dt}\right)^2,$$

where θ is as in Fig. 10.69 and L is the length.

(b) Show that the potential energy is given by $mgL(1 - \cos\theta)$.

(c) By differentiating the equation

$$\text{kinetic energy} + \text{potential energy} = C$$

with respect to t, obtain equation (10.96).

11. The motion of an undamped pendulum is described by the equation $\theta'' + \frac{g}{L}\sin\theta = 0$. Any equation of the form $\theta'' + f(\theta) = 0$ where f is a continuously differentiable function satisfying

$$f(0) = 0, \quad f > 0 \text{ on some interval } (0, \alpha), \quad \text{and} \quad f < 0 \text{ on } (-\alpha, 0)$$

is called a generalized pendulum equation. The substitution $v = d\theta/dt$ converts this to the system

$$\text{(10.97)} \qquad \frac{d\theta}{dt} = v, \quad \frac{dv}{dt} = -f(\theta).$$

(a) Show that if

$$F(\theta) = \int_0^\theta f(u)\,du,$$

then the energy function
$$E(\theta, v) = \frac{1}{2}v^2 + F(\theta)$$
is a conserved quantity for the system, and hence the orbits lie in the level curves of $E(\theta, v)$.

(b) Show that the equilibrium points of the system are the points $(\theta^*, 0)$ where θ^* is a critical point of F.

(c) Show that if θ^* is a critical point of F and $f'(\theta^*) < 0$, then $(\theta^*, 0)$ is a saddle point of (10.97).

(d) Show that the hypothesis on $f(\theta)$ says that $F(0) = 0$, $F(\theta)$ is strictly increasing for $0 < \theta < \alpha$ and $F(\theta)$ is strictly decreasing for $-\alpha < \theta < 0$. Thus F has a strict relative minimum at $\theta = 0$ and $f'(0) \geq 0$.

(e) Show that if $f'(0) > 0$, the linear approximation near $(0,0)$ to the system in (10.97) has a center there (a borderline case).

(f) Use the following argument to show that the nonlinear system has a center at $(0,0)$: Show that if k is a sufficiently small positive value, then we can find two values a, b with $a < 0 < b$, $F(a) = F(b) = k$, and $F(\theta) < k$ on (a, b). Explain why for each value $\tilde{\theta}$ between a and b the level curve $E(v, \theta) = k$ will contain exactly two points $(\tilde{\theta}, \tilde{v})$ and $(\tilde{\theta}, -\tilde{v})$, and as $\tilde{\theta}$ approaches either a or b, $\tilde{v}$ will approach 0. Thus the orbits of (10.97) near $(0,0)$ are closed curves that encircle the origin, so that $(0,0)$ is indeed a center.

10.8. Cycles and limit cycles

A periodic solution of an autonomous system

(10.98) $$\frac{dx}{dt} = f(x, y), \quad \frac{dy}{dt} = g(x, y)$$

is a solution $x = x(t), y = y(t)$ with the property that $x(t+T) = x(t)$ and $y(t+T) = y(t)$ for some $T > 0$ and all t. We also require that $x(t)$ and $y(t)$ be nonconstant functions. A periodic solution has an orbit which is a closed curve, traversed over and over as t varies. We'll call such an orbit a **cycle**. An important part of the global analysis of (10.98) is to determine if the system has any cycles. Throughout this section we assume that the functions $f(x, y)$ and $g(x, y)$ are defined and continuously differentiable at every point in the xy-plane.

We've seen numerous examples of systems with cycles. In any planar linear system $\mathbf{X}' = \mathbf{AX}$ where the eigenvalues of $\mathbf{A}$ are a pair of purely imaginary conjugates $\pm i\beta$, $\beta \neq 0$, all nonconstant orbits are cycles (in fact, they are ellipses, by Exercise 13 of Section 8.9). The Lotka-Volterra predator-prey system of Section 10.4 has orbits in the first quadrant that are cycles enclosing the equilibrium point $(\frac{d}{c}, \frac{a}{b})$.

A more exotic example of a cycle was seen in Example 10.3.3 of Section 10.3. We saw there that the system

$$\frac{dx}{dt} = -y + x(1 - x^2 - y^2), \quad \frac{dy}{dt} = x + y(1 - x^2 - y^2)$$

has the periodic solution $x(t) = \cos t$, $y(t) = \sin t$ whose orbit is the unit circle $\mathbb{T}$. Moreover we saw that orbits that start inside the unit circle, as well as orbits that start outside $\mathbb{T}$, spiral towards $\mathbb{T}$ as $t \to \infty$. Recall that we called $\mathbb{T}$ a **limit cycle**, to capture the fact that it is approached by other nearby trajectories as $t \to \infty$. In fact, it is approached by *every* nearby trajectory as $t \to \infty$, and we will call it an **attracting limit cycle**. Limit cycles can also be **repelling**; this means that nearby trajectories approach it as $t \to -\infty$. The cycles in the Lotka-Volterra system are not limit cycles, since nearby trajectories are cycles themselves and so remain at a positive distance from the given cycle.

10.8. Cycles and limit cycles

Cycles and equilibrium points. Cycles must always enclose one or more equilibrium points. This fact can sometimes be used to rule out the presence of cycles. For example, the competing species model

$$\frac{dx}{dt} = x(2 - x - y), \quad \frac{dy}{dt} = y(1 - x - 4y)$$

cannot have any cycles in the biologically relevant first quadrant $x > 0$, $y > 0$, since its equilibrium points are $(0,0)$, $(0, \frac{1}{4})$, $(2,0)$, and $(\frac{7}{3}, -\frac{1}{3})$, none living in the first quadrant. The first quadrant is an example of a **simply connected region**, which roughly means a region with no holes. Other examples of simply connected regions include any rectangular region or any disk. A region that is not simply connected is the ring-shaped region between two concentric circles. A region R is simply connected if every **simple closed curve** in R encloses only points of R. A **simple closed curve** begins and ends at the same point but otherwise does not cross itself. Figs. 10.84–10.89 illustrate these ideas.

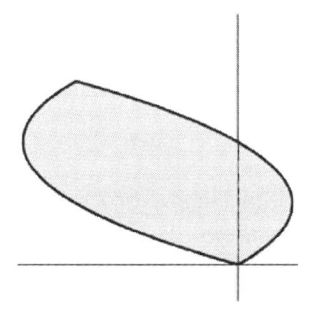

Figure 10.84. A simply connected region.

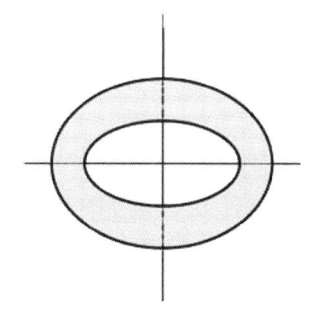

Figure 10.85. A region that is not simply connected.

Figure 10.86. Another simply connected region—the first quadrant.

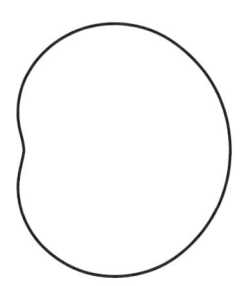

Figure 10.87. A simple closed curve.

Figure 10.88. A curve that is neither simple nor closed.

Figure 10.89. A closed curve that is not simple.

Cycles or no cycles? In a simply connected region, the absence of equilibrium points means no cycles:

> If a region R is simply connected and contains no equilibrium point of the system (10.98), then there is no cycle in R.

On the other hand, there are tools that sometimes allow us to know that a cycle exists in a certain region. The most important of these is the **Poincaré-Bendixson** Theorem, which we will describe next.

Our version of the Poincaré-Bendixson Theorem deals with the annular, or ring-shaped, region between two simple closed curves, one lying inside the other, as shown in Fig. 10.90.

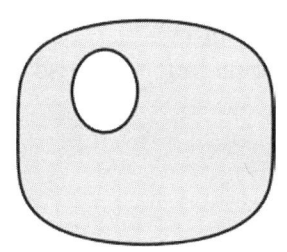

Figure 10.90. Annular region between two simple closed curves.

Theorem 10.8.1 (Poincaré-Bendixson). *Let R be the region consisting of two simple closed curves C_o and C_i, where C_i is inside C_o, and the annular region between them. Suppose that R contains no equilibrium points of the system (10.98), where f and g are continuously differentiable in $\mathbb{R}^2$, and suppose we have a trajectory $(x(t), y(t))$ which stays in R for all $t \geq 0$. Then either this trajectory is itself a cycle in R or it spirals towards some limit cycle in R as $t \to \infty$.*

Notice that this result tells you, in particular, that when the hypotheses are satisfied, *the region R must contain a cycle*.

The work in applying Theorem 10.8.1 is identifying curves C_o and C_i, and thus the region R, so that the hypotheses are satisfied. Roughly speaking, you want to look for a region such that the vector field for (10.98) points *into* the region at every point of the boundary curves C_o and C_i. This will guarantee that a trajectory that starts in R will stay in R for all later times.

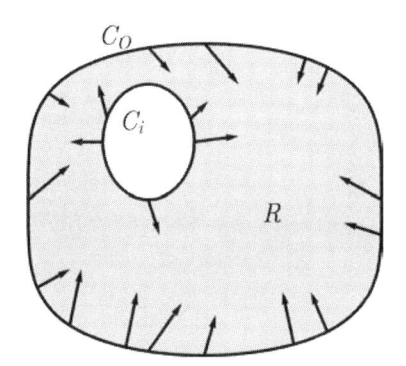

Figure 10.91. Vector field points into R at each point of C_o and C_i.

We'll illustrate Theorem 10.8.1 with some examples. The first one is a "warm-up".

10.8. Cycles and limit cycles

Example 10.8.2. Consider again the system

$$\frac{dx}{dt} = -y + x(1 - x^2 - y^2), \quad \frac{dy}{dt} = x + y(1 - x^2 - y^2)$$

of Example 10.3.3 in Section 10.3. A portion of its vector field, with the vectors scaled to all have the same length, is shown in Fig. 10.92. We aim to use the Poincaré-Bendixson Theorem to show this system has a cycle, which must enclose the only equilibrium point, located at $(0, 0)$. We'll use an idea suggested by our work on Lyapunov functions in the last section. If $E(x, y) = x^2 + y^2$, then the derivative of E following the motion of the trajectories is

$$\begin{aligned}\dot{E}(x,y) &= \frac{\partial E}{\partial x}\frac{dx}{dt} + \frac{\partial E}{\partial y}\frac{dy}{dt} \\ &= 2x(-y + x(1-x^2-y^2)) + 2y(x + y(1-x^2-y^2)) \\ &= 2(x^2+y^2)(1-(x^2+y^2)).\end{aligned}$$

The level curves of E are circles centered at the origin. Since $\dot{E}$ is negative at points of the circle $x^2 + y^2 = r$ for any $r > 1$, this tells us that trajectories crossing circles with radius greater than one do so moving in the direction of circles with smaller radii; i.e., they cross the circle $x^2 + y^2 = r, r > 1$, pointing inwards. Similarly, since $\dot{E}$ is positive at points of the circle $x^2 + y^2 = r, r < 1$, as orbits cross this circle they are moving outwards, towards circles with larger radii. Thus if we choose C_i to be the circle $x^2 + y^2 = r_1$ where $r_1 < 1$, C_o to be the circle $x^2 + y^2 = r_2$ where $r_2 > 1$, and R to be the region between two circles C_i and C_o, together with the circles themselves, then an orbit that starts in R must stay in R. By the Poincaré-Bendixson Theorem, R must contain a cycle. This is true for any choice $r_1 < 1 < r_2$, which leaves us with the only possible conclusion that the unit circle is a cycle! We can verify this directly, by checking that $x(t) = \cos t, y(t) = \sin t$ is a solution; its orbit is included in Fig. 10.93.

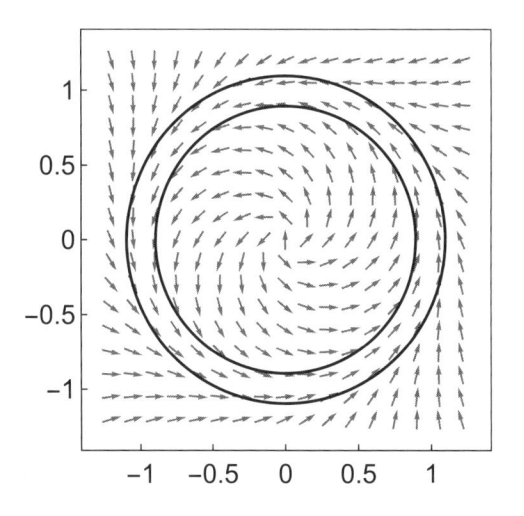

Figure 10.92. Vector field and circles $x^2 + y^2 = r_1$ and $x^2 + y^2 = r_2$ for $r_1 < 1 < r_2$.

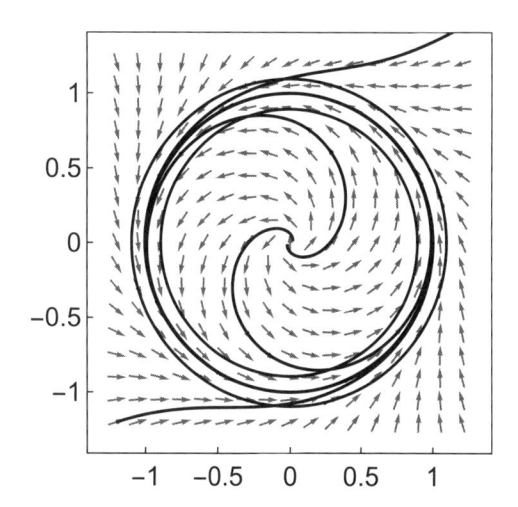

Figure 10.93. Some trajectories for Example 10.8.2.

Example 10.8.3. Consider the system

$$\frac{dx}{dt} = y, \quad \frac{dy}{dt} = -x - y\ln\left(9x^2 + 4y^2 + \frac{1}{2}\right).$$

We will show how to apply the Poincaré-Bendixson Theorem in the region R consisting of the circles $x^2 + y^2 = 1$ and $x^2 + y^2 = \frac{1}{25}$ and the annular ring between them (see Fig. 10.94), to conclude this region contains a cycle. The only equilibrium point of our system is $(0,0)$, and this lies outside of R.

Conversion to polar coordinates will make our calculations simpler. Note that in polar coordinates the boundary circles for R have equations $r = 1$ and $r = \frac{1}{5}$. Since $r^2 = x^2 + y^2$ and $x = r\cos\theta$, $y = r\sin\theta$, we have (as in equation (10.42)) $\frac{dr}{dt} = \frac{1}{r}(x\frac{dx}{dt} + y\frac{dy}{dt})$ and using the equations for the system,

$$\frac{dr}{dt} = \frac{xy + y(-x - y\ln\left(9x^2 + 4y^2 + \frac{1}{2}\right))}{r} = \frac{-y^2\ln(9x^2 + 4y^2 + \frac{1}{2})}{r}$$

$$= -r\sin^2\theta \ln\left(4r^2 + 5r^2\cos^2\theta + \frac{1}{2}\right).$$

On the circle $r = 1$ we have

$$\frac{dr}{dt} = -\sin^2\theta \ln\left(4 + 5\cos^2\theta + \frac{1}{2}\right) \leq 0$$

for all θ, and r is nonincreasing. This tells us that no trajectory can leave the region R by crossing over the outer boundary $r = 1$ and moving farther away from the origin.

On the circle $r = \frac{1}{5}$ we have

$$\frac{dr}{dt} = -\frac{1}{5}\sin^2\theta \ln\left(\frac{4}{25} + \frac{5}{25}\cos^2\theta + \frac{1}{2}\right) \geq 0$$

for all θ, and r is nondecreasing. Thus no trajectory can leave the region R by crossing over the inner boundary $r = \frac{1}{5}$ and moving towards the origin. Thus a trajectory which is in R at $t = 0$ must stay in R for $t > 0$. By the Poincaré-Bendixson Theorem, there must be a cycle in R. Fig. 10.95 is a sketch of the phase portrait for this system.

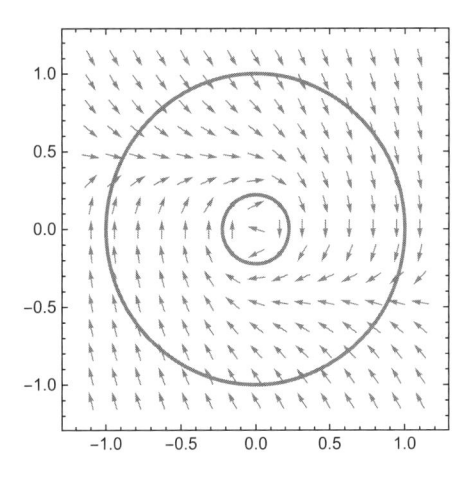

Figure 10.94. Scaled vector field for Example 10.8.3 and circles $x^2 + y^2 = 1$, $x^2 + y^2 = \frac{1}{25}$.

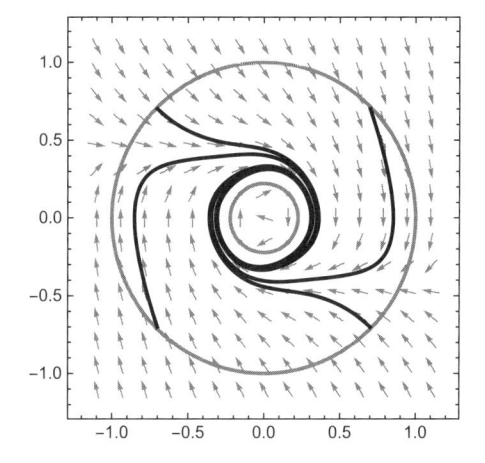

Figure 10.95. Phase portrait in the region R for Example 10.8.3.

Example 10.8.4. The nonlinear system

(10.99) $$\frac{dx}{dt} = f(x, y) = y - x^3 + x, \quad \frac{dy}{dt} = g(x, y) = -x$$

is an example of a van der Pol equation. Variations of the van der Pol equation are used in the study of nonlinear circuits, as well as in modeling the pumping action of the human heart. There is a single equilibrium point at the origin, and linear approximation shows that $(0,0)$ is an unstable spiral point. We look for a cycle using the Poincaré-Bendixson Theorem. Fig. 10.96 suggests how

10.8. Cycles and limit cycles

this might be done. The vector field (with all vectors scaled to have the same length) is shown there. Focus on the region R, consisting of the six-sided polygon C_o with vertices at $(\frac{1}{2}, 3)$, $(3, 3)$, $(\frac{1}{2}, -2)$, $(-\frac{1}{2}, -3)$, $(-3, -3)$, and $(-\frac{1}{2}, 2)$, the circle C_i with equation $x^2 + y^2 = \frac{1}{4}$, and the annular region between them. We will show that any orbit which starts in R at $t = 0$ must remain in R for all later times $t > 0$. Theorem 10.8.1 then says that R must contain a cycle (which encircles the origin). Fig. 10.97, which shows several orbits of the van der Pol system, suggests correctly that there is a *unique* cycle, which attracts all nonconstant orbits.

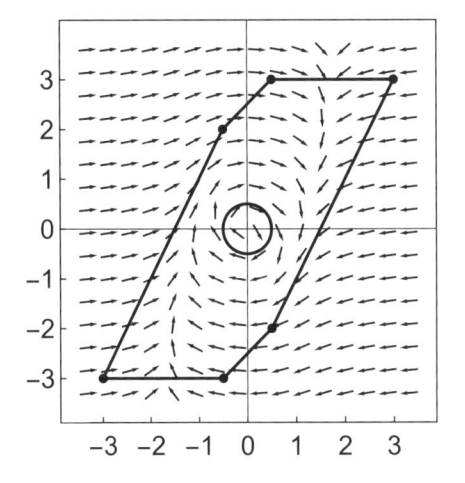

Figure 10.96. Scaled vector field for equation (10.99) with region R bounded by the circle and polygon.

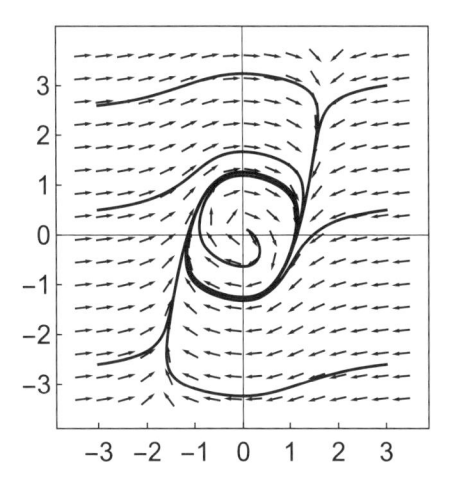

Figure 10.97. Phase portrait for Example 10.8.4.

To verify that no orbit that starts in R can leave R, either by escaping outward across C_o or inward across C_i, we will first show that at every point of C_o the vectors in the vector field point *into* the region R. Since C_o is a polygon, we will work on each of its six edges, as well as the six vertices, in turn. The basic principle we use is: *If the dot product of an outward pointing normal vector at a point of C_o and the vector field vector at that point is negative, then the vector field vector points into the region R at that point.* This follows since if the dot product is negative, the angle between the two vectors is greater than $\pi/2$; see Fig.10.98.

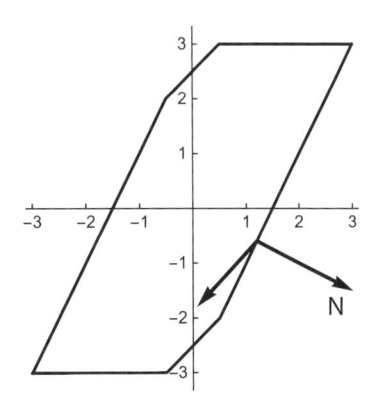

Figure 10.98. Outward normal vector and field vector at a point of C_o.

The top edge of C_o, between the vertices $(\frac{1}{2}, 3)$ and $(3, 3)$, is a horizontal line segment $y = 3$, $\frac{1}{2} < x < 3$. The vector $(0, 1)$ is perpendicular to this line segment and points outward from the region R; this serves as an outward normal vector at each point of this edge. The dot product of

this outward normal with a vector in the vector field is
$$(y - x^3 + x, -x) \cdot (0, 1) = -x$$
and for all x with $\frac{1}{2} < x < 3$ this is negative, as desired. At the vertex $(\frac{1}{2}, 3)$ the field vector is $(3 + \frac{3}{8}, -\frac{1}{2})$, which clearly points into the region R. Similarly, at the vertex $(3, 3)$ the field vector is $(-21, -3)$, which points into R as well.

Next we consider the edge of the C_o joining $(3, 3)$ to $(\frac{1}{2}, -2)$. This is the portion of the line $y = 2x - 3$ between $x = \frac{1}{2}$ and $x = 3$. Since this line has direction vector $(1, 2)$, we can choose $N = (2, -1)$ as an outward normal vector at any point on this line. The dot product of this normal vector and a field vector is
$$(y - x^3 + x, -x) \cdot (2, -1) = 2y - 2x^3 + 3x = 2(2x - 3) - 2x^3 + 3x = 7x - 6 - 2x^3.$$
A calculus argument shows that the maximum value of $7x - 6 - 2x^3$ for $\frac{1}{2} < x < 3$ occurs at $x = \sqrt{7/6}$ and is approximately -0.96. Thus the dot product is negative for $-\frac{1}{2} < x < 3$ and the field vectors point into the region R along this segment. Checking the vertex $(\frac{1}{2}, -2)$ we compute the field vector there to be $(-\frac{13}{8}, -\frac{1}{2})$ and this is easily seen to point into the region R (notice that the segment from $(\frac{1}{2}, -2)$ to $(-\frac{1}{2}, -3)$ has slope 1).

The edge of C_o which joins the vertex $(\frac{1}{2}, -2)$ to the vertex $(-\frac{1}{2}, -3)$ has equation $y = x - \frac{5}{2}$ for $-\frac{1}{2} < x < \frac{1}{2}$. We can choose $(1, -1)$ as an outward normal vector at points along this segment. Computing the dot product
$$(y - x^3 + x, -x) \cdot (1, -1) = y - x^3 + 2x = x - \frac{5}{2} - x^3 + 2x = 3x - \frac{5}{2} - x^3.$$
The function $3x - \frac{5}{2} - x^3$ has no critical points in the interval $-\frac{1}{2} < x < \frac{1}{2}$ and is increasing in this interval, with value $-\frac{9}{8}$ at $x = \frac{1}{2}$. Thus $3x - \frac{5}{2} - x^3$ is negative for all $-\frac{1}{2} < x < \frac{1}{2}$, giving the desired conclusion that the field vectors point into R along this edge. We check the vertex $(-\frac{1}{2}, -3)$ separately, obtaining the field vector $(-\frac{27}{8}, \frac{1}{2})$ there; this clearly points into the region R.

The remaining three edges (and two vertices) of C_o are checked similarly. We leave the details to you in Exercise 7. Summarizing our work so far, we see that no orbit starting in R can exit R through C_o.

To finish, we show that no orbit which begins in R at $t = 0$ can exit R by crossing over the circle C_i. Set $E(x, y) = x^2 + y^2$ and compute the derivative of E along orbits:
$$\dot{E}(x, y) = 2x(y - x^3 + x) + 2y(-x) = 2x^2(1 - x^2).$$
Since this is nonnegative at every point (x, y) with $x^2 + y^2 < 1$, no orbit can cross the circle C_i moving towards the origin, giving the desired conclusion. Thus the Poincaré-Bendixson Theorem applies to show that R contains a cycle (pictured in Fig. 10.97) of the van der Pol system.

The Poincaré-Bendixson Theorem gives conditions under which you can conclude a region contains a cycle. The next result gives conditions under which a region is guaranteed to *not* contain a cycle.

Theorem 10.8.5. *Consider the system*
$$\frac{dx}{dt} = f(x, y), \quad \frac{dy}{dt} = g(x, y),$$
where f and g are continuously differentiable. If there is a rectangle R in which the function
$$\frac{\partial f}{\partial x} + \frac{\partial g}{\partial y}$$
is either always strictly positive or always strictly negative, then there are no cycles in R.

10.8. Cycles and limit cycles

The rectangle in this result is allowed to be the entire xy-plane (an infinite rectangle).[16] The next example shows such an application.

Example 10.8.6. We revisit the damped pendulum equations
$$\frac{d\theta}{dt} = v, \quad \frac{dv}{dt} = -\frac{c}{m}v - \frac{g}{L}\sin\theta.$$

Here our dependent variables are θ and v instead of x and y. Adapting the statement of Theorem 10.8.5, we then consider
$$\frac{\partial}{\partial \theta}(v) + \frac{\partial}{\partial v}\left(-\frac{c}{m}v - \frac{g}{L}\sin\theta\right) = -\frac{c}{m}.$$

Since c and m are positive constants, this is negative at every point in the θv-plane. We conclude that the phase portrait for the damped pendulum system contains no cycles.

Theorem 10.8.5 can be proved in a few lines as an application of Green's Theorem from multivariable calculus; see Exercise 8.

10.8.1. Exercises.

1. Show that the system
$$\frac{dx}{dt} = -y + x(1 - 2x^2 - 3y^2), \quad \frac{dy}{dt} = x + y(1 - 2x^2 - 3y^2)$$
has a cycle in the region between the circles $x^2 + y^2 = 1$ and $x^2 + y^2 = \frac{1}{4}$.

2. Use the Poincaré-Bendixson Theorem to show that the system
$$\frac{dx}{dt} = 3x + 3y - x(x^2 + y^2), \quad \frac{dy}{dt} = -3x + 3y - y(x^2 + y^2)$$
has a cycle in the region R between the two circles $x^2 + y^2 = 2$ and $x^2 + y^2 = 4$. Then verify this by finding values of a and b so that
$$x = -a\cos(bt), \quad y = a\sin(bt)$$
is a solution whose orbit lies in R.

3. Find values r_1 and r_2 such that
$$\frac{dx}{dt} = 2x + y - xe^{x^2+y^2}, \quad \frac{dy}{dt} = -x + 2y - ye^{x^2+y^2}$$
has a cycle in the region between the two circles $x^2 + y^2 = r_1$ and $x^2 + y^2 = r_2$.

4. Modify the argument in Example 10.8.3 to show that there is a cycle in the region between the ellipses $36x^2 + 16y^2 = 1$ and $\frac{9}{2}x^2 + 2y^2 = 1$.

5. Show that the system
$$\frac{dx}{dt} = -xy + x, \quad \frac{dy}{dt} = xy + y$$
has no cycles in the right half-plane $\{(x,y) : x > 0\}$.

6. Show that the system (10.59) used in Example 10.5.2 to model the spread of a sexually transmitted disease has no cycles in the open first quadrant $\{(x,y) : x > 0, y > 0\}$.

7. In this problem you will finish the analysis of the polygon C_o in Example 10.8.4.
   (a) Find an outward normal vector at each point of the edge joining $(-\frac{1}{2}, -3)$ to $(-3, -3)$ and verify that the dot product of this normal vector with the vector field is negative for $-3 < x < -\frac{1}{2}$, so that the vector field points into R at points of this edge.
   (b) Show that the vector field at the vertex $(-3, -3)$ points into R.

---
[16] In fact, the rectangle R in Theorem 10.8.5 can be replaced by a simply connected region.

(c) Find an outward normal vector at each point of the edge joining $(-3,-3)$ to $(-\frac{1}{2}, 2)$ and verify that the dot product of this normal vector with the vector field is negative for $-3 < x < -\frac{1}{2}$.

(d) Show that the vector field at the vertex $(-\frac{1}{2}, 2)$ points into R.

(e) Find an outward normal vector at each point of the edge joining $(-\frac{1}{2}, 2)$ to $(\frac{1}{2}, 3)$ and verify that the dot product of this normal vector with the vector field is negative for $-\frac{1}{2} < x < \frac{1}{2}$.

8. For a simple closed curve C enclosing a region Ω as shown in Fig. 10.99, Green's Theorem says that

$$\int_C f(x,y)dy - g(x,y)dx = \int\int_\Omega \left(\frac{\partial f}{\partial x} + \frac{\partial g}{\partial y}\right) dxdy.$$

The left side of this identify is a line integral over the curve C and the right side is a double integral over the region Ω bounded by C.

(a) Suppose that the hypothesis of Theorem 10.8.5 hold for some rectangle R and suppose that C is a cycle in R. Use Green's Theorem to explain why

$$\int_C f(x,y)dy - g(x,y)dx \neq 0.$$

(b) Directly compute the line integral

$$\int_C f(x,y)dy - g(x,y)dx$$

using the fact that C is a cycle and therefore parametrized by a solution $x = x(t), y = y(t)$ of the system featured in Theorem 10.8.5. You should have reached a contradiction to the result in (a); hence there can be no cycle in R.

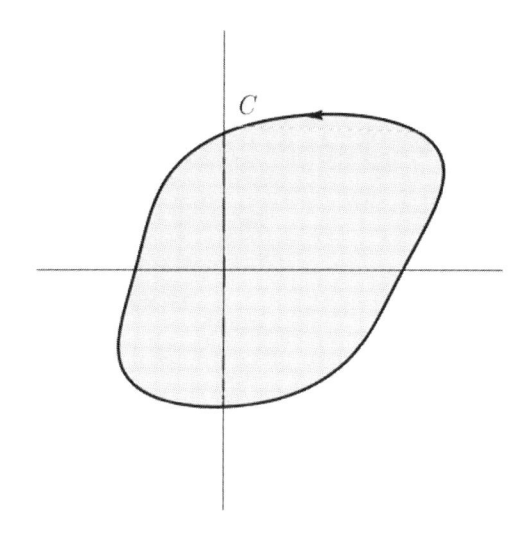

Figure 10.99. The region Ω bounded by the simple closed curve C.

9. (a) Write the second-order equation

$$x'' + p(x)x' + q(x) = 0$$

as a system by setting $y = x'$.

(b) Show that if $p(x) > 0$ for all x, then the system in (a) cannot have any periodic solution. Assume $p(x)$ and $q(x)$ are continuously differentiable.

(c) Explain why the damped pendulum is a special case of (a)–(b).

Chapter 11

Partial Differential Equations and Fourier Series

11.1. Introduction: Three interesting partial differential equations

So far in this book we have considered exclusively ordinary differential equations, those whose solutions are functions of a single independent variable. In this chapter we present an introduction to partial differential equations, which are distinguished by the fact that their solutions are functions of two or more independent variables. We concentrate mainly on three justly famous equations having great importance in both the theory of the subject and physical applications. For simplicity we restrict our attention to the case of two independent variables, with solutions $u = u(x,t)$ (one space variable x and one time variable t) or $u = u(x,y)$ (two space variables x and y). Our three equations are

- the **heat** equation
$$\frac{\partial u}{\partial t} = \kappa \frac{\partial^2 u}{\partial x^2}, \quad \kappa \text{ a positive constant,}$$
- the **wave** equation
$$\frac{\partial^2 u}{\partial t^2} = \alpha^2 \frac{\partial^2 u}{\partial x^2}, \quad \alpha \text{ a positive constant,}$$
- and **Laplace's** equation
$$\frac{\partial^2 u}{\partial x^2} + \frac{\partial^2 u}{\partial y^2} = 0.$$

Notice that each of these partial differential equations has order two, and any solution will be a function of two independent variables ($u = u(x,t)$ in the case of the heat or wave equation or $u = u(x,y)$ for Laplace's equation). Each of these three equations is linear, and each is a prototype for a whole class of second-order equations, called parabolic, hyperbolic, or elliptic equations, respectively. Many of the natural laws of physics can be formulated in terms of partial differential equations, which is one reason for their importance. The wave equation is used, for example, to model the motion of a vibrating string such as a guitar string, so that $u(x,t)$ gives the guitar string's displacement from equilibrium, x units from one of it's ends at time t. The heat equation can be used to model the temperature in a uniform heated rod, so that $u(x,t)$ is the temperature of the rod at a point x units from one end at time t. Laplace's equation has solutions which depend on two space variables but not on time. Thus Laplace's equation, which has possibly the broadest applications of any second-order partial differential equation, appears in steady-state problems. One simple application is to model the steady-state temperature (not varying with time) of an insulated metal plate with prescribed temperature along the edges.

For a preview of some new ideas we will encounter, consider the following instance of the heat equation. Imagine a uniform metal rod of length L occupying the line segment $[0, L]$ on the x-axis. Suppose that the rod is insulated along its length but that both ends are held at constant temperature $0°$. Let $u(x, t)$ denote the temperature of the rod at position x, $0 \leq x \leq L$, and time t. The function $u(x, t)$ satisfies the heat equation with the constant κ taken to be the "thermal diffusivity" of the particular metal from which the rod is made. Our assumption that the ends of the rod are held at $0°$ imposes two **boundary conditions** on $u(x,t)$: $u(0, t) = 0$ and $u(L, t) = 0$ for all times t. Note that these conditions are something different from the "initial conditions" that we use with ordinary differential equations since they specify $u(x, t)$ at **two different** x-values, $x = 0$ and $x = L$. There is also an actual **initial condition**

$$u(x, 0) = f(x), \quad 0 < x < L,$$

which represents the temperature of the rod at every position x at time $t = 0$. This initial temperature function is taken to be known. Note that it specifies much more information (a whole function's worth!) than the initial conditions encountered in ordinary differential equations, which typically prescribe only a single value of the solution and possibly some of its derivatives at that same single point. Our goal is then to find the temperature $u(x, t)$ of the rod at any position x and any future time $t > 0$. Mathematically this amounts to solving the heat equation for $u(x, t)$ subject to the given boundary and initial conditions. This may seem a challenging task, and it is not surprising that several new ideas are needed. In particular we introduce **separation of variables** in Section 11.3 (here this term has a meaning different from that used in Section 2.2), which enables us to first find all simple "product solutions" $u(x, t) = X(x)T(t)$ of the heat equation, that is, the solutions that can be separated into two factors, one of which is a function of the single variable x and the other of which is a function of the single variable t. In Section 11.2, using the elementary theory of **boundary value problems**, we identify exactly those product solutions that satisfy the boundary conditions $u(0, t) = 0$ and $u(L, t) = 0$. We call these the "building-block solutions" for our problem. (Other boundary conditions for our heated rod are possible and will lead to different building-block solutions.)

As we will see in Section 11.3, the building-block solutions for the heat equation with boundary conditions $u(0, t) = 0$ and $u(L, t) = 0$ are the functions

$$u_n(x, t) = e^{-\kappa \frac{n^2 \pi^2}{L^2} t} \sin \frac{n \pi x}{L}$$

where n is any positive integer. You should verify by substitution that these solutions satisfy the heat equation. Since the heat equation is *linear* and *homogeneous*, any finite linear combination of these building-block solutions is still a solution satisfying the boundary conditions. That is, functions of the form

(11.1) $$u(x, t) = \sum_{n=1}^{k} c_n e^{-\kappa \frac{n^2 \pi^2}{L^2} t} \sin \frac{n \pi x}{L}$$

satisfy both the heat equation and the boundary conditions $u(0, t) = 0$ and $u(L, t) = 0$.

11.2. Boundary value problems

So far, we have yet to bring the initial condition $u(x,0) = f(x)$ into play. Notice that if we set $t = 0$ in equation (11.1) we have

$$u(x,0) = \sum_{n=1}^{k} c_n \sin \frac{n\pi x}{L}.$$

Roughly speaking, we are hoping to choose coefficients c_n so that $u(x,0)$ is our desired initial value $f(x)$ for all $0 < x < L$. If we are limited to finite sums in equation (11.1) there will be many choices of $f(x)$ which cannot be so obtained. However, there are infinitely many building-block functions, and it is natural to ask if by passing to an infinite sum in (11.1) we can accommodate all "sufficiently nice" choices of $f(x)$ for the initial temperature and write

(11.2) $$f(x) = \sum_{n=1}^{\infty} c_n \sin \frac{n\pi x}{L}$$

for some choice of constants c_n. This is precisely the question that French mathematician and physicist Joseph Fourier was led to in the early nineteenth century as part of his study of heat flow. The tools for answering this and related questions come under the heading of **Fourier analysis**, which we introduce in Section 11.6. In particular, we will see that if our desired initial condition function f is differentiable with a continuous derivative on $[0, L]$, then for each x with $0 < x < L$ equation (11.2) holds for constants c_n whose values we can determine. This is an example of a **Fourier series**. We will also use similar results involving infinite sums of cosine functions or mixed infinite sums of both sine and cosine functions. With our Fourier series in hand, it will be a short step to finish constructing the desired solution $u(x,t)$ to our heat equation problem as an explicit infinite linear combination of building-block solutions.

We will see in this chapter that the solutions of the heat equation, wave equation, and Laplace equation all have rather different characters. Nevertheless the general scheme outlined above works to solve all three of these second-order equations. In addition, we will solve and apply certain *first-order* partial differential equations such as the **advection-decay** equation

$$\frac{\partial u}{\partial t} = -\kappa \frac{\partial u}{\partial x} - ru$$

where κ and r are constants (see Section 11.4).

In solving the heat and wave equations and Laplace's equation, we will begin with the separation of variables procedure. This will lead to two familiar *ordinary* differential equations. The boundary conditions on the original partial differential equation will translate into boundary conditions on one of these ordinary differential equations. To prepare for this, in the next section we will consider boundary value problems for ordinary differential equations, focusing especially on those problems that arise in our study of the heat, wave, and Laplace equations.

11.2. Boundary value problems

We begin by taking a new look at the familiar ordinary differential equation

(11.3) $$\frac{d^2 y}{dx^2} + \lambda y = 0,$$

where λ is a real constant. Now we will be interested in solving this equation as part of a **boundary value problem**, rather than in an initial value problem. Recall that initial conditions for a second-order equation like (11.3) specifies y and $\frac{dy}{dx}$ at *one* value of x. By contrast, in a boundary value problem for an ordinary differential equation, we give information about y (and/or y') at two *different* values of x. Think of x as a space variable, rather than time. We seek a solution to the differential equation on an interval $[x_0, x_1]$ having the specified values at the endpoints of this interval.

Recall that the Existence and Uniqueness Theorem (Theorem 4.1.4) guarantees that the initial value problem
$$y'' + p(x)y' + q(x)y = g(x), \qquad y(x_0) = \alpha, \quad y'(x_0) = \beta,$$
has a unique solution on any open interval containing x_0 on which p, q, and g are continuous. When the initial conditions are replaced by the boundary conditions
$$y(x_0) = \alpha, \quad y(x_1) = \beta$$
or more generally by boundary conditions of the form
$$ay(x_0) + by'(x_0) = \alpha, \quad cy(x_1) + dy'(x_1) = \beta$$
there may be one solution or many solutions or no solution. Let's look at a few examples that illustrate this.

Example 11.2.1. Solve $y'' + 2y = 0$ with boundary conditions $y(0) = 1$ and $y(\pi) = 0$. The differential equation has characteristic equation $z^2 + 2 = 0$, with roots $\pm\sqrt{2}\,i$. Thus the general solution to the differential equation is
$$y = c_1 \cos(\sqrt{2}\,x) + c_2 \sin(\sqrt{2}\,x).$$
What do the boundary conditions tell us about c_1 and c_2? Since $y(0) = 1$, we have $1 = c_1$. Using this and the requirement that $y(\pi) = 0$, we see that $0 = \cos(\sqrt{2}\,\pi) + c_2 \sin(\sqrt{2}\,\pi)$, which tells us that
$$c_2 = -\frac{\cos(\sqrt{2}\,\pi)}{\sin(\sqrt{2}\,\pi)} = -\cot(\sqrt{2}\,\pi) \approx -0.2762.$$
Thus, this boundary value problem has a unique solution
$$y = \cos(\sqrt{2}\,x) - \cot(\sqrt{2}\,\pi) \sin(\sqrt{2}\,x).$$

Example 11.2.2. Solve $y'' + y = 0$ with the boundary conditions $y(0) = 1$ and $y(\pi) = a$, where a is a real constant to be specified. Since the characteristic equation is $z^2 + 1 = 0$, with roots $\pm i$, the general solution to $y'' + y = 0$ is $y = c_1 \cos x + c_2 \sin x$. The condition $y(0) = 1$ tells us that $c_1 = 1$. The condition $y(\pi) = a$ says $a = -c_1 = -1$ and puts no restriction on c_2. Here is what we conclude:

- When $a = -1$ there are infinitely many solutions
$$y = \cos x + c \sin x$$
for arbitrary choice of c.
- When $a \neq -1$ there is **no** solution.

Homogeneous boundary value problems. The boundary value problem
(11.4) $$y'' + p(x)y' + q(x)y = 0,$$
(11.5) $$ay(0) + by'(0) = 0, \quad cy(L) + dy'(L) = 0,$$
is said to be **homogeneous**. Notice that the two boundary conditions, as well as the differential equation itself, have zeros on the right-hand side. A homogeneous boundary value problem has the following useful and familiar property:

Superposition Principle. If y_1 and y_2 are solutions to (11.4)–(11.5), then so is $c_1 y_1 + c_2 y_2$ for any choice of constants c_1 and c_2. Note also that a homogeneous boundary value problem always

11.2. Boundary value problems

has (at least) the trivial solution $y = 0$. You are asked to verify the Superposition Principle in Exercise 9.

An important boundary value problem. For solving certain partial differential equations it will be important to understand the solutions to the particular equation

(11.6) $$y'' + \lambda y = 0$$

with various boundary conditions, where λ is a real constant. Let's first impose the homogeneous boundary conditions

(11.7) $$y(0) = 0, \quad y(L) = 0$$

for some $L > 0$. We consider three cases: $\lambda = 0$, $\lambda > 0$, and $\lambda < 0$.

Case (i): $\lambda = 0$. When $\lambda = 0$ in (11.6), the general solution to $y'' = 0$ is $y = c_1 + c_2 x$. The condition $y(0) = 0$ tells us that $c_1 = 0$, and then the condition $y(L) = 0$ forces $c_2 = 0$. So with $\lambda = 0$, there is only one solution, $y = 0$ (the trivial solution) to (11.6) satisfying the boundary conditions in (11.7).

Case (ii): $\lambda < 0$. When $\lambda < 0$, the roots of the characteristic equation $z^2 + \lambda = 0$ are $\pm\sqrt{-\lambda}$. Since $\lambda < 0$, these are two real numbers, one positive and one negative. A fundamental set of solutions to $y'' + \lambda y = 0$ is thus

$$\left\{ e^{\sqrt{-\lambda}\,x}, e^{-\sqrt{-\lambda}\,x} \right\}.$$

We prefer to work with a *different* fundamental set of solutions. Recall (cf. Section 4.2, Exercise 15) the functions

$$\sinh(\sqrt{-\lambda}\,x) = \frac{e^{\sqrt{-\lambda}\,x} - e^{-\sqrt{-\lambda}\,x}}{2}$$

and

$$\cosh(\sqrt{-\lambda}\,x) = \frac{e^{\sqrt{-\lambda}\,x} + e^{-\sqrt{-\lambda}\,x}}{2}.$$

Since $\sinh(\sqrt{-\lambda}\,x)$ and $\cosh(\sqrt{-\lambda}\,x)$ are just linear combinations of

$$e^{\sqrt{-\lambda}\,x} \quad \text{and} \quad e^{-\sqrt{-\lambda}\,x}$$

they are also solutions to $y'' + \lambda y = 0$ when $\lambda < 0$. Since $\sinh(\sqrt{-\lambda}\,x)$ and $\cosh(\sqrt{-\lambda}\,x)$ are linearly independent,

(11.8) $$\left\{ \cosh(\sqrt{-\lambda}\,x), \sinh(\sqrt{-\lambda}\,x) \right\}$$

is a fundamental set of solutions to the equation $y'' + \lambda y = 0$ (when $\lambda < 0$). Taking

$$y = c_1 \cosh(\sqrt{-\lambda}\,x) + c_2 \sinh(\sqrt{-\lambda}\,x)$$

as the general solution, we see that to meet the boundary condition $y(0) = 0$ we need $c_1 = 0$, and then to satisfy $y(L) = 0$ we must have

$$c_2 \sinh(\sqrt{-\lambda}\,L) = 0.$$

Since $L \neq 0$, a look at the graph of the hyperbolic sine function tells us that $c_2 = 0$ also, and there are no nontrivial solutions to (11.6)–(11.7) with $\lambda < 0$.

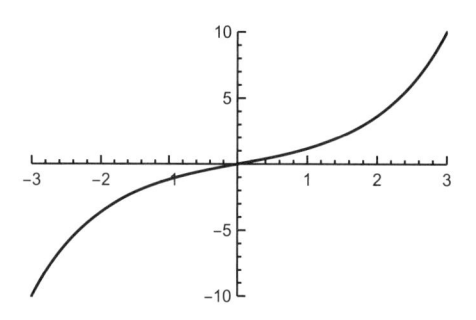

Figure 11.1. Graph of $y = \sinh x$.

Case (iii): $\lambda > 0$. When $\lambda > 0$, the general solution to $y'' + \lambda y = 0$ is
$$y = c_1 \cos(\sqrt{\lambda}\, x) + c_2 \sin(\sqrt{\lambda}\, x).$$
Since we want $y(0) = 0$, we must have $c_1 = 0$. The requirement $y(L) = 0$ says
$$(11.9) \qquad c_2 \sin(\sqrt{\lambda}\, L) = 0.$$
We're not so interested in the choice $c_2 = 0$, since that gives the trivial solution $y = 0$. For certain values of $\lambda > 0$ there **will** be a nontrivial solution: If $\sqrt{\lambda}\, L$ is a positive number whose sine is 0, then **any** value of c_2 will satisfy (11.9). In other words if
$$\sqrt{\lambda}\, L = n\pi$$
for some positive integer n, then we will have infinitely many solutions. More precisely, if
$$\lambda = \frac{n^2 \pi^2}{L^2}$$
for any positive integer n, then for every constant c
$$y = c \sin \frac{n\pi x}{L}$$
is a solution to
$$y'' + \frac{n^2 \pi^2}{L^2} y = 0, \qquad y(0) = 0, \quad y(L) = 0.$$

The language of eigenvalues and eigenfunctions. In this last example, we say that the numbers
$$\lambda_n = \frac{n^2 \pi^2}{L^2}, \quad n = 1, 2, 3, \ldots,$$
are **eigenvalues** of our boundary value problem $y'' + \lambda y = 0$, $y(0) = 0$, $y(L) = 0$, with corresponding **eigenfunctions**
$$c \sin \frac{n\pi x}{L}, \quad c \neq 0.$$
To see where this terminology comes from, write the differential equation as $-y'' = \lambda_n y$, and think of the linear operator pictured below, which produces the output $-y''$ when the input is a function y:

11.2. Boundary value problems

The (nontrivial) functions
$$y = c\sin\frac{n\pi x}{L}, \quad c \neq 0,$$
meet our boundary conditions and also satisfy $y'' + \lambda_n y = 0$. They are transformed by this operator $(-D^2)$ into scalar multiples of themselves, $\lambda_n y$, with the scalar being $\lambda_n = \frac{n^2\pi^2}{L^2}$. So this generalizes the notion of eigenvalue and eigenvector for the linear operation of multiplication by a matrix. Just as the zero vector is not allowed to be an eigenvector, the zero function is not allowed as an eigenfunction. Our work on this example may be summarized as follows:

> The boundary value problem
> $$y'' + \lambda y = 0, \quad y(0) = 0, \quad y(L) = 0,$$
> has no nonpositive eigenvalues, and it has positive eigenvalues
> $$\lambda = \frac{n^2\pi^2}{L^2}, \quad n = 1, 2, 3, \ldots,$$
> with corresponding eigenfunctions
> $$y = c\sin\frac{n\pi x}{L}, \quad c \neq 0.$$

Our next example is also phrased in this language.

Example 11.2.3. Does the boundary value problem
$$y'' + \lambda y = 0, \quad y(0) + y'(0) = 0, \quad y(1) = 0,$$
have any positive eigenvalues? In other words, are there positive values of λ and nontrivial functions y that solve $y'' + \lambda y = 0$ and satisfy $y(0) + y'(0) = 0$, $y(1) = 0$? When $\lambda > 0$, the general solution of $y'' + \lambda y = 0$ is
$$y = c_1 \cos(\sqrt{\lambda}\,x) + c_2 \sin(\sqrt{\lambda}\,x).$$
The condition $y(0) + y'(0) = 0$ says

(11.10) $$c_1 + c_2\sqrt{\lambda} = 0,$$

and the condition $y(1) = 0$ says

(11.11) $$c_1 \cos\sqrt{\lambda} + c_2 \sin\sqrt{\lambda} = 0.$$

We want to know if there are any positive values of λ for which equations (11.10) and (11.11) have a solution other than $c_1 = c_2 = 0$. Substituting $c_1 = -c_2\sqrt{\lambda}$ from equation (11.10) into equation (11.11) gives
$$c_2(\sin\sqrt{\lambda} - \sqrt{\lambda}\cos\sqrt{\lambda}) = 0.$$
This equation holds if $c_2 = 0$, but then $c_1 = 0$ too, leaving us with the trivial solution. Thus we look for values of λ with
$$\sin\sqrt{\lambda} - \sqrt{\lambda}\cos\sqrt{\lambda} = 0.$$
Rewrite this as $\tan\sqrt{\lambda} = \sqrt{\lambda}$. To analyze this last equation, set $w = \sqrt{\lambda}$ and consider the graphs of the equations $v = \tan w$ and $v = w$ in the wv-plane, as shown in Fig. 11.2. There are infinitely many points of intersection of these two graphs with $w > 0$ as shown, denoted $w = w_1, w_2, w_3, \ldots$. Notice that the smallest value w_1 is between $\frac{\pi}{2}$ and $\frac{3\pi}{2}$ and that there is a natural approximation for w_n for large n; see Exercises 12 and 14. These correspond to infinitely many positive eigenvalues $\lambda_1 = w_1^2, \lambda_2 = w_2^2, \ldots$. The corresponding eigenfunctions are
$$c(\sin(\sqrt{\lambda_n}\,x) - \sqrt{\lambda_n}\cos(\sqrt{\lambda_n}\,x)), \quad c \neq 0.$$

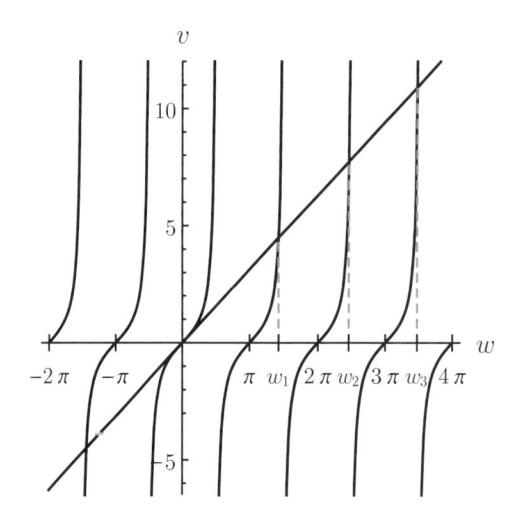

Figure 11.2.

In Exercise 10 you are asked to solve the same equation $y'' + \lambda y = 0$ with other kinds of boundary conditions. In each case, the goal is to determine, for $\lambda = 0, \lambda < 0$, and $\lambda > 0$ in turn, whether or not the boundary value problem has any nontrivial solutions and to find all nontrivial solutions when they exist.

When we solve a boundary value problem such as that of Example 11.2.3, which features a constant coefficient linear differential equation together with boundary conditions at a and b (with $a < b$), solutions, if they exist, are necessarily continuous on $[a, b]$ and have derivatives that extend continuously to $[a, b]$. The reason is that our fundamental sets of solutions, derived in Chapter 4, for such constant coefficient equations consist of functions continuously differentiable on $(-\infty, \infty)$. In general, solutions to boundary value problems on $[a, b]$ should always be at least continuous on $[a, b]$, satisfying the specified boundary conditions at a and b. See Exercise 13.

11.2.1. Exercises.

In Exercises 1–8 solve the given boundary value problem or show that it has no solution.

1. $y'' + y = 0$, $y(0) = 0, y'(\pi) = 1$.
2. $y'' + 3y = 0$, $y'(0) = 1, y'(\pi) = 0$.
3. $y'' + y = x$, $y(0) = 0, y(\pi) = 0$.
4. $y'' + 2y = x$, $y(0) = 0, y(\pi) = 0$.
5. $y'' + y = 0$, $y(-\pi/2) = 0, y(\pi/4) + y'(\pi/4) = 0$.
6. $y'' - y = 0$, $y(0) = 0, y(1) = 0$.
7. $y'' + 2y' + 2y = 0$, $y(0) = 0, y(\pi) = 0$.
8. $y'' - 2y' + y = 0$, $y(0) = 0, y(1) = 1$,

9. Show that if y_1 and y_2 are solutions to the homogeneous boundary value problem
$$y'' + p(x)y' + q(x)y = 0, \quad ay(0) + by'(0) = 0, \quad cy(L) + dy'(L) = 0,$$
then so is $c_1 y_1 + c_2 y_2$ for any choice of constants c_1 and c_2.

10. Find the eigenvalues and eigenfunctions of each of the following boundary value problems. Your answer will depend on whether $\lambda = 0$, $\lambda > 0$, or $\lambda < 0$. When $\lambda < 0$ you may want to describe a fundamental set of solutions to $y'' + \lambda y = 0$ using the hyperbolic sine and cosine functions, as in equation (11.8). For future reference, you may want to summarize your results in the table below, showing the eigenvalues and eigenfunctions (or write "none").
    (a) $y'' + \lambda y = 0$, $y(0) = 0$, $y'(L) = 0$.
    (b) $y'' + \lambda y = 0$, $y'(0) = 0$, $y'(L) = 0$.
    (c) $y'' - \lambda y = 0$, $y'(0) = 0$, $y'(L) = 0$. Hint: Don't work too hard if you've already done (b).

11.2. Boundary value problems

(d) $y'' - \lambda y = 0$, $y(0) = 0$, $y'(L) = 0$.
(e) $y'' + \lambda y = 0$, $y(0) = 0$, $y(\pi) - y'(\pi) = 0$.
(f) $y'' + \lambda y = 0$, $y(0) - y'(0) = 0$, $y(\pi) - y'(\pi) = 0$.

boundary value problem	where $\lambda = 0$	where $\lambda > 0$	where $\lambda < 0$
$y'' + \lambda y = 0$, $y(0) = 0$, $y'(L) = 0$			
$y'' + \lambda y = 0$, $y'(0) = 0$, $y'(L) = 0$			
$y'' - \lambda y = 0$, $y'(0) = 0$, $y'(L) = 0$			
$y'' - \lambda y = 0$, $y(0) = 0$, $y'(L) = 0$			
$y'' + \lambda y = 0$, $y(0) = 0$, $y(\pi) - y'(\pi) = 0$			
$y'' + \lambda y = 0$, $y(0) - y'(0) = 0$, $y(\pi) - y'(\pi) = 0$			

11. (a) By considering the three cases $\lambda = 0$, $\lambda > 0$, and $\lambda < 0$, find all solutions to

    (11.12) $$y'' + \lambda y = 0$$

    that satisfy

    (11.13) $$y(L) = y(-L), \quad y'(L) = y'(-L)$$

    for some positive constant L.
    (b) If y_1 and y_2 solve (11.12)–(11.13) for a specific λ, are linear combinations of y_1 and y_2 also solutions?

12. Show that in Example 11.2.3, the graphs of $v = \tan w$ and $v = w$ are as shown there and *not* as in Fig. 11.3, with the first positive intersection between 0 and $\pi/2$.

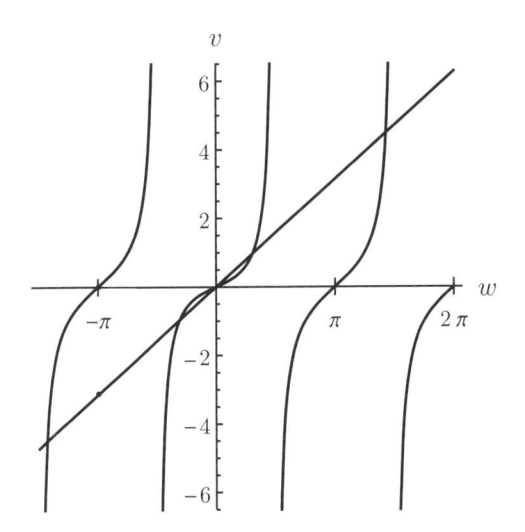

Figure 11.3.

13. Boundary value problems featuring second-order ordinary differential equations arise frequently in applications. (See, for example, Exercise 30 in Section 7.3 or Exercise 14 in Section 7.6.) Our general form for such a problem is

    (11.14) $\quad F(x, y, y', y'') = 0, \quad ay(x_0) + by'(x_0) = \alpha, \quad cy(x_1) + dy'(x_1) = \beta,$

    where $x_0 < x_1$ and F is some real-valued function of four real variables. The function $y = \varphi(x)$, defined on the interval $[x_0, x_1]$, is considered to be a solution of the preceding problem provided that
    (i) for each x in the open interval (x_0, x_1), φ satisfies the differential equation of (11.14), meaning $F(x, \varphi(x), \varphi'(x), \varphi''(x)) = 0$ for all x in (x_0, x_1),
    (ii) the boundary conditions of (11.14) are satisfied, and
    (iii) φ is continuous on $[x_0, x_1]$ while φ' is continuous on $[x_0, x_1)$ if $b \neq 0$ and continuous on $(x_0, x_1]$ if $d \neq 0$.

    (a) Solve the boundary value problem
    $$2x^2 y'' + 3xy' - y = 0, \quad y(0) = 0, \quad y(1) = 1.$$

    (b) Show that the boundary value problem
    $$2x^2 y'' + 3xy' - y = 0, \quad y(0) + y'(0) = 0, \quad y(1) = 1,$$
    has no solution.

14. (CAS) Recall that in Example 11.2.3 the positive roots of $\tan w = w$, in ascending order, were labeled $w_1, w_2, w_3, \ldots$. What is the approximate value of w_n for large n? Use any technology at your disposal to compute (numerically) w_{10} and compare the result to your approximation for w_{10}.

15. A 10-foot long piece of PEX plastic pipe is suspended between two metal joists 8 feet above a basement floor. Its left and right endpoints are held horizontal by the joists. Let $y(x)$ be the height of the pipe x feet from its left endpoint. We have $y(0) = 8 = y(10)$. The Euler beam equation predicts that y satisfies the following differential equation:

    $$\frac{d^4 y}{dx^4} = -\rho,$$

    where ρ is a positive constant.

(a) Provide two additional boundary conditions satisfied by y based on the pipe's being held horizontal at both endpoints.

(b) Assume that the pipe sags 1 inch, i.e., 1/12 ft, at its center. Solve the Euler beam equation subject to the boundary conditions $y(0) = 8 = y(10)$ as well as your conditions from part (a). Find ρ using the sag information, and then plot your solution, which is the pipe's "profile" function.

(c) Suppose that before being shortened for installation, the pipe is slightly longer than 10 feet and it is resting between two saw horses spread 10 feet apart. In this situation the pipe is said to be "simply supported" at its endpoints. Again, let $y(x)$ be the height of the pipe x feet from its left endpoint. Assuming the saw horses are 3 feet high, we have

$$\frac{d^4y}{dx^4} = -\rho, \qquad y(0) = 3 = y(10), \quad y''(0) = 0 = y''(10).$$

  (i) Using the value of ρ from part (b) above, solve the preceding boundary value problem and plot the resulting profile function.
  (ii) Can you explain the boundary conditions $y''(0) = 0 = y''(10)$? (Hint: Consider the following picture of a very flexible pipe simply supported by two vertical posts.)

11.3. Partial differential equations: A first look

The boundary value problems we studied in the last section play a key role in solving certain important partial differential equations. To illustrate this, here we will consider the second-order partial differential equation

(11.15) $$\frac{\partial u}{\partial t} = \kappa \frac{\partial^2 u}{\partial x^2}$$

for an unknown function $u(x, t)$ of two variables, where κ is a positive constant. You might prefer to write this equation a bit more compactly as

$$u_t(x, t) = \kappa u_{xx}(x, t)$$

using alternate notation for the partial derivatives. In different contexts, equation (11.15) is called the **heat equation** or the **diffusion equation**. Later we will discuss some applications of this equation which will help explain this terminology, as well as its derivation. There, x will be a space variable (giving location along a line) and t will be time.

To get oriented, we begin with two facts about equation (11.15).

- The **Superposition Principle** applies to equation (11.15). That is, if $u_1(x,t)$ and $u_2(x,t)$ are both solutions, so is $c_1 u_1(x,t) + c_2 u_2(x,t)$ for any constants c_1 and c_2. You are asked to check this in Exercise 1. In general, the Superposition Principle applies to any homogeneous linear partial differential equation—see Exercises 7–10.
- Under suitable conditions, if the function $u(x,t)$ is specified for $t = 0$ as some function $g(x)$, this determines $u(x,t)$ for all $t > 0$.

The second of these facts alludes to a "uniqueness theorem" for equation (11.15). We'll be a bit more precise about this later, when we solve (11.15) on an interval $0 \leq x \leq L$, with boundary conditions specified at the endpoints of this interval, as well as a given initial value $u(x, 0)$ for all x in $(0, L)$.

A powerful technique for solving equation (11.15), as well as other important partial differential equations, is separation of variables . We describe this next.

Separation of variables. To produce *some* solutions to equation (11.15) we begin by looking for solutions that have a particularly simple form

$$u(x,t) = X(x)T(t).$$

In other words, we will look for a solution function of two variables that is a product of two one-variable functions (one a function of x and the other a function of t). This may seem like a big restriction to start with (because many functions of two variables don't have this particularly simple form), but as we will see, we can get a lot of mileage out of it. If $u(x,t) = X(x)T(t)$, then

$$u_t(x,t) = \frac{\partial u}{\partial t} = X(x)T'(t)$$

and

$$u_{xx}(x,t) = \frac{\partial^2 u}{\partial x^2} = X''(x)T(t).$$

The function $X(x)T(t)$ will satisfy equation (11.15) if

$$X(x)T'(t) = \kappa X''(x)T(t).$$

Dividing by $\kappa T(t)X(x)$ (which we assume is nonzero for the points (x,t) of interest), we can write this as

(11.16) $$\frac{T'(t)}{\kappa T(t)} = \frac{X''(x)}{X(x)}.$$

Now something surprising happens. The left-hand side of equation (11.16) is a function of t only, and the right-hand side is a function of x only. In other words, the left side doesn't change as x varies, and the right side doesn't change as t varies. Since the two sides are equal, neither can change as either x or t varies. In other words, if $X(x)T(t)$ solves equation (11.15), both

$$\frac{T'(t)}{\kappa T(t)} \quad \text{and} \quad \frac{X''(x)}{X(x)}$$

must equal the **same** constant. We will denote this common constant value $-\lambda$, where the minus sign is for later convenience. The parameter $-\lambda$ is called the **separation constant**. Thus for a function of the form $u(x,t) = X(x)T(t)$ to solve $u_t = \kappa u_{xx}$, we must have

$$\frac{X''(x)}{X(x)} = -\lambda = \frac{T'(t)}{\kappa T(t)},$$

or equivalently,

(11.17) $$X''(x) + \lambda X(x) = 0$$

and

(11.18) $$T'(t) = -\kappa \lambda T(t).$$

Effectively, we have replaced the original *partial* differential equation (11.15) by the two *ordinary* differential equations (11.17) and (11.18).

The first-order equation in (11.18) is easily solved to obtain

(11.19) $$T(t) = Ce^{-\kappa \lambda t}.$$

Equation (11.17) is one that we solved with various boundary conditions in Section 11.2. Moreover, if you completed the table in Exercise 10 of the previous section, you have a nice summary of the solutions to equation (11.17) for a number of important boundary conditions.

The heat equation with boundary conditions. Typically, physical problems involve boundaries of some kind, and their mathematical models correspondingly include boundary conditions. Next we discuss solving the heat equation $u_t = \kappa u_{xx}$ with the **boundary conditions**
$$u(0,t) = 0, \quad u(L,t) = 0$$
for some positive value L and all $t > 0$. We'll see in later applications that these are natural conditions to invoke. With $u(x,t) = X(x)T(t)$ as above, they say $X(0)T(t) = 0$ and $X(L)T(t) = 0$ for all $t > 0$. Thus either
$$X(0) = 0 \quad \text{and} \quad X(L) = 0$$
or $T(t) = 0$ for all $t > 0$. The latter option is not so interesting, since then $u(x,t) = X(x)T(t) = 0$. So to meet the boundary conditions $u(0,t) = 0$ and $u(L,t) = 0$ (for all $t > 0$) in a nontrivial way, with a function of the form $u(x,t) = X(x)T(t)$, we must have $X(0) = 0$ and $X(L) = 0$, where X is a solution to equation (11.17).

The boundary value problem
$$(11.20) \qquad X''(x) + \lambda X(x) = 0, \quad X(0) = 0, \quad X(L) = 0,$$
was analyzed in Section 11.2. Importing our work from there into the current notation, we see that (11.20) has a nontrivial solution only if $\lambda = (n^2\pi^2)/L^2$ for some positive integer n, and for such λ, we have a solution
$$X(x) = \sin\frac{n\pi x}{L}$$
for $n = 1, 2, 3, \ldots$.

Using these same values of λ, $\lambda = (n^2\pi^2)/L^2$, we obtain (choosing $C = 1$ in equation (11.19))
$$T(t) = e^{-\kappa\frac{n^2\pi^2}{L^2}t}.$$
Assembling the ingredients, we now know that the functions
$$(11.21) \qquad u(x,t) = e^{-\kappa\frac{n^2\pi^2}{L^2}t}\sin\frac{n\pi x}{L}, \quad \text{for } n = 1, 2, 3, \ldots,$$
solve the heat equation $u_t = \kappa u_{xx}$ and satisfy the boundary conditions $u(0,t) = 0$ and $u(L,t) = 0$, $t > 0$. Thus we have produced a family of "building-block solutions":

> The problem
> $$u_t = \kappa u_{xx}, \quad u(0,t) = 0, \quad u(L,t) = 0,$$
> has building-block solutions
> $$u_n(x,t) = e^{-\kappa\frac{n^2\pi^2}{L^2}t}\sin\frac{n\pi x}{L}$$
> for $n = 1, 2, 3, \ldots$.

Superposition allows us to generate more solutions by linear combinations of our building block solutions:
$$(11.22) \qquad u(x,t) = c_1 e^{-\kappa\frac{\pi^2}{L^2}t}\sin\frac{\pi x}{L} + c_2 e^{-\kappa\frac{2^2\pi^2}{L^2}t}\sin\frac{2\pi x}{L} + \cdots + c_n e^{-\kappa\frac{n^2\pi^2}{L^2}t}\sin\frac{n\pi x}{L}$$
for some positive integer n. Any such linear combination still will satisfy the boundary conditions $u(0,t) = u(L,t) = 0$.

In the next example, we will also impose an **initial condition** that specifies $u(x,0)$, the value of u at each x, $0 \le x \le L$, at time $t = 0$.

Example 11.3.1. We find a function $u(x,t)$ that solves

(11.23) $$u_t = u_{xx}$$

(equation (11.15) with $\kappa = 1$), satisfies the boundary conditions

(11.24) $$u(0,t) = 0 \quad \text{and} \quad u(10,t) = 0$$

for all t, and also meets the initial condition

(11.25) $$u(x,0) = 4\sin\frac{\pi x}{5} - 3\sin(\pi x), \quad 0 \le x \le 10.$$

Our work above, with $\kappa = 1$ and $L = 10$, gives, for positive integers n, building-block solutions

$$u_n(x,t) = e^{-\frac{n^2\pi^2}{100}t}\sin\frac{n\pi x}{10}$$

satisfying the boundary conditions (11.24). Linear combinations of these functions will also solve our partial differential equation and satisfy the desired boundary conditions. Can we find a linear combination that will in addition satisfy the given initial condition? We want to determine coefficients c_j so that with

(11.26) $$u(x,t) = c_1 e^{-\frac{\pi^2}{100}t}\sin\frac{\pi x}{10} + c_2 e^{-\frac{2^2\pi^2}{100}t}\sin\frac{2\pi x}{10} + \cdots + c_n e^{-\frac{n^2\pi^2}{100}t}\sin\frac{n\pi x}{10},$$

we will have

$$u(x,0) = 4\sin\frac{\pi x}{5} - 3\sin(\pi x).$$

Setting $t = 0$ in equation (11.26) we have

$$u(x,0) = c_1 \sin\frac{\pi x}{10} + c_2 \sin\frac{2\pi x}{10} + \cdots + c_n \sin\frac{n\pi x}{10}.$$

Comparing this with the desired initial value specified in equation (11.25), we see that we should choose

$$c_2 = 4, \quad c_{10} = -3, \quad \text{and all other } c_j\text{'s} = 0.$$

Summarizing, the function

$$u(x,t) = 4e^{-\frac{2^2\pi^2}{100}t}\sin\frac{\pi x}{5} - 3e^{-\frac{10^2\pi^2}{100}t}\sin(\pi x)$$

solves equation (11.23) and satisfies the boundary conditions in (11.24) as well as the initial condition (11.25).

The initial condition in the last example was rather special. It allowed us to find a *finite* linear combination of the building-block solutions that meets the initial condition. Eventually we will see how to go further, beyond the scope of finite linear combinations, to produce even more solutions from our building blocks. This will allow us to cope with more general initial conditions.

The building-block solutions in equation (11.21), which are solutions of the heat equation (11.15), have continuous partial derivatives of all orders. In general, solutions of partial differential equations need not be so nice. Consider, for example, the function $u(x,y) = x + |y|$ which solves the partial differential equation $u_x(x,y) = 1$ on the whole plane but fails to have a partial derivative with respect to y at the origin. However, usually our interest will be in finding classical solutions to our partial differential equations, defined as follows.

> **A classical solution** of a kth-order partial differential equation is one that has continuous partial derivatives through order k (at least).

11.3. Partial differential equations: A first look

Our building-block solutions of the heat equation are thus classical solutions. The separation of variables technique, employed above, as well as in Sections 11.7, 11.8, and 11.10, will lead us to classical solutions. However, methods employed in the next section to solve certain first-order partial differential equations require only the assumption that solutions be differentiable (rather than continuously differentiable). A function $g = g(x, y)$ is differentiable at (x_0, y_0) precisely when g has a tangent plane approximation near (x_0, y_0) described by (10.18) and (10.19) in Section 10.2.

11.3.1. Exercises.

1. Verify that if $u_1(x,t)$ and $u_2(x,t)$ solve equation (11.15), so does $c_1 u_1(x,t) + c_2 u_2(x,t)$ for any constants c_1 and c_2.

2. Solve the partial differential equation $u_t = u_{xx}$ with boundary conditions $u(0,t) = 0$, $u(2,t) = 0$ and each of the following initial conditions.
   (a) $u(x,0) = 3\sin\frac{5\pi x}{2} - 7\sin\frac{9\pi x}{2}$.
   (b) $u(x,0) = \sin(\pi x) - 5\sin(4\pi x)$.
   (c) $u(x,0) = -\sin(\pi x) + 3\sin(2\pi x)\cos(2\pi x)$. Hint: Use a trigonometric identity for $\sin(2\theta)$.

In Exercises 3–6, if we use the method of separation of variables to solve the given partial differential equation, what pair of ordinary differential equations (analogous to equations (11.17) and (11.18) for the heat equation) are obtained?

3. $u_t = u_{xx} - u_{tt}$.

4. $u_{xx} + u_{yy} = 0$.

5. $\frac{\partial u}{\partial t} + \frac{\partial^4 u}{\partial x^4} = 0$.

6. $tu_{tt} + u_x = 0$.

Linearity of partial differential equations. Exercises 7–10 below explore the property of linearity for partial differential equations.

7. The operator depicted below takes an input function $u(x,t)$ and produces as the output
$$\frac{\partial u}{\partial t} - \kappa \frac{\partial^2 u}{\partial x^2}.$$
(Here κ is a positive constant.) Is this operator linear? In other words, is the output function for $u_1(x,t) + u_2(x,t)$ the same as the sum of the outputs for u_1 and u_2? Is the output for $cu_1(x,t)$ equal to c times the output for u_1?

$$u(x,t) \longrightarrow \boxed{\frac{\partial}{\partial t} - \kappa \frac{\partial^2}{\partial x^2}} \longrightarrow \frac{\partial u}{\partial t} - \kappa \frac{\partial^2 u}{\partial x^2}$$

8. An operator takes an input function $u(x,t)$ and produces as the output $u_{xx} - u_{tt}$. Is this operator linear? Does the Superposition Principle hold for solutions to $u_{xx} - u_{tt} = 0$?

9. An operator takes an input function $u(x,y)$ and produces as the output $u_x + (u_y)^2$. Is this operator linear? Does the Superposition Principle hold for solutions to $u_x + (u_y)^2 = 0$?

10. In general a **linear partial differential operator** acting on functions of x and t is a sum of terms of the form $g_{ij}\frac{\partial^{i+j}}{\partial t^i \partial x^j}$ over some finite collection of pairs of nonnegative integers i and j, where g_{ij} is a function of x and t for each i and j. For example,
$$L = \cos(xt)\frac{\partial^5}{\partial t^3 \partial x^2} + x^3 t^2 \frac{\partial^2}{\partial t \partial x} + 5\frac{\partial}{\partial x}$$
is a linear partial differential operator with $g_{32}(x,t) = \cos(xt)$, $g_{11}(x,t) = x^3 t^2$, and $g_{01}(x,t) = 5$. A partial differential equation with dependent variable u and independent variables x and t is

linear provided it may be written in the form

(11.27) $$L[u] = g(x,t)$$

where L is a linear partial differential operator acting on functions of x and t. The equation (11.27) is **homogeneous** provided that the function g on the right-hand side is the zero function. Show that the Superposition Principle holds for homogeneous linear partial differential equations: If $u_1 = u_1(x,t), u_2 = u_2(x,t), \ldots, u_n = u_n(x,t)$ are solutions to $L[u] = 0$, then $u(x,t) = \sum_{j=1}^{n} c_j u_j(x,y)$, for any constants $c_1, \ldots, c_n$, will satisfy $L[u] = 0$.

11. In this problem, we'll use separation of variables to solve

    (11.28) $$u_t = \kappa u_{xx}$$

    with the boundary conditions

    (11.29) $$u(0,t) = 0, \quad u_x(L,t) = 0.$$

    (a) Does the Superposition Principle hold for solutions to equation (11.28) together with (11.29)? In other words, is a linear combination of solutions still a solution?
    (b) Show that if we look for a solution of the form
    $$u(x,t) = X(x)T(t),$$
    we'll want $X(x)$ to solve the boundary value problem

    (11.30) $$X'' + \lambda X = 0, \quad X(0) = 0, \quad X'(L) = 0,$$

    for some constant λ, and we'll want T to satisfy
    $$\frac{1}{\kappa}\frac{T'}{T} = -\lambda$$
    for the same λ. Determine the values of λ and the corresponding functions $X(x)$ that give nontrivial solutions to (11.30). If you did Exercise 10 in Section 11.2, you can use your answer from there.
    (c) For each λ from (b), determine the corresponding function $T(t)$.
    (d) With $\kappa = 1$ and $L = \pi$, find the solution to equation (11.28) satisfying the boundary conditions in (11.29) and the initial condition
    $$u(x,0) = 5\sin\frac{x}{2} - 3\sin\frac{3x}{2}.$$

12. (a) Use the separation of variables technique to find some solutions of $u_x(x,y) = u_y(x,y)$.
    (b) Can your solution be written in the form $u(x,y) = f(x+y)$ where f is some real-valued function of a real variable?
    (c) Show that if f is any differentiable, real-valued function of a real variable, then $u(x,y) = f(x+y)$ solves $u_x(x,y) = u_y(x,y)$.

13. (a) Show that $u_1(x,t) = \cos(2t)\sin x$ and $u_2(x,t) = \cos(4t)\sin(2x)$ satisfy the partial differential equation
    $$\frac{\partial^2 u}{\partial t^2} - 4\frac{\partial^2 u}{\partial x^2} = 0.$$
    (b) Using u_1 and u_2, build at least two more solutions to the equation in (a), neither equal to u_1 or u_2.

14. How must r and s be related if $u(x,t) = e^{rx+st}$ solves $u_t(x,t) = u_{xx}(x,t)$?

15. The vibrations of a hanging chain (see Exercise 14 in Section 7.6) are modeled by the partial differential equation
    $$u_{tt} = g(xu_{xx} + u_x),$$

where $u(x,t)$ is the displacement of the point x units up the chain at time t (see Fig. 7.16 in Section 7.6) and g is the acceleration due to gravity. If the chain has length L and the top is fixed, $u(L,t) = 0$ for all $t > 0$.
  (a) Show that a separated solution $u(x,t) = y(x)s(t)$ must satisfy the ordinary differential equations
$$xy''(x) + y'(x) + ky(x) = 0 \quad \text{and} \quad s''(t) + kgs(t) = 0$$
  for k a constant.
  (b) What does the boundary condition $u(L,t) = 0$ tell you about $y(L)$ if you want a nontrivial solution for $u(x,t)$?
  (c) Suppose we seek a solution $u(x,t) = y(x)s(t)$ that is periodic in t. Show that the constant k must then be positive.

16. Suppose we change equation (11.15) to
$$\frac{\partial u}{\partial t} = -\frac{\partial^2 u}{\partial x^2}$$
and impose the boundary conditions $u(0,t) = 0 = u(1,t)$.
  (a) Explain how you know that this has building-block solutions
$$u_n(x,t) = e^{n^2\pi^2 t}\sin(n\pi x)$$
  for $n = 1, 2, 3, \ldots$.
  (b) Suppose $U(x,t)$ solves the above partial differential equation and the given boundary conditions and also satisfies the initial condition $U(x,0) = f(x)$ for some function $f(x)$. Let n be a positive integer. Give a solution (in terms of $U(x,t)$) to the new boundary-initial value problem
$$u_t = -u_{xx}, \quad u(0,t) = 0 = u(1,t), \quad u(x,0) = f(x) + \frac{1}{n}\sin(n\pi x).$$
  for any positive integer n. Call the new solution $U_n^*(x,t)$. Hint: Use the Superposition Principle.
  (c) Show that $|U(x,0) - U_n^*(x,0)| \leq \frac{1}{n}$ for all x, $0 < x < 1$, and show that
$$|U(x,1) - U_n^*(x,1)| = \left|\frac{1}{n}e^{n^2\pi^2}\sin(n\pi x)\right|,$$
  and that for $x = \frac{1}{2}$ and n odd this is $\frac{1}{n}e^{n^2\pi^2}$. Explain why by choosing n large, you can make $\frac{1}{n}e^{n^2\pi^2}$ as large as you want.
  (d) Why is the phenomena in (c) undesirable if you are modeling a physical application?

17. **Classifying a partial differential equation.** A homogeneous constant-coefficient second-order partial differential equation with dependent variable u and independent variables x and y is linear provided it may be written in the form

(11.31) $$Au_{xx} + Bu_{xy} + Cu_{yy} + Du_x + Eu_y + Fu = 0,$$

where $A, B, C, D, E,$ and F are constants. The **characteristic equation** of (11.31) is
$$Ar^2 + Brs + Cs^2 + Dr + Es + F = 0.$$

The equation in (11.31) is classified as follows:
  - **elliptic** if $B^2 - 4AC < 0$,
  - **parabolic** if $B^2 - 4AC = 0$,
  - **hyperbolic** if $B^2 - 4AC > 0$.
  (a) Show that $u(x,y) = e^{rx+sy}$ satisfies (11.31) if and only if r and s satisfy the characteristic equation of (11.31).
  (b) Letting t play the role of y, classify $u_t = \kappa u_{xx}$ (elliptic, parabolic, or hyperbolic).

(c) Classify $u_{xx}(x,y) + u_{yy}(x,y) = 0$.

(d) Again, letting t play the role of y, classify $u_{tt}(x,t) = \alpha^2 u_{xx}(x,t)$.

18. Consider the boundary-initial value problem of Example 11.3.1 with a change of its initial condition: $u_t = u_{xx}$ and

(11.32) $\quad u(0,t) = 0, \quad u(10,t) = 0$ for $t > 0$; $\quad u(x,0) = 25 - (x-5)^2$ for $0 \leq x \leq 10$.

At first glance, it may seem that the building-block solutions to $u_t = u_{xx}$,

(11.33) $$u_n(x,t) = e^{-\frac{n^2\pi^2}{100}t} \sin \frac{n\pi x}{10}, \quad n = 1, 2, \ldots,$$

cannot help us solve (11.32). This problem should convince you that this first impression is incorrect.

(a) Using the building blocks in (11.33), solve

$$u_t = u_{xx}; \quad u(0,t) = 0, \quad u(10,t) = 0 \text{ for } t > 0; \quad u(x,0) = f(x) \text{ for } 0 \leq x \leq 10,$$

where

$$f(x) = \frac{800}{\pi^3} \sin \frac{\pi x}{10} + \frac{800}{27\pi^3} \sin \frac{3\pi x}{10}.$$

(b) (CAS) Let u be your solution from part (a). Using any technology at your disposal, plot, on the same coordinate grid, both the graphs of $u(x,0)$ and $g(x) = 25 - (x-5)^2$, with $0 \leq x \leq 10$. Now, plot $|u(x,0) - g(x)|$, $0 \leq x \leq 10$, and record a rough estimate of its maximum value over $[0,10]$. This work should convince you that your solution $u(x,t)$ may be regarded as an approximate solution of (11.32).

(c) (CAS) The accuracy of the approximate solution of (11.32) that you found in part (a) can be improved. Let

$$f(x) = \frac{800}{\pi^3} \sin \frac{\pi x}{10} + \frac{800}{27\pi^3} \sin \frac{3\pi x}{10} + \frac{800}{125\pi^3} \sin \frac{5\pi x}{10} + \frac{800}{343\pi^3} \sin \frac{7\pi x}{10}.$$

With g as in part (b), plot $|f(x) - g(x)|$ over $[0,10]$ and estimate its maximum value. Note: It is possible to show—see Section 11.6—that for each x in $[0,10]$,

$$25 - (x-5)^2 = \sum_{n=0}^{\infty} \frac{800}{(2n+1)^3 \pi^3} \sin \frac{(2n+1)\pi x}{10}$$

and that

$$u(x,t) = \sum_{n=0}^{\infty} \frac{800}{(2n+1)^3 \pi^3} e^{-\frac{(2n+1)^2 \pi^2}{100}t} \sin \frac{(2n+1)\pi x}{10}$$

solves (11.32). This problem raises the question of which functions can be similarly expressed as an infinite sum of sine and/or cosine functions. Such series, called **Fourier series**, will be discussed in detail in Section 11.6.

11.4. Advection and diffusion

11.4.1. Some first-order partial differential equations. We'll begin by describing how to solve first-order linear equations of the form

(11.34) $$\frac{\partial u}{\partial x} + p(x,y) \frac{\partial u}{\partial y} = 0$$

for $u = u(x,y)$ a function of two variables. We will assume throughout this section that $u = u(x,y)$ is everywhere differentiable, which means its partial derivatives exist everywhere and provide a tangent plane approximation (see equation (10.18) in Section 10.2) at each point (x_0, y_0) satisfying equation (10.19).

11.4. Advection and diffusion

We'll see that the solutions to equation (11.34) can be described in terms of the solutions to the *ordinary* differential equation

(11.35) $$\frac{dy}{dx} = p(x,y).$$

To see how this goes, we first review the notion of the direction field for equation (11.35), discussed in Chapter 1. To sketch the direction field for (11.35), we choose points (x,y), compute $p(x,y)$, and draw the corresponding field mark—a short line segment centered at the point (x,y) having slope $p(x,y)$. Graphically, a solution curve to (11.35) is tangent to the field mark at each point it passes through. In other words, if we think of a solution to (11.35) given implicitly by $\varphi(x,y) = C$ for a constant C, then at any point (x,y) on this solution curve, the tangent line to the curve has slope $p(x,y)$. The vector $(1, p(x,y))$ gives a direction vector for this tangent line.

An illustrative example. Fig. 11.4 shows the direction field for the equation

(11.36) $$\frac{dy}{dx} = p(x,y) = x^2,$$

along with several solution curves. Since the equation $\frac{dy}{dx} = x^2$ is simple to solve, we see that the solution curves sketched in Fig. 11.4 have equations

$$y - \frac{x^3}{3} = C$$

for any constant C. Different values of C correspond to different solution curves.

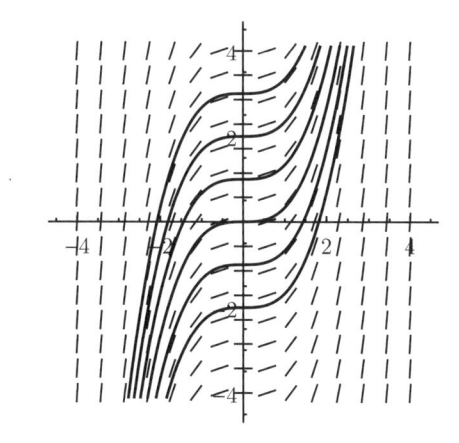

Figure 11.4. Direction field for $dy/dx = x^2$ with some solution curves.

Let's see how this helps us to solve the partial differential equation

(11.37) $$\frac{\partial u}{\partial x} + x^2 \frac{\partial u}{\partial y} = 0$$

(this is equation (11.34) with $p(x,y) = x^2$). The gradient of a function $u(x,y)$ is

$$\text{grad } u(x,y) = \left(\frac{\partial u}{\partial x}, \frac{\partial u}{\partial y}\right).$$

Equation (11.37) says that the dot product of the gradient vector grad u with the vector $(1, p(x,y))$ $= (1, x^2)$ must be zero. In a multivariable calculus course you might interpret this as saying "at the point (x,y), the directional derivative of u in the direction of the vector $(1, x^2)$ is zero". Thus at any point (x,y), the function u doesn't change in the direction given by $(1, x^2)$. This direction coincides with that of the line tangent to the solution curve at (x,y), and therefore u **is constant** along the solution curves of equation (11.36) (the curves $y - \frac{x^3}{3} = C$). The value of $u(x,y)$ only changes when we change which of the solution curves $y - x^3/3 = C$ we are on; in other words it only changes when we change the value of C. Thus u is a function of $C = y - \frac{x^2}{2}$. Here's how to say this in symbols: If $u(x,y)$ is a solution of (11.37), then

$$u(x,y) = f(C) = f\left(y - \frac{x^3}{3}\right)$$

for some one-variable differentiable function $f = f(s)$.

This gives us an enormous collection of solutions to (11.37). Choose any differentiable function $f(s)$ and set

$$u(x,y) = f\left(y - \frac{x^3}{3}\right)$$

(this just means replace s by $y - \frac{x^3}{3}$) to get a solution to (11.37). So, for example,

$$u(x,y) = e^{y - \frac{x^3}{3}}, \quad u(x,y) = \sin\left(y - \frac{x^3}{3}\right), \quad \text{and}$$

$$u(x,y) = 5\left(y - \frac{x^3}{3}\right)^3 - 3\left(y - \frac{x^3}{3}\right)$$

are each solutions to (11.37), coming from the choices $f(s) = e^s$, $f(s) = \sin s$, and $f(s) = 5s^3 - 3s$, respectively.

We can generalize this example as follows:

> Suppose $\varphi(x,y) = C$ solves the ordinary differential equation
>
> (11.38) $$\frac{dy}{dx} = p(x,y)$$
>
> and that $u = u(x,y)$ is differentiable. Then u is a solution to the partial differential equation
>
> (11.39) $$\frac{\partial u}{\partial x} + p(x,y)\frac{\partial u}{\partial y} = 0$$
>
> if and only if
>
> (11.40) $$u(x,y) = f(\varphi(x,y))$$
>
> for some differentiable function f.

The reasoning is just as above—if u is a solution of (11.39), then the dot product

$$\text{grad } u \cdot (1, p(x,y)) = 0,$$

and u is constant along any solution curve $\varphi(x,y) = C$ of (11.38). Thus u is a function of $C = \varphi(x,y)$ and we have

$$u(x,y) = f(\varphi(x,y))$$

for arbitrary (differentiable) functions f. Conversely, you can verify by substitution that (11.40) provides a solution to (11.39) when $\varphi(x,y) = C$ solves (11.38); see Exercise 3.

Example 11.4.1. Traveling waves. If you solve the first-order partial differential equation

$$\frac{\partial u}{\partial x} + k\frac{\partial u}{\partial y} = 0$$

11.4. Advection and diffusion

for constant k by the method just described, you will obtain the solutions $u(x,y) = f(y - kx)$, where f is any one-variable differentiable function (see Exercise 2). For the point we want to make about these solutions, it's helpful to think of one of the variables as representing time, so let's rewrite this using t instead of x. The equation is then $u_t + ku_y = 0$, and the solutions are $u(t,y) = f(y - kt)$. These solutions are called **traveling waves**. To see why this language is used, notice that at time $t = 0$ we have

$$u(0,y) = f(y),$$

and at a later time $t = t_1$ we have

$$u(t_1, y) = f(y - kt_1).$$

As a function of y, the graph of $f(y - kt_1)$ is just a translate of the graph of $f(y)$ (translated kt_1 units to the right if $k > 0$). Fig. 11.5 illustrates this by showing the graph of a function f and then "snapshot" pictures of $u(t,y) = f(y - kt)$ for times $0 < t_1 < t_2 < t_3$. The shape of f propagates unchanged as time goes on, with velocity of propagation k. (Note that in order for the units of the equation $u_t + ku_y = 0$ to balance, k must have units of velocity, that is, distance/time.)

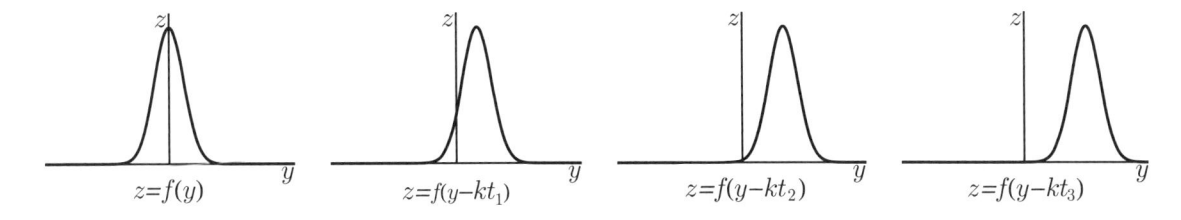

Figure 11.5. Traveling wave.

11.4.2. Applications: Advection and diffusion.

Example 11.4.2. An accident causes a factory to dump a large amount of a pollutant into a river. The river current carries the pollution downstream. Additionally, some of the pollutant decomposes due to the action of bacterial agents. How do we model this situation to determine the concentration of the pollutant at points on the river as time goes on?

To get started, we make an assumption that allows us to propose a model with only one space variable. We assume that the concentration of the pollutant depends only on time and how far down river we are (so that the pollution is assumed to be homogeneous in the other directions). Thus we visualize the river as a one-dimensional line, with the factory at the origin and "downstream" in the positive direction. We ultimately hope to determine the function $u(x,t)$ that gives the concentration of pollutant at time t and at point x downstream from the factory. To do this, we first develop a partial differential equation for $u(x,t)$. By our one-dimensional set-up, the units on $u(x,t)$ are units of mass/distance.

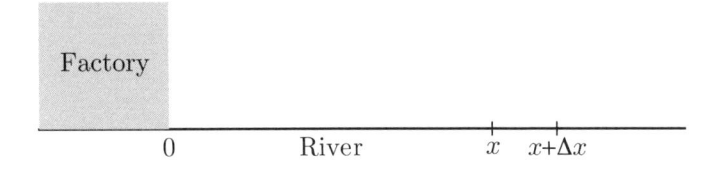

Figure 11.6. River with polluting factory at the origin.

At time t, the amount of pollutant present in a short stretch of river, from point x to point $x + \Delta x$, is *approximately* [1]
$$u(x,t)\Delta x.$$
How is this amount changing with time? Several factors come into play: Pollution is crossing over into the interval $[x, x + \Delta x]$ at the left endpoint and leaving the interval at the right endpoint (all due to the river current) and there is loss of pollutant in the interval due to bacterial decomposition.

If we assume the river carries the pollutant along at constant velocity κ (with units distance/time), then the rate at which pollutant flows into the stretch from x to $x + \Delta x$ is $\kappa u(x,t)$ (with units of mass/time), and the rate at which the pollutants flow out of this stretch of river is $\kappa u(x + \Delta x, t)$. If we assume that the pollutant is decomposed at a rate proportional to the amount of pollutant, then the amount lost to decomposition in the interval from x to $x + \Delta x$ is (approximately) $ru(x,t)\Delta x$ for some positive constant r. The units on r are 1/time. This leads us to an (approximate) equation for how the amount of pollution in the short stretch of river is changing with time:
$$\frac{\partial}{\partial t}(u(x,t)\Delta x) = \kappa u(x,t) - \kappa u(x + \Delta x, t) - ru(x,t)\Delta x.$$

Check that all terms in this equation have units of mass/time. If we divide by Δx, we have
$$\frac{\partial}{\partial t}(u(x,t)) = \kappa \frac{u(x,t) - u(x + \Delta x, t)}{\Delta x} - ru(x,t).$$
Taking the limit as $\Delta x \to 0$ we obtain the partial differential equation

(11.41)
$$\frac{\partial u}{\partial t} = -\kappa \frac{\partial u}{\partial x} - ru.$$

Equation (11.41) is called an **advection-decay equation**[2]. Notice how we have used the definition
$$\frac{\partial u}{\partial x} = \lim_{\Delta x \to 0} \frac{u(x + \Delta x, t) - u(x,t)}{\Delta x}$$
to obtain the first term in the right-hand side of this equation.

In Exercise 9 you are asked to show that if f is an arbitrary one-variable differentiable function, then

(11.42)
$$u(x,t) = e^{-rt} f(x - \kappa t)$$

solves (11.41). Notice that when $r = 0$, this reduces to a traveling wave solution discussed earlier. Moreover, we can choose f so that $u(x,0)$ has a prescribed value for all x, since for $u(x,t)$ as in equation (11.42) we have $u(x,0) = f(x)$. For example, if we want
$$u(x,0) = \frac{1}{1 + x^2},$$
then our solution is obtained by choosing
$$f(s) = \frac{1}{1 + s^2},$$
so that for $t > 0$
$$u(x,t) = e^{-rt} \frac{1}{1 + (x - \kappa t)^2}.$$

---

[1] The exact value would be $\int_x^{x+\Delta x} u(s,t)ds$, but for small values of Δx we can approximate this by $u(x,t)\Delta x$, by taking the value of the integrand at the left endpoint of the interval of integration and multiplying by the length of the interval. The smaller the value of Δx, the better this approximation.

[2] You can find a similar but more general version of the advection-decay equation discussed in *Mathematics for Dynamic Modeling* by Edward Beltrami; see [5]. Further applications to modeling pollution in a river can also be found there.

11.4. Advection and diffusion

Alternately, we can determine $u(x,t)$ if we know $u(x_0, t)$ for all $t > 0$ and some particular point at distance x_0 downstream from the factory; see Exercise 10. This corresponds to the situation where pollution levels are being monitored at one particular place on the river.

In the above discussion, the motion of the "stuff" that we were measuring was due to the motion of the water in the river; the pollutant was carried along by the current, a typical example of the process called advection. In other situations we might want to measure the concentration of a substance that moves "randomly". We'll look at an example of this next.

Example 11.4.3. Suppose that adjacent to a long straight stretch of coastline a no-fishing zone is to be created. Within that zone, a species of desirable fish has an optimal habitat so that it can grow exponentially (according to a Malthusian model). Beginning at the edge of the protected zone, fishing is so heavy that all members of this fish species are harvested. The fish move randomly, so that some move outside the zone and are lost. Is it possible that if the width of the protected zone is too small, the species will be driven to extinction?

Just as with our previous example, we will use one space dimension x in our model (see Fig. 11.7) which measures distance from shore and assume that the density of fish depends only on this distance x. We'll denote the density of fish at point x and time t by $u(x,t)$. Our first goal is to find a partial differential equation for u.

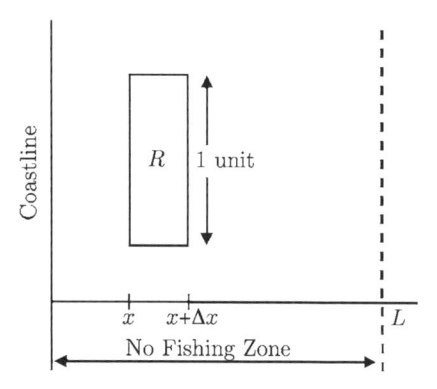

Figure 11.7. A no-fishing zone adjacent to a straight coastline.

We are thinking of u as "area density", with units of number of fish/unit area as seen from above. Fig. 11.7 presents a view of the water's surface from directly overhead. The vertical axis represents the shoreline, and the distance from shore is measured to the right along the positive x-axis. Image a thin rectangle R lying on the water's surface, with sides parallel to the axes, as shown in Fig. 11.7. The x-coordinates of the left and right edges of R are x and $x+\Delta x$, respectively. For simplicity, suppose the height of R is one unit (say, one kilometer), so that the area of R is $(\Delta x)(1) = \Delta x$. How many fish are in the solid block of water below R and above the sea floor? (Thinking two-dimensionally via our overhead view, we will call this the number of fish **in** R.) If the width Δx is very small, the value of u at time t and any point in R is approximately $u(x,t)$, the value of u on the left edge. Thus the number of fish in R at time t is approximately

$$u(x,t) \times \text{ area } R = u(x,t)\Delta x.$$

How is this changing with time? Temporarily ignoring the reproduction of the fish, we need to account for fish crossing into or out of R along its left or right edges (the top and bottom of R are so short that they can be ignored). In contrast to our earlier advection example, random motion means that fish can move in any direction, not just the direction of increasing x. It's just as likely that a fish just to the left of the left edge of R will cross into R as not (and just as likely that a fish just to the right of the left edge of R will cross out as not). Whether there is a net flow into or out of R along its left edge depends on whether there are more fish just to the left or just to the right of this edge. The density u expresses the variability in the distribution of fish, and we expect the net flow of fish to be from areas of higher to lower density. A precise mathematical statement in the current context is the following: The net rate of flow of fish into R along its left edge (in number of fish per unit of length, measured vertically along the left edge of R, per unit of time) at time t is

$$(11.43) \qquad -\kappa \frac{\partial u}{\partial x}(x,t).$$

Here κ is a positive constant with units distance^2/time, called the **diffusion coefficient**. Since R has unit height, the expression (11.43) is actually just the number of fish entering R though its left edge, per unit of time, at time t. This appealing fact is known as Fick's law. Although its full mathematical justification requires some careful thought about random motion, we can easily explain the presence of the minus sign in the constant of proportionality in (11.43). To see how this goes, note that if $\frac{\partial u}{\partial x}(x,t) > 0$, there are more fish just to the right of x than just to the left, and we expect a net flow of fish *out* of R along its left edge, so the overall sign of the flow rate here should be negative. Conversely, if $\frac{\partial u}{\partial x}(x,t) < 0$, we expect a net flow *into* R along its left edge, so the sign of the flow rate should be positive here. In either case, the minus sign in (11.43) guarantees that the whole quantity has the correct sign. We can also apply Fick's law to the right edge of R to find that the net rate of flow of fish into R on the right is

$$(11.44) \qquad \kappa \frac{\partial u}{\partial x}(x + \Delta x, t).$$

Here, the sign is positive because entering R means going left rather than right.

So without including the effect of fish reproduction in R, we can add (11.43) and (11.44) to find

$$(11.45) \qquad \frac{\partial}{\partial t}(u(x,t)\Delta x) = -\kappa \frac{\partial u}{\partial x}(x,t) + \kappa \frac{\partial u}{\partial x}(x + \Delta x, t)$$

for the time rate of change of the number of fish in R. To include the effect of reproduction, remember that we have approximately $u(x,t)\Delta x$ fish in R at time t, reproducing at a rate proportional to their number. Thus we add a term $ru(x,t)\Delta x$ to the right side of equation (11.45), where r is a positive constant. We obtain

$$\frac{\partial}{\partial t}(u(x,t)\Delta x) = -\kappa \frac{\partial u}{\partial x}(x,t) + \kappa \frac{\partial u}{\partial x}(x + \Delta x, t) + ru(x,t)\Delta x.$$

Dividing by Δx and letting $\Delta x \to 0$ we see that

$$(11.46) \qquad \frac{\partial u}{\partial t} = \kappa \frac{\partial^2 u}{\partial x^2} + ru.$$

This is a **diffusion** equation. Note that we used the fact that

$$\frac{\partial^2 u}{\partial x^2} = \lim_{\Delta x \to 0} \frac{\frac{\partial u}{\partial x}(x+\Delta x, t) - \frac{\partial u}{\partial x}(x,t)}{\Delta x}$$

when letting $\Delta x \to 0$. The second term on the right-hand side of (11.46) has the same form as that in equation (11.41), but it now appears with a positive sign since fish are being "created" by reproduction.

We want to attach some boundary conditions to this equation too. If the width of the protected zone is L, $u(L,t) = 0$ by our assumption that all fish which stray outside the protected zone are harvested. On the other hand, since there is no movement of fish onto land at $x = 0$, it is reasonable to require that $\frac{\partial u}{\partial x}(0,t) = 0$.

11.4. Advection and diffusion

Our question is: Are there values of L for which the problem

(11.47) $$\frac{\partial u}{\partial t} = \kappa \frac{\partial^2 u}{\partial x^2} + ru, \qquad \frac{\partial u}{\partial x}(0,t) = 0, \quad u(L,t) = 0$$

cannot lead to a sustainable fish population? Let's see what separation of variables has to tell us about solving (11.47). Setting $u(x,t) = X(x)T(t)$ in (11.47) gives us

$$X(x)T'(t) = \kappa X''(x)T(t) + rX(x)T(t).$$

Dividing this by $X(x)T(t)$ we have

$$\frac{T'(t)}{T(t)} = \kappa \frac{X''(x)}{X(x)} + r,$$

in which the left side is a function of t only and the right side is a function of x only (remember r is a constant). Thus

$$\frac{T'(t)}{T(t)} = \lambda = \kappa \frac{X''(x)}{X(x)} + r$$

for some constant λ.

We know that the equation

$$\frac{T'(t)}{T(t)} = \lambda$$

has solutions $T(t) = ce^{\lambda t}$ for constant c. Rewrite

$$\lambda = \kappa \frac{X''(x)}{X(x)} + r$$

as

(11.48) $$X'' + \frac{r-\lambda}{\kappa} X = 0.$$

This has characteristic equation $\kappa z^2 + (r - \lambda) = 0$. Since $\kappa > 0$, the solutions to this equation depend on three cases: $\lambda = r$, $\lambda > r$, or $\lambda < r$. Considering each case in turn, we solve (11.48) for $X(x)$ and obtain

Case (i), $\lambda = r$. $X(x) = c_1 + c_2 x$.

Case (ii), $\lambda > r$. $X(x) = c_1 \sinh(\alpha x) + c_2 \cosh(\alpha x)$, where $\alpha = \sqrt{(\lambda - r)/\kappa}$.

Case (iii), $\lambda < r$. $X(x) = c_1 \cos(\beta x) + c_2 \sin(\beta x)$, where

$$\beta = \sqrt{(r - \lambda)/\kappa}.$$

To meet the requirements $\frac{\partial u}{\partial x}(0,t) = 0$ and $u(L,t) = 0$ with a nontrivial solution $u(x,t) = X(x)T(t)$, we must have $X'(0) = 0$ and $X(L) = 0$. In Exercise 12 you are asked to show that this forces us to be in Case (iii), with $X(x) = \cos(\beta x)$, or a constant multiple of this, where

(11.49) $$\beta L = \frac{n\pi}{2}$$

for some *odd* integer n. Building-block solutions to (11.47) have the form

$$u(x,t) = e^{\lambda t} \cos(\beta x)$$

where
$$\beta = \frac{n\pi}{2L}$$
for odd n. Since $\beta = \sqrt{\frac{r-\lambda}{\kappa}}$, equation (11.49) says that
$$L^2 = \frac{n^2\pi^2\kappa}{4(r-\lambda)}$$
for some odd positive integer n. Taking sums of the building-block solutions, we have
(11.50) $$u(x,t) = \sum_{n \text{ odd}} c_n e^{\lambda_n t} \cos \frac{n\pi x}{2L}$$
where λ_n, for n odd, is determined from
$$\sqrt{\frac{r-\lambda_n}{\kappa}} = \frac{n\pi}{2L},$$
so that
(11.51) $$\lambda_n = r - \frac{\kappa n^2 \pi^2}{4L^2}.$$

If all the λ_n are negative, the solution $u(x,t)$ given in (11.50) tends to 0 as $t \to \infty$, which is the situation we want to avoid if we desire a sustainable fish population. The formula for λ_n in (11.51) shows that $\lambda_n < 0$ for all n precisely when $\lambda_1 < 0$, which is equivalent to
$$L < \frac{\pi}{2}\sqrt{\frac{\kappa}{r}}.$$

Thus our model gives the biological conclusion that if L, the width of the protected strip, is less than
$$\frac{\pi}{2}\sqrt{\frac{\kappa}{r}},$$
the fish population will not be sustainable.

11.4.3. Exercises.

1. For which of the following first-order partial differential equations does the Superposition Principle apply? To answer this, you must decide if $u_1(x,y) + u_2(x,y)$ is a solution whenever $u_1(x,y)$ and $u_2(x,y)$ are and if $cu(x,y)$ is a solution for arbitrary constant c whenever $u(x,y)$ is a solution.
   (a) $u_x + ku_y = 0$.
   (b) $u_x + xu_y = 0$.
   (c) $u_x + yu_y = 0$.
   (d) $u_x + uu_y = 0$.
   (e) $u_x + p(x,y)u_y = 0$.

2. Give a solution $u(t,y)$ to $u_t + ku_y = 0$ (where k is a constant) that satisfies the given initial condition.
   (a)
   $$u(0,y) = 4\sin\frac{y}{2} \quad \text{for all } y.$$
   (b)
   $$u(0,y) = e^{y^2} \quad \text{for all } y.$$

3. Show by substitution that for each one-variable (differentiable) function f, $u(x,y) = f(\varphi(x,y))$ is a solution to (11.39) when $\varphi(x,y) = C$ solves (11.38). Hint: You'll need the Chain Rule. Also, recall equation (2.123) from Section 2.8.

11.4. Advection and diffusion

In Exercises 4–7, use equations (11.38)–(11.39) to solve the given equation.

4. $2\dfrac{\partial u}{\partial x} + 3\dfrac{\partial u}{\partial y} = 0.$

5. $x\dfrac{\partial u}{\partial x} + (x+1)\dfrac{\partial u}{\partial y} = 0.$

6. $-x\dfrac{\partial u}{\partial x} + y\dfrac{\partial u}{\partial y} = 0.$

7. $y\dfrac{\partial u}{\partial x} + x\dfrac{\partial u}{\partial y} = 0.$

8. In Exercise 4, give a solution that meets the initial condition $u(0,y) = \sin y$.

9. Consider the advection-decay equation
$$\dfrac{\partial u}{\partial t} = -\kappa \dfrac{\partial u}{\partial x} - ru,$$
for constants κ and r.
   (a) Show by substitution that
   $$u(x,t) = e^{-rt}\sin(x - \kappa t)$$
   solves the advection-decay equation.
   (b) Show that
   $$u(x,t) = e^{-rt}(x - \kappa t)^3$$
   solves the advection-decay equation.
   (c) Show that
   $$u(x,t) = e^{-rt}e^{x-\kappa t} = e^x e^{-(r+\kappa)t}$$
   solves the advection-decay equation.
   (d) Show that for any differentiable function $f(s)$ of a single variable, the function $u(x,t) = e^{-rt}f(x - \kappa t)$ is a solution of the advection-decay equation. Hint: Compute $\frac{\partial u}{\partial t}$ and $\frac{\partial u}{\partial x}$ using the Chain Rule as needed, and then check that u is a solution by substitution.

10. Suppose in our example of pollution downstream from a factory (equation (11.41)) we know $u(x_0, t) = g(t)$ for all $t > 0$ and some x_0 representing a point downstream from the factory. Show that if you choose
$$f(s) = e^{r(x_0 - s)/\kappa} g\left(\dfrac{x_0 - s}{\kappa}\right),$$
then
$$u(x,t) = e^{-rt}f(x - \kappa t)$$
solves equation (11.41) and satisfies the desired boundary conditions.

11. Suppose that a clump of vegetation floats downstream at a rate of 500 meters/hr along a straight portion of a river. Suppose that at time $t = 0$ the vegetation is centered at position 0 and extends a total of 4 meters as pictured below.

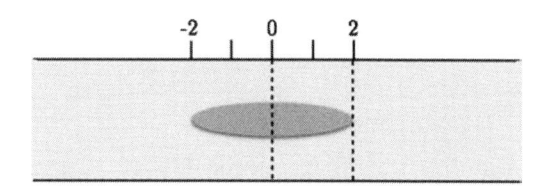

The mass density of the vegetation at time $t = 0$ per meter swath along the river is given by
$$f(x) = 4 - x^2 \text{ kg/meter}, \quad -2 \le x \le 2;$$

hence, e.g., at $t = 0$ the total mass of vegetation in the shaded portion of the stream, between $x = 0$ and $x = 2$, is
$$\int_0^2 \left(4 - s^2\right) ds = 16/3 \text{ kg}.$$
Hungry fish eat the vegetation as it floats down the river so that the equation modeling its linear mass density is

(11.52) $$\frac{\partial u}{\partial t} + 500 \frac{\partial u}{\partial x} + 0.01 u = 0,$$

where $u(x,t)$ is the mass density of the vegetation per meter swath at position x along the river at time t, where x is measured in meters and t is measured in hours.

(a) Solve (11.52) subject to the initial condition $u(x,0) = f(x) = 4 - x^2$.

(b) After $t = 5$ hours, what is the mass of the vegetation in the swath of the river from $x = 2{,}501$ to $x = 2{,}502$?

12. In Example 11.4.3, we found the general solution to
$$X''(x) + \frac{r - \lambda}{\kappa} X(x) = 0 \quad (\kappa > 0)$$
in the three cases $\lambda = r$, $\lambda > r$, and $\lambda < r$. Show that if we want to satisfy the boundary conditions
$$X'(0) = 0, \quad X(L) = 0,$$
with a nontrivial function $X(x)$, we must have $\lambda < r$. Verify that with $\lambda < r$, the solutions satisfying these boundary conditions are $X(x) = c \cos(\beta x)$ where
$$\beta = \sqrt{\frac{r - \lambda}{\kappa}}$$
and
$$\beta L = \frac{n\pi}{2}$$
for some odd integer n.

13. In the protected fishing setting of Example 11.4.3, does the minimum width of the strip for a sustainable population increase or decrease as the reproduction rate r increases?

14. Suppose some limited fishing is permitted in the "protected strip" of Example 11.4.3, so that
$$u_t = \kappa u_{xx} + ru - Eu$$
for some small positive constant E, reflecting that the fishing is permitted "at a rate proportional to the population size". As in Example 11.4.3, $u_x(0,t) = 0$, $u(L,0) = 0$. What is the minimal width L necessary to avoid a collapse of the fish population? Your answer should involve r and E.

15. Suppose in Example 11.4.3 the protected strip is in the open ocean, bounded at $x = 0$ and $x = L$ by open water into which the fish can cross. This changes the boundary conditions in equation (11.47). What are the new conditions? Show that in this case the minimum width of the protected strip for a sustainable fish population is
$$\pi \sqrt{\frac{\kappa}{r}},$$
twice the value of the minimum width in Example 11.4.3.

16. Suppose the density $u(x,t)$ of fish in Example 11.4.3 satisfies not only the partial differential equation and boundary conditions in equation (11.47) but also the initial condition $u(x,0) = \cos \frac{\pi x}{2L}$.

(a) Find $u(x,t)$ for all $t > 0$.

11.5. Functions as vectors

(b) We know that if $L < \frac{\pi}{2}\sqrt{\frac{\kappa}{r}}$, $u(x,t) \to 0$ as $t \to \infty$. If $L > \frac{\pi}{2}\sqrt{\frac{\kappa}{r}}$, what happens to $u(x,t)$ as t increases?

(c) Answer the same question if $L = \frac{\pi}{2}\sqrt{\frac{\kappa}{r}}$.

17. In this problem we compare the equation

(11.53) $$u_t = \kappa u_{xx} + ru,$$

where r is a constant, and the equation

(11.54) $$u_t = \kappa u_{xx}.$$

(a) Show that if $u_1(x,t)$ solves (11.53), then $e^{-rt}u_1(x,t)$ solves (11.54).
(b) Show that if $u_2(x,t)$ solves (11.54), then $e^{rt}u_2(x,t)$ solves (11.53).
(c) Using (b) and the building-block solutions in equation (11.21), solve

$$u_t = u_{xx} + 2u, \qquad 0 < x < 1, \quad t > 0,$$

with boundary conditions

$$u(0,t) = 0 = u(1,t), \quad t > 0$$

and initial condition

$$u(x,0) = 3\sin(2\pi x) - 4\sin(5\pi x), \quad 0 \le x \le 1.$$

11.5. Functions as vectors

In Sections 11.3 and 11.4 we used separation of variables to produce "building-block" solutions to certain partial differential equations. Since these equations also obey the Superposition Principle, we could take finite linear combinations of these building blocks to obtain yet more solutions. In Sections 11.6 and 11.7 we will see how to pass from finite linear combinations to infinite series, in the process obtaining many more solutions. In so doing, an analogy between functions and vectors will help us understand what is going on. We turn to this vector analogy in Section 11.5.2 but begin here with some basic properties of functions.

11.5.1. Some concepts for functions. A function $f(x)$ defined on the real line is **even** if $f(x) = f(-x)$ for all x. Its graph will be symmetric about the y-axis, so that if (a,b) is on the graph of an even function, so is $(-a,b)$. If instead $f(-x) = -f(x)$ for all x, we say f is **odd**. In this case the graph of f will be symmetric with respect to the origin; if (a,b) is a point on the graph of an odd function, then so is $(-a,-b)$.

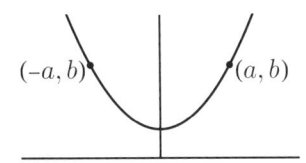

Figure 11.8. An even function.

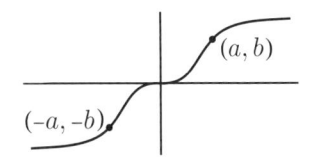

Figure 11.9. An odd function.

Even functions include $\cos x$ and polynomials like $p(x) = x^4 + 3x^2 - 5$ for which all powers of x are even. The function $\sin x$ is odd, as are polynomials like $p(x) = x^9 - 4x^5 + x^3 + 2x$ for which all powers of x are odd.

We will find it natural in this chapter to work with functions that have certain kinds of "mild" discontinuities. The main ideas are contained in the following definitions.

Definition 11.5.1. A function $f(x)$ defined on an open interval in the real line is said to be **piecewise continuous** on that interval if it has at most a finite number of discontinuities and at each discontinuity the right-hand and left-hand limits of $f(x)$ exist as finite numbers.[3]

If the interval in question is a closed interval $[a, b]$, then at $x = a$ we only require the existence of a finite *right-hand* limit for f, and at $x = b$, only a finite *left-hand* limit is required to exist. Moreover, in Definition 11.5.1, we don't actually need to require that f even be defined at the (finitely) many points of discontinuity.

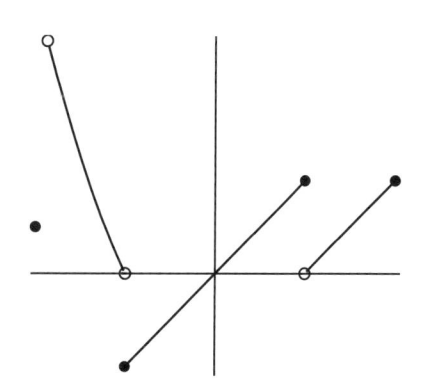

Figure 11.10. Piecewise continuous on $[-2, 2]$.

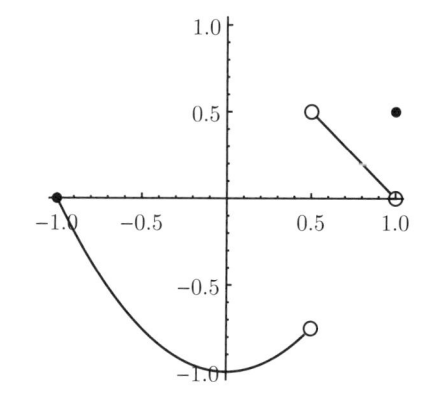

Figure 11.11. Piecewise continuous on $[-1, 1]$, not defined at $x = 1/2$.

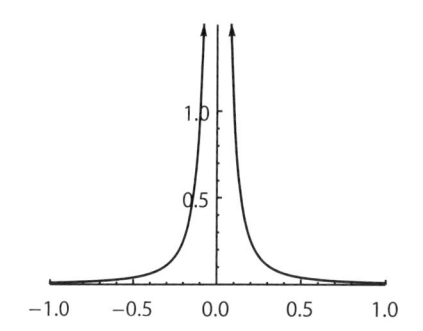

Figure 11.12. Not piecewise continuous on $[-1, 1]$; $\lim_{x \to 0^+} f(x) = \lim_{x \to 0^-} f(x) = \infty$.

Definition 11.5.2. A function $f(x)$ is said to be **piecewise continuously differentiable** on an interval if f and f' have at most a finite number of discontinuities in the interval and at each discontinuity both f and f' have finite right- and left-hand limits (if the interval is closed, then we only require the appropriate one-sided limits at the endpoints of the interval).

So a function f is piecewise continuously differentiable if both f and f' are piecewise continuous. The functions shown in Figs. 11.10 and 11.11 are piecewise continuously differentiable.

Recall that a function $f(x)$ is said to be **periodic** if there is a number $T > 0$ so that

(11.55) $$f(x + T) = f(x)$$

for all x in the domain of f. Note that for this to make sense, it must be the case that $x + T$ is in the domain of f whenever x is in the domain of f. Such a number T is called a period of f. If f is continuous and periodic but not constant, then we sometimes distinguish the *smallest $T > 0$*

---

[3]If the open interval in the definition is infinite, we can allow infinitely many discontinuities provided that there are only finitely many discontinuities in each finite subinterval.

11.5. Functions as vectors

for which (11.55) holds and call this the fundamental period. So while $f(x) = \cos x$ has periods $2\pi, 4\pi, 6\pi$, and so on, its fundamental period is 2π.

For a function defined on an interval $[0, L]$, we will consider two ways of extending f to be periodic, with period $2L$, on the whole real line, the *even* and *odd* periodic extension of f.

Even periodic extension. Starting with a function $f(x)$ defined on $[0, L]$, we first extend f to be an even function on $[-L, L]$ by setting $f(-x) = f(x)$ for $0 \leq x \leq L$. This produces a function defined on $[-L, L]$ whose graph is symmetric about the y-axis. As the second step, we extend the function (now defined on $[-L, L]$) to all of $\mathbb{R}$ by making it periodic, with period $2L$. In other words, define f for all x by the requirement $f(x) = f(x + 2L)$.

For example, suppose we start with the function

$$f(x) = \begin{cases} 1 & \text{if } 0 \leq x \leq 1, \\ 2 & \text{if } 1 < x \leq 2. \end{cases}$$

The graph of f on $[0, 2]$ is shown in Fig. 11.13.

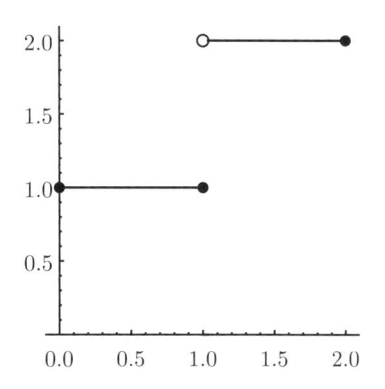

Figure 11.13. Graph of f on $[0, 2]$.

Its even extension to $[-2, 2]$, which we still call f, is given by

$$f(x) = \begin{cases} 1 & \text{if } -1 \leq x \leq 1, \\ 2 & \text{if } -2 \leq x < -1 \text{ or } 1 < x \leq 2. \end{cases}$$

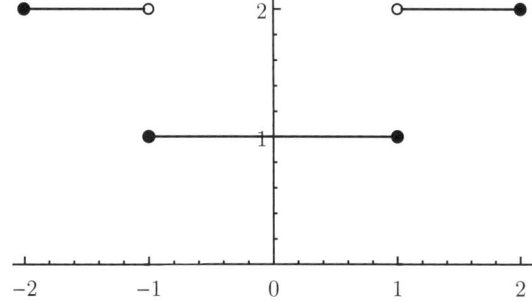

Figure 11.14. Even extension of f to $[-2, 2]$.

Finally, Fig. 11.15 shows the resulting graph on $[-6, 6]$ when the periodic extension is carried out to extend the function graphed in Fig. 11.14.

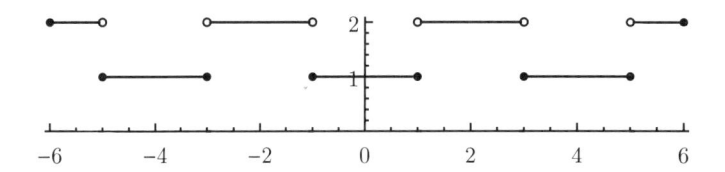

Figure 11.15. Periodic extension of f pictured on $[-6, 6]$.

Example 11.5.3. We'll sketch the graph of the even periodic extension of the function

$$g(x) = \begin{cases} x & \text{if } 0 \leq x \leq 2, \\ 4 - x & \text{if } 2 < x \leq 4, \end{cases}$$

which is shown in Fig 11.16. Its even extension to $[-4, 4]$ is shown in Fig. 11.17.

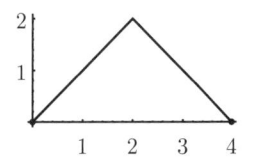

Figure 11.16. Graph of $y = g(x)$.

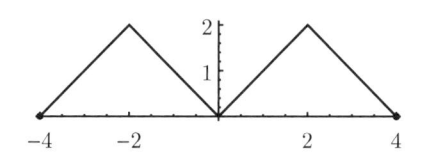

Figure 11.17. Even extension of $y = g(x)$ to $[-4, 4]$.

The periodic extension to $\mathbb{R}$ is shown in Fig. 11.18.

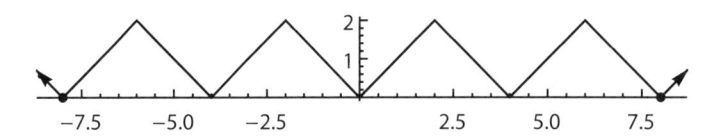

Figure 11.18. Even periodic extension of $y = g(x)$.

It has period 8, but fundamental period 4.

Conflict points. Starting with a (not necessarily even) function defined on $[-L, L]$, the prescription to "create its periodic extension" to all of $\mathbb{R}$ can create conflicts at a few points, arising from the endpoints $x = L$ and $x = -L$. For example, suppose we want to give a periodic extension of the function f defined on $[-2, 2]$ whose graph there is as shown in Fig. 11.19:

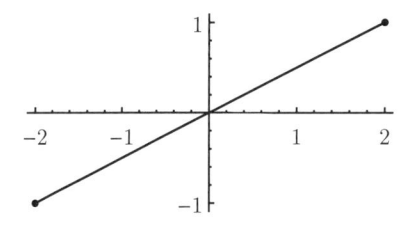

Figure 11.19. A function f on $[-2, 2]$.

Extending it periodically, with period 4, we already have an issue at $x = 2$. Its original value there is 1, but periodicity says that we should have $f(2) = f(-2 + 4) = f(-2) = -1$. The same issue arises at $x = 6, 10, \ldots$. For our purposes we can resolve these "endpoint" conflicts any way we want. Figs. 11.20–11.22 show three possible "resolutions".

11.5. Functions as vectors

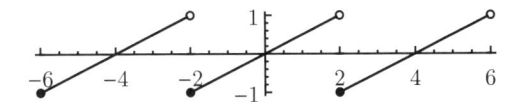

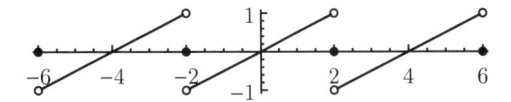

Figure 11.20. One way to resolve endpoint conflicts.

Figure 11.21. Another way to resolve endpoint conflicts.

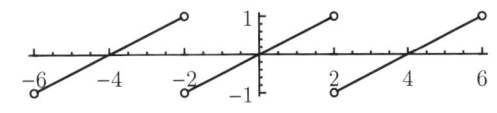

Figure 11.22. Another resolution of conflicts; extension not defined at $x = \pm 2, \pm 6, \pm 10, \ldots$.

If we initially think of f as defined on either half-open interval $[-L, L)$ or $(-L, L]$, then no conflict points will arise when we extend it $2L$-periodically to $\mathbb{R}$.

Odd periodic extension. Consider again the function $g(x)$ of Example 11.5.3. We extend this function, initially defined on $[0, 4]$, to the interval $[-4, 4]$ so as to obtain an *odd* function. The graph will look like this:

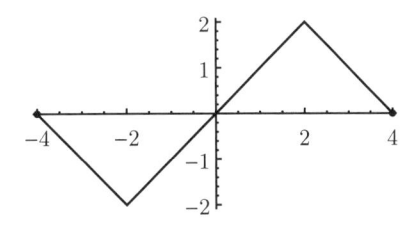

Figure 11.23. Odd extension of g to $[-4, 4]$.

We can describe this extension to $[-4, 0)$ by

$$g(x) = \begin{cases} x & \text{if } -2 \leq x < 0, \\ -4 - x & \text{if } -4 \leq x \leq -2. \end{cases}$$

We'll continue to call the extended function $g(x)$. Next, we extend g from $[-4, 4]$ to all of $\mathbb{R}$ periodically. It will have period 8, and thus $g(x + 8) = g(x)$ for all real x.

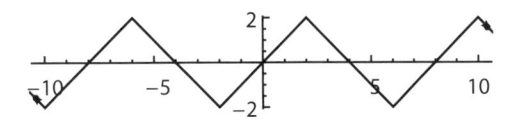

Figure 11.24. Periodic extension of g.

In generalizing this two-step process to create the odd periodic extension to all of $\mathbb{R}$ of an arbitrary function defined initially on $[0, L]$, there can be potential conflicts at a few points. The next example illustrates this.

Example 11.5.4. Begin with the function defined on $[0, 2]$ described by

$$f(x) = \begin{cases} 1 & \text{if } 0 \leq x \leq 1, \\ 2 & \text{if } 1 < x \leq 2. \end{cases}$$

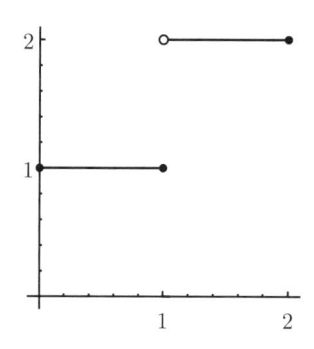

Figure 11.25. Graph of f.

We'll first do the odd extension to $[-2, 0)$. Graphically this means that whenever (a, b) is on the graph of f, then $(-a, -b)$ should also be on the graph. This tells us clearly that the points $(-x, -1)$ for $0 < x \leq 1$ belong to the graph, as do the points $(-x, -2)$ for $1 < x \leq 2$. However at $x = 0$ we have a conflict: We started with the point $(0, 1)$ on the graph of f, and we can't also have $(0, -1)$ on the graph of the extended function. We can resolve this conflict by changing the value of f at $x = 0$ to zero ($f(0) = 0$). In fact, *any* odd function defined at 0 must satisfy $f(0) = 0$. On $[-2, 2]$ the graph now looks like this:

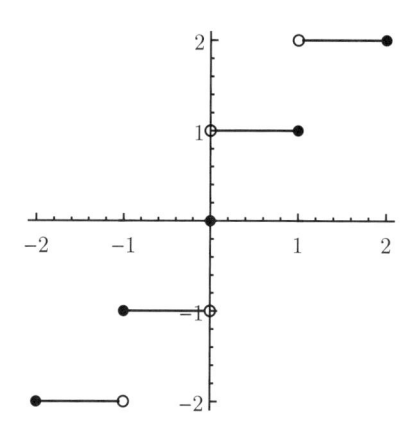

Figure 11.26. Odd extension of f to $[-2, 2]$.

Our last step is to give the periodic extension to all of $\mathbb{R}$, with period 4. Again, there will be conflicts at a few points. For example, we have $f(-2) = -2$ and $f(2) = 2$ (see Fig. 11.26). Periodicity says $f(x) = f(x+4)$ for all x, and so $f(-2) = f(-2+4) = f(2) = 2$ which is in conflict with $f(-2) = -2$. Again, we can resolve this conflict as we want. Our choice here is to change $f(-2)$ and $f(2)$ to both be 0 and use periodicity to determine its values everywhere else.

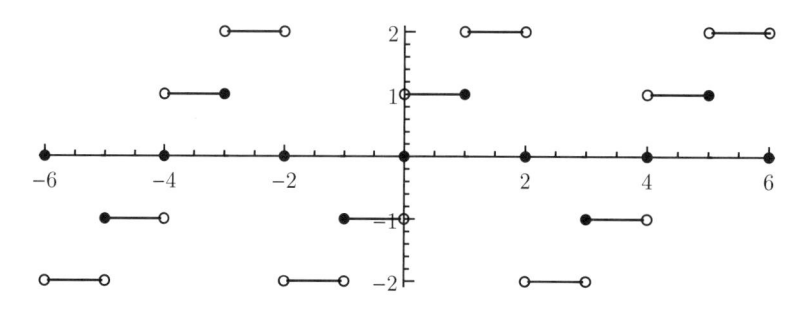

Figure 11.27. Periodic extension of f.

11.5.2. Some vector concepts. We will be extending some ideas from the study of vectors in $\mathbb{R}^2$ or $\mathbb{R}^3$; so we begin with a brief review. Consider, for example, two vectors in $\mathbb{R}^3$. Let $\mathbf{A} = (a_1, a_2, a_3)$ and $\mathbf{B} = (b_1, b_2, b_3)$. The **dot product** of $\mathbf{A}$ and $\mathbf{B}$ is

$$\mathbf{A} \cdot \mathbf{B} = a_1 b_1 + a_2 b_2 + a_3 b_3.$$

Recall that we say $\mathbf{A} \perp \mathbf{B}$ ($\mathbf{A}$ is **orthogonal**, or **perpendicular**, to $\mathbf{B}$) if and only if $\mathbf{A} \cdot \mathbf{B} = 0$.

The **norm** or **length** of $\mathbf{A}$ is

$$\|\mathbf{A}\| = \sqrt{\mathbf{A} \cdot \mathbf{A}} = \sqrt{a_1^2 + a_2^2 + a_3^2}.$$

In general, the dot product of two vectors is related to the lengths of the vectors and the angle θ between the vectors by the formula $\mathbf{A} \cdot \mathbf{B} = \|\mathbf{A}\| \|\mathbf{B}\| \cos \theta$, illustrated in Fig. 11.28.

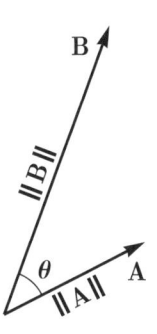

Figure 11.28. $\mathbf{A} \cdot \mathbf{B} = \|\mathbf{A}\| \|\mathbf{B}\| \cos \theta$.

The dot product can be just as easily defined for vectors in $\mathbb{R}^n$. If $\mathbf{A} = (a_1, a_2, \ldots, a_n)$ and $\mathbf{B} = (b_1, b_2, \ldots, b_n)$, then

$$\mathbf{A} \cdot \mathbf{B} = a_1 b_1 + a_2 b_2 + \cdots + a_n b_n.$$

The vectors $\mathbf{A}$ and $\mathbf{B}$ are orthogonal if this dot product is zero. The length of $\mathbf{A}$ is again

$$\|\mathbf{A}\| = \sqrt{\mathbf{A} \cdot \mathbf{A}} = \sqrt{a_1^2 + a_2^2 + \cdots + a_n^2}.$$

The dot product has a number of properties that follow easily from the definition.

Theorem 11.5.5. *For vectors $\mathbf{A}$, $\mathbf{B}$, and $\mathbf{C}$ in $\mathbb{R}^n$ and a real number s, we have:*

(a) $\mathbf{A} \cdot \mathbf{A} \geq 0$ *with* $\mathbf{A} \cdot \mathbf{A} = 0$ *if and only if* $\mathbf{A} = (0, 0, \ldots, 0)$.

(b) $\mathbf{A} \cdot \mathbf{B} = \mathbf{B} \cdot \mathbf{A}$ *(commutativity)*.

(c) $\mathbf{A} \cdot (\mathbf{B} + \mathbf{C}) = \mathbf{A} \cdot \mathbf{B} + \mathbf{A} \cdot \mathbf{C}$ *(distributivity)*.

(d) $\mathbf{A} \cdot (s\mathbf{B}) = s(\mathbf{A} \cdot \mathbf{B}) = (s\mathbf{A}) \cdot \mathbf{B}$.

A **unit vector** is a vector of length one. In $\mathbb{R}^2$ we cannot have a family of *mutually orthogonal unit vectors* (the terminology "pairwise orthogonal" is sometimes used instead of "mutually orthogonal") containing more than two vectors. In $\mathbb{R}^3$ a family of mutually orthogonal unit vectors cannot contain more than three vectors, and in $\mathbb{R}^n$ the maximum size of a family of mutually orthogonal unit vectors is n. Perhaps the most natural family of n mutually orthogonal unit vectors in $\mathbb{R}^n$ has as its jth member the vector with 1 as its jth component and zeros elsewhere (where j is in the set $\{1, 2, \ldots, n\}$).

We are going to investigate writing a vector in $\mathbb{R}^n$ as a linear combination of mutually orthogonal unit vectors. Let's start with $\mathbb{R}^2$ where it will be easy to draw pictures. Suppose we have any pair of orthogonal unit vectors $\mathbf{E}_1$ and $\mathbf{E}_2$ in $\mathbb{R}^2$ and an arbitrary vector $\mathbf{A}$ in $\mathbb{R}^2$ as well. Suppose we are able to write $\mathbf{A}$ as a linear combination of $\mathbf{E}_1$ and $\mathbf{E}_2$:

(11.56) $$\mathbf{A} = c_1\mathbf{E}_1 + c_2\mathbf{E}_2.$$

What must the constants c_1 and c_2 be? In equation (11.56), take the dot product on both sides with the vector $\mathbf{E}_1$:

$$\mathbf{A} \cdot \mathbf{E}_1 = (c_1\mathbf{E}_1 + c_2\mathbf{E}_2) \cdot \mathbf{E}_1 = c_1(\mathbf{E}_1 \cdot \mathbf{E}_1) + c_2(\mathbf{E}_2 \cdot \mathbf{E}_1).$$

Note how we have used properties from Theorem 11.5.5. Since $\mathbf{E}_1 \perp \mathbf{E}_2$ this simplifies to

$$\mathbf{A} \cdot \mathbf{E}_1 = c_1(\mathbf{E}_1 \cdot \mathbf{E}_1).$$

But $\mathbf{E}_1$ is a *unit* vector, so $\mathbf{E}_1 \cdot \mathbf{E}_1 = 1$ and

$$c_1 = \mathbf{A} \cdot \mathbf{E}_1.$$

Similarly, you should be able to verify (see Exercise 14) that

$$c_2 = \mathbf{A} \cdot \mathbf{E}_2.$$

In a calculus class you might have called the vectors

$$(\mathbf{A} \cdot \mathbf{E}_1)\mathbf{E}_1$$

and

$$(\mathbf{A} \cdot \mathbf{E}_2)\mathbf{E}_2$$

the "projections of $\mathbf{A}$ onto $\mathbf{E}_1$ and $\mathbf{E}_2$", respectively. It is indeed the case that the vector $\mathbf{A}$ is the sum of these two projections, so that equation (11.56) holds with $c_1 = \mathbf{A} \cdot \mathbf{E}_1$ and $c_2 = \mathbf{A} \cdot \mathbf{E}_2$; see the figure below.

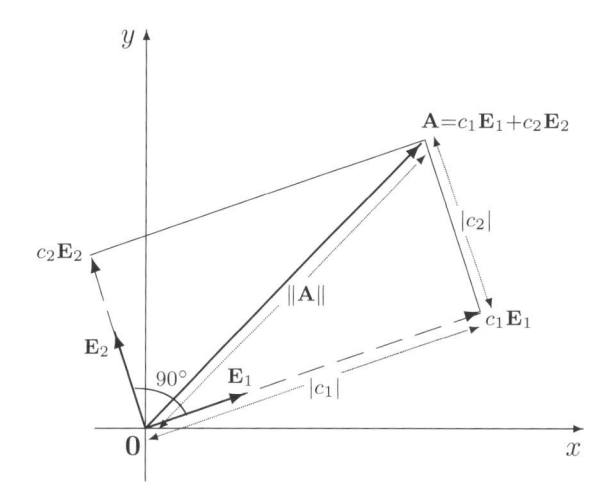

Regarding norms, we have

$$\begin{aligned}\|\mathbf{A}\|^2 &= \mathbf{A} \cdot \mathbf{A} = \mathbf{A} \cdot (c_1\mathbf{E}_1 + c_2\mathbf{E}_2) \\ &= c_1\mathbf{A} \cdot \mathbf{E}_1 + c_2\mathbf{A} \cdot \mathbf{E}_2 = c_1^2 + c_2^2 \quad \text{(Pythagorean Theorem!!)}.\end{aligned}$$

The analogue in $\mathbb{R}^3$ is as follows. If $\mathbf{E}_1$, $\mathbf{E}_2$, and $\mathbf{E}_3$ are any three mutually orthogonal unit vectors in $\mathbb{R}^3$ and $\mathbf{A}$ is an arbitrary vector in $\mathbb{R}^3$, then

(11.57) $$\mathbf{A} = c_1\mathbf{E}_1 + c_2\mathbf{E}_2 + c_3\mathbf{E}_3$$

11.5. Functions as vectors

where
$$c_1 = \mathbf{A} \cdot \mathbf{E}_1, \quad c_2 = \mathbf{A} \cdot \mathbf{E}_2, \quad \text{and} \quad c_3 = \mathbf{A} \cdot \mathbf{E}_3.$$
You can use equation (11.57) to see that $\|\mathbf{A}\|^2 = \mathbf{A} \cdot \mathbf{A} = c_1^2 + c_2^2 + c_3^2$; see Exercise 14. Exercise 14 also asks you to formulate the analogue of these results in $\mathbb{R}^n$.

We want to expand some of these ideas beyond $\mathbb{R}^n$. Let's mentally replace $\mathbb{R}^n$ by the collection of all piecewise continuous real-valued functions on some interval $[-L, L]$ in the real line. Define a dot product-like operation for this collection of functions by setting

$$(11.58) \qquad \langle f, g \rangle = \frac{1}{L} \int_{-L}^{L} f(x) g(x) dx.$$

This is called the **inner product**[4] of f and g. Note that $\langle f, g \rangle$ is a *number*. The inner product has the following properties, which should remind you of the properties of the dot product in Theorem 11.5.5.

-
$$\langle f, f \rangle = \frac{1}{L} \int_{-L}^{L} f(x)^2 dx \geq 0.$$

-
$$\langle f, g \rangle = \langle g, f \rangle.$$

-
$$\langle f, g + h \rangle = \langle f, g \rangle + \langle f, h \rangle$$

  since
$$\frac{1}{L} \int_{-L}^{L} f(x)(g(x) + h(x))dx = \frac{1}{L} \int_{-L}^{L} f(x)g(x)dx + \frac{1}{L} \int_{-L}^{L} f(x)h(x)dx.$$

- For any constant s,
$$\langle sf, g \rangle = \frac{1}{L} \int_{-L}^{L} sf(x)g(x)dx = \langle f, sg \rangle = s\langle f, g \rangle.$$

The last equality follows since we can move the constant s outside of the integral sign. Notice that, for a *continuous* function f, $\langle f, f \rangle = 0$ if and only if $f(x) \equiv 0$ for all x in $[-L, L]$.

Again by analogy, let's call $\sqrt{\langle f, f \rangle}$ the **norm** of the function f on the interval $[-L, L]$. Thus we define
$$\|f\| = \sqrt{\langle f, f \rangle} = \left(\frac{1}{L} \int_{-L}^{L} |f(x)|^2 dx \right)^{1/2}.$$

We will say that the functions f and g are **orthogonal on** $[-L, L]$ if $\langle f, g \rangle = 0$, that is, if
$$\frac{1}{L} \int_{-L}^{L} f(x)g(x)dx = 0.$$

Caution: *Don't* expect to see this kind of orthogonality as some kind of right angle in the graphs of f and g. But *do* consider this exchange when University of Michigan law professor Richard Friedman argued before the Supreme Court on January 11, 2010:[5]

> Friedman: "I think that issue is entirely orthogonal to the issue here..." Chief Justice Roberts: "I'm sorry. Entirely what?" Friedman: "Orthogonal. Right angle. Unrelated. Irrelevant." Roberts: "Oh." Justice Scalia: "What was that adjective? I like that." Friedman: "Orthogonal." Justice Scalia: "Orthogonal?" Justice

---

[4] Technically, this is a semi-inner product, but we will use the briefer terminology.
[5] *Briscoe v. Virginia*, Number 07-11191.

Scalia: "Ooh... I think we should use that in the opinion." Justice Roberts: "Or the dissent."

Since the inner product involves an integral over $[-L, L]$, this is a good place to recall from calculus the relationship of such integrals to even and odd functions: If $f(x)$ is even on $[-L, L]$, then
$$\int_{-L}^{L} f(x)\,dx = 2\int_{0}^{L} f(x)\,dx.$$
On the other hand, if $g(x)$ is odd on $[-L, L]$, then
$$\int_{-L}^{L} g(x)\,dx = 0.$$
It's also helpful to keep in mind that the product of two even functions and the product of two odd functions are both even, while the product of an odd function and an even function is odd. In shorthand, **(even)(even) = even**, **(odd)(odd) = even**, and **(odd)(even) = odd**. You are invited to verify these rules in Exercise 1. These facts can sometimes simplify integrations of such products.

The next example shows that we can have *infinitely* many mutually orthogonal functions on $[-L, L]$, each having norm 1.

Example 11.5.6. On any interval $[-L, L]$, consider the functions
$$\{\sin \frac{n\pi x}{L}\}_{n=1}^{\infty} \quad \text{and} \quad \{\cos \frac{m\pi x}{L}\}_{m=1}^{\infty}$$
with the inner product as defined by (11.58). We claim that

(11.59) $$\left\langle \sin \frac{n\pi x}{L}, \sin \frac{m\pi x}{L} \right\rangle = \frac{1}{L}\int_{-L}^{L} \sin \frac{n\pi x}{L} \sin \frac{m\pi x}{L}\,dx = \begin{cases} 0 & \text{if } n \neq m, \\ 1 & \text{if } n = m, \end{cases}$$

(11.60) $$\left\langle \cos \frac{n\pi x}{L}, \cos \frac{m\pi x}{L} \right\rangle = \frac{1}{L}\int_{-L}^{L} \cos \frac{n\pi x}{L} \cos \frac{m\pi x}{L}\,dx = \begin{cases} 0 & \text{if } n \neq m, \\ 1 & \text{if } n = m, \end{cases}$$

and

(11.61) $$\left\langle \sin \frac{n\pi x}{L}, \cos \frac{m\pi x}{L} \right\rangle = \frac{1}{L}\int_{-L}^{L} \sin \frac{n\pi x}{L} \cos \frac{m\pi x}{L}\,dx = 0 \quad \text{for all } n, m = 1, 2, 3, \ldots.$$

Once verified, this claim tells us that the infinitely many functions
$$\{\sin \frac{n\pi x}{L}\}_{n=1}^{\infty} \quad \text{and} \quad \{\cos \frac{m\pi x}{L}\}_{m=1}^{\infty}$$
are mutually orthogonal and that each of these functions has norm 1. One way to verify equations (11.59)–(11.60) is to use the trigonometric identities

(11.62) $$\cos u \cos v = \frac{1}{2}(\cos(u+v) + \cos(u-v)),$$

(11.63) $$\sin u \sin v = \frac{1}{2}(\cos(u-v) - \cos(u+v))$$

to compute the integrals; see Exercise 15. Notice that equation (11.61) follows with no computation at all simply by observing that the integrand
$$F(x) = \sin \frac{n\pi x}{L} \cos \frac{m\pi x}{L}$$
is an odd function, so that when we integrate over the symmetric interval from $-L$ to L we must get 0.

11.5.3. An eigenfunction approach to orthogonality.

There is another approach to the $n \neq m$ cases of equations (11.59)–(11.60) which connects them to work we did in Section 11.2. Recall that the functions
$$f_n(x) = \sin\frac{n\pi x}{L}$$
on the interval $[0, L]$ are solutions to the boundary value problem

(11.64) $$y'' + \frac{n^2\pi^2}{L^2}y = 0, \quad y(0) = 0, \quad y(L) = 0.$$

In other language, we said that these functions f_n are eigenfunctions corresponding to the eigenvalues $\lambda_n \equiv \frac{n^2\pi^2}{L^2}$. The principle we want to verify is that *eigenfunctions of* (11.64) *corresponding to distinct eigenvalues are orthogonal*. For equation (11.59), we want to show that $\int_{-L}^{L} f_n(x)f_m(x)dx = 0$ whenever $n \neq m$. Since for every n and m the product of the two odd functions $f_n(x) = \sin\frac{n\pi x}{L}$ and $f_m(x) = \sin\frac{m\pi x}{L}$ is an even function,
$$\int_{-L}^{L} f_n(x)f_m(x)dx = 2\int_0^L f_n(x)f_m(x)dx$$
and if our goal is to show that the left-hand side is 0, it suffices to show that $\int_0^L f_n(x)f_m(x)dx = 0$.

With $\lambda_n = \frac{n^2\pi^2}{L^2}$ and $\lambda_m = \frac{m^2\pi^2}{L^2}$, we know that

(11.65) $$f_n'' + \lambda_n f_n = 0$$

and

(11.66) $$f_m'' + \lambda_m f_m = 0$$

since the functions solve the differential equation $y'' + \lambda y = 0$ for the appropriate values of λ. Multiply equation (11.65) by f_m and equation (11.66) by f_n, and subtract the resulting equations to obtain

(11.67) $$f_m f_n'' - f_n f_m'' + (\lambda_n - \lambda_m)f_m f_n = 0.$$

At this point we can invoke a useful fact about the Wronskian $W(u, v) = uv' - vu'$ of two functions u and v, namely that
$$\frac{d}{dx}W(u, v) = uv'' - vu''.$$
This follows from differentiating $uv' - vu'$ using the product rule and observing that two of the four resulting terms cancel. Thus we can rewrite equation (11.67) as
$$\frac{d}{dx}W(f_m, f_n) + (\lambda_n - \lambda_m)f_m f_n = 0.$$
Rearranging this equation and integrating from 0 to L tells us that
$$(\lambda_m - \lambda_n)\int_0^L f_m f_n\, dx = \int_0^L \frac{d}{dx}W(f_m, f_n)\, dx$$
$$= W(f_m, f_n)\big|_0^L$$
$$= W(f_m, f_n)(L) - W(f_m, f_n)(0).$$
Since $f_m(x)$ and $f_n(x)$ are both zero when $x = 0$ or $x = L$, both $W(f_m, f_n)(L)$ and $W(f_m, f_n)(0)$ are zero, and we have
$$(\lambda_m - \lambda_n)\int_0^L f_m f_n\, dx = 0.$$

Since $\lambda_m - \lambda_n$ is nonzero, we can divide by it to obtain
$$\int_0^L f_m f_n \, dx = 0$$
whenever $m \neq n$, as desired.

The $n \neq m$ case of equation (11.60) can be obtained in a similar way, starting with the fact that the boundary value problem
$$y'' + \lambda y = 0, \quad y'(0) = 0, \quad y'(L) = 0,$$
has eigenvalues
$$\lambda_n = \frac{n^2 \pi^2}{L^2}$$
and eigenfunctions
$$g_n(x) = \cos \frac{n\pi x}{L}.$$
The argument just given can be modified to show that the functions
$$\cos \frac{n\pi x}{L}$$
are orthogonal on $[0, L]$ and thus also on $[-L, L]$, since
$$\int_{-L}^{L} g_n(x) g_m(x) \, dx = 2 \int_0^L g_n(x) g_m(x) \, dx.$$
Exercise 17 asks for the details.

11.5.4. Using complex exponentials. Our experience has shown us that sometimes we can make things simpler by calculating with complex exponentials rather than sines and cosines. This is possible because of Euler's formula and its close relations:
$$e^{i\theta} = \cos\theta + i\sin\theta, \quad e^{-i\theta} = \cos\theta - i\sin\theta$$
and
$$\cos\theta = \frac{e^{i\theta} + e^{-i\theta}}{2}, \quad \sin\theta = \frac{e^{i\theta} - e^{-i\theta}}{2i}.$$
In particular, we have $e^{2\pi i} = 1$ and, more generally, $e^{2\pi i k} = 1$ for any integer k. In the next result, we integrate the complex exponential function $e^{ik\pi x/L}$.

Lemma 11.5.7. *Whenever k is an integer not equal to 0,*

(11.68)
$$\int_{-L}^{L} e^{ik\pi x/L} \, dx = 0.$$

Proof. Since $\frac{d}{dx}(e^{rx}) = re^{rx}$ for any real or complex constant r, an antiderivative of $e^{ik\pi x/L}$ (if $k \neq 0$) is
$$\frac{L}{ik\pi} e^{ik\pi x/L}.$$
Thus
$$\begin{aligned}
\int_{-L}^{L} e^{ik\pi x/L} \, dx &= \frac{L}{ik\pi} e^{ik\pi x/L} \Big|_{-L}^{L} = \frac{L}{ik\pi} \left(e^{ik\pi} - e^{-ik\pi} \right) \\
&= \frac{L}{ik\pi} e^{-ik\pi} \left(e^{2\pi i k} - 1 \right) = 0,
\end{aligned}$$
since $e^{2\pi i k} = 1$. □

11.5. Functions as vectors

We'll use this lemma to verify in a new way the orthogonality of the functions $\cos \frac{m\pi x}{L}$ and $\sin \frac{n\pi x}{L}$ discussed in Example 11.5.6. We'll include the constant function 1 by allowing $m = 0$ and state our results as an official theorem.

Theorem 11.5.8. *For $m = 0, 1, 2, \ldots$ and $n = 1, 2, 3, \ldots$, the functions $\cos \frac{m\pi x}{L}$ and $\sin \frac{n\pi x}{L}$ are mutually orthogonal in the inner product given by (11.58). That is,*

$$\left\langle \sin \frac{n\pi x}{L}, \cos \frac{m\pi x}{L} \right\rangle = 0 \quad \text{for all } m, n,$$

$$\left\langle \sin \frac{m\pi x}{L}, \sin \frac{n\pi x}{L} \right\rangle = 0 \quad \text{if } m \neq n,$$

$$\left\langle \cos \frac{m\pi x}{L}, \cos \frac{n\pi x}{L} \right\rangle = 0 \quad \text{if } m \neq n.$$

Moreover, for $m \geq 1$

$$\left\| \cos \frac{m\pi x}{L} \right\| = \left\| \sin \frac{m\pi x}{L} \right\| = 1,$$

and (with $m = 0$)

$$\| \cos 0 \| = \| 1 \| = \sqrt{2},$$

so they are all "unit vectors" except for $\cos 0 = 1$.

Proof. Consider

(11.69) $$\left\langle \cos \frac{m\pi x}{L}, \cos \frac{n\pi x}{L} \right\rangle = \frac{1}{L} \int_{-L}^{L} \cos \frac{m\pi x}{L} \cos \frac{n\pi x}{L} \, dx.$$

Writing $\cos \frac{m\pi x}{L} = \frac{1}{2}\left(e^{im\pi x/L} + e^{-im\pi x/L}\right)$ and plugging into (11.69) gives us

$$\left\langle \cos \frac{m\pi x}{L}, \cos \frac{n\pi x}{L} \right\rangle = \frac{1}{L} \int_{-L}^{L} \frac{\left(e^{\frac{im\pi x}{L}} + e^{\frac{-im\pi x}{L}}\right)}{2} \frac{\left(e^{\frac{in\pi x}{L}} + e^{\frac{-in\pi x}{L}}\right)}{2} dx$$

$$= \frac{1}{4L} \int_{-L}^{L} \left[e^{\frac{i(m+n)\pi x}{L}} + e^{\frac{i(m-n)\pi x}{L}} + e^{\frac{i(n-m)\pi x}{L}} + e^{\frac{-i(m+n)\pi x}{L}} \right] dx$$

$$= \frac{1}{4L} \int_{-L}^{L} e^{\frac{i(m+n)\pi x}{L}} dx + \frac{1}{4L} \int_{-L}^{L} e^{\frac{i(m-n)\pi x}{L}} dx$$

$$+ \frac{1}{4L} \int_{-L}^{L} e^{\frac{i(n-m)\pi x}{L}} dx + \frac{1}{4L} \int_{-L}^{L} e^{\frac{-i(m+n)\pi x}{L}} dx.$$

If $m \neq n$, all the integrals are 0, by Lemma 11.5.7. Therefore $\left\langle \cos \frac{m\pi x}{L}, \cos \frac{n\pi x}{L} \right\rangle = 0$ for $n \neq m$. If

$m = n > 0$, the same calculations show

$$\left\|\cos\frac{m\pi x}{L}\right\|^2 = \left\langle \cos\frac{m\pi x}{L}, \cos\frac{m\pi x}{L}\right\rangle$$

$$= \frac{1}{4L}\int_{-L}^{L}\left[e^{\frac{i(m+m)\pi x}{L}} + 1 + 1 + e^{\frac{-i(m+m)\pi x}{L}}\right]dx$$

$$= \frac{1}{4L}(0 + 2L + 2L + 0)$$

$$= 1.$$

When $m = 0$, we have

$$\left\|\cos\frac{0\pi x}{L}\right\| = \|1\| = \left(\frac{1}{L}\int_{-L}^{L}1\,dx\right)^{1/2} = \sqrt{2}.$$

For the remaining inner product calculations, proceed similarly using

$$\sin\frac{m\pi x}{L} = \frac{1}{2i}\left(e^{\frac{im\pi x}{L}} - e^{\frac{-im\pi x}{L}}\right).$$
□

11.5.5. Exercises.

1. Show that the product of two odd functions and the product of two even functions are both even. Show that the product of an odd function with an even function is odd.

2. (a) Show that if f and g are both periodic with period T, then any linear combination $c_1 f + c_2 g$ also has period T.
   (b) Is $f(x) = \cos x + \cos(\pi x)$ a periodic function? (Notice $\cos x$ has period 2π and $\cos \pi x$ has period 2.) Hint: What are the solutions to $\cos x + \cos \pi x = 2$?

3. Suppose $f(x) = e^{2x}\sin x$ for $0 \leq x \leq \pi$. If f is extended to be a 2π-periodic odd function, what is $f(-\pi/2)$? $f(8)$? $f(-2-\pi)$?

In Exercises 4–9 sketch the graphs of the even and odd periodic extensions of the given function to $\mathbb{R}$, showing at least 3 periods.

4. $f(x) = 2x$ on $[0, 3]$.
5. $f(x) = 1 - x$ on $[0, 2]$.
6. $f(x) = \sin x$ on $[0, \pi]$.
7. $f(x) = |\cos x|$ on $[0, \pi]$.

8. $f(x) = \begin{cases} 0 & \text{if } 0 \leq x < 1, \\ x - 1 & \text{if } 1 \leq x \leq 2. \end{cases}$

9. $f(x) = \begin{cases} 0 & \text{if } 0 \leq x < 1, \\ x - 2 & \text{if } 1 \leq x \leq 2. \end{cases}$

10. If $f(x)$ is continuous on the interval $[0, L]$, must the even periodic extension of f to $\mathbb{R}$ be continuous everywhere?

11. If $f(x)$ is continuous on the interval $[0, L]$, show by example that the odd periodic extension of f to $\mathbb{R}$ need not be continuous everywhere. Can you give additional conditions on f that will guarantee that its odd periodic extension is everywhere continuous?

12. Find the fundamental period of each of the following periodic functions or else show that the function does not have a smallest positive period.
    (a) $f(x) = \sin(5x)$.
    (b) $f(x) = 1 + \cos(4x)$.
    (c) $f(x) = \tan x$.

11.5. Functions as vectors

(d) $f(x) = \begin{cases} 1 & \text{if } x \text{ is a rational number,} \\ 0 & \text{if } x \text{ is an irrational number.} \end{cases}$

13. (a) Suppose that f is periodic with period T. Using mathematical induction, show that f is periodic with period nT for every positive integer n.
    (b) Suppose that f is periodic with period T and that f has domain $(-\infty, \infty)$. Show that $f(x - T) = f(x)$ for all x.
    (c) Continue to assume the domain of f is all real numbers. Combine the results of parts (a) and (b) to conclude that if f is periodic with period T, then $f(x + kT) = f(x)$ for all integers k and all real numbers x.

14. (a) Suppose that $\mathbf{E}_1$, $\mathbf{E}_2$, and $\mathbf{E}_3$ are any three mutually orthogonal unit vectors in $\mathbb{R}^3$ and that $\mathbf{A}$ is any vector in $\mathbb{R}^3$. Suppose we have written
$$\mathbf{A} = c_1 \mathbf{E}_1 + c_2 \mathbf{E}_2 + c_3 \mathbf{E}_3$$
for some constants c_1, c_2, and c_3. Show that
$$c_1 = \mathbf{A} \cdot \mathbf{E}_1, \quad c_2 = \mathbf{A} \cdot \mathbf{E}_2, \quad \text{and} \quad c_3 = \mathbf{A} \cdot \mathbf{E}_3.$$
Also verify that $\|\mathbf{A}\|^2 = c_1^2 + c_2^2 + c_3^2$.
    (b) Suppose the $\mathbf{E}_1, \mathbf{E}_2, \ldots, \mathbf{E}_n$ are n mutually orthogonal unit vectors in $\mathbb{R}^n$. It is a fact that an arbitrary vector $\mathbf{A}$ in $\mathbb{R}^n$ can be written as a linear combination of the vectors $\mathbf{E}_j$:
$$\mathbf{A} = c_1 \mathbf{E}_1 + c_2 \mathbf{E}_2 + \cdots + c_n \mathbf{E}_n.$$
Find formulas for the c_j's. What is $\|\mathbf{A}\|$ in terms of the c_j's?

15. Use the trigonometric identities (11.62)–(11.63) given in Example 11.5.6 to verify equations (11.59)–(11.60).

16. Using the result of Example 11.5.6, show that for $n = 1, 2, 3, \ldots$ and $m = 1, 2, 3, \ldots$
$$\frac{2}{L} \int_0^L \sin \frac{n\pi x}{L} \sin \frac{m\pi x}{L} \, dx = \begin{cases} 0 & \text{if } n \neq m, \\ 1 & \text{if } n = m, \end{cases}$$
and
$$\frac{2}{L} \int_0^L \cos \frac{n\pi x}{L} \cos \frac{m\pi x}{L} \, dx = \begin{cases} 0 & \text{if } n \neq m, \\ 1 & \text{if } n = m. \end{cases}$$
Hint: No new integral computations are needed if you use the fact that the product of two odd functions or of two even functions is even.

17. The functions
$$g_n(x) = \cos \frac{n\pi x}{L}, \quad n = 0, 1, 2, 3, \ldots,$$
solve the boundary value problem
$$y'' + \lambda_n y = 0, \quad y'(0) = 0, \quad y'(L) = 0,$$
when
$$\lambda_n = \frac{n^2 \pi^2}{L^2}.$$
Use the eigenfunction approach of Section 11.5.3 to show that the functions g_n, $n = 0, 1, 2, 3 \ldots$, are orthogonal on $[0, L]$ and thus also on $[-L, L]$. Hint: The situation here differs from that discussed in Section 11.5.3 only in replacing the sine functions f_n by the cosine functions g_n and accordingly altering the boundary conditions.

11.6. Fourier series

Let's consider an expression built as a (possibly infinite) sum of constant multiples of the functions

(11.70) $$\cos \frac{n\pi x}{L}, \ n = 0, 1, 2, \ldots, \quad \text{and} \quad \sin \frac{n\pi x}{L}, \ n = 1, 2, 3, \ldots.$$

Each of these functions has period $2L$. Notice that $\cos \frac{0\pi x}{L} = 1$, so our list in (11.70) includes the constant function 1. In other words, given numbers $a_0, a_1, a_2, \ldots$ and $b_1, b_2, b_3, \ldots$, we form the expression

(11.71) $$\frac{a_0}{2} + \sum_{n=1}^{\infty} \left(a_n \cos \frac{n\pi x}{L} + b_n \sin \frac{n\pi x}{L} \right).$$

This is called a **Fourier series**. We'll see later that writing the constant term as $\frac{a_0}{2}$ (instead of just a_0) has certain notational conveniences. At any given value of x, the infinite sum (11.71) may or may not converge. If it does, let's call the value of the sum $f(x)$. Notice that if the sum in (11.71) converges at some x, then it must also converge at $x + 2L$ and take the same value, since each term in the sum has period $2L$. Thus $f(x)$ is periodic with period $2L$.

Starting with an infinite sum as in equation (11.71), we can ask, "For what values of x does this sum converge?" More important for us, though, will be to **start with some ($2L$-periodic) function** $f(x)$ and ask whether it can be written as a Fourier series

(11.72) $$f(x) = \frac{a_0}{2} + \sum_{n=1}^{\infty} \left(a_n \cos \frac{n\pi x}{L} + b_n \sin \frac{n\pi x}{L} \right)$$

for "most" values of x and, if the answer to this question is yes, to determine the values of a_n and b_n.

Computing the Fourier coefficients. Let's start with the second issue—determining the coefficients a_n and b_n if we have

$$f(x) = \frac{a_0}{2} + \sum_{n=1}^{\infty} \left(a_n \cos \frac{n\pi x}{L} + b_n \sin \frac{n\pi x}{L} \right).$$

Here's the idea, which uses the orthogonality of the functions in (11.70). Fix a positive integer n. We have

$$\frac{1}{L} \int_{-L}^{L} f(x) \sin \frac{n\pi x}{L} dx = \frac{1}{L} \int_{-L}^{L} \left[\frac{a_0}{2} + \sum_{m=1}^{\infty} \left(a_m \cos \frac{m\pi x}{L} + b_m \sin \frac{m\pi x}{L} \right) \right] \sin \frac{n\pi x}{L} dx$$

$$= \frac{1}{L} \int_{-L}^{L} \left[\frac{a_0}{2} \sin \frac{n\pi x}{L} + \sum_{m=1}^{\infty} \left(a_m \cos \frac{m\pi x}{L} \sin \frac{n\pi x}{L} + b_m \sin \frac{m\pi x}{L} \sin \frac{n\pi x}{L} \right) \right] dx$$

$$= \frac{a_0}{2} \frac{1}{L} \int_{-L}^{L} \sin \frac{n\pi x}{L} dx$$

$$+ \sum_{m=1}^{\infty} \left[a_m \frac{1}{L} \int_{-L}^{L} \cos \frac{m\pi x}{L} \sin \frac{n\pi x}{L} dx + b_m \frac{1}{L} \int_{-L}^{L} \sin \frac{m\pi x}{L} \sin \frac{n\pi x}{L} dx \right]$$

$$= b_n.$$

The first equality in this calculation comes from replacing the function $f(x)$ by its Fourier series representation as in (11.72). For the second equality, we multiply $\sin \frac{n\pi x}{L}$ through the sum using the distributive law. For the third equality, take the integral *inside* the sum sign. The integral of a sum is the sum of the integrals; of course, there's more to it than that here, because the sum is

11.6. Fourier series

an *infinite series*. The interchange of integral and sum can be justified in this setting, however, provided the function f is sufficiently nice (for example, if f is continuously differentiable). For the final equality, apply Theorem 11.5.8 on mutual orthogonality to find that all terms in the sum are zero *except* for the one that comes from $m = n$, and *this* one is $a_n \cdot 0 + b_n \cdot 1 = b_n$.

Thus we have shown that

$$(11.73) \qquad b_n = \frac{1}{L} \int_{-L}^{L} f(x) \sin \frac{n\pi x}{L} \, dx = \left\langle f, \sin \frac{n\pi x}{L} \right\rangle,$$

for $n = 1, 2, 3, \ldots$. Similarly (see Exercise 15),

$$(11.74) \qquad a_n = \left\langle f, \cos \frac{n\pi x}{L} \right\rangle$$

for $n = 1, 2, 3, \ldots$. What about a_0? Using orthogonality again, with the expression in (11.72) for $f(x)$ and following the steps in the above calculation, we see that

$$\begin{aligned}
\langle f, \cos 0 \rangle &= \frac{1}{L} \int_{-L}^{L} f(x) \cdot \cos 0 \, dx = \frac{1}{L} \int_{-L}^{L} f(x) \, dx \\
&= \frac{1}{L} \int_{-L}^{L} \left[\frac{a_0}{2} + \sum_{m=1}^{\infty} \left(a_m \cos \frac{m\pi x}{L} + b_m \sin \frac{m\pi x}{L} \right) \right] dx \\
&= \frac{1}{L} \int_{-L}^{L} \frac{a_0}{2} \, dx = \frac{a_0}{2} \frac{1}{L} \cdot 2L = a_0,
\end{aligned}$$

and equation (11.74) is correct for $n = 0$ too. To summarize:

If f is a sufficiently nice function with

$$(11.75) \qquad f(x) = \frac{a_0}{2} + \sum_{n=1}^{\infty} \left(a_n \cos \frac{n\pi x}{L} + b_n \sin \frac{n\pi x}{L} \right),$$

then for $n = 0, 1, 2, \ldots$

$$(11.76) \qquad a_n = \frac{1}{L} \int_{-L}^{L} f(x) \cos \frac{n\pi x}{L} \, dx = \left\langle f, \cos \frac{n\pi x}{L} \right\rangle,$$

and for $n = 1, 2, 3, \ldots$

$$(11.77) \qquad b_n = \frac{1}{L} \int_{-L}^{L} f(x) \sin \frac{n\pi x}{L} \, dx = \left\langle f, \sin \frac{n\pi x}{L} \right\rangle.$$

These are **Euler-Fourier** formulas for the Fourier coefficients a_n and b_n. Writing the constant term in equation (11.72) as $\frac{a_0}{2}$ instead of as just a_0 allows us to use the formula $a_n = \langle f, \cos \frac{n\pi x}{L} \rangle$ even for $n = 0$.

An analogy with $\mathbb{R}^n$. Notice the strong analogy with our discussion about $\mathbb{R}^n$. For example, with $n = 3$, we saw that given any three mutually orthogonal unit vectors $\mathbf{E}_1, \mathbf{E}_2$, and $\mathbf{E}_3$ in $\mathbb{R}^3$ and an arbitrary vector $\mathbf{A}$, we have

$$\mathbf{A} = c_1 \mathbf{E}_1 + c_2 \mathbf{E}_2 + c_3 \mathbf{E}_3$$

where the coefficients c_j are determined by

$$c_1 = \mathbf{A} \cdot \mathbf{E}_1, \quad c_2 = \mathbf{A} \cdot \mathbf{E}_2, \quad \text{and} \quad c_3 = \mathbf{A} \cdot \mathbf{E}_3.$$

In the Euler-Fourier equations, the role of the $\mathbf{E}_j$'s is played by the family of mutually orthogonal functions $\sin \frac{n\pi x}{L}$ and $\cos \frac{m\pi x}{L}$, the vector $\mathbf{A}$ is replaced by the function f, and the dot product is replaced by the inner product.

The Fourier Convergence Theorem. We have just seen that if you start with a sufficiently nice function $f(x)$ which can be represented as the sum of a Fourier series as in (11.75) above, then the Fourier coefficients a_n and b_n are determined by the function f via the formulas (11.76) and (11.77). This motivates the following definition of the **Fourier series** of *any* function f for which the coefficients a_n and b_n given by the formulas in (11.76) and (11.77) make sense.

> Given any function f such that the integrals defining a_n and b_n in equations (11.76)–(11.77) exist as finite numbers, the **Fourier series of** f is the expression
> $$\frac{a_0}{2} + \sum_{n=1}^{\infty}\left(a_n \cos \frac{n\pi x}{L} + b_n \sin \frac{n\pi x}{L}\right).$$

For example, if f is merely piecewise continuous on $[-L, L]$, the integrals in (11.76)–(11.77) will exist as finite numbers, and we can define its Fourier series. It is natural to then ask:

Question: Does the Fourier series having these coefficients a_n and b_n actually converge at every x, and if so, is its sum equal to the function $f(x)$ we started with? To answer this, we require the **Fourier Convergence Theorem**. The answer is a qualified **yes** (at least for most x), provided f is nice enough.

Theorem 11.6.1 (Fourier Convergence Theorem). *Suppose f is piecewise continuously differentiable on $-L \leq x < L$ and extended to $\mathbb{R}$ to be $2L$-periodic. For $n = 0, 1, 2, \ldots$, set*

$$a_n = \left\langle f, \cos \frac{n\pi x}{L}\right\rangle = \frac{1}{L}\int_{-L}^{L} f(x) \cos \frac{n\pi x}{L} dx,$$

and for $n = 1, 2, 3 \ldots$, set

$$b_n = \left\langle f, \sin \frac{n\pi x}{L}\right\rangle = \frac{1}{L}\int_{-L}^{L} f(x) \sin \frac{n\pi x}{L} dx.$$

Then the Fourier series

$$\frac{a_0}{2} + \sum_{n=1}^{\infty}\left(a_n \cos \frac{n\pi x}{L} + b_n \sin \frac{n\pi x}{L}\right)$$

converges to $f(x)$ at every point x of continuity of f. At other points, the Fourier series converges to

$$\frac{f(x+) + f(x-)}{2},$$

where we define

$$f(x+) = \lim_{t \to x^+} f(t)$$

and

$$f(x-) = \lim_{t \to x^-} f(t),$$

the right-hand and left-hand limits of f at x.

11.6. Fourier series

In other words, the Fourier series converges to the average value of the right-hand and left-hand limits of f at the points of discontinuity, and to f at each point of continuity. Notice that the value of the one-sided *limits* $f(x+)$ and $f(x-)$ are not affected by the value of f at the point $t = x$. This explains why in creating the odd periodic or even periodic extension of a function we were able to resolve any endpoint conflicts that occurred however we wanted. Notice also that x is a point of continuity of f exactly when $f(x)$, $f(x-)$, and $f(x+)$ all coincide; in this case the average value of $f(x+)$ and $f(x-)$ is actually $f(x)$ itself.

Example 11.6.2. Consider the function
$$f(x) = \begin{cases} x & \text{if } 0 \leq x < 1, \\ 2 & \text{if } -1 \leq x < 0. \end{cases}$$

This is a piecewise continuously differentiable function initially defined for $-1 \leq x < 1$. Extend f to $\mathbb{R}$ to get a periodic function with period 2. The resulting function has discontinuities at each integer value of x and nowhere else. A graph of the extended f on $[-3, 3]$ is shown in Fig. 11.29.

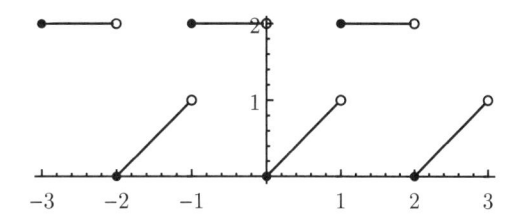

Figure 11.29. Graph of f (extended).

By the Fourier Convergence Theorem, the Fourier series for f,
$$\frac{a_0}{2} + \sum_{n=1}^{\infty} [a_n \cos(n\pi x) + b_n \sin(n\pi x)]$$

where a_n and b_n are computed by equations (11.76) and (11.77) with $L = 1$, converges to (the extended) f at each x that is not an integer. At $x = 0$, the Fourier series converges to 1, the average value of $\lim_{x \to 0+} f(x) = 0$ and $\lim_{x \to 0-} f(x) = 2$. Similarly, the Fourier series converges to 1 at $x = \pm 2, \pm 4, \pm 6, \ldots$.

At $x = 1$ the Fourier series converges to $\frac{3}{2}$, the average value of $\lim_{x \to 1+} f(x) = 2$ and $\lim_{x \to 1-} f(x) = 1$. Similarly, the Fourier series converges to $\frac{3}{2}$ at $x = \pm 1, \pm 3, \pm 5, \ldots$.

Fig. 11.30 shows the graph of the Fourier series. This is just a pictorial answer to the question, "What does the Fourier series of f converge to?"

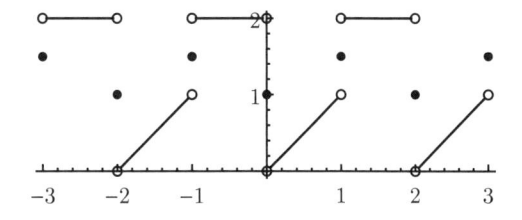

Figure 11.30. Graph of the Fourier series of f.

Example 11.6.3. Let's consider a **square wave**. Begin with

(11.78) $$f(x) = \begin{cases} 1, & 0 < x < 1, \\ -1, & -1 < x < 0. \end{cases}$$

In our previous notation we have $L = 1$. Set $f(-1) = f(1) = 0$ and $f(0) = 0$, so that f, now defined on $[-1, 1]$, is odd. Next extend f to be periodic, with period 2, on the whole real line.

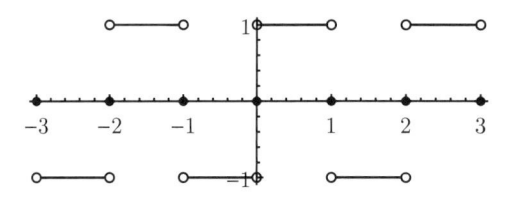

Figure 11.31. Graph of the square wave f on $[-3, 3]$.

We use the Euler-Fourier formulas to compute the Fourier coefficients of f. Since the product of the odd function f times the even function $\cos(n\pi x)$ is odd, we see by equation (11.76), with $L = 1$, that

$$a_n = \int_{-1}^{1} f(x) \cos(n\pi x) \, dx = 0$$

for all $n = 0, 1, 2, 3, \ldots$, since the integral of an odd function over the symmetric interval $[-L, L]$ must be 0.

Since the odd function $f(x)$ times the odd function $\sin(n\pi x)$ is even, we have (by equation (11.77) with $L = 1$)

$$\begin{aligned} b_n &= \int_{-1}^{1} f(x) \sin(n\pi x) \, dx = 2 \int_{0}^{1} f(x) \sin(n\pi x) \, dx \\ &= 2 \int_{0}^{1} (1) \sin(n\pi x) \, dx = 2 \left(-\frac{1}{n\pi} \right) \cos(n\pi x) \Big|_{0}^{1} \\ &= -\frac{2}{n\pi} (\cos(n\pi) - \cos 0) = \frac{2}{n\pi} (1 - (-1)^n) \\ &= \begin{cases} \dfrac{4}{n\pi} & \text{if } n \text{ is odd,} \\ 0 & \text{if } n \text{ is even.} \end{cases} \end{aligned}$$

Thus $f(x)$ has Fourier series

$$(11.79) \qquad \frac{4}{\pi} \left(\sin(\pi x) + \frac{1}{3} \sin(3\pi x) + \frac{1}{5} \sin(5\pi x) + \frac{1}{7} \sin(7\pi x) + \cdots \right).$$

For what x does this Fourier series converge, and what does it converge to? We use the Fourier Convergence Theorem to answer this, noting that f is piecewise continuously differentiable. Since f is continuous at every noninteger x, the Fourier Convergence Theorem tells us that the sum in (11.79) converges to $f(x)$ for each such x. On the other hand $f(x) = 0$ whenever x is an integer (this follows from periodicity of period 2 and the fact that $f(0) = f(1) = f(-1) = 0$), and 0 is exactly the average value of 1 and -1, which are the one-sided limits of f at any integer. Thus

$$(11.80) \qquad f(x) = \frac{4}{\pi} \left(\sin(\pi x) + \frac{1}{3} \sin(3\pi x) + \frac{1}{5} \sin(5\pi x) + \frac{1}{7} \sin(7\pi x) + \cdots \right)$$

for every value of x, integer as well as noninteger. (Note the validity of (11.80) for integers x can be checked directly; each summand on the right-hand side is 0 in this case.) The graph in Fig. 11.31 is thus the graph of the Fourier series of f as well.

11.6. Fourier series

An equation such as (11.80) is a remarkable statement. For example, setting $x = \frac{1}{2}$ it tells us that
$$1 = f(\tfrac{1}{2}) = \tfrac{4}{\pi}\left(1 - \tfrac{1}{3} + \tfrac{1}{5} - \tfrac{1}{7} + \tfrac{1}{9} - \cdots\right),$$
or
$$1 - \frac{1}{3} + \frac{1}{5} - \frac{1}{7} + \frac{1}{9} - \cdots = \frac{\pi}{4}.$$

In calculus one learns why such an alternating series converges, but it is a higher-order achievement to actually determine the sum of this series.

Let's now derive **Parseval's equality**, an analogue of the Pythagorean Theorem in Euclidean space. Suppose that f is piecewise continuously differentiable on $[-L, L]$, so that
$$f(x) = \frac{a_0}{2} + \sum_{n=1}^{\infty}\left(a_n \cos\frac{n\pi x}{L} + b_n \sin\frac{n\pi x}{L}\right)$$
for all except (at most) a finite number of points x in $[-L, L]$ (the points of discontinuity of f). We have
$$\frac{1}{L}\int_{-L}^{L} f(x)^2\, dx = \frac{1}{L}\int_{-L}^{L}\left[\frac{a_0}{2} + \sum_{n=1}^{\infty}\left(a_n \cos\frac{n\pi x}{L} + b_n \sin\frac{n\pi x}{L}\right)\right] f(x)\, dx$$

$$= \frac{a_0}{2}\frac{1}{L}\int_{-L}^{L} f(x)\, dx$$

$$+ \sum_{n=1}^{\infty}\left[a_n \frac{1}{L}\int_{-L}^{L} f(x)\cos\frac{n\pi x}{L}\, dx + b_n\frac{1}{L}\int_{-L}^{L} f(x)\sin\frac{n\pi x}{L} dx\right]$$

$$= \frac{a_0^2}{2} + \sum_{n=1}^{\infty}(a_n^2 + b_n^2).$$

The first equality holds because two functions that differ at (at most) a finite number of points have the same integrals over $[-L, L]$. Note that the second equality follows from again taking the integral inside the (infinite!) sum, while the third uses the Euler-Fourier formulas for a_n and b_n. This is **Parseval's equality**:
$$\frac{1}{L}\int_{-L}^{L} f(x)^2\, dx = \frac{a_0^2}{2} + \sum_{n=1}^{\infty}(a_n^2 + b_n^2).$$

Although our derivation of Parseval's equality depended on f being nice enough to invoke the Fourier Convergence Theorem and to justify interchanging an integral with an infinite sum, it is possible to show the equality holds for any f such that $\int_{-L}^{L} f(x)^2\, dx$ is finite.

Let's apply Parseval's equality to the square wave in Example 11.6.3, with $L = 1$ and $f(x)^2 = 1$ everywhere except at the integer points $x = 0, \pm 1, \pm 2,\ldots$ (where $f(x) = 0$). Using Parseval's equality we get
$$2 = \int_{-1}^{1} 1\, dx = \left(\frac{4}{\pi}\right)^2\left(1 + \frac{1}{3^2} + \frac{1}{5^2} + \frac{1}{7^2} + \cdots\right),$$
or
$$\frac{\pi^2}{8} = 1 + \frac{1}{3^2} + \frac{1}{5^2} + \frac{1}{7^2} + \cdots,$$
another remarkable evaluation of a numerical infinite series.

We summarize with a table comparing Euclidean space $\mathbb{R}^n$ with our "vector space" of functions. Notice in the last entry that Parseval's equality above plays the role of the Pythagorean Theorem in Euclidean space.

Table 11.1. Comparison Chart	
n-dimensional Euclidean space $\mathbb{R}^n$	the space of piecewise continuously differentiable functions, periodic of period $2L$
vectors $\mathbf{A} = (a_1, \ldots, a_n)$ in $\mathbb{R}^n$	**functions** $f(x)$
dot product of two vectors in $\mathbb{R}^n$: $\mathbf{A} \cdot \mathbf{B} = a_1 b_1 + \cdots + a_n b_n,$ where $\mathbf{B} = (b_1, \ldots, b_n)$	**inner product** of two functions: $\langle f, g \rangle = \dfrac{1}{L} \int_{-L}^{L} f(x) g(x) dx$
norm, or **length**, of a vector $\mathbf{A}$: $\|\mathbf{A}\| = \sqrt{\mathbf{A} \cdot \mathbf{A}} = \sqrt{a_1^2 + \cdots + a_n^2}$	**norm** of a function f: $\|f\| = \sqrt{\langle f, f \rangle} = \sqrt{\dfrac{1}{L} \int_{-L}^{L} f(x)^2 dx}$
$\mathbf{A}$ and $\mathbf{B}$ are **orthogonal**: $\mathbf{A} \cdot \mathbf{B} = 0$.	f and g are **orthogonal**: $\langle f, g \rangle = 0$.
any mutually orthogonal vectors of norm 1 (**unit** vectors) $\mathbf{E}_1, \mathbf{E}_2, \ldots, \mathbf{E}_n$ in $\mathbb{R}^n$	mutually orthogonal functions of norm 1: $\sin \dfrac{n\pi x}{L}, \ n = 1, 2, 3, \ldots;$ $\cos \dfrac{n\pi x}{L}, \ n = 0, 1, 2, \ldots$ **Exception:** $\|\cos 0\| = \|1\| = \sqrt{2}$
representing a vector $\mathbf{A}$ as a linear combination of $\mathbf{E}_1, \ldots, \mathbf{E}_n$: $\mathbf{A} = c_1 \mathbf{E}_1 + \cdots + c_n \mathbf{E}_n$	representing a function f as a Fourier series: $f(x) \sim \dfrac{a_0}{2} + \sum_{n=1}^{\infty} \left(a_n \cos \dfrac{n\pi x}{L} + b_n \sin \dfrac{n\pi x}{L} \right)$
formulas for coefficients: $c_m = \mathbf{A} \cdot \mathbf{E}_m, \ m = 1, \ldots, n$	Euler-Fourier formulas for the Fourier coefficients: $a_n = \left\langle f, \cos \dfrac{n\pi x}{L} \right\rangle, \ n = 0, 1, 2, \ldots,$ $b_n = \left\langle f, \sin \dfrac{n\pi x}{L} \right\rangle, \ n = 1, 2, \ldots$
Pythagorean Theorem: If $\mathbf{A} = c_1 \mathbf{E}_1 + \cdots + c_n \mathbf{E}_n,$ then $\|\mathbf{A}\|^2 = c_1^2 + \cdots + c_n^2.$	Parseval's equality: If $f(x) \sim$ $\dfrac{a_0}{2} + \sum_{n=1}^{\infty} \left(a_n \cos \dfrac{n\pi x}{L} + b_n \sin \dfrac{n\pi x}{L} \right),$ then $\|f\|^2 = \dfrac{a_0^2}{2} + \sum_{n=1}^{\infty} (a_n^2 + b_n^2).$

11.6.1. Fourier sine and cosine series. Starting with a (piecewise continuously differentiable) function f defined on an interval $[0, L]$, we have considered two ways to extend f to a $2L$-periodic function on $\mathbb{R}$: the even periodic extension of f and the odd periodic extension of f. The Fourier

11.6. Fourier series

series of the *odd* periodic extension of f will contain only sine terms

(11.81) $$\sum_{n=1}^{\infty} b_n \sin \frac{n\pi x}{L}$$

since the "oddness" of the integrand $f(x)\cos\frac{n\pi x}{L}$ in equation (11.76) tells you that all of the coefficients a_n there are 0. We call (11.81) the **Fourier sine series** of the function f (initially defined on $[0, L]$). The coefficients b_n are computed by

$$b_n = \frac{2}{L} \int_0^L f(x) \sin \frac{n\pi x}{L} \, dx.$$

This follows from equation (11.77), since we are using the odd extension of f, so the product $f(x)\sin\frac{n\pi x}{L}$ is even, and

$$b_n = \frac{1}{L} \int_{-L}^L f(x) \sin \frac{n\pi x}{L} \, dx = \frac{2}{L} \int_0^L f(x) \sin \frac{n\pi x}{L} \, dx.$$

The Fourier series of the *even* periodic extension of f will contain only cosine terms (and a constant term)

(11.82) $$\frac{a_0}{2} + \sum_{n=1}^{\infty} a_n \cos \frac{n\pi x}{L}$$

for a similar reason: All of the coefficients b_n given by equation (11.77) are 0 since the integrand is odd. We call (11.82) the **Fourier cosine series** of f. The coefficients a_n in (11.82) are computed by

$$a_n = \frac{2}{L} \int_0^L f(x) \cos \frac{n\pi x}{L} \, dx.$$

This follows from equation (11.76), since we now take the even extension of f, and $f(x)\cos\frac{n\pi x}{L}$ is even.

Since the Fourier sine and Fourier cosine series of f are just the Fourier series for a particular (odd or even) extension of f, the convergence of the Fourier sine or cosine series is still described by Theorem 11.6.1.

Example 11.6.4. Consider the function f defined on $[0, \pi]$ by $f(x) = x^2$ for $0 \le x \le \pi$. To determine the Fourier cosine series of f, we work with the even periodic extension of f, whose graph is shown in Fig. 11.32.

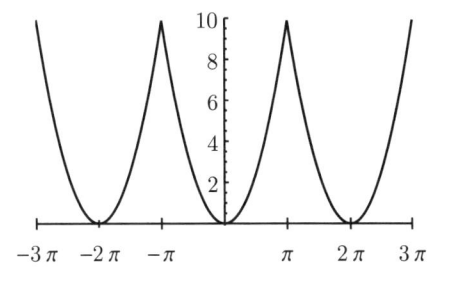

Figure 11.32. Even extension of f.

The Fourier cosine series is

$$\frac{a_0}{2} + \sum_{n=1}^{\infty} a_n \cos(nx)$$

where

$$a_n = \frac{2}{\pi} \int_0^{\pi} x^2 \cos(nx)\,dx.$$

Computing these integrals we find that

$$a_0 = \frac{2\pi^2}{3}$$

and for $n \geq 1$,

$$a_n = (-1)^n \frac{4}{n^2}.$$

Thus f has Fourier cosine series

(11.83) $$\frac{\pi^2}{3} + \sum_{n=1}^{\infty} (-1)^n \frac{4}{n^2} \cos(nx).$$

Since there are no points of discontinuity for the even periodic extension of f, this Fourier cosine series converges at every point to the function graphed in Fig. 11.32.

The partial sums of the Fourier cosine series in (11.83) converge *rapidly* to the even 2π-periodic extension of f; see Figs. 11.33–11.35, which show the partial sums obtained by taking the first 3, 5, and 9 terms of (11.83), respectively.

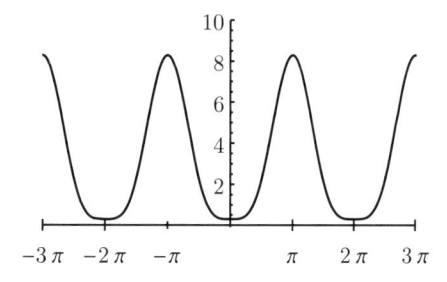

Figure 11.33. $\dfrac{\pi^2}{3} + \sum_{n=1}^{2} (-1)^n \dfrac{4}{n^2} \cos(nx).$

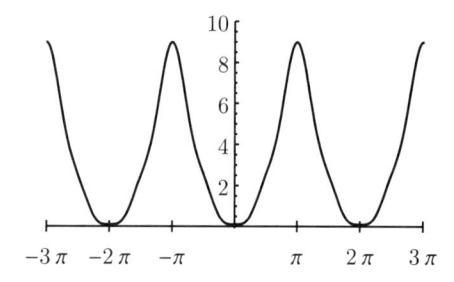

Figure 11.34. $\dfrac{\pi^2}{3} + \sum_{n=1}^{4} (-1)^n \dfrac{4}{n^2} \cos(nx).$

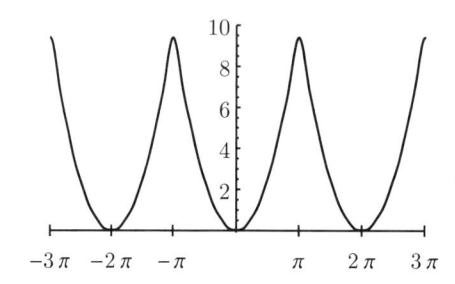

Figure 11.35. $\dfrac{\pi^2}{3} + \sum_{n=1}^{8} (-1)^n \dfrac{4}{n^2} \cos(nx).$

For the Fourier sine series of f, we use the odd periodic extension of f, whose graph is shown in Fig. 11.36.

11.6. Fourier series

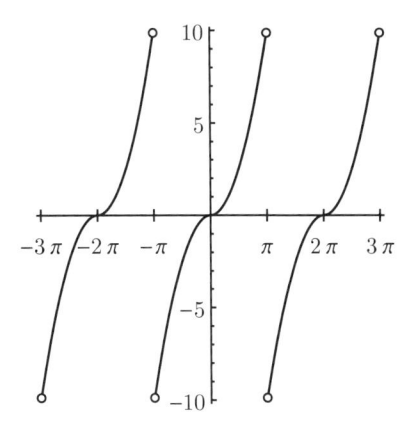

Figure 11.36. Odd extension of f (undefined at conflict points).

The Fourier sine series of f has the form

$$\sum_{n=1}^{\infty} b_n \sin(nx)$$

where

$$b_n = \frac{2}{\pi} \int_0^{\pi} x^2 \sin(nx)\, dx.$$

Computing these integrals for the first few values $n = 1, 2, 3$, and 4 gives the sine series with the first four nonzero terms shown explicitly:

$$\left(2\pi - \frac{8}{\pi}\right) \sin x - \pi \sin(2x) + \left(\frac{2\pi}{3} - \frac{8}{27\pi}\right) \sin(3x) - \frac{\pi}{2} \sin(4x) + \cdots$$

for f. This series converges to the odd periodic extension of f at every point where this extension is continuous. At the points of discontinuity (for example, $x = \pm\pi$ or $x = \pm 3\pi$), the series converges to 0, which is the average of the right- and left-hand limits at the points of discontinuity. Pictorially, the series converges to the function whose graph is shown in Fig. 11.37.

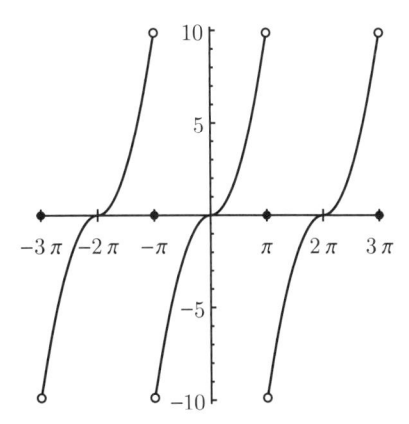

Figure 11.37. Convergence of the Fourier sine series.

In comparison with the Fourier cosine series in (11.83), the Fourier sine series for $f(x) = x^2$ on $[0, \pi]$ converges rather slowly. Fig. 11.38 shows the partial sums obtained, respectively, from the first 2, 8, and 20 terms of the Fourier sine series.

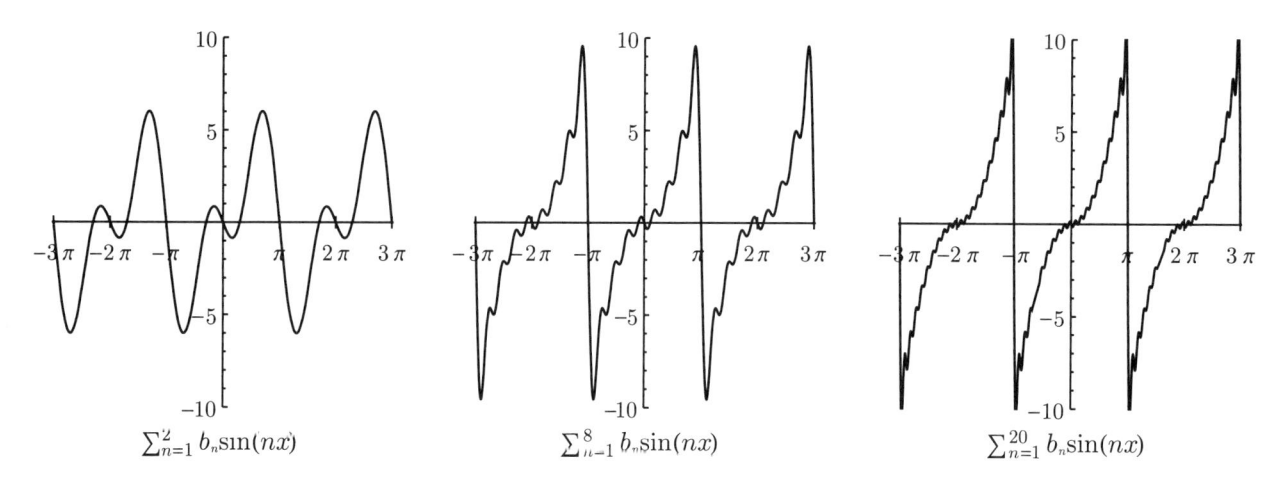

$\sum_{n=1}^{2} b_n \sin(nx)$ $\sum_{n=1}^{8} b_n \sin(nx)$ $\sum_{n=1}^{20} b_n \sin(nx)$

Figure 11.38. Some partial sums of the Fourier sine series.

Note that the graph of the 20th partial sum in Fig. 11.38 (on the right) still hasn't settled down near the jump discontinuities of the odd 2π-periodic extension of f. In particular, as we get close to the jump discontinuity of f at $x = \pi$, the various partial sums consistently "overshoot" the value $f(\pi) = \pi^2 \approx 9.869$. This "wiggling" as the (continuously differentiable) partial sums try to approximate a function near one of its jump discontinuities is called the **Gibbs phenomena**, discussed further in Exercise 22. There you are asked to test the following "rule of thumb": If, for large n, the nth partial sum of a Fourier series for a piecewise continuously differentiable function f is used to approximate f near one of its jump discontinuities, then the maximum overshoot/undershoot error in the approximation is about 0.09 times the size of the jump.

11.6.2. A brief historical note: Fourier provokes a crisis. In the latter part of the eighteenth century and the beginning of the nineteenth century, the physical problems of heat conduction and vibrating strings were under investigation, and in connection with these problems, infinite sums of sines and cosines (what we now call Fourier series) started to appear. In 1807 Joseph Fourier, who was modeling heat flow, submitted a paper to the French Academy of Sciences that derived the heat equation and developed methods for its solution. In particular, he claimed that *any* function $f(x)$ could be written as an infinite series of sines and cosines, that is, as a Fourier series. To say that this announcement was greeted with skepticism and suspicion is an understatement. Fourier provided no proof for his claim, although he did provide a great deal of empirical evidence—in the form of verifiable solutions to concrete physical problems—and thus his work could not easily be just dismissed. Leading mathematicians, including those on the committee charged with evaluating Fourier's submission, understood on some level that Fourier's claim called into question the understanding of calculus at the time, including such basic concepts as the nature of "function" and "continuity". Subsequently, several well-known mathematicians of the nineteenth century, including Cauchy, provided incorrect proofs of Fourier's assertion. Eventually, in 1829, Dirichlet provided a rigorous justification of a result along the lines of Theorem 11.6.1. Dirichlet believed his theorem should be true more generally for any piecewise continuous function (as did such other mathematical luminaries as Riemann and Weierstrass) and it was a shock when forty years later a mathematician named DuBois-Reymond produced an example of a continuous function whose Fourier series failed to converge at a point. Work on understanding the pointwise convergence of the Fourier series of a continuous function continued well into the twentieth century.

In 1811 the French Academy announced a competition for the best work in the theory of heat conduction. Fourier presented an expanded and revised version of his 1807 work, and it was awarded the grand prize. However, objections, particularly by Laplace and Lagrange, were still

11.6. Fourier series

strongly expressed and publication of Fourier's paper was delayed until 1822. In retrospect, the investigation of the convergence of Fourier series initiated by Fourier's 1807 paper could reasonably be credited with bringing about the birth of a new field of mathematics—now called real analysis—which, among other developments, worked out a thorough understanding of continuity, convergence in all of its subtleties, and integration. Fourier died in 1830, living just long enough to see the publication of Dirichlet's work establishing the validity of Fourier series. Fittingly, Dirichlet's work demonstrates the standards of clarity and rigor which became a characteristic feature of the new real analysis.

11.6.3. Exercises.

1. Consider the function
$$f(x) = \begin{cases} 1 & \text{if } 0 \leq x \leq 1, \\ 3 & \text{if } 1 < x \leq 2. \end{cases}$$

   Extend f to be even on $[-2, 2]$, and then extend to all of $\mathbb{R}$ to obtain a periodic function with period 4.
   (a) Sketch the graph of this even periodic extension, showing at least three periods.
   (b) Explain why in the Fourier series for (the extended) f, you expect all coefficients b_n in the sine terms to be 0.
   (c) Find the Fourier (cosine) series expansion for f.
   (d) Describe the convergence of the Fourier series you found in (c). Where does it converge and to what?
   (e) How do your answers to (c) and (d) change if the definition of f is slightly changed to
$$f(x) = \begin{cases} 1 & \text{if } 0 \leq x < 1, \\ 3 & \text{if } 1 \leq x \leq 2. \end{cases}$$

2. Let f be defined on $[0, 2]$ by
$$f(x) = \begin{cases} 1 & \text{if } 0 \leq x \leq 1, \\ 0 & \text{if } 1 < x \leq 2. \end{cases}$$

   Extend f to be periodic of period 4 and even on all of $\mathbb{R}$.
   (a) Sketch the graph of the extended f for $-6 \leq x \leq 6$.
   (b) Find the Fourier coefficients a_n and b_n for f.
   (c) Does the Fourier series of f converge at $x = 0$? If so, what does it converge to? Does it converge at $x = 1$ and at $x = 2$? If so, to what?

3. Consider the function
$$f(x) = \begin{cases} x & \text{if } 0 \leq x \leq 2, \\ 2 - x & \text{if } 2 < x \leq 4. \end{cases}$$

   (a) Extend f to $[-4, 4]$ so as to have an odd function and then sketch the periodic extension of f (with period 8) from $[-4, 4]$ to $\mathbb{R}$, showing at least two periods.
   (b) Explain why, in the Fourier series for f, you expect all coefficients a_n in the cosine terms to be 0.
   (c) In the Fourier (sine) series expansion $\sum_{n=1}^{\infty} b_n \sin \frac{n\pi x}{4}$ for f, find b_1, b_2, and b_3.
   (d) Describe the convergence of the Fourier series in (c). Where does it converge and to what?

4. Match the following functions $f, g,$ and h with the form of their Fourier series, (A), (B), or (C) below:

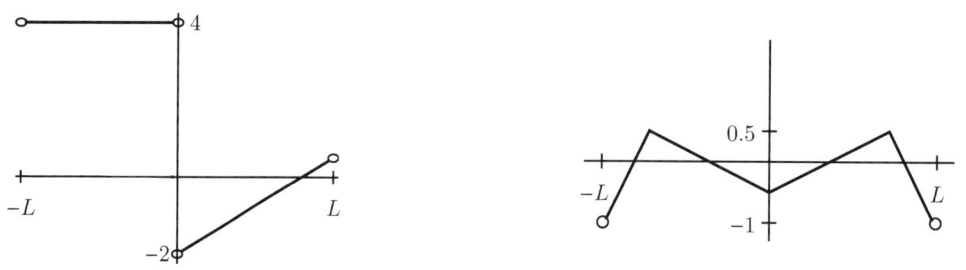

Figure 11.39. Graph of f.

Figure 11.40. Graph of g.

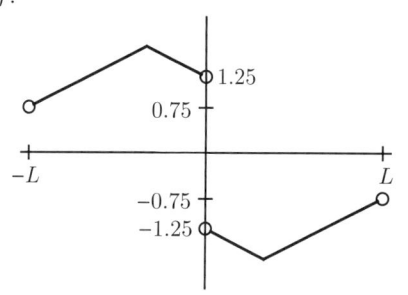

Figure 11.41. Graph of h.

(A) $\dfrac{a_0}{2} + \sum_{n=1}^{\infty}\left(a_n \cos \dfrac{n\pi x}{L} + b_n \sin \dfrac{n\pi x}{L}\right).$ (B) $\dfrac{a_0}{2} + \sum_{n=1}^{\infty} a_n \cos \dfrac{n\pi x}{L}.$ (C) $\sum_{n=1}^{\infty} b_n \sin \dfrac{n\pi x}{L}.$

5. (a) For the function f in Exercise 4, does the Fourier series of f converge at $x = 0$? If so, to what?
   (b) For the function g in Exercise 4, does the Fourier series of g converge at $x = L$? If so, to what?
   (c) At what points x in $[-L, L]$ does the Fourier series of the function h in Exercise 4 converge to 0?

6. Show that the Fourier series for $f(x) = x$ on $[-1, 1]$ is
$$\frac{2}{\pi} \sum_{n=1}^{\infty} (-1)^{n+1} \frac{1}{n} \sin(n\pi x).$$

7. (a) (CAS) Plot the 10th, 20th, and 50th partial sums of the Fourier series in Exercise 6 over the interval $[-1, 3]$, comparing your plot with that of the period-2 extension of $f(x) = x$ from $[-1, 1]$ to $[-1, 3]$.
   (b) (CAS) Letting $S(x) = \frac{2}{\pi} \sum_{n=1}^{50} (-1)^{n+1} \frac{1}{n} \sin(n\pi x)$ be the 50th partial sum, compute, with 3-decimal place accuracy, the maximum of $|S(x) - x|$ on $[-0.7, 0.7]$, $[-0.8, 0.8]$, $[-0.9, 0.9]$, and $[-0.99, 0.99]$.

8. We wish to write the function
$$f(x) = \begin{cases} 2 & \text{if } 0 \leq x \leq 2, \\ 1 & \text{if } 2 < x \leq 4, \end{cases}$$
as a Fourier cosine series
$$\frac{a_0}{2} + \sum_{n=1}^{\infty} a_n \cos \frac{n\pi x}{4}.$$

11.6. Fourier series

(a) Calculate the a_n's.

(b) For what values of x between $x = 0$ and $x = 4$ is

$$f(x) = \frac{a_0}{2} + \sum_{n=1}^{\infty} a_n \cos \frac{n\pi x}{4},$$

when the a_n's are as determined in (a)? That is, for what values of x in $[0, 4]$ does the Fourier cosine series converge to $f(x)$?

(c) For what values of x in $[0, 4]$ does the Fourier cosine series not converge to $f(x)$? For these x, what does it converge to?

9. We wish to write the function f in Exercise 8 as a Fourier sine series

$$\sum_{n=1}^{\infty} b_n \sin \frac{n\pi x}{4}.$$

(a) Determine b_1, b_2, b_3, b_4, and b_5.

(b) For what values of x between $x = 0$ and $x = 4$ does the Fourier sine series converge to $f(x)$?

10. For each of the functions shown below, sketch the graph of the 2π-periodic extension on the interval $[-3\pi, 3\pi]$, and then sketch the graph of the Fourier series of the given function on $[-3\pi, 3\pi]$. You can do this without computing the Fourier series if you use the Fourier Convergence Theorem.

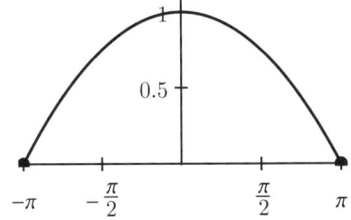

Figure 11.42. (a) $f(x) = 1 - (x/\pi)^2$.

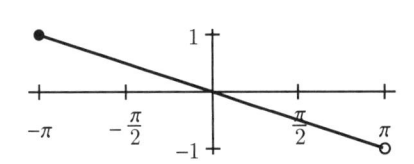

Figure 11.43. (b) $g(x) = -x/\pi$, $-\pi \leq x < \pi$.

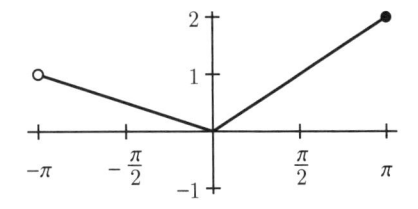

Figure 11.44. (c) $h(x) = (2/\pi)x$ if $0 \leq x \leq \pi$ and $-x/\pi$ if $-\pi < x < 0$.

11. Without doing any integral computations, give the Fourier series for $f(x) = \sin^2 x$ and $g(x) = \cos^2 x$, viewing f and g to be periodic of period 2π. Hint: Decide first what *form* these Fourier series should have, and then use a trigonometric identity to write f and g in the desired form.

12. Suppose $f(x) = x \sin x$ on $-\pi \leq x \leq \pi$, and extend f to be periodic with period 2π. Find the Fourier series $\frac{a_0}{2} + \sum_{n=1}^{\infty} (a_n \cos(nx) + b_n \sin(nx))$, determining all coefficients a_n and b_n explicitly up to $n = 5$. The trigonometric identity

$$\sin u \cos v = \frac{1}{2} [\sin(u+v) + \sin(u-v)]$$

may be helpful.

13. (a) Show that the Fourier sine series expansion of $g(x) = x(\pi - x)$ on $[0, \pi]$ is
$$\frac{8}{\pi} \sum_{k=0}^{\infty} \frac{1}{(2k+1)^3} \sin((2k+1)x).$$

(b) Use the series from (a) and the Fourier Convergence Theorem to find
$$\sum_{k=0}^{\infty} \frac{(-1)^k}{(2k+1)^3}.$$

14. (CAS) **Fourier sine series of a triangle function.** Define the triangle function
$$f(x) = \begin{cases} \frac{2h}{L}x & \text{if } 0 \leq x \leq \frac{L}{2}, \\ \frac{2h}{L}(L-x) & \text{if } \frac{L}{2} < x \leq L. \end{cases}$$

Graph of $f(x)$ on $[0, L]$.

(a) Sketch the odd $2L$-periodic extension of f, showing at least two periods.
(b) (CAS) Using a computer algebra system to help with the integration, show that f has Fourier sine series
$$\frac{8h}{\pi^2} \sum_{k=0}^{\infty} (-1)^k \frac{1}{(2k+1)^2} \sin \frac{(2k+1)\pi x}{L}.$$

15. Verify equation (11.74); that is, show that if
$$f(x) = \frac{a_0}{2} + \sum_{n=1}^{\infty} \left(a_n \cos \frac{n\pi x}{L} + b_n \sin \frac{n\pi x}{L} \right)$$

for $-L \leq x \leq L$, then for $n \geq 1$,
$$a_n = \frac{1}{L} \int_{-L}^{L} f(x) \cos \frac{n\pi x}{L} \, dx = \left\langle f, \cos \frac{n\pi x}{L} \right\rangle.$$

Hint: Mimic the calculations that led to equation (11.73) for the coefficients b_n.

The purpose of Exercises 16–17 is to add two more rows to the comparison table, Table 11.1.

16. (a) For vectors $\mathbf{A}$ and $\mathbf{B}$ in $\mathbb{R}^n$, the **Cauchy-Schwarz inequality** says that $|\mathbf{A} \cdot \mathbf{B}| \leq \|\mathbf{A}\| \|\mathbf{B}\|$. Show how this follows easily from the equation $\mathbf{A} \cdot \mathbf{B} = \|\mathbf{A}\| \|\mathbf{B}\| \cos \theta$, where θ is the angle between $\mathbf{A}$ and $\mathbf{B}$.
(b) Conjecture the form of the Cauchy-Schwarz inequality for real-valued piecewise continuously differentiable functions f and g on $[-L, L]$. Your answer should be a statement about integrals.

17. (a) For vectors $\mathbf{A}$ and $\mathbf{B}$ in $\mathbb{R}^n$, the **triangle inequality** says that $\|\mathbf{A} + \mathbf{B}\| \leq \|\mathbf{A}\| + \|\mathbf{B}\|$. Draw a picture in $\mathbb{R}^2$ to explain the name "triangle inequality".
(b) Conjecture the form of the triangle inequality for real-valued piecewise continuously differentiable functions f and g on $[-L, L]$. Your answer should be a statement about integrals.

18. Suppose f is a $2L$-periodic piecewise continuously differentiable function on $[-L, L]$ with Fourier series
$$\frac{a_0}{2} + \sum_{n=1}^{\infty} \left(a_n \cos \frac{n\pi x}{L} + b_n \sin \frac{n\pi x}{L} \right).$$

11.6. Fourier series

What does Parseval's equality tell you about $\lim_{n\to\infty} a_n$ and $\lim_{n\to\infty} b_n$?

19. If we know the Fourier series (or sine or cosine series) for $f(x)$, can we obtain the Fourier series for $f'(x)$ by "term-by-term" differentiation? This problem illustrates that the answer to this is "sometimes".

    (a) Show that the Fourier series for $f(x) = x$ on $-\pi < x < \pi$ is
    $$2\sum_{n=1}^{\infty} \frac{(-1)^{n+1}}{n} \sin(nx)$$
    but that the Fourier series for $f'(x) = 1$ on $-\pi < x < \pi$ is **not**
    $$2\sum_{n=1}^{\infty} (-1)^{n+1} \cos(nx),$$
    the series that we get by differentiating term by term the Fourier series for f.

    (b) In this part you will show that if f is "nice enough", term-by-term differentiation of the Fourier series for f will produce the Fourier series for f'. To keep things simple, we assume that
       (i) f and f' are both continuous on $-L \le x \le L$,
       (ii) $f(-L) = f(L)$.
    Use integration by parts (choosing "$dv = f'(x)\,dx$") to compute
    $$\int_{-L}^{L} f'(x) \sin\frac{n\pi x}{L}\,dx \quad \text{and} \quad \int_{-L}^{L} f'(x) \cos\frac{n\pi x}{L}\,dx$$
    and use your answer to show that if f has Fourier series
    $$\frac{a_0}{2} + \sum_{n=1}^{\infty} \left(a_n \cos\frac{n\pi x}{L} + b_n \sin\frac{n\pi x}{L}\right),$$
    then f' has Fourier series
    $$\sum_{n=1}^{\infty} \left(-na_n \frac{\pi}{L} \sin\frac{n\pi x}{L} + nb_n \frac{\pi}{L} \cos\frac{n\pi x}{L}\right).$$
    Observe that the last series expression is exactly what you would obtain if you differentiated the Fourier series for f term by term.

    (c) Use the result of (b) to compute the Fourier series of $g(x) = x$ on $-L \le x \le L$, given that the Fourier series of $f(x) = 1 - (\frac{x}{L})^2$ is
    $$\frac{2}{3} + 4\sum_{n=1}^{\infty} \frac{(-1)^{n+1}}{(n\pi)^2} \cos\frac{n\pi x}{L}.$$

20. Suppose that f, f', and f'' are continuous on $[0, L]$ and that $f(0) = 0$ and $f(L) = 0$. In this problem we will show how, under these hypotheses, we can obtain the Fourier sine series of f from the Fourier sine series for f''. This is useful if f'' is simpler than f.

    (a) Show that with an integration by parts (using $u = f$ and $dv = \sin\frac{n\pi x}{L}\,dx$) we have
    $$\int_0^L f(x) \sin\frac{n\pi x}{L}\,dx = \frac{L}{n\pi} \int_0^L f'(x) \cos\frac{n\pi x}{L}\,dx.$$

    (b) Use a second integration by parts to show that
    $$\frac{L}{n\pi} \int_0^L f'(x) \cos\frac{n\pi x}{L}\,dx = -\left(\frac{L}{n\pi}\right)^2 \int_0^L f''(x) \sin\frac{n\pi x}{L}\,dx.$$

(c) Using the results of (a) and (b), show that if f has Fourier sine series $\sum_1^\infty c_n \sin \frac{n\pi x}{L}$ and if f'' has Fourier sine series $\sum_1^\infty d_n \sin \frac{n\pi x}{L}$, then the coefficients c_n and d_n are related by

$$c_n = -d_n \left(\frac{L}{n\pi}\right)^2.$$

(d) Using the result of Exercise 6, find the first four nonzero terms of the Fourier sine series for $f(x) = x^3 - x$ on the interval $0 \le x \le 1$.

21. Let $f(x)$ be the periodic function defined on $\mathbb{R}$ with period 2π and with $f(x) = x^2$ for $-\pi \le x \le \pi$. The Fourier series of f was computed in Example 11.6.4.
    (a) Using the Fourier convergence theorem with well-chosen values of x, compute

    $$\sum_{n=1}^\infty \frac{1}{n^2} \quad \text{and} \quad \sum_{n=1}^\infty \frac{(-1)^{n+1}}{n^2}.$$

    (b) Use Parseval's equality to evaluate the sum

    $$\frac{1}{1^4} + \frac{1}{2^4} + \frac{1}{3^4} + \cdots.$$

22. Let

$$f(x) = \begin{cases} -1 & \text{if } -2 < x < 0, \\ 0 & \text{if } 0 < x < 1, \\ 2 & \text{if } 1 < x < 2, \end{cases}$$

and extend f to be periodic on $\mathbb{R}$ of period 4, calling the extension f as well.
(a) Find formulas for the Fourier series coefficients of f.
(b) (CAS) Let s_{50} be the 50th partial sum of the Fourier series of f. Plot s_{50} over the interval $-5.5 \le x \le 5.5$ and observe how the plot oscillates away from the value of f near its jump discontinuities. This is the Gibbs phenomenon.
(c) (CAS) Plot $s_{50}(x)$ over the interval $I_0 = (1.95, 2.05)$ and estimate its maximum value M and minimum value m. The "overshoot error" of the Fourier series approximation s_{50} of f near its jump discontinuity 2 is $M - 2$ and the "undershoot error" is $-1 - m$. Similarly, obtain estimates for the overshoot and undershoot errors resulting from using s_{50} to approximate f near its jump discontinuities at 0 and 1.
(d) (CAS) Test the rule of thumb described in the text for predicting Fourier series overshoot/undershoot errors near jump discontinuities. Near jump discontinuities $j = 0, 1,$ and 2 of f, does

$$(0.09) \cdot \left| \lim_{x \to j^+} f(x) - \lim_{x \to j^-} f(x) \right|$$

provide a rough approximation for overshoot/undershoot errors estimated in part (c)?

23. Since

$$\cos\theta = \frac{e^{i\theta} + e^{-i\theta}}{2} \quad \text{and} \quad \sin\theta = \frac{e^{i\theta} - e^{-i\theta}}{2i},$$

the Fourier series of a $2L$-periodic piecewise continuously differentiable function f can be written as

$$\sum_{n=-\infty}^\infty c_n e^{i\frac{n\pi x}{L}}$$

for some (possibly complex) constants c_n. Find a formula (analogous to the Euler-Fourier equations) for the c_n's.

24. **A specialty Fourier series.** In this exercise and the next, we look at constructing a Fourier series having a special form. This will be used in some later application exercises (see Exercise 6 in Section 11.7 and Exercise 14 in Section 11.8).
   (a) (CAS) Using a computer algebra system to help with the integration, find the Fourier sine series
$$\sum_{n=1}^{\infty} b_n \sin \frac{n\pi x}{2\pi}$$
   for the function $f(x) = |\sin x|$ on the interval $[0, 2\pi]$, showing the first three nonzero values of b_n explicitly.
   (b) Using your result in (a), explain why
$$\sin x = \frac{8}{3\pi} \sin \frac{x}{2} + \frac{8}{5\pi} \sin \frac{3x}{2} - \frac{8}{21\pi} \sin \frac{5x}{2} + \cdots + b_n \sin \frac{nx}{2} + \cdots$$
   for $0 \leq x \leq \pi$, where $b_n = 0$ if n is even and
$$b_n = \frac{8}{(4-n^2)\pi} \sin \frac{n\pi}{2}$$
   for n odd.

25. Suppose a piecewise continuously differentiable function f is defined initially on $[0, L]$. Extend f to $[L, 2L]$ by making it symmetric across the line $x = L$. This means $f(2L - x) = f(x)$ for $0 \leq x \leq L$; we'll still call this extended function f. The Fourier sine series for f on $[0, 2L]$ is $\sum_{n=1}^{\infty} b_n \sin \frac{n\pi x}{2L}$ where
$$b_n = \frac{2}{L} \int_0^{2L} f(x) \sin \frac{n\pi x}{2L} \, dx.$$
Show that $b_n = 0$ for every even integer n. Hint:
$$\int_0^{2L} f(x) \sin \frac{n\pi x}{2L} \, dx = \int_0^L f(x) \sin \frac{n\pi x}{2L} \, dx + \int_L^{2L} f(x) \sin \frac{n\pi x}{2L} \, dx$$
$$= \int_0^L f(x) \sin \frac{n\pi x}{2L} \, dx + \int_L^{2L} f(2L - x) \sin \frac{n\pi x}{2L} \, dx.$$
Make the substitution $w = 2L - x$ in the last integral, and use the identity
$$\sin\left(n\pi - \frac{n\pi w}{2L}\right) = \sin(n\pi) \cos \frac{n\pi w}{2L} - \cos(n\pi) \sin \frac{n\pi w}{2L}.$$

11.7. The heat equation

In this section, we return to the heat equation

(11.84) $$u_t(x, t) = \kappa u_{xx}(x, t),$$

with κ a positive constant. We can give a physical meaning to this as follows: Imagine a thin rod or wire of homogeneous material with uniform cross sections, stretched out along the x-axis, with its left end at $x = 0$ and its right end at $x = L$:

```
0                                    L
```

The rod is laterally insulated, so no heat passes through its cylindrical surface; however its ends are potentially exposed. If the function $u(x, t)$ gives the temperature of the rod at position x and time t, then the temperature function satisfies the equation

$$u_t(x, t) = \kappa u_{xx}(x, t)$$

where the positive constant κ (the "thermal diffusivity") depends on the material of the rod. If u is measured in degrees Celsius while x is measured in centimeters and t in seconds, then κ has units of cm^2/sec. Some values of κ for various substances are given in the following table:

Substance	κ in cm^2/sec
Silver	1.66
Copper	1.11
Iron	0.23
Stainless Steel	0.042
Glass	0.0034

You should convince yourself that the heat equation is a plausible model for temperature variation of a laterally insulated rod by considering the relationship the equation provides between the concavity of the graph of $g(x) = u(x,t)$, $0 < x < L$, at a fixed time t and the time rate of change of $u(x,t)$ at that time. (See Exercise 7.) Observe that for a given shape of the graph of $g(x)$, the larger κ is, the more rapid the temperature change will be. Looking at values of κ tabulated above, you can understand why stainless steel pans might have copper incorporated into their bottoms.

Deriving the heat equation. The formal derivation of the heat equation (11.84) as a model for temperature of a laterally insulated rod is similar to our derivation of the diffusion equation model of Example 11.4.3 for fish population density in a no-fishing zone. In deriving the heat equation, one uses Fourier's law of heat conduction to describe heat transfer into a short segment of the rod though its left and right ends in the same way that Fick's law was used in Example 11.4.3 to describe fish transfer through the left and right edges of a small slice of the no-fishing zone. The one-dimensional form of Fourier's law matches that of Fick's law from Example 11.4.3. Suppose S is a short segment of the rod extending from x to $x + \Delta x$. The rate at which heat energy flows, at time t, into S through its left endpoint is proportional to

$$(11.85) \qquad -\frac{\partial u}{\partial x}(x,t)$$

while the rate of flow into S through its right endpoint is proportional to

$$(11.86) \qquad \frac{\partial u}{\partial x}(x + \Delta x, t)$$

(and the common constant of proportionality, call it b, is positive). Focusing on the left endpoint, imagine time is fixed and the slope of $f(x) = u(x,t)$ is upward; i.e., $\frac{\partial u}{\partial x}(x,t)$ is positive. Then the temperature is increasing with x near the left endpoint of S and heat should be flowing to the left—from hot to cold—which is exactly what the sign in (11.85) tells us. A similar analysis confirms that the sign of (11.86) yields the correct direction for heat flow into S through its right endpoint. On the other hand, the total rate at which heat energy flows into (or out of) S is proportional to

$$(11.87) \qquad \text{(the time rate of change of the average temperature of } S) \times (\text{length of } S)$$

with positive constant of proportionality a. If S is very short (that is, it's length Δx is very small), we can approximate the left-hand factor in equation (11.87) by $\frac{\partial u}{\partial t}(x,t)$, the time rate of change of u at the left endpoint x of S. This approximation can be justified with a bit more care. Since the sides of S are insulated, the total flow of heat energy into (or out of) S results only from flow through the left and right endpoints. Thus combining the above we have, to a good approximation,

$$a\frac{\partial u}{\partial t}(x,t)\Delta x = b\frac{\partial u}{\partial x}(x + \Delta x, t) - b\frac{\partial u}{\partial x}(x,t).$$

11.7. The heat equation

On dividing by $a\Delta x$, labeling b/a as κ, and letting $\Delta x \to 0$, we arrive at the heat equation

$$\frac{\partial u}{\partial t}(x,t) = \kappa \frac{\partial^2 u}{\partial x^2}(x,t).$$

The heat equation with boundary and initial conditions. In solving the heat equation, we assume that there is some initial temperature distribution in the rod, so that at time $t = 0$ we know the temperature at each position x in the rod:

(11.88) $$u(x,0) = f(x).$$

We also know something about the ends of the rod for all times. For example, we might assume that the ends of the rod are held at fixed temperature 0:

(11.89) $$u(0,t) = 0, \quad u(L,t) = 0, \qquad t > 0,$$

or that the ends of the rod are insulated, which gives different boundary conditions:

(11.90) $$u_x(0,t) = 0, \quad u_x(L,t) = 0, \qquad t > 0.$$

Notice that the insulated ends boundary conditions are consistent with Fourier's law: Because the ends are insulated, no heat is flowing through the endpoint surfaces and u_x must be zero at each endpoint. Both (11.89) and (11.90) are homogeneous boundary conditions, and any linear combination of functions solving (11.84) and satisfying either the boundary conditions in (11.89) or in (11.90) will also meet the same boundary conditions.

In Section 11.3 we used the technique of separation of variables to show that with the **zero-temperature-ends boundary conditions** (11.89), the heat equation has building-block solutions

$$u_n(x,t) = e^{-\kappa\left(\frac{n\pi}{L}\right)^2 t} \sin \frac{n\pi x}{L}$$

for $n = 1, 2, 3\ldots$. Using Fourier series, we can combine the building-block solutions to find the solution that also meets the desired initial condition (11.88), provided the initial temperature function f is piecewise continuously differentiable. The next example shows how this goes.

Example 11.7.1. At time $t = 0$ the temperature at every point $0 < x < 2$ in a copper rod of length 2 cm is $1°$. The thermal diffusivity of copper is $\kappa = 1.11$ cm^2/sec. The ends of the rod are packed in ice in order to maintain them at $0°$. We will find the temperature $u(x,t)$ for any point in the rod for any $t > 0$.

For each $n = 1, 2, 3\ldots$, the functions

$$u_n(x,t) = e^{-\frac{1.11 n^2 \pi^2}{4} t} \sin \frac{n\pi x}{2}$$

satisfy the heat equation and meet the boundary conditions

$$u(0,t) = 0, \quad u(2,t) = 0, \qquad t > 0.$$

We seek a function in the form

(11.91) $$u(x,t) = \sum_{n=1}^{\infty} b_n e^{-\frac{1.11 n^2 \pi^2}{4} t} \sin \frac{n\pi x}{2}$$

that will satisfy

$$u(x,0) = 1$$

for all $0 < x < 2$. Setting $t = 0$ in (11.91), we have

$$u(x,0) = \sum_{n=1}^{\infty} b_n \sin \frac{n\pi x}{2},$$

so we should choose the coefficients b_n to be the Fourier sine series coefficients of $f(x) = 1$ on the interval $(0, 2)$. From Section 11.6.1, we know that these sine series coefficients are given by

$$b_n = \frac{2}{2}\int_0^2 f(x) \sin \frac{n\pi x}{2} dx = \int_0^2 \sin \frac{n\pi x}{2} dx = -\frac{2}{n\pi}\cos \frac{n\pi x}{2} \Big|_0^2 = -\frac{2}{n\pi}\cos(n\pi) + \frac{2}{n\pi}.$$

Thus

$$b_n = \begin{cases} \frac{4}{n\pi} & \text{if } n \text{ is odd}, \\ 0 & \text{if } n \text{ is even}. \end{cases}$$

The Fourier Convergence Theorem tells us that for every x, $0 < x < 2$, it is correct to write

$$1 = \frac{4}{\pi}\sin\frac{\pi x}{2} + \frac{4}{3\pi}\sin\frac{3\pi x}{2} + \frac{4}{5\pi}\sin\frac{5\pi x}{2} + \cdots.$$

At the endpoints $x = 0$ and $x = 2$, the sine series converges to 0 (the average value of the right- and left-hand limits at $x = 0$, and at $x = 2$, when we consider the odd periodic extension of f).

Putting the pieces together, we find that

$$u(x, t) = \frac{4}{\pi}e^{-\frac{1.11\pi^2}{4}t}\sin\frac{\pi x}{2} + \frac{4}{3\pi}e^{-\frac{(1.11)(3^2)\pi^2}{4}t}\sin\frac{3\pi x}{2} + \frac{4}{5\pi}e^{-\frac{(1.11)(5^2)\pi^2}{4}t}\sin\frac{5\pi x}{2} + \cdots$$

solves the heat equation with the desired boundary and initial conditions. We can write this infinite sum more compactly as

$$u(x, t) = \sum_{m=0}^{\infty} \frac{4}{(2m+1)\pi} e^{-\frac{1.11(2m+1)^2\pi^2}{4}t} \sin\frac{(2m+1)\pi x}{2}.$$

Figs. 11.45–11.47 show three snapshots of the approximate temperature at times $t = 1/50$, $t = 1/2$, and $t = 1$, respectively. Fig. 11.48 shows the approximate graph of the two-variable function $u(x, t)$ for $0 \le x \le 2$ and $0 < t < 1.9$. We used the first four nonzero terms in the infinite sum for $u(x, t)$ in constructing Figs. 11.45–11.48. Because the exponential factors get small very quickly for $t > 0$, this gives a good approximation. In Fig. 11.49 we used the first 20 nonzero terms in the infinite sum for $u(x, t)$ to produce an more accurate approximation of the graph of $u(x, t)$.

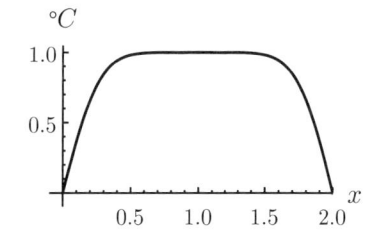

Figure 11.45. Graph of $u(x, 0.02)$.

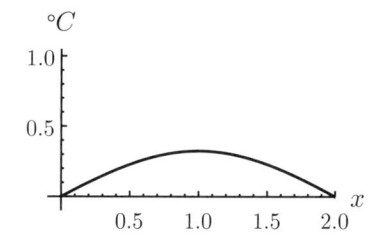

Figure 11.46. Graph of $u(x, 0.5)$.

11.7. The heat equation

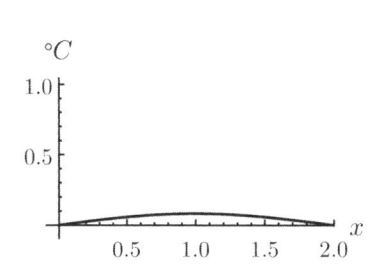

Figure 11.47. Graph of $u(x,1)$.

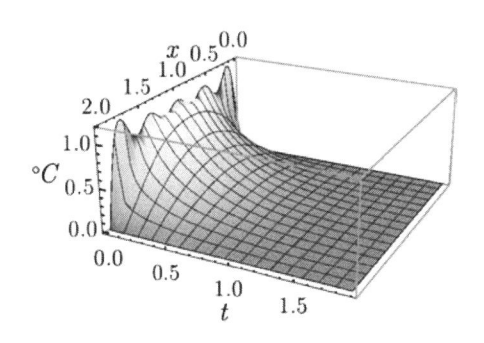

Figure 11.48. Approximate graph of $u(x,t)$, $0 \leq x \leq 2$, $0 < t < 1.9$.

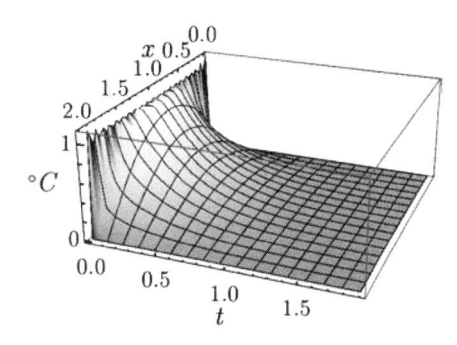

Figure 11.49. Another approximate graph of $u(x,t)$, $0 \leq x \leq 2$, $0 < t < 1.9$.

The t-axis runs left to right in Figs. 11.48 and 11.49, and the x-axis runs back to front. The snapshots are cross sections of the three-dimensional picture in Fig. 11.48 for the values $t = 1/50, 1/2$, and $t = 1$.

We summarize our work on solving the heat equation with the zero-temperature-ends boundary conditions as follows.

Zero-temperature-ends: For the boundary-initial value problem
$$u_t = \kappa u_{xx},$$
$$u(0,t) = 0, \ u(L,t) = 0,$$
we have building-block solutions
$$e^{-\kappa\left(\frac{n\pi}{L}\right)^2 t} \sin\frac{n\pi x}{L}$$
for $n = 1, 2, 3 \ldots$.
We satisfy the initial condition
$$u(x,0) = f(x), \quad 0 < x < L,$$
for f piecewise continuously differentiable, with solution
$$u(x,t) = \sum_{n=1}^{\infty} b_n e^{-\kappa\left(\frac{n\pi}{L}\right)^2 t} \sin\frac{n\pi x}{L}$$
where
$$b_n = \frac{2}{L}\int_0^L f(x) \sin\frac{n\pi x}{L}\,dx.$$

Insulated ends. Next, we consider the heat equation $u_t = \kappa u_{xx}$ with the **insulated-ends boundary conditions**

$$u_x(0, t) = 0, \quad u_x(L, t) = 0, \quad t > 0.$$

To find building-block solutions using separation of variables, we set $u(x, t) = X(x)T(t)$ and substitute into the heat equation to see that

$$\frac{1}{\kappa} \frac{T'(t)}{T(t)} = \frac{X''(x)}{X(x)}.$$

For this to hold, both sides of the last equation must equal the same constant value, which we denote $-\lambda$:

(11.92) $$\frac{1}{\kappa} \frac{T'(t)}{T(t)} = -\lambda, \quad \frac{X''(x)}{X(x)} = -\lambda.$$

To satisfy the boundary conditions in a nontrivial way we must have $X'(0) = 0$ and $X'(L) = 0$. From Exercise 10 in Section 11.2, we know that to solve the boundary value problem

$$\frac{X''(x)}{X(x)} = -\lambda, \quad X'(0) = 0 = X'(L),$$

nontrivially, we must have

(11.93) $$\lambda = \left(\frac{n\pi}{L}\right)^2, \quad n = 0, 1, 2, \ldots,$$

and $X(x) = \cos \frac{n\pi x}{L}$ (or some nonzero constant multiple of this function). Notice that the choice $n = 0$ gives $\lambda = 0$ and $X(x) = 1$.

With any of these λ values, we next solve the first-order equation in (11.92) and obtain

$$T(t) = Ce^{-\kappa\lambda t}, \quad \text{where } \lambda = \left(\frac{n\pi}{L}\right)^2 \text{ for } n = 0, 1, 2, \ldots$$

for constant C. Choosing $C = 1$, we assemble our building-block solutions for the heat equation with the insulated-ends boundary conditions:

$$u_n(x, t) = e^{-\kappa\left(\frac{n\pi}{L}\right)^2 t} \cos \frac{n\pi x}{L} \quad \text{for } n = 0, 1, 2, \ldots.$$

Linear combinations of the building blocks give further solutions, and passing to infinite sums will allow us to also satisfy a general initial condition $u(x, 0) = f(x)$:

> **Insulated-ends**: For the boundary-initial value problem
>
> $$u_t = \kappa u_{xx}, \quad u_x(0, t) = 0, \quad u_x(L, t) = 0,$$
>
> we have building-block solutions
>
> $$e^{-\kappa\left(\frac{n\pi}{L}\right)^2 t} \cos \frac{n\pi x}{L} \quad \text{for } n = 0, 1, 2, 3, \ldots.$$
>
> We satisfy the initial condition $u(x, 0) = f(x)$, $0 < x < L$, for f piecewise continuously differentiable, with solution
>
> $$u(x, t) = \frac{a_0}{2} + \sum_{n=1}^{\infty} a_n e^{-\kappa\left(\frac{n\pi}{L}\right)^2 t} \cos \frac{n\pi x}{L}$$
>
> where
>
> $$a_n = \frac{2}{L} \int_0^L f(x) \cos \frac{n\pi x}{L} \, dx.$$

Notice that it is the Fourier cosine series of $f(x)$ that is used to determine the coefficients of our solution $u(x,t)$.

Example 11.7.2. A wire has length $L = \pi$ and thermal diffusivity $\kappa = 1$. Both ends of the wire are insulated and the initial temperature in the wire is given by $u(x,0) = 10x$. To find the temperature $u(x,t)$ in the wire for all future times, we solve the heat equation

$$u_t = u_{xx}$$

with boundary conditions

$$u_x(0,t) = 0, \quad u_x(\pi,t) = 0, \quad t > 0,$$

and initial condition

$$u(x,0) = 10x$$

for $0 < x < \pi$ and $t > 0$.

The building-block solutions are

$$u_n(x,t) = e^{-n^2 t} \cos(nx)$$

for $n = 0, 1, 2, 3, \ldots$. We want a function of the form

$$u(x,t) = \frac{a_0}{2} + \sum_{n=1}^{\infty} a_n e^{-n^2 t} \cos(nx)$$

that will satisfy

$$10x = u(x,0) = \frac{a_0}{2} + \sum_{n=1}^{\infty} a_n \cos(nx).$$

This means that the coefficients a_n should be chosen to be the Fourier cosine coefficients of the initial temperature distribution $f(x) = 10x$. From Section 11.6.1, this tells us that

$$a_n = \frac{2}{\pi} \int_0^{\pi} 10x \cos(nx) \, dx$$

for $n = 0, 1, 2, 3 \ldots$. Computing the integrals, we see that $a_0 = 10\pi$, and for $n \geq 1$,

$$a_n = \begin{cases} -\frac{40}{n^2 \pi} & \text{if } n \text{ is odd}, \\ 0 & \text{if } n \text{ is even}. \end{cases}$$

Thus the temperature at time $t > 0$ is given by

$$u(x,t) = 5\pi + \sum_{m=0}^{\infty} -\frac{40}{(2m+1)^2 \pi} e^{-(2m+1)^2 t} \cos((2m+1)x).$$

In Fig. 11.50 we have used the first five nonzero terms in this sum to give an approximate graph of $u(x,t)$ for $0 < x < \pi$ and $0 < t < 4$. Cross sections of this three-dimensional picture give snapshots of the (approximate) temperature function at $t = 0.05$, $t = 0.2$, and $t = 2$ as shown in Figs. 11.51–11.53.

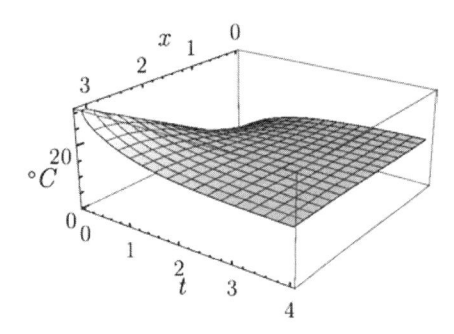

Figure 11.50. Graph of $u(x,t)$, $0 < x < \pi$ (back to front), $0 < t < 4$ (left to right).

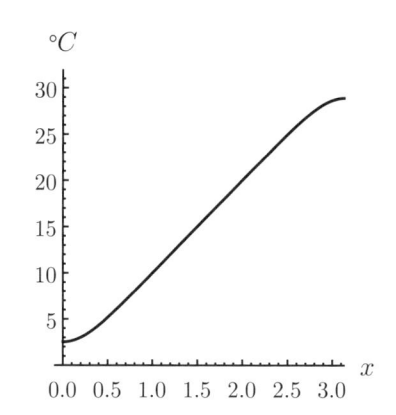

Figure 11.51. Graph of $u(x, 0.05)$.

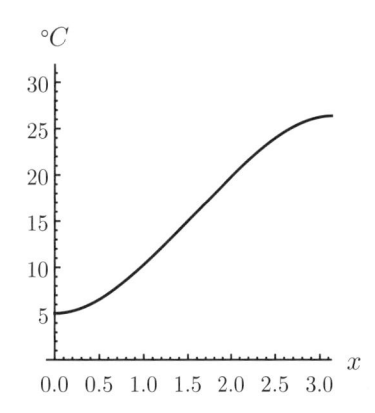

Figure 11.52. Graph of $u(x, 0.2)$.

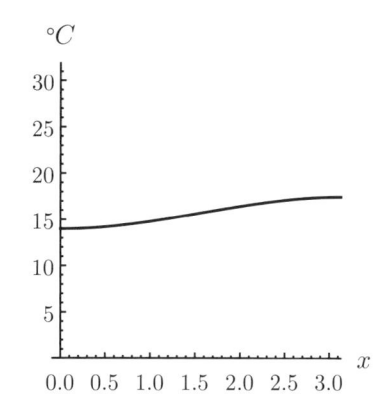

Figure 11.53. Graph of $u(x, 2)$.

Our last example involves *nonhomogeneous boundary conditions*.

Example 11.7.3. Our goal in this example is to describe how to solve

(11.94) $$u_t(x,t) = \kappa u_{xx}(x,t), \qquad u(0,t) = T_1, \quad u(L,t) = T_2,$$

with the initial condition

(11.95) $$u(x,0) = f(x).$$

Physically, we are thinking of a laterally insulated rod with one end held at temperature T_1 and the other at temperature T_2.

First we find the **steady-state** solution to

$$u_t(x,t) = \kappa u_{xx}(x,t), \qquad u(0,t) = T_1, \quad u(L,t) = T_2.$$

A steady-state solution is one that is constant in time; that is, it depends only on x. Denoting such a solution by $s(x)$ and substituting into the heat equation gives $0 = \kappa s''(x)$, or simply

$$s''(x) = 0.$$

Thus a steady-state solution looks like $s(x) = ax + b$ for some constants a and b. We use the boundary conditions $u(0,t) = T_1$ and $u(L,t) = T_2$ to determine

$$b = T_1, \quad a = \frac{T_2 - T_1}{L}.$$

11.7. The heat equation

So the steady-state solution is

$$s(x) = \frac{T_2 - T_1}{L}x + T_1.$$

Next we solve the zero-temperature-ends homogeneous problem

$$u_t(x,t) = \kappa u_{xx}(x,t), \quad u(0,t) = 0, \quad u(L,t) = 0,$$

with the initial condition

$$u(x,0) = f(x) - \frac{T_2 - T_1}{L}x - T_1.$$

Note that on the right-hand side, we have subtracted the steady-state solution from $f(x)$. Call this solution $v(x,t)$. The sum of $v(x,t)$ with the steady-state solution, namely

(11.96) $$v(x,t) + \frac{T_2 - T_1}{L}x + T_1,$$

is our desired solution to (11.94)–(11.95). You should verify this by checking three things:

- The function in (11.96) solves the partial differential equation

$$u_t(x,t) = \kappa u_{xx}(x,t).$$

- It has value T_1 when $x = 0$ and value T_2 when $x = L$.
- It has value $f(x)$ when $t = 0$.

As a special case of the last example, if the two ends of a wire are held at the same constant (but nonzero) temperature T_1 and if the initial temperature in the wire is given by $f(x)$, then the temperature at any point x and time t is given by

$$u(x,t) = v(x,t) + T_1,$$

where $v(x,t)$ is the solution to

$$v_t = \kappa v_{xx}, \quad v(0,t) = v(L,t) = 0, \quad v(x,0) = f(x) - T_1.$$

11.7.1. Exercises.

1. A thin rod with length π, composed of a metal with thermal diffusivity $\kappa = 1$, is completely insulated (laterally and at both ends). Denote the temperature distribution in the rod by $u(x,t)$ for all times $t > 0$.
   (a) What is the temperature distribution $u(x,t)$ if $u(x,0) = 2 + 3\cos x$ for $0 < x < \pi$?
   (b) What is the temperature distribution $u(x,t)$ if $u(x,0) = 2 - \sin^2 x$ for $0 < x < \pi$? (Hint: You can use a trigonometric identity to avoid any integral computations).

2. Consider the function defined on the interval $[-20, 20]$ by

$$f(x) = \frac{2}{3}\sin\frac{\pi x}{10} - 4\sin\frac{\pi x}{5} + 7\sin\frac{\pi x}{4} - \sin \pi x.$$

   (a) Find all of the Fourier coefficients $a_0, a_1, a_2 \ldots$ and $b_1, b_2, b_3, \ldots$ of f. Hint: This shouldn't require any integral computations!
   (b) Suppose the temperature $u(x,t)$ in a rod of length 20 constructed of a metal with thermal diffusivity $\kappa = 1$ obeys the heat equation $u_t = u_{xx}$. If at time $t = 0$ we have $u(x,0) = f(x)$ for f as in (a) and the ends of the rod are held at temperature 0 for all t, find $u(x,t)$ for $0 < x < 20$ and $t > 0$.

3. A metal rod extends along the x-axis from $x = 0$ to $x = 2$. The two ends of the rod are kept at temperature 0 for all time. The temperature $u(x,t)$ of the rod at position x and time t satisfies the heat equation $9u_{xx} = u_t$. If the initial temperature of the rod is given by

$$f(x) = \begin{cases} 30x & \text{if } 0 < x < 1, \\ 60 - 30x & \text{if } 1 < x < 2, \end{cases}$$

find $u(x,t)$ for all $t > 0$. Hint: The result in Exercise 14 of Section 11.6 will shorten the computations.

4. A thin rod with length π, composed of a metal with thermal diffusivity $\kappa = 1$, is heated to $100°$. The lateral sides are insulated, and beginning at time $t = 0$ the ends are kept at a constant temperature of $0°$.
   (a) Find the temperature distribution $u(x,t)$ for $t > 0$.
   (b) What is $\lim_{t \to \infty} u(x,t)$ for any point x in the rod? Does your answer make sense intuitively?
   (c) Let $s_2(x,t)$ be the sum of the first two nonzero terms of your series solution to part (a). Compute $s_2(\pi/2, 1/2)$ and then use the alternating series error estimate (if needed, consult your favorite calculus text) to compute a bound on the error that results when $s_2(\pi/2, 1/2)$ is used to approximate $u(\pi/2, 1/2)$.
   (d) Let s_2 be as in part (c). Show that for $t \geq 0$,

$$|u(x,t) - s_2(x,t)| \leq \frac{400}{5\pi} \sum_{n=2}^{\infty} \left(e^{-4t}\right)^n.$$

   Then calculate the preceding error bound given $t = 1$; that is, find the sum of the geometric series

$$\frac{400}{5\pi} \sum_{n=2}^{\infty} \left(e^{-4}\right)^n.$$

5. Consider the heat equation $u_t = \kappa u_{xx}$ with the boundary conditions

$$u(0,t) = 0, \quad u_x(L,t) = 0, \quad t > 0.$$

   We can interpret these conditions as requiring the left end of the rod to be held at $0°$ and the right end of the rod to be insulated.
   (a) Use the method of separation of variables to find building-block solutions of the form $X(x)T(t)$.
   (b) Will a linear combination of building-block solutions still satisfy the boundary conditions?
   (c) With $L = \pi$ and $\kappa = 1$, find the solution that satisfies the boundary conditions and the initial condition

$$u(x,0) = 5 \sin \frac{3x}{2} - 2 \sin \frac{7x}{2}$$

   for $0 < x < \pi$.

6. Suppose a thin laterally insulated metal rod of length π has its left end held at $0°$ and the other end insulated. At time $t = 0$, the temperature at point x units from the left end of the rod is given by a function $f(x) = \sin x$.
   (a) Give a boundary-initial value problem to model this set-up.
   (b) Solve the problem in (a) by separation of variables, using the result of Exercise 24 of Section 11.6, as well as the solution from Exercise 5(a) above.
   (c) What is the temperature at the middle of the rod at time t? Give your answer as an infinite sum.

7. Suppose the function $u(x,t)$ satisfies the heat equation $u_t = u_{xx}$ for $t > 0$ and $0 < x < L$. For some time $t_1 > 0$ the graph $u(x, t_1)$ is as shown below.

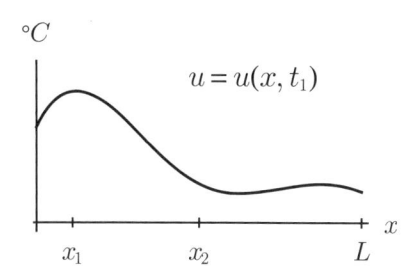

11.7. The heat equation

(a) Thinking of $u(x,t)$ as temperature distribution, do you expect $u_t(x_1, t_1)$ to be positive or negative? Confirm your answer using the heat equation and the concavity of the graph of $u(x, t_1)$ at $x = x_1$.

(b) Is $u_t(x_2, t_1)$ positive or negative? Explain.

8. (a) Consider a uniform, laterally insulated iron rod ($\kappa = 0.23$ cm^2/sec) with length 50 cm. Suppose its left end is held at 20 degrees and its right end at 70 degrees for all $t > 0$. Its initial temperature at point x, $0 < x < 50$, is given by $f(x) = x$. Determine the function $u(x,t)$ modeling the temperature at position x and time t.

(b) (CAS) Approximate your solution $u(x,t)$ from (a) by using the first 50 terms of your series solution. Plot the snapshot pictures of this approximate solution for $0 \le x \le 50$ and $t = 60, 120, 500$, and $1{,}000$. Has the rod (essentially) reached its steady-state temperature at $t = 1{,}000$?

9. Suppose that $u(x,t)$ is the temperature distribution in a thin laterally insulated rod of length L. For each $t > 0$, let

$$E(t) = \int_0^L u(x,t)\,dx.$$

The derivative of $E(t)$ is obtained by differentiation (with respect to t) inside the integral sign:

$$\frac{dE}{dt} = \int_0^L u_t(x,t)\,dx.$$

(a) Suppose that the ends of the rod are insulated. Using the heat equation, show that $\frac{dE}{dt} = 0$.

(b) The "total thermal energy" contained in the rod is a constant multiple of $E(t)$. Explain why (a) tells you that when the ends of the rod are insulated, the total thermal energy is constant. Does this make intuitive sense?

10. **Average temperature of a completely insulated rod.** Let $u(x,t)$ be the solution of the insulated-ends boundary-initial value problem of Example 11.7.2.

(a) Compute the average temperature of the rod at time $t = 0$.

(b) Assume you can integrate term by term in the infinite series defining $u(x,t)$. What is the average temperature of the rod at time t for $t > 0$? How does this tie in with the results in Exercise 9?

(c) For x in $(0, \pi)$, find $\lim_{t \to \infty} u(x,t)$.

11. Suppose a thin metal rod of length 2π is formed into a circular ring. Assume it is insulated along its sides, and let $u(x,t)$ represent the temperature at point x, for $-\pi < x \le \pi$ (as shown below) and time t. Because of the circular shape, we require that the ends "match up":

(11.97) $\qquad u(-\pi, t) = u(\pi, t), \qquad u_x(-\pi, t) = u_x(\pi, t), \qquad t > 0.$

Assume $u(x,t)$ satisfies the heat equation $u_t = \kappa u_{xx}$.

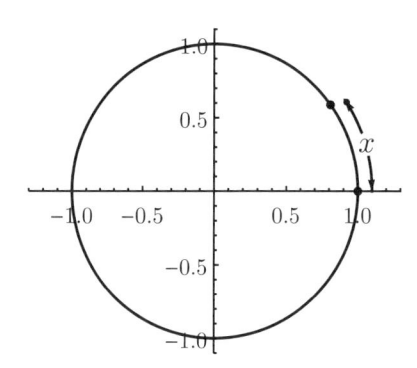

(a) Show that the method of separation of variables leads to the equations
$$T'(t) = -\lambda\kappa T(t),$$
$$X''(x) = -\lambda X(x),$$
$$X(-\pi) = X(\pi), \ X'(-\pi) = X'(\pi)$$
for $u(x,t) = X(x)T(t)$, where $-\lambda$ is the separation constant.

(b) If $\lambda = 0$, are there any nontrivial solutions in (a) for $X(x)$? If so, find them, and find the corresponding solutions for $T(t)$.

(c) If $\lambda < 0$, are there any nontrivial solutions for $X(x)$? If so, find them.

(d) Show that when $\lambda > 0$ there are nontrivial solutions for $X(x)$ precisely when $\lambda = n^2$ for n a positive integer. Find them, and find the corresponding solutions for $T(t)$.

(e) Give building-block solutions satisfying the boundary conditions (11.97).

(f) If the initial temperature is given by
$$u(x,0) = f(x), \quad -\pi < x < \pi,$$
where f has Fourier series
$$f(x) = \frac{a_0}{2} + \sum_{n=1}^{\infty}(a_n\cos(nx) + b_n\sin(nx)),$$
what is the solution $u(x,t)$?

(g) Intuitively, what happens to $u(x,t)$ as $t \to \infty$? Can you reconcile this with your answer in (f)?

12. **Reversing the seasons.** The temperature at ground level in North Ontario (ignoring daily variations) is $T(t) = 40\cos(2\pi t)$ degrees Fahrenheit, where t has units of years/2π and $t = 0$ corresponds to August 1. Let $u(x,t)$ be the temperature at time t and depth x feet **below** ground level. Assume $u(x,t)$ satisfies the heat equation $u_t = \kappa u_{xx}$ for some positive constant κ. Suppose we start with a complex trial solution
$$v(x,t) = f(x)e^{2\pi it}$$
where f is possibly complex valued.

(a) Show that v solves the heat equation exactly when $f(x)$ satisfies the second-order equation
$$f''(x) = \frac{2\pi i}{\kappa}f(x).$$

(b) The characteristic equation for the differential equation in (a) is $z^2 - \frac{2\pi i}{\kappa} = 0$. Solve this, using the fact that
$$\sqrt{i} = \pm\frac{1}{2}\left(\sqrt{2} + i\sqrt{2}\right).$$

Notice that the roots w_1 and w_2 are not complex conjugates (the characteristic equation does not have real coefficients). One of the roots w gives a complex solution $f(x) = ce^{wx}$ which remains bounded as x increases. Find it, and then write down the trial solution
$$v(x,t) = f(x)e^{2\pi it} = ce^{wx}e^{2\pi it}$$
where you can assume c is real.

(c) Find $u(x,t) \equiv$ the real part of $v(x,t)$ (for v as in (b)), and explain why this also solves the heat equation. Determine c so that $u(0,t) = T(t) = 40\cos(2\pi t)$.

(d) Show from your solution that at a depth $x_0 = \sqrt{\kappa\pi}$ feet the seasons are exactly reversed (hottest in winter and coldest in summer). At ground level, the amplitude of the temperature swings is 40 degrees Fahrenheit. What is the amplitude at depth x_0? A reasonable value for κ in the units we are using is $\kappa = 55$. What is the numerical value of the depth x_0?

13. **A noninsulated rod.** We have a noninsulated uniform rod of length L cm lying along the x-axis from 0 to L. Let $u(x,t)$ be the temperature of the rod at position x and time t. We assume the rod's left endpoint is held at temperature T_1 and its right endpoint is held at T_2. We assume that the thermal diffusivity of the rod is κ, that the rod's initial temperature is given by a piecewise continuously differentiable function $f(x)$ on $[0, L]$, and that the ambient temperature is A and its influence on u_t is proportional to $(A - u)$ with positive constant of proportionality h. (One might call h a "heat-transfer coefficient".) With the preceding assumptions, u satisfies the following boundary-initial value problem:

$$(\dagger) \quad \begin{array}{l} \text{(i)} \quad u_t - \kappa u_{xx} + h(u - A) = 0, \ 0 < x < L, \ t > 0, \\ \text{(ii)} \quad u(0,t) = T_1; \ u(L,t) = T_2, \ t > 0, \\ \text{(iii)} \quad u(x,0) = f(x), \ 0 < x < L. \end{array}$$

Let s denote the steady-state solution of $(\dagger)$(i) so that

(11.98) $$-\kappa s''(x) + h(s(x) - A) = 0, \quad 0 < x < L,$$

and $s(0) = T_1$ while $s(L) = T_2$. (Note that (11.98) results when $s = s(x)$ replaces u in $(\dagger)$(i) since $s_t = 0$.) Let $v = v(x,t)$ be a solution to the following zero-temperature-ends boundary-initial problem *for a laterally insulated rod*:

$$(\ddagger) \quad \begin{array}{l} \text{(i)} \quad v_t - \kappa v_{xx} = 0, \ 0 < x < L, \ t > 0, \\ \text{(ii)} \quad v(0,t) = 0, \ v(L,t) = 0, \ t > 0, \\ \text{(iii)} \quad v(x,0) = f(x) - s(x), \ 0 < x < L. \end{array}$$

(a) Prove that $u(x,t) = s(x) + e^{-ht}v(x,t)$ solves $(\dagger)$ (assuming v solves $(\ddagger)$ and s solves equation (11.98)).

(b) (CAS) Use a computer algebra system and the procedure outlined above to solve $(\dagger)$, where $L = 100$ cm, $\kappa = 1.1$ cm²/sec, $h = 0.01$ sec^{-1}, $T_1 = 10°$, $T_2 = 60°$, $A = 20°$, and $f(x) = 0°, 0 < x < 100$. Plot the steady-state solution as well as the following snapshots of your solution: $u(x, 10)$, $u(x, 100)$, and $u(x, 500)$.

14. **An endpoint exposed to the environment.** Suppose a laterally insulated rod extending from $x = 0$ to $x = L$ cm has its right endpoint exposed to an environment with ambient temperature A degrees Celsius. Assuming that at this right endpoint the flow of heat between the rod to the environment is proportional to the difference between the rod's right-endpoint temperature and that of the environment, we obtain from Fourier's law of heat conduction the following right-endpoint boundary condition:

$$u_x(L,t) = -h(u(L,t) - A),$$

where $h > 0$ is the proportionality constant.

(a) Suppose that a laterally insulated rod extending from $x = 0$ to $x = L$ cm has its right endpoint exposed to an environment having temperature 0 degrees Celsius and its left endpoint is held at $0°$. Use separation of variables to find building-block solutions of the heat equation $u_t(x,t) = \kappa u_{xx}(x,t)$, $0 < x < L$, $t > 0$, such that $u(0,t) = 0$ for $t > 0$ and $u_x(L,t) = -hu(L,t), t > 0$. (Hint: Review Example 11.2.3.)

(b) (CAS) Let $u_n = u_n(x,t)$, $n = 1, 2, 3, \ldots$, be the building-block solutions you found in part (a). A consequence of a result due to Jacques Sturm and Joseph Liouville[6] is that the boundary value problem eigenfunctions $g_n(x) = u_n(x, 0)$ found in (a) are orthogonal in the following sense:

(11.99) $$\int_0^L g_n(x) g_m(x)\, dx = 0 \quad \text{whenever } n \neq m.$$

Assume that $L = 100$, $h = 0.01$, and as above, let $g_n = u_n(x,0)$. To provide evidence for (11.99) compute, using a CAS,

$$\int_0^{100} g_1(x) g_2(x)\, dx \quad \text{and} \quad \int_0^{100} g_2(x) g_4(x)\, dx.$$

Are your results consistent with the orthogonality of the family $\{g_n : n = 1, 2, 3, \ldots\}$ of functions?

(c) Again, for $n = 1, 2, 3, \ldots$, let $g_n(x) = u_n(x, 0)$, where $u_n = u_n(x,t)$ are your building-block solutions obtained in (a). Sturm-Liouville theory not only tells us that the family of eigenfunctions $\{g_n : n = 1, 2, 3, \ldots\}$ is orthogonal, but it also tells us that the family is **complete** in the sense that if f is any piecewise continuously differentiable function defined on $[0, L]$, then f has a **generalized Fourier representation**

(11.100) $$f(x) = \sum_{n=1}^{\infty} c_n g_n(x),$$

which holds at every point of continuity of f. Using as a model our derivation of the Fourier coefficient formulas (11.76) and (11.77), which relied on the orthogonality of the functions $\sin\left(\frac{n\pi x}{L}\right)$ and $\cos\left(\frac{n\pi x}{L}\right)$, find a formula for the coefficients c_n in equation (11.100) relying on the orthogonality relationship (11.99). Your formula for c_n will be a quotient of integrals both of which will involve g_n.

(d) (CAS) Consider the boundary-initial value problem

$$u_t = u_{xx}; \; u(0,t) = 0, \; t > 0; \; u_x(L,t) - hu(L,t) = 0, \; t > 0; \; u(x) = f(x), \; 0 < x < L,$$

where $\kappa = 1$, $L = 100$, $h = 0.01$, and $f(x) = x(100-x)/100$ (degrees), $0 \le x \le 100$. Using your building-block solutions of (a) and the coefficient formula derived in (c), find the following approximate solution of this boundary-initial value problem: $u(x,t) \approx \sum_{n=1}^{40} c_n u_n(x,t)$. Plot snapshots of your solution at times $t = 10, 100, 1{,}000$, and $3{,}000$. What are the predicted right-hand endpoint temperatures $u(100, 10)$, $u(100, 100)$, $u(100, 1{,}000)$, and $u(100, 3{,}000)$? With $u(x,t) \approx \sum_{n=1}^{80} c_n u_n(x,t)$, what are the predicted endpoint temperatures $u(100, 10)$, $u(100, 100)$, $u(100, 1{,}000)$, $u(100, 3{,}000)$? There should be essentially no difference between the endpoint-temperature approximations with 40 summands versus 80 summands, which suggests these approximations are accurate.

(e) Using Example 11.7.3 and the preceding exercise as models, explain how to solve the following boundary-initial value problem in which the ambient temperature A is not zero:

$$u_t = \kappa u_{xx}; \; u(0,t) = 0 \text{ for } t > 0; \; u_x(L,t) = -h(u(L,t) - A) \text{ for } t > 0,$$

---

[6]You can read more about Sturm-Liouville theory in [37].

11.7. The heat equation

and
$$u(x,0) = f(x) \quad \text{for } 0 < x < L.$$

15. (a) Suppose that a laterally insulated rod extending from $x = 0$ to $x = L$ cm has its right endpoint insulated and its left endpoint exposed to an environment at temperature A degrees Celsius. Let the initial temperature of the rod be given by $f = f(x)$. As in Exercise 14, assume that at its exposed endpoint the flow of heat between the rod to the environment is proportional to the difference between the rod's endpoint temperature and that of the environment. Set up a boundary-initial value problem modeling the temperature $u(x,t)$ of the rod at position x and time t. Assume κ is the thermal diffusivity of the rod.
    (b) For the problem in (a), what initial condition would result in an equilibrium (i.e., constant) solution?
    (c) Assuming that the ambient temperature A is zero in the problem of (a), use separation of variables to find building-block solutions of the problem that satisfy both the heat equation and the boundary conditions.

16. **Uniqueness of solutions.** Suppose that $u_1(x,t)$ and $u_2(x,t)$ are two (twice continuously differentiable) solutions to the zero-temperature-ends boundary-initial value problem
$$u_t = \kappa u_{xx}, \quad u(0,t) = u(L,t) = 0, \quad u(x,0) = f(x)$$
where f is continuously differentiable on $[0, L]$. The goal of this problem is to show $u_1(x,t) = u_2(x,t)$; i.e., solutions are *unique*.
    (a) Show that $v(x,t) \equiv u_1(x,t) - u_2(x,t)$ solves the zero-temperature-ends boundary value problem with a different initial condition, namely $v(x,0) = 0$ for all $0 \le x \le L$.
    (b) For each fixed $t \ge 0$ define
$$E(t) = \int_0^L v(x,t)^2 \, dx.$$

    Notice that $E(t) \ge 0$. Check that $E(0) = 0$ and then explain why it's enough to show that $E(t) = 0$ for all $t \ge 0$ to be able to conclude that $u_1(x,t) = u_2(x,t)$.
    (c) Using the fact that $v_t = \kappa v_{xx}$ (from (a)), show that
$$\frac{d}{dt}(E(t)) = 2\kappa \int_0^L v v_{xx} \, dx.$$

    You may assume that the differentiation with respect to t can be moved "inside" the integral sign, so that
$$\frac{d}{dt}(E(t)) = \int_0^L \frac{\partial}{\partial t}(v(x,t)^2) \, dx.$$

    (d) Show that an integration by parts (with $U = v$ and $dV = v_{xx}\,dx$) gives
$$\int_0^L v v_{xx} \, dx = v v_x \Big|_0^L - \int_0^L (v_x)^2 \, dx.$$

    Show this simplifies to
$$-\int_0^L (v_x)^2 \, dx$$

    using the boundary conditions $v(0,t) = v(L,t) = 0$.
    (e) Assemble the pieces to complete the argument that $E(t) = 0$ for all $t \ge 0$. Hint: By part (d), $E(t)$ is nonincreasing for $t \ge 0$.

11.8. The wave equation: Separation of variables

Imagine a homogeneous string, held taut and fastened at each end at a point on the x-axis as in Fig. 11.54. The string is plucked or struck to put it in motion so that it vibrates up and down in a plane. Our mathematical model assumes each point on the string just moves up or down in a line perpendicular to the x-axis. The vertical displacement of a point x on the string at time t is given by a function $u(x,t)$. A positive value of u means that at that moment the point is above the x-axis, while a negative value of u indicates the point is below the x-axis. The displacement $u(x,t)$ is governed by the wave equation

(11.101) $$u_{tt} = \alpha^2 u_{xx}$$

for α a constant.[7] The meaning of the constant α will be revealed later (see Exercise 12 and Section 11.9), but for now note that in order for the units in the wave equation to balance, α must have units of distance/time = velocity.

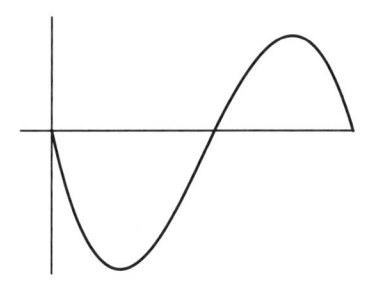

Figure 11.54. A vibrating string.

To think about equation (11.101) intuitively, notice that the derivative u_{xx} describes the concavity of u considered as a function of x, while u_{tt} is the acceleration (in the vertical direction). The larger u_{xx} is (in either the positive or negative sense), the greater the concavity and the stronger the force (due to tension), hence the larger the acceleration. When $u_{xx} > 0$, so the string is concave up, we expect the acceleration to be upward, so that u_{tt} should be positive too, as (11.101) says. When $u_{xx} < 0$, the string is concave down, and the acceleration should be downward.

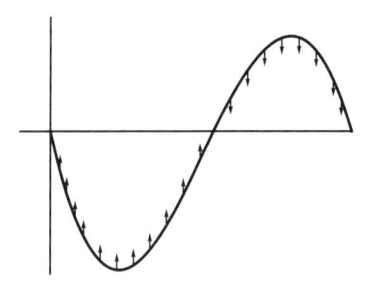

We will be solving equation (11.101) with boundary and initial conditions. Fixing the endpoints of the string on the x-axis at points $x = 0$ and $x = L$ gives us the boundary conditions

(11.102) $$u(0,t) = 0, \quad u(L,t) = 0.$$

Other types of boundary conditions are possible too, and some will be considered in the exercises. Because equation (11.101) has a second-order time derivative, we expect the appropriate initial

---

[7]Various assumptions are made in developing the wave equation, for example, that the string is perfectly flexible with constant linear density and that the effect of gravity as well as other possible outside forces is ignored. The model also applies only to "small" vibrations on the string.

conditions will specify both $u(x, 0)$, the initial position, and $u_t(x, 0)$, the initial velocity:

(11.103) $$u(x, 0) = f(x), \quad u_t(x, 0) = g(x)$$

for some continuous functions $f(x)$ and $g(x)$ on $[0, L]$. Generally we will want the boundary and initial conditions to be consistent at $x = 0$ and $x = L$; this requires that

$$f(0) = f(L) = 0 \quad \text{and} \quad g(0) = g(L) = 0.$$

The choice $g(x) \equiv 0$, corresponding to zero initial velocity, is sometimes called a "plucked string" problem. If by contrast, $f(x) \equiv 0$, so we have initial position zero (but nonzero initial velocity), we call this a "struck string" problem.

Solving the wave equation by separation of variables. Let's see how we can use separation of variables to solve equation (11.101) with the boundary conditions in (11.102) and initial conditions as in (11.103). We seek solutions in the form

$$u(x, t) = X(x)T(t).$$

Since then $u_{xx} = X''(x)T(t)$ and $u_{tt} = X(x)T''(t)$, substituting into the wave equation (11.101) gives

$$X(x)T''(t) = \alpha^2 X''(x)T(t).$$

We divide by $\alpha^2 X(x)T(t)$ to obtain

$$\frac{T''(t)}{\alpha^2 T(t)} = \frac{X''(x)}{X(x)}.$$

Since the left-hand side is a function of t only and the right-hand side is a function of x only, both must be a (common) constant value, which we call $-\lambda$ (the minus sign for later convenience):

$$\frac{T''(t)}{\alpha^2 T(t)} = -\lambda \quad \text{and} \quad \frac{X''(x)}{X(x)} = -\lambda.$$

Also notice that with $u(x, t) = X(x)T(t)$, the boundary conditions in (11.102) force $X(0) = 0$ and $X(L) = 0$ if we want to avoid the trivial solution $T(t) \equiv 0$, which just gives $u(x, t) \equiv 0$.

Now the boundary value problem

$$X''(x) + \lambda X(x) = 0, \quad X(0) = 0, \quad X(L) = 0,$$

is one we have solved already (see Section 11.2). The only nontrivial solutions come from the values

$$\lambda = \frac{n^2 \pi^2}{L^2}$$

for $n = 1, 2, 3\ldots$, and corresponding to each such λ we have the solution

$$X(x) = \sin \frac{n\pi x}{L}$$

or any constant multiple of this. Keep this on hold, and turn to the equation

$$\frac{T''(t)}{\alpha^2 T(t)} = -\lambda = -\frac{n^2 \pi^2}{L^2},$$

or

$$T''(t) + \frac{\alpha^2 n^2 \pi^2}{L^2} T(t) = 0.$$

This second-order ordinary differential equation has general solution
$$T(t) = a_n \cos \frac{\alpha n \pi t}{L} + b_n \sin \frac{\alpha n \pi t}{L}$$
for arbitrary constants a_n and b_n. Building-block solutions $u(x,t) = X(x)T(t)$ to the wave equation (11.101) with the boundary conditions (11.102) thus come in two types:
$$\sin \frac{n\pi x}{L} \cos \frac{\alpha n \pi t}{L}$$
and
$$\sin \frac{n\pi x}{L} \sin \frac{\alpha n \pi t}{L}$$
as n ranges over the positive integers. Since the Superposition Principle holds both for the wave equation (see Exercise 1) and our chosen boundary conditions (11.102), we can build more solutions by taking linear combinations of the building blocks. This leads us to solutions written as an infinite sum

$$\begin{aligned} u(x,t) &= \sum_{n=1}^{\infty} \left(a_n \sin \frac{n\pi x}{L} \cos \frac{\alpha n \pi t}{L} + b_n \sin \frac{n\pi x}{L} \sin \frac{\alpha \pi n t}{L} \right) \\ &= \sum_{n=1}^{\infty} \sin \frac{n\pi x}{L} \left(a_n \cos \frac{\alpha n \pi t}{L} + b_n \sin \frac{\alpha n \pi t}{L} \right). \end{aligned}$$

(11.104)

So far, we haven't used our initial conditions (11.103). To finish, we want to see if we can choose the constants a_n and b_n in (11.104) so that the initial conditions (11.103) are satisfied. With $u(x,t)$ as in (11.104), we have
$$u(x,0) = \sum_{n=1}^{\infty} a_n \sin \frac{n\pi x}{L}.$$
We want this to be the function $f(x)$, so we should choose the a_n's so that
$$\sum_{n=1}^{\infty} a_n \sin \frac{n\pi x}{L} = f(x).$$
We're looking at the Fourier sine series expansion for $f(x)$ on the interval $[0, L]$, so we must have

(11.105) $$a_n = \frac{2}{L} \int_0^L f(x) \sin \frac{n\pi x}{L} \, dx,$$

the Fourier sine series coefficients for f.

Now return to (11.104), differentiate with respect to t (we do this "formally", differentiating each term inside the sum), and then set $t = 0$. The result is
$$u_t(x,0) = \sum_{n=1}^{\infty} b_n \frac{\alpha n \pi}{L} \sin \frac{n\pi x}{L}.$$
We want this to be $g(x)$, so we should choose b_n so that the Fourier sine series coefficients of $g(x)$ are equal to $b_n \alpha n \pi / L$:
$$\frac{2}{L} \int_0^L g(x) \sin \frac{n\pi x}{L} \, dx = b_n \frac{\alpha n \pi}{L}.$$
Solving for b_n we see that b_n should be the product of $L/(\alpha n \pi)$ and the nth Fourier sine series coefficient for g. Summarizing our work we have:

11.8. The wave equation: Separation of variables

> A formal solution to the wave equation with boundary conditions $u(0,t) = 0, u(L,t) = 0$ and initial conditions $u(x,0) = f(x)$, $u_t(x,0) = g(x)$ is given by
>
> (11.106) $$u(x,t) = \sum_{n=1}^{\infty} \sin\frac{n\pi x}{L}\left(a_n \cos\frac{\alpha\pi n t}{L} + b_n \sin\frac{\alpha\pi n t}{L}\right),$$
>
> where
> $$a_n = \frac{2}{L}\int_0^L f(x)\sin\frac{n\pi x}{L}\,dx,$$
> the nth Fourier sine series coefficient for f, and
>
> (11.107) $$b_n = \frac{2}{\alpha n \pi}\int_0^L g(x)\sin\frac{n\pi x}{L}\,dx = \frac{L}{\alpha n \pi}\cdot n\text{th Fourier sine series coefficient for } g.$$

The terminology "formal solution" is used here because we haven't addressed the question of convergence of the infinite sum in this formula. If it does converge and if the resulting sum $u(x,t)$ is twice continuously differentiable by term-by-term differentiation, it will be the true desired solution. A proof that the infinite series in the formal solution does indeed converge when the initial position function f is continuously differentiable and the initial velocity function g is continuous is outlined in Exercise 20. This exercise also shows that the series converges to the desired solution on any interval when g is differentiable and f is twice differentiable. In the next section we will see how to circumvent the issue of convergence altogether, by obtaining the solution in a form that does not involve an infinite series.

> **Applying series solutions.** A good perspective to have on our series solutions of the heat and wave equations is that for applied problems we always use a partial sum of the series solution for modeling. For the heat and wave equations, this partial sum is clearly a classical solution (in fact infinitely differentiable) that satisfies the boundary conditions (at $x = 0$ and $x = L$). How well the partial sum serves to approximate the exact solution thus reduces to how well the partial sum at $t = 0$ approximates the initial temperature function (heat equation) or the initial position and initial velocity functions (wave equation).

Example 11.8.1. A string of length $L = \pi$ has both ends fixed on the x-axis. It is put into the position of a triangle with height $\frac{1}{10}$ as shown in Fig. 11.55 and released. We'll find the displacement $u(x,t)$ for $t > 0, 0 \leq x \leq \pi$. By our discussion above we seek $u(x,t)$ in the form

$$u(x,t) = \sum_{n=1}^{\infty} \sin nx[a_n\cos(\alpha n t) + b_n\sin(\alpha n t)],$$

where the coefficients a_n and b_n are to be determined from the initial conditions.

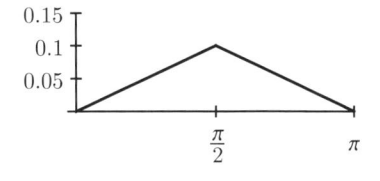

Figure 11.55. Initial position.

The initial position is

$$u(x,0) = \begin{cases} \frac{1}{5\pi}x & \text{if } 0 \leq x \leq \frac{\pi}{2}, \\ -\frac{1}{5\pi}x + \frac{1}{5} & \text{if } \frac{\pi}{2} \leq x \leq \pi, \end{cases}$$

and the initial velocity is $u_t(x,0) = 0$ for $0 \leq x \leq \pi$. Equation (11.105) says that the a_n's are the Fourier sine series coefficients of $f(x) = u(x,0)$.

Either computing these directly or using Exercise 14 in Section 11.6, we have

$$a_n = \begin{cases} \frac{4}{5\pi^2 n^2} & \text{if } n = 1, 5, 9, \ldots, \\ 0 & \text{if } n = 2, 4, 6, 8, \ldots, \\ -\frac{4}{5\pi^2 n^2} & \text{if } n = 3, 7, 11, \ldots. \end{cases}$$

Since $g(x) \equiv 0$, equation (11.107) says that $b_n = 0$ for all n. Thus

$$\begin{aligned} u(x,t) &= \sum_{n=1}^{\infty} a_n \sin(nx)\cos(\alpha nt) \\ (11.108) \quad &= \frac{4}{5\pi^2}\sin x \cos(\alpha t) - \frac{4}{45\pi^2}\sin(3x)\cos(3\alpha t) + \frac{4}{125\pi^2}\sin(5x)\cos(5\alpha t) - \cdots \end{aligned}$$

is the displacement function for our vibrating string. With $\alpha = 1$, Figs. 11.56–11.59 give a snapshot of the string at times $t = 0.2, 1, 2,$ and 3. To produce these graphs, we used a partial sum approximation to $u(x,t)$, keeping just the first three nonzero terms in the infinite sum expression for $u(x,t)$, with the chosen values of t. This gives pictures of the approximate displacement at the various times. We'll be able to compare these with pictures of the exact displacement in the next section.

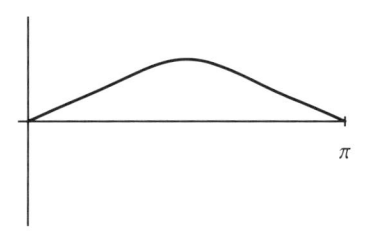

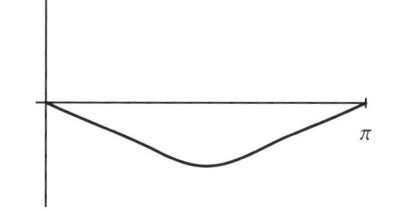

Figure 11.56. Approximate displacement at $t = 0.2$.

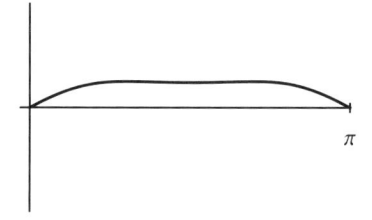

Figure 11.57. Approximate displacement at $t = 1$.

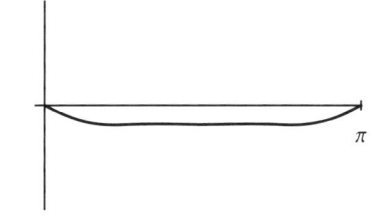

Figure 11.58. Approximate displacement at $t = 2$.

Figure 11.59. Approximate displacement at $t = 3$.

Harmonics. We've seen that among the building-block solutions to

$$u_{tt} = \alpha^2 u_{xx}, \quad u(0,t) = 0, \quad u(L,t) = 0,$$

11.8. The wave equation: Separation of variables

we have the functions

$$u(x,t) = \sin\frac{n\pi x}{L}\sin\frac{\alpha n\pi t}{L}$$

and

$$v(x,t) = \sin\frac{n\pi x}{L}\cos\frac{\alpha n\pi t}{L}.$$

More generally, for each n the functions

(11.109) $$u_n(x,t) = \sin\frac{n\pi x}{L}\left[a_n\sin\frac{\alpha n\pi t}{L} + b_n\cos\frac{\alpha n\pi t}{L}\right]$$

are solutions for constants a_n and b_n. For a fixed value of n, this is called an nth **harmonic**. It is a product of the function

$$\sin\frac{n\pi x}{L}$$

(a function of x alone) with a *time-varying amplitude*

$$a_n\sin\frac{\alpha n\pi t}{L} + b_n\cos\frac{\alpha n\pi t}{L}$$

(a function of t alone). Just as in Section 4.4.2, we know we can write

$$a_n\sin\frac{\alpha n\pi t}{L} + b_n\cos\frac{\alpha n\pi t}{L} = R_n\cos\left(\frac{\alpha n\pi t}{L} - \delta_n\right)$$

for some amplitude $R_n > 0$ and phase shift δ_n. Thus the displacement function in (11.109) is

$$u_n(x,t) = R_n\sin\frac{n\pi x}{L}\cos\left(\frac{\alpha n\pi t}{L} - \delta_n\right).$$

Each point on the string vibrates up and down with period $\frac{2L}{\alpha n}$ and circular frequency

(11.110) $$\frac{\alpha n\pi}{L}.$$

The amplitude of this vibration at position x is

$$R_n\left|\sin\frac{n\pi x}{L}\right|.$$

A point x on the string for which $\sin\frac{n\pi x}{L} = 0$ never moves; it is called a **node**. A function as in (11.109) is sometimes called a *standing wave*; it represents vertical vibrations of the string without apparent side-to-side motion of the waveform.

For an alternate perspective, fix a value of t. At this moment in time, the displacement $u_n(x,t)$ is given by $\sin\frac{n\pi x}{L}$ multiplied by the constant

$$R_n\cos\left(\frac{\alpha n\pi t}{L} - \delta_n\right).$$

Graphically, the position of the string at this instant is an amplitude-scaled version of the sine curve $\sin\frac{n\pi x}{L}$.

Fig. 11.60 shows that graph of

$$u(x,t) = \sin x\left[\frac{\sqrt{2}}{2}\sin t - \frac{\sqrt{2}}{2}\cos t\right] = \sin x\left[\cos\left(t - \frac{3\pi}{4}\right)\right]$$

(equation (11.109) with $L = \pi, n = 1, \alpha = 1, a_n = \sqrt{2}/2, b_n = -\sqrt{2}/2$) at six different times, $t = 0, \frac{\pi}{16}, \frac{\pi}{4}, \frac{\pi}{3}, \frac{\pi}{2}$, and $t = \frac{3\pi}{4}$. Every point x on the string is vibrating up and down with period 2π and circular frequency 1 radian/sec.

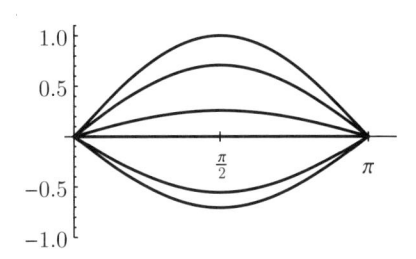

Figure 11.60. $u(x,t)$ at $t = 0, \pi/16, \pi/4, \pi/3, \pi/2, 3\pi/4$ (from bottom to top).

With $n = 1$, the value in (11.110) is $\frac{\alpha\pi}{L}$; dividing by 2π we obtain the **fundamental frequency** $\nu_1 = \frac{\alpha}{2L}$ cycles/sec. Values of $n > 1$ give rise to frequencies that are *integer* multiples of this fundamental frequency. The frequencies corresponding to larger values of n in (11.110) are called **overtones**[8] in acoustics. A vibrating piano or guitar string gives multiple tones, corresponding to the fundamental frequency and several overtones. You can demonstrate this if you, say, pluck a guitar string near one end while lightly touching the string at its midpoint. This will damp out the fundamental frequency, and you will find that you now hear the second harmonic. Touching the string 1/3 of the way down its length instead will bring the third harmonic into prominence. If you actually try this, you may discover that you can now hear these overtones more clearly even when you just pluck the string.

The difference in the quality of sound (or timbre) when a guitar or violin or piano string is struck (to give a note at the same pitch and loudness) depends on the relative contribution of the higher n-value harmonics to the vibrations. Various musical attributes—brilliance, richness, clearness, hollowness, etc.—are associated to the contributions of particular overtones. The fact that the overtones of a vibrating string are integer multiples of the fundamental frequency gives musical quality. By contrast, the vibrations of a drumhead (described by a two-dimensional wave equation) don't have this property and are not "melodic" in the same way that a vibrating violin or guitar string is.

Similar comments apply to wind instruments such as clarinets, flutes, and oboes, where a column of air vibrates. Two instruments playing the same note produce vibrations at the same fundamental frequency but have different sound quality owing to the relative intensities of the various harmonics. The relative contribution of the higher harmonics to the overall sound depends on the construction of the instrument and to a lesser degree the person playing it.

11.8.1. Exercises.

1. Show that a linear combination of solutions to $u_{tt} = \alpha^2 u_{xx}$ is again a solution.

2. For $0 \leq x \leq \pi$ and $t > 0$, solve the wave equation $u_{tt} = u_{xx}$ with boundary conditions
$$u(0,t) = u(\pi,t) = 0, \quad t > 0$$
and initial conditions
$$u(x,0) = \sin(3x) - \frac{1}{2}\sin(5x), \quad u_t(x,0) = \sin(2x) + \sin(4x).$$

In Exercises 3–6, solve the wave equation $u_{tt} = u_{xx}$ with boundary conditions
$$u(0,t) = 0, \quad u(L,t) = 0 \quad \text{for } t > 0$$
and initial conditions
$$u(x,0) = f(x), \quad u_t(x,0) = g(x) \quad \text{for } 0 < x < L$$

---
[8]The first overtone is the second harmonic, the second overtone is the third harmonic, and so on.

11.8. The wave equation: Separation of variables

for the given value of L and functions f and g. Exercises 13 and 14 in Section 11.6 may be useful to simplify the computational work in some parts.

3. $L = 1$, $f(x) = \frac{1}{10}\sin(\pi x)$, $g(x) = 0$.
4. $L = \pi$, $f(x) = x(\pi - x)$, $g(x) = 0$.
5. $L = \pi$, $f(x) = 0$, $g(x) = \frac{1}{10}\sin(2x)$.
6. $L = \pi$, $f(x) = 0$,
$$g(x) = \begin{cases} -\frac{\pi}{10}x & \text{if } 0 < x < \frac{\pi}{2}, \\ \frac{\pi}{10}(x - \pi) & \text{if } \frac{\pi}{2} \le x < \pi. \end{cases}$$

7. (CAS) Consider the wave equation $u_{tt} = u_{xx}$ with boundary conditions
$$u(0, t) = 0, \quad u(3, t) = 0 \quad \text{for } t > 0$$
and initial conditions
$$u(x, 0) = f(x), \quad u_t(x, 0) = g(x) \quad \text{for } 0 < x < 3.$$
If $g(x) = 0$ and
$$f(x) = \begin{cases} x & \text{if } 0 \le x < 1, \\ 1 & \text{if } 1 \le x < 2, \\ 3 - x & \text{if } 2 \le x \le 3, \end{cases}$$
find the solution $u(x, t)$ explicitly up to the $n = 5$ term. You may use a computer algebra system to help with integral computations.

8. The position $u(x, t)$ of a vibrating string of length π is described by the boundary-initial value problem
$$u_{tt} = \frac{1}{9}\pi^2 u_{xx},$$
$$u(0, t) = u(\pi, t) = 0, \quad t > 0,$$
and
$$u(x, 0) = \sin x - \frac{1}{2}\sin(3x), \quad u_t(x, 0) = 0, \quad 0 \le x \le \pi.$$
(a) Determine $u(x, t)$ for $0 < x < \pi$ and $t > 0$.
(b) Is there any time $t > 0$ for which $u(x, t) \equiv 0$? If so, find such a time.
(c) Is there any time $t > 0$ when the string is back in its original shape? If so, find the first such time.

9. Suppose that for $t > 0$, $0 \le x \le L$, $u_1(x, t)$ solves the boundary-initial value problem
$$u_{tt} = \alpha^2 u_{xx}, \quad u(0, t) = u(L, t) = 0, \quad u(x, 0) = f(x), \quad u_t(x, 0) = 0$$
(zero initial velocity) and $u_2(x, t)$ similarly solves
$$u_{tt} = \alpha^2 u_{xx}, \quad u(0, t) = u(L, t) = 0, \quad u(x, 0) = 0, \quad u_t(x, 0) = g(x)$$
(zero initial position). Show that $u_1(x, t) + u_2(x, t)$ solves
$$u_{tt} = \alpha^2 u_{xx}, \quad u(0, t) = u(L, t) = 0, \quad u(x, 0) = f(x), \quad u_t(x, 0) = g(x)$$
(general initial conditions).

10. For the solution (11.108) of Example 11.8.1, how does the position at time $t = \pi/\alpha$ compare to the position at time $t = 0$? How does the position at time $t = 2\pi/\alpha$ compare to the initial position?

11. (CAS) For your solution to Exercise 4, sketch (approximately) the graph of the solution $u(x, 1)$, the displacement at time $t = 1$, and the graph of the solution $u(x, 2)$, the displacement at time $t = 2$. Use the partial sum with two nonzero terms to approximate $u(x, 1)$ and $u(x, 2)$.

12. The constant α in the wave equation is related to the tension force τ in the string (with units mass-distance/time^2) and linear density ρ of the string (with units mass/distance) by $\alpha^2 = \tau/\rho$.
    (a) Verify that the given information on the units of τ and ρ is consistent with α having units of velocity.
    (b) If the tension τ is increased, does the fundamental frequency (equation (11.110) with $n = 1$) increase or decrease?
    (c) If the string density ρ is increased, does the fundamental frequency increase or decrease?
    (d) If the length of the string is increased, does the fundamental frequency increase or decrease?

13. In this problem we consider the wave equation with different boundary conditions ("free ends"). Physically, you can picture this as a string where both ends $x = 0$ and $x = L$ of the string are free to slide vertically (along the lines $x = 0$ and $x = L$ in the xu-plane) without friction.

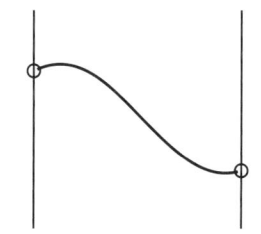

    (a) Find building-block solutions $X(x)T(t)$ to
    $$u_{tt} = \alpha^2 u_{xx}, \qquad 0 < x < L, \quad t > 0,$$
    if
    $$u_x(0, t) = u_x(L, t) = 0, \quad t > 0,$$
    using separation of variables. Don't forget to include the solutions that arise from the now-allowable value $\lambda = 0$ for the separation constant.
    (b) Using your building-block solutions from (a), describe how you would construct a solution that also satisfies
    $$u(x, 0) = f(x), \quad u_t(x, 0) = g(x)$$
    for $0 \leq x \leq L$.
    (c) If the average initial velocity $\frac{1}{L}\int_0^L g(x)dx$ is nonzero, what is happening to the overall motion of the string as time goes on? Hint: Focus on the term in the solution $u(x, t)$ that arises from the choice $\lambda = 0$ for a building block.

14. Consider the wave equation $u_{tt} = \alpha^2 u_{xx}$ with boundary conditions $u(0, t) = 0$ and $u_x(L, t) = 0$ ("one free end").
    (a) Use separation of variables to show that building-block solutions to this are
    $$\sin\frac{(2n-1)\pi x}{2L}\cos\frac{(2n-1)\pi\alpha t}{2L} \quad \text{and} \quad \sin\frac{(2n-1)\pi x}{2L}\sin\frac{(2n-1)\pi\alpha t}{2L}$$
    for $n = 1, 2, 3, \ldots$.
    (b) Suppose we also want to satisfy the initial conditions
    $$u(x, 0) = f(x), \quad u_t(x, 0) = 0$$
    for $0 < x < L$. What issue arises in trying to do this with a solution of the form
    $$u(x, t) = \sum \left(a_n \sin\frac{(2n-1)\pi x}{2L}\cos\frac{(2n-1)\pi\alpha t}{2L} + b_n \sin\frac{(2n-1)\pi x}{2L}\sin\frac{(2n-1)\pi\alpha t}{2L}\right)?$$
    Hint: The sine series for $f(x)$ on $[0, L]$ has the form
    $$\sum_{n=0}^{\infty} A_n \sin\frac{n\pi x}{L}.$$

11.8. The wave equation: Separation of variables

(c) Use the result of Exercise 24 in Section 11.6 to find $u(x,t)$ if $L = \pi$ and $f(x) = \sin x$.

15. **The wave equation with a damping term.** Here we consider the equation $u_{tt} + u_t - u_{xx} = 0$ with boundary conditions
$$u(0,t) = u(\pi, t) = 0, \quad t > 0.$$

(a) Using separation of variables, show that this has building-block solutions
$$e^{-t/2} \cos \frac{\sqrt{4n^2 - 1}\, t}{2} \sin(nx)$$
and
$$e^{-t/2} \sin \frac{\sqrt{4n^2 - 1}\, t}{2} \sin(nx).$$

(b) Use (a) to give the solution that also satisfies the initial conditions
$$u(x,0) = \sin(2x), \quad u_t(x,0) = 0$$
for $0 \le x \le \pi$.

(c) Consider any initial conditions with zero-initial-velocity:
$$u(x,0) = f(x), \quad u_t(x,0) = 0$$
for $0 \le x \le \pi$. Show that the solution in this case is
$$\sum_{n=1}^{\infty} \frac{b_n}{\sqrt{4n^2 - 1}} e^{-\frac{t}{2}} \left(\sqrt{4n^2 - 1} \cos \frac{\sqrt{4n^2 - 1}\, t}{2} + \sin \frac{\sqrt{4n^2 - 1}\, t}{2} \right) \sin(nx)$$
if f has Fourier sine series $\sum_{n=1}^{\infty} b_n \sin(nx)$ on $[0, \pi]$. (Assume f is continuous and piecewise continuously differentiable.)

16. The air pressure $p(x,t)$ in an organ pipe of length L satisfies the wave equation $p_{tt} = \alpha^2 p_{xx}$. If the pipe is open, it also satisfies the boundary conditions
$$p(0,t) = c, \quad p(L,t) = c, \quad t > 0,$$
for some constant c. If the pipe is closed at $x = L$ instead, it satisfies the boundary conditions
$$p(0,t) = c, \quad p_x(L,t) = 0, \quad t > 0.$$

The purpose of this problem is to decide which sounds higher—an open pipe or a closed pipe?

(a) Show if $P(x,t)$ is a solution with $c = 0$ for either set of boundary conditions, then $p(x,t) = P(x,t) + c$ is a solution satisfying either set of boundary conditions.

(b) With $c = 0$, find the fundamental frequency for both the closed and open organ pipes. Which sounds higher? Whenever the frequency of an audible vibration doubles, the pitch goes up an octave.

17. A string of length 1 is plucked so that its initial displacement and velocity are
$$u(x,0) = f(x), \quad u_t(x,0) = 0$$
where $f(x)$ is the "triangle function"
$$f(x) = \begin{cases} \frac{1}{10b} x & \text{if } 0 \le x < b, \\ -\frac{1}{10(1-b)}(x - 1) & \text{if } b < x \le 1. \end{cases}$$

Do you hear the second harmonic more clearly when $b = \frac{1}{4}$ or when $b = \frac{1}{8}$? The Fourier sine series for $f(x)$ is
$$f(x) = \frac{2}{10b(1-b)\pi^2} \sum_{n=1}^{\infty} \frac{\sin(bn\pi)}{n^2} \sin(n\pi x).$$

18. **Uniqueness of solutions.** For a string of length L whose displacement is given by $u(x,t)$, the energy at time t is defined to be

$$E(t) = \int_0^L \left(\frac{1}{2}\rho \left[\frac{\partial u}{\partial t}(x,t)\right]^2 + \frac{1}{2}\tau \left[\frac{\partial u}{\partial x}(x,t)\right]^2 \right) dx$$

where ρ and τ are related to the constant α in the wave equation by $\alpha^2 = \tau/\rho$ (see Exercise 12). The first term in the expression for $E(t)$ represents kinetic energy and the second represents potential energy.

(a) Show that

$$\frac{dE}{dt} = \tau \int_0^L \left(\frac{\partial u}{\partial t} \frac{\partial^2 u}{\partial x^2} + \frac{\partial u}{\partial x} \frac{\partial^2 u}{\partial t \partial x} \right) dx.$$

You may assume that the differentiation can be moved inside the integral sign, that is, that

$$\frac{d}{dt}\left(\int_0^L \left(\frac{1}{2}\rho \left(\frac{\partial u}{\partial t}\right)^2 + \frac{1}{2}\tau \left(\frac{\partial u}{\partial x}\right)^2 \right) dx \right) = \int_0^L \frac{\partial}{\partial t}\left(\frac{1}{2}\rho \left(\frac{\partial u}{\partial t}\right)^2 + \frac{1}{2}\tau \left(\frac{\partial u}{\partial x}\right)^2 \right) dx.$$

You'll also need to use the wave equation

$$\frac{\partial^2 u}{\partial t^2} = \alpha^2 \frac{\partial u^2}{\partial x^2}.$$

(b) Using the result from (a), show that

$$\frac{dE}{dt} = \tau \int_0^L \frac{\partial}{\partial x}\left(\frac{\partial u}{\partial x}\frac{\partial u}{\partial t} \right) dx.$$

(c) Explain why the boundary conditions $u(0,t) = 0$, $u(L,t) = 0$ imply that $u_t(0,t) = 0 = u_t(L,t)$ for all $t > 0$.

(d) Use (b) and (c) to show that $\frac{dE}{dt} = 0$. This is a "conservation of energy" principle for free vibrations of a string. It says that $E(t)$ is constant.

(e) Suppose that $u(0,t) = 0 = u(L,t)$ and that the string has initial position and velocity both zero. Show that $E(0) = 0$ and hence $E(t) = 0$ for all $t > 0$. Conclude that the only solution to the wave equation with these boundary and initial conditions is $u(x,t) = 0$, $0 \le x \le L$, $t > 0$.

(f) Consider the wave equation with boundary conditions $u(0,t) = 0 = u(L,t)$. Show that for a given $f(x)$ and $g(x)$ there is at most one solution that also satisfies the initial conditions

$$u(x,0) = f(x), \quad u_t(x,0) = g(x), \quad 0 < x < L.$$

Hint: If $u_1(x,t)$ and $u_2(x,t)$ are both solutions, what boundary-initial value problem does $u_1 - u_2$ solve?

19. An old guitar is strapped down in the back of an RV, which slams into a brick wall at a speed of 10 meters per second (about 22 miles per hour). We model the motion of one of the guitar strings of that old guitar via the following boundary-initial value problem, with u and x in centimeters:

$$u_{tt} = (110)^2(130)^2 u_{xx}, \quad 0 < x < 65, \quad t > 0,$$
$$u(0,t) = 0, \quad u(65,t) = 0, \quad t > 0,$$
$$u(x,0) = 0, \quad u_t(x,0) = 1{,}000, \quad 0 < x < 65.$$

(a) Find a formal series solution of the preceding problem.
(b) What is the fundamental frequency of the string's oscillation in hertz?
(c) Can you identify the note corresponding to the string's fundamental frequency?

(d) (CAS) Using any technology at your disposal create a slow-motion animation of the string's motion consisting of 30 frames per 1 period of motion. In creating the animation, approximate your series solution from (a) using its first 40 nonzero terms.

20. **Convergence of the formal series solution of the wave equation.** Suppose that f is a continuously differentiable function on $[0, L]$ with $f(0) = f(L) = 0$ and that g is a continuous function on $[0, L]$ with $g(0) = g(L) = 0$. Let G be defined on $[0, L]$ by $G(x) = \int_0^x g(t)\,dt$, so that G is an antiderivative of g on $[0, L]$ (i.e., G is differentiable on $[0, L]$ and $G'(x) = g(x)$ for all x in $[0, L]$). Let $\tilde{f}$ denote the odd period-$2L$ extension of f to $\mathbb{R}$ and let G_e denote the even period-$2L$ extension of G to $\mathbb{R}$. It is easy to see that $\tilde{f}$ and G_e are continuously differentiable on all of $\mathbb{R}$. The interested reader can check this, but for the purposes of this problem, assume that it is true. Recall that

$$\sum_{n=1}^{\infty} \sin\frac{n\pi x}{L}\left(a_n \cos\frac{\alpha\pi n t}{L} + b_n \sin\frac{\alpha\pi n t}{L}\right)$$

is our formal series solution of the general wave equation problem (with boundary conditions as in (11.102) and initial conditions as in (11.103)), where

$$a_n = \frac{2}{L}\int_0^L f(x)\sin\frac{n\pi x}{L}\,dx$$

are the Fourier sine series coefficients of f and

$$b_n = \frac{2}{\alpha n\pi}\int_0^L g(x)\sin\frac{n\pi x}{L}\,dx,$$

which you will show in part (b) of this exercise are closely related to the Fourier cosine coefficients of G. Note that for any positive integer N, we can write the Nth partial sum of the formal series solution as

$$\sum_{n=1}^{N} a_n \sin\frac{n\pi x}{L}\cos\frac{\alpha\pi n t}{L} + \sum_{n=1}^{N} b_n \sin\frac{n\pi x}{L}\sin\frac{\alpha\pi n t}{L}.$$

In parts (a)–(c) below you will show that the limit of each of the two partial sums in the preceding expression exists as $N \to \infty$ for all choices of x and t. In part (d) you will show that the series converges to a solution of the general boundary-initial value wave equation problem if f is actually twice differentiable and g is once differentiable on $[0, L]$.

(a) Use the trigonometric identity $\sin(A)\cos(B) = \frac{1}{2}\Big(\sin(A-B) + \sin(A+B)\Big)$ to show

$$\lim_{N\to\infty}\sum_{n=1}^{N} a_n \sin\frac{n\pi x}{L}\cos\frac{\alpha\pi n t}{L} = \frac{1}{2}\Big(\tilde{f}(x-\alpha t) + \tilde{f}(x+\alpha t)\Big)$$

for all real numbers x and t. (Don't forget the Fourier Convergence Theorem.)

(b) Use $b_n = \frac{2}{\alpha n\pi}\int_0^L g(x)\sin\frac{n\pi x}{L}\,dx$ and integration by parts (with "$u = \sin\frac{n\pi x}{L}$" and "$dv = g(x)\,dx$") to show that for $n = 1, 2, 3, \ldots$,

$b_n = -\frac{1}{\alpha}\cdot$ the nth Fourier cosine series coefficient of G.

(c) Use the trigonometric identity $\sin(A)\sin(B) = \frac{1}{2}\Big(\cos(A-B) - \cos(A+B)\Big)$ and the result of part (b) above to show that

$$\lim_{N\to\infty}\sum_{n=1}^{N} b_n \sin\frac{n\pi x}{L}\sin\frac{\alpha\pi n t}{L} = \frac{1}{2\alpha}\Big(G_e(x+\alpha t) - G_e(x-\alpha t)\Big)$$

for all real numbers x and t.

(d) Let $u(x,t)$ be defined by

$$u(x,t) = \frac{1}{2}\left(\tilde{f}(x-\alpha t) + \tilde{f}(x+\alpha t)\right) + \frac{1}{2\alpha}\left(G_e(x+\alpha t) - G_e(x-\alpha t)\right).$$

Also let V be the diamond-shaped region in the xt-plane defined by the inequalities $0 < x - \alpha t < L$ and $0 < x + \alpha t < L$. If (x,t) lies in V, $\tilde{f}$ becomes f and G_e becomes G in the above formula for $u(x,t)$. Show directly from this formula that if we now require f to be twice differentiable and g to be once differentiable on $[0,L]$, then $u(x,t)$ satisfies the wave equation for (x,t) in V, as well as the boundary conditions $u(0,t) = 0$, $u(L,t) = 0$, for all t, and the initial conditions $u(x,0) = f(x)$, $u_t(x,0) = g(x)$, for all x in $[0,L]$. You have already shown in parts (a) and (c) that

$$u(x,t) = \lim_{N \to \infty} \sum_{n=1}^{N} \sin \frac{n\pi x}{L} \left(a_n \cos \frac{\alpha \pi n t}{L} + b_n \sin \frac{\alpha \pi n t}{L}\right)$$

for all x and t. This is the desired conclusion.

11.9. The wave equation: D'Alembert's method

In this section we are going to focus on the wave equation

(11.111) $$u_{tt} = \alpha^2 u_{xx}$$

with initial conditions

$$u(x,0) = f(x), \quad u_t(x,0) = g(x)$$

for $-\infty < x < \infty$, but **no** boundary conditions. This can be thought of as describing the motion of an *infinite* string. We'll see an ingenious way, quite different from the method of separation of variables, to solve any such problem. As we will see, the role of the "data" f and g in the initial conditions will be rather simple, and our solution will be easy to interpret intuitively to give more geometric insight into the motion of the string.

The basic idea of the D'Alembert method is to make a particular kind of substitution at the outset. The original variables x and t will be replaced by new variables we call ξ and η, defined by

$$\xi = x + \alpha t$$

and

$$\eta = x - \alpha t.$$

To see how to rewrite equation (11.111) in terms of our new variables, we first compute

$$\frac{\partial \xi}{\partial x} = 1, \quad \frac{\partial \xi}{\partial t} = \alpha \quad \text{and} \quad \frac{\partial \eta}{\partial x} = 1, \quad \frac{\partial \eta}{\partial t} = -\alpha.$$

We are temporarily using our more long-winded notation for partial derivatives for greater clarity. Next, we use the Chain Rule to see that

(11.112) $$\frac{\partial u}{\partial x} = \frac{\partial u}{\partial \xi}\frac{\partial \xi}{\partial x} + \frac{\partial u}{\partial \eta}\frac{\partial \eta}{\partial x} = \frac{\partial u}{\partial \xi} + \frac{\partial u}{\partial \eta}.$$

Similarly you should verify (see Exercise 5)

(11.113) $$\frac{\partial u}{\partial t} = \alpha\left(\frac{\partial u}{\partial \xi} - \frac{\partial u}{\partial \eta}\right),$$

(11.114) $$\frac{\partial^2 u}{\partial x^2} = \frac{\partial^2 u}{\partial \xi^2} + 2\frac{\partial^2 u}{\partial \xi \partial \eta} + \frac{\partial^2 u}{\partial \eta^2},$$

11.9. The wave equation: D'Alembert's method

and

(11.115) $$\frac{\partial^2 u}{\partial t^2} = \alpha^2 \left(\frac{\partial^2 u}{\partial \xi^2} - 2 \frac{\partial^2 u}{\partial \xi \partial \eta} + \frac{\partial^2 u}{\partial \eta^2} \right).$$

Now substitute equations (11.114) and (11.115) into the wave equation (11.111). After the dust settles, we obtain

$$\frac{\partial^2 u}{\partial \xi \partial \eta} = 0,$$

which we think of as

(11.116) $$\frac{\partial}{\partial \xi} \left(\frac{\partial u}{\partial \eta} \right) = 0.$$

Any function of the form

$$u(\xi, \eta) = \varphi(\eta) + \psi(\xi)$$

for arbitrary (twice differentiable) one-variable functions φ and ψ is a solution to (11.116). You can verify this immediately by substitution. Moreover, every solution to (11.116) has this form; see Exercise 6.

Returning to our original variables x and t, we see that the wave equation $u_{tt} = \alpha^2 u_{xx}$ has general solution

(11.117) $$u(x,t) = \varphi(x - \alpha t) + \psi(x + \alpha t)$$

for our arbitrary functions φ and ψ. We've seen expressions like $\varphi(x - \alpha t)$ and $\psi(x + \alpha t)$ before (for example, in Section 11.4) and we've called them **traveling waves**. With $\alpha > 0$, the first has positive velocity α and is therefore moving to the right. The second, $\psi(x + \alpha t)$, is moving to the left, with velocity $-\alpha$. The function in (11.117) is the sum of these two oppositely moving waves. So the constant α in the wave equation is the speed of such waves as they move along the string. This is different from the velocity u_t, which describes the rate of change of the vertical displacement of a particular point on the string.

Imposing the initial conditions. Starting from our general solution

(11.118) $$u(x,t) = \varphi(x - \alpha t) + \psi(x + \alpha t)$$

to the wave equation, can we choose φ and ψ so that the initial conditions

(11.119) $$u(x, 0) = f(x), \quad u_t(x, 0) = g(x)$$

are satisfied?

A special case. As a warm-up calculation, let's first assume $g(x) \equiv 0$, so that our string has zero initial velocity at every point. In this case we want

(11.120) $$u(x, 0) = \varphi(x) + \psi(x) = f(x)$$

and (using the Chain Rule in equation (11.118))

(11.121) $$u_t(x, 0) = -\alpha \varphi'(x) + \alpha \psi'(x) = g(x) = 0.$$

Our task is to determine φ and ψ from equations (11.120) and (11.121). From equation (11.121) we see that

$$(\psi - \varphi)'(x) = 0$$

so that

(11.122) $$\psi(x) - \varphi(x) = k$$

for some constant k. Adding equations (11.122) and (11.120) gives
$$\psi(x) = \frac{f(x) + k}{2},$$
while subtracting equation (11.122) from (11.120) gives
$$\varphi(x) = \frac{f(x) - k}{2}.$$
By equation (11.118) we have
$$u(x,t) = \frac{f(x - \alpha t) - k}{2} + \frac{f(x + \alpha t) + k}{2} = \frac{1}{2}f(x - \alpha t) + \frac{1}{2}f(x + \alpha t).$$
Let's summarize our work so far:

> The D'Alembert solution to the zero-initial-velocity infinite string problem
> $$u_{tt} = \alpha u_{xx}, \quad u(x,0) = f(x), \quad u_t(x,0) = 0$$
> is
> $$u(x,t) = \frac{1}{2}f(x - \alpha t) + \frac{1}{2}f(x + \alpha t).$$

The general case. Now suppose that the initial velocity $u_t(x,0)$ is not necessarily zero, say, $u_t(x,0) = g(x)$. As before, we have

(11.123) $$-\alpha \varphi'(x) + \alpha \psi'(x) = g(x),$$

so that
$$(\psi - \varphi)'(x) = \frac{1}{\alpha} g(x).$$

Integrating this gives

(11.124) $$\psi(x) - \varphi(x) = \frac{1}{\alpha} \int_0^x g(w)\, dw + k$$

for some constant k. Adding equations (11.124) and (11.120) yields
$$\psi(x) = \frac{1}{2}f(x) + \frac{1}{2\alpha} \int_0^x g(w)\, dw + \frac{k}{2},$$
while subtracting equation (11.124) from (11.120) gives us
$$\varphi(x) = \frac{1}{2}f(x) - \frac{1}{2\alpha} \int_0^x g(w)\, dw - \frac{k}{2}.$$

Our solution is then
$$\begin{aligned} u(x,t) &= \varphi(x - \alpha t) + \psi(x + \alpha t) \\ &= \frac{1}{2}f(x - \alpha t) - \frac{1}{2\alpha} \int_0^{x-\alpha t} g(w)\, dw + \frac{1}{2}f(x + \alpha t) + \frac{1}{2\alpha} \int_0^{x+\alpha t} g(w)\, dw. \end{aligned}$$

The sum of the integrals,
$$-\frac{1}{2\alpha} \int_0^{x-\alpha t} g(w)\, dw + \frac{1}{2\alpha} \int_0^{x+\alpha t} g(w)\, dw = \frac{1}{2\alpha} \left(\int_{x-\alpha t}^0 g(w)\, dw + \int_0^{x+\alpha t} g(w)\, dw \right),$$

can be written as
$$\frac{1}{2\alpha} \int_{x-\alpha t}^{x+\alpha t} g(w)\, dw.$$

11.9. The wave equation: D'Alembert's method

This simplifies our solution to

$$u(x,t) = \frac{1}{2}f(x-\alpha t) + \frac{1}{2}f(x+\alpha t) + \frac{1}{2\alpha}\int_{x-\alpha t}^{x+\alpha t} g(w)\,dw.$$

This is called **D'Alembert's solution**, summarized as follows:

> The D'Alembert solution to the infinite string problem
> $$u_{tt} = \alpha u_{xx}, \quad u(x,0) = f(x), \quad u_t(x,0) = g(x)$$
> is
> $$u(x,t) = \frac{1}{2}f(x-\alpha t) + \frac{1}{2}f(x+\alpha t) + \frac{1}{2\alpha}\int_{x-\alpha t}^{x+\alpha t} g(w)\,dw.$$

Ideally, we would want f to be twice continuously differentiable (and g to be at least one time continuously differentiable) so that the D'Alembert solution $u(x,t)$ has continuous second-order partial derivatives. In practice, though, we may want to consider functions f and g which have isolated points at which they fail to be differentiable, as the next example shows.

Example 11.9.1. We'll find the D'Alembert solution to

$$u_{tt} = u_{xx}, \quad u(x,0) = f(x), \quad u_t(x,0) = 0, \quad -\infty < x < \infty, \quad t > 0,$$

where

$$f(x) = \begin{cases} 2x+2 & \text{if } -1 \le x \le -\frac{1}{2}, \\ 1 & \text{if } -\frac{1}{2} < x < \frac{1}{2}, \\ -2x+2 & \text{if } \frac{1}{2} \le x \le 1, \\ 0 & \text{for all other } x. \end{cases}$$

The graph of the initial position $f(x)$ is shown in Fig. 11.61. Keep in mind that we are imposing **no** boundary conditions, just initial conditions.

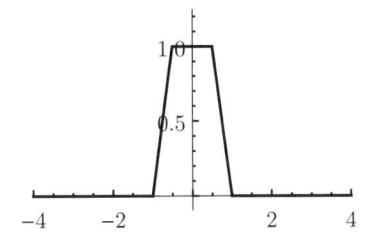

Figure 11.61. Initial position $f(x)$.

Since $g(x) \equiv 0$, the D'Alembert solution is

$$u(x,t) = \frac{1}{2}f(x-t) + \frac{1}{2}f(x+t)$$

for all x and all $t \ge 0$. Figs. 11.63–11.65 show snapshots of the solution at three values of t: $t = 1, 2, 3$. To see how these are visually produced, we first sketch the function $\frac{1}{2}f(x)$, as in Fig. 11.62. A sketch of the function $u(x,1) = \frac{1}{2}f(x-1) + \frac{1}{2}f(x+1)$ is obtained by translating the graph of $\frac{1}{2}f$ in Fig. 11.62 one unit to the right (this is $\frac{1}{2}f(x-1)$) and one unit to the left (this is $\frac{1}{2}f(x+1)$) and then adding the results (Fig. 11.63). Fig. 11.64, which shows $u(x,2)$, is similarly obtained as the sum of two translations of $\frac{1}{2}f(x)$: one to the right two units and the other to the left two units. For $u(x,3)$, the translates are three units, right and left.

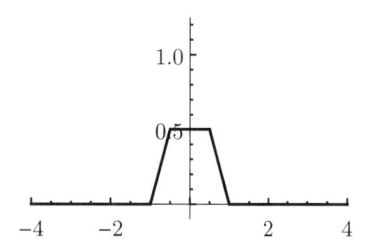

Figure 11.62. $\frac{1}{2}f(x)$.

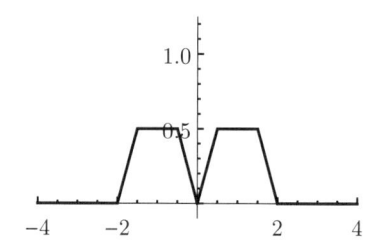

Figure 11.63. Position at $t = 1$, which is $\frac{1}{2}f(x-1) + \frac{1}{2}f(x+1)$.

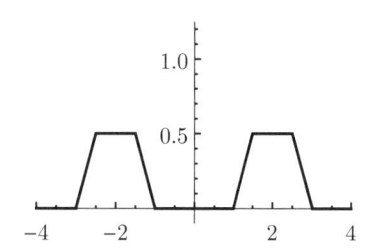

Figure 11.64. Position at $t = 2$, which is $\frac{1}{2}f(x-2) + \frac{1}{2}f(x+2)$.

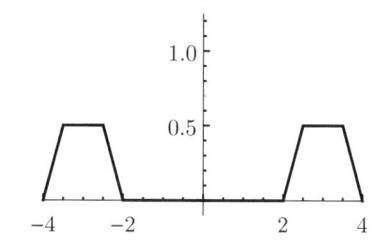

Figure 11.65. Position at $t = 3$, which is $\frac{1}{2}f(x-3) + \frac{1}{2}f(x+3)$.

Exercise 20 in Section 11.8 reconciles D'Alembert's solution with the "infinite series" solution (equation (11.106)) we obtained by separation of variables in Section 11.8. If you look back at this exercise, you will see, for example, that it shows that if f is twice differentiable on $[0, L]$ with $f(0) = f(L) = 0$, then the infinite series solution $u(x,t)$ of the wave equation $u_{tt} = \alpha^2 u_{xx}$, with boundary conditions $u(0,t) = u(L,t) = 0$, $t \geq 0$ and initial conditions $u(x,0) = f(x), u_t(x,0) = 0$, $0 \leq x \leq L$, is equal to the D'Alembert solution $\frac{1}{2}\tilde{f}(x-\alpha t) + \frac{1}{2}\tilde{f}(x+\alpha t)$, where $\tilde{f}$ is the odd periodic extension of f to $(-\infty, \infty)$.

Example 11.9.2. We'll revisit Example 11.8.1 from Section 11.8, now with the D'Alembert perspective. The initial position of the string (for $0 \leq x \leq \pi$) is

$$u(x,0) = f(x) = \begin{cases} \frac{1}{5\pi}x & \text{if } 0 \leq x \leq \frac{\pi}{2}, \\ -\frac{1}{5\pi}x + \frac{1}{5} & \text{if } \frac{\pi}{2} \leq x \leq \pi, \end{cases}$$

and the initial velocity is $u_t(x,0) = g(x) = 0$. The odd periodic extension $\tilde{f}(x)$ of $f(x)$ is sketched in Fig. 11.66. Using $\tilde{f}$ as the initial position for $-\infty < x < \infty$ and initial velocity zero, we have the D'Alembert solution

$$u(x,t) = \frac{1}{2}\tilde{f}(x-\alpha t) + \frac{1}{2}\tilde{f}(x+\alpha t)$$

for $t \geq 0$ and all x. While for $0 \leq x \leq \pi$, this is the same as the separation of variables solution we produced in Example 11.8.1, it gives the solution in a quite different form.

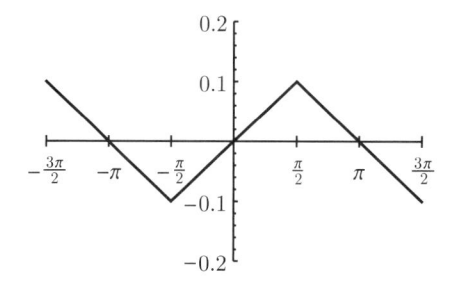

Figure 11.66. Odd periodic extension $\tilde{f}$ of f.

11.9. The wave equation: D'Alembert's method

Here is one way we can visually produce a snapshot of the solution at any chosen time. Again, let's set $\alpha = 1$ for convenience. Suppose we want to see $u(x,1)$, the solution at time $t = 1$:

$$u(x,1) = \frac{1}{2}\left(\tilde{f}(x-1) + \tilde{f}(x+1)\right).$$

Thus at each point x, $u(x,1)$ is the average value of $\tilde{f}(x-1)$ and $\tilde{f}(x+1)$. The graph of the function $\tilde{f}(x-1)$ is a translation of the graph of $\tilde{f}(x)$, by one unit to the right. The graph of $\tilde{f}(x+1)$ is a translation of the graph of $\tilde{f}$ one unit to the left. We sketch these in Figs. 11.67 and 11.68. The function $u(x,1)$ is obtained by averaging the values $\tilde{f}(x-1)$ and $\tilde{f}(x+1)$. We can do this averaging visually to sketch the graph of $u(x,1)$. In doing so we're helped by the fact that the average of two piecewise linear functions is also piecewise linear, so we can average the values of $\tilde{f}(x-1)$ and $\tilde{f}(x+1)$ at each point x where one or the other graphs has a corner and then join these points by line segments; see Fig. 11.69.

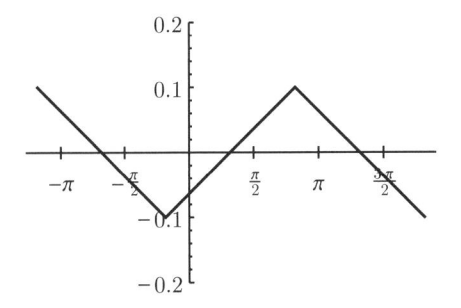

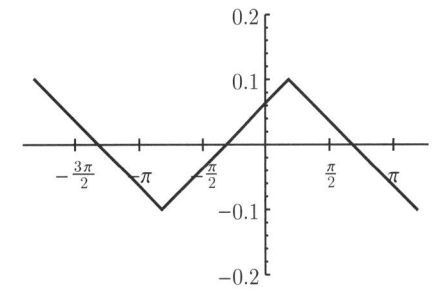

Figure 11.67. The graph of $\tilde{f}(x-1)$.

Figure 11.68. The graph of $\tilde{f}(x+1)$.

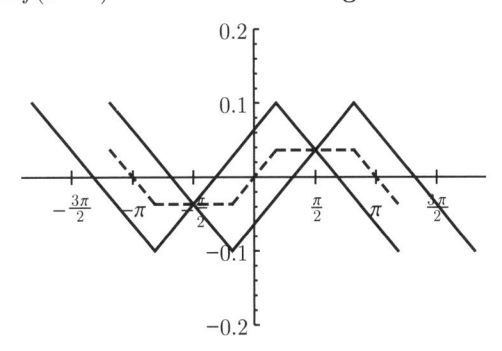

Figure 11.69. $u(x,1) = \frac{1}{2}[\tilde{f}(x-1) + \tilde{f}(x+1)]$ (shown dashed).

You might compare the graph in Fig. 11.69 (restricted to the interval $0 \le x \le \pi$) with that in Fig. 11.57 of Example 11.8.1. There, we were solving the same problem by separation of variables, and we produced the sketch in Fig. 11.57 by taking only the first three nonzero terms in the infinite sum solution for $u(x,1)$. This approximation smooths out the corners we see in the graph of the D'Alembert solution in Fig. 11.69.

11.9.1. Exercises.

1. At time $t = 0$ the position of an infinite string is given by

$$u(x,0) = e^{-x^2}.$$

(a) Give the D'Alembert solution to the wave equation $u_{tt} = u_{xx}$ for this string if the initial velocity is equal to zero everywhere.

(b) The graph of e^{-x^2} is shown below. Using this and your answer to (a) sketch the position of the string at $t = 2$.

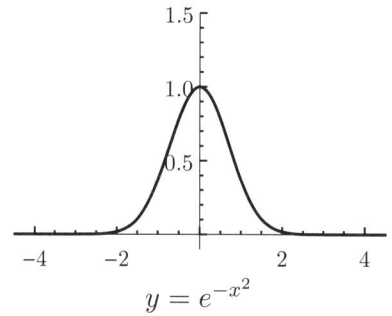

$$y = e^{-x^2}$$

2. Suppose that at time $t = 0$ the position of an infinite string is given by $u(x, 0) = f(x)$ where

$$f(x) = \begin{cases} \cos x & \text{if } -\frac{\pi}{2} \le x \le \frac{\pi}{2}, \\ 0 & \text{for all other } x, \end{cases}$$

and the initial velocity is zero.

(a) Give the D'Alembert solution to $u_{tt} = u_{xx}$ with this initial position and velocity, and determine the value of $u(2, \pi)$.

(b) Sketch the position of the string at times $t = \frac{\pi}{2}$ and $t = \pi$, using the method of Example 11.9.1.

(c) How long does it take before the point $x = 3\pi$ begins to vibrate?

3. Consider the wave equation $u_{tt} = u_{xx}$ with initial conditions

$$u(x, 0) = \sin x, \quad u_t(x, 0) = 0$$

for $-\infty < x < \infty$ and $t > 0$.

(a) Find the D'Alembert solution and show that it is $u(x, t) = \cos t \sin x$.

(b) Sketch the solution $u(x, t)$ for $t = 0$, $t = \frac{\pi}{4}$, and $t = \frac{\pi}{2}$.

4. Solve the boundary-initial value problem

$$u_{tt} = 100 u_{xx},$$
$$u(0, t) = u(10, t) = 0, \quad t > 0,$$
$$u(x, 0) = 2 \sin \frac{\pi x}{10} + \sin \frac{3 \pi x}{10}, \quad u_t(x, 0) = 0, \quad 0 < x < 10,$$

(a) using "building-block solutions" produced by separation of variables and
(b) using the D'Alembert method.
(c) Show your solutions to (a) and (b) agree.

5. Verify equations (11.113)–(11.115) using the Chain Rule. Assume that u is sufficiently nice that the mixed second-order partial derivatives $\frac{\partial^2 u}{\partial \eta \partial \xi}$ and $\frac{\partial^2 u}{\partial \xi \partial \eta}$ are equal. Hint: For $\frac{\partial^2 u}{\partial x^2}$, start with

$$\frac{\partial^2 u}{\partial x^2} = \frac{\partial}{\partial x}\left(\frac{\partial u}{\partial x}\right) = \frac{\partial}{\partial \xi}\left(\frac{\partial u}{\partial x}\right)\frac{\partial \xi}{\partial x} + \frac{\partial}{\partial \eta}\left(\frac{\partial u}{\partial x}\right)\frac{\partial \eta}{\partial x}$$

and then use (11.112).

6. Suppose that $u(\xi, \eta)$ is a solution of the partial differential equation (11.116). Show that there exist functions $\varphi(\xi)$ and $\psi(\eta)$ such that

$$u(\xi, \eta) = \varphi(\xi) + \psi(\eta).$$

Hint: Integrate twice, first with respect to ξ, then with respect to η. If $v(\xi, \eta)$ is a two-variable function such that $\frac{\partial v}{\partial \xi} = 0$, then $v(\xi, \eta)$ is constant with respect to the variable ξ, but this "constant" can depend of the other variable η, so that $v(\xi, \eta)$ has the form $\Psi(\eta)$ for some one-variable function Ψ. Apply this with $v = \frac{\partial u}{\partial \eta}$.

7. Suppose we make the substitutions $X = x/L$ and $\tau = \alpha t/L$ in the wave equation
$$u_{tt} = \alpha^2 u_{xx}.$$
   (a) What is the resulting partial differential equation in terms of X and τ?
   (b) Show that the new variables X and τ are dimensionless. Hint: The units on α are distance/time.

8. In obtaining the D'Alembert solution to the wave equation $u_{tt} = \alpha^2 u_{xx}$ we made the substitutions $\xi = x - \alpha t$, $\eta = x + \alpha t$.
   (a) Similarly, make the substitution $\xi = x$, $\eta = x + t$ in $u_{xt} - u_{tt} = 0$ and rewrite the equation in terms of ξ and η.
   (b) Solve the equation you obtained in (a), and then give the solution to $u_{xt} - u_{tt} = 0$.
   (c) Solve the partial differential equation
$$u_{xx} + u_{xt} - 2u_{tt} = 0$$
   for unknown function $u(x,t)$ by first making the substitution $\xi = x + t$, $\eta = 2x - t$.

9. (a) Make the substitution $\zeta = x - \alpha t$, $\eta = x + \alpha t$ in the first-order partial differential equation
$$u_t - \alpha u_x = 0.$$
   What new equation results?
   (b) Solve the equation you obtained in (a), and then give the solution to $u_t - \alpha u_x = 0$. Compare your answer to what you would obtain by applying the method of Section 11.4 to solve $u_t - \alpha u_x = 0$.

10. Show that if $u(x,t)$ solves the equation
$$u_{tt} + 2bu_t + b^2 u = \alpha^{-2} u_{xx}$$
    for positive constants b and α, then
$$w(x,t) = e^{bt} u(x,t)$$
    solves the wave equation
$$w_{xx} = \alpha^2 w_{tt}.$$

11. Consider the D'Alembert solution
$$u(x,t) = \frac{1}{2}\left(f(x+\alpha t) + f(x-\alpha t)\right)$$
    to the wave equation with zero initial velocity. Show that:
    (a) If f is an odd function, then $u(x,t)$ is an odd function of x for each fixed t.
    (b) If f is an even function, then $u(x,t)$ is an even function of x for each fixed t.
    (c) If f is periodic, then for each fixed t, $u(x,t)$ is periodic in x, and for each fixed x, $u(x,t)$ is periodic in t. If f has period p, give the period of u (in x and in t).

12. Starting with functions $f(x)$ and $g(x)$ defined on $[0,L]$ with $f(0) = f(L) = 0$ and $g(0) = g(L) = 0$, let $\tilde{f}(x)$ and $\tilde{g}(x)$ denote their odd $2L$-periodic extensions to $(-\infty, \infty)$. Show that the D'Alembert solution
$$u(x,t) = \frac{1}{2}\left[\tilde{f}(x-\alpha t) + \tilde{f}(x+\alpha t)\right] + \frac{1}{2\alpha}\int_{x-\alpha t}^{x+\alpha t} \tilde{g}(w)\, dw$$
    to $u_{tt} = \alpha^2 u_{xx}$ satisfies $u(0,t) = 0$ and $u(L,t) = 0$ for all $t > 0$.

13. **Solving the wave equation via operator factorization.** Let α be a positive constant.
    (a) Let $v = v(x,t)$ be a function with continuous second-order partial derivatives. Show that
    $$\frac{\partial^2 v}{\partial t^2} - \alpha^2 \frac{\partial^2 v}{\partial x^2} = \left(\frac{\partial}{\partial t} - \alpha \frac{\partial}{\partial x}\right)\left(\frac{\partial}{\partial t} + \alpha \frac{\partial}{\partial x}\right) v,$$
    establishing that the second-order operator corresponding to the wave equation is a product of two first-order operators:

    (11.125) $$\frac{\partial^2}{\partial t^2} - \alpha^2 \frac{\partial^2}{\partial x^2} = \left(\frac{\partial}{\partial t} - \alpha \frac{\partial}{\partial x}\right)\left(\frac{\partial}{\partial t} + \alpha \frac{\partial}{\partial x}\right).$$

    Remark: Also note the same reasoning yields

    (11.126) $$\frac{\partial^2}{\partial t^2} - \alpha^2 \frac{\partial^2}{\partial x^2} = \left(\frac{\partial}{\partial t} + \alpha \frac{\partial}{\partial x}\right)\left(\frac{\partial}{\partial t} - \alpha \frac{\partial}{\partial x}\right).$$

    (b) Assume that $v = v(x,t)$ has continuous partial derivatives through at least second order. Use (11.125) and (11.126) from (a) to show that if $v = v(x,t)$ satisfies *either*
    $$\frac{\partial v}{\partial t} + \alpha \frac{\partial v}{\partial x} = 0 \quad \text{or} \quad \frac{\partial v}{\partial t} - \alpha \frac{\partial v}{\partial x} = 0,$$
    then v must also satisfy the wave equation: $\frac{\partial^2 v}{\partial t^2} - \alpha^2 \frac{\partial^2 v}{\partial x^2} = 0$.

    (c) Use part (b) and the technique of solution of first-order equations described in Section 11.4 (see equations (11.38)–(11.40), as well as Example 11.4.1) to show that the wave equation $v_{tt} = \alpha^2 v_{xx}$ has solutions of the form $v(x,t) = F(x + \alpha t)$ and $v(x,t) = G(x - \alpha t)$ where F and G are twice continuously differentiable real-valued functions of one real variable. Apply the Superposition Principle to conclude that

    (11.127) $$u(x,t) = F(x + \alpha t) + G(x - \alpha t)$$

    solves the wave equation.

    (d) Show that every solution of the wave equation takes the form (11.127) by following the steps below.

    Let u be a classical solution of the wave equation $u_{tt} = \alpha^2 u_{xx}$ (so u is twice continuously differentiable). Set $v(x,t) = \frac{\partial u}{\partial t} + \alpha \frac{\partial u}{\partial x}$.
      (i) Show that $v(x,t) = h(x + \alpha t)$ for some continuously differentiable function h.
      (ii) Because h is continuous, the function $\frac{1}{2\alpha} h$ is also continuous; hence, it has an antiderivative which you should label F: $F'(s) = \frac{1}{2\alpha} h(s)$ for all real s. Show that $w(x,t) = u(x,t) - F(x + \alpha t)$ is a solution of $\frac{\partial w}{\partial t} + \alpha \frac{\partial w}{\partial x} = 0$. Conclude that $u(x,t) = F(x + \alpha t) + G(x - \alpha t)$, where F and G are twice continuously differentiable real-valued functions of a real variable.

11.10. Laplace's equation

Laplace's equation for a function $u(x,y)$ of two (space) variables is

(11.128) $$u_{xx} + u_{yy} = 0.$$

Laplace's equation is one of the most important partial differential equations of mathematical physics. Functions that satisfy this equation are called **harmonic** functions. Harmonic functions arise in a wide variety of applications, including steady-state temperature distributions, electrostatics, and two-dimensional fluid flow.

If $\mathcal{R}$ is a region in the xy-plane, then we say a point (x_0, y_0) is *inside* $\mathcal{R}$ provided that it is the center of some disk of positive radius that lies entirely in $\mathcal{R}$. A region $\mathcal{R}$ in the xy-plane is **open** provided that every point of $\mathcal{R}$ is inside $\mathcal{R}$. Recall, for example, that the rectangular region

11.10. Laplace's equation

$\{(x,y) : 2 < x < 5, 1 < y < 2\}$ is open but $\{(x,y) : 2 \leq x < 5, 1 < y < 2\}$ is not open. Intuitively speaking, you can think of an open region as one that includes none of its "boundary points".

Figure 11.70. An open region.

Figure 11.71. Not open.

A function $u(x,y)$ that solves Laplace's equation on an open region $\mathcal{R}$ has the following interesting properties:

Mean Value Property. The value of u at a point (x_0, y_0) inside $\mathcal{R}$ is equal to the average value of u over any disk centered at (x_0, y_0) and contained in $\mathcal{R}$.

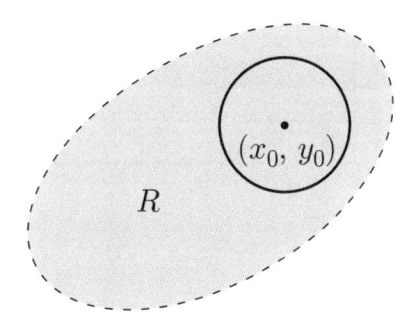

Figure 11.72. Value at $P = (x_0, y_0)$ is the average value over any disk centered at P and contained in $\mathcal{R}$.

Maximum Principle. The function $u(x,y)$ cannot have a *strict* local maximum at a point (x_0, y_0) in $\mathcal{R}$. This means that we cannot have $u(x_0, y_0) > u(x,y)$ for all (x,y) different from (x_0, y_0) in some disk centered at (x_0, y_0) and lying in $\mathcal{R}$.

The Maximum Principle is a consequence of the Mean Value Property, since if $u(x_0, y_0)$ is strictly larger than all the nearby values $u(x,y)$, then $u(x_0, y_0)$ can't be equal to the average value of u on a small disk centered at (x_0, y_0).

These properties often make intuitive sense for solutions to Laplace's equation arising in physical problems. As an example, let's look at steady-state temperature distributions. The two-dimensional version of the heat equation

(11.129) $$u_t = \kappa(u_{xx} + u_{yy})$$

describes the temperature $u(t, x, y)$ at time t at a point (x, y) in a two-dimensional plate (with insulated faces). A *steady-state* solution to the equation is one which does not vary with time, so that $u_t = 0$, and we can think of u as a function of the two spatial variables x and y. Setting $u_t = 0$ in (11.129) gives the equation $0 = \kappa(u_{xx} + u_{yy})$, or just

$$u_{xx} + u_{yy} = 0,$$

which is Laplace's equation. Does it make sense that a steady-state temperature function can't have a local maximum at a point (x_0, y_0) in the plate? If $u(x_0, y_0)$ is *larger* than the surrounding values of u, then since heat flows from warmer to cooler areas, the temperature at $u(x_0, y_0)$ would decrease over time, contrary to our assumption that u is *not* changing with time. So it is intuitively plausible that the maximum principle holds for steady-state temperature.

Solving Laplace's equation on a rectangle. We'll first discuss solving equation (11.128) for (x, y) in the rectangle $0 < x < a$, $0 < y < b$, when the value of $u(x, y)$ is specified along the edges of the rectangle:

$$u(x, 0) = f_1(x), \qquad u(x, b) = f_2(x) \quad \text{for } 0 < x < a$$

and

$$u(0, y) = g_1(y), \qquad u(a, y) = g_2(y) \quad \text{for } 0 < y < b;$$

see Fig. 11.73. Solving Laplace's equation in a particular region, with the values of the solution specified along the boundary of that region, is called a **Dirichlet problem** for that region.

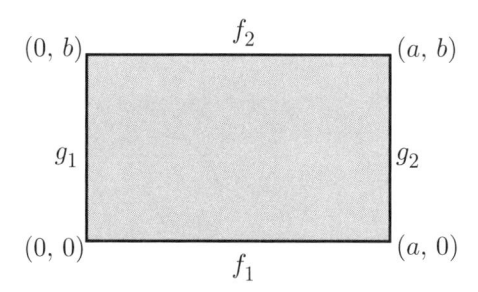

Figure 11.73. Dirichlet boundary condition for the rectangle $0 \leq x \leq a$, $0 \leq y \leq b$.

In the following example, we will take the boundary values on three sides of the rectangle to be zero.

Example 11.10.1. Let's use the technique of separation of variables to find solutions to the Dirichlet problem depicted in Fig. 11.74:

(11.130) $\quad u_{xx} + u_{yy} = 0, \quad u(x, 0) = 0, \quad u(0, y) = 0, \quad u(a, y) = 0, \quad u(x, b) = f_2(x)$

for $0 < x < a, 0 < y < b$.

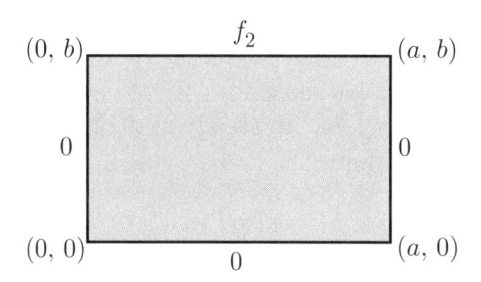

Figure 11.74. Boundary value 0 on three sides.

We look for solutions in the form

$$u(x, y) = X(x)Y(y).$$

For such u, $u_{xx}(x, y) = X''(x)Y(y)$ and $u_{yy}(x, y) = X(x)Y''(y)$. To satisfy Laplace's equation, we must have

$$X''(x)Y(y) + X(x)Y''(y) = 0.$$

11.10. Laplace's equation

Dividing by $X(x)Y(y)$, rewrite this as

$$\frac{X''(x)}{X(x)} = -\frac{Y''(y)}{Y(y)},$$

where the left-hand side depends on x only and the right-hand side depends on y only. We see that both $\frac{X''}{X}$ and $-\frac{Y''}{Y}$ must have the same constant value which we call $-\lambda$:

$$\frac{X''(x)}{X(x)} = -\lambda \quad \text{and} \quad \frac{Y''(y)}{Y(y)} = \lambda.$$

Our conditions on the three "zero" sides of the boundary say

$$X(x)Y(0) = 0, \quad X(0)Y(y) = 0, \quad X(a)Y(y) = 0$$

for $0 < x < a$ and $0 < y < b$. To find a nontrivial solution (i.e., a solution other than $u(x, y) \equiv 0$) we must have $Y(0) = 0$, $X(0) = 0$, and $X(a) = 0$. So we have two boundary value problems:

(11.131) $$X'' + \lambda X = 0, \quad X(0) = X(a) = 0,$$

and

(11.132) $$Y'' - \lambda Y = 0, \quad Y(0) = 0.$$

From Section 11.2, we know that to get a nontrivial solution to (11.131) we must have $\lambda = \left(\frac{n\pi}{a}\right)^2$ for $n = 1, 2, 3 \ldots$. For such λ, the corresponding solutions for $X(x)$ are

$$X_n(x) = \sin \frac{n\pi x}{a}$$

or any nonzero constant multiple thereof. Using $\lambda = \left(\frac{n\pi}{a}\right)^2$ we next solve $Y'' - \lambda Y = 0$ to obtain

$$Y_n(y) = A_n \cosh \frac{n\pi y}{a} + B_n \sinh \frac{n\pi y}{a}$$

for constants A_n and B_n (see Section 11.2). To meet the condition $Y(0) = 0$ we must have $A_n = 0$, so that $Y(y) = B_n \sinh \frac{n\pi y}{a}$. Choosing $B_n = 1$ for convenience and putting the pieces together, we see that for each integer $n = 1, 2, 3, \ldots$, the function

(11.133) $$u_n(x, y) = \sin \frac{n\pi x}{a} \sinh \frac{n\pi y}{a}$$

solves the problem (11.130). Superposition allows linear combinations of these building blocks as well (check!). Passing to an infinite sum gives

(11.134) $$u(x, y) = \sum_{n=1}^{\infty} c_n \sin \frac{n\pi x}{a} \sinh \frac{n\pi y}{a},$$

and we hope to be able to meet our final boundary condition

$$u(x, b) = f_2(x)$$

by an appropriate choice of coefficients c_n. How does this go? Setting $y = b$ in (11.134), we see that we want

(11.135) $$f_2(x) = \sum_{n=1}^{\infty} c_n \sinh \frac{n\pi b}{a} \sin \frac{n\pi x}{a}$$

for $0 < x < a$. If we write $f_2(x)$ as a Fourier sine series on the interval $0 < x < a$,

$$f_2(x) = \sum_{n=1}^{\infty} k_n \sin \frac{n\pi x}{a},$$

we know the sine series coefficients of f_2 are given by the Fourier-Euler formula
$$k_n = \frac{2}{a} \int_0^a f_2(x) \sin \frac{n\pi x}{a} dx.$$

Thus to satisfy (11.135) we should choose the coefficients c_n so that
$$c_n \sinh(n\pi b/a) = k_n = \frac{2}{a} \int_0^a f_2(x) \sin \frac{n\pi x}{a} dx.$$

This says c_n should be the product of $(\sinh \frac{n\pi b}{a})^{-1}$ and the nth Fourier sine series coefficient of f_2. Summarizing, we have:

> **The Dirichlet problem**
> $$u_{xx} + u_{yy} = 0, \; u(x,0) = 0, \; u(0,y) = 0, \; u(a,y) = 0, \; u(x,b) = f_2(x),$$
> for the rectangle $0 < x < a$, $0 < y < b$, has solution
> $$u(x,y) = \sum_{n=1}^{\infty} c_n \sin \frac{n\pi x}{a} \sinh \frac{n\pi y}{a},$$
> where
> $$c_n = \frac{2}{a \sinh(n\pi b/a)} \int_0^a f_2(x) \sin \frac{n\pi x}{a} dx.$$

Example 11.10.2. Let's find the steady-state temperature $u(x,y)$ in a 1×1 plate if the temperature along three edges is $0°$ and along the fourth edge it is given by
$$f(x) = \begin{cases} x & \text{if } 0 \leq x \leq \frac{1}{2}, \\ 1-x & \text{if } \frac{1}{2} \leq x \leq 1, \end{cases}$$
whose graph is shown in Fig. 11.75.

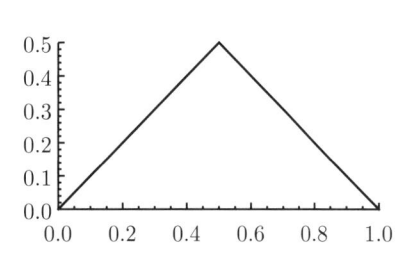

Figure 11.75. Graph of f.

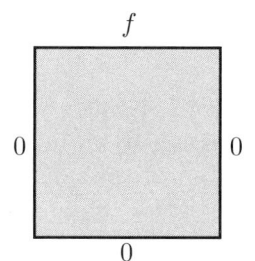

Figure 11.76. Solving a Dirichlet problem.

A computation (see Exercise 14 in Section 11.6) shows that $f(x)$ has Fourier sine series on $[0,1]$ given by

(11.136)
$$f(x) = \sum_{k=0}^{\infty} \frac{4}{\pi^2} \frac{(-1)^k}{(2k+1)^2} \sin[(2k+1)\pi x].$$

By our work in the last example, with $a = b = 1$,
$$u(x,y) = \sum_{n=1}^{\infty} c_n \sin(n\pi x) \sinh(n\pi y)$$

11.10. Laplace's equation   817

where c_n is the product of $(\sinh(n\pi))^{-1}$ and the nth Fourier sine series coefficient of f. By (11.136), this nth coefficient is 0 if n is even, and for $n = 2k + 1$ odd it is

$$\frac{4}{\pi^2}\frac{(-1)^k}{(2k+1)^2}.$$

Thus

$$u(x,y) = \sum_{k=0}^{\infty} \frac{4(-1)^k}{\pi^2(2k+1)^2 \sinh((2k+1)\pi)} \sin((2k+1)\pi x)\sinh((2k+1)\pi y).$$

Fig. 11.77 shows the curves in the plate along which the temperature $u(x,y)$ is constant. These are called **isotherms**.

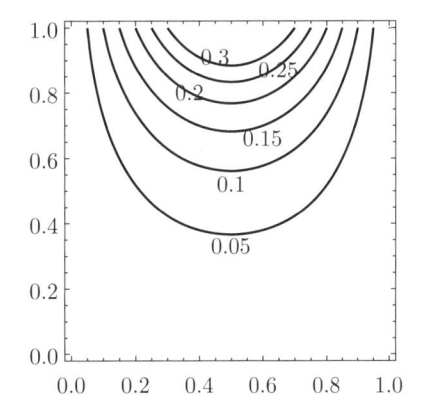

Figure 11.77. Isotherms $u(x,y) = K$ for various constants K.

Once you can solve the Dirichlet problem for a rectangle with boundary values 0 on three sides of the rectangle, it is not hard to solve the Dirichlet problem with arbitrary prescribed values f_1, f_2, f_3, and f_4 on the four sides of the rectangle; see Exercise 6.

11.10.1. Laplace's equation in a disk. To work with Laplace's equation in a disk or similar shaped regions like an annulus or a pie-shaped wedge, it's much more convenient to use *polar coordinates* (r, θ) in place of rectangular coordinates (x, y). Just as a rectangle is described in rectangular coordinates by $c_1 \le x \le d_1, c_2 \le y \le d_2$ for constants c_1, d_1, c_2, and d_2, the regions we'll be considering in this section are described in polar coordinates by

$$c_1 \le r \le d_1, \quad c_2 \le \theta \le d_2$$

for constants c_1, d_1, c_2, and d_2. Choosing $c_1 = 0, d_1 = s, c_2 = -\pi, d_2 = \pi$ we have the disk of radius s. Annuli or wedges as shown below are described by making other choices for these constants.

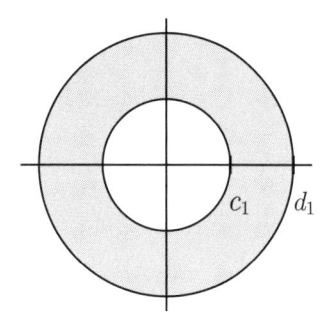

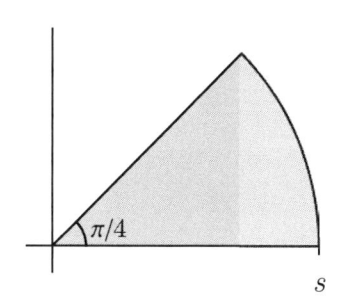

Figure 11.78. Annulus $c_1 \leq r \leq d_1$, $-\pi \leq \theta \leq \pi$.

Figure 11.79. Wedge $0 \leq r \leq s$, $0 \leq \theta \leq \pi/4$.

The basic conversion equations from polar to rectangular coordinates are

$$x = r\cos\theta, \quad y = r\sin\theta,$$

and the equations

$$r^2 = x^2 + y^2, \quad \tan\theta = \frac{y}{x}$$

convert from rectangular to polar coordinates. Laplace's equation $u_{xx} + u_{yy} = 0$ in polar coordinates is

(11.137) $$u_{rr} + \frac{1}{r}u_r + \frac{1}{r^2}u_{\theta\theta} = 0.$$

Exercise 9 outlines how to use the conversion equations and the Multivariable Chain Rule to obtain (11.137).

Dirichlet's problem for a disk. We'll solve (11.137) in a disk $0 \leq r < s$, $-\pi \leq \theta \leq \pi$, given the value of the solution at each point on the (circle) boundary of the disk. *Mathematically*, our problem is to solve

(11.138) $$u_{rr} + \frac{1}{r}u_r + \frac{1}{r^2}u_{\theta\theta} = 0, \quad u(s,\theta) = f(\theta),$$

for a given function $f(\theta)$. *Physically*, we know we can interpret this as the steady-state temperature in a circular plate, given the temperature $f(\theta)$ at points (s, θ) on the boundary of the plate. Here's a different physical interpretation which is intuitively appealing. If we form a wire into a circle with radius s and dip it into a soap solution, a soap film in the shape of a flat disk forms. Next, suppose we distort the wire a bit, bending it up or down from its original shape by a small amount $f(\theta)$ at each point (s, θ). If the resulting frame is dipped into a soap solution, the soap film that forms can be described by giving the height $u(r, \theta)$ of the film at each point (r, θ), $0 \leq r \leq s, -\pi \leq \theta \leq \pi$ (see Fig. 10.56). This function $u(r, \theta)$ is approximately[9] the solution to the Dirichlet problem (11.138). Notice that with the original circular frame, $f(\theta) \equiv 0$, and the solution is $u(r, \theta) \equiv 0$ — the "flat disk" soap film.

---

[9] The exact solution $u(r, \theta)$ is found by solving a more complicated nonlinear partial differential equation called the minimal surface equation. When we only mildly distort the wire, Laplace's equation serves as a good approximation.

11.10. Laplace's equation

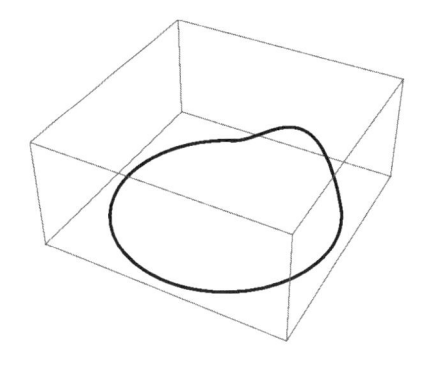

Figure 11.80. Circular frame with slight deformation.

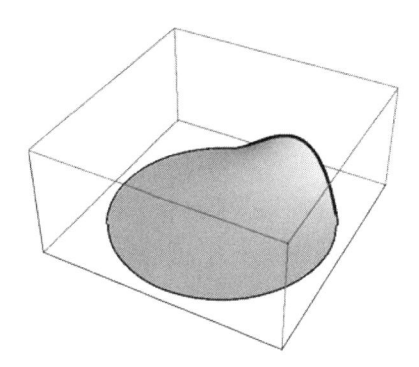

Figure 11.81. Soap film.

Our strategy for solving (11.138) is a familiar one: Use separation of variables to find building-block solutions to (11.137) in the disk, and choose the appropriate (possibly infinite) combinations of these building-block solutions to meet the desired condition on the boundary of the disk. We will require that our building-block solutions be continuous on the disk, as well as twice differentiable in each of the variables r and θ.

Separation of variables. We begin by seeking continuous solutions of (11.137) in the form

$$u(r, \theta) = R(r)\Theta(\theta),$$

a product of two one-variable functions. Computing the various derivatives and substituting into (11.137) we obtain

$$R''(r)\Theta(\theta) + \frac{1}{r}R'(r)\Theta(\theta) + \frac{1}{r^2}R(r)\Theta''(\theta) = 0.$$

Dividing by $R(r)\Theta(\theta)$ gives

$$\frac{R''(r)}{R(r)} + \frac{1}{r}\frac{R'(r)}{R(r)} + \frac{1}{r^2}\frac{\Theta''(\theta)}{\Theta(\theta)} = 0.$$

Multiplying by r^2 we can rewrite this as

(11.139) $$\frac{r^2 R''(r) + r R'(r)}{R(r)} = -\frac{\Theta''(\theta)}{\Theta(\theta)}.$$

The next step is a familiar one: The left-hand side of (11.139) is a function of r only, while the right-hand side is a function of θ only. For equation (11.139) to hold, both

$$\frac{r^2 R'' + r R'}{R} \quad \text{and} \quad -\frac{\Theta''}{\Theta}$$

must have a common constant value, which we'll call λ. Thus we have

(11.140) $$\Theta'' + \lambda \Theta = 0$$

and

(11.141) $$r^2 R'' + r R' = \lambda R.$$

Turning first to equation (11.140), recall that we want $u(r, \theta) = R(r)\Theta(\theta)$ to be continuous. The function $\Theta(\theta)$ must be periodic, with period 2π, and in particular we must have $\Theta(-\pi) = \Theta(\pi)$. Keeping this in mind, in Exercise 10 you are asked to show that if a 2π-periodic function Θ

solves (11.140), then either

(i) $\lambda = n^2$ for some $n = 1, 2, 3, \ldots$ and $\Theta(\theta) = c_1 \cos(n\theta) + c_2 \sin(n\theta)$ for some constants c_1 and c_2 or

(ii) $\lambda = 0$ and $\Theta(\theta) = c$ for some constant c.

Option (ii) could be included in the list in (i) if we allow $n = 0$ there, but we'll find it convenient to list it separately.

Using our options $\lambda = n^2$, $n = 1, 2, 3, \ldots$, and $\lambda = 0$, we next solve equation (11.141). Both

(11.142) $$r^2 R'' + rR' = n^2 R$$

and

(11.143) $$r^2 R'' + rR' = 0$$

are Cauchy-Euler equations (the independent variable is r, and the dependent variable is R). By our work in Section 7.5, equations (11.142)–(11.143) have solutions of the form $R = r^w$ where w is a root of the indicial equation. The indicial equation for (11.142) is

$$w(w-1) + w - n^2 = 0,$$

which has roots $w = \pm n$, while the indicial equation for (11.143) is

$$w(w-1) + w = 0,$$

which has a single repeated root $w = 0$. From these calculations we see that:

(i) When $n = 1, 2, 3, \ldots$, the equation $r^2 R'' + rR' = n^2 R$ has general solution $R(r) = c_1 r^n + c_2 r^{-n}$.

(ii) The equation $r^2 R'' + rR' = 0$ has general solution $R(r) = c_1 + c_2 \ln r$.

If c_2 is not zero in either (i) or (ii), the function $R(r)$ becomes unbounded (in either a positive or negative sense) as r approaches zero through positive values. This will not give us a (nontrivial) solution $u(r, \theta) = R(r)\Theta(\theta)$ that is continuous on the disk. Thus we must choose $c_2 = 0$ in the solutions for $R(r)$ in either (i) or (ii).

Let's summarize our work so far by specifying building-block solutions to Laplace's equation in the disk that arise from separation of variables. These are reconstituted from our computations of "allowable" choices of $R(r)$ and corresponding $\Theta(\theta)$.

> In the disk $0 \leq r < s$, $-\pi \leq \theta \leq \pi$, Laplace's equation
> $$u_{rr} + \frac{1}{r} u_r + \frac{1}{r^2} u_{\theta\theta} = 0$$
> has building-block solutions of three types:
> $$r^n \cos(n\theta), \quad n = 1, 2, 3, \ldots,$$
> and
> $$r^n \sin(n\theta), \quad n = 1, 2, 3, \ldots,$$
> and
> the constant function 1.

Taking combinations of these building blocks in the usual way we have solutions

(11.144) $$u(r, \theta) = \frac{A_0}{2} + \sum_{n=1}^{\infty} (A_n r^n \cos(n\theta) + B_n r^n \sin(n\theta)).$$

11.10. Laplace's equation

Here, we've written the constant term as $\frac{A_0}{2}$ (instead of just A_0) in anticipation of the role that Fourier series will soon play.

Imposing the boundary condition. Now we specify $u(s, \theta)$, $-\pi \leq \theta \leq \pi$, as

$$u(s, \theta) = f(\theta), \quad -\pi \leq \theta \leq \pi,$$

for some (piecewise continuously differentiable) 2π-periodic function f. Setting $r = s$ in (11.144), how should the coefficients A_n and B_n be chosen so that

$$\frac{A_0}{2} + \sum_{n=1}^{\infty} (A_n s^n \cos(n\theta) + B_n s^n \sin(n\theta)) = f(\theta)?$$

Since the sum on the left-hand side has the form of the Fourier series expansion on the interval $[-\pi, \pi]$, the values of A_n, $n = 0, 1, 2, \ldots$, and B_n, $n = 1, 2, 3, \ldots$, are related to the Fourier coefficients of f by

(11.145) $$A_0 = \frac{1}{\pi} \int_{-\pi}^{\pi} f(\theta) \, d\theta,$$

and for $n \geq 1$,

(11.146) $$A_n s^n = \frac{1}{\pi} \int_{-\pi}^{\pi} f(\theta) \cos(n\theta) \, d\theta$$

and

(11.147) $$B_n s^n = \frac{1}{\pi} \int_{-\pi}^{\pi} f(\theta) \sin(n\theta) \, d\theta.$$

These three equations tell us how to determine A_0, A_n, and B_n.

Example 11.10.3. We'll solve the Dirichlet problem

$$u_{rr} + \frac{1}{r} u_r + \frac{1}{r^2} u_{\theta\theta} = 0, \quad u(1, \theta) = f(\theta),$$

for the unit disk $r < 1$, $-\pi \leq \theta \leq \pi$, where

$$f(\theta) = \begin{cases} 100 & \text{if } 0 < \theta < \pi, \\ 0 & \text{if } -\pi < \theta < 0. \end{cases}$$

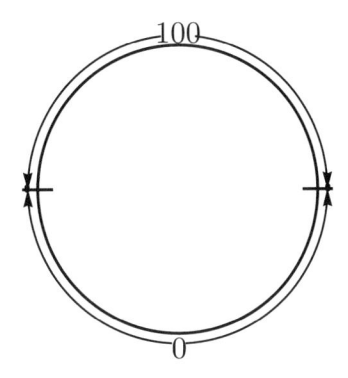

Figure 11.82. Boundary condition in Example 11.10.3.

The Fourier series for $f(\theta)$ on $[-\pi, \pi]$ is

$$f(\theta) = \frac{a_0}{2} + \sum_{n=1}^{\infty}(a_n \cos(n\theta) + b_n \sin(n\theta))$$

where

$$a_0 = \frac{1}{\pi}\int_{-\pi}^{\pi} f(\theta)\, d\theta = \frac{1}{\pi}(100\pi) = 100,$$

$$a_n = \frac{1}{\pi}\int_{-\pi}^{\pi} f(\theta)\cos(n\theta)\, d\theta = \frac{1}{\pi}\int_{0}^{\pi} 100\cos(n\theta)\, d\theta = 0$$

for $n \geq 1$, and

$$b_n = \frac{1}{\pi}\int_{-\pi}^{\pi} f(\theta)\sin(n\theta)\, d\theta = \frac{1}{\pi}\int_{0}^{\pi} 100\sin(n\theta)\, d\theta = -\frac{100}{n\pi}(\cos(n\pi) - 1)$$

for $n \geq 1$. Since $\cos(n\pi) = 1$ for all even integers n and $\cos(n\pi) = -1$ when n is an odd integer, $b_n = 0$ when n is even and $b_n = \frac{200}{n\pi}$ when n is odd. Since the radius of our disk is $s = 1$, we see from equations (11.145)–(11.147) that $A_0 = a_0 = 100$, and for $n \geq 1$,

$$A_n = a_n = 0 \quad \text{while} \quad B_n = b_n = \begin{cases} 0 & \text{if } n \text{ is even,} \\ \frac{200}{n\pi} & \text{if } n \text{ is odd.} \end{cases}$$

Using these values in (11.144), the harmonic function that solves our Dirichlet problem is

$$\begin{aligned} u(r,\theta) &= \frac{100}{2} + \sum_{n=1}^{\infty} B_n r^n \sin(n\theta) \\ &= 50 + \sum_{k=0}^{\infty} \frac{200}{(2k+1)\pi} r^{2k+1} \sin((2k+1)\theta). \end{aligned}$$

The Poisson Integral Formula. We have a series solution to the Dirichlet problem for the disk $0 \leq r < s$, $-\pi \leq \theta \leq \pi$, with prescribed values $f(\theta)$ on the boundary circle given by

(11.148) $$u(r,\theta) = \frac{A_0}{2} + \sum_{n=1}^{\infty}(A_n r^n \cos(n\theta) + B_n r^n \sin(n\theta))$$

where A_n and B_n are determined from the boundary function $f(\theta)$ via the equations

(11.149) $$A_0 = \frac{1}{\pi}\int_{-\pi}^{\pi} f(\varphi)\, d\varphi$$

and for $n \geq 1$,

(11.150) $$A_n = \frac{1}{s^n \pi}\int_{-\pi}^{\pi} f(\varphi)\cos(n\varphi)\, d\varphi, \quad B_n = \frac{1}{s^n \pi}\int_{-\pi}^{\pi} f(\varphi)\sin(n\varphi)\, d\varphi.$$

(These are equations (11.145)–(11.147), now written with the variable of integration as φ.) We'll finish this section by showing how we can obtain an alternate form of this solution, called the **Poisson Integral Formula**. It has advantages in certain situations. Obtaining this alternate form involves some algebraic and trigonometric manipulations. In particular, we'll be using the identity

$$\cos\alpha = \frac{e^{i\alpha} + e^{-i\alpha}}{2}.$$

11.10. Laplace's equation

To see how this goes, start with the solution in (11.148)

$$u(r,\theta) = \frac{A_0}{2} + \sum_{n=1}^{\infty} r^n (A_n \cos(n\theta) + B_n \sin(n\theta))$$

and rewrite the expression $A_n \cos(n\theta) + B_n \sin(n\theta)$, using equation (11.150), as

$$
\begin{aligned}
A_n \cos(n\theta) + B_n \sin(n\theta) &= \left(\frac{1}{s^n \pi} \int_{-\pi}^{\pi} f(\varphi) \cos(n\varphi)\, d\varphi\right) \cos(n\theta) \\
&\quad + \left(\frac{1}{s^n \pi} \int_{-\pi}^{\pi} f(\varphi) \sin(n\varphi)\, d\varphi\right) \sin(n\theta) \\
&= \frac{1}{s^n \pi} \int_{-\pi}^{\pi} f(\varphi)[\cos(n\varphi)\cos(n\theta) + \sin(n\varphi)\sin(n\theta)]\, d\varphi \\
&= \frac{1}{s^n \pi} \int_{-\pi}^{\pi} f(\varphi) \cos(n(\varphi - \theta))\, d\varphi.
\end{aligned}
$$

(11.151)

We used the trigonometric identity for the cosine of a difference to obtain the last line. Inserting (11.151) and formula (11.149) for A_0 into equation (11.148) gives

$$u(r,\theta) = \frac{1}{\pi}\left(\frac{1}{2}\int_{-\pi}^{\pi} f(\varphi)\, d\varphi + \sum_{n=1}^{\infty} \frac{r^n}{s^n} \int_{-\pi}^{\pi} f(\varphi) \cos(n(\varphi - \theta))\, d\varphi\right).$$

Next we interchange the sum and integral in the second term inside the big parentheses above to obtain

$$\sum_{n=1}^{\infty} \frac{r^n}{s^n} \int_{-\pi}^{\pi} f(\varphi) \cos(n(\varphi - \theta))\, d\varphi = \int_{-\pi}^{\pi} f(\varphi) \left(\sum_{n=1}^{\infty} \frac{r^n}{s^n} \cos(n(\varphi - \theta))\right) d\varphi.$$

Strictly speaking this interchange requires justification, since the sum is an *infinite* sum. We will omit this justification. Proceeding further hinges on getting a closed form expression for the infinite sum

$$\sum_{n=1}^{\infty} \frac{r^n}{s^n} \cos(n(\varphi - \theta)).$$

We'll do this using the identity

(11.152) $$\cos(n(\varphi - \theta)) = \frac{e^{in(\varphi - \theta)} + e^{-in(\varphi - \theta)}}{2}.$$

This gives

(11.153) $$\sum_{n=1}^{\infty} \frac{r^n}{s^n} \cos(n(\varphi - \theta)) = \frac{1}{2} \sum_{n=1}^{\infty} \left(\frac{r}{s}\right)^n \left(e^{in(\varphi - \theta)} + e^{-in(\varphi - \theta)}\right).$$

In Exercise 18 you are asked to show that you can use the geometric series formula to rewrite the right-hand side of (11.153) as

$$\frac{1}{2}\left(\frac{re^{i(\varphi-\theta)}}{s - re^{i(\varphi-\theta)}} + \frac{re^{-i(\varphi-\theta)}}{s - re^{-i(\varphi-\theta)}}\right).$$

Putting our calculations together, we have

$$\begin{aligned} u(r,\theta) &= \frac{1}{2\pi}\int_{-\pi}^{\pi} f(\varphi)d\varphi + \frac{1}{2\pi}\int_{-\pi}^{\pi} f(\varphi)\left(\frac{re^{i(\varphi-\theta)}}{s - re^{i(\varphi-\theta)}} + \frac{re^{-i(\varphi-\theta)}}{s - re^{-i(\varphi-\theta)}}\right)d\varphi \\ &= \frac{1}{2\pi}\int_{-\pi}^{\pi} f(\varphi)\left(1 + \frac{re^{i(\varphi-\theta)}}{s - re^{i(\varphi-\theta)}} + \frac{re^{-i(\varphi-\theta)}}{s - re^{-i(\varphi-\theta)}}\right)d\varphi \\ &= \frac{1}{2\pi}\int_{-\pi}^{\pi} f(\varphi)\frac{s^2 - r^2}{s^2 - 2sr\cos(\varphi - \theta) + r^2}\,d\varphi. \end{aligned}$$

For the last step, we combined the terms

$$1 + \frac{re^{i(\varphi-\theta)}}{s - re^{i(\varphi-\theta)}} + \frac{re^{-i(\varphi-\theta)}}{s - re^{-i(\varphi-\theta)}}$$

with a common denominator of

$$\begin{aligned} (s - re^{i(\varphi-\theta)})(s - re^{-i(\varphi-\theta)}) &= s^2 - rse^{i(\varphi-\theta)} - rse^{-i(\varphi-\theta)} + r^2 \\ &= s^2 - rs\left(e^{i(\varphi-\theta)} + e^{-i(\varphi-\theta)}\right) + r^2 \\ &= s^2 - 2rs\cos(\varphi - \theta) + r^2, \end{aligned}$$

where we have used (11.152) with $n = 1$. We've arrived at the desired result:

$$(11.154) \qquad u(r,\theta) = \frac{1}{2\pi}\int_{-\pi}^{\pi} f(\varphi)\frac{s^2 - r^2}{s^2 - 2sr\cos(\varphi - \theta) + r^2}\,d\varphi.$$

This is called the Poisson Integral Formula. It gives the solution to the Dirichlet problem for the disk of radius s, with boundary value $f(\theta)$, as an integral, rather than a series as in (11.148).

The integrand in (11.154) is the product of the boundary function $f(\varphi)$ and the **Poisson kernel**

$$\frac{s^2 - r^2}{s^2 - 2rs\cos(\varphi - \theta) + r^2}.$$

In Exercise 20 you are asked to show that the Poisson kernel is *large* when the point with polar coordinates (r, θ) (inside the disk) is close the point (s, φ) on the boundary of the disk. Roughly speaking, this means that when (r, θ) is close to a boundary point (s, φ), the value of $u(r, \theta)$ is highly influenced by the value $f(\varphi)$, a fact that shouldn't be surprising.

Example 11.10.4. Suppose we have a Dirichlet problem

$$u_{rr} + \frac{1}{r}u_r + \frac{1}{r^2}u_{\theta\theta} = 0, \quad u(s, \theta) = f(\theta),$$

in a disk of radius s. We'll use the Poisson Integral Formula to relate the value of u at the center of the disk with its values on the boundary circle. The center of the disk has polar coordinates $r = 0$ and θ arbitrary, so by the Poisson Integral Formula we have

$$u(0, \theta) = \frac{1}{2\pi}\int_{-\pi}^{\pi} f(\varphi)d\varphi = \frac{1}{2\pi}\int_{-\pi}^{\pi} u(s, \varphi)d\varphi.$$

This says that the value of u at the center of the disk is the *average value* of the boundary values $f(\varphi)$. This might be called a "boundary mean-value property".

11.10. Laplace's equation

The Poisson Integral Formula can be useful both computationally and in deriving some more theoretical properties of the solutions of Dirichlet problems for a disk. See Exercise 21 for an example of the latter.

11.10.2. Exercises.

1. Solve Laplace's equation in the square $0 < x < 1$, $0 < y < 1$ with
$$u(x,0) = 0, \qquad u(x,1) = \sin(\pi x) \quad \text{for } 0 < x < 1$$
and
$$u(0,y) = 0, \qquad u(1,y) = 0 \quad \text{for } 0 < y < 1.$$

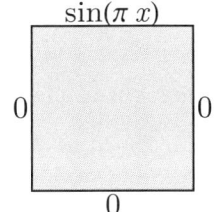

2. A square metal sheet has dimensions $\pi \times \pi$. The top edge is held at $10°$ and the other 3 edges at $0°$. Find the steady-state temperature distribution $u(x,y)$ at any point in the sheet.

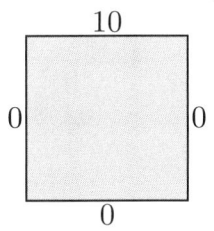

3. In this problem we will solve the "three 0 sides" Dirichlet problem for the rectangle $0 \le x \le a$, $0 \le y \le b$ depicted below.

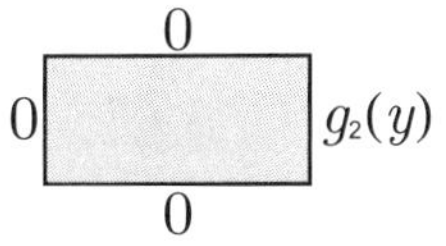

   (a) If we use separation of variables, how do equations (11.131) and (11.132) change to reflect the new boundary conditions?
   (b) Solve the boundary value problem in (a) to show that building-block solutions to the problem at hand are
   $$u_n(x,y) = \sin\frac{n\pi y}{b} \sinh\frac{n\pi x}{b}.$$
   Remark: Notice these building blocks could be obtained from equation (11.133) by interchanging x and y and interchanging a and b. This should make sense.
   (c) To meet the condition $u(a,y) = g_2(y)$ with a function of the form
   $$\sum_{n=1}^{\infty} c_n \sin\frac{n\pi y}{b} \sinh\frac{n\pi x}{b},$$
   how should the coefficients c_n be chosen?

4. (a) Show that the "three 0 sides" Dirichlet problem for the rectangle $0 \le x \le a$, $0 \le y \le b$ depicted below has solution
$$u(x,y) = \sum_{n=1}^{\infty} \left(a_n \cosh\frac{n\pi y}{a} + b_n \sinh\frac{n\pi y}{a} \right) \sin\frac{n\pi x}{a},$$

where the a_n's are the Fourier sine series coefficients of f on $[0,a]$ and $b_n = -a_n \dfrac{\cosh(n\pi b/a)}{\sinh(n\pi b/a)}$.

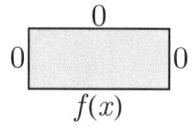

(b) Solve the Dirichlet problem depicted by

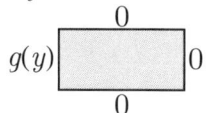

[Rectangle with $g(y)$ on left side, 0 on top, right, and bottom]

Hint: You can do this in a computation-free way if you make use of the remark in (b) of Exercise 3.

5. If $u_2(x,y)$ is a solution to Laplace's equation in the rectangle $0 < x < a$, $0 < y < b$ satisfying

$$u_2(x,0) = 0, \qquad u_2(x,b) = f_2(x) \quad \text{for } 0 < x < a$$

and

$$u_2(0,y) = 0, \qquad u_2(a,y) = 0 \quad \text{for } 0 < y < b,$$

[Rectangle with $f_2(x)$ on top, 0 on other three sides]

and $u_1(x,y)$ solves Laplace's equation in the same rectangle and satisfies

$$u_1(x,0) = f_1(x), \qquad u_1(x,b) = 0 \quad \text{for } 0 < x < a$$

and

$$u_1(0,y) = 0, \qquad u_1(a,y) = 0 \quad \text{for } 0 < y < b,$$

[Rectangle with $f_1(x)$ on bottom, 0 on other three sides]

give the solution to

$$u_{xx} + u_{yy} = 0, \quad u(x,0) = f_1(x), \quad u(x,b) = f_2(x), \quad u(0,y) = 0, \quad u(a,y) = 0,$$

for $0 < x < a$, $0 < y < b$.

6. Describe how to obtain the solution to the Dirichlet problem depicted below in terms of the solutions to four other Dirichlet problems, each having three "0" sides.

[Rectangle with f_2 on top, f_1 on bottom, g_1 on left, g_2 on right]

7. To find the steady-state temperature in a rectangle whose bottom edge is held at $0°$, top edge at $f(x)°$, and has the two other edges insulated, we must solve the boundary-initial value problem

$$u_{xx} + u_{yy} = 0, \qquad 0 < x < a, \quad 0 < y < b,$$

$$u(x,0) = 0, \quad u(x,b) = f(x),$$

$$u_x(0,y) = 0, \quad u_x(a,y) = 0.$$

(a) Using separation of variables, find building-block solutions to Laplace's equation with homogeneous boundary conditions
$$u(x,0) = 0 \quad \text{for } 0 < x < a$$
and
$$u_x(0,y) = 0, \quad u_x(a,y) = 0 \quad \text{for } 0 < y < b.$$

(b) Using the building blocks from (a), describe how to find the steady-state temperature so that the nonhomogeneous boundary condition $u(x,b) = f(x)$ is also satisfied.

8. (a) (CAS) Use the first three terms of your steady-state temperature series for Exercise 2 to approximate the steady-state temperature at the center $(\pi/2, \pi/2)$ of the square there.

(b) What is the exact steady-state temperature at the center? Prove you are right without relying on the series solution of Exercise 2. Hint: Consider the sum $u_1 + u_2 + u_3 + u_4$ where u_j solves the Dirichlet problem for the square with boundary temperature 10 on the jth side and 0 elsewhere.

9. The purpose of this problem is to show how to obtain Laplace's partial differential equation in polar coordinates.

(a) Using the conversion equations $x = r\cos\theta$ and $y = r\sin\theta$, show that
$$rx_r = x, \quad ry_r = y, \quad x_\theta = -y, \quad \text{and} \quad y_\theta = x.$$

(b) The Chain Rule says that for a function $w = w(x,y)$,
$$w_r = w_x \cdot x_r + w_y \cdot y_r \quad \text{and} \quad w_\theta = w_x \cdot x_\theta + w_y \cdot y_\theta.$$
Use these, with the choices $w = u$, $w = u_x$, and $w = u_y$, to obtain the formulae
$$ru_r = xu_x + yu_y, \quad u_\theta = -yu_x + xu_y,$$
$$r(u_x)_r = xu_{xx} + yu_{xy}, \quad (u_x)_\theta = -yu_{xx} + xu_{xy},$$
and
$$r(u_y)_r = xu_{yx} + yu_{yy}, \quad (u_y)_\theta = -yu_{yx} + xu_{yy}.$$

(c) With the results of (b) in hand, differentiate the expression there for ru_r (using the product rule as needed) to show that
$$ru_{rr} + u_r = \frac{x^2}{r}u_{xx} + \frac{xy}{r}u_{xy} + \frac{xy}{r}u_{yx} + \frac{y^2}{r}u_{yy} + \frac{x}{r}u_x + \frac{y}{r}u_y.$$
Similarly, differentiate the formula for u_θ to obtain
$$u_{\theta\theta} = -xu_x + y^2 u_{xx} - xyu_{xy} - yu_y - xyu_{yx} + x^2 u_{yy}.$$

(d) Using the results from (c), show that
$$r^2 u_{rr} + ru_r + u_{\theta\theta} = (x^2 + y^2)[u_{xx} + u_{yy}].$$
Conclude that the Laplacian
$$u_{xx} + u_{yy}$$
is written in polar coordinates as
$$u_{rr} + \frac{1}{r}u_r + \frac{1}{r^2}u_{\theta\theta}.$$

10. By considering $\lambda = 0$, $\lambda < 0$, and $\lambda > 0$, find all solutions $\Theta(\theta)$ to
$$\Theta'' + \lambda\Theta = 0$$
that are 2π-periodic and not identically 0.

In Exercises 11–13 solve the Dirichlet problem

$$u_{rr} + \frac{1}{r}u_r + \frac{1}{r^2}u_{\theta\theta} = 0, \quad u(1,\theta) = f(\theta),$$

for the unit disk $0 \le r < 1$, $-\pi \le \theta \le \pi$ for the given choice of boundary data $f(\theta)$.

11. $f(\theta) = \sin\theta$.

12. $f(\theta) = \cos(2\theta)$.

13.
$$f(\theta) = \begin{cases} 100 & \text{if } 0 < \theta < \frac{\pi}{2}, \\ 0 & \text{if } \frac{\pi}{2} \le \theta \le \pi \text{ or } -\pi \le \theta \le 0. \end{cases}$$

14. Solve the Dirichlet problem for the disk $0 \le r < 3$, $-\pi \le \theta \le \pi$ with boundary data $f(\theta) = \sin(6\theta)$.

15. In this problem we'll solve some Dirichlet problems for the pie-shaped wedge shown below, with angle opening $\frac{\pi}{4}$ and radius 1. The boundary data is as shown: 0 along the two line segments and $f(\theta)$ along the arc.

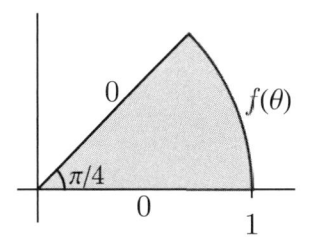

(a) Show that building-block solutions for solving

$$u_{rr} + \frac{1}{r}u_r + \frac{1}{r^2}u_{\theta\theta} = 0, \quad u(r,0) = 0, \quad u\left(r, \frac{\pi}{4}\right) = 0,$$

on the wedge $0 \le r \le 1$, $0 \le \theta \le \frac{\pi}{4}$ are

$$r^{4n}\sin(4n\theta) \quad \text{for } n = 1, 2, 3, \ldots.$$

Hint: The separation of variable computation that we used in the disk still applies to obtain equations (11.140) and (11.141). You now want to solve (11.140) with the boundary conditions $\Theta(0) = 0$, $\Theta(\frac{\pi}{4}) = 0$ (why?).

(b) Now we will prescribe the boundary values $f(\theta)$ along the arc to be

$$f(\theta) = \frac{1}{2}\sin(12\theta) - \frac{3}{2}\sin(24\theta).$$

With

$$u(r,\theta) = \sum_{n=1}^{\infty} B_n r^{4n}\sin(4n\theta),$$

how should the coefficients be chosen so that $u(1,\theta) = f(\theta)$?

(c) Next let $f(\theta) = 1$. How should the coefficients B_n be chosen so that $u(1,\theta) = 1$?

(d) For general $f(\theta)$, how do you choose the coefficients B_n so that $u(1,\theta) = f(\theta)$?

(e) How do the results in (a)–(d) change if the wedge has angle opening $\pi/3$?

16. Show that the function $u(r,\theta) = c_1 + c_2\theta$ with the appropriate choice of c_1 and c_2 will solve the Dirichlet problem schematically indicated below (with T_1 and T_2 constant). What are c_1 and c_2?

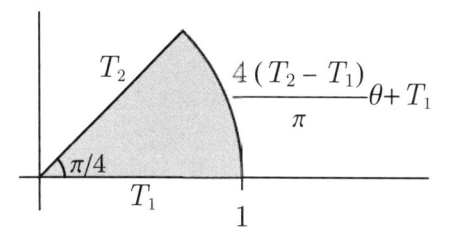

11.10. Laplace's equation

17. The Dirichlet problems depicted below were solved in Exercises 15 and 16.

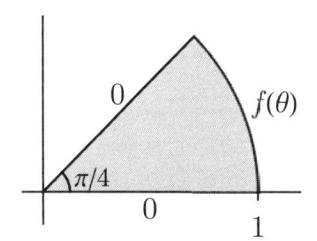

Figure 11.83. 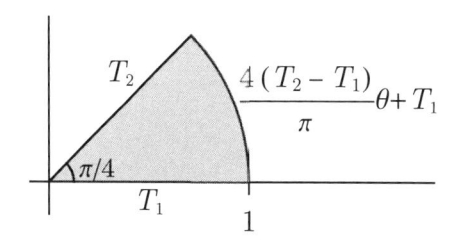   **Figure 11.84.**

Explain how to use these two solutions to solve the Dirichlet problem depicted by

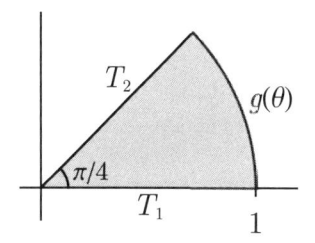

In particular, how should $f(\theta)$ be chosen if you are given $g(\theta)$, T_1, and T_2?

18. The geometric series formula
$$\sum_{n=1}^{\infty} z^n = \frac{z}{1-z}$$
holds for any real or complex number z with $|z| < 1$. Using this, show that if $0 \leq r < s$,
$$\sum_{n=1}^{\infty} \left(\frac{r}{s}\right)^n \left(e^{in(\varphi-\theta)} + e^{-in(\varphi-\theta)}\right) = \frac{re^{i(\varphi-\theta)}}{s - re^{i(\varphi-\theta)}} + \frac{re^{-i(\varphi-\theta)}}{s - re^{-i(\varphi-\theta)}}.$$

19. Suppose the temperature at each point on the boundary of the unit disk is as shown below, with value 100 on a quarter circle and 0 elsewhere. What is the (steady-state) temperature at the center of the disk?

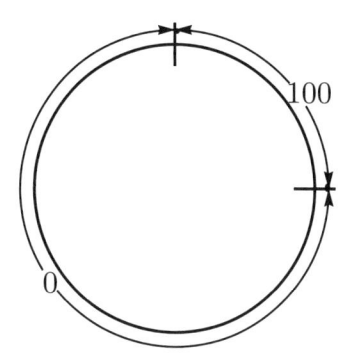

20. (a) Show that the denominator of the Poisson kernel

$$s^2 - 2rs\cos(\varphi - \theta) + r^2$$

is equal to the square of the distance from the point with polar coordinates (r, θ) to the point with polar coordinates (s, φ).

(b) Let the point with polar coordinates (s, φ) be fixed on the boundary circle and fix r with $0 < r < s$. Use your result in (a) to find the maximum and minimum values of the Poisson kernel on the circle of radius r; that is, find the maximum and minimum of the function

$$F(\theta) = \frac{s^2 - r^2}{s^2 - 2rs\cos(\varphi - \theta) + r^2}.$$

Where are they located? Hint: If you use part (a) and think geometrically, you don't need to use calculus.

21. (a) Solve the Dirichlet problem

$$u_{rr} + \frac{1}{r}u_r + \frac{1}{r^2}u_{\theta\theta} = 0, \quad u(s, \theta) = 1,$$

for the disk of radius s. Do this by either computing the coefficients A_n and B_n in the series solution given by (11.144) or simply by thinking about the problem in physical terms. ("What is the steady-state temperature in a disk of radius s if the temperature at every point of the boundary is 1?")

(b) Use your answer to (a) to compute

$$\frac{1}{2\pi}\int_{-\pi}^{\pi} \frac{s^2 - r^2}{s^2 - 2rs\cos(\varphi - \theta) + r^2} d\varphi$$

for any $0 \le r < s$ and $-\pi \le \theta \le \pi$.

(c) Explain why

$$\left|\frac{1}{2\pi}\int_{-\pi}^{\pi} f(\varphi) \frac{s^2 - r^2}{s^2 - 2sr\cos(\varphi - \theta) + r^2} d\varphi\right|$$

can never be larger (for any choice of r, $0 \le r < s$) than the maximum value of $|f(\varphi)|$ for $-\pi \le \varphi \le \pi$. You may use the fact that

$$\left|\frac{1}{2\pi}\int_{-\pi}^{\pi} f(\varphi) \frac{s^2 - r^2}{s^2 - 2sr\cos(\varphi - \theta) + r^2} d\varphi\right| \le \max|f(\varphi)| \int_{-\pi}^{\pi} \frac{s^2 - r^2}{s^2 - 2sr\cos(\varphi - \theta) + r^2} d\varphi.$$

22. (CAS) In this problem you will use a geometric series to bound the error that results when a partial sum of the series solution of Example 11.10.3 is used to approximate the solution at the point $(1/2, 1/2)$ or, in polar coordinates, $r = 1/\sqrt{2}$, $\theta = \pi/4$.

(a) Use

$$u(r, \theta) \approx 50 + \sum_{k=0}^{10} \frac{200}{\pi(2k+1)} r^{2k+1} \sin((2k+1)\theta)$$

to approximate $u\left(\frac{1}{\sqrt{2}}, \frac{\pi}{4}\right)$.

(b) Observe that

$$\left| u\left(\frac{1}{\sqrt{2}}, \frac{\pi}{4}\right) - 50 - \sum_{k=0}^{10} \frac{200}{\pi(2k+1)} \left(\frac{1}{\sqrt{2}}\right)^{2k+1} \sin\frac{(2k+1)\pi}{4} \right|$$

$$= \left| \sum_{k=11}^{\infty} \frac{200}{\pi(2k+1)} \left(\frac{1}{\sqrt{2}}\right)^{2k+1} \sin\frac{(2k+1)\pi}{4} \right|$$

$$\leq \frac{200}{23\pi\sqrt{2}} \sum_{k=11}^{\infty} \left(\frac{1}{2}\right)^k.$$

Find the sum of the geometric series on the preceding line to obtain a bound on the error in the approximation of part (a).

23. The two-dimensional wave equation is used to model the vibrations of a drumhead. For a circular drumhead, it is best to use polar coordinates. In polar coordinates the two-dimensional wave equation is

(11.155) $$u_{tt} = \alpha^2 \left(u_{rr} + \frac{1}{r} u_r + \frac{1}{r^2} u_{\theta\theta} \right)$$

for the deflection $u(r, \theta, t)$ of the drumhead at position (r, θ) and time t.

(a) A solution of (11.155) is radially symmetric if it depends on r and t but not θ. How does equation (11.155) simplify if $u(r, \theta)$ is radially symmetric?

(b) Show that for a separated solution $u(r, t) = y(r)s(t)$ of

$$u_{tt} = \alpha^2 \left(u_{rr} + \frac{1}{r} u_r \right)$$

we must have

$$ry''(r) + y'(r) + kry(r) = 0$$

and

$$s''(t) + k\alpha^2 s(t) = 0,$$

where the constant k arises from the separation constant. Solving these two ordinary differential equations (under appropriate boundary conditions) is discussed in Exercise 13 of Section 7.6.

24. In this exercise you will derive the Mean Value Property for harmonic functions stated at the beginning of this section. This exercise assumes familiarity with double integrals in both rectangular and polar coordinates.

Suppose we have a harmonic function $v(x, y)$ in an open region $\mathcal{S}$ in the xy-plane, and suppose also that D is a disk with center (x_0, y_0) which (together with its boundary circle) lies in $\mathcal{S}$. The Mean Value Property for v and this disk states that

(11.156) $$v(x_0, y_0) = \frac{1}{\text{area}(D)} \iint_D v(x, y)\, dx\, dy.$$

The expression on the right-hand side is the "area average" of $v(x, y)$ over D. For simplicity we will assume (x_0, y_0) is the origin $(0, 0)$, so that D is a disk centered at the origin, having some radius $s > 0$, and we can put $v(x, y)$ in polar coordinates by defining $u(r, \varphi) = v(r\cos\varphi, r\sin\varphi)$. We can then rewrite the integral in (11.156) to obtain

$$\iint_D v(x, y)\, dx\, dy = \int_{-\pi}^{\pi} \int_0^s u(r, \varphi)\, r\, dr\, d\varphi.$$

Show that this integral is equal to area$(D)v(0,0)$, thus proving (11.156) in the case where (x_0, y_0) is the origin. Hint: Evaluate the interated integral above by doing the $d\varphi$ integral first using the result in Example 11.10.4.

Notes and Further Reading

Chapter 1

Interesting discussions of the current role of mathematical modeling in the biological sciences can be found in the preface to [**31**] and Chapter 1 of [**51**].

Chapter 2

You can read more on the Darden Business School's decision analysis course described in Exercise 28 of Section 2.2 in Robert Bruner's blog "The World's Toughest Interview Question" from February 17, 2008, archived at https://blogs.darden.virginia.edu/brunerblog. References for further study of utility functions and the notion of expected utility are [**25**] and the more advanced article [**23**].

For more on using mathematics to construct scuba diving tables, including a detailed derivation of the differential equation we used for tissue pressure, see [**56**]. You can also find there more information on constructing dive tables for dives that require one or more decompression stops. The recent book [**33**] tells the fascinating story of the sport of breath-hold diving and "deepest man" Herbert Nitsch from Exercise 12 in Section 2.3.

A good reference for a proof of both the existence and uniqueness parts of Theorem 2.4.5 under the stronger hypotheses of part (b), namely that both $f(x,y)$ and $\frac{\partial f}{\partial y}$ exist and are continuous functions on the rectangle in question, is Section 1.10 in [**11**]. The existence theorem as stated in part (a), with only the hypothesis that $f(x,y)$ is continuous on the rectangle in question, is known as Peano's Theorem. Its proof requires a bit more work and can be found in [**14**, pp. 1–7].

The cow-grazing model in Exercise 24 of Section 2.6 is taken from [**26**].

The spruce budworm model of Exercise 26 in Section 2.7 has been much studied. The original reference for this is [**22**]. There are several sources for interesting guided projects on this model, including "Project 3.5: The Spruce Budworm" in [**36**] and Project 1.1 in [**10**]. Section 5.6 of [**10**] contains a more elaborate model of the spruce budworm, incorporating additional variables to describe the health and dynamics of the forest.

Chapter 3

You can read further about numerical methods in [42], including a discussion of multistep methods and Runge-Kutta formulas in general. For a nice expository discussion of choosing a numerical method and things that can go wrong with numerical methods, see Section 6.4: "Practical Use of Solvers" in [36]. A good general reference for the numerical solution of differential equations is [45].

Exercise 7 in Section 3.3 is adapted from [13]; see also Section 1.6 in [10] for further discussion of the issues relevant to this exercise.

Chapter 4

For further reading on falling cats and other objects, see [28], [54], and the websites of Paul Tavilla or the Free Fall Research Page.

Chapter 5

Mesmerizing videos of the Tacoma Narrows Bridge collapse can be found on YouTube. The fiftieth anniversary (in 1990) of the collapse of the Tacoma Narrows Bridge seems to have renewed debate on the exact causes of the catastrophic failure. The articles [7] and [35] discuss some of the continuing controversy.

Chapter 6

A very readable discussion of "combat models" in general can be found in the chapter by Courtney Coleman in [12]. Our presentation of a differential equations model for the Battle of Iwo Jima is taken from there. The data for casualty figures at Iwo Jima is from [30], although this reference is not easy to find.

Chapter 7

Clever substitutions will allow you to solve various (non-Bessel) second-order linear differential equations in terms of Bessel functions; see Section 8.6 in [16] for a discussion of this. The problem of "bending of a flagpole" can also be found there.

For more on finding a second linearly independent solution near a regular singular point, in particular when the roots of the indicial equation differ by an integer, see [36].

Chapter 8

A captivating book on the history of the Battle of Trafalgar is [1]. A readable brief discussion of both the history of the battle and the modeling of it by a system of differential equations can be found in [32]. As previously mentioned, [12] contains a general discussion of mathematical combat models.

Fun articles to read related to the "Romeo and Juliet" exercises in Section 8.1 and Section 8.9 are Steven Strogatz's paper [48], the guest column "Loves me, loves me not (do the math)" by Strogatz in [49], and Section 5.3 in [50]. Inspired in part by the Strogatz article, Sergio Rinaldi [43] used a three differential equation model to analyze the love dynamics between the fourteenth-century poet Petrarch and his platonic mistress Laura (to whom he wrote in excess of 200 poems over more than 20 years). The three dependent variables in Rinaldi's paper represent Laura's love for Petrarch, his love for her, and his "poetic inspiration".

Notes and Further Reading 835

The classic reference for data about cleaning out pollution in the Great Lakes is [41]. See also Chapter 9 in [29].

Exercises 16 in Section 8.8 and 19 in Section 9.4 are based on the articles [39] and [40].

For more information on compartment models, see [3] or [15].

A good general reference for further reading on the topics of Chapters 8–10 is [20]. In addition to more on the theory of linear and nonlinear systems, there are several chapters devoted to applications of systems of differential equations.

Chapter 9

Putzer's original article [38] is quite readable; see especially Theorem 2 there. Our proof of Theorem 9.5.1 on periodic forcing is taken from [20].

Chapter 10

There is much literature available on interacting population models. Two good sources for additional reading are [10] and [31]. This latter book also has an extensive chapter on modeling infectious diseases, with examples supported by much interesting public health data.

Section 10.4.1 on competition for space and habitat destruction is based on the articles [47], [52], and [53].

References for Exercises 8 and 11 in Section 10.4 include [19] and [27].

Example 10.5.2 is based in part on the article "A model for the spread of gonorrhea" by M. Braun in [24].

The original source for Exercise 4 in Section 10.5 is David Bernoulli's 1760 paper in [6]. Our approach to solving Bernoulli's model follows that in [9]. You can find a slightly different approach in Section 3.1 of [42].

A good reference for further reading on nonlinear systems, at a somewhat more advanced level, is [34].

Chapter 11

One of our favorite texts for more information on Fourier series and partial differential equations is [17]. This book also contains very readable treatments of Bessel functions, Hermite polynomials, and other special functions, including their orthogonality properties and applications to solving the heat equation, the wave equation, and Laplace's equation in certain situations we have not discussed here. Books with more emphasis on partial differential equations that we like are the classic [55], which begins with a derivation of the wave equation from physical principles and includes introductions to complex analysis and transform techniques, and [57], a slightly more advanced text (but still suitable for undergraduates) that presents an extended look at the geometry and solutions of first-order partial differential equations using ideas from multivariable calculus.

Example 11.4.3 is inspired by the related models in Chapter 16 of [51] on trawl-fishing and lobsters and Section 5.2 of [5] on algae blooms.

You can find more details on the derivation of the heat and wave equations in Appendices A and B to Chapter 10 in [9].

For more information on the life of Joseph Fourier, who made equally important contributions to the field of Egyptology as he did to mathematics, see the biographies in [21] and [44]. The text [8] is a good place to read more about the mathematical problems and questions created by Fourier's work on trigonometric series and the role that these played in the development of real analysis.

Selected Answers to Exercises

Section 1.2

1. For $y = ce^{2x}$, $\frac{dy}{dx} = 2ce^{2x} = 2y$. Through the point $(1,3)$ if $c = 3/(e^2)$, through $(1,0)$ if $c = 0$.

3. (b) $y = e^x/(1 + e^x)$
   (c) $y = e^c e^x/(1 + e^c e^x)$

5. $c > 0$

7. (b) $c = 4$
   (c) $(-2, 2)$

9. (a) (i) (D), (ii) (B), (iii) (C), (iv) (A)
   (b) Along any horizontal line, all field marks have the same slope.

11. (a) $\frac{dy}{dx} = y$

13. (c) Right-hand endpoint ≈ 1.7

15. $y = 0$ and $y = 1$

17. $y = 2$

19. $y = \frac{1}{2}$

21. $y = 1$

23. (a) $y = 0$ and $y = 3$
    (b) $y = \frac{1}{2}$ and $y = \frac{5}{2}$
    (c) None if $k > \frac{9}{4}$

29. $\frac{dP}{dt} = cP$; independent variable t, dependent variable P, proportionality constant c

31. (a) Sally's is larger
    (b) $L(0) = 0$
    (c) $\frac{dL}{dt} = k_1(M - L) - k_2 L$ for positive constants k_1 and k_2

Section 1.3

1. (b) Isoclines are the horizontal lines $y = k$. Field marks along the line $y = k$ all have slope k.
   (d) Yes; $y = 0$

3. (a) Isoclines $x - 2y = k$ for k a constant (lines with slope $\frac{1}{2}$).
   (c) The isocline $y = \frac{x}{2} - \frac{1}{4}$ is also a solution curve.

5. (a) Isocline B, (b) isocline C, (c) isocline A

7. For each integer n, $y = x + n\pi$ is a solution.

9. (b) 3,000 bacteria
   (c) $P(t) = 25{,}165{,}824 e^{3t}/(8{,}388{,}608 e^{3t} + (4 + t)^{12})$
   (d) $\lim_{t \to \infty} P(t) = 3$

Section 2.1

1. $y = \frac{3}{7}e^{2t} + ce^{-5t}$
3. $y = \frac{x^4+2x^2}{4(x^2+1)} + \frac{c}{x^2+1}$
5. $y = -t\cos t + ct$
7. $y = (t+c)/\sqrt{t^2+1}$
9. $y = (\ln|x-2| - \ln|x+2|)/(4x) + \frac{c}{x}$
11. $y = \frac{e^{2t}}{2t} + \frac{8-e^2}{2t}$; interval of validity $(0, \infty)$
13. $y = \frac{1}{2}x^2\cos x + \cos x$; interval of validity $(-\frac{\pi}{2}, \frac{\pi}{2})$
15. $y = \frac{1}{2} - \frac{3}{2}e^{-x^2}$; interval of validity $(-\infty, \infty)$
19. (a) Solutions are $y = ce^{-at}$ and $y = bte^{-at} + ce^{-at}$; for the second you can use l'Hôpital's Rule.
    (b) Solution is $y = be^{ct}/(a+c) + ke^{-at}$
21. (c) $A = -1$ and $r = 3$
23. $p(t) = 5 - 3e^{t/10}$; population tends to 0 as t approaches $10\ln(5/3)$
25. (a) Approximately 100.65 meters
    (b) Approximately 100.001 meters

Section 2.2

1. $y = ((x^2+c)/3)^3$
3. $\frac{1}{2}e^{x^2} = ye^y - e^y + c$
5. $y + \ln|y| = -t + c$
7. $y = -\frac{1}{2}\ln(c - 2x)$
9. $y = (\tan x + c)^3$
11. $y = (t^4 + 2t^2 + 13)^{1/4}$
13. $y = \ln(e^x + e^2 - 1)$
15. (c) only
17. (a) Equilibrium solution $y = 0$
    (b) $y^{3/2}$ only defined for $y \geq 0$
19. (a) $y = 2$
    (b) $y = 2$
    (c) No equilibrium solution
21. Check that $G'(x) = 0$ on (a,b).
23. (a) Equilibrium solutions $y = 2/\sqrt{b}$, $y = -2/\sqrt{b}$ if $b > 0$
    (b) (i) $-\infty$, (ii) ∞, (iii) $-2/\sqrt{b}$, (iv) $-2/\sqrt{b}$
25. (a) Surface area $= 4\pi\left(\frac{3V}{4\pi}\right)^{2/3}$, $\frac{dV}{dt} = CV^{2/3}$ for some negative constant C
    (b) After 6 hours

Section 2.3

1. About 4.87 hours after death
3. Approximately 100.33 degrees F
7. $k \approx \ln(28.6/0.7)/18 \approx 0.2061$ with units 1/hr
9. At most 24.6 minutes at 80 feet
11. (a) 1.7988 atm
    (b) 2.478 atm
    (c) $\frac{dP}{dt} = 0.1(1 + \frac{150}{33} - \frac{30}{33}t - P)$, $P(0) = 2.478$
    (d) 2.716; not safe
13. (a) $\ln 2/k$
    (b) Approximately 27.7 minutes, 69.3 minutes, and 115.5 minutes, respectively
15. Yes, due to the tissues with $k = 0.01$.
17. (a) 12.7 minutes; pressures are 2.26 and 1.89 in the other tissues
    (b) 22 minutes
19. (a) $S(t) = (40-t)/2 - 0.01(40-t)^2$
    (b) At $t = 15$
    (c) $\frac{1}{2} - 0.01(40-t)$
    (d) $\frac{1}{2}$ lbs/gal
21. Approximately 74 days
23. (a) $p(t) = -\frac{1}{2}gt^2 + h$
    (b) $t = \sqrt{2h/g}$
    (c) $v = -\sqrt{2gh}$
25. (a) Both rates are approximately -2.5×10^{-4} in/sec
    (b) About 5.7 inches
    (c) The more nearly constant dh/dt is, the better it functions, so the graph that is closer to a horizontal line corresponds to the better clock.
27. $2(0.00997)$ inches
29. (a) $W(t) = \left(-\frac{0.03}{4}t + W_0^{1/4}\right)^4$

(b) $t \approx 21.2 W_0^{1/4}$

31. $k = \ln(10)/21$; about 6.32 years

Section 2.4

1. $(-\frac{\pi}{2}, \frac{\pi}{2})$

3. $(-2, 0)$

5. $f(x,y) = x^2 + y^2$, $\frac{\partial f}{\partial y} = 2y$ are everywhere continuous, so existence and uniqueness are guaranteed on some interval containing 1.

7. Since $f(x,y) = \sqrt{y}$ is not continuous on an open rectangle containing $(1,0)$, existence is not guaranteed by Theorem 2.4.5.

9. Since $f(x,y) = x/(y^2 - 4)$ is not continuous on an open rectangle containing $(1,2)$, existence is not guaranteed by Theorem 2.4.5.

11. $y = cx^2$ is a solution for all constants c. The function $p(x) = -2/x$ is not continuous at $x = 0$.

19. (a) $f(x,y) = g(x) - p(x)y$
    (b) $g(x) - p(x)y$ and $\frac{\partial f}{\partial y} = -p(x)$ are continuous on $a < x < b$, $-\infty < y < \infty$

(c) If $a < x_0 < b$, then Theorem 2.4.5 says a unique solution exists on some open interval containing x_0 and contained in (a,b). Theorem 2.4.4 gives the same conclusion on the entire interval (a,b).

23. (a) No equilibrium solutions
    (b) $e^y = \frac{1}{2}e^{2x} + c$; $y = \ln(\frac{1}{2}e^{2x} + c)$
    (c) Yes, the theorem applies.

29. One solution is the function $y = (1-x^2)^3$ for $x \leq 1$ and $y = 0$ for $x > 1$.

31. (a) Tangent line at (x_0, x_0^2) is $y - x_0^2 = 2x_0(x - x_0)$ which is $y = 2cx - c^2$ for $c = x_0$.
    (c) Yes a solution, but no, not obtained by any choice of c.

33. (a) If $a = b$, we have $|y_1(x) - y_2(x)| \leq 0$.
    (b) $y_1(x) = ae^{3x}$ and $y_2(x) = be^{3x}$ so $y_1(x) - y_2(x) = |a-b|e^{3x}$. Since $\frac{\partial f}{\partial y} = 3$, we may choose $M = 3$.
    (c) $t \geq \ln(100)/3$

Section 2.5

1. $k = \ln(503)/27 \approx 0.2304$. Approximately 15,970 fish.

3. Approximately 4.3×10^{-5} percent of the original

5. $\frac{dP}{dt} = 0.015P - 0.2$. Implicit solution $\ln|0.015P - 0.2|/0.015 = t + c$ where $c \approx -174.945$. Approximately 8.045 million.

7. (a) Payments of about $2,705.33 per year, so a total of about $10,821 or $821 in interest.
   (b) Yes
   (c) No

9. (a) About $1,492
   (b) About $12,131
   (c) About $17,296
   (d) About 8.1 years. $k \leq 250$

11. Projected population for 2020 is about 221,000. Population of 100 in mid-1981.

15. (d) $x \approx 0.0370$, $s \approx 1.689 \times 10^{-5}$, $a \approx 3.297$, $b \approx 0.999 \times 10^{-5}$, $r \approx 0.303 \times 10^{-5}$

(e) About 329,000
(f) About 330,000

17. $a_2 > a_1$

19. (a) $P(t) = 1/(1 - Ce^t)$
    (b) $P(t) = 1/(1 - 0.5e^t)$; explodes as $t \to \ln(2)$
    (c) $P(t) = 1/(1 + e^t)$; tends to 0 as $t \to \infty$

21. $r(P) = a - b\ln(P)$

23. Equilibrium solution $P = 30,000$

25. (a) $N = (\frac{kt}{3} + N_0^{1/3})^3$
    (b) The number of bacteria is proportional to the volume and should grow at a rate proportional to the surface area. Use this to show $\frac{dN}{dt} = cN^{2/3}$ for some constant c.

27. (a) A normal k value is greater than 0.231049, so not in normal limits.
    (b) Yes, in normal limits.
    (c) Only the last two measurements are in normal limits.

Section 2.6

1. (a) and (c): Equilibrium solutions are $y = 4$ (unstable) and $y = -1$ (stable).
   (b) For the phase line, note $\frac{dy}{dx} > 0$ for $y > 4$ and $y < -1$ and negative for $-1 < y < 4$.
   (d) $y(t) \to -1$

3. (a) and (c): Equilibrium solutions are $y = 0$ (semistable) and $y = 3$ (stable).
   (b) For the phase line, note $\frac{dy}{dx} > 0$ for $y < 0$ and $0 < y < 3$ and negative for $y > 3$.
   (d) $y(t) \to 3$

5. (a) and (c): Equilibrium solutions are $y = 2$ (unstable) and $y = 0$ (stable).
   (b) For the phase line, note $\frac{dy}{dx} > 0$ for $y > 2$ and $y < 0$ and negative for $0 < y < 2$.
   (d) $y(t) \to 0$

7. (a) and (c): Equilibrium solutions are $y = 0$ (semistable) and $y = 2$ (stable).
   (b) For the phase line, note $\frac{dy}{dx} > 0$ for $y < 0$ and $0 < y < 2$ and negative for $y > 2$.
   (d) $y(t) \to 2$

9. Implicit solution $\ln|y| - \ln|1 + y| = t + c$. Exponentiate to find the explicit solution.

11. (a) and (e): $y = -2$ (stable) and $y = 3$ (unstable).

    (b) For the phase line, note $\frac{dy}{dx} > 0$ for $y > 3$ and $y < -2$ and negative for $-2 < y < 3$.
    (c) $\frac{d^2y}{dx^2} = (y+2)(y-3)(2y-1)$, which is 0 when $y = -2$, $y = 3$, $y = \frac{1}{2}$. Concave up for $y > 3$ or $-2 < y < \frac{1}{2}$ and concave down for $y < -2$ or $\frac{1}{2} < y < 3$.

13. (a) Equilibrium solutions $P = 0$, $P = a$, and $P = b$.
    (b) $P = b$ is stable.
    (c) $P(t) \to b$ if $a < P(0)$ and $P(t) \to 0$ if $0 < P(0) < a$.

15. (a) True, (b) True, (c) True

17. (a) is the correct statement.

19. (a) $P = 2$, $P = 1$.
    (b) Increasing for $1 < P < 2$ and decreasing for $P > 2$ and $P < 1$.
    (c) Yes if $P(0) = 1.5$ and no if $P(0) = 0.8$.
    (e) $P(t) = (1 + 2e^t)/(1 + e^t)$
    (f) $\ln(4)$ years

21. (b) $P(t) \to (a - c)/b$
    (c) $Y = (ac - c^2)/b$ which is maximized when $c = a/2$.
    (d) $(a + \sqrt{a^2 - 4bh})/(2b)$ is unstable and $(a - \sqrt{a^2 - 4bh})/(2b)$ is stable.

Section 2.7

1. $y = (\frac{1}{2} + ce^{-8t})^{-1/2}$
3. $-\frac{3}{5}\ln|1 - \frac{y}{t}| - \frac{2}{5}\ln|2 + 3\frac{y}{t}| = \ln|t| + c$
5. $y = (c\cos^2 x - 6x\cos^2 x)^{-1/2}$
7. $\arctan\frac{y}{x} - \frac{1}{2}\ln(1 + \frac{y^2}{x^2}) = \ln|x| + c$
9. $\frac{1}{4}\ln(1+(y/x)^4) = \ln|x|+c$ or $y = (cx^8 - x^4)^{1/4}$
15. (a) Equilibrium solutions $W = 0$ and $W = (a/b)^3$, $W^{1/3} = \frac{a}{b} + ce^{-bt/3}$
17. $y = -\frac{3}{4} - \frac{1}{2}t + ce^{2t}$ with $t = x^2$

19. (a) $\frac{dv}{dt} = (v - 2)(v + 1)$
    (b) Equilibrium solutions are $v = 2$ and $v = -1$
    (c) $y = t + 2$ and $y = t - 1$

21. (c) $\beta = \frac{c}{m}$, $\alpha = \frac{c}{mg}$

23. $\frac{1}{P} = \frac{b}{a} + \frac{b\beta}{a^2+\gamma^2}(a\cos(\gamma t) + \gamma\sin(\gamma t)) + ce^{-at}$; $P(0) = P_0$ gives $c = \frac{1}{P_0} - \frac{b}{a} - \frac{ab\beta}{a^2+\gamma^2}$

25. (b) $y = x\sinh(\alpha \ln(100) - \alpha \ln x)$
    (d) $(0, 50)$

Section 2.8

1. Exact; $x^2 y^3 = c$
3. Not exact
5. Not exact
7. Exact; $\ln(1 + x^2 + y^2) = c$
9. $x^2 y^3 + y^2/2 = 9/2$
11. $\sin(xy) + y = 3/2$
13. (a) $5xy + \sin y + h(x)$
    (b) $h(x) = e^x (+c)$

15. $\ln x + y^3/x + y = c$

17. (a) $e^{\int p(x)dx}$
    (b) $p(x)y - g(x)$

19. $f(x) = \frac{x}{2} + \frac{c}{x}$

21. (b) $s = \frac{4}{5}r^2 + \frac{c}{r^3}$
    (c) $\ln|y| = -\ln(1 + x^2/y^2) + c$

23. $\frac{x}{y} = \frac{(xy)^5}{5} + c$

25. (a) $\frac{x^2y^2}{2} - xy = c$
    (b) $xy - \ln|x| - \ln|y| = c$

27. $\sqrt{x^2 + y^2} = x + c$ or $x = (y^2 - c^2)/(2c)$, the equation of a parabola

Section 3.1

1. $y(2) \approx 1.14642$. The exact solution to $\frac{dy}{dt} = 1/(1 + t^2)$ with initial condition $y(1) = \frac{\pi}{4}$ is $y(t) = \arctan t$, so the exact value of $y(2)$ is $\arctan 2$.

3. (a) $y(2) \approx 1.621$
   (b) $y(2) \approx 1.93045$
   (c) $\ln y = \frac{t^2}{2} + c$ with $c = \ln\frac{1}{2} - \frac{1}{2}$. Exact value of $y(2)$ is obtained from $\ln y = 1.5 + \ln\frac{1}{2}$ and is $y(2) = \frac{1}{2}e^{1.5} \approx 2.2408$.
   (d) Relative percent error in (a) is approximately 27.7% and in (b) is approximately 13.8%.
   (e) $y(\frac{1}{2}) \approx 0.329346$

5. $P(100) \approx 80.831$ million

9. (a) $y(1) \approx 3.18748$
   (b) $y(1) = 2e - 2$; relative percent error in (a) is $\frac{2e-2-3.18748}{2e-2} \times 100 \approx 7.2\%$

(c) Step size of $\frac{1}{100}$ gives $y(1) \approx 3.40963$ and step size of $\frac{1}{200}$ gives $y(1) \approx 3.42303$, with relative percent errors approximately 0.78% and 0.39%

11. (a) $y(1) \approx 6.1289$
    (b) $y(1) \approx 30.3897$
    (c) $y = 1/(1 - t)$. This is undefined when $t = 1$.

13. (a) Any root of the polynomial gives an equilibrium solution.
    (b) From the direction field it appears that there are equilibrium solutions of approximately $y = -1$, $y = 0.5$, and $y = 1.5$. Thus the polynomial $p(y)$ has roots at approximately these values: -1, 0.5, and 1.5.
    (c) $y_{1000} \approx 0.482856$, which is an accurate estimation of the root of p that is near 0.5.

Section 3.2

1. $y = \left(\frac{t^3}{2} - \frac{6}{t}\right)^2 = \frac{(t^4 - 12)^2}{4t^2}$

3. Rounding computations to six decimal places, we obtain $y(1) \approx 2.694856$ by the Heun method and $y(1) \approx 2.718210$ via the Runge-Kutta method.

5. $j = 7$

7. (a) $y(0.5) \approx y_5 = 1.386133$
   (b) $y(0.5) \approx y_5 = 1.386565$

9. A natural choice is $f(t, y) = 1/t$ and $y(1) = 0$, so that $\ln 2$ is the exact solution of $\frac{dy}{dt} = f(t, y)$, $y(1) = 0$. Using Runge-Kutta with step size $\frac{1}{10}$ we find $\ln 2 \approx y_{10} = 0.06931473747$.

11. (a) $y(1) \approx y_4 = 0.519763$ (rounded to six decimal places)

13. The Heun method requires two evaluations; the Runge-Kutta method requires four.

15. (a) The Heun method with step size $\frac{1}{10}$ gives $y(30) \approx y_{300} = 0.3728436787$ while $e^{\sin(30)} \approx 0.3723088139$.

17. (a) The interval $[t_0, t_1]$ should be longer than the interval $[t_3, t_4]$.

Section 3.3

1. (a) $y^{1+\epsilon}\left(x+\sqrt{x^2+y^2}\right)^{-\epsilon}=C$

3. (a) At a point (x,y) on a circle centered at the origin, the tangent vector to the circle is perpendicular to the radial segment ending at (x,y). Since the dot product of the vector (x,y) and the vector $(-y,x)$ is 0, the vector $T=(-y,x)$ is tangent to the circle centered at $(0,0)$ and passes through the point (x,y).
(b) The apparent normal $(0,1)+\epsilon(-y,x)=(-\epsilon y,1+\epsilon x)$ has slope $-(1+\epsilon x)/(\epsilon y)$. The slope of the apparent curve is the negative reciprocal of this value; thus $\dfrac{dy}{dx}=\dfrac{\epsilon y}{1+\epsilon x}$ is the equation for the apparent curve.

7. (a) Equilibrium solutions are found from the points of intersection of the two graphs; there are three, which we denote as $0<e_1<e_2<e_3$. A phase line sketch determines that e_1 and e_3 are stable, while e_2 is unstable.

(b) Translating the graph a sufficient amount downward (smaller L) we can obtain one stable equilibrium point. Translating upward a sufficient amount (larger L) we can obtain one stable equilibrium point.
(c) The smallest equilibrium is about 4,900, the intermediate value equilibrium is about 86,850, and the largest equilibrium is about 900,000.
(e) Changing q from the value 7.88 to the value 2 eliminates all but the largest equilibrium point, which is stable. Thus in the long run, the practical effect is an unhealthier lake.

8. (a) Approximate values of the equilibrium solutions are $e_1\approx 1.99$, $e_2\approx 7.13$, and $e_3\approx 15.88$.
(d) We have $V(31.25)\approx 14.99$ while $V(31.5)\approx 15.01$. Thus the vegetation level reaches 15 tons after about 31 days.

Section 4.1

1. (a), (b), (c), (d), and (f) are linear; of these, (b), (d), and (f) are homogeneous

3. (a) $(0,\frac{\pi}{2})$
(b) $(-3,0)$
(c) $(0,\infty)$
(d) $(-\infty,\infty)$

7. $c>42.3$ kg/sec

9. (a) Terminal velocity is
$$\frac{gm}{c}=\frac{(980)(\frac{4}{3}\pi(.05)^3)}{1.2\times 10^{-3}}\approx 427.6$$
with units cm/sec
(b) Within 10 percent of its terminal velocity when $t\approx 1.0037$ sec. At this point it has only fallen about 2.6 meters.

11. (c) $|1+w|/|1-w|=e^{2s}$; as $s\to\infty$ we must have $w\to 1$

13. (a) $k=32/6$
(b) Terminal velocity is $32/6$ (not reached in finite time)

15. d tends to $\frac{1}{2}$ as $x\to 0$ (i.e., as $t\to\infty$), but for finite t the smuggler is never in range.

17. Graph (C)

19. (e)

21. $E'(t)=my'y''+kyy'=my'(-\frac{b}{m}y'-\frac{k}{m}y)+kyy'=-b(y')^2$, where primes denote differentiation with respect to t

23. (b) $y=\frac{1}{t+1}$ and $y\to 0$ as $t\to\infty$
(c) $y=\frac{e^{t/2}}{2e^{t/2}-1}$. As $t\to\infty$, $y\to\frac{1}{2}$.

25. (a) No, by the No Intersection Theorem, since both $f(t,y)=y\sin y+t$ and $\frac{\partial f}{\partial y}=y\cos y+\sin y$ are continuous everywhere.

26. True

Section 4.2

1. (a) Linear, (b) nonlinear, (c) linear, (d) linear, (e) nonlinear, (f) linear

3. (a) $6t+7t^4$
(b) $-\sin t+3t^2\cos t-2t\sin t$
(c) $6t+7t^4-\sin t+3t^2\cos t-2t\sin t$

5. $y = c_1 + c_2 e^{-3t}$

7. $y = c_1 e^{3t/2} + c_2 e^{-t}$

9. $y = e^{4t} - 2e^{-t}$

11. $y = e^{2t}$

15. (d) $y = 8 \sinh t + 5 \cosh t$

Section 4.3

1. $(1) 2 \sin^2 t + (-2)(1 - \cos^2 t) = 0$

3. $(2) \sin(t + \frac{\pi}{2}) + (-1)(2 \cos t) = 0$

4. (a) Dependent
   (b) Independent
   (c) Dependent
   (d) Dependent

17. Yes, $y_1(t) + y_2(t)$

18. (a) $y'' - y' - 2y = 0$

32. (a) $6D^2 + D - 2 = (3D + 2)(2D - 1)$

41. $W(y_1, y_2)(t) = 3e^{-t^2}$

7. $\cos(t - \frac{\pi}{3}) = \cos t \cos \frac{\pi}{3} + \sin t \sin \frac{\pi}{3} = \frac{1}{2} \cos t + \frac{\sqrt{3}}{2} \sin t$

9. (a) True, (b) False

11. (a) Yes, (b) Yes, (c) No, (d) Yes

13. Yes

Section 4.4

1. $y = c_1 e^{-\frac{1}{2}t} \cos(\frac{\sqrt{3}}{2}t) + c_2 e^{-\frac{1}{2}t} \sin(\frac{\sqrt{3}}{2}t)$

3. $y = c_1 e^{3t} + c_2 e^{-2t}$

5. $y = c_1 e^{-2t} \cos(\sqrt{3}\, t) + c_2 e^{-2t} \sin(\sqrt{3}\, t)$

7. $y = c_1 \cos(\sqrt{2}\, t) + c_2 \sin(\sqrt{2}\, t)$

10. $\sqrt{2} \cos(2t + \frac{\pi}{4})$; amplitude $\sqrt{2}$, period π, frequencey $\frac{1}{\pi}$

12. $5 \cos(2t - \delta)$ where $\cos \delta = \frac{3}{5}$, $\sin \delta = -\frac{4}{5}$ so $\delta \approx -0.9273$; amplitude 5, period π, frequency $\frac{1}{\pi}$

17. (a) $t = \frac{\pi}{12}$
    (b) Yes, two successive zeros are separated by $\frac{\pi}{3}$.

19. (a) $y = c_1 e^{kt} + c_2 e^{-kt}$ and $y = c_1 \cos(kt) + c_2 \sin(kt)$, respectively
    (b) $y = \frac{3}{2} e^{kt} - \frac{1}{2} e^{-kt}$ and $y = \cos(kt) + 2 \sin(kt)$
    (c) They solve $y'' - k^2 y = 0$.

21. (b) $y_1 = e^{3it}$ and $y_2 = e^{-it}$; no
    (c) $y = c_1 e^{3it} + c_2 e^{-it}$
    (d) $c_1 e^{-it} + c_2 e^{2it}$

Section 4.5

1. $y = c_1 e^{2t} + c_2 t e^{2t}$

3. $y = c_1 e^{\sqrt{2}\, t} + c_2 e^{-\sqrt{2}\, t}$

5. $y = c_1 e^{-t} + c_2 t e^{-t}$

7. $y = 3e^{-3t} + 4t e^{-3t}$

9. $y = e^{-t} - 2t e^{-t}$

11. $y'' - y' - 2y = 0$, $y(0) = -5, y'(0) = 8$ (other initial conditions are possible)

13. $y'' + 4y = 0$, $y(0) = 1, y'(0) = -2$ (other initial conditions are possible)

19. (a) $y = c_1 t^{1/3} + c_2 t^2$
    (c) $y = c_1 t^{-\frac{1}{2}} + c_2 t^{-2}$

21. $y = c_1 + c_2 t^3$

23. (a) $y_1(t) = t^7$

(b) $y_2(t) = t^7 \ln t$

25. (c) $c_1 t + c_2 t^2$

27. $y = c_1 \sin t + c_2 t \sin t$

31. $y = c_1 t + c_2 (t^2 + 1)$

36. (a) $(D + \frac{1}{2t})(D + \frac{3}{t})$ and $(D + \frac{4}{t})(D - \frac{1}{2t})$

Section 4.6

1. (b) $(-3e)e^t+(-e^{-4})e^{2t}+(1)(3e^{t+1}+e^{2t-4}) = 0$
3. (a) $t = \frac{1}{8}(t^2 + 5t) - \frac{1}{8}(t^2 - 3t)$
   (b) $(-1)t + (\frac{1}{8})(t^2 + 5t) + (-\frac{1}{8})(t^2 - 3t) = 0$
5. Dependent
7. Independent
9. Independent
11. Dependent
13. (a) True, (b) False
14. (a) $(-3, 2)$
15. Yes

Section 4.7

1. (a) $z = 2$, $z = 2i$, $z = -2$, and $z = -2i$
   (b) $w = \sqrt{8}e^{(\frac{\pi}{4}+2k\pi)i}$ for k any integer;
   $z = 8^{1/6}e^{i\frac{\pi}{12}} = 8^{1/6}(\cos\frac{\pi}{12} + i\sin\frac{\pi}{12})$, $z = 8^{1/6}e^{i\frac{9\pi}{12}} = 8^{1/6}(\cos\frac{9\pi}{12} + i\sin\frac{9\pi}{12})$, and $z = 8^{1/6}e^{i\frac{17\pi}{12}} = 8^{1/6}(\cos\frac{17\pi}{12} + i\sin\frac{17\pi}{12})$
   (c) $w = 2e^{i(-\frac{\pi}{3}+2k\pi)}$ for k any integer;
   $z = 2^{1/4}e^{-i\frac{\pi}{12}} = 2^{1/4}(\cos\frac{\pi}{12} - i\sin\frac{\pi}{12})$, $z = 2^{1/4}e^{i\frac{5\pi}{12}} = 2^{1/4}(\cos\frac{5\pi}{12} + i\sin\frac{5\pi}{12})$, $z = 2^{1/4}e^{i\frac{11\pi}{12}} = 2^{1/4}(\cos\frac{11\pi}{12} + i\sin\frac{11\pi}{12})$, and $z = 2^{1/4}e^{i\frac{17\pi}{12}} = 2^{1/4}(\cos\frac{17\pi}{12} + i\sin\frac{17\pi}{12})$
   (d) $w = 2e^{i(\frac{5\pi}{6}+2k\pi)}$ for k any integer;
   $z = 2^{1/4}e^{i\frac{5\pi}{24}} = 2^{1/4}(\cos\frac{5\pi}{24} + i\sin\frac{5\pi}{24})$, $z = 2^{1/4}e^{i\frac{17\pi}{24}} = 2^{1/4}(\cos\frac{17\pi}{24} + i\sin\frac{17\pi}{24})$, $z = 2^{1/4}e^{i\frac{29\pi}{24}} = 2^{1/4}(\cos\frac{29\pi}{24} + i\sin\frac{29\pi}{24})$, and $z = 2^{1/4}e^{i\frac{41\pi}{24}} = 2^{1/4}(\cos\frac{41\pi}{24} + i\sin\frac{41\pi}{24})$

3. $e^{i\frac{\pi}{8}} = \cos\frac{\pi}{8} + i\sin\frac{\pi}{8}$, $e^{i\frac{5\pi}{8}} = \cos\frac{5\pi}{8} + i\sin\frac{5\pi}{8}$, $e^{i\frac{9\pi}{8}} = \cos\frac{9\pi}{8} + i\sin\frac{9\pi}{8}$, and $e^{i\frac{13\pi}{8}} = \cos\frac{13\pi}{8} + i\sin\frac{13\pi}{8}$
5. $y = c_1 e^{-t} + c_2 t e^{-t} + c_3 t^2 e^{-t}$
7. $y = c_1 e^{2t} + c_2 \cos t + c_3 \sin t$
9. $y = c_1 e^t + c_2 e^{(2+\sqrt{5})t} + c_3 e^{(2-\sqrt{5})t}$
11. $y = c_1 e^{\sqrt{3}t}\cos t + c_2 e^{\sqrt{3}t}\sin t + c_3 \cos(2t) + c_4 \sin(2t) + c_5 e^{-\sqrt{3}t}\cos t + c_6 e^{-\sqrt{3}t}\sin t$
17. $y = k_1(e^t - e^{-t}) + k_2 \cos t$

Section 4.8

1. (a) Above and moving up
   (b) $\frac{1}{5}$ meters
3. $y = \frac{1}{2}e^{-16t} + 8te^{-16t}$; never for $t > 0$
5. (b) $t = \frac{1}{2}\ln(5/3)$
7. (a) No
   (b) Yes only for $v_0 = 3$
   (c) $b = 2$
9. $c = 2\sqrt{4.9}$
11. Since $m \approx \frac{8}{32}$ and the period is $2\pi\sqrt{m/k}$ we have $k = 16m \approx 4$ lb/ft.
15. (b) Graph (A)

17. $y = \alpha\cos(\mu t) + \frac{\beta}{\mu}\sin(\mu t)$, where $\mu = \sqrt{k/m}$. This is $y = R\cos(\mu t - \delta)$ for $R = \sqrt{\alpha^2 + m\beta^2/k}$, which gets larger as m is increased.
19. Overdamped
21. Period is $2\pi\sqrt{m/k} \approx 2\pi\sqrt{d/9.8}$ sec.
23. (a) Left: $8/20 = 2/5$, right: $12/20 = 3/5$
    (b) Left: $(8-y)/20$, right: $(12+y)/20$
    (c) $y(t) = c_1 e^{t\sqrt{3.2}} + c_2 e^{-t\sqrt{3.2}} - 2$
    (d) $t = \frac{1}{\sqrt{3.2}}\ln(5 + 2\sqrt{6}) \approx 1.282$ sec.

Section 5.1

1. $y = c_1 e^t + c_2 e^{-t} + te^t$
3. $y = c_1 \cos t + c_2 \sin t + \frac{1}{3}\cos^4 t + \sin^2 t - \frac{1}{3}\sin^4 t$
5. $y = c_1 t^2 + c_2 t + \frac{t^4}{6}$

7. $y = c_1 e^{-t} + c_2 t e^{-t} - \frac{2}{5} e^{-t}(1+t)^{5/2} + \frac{2}{3} e^{-t}(1+t)^{3/2} + \frac{2}{3} t e^{-t}(1+t)^{3/2} = c_1 e^{-t} + c_2 t e^{-t} + \frac{4}{15} e^{-t}(1+t)^{5/2}$

9. $y = \frac{5}{2} t + \frac{1}{2} t^{-1} + (\ln t)^2 t - t \ln t$

Section 5.2

1. $D-1$ annihilates e^t, $(D-1)^2$ annihilates $t e^t$, $(D-1)^3$ annihilates $t^2 e^t$, and $(D-1)^{m+1}$ annihilates $t^m e^t$.

3. $(D^2 - 2D + 5)^2$

6. (a) $z^2 - 4z + 5 = 0$
   (b) $D^2 - 4D + 5$

9. $y(t) = c_1 \cos t + c_2 \sin t - 2 + t^2$

11. $y = c_1 \cos(4t) + c_2 \sin(4t) - \frac{1}{4} t \cos(4t)$

13. $y(t) = c_1 e^{2t} + c_2 e^{-2t} - \frac{8}{3} e^t$

15. $y(t) = c_1 e^{2t} + c_2 e^{-2t} + c_3 \cos(2t) + c_4 \sin(2t) + \frac{1}{32} t e^{2t}$

17. (a) $y(t) = -\frac{1}{8} t^2 + c_1 + c_2 e^{2t} + c_3 e^{-2t}$
    (b) $y(t) = -\frac{3}{5} \sin t + c_1 + c_2 e^{2t} + c_3 e^{-2t}$
    (c) $y(t) = -\frac{1}{8} t^2 - \frac{3}{5} \sin t + c_1 + c_2 e^{2t} + c_3 e^{-2t}$

19. $y(t) = c_1 + c_2 e^{-t} + t - t^2 + \frac{1}{3} t^3$

15. The general solution may be written as $y = c_1 (3 - 7t) + c_2 e^{2t} + 2t$.

17. The general solution may be written as $y = t^2 + c_1 (t - t^2) + c_2 (1 - t^2)$.

21. $\frac{1}{3} e^{3t} - \cos(3t) \ln|\sec(3t) + \tan(3t)|$

23. $y = c_1 e^{-2t} + c_2 e^{-t} + \frac{5}{4} - \frac{1}{8} \cos(2t) + \frac{3}{8} \sin(2t)$

25. $y(t) = \frac{1}{4} e^t + \frac{1}{4} e^{-t} - \frac{1}{2} \cos t$

27. $y = 3 y_{p_1} - 2 y_{p_2}$

29. $y = c_1 e^{2t} + c_2 t e^{2t} + t \ln t \, e^{2t}$

31. $y = c_1 e^{-2t} + c_2 e^{-t} - \frac{9}{10} \cos t + \frac{3}{10} \sin t$

33. (b) $c = \frac{6}{-6 + 12i} = \frac{1}{-1 + 2i} = \frac{-1 - 2i}{5}$
    (c) $y = -\frac{1}{5} \cos(3t) + \frac{2}{5} \sin(3t)$ and $v = -\frac{2}{5} \cos(3t) - \frac{1}{5} \sin(3t)$

35. (b) $\frac{1}{4+4i} e^{(1+i)t} = \frac{1}{\sqrt{32}} e^{t + i(t - \pi/4)}$ since $4 + 4i = \sqrt{32} e^{i\pi/4}$
    (c) $\frac{1}{\sqrt{32}} e^t \cos(t - \frac{\pi}{4})$

37. (a) $-e^{-2t}$
    (b) $-3 e^{-2t}$
    (c) $-\frac{1}{5} t e^{-2t}$

Section 5.3

1. Resonance will occur for solutions to $y'' + 4y = 5 \cos(2t)$ and for solutions to $y'' + 4y = \sin(2t)$ but not for $y'' + 4y = -\cos \frac{t}{2}$.

3. For the third equation, (c)

5. All solutions will exhibit resonance.

7. $y(t) = -20 e^{-t} - 20 t e^{-t} + 20$

11. He should sing a note 13 half-steps above middle C; this is the $C\sharp$ above the C which is one octave up from middle C.

15. The maximum blood level is approximately 252 mg.

18. (b) $x(t) = \frac{c_1}{2} e^{3t} - \frac{c_2}{2} e^{-t} - \frac{1}{3} e^{2t}$, $y(t) = c_1 e^{3t} + c_2 e^{-t} - \frac{4}{3} e^{2t}$

Section 5.4

1. (a) $a = L$, $b = R$, and $c = \frac{1}{C}$

3. (a) $b = (V(0) - R I_0 - \frac{1}{C} Q_0)/L$
   (b) $I(t) = -e^{-75t} \sin(25\sqrt{71} t)/(5\sqrt{71})$

6. $V_{ab}(t) = I_s(t) R = \frac{\omega V_0 R}{r} \cos(\omega t - \delta)$, $V_{cd}(t) = L I_s'(t) = -\frac{\omega^2 L V_0}{r} \sin(\omega t - \delta)$, $V_{bc}(t) = V_0 \sin(\omega t) - \frac{\omega V_0 R}{r} \cos(\omega t - \delta) + \frac{\omega^2 L V_0}{r} \sin(\omega t - \delta) = \frac{V_0}{Cr} \sin(\omega t - \delta)$.

8. (a) Solutions oscillate in the underdamped case $R^2 < \frac{4L}{C}$, corresponding to the graphs of I vs. t and Q vs. t on the top.

Section 6.1

1. $\frac{1}{s-3}$ for $s > 3$
3. $\frac{2}{s^3}$ for $s > 0$
5. $\frac{1}{s^2+1}$ for $s > 0$.
7. $\frac{1}{1+s} - e^{-1}\frac{e^{-s}}{1+s}$
9. $-\frac{2}{s}e^{-2s} - \frac{1}{s^2}e^{-2s} + \frac{1}{s^2}$
11. $2\frac{s}{s^2+9} - \frac{10}{s^2+4}$
13. $\frac{s}{s^2+4} + \frac{s^2-4}{(s^2+4)^2}$
15. $\frac{2}{(s-1)^3} + \frac{2}{(s-1)^2} + \frac{1}{s-1}$
17. $\frac{s^2+8}{s(s^2+16)}$
19. $\frac{6}{s^2-9}$
27. $(e^{-sa} - e^{-sb})/s$
29. (a) $g(t) = t(u_0(t) - u_3(t)) + e^{-t}u_3(t)$

(b) $g(t) = 2(u_0(t) - u_2(t)) + 3t(u_2(t) - u_3(t)) + 5u_3(t)$
31. $\mathcal{L}[f] = \frac{1+e^{-\pi s}}{(1-e^{-2\pi s})(s^2+1)} = \frac{1}{(1-e^{-\pi s})(s^2+1)}$
33. (a) $\int_0^4 f(t)\,dt = \frac{17}{3}$
(b) Yes, piecewise continuous with discontinuities at $t = 1, 2$ and $t = 3$
35. (c) 1
36. (a) Yes, for example with $A = 1$, $b = 0$.
(b) Yes, for example with $A = \frac{10}{e}$, $b = 1$.
(f) Yes; $|\sin(e^{t^2})| \leq 1$ so we can choose $A = 1$, $b = 0$.
37. (b) $\mathcal{L}[t^n e^{at}] = \frac{n!}{(s-a)^{n+1}}$
43. (c) No; we would have to have $\lim_{s \to \infty} \sin s = 0$, which is not true, since $\lim_{s \to \infty} \sin s$ does not exist.

Section 6.2

1. $\frac{1}{s^2-5s+6} = \frac{1}{s-3} - \frac{1}{s-2}$ and $\mathcal{L}^{-1}\left[\frac{1}{s^2-5s+6}\right] = e^{3t} - e^{2t}$
3. $\frac{s+1}{(s+2)(s^2+1)} = -\frac{1/5}{s+2} + \frac{(s+3)/5}{s^2+1}$ and $\mathcal{L}^{-1}\left[\frac{s+1}{(s+2)(s^2+1)}\right] = -\frac{1}{5}e^{-2t} + \frac{1}{5}\cos t + \frac{3}{5}\sin t$
5. $5\cos t$
7. $-\frac{1}{8}\cos(2t) + \frac{1}{8}\sin(2t) + \frac{1}{8}e^{-2t}$
9. $\frac{1}{2}t^2, \frac{5}{6}t^3$
11. (a) $\frac{s+2}{(s-3)(s-1)} = \frac{5/2}{s-3} - \frac{3/2}{s-1}$ and $\mathcal{L}^{-1}\left[\frac{s+2}{s^2-4s+3}\right] = \frac{5}{2}e^{3t} - \frac{3}{2}e^t$
(b) $\frac{s+2}{s^2-4s+3} = \frac{s-2}{(s-2)^2-1} + \frac{4}{(s-2)^2-1}$ and $\mathcal{L}^{-1}\left[\frac{s+2}{s^2-4s+3}\right] = e^{2t}\cosh t + 4e^{2t}\sinh t$
(c) Use $\cosh t = (e^t + e^{-t})/2$ and $\sinh t = (e^t - e^{-t})/2$ to see that answers in (a) and (b) are the same.

Section 6.3

1. $\frac{s^2 - 3}{(s^2 + 3)^2}$
3. $\frac{1}{(s+2)^2}$
5. $\frac{5}{2}\frac{1}{s} + \frac{5}{2}\frac{s}{s^2+16}$
9. $\frac{(s-2)^2-9}{((s-2)^2+9)^2}$
11. $\frac{1}{s} - \frac{2}{s^2} + e^{-\frac{1}{2}s}\frac{4}{s^2}$
13. $y = 2e^t + e^{-t}$
15. $y = \frac{3}{10}\sin(2t) - \frac{1}{5}\sin(3t)$
17. $y = \frac{1}{14}e^{-4t} + \frac{1}{35}e^{3t} - \frac{1}{10}e^{-2t}$
19. $y = \frac{3}{8} - \frac{1}{2}t + \frac{1}{4}t^2 - \frac{3}{8}e^{-2t} - \frac{1}{4}te^{-2t}$
21. $y = -\frac{2}{36} - \frac{1}{34}t^2 - \frac{1}{108}t^4 + \frac{1}{36}e^{3t} + \frac{1}{36}e^{-3t}$
23. $c_1 = (2b^3)^{-1}$, $c_2 = -(2b^2)^{-1}$
28. (a) Graph II
(b) Graph III
29. (a) $u_2(1)f(1) = 0$
(b) 6
(c) 2
(d) 11
31. $t^7/7!$
33. $\frac{1}{2}\sinh(2t)$
35. $t^2 e^{2t}$
37. $e^{2t} - \frac{3}{2}e^t + \frac{1}{2}e^{-t}$
39. $\frac{3}{2}e^{3t} - \frac{1}{2}e^{-t}$

Section 6.4

1. $h \leq 50e^{0.6}/(e^{0.6} - 1) \approx 110.8$

3. (a) $T(t) = 9e^{-kt} + 70$, $k = \ln(9/5)$, time of death approximately 10:02 PM
   (b) $T(t) = 98.6e^{-kt} + 50(1 - e^{-kt}) + 20\left(u_{1/2}(t) - u_{1/2}(t)e^{-k(t-\frac{1}{2})}\right)$, $k = \ln(9/5)$, time of death approximately 10:30 PM

5. $r = 2.4$ ounces per hour. So between 9:00 PM and 1:00 AM, $2.4 \times 4 = 9.6$ ounces of liquor are drunk.

7. (a) Regimen I: $\frac{dy}{dt} = -0.34y + 500(u_0(t) - u_1(t)) + 500(u_6(t) - u_7(t)) + 500(u_{12}(t) - u_{13}(t)) + 500(u_{18}(t) - u_{19}(t))$, $y(0) = 0$. Regimen II: $\frac{dy}{dt} = -0.34y + \frac{1000}{1.5}(u_0(t) - u_{3/2}(t)) + \frac{1000}{1.5}(u_{12}(t) - u_{13.5}(t))$, $y(0) = 0$.
   (b) $y(2) = \frac{500}{0.34}(-e^{-0.68} + e^{-0.34}) \approx 302$ mg under regimen I and $y(2) = \frac{1000}{(1.5)(0.34)}(-e^{-0.68} + e^{-0.17}) \approx 661$ mg under regimen II.

9. (a) $\frac{dT}{dt} = -k(T - 450)(u_0(t) - u_{15}(t)) - k(T - 350)u_{15}(t)$, $T(0) = 55°$ where T is the temperature after t minutes
   (b) $\frac{dT}{dt} = -k(T - 450)(u_0(t) - u_{15}(t)) - k(T - 600 + 10t)(u_{15}(t) - u_{25}(t)) - k(T - 350)u_{25}(t)$, $T(0) = 55°$

Section 6.5

1. $x(t) = \frac{3}{4}e^{3t} + \frac{1}{4}e^{-t}$ and $y(t) = \frac{3}{2}e^{3t} - \frac{1}{2}e^{-t} - e^t$

3. $x(t) = -\cosh(2t) - \sinh(2t) = -e^{2t}$, $y = 2\cosh(2t) + 2\sinh(2t) = 2e^{2t}$

7. $(s + k_1)\mathcal{L}[g] = g_0$ and $g(t) = g_0 e^{-k_1 t}$; $\mathcal{L}[b] = \frac{k_1 g_0}{(s+k_1)(s+k_2)}$ and $b(t) = \frac{k_1 g_0}{k_1 - k_2}\left(e^{-k_2 t} - e^{-k_1 t}\right)$

9. (a) $U_1(t) = \frac{2}{3}(\cos t - \cos(2t))$ and $U_2(t) = \frac{2}{3}\cos(2t) + \frac{4}{3}\cos t$

Section 6.6

1. $t^5/20$

3. $\frac{t}{4} + \frac{1}{16}e^{-4t} - \frac{1}{16}$

5. (b) te^{bt}

9. $e^{at} * e^{bt} = \frac{1}{a-b}(e^{at} - e^{bt})$

11. $-\frac{2}{5}(2\cos(2t) + \sin(2t) - 2e^t)$

14. $f(t) = 2te^{-t}$

Section 6.7

1. $\mathcal{L}[y] = 4/s$ and $y(t) = 4$

3. $s\mathcal{L}[y] = 1 - e^{-2s}$ and $y(t) = 1 - u_2(t)$

5. $s^2\mathcal{L}[y] = 2e^{-\pi s}$ and $y(t) = 2u_\pi(t)(t - \pi)$

7. $s^2\mathcal{L}[y] - 1 + \mathcal{L}[y] = 3e^{-\pi s/2}$ and $y(t) = \sin t + 3u_{\frac{\pi}{2}}(t)\sin(t - \frac{\pi}{2}) = \sin t - 3u_{\frac{\pi}{2}}(t)\cos t$

9. $s^2\mathcal{L}[y] + 2s\mathcal{L}[y] + 2\mathcal{L}[y] = e^{-4s}$ and $y(t) = u_4(t)e^{-(t-4)}\sin(t - 4)$

11. $\mathcal{L}[y] = \frac{1}{s^2+1} + \frac{e^{-\pi s}}{s^2+1} + \frac{e^{-2\pi s}}{s^2+1} + \frac{e^{-3\pi s}}{s^2+1}$ and $y(t) = \sin t + u_\pi(t)\sin(t - \pi) + u_{2\pi}(t)\sin(t - 2\pi) + u_{3\pi}(t)\sin(t - 3\pi) = \sin t - u_\pi(t)\sin t + u_{2\pi}(t)\sin t - u_{3\pi}(t)\sin t$

13. $y = -\frac{1}{2}e^{-2t} + \frac{1}{2}e^{-t} - u_5(t)e^{-2(t-5)} + u_5(t)e^{-(t-5)} + \frac{1}{2}u_8(t) + \frac{1}{2}u_8(t)e^{-2(t-8)} - u_8(t)e^{-(t-8)}$

15. Sally is correct; $y = 4u_2(t)\left((t - 2)e^{-(t-2)}\right)$.

17. In both parts, the initial conditions for the second initial value problem are satisfied, but the initial conditions for the first are not.

19. (a) $1/h$
    (b) $0, 0, 1/h$

21. (b) $y(t) = u_2(t)e^{-2(t-2))}\sin(t - 2)$

23. (a) $y(t) = a\cos t + b\sin t$. So long as a and b are not both zero, we can rewrite this as $y(t) = R\cos(t - \delta)$ for some $R > 0$.

25. (a) $y(t) = y_0 e^{-0.1t}$ which tends to 0 as $t \to \infty$
    (b) $A = y_0(1 - e^{-0.1})$; yes

27. $y(t) = e^{-t}$

29. (a) $\sin t$
    (c) $f(t)$

Section 7.2

1. (a) $\sum_{n=3}^{\infty} n a_{n-1}(x-2)^n$
   (b) $\sum_{n=4}^{\infty} (n-1) a_{n-2}(x-2)^{n-1}$

5. $1 + \sum_{n=1}^{\infty} \left(\dfrac{2^{n-1}}{(n-1)!} + \dfrac{2^n}{n!} \right)(x-1)^n$

7. $\sum_{n=0}^{\infty} a_n (x-1)^n$ for $a_0 = 1$, $a_1 = 0$, and
$a_n = \dfrac{2^{n-2}}{(n-2)!} - \dfrac{2^n}{(n-1)!} + \dfrac{2^n}{n!}$ for $n \geq 2$

9. $a_2 b_0 + a_1 b_1 + a_0 b_2$

11. (a) $(1-x^2)^{-1}$; $(-1, 1)$, (b) $2/(2-x)$; $(-2, 2)$,
    (e) $-\ln(1-x)$; $(-1, 1)$

13. $R = 3$, $(-3, 3)$

15. $R = \infty$, $(-\infty, \infty)$

17. (a) Since $a_n = 0$ for every odd value of n, the ratio $|a_{n+1}/a_n|$ is not defined for n odd. (d) The series in x has radius of convergence $\sqrt{2}$.

19. $a_n = e/n!$

Section 7.3

1. For n odd, $a_n = 0$ and for $n = 2k$ even, $a_n = a_{2k} = (-1)^{k+1} \dfrac{1}{n-1}$.

3. Recursion relation is $a_{n+2} = \frac{1}{4} a_n$. With $a_0 = 1$ and $a_1 = 0$ we have the solution
$y_1 = 1 + \tfrac{1}{4} x^2 + \tfrac{1}{4^2} x^4 + \cdots = \sum_{n=0}^{\infty} \left(\dfrac{x}{2}\right)^{2n}$.
Using the geometric series formula, $y_1 = 4/(4-x^2)$. The choice $a_0 = 0$, $a_1 = 1$ gives a second linearly independent solution $y_2 = \sum_{k=0}^{\infty} \dfrac{1}{4^k} x^{2k+1} = 4x/(4-x^2)$. The general solution is $a_0 y_1 + a_1 y_2$ for arbitrary a_0 and a_1. The minimum guaranteed radius of convergence of our series solutions is 2.

5. Recursion relation is $a_{n+2} = -\dfrac{n+3}{9(n+1)} a_n$. With $a_0 = 1$ and $a_1 = 0$ we have the solution
$y_1 = \sum_{k=0}^{\infty} (-1)^k \dfrac{2k+1}{9^k} x^{2k}$. The choice $a_0 = 0$, $a_1 = 1$ gives a second linearly independent solution $y_2 = \sum_{k=0}^{\infty} (-1)^k \dfrac{k+1}{9^k} x^{2k+1}$. The general solution is $a_0 y_1 + a_1 y_2$ for arbitrary a_0 and a_1. The minimum guaranteed radius of convergence of our series solutions is 3.

7. (a) $\sum_{k=0}^{\infty} (-1)^k \dfrac{1}{(2k+1)!} x^{2k+1}$
   (b) Converges on $(-\infty, \infty)$
   (c) $\sum_{k=0}^{\infty} (-1)^k \dfrac{1}{(2k)!} x^{2k}$; converges on $(-\infty, \infty)$

9. $y_1(x) = e^{-x^2/2}$

11. (a) $y_2(x) = \sum_{k=0}^{\infty} a_{2k+1} x^{2k+1}$ where $a_{2k+1} = \dfrac{k!}{3 \cdot 5 \cdot 7 \cdots (2k+1)}$
    (b) $y = 2 y_1 + 3 y_2$ where y_1 is the solution of $(x^2 - 2) y'' + 3 x y' + y = 0$ obtained in Example 7.3.3, and y_2 is from (a).

13. $y = 1 - x - \tfrac{1}{2} x^2 - \tfrac{1}{6} x^3 - \tfrac{1}{8} x^4 - \tfrac{1}{24} x^5 - \cdots$

15. (a) Radius of convergence is at least $\sqrt{3}$
    (b) Radius of convergence is at least 2

19. (b) $u(t) = 1 - \tfrac{1}{4} t^2 + \tfrac{1}{12} t^3 - \tfrac{1}{48} t^4 + \cdots$

23. (a) $(-3, 3)$, (b) $(1 - \sqrt{10}, 1 + \sqrt{10})$

25. (b) $a_2 = 0$, $a_3 = -\tfrac{1}{6}$, $a_4 = \tfrac{1}{24}$, $a_5 = -\tfrac{1}{120}$

27. (c) $y = 1 - \tfrac{1}{2} x^2 - \tfrac{1}{6} x^3 - \tfrac{1}{8} x^4 - \cdots$

29. (b) $y = \tfrac{1}{2} - \tfrac{19}{4} t^2 + \tfrac{19}{12} t^3 + \tfrac{95}{12} t^4 + \cdots$

31. (c) $H_2(x) = 4x^2 - 2$, $H_3(x) = 8x^3 - 12x$

Section 7.4

1. $a_5 - \dfrac{a_3}{3!} + \dfrac{a_1}{5!}$

4. $a_0 = 1$, $a_1 = 0$, $a_2 = \tfrac{1}{2}$, $a_3 = -\tfrac{1}{6}$, $a_4 = \tfrac{1}{12}$, $a_5 = -\tfrac{1}{60}$

7. $1 + x - \tfrac{3}{2} x^2 + \tfrac{1}{6} x^3 + \tfrac{2}{9} x^4 + \cdots$

9. (a) $y = 1 - \dfrac{k}{2m} x^2 - \dfrac{k\epsilon}{6m} x^3 + \left(\dfrac{k^2}{24 m^2} - \dfrac{k\epsilon^2}{24m} \right) x^4 + \cdots$

11. $y = \tfrac{1}{2} x^2 + \tfrac{1}{6} x^3 - \tfrac{7}{24} x^4 - \tfrac{3}{40} x^5 + \cdots$

Selected Answers 849

Section 7.5

1. $y = c_1 x^2 + c_2 x^2 \ln x$
3. $y = c_1 \cos(\ln x) + c_2 \sin(\ln x)$
5. $y = c_1 x^{-1} \cos(2 \ln x) + c_2 x^{-1} \sin(2 \ln x)$
7. $y = c_1 x^{5/3} + c_2 x^{1/3}$
10. (a) $y = c_1 x^3 + c_2 x^{-2}$
    (b) A particular solution is $y = -x^2$.
13. (a) 0 is a regular singular point.
    (b) 0 is an irregular singular point.
    (c) 1 and -1 are regular singular points.
15. $y(x) = c_1(x-4)\cos(2\ln(x-4)) + c_2(x-4)\sin(2\ln(x-4))$ in either case.

17. $y = c_1(x+1) + c_2(x+1)^{-1}$
21. (b) $c_8 = \frac{(3)(10)}{11!} c_0$ and $c_{10} = \frac{(3)(12)}{13!} c_0$
25. (a) The indicial equation is $r(r-1)+r-1 = 0$, with roots $r_1 = 1$ and $r_2 = -1$.
27. (a) The indicial equation is $r^2 - 1 = 0$ with roots $r_1 = 1$ and $r_2 = -1$.
29. (b) $y_1(x) = x^{\frac{1}{2}} \sum_{n=0}^{\infty} \frac{1}{(2n+1)!} x^n$ and $y_2(x) = \sum_{n=0}^{\infty} \frac{1}{(2n)!} x^n$
33. (c) $(x+1)^2 \left(1 + \sum_{k=1}^{\infty} \frac{3(2k+2)}{(2k+3)!}(x+1)^{2k}\right)$ and $\frac{1}{(x+1)^{-1}}\left(1 - \frac{(x+1)^2}{2} - \frac{(x+1)^4}{8} - \frac{(x+1)^6}{144} - \cdots\right)$

Section 7.6

2. $y = c_1 x^{-\frac{1}{2}} \cos x + c_2 x^{-\frac{1}{2}} \sin x$
7. (b) $\Gamma(\frac{3}{2}) = \frac{1}{2}\Gamma(\frac{1}{2}) = \frac{\sqrt{\pi}}{2}$, $\Gamma(\frac{5}{2}) = \frac{3\sqrt{\pi}}{4}$, $\Gamma(\frac{7}{2}) = \frac{15\sqrt{\pi}}{8}$
   (c) $\Gamma(-\frac{3}{2}) = \frac{4}{3}\sqrt{\pi}$, $\Gamma(-\frac{5}{2}) = -\frac{8}{15}\sqrt{\pi}$
9. (b) $d_{2k+1} = 0$

(c) $d_4 = -\frac{1}{8}d_0 = -\frac{d_0}{4(2!)}$, $d_6 = \frac{1}{144}d_0 = \frac{d_0}{6(4!)}$, $d_8 = -\frac{d_0}{8(6!)}$

13. (d) $k = (\gamma_j/b)^2$, natural mode frequencies are $\frac{\gamma_j \alpha}{2\pi b}$ for positive integer j
    (e) $f_j/f_k = \gamma_j/\gamma_k$

Section 8.1

1. (a) $\frac{dy}{dx} = \frac{dy/dt}{dx/dt} = \frac{-bx}{-axy} = \frac{b}{ay}$, with solutions $\frac{a}{2}y^2 = bx + c$. These are equations of parabolas which open to the right and have vertex at $(-k, 0)$, where $k = c/b$.
   (c) If $y_0^2 - \frac{2b}{a}x_0 < 0$, the x-force wins; if $y_0^2 - \frac{2b}{a}x_0 > 0$, the y-force wins, and if $y_0^2 - \frac{2b}{a}x_0 = 0$, it is a tie.
5. (a) $(0,0)$ is the only equilibrium point.
   (b) The points $(k, -k)$ for any constant k are the equilibrium points.
7. (a) The orbits are hyperbolas. If the x-force is the losing side, we have $ay_0^2 - bx_0^2 > 0$.
   (b) Double the initial force
9. $q(t)$ is the predator population, and $p(t)$ is the prey population.
13. (a) $y_1' = y_2$, $y_2' = y_3$, $y_3' = 2y_1 - 3y_2 - 2y_3$ with initial conditions $y_1(0) = 1$, $y_2(0) = -2$, and $y_3(0) = 5$.
    (b) The system is linear and homogeneous.

17. $y'' - 4y' + 3y = 0$, which has general solution $y(t) = c_1 e^{3t} + c_2 e^t$. Then $x(t) = \frac{dy}{dt} - 2y = 3c_1 e^{3t} + c_2 e^t - 2(c_1 e^{3t} + c_2 e^t) = c_1 e^{3t} - c_2 e^t$.
21. (a) $\frac{dx}{dt} = -\frac{1}{5}x + \frac{2}{5}y$, $\frac{dy}{dt} = \frac{1}{5}x - \frac{2}{5}y$ with initial conditions $x(0) = 15$, $y(0) = 0$.
    (b) $x(t) = 10 + 5e^{-\frac{3}{5}t}$, $y(t) = 5 - 5e^{-\frac{3}{5}t}$
    (c) As $t \to \infty$, $x(t) \to 10$ lbs and $y(t) \to 5$ lbs.
22. (a) $\frac{dS}{dt} = -\alpha SI$, $\frac{dI}{dt} = \alpha SI - \beta I$, $\frac{dR}{dt} = \beta I$ for positive constants α and β
    (c) $\frac{dI}{dS} = \frac{(\alpha S - \beta)I}{-\alpha SI} = -1 + \frac{\beta}{\alpha}\frac{1}{S}$; $I = -S + \frac{\beta}{\alpha}\ln S + (I_0 + S_0 - \frac{\beta}{\alpha}\ln S_0)$
    (d) No one is left in the infective compartment.
    (e) $S = \frac{\beta}{\alpha}$. This value of S gives the maximum value of I by, for example, the Second Derivative Test.
24. (a) $y_1' = y_2$, $y_2' = -\frac{k_1 + k_2}{m_1}y_1 + \frac{k_2}{m_1}y_3$, $y_3' = y_4$, $y_4' = \frac{k_2}{m_2}y_1 - \frac{k_2}{m_2}y_3$
    (c) $u_1(t) = -\frac{2}{3}\cos(2t) + \frac{2}{3}\cos t$ and $u_2(t) = \frac{2}{3}\cos(2t) + \frac{4}{3}\cos t$

Section 8.2

1. (a)
$$\begin{pmatrix} 4 & -1 & 5 \\ 10 & -2 & 6 \\ -6 & -3 & -2 \end{pmatrix}$$

(b)
$$\begin{pmatrix} 13 & 7 & 12 \\ -18 & -12 & -6 \\ 13 & 2 & -5 \end{pmatrix}$$

3. (a)
$$\mathbf{A}^{-1} = -\frac{1}{14}\begin{pmatrix} -2 & -4 \\ -3 & 1 \end{pmatrix} = \begin{pmatrix} \frac{1}{7} & \frac{2}{7} \\ \frac{3}{14} & -\frac{1}{14} \end{pmatrix}.$$

(b) Since the determinant of the given matrix is 0, the matrix is not invertible.

5. (a) $\begin{pmatrix} -9 \\ 5 \end{pmatrix}$

(b) $c_1 = 2$, $c_2 = \frac{3}{2}$, $\begin{pmatrix} \frac{1}{2} \\ 12 \end{pmatrix}$

9. Invertible for all t except $t = 0$ and $t = -\frac{2}{3}$. The inverse matrix for any other value of t is

$$\begin{pmatrix} 3t^3/(3t^4 + 2t^3) & -t^2/(3t^4 + 2t^3) \\ 2t/(3t^4 + 2t^3) & t/(3t^4 + 2t^3) \end{pmatrix}.$$

Section 8.3

1. One repeated eigenvalue 2 with eigenvector $\begin{pmatrix} 1 \\ -1 \end{pmatrix}$ or any nonzero multiple

3. Eigenvalues $z = 1 + 2i$ and $z = 1 - 2i$ with eigenvectors $\begin{pmatrix} 1 \\ 1-i \end{pmatrix}$ (or any nonzero multiple) and $\begin{pmatrix} 1 \\ 1+i \end{pmatrix}$, respectively

5. Eigenvalues $z = 2$ and $z = 0$ with eigenvectors $\begin{pmatrix} i \\ 1 \end{pmatrix}$ (or any nonzero multiple) and $\begin{pmatrix} -i \\ 1 \end{pmatrix}$, respectively.

11. (a) Rotation is through an angle $\theta = 0, \pm 2\pi, \pm 4\pi, \ldots$ (and every vector rotates to itself) or through an angle $\theta = \pm\pi, \pm 3\pi, \pm 5\pi$ (and every vector rotates to its negative). In the first case the eigenvalue is 1, and in the second case it is -1.

17. (a) The $n \times n$ identity matrix $\mathbf{I}$ has determinant 1. The $n \times n$ matrix $s\mathbf{I}$ has determinant s^n.

19. (b) Yes

23. -126

Section 8.4

1. $\mathbf{X}(t) = c_1 e^{3t}\begin{pmatrix} 1 \\ 1 \end{pmatrix} + c_2 e^{-t}\begin{pmatrix} -1 \\ 1 \end{pmatrix}$, where c_1 and c_2 are arbitrary constants.

3. $\mathbf{X}(t) = c_1 e^{-t}\begin{pmatrix} 1 \\ 0 \end{pmatrix} + c_2 e^{t}\begin{pmatrix} -3 \\ 2 \end{pmatrix}$, where c_1 and c_2 are arbitrary constants.

5. $\mathbf{X}(t) = c_1 e^{4t}\begin{pmatrix} 1 \\ 3 \end{pmatrix} + c_2 e^{-2t}\begin{pmatrix} -1 \\ 3 \end{pmatrix}$, where c_1 and c_2 are arbitrary constants.

9. $c_1 e^{5t}\begin{pmatrix} -1 \\ -6 \\ 1 \end{pmatrix} + c_2 e^{-3t}\begin{pmatrix} -1 \\ 2 \\ 1 \end{pmatrix} + c_3\begin{pmatrix} 4 \\ 4 \\ 1 \end{pmatrix}$, for arbitrary constants c_1, c_2, and c_3

11. $x(t) = -e^t$, $y(t) = 3e^t + 3e^{2t}$

24. $x = c_1 e^t + 2c_2 e^{-t} + 2$, $y = c_1 e^t + c_2 e^{-t} - 1$

28. $c_1 e^{-5t} \begin{pmatrix} 1 \\ -1 \end{pmatrix} + c_2 e^t \begin{pmatrix} 1 \\ 1 \end{pmatrix} + \begin{pmatrix} 1 \\ -2 \end{pmatrix}$

29. (a) $r = 3$ and $r = -2$ with $b = -\frac{r}{2}$

(b) $\begin{pmatrix} x(t) \\ y(t) \end{pmatrix} = c_1 \begin{pmatrix} t^3 \\ -\frac{3}{2}t^3 \end{pmatrix} + c_2 \begin{pmatrix} t^{-2} \\ t^{-2} \end{pmatrix}$,

for arbitrary constants c_1 and c_2

Section 8.5

1. (b) $(1, 1)$

3. (a) Figure (A)
   (b) The line $y = -x$

5. (B) corresponds to one positive and one negative eigenvalue, (C) to two positive eigenvalues, and (A) to two negative eigenvalues.

7. (a) Two negative eigenvalues
   (b) The lines $y = 2x$ and $y = -3x$ are the nodal tangents and the preferred nodal tangent is $y = 2x$.

Section 8.6

1. (a) $x(t) = -\frac{4}{7}e^{5t} - \frac{3}{7}e^{-2t}$, $y(t) = -\frac{2}{7}e^{5t} + \frac{9}{7}e^{-2t}$
   (c) Does not cross the y-nullcline $y = 3x$ but it does cross the x-nullcline $y = -2x$ at approximately $(-0.846, 1.692)$

3. (a) The x-nullcline: $y = 11x$; y-nullcline: $x = 3y$
   (c) Eigenvectors

   $\begin{pmatrix} 1 \\ -1 \end{pmatrix}$ and $\begin{pmatrix} 1 \\ 3 \end{pmatrix}$

   with eigenvalues 3 and 2, respectively. General solution:

   $c_1 e^{3t} \begin{pmatrix} 1 \\ -1 \end{pmatrix} + c_2 e^{2t} \begin{pmatrix} 1 \\ 3 \end{pmatrix}$

   (d) $x = e^{2t} + 2e^{3t}$, $y = 3e^{2t} - 2e^{3t}$. Tangent vectors on trajectory are nearly parallel to the vector $\begin{pmatrix} 1 \\ -1 \end{pmatrix}$ as $t \to \infty$. The trajectory approaches the origin tangent to the line $y = 3x$.

5. (b) $y = x/2$ and $y = -3x$. Two ray solutions of the system are

   $e^{5t} \begin{pmatrix} 2 \\ 1 \end{pmatrix}$ and $e^{-2t} \begin{pmatrix} -1 \\ 3 \end{pmatrix}$,

   which lie along the lines $y = x/2$ and $y = -3x$.

7. (a) (A): both eigenvalues positive; (B): both eigenvalues negative; (C): one eigenvalue positive, one negative.

11. (b) $x(t) = \frac{1}{7}e^{-4t} + \frac{6}{7}e^{3t}$, $y = \frac{-1}{7}e^{-4t} + \frac{1}{7}e^{3t}$
    (c) Slant asymptotes: $y = -x$ and $y = \frac{1}{6}x$. Points on the trajectory along the vertical line $x = 48$ lie close to the points where the line $x = 48$ intersects the slant asymptotes, namely $(48, 8)$ and $(48, -48)$. The actual y-coordinates should be a little less than 8 for the first trajectory point and a little larger than -48 for the second.
    (d) Solving numerically $\frac{1}{7}e^{-4t} + \frac{6}{7}e^{3t} = 48$, we find $t \approx 1.34178$ and $t \approx -1.45422$ and $y(1.34178) \approx 7.99924$ and $y(-1.45222) \approx -47.9871$, consistent with the estimates of (c).

13. (c) The equilibrium points are points of the form $c(1, 0)$ for c any constant.

Section 8.7

1. $\mathbf{A} = \begin{pmatrix} 1 & -1 \\ 5 & -3 \end{pmatrix}$ and $\mathbf{X} = \begin{pmatrix} x \\ y \end{pmatrix}$.
The general real-valued solution is
$c_1 e^{-t} \begin{pmatrix} \cos t \\ 2\cos t + \sin t \end{pmatrix} + c_2 e^{-t} \begin{pmatrix} \sin t \\ 2\sin t - \cos t \end{pmatrix}$.

3. $\mathbf{A} = \begin{pmatrix} 1 & -2 \\ 2 & 1 \end{pmatrix}$ and $\mathbf{X} = \begin{pmatrix} x \\ y \end{pmatrix}$.
The general real-valued solution is
$c_1 e^t \begin{pmatrix} -\sin(2t) \\ \cos(2t) \end{pmatrix} + c_2 e^t \begin{pmatrix} \cos(2t) \\ \sin(2t) \end{pmatrix}$.

5. $c_1 e^t \begin{pmatrix} \cos t \\ -\sin t \end{pmatrix} + c_2 e^t \begin{pmatrix} \sin t \\ \cos t \end{pmatrix}$

7. $x(t) = 3e^{-3t} + 4te^{-3t}$, $y(t) = 2e^{-3t} + 4te^{-3t}$

13. (a) In the first quadrant, x and y are both positive, so $2x+5y$ will be positive also. Thus $\frac{dx}{dt}$ is positive in the first quadrant.
(b) Since $-x - 2y$ will be negative at every point in the first quadrant, $\frac{dy}{dt}$ will be negative there.
(c) In the first quadrant, an orbit will be moving right and moving down. The closed loop orbits must be clockwise.

20. (a) $v_1(t) = c_1 \cos(\alpha t) + c_2 \sin(\alpha t)$, $v_2(t) = -c_1 \sin(\alpha t) + c_2 \cos(\alpha t)$
(b) $x(t) = \frac{c_1}{\alpha} \sin(\alpha t) - \frac{c_2}{\alpha} \cos(\alpha t) + k_1$, $y(t) = \frac{c_1}{\alpha} \cos(\alpha t) + \frac{c_2}{\alpha} \sin(\alpha t) + k_2$. Notice that $(x - k_1)^2 + (y - k_2)^2 = \left(\frac{c_1}{\alpha}\right)^2 + \left(\frac{c_2}{\alpha}\right)^2$, the equation of a circle.

Section 8.8

1. $x(t) = 5 + 5e^{-\frac{1}{5}t}$, $y(t) = 5 - 5e^{-\frac{1}{5}t}$

3. $a_{31} = \frac{1}{2}$, $a_{12} = 2$, $a_{21} = 1$, and all other $a_{ij} = 0$

4. (a) $c_1 \begin{pmatrix} 1 \\ 2 \\ \frac{9}{2} \end{pmatrix} + c_2 e^{-10t} \begin{pmatrix} -2 \\ 1 \\ 1 \end{pmatrix} + c_3 e^{-6t} \begin{pmatrix} 2 \\ 1 \\ -3 \end{pmatrix}$

(b) $\begin{pmatrix} 1 \\ 2 \\ \frac{9}{2} \end{pmatrix}$

7. (a) $\frac{dy_1}{dt} = -(a_{21}+a_{31})y_1 + a_{12}y_2$, $\frac{dy_2}{dt} = a_{21}y_1 - (a_{12}+a_{32})y_2$, $\frac{dy_3}{dt} = a_{31}y_1 + a_{32}y_2 - a_{43}y_3$, $\frac{dy_4}{dt} = a_{43}y_3$

11. (a) Eigenvalues $-a_{21}$, $-a_{32}$, and 0. Except for 0, these are the same as the eigenvalues of the given 2×2 matrix.

(b) Equilibrium solutions: $x = 0$ and $y = 0$, z arbitrary

13. (c) Equilibrium solutions $x = y = z = c$ for any constant c

15. (a) $x_S(0) = 3 \times 10^7$, $x_M(0) = 17.5 \times 10^6$, $x_H(0) = 7 \times 10^6$, $x_E(0) = 6.345 \times 10^6$, $x_O(0) = 6.4 \times 10^6$
(b) $c_1 = 3 \times 10^7$, $c_2 = 21.6 \times 10^6$, $c_3 = -7.587 \times 10^6$, $c_4 = 10.3 \times 10^6$, $c_5 = 13.2 \times 10^6$
(d) only Lake Erie

17. For the number of adults $a(t)$ and children $c(t)$ at time t: $\frac{da}{dt} = -\delta_1 a(t) + gc(t)$, $\frac{dc}{dt} = ba(t) - (g + \delta_2)c(t)$

19. (b) $u' = -0.2u$, $v' = 0.2u - 0.05v$ with initial condition $u(0) = -5I$, $v(0) = -20I$

21. $x(t) = -\frac{1}{2}t^2$, $y(t) = t + \frac{1}{2}t^2$

Section 8.9

1. $\Delta = 3$, $\tau = -4$, and $\tau^2 - 4\Delta = 4$; $(0,0)$ is a stable node.

3. $\Delta = 4$, $\tau = 4$, and $\tau^2 - 4\Delta = 0$; $(0,0)$ is an unstable degenerate node (improper).

5. $\Delta = 32$, $\tau = 12$, and $\tau^2 - 4\Delta = 16 > 0$; $(0,0)$ is an unstable node.

8. Stable node when $0 < c < 1$, saddle point when $c < 0$, stable spiral point when $c > 1$

Selected Answers

11. (a) In case (i) $(0,0)$ is a stable node ($b \neq 0$) or a stable degenerate (proper) node ($b = 0$). In cases (ii) and (iii) $(0,0)$ is a saddle point. In case (iv) $(0,0)$ is an unstable node ($b \neq 0$) or an unstable degenerate (proper) node ($b = 0$).
    (b) For a stable node or stable degenerate (proper) node (case (i)), all orbits in the phase portrait tend to the origin as $t \to \infty$. This corresponds to Romeo and Juliet ending up in a state of mutual indifference.
    (c) $a + b$ and $a - b$, respectively
    (d) Mutual love

13. (a) $-\frac{a_2}{2}x^2 + a_1 xy + \frac{b_1}{2}y^2 = c$ for any constant c.

Section 9.1

1. $e^{t\mathbf{A}} = \begin{pmatrix} \cos t & \sin t \\ -\sin t & \cos t \end{pmatrix}$,

   $\mathbf{X}(t) = \begin{pmatrix} \cos t + \sin t \\ -\sin t + \cos t \end{pmatrix}$

3. $e^{t\mathbf{A}} = e^t \mathbf{M}$ where $\mathbf{M} =$

   $\begin{pmatrix} \cos(3t) - \frac{1}{3}\sin(3t) & -\frac{2}{3}\sin(3t) \\ \frac{5}{3}\sin(3t) & \cos(3t) + \frac{1}{3}\sin(3t) \end{pmatrix}$,

   $\mathbf{X}(t) = e^t \begin{pmatrix} -\frac{2}{3}\sin(3t) \\ \cos(3t) + \frac{1}{3}\sin(3t) \end{pmatrix}$

5. $e^{t\mathbf{A}} = e^{-\frac{3}{2}t}\mathbf{M}$ where $\mathbf{M} =$

   $\begin{pmatrix} \cos\frac{\sqrt{7}t}{2} + \frac{1}{\sqrt{7}}\sin\frac{\sqrt{7}t}{2} & \frac{2}{\sqrt{7}}\sin\frac{\sqrt{7}t}{2} \\ -\frac{4}{\sqrt{7}}\sin\frac{\sqrt{7}t}{2} & \cos\frac{\sqrt{7}t}{2} - \frac{\sqrt{7}}{7}\sin\frac{\sqrt{7}t}{2} \end{pmatrix}$,

   $\mathbf{X}(t) = e^{-\frac{3}{2}t}\begin{pmatrix} \frac{2}{\sqrt{7}}\sin\frac{\sqrt{7}t}{2} \\ \cos\frac{\sqrt{7}t}{2} - \frac{\sqrt{7}}{7}\sin\frac{\sqrt{7}t}{2} \end{pmatrix}$

7. $e^{t\mathbf{A}} = \begin{pmatrix} \cos(2t) + \sin(2t) & -\frac{1}{2}\sin(2t) \\ 4\sin(2t) & \cos(2t) - \sin(2t) \end{pmatrix}$,

   $\mathbf{X}(t) = \begin{pmatrix} \cos(2t) + \frac{1}{2}\sin(2t) \\ 3\sin(2t) + \cos(2t) \end{pmatrix}$

9. $\sqrt{2}$

13. (a) $\begin{pmatrix} 1 & t & \frac{t^2}{2} \\ 0 & 1 & t \\ 0 & 0 & 1 \end{pmatrix}$

14. (a) $\mathbf{A}^2 = \begin{pmatrix} -1 & 0 \\ 0 & -1 \end{pmatrix}$, $\mathbf{A}^3 = \begin{pmatrix} 0 & 1 \\ -1 & 0 \end{pmatrix}$

    and $\mathbf{A}^4 = \begin{pmatrix} 1 & 0 \\ 0 & 1 \end{pmatrix}$ Since $\mathbf{A}^4 = \mathbf{I}$, we have $\mathbf{A}^5 = \mathbf{A}$, $\mathbf{A}^6 = \mathbf{A}^2$, $\mathbf{A}^7 = \mathbf{A}^3$, $\mathbf{A}^8 = \mathbf{A}^4 = \mathbf{I}$, $\mathbf{A}^9 = \mathbf{A}$ and the cycle repeats.

    (b) $\begin{pmatrix} \cos(bt) & -\sin(bt) \\ \sin(bt) & \cos(bt) \end{pmatrix}$

    (c) $e^{at}\begin{pmatrix} \cos(bt) & -\sin(bt) \\ \sin(bt) & \cos(bt) \end{pmatrix}$

15. $\mathbf{AB} = \begin{pmatrix} 2 & -2 \\ 0 & 0 \end{pmatrix}$, $\mathbf{BA} = \begin{pmatrix} 2 & 2 \\ 0 & 0 \end{pmatrix}$,

    $e^{\mathbf{A}}e^{\mathbf{B}} = \begin{pmatrix} e^3 & 2e^2 - e^3 - 1 \\ 0 & 1 \end{pmatrix}$, $e^{\mathbf{B}}e^{\mathbf{A}} = \begin{pmatrix} e^3 & e^3 - 2e + 1 \\ 0 & 1 \end{pmatrix}$, and $e^{\mathbf{A}+\mathbf{B}} = \begin{pmatrix} e^3 & (e^3 - 1)/3 \\ 0 & 1 \end{pmatrix}$.

17. No, correct formula is $\frac{d}{dt}(\mathbf{A}(t)^2) = \frac{d}{dt}(\mathbf{A}(t)\mathbf{A}(t)) = \mathbf{A}(t)\frac{d}{dt}(\mathbf{A}(t)) + \frac{d}{dt}(\mathbf{A}(t))\mathbf{A}(t)$

19. $\frac{d}{dt}(e^{t\mathbf{A}}) = \begin{pmatrix} e^t & 0 \\ e^{-t} + e^t & -e^{-t} \end{pmatrix}$. Evaluating this at $t = 0$ gives $\mathbf{A}$, so $\mathbf{A} = \begin{pmatrix} 1 & 0 \\ 2 & -1 \end{pmatrix}$.

Section 9.2

1. (a) Eigenvalues ± 1 with eigenvectors $\begin{pmatrix} 1 \\ 1 \end{pmatrix}$ and $\begin{pmatrix} -1 \\ 1 \end{pmatrix}$, respectively. Fundamental matrix $\mathbf{Y}(t) = \begin{pmatrix} e^t & -e^{-t} \\ e^t & e^{-t} \end{pmatrix}$

3. $\begin{pmatrix} \frac{1}{2}e^{-t} + \frac{1}{2}e^{5t} & 0 & -\frac{3}{2}e^{-t} + \frac{3}{2}e^{5t} \\ 0 & e^{3t} & 0 \\ -\frac{1}{6}e^{-t} + \frac{1}{6}e^{5t} & 0 & \frac{1}{2}e^{-t} + \frac{1}{2}e^{5t} \end{pmatrix}$

5. Matrix is not invertible for every t.

6. $e^{t\mathbf{A}} = e^{-t}\begin{pmatrix} 1 - 3t & -t \\ 9t & 1 + 3t \end{pmatrix}$. The general solution of $\mathbf{X}' = \mathbf{A}\mathbf{X}$ is

$e^{t\mathbf{A}}\begin{pmatrix} c_1 \\ c_2 \end{pmatrix} = c_1 e^{-t}\begin{pmatrix} 1 - 3t \\ 9t \end{pmatrix} + c_2 e^{-t}\begin{pmatrix} -t \\ 1 + 3t \end{pmatrix}$.

Section 9.3

1. $\begin{pmatrix} e^t & 0 & 0 \\ 0 & e^{2t} & 0 \\ 0 & 2te^{2t} & e^{2t} \end{pmatrix}$, $\mathbf{X}(t) = \begin{pmatrix} e^t \\ e^{2t} \\ 2te^{2t} \end{pmatrix}$

3. $\mathbf{X}(t) = \begin{pmatrix} e^t \\ (2te^t + e^{-t} + 3e^t)/4 \\ (2te^t + e^t - e^{-t})/4 \end{pmatrix}$

7. $x(t) = \frac{16}{3}e^{2t} - 6e^t + \frac{5}{3}e^{-t}$, $y(t) = \frac{16}{3}e^{2t} - 4e^t + \frac{2}{3}e^{-t}$, and $z(t) = \frac{16}{3}e^{2t} - 6e^t - \frac{1}{3}e^{-t}$

11. (a) $e^{t\mathbf{A}} = \begin{pmatrix} 1 & t & 2t + t^2/2 \\ 0 & 1 & t \\ 0 & 0 & 1 \end{pmatrix}$

(b) $\mathbf{A}^2 = \begin{pmatrix} 0 & 0 & ac \\ 0 & 0 & 0 \\ 0 & 0 & 0 \end{pmatrix}$ and for $n \geq 3$

$\mathbf{A}^n = \begin{pmatrix} 0 & 0 & 0 \\ 0 & 0 & 0 \\ 0 & 0 & 0 \end{pmatrix}$. Thus

$e^{t\mathbf{A}} = \mathbf{I} + t\mathbf{A} + \frac{t^2}{2!}\mathbf{A}^2 = \begin{pmatrix} 1 & at & bt + \frac{ac}{2}t^2 \\ 0 & 1 & ct \\ 0 & 0 & 1 \end{pmatrix}$.

Section 9.4

1. $\mathbf{X}(t) = \begin{pmatrix} 1 + t - \cos t \\ 1 - t + \sin t \end{pmatrix}$

2. (a) $\begin{pmatrix} \frac{t^2}{2}\cos(2t) \\ \frac{t^2}{2}\sin(2t) \end{pmatrix}$

3. $x(t) = -\frac{15}{35}e^{-2t} - \frac{21}{35} + \frac{36}{35}e^{5t}$

5. $x(t) = \frac{1}{15}e^{4t} + \frac{9}{15}e^{-t} - \frac{2}{3}e^{-2t}$

7. $\begin{pmatrix} -te^{-t} \\ te^{-t} \end{pmatrix}$

9. $\begin{pmatrix} -e^{-t} - 2e^{2t} + 2e^{3t} \\ e^{-t} + e^{2t} - 2e^{3t} \end{pmatrix}$

10. (a) $x = 2$, $y = -\frac{5}{2}$

Section 9.5

3. If $\mathbf{AB} = \mathbf{BA}$ where $\mathbf{B}$ is invertible, then $\mathbf{ABB}^{-1} = \mathbf{BAB}^{-1}$. This says $\mathbf{A} = \mathbf{BAB}^{-1}$. Multiplying in the last line by $\mathbf{B}^{-1}$ (on the left!) we obtain $\mathbf{B}^{-1}\mathbf{A} = \mathbf{B}^{-1}\mathbf{BAB}^{-1}$ which simplifies to the desired conclusion.

5. (a) $-\frac{2}{10}\cos(2t)\begin{pmatrix} 6 \\ 3 \end{pmatrix} - \frac{1}{10}\sin(2t)\begin{pmatrix} -9 \\ 3 \end{pmatrix}$

   (b) $-\frac{1}{10}\cos(2t)\begin{pmatrix} 1 \\ 3 \end{pmatrix} + \frac{2}{10}\sin(2t)\begin{pmatrix} 1 \\ -2 \end{pmatrix}$

6. (b) Eigenvalues of $\mathbf{A}$ are $\pm 4i$ and these are integer multiples of $2\pi i/T$, where $T = \pi$ is the period of the forcing term $\cos(2t)$.
   (c) There is one periodic solution, with period π.
   (d) No periodic solutions.

Section 10.1

1. Equilibrium points are $(0,0)$, $(0,2)$, $(5,0)$, and $(1,1)$. Long term, we expect one population to be driven to extinction (which one depends on the initial conditions). Rarely, there is peaceful coexistence at $(1,1)$.

3. Equilibrium points are $(0,0)$, $(0,7/3)$, $(5/2,0)$, and $(13/10, 3/5)$ Long term, we expect one population or the other, depending on initial conditions, to be driven to extinction. Rarely, there is peaceful coexistence at $(13/10, 3/5)$.

5. (a) $(0,0)$ and $(1,1)$

7. (a) $x = Ke^{-at^2/2}$
   (b) $y = -Ke^{-at^2/2} + K$

(c) $t = \sqrt{36\ln 10/\ln 1.5} \approx 14.3$ years

11. (a) $x(t) = \frac{100}{49e^{-5t}+1}$. As $t \to \infty$ this has limit 100 and as $t \to -\infty$, the limit is 0. As t increases from $-\infty$ to ∞, x increases from 0 to 100.
    (b) $x(t) = \frac{200e^{5t}}{2e^{5t}-1}$. As $t \to \infty$, x decreases to 100. As $t \to -\frac{\ln 2}{5}$ from the right, the denominator $2e^{5t} - 1$ tends to 0 from above. As t increases from $-\frac{\ln 2}{5}$ to ∞, x decreases from ∞ to 100

13. (b) $\mathbf{Y}_1(t) = \mathbf{X}(t-\frac{1}{3}\ln 3)$ and $\mathbf{Y}_3(t) = \mathbf{X}(t-1)$ are time translates of $\mathbf{X}$; $\mathbf{Y}_2(t)$ is not.
    (c) $\mathbf{Y}_2$ cannot be a solution.
    (d) $(0,0)$ must be an equilibrium point

Section 10.2

1. Equilibrium points $(0,0)$ and $(1,1)$. At $(0,0)$ the Jacobian matrix is

$$\mathbf{J}_1 = \begin{pmatrix} 1 & 0 \\ 1 & -1 \end{pmatrix},$$

a matrix with eigenvalues 1 and -1. Linear approximation system $\mathbf{X}' = \mathbf{J}_1\mathbf{X}$ has an (unstable) saddle at $(0,0)$, as does the original nonlinear system. Jacobian matrix at $(1,1)$ is

$$\mathbf{J}_2 = \begin{pmatrix} 1 & -2 \\ 1 & -1 \end{pmatrix}$$

with eigenvalues $\pm i$ so linear system $\mathbf{X}' = \mathbf{J}_2\mathbf{X}$ has a center at $(0,0)$. Theorem 10.2.3 does not apply.

3. Equilibrium points: $(0,0)$, $(\sqrt{10},0)$, $(-\sqrt{10},0)$, $(3,1)$, $(-3,-1)$, $(1,3)$, and $(-1,-3)$. The Jacobian matrix is

$$\begin{pmatrix} 3x^2 + y^2 - 10 & 2xy \\ y^2 & 2xy - 3 \end{pmatrix}.$$

Evaluating the Jacobian matrix at equilibrium points shows $(0,0)$ is a stable node and $(\pm\sqrt{10},0)$ are saddle points, $(3,1)$ and $(-3,-1)$ are unstable nodes, and $(1,3)$ and $(-1,-3)$ are saddle points.

5. One equilibrium point, at $(0,0)$. The Jacobian matrix evaluated at $(0,0)$ is

$$\mathbf{J} = \begin{pmatrix} 0 & 1 \\ -1 & 1 \end{pmatrix}.$$

Compute $\Delta = 1, \tau = 1, \tau^2 - 4\Delta < 0$, to see the eigenvalues of $\mathbf{J}$ must be complex conjugates with a positive real part. $(0,0)$ is an unstable spiral point for the linear approximation system and original nonlinear system.

7. One equilibrium point, at $(0,0)$. The Jacobian matrix evaluated at $(0,0)$ is

$$\mathbf{J} = \begin{pmatrix} 0 & 1 \\ -1 & 3 \end{pmatrix}.$$

Since $\Delta = 1, \tau = 3, \tau^2 - 4\Delta > 0$, the eigenvalues of **J** must be two positive real values. For the linear approximation system $\mathbf{X}' = \mathbf{JX}$, $(0,0)$ is an unstable node and $(0,0)$ is also an unstable node for the nonlinear system.

9. Equilibrium points $(0, n\pi)$ where n is any integer. The Jacobian matrix is
$$\begin{pmatrix} \cos(x+y) & \cos(x+y) \\ e^x & 0 \end{pmatrix}.$$
Evaluate this at $(0, n\pi)$ to see that when n is even, $(0, n\pi)$ is an (unstable) saddle point and when n is odd, $(0, n\pi)$ is a stable spiral point for the linear approximation and nonlinear systems.

11. Equilibrium point is $(0,0)$. Jacobian matrix evaluated at $(0,0)$ is
$$\mathbf{J} = \begin{pmatrix} 0 & 1 \\ -1 & 0 \end{pmatrix},$$
whose eigenvalues are $\pm i$. Theorem 10.2.3 does not apply. Theorem 10.2.4 gives no stability conclusion for $(0,0)$ and indicates that the point might be a center, spiral, or hybrid center/spiral.

13. Equilibrium point is $(1,1)$. Jacobian matrix evaluated at $(1,1)$ is
$$\mathbf{J} = \begin{pmatrix} -3 & 3 \\ 0 & -3 \end{pmatrix},$$
which has -3 as a repeated eigenvalue. Theorem 10.2.3 does not apply. Theorem 10.2.4 tells us that $(1,1)$ is asymptotically stable but does not identify the type of $(1,1)$; it might be a node or a spiral.

15. Equilibrium point is $(0,0)$. Jacobian matrix evaluated at $(0,0)$ is
$$\mathbf{J} = \begin{pmatrix} 1 & 1 \\ 0 & 1 \end{pmatrix},$$
which has 1 as a repeated eigenvalue. Theorem 10.2.3 does not apply. Theorem 10.2.4 tell us that $(0,0)$ is unstable but does not identify the type of $(0,0)$; it might be a node or a spiral.

19. (a) Equilibrium points: $(0,0)$, $(4,4)$, and $(1,1)$. Jacobian matrices at these points are, respectively, $\mathbf{J}_1 = \begin{pmatrix} -1 & 1 \\ 0 & -1 \end{pmatrix}$, $\mathbf{J}_2 = \begin{pmatrix} -1 & 1 \\ \frac{2}{5} & -1 \end{pmatrix}$, $\mathbf{J}_3 = \begin{pmatrix} -1 & 1 \\ \frac{8}{5} & -1 \end{pmatrix}$. The matrix $\mathbf{J}_1$ has one repeated eigenvalue, -1, and Theorem 10.2.3 does not apply to classify $(0,0)$.
For $\mathbf{J}_2$, we have $\Delta > 0$, $\tau < 0$, and $\tau^2 - 4\Delta > 0$, so by Theorem 10.2.3, $(4,4)$ is a stable node.
For $\mathbf{J}_3$, we have $\Delta < 0$, by Theorem 10.2.3 $(1,1)$ is a saddle point.

21. All points in the first quadrant that lie strictly to the right of the stable line S.

23. (c) Any point (x, y) in the plane with $y < -3$ cannot be in the basin of attraction of $(0,0)$.

25. (a) Nodal tangents for the nonlinear system are $y = -2x + 2$ and $x = 0$ with $y = -2x + 2$ the preferred nodal tangent.

27. (b) Jacobian at $(2, 1)$ is
$$\mathbf{J} = \begin{pmatrix} 0 & -2 \\ 1 & 0 \end{pmatrix},$$
which has eigenvalues $\pm i\sqrt{2}$. Theorem 10.2.3 does not apply; Theorem 10.2.4 gives no stability conclusion for $(2, 1)$ and the point might be a center, spiral, or hybrid center/spiral.

Section 10.3

1. $\frac{dy}{dx} = -\frac{x}{y}$, so that $\frac{y^2}{2} = -\frac{x^2}{2} + c$. The orbits are circles centered at the origin, and $(0,0)$ is a center.

3. The orbit starts outside the circle $x^2 + y^2 = 1$ for $t = 0$. As $t \to \infty$, $r(t) \to 1$, and the orbit spirals towards the circle $x^2 + y^2 = 1$.

5. (a) $(0,0)$ and $(1,0)$.

(b) At $(0,0)$, $\mathbf{J} = \begin{pmatrix} 1 & 0 \\ 0 & 1 \end{pmatrix}$, and the linear system $\mathbf{X}' = \mathbf{JX}$ has an unstable, degenerate, proper node (star point) at $(0,0)$.
(c) $y = C\frac{x}{1-x}$ for C a constant.
(d) True (note there are trajectories on the y-axis.
(e) We might classify $(0,0)$ as a proper or "star-like node". It is unstable.

7. (e) For $c > 0$ then if $0 < r(0) < \sqrt{c}$, r increases to $\sqrt{c}$ as $t \to \infty$. These orbits spin clockwise into the circle $x^2 + y^2 = c$. If $r(0) > \sqrt{c}$, then r decreases to $\sqrt{c}$ as $t \to \infty$. If $c < 0$ and $0 < r(0)$, then r decreases to 0 as $t \to \infty$. The orbits spin clockwise into the origin.
(f) Stable if $c < 0$ and unstable if $c > 0$.

9. (b) Evaluating the Jacobian matrix at $(0,0)$ gives $\mathbf{J}_1 = \begin{pmatrix} 0 & 0 \\ 0 & 0 \end{pmatrix}$. For the linear system $\mathbf{X}' = \mathbf{J}_1\mathbf{X}$, *every* point is an equilibrium point. In particular, $(0,0)$ is a nonisolated equilibrium point for the linear system.

Section 10.4

1. The first graph.

3. Average aphid population is increased by the use of the pesticide.

5. (a) y is the predator and x is the prey.
(b) Equilibrium point in open first quadrant is $(4,1)$. The Jacobian matrix at $(4,1)$ is $\begin{pmatrix} 1/2 & -4 \\ 1/4 & 0 \end{pmatrix}$. We have $\Delta = 1$, $\tau = \frac{1}{2}$, and $\tau^2 - 4\Delta < 0$, so the equilibrium point $(4,1)$ is an unstable spiral point.

7. (a) Equilibrium points are $(0,0)$, $(r/a, 0)$, $(0, s/d)$ (and a fourth point which is not in the first quadrant if $s/d > r/b$).
(b) Evaluating the Jacobian matrix at $(0,0)$, $(r/a, 0)$, and $(0, s/d)$ gives $\mathbf{J}_1 = \begin{pmatrix} r & 0 \\ 0 & s \end{pmatrix}$, $\mathbf{J}_2 = \begin{pmatrix} -r & -br/a \\ 0 & s + cr/a \end{pmatrix}$, and $\mathbf{J}_3 = \begin{pmatrix} r - sb/d & 0 \\ cs/d & -s \end{pmatrix}$. $(0,0)$: unstable node; $(r/a, 0)$: saddle point; $(0, s/d)$: stable node.
(c) x-population tends to 0 and y-population tends to s/d.

Section 10.5

1. (c) As $t \to \infty$, $S \to K$ and $I \to 0$.

3. At least $\approx 79.6\%$ of the population should be immunized.

5. (a) a_1 and a_2 have units of $1/$(people-day), b_1 and b_2 have units of $1/$day, and ρ_1 and ρ_2 have units of number of people.
(b)
$$y' = \frac{b_2 W a_2}{(b_2 + a_2 x)^2} > 0, \quad y'' = -\frac{2W b_2 a_2^2}{(b_2 + a_2 x)^3} < 0,$$
and y is increasing and concave down for all x's of interest. Moreover, $y \to W$ as $x \to \infty$, and $y \to -\infty$ as $x \to -\frac{b_2}{a_2}$.
(c) $F_1'(0) > F_2'(0)$ precisely when $b_1 b_2 > MW a_1 a_2$.

6. (a) The total number of infected people at equilibrium is about 895.
(b) With $W = 500$, $M = 1{,}000$, and the parameter values as in (a), the new equilibrium point is approximately $(416, 71)$. The total number of infected persons is about 487.
(c) With $W = 1{,}000$, $M = 500$, and the parameter values as in (a), the new equilibrium point is approximately $(227, 83)$. The total number of infected persons is about 310.
(d) Halving the number of at-risk men is the more effective policy.

11. (b) Biologically relevant equilibrium points at $(0,0)$ and approximately $(0.541, 0.032)$; $(0,0)$

is a saddle point and $(0.541, 0.032)$ is a stable node.
(c) Long term, we expect $S(t) \to 0.541$ and $Q(t) \to 0.032$ (approximately).

13. (a) $\frac{dx}{dt} = gN - bx\frac{y}{N} - gx$, $\frac{dy}{dt} = bx\frac{y}{N} - (g+c)y + rz$, $\frac{dz}{dt} = cy - gz - rz$

(b) $\frac{dS}{dt} = b(1-S-Q)S - (g+c)S + rQ$, $\frac{dQ}{dt} = cS - gQ - rQ$. The point $(0,0)$ is unstable if $g(g+r+c) - b(g+r) < 0$ (in which case $(0,0)$ is a saddle point) or if $r < b - 2g - c$ (so that the trace $\tau > 0$).

Section 10.6

1. $H(x,y) = axy + \frac{by^2}{2} - \frac{cx^2}{2}$ is a Hamiltonian function.

3. The system is Hamiltonian with Hamiltonian function $H(x,y) = \frac{y^2}{2} + \int f(x)\,dx$, where $\int f(x)\,dx$ is any antiderivative of $f(x)$.

5. (a) Equilibrium points are $R = 0$, $J = 0$ and $R = 0$, $J = 1$.
(c) As $t \to \infty$, $R \to -\infty$ and $J \to -\infty$, so the relationship ends in mutual dislike.

7. (b) The quantity $E(x,y) = \frac{y^2}{2} + x^{-1} + x^{-2}$ is conserved for the second system, and this quantity is also conserved for the first system.

9. With $E(x,y) = x^2 + y^2$ we have $E(0,0) = 0$, $E(x,y) > 0$ for $(x,y) \neq (0,0)$ and $\dot{E} < 0$ for all $(x,y) \neq (0,0)$, so that $(0,0)$ is asymptotically stable.

11. If a and b are both positive and $a = b > 0$ (for example, $a = b = 1$), then $E(x,y)$ is a Lyapunov function. $(0,0)$ is asymptotically stable.

21. (b) Level curves are ellipses. Trajectories are perpendicular to the elliptical level curves.

23. Hamiltonian system, with Hamiltonian function $H(x,y) = y\sin x$

25. This is both a gradient system, with $F(x,y) = -2xy + 4x + 2y$, and a Hamiltonian system, with Hamiltonian function $H(x,y) = y^2 - 4y - x^2 + 2x$.

Section 10.7

1. $\dot{E}(\theta, v) = -cL^2 v^2$
6. $E(n\pi, 0) \neq 0$ for n odd.
7. Orbit 1 corresponds to graph (A). Orbit 2 corresponds to graph (C).

9. If $\frac{c^2}{m^2} - \frac{4g}{L} > 0$, $(n\pi, 0)$ is a stable node. If $\frac{c^2}{m^2} - \frac{4g}{L} < 0$, $(n\pi, 0)$ is a stable spiral.

Section 10.8

2. $a = \sqrt{3}$ and $b = 3$
3. Any $0 < r_1 < \ln 2 < r_2$; for example $r_1 = \frac{1}{2}$ and $r_2 = 1$.
5. Equilibrium points are $(0,0)$ and $(-1,1)$. Since a cycle must enclose an equilibrium point, there can be no cycles in the right half-plane.

7. (a) Outward normal vector $(0,-1)$ or any positive scalar multiple
(c) Outward normal vector $(-2,1)$ or any positive scalar multiple
(e) Outward normal vector $(-1,1)$ or any positive scalar multiple

9. (a) $x' = y$, $y' = -p(x)y - q(x)$

Section 11.2

1. $y = -\sin x$
3. No solution

5. $y = c\cos x$ for any constant c
7. $y = ce^{-x}\sin x$ for any constant c

10. (a) No eigenfunctions when $\lambda = 0$; positive eigenvalues $\lambda = \left(\frac{n\pi}{2L}\right)^2$ for n odd, with eigenfunctions $c\sin\frac{n\pi x}{2L}$; no negative eigenvalues
(b) Eigenvalue $\lambda = 0$ with eigenfunctions $y = c$, $c \neq 0$; positive eigenvalues $\lambda = (n\pi/L)^2$ for n any integer, and corresponding eigenfunctions $c\cos(n\pi x/L)$; no negative eigenvalues
(c) Eigenvalue $\lambda = 0$ with eigenfunctions $y = c$, $c \neq 0$; no positive eigenvalues; negative eigenvalues $-\left(\frac{n\pi}{L}\right)^2$ for each integer n, with eigenfunctions $c\cos\frac{n\pi x}{L}$.

11. (a) $y = c$, c any constant, when $\lambda = 0$; $\lambda = \left(\frac{n\pi}{L}\right)^2$ and solutions $y = c_1\cos\frac{n\pi x}{L} + c_2\sin\frac{n\pi x}{L}$; no nontrivial solutions with $\lambda < 0$

13. (a) $y = x^{1/2}$

15. (a) $y'(0) = 0$ and $y'(10) = 0$
(b) $y(x) = 8 - \frac{25\rho}{6}x^2 + \frac{5\rho}{6}x^3 - \frac{\rho}{24}x^4$ where $\rho = \frac{2}{625}$
(c) $y(x) = 3 - \frac{125}{3}\rho x + \frac{5\rho}{6}x^3 - \frac{\rho}{24}x^4$ with ρ as in (b)

Section 11.3

2. (a) $3e^{-\frac{25\pi^2}{4}t}\sin\frac{5\pi x}{2} - 7e^{-\frac{81\pi^2}{4}t}\sin\frac{9\pi x}{2}$

3. $X''(x) + \lambda X(x) = 0$, $T''(t) + T'(t) + \lambda T(t) = 0$

5. $X^{(4)}(x) - \lambda X(x) = 0$, $T'(t) + \lambda T(t) = 0$

7. Yes

9. No

11. (b) $\lambda = \left(\frac{n\pi}{2L}\right)^2$ for n odd, and $X(x) = c\sin\frac{n\pi x}{2L}$, $c \neq 0$
(c) $T = ce^{-\left(\frac{n\pi}{2L}\right)^2\kappa t}$
(d) $u(x,t) = 5e^{-\frac{1}{4}t}\sin\frac{x}{2} - 3e^{-\frac{9}{4}t}\sin\frac{3x}{2}$

15. (b) $y(L) = 0$

17. (b) Parabolic
(d) Hyperbolic

Section 11.4

1. All except (d)

4. $u(x,y) = f(y - \frac{3}{2}x)$ for any differentiable function f

7. $u(x,y) = f(y^2 - x^2)$ for any differentiable function f

11. (a) $u(x,t) = e^{-0.01t}(4 - (x - 500t)^2)$

13. As r increases, the minimum width decreases.

15. New boundary condition is $u(0,t) = 0$, $u(L,t) = 0$

17. (c) $u(x,t) = 3e^{(2-4\pi^2)t}\sin(2\pi x) - 4e^{(2-25\pi^2)t}\sin(5\pi x)$

Section 11.5

2. (b) No; the only solution to $\cos x + \cos(\pi x) = 2$ is $x = 0$

3. $f(-\frac{\pi}{2}) = -e^\pi$, $f(8) = f(8 - 2\pi) = e^{16-4\pi}\sin(8 - 2\pi) = e^{16-4\pi}\sin 8$, and $f(-2 - \pi) = e^{2\pi-4}\sin 2$

11. If $f(0) = 0$ and $f(L) = 0$, then the odd periodic extension of f from $[0, L]$ to the real line will be everywhere continuous.

12. (b) The fundamental period is $\pi/2$.
(d) There is no smallest positive period. For each rational number r, $f(x) = f(x + r)$ for all real numbers x. Since there is no smallest positive rational number, there is no smallest positive period.

14. (b) $c_j = \mathbf{A} \cdot \mathbf{E}_j$, $\|\mathbf{A}\| = \sqrt{c_1^2 + c_2^2 + \cdots + c_n^2}$

Section 11.6

1. (b) In the formula for b_n, the integrand is odd.
   (c) Fourier series $\frac{4}{2} - \frac{4}{\pi}\cos\frac{\pi x}{2} + \frac{4}{3\pi}\cos\frac{3\pi x}{2} - \frac{4}{5\pi}\cos\frac{5\pi x}{2} + \cdots$
   (d) Converges to $f(x)$ at every $x \neq \pm 1, \pm 3, \pm 5, \pm 7$, etc. At $x = \pm 1, \pm 3, \pm 5, \pm 7$, etc., converges to 2.
   (e) Neither one changes.

3. (b) Since f is odd, so is the integrand $f(x)\cos\frac{n\pi x}{4}$, and the integral of an odd function over the symmetric interval $[-4, 4]$ is 0.
   (c) $b_1 = \frac{16-4\pi}{\pi^2}$, $b_2 = \frac{4}{\pi}$, $b_3 = \frac{-16-12\pi}{9\pi^2}$
   (d) At all x except $x = \pm 2, \pm 4, \pm 6, \pm 10$, etc., the Fourier series converges to (the extended) $f(x)$. At $x = 2$ the Fourier series converges to 1, at $x = -2$ it converges to -1, at $x = \pm 4$ it converges to 0, at $x = 6$ it converges to -1, and at $x = 10$ it converges to 1, etc.

5. (a) Yes, to 1
   (b) Yes, to -1
   (c) At $x = 0$, $x = \pm L$.

7. (b) $\text{Max}\{|S(x) - x| : -0.7 \le x \le 0.7\} = 0.013$, $\text{Max}\{|S(x) - x| : -0.8 \le x \le 0.8\} = 0.020$, $\text{Max}\{|S(x) - x| : -0.9 \le x \le 0.9\} = 0.040$, and $\text{Max}\{|S(x) - x| : -0.99 \le x \le 0.99\} = 0.179$

9. (a) $b_1 = \frac{6}{\pi}$, $b_2 = \frac{2}{\pi}$, $b_3 = \frac{2}{\pi}$, $b_4 = 0$, and $b_5 = \frac{6}{5\pi}$
   (b) Converges to f at every $0 \le x \le 4$ except at $x = 0$, $x = 2$, and $x = 4$

11. $\sin^2 x = \frac{1}{2} - \frac{1}{2}\cos(2x)$, $\cos^2 x = \frac{1}{2} + \frac{1}{2}\cos(2x)$

13. (b) $\frac{\pi^3}{32}$

20. (d) $-\frac{12}{\pi^3}\sin(\pi x) + \frac{3}{2\pi^3}\sin(2\pi x) - \frac{4}{9\pi^3}\sin(3\pi x) + \frac{3}{16\pi^3}\sin(4\pi x) + \cdots$

Section 11.7

1. (a) $u(x, t) = 2 + 3e^{-t}\cos x$
   (b) $u(x, t) = \frac{3}{2} + \frac{1}{2}e^{-4t}\cos(2x)$

3. $\frac{8(30)}{\pi^2}\sum_{k=0}^{\infty}\frac{(-1)^k}{(2k+1)^2}e^{-9((2k+1)\pi/2)^2 t}\sin\frac{(2k+1)\pi x}{2}$

5. (a) $e^{-\kappa(\frac{n\pi}{2L})^2 t}\sin\frac{n\pi x}{2L}$ for n any odd integer
   (b) Yes
   (c) $u(x, t) = 5e^{-9t/4}\sin\frac{3x}{2} - 2e^{-49t/4}\sin\frac{7x}{2}$

Section 11.8

3. $u(x, t) = \frac{1}{10}\sin(\pi x)\cos(\pi t)$

5. $u(x, t) = \frac{1}{20}\sin(2x)\sin(2t)$

7. $\frac{6\sqrt{3}}{\pi^2}\sin\frac{\pi x}{3}\cos\frac{\pi t}{3} - \frac{6\sqrt{3}}{25\pi^2}\sin\frac{5\pi x}{3}\cos\frac{5\pi t}{3} + \cdots$

8. (a) $u(x, t) = \sin x \cos\frac{\pi t}{3} - \frac{1}{2}\sin(3x)\cos(\pi t)$
   (b) $t = \frac{3}{2}$; other answers are possible
   (c) $t = 6$

10. $u(x, \pi/\alpha) = -u(x, 0)$ and $u(x, 2\pi/\alpha) = u(x, 0)$, for all x

13. (a) Building-block solutions are
$$at + b, \quad \cos\frac{n\pi x}{L}\cos\frac{\alpha n\pi t}{L}, \quad \cos\frac{n\pi x}{L}\sin\frac{\alpha n\pi t}{L}.$$

15. (b) $e^{-t/2}\sin(2x)\left(\cos\frac{\sqrt{15}t}{2} + \frac{1}{\sqrt{15}}\sin\frac{\sqrt{15}t}{2}\right)$

17. When $b = \frac{1}{8}$

Section 11.9

1. (a) $u(x, t) = \frac{1}{2}e^{-(x-t)^2} + \frac{1}{2}e^{-(x+t)^2}$

3. (a) D'Alembert solution is $u(x, t) = \frac{1}{2}\sin(x + t) + \frac{1}{2}\sin(x - t)$. The identity $2\sin w \cos v = \sin(w + v) + \sin(w - v)$ shows this is the same as $u(x, t) = \sin x \cos t$.

4. (a) $u(x, t) = 2\sin\frac{\pi x}{10}\cos(\pi t) + \sin\frac{3\pi x}{10}\cos(3\pi t)$

7. (a) $u_{\tau\tau} = u_{XX}$

9. (a) $-\alpha u_\zeta + \alpha u_\eta - \alpha(u_\zeta + u_\eta) = 0$ or $u_\zeta = 0$
   (b) $u(x, t) = f(x + \alpha t)$

11. (c) For fixed t, u is periodic in x with period p. For fixed x, u is periodic in t with period p/α.

Section 11.10

1. $u(x,y) = \frac{1}{\sinh(\pi)} \sinh(\pi y) \sin(\pi x)$

3. (a) If $u(x,y) = X(x)Y(y)$, we want $X''(x) + \lambda X(x) = 0$, $X(0) = 0$, and $Y''(y) - \lambda Y(y) = 0$, $Y(0) = Y(b) = 0$.
   (c) $c_n = \frac{2}{b \sinh(n\pi a/b)} \int_0^b g_2(y) \sin \frac{n\pi y}{b} \, dy$

5. $u(x,y) = u_1(x,y) + u_2(x,y)$

7. (a) Building-block solutions are $u(x,y) = y$ and $u(x,y) = \cos \frac{n\pi x}{a} \sinh \frac{n\pi y}{a}$.

10. For $\lambda = 0$, only for Θ a nonzero constant. There are no (nontrivial) periodic solutions with $\lambda < 0$.
    If $\lambda > 0$, the nontrivial 2π-periodic solutions are $\Theta(\theta) = c_1 \cos(n\theta) + c_2 \sin(n\theta)$, n an integer.

11. $r \sin \theta$

14. $\frac{1}{729} r^6 \sin(6\theta)$

16. $c_1 = T_1$ $c_2 = 4(T_2 - T_1)/\pi$

Bibliography

[1] Mark Adkin, *The Trafalgar Companion: A Guide to History's Most Famous Sea Battle and the Life of Admiral Lord Nelson*, Aurum Press, London, 2005.

[2] P. A. Assimakopoulos, K. G. Ioannides, and A. A. Pakou, *The propagation of the Chernobyl ^{131}I impulse through the air-grass-animal-milk pathway in Northwestern Greece*, The Science of the Total Environment, 85(1989), 295–305.

[3] Ron Barnes, *Compartment Models in Biology*, UMAP Module 676, COMAP, Arlington, MA, 1987.

[4] P. Blanchard, R. Devaney, and G. Hall, *Differential Equations*, 3rd ed., Thomson Brooks/Cole, 2002.

[5] Edward Beltrami, *Mathematics for Dynamic Modeling*, Academic Press, Orlando, FL, 1987.

[6] D. Bernoulli, *Essai d'une nouvelle analyse de la mortalité causée par la petite vérole, et des avantages de l'inoculation pour la prévenir*, Histoire de L'Acad. Roy. Sci. (Paris) avec Mém. des Math. et Phys and Mém, 1760, 1–45.

[7] David Berreby, *The great bridge controversy*, Discovery, February 1992, 26–33.

[8] David Bressoud, *A Radical Approach to Real Analysis*, 2nd ed., Mathematical Association of America, 2007.

[9] W. Boyce and R. DiPrima, *Elementary Differential Equations and Boundary Value Problems*, 10th ed., Wiley, 2012.

[10] Fred Brauer and Carlos Castillo-Chávez, *Mathematical Models in Population Biology and Epidemiology*, Texts in Applied Mathematics 40, Springer, New York, 2001.

[11] Martin Braun, *Differential Equations and Their Applications*, 3rd ed., Springer-Verlag, New York, 1983.

[12] Martin Braun, Courtney Coleman, and Donald Drew, eds., *Differential Equation Models*, Modules in Applied Mathematics, Vol. 1, Springer-Verlag, New York, 1983.

[13] S. Carpenter, D. Ludwig, and W. Brock, *Management of eutrophication for lakes subject to potentially irreversible change*, Ecological Applications, 9(1999), 751–771.

[14] Earl Coddington and Norman Levinson, *Theory of Ordinary Differential Equations*, McGraw-Hill, New York, 1955.

[15] M. R. Cullen, *Linear Models in Biology: Linear Systems Analysis with Biological Applications*, Ellis Horwood Limited, Chichester, 1985.

[16] C. H. Edwards, Jr. and David Penney, *Differential Equations and Boundary Value Problems: Computing and Modeling*, Prentice-Hall, Englewood Cliffs, New Jersey, 1996.

[17] Gerald Folland, *Fourier Analysis and Its Applications*, Wadsworth and Brooks/Cole, Pacific Grove, CA, 1992.

[18] Justin Gillis and Celia Dugger, *U.N. sees rise for the world population to 10.1 billion*, The New York Times, May 4, 2011, p. A1.

[19] Raymond Greenwell, *Whales and Krill: A Mathematical Model*, UMAP Module 610, COMAP, Arlington, MA, reprinted in UMAP Modules: Tools for Teaching, 1982.

[20] Morris Hirsch, Stephen Smale, and Robert Devaney, *Differential Equations, Dynamical Systems, and an Introduction to Chaos*, 3rd ed., Academic Press, Elsevier, Boston, 2013.

[21] T. W. Körner, *Fourier Analysis*, Cambridge University Press, 1988.

[22] D. Ludwig, D. D. Jones, and C. S. Holling, *Qualitative analysis of insect outbreak systems: The spruce budworm and forest*, J. Anim. Ecol., 47(1978), 315–332.

[23] Mark J. Machina, *Decision-making in the presence of risk*, Science, 236(1987), 537–543.

[24] Helen Marcus-Roberts and Maynard Thompson, eds., *Life Science Models*, Modules in Applied Mathematics, Vol. 4, Springer-Verlag, New York, 1983.

[25] A. Mas-Colell, M. Whinston, and J. Green, *Microeconomic Theory*, Oxford University Press, Oxford, 1995.

[26] Robert M. May, *Stability and Complexity in Model Ecosystems*, Princeton University Press, Princeton NJ, 1974.

[27] Robert May, John Beddington, Colin Clark, Sidney Holt and Richard Laws, *Management of multispecies fisheries*, Science, 205(1979), 267–277.

[28] Mark Memmott, *Cat falls 19 floors, lands purrfectly*, NPR's News Blog, March 22, 2012.

[29] Michael Mesterton-Gibbons, *A Concrete Approach to Mathematical Modeling*, Addison-Wesley, Redwood City, CA, 1989.

[30] Clifford Morehouse, *The Iwo Jima Operations*, Historical Division, Headquarters U.S. Marine Corps, 1946.

[31] J. D. Murray, *Mathematical Biology I: An Introduction*, 3rd ed., Interdisciplinary Applied Mathematics, 17, Springer-Verlag, New York, 2002.

[32] David H. Nash, *Differential equations and the Battle of Trafalgar*, The College Mathematics Journal, 16(1985), No. 2, 98–102.

[33] James Nestor, *Deep*, Houghton Mifflin Harcourt, Boston, 2015.

[34] Lawrence Perko, *Differential Equations and Dynamical Systems*, 3rd ed., Springer, New York, 2001.

[35] Henry Petroski, *Still twisting*, American Scientist, 79(5), 1991, 398–401.

[36] John Polking, Albert Boggess, and David Arnold, *Differential Equations with Boundary Value Problems*, 2nd ed., Pearson Education, Upper Saddle River, New Jersey, 2006.

[37] David L. Powers, *Boundary Value Problems*, Academic Press, New York, 1972.

[38] E. J. Putzer, *Avoiding the Jordan canonical form in the discussion of linear systems with constant coefficients*, Amer. Math. Monthly, 73(1966), 2–5.

[39] Michael Rabinowitz, *Toxicokinetics of bone lead*, Environmental Health Perspectives, 91(1991), 33–37.

[40] Michael Rabinowitz, George Wetherill, and Joel Kopple, *Lead metabolism in the normal human: Stable isotope studies*, Science, 182(1973), 725–727.

[41] Robert Rainey, *Natural displacement of pollution from the Great Lakes*, Science, 155(1967), 1242–1243.

[42] Ray Redheffer, *Differential Equations Theory and Applications*, Jones and Bartlett, Boston, 1991.

[43] Sergio Rinaldi, *Laura and Petrarch: An intriguing case of cyclical love dynamics*, Siam. J. Appl. Math., 58(1998), 1205–1221.

[44] Karen Saxe, *Beginning Functional Analysis*, Springer-Verlag, New York, 2002.

[45] Lawrence Shampine, *Numerical Solution of Ordinary Differential Equations*, Chapman and Hall, New York, 1994.

[46] David A. Smith, *Descriptive Models for Perception of Optical Illusions I, II*, UMAP Module 534–535, COMAP, Arlington, MA, 1986.

[47] Lewi Stone, *Biodiversity and habitat destruction: A comparative study of model forest and coral reef ecosystems*, Proc. R. Soc. Lond. B, 261(1995), 381–388.

[48] Steven Strogatz, *Love affairs and differential equations*, Math. Magazine, 61(1998), 35.

[49] Steven Strogatz, *Loves me, loves me not (do the math)*, Olivia Judson blog, New York Times, May 26, 2009.

[50] Steven Strogatz, *Nonlinear Dynamics and Chaos*, 2nd ed., CRC Press, Taylor and Francis Group, Boca Raton, FL, 2018.

[51] Clifford Taubes, *Modeling Differential Equations in Biology*, 2nd ed., Cambridge University Press, New York, 2008.

[52] David Tilman, *Competition and biodiversity in spatially structured habitats*, Ecology, 75(1994), 2–16.

[53] David Tilman, Robert May, Clarence Lehman, and Martin Nowak, *Habitat destruction and the extinction debt*, Nature, 371, September 1994, 65–66.

[54] D. Vnuk, B. Pirkic, D. Maticic, B. Radisic, S. Stegskal, T. Babic, M. Kreszinger, and N. Lemo, *Feline high-rise syndrome: 119 cases (1988 1998)*, J. Feline Medicine and Surgery, 6(2004), 305–312.

[55] Hans F. Weinberger, *A First Course in Partial Differential Equations with Complex Variables and Transform Methods*, Dover Publications, New York, 1965.

[56] D. R. Westbrook, *The Mathematics of Scuba Diving*, UMAP Module 767, COMAP, reprinted in UMAP Modules: Tools for Teaching, 1997.

[57] E. C. Zachmanoglou and Dale W. Thoe, *Introduction to Partial Differential Equations with Applications*, Dover Publications, New York, 1986.

Index

Abel's formula, 200
Abel's Theorem, 200
Aberdeen Proving Ground, 58
adaptive methods, 158
advection, 738, 739
advection-decay equation, 738, 743
AIDS, 623
Airy's equation, 398, 403
albumin, 544
amplitude modulation, 291
analytic function, 387
annihilating operator, 276
    complexified, 283, 287
    table of, 278
antiperiodic function, 327
apparent curve, 162
Arafat, Yasir, 91
argument, 249
aspirin poisoning, 294
asymptote, 513
asymptotically stable, 101, 555, 556
autonomous equation, 85, 98
autonomous system, 181, 447, 616
auxiliary condition, 267, 274

basic region, 509
basic reproduction number, 671, 680
basin of attraction, 642, 703
Battle of Trafalgar, 443
beam equation, 254, 727
beats, 300, 301
Bendixson Theorem, 710
Bernoulli, 668
    smallpox model, 668, 680
Bernoulli equation, 114, 119, 649, 652
Bessel function, 430
    order 0, 432
    order 1, 432
    order p, 432
Bessel's equation, 430
    parametric, 439
bifurcation, 105
    diagram, 105
    value, 42, 105
Binomial Theorem, 220, 405
blood alcohol concentration, 167, 355
Bolt, Usain, 34
borderline case, 647
bounce dive, 47
boundary conditions, 407, 718, 720
    and heat equation, 729, 779, 781, 782
    Dirichlet, 814
    insulated-ends, 782
    nonhomogeneous, 784
    wave equation, 792
    zero-temperature-ends, 779, 781
boundary value problem, 189, 264, 719, 721, 724, 726
    homogeneous, 720
breath-hold diving, 58
building-block solutions, 718, 741
    heat equation, 729, 779, 781, 782, 786, 790
    Laplace's equation, 820
    wave equation, 794

capacitor, 305
carrying capacity, 88, 94, 118, 615
Cauchy-Euler equation, 233, 413
Cauchy-Schwarz, 774
Cayley-Hamilton Theorem, 535, 566, 585, 586
center, 525, 555, 647
certainty equivalent, 44

change in time scale, 115
change of variable, 112
characteristic equation, 216, 475, 733
   complex roots, 219, 246
   repeated roots, 229, 245
characteristic polynomial, 245, 474
charge, 304
Chernobyl, 599
circuit
   elements, 304
   tuning, 312
   water analogy, 305, 306
Clairaut equation, 81
clepsydra, 52
   perfect, 55
closed system, 546
coefficient function, 25, 240, 447
colonization-rate constant, 658
combat model, 357, 443, 451
compartment model, 295, 536, 669
   two-way transfer, 542
competing species, 618, 622, 658
competition for space, 658
competitive exclusion, 615
complementary
   equation, 265, 279
   solution, 266
complex exponential, 219, 756
   derivative of, 222
complex number, 217
   conjugate, 218
   imaginary part, 217
   modulus, 218
   polar form, 249
   real part, 217
complex plane, 217
complex ray, 524
complex sinusoid, 610
concavity chart, 511
conflict point, 748
conjugate, 218
conservation of energy, 188, 707, 802
conservative system, 684
conservative vector field, 122
conserved quantity, 684, 686
continuously differentiable, 628
convergence
   absolute, 385
   tests for, 385
convolution, 363
Convolution Theorem, 363, 365
Cooper, D. B., 184
correction curve, 165

cow-grazing model, 111, 169
Cramer's Rule, 208, 274, 471
critical point, 450, 693
critically damped, 259
cumulative error, 144
current, 304
cycle, 708

D'Alembert
   method, 804
   solution, 807
D'Ancona, 653
damped oscillations, 226
damping force, 187, 256
Darwin, 615
decompression sickness, 46
deepest man on earth, 58
degenerate node, 531, 552
delayed function, 343
delta function, 370
demand step, 280, 286
dependent variable, 1, 174, 448
determinant, 196, 244, 465, 478, 550
diagonal matrix, 477
   eigenvalues, 481
   powers, 481
differential, 130
   operator, 194
differential equation, 1
diffusion
   coefficient, 740
   equation, 727, 740
dimensionless, 118
Dirac delta function, 370
   Laplace transform of, 370
direction field, 6, 139, 461, 735
   and isoclines, 16
   autonomous equation, 98
   computer generated, 18
direction-indicator arrow, 509
Dirichlet problem, 814
   in a disk, 818
   in a wedge, 828
disease model, 456, 669
dissipative system, 591, 609
distortion pattern, 162
doomsday model, 95
dot product, 690, 751
   properties, 751
doubling time, 84, 97
drag, 171, 256
drag coefficient, 171
drug infusion, 352
drug overdose, 294

drum, 439, 798

eigenfunction, 722
    orthogonality, 755, 759
eigenvalue, 473, 722
eigenvector, 473
eigenvector line, 484
electric circuit, 303
elementary function, 15
elimination
    method of, 451
endemic, 671
energy, 707, 787, 802
equilibrium point, 450, 616, 621, 627
    center, 647
    node, 632
    saddle, 632
    spiral, 632
equilibrium position, 255
equilibrium solution, 7, 74, 99, 104
    stability, 101
Euler's formula, 222
Euler's method, 139, 152
    error estimates, 143
    second-order equation, 192
Euler-Fourier formulas, 761
even function, 79, 745, 754
even periodic extension, 747
exact equation, 122, 686
    integrating factor, 125
    Test for Exactness, 123
existence and uniqueness, 68, 69, 78, 180, 190, 208, 212, 240, 448
    systems, 448, 580, 620
expected utility, 44
expected value, 44
explicit solution, 2
exponentially bounded, 322

falling objects, 171
feline high-rise syndrome, 184
Fick's law, 740, 778
field mark, 6
financial model, 89
First Differentiation Formula, 336
First Shift Formula, 339
first-order equation
    autonomous, 98
    Bernoulli, 114
    change of variable, 112
    Clairaut, 81
    comparing linear and nonlinear, 73
    exact, 122
    existence and uniqueness, 68, 69
    explicit solution, 2
    implicit solution, 4
    linear, 25
    lost solution, 37, 136
    nonlinear, 25
    normal form, 2
    qualitative solution, 6, 98
    separable, 36
first-order kinetics, 295
forced vibration, 289
Fourier coefficients, 760
Fourier Convergence Theorem, 762
Fourier cosine series, 767
Fourier series, 719, 760, 762
    specialty, 777
Fourier sine series, 767
Fourier, Joseph, 719, 770
free end, 800
free vibration, 256
frequency, 257
Frobenius equation, 416, 424
Frobenius method, 420
Frobenius solution, 417
fundamental frequency, 798
fundamental matrix, 581
fundamental period, 747
fundamental set of solutions, 197, 213, 241, 242, 246, 489
Fundamental Theorem of Algebra, 245
Fundamental Theorem of Calculus, 15, 29, 53, 80, 176
    vector version, 596

Gamma function, 436
general solution, 3, 27, 37, 41, 73, 198, 200, 242, 246, 266, 483, 487
General Solutions Theorem, 200
generalized function, 374
geometric series, 384, 386, 388, 390
Gibbs phenomena, 770
Gompertz equation, 89, 95, 96, 117
gradient, 122, 693, 735
gradient system, 693
gradient vector, 690
Great Lakes, 60, 539
Green's Theorem, 715, 716

habitat destruction, 661, 667
half-time, 59
Hamiltonian
    function, 685
    system, 685, 696
hanging chain, 440, 732
harmonic, 797

harmonic function, 812
harmonic oscillator, 178
harvesting renewable resources, 109
HCG, 96
heat equation, 264, 717, 727, 777, 779
    uniqueness, 791
herd immunity, 672
Hering illusion, 79, 162
Hering models
    derivation, 165
Hermite polynomial, 408
Hermite's equation, 239, 408
hertz, 257
Heun method, 150, 153
    error estimates, 155
homogeneous, 25, 113, 174, 240, 447, 448
Hooke's law, 178, 255
House, on Fox TV, 59
Hurricane Katrina, 93
hyperbolic cosine, 203, 429
hyperbolic sine, 203
hyperbolic tangent, 327

identity matrix, 465
imaginary axis, 217
Implicit Function Theorem, 5
implicit solution, 4
improper integral, 319
improper node, 530, 531, 554, 634
impulse response function, 379
in-vitro fertilization, 97
incoming ray, 484
independent variable, 1, 448
indicial equation, 414, 418, 420, 426
inductor, 306
infective, 669
infinite string, 804
initial condition, 8, 180, 240, 443, 448
    heat equation, 729, 779
    wave equation, 792
initial value problem, 8, 38, 449
    Laplace transform method, 336
inner product, 753
insulated ends, 782
integral equation, 380
integral transform, 329
integrating factor, 27, 125, 127, 137
Intermediate Value Theorem, 173
interval of existence, *see also* interval of validity, 2
interval of validity, 2, 8, 70, 73
intrinsic growth rate, 85
inverse, 465
inverse Laplace transform, 330

    linearity, 331
iodine tracer, 536
iPod sales, 94
irregular singular point, 416
isocline, 16
isotherm, 817
Iwo Jima, 357

Jacobian matrix, 629, 633

kinetic energy, 188
Kirchhoff's current law, 308
Kirchhoff's voltage law, 307

lake eutrophication, 168
Lanchester, 443
Laplace differentiation formula, 335
Laplace transform, 319
    and pharmacology, 351
    and systems, 357, 541
    as an operator, 319
    computational rules, 343
    inverse, 330
    linearity, 321
    of antiperiodic function, 327
    of Dirac delta function, 370
    of periodic function, 327
    of square wave, 327
    of unit step function, 323
    shift formulas, 339
    solving initial value problem, 336
    table, 343
Laplace's equation, 717, 812
    in disk, 817
    in polar coordinates, 818, 827
    in rectangle, 814
Law of Reflection, 132
lead poisoning, 547, 605
level curve, 684
limit cycle, 649, 708
    attracting, 708
    repelling, 708
linear approximation, 627
linear combination, 197, 209, 241, 242, 468
    nontrivial, 214
linear differential equation, 25, 174, 240
linear independence, 488
linear operator, 194, 276
linearly dependent, 209, 210, 241
linearly independent, 209, 210, 241, 468
lionfish, 663
logistic model, 85, 119, 615
long-term memory, 96
Lord Nelson, 443

Index 871

Lotka-Volterra system, 653, 663, 708
love affairs, 453, 695
Lyapunov function, 688
   strong, 689

major type, 632
Malthusian model, 83, 84, 653
mass-spring, 178, 255, 289
   critically damped, 259
   double, 362, 457
   overdamped, 259
   underdamped, 258
mathematical modeling, 8, 45, 134, 180, 254, 303
matrix, 462
   addition, 463
   diagonal, 477
   identity, 465
   inverse, 465
   multiplication, 462, 463
   symmetric, 697
   triangular, 481
matrix exponential, 563, 584, 586
   distinct eigenvalues, 585
   formulas for 2×2 case, 574
   Putzer's Algorithm, 587
matrix function, 482, 564
maximal sustainable yield, 109
Maximum Principle, 813
Mean Value Property
   for harmonic functions, 813, 824
Mean Value Theorem, 67, 437
measles, 669
memorization, 96
metabolism, 64
modulus, 218
mortality rate, 658
multiplicity, 217, 245, 477
multistep methods, 158
musical attributes, 798

nth-order linear equation, 240
   constant coefficient, 245, 275
   existence and uniqueness, 240
   fundamental set of solutions, 242
   general solution, 242
   homogeneous, 240
   standard form, 240
   Wronskian, 243
nth roots, 250
Navy dive table, 48
Neanderthal, 623
negative definite, 692
negative semidefinite, 692
Newton's law of heating and cooling, 45

Newton's second law, 172, 178, 256
Nitsch, Herbert, 58
No Intersection Theorem, 72, 100
nodal tangent, 502, 634
   preferred, 502, 554, 635
node, 499, 529, 551, 632, 797
   degenerate, 529
   improper, 529, 554, 634
   nondegenerate, 531
   nonlinear, 633
   proper, 531, 553, 634
   stable, 519, 634
   unstable, 634
nondegenerate node, 531
nonhomogeneous system, 595
nonlinear, 25
nonlinear system, 616
norm, 751, 753
normal form, 174
   first-order equation, 2
normal vector, 690
norovirus, 682
nullcline, 508, 616
nullcline arrow, 509
nullcline-and-arrow diagram, 507, 616
nullspace, 467
numerical solution, 139, 192

odd function, 745, 754
odd periodic extension, 749
Ohm's law, 304
one-parameter family, 3, 73
one-step methods, 158
open region, 812
operator, 193
   annihilating, 277
   differential, 193, 194
   factoring, 201, 206, 231, 235, 239, 270, 277, 812
   linear, 194
   linear partial differential, 731
optical illusion, 79, 162
Orbison illusion, 164
orbit, 448, 496
   elliptical, 533
order, 1
   annihilating operator, 278
   linear differential operator, 275
ordinary differential equation, 1
ordinary point, 392, 409
organ pipe, 801
orthogonal, 751, 753
   eigenfunctions, 755
   Supreme Court arguments, 753
outgoing ray, 484

overdamped, 259
overtones, 798

parakeet, 92
parallelogram law, 458
Parseval's equality, 765
partial derivative, 69, 128
partial differential equation, 1, 127
   elliptic, 717, 733
   first-order, 734
   hyperbolic, 717, 733
   linear, 732
   parabolic, 717, 733
   second-order linear, 733
partial fraction decomposition, 87, 331
partial integration, 124
particular solution, 265
pendulum
   damped, 702, 715
   generalized, 707
   undamped, 700
periodic, 608, 746
   extension, 747
   fundamental period, 747
periodic forcing, 289, 608
periodic solution, 708
pharmacokinetics, 294
phase line, 100
phase plane, 181, 448
phase portrait, 181, 445, 448, 496
phase shift, 225
piecewise
   continuous, 322, 746
   continuously differentiable, 746
   defined function, 323
pitchfork, 105
plucked string, 793
Poincaré-Bendixson Theorem, 710
Poisson Integral Formula, 822
Poisson kernel, 824
polar coordinates, 225, 249, 648
polar form, 249, 250
pollution
   Great Lakes, 60, 539
   lake, 168
   river, 737
population
   New Orleans, 93
   seasonally varying, 89, 119, 167
   United States, 83, 146
   world, 88
population model, 83
   logistic, 85
   Malthusian, 83

positive definite, 692
potential energy, 188
potential function, 122
power series, 382, 383
   center, 383
   computational properties, 386
   convergence of, 383, 395
   radius of convergence, 384
   uniqueness, 386
predator-prey, 653, 664
predictor-corrector, 153
preferred nodal tangent, 554, 635
principal operator, 279
proper node, 531, 553, 568
proportional growth rate, 84
pursuit curve, 174
Putzer's Algorithm, 587

quadratic form, 697
qualitative solution, 6
   autonomous equation, 98

radioactive decay, 91
radius of convergence, 384
Ratio Test, 383, 385
ray solution, 484
recognizing differentials, 130
recursion relation, 393
reduction of order, 229
regular point, 693
regular singular point, 416
relative percent error, 141
repeated eigenvalue, 524
repeated root, 229
resistor, 304
resonance, 289, 290, 300
rest condition, 336
restoring force, 178
reversal of seasons, 789
Riccati equation, 240
risk-averse, 44
risk-loving, 45
RK4, 155
Romeo and Juliet, 453, 559, 695
Root Test, 385
round-off error, 143, 159
Runge-Kutta method, 155, 157, 163
   error estimates, 157

saddle point, 498, 515, 550, 554, 632
   nonlinear, 635
scalar multiplication, 458
scuba diving, 46
Second Differentiation Formula, 337, 338

Index

Second Shift Formula, 340
second-order equation
    Cauchy-Euler, 233
    constant coefficient, 216
    Euler's method, 192
    existence and uniqueness, 180
    initial conditions, 180
    linear, 174
    normal form, 174
    reduction of order, 229
    standard form, 268
    superposition principle, 195
semistable, 101
separable equation, 36, 74
separation constant, 728
separation of variables, 35, 718
    diffusion equation, 728
    heat equation, 728
    Laplace's equation, 814, 819
    wave equation, 793
separatrix, 554, 638
series solution
    convergence of, 395, 409
    near ordinary point, 392, 409
    nonhomogeneous equation, 411
sexually transmitted disease, 672
shifting index of summation, 388
similar, 477, 490
simple closed curve, 709
simple harmonic motion, 257
simply connected, 709
Simpson's Rule, 30, 155
singular point, 416
    irregular, 416
    regular, 416
singular solution, 81
SIR model, 456, 669, 676
slope field, 6
slow sine wave, 291
smallpox, 680
smoking model, 675
snakehead, 91
soap film, 818
special function, 381
speed, 459
spiral point, 551, 554, 632
    nonlinear, 645
spotlight, 132
spring constant, 255
sprinting, 33
spruce budworm, 120
square wave, 327, 763
stable, 86, 462, 499, 555
    asymptotically, 556
    curve, 636
    line, 554, 636
    node, 634
standard form
    first-order linear, 25
    linear differential equation, 240
    second-order linear, 268
standing wave, 797
star node, 531
steady-state, 293, 609, 718, 784, 813, 816
step size, 139
struck string, 793
Sturm-Liouville theory, 790
Superposition Principle, 195, 223, 240, 273, 483, 604, 720, 721, 727
susceptible, 669
swine flu, 456, 668
symmetric form, 130
system
    autonomous, 447, 494, 616
    constant coefficient, 457, 484, 485, 489
    differential equations, 443
    from higher-order equation, 450
    homogeneous, 447
    linear, 447, 448, 487
    method of elimination, 451, 455
    nonhomogeneous, 595
    nonlinear, 616

Tacoma Narrows Bridge, 289
tangent plane, 129
Tavilla, Paul, 171
Taylor polynomial, 382
Taylor series, 219, 387
tension, 800
terminal velocity, 174
Test for Exactness, 123
thermal diffusivity, 778
threshold, 671
time derivative, 688
time translate, 449, 625
Torricelli's law, 53, 61
trace, 549
trace-determinant plane, 551
trajectory, 444, 448, 496
transfer coefficient, 536
transient solution, 293, 610
translation in time, 343
trapezoid rule, 30, 153
traveling waves, 737, 805
trial solution, 233
triangle function, 774
triangle inequality, 774

triangular matrix, 481
Trump, Donald, 57
truncation error, 141, 144

underdamped, 258
uniqueness
   heat equation, 791
   wave equation, 802
Uniqueness Property, 386
unit step function, 323, 342
   derivative of, 375, 377
   Laplace transform, 323
unit vector, 751
units, 172, 187, 257, 314
unstable, 86, 101, 498, 515, 555
   curve, 636
   line, 554, 636
   node, 634
upper triangular, 481
utility function, 44

vaccination, 671
van der Pol equation, 712
variation of parameters, 43, 119, 266, 273
vector, 458
   addition, 458
   dot product, 751
   norm, 751
   orthogonal, 751
   unit, 751
vector field, 461, 619
vector function, 482
velocity vector, 459
Vendera, Jamie, 301
Verhulst, 88
vibration
   beam, 254
   drum, 798, 831
   mass-spring, 255
   string, 792
voltage, 303
Volterra, 653
von Bertalanffy model, 64, 65, 116

wave equation, 717, 792
   D'Alembert method, 804
   separation of variables, 793
   two-dimensional, 831
   with damping, 801
Wronskian, 198, 211, 243, 488, 492
   dichotomy, 200, 243

yield
   maximal sustainable, 109

zero initial state solution, 610
zero matrix, 465
zero-order kinetics, 297
zero-temperature-ends, 781